José R. Correa Alejandro Hevia
Marcos Kiwi (Eds.)

LATIN 2006: Theoretical Informatics

7th Latin American Symposium
Valdivia, Chile, March 20-24, 2006
Proceedings

Volume Editors

José R. Correa
Universidad Adolfo Ibáñez, School of Business
Av. Presidente Errázuiz 3485, Las Condes, Santiago, Chile
url: http://www.jcorea.uai.cl

Alejandro Hevia
Universidad de Chile, Department of Computer Science
Blanco Encalada 2120, Santiago Centro, Santiago, Chile
url: http://www.dcc.uchile.cl/~ahevia

Marcos Kiwi
Universidad de Chile
Center for Mathematical Modelling and Department of Mathematical Engineering
Blanco Encalada 2120, Santiago Centro, Santiago, Chile
url: http://www.dim.uchile.cl/~mkiwi

Library of Congress Control Number: Applied for

CR Subject Classification (1998): F.2, F.1, E.1, E.3, G.2, G.1, I.3.5, F.3, F.4

LNCS Sublibrary: SL 1 – Theoretical Computer Science and General Issues

ISSN 0302-9743
ISBN-10 3-540-32755-X Springer Berlin Heidelberg New York
ISBN-13 978-3-540-32755-4 Springer Berlin Heidelberg New York

Springer is a part of Springer Science+Business Media

springer.com

Printed in Germany

Typesetting: Camera-ready by author, data conversion by Scientific Publishing Services, Chennai, India
Printed on acid-free paper SPIN: 11682462 06/3142 5 4 3 2 1 0

Preface

This volume contains the papers accepted for publication at LATIN 2006, the 7th Latin American Theoretical Informatics Symposium held in Valdivia, Chile, March 20-24, 2006. The LATIN series of conferences presents recent results in theoretical computer science. It was launched in 1992 to foster the interaction between the Latin American community and computer scientists around the world. LATIN 2006 was the seventh of a series, after São Paulo, Brazil (1992); Valparaiso, Chile (1995); Campinas, Brazil (1998), Punta del Este, Uruguay (2000), Cancún, Mexico (2002), and Buenos Aires, Argentina (2004).

In response to the call for papers, a record number of 224 submissions were received. The Program Committee accepted 66 papers in order to meet the goal of having five days of talks with no parallel sessions. Therefore, many good papers could not be accepted. The Program Committee met electronically from October 25 to November 10, 2005. The selection of papers was based on originality, quality, and relevance to theoretical computer science. It is expected that most of these papers will appear in a more complete and polished form in scientific journals in the future.

In addition to the contributed papers, this volume contains the abstracts of seven invited plenary talks given at the conference by Ricardo Baeza-Yates, Anne Condon, Ferran Hurtado, R. Ravi, Madhu Sudan, Sergio Verdú and Avi Wigderson.

The Program Committee thanks all authors of submitted manuscripts for their support of LATIN, and the many colleagues listed in pages VIII-X, who helped reviewing the submissions.

The LATIN proceedings have been published by Springer since the first edition. We are grateful to Springer for their continuous support.

January 2006

José R. Correa (Co-organizer)
Alejandro Hevia (Co-organizer)
Marcos Kiwi (Program Chair)
LATIN 2006

The Conference

Program Committee

Argimiro Arratia Quesada	U. Valladolid, Spain
Marcelo Arenas	U. Toronto, Canada
Jose Correa	U. Adolfo Ibáñez, Chile
Luc Devroye	McGill U., Canada
Guillermo Durán	U. Chile, Chile / U. Bs. Aires, Argentina
Antonio Fernández	U. Rey Juan Carlos, Spain
David Fernández-Baca	Iowa State U., USA
Afonso Ferreira	CNRS & COST Office, France
Fedor Fomin	U. Bergen, Norway
Joachim von zur Gathen	Bonn U., Germany
Cristina Gomes Fernandes	U. de São Paulo, Brazil
Mordecai Golin	Hong Kong UST, Hong Kong, China
Alejandro Hevia	U. Chile, Chile
Marcos Kiwi (Chair)	U. Chile, Chile
Hugo Krawczyk	IBM Research, USA
Eduardo Laber	PUC-Rio, Brazil
Martin Matamala	U. Chile, Chile
Jiří Matoušek	Charles U., Czech Republic
Gonzalo Navarro	U. Chile, Chile
Daniel Panario	Carleton U., Canada
Andreas Schulz	MIT, USA
Jacques Sakarovitch	CNRS/ENST, France
Amin Shokrollahi	EPFL, Switzerland
Mona Singh	Princeton U., USA
Howard Straubing	Boston College, USA
Shang-Hua Teng	Boston U., USA
Luca Trevisan	UC Berkeley, USA
Jorge Urrutia	UNAM, Mexico
Umesh Vazirani	UC Berkeley, USA
Santosh Vempala	MIT, USA
Alfredo Viola	U. de la República, Uruguay
Yoshiko Wakabayashi	U. de São Paulo, Brazil

Organizing Committee

José Rafael Correa (Co-organizer)	U. Adolfo Ibáñez, Chile
Alejandro Hevia (Co-organizer)	U. Chile, Chile
Michael Vielhaber (Local liaison)	U. Austral, Chile

Sterring Committee

Ricardo Baeza-Yates (Chair)	U. Chile, Chile
Martin Farach-Colton	Rutgers U., USA
Gaston Gonnet	ETH Zürich, Switzerland
Yoshiharu Kohayakawa	U. de São Paulo, Brazil
Sergio Rajsbaum	U. Nacional Autónoma de México, Mexico
Gadiel Seroussi	MSRI, USA

Plenary Speakers

Ricardo Baeza-Yates	U. Chile, Chile
Anne Condon	U. British Columbia, Canada
Ferran Hurtado	U. Politècnica de Catalunya, Spain
R. Ravi	Carnegie Mellon U., USA
Madhu Sudan	MIT, USA
Sergio Verdú	Princeton U., USA
Avi Wigderson	Institute for Advanced Study, USA

Sponsors

Centro Latinoamericano de Estudios en Informática (CLEI)
Centro de Modelamiento Matemático (CMM), U. Chile
CONICYT via grant *Anillo en Redes*
International Federation for Information Processing (IFIP)

Referees

Scott Aaronson
Laila El Aimani
Tatsuya Akutsu
Liliana Alcon
Sergio Alvarez
Andris Ambainis
Arvind Arasu
Sergio Arévalo
Ofer Arieli
Esther Arkin
Sunil Arya
Albert Atserias
Jean-Michel Autebert
Nikhil Bansal
Denilson Barbosa
Pablo Barcelo
Amotz Barnoy
David Mix Barrington
Michael Bender
Leopoldo Bertossi
Gustavo Betarte
Silvia Bianchi
Ernesto G. Birgin
Hans L. Bodlaender
Alexandra Boldyreva
Maria Luisa Bonet
Flavia Bonomo
Claudson Bornstein
Endre Boros
Jeremie Bourdon
Xavier Boyen
Richard Brent
Hajo Broersma
Andrei Bulatov
Samuel Buss
Benjamin Bustos
Leizhen Cai
Alberto Caprara
Rafael Carrasco

Carlos Castillo
Marcia Cerioli
Melissa Chase
Frédéric Chataigner
Kamalika Chaudhuri
Edgar Chávez
Bernard Chazelle
Siu-Wing Cheng
Joseph Cheriyan
Francis Y.L. Chin
Vicent Cholvi
Rezaul Alam Chowdhury
Ferdinando Cicalese
Julien Clement
Peter Clote
Anne Condon
William Cook
Graham Cormode
Derek Corneil
Bruno Courcelle
Ricardo Dahab
Josep Diaz
Volker Diekert
Ajit Diwan
Yevgeniy Dodis
Gerhard Dorfer
Feodor Dragan
Michael Drmota
Georges Dupret
Mariana Escalante
Daniel Espinoza
Luerbio Faria
Uriel Feige
Jose Fernandez
Esteban Feuerstein
Jiri Fiala
Celina Figueiredo
Samuel Fiorini
Marc Fischlin
David Flores
Pierre Fraigniaud
Matthew Franklin
Michael J. Freedman
Armin Fügenschuh
Henryk Fuks
Martin Fürer
Peter Gacs
David Gamarnik
Theo Garefalakis
Paul Gastin
Mark Giesbrecht
Michael Goldwasser
Jovan Golic
Martin Golumbic
Marcos Goycoolea
Eduardo Grampin
Lov K. Grover
Irene Guessarian
Gregory Gutin
András Gyárfás
Michel Habib
Yosub Han
Pierre Hansen
Meng He
Pinar Heggernes
Maurice Herlihy
Miki Hermann
Celina Herrera
Shlomo Hershkop
Susan Hohenberger
Stefan Hougardy
Ferran Hurtado
Hsien-Kuei Hwang
John Iacono
Costas Iliopoulos
Neil Immerman
Nicole Immorlica
Daniel Alejandro Jaume
Emmanuel Jeandel
Shaoquan Jiang
Charanjit Jutla
Bala Kalyanasundaram
Sampath Kannan
Juhani Karhumaki
Marek Karpinski
Eike Kiltz
Kwangjo Kim
Ekkehard Koehler
Yoshiharu Kohayakawa
Michael Köhler
Jochen Könemann
Daniel Kral
Dieter Kratsch
Manfred Kufleitner
Peter Kugel
Alberto Lafuente
Alair Pereira Do Lago
Tanja Lange
John Langford
Federico Lecumberry
Orlando Lee
Xiang-Yang Li
Leonid Libkin
Min Chih Lin
Claudia Linhares
Andrea Lodi
Sylvain Lombardy
Alex Lopez-Ortiz
Florent Madelaine
Frédéric Magniez
Mohammad Mahdian
Veli Makinen
Arnaldo Mandel
Yishay Mansour
Javier Marenco
Maurice Margenstern
Alvaro Martín
Conrado Martínez
Fábio Viduani Martinez
Guillermo Matera
Jacques Mazoyer
Ross McConnell
Luis A. A. Meira
Donatella Merlini
Ramgopal Mettu
Gatis Midrijanis
Peter Bro Miltersen
Lorenz Minder
Joseph Mitchell
F. K. Miyazawa
Manal Mohamed
Cristopher Moore
Ruggero Morselli
J. Ian Munro

Ashwin Nayak
Sergio Nesmachnow
Gregory Neven
Alantha Newman
Dejan Nickovic
Harald Niederreiter
Johannes Nowak
Juha Nurmonen
Michael Nüsken
José Oncina-Carratalá
Alan Oppenheim
Gianpaolo Oriolo
Luis Ortiz
Rafail Ostrovsky
Alberto Pardo
Alvaro Pardo
Rodrigo Paredes
Ojas Parekh
David Parkes
Rene Peralta
Gonzalo Perera
Dominique Perrin
Artur Pessoa
Benny Pinkas
Valentin Polishchuk
Helmut Prodinger
Fábio Protti
Mihai Prunescu
Artem Pyatkin
Charles Rackoff
Rajmohan Rajaraman
Rajeev Raman
Michael Rao
Ivan Rapaport
André Raspaud
Bruce Reed
Omer Reingold
Pablo Rey
Franco Robledo
Virgínia M. Rodrigues
Andrea Rodriguez
Frank Ruskey
Alexander Russell
Michel Saks
Juan Jose Salazar
Matias Salibian-Barrera
Nicolas Schabanel
Jay Sethuraman
Ronen Shaltiel
David Shmoys
Jamshid Shokrollahi
Igor Shparlinksi
Martin Skutella
Adam Smith
Siang W. Song
Gregory Sorkin
Jeremy Spinrad
Rob van Stee
Lorna Stewart
Nicolas E. Stier-Moses
Paul K. Stockmeyer
Maxim Sviridenko
Mario Szegedy
Jayme L. Szwarcfiter
Arie Tamir
Dimitrios Thilikos
Christopher Thraves
Ioan Todinca
Vilmar Trevisan
Akio Tsuneda
Mario Valencia-Pabon
Gabriel Valiente
Brigitte Vallee
Eric Vigoda
Jesús Villadangos
Yngve Villanger
Jan Vondrák
Gustavo Vulcano
Michael Wagner
John Watrous
Marcelo Weinberger
Renato Werneck
Jiri Wiedermann
Thomas Wilke
Ryan Williams
Erik Winfree
Prudence Wong
Derik Wood
Moti Yung
Paula Zabala
Fangguo Zhang
Yong Zhang

Table of Contents

Keynotes

Regular Contributions

Algorithmic Challenges in Web Search Engines

Ricardo Baeza-Yates[1,2]

[1] Center for Web Research, Dept. of Computer Science,
Universidad de Chile, Santiago, Chile

[2] ICREA Professor, Dept. of Technology, Universitat Pompeu Fabra,
Barcelona, Catalunya, Spain

ricardo@baeza.cl

Abstract. In this paper we present the main algorithmic challenges that large Web search engines face today. These challenges are present in all the modules of a Web retrieval system, ranging from the gathering of the data to be indexed (crawling) to the selection and ordering of the answers to a query (searching and ranking). Most of the challenges are ultimately related to the quality of the answer or the efficiency in obtaining it, although some are relevant even to the existence of search engines: context based advertising.

1 Introduction

The Web is the largest public collection of data, and therefore, Web search has become one of the main challenges in the field of information retrieval. The complexity is not only due to its volume, but also because of its dynamics and heterogeneity. In addition, as search engines are (still) free as a consequence of Web advertising, the choice and placement of advertisements in the answer page is crucial to their survival. For all these reasons, we believe that Web retrieval is one of the main sources for interesting and challenging applied algorithmic problems.

A Web search engine has basically four software modules around an index, as shown in Figure 1. We know detail this simplified software architecture. The Crawling module brings new or updated pages to the Indexing module. The Indexing module creates a compact searchable index with key preprocessed information for page ranking. The Searching module, using the index, finds a ranked answer to a stream of queries from remote users. Finally, the Answering module creates the answer page and places the right advertisement related to a query. Each of these modules presents a different set of challenges, which motivate this paper.

The next sections summarize the main algorithmic challenges of the software modules mentioned before, using simplified versions of each problem so that we can (more or less) formalize them. We include recent results, although the bibliography is by no means exhaustive. The final section mentions additional problems related to the Web[1].

[1] Disclaimer: The choice of problems is biased to the preferences of the author who is now at Yahoo! Research.

J.R. Correa, A. Hevia, and M. Kiwi (Eds.): LATIN 2006, LNCS 3887, pp. 1–7, 2006.

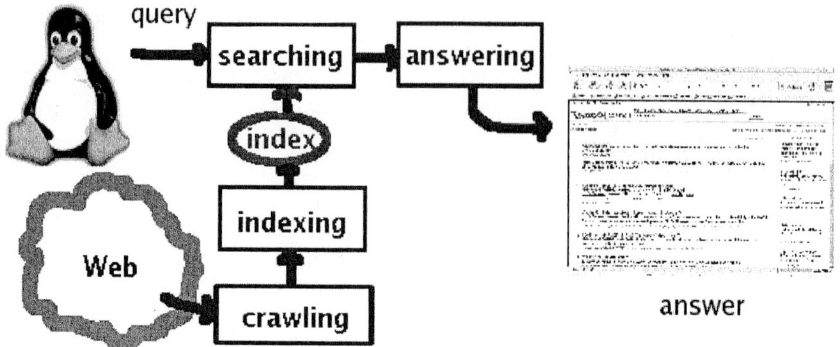

Fig. 1. Simplified software architecture of a Web search engine

2 Crawling

A crawler is a software module that gathers Web pages to create an index, typically using a parallel and distributed architecture. In practice a crawler never stops as the Web keeps growing, but we simplify the problem by limiting the crawling time and scope.

Given a set of Web sites with their bandwidth W, a period of time T, a set of politeness rules P [14], a set of resources R (computers, crawler bandwidth, etc.), and three functions V, Q and F (over a set of pages), a crawler has to bring a collection of pages $C \in W$, achieving the following four main goals:

- maximize the volume $V(C)$ of the pages,
- maximize the quality $Q(C)$ of the pages,
- minimize the freshness $F(C)$ of the pages, and
- maximize the use of R while satisfying the politeness rules P.

Part of the problem is how to define the three optimization functions and how to combine them to obtain the best possible collection C that should imply the best possible index. Each possible choice brings new problems. It is out of the scope of this paper to present an exhaustive list of possibilities, but here are a few examples:

- V could be the number of different pages or the total number of bytes of text. The former brings another problem, detection of duplicated pages, while the latter raises the question of how many bytes (or percentage) are needed to have a good textual description of a page;
- Q could be based on the distribution of words or the link structure of the Web (e.g. Pagerank [17]); and
- F could be the absolute time difference between the time when the page was crawled and the last modification time. However this raises the problem of measuring such function, as we cannot know its value until the end. Hence, usually F is an estimation of the freshness.

Considering the above functions as given, we could define an overall function to optimize such as $f(C) = \alpha V(C) + \beta Q(C) - \gamma F(C)$, for some $\alpha, \beta, \gamma > 0$.

Regarding R and P, we can simplify R as the maximum bandwidth available, B, and P as a minimum number of seconds, s, the wait time between any two page requests to the same site. Notice that being polite to Web sites opposes the goal of using all the bandwidth available at the search engine side.

Then we have a formal scheduling problem: find a sequence of requests for complete pages[2] at given times $(p_1, t_1), \cdots, (p_n, t_n)$, $n > 0$, such that we maximize $f(C)$ $(C = \cup_{i=1}^{n} p_i)$, satisfying the following:

- Crawling period: $t_n - t_1 \leq T$,
- Politeness: $|t_i - t_j| \geq s$ for any pair of pages p_i, p_j that belong to the same Web site, and
- Overall bandwidth: for any time τ $(t_1 \leq \tau \leq t_n)$, $b(W_\tau) \leq B$; where $b(W_\tau)$ is the bandwidth of the set of active requests at a given time τ. A request (p_i, t_i) is active if p_i belongs to site $w_i \in W_\tau$, and if $t_i \leq \tau \leq t_i + size(p_i)/b(w_i)$ ($b(w_i)$ is the bandwidth of the site w_i[3]).

In the open scope case, finding new pages implies to find modified pages having new links, and that implies wasting time revisiting known pages. Hence, freshness opposes volume, given that wasted time cannot be used to crawl new pages. But, paradoxically, the number of new pages only increase by wasting that time. Another practical issue is dynamic pages, which can be unbounded.

Several heuristics have been used, from breadth-first to ordering pages based on quality. Recently, a strategy based on Web site sizes (largest sites first, LSF) has shown to be competitive with strategies that use more information [3]. One problem is that there is no standard benchmark to compare crawling strategies, given that we would need the same Internet location for all the experiments.

3 Indexing and Searching

These two modules are interdependent, as the search time and the quality of the ranking will depend on the information stored in the index. Hence we present them as one integrated problem.

The best index for searching words up to now is an inverted index, one of the oldest data structures [1]. An inverted index basically consists of a set of unique words (vocabulary) and a set of corresponding lists of pages where each word occurs. However, better solutions may exist, in particular given the new conditions imposed by the Web: smaller but distributed indexes, frequent updates, and fast answer time, to name a few.

The index has to contain pre-calculated information that will be useful when ranking the answer. This information depends on the document similarity model

[2] In practice could be partial pages, and in that case we also have to add to the request how many bytes to bring after we know the total size of the page.

[3] We assume that the bandwidth for each site is constant, which in practice is not true.

used by the search engine and the query language used (they are not independent). Examples would be the vector model [1] that uses information on term distribution or a link based model such as Pagerank [17] or global Authorities [12]. Nowadays, search engines use many sources of information for ranking: text content, link structure, search engine usage, etc. One of the main problems is the evaluation of the quality of the ranking, as we do not know which are the best answers. Current evaluation techniques are based on click-through analysis (that is, how people click on the answer pages).

Another restriction is the current Web volume, which implies that the index must be distributed across many machines. It also implies a partial evaluation of the query to give a fast response (in addition, as people on average looks at two answer pages, does not make sense to do additional work). If we add to this the current query volume, we need a parallel processing of the queries to increase the concurrency level of the overall retrieval system.

Hence, we have a variant of the dictionary problem: design a dynamic data structure (the index) that, given a maximum space available M, achieves at least a throughput T of queries per second, satisfying the following requirements:

- any query must be solved using only the index (accessing secondary memory or a remote page is too slow);
- the index should be easy to distribute across many processors without making the search more difficult;
- the index should be easy to update frequently, maintaining its quality and speed; and
- the query can be partially evaluated to find the best B ranked answers.

Almost all the requirements just mentioned imply some amount of extra information to be stored in the index. Hence, compression is an important element of current solutions. This can be approached from two directions: design a compressed searchable index, or, design a compression technique that allows fast searching [15]. In both cases, being able to search without the need of decompressing the index, improves the search time. In fact, these is one of the few cases where we can do faster search by using less space. Fast querying implies other interesting subproblems, such as fast computation of set operations (e.g. see [5]).

Another of the most interesting subproblems are distributed indexes. For inverted indexes, there are two ways to distribute the index:

- *document partitioning* splits the document collection in pieces and builds one partial index for each piece. Searching is achieved by merging partial answers from the processors that store the partial indexes.
- *term partitioning* splits the vocabulary in pieces, and each processor holds one subset of the index, and hence, one part of a global inverted index. Searching is achieved by merging the answers for every word in the query.

Current search engines use the former, as the main problem of the latter is that of building and maintaining a global index. However, new ideas may change in the future this choice, as term partitioning allows higher concurrency.

4 Advertising

Deciding the right advertising is one of the main tasks of the Answering module. Context based advertising comes in two flavors: advertising shown after a query in the answer page of a search engine (e.g. Google's AdWords) or the advertising shown in a syndicated page (e.g. Yahoo's ContentMatch). The differences are the data available for matching the advertising: in one case a query and all its attributes, and in the second case the content of a page and the referrer information of the visitor to that page (which in some cases could include a query); and the number of places available (around ten in the former, two or three in the latter). In Figure 2 we present an example for the first case.

Advertisers pay for each visit to a given site (this is called pay per click, PPC), so the search engine is interested in maximizing the future income. Clicks depend on the position of the ad, so the advertisers should be ranked. However, the choice of advertisers it is not as simple as choosing the ones that match and that would pay more for a click, as some advertisers are more clicked than others. Also, we cannot use click frequency as a definite rule, because advertisers that have had more exposition time, would have more clicks and new advertisers would then never appear, without having the chance to become popular and profitable.

Then we have an on-line problem: given a set of evidences E regarding a page p, and a database of advertisers A, find a ranked subset $a \in A$ that maximizes the expected income from clicks in p, such that $|a| = n$, where n is the number of places available in p.

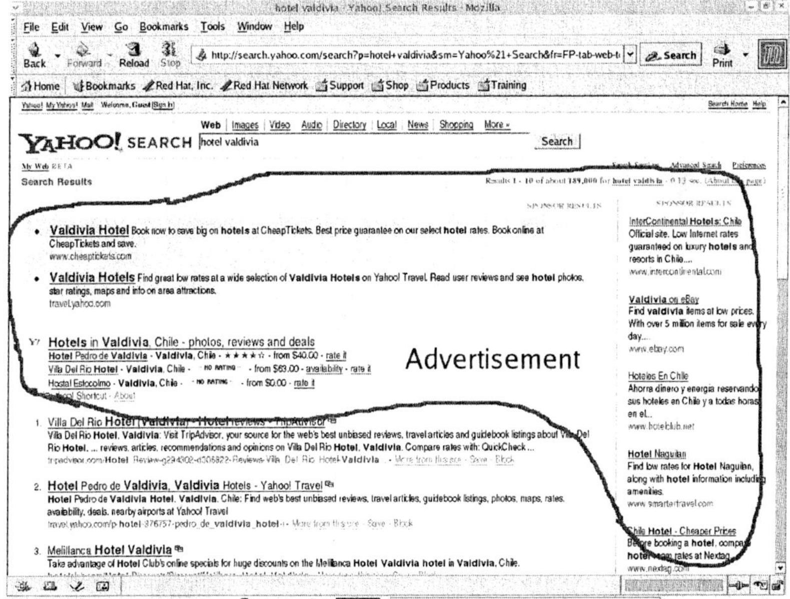

Fig. 2. Example of keyword-based advertisement for the query "hotel valdivia"

Notice that we do not separate the matching part (find advertisers that satisfy the evidence) from the ranking part (find advertisers that will maximize the future income), so E includes the data available for the match, past history, etc. There are few results for this problem. For example [18] explores the matching part, while [9, 16] the placement part.

Another problem in the Answering module is the fast generation of the text summary (*snippets*) for each result in the answer page.

5 Concluding Remarks

We have briefly surveyed the algorithmic challenges of Web search. Solutions to these theoretical problems can help to find solutions to real ones, as in practice the problems are much more complex. For example, currently the Web has many forms of spam, including content, links and usage. So, finding the bad guys (e.g. pages that have misleading content, links that are used just to improve the ranking of the linked page, or clicks that come from malicious software agents) is an interesting dynamic problem. A possible easier solution could be to help the good guys, that is, the Web sites that do have good content and good links. Still, we have the problem of recognizing the good from the bad, which is related to the problem of information trust in the Semantic Web. This area is now called adversarial information retrieval(e.g. [11]).

When we search we know what we are looking for. However, in the Web there could be interesting answers waiting to be discovered or interesting usage patterns that could help to improve a search engine. These are two examples of Web mining [10], a field that is still in its infancy. One interesting case is queries and the user actions after a query, or query mining [6]. When people formulate queries and click on answers, they are giving away "semantic information" for free, and with the current volume of queries per day in a search engine, the potential of this data is still unknown. For example, it could be used for better index design, better ranking, query optimization, query recommendation [4], generation of pseudo-semantic resources, Web site design [7], to name a few. We are currently building a platform to formalize Web mining tasks [8].

Finally, advertising is related to two newer fields: social networks [19] and Internet economy [20]. The intersection of these two fields will sparkle many interesting problems such as query incentive networks [13] or auction pricing.

Acknowledgments

We thank the helpful comments of David Nettleton.

References

1. R. Baeza-Yates and B. Ribeiro-Neto, *Modern Information Retrieval*, Addison-Wesley, England, 513 pages, 1999.
2. R. Baeza-Yates. Information Retrieval in the Web: beyond current search engines, Int. Journal of Approximate Reasoning 34 (2-3), 97-104, 2003.

3. R. Baeza-Yates, C. Castillo, M. Marin, A. Rodriguez. Crawling a Country: Better Strategies than Breadth-First for Page Ordering. In *WWW 2005, Industrial Track*, ACM Press, Chiba, Japan, May 2005.
4. Ricardo A. Baeza-Yates, Carlos A. Hurtado, Marcelo Mendoza. Query Recommendation Using Query Logs in Search Engines, Current Trends in Database Technology - EDBT 2004 Workshops, Workshop on Clustering Information over the Web, Heraklion, Crete, Greece, March 14-18, 2004, Revised Selected Papers. LNCS 3268, Springer, p. 588-596.
5. R. Baeza-Yates, A Fast Set Intersection Algorithm for Sorted Sequences, In *15th Combinatorial Pattern Matching 2004*, LNCS, Springer, Istanbul, Turkey, July 2004.
6. Ricardo Baeza-Yates. Applications of Web Query Mining. In *European Conference on Information Retrieval (ECIR'05)*, D. Losada, J. Fernández-Luna (editors), Springer LNCS 3408, Santiago de Compostela, Spain, March 2005, 7–22.
7. Ricardo Baeza-Yates, Barbara Poblete, A Website Mining Model Centered on User Queries, European Web Mining Forum, B. Berendt *et al*, editors. October 2005, Oporto, Portugal, p. 3-15.
8. R. Baeza-Yates, A. Pereira and N. Ziviani, WIM: A Web Information Mining Model for the Web, In *LA-WEB 2005*, IEEE CS Press, Oct 2005, 233-241.
9. H. K. Bhargava and J. Feng. Paid placement strategies for internet search engines. In Proceedings of the eleventh international conference on World Wide Web, pages 117-123. ACM Press, 2002.
10. S. Chakrabarti. Mining the Web: Discovering knowledge from hypertext data, Morgan Kaufmann, 2003.
11. B. Davison, editor. Workshop on Adversarial Information Retrieval on the Web. URL: http://airweb.cse.lehigh.edu/2005/. Chiba, Japan, May 2005.
12. J. Kleinberg. Authoritative sources in a hyperlinked environment. *Journal of the ACM*, 46(5):604–632, 1999. Preliminary version presented at SODA 1998.
13. J. Kleinberg and P. Raghavan. Query Incentive Networks. Proc. 46th IEEE Symposium on Foundations of Computer Science, 2005.
14. M. Koster. A standard for robot exclusion. http://www.robotstxt.org/wc/exclusion.html, 1996.
15. V. Makinen and G. Navarro. Compressed Full Text Indexes. Technical Report TR/DCC-2005-7, Dept. of Computer Science, University of Chile, June 2005. Available at http://pizzachili.dcc.uchile.cl/biblio.html
16. S. Nicholson, T. Sierra, U.Y. Eseryel, J.H. Park, P. Barkow, E.J. Pozo, J. Ward. How Much of It is Real? Analysis of Paid Placement in Web Search Engine Results, JASIST, 2005.
17. L. Page, S. Brin, R. Motwani, and T. Winograd. The Pagerank citation algorithm: bringing order to the web. Technical report, Stanford Digital Library Technologies Project, 1998.
18. B. Ribeiro-Neto, M. Cristo, P. Golgher, and E. Silva de Moura. Impedance coupling in content-targeted advertising. In Proceedings of the 28th Annual international ACM SIGIR Conference on Research and Development in information Retrieval (Salvador, Brazil, August 15 - 19, 2005). SIGIR '05. ACM Press, New York, NY, 496-503, 2005.
19. Barry Wellman. Computer Networks As Social Networks, *Science* 293 (5537), p. 2031-2034, 2001.
20. A.C-C. Yao, editor. First Workshop on Internet and Networks Economy, WINE 2005. URL: http://www.cs.cityu.edu.hk/ wine2005/, Hong-Kong, 2005.

RNA Molecules: Glimpses Through an Algorithmic Lens

Anne Condon

The Department of Computer Science,
U. British Columbia, Canada
condon@cs.ubc.ca

Dubbed the "architects of eukaryotic complexity" [8], RNA molecules are increasingly in the spotlight, in recognition of the important catalytic and regulatory roles they play in our cells and their promise in therapeutics. Our goal is to describe the ways in which algorithms can help shape our understanding of RNA structure and function.

Computational means for prediction of the structure of RNA and DNA molecules – collectively known as nucleic acids – are invaluable in determining the functions of molecules in the cell. Structure prediction problems, as well as the inverse problem of designing nucleic acids with specific structural properties, also arise in biological research aimed at creating new catalysts and biosensors, in nanotechnology, and in efforts to recreate an RNA world that may have preceded modern life [11].

Put simply, a DNA or RNA molecule is a sequence of units, called bases, over a four-letter alphabet. Prediction of nucleic acid structure is easier than prediction of protein structure, because the primary forces that determine nucleic acid structure are pairings (bonds) between individual bases of the molecule, with each base in at most one pair. This set of base pairs is called the *secondary structure* of the molecule. One premise is that, of the exponentially many possibilities, an RNA molecule folds into that secondary structure which has *minimum free energy* (MFE) [7]. Finding the MFE structure for a given RNA molecule is NP-hard [6]. However, the range of structures that arise in nature is relatively limited, making it feasible to find MFE predictions of almost all naturally-occurring secondary structures in polynomial time [10]. There is still much potential to advance the state of the art in MFE secondary structure prediction of nucleic acids.

- The algorithm of Rivas and Eddy [10] is very general in terms of types of structures that it can predict [3], but with $\Theta(n^6)$ running time is limited to relatively short inputs. One challenge is to find the sweet spot in the trade-off between algorithmic generality and efficiency. A concrete goal is to find MFE "kissing hairpin" structures in less than $O(n^6)$ time.
- MFE prediction algorithms can only be as good as their underlying energy models. While experimental wet-lab work has provided hundreds of high-quality parameters for use in RNA structure prediction [7], some model features, including multi-loops and pseudoknots, have been parameterized in

J.R. Correa, A. Hevia, and M. Kiwi (Eds.): LATIN 2006, LNCS 3887, pp. 8–10, 2006.

a somewhat ad-hoc way, and indeed the model features themselves may have been chosen in part for algorithmic efficiency, rather than accuracy. Thus, we believe that a reassessment of the energy model, informed by physical principles, known nucleic acid structures, and algorithmic complexity, should be fruitful in improving the quality of prediction algorithms.

Stepping back from our focus on MFE prediction from a single sequence, we note that there are many other interesting computational problems relating to nucleic acid structure prediction. Minimum free energy prediction of complexes of two or more molecules is an important goal that arises, for example, in determining the targets of anti-sense RNA's [1]. Moreover, partition function prediction provides additional useful information, including base pairing probabilities [5]. In a different direction, the premise that molecules fold into their MFE structures may be false for some structures, when "kinetic traps" - low energy structures with no low-energy paths to a MFE structure - exist, or when folding occurs co-transcriptionally [9]. Thus, also important are alternative structure prediction approaches [9] and efficient simulation of folding kinetics [12, 13]. The problem of predicting three-dimensional structure is largely unsolved. Finally, although good heuristics have been developed for the design problem, which is to determine a sequence that folds into a given input secondary structure [2, 4], its computational complexity is still open.

References

1. C. Alkan, E. Karakoç, J. Nadeau, S.C. Sahinalp, and K. Zhang. "RNA-RNA interaction prediction and antisense RNA target search", Proc. Ninth Annual Intl. Conf. on Research in Computational Molecular Biology (RECOMB), 152–171, 2005.
2. M. Andronescu, A.P. Fejes, F. Hutter, H.H. Hoos, and A. Condon. "A new algorithm for RNA secondary structure design", *J. Mol. Biol.*, 336(3):607–624, 2004.
3. A. Condon, B. Davy, B. Rastegari, F. Tarrant, and S. Zhao. "Classifying RNA pseudoknotted structures", *Theoretical Computer Science*, 320(1):35–50, 2004.
4. R.M. Dirks, M. Lin, E. Winfree and N.A. Pierce. "Paradigms for computational nucleic acid design", *Nucleic Acids Res.*, 32(4):1392–1403, 2004.
5. R.M. Dirks and N.A. Pierce. "A partition function algorithm for nucleic acid secondary structure including pseudoknots", *J. Comput. Chem.*, 24(13):1664–1677, 2004.
6. R.B. Lyngsø and C.N.S. Pedersen. "Pseudoknot prediction in energy based models", *J. Comput. Biol.*, 7:3409–427, 2000.
7. D.H. Mathews, J. Sabina, M. Zuker and D.H. Turner. "Expanded sequence dependence of thermodynamic parameters improves prediction of RNA secondary structure", *J. Mol. Biol.*, 288:911–940, 1999.
8. J.S. Mattick. "Non-coding RNAs: the architects of eukaryotic complexity", EMBO (European Molecular Biology Organization) Reports, 2(11):986–991, 2001.
9. I.M. Meyer and I. Miklos. "Co-transcriptional folding is encoded within RNA genes", *BMC Molecular Biology*, 5:10, 2004.
10. E. Rivas and S. R. Eddys. "A dynamic programming algorithm for RNA structure prediction including pseudoknots", *J. Mol. Biol.*, 225:2053–2068, 1999.

11. J.W. Szostak, D.P. Bartel, and L. Luisi. “Synthesizing life”, *Nature*, 409:387–389, 2001.
12. X. Tang, B. Kirkpatrick, S. Thomas, G. Song, and N.M. Amato. “Using Motion Planning to Study RNA Folding Kinetics”, *J. Comput. Biol.*, 12(6):862–881, 2005.
13. A. Xayaphoummine, T. Bucher, and H. Isambert. “Kinefold web server for RNA/DNA folding path and structure prediction including pseudoknots and knots”, *Nucleic Acid Res.*, 33:605–610, 2005.

Squares*

Ferran Hurtado

Departament de Matemàtica Aplicada II,
Universitat Politècnica de Catalunya

In this talk we present several results and open problems having squares, the basic geometric entity, as a common thread. These results have been gathered from various papers; coauthors and precise references are given in the descriptions that follow.

Disassembling Sets of Colored Squares

Given a set of disjoint convex objects in the plane, it is well known that they can be moved to infinity without collision, one at a time, using only translations in the direction given by any vector $\boldsymbol{v}$. If we have convex objects that have two different colors it is always possible to obtain two vectors $\boldsymbol{v}_1$ and $\boldsymbol{v}_2$, forming possibly an infinitesimally small angle, such that each direction is used for one of the colors and the objects in their final far away positions are well separated, say by a line. An infinitesimally small angle is not quite satisfactory and it is natural to wonder whether a larger *separating angle* independent of n may always be obtained. This is precisely the problem we have considered in [5]: we study in which cases the separating angle can be guaranteed to be bounded by below, and consider also the similar problem for c colors.

Somehow surprisingly, the shape of the objects happens to be a crucial issue; for example, any c-colored set of isothetic squares can be separated using separating vectors such that the angle between any two of them is at least $\pi/(2c-2)$, while for disks the situations is quite different, as the angle between separating vectors may be required to be arbitrarily small. We will discuss as well some algorithmic issues on these problems.

Matching with Squares

Let $\mathcal{C}$ be a class of geometric objects and let P be a point set with n elements $p_1, \ldots, p_n$ in general position, n even. A $\mathcal{C}$-*matching* of P is a set $M = \{C_1, \ldots, C_k\}$ of elements of $\mathcal{C}$, such that every C_i contains exactly two elements of P. If all the elements of P belong to some C_i, M is called a *perfect matching.* If in addition all the elements of M are pairwise disjoint we say that the matching M is *strong.* If we define a graph $G_{\mathcal{C}}(P)$ in which the vertices are the elements of P, two of which are adjacent if there is an element of $\mathcal{C}$ containing them and no other element from P, a perfect matching in $G_{\mathcal{C}}(P)$ in the graph theory sense corresponds naturally with our definition of $G_{\mathcal{C}}(P)$-matchings.

* Research partially supported by Projects MCYT-FEDER BFM20033-0368 and Gen. Cat 2005SGR00692.

J.R. Correa, A. Hevia, and M. Kiwi (Eds.): LATIN 2006, LNCS 3887, pp. 11–12, 2006.

If C is the set of all isothetic squares, M will be called a *square-matching* and the graph $G_C(P)$ is the Delaunay graph for the L_1 (or the L_∞) metric. We prove that there is always a square-matching, which translates to these graphs to contain a perfect matching. In fact we have obtained a stronger result, namely that these graphs contain a Hamiltonian path, a question that remained unsolved since it was posed by Dillencourt [4], who proved the existence of perfect matchings in Delaunay triangulations for the Euclidean metric, i.e., that point sets always admit circle-matchings. On the other hand, the problem of deciding whether a point set admits a strong square-matching has recently been proved to be NP-hard [2].

The results we present were developed in [1], where this class of problems was introduced and studied on the light of *geometric matchings*.

Tiny Squares

We call a function $I : \mathbb{Z}^2 \to \{0, 1\}$ a *binary image*. We call the elements of $\mathbb{Z}^2$ *pixels* and we say that a pixel p is *black* (respectively, *white*) if $I(p) = 1$ (respectively, $I(p) = 0$). Taking the pixels as vertex set and the natural definitions of 4-*neighbourhood* and 8-*neighbourhood* the graphs G_4 and G_8 are obtained. For a given image I, its black and white pixels induce subgraphs that we denote $B_4(I)$ and $W_4(I)$, and $B_8(I)$ and $W_8(I)$ respectively. For $a, b \in \{4, 8\}$ we say that an image I is $\mathsf{B}_a, \mathsf{W}_b$*-connected* if the graphs $B_a(I)$ and $W_b(I)$ are each connected, that is, each has a single connected component.

In the paper [3] we consider for both graphs a local modification operation on binary images in which a black pixel p and a white pixel q can interchange their colours when they are neighbours, and we prove that, for any $(a, b) \in \{(4, 4), (4, 8), (8, 4), (8, 8)\}$, any two $\mathsf{B}_a, \mathsf{W}_b$-connected images I and J each with n black pixels can be converted into the other with a sequence of $O(n^2)$ 8-local interchanges if $(a, b) \in \{(4, 8), (8, 4), (8, 8)\}$ and $O(n^4)$ 8-local interchanges if $(a, b) = (4, 4)$.

References

1. Bernardo Ábrego, Esther Arkin, Silvia Fernández, Ferran Hurtado, Mikio Kano, Joseph Mitchell and Jorge Urrutia *Matching points with geometric objects*, Proc. Japan Conf. on Discrete and Computational Geometry 2004, LNCS Springer-Verlag 3742, pp.1-15, 2005.
2. Sergey Bereg, Nikolaus Mutsanas and Alexander Wolff, *Matching points with rectangles and squares*, Proc. 32nd Int. Conf. on Current Trends in Theory and Practice of Computer Science, SOFSEM'06, January 2006, Merin, Czech Republic.
3. Prosenjit Bose, Vida Dujmovic, Ferran Hurtado and Pat Morin, *Connectivity-Preserving Transformations of Binary Images*, submitted for publication.
4. Michael Dillencourt, *Toughness and Delaunay Triangulations*, Discrete and Computational Geometry, 5(6), 1990, 575-601. Preliminary version in Proc. of the 3rd Ann. Symposium on Computational Geometry, Waterloo, 1987, 186-194.
5. Ferran Hurtado and Mikio Kano, *On the translational separation of colored convex objects*, Proc. XI Encuentros de Geometría Computacional, Santander, 2005, pp. 225-230.

Matching Based Augmentations for Approximating Connectivity Problems*

R. Ravi

Tepper School of Business, Carnegie Mellon University,
Pittsburgh, PA 15213
ravi@cmu.edu

Abstract. We describe a very simple idea for designing approximation algorithms for connectivity problems: For a spanning tree problem, the idea is to start with the empty set of edges, and add matching paths between pairs of components in the current graph that have desirable properties in terms of the objective function of the spanning tree problem being solved. Such matching augment the solution by reducing the number of connected components to roughly half their original number, resulting in a logarithmic number of such matching iterations. A logarithmic performance ratio results for the problem by appropriately bounding the contribution of each matching to the objective function by that of an optimal solution.

In this survey, we trace the initial application of these ideas to traveling salesperson problems through a simple tree pairing observation down to more sophisticated applications for buy-at-bulk type network design problems.

1 Introduction

Approximation algorithms have been traditionally designed and taught on a problem-by-problem basis; Surveys (e.g., [6]) and recent courses and books (e.g., [13]) have approached the area in this way by mainly classifying key results based on a problem-specific basis. As the field matures to provide a rich variety of results, commonalities can be identified to highlight key techniques that become repeatedly useful.

In this survey, we point to one such extremely simple technique that we term MBA, an acronym for Matching Based Augmentation. The two salient features that determine the applicability of the method are that the problem at hand must be a connectivity problem where one tries to connect up various demands (either among themselves or to a common root) in a network, and that the optimal solution can be used to identify an appropriate polynomial-time solvable augmenting subproblem that is a variant of matching. Since the method proceeds by finding such matching iteratively and adding them to augment the solution, the approximation ratio is typically bounded by the number of iterations of the process; Furthermore, since the cost paid by the augmentation

* Supported in part by NSF grant CCF-0430751 and ITR grant CCR-0122581 (The ALADDIN project).

J.R. Correa, A. Hevia, and M. Kiwi (Eds.): LATIN 2006, LNCS 3887, pp. 13–24, 2006.

in each iteration is bounded with respect to the optimal, the method of proof of the performance ratio is primal-based, without relying on any new (lower) bounds on the (minimization) problem to argue the guarantee. Finally, since each augmentation proceeds by matching up components of the current solution, the number of iterations before there is one single component and hence a feasible solution, is logarithmic in the number of demand points that need to be connected. This explains why most approximation algorithms based on this method have logarithmic performance ratio.

We have structured this survey chronologically by describing the applications of the method in the order of their first (typically conference or technical report) publication. In this order, the classic paper of Christofides [2] is the first paper of the sequence to contain most features of the MBA idea: the missing idea is the iterative augmentation. The ATSP approximation of Frieze et al. [3] uses the MBA idea in it complete form to obtain a logarithmic approximation for metric ATSPs. We review these two results in the next section. In the following section, we trace our own work in a series of papers [7, 10, 12, 8, 11] that use this idea in various contexts for NP-hard undirected spanning tree problems. In the next section, we review some more sophisticated uses of the method to solve generalizations of basic connectivity problems so as to route flow under concave cost functions [9, 1, 5]. We close by summarizing the method.

2 The Early Applications

The roots of the matching based augmentation method can be traced back to Christofides' $\frac{3}{2}$-approximation algorithm for the traveling salesperson problem on undirected graphs with metric costs. Recall that in this problem, we are given an undirected (without loss of generality, complete) graph with nonnegative costs obeying the triangle inequality on the edges, and the goal is to find a TSP tour (Hamiltonian cycle that visits every vertex exactly once) of minimum total edge cost.

2.1 Christofides' Algorithm for Metric TSP

Christofides' heuristic [2] first computes a spanning tree T on the graph G. Next, we observe by a simple parity argument on the sum of all degrees in any graph that the number of odd-degree nodes is even. Applying this to the tree T, we see that the number of nodes of odd degree in T is even. We now consider the induced (complete) subgraph on only the odd-degree nodes of T and compute a perfect matching M on this (even-sized) set. Now $T \cup M$ is a connected graph of even degree, which implies that it is Eulerian. An Euler tour of this graph can be shortcut to yield a TSP solution of no higher value (using the triangle inequality property of the metric costs).

While it is clear that the MST T has cost at most that of an optimal tour, bounding the cost of M with respect to an optimal TSP tour requires a little work. Consider an optimal tour and induce it on the odd-degree nodes of T

(short-cutting over the even degree nodes of T). This tour, by the triangle inequality, has cost no more than that of the optimal solution. This induced tour is on an even sized set (by the earlier observation) and hence can be exactly decomposed into two disjoint matchings. The cheaper of these two matchings has cost at most half that of an optimal solution. This in turn upper bounds the cost of the minimum-cost matching M we found on the odd-degree nodes. Overall, the $\frac{3}{2}$ performance ratio is proved.

The key step in the algorithm is to augment the initial tree T by a matching M which can be appropriately bounded by a fraction of the cost of the optimal tour solution. In this way, this algorithm lays out the idea of augmenting a current solution with a matching the cost of which can be bounded by comparing it with an optimal solution. As we shall see, this is the underlying idea of the MBA method.

2.2 The FGM Algorithm for Metric ATSPs

Next, we consider an algorithm due to Frieze, Galbiati and Maffioli [3], henceforth referred to as the FGM algorithm for the asymmetric version of the TSP problem. In this version, a complete directed graph is given with arc costs that are not necessarily symmetric but obey the triangle inequality, and the goal is to find a traveling salesperson directed tour (that visits each vertex exactly once) of minimum total arc cost.

The FGM algorithm is a "greedy" augmentation algorithm that adds arcs to the solution in iterations. It starts with an empty graph in which each node is a singleton component. In each iteration, it adds a collection of cycles that merge these components into larger components. In particular, in the first iteration, it computes a minimum cost directed cycle cover of the nodes and adds it to the solution. This merges the nodes into cycles, and for each cycle a representative node is chosen. In the next iteration, only the induced complete digraph on the representative nodes is considered and a minimum cost cycle cover on the representative nodes is computed and added to the solution. This merges the set of representative nodes (and hence their respective components) in a cycle into a larger component. Notice that every component is strongly connected and Eulerian (every node has indegree equal to outdegree). This proceeds in every iteration by first identifying a representative node in each Eulerian component and computing a minimum cost cycle cover on these representatives to merge components into larger Eulerian components. Finally, when all nodes are in one Eulerian component, we can shortcut an Eulerian tour on all the edges into a Hamiltonian tour of no higher cost using the triangle inequality on the costs.

Two simple observations prove the performance guarantee of $\log_2 n$ for the FGM algorithm on a graph with n nodes: (i) In each iteration the Eulerian components at least halve in number; This is a simple consequence of the fact that every cycle in a cycle cover has length at least two leading to every Eulerian component merging with at least one other such component. (ii) The cost of the cycle cover added in any iteration is at most that of a minimum TSP tour; This follows as a simple consequence of the fact that the minimum TSP tour induced on the representative nodes in any iteration (and shortcut over the other nodes) is a feasible solution to the cycle cover problem for that iteration, and hence the minimum cover computed has cost no more than that of this optimal TSP solution. Putting these two observations together, we see that the approximation ratio of the FGM algorithm is bounded by the number of iterations, which in turn is at most $\log_2 n$.

The FGM algorithm has all the salient features of the MBA idea: (i) Construct the solution by iterative augmentation using a matching based routine in each iteration (Note that a cycle cover problem on a digraph $G = (V, A)$ is solved by an assignment problem on an auxiliary bipartite graph with node bipartition (V_1, V_2), each of the parts being a copy of V, and edges u_1, v_2 for every arc u, v in A). (ii) The cost of the augmenting solution in each iteration is bounded by that of the optimal by identifying the appropriate matching subproblem to solve the augmentation problem. The overall performance ratio is then proportional to the number of iterations.

3 A Tree Pairing Lemma and Its Applications

In our own work, the MBA method took shape in an unintended context, namely in deriving an approximation algorithm for the node-weighted Steiner tree problem. The conference version of our work [7] proved the performance ratio of the greedy algorithm therein via a simple pairing argument on an even number of nodes in a tree. We recall that here.

Lemma 1. *Let T be a tree and M be an even subset of the nodes of T. There exists a pairing (loosely a "matching") of the nodes of M such that the paths between the pairs in T are edge-disjoint.*

Proof: For a pair (u, v) define the length of the pair to be the number of edges (hops) in T between u and v, The pairing that minimizes the total length has the claimed property. Suppose for a contradiction, two pairs in such a pairing, say (u_1, v_1) and (u_2, v_2) have a common edge e in their paths in T: Breaking up the pairing and re-pairing them using only the paths until e results in a new pairing that reduces the total length of the resulting pairing, contradicting our choice of the pairing.

While being immaterial to our subsequent application of the above lemma, the above proof suggests a constructive method for finding such a pairing: Start with any pairing and repeatedly pick any two pairs that overlap and re-pair them until there are no more such pairs. Since the total length of the pairing reduces at each re-pairing, it is not hard to argue polynomial time termination. Other alternate algorithmic approaches that work include using a minimum length perfect matching procedure on the marked nodes.

3.1 A logarithmic Approximation for MST

We can now use the above lemma to design a simple (but somewhat ridiculous) algorithm for approximating the cost of a minimum spanning tree in an undirected graph. While there are simple (Kruskal' and Prim's) and even linear time exact algorithms for this problem, the approximation algorithm illustrates some general principles that will explain our subsequent algorithms.

The idea for the approximation is to build the spanning tree in iterations starting with the empty set of edges; The aim is to reduce the number of connected components at the end of each iteration to a constant fraction (typically half) of the number at the beginning of the iteration. For spanning trees, the simplest way to accomplish this is to ensure that every component connects with at least one other component via the edges added in a typical iteration.

How can one arrive at the polynomial time subproblem that accomplishes the component reduction but whose solution can be bounded against an optimal solution? This is the crux of applying the MBA method and the answer depends on the problem at hand.

Let's develop some common notation that will be useful for the rest of this section. Let the total number of iterations for the MBA based algorithm be denoted by τ (typically, $\tau = O(\log n)$). In iteration $t \in \{1, 2, \ldots, \tau\}$, let the set of edges added to augment the solution be denoted E_t, and let the set of connected components at the end of this iteration be denoted $\mathcal{C}_t$ with the connected components labeled $C_t(1), C_t(2), \ldots, C_t(k_t)$, where k_t is the number of connected components in $\mathcal{C}_t$. For example, $\mathcal{C}_0 = V$ with $k_0 = |V| = n$, while $\mathcal{C}_\tau$ is one single connected component with $k_\tau = 1$.

To return to the question about the subroutine to employ at each iteration, we reason as follows: Consider an optimal MST, T^* say, and at the start of iteration $t+1$, we look at the components of $\mathcal{C}_t$ and contract them to supernodes in T^*. The edges of T^* now form a potentially cyclic set of edges with some self loops and multiedges on the node set $\mathcal{C}_t$. We can remove cycles (and self-loops) to finally get a tree (call it $T^*(t)$) on this set of supernodes that use only edges of T^* and hence of total cost no more than the optimum. Now we can apply the tree pairing lemma to $T^*(t)$ (Assume for now that the number of supernodes in $T^*(t)$ is even for otherwise we can omit an arbitrary supernode). The tree-pairing lemma shows how the supernodes can be paired off using edges of T^* and be connected between these pairs. The resulting matching problem that can be used to solve the resulting connection problem is to connect each component of $\mathcal{C}_t$ with another at minimum total cost of all such pairwise connections. Note that even though the original costs may not be metric, we can use a metric completion between supernodes in solving this matching problem: Indeed, if an edge used in the matching is not a direct edge but one in the metric completion, we can use the path of this cost to connect the two endpoints, satisfying the connectivity feasibility requirement of this iteration.

To summarize, in iteration $t+1$, we compute the metric completion of the supernodes in $T^*(t)$ and solve a minimum cost perfect matching problem (assuming the number of supernodes in it is even). For every edge in the matching, we add the path in the graph of this cost during this iteration. The following two lemmas are now immediate.

Lemma 2. *The number of iterations of the MBA-based algorithm is* $O(\log n)$.

The proof follows from the observation that all but one component are paired off in every iteration this reducing the number of components in any iteration by at least a fraction of $\frac{2}{3}$. Starting with $|V| = n$ components, the number of iterations is bounded as above.

Lemma 3. *The cost of edges added at every iteration at most that of an optimal solution T^*.*

The proof of this lemma uses the metric completion on the components of $\mathcal{C}_t$ and using the induced solution $T^*(t)$ and the tree-pairing lemma on it, identifies a matching of cost at most $T^*(t)$ that pairs up the components. Since a minimum cost perfect matching subroutine finds such a pairing of minimum cost, its cost is no more that that of $T^*(t)$ as stated.

Putting the above two lemmas together and observing that at the end of the last iteration, we have added a set of edges that form a single connected component, we can delete edges as required to get a final spanning tree of cost no more than the number of iterations times that of T^*. Along with the observation that the subproblem we set up at each iteration is polynomial-time solvable we have the following theorem.

Theorem 1. *The MBA-based algorithm using a minimum cost perfect matching subroutine at each iteration outputs a spanning tree of total cost $O(\log n)$ times the minimum.*

Since the tree pairing lemma works only on a subset of nodes, the results in the following sections all apply to finding Steiner trees that connect a subset of the nodes (called terminals) rather than the whole node set as in a spanning tree. We restrict our discussion to spanning trees for the sake of simplicity and reduced notation, but note that the $O(\log n)$ factors in the treatment below is typically reduced to $O(\log k)$ where k is the number of terminals in the Steiner tree problem.

3.2 Degree Bounded MSTs

The first problem using the MBA framework is the degree-constrained minimum spanning tree problem: Given integer degree budgets $B_v > 0$ for every vertex v of an undirected graph with nonnegative edge costs, the goal is to find a spanning tree of minimum total cost obeying all the degree bounds (if it exists), i.e., the degree of node v in the tree is at most B_v. This problem generalizes minimum-cost TSP paths by setting the budget to one at the endpoints and two elsewhere. Furer and Raghavachari [4] used a matching based approach to derive the first approximation algorithm for a special case of the problem with all edge costs being either one or infinity (the unweighted graph case), and the solution output by their method used a degree-constrained subgraph subroutine to get an $O(\log n)$ approximation ratio for all the degree budgets simultaneously (i.e., if B_v is feasible for all v, their solution has degree $O(\log n \cdot B_v)$ at v for all v. Their algorithm can be seen as an early application of the MBA method.

The degree constrained MST problem was first addressed in our work [10] where the tree-pairing lemma was used to identify a matching subproblem to connect up components in each iteration [1].

The algorithm in [10] follows the same outline as that for MSTs in the previous subsection. The subroutine at each iteration must be tailored to add a subgraph that induces degree no more than about B_v at any node v, and has cost no more than that of an optimum solution, while merging components in pairs. The resulting matching problem turns out to be a bit more sophisticated than that for MST as expected since it handles two different objectives, namely node degrees and edge costs. The subroutine builds a bipartite graph with the original nodes on the left part and the current connected components $\mathcal{C}_t$ on the right part. Original graph edges are duplicated to go between each vertex endpoint on the left part to the component on the right part containing the other endpoint. The subroutine is now to choose a minimum cost set of edges that have at least one edge leaving every component (on the right part) but have degree at most say $2B_v$ at any node v on the left part. The tree pairing lemma guarantees that the paths between the pairings induce degree at most twice the original degree. While the counterpart of Lemma 2 is immediate, we have the following version of Lemma 3.

Lemma 4. *The cost of edges added at every iteration at most that of an optimal solution T^*, while the degree added to any node v in any iteration is at most $2B_v$.*

Finally we get the following theorem.

Theorem 2. *[10] The MBA-based algorithm using a minimum-cost degree-constrained subgraph subroutine at each iteration outputs a spanning tree of total cost $O(\log n)$ times that of a minimum cost tree obeying the degree bounds; Moreover, the spanning tree output has degree at most $O(\log n \cdot B_v)$ at node v for all vertices v.*

3.3 Diameter Bounded MSTs

Next, we turn to a "cost-diameter" version of the MST problem: Given a nonnegative length l_e and a nonnegative cost c_e for every edge e of an undirected complete graph, the goal is to find a cheap tree (in terms of total cost) and also low diameter (in terms of lengths). In a particular budgeted version of the problem, we are given a bound L on the total (length) diameter of the spanning tree to be output and the goal is to find such a spanning tree of minimum total edge cost. This minimum cost-diameter spanning tree problem can be easily shown to be NP-hard [8], as is a cost-radius version of the problem. In the cost-radius version, we are given a root node r, and a bound R on the total length of any path in the output tree from r to any node (hence the name radius, in terms of

[1] While this treatment has been completely worked out in the conference version of our paper [10], the journal version [12] uses a different greedy approach that can also handle node weights in a generalized version of the basic problem.

the length function). We will relate this cost-radius spanning tree problem to a cost-distance version in the next section.

Let us apply the MBA based method to find a minimum cost-diameter spanning tree. At each iteration, we must merge components but with two different goals in mind: the total cost of the matching paths added in the iteration must be at most that of an optimal solution, and the diameter of every path added in the matching should also not exceed the bound L. A further complication is introduced in keeping the total diameter of the final solution bounded with respect to L. For this reason, we simple promote one of the two endpoints of the matched pairs as a representative for its connected component in the next iteration to control the growing radius of the component. Applying the tree pairing lemma to the set of representatives in an optimal tree (of diameter L and total cost C^* say), we can pair the representatives using paths of length at most L each and of total cost at most C^*.

This leads to the following matching subroutine in each iteration. We have a set of connected components, each with a representative. We build an auxiliary graph only on the representatives connecting every pair of representatives by an edge that represents paths of length at most L. Furthermore, we want the cost of these paths to be minimum under the length constraint. For this, we solve a constrained shortest-path problem between this pair of representatives: in particular, we find the minimum cost of a path of total length at most L between these representatives. This problem is itself weakly NP-hard but a scaled adaptation of Djikstra's algorithm gives a PTAS for this path cost computation (i.e, we can get a $(1 + \epsilon)$-approximation to the minimum cost path of length at most L in polynomial time for any fixed $\epsilon > 0$). After filling in all these path costs between representatives, we find a minimum cost perfect matching under these costs. Note that this pairs up components via their representatives using paths of length no more than L^* and nearly minimum total cost.

As in Lemma 2, the guarantee on the number of iterations follows from the pairing property of the paths added in every iteration. We also have the following guarantee on the cost and diameter of components at the end of every iteration.

Lemma 5. *The cost of edges added at every iteration at most $(1 + \epsilon)$ times that of an optimal solution T^* for some fixed $\epsilon > 0$, while the diameter of any connected component at the end of iteration t under the length function is at most $2tL$.*

The bound on the diameter follows from an inductive argument while the cost guarantee is a consequence of the tree-pairing lemma. To obtain the final solution, we observe that even though the set of edges we may have added may form cycles, we can choose a minimum radius tree (under the length function) rooted at the representative of the final component. This tree obeys the bounds in the next theorem.

Theorem 3. *[8] The MBA-based algorithm using a minimum-cost length constrained subgraph subroutine at each iteration outputs a spanning tree of total cost $O(\log n)$ times that of a minimum cost tree obeying the diameter bounds; Moreover, the spanning tree output has diameter at most $O(\log n \cdot L)$.*

3.4 Degree and Diameter Bounded Trees

A third application of the tree-pairing lemma to formulate an MBA based approximation came in the unlikely guise of minimizing the broadcast time in an undirected graph. In this problem, we are given a undirected graph with a root node r containing a message to be broadcast to all the nodes in the graph. At each time step, every node that has a copy of the message can transmit it to one of the (uninformed) neighbors, in the so-called telephone model. The goal is to find a scheme for broadcasting the message to all nodes in the minimum number of time steps. In the first poly-logarithmic approximation algorithm for this problem [11], we showed how to reduce this problem to one of finding a spanning tree with simultaneously low diameter and low maximum node degree. The *poise* of a spanning tree in an undirected graph captures this notion and is defined as the sum of the diameter and the maximum degree. A spanning tree of an undirected graph on n nodes with poise ρ can be used to broadcast a message from any root node within $O(\frac{\log n}{\log\log n} \cdot \rho)$ time steps.

The problem of finding spanning trees with minimum poise can be attacked using the MBA method. At each iteration, the matching based subroutine is require to add paths between matched components that have low diameter (number of hops) as well as induce low degree on any node in the graph. We can use the idea of promoting representatives from the previous subsection (for minimum cost-diameter spanning trees) to control the diameter of the connected components at each iteration. Applying the tree-pairing lemma to the representatives on an optimal tree, we can infer that there is a matching between them using paths of length at most the optimal poise such that the maximum degree induced by these paths at node is also at most the optimal poise. This motivates a corresponding matching problem of pairing up the representatives using short paths with low congestion at any node.

To set up this problem so as to control for the maximum degree of any node induced by the set of matching paths, we use ideas from minimizing congestion in routing integral multicommodity flow, and formulate a linear programming problem to which we can apply randomized rounding. To summarize, the set of representatives from the connected components at each iteration are the sources of multicommodity flow that sinks at any of the other representatives. Furthermore, the length of any of these flow paths is bounded by a given budget (on the poise). An LP solution to the resulting problem of minimizing the node congestion can be rounded randomly to get a near-optimal integral solution. The tree-pairing lemma again provides a proof that there is an integral (and hence LP) solution for the right guess value of the poise with maximum node congestion also at most this poise. The integral rounded solutions can be used to find appropriate matching paths between components in a way that the diameter only increases linearly with the number of iterations. A slightly more careful choice of pairing paths still guarantees the bounds of Lemma 2 while we can get the following analogue of the cost bounding lemma.

Lemma 6. *If there is a tree of poise ρ in the input graph, the LP rounding method with subsequent careful choice of matching paths induces degree at most*

$O(\rho + \log n)$ at any node in each iteration, while the diameter of any connected component at the end of iteration t under the length function is at most $2t\rho$.

Noting that the minimum poise of any spanning tree of an n-node graph is $\Omega(\frac{\log n}{\log\log n})$, we get the following result.

Theorem 4. *[11] The MBA-based algorithm using randomized rounding of a length-constrained node-congestion minimizing LP at each iteration outputs a spanning tree of poise $O(\frac{\log^2 n}{\log logn})$ times the minimum.*

This subsequently leads to the same performance guarantee for the minimum broadcast time problem as shown in [11].

4 Algorithms Inspired by MBA

In this section, we briefly review two lines of work that have used the MBA technique but pushed it to a whole new level. While the underlying matching is recognized as a vehicle to argue the cost incurred by the algorithm by charging it against an optimal solution, these methods typically employ randomization (in their simplest versions) to show expected guarantees on the cost of one iteration. Logarithmic guarantees follow using the same basic line of argument as for the MBA method.

4.1 Cost-Distance Network Design

The cost-distance network design problem is a variant of the set of distance-constrained minimum-cost spanning tree problems introduced in Section 3.3. In this problem, we are given a nonnegative length l_e and a nonnegative cost c_e for every edge e of an undirected complete graph as well as a root node r. In the simplest version, the goal is to find a spanning tree that minimizes the sum of the costs of the edges in the tree (under the c-function) and the distances in the tree (under the l-function) from the root to all the nodes.

The algorithm given by Myerson et al. [9] for this simple version is to define a composite weight function that is the sum of the cost and length for each edge. The algorithm then finds a near-perfect minimum weight perfect matching (ignoring the root and connecting to it only in the last iteration) and chooses one of the two endpoints to be a representative for the whole component randomly. As in the MBA algorithms, these paths are added and the process continues until a tree is obtained.

As in the MBA method, the proof of performance ratio proceeds by showing that the expected cost of eventually connecting all the vertices to the root via the matching added in one iteration is bounded by a constant factor times that of the optimal solution. The randomization allows one to argue that as the iterations proceed that the subproblems on the representatives (which can be thought of as aggregating the demand of all nodes in its component) has expected cost at most that of an optimal solution. Using a similar line of proof as in the MBA method, a performance ratio of $O(\log n)$ follows for this algorithm.

A derandomization of the method along with links to the integrality gap of a natural LP relaxation was provided in [1]. This derandomizing procedure proceeds via the method of conditional probabilities using an LP relaxation; The underlying matching problem is solved motivated by an argument that can be viewed as a more sophisticated matching version of the tree pairing lemma arising in the context of the new composite cost function.

4.2 Simultaneous Optimization for Concave Costs

A further generalization was studies by Goel and Estrin in [5]. In this version, we are given an undirected graph with a root r and a nonnegative cost c_e for every edge e. The goal is to find an "aggregation" tree that collects information from all the nodes to the root. The cost of the tree depends on the aggregation functions on the edges. Let f be a real-valued function defined on non-negative real numbers that is concave and nondecreasing. The cost of an edge e is then $c_e f(flow_e)$ where $flow_e$ is the flow routed through e, in this case the total number of nodes in the subtree under e, when the solution is rooted at r.

Goel and Estrin use a variant of the MBA method to prove a surprising result: There is a tree that is simultaneously near-optimal for all concave aggregating functions for a given undirected graph with costs. This tree is none other than a MBA-based tree constructed in iterations based on the cost function c on the edges. Assuming that the number of nodes n is a power of two. This method simply finds a minimum-cost perfect matching on the nodes and chooses one endpoint as a representative with probability half, and continues until all nodes are connected in a spanning tree.

The proof of performance of this aggregating tree for any fixed concave aggregating function proceeds in a similar way as for the cost-distance problem. First, the expected cost of the rerouted instances is bounded by that of the optimal. Second, the expected aggregated routing cost of the matching edges added in each iteration is bounded by the cost of the optimal solution. To prove the result for general functions, the method employed is to carry out the analysis in terms of some basis aggregation functions (also called "atomic" functions in [5]) that aggregate linearly up to some power of two. Any concave aggregating function's cost is written as a scaled contribution from an appropriate basis function, which are then used in a style similar to that for a fixed function to argue the final result. At this level, while the basic algorithm and outline of the proof technique (using an optimal solution to bound expected cost of the current augmentation) are as in the MBA based methods, this application requires a much more involved argument.

5 Summary

We have reviewed various applications of a simple construction heuristic idea with the augmentations coming from a matching-like subroutine that is inspired by a very simple tree-pairing lemma. Recent refinements replace the tree-pairing with a randomized demand redistribution for reallocation of the cost of the current iteration to that of an optimal solution. The simple idea of using an optimal solution appropriately to derive an augmentation of the solution has been effectively used in a variety of contexts, but we hope the reader is left with a sense of commonality in these applications for network design problems.

References

1. C. Chekuri, S. Khanna, and S, Naor. A deterministic algorithm for the COST-DISTANCE problem. *Proceedings of the 12th Annual ACM-SIAM Symposium on Discrete Algorithms*, 232-233, 2001.
2. N. Christofides. Worst-case analysis of a new heuristic for the travelling salesman problem. Report 388, Graduate School of Industrial Administration, CMU, 1976.
3. A.M. Frieze, G. Galbiati, and F. Maffioli. On the worst-case performance of some algorithms for the asymmetric traveling salesman problem. *Networks* 12 (1982) 23-39.
4. M. Fürer and B. Raghavachari. An NC approximation algorithm for the minimum-degree spanning tree problem. *Proceedings of the 28th Annual Allerton Conference on Communication, Control and Computing* 274-281, 1990.
5. A. Goel and D. Estrin. Simultaneous Optimization for Concave Costs: Single Sink Aggregation or Single Source Buy-at-Bulk. *Proceedings of the 14th Annual ACM-SIAM Symposium on Discrete Algorithms*, 2003.
6. D. Hochbaum, editor. *Approximation algorithms for NP-hard problems.* P.W.S, 1997.
7. Philip N. Klein and R. Ravi. A nearly best-possible approximation algorithm for node-weighted Steiner trees. *J. Algorithms*, 19(1):104–115, 1995. An early version appeared in the *Proceedings of the Annual MPS conference on Integer Programming and Combinatorial Optimization*, 1992.
8. M. V. Marathe, R. Ravi, R. Sundaram, S. S. Ravi, D. J. Rosenkrantz, and H. B. Hunt. Bicriteria network design problems. *J. of Algorithms*, 28, 142171, 1998.
9. A. Meyerson, K. Munagala, and S. Plotkin. Cost-Distance: Two Metric Network Design. *Proceedings of the 41st Annual IEEE Symposium on the Foundations of Computer Science*, 2000.
10. Many birds with one stone: Multi-objective approximation algorithms. R. Ravi, M. V. Marathe, S. S. Ravi, D. J. Rosenkrantz, and H. B. Hunt III. *Proceedings of the ACM Symposium on the Theory of Computing*, 438-447, 1993.
11. R. Ravi. Rapid Rumor Ramification: Approximating the minimum broadcast time. *Proceedings of the 35th Annual IEEE Symposium on the Foundations of Computer Science*, 202–213, 1994.
12. Approximation Algorithms for Degree-Constrained Minimum-Cost Network-Design Problems. R. Ravi, M. V. Marathe, S. S. Ravi, D. J. Rosenkrantz, H. B. Hunt III. *Algorithmica* 31(1), 58-78, 2001.
13. Vijay V. Vazirani. *Approximation Algorithms.* Springer-Verlag, Berlin, 2001.

Modelling Errors and Recovery for Communication

Madhu Sudan

MIT, Cambridge, USA

The theory of error-correction has had two divergent schools of thought, going back to the works of Shannon and Hamming. In the Shannon school, error is presumed to have been effected probabilistically. In the Hamming school, the error is modeled as effected by an all-powerful adversary. The two schools lead to drastically different limits. In the Shannon model, a binary channel with error-rate close to, but less than, 50% is useable for effective communication. In the Hamming model, a binary channel with an error-rate of more than 25% prohibits unique recovery of the message.

In this talk, we describe the notion of list-decoding, as a bridge between the Hamming and Shannon models. This model relaxes the notion of recovery to allow for a "list of candidates". We describe results in this model, and then show how these results can be applied to get unique recovery under "computational restrictions" on the channel's ability, a model initiated by R. Lipton in 1994.

Based on joint works with Venkatesan Guruswami (U. Washington), and with Silvio Micali (MIT), Chris Peikert (MIT) and David Wilson (MIT).

J.R. Correa, A. Hevia, and M. Kiwi (Eds.): LATIN 2006, LNCS 3887, p. 25, 2006.

Lossless Data Compression Via Error Correction

Sergio Verdú

Dept. Electrical Engineering, Princeton University,
Princeton, New Jersey 08540, USA
verdu@princeton.edu

Abstract. This plenary talk gives an overview of recent joint work with G. Caire and S. Shamai on the use of linear error correcting codes for lossless data compression, joint source/channel coding and interactive data exchange.

1 Summary

Over the last five decades, significant inventions have led to data compression and data transmission systems whose efficiency approaches Shannon's fundamental limits [1]. Error-correcting codes now exist (i.e. sparse-graph linear codes) that can achieve performance close to channel capacity with complexity and delay that are tolerable for many applications. Similarly, lossless data compression algorithms exist (most notably the Lempel-Ziv algorithm) that can provably achieve the entropy rate of a wide class of sources with very low complexity. Curiously, although Shannon's development of the theories of fundamental limits for data compression and transmission shared very strong commonalities, there has been essentially no intercourse between the respective constructive theories throughout their long histories.

While Shannon's separation principle establishes no loss in asymptotic performance when compression and transmission are performed separately, it has long been expected (but not fully realized) that, in the nonasymptotic regime, gains may accrue by joint design. Furthermore, in systems such as packet-oriented wireless high data rate systems, it is sometimes cumbersome to design systems based on the separation principle.

Lossless data compression algorithms find numerous applications in information technology, such us packing utilities (e.g. `gzip`), modem standards, fax standards, back-end of lossy compression algorithms (e.g. JPEG and MPEG), and compression of headers of TCP/IP packets in wireless networks.

Indeed, the field of lossless data compression has achieved a state of maturity, with algorithms that admit fast (linear-complexity) implementations and achieve asymptotically the fundamental information theoretic limits.

The availability of linear codes (such as the low-density parity check codes) that allow for very efficient encoding/decoding algorithms while operating near the Shannon limit makes their application in data compression competitive with state-of-the-art methods while not suffering from some of their shortcomings.

J.R. Correa, A. Hevia, and M. Kiwi (Eds.): LATIN 2006, LNCS 3887, pp. 26–27, 2006.

A series of recent papers [2, 3, 4, 5] presents a new approach to universal noiseless compression based on error correcting codes. The scheme is based on the concatenation of the Burrows-Wheeler block sorting transform (BWT) with the syndrome former of a Low-Density Parity-Check (LDPC) code. The proposed scheme has linear encoding and decoding times and uses a new closed-loop iterative doping (CLID) algorithm that works in conjunction with belief-propagation decoding.

Alternatively, fountain codes can replace the LDPC codes [6] to provide a streamlined design which is ideally suited for variable-length lossless compression.

One of the incentives to use error correcting codes for data compression is the natural extension of the schemes to joint source/channel encoding and decoding. Schemes for that purpose are explored in [7].

Building upon Slepian-Wolf coding [8] , sparse-graph codes, belief propagation, and closed-loop iterative doping, new schemes for interactive data exchange between two agents who want to communicate losslessly their respective information via several rounds of communication are proposed in [9].

References

1. C. E. Shannon, "A mathematical theory of communication," *Bell Sys. Tech. J.*, vol. 27, pp. 379–423, 623–656, Jul.-Oct. 1948.
2. G. Caire, S. Shamai, and S. Verdú, "A new data compression algorithm for sources with memory based on error correcting codes," *2003 IEEE Workshop on Information Theory*, pp. 291–295, Mar. 30- Apr. 4, 2003.
3. G. Caire, S. Shamai, and S. Verdú, "Lossless data compression with error correction codes," *2003 IEEE Int. Symp. on Information Theory*, p. 22, June 29- July 4, 2003.
4. G. Caire, S. Shamai, and S. Verdú, "Universal data compression with LDPC codes," *Third International Symposium On Turbo Codes and Related Topics*, pp. 55–58, Brest, France, September 1-5, 2003.
5. G. Caire, S. Shamai, and S. Verdú, "Noiseless data compression with low density parity check codes," in *DIMACS Series in Discrete Mathematics and Theoretical Computer Science*, P. Gupta and G. Kramer, Eds., pp. vol. 66, pp. 263–284. American Mathematical Society, 2004.
6. G. Caire, S. Shamai, A. Shokrollahi and S. Verdú, "Fountain Codes for Lossless Data Compression," in *DIMACS Series in Discrete Mathematics and Theoretical Computer Science: Algebraic Coding Theory and Information Theory*, A. Ashikhmin, A. Barg, I. Duursma, Eds., pp. vol. 68, pp. 1-20. American Mathematical Society, 2005.
7. G. Caire, S. Shamai, and S. Verdú, "Almost-noiseless joint source-channel coding-decoding of sources with memory," *Proc. Fifth International ITG Conference on Source and Channel Coding (SCC)*, pp. 295–304, Jan 14-16, 2004.
8. D. Slepian and J. K. Wolf. Noiseless coding of correlated information sources. *IEEE Trans. Information Theory*, IT-19:471–480.
9. G. Caire, S. Shamai, and S. Verdú, "Practical schemes for interactive data exchange," *Proc. 2004 Int. Symp. Information Theory and its Applications*, Parma, Italy, Oct. 2004.

The Power and Weakness of Randomness in Computation

Avi Wigderson

Institute for Advanced Study, Princeton

Humanity has grappled with the meaning and utility of randomness for centuries. Research in the Theory of Computation in the last thirty years has enriched this study considerably. We describe two main aspects of this research on randomness, demonstrating its power and weakness respectively.

Randomness is Paramount to Computational Efficiency. The use of randomness can dramatically enhance computation (and do other wonders) for a variety of problems and settings. In particular, examples will be given of probabilistic algorithms (with tiny error) which are exponentially faster than their (best known) deterministic counterparts, and probabilistic algorithms which achieve significant space savings over deterministic ones. Other settings include distributed algorithms where randomness (provably) achieves exponentially smaller congestion than deterministic ones. Finally we'll show that using randomness, proof systems can be enhanced to allow properties unattainable without it. Letting the verifier and prover toss coins, proof systems can allow spot checking of proofs (PCPs - a central tool in the theory of approximation), as well as zero-knowledge proofs (proofs revealing nothing except their validity - a central tool in cryptography).

Computational Efficiency is Paramount to Understanding Randomness. We explain the computationally-motivated definition of randomness, and try to argue its merits as the "right" definition. The central idea is "computational indistinguishability" - declaring a distribution pseudorandom if it cannot be distinguished from the uniform distribution by any efficient procedure (in a given class, say time or space bounded algorithms). It is evident, almost by definition, that such pseudorandom distributions are as good as uniform as sources of randomness for probabilistic algorithms in the given class. We then demonstrate the remarkable fact, known as the "hardness vs. randomness paradigm" that such pseudorandomness may be generated deterministically and efficiently, from (appropriate) computationally difficult problems. This leads to a deterministic "derandomization" of any given probabilistic algorithm, which is not much slower. Consequently, randomness is probably not as powerful as it seems above.

For a comprehensive text on probabilistic algorithms the reader is refered to [MR]. For a thorough discussion of both probabilistic proof systems, as well as pseudorandomness, the reader is refered to [G].

J.R. Correa, A. Hevia, and M. Kiwi (Eds.): LATIN 2006, LNCS 3887, pp. 28–29, 2006.

References

[MR] Motwani, Rajeev and Raghavan, Prabhakar Randomized Algorithms, Cambridge University Press, 1995.

[G] Goldreich, Oded Modern Cryptography, Probabilisitc Proofs and Pseudorandomness, Springer Verlag, Algorithms and Combinatorics, Vol 17, 1998.

A New GCD Algorithm for Quadratic Number Rings with Unique Factorization

Saurabh Agarwal and Gudmund Skovbjerg Frandsen

BRICS*, Department of Computer Science, University of Aarhus,
IT-Parken, Åbogade 34, DK-8200, Aarhus N, Denmark
{saurabh, gudmund}@daimi.au.dk

Abstract. We present an algorithm to compute a greatest common divisor of two integers in a quadratic number ring that is a unique factorization domain. The algorithm uses $O(n \log^2 n \log\log n + \Delta^{\frac{1}{2}+\epsilon})$ bit operations in a ring of discriminant Δ. This appears to be the first gcd algorithm of complexity $o(n^2)$ for any fixed non-Euclidean number ring. The main idea behind the algorithm is a well known relationship between quadratic forms and ideals in quadratic rings. We also give a simpler version of the algorithm that has complexity $O(n^2)$ in a fixed ring. It uses a new binary algorithm for reducing quadratic forms that may be of independent interest.

Keywords: gcd, quadratic number ring, quadratic form reduction.

1 Introduction

Given a squarefree integer $d \neq 1$, let $\mathcal{Z}_d$ denote the ring of integers in the quadratic number field $\mathbb{Q}(\sqrt{d})$. The prototypical example is the Gaussian integers $\mathcal{Z}_{-1} = \{a + bi \mid a, b \in \mathbb{Z}\}$. When $\mathcal{Z}_d$ is a unique factorization domain (UFD), the greatest common divisor (gcd) of two elements in $\mathcal{Z}_d$ always exists and is unique up to multiplication with a unit. $\mathcal{Z}_d$ is known to be a UFD for precisely 9 values of $d < 0$ (complex quadratic rings), but it is unresolved whether $\mathcal{Z}_d$ is a UFD for infinitely many $d > 0$ (real quadratic rings) [7].

We consider the following problem. On input $\alpha, \beta \in \mathcal{Z}_d$, where $\mathcal{Z}_d$ is a UFD, compute a gcd of α and β. We consider the problem for both a fixed ring, and when Δ the discriminant of the ring is given as part of the input (Δ is d or $4d$).

1.1 Earlier Work

Greatest common divisor is a basic concept in number theory. The problem of computing the gcd is as old as number theory since many computational number theory problems require gcd or extended gcd. Euclid presented an algorithm to compute the gcd of rational integers in 300 B.C. [9]. The rings in which one can construct a similar algorithm are called Euclidean rings. These do not include

* Basic Research in Computer Science (www.brics.dk), funded by the Danish National Research Foundation.

J.R. Correa, A. Hevia, and M. Kiwi (Eds.): LATIN 2006, LNCS 3887, pp. 30–42, 2006.

all quadratic number rings that are UFD's, but a fairly complete list of all known Euclidean number rings can be found in [13]. However, for some of the Euclidean rings it is not clear if the Euclidean algorithm runs in polynomial time. For the integers, Lehmer presented an improved version of complexity $O(n^2)$ [12], and Schönhage gave a version of complexity $O(n \log^2 n \log \log n)$ [15]. The latter algorithm was generalized to Euclidean complex quadratic number rings by Weilert [21].

A different class of algorithms to compute gcd comes from the binary gcd algorithm [19]. Simple and practical generalisations of the binary gcd algorithms are known for some complex quadratic number rings including a non-Euclidean ring [20, 5, 1]. The running time of all these algorithms is $O(n^2)$ with small constants hidden under the big-oh notation. Wikström [22] has shown that one can extend the binary gcd algorithm to all number rings that are UFD's. The complexity of the algorithm is $O(n^2)$ in a fixed ring. The dependence of the ring is not made explicit in the runtime analysis.

Kaltofen and Rolletschek [8] gave an $O(n^2)$ algorithm to compute gcd in any fixed quadratic ring This algorithm appears difficult to implement and the size of constants under the big-oh notation is not small [8, 20].

1.2 Results

We present a gcd algorithm applicable for all quadratic number rings that are UFD's. The algorithm has complexity $O(n \log^2 n \log \log n)$ assuming a fixed ring. If the ring is not fixed, we assume that the discriminant Δ is given as part of the input. The complexity of the algorithm is still $O(n \log^2 n \log \log n)$ for complex quadratic rings, but for real quadratic rings it takes $O(n \log^2 n \log \log n + \Delta^{\frac{1}{2}+\epsilon})$ time.

For the complexity bound of $O(n \log^2 n \log \log n)$, the algorithm needs to use similar bounds for integer multiplication, extended integer gcd and reduction of quadratic forms [16, 15, 17]. Though the corresponding algorithms are the best known with respect to asymptotic complexity, they may be impractical for moderate input sizes. We give an alternative version of our algorithm of complexity $O(n^2 + \Delta^{1+\epsilon})$. We believe it may be more practical than the known general algorithms of complexity $O(n^2)$ [8, 22]. It uses a "binary" algorithm of complexity $O(n^2)$ for reducing quadratic forms, which may be of independent interest.

1.3 Main Idea of Algorithm

Let α and β be integers in some quadratic ring that is a UFD. Computing a gcd of α and β is the same as computing a generator of the ideal generated by α and β. Every ideal can be viewed as a module and for every ideal there exists a module basis such that one of the elements in the basis is a generator of the ideal. The idea of the algorithm is to compute such a basis. All (ordered) bases are equivalent up to multiplication by an $\mathbb{SL}_2(\mathbb{Z})$ matrix, so we simply start by some basis and look for a transformation that maps our initial basis into one containing a generator. To guide this search, we use that each (ordered) basis for the ideal has an associated quadratic form of the same discriminant as the ring. We start by finding the form corresponding to our initial basis. It turns out that we can then easily find an $\mathbb{SL}_2(\mathbb{Z})$ matrix U taking this form into a reduced principal form. The same U is then applied to the initial module basis and will give a basis containing a gcd for α and β.

For complex quadratic rings that are UFD's, there is only one reduced form of the corresponding discriminant. Therefore a standard algorithm for reducing binary quadratic forms suffices to find U. In the case of real quadratic rings there are in general many reduced forms of a given discriminant. Hence, for such rings, one must in addition find an $\mathbb{SL}_2(\mathbb{Z})$ matrix that takes the encountered reduced form into a principal form.

We illustrate the idea of the algorithm by considering a concrete simple example, namely computing $\gcd(\alpha, \beta)$ in the ring of Gaussian integers $\mathbb{Z}[i]$ for $\alpha = 3 - i$ and $\beta = 4 - 2i$. First, we compute an ordered module basis $[\alpha_1, \beta_1]$ for the ideal $I = (\alpha, \beta)$. This may result in $\alpha_1 = 3 - i$ and $\beta_1 = 2$. The associated quadratic form is $Q_1 = 5x^2 + 6xy + 2y^2$. Using a standard reduction algorithm one may find that $U = \begin{bmatrix} -1 & 1 \\ 1 & -2 \end{bmatrix} \in \mathbb{SL}_2(\mathbb{Z})$ takes Q_1 into the reduced (principal) form $Q_2 = x^2 + y^2$. When applying the same transformation U to $[\alpha_1, \beta_1]$, one obtains $[\alpha_2, \beta_2]$, where $\alpha_2 = -1 + i$ and $\beta_2 = -1 - i$, both of which are gcd's of α and β (and associates).

2 Preliminaries

The definitions/facts in this section are found in most books on algebra and/or algebraic number theory (for example see [11, 6, 7, 4, 18]). Most of the concepts are also covered in [14].

In the following, the letters $\mathbb{Q}$ and $\mathbb{Z}$ denote the set of rational numbers and rational integers. The notation $\mathbb{SL}_2(\mathbb{Z})$ is used to denote the set of all 2×2 matrices with entries from $\mathbb{Z}$ and determinant 1.

2.1 Quadratic Fields and Rings

Quadratic number fields are of the form $\mathcal{Q} = \mathbb{Q}(\sqrt{d})$ where $d \in \mathbb{Z}$ is square-free. If $\mathcal{Z}$ is the ring of integers in $\mathcal{Q}$, then $\mathcal{Z} = \mathbb{Z}[\omega]$ where

$$\omega = \begin{cases} \sqrt{d} & \text{if } d \equiv 2, 3 \pmod 4 \\ \frac{1+\sqrt{d}}{2} & \text{if } d \equiv 1 \pmod 4 \end{cases}$$

Any $\theta \in \mathcal{Q}$ is of the form $q_1 + q_2\sqrt{d}$ where $q_1, q_2 \in \mathbb{Q}$. If θ is also an element of $\mathcal{Z}$, then θ can be written as $a_1 + a_2\omega$ where $a_1, a_2 \in \mathbb{Z}$. The *discriminant* Δ of the field $\mathcal{Q}$ (or the ring $\mathcal{Z}$) is,

$$\Delta = \begin{cases} 4d & \text{if } d \equiv 2, 3 \pmod 4 \\ d & \text{if } d \equiv 1 \pmod 4 \end{cases}$$

Let $\theta = q_1 + q_2\sqrt{d} \in \mathcal{Q}$. The number, $\bar{\theta} = q_1 - q_2\sqrt{d}$ is the *conjugate* of θ. The *norm* of θ is $N(\theta) = \theta\bar{\theta} = q_1^2 - dq_2^2$. An element $\nu \in \mathcal{Z}$ is a *unit* if $\frac{1}{\nu} \in \mathcal{Z}$. A unit $\nu \in \mathcal{Z}$ is also characterized by $N(\nu) \in \{-1, 1\}$.

In the following the term "quadratic ring" will always refer to the ring of integers in a quadratic number field.

2.2 Modules

Let $\mathcal{Z}$ be a quadratic ring. Any $M \subseteq \mathcal{Z}$ which is closed under addition and subtraction is a module in $\mathcal{Z}$. A collection of elements $\{\alpha_1, \ldots, \alpha_k\} \in M$ *spans* M if for all $\alpha \in M$ there exist $x_1, \ldots, x_k \in \mathbb{Z}$ such that $\alpha = x_1\alpha_1 + \cdots + x_k\alpha_k$ and we write

$$M = [\alpha_1, \alpha_2, \ldots, \alpha_k] = \begin{bmatrix} m_{11} & m_{12} & \cdots & m_{k1} \\ m_{21} & m_{22} & \cdots & m_{k2} \end{bmatrix} \quad \text{where } \alpha_j = m_{1j} + m_{2j}\omega \ .$$

A collection of elements $\{\alpha_1, \ldots, \alpha_k\} \in M$ is *linearly independent* (over $\mathbb{Z}$) if $x_1\alpha_1 + \cdots + x_k\alpha_k = 0$ and $x_1, \ldots, x_k \in \mathbb{Z}$ imply that all x_i are zero. A collection of elements $\{\alpha_1, \ldots, \alpha_k\} \in M$ is a *basis* for M if $\alpha_1, \ldots, \alpha_k$ span M and are linearly independent. All bases for a module has the same number of elements and that number is called the *dimension* of the module.

2.3 Ideals and Bases

Let $\mathcal{Z}$ be a quadratic ring. A set $I \subseteq \mathcal{Z}$ is an ideal of $\mathcal{Z}$ if I is a module and $\alpha I \subseteq I$ for all $\alpha \in \mathcal{Z}$. A collection of elements $\{\alpha_1, \ldots, \alpha_m\} \in I$ *generates* I, if for all $\alpha \in I$ there exist $\beta_1, \ldots, \beta_m \in \mathcal{Z}$, such that $\alpha = \beta_1\alpha_1 + \cdots + \beta_m\alpha_m$ and we write

$$I = (\alpha_1, \ldots, \alpha_m) \ .$$

A *principal* ideal is an ideal generated by a single element. When the quadratic ring $\mathcal{Z}$ is a UFD then all ideals in $\mathcal{Z}$ are principal.

Proposition 1. *An ideal has dimension two when regarded as a module. Thus if I is an ideal in $\mathcal{Z}$, then there exists $\alpha = a_1 + a_2\omega$ and $\beta = b_1 + b_2\omega$ in I such that,*

$$I = \mathbb{Z}\alpha + \mathbb{Z}\beta = [\alpha, \beta] = \begin{bmatrix} a_1 & b_1 \\ a_2 & b_2 \end{bmatrix} \ .$$

Let $I = [\alpha, \beta]$ be an ideal in $\mathcal{Z}$. The module basis $[\alpha, \beta]$ is an *ordered basis* of I if $\det([\alpha, \beta]) > 0$. The norm of I is the number of elements in $\mathcal{Z}/I$ and is denoted by $N(I)$ and when $[\alpha, \beta]$ forms an ordered basis for I,

$$N(I) = \frac{\bar{\alpha}\beta - \alpha\bar{\beta}}{\omega - \bar{\omega}} = \det[\alpha, \beta].$$

If I is a principal ideal generated by γ, then $N(I) = |N(\gamma)|$ and $[\gamma, \omega\gamma]$ forms a module basis for I.

There is a natural $\mathbb{SL}_2(\mathbb{Z})$-action on ordered module-bases, given by $[\alpha, \beta]U = [t\alpha + v\beta, u\alpha + w\beta]$ for $U = \begin{bmatrix} t & u \\ v & w \end{bmatrix} \in \mathbb{SL}_2(\mathbb{Z})$. The action of $\mathbb{SL}_2(\mathbb{Z})$ does not change the ideal, and all ordered bases for a specific ideal are equivalent under the action. In particular

Proposition 2. *Let $[\alpha, \beta]$ be an ordered module basis for a principal ideal $I = (\gamma)$ in $\mathcal{Z}$. There exists $U \in \mathbb{SL}_2(\mathbb{Z})$ such that $[\alpha, \beta]U$ is the ordered basis among $[\gamma, \pm\omega\gamma]$.*

2.4 GCD and Principal Ideals

Let $\mathcal{Z}$ be a quadratic ring, and let $\alpha, \beta \in \mathcal{Z}$. If $\alpha\beta \neq 0$ then a non-zero element $\gamma \in \mathcal{Z}$ is a *greatest common divisor* (gcd) of α and β if

i. $\gamma|\alpha$ and $\gamma|\beta$, and
ii. for any $\delta \in \mathcal{Z}\backslash\{0\}$, if $\delta|\alpha$ and $\delta|\beta$, then $\delta|\gamma$.

For any $\alpha \neq 0$, gcd of α and 0 is defined to be α.

Proposition 3. *Let $\mathcal{Z}$ be any quadratic number ring that is a UFD and let $\alpha, \beta \in \mathcal{Z}$. Then γ is a gcd of α and β iff γ is a generator of the ideal generated by α and β.*

2.5 Quadratic Forms

Let $\Delta \in \mathbb{Z} \setminus \{0\}$. A *primitive integral binary quadratic form of discriminant Δ* (henceforth forms of discriminant Δ or simply forms) is a polynomial $\mathfrak{Q}(A, B, C) = Ax^2 + Bxy + Cy^2 \in \mathbb{Z}[x, y]$, for which $\gcd(A, B, C) = 1$, $B^2 - 4AC = \Delta$ and if $\Delta < 0$ then $A > 0$. The form $\mathfrak{Q}(A, B, C)$ is said to be reduced if

$$|\sqrt{\Delta} - 2|A|| < B < \sqrt{\Delta} \text{ if } \Delta > 0$$

$$\left.\begin{array}{c} |B| \leq A \leq C \\ B \geq 0 \text{ if } |B| = A \text{ or} A = C \end{array}\right\} \text{ if } \Delta < 0$$

The group $\mathbb{SL}_2(\mathbb{Z})$ acts on the right on $\mathbb{Z}[x, y]$ as a group of ring automorphisms given by $xU = tx + uy$ and $yU = vx + wy$ for $U = \begin{bmatrix} t & u \\ v & w \end{bmatrix} \in \mathbb{SL}_2(\mathbb{Z})$ transforming the set of forms of discriminant Δ into itself, i.e.

$$(Ax^2 + Bxy + Cy^2)\,U = A(xt + uy)^2 + B(xt + uy)(xv + yw) + C(xv + wy)^2.$$

Two forms are said to be equivalent if they can be transformed into each other by elements of $\mathbb{SL}_2(\mathbb{Z})$. Every form is equivalent to a reduced form.

3 Overview of the Algorithm

To compute the gcd of α and β, it suffices by Proposition 3 to compute a generator γ for the ideal generated by α and β. It turns out to be easy to find an ordered module basis $[\alpha_1, \beta_1]$ for the ideal (α, β). We would be done, if we could find some $U \in \mathbb{SL}_2(\mathbb{Z})$ such that $[\alpha_1, \beta_1]U = [\alpha_2, \beta_2]$, where α_2 alone is a generator of the ideal. Proposition 2 guarantees that such a U always exists. The basic idea of the algorithm is to find a suitable U by using a well studied relationship between ideals and quadratic forms.

Proposition 4. *[14–sect. 3] or [4–ch.12] Let $\mathcal{Z} = \mathbb{Z}[\omega]$ be the quadratic ring of integers with discriminant Δ. Let $[\alpha, \beta]$ be an ordered basis for an ideal I in $\mathcal{Z}$. The form*

$$\mathfrak{Q}\big([\alpha,\beta]\big) = \frac{N(x\alpha + y\beta)}{N(I)} = \frac{N(\alpha)}{N(I)}x^2 + \frac{N(\alpha+\beta) - N(\alpha) - N(\beta)}{N(I)}xy + \frac{N(\beta)}{N(I)}y^2$$

is a primitive integral form of discriminant Δ.

To make use of this relationship, we need several facts. Firstly, the action of $\mathbb{SL}_2(\mathbb{Z})$ commutes with the mapping from ideals to forms.

Lemma 1. *Let $[\alpha, \beta]$ be an ordered basis for an ideal I in $\mathcal{Z}$, the quadratic ring of integers with discriminant Δ. Let $U \in \mathbb{SL}_2(\mathbb{Z})$ be arbitrary. Then,*

$$\mathfrak{Q}\big([\alpha,\beta]\big)U = \mathfrak{Q}([\alpha,\beta]U)\ .$$

Proof. Let $U = \begin{bmatrix} t & u \\ v & w \end{bmatrix} \in \mathbb{SL}_2(\mathbb{Z})$. It suffices to note that

$$\begin{aligned}\mathfrak{Q}([\alpha,\beta])U &= \frac{N((xt+uy)\alpha + (xv+yw)\beta)}{N(I)} \\ &= \frac{N(x(t\alpha+v\beta) + y(u\alpha+w\beta))}{N(I)} \\ &= \mathfrak{Q}([\alpha,\beta]U)\end{aligned}$$

Secondly, we can recognize a form corresponding to a module basis containing a generator for the ideal.

Lemma 2. *Let $[\alpha, \beta]$ be an ordered basis for a principal ideal I in the quadratic ring of integers $\mathcal{Z}$. Let $\mathfrak{Q}\big([\alpha,\beta]\big)$ be the form corresponding to this basis as given by Proposition 4. Then α is a generator of I iff the coefficient of x^2 in $\mathfrak{Q}\big([\alpha,\beta]\big)$ is ± 1.*

Proof. If α is a generator of I, then $N(\alpha) = \pm N(I)$. Thus the coefficient of x^2 will be ± 1. Conversely if the coefficient of x^2 is ± 1, then $N(\alpha) = \pm N(I)$. If δ is any generator of I, then $N(\delta) = \pm N(I)$. Since $\alpha \in I$, $\alpha = \gamma\delta$ for some $\gamma \in \mathcal{Z}$. But as $N(\alpha) = \pm N(\delta)$, $N(\gamma) = \pm 1$ and hence γ is a unit. Thus α is also a generator of I.

To simplify the rest of the article, a form is a *principal* form if the coefficient of x^2 is ± 1. Proposition 2, Lemma 1 and Lemma 2 imply that given a module basis $[\alpha_1, \beta_1]$ for a (principal) ideal, there exist a transformation $U_p \in \mathbb{SL}_2(\mathbb{Z})$ such that the form $\mathfrak{Q}([\alpha, \beta])U_p$ is a principal form, and the basis $[\alpha, \beta]U_p$ contains a generator of the ideal, i.e. a gcd. To find such a principal form, we need only look among reduced forms.

Lemma 3. *A principal form is equivalent to a form that is both principal and reduced.*

Proof. Given a principal form $\mathfrak{Q}(A, B, C)$, ie. $A = \pm 1$ for $\Delta > 0$ and $A = 1$ for $\Delta < 0$, it suffices to argue that we can find $U = \begin{bmatrix} 1 & m \\ 0 & 1 \end{bmatrix}$ such that $\mathfrak{Q}(A, B, C)U = \mathfrak{Q}(A, B + 2Am, C')$ is principal and reduced. $\mathfrak{Q}(A, B + 2Am, C')$ is clearly principal, and using the definition of a reduced form, it follows that there is a unique integral m such $\mathfrak{Q}(A, B + 2Am, C')$ is reduced.

Algorithm 1. Compute gcd in a quadratic number ring $\mathcal{Z}$ of discriminant Δ

Require: $\alpha, \beta \in \mathcal{Z}$ with $\alpha, \beta \neq 0$
Ensure: $\gamma = \gcd(\alpha, \beta)$.

1: Compute an ordered basis, $[\alpha_1, \beta_1]$ for the ideal I generated by α and β.
2: Compute a quadratic form $\mathfrak{Q}(A_1, B_1, C_1)$ corresponding to the basis $[\alpha_1, \beta_1]$ using Proposition 4.
3: Compute a reduced form $\mathfrak{Q}(A_2, B_2, C_2)$ and a corresponding transformation $U_1 \in \mathbb{SL}_2(\mathbb{Z})$ such that $\mathfrak{Q}(A_1, B_1, C_1)U_1 = \mathfrak{Q}(A_2, B_2, C_2)$.
4: **if** $\Delta < 0$ let $U_2 = I$.
5: **if** $\Delta > 0$ Let U_2 be a transformation arising from applying ϱ of Proposition 5 repeatedly such that $\mathfrak{Q}(A_2, B_2, C_2)U_2$ is a principal form.
6: Compute $[\alpha_2, \beta_2] = [\alpha_1, \beta_1]U_1U_2$.
7: **return** $\gamma = \alpha_2$.

As a first step towards finding U_p, we apply a (standard) reduction algorithm to the initial quadratic form $\mathfrak{Q}([\alpha_1, \beta_1])$ in order to obtain a transformation U_r which will reduce it. Lets say that the reduced form is $\mathfrak{Q}(A, B, C)$. We then use the following result to find a principal reduced form among the reduced forms that are equivalent to $\mathfrak{Q}(A, B, C)$.

Proposition 5. *[14–sect. 5] A form of discriminant $\Delta < 0$ is equivalent to precisely one reduced form.*

Let $\mathfrak{Q}(A, B, C)$ be a reduced form of discriminant $\Delta > 0$. If one applies the transformation $U = \begin{bmatrix} 0 & -1 \\ 1 & m \end{bmatrix}$ where m is chosen such that $\sqrt{\Delta} - 2|C| \leq -B + 2Cm < \sqrt{\Delta}$ then $\mathfrak{Q}(A, B, C)U = \mathfrak{Q}(C, -B + 2Cm, A - mB + m^2C)$ is also a reduced form. If we denote by ϱ the action of applying such a U on a reduced quadratic form, then ϱ is a permutation on the set of all reduced quadratic forms of discriminant Δ. Two reduced forms are equivalent precisely when they lie on the same cycle of this permutation.

Let $\mathfrak{Q}(A, B, C)$ be a reduced form that is equivalent to a principal form. If $\Delta < 0$, then Proposition 5 and Lemma 3 imply that $\mathfrak{Q}(A, B, C)$ is necessarily also principal. If $\Delta > 0$, then Proposition 5 and Lemma 3 imply that $\mathfrak{Q}(A, B, C)$ is on the same cycle as a principal form. Thus if one applies ϱ repeatedly on $\mathfrak{Q}(A, B, C)$, one will eventually encounter a principal form.

Broadly the algorithm is as shown in Algorithm 1. In the next section we consider step 1 and step 3 of the algorithm and the complexity analysis.

4 Implementation and Complexity Analysis

All complexity bounds refer to the number of bit operations. We define the size of numbers as the number of bits needed for their representation (disregarding signs). For integer $m \neq 0$ let $\text{size}(m) = 1 + \lfloor \log |m| \rfloor$, for $\alpha = a + b\omega$ let $\text{size}(\alpha) = \text{size}(a) + \text{size}(b)$ and for $\mathfrak{Q} = Ax^2 + Bxy + Cy^2$ define $\text{size}(\mathfrak{Q}) = \text{size}(A) + \text{size}(B) + \text{size}(C)$. For $M = \begin{bmatrix} p & q \\ r & s \end{bmatrix}$, define $\text{size}(M) = \text{size}(p) + \text{size}(q) + \text{size}(r) + \text{size}(s)$.

4.1 Computing the Module Basis of an Ideal

If I is an ideal generated by α and β, then as a module I is spanned by α, $\omega\alpha$, β and $\omega\beta$, i.e.

$$I = [\alpha, \omega\alpha, \beta, \omega\beta] \ .$$

One can use a standard basis extraction method [4–sect.4.9] to get a basis for this module. To make the time analysis explicit we present Algorithm 2.

Lemma 4. *Given α and β in the quadratic ring $\mathbb{Z}[\omega]$ of discriminant Δ, let $n = size(\alpha) + size(\beta) + size(\Delta)$. Algorithm 2 outputs an ordered basis $[\alpha_1, \beta_1]$ for the ideal generated by α and β. Algorithm 2 uses time corresponding to $O(1)$ multiplications, divisions and extended gcd computations on numbers with $O(n)$ bits.*

Algorithm 2. Computing Module Basis of an Ideal

Require: $\alpha, \beta \in \mathcal{Z}$
Ensure: α_1, β_1 satisfies that $[\alpha_1, \beta_1]$ is an ordered basis for the ideal (α, β)
Given $\alpha = a_1 + a_2\omega$ and $\beta = b_1 + b_2\omega$.
Compute $\omega\alpha = j_1 + j_2\omega$ and $\omega\beta = k_1 + k_2\omega$.
Let $M = \begin{bmatrix} a_1 & j_1 & b_1 & k_1 \\ a_2 & j_2 & b_2 & k_2 \end{bmatrix} = [m_{ij}]_{1 \le i \le 2, 1 \le j \le 4}$
Assert $m_{21} \neq 0$. Swap columns of M if needed.
for $k = 2$ to 4 **do**
  Compute s, t, g such that $g = \gcd(m_{21}, m_{2k}) = sm_{21} + tm_{2k}$
  Let $\begin{bmatrix} m_{11} & m_{1k} \\ m_{21} & m_{2k} \end{bmatrix} = \begin{bmatrix} m_{11} & m_{1k} \\ m_{21} & m_{2k} \end{bmatrix} \begin{bmatrix} s & -m_{2k}/g \\ t & m_{21}/g \end{bmatrix}$
let $\alpha_1 = m_{11} + m_{21}\omega$ and $g_1 = \gcd(m_{12}, m_{13}, m_{14})$
if $m_{21}g_1 > 0$ let $\beta_1 = -g_1$ else let $\beta_1 = g_1$
return α_1, β_1

Proof. Let us first argue correctness of the algorithm. Observe that since

$$U = \begin{bmatrix} s & -m_{2k}/g \\ t & m_{21}/g \end{bmatrix}$$

has determinant 1, then the multiplication with U in step 7 does not change the span of the first and k'th column of M. Thus M spans the same ideal after each iteration of the for-loop. After the for-loop, M has the form

$$M = \begin{bmatrix} m_{11} & m_{12} & m_{13} & m_{14} \\ m_{21} & 0 & 0 & 0 \end{bmatrix}$$

Since $m_{12}\mathbb{Z}+m_{13}\mathbb{Z}+m_{14}\mathbb{Z} = \gcd(m_{12}, m_{13}, m_{14})\mathbb{Z} = g_1\mathbb{Z}$, the module M is also spanned by

$$M = \begin{bmatrix} m_{11} & g_1 \\ m_{21} & 0 \end{bmatrix}$$

By Proposition 1 an ideal has dimension 2 when regarded as a module. Thus g_1 and $m_{11} + m_{12}\omega$ form a module basis for the ideal generated by α and β. Step 9 ensures that the algorithm outputs an ordered basis.

The complexity bound follows by inspection of the algorithm.

4.2 "Binary" Algorithm for Reducing a Quadratic Form

Schönhage has shown how to reduce quadratic forms in time $O(n \log^2 n \log\log n)$. For moderate input sizes a simpler algorithm my be faster. Buchmann and Biehl [2] has shown that the classical reduction algorithm for quadratic forms (see e.g. Lagarias [10]) is of complexity $O(n^2)$.

We present an alternative "binary" algorithm also of complexity $O(n^2)$. In this alternative version we seek to replace multiplications/divisions by additions, subtractions and binary shifts. We believe the resulting algorithm is quite practical for moderate input sizes and may be of independent interest.

Lemma 5. *Given a quadratic form $\mathfrak{Q}(A, B, C)$, let $n = size(\mathfrak{Q}(A, B, C))$. Algorithm 3 reduces the quadratic form and computes a corresponding transformation U satisfying that $size(U) = O(n)$ in time $O(n^2)$.*

Proof. Assume for the moment that the algorithm terminates. To show that the algorithm is correct, we just need to show that the final form returned is reduced.

Consider the case when $\Delta > 0$. Following the outer while-loop it holds that $|B| \leq 2|A| \leq 2|C|$. In addition it holds that $2|A| \leq \sqrt{\Delta}$. To see this observe that $B^2 - 4AC = \Delta > 0$ combined with $|B| \leq 2|A| \leq 2|C|$ implies that $B^2 + 4|A||C| = \Delta$. If the form is not reduced, then the choice of m in step 13 gives a reduced form.

Consider similarly the case of $\Delta < 0$. By our definition of quadratic form, $A > 0$ and from $B^2 - 4AC = \Delta < 0$ we deduce that also $C > 0$. Each application of S and T_m will preserve the signs of A and C, and at step 16 it holds that $|B| \leq 2A \leq 2C$. If $A < |B|$ then steps 18-22 will transform $\mathfrak{Q}(A, B, C)$ to

Algorithm 3. Reducing a Quadratic Form ("binary" version)

Require: A quadratic form $\mathfrak{Q}(A_1, B_1, C_1)$ of discriminant Δ.
Ensure: $\mathfrak{Q}(A, B, C)$ is reduced and $U \in \mathbb{SL}_2(\mathbb{Z})$ such that $\mathfrak{Q}(A_1, B_1, C_1)U = \mathfrak{Q}(A, B, C)$

1: Let $\mathfrak{Q}(A, B, C) = \mathfrak{Q}(A_1, B_1, C_1)$.
2: Let $U = I$.
3: Let $S = \begin{bmatrix} 0 & -1 \\ 1 & 0 \end{bmatrix}$. (note that $\mathfrak{Q}(A, B, C)S$ is $\mathfrak{Q}(C, -B, A)$)
4: Let $T_m = \begin{bmatrix} 1 & m \\ 0 & 1 \end{bmatrix}$. (note that $\mathfrak{Q}(A, B, C)T_m$ is $\mathfrak{Q}(A, B + 2mA, m^2A + mB + C)$)
5: **while** $\neg(|B| \leq 2|A| \leq 2|C|)$ **do**
6:     **while** $|B| > 2|A|$ **do**
7:         Let $j = \text{size}(B) - \text{size}(A) - 1$.
8:         **if** $AB > 0$ **then** $m = -2^j$ **else** $m = 2^j$
9:         Let $\mathfrak{Q}(A, B, C) = \mathfrak{Q}(A, B, C)T_m$ and $U = UT_m$
10:     **if** $|A| > |C|$ **then**
11:         Let $\mathfrak{Q}(A, B, C) = \mathfrak{Q}(A, B, C)S$ and $U = US$.
12: **if** $\Delta > 0$ **then**
13:     Let m be such that $\sqrt{\Delta} - 2|A| \leq B + 2Am \leq \sqrt{\Delta}$
14:     Let $\mathfrak{Q}(A, B, C) = \mathfrak{Q}(A, B, C)T_m$ and $U = UT_m$
15: **if** $\Delta < 0$ **then**
16:     Assert $|B| \leq 2A \leq 2C$.
17:     **if** $|B| > A$ **then**
18:         **if** $B > 0$ **then** $m = -1$ **else** $m = 1$.
19:         Let $\mathfrak{Q}(A, B, C) = \mathfrak{Q}(A, B, C)T_m$ and $U = UT_m$
20:         Assert $|B| \leq \min\{A, C\}$.
21:         **if** $A > C$ **then**
22:             Let $\mathfrak{Q}(A, B, C) = \mathfrak{Q}(A, B, C)S$ and $U = US$.
23:     Assert $|B| \leq A \leq C$.
24:     **if** $B < 0$ and $A = C$ **then** Let $\mathfrak{Q}(A, B, C) = \mathfrak{Q}(A, B, C)S$ and $U = US$.
25:     **if** $B < 0$ and $A = -B$ **then** Let $\mathfrak{Q}(A, B, C) = \mathfrak{Q}(A, B, C)T_1$ and $U = UT_1$.
26: return $Q(A, B, C)$ and U.

ensure $|B| \leq A \leq C$. The form is reduced now except possibly for the sign of B. This part is handled in steps 24-25. Thus the form returned by the algorithm is reduced.

Finally, consider termination and complexity. Assume for the moment that $\text{size}(U) = O(n)$ throughout the algorithm. Consider the while-loops. Each application of T_m in step 9 strictly decreases $\text{size}(B)$, and an application of S does not change $|B|$. Hence, there are at most $\text{size}(B)$ executions of step 9. Note also that $|A|$ never increases and $|C|$ is bounded by the equation $B^2 - 4AC = \Delta$. Since each application of T_m in step 9 can be done in time $O(n)$, the total time spent in the while-loops is $O(n^2)$. The same time bound clearly applies to the remaining part of the algorithm.

We still need to argue that $\text{size}(U) = O(n)$ through the entire algorithm. Let $U = \{u_{ij}\}$. It will be enough to bound $\max\{\text{size}(u_{ij})\}$. Let t be the time interval used on a specific execution of the inner while loop. Let $j_0 = \text{size}(B) - \text{size}(A) - 1$

be the value of j in the first iteration of the specific execution of the while-loop. One may verify that $\max\{\text{size}(u_{ij})\}$ increases by at most $j_0 + 2$ during all the iterations in the time interval t, and clearly size(B) decreases by at least j_0+1 in the same time interval. Since there are at most $O(n)$ executions of the inner while loop, and size(B) never increases, we find that $\max\{\text{size}(u_{ij})\}$ and therefore also size(U) remains $O(n)$ throughout.

4.3 Complexity Analysis

Let us first consider steps 1,2 and 3 of Algorithm 1. By using asymptotically fast algorithms for integer multiplication, extended integer gcd and reduction of quadratic forms [16, 15, 17], it may be done in time $O(n \log^2 n \log\log n)$ by Lemma 4. Though asymptotically fast, this implementation may be impractical for moderate input sizes.

By using a simpler implementation such as naive multiplication/division, the binary algorithm for (extended) gcd, and our binary reduction algorithm, one may execute steps 1, 2 and 3 of Algorithm 1 in time $O(n^2)$ by Lemma 4 and Lemma 5.

If $\Delta < 0$ then steps 4 and 6 and hence the entire Algorithm 1 runs within the same time bound as steps 1-3.

If $\Delta > 0$ then we can upper bound the time for steps 5 and 6 as follows. It is known that the number of reduced forms of discriminant Δ is $O(\Delta^{\frac{1}{2}+\epsilon})$ for every $\epsilon > 0$ [14]. This implies that we need to apply ϱ repeatedly at most $O(\Delta^{\frac{1}{2}+\epsilon})$ times in step 5. The bit-size of a reduced form is $O(\log \Delta)$ by definition. So the matrix U corresponding to a single application of ϱ of Proposition 5 has also bit-length bounded by $O(\log \Delta)$. However, we can only bound the bit-length of the matrix U_2 computed in step 5 by $O(\Delta^{\frac{1}{2}+\epsilon})$. Hence, using naive arithmetic for step 5-6 and the simple implementation for steps 1-3, Algorithm 1 runs within time $O(n^2 + \Delta^{1+\epsilon})$ and when using asymptotically fast arithmetic throughout, Algorithm 1 runs within time $O(n \log^2 n \log\log n + \Delta^{\frac{1}{2}+\epsilon})$.

There seems to be no simple way to improve this analysis. Buchmann, Thiel and Williams [3] state "it can be shown under reasonable assumptions that there cannot be a polynomial time algorithm that on input of Δ and the norm of a principal ideal in $\mathcal{O}_\Delta$ computes the standard representation of a generator of such an ideal because the length of this representation is too big." If this also holds in our context, which is not general quadratic number rings, but only UFD's, we have that size(gcd(α, β)) may be super polynomial in size(α) + size(β) + $\log \Delta$.

We can summarize the complexity analysis as follows.

Theorem 1. *Let n = size(α) + size(β) + $\log \Delta$.*

For $\Delta < 0$, Algorithm 1 runs in time $O(n \log^2 n \log\log n)$.

For $\Delta > 0$, Algorithm 1 runs in time $O(n \log^2 n \log\log n + \Delta^{\frac{1}{2}+\epsilon})$ for every $\epsilon > 0$.

Remark. Our algorithms can be augmented to compute an extended gcd. One may also apply our algorithm in quadratic rings that are not UFD's, provided the inputs α, β generate a principal ideal. In the case of $\Delta > 0$ the matrix

U_2 of step 5 of Algorithm 1 may be precomputed for all distinct reduced forms of discriminant Δ, allowing the actual gcd algorithm to benefit from a table look-up. These topics will be elaborated in the full version of the paper.

Acknowledgment. The first author wishes to thank Hendrik Lenstra for suggesting the use of the relation between ideals and quadratic forms for gcd computation.

References

1. Saurabh Agarwal and Gudmund Frandsen. Binary GCD like algorithms in some complex quadratic rings. In *Proceedings of the sixth Algorithmic Number Theory Symposium*, volume 3076 of *Lecture Notes in Comput. Sci.*, pages 57–71, 2004.
2. Johannes Buchmann and Ingrid Biehl. An analysis of the reduction algorithms for binary quadratic forms. In *Voronoi's impact on Modern Science*, pages 71–98, 1998.
3. Johannes Buchmann, Christoph Thiel, and Hugh Williams. Short representation of quadratic integers. In *Computational algebra and number theory (Sydney, 1992)*, volume 325 of *Math. Appl.*, pages 159–185. Kluwer Acad. Publ., Dordrecht, 1995.
4. Harvey Cohn. *Advanced number theory.* Dover Publications Inc., New York, 1980. Reprint of *A second course in number theory*, 1962, Dover Books on Advanced Mathematics.
5. Ivan Bjerre Damgård and Gudmund Skovbjerg Frandsen. Efficient algorithms for gcd and cubic residuosity in the ring of Eisenstein integers. *J. Symb. Comput.*, 39(6):643–652, 2005.
6. Thomas W. Hungerford. *Algebra*, volume 73 of *Graduate Texts in Mathematics.* Springer-Verlag, New York, 1980. Reprint of the 1974 original.
7. Kenneth Ireland and Michael Rosen. *A classical introduction to modern number theory*, volume 84 of *Graduate Texts in Mathematics.* Springer-Verlag, New York, second edition, 1990.
8. Erich Kaltofen and Heinrich Rolletschek. Computing greatest common divisors and factorizations in quadratic number fields. *Math. Comp.*, 53(188):697–720, 1989.
9. Donald E. Knuth. *The art of computer programming. Vol. 2.* Addison-Wesley Publishing Co., Reading, Mass., second edition, 1981. Seminumerical algorithms.
10. J. C. Lagarias. Worst-case complexity bounds for algorithms in the theory of integral quadratic forms. *J. Algorithms*, 1(2):142–186, 1980.
11. Serge Lang. *Algebra.* Addison-Wesley Publishing Company, third edition, 1993.
12. D. H. Lehmer. Euclid's algorithm for large numbers. *American Mathematical Monthly*, 45:227–233, 1938.
13. Franz Lemmermeyer. The Euclidean algorithm in algebraic number fields. *Expo-sition. Math.*, 13(5):385–416, 1995. Updated version (Feb. 2004) at `http://www.fen.bilkent.edu.tr/~franz/publ/survey.pdf`.
14. H. W. Lenstra, Jr. On the calculation of regulators and class numbers of quadratic fields. In *Number theory days, 1980 (Exeter, 1980)*, volume 56 of *London Math. Soc. Lecture Note Ser.*, pages 123–150. Cambridge Univ. Press, Cambridge, 1982.
15. A. Schonhage. Schnelle berechnung von kettenbruchentwicklungen. *Acta Informatica*, 1:139–144, 1971.
16. A. Schönhage and V. Strassen. Schnelle Multiplikation grosser Zahlen. *Computing (Arch. Elektron. Rechnen)*, 7:281–292, 1971.

17. Arnold Schönhage. Fast reduction and composition of binary quadratic forms. In *ISSAC '91: Proceedings of the 1991 international symposium on Symbolic and algebraic computation*, pages 128–133, New York, NY, USA, 1991. ACM Press.
18. Victor Shoup. *A Computational Introduction to Number Theory and Algebra.* Cambridge University Press, 2005.
19. J. Stein. Computational problems associated with Racah algebra. *J. Comput. Phys.*, (1):397–405, 1967.
20. André Weilert. $(1+i)$-ary GCD computation in $\mathbf{Z}[i]$ as an analogue to the binary GCD algorithm. *J. Symbolic Comput.*, 30(5):605–617, 2000.
21. André Weilert. Asymptotically fast GCD computation in $\mathbb{Z}[i]$. In *Algorithmic number theory (Leiden, 2000)*, volume 1838 of *Lecture Notes in Comput. Sci.*, pages 595–613. Springer, Berlin, 2000.
22. Douglas Wikström. On the l-ary gcd-algorithm in rings of integers. In *Proceedings 32nd International Colloquium Automata, Languages and Programming*, volume 3580 of *Lecture Notes in Comput. Sci.*, pages 1189–1201, 2005.

On Clusters in Markov Chains

Nir Ailon[1,*], Steve Chien[2], and Cynthia Dwork[2]

[1] Princeton University,
Princeton NJ
nailon@cs.princeton.edu
[2] Microsoft Research,
Mountain View, CA 94043
{schien, dwork}@microsoft.com

Abstract. Motivated by the computational difficulty of analyzing very large Markov chains, we define a notion of clusters in (not necessarily reversible) Markov chains, and explore the possibility of analyzing a cluster "in vitro," without regard to the remainder of the chain. We estimate the stationary probabilities of the states in the cluster using only transition information for these states, and bound the error of the estimate in terms of parameters measuring the quality of the cluster. Finally, we relate our results to searching in a hyperlinked environment, and provide supporting experimental results.

1 Introduction

Motivated by the computational difficulty of analyzing very large Markov chains, we define a notion of clusters in (not necessarily reversible) Markov chains, and explore the possibility of analyzing a cluster "in vitro," without regard to the remainder of the chain. Given a cluster in an aperiodic and irreducible Markov chain, our goal is to approximate the relative stationary probabilities of the states within the cluster; that is, while we cannot know the total probability mass of the cluster at stationarity – this depends heavily on the rest of the chain – we may hope to learn, for each state in the cluster, the fraction of the cluster's mass at stationarity held by the given state. If the cluster is much smaller than the whole chain, then this analysis can be dramatically less expensive than, say, running power iteration on the whole chain to find the complete stationary distribution.

Although to our knowledge we are the first to explicitly define a notion of clusters for Markov chains, much previous work has noted a correlation between clusters in hyperlinked media and semantic topics (see [2] for a nice summary), and the interpretation of (a slight modification of) the WWW graph as a Markov chain is the basis for PageRank [1][1]. Given these precedents, it is a small step

* Work performed while author was visiting Microsoft Research.

[1] The elegant work of Madras and Randall [7], while explicitly decomposing a Markov chain into (not necessarily disjoint) pieces, deals with a converse problem: examine the pieces and a crude model of their interactions to analyze the rate at which the full chain mixes.

J.R. Correa, A. Hevia, and M. Kiwi (Eds.): LATIN 2006, LNCS 3887, pp. 43–55, 2006.

to defining Markov chain clusters. In the context of web search or analysis of other hyperlinked media, being able to analyze a cluster in isolation should give an inexpensive method of ranking the utility of different web pages on a given topic.

We have three contributions: the definition of a cluster, a theoretical analysis of the value of the definition, and related experimental results. Our formal measure of cluster in a Markov chain is based on the bicriteria measure used by Kannan, Vempala, and Vetta[3]; a similar intuition to ours underlies the definition of *community* of Flake, Lawrence, and Giles [2]. We show that our measure does in fact capture at least one desirable property that a cluster should intuitively have–namely, that the stationary distribution of a good cluster viewed as its own self-contained small Markov chain is close to that of its induced stationary distribution in the larger chain. Finally, we conduct experiments on both synthetic graphs and a large scale section of the web to test the applicability of our measure. Our results show, perhaps surprisingly, that the PageRank Markov chain is initially ill-suited for study by clusters, as its distinctive ϵ-reset parameter "blurs" clusters, making it difficult to isolate any one set of pages from the web at large. However, we show that we can still accurately estimate of a cluster's relative stationary distribution at a fraction of the cost of computing the global stationary distribution.

We now describe our results more fully.

A Formal Definition of Clusters in Markov Chains. There is a rich literature on measures of clusterings for *graphs*. Kannan, Vempala, and Vetta[3], proposed a measure for clustering in weighted similarity graphs that seeks to maximize the smallest conductance[2] (roughly, the flow) within the individual clusters while minimizing the fraction of total edge weight that crosses between clusters. (This generalizes the Flake, Lawrence, and Giles definition of a web community as "a set of pages that link (in either direction) to more pages within a community than to pages outside the community" [2].) A partitioning of the vertices into clusters is considered an (α, ϵ)-clustering if the conductance of each cluster is at least α and the combined weight of the inter-cluster edges is at most ϵ of the total edge weight.

In our applications, we will be more interested in individual clusters of states in a large Markov chain, and not necessarily a full partition of all states into clusters. However, we can still adapt the above definition of clustering to Markov chains in a natural way. In particular, we say that a set of states C in a Markov chain forms an (α, β)-cluster if the conductance *within* C is at least α, and the conductance from C to the rest of the chain is at most β. Intuitively, then, a cluster is a set of vertices within which a Markov chain mixes rapidly (due to α), but from which it is difficult to escape (because of β). The exact definition of an (α, β)-cluster will be given in Sect. 2.

---

[2] Both the measure in [3] for graphs and our measure for Markov chains refer to conductance. The concept is slightly different in the two settings, but intuitively similar.

Properties of Markov Chain Clusters. For our definition of clusters in Markov chains to be interesting, we must now show that an (α, β) cluster has useful non-trivial properties. Our main theoretical result is that clusters are well-isolated from the larger Markov chain in terms of their stationary distributions, in that we can obtain a relatively accurate approximation of the induced stationary distribution for a small cluster in a much larger Markov chain by examining only the cluster itself.

More precisely, given a set of states C in a large Markov chain P with stationary distribution π, let π^{real} be the stationary probability on C induced by π; namely $\pi_i^{real} = \pi_i/\pi(C)$, where $\pi(C) = \sum_{j \in C} \pi_j$. Let π^{est} (for "estimated") denote the stationary probability of C, treated as its own self-contained Markov chain[3]. Our main result (Theorem 4) is that if C is an (α, β)-cluster then the ℓ_1 difference between π^{real} and π^{est} is bounded by $c\frac{\beta}{\alpha^2} \log \frac{1}{\pi_{min}^{est}}$ for some global constant $c > 0$, where π_{min}^{est} is the minimum over all π_i^{est} for $i \in C$. This result shows that π^{real} and π^{est} will not be too far apart, so long as α is large and β is small, as we might intuitively expect.

Experimental Results. To test the applicability of our results in real world settings, we conducted experiments on two types of graphs. To test the basic feasibility of our approach, we first generated a series of random Markov chains with planted clusters. The underlying random graph model we used is well-behaved and represents a favorable situation for our approach, and we find that it is indeed possible to obtain a good approximation for the induced stationary distribution of a cluster in this setting.

We next performed tests on the PageRank Markov chain applied to a large crawl of the web with over 90 million pages and 2.4 billion links. Here we use individual domains (such as corporate and university web sites) as clusters, and find that the cluster's own stationary distribution π^{est} is a poor approximation for the induced stationary distribution π^{real}. It turns out that PageRank's ϵ-reset feature has the side effect of obscuring natural clusters in the underlying web graph by virtually guaranteeing that β will be at least ϵ for any set of pages. However, we show that with a small amount of preprocessing, we can still obtain good estimates in this setting.

2 Background and Definitions

2.1 Facts About Markov Chains

If P is a finite Markov chain over a set of n states V, we will write P as an $n \times n$ transition matrix in which p_{ij} is the probability of moving to state j given that the chain is in state i. We consider only chains that are finite-state and *regular* i.e., the transition matrix P satisfies $\exists k\, P^k > 0$. The stationary distribution (usually denoted π) is then principal left eigenvector of P. We will use several well-known facts about Markov chains, summarized next.

[3] The exact definition of the Markov chain associated with C appears in Sect. 3.

The *fundamental matrix* $Z = \{z\}_{ij}$ of a Markov chain P is the matrix

$$Z = \{z\}_{ij} = (I - (P - B))^{-1},$$

where $B = \lim_{k\to\infty} P^k$. The fundamental matrix captures effects of the choice of starting state: as $t \to \infty$, $\#_{ij}^{(t)} - t\pi_j = z_{ij} - \pi_j$, where $\#_{ij}^{(t)}$ is the expected number of times the chain starting in state i will visit state j in the first t steps (including the initial state as one step). The entry z_{ij} thus helps measure how many extra times the chain reaches j in the first t steps when started at i. This quantity may be negative. See [4] for a beautiful treatment of the fundamental matrix.

Fact 1. *Let P be the transition matrix of a Markov chain with fundamental matrix Z and stationary distribution π. Then $\pi Z = \pi$.*

Definition 1. *The* discrepancy[4] *of a Markov chain is the quantity*

$$\mathcal{Z} = \max_i \sum_j |z_{ij} - \pi_j|.$$

The mixing time $\mathcal{H}$ of a Markov chain measures how long it takes for the chain to converge to its stationary distribution from a worse-case start state. As is well known (see, e.g. [9]), the mixing time $\mathcal{H}$ is governed by the *conductance*, which is defined as follows.

Let V be the state space of the Markov chain. For any disjoint subsets $A, B \subset V$, define $Q(A, B) = \sum_{i\in A, j\in B} \pi_i p_{ij}$. For any $C \subset V$ we let $\Phi_V(C)$ denote the conductance (out of S to its complement in V), ie, $\Phi_V(C) = \frac{Q(C, V\setminus C)}{\pi(C)}$. We define the conductance (within V) to be $\Phi_V = \min_{C\subseteq V:\pi(C)\leq\frac{1}{2}} \Phi_V(C)$. It is common to omit the subscript V; however, since we will be talking about Markov chains induced by subsets of V, we sometimes explicitly name the state space for clarity.

Letting π_{min} denote the minimum stationary probability of any state, the conductance, mixing time, and discrepancy enjoy the following relationships (see [5, 9]):

Fact 2. $\mathcal{Z} \leq 4\mathcal{H} \leq \frac{64}{\Phi^2} \log \frac{1}{\pi_{min}}$.

It follows that sets C from which it is difficult to escape limit the rate of convergence. Markov chains that do not have such sets therefore mix rapidly. Note that if there is a very well isolated cluster then the conductance is low and so the mixing time is high. Thus, not only will our results be more meaningful when clusters are well isolated, but they will also be more useful (because power iteration necessarily must be run for more steps).

We will use the following deep theorem of Schweitzer.

Theorem 3 [Schweitzer]. *Let $P^{(1)}$ and $P^{(2)}$ be Markov chains on the same state space and with respective stationary distributions $\pi^{(1)}$ and $\pi^{(2)}$. Then $\pi^{(1)} - \pi^{(2)} = \pi^{(1)} E Z^{(2)}$, where $E = P^{(1)} - P^{(2)}$ and $Z^{(2)}$ is the fundamental matrix of $P^{(2)}$.*

[4] See [5]. Discrepancy is usually defined in terms of *hitting times*. The definition here is equivalent and simplifies our proof in Sect. 3.

2.2 The PageRank Markov Chain

Consider the Web as a very large graph, in which each page is a vertex and each link is a directed edge from the source page to the target page. Let M be the natural transition matrix associated with the web graph; namely, if a page i contains d_i links, then for each link (i, j), we have that $m_{ij} = 1/d_i$. We assume that the Markov chain is ergodic and aperiodic. (If $d_i = 0$, we can somewhat arbitrarily say that $d_{ii} = 1$, or alternatively, that $d_{ij} = 1/n$ for all j.)

The PageRank Markov chain is then defined as $P = (1-\epsilon)M + \epsilon U$, where U is the uniform matrix ($u_{ij} = 1/n$ for all i,j), and ϵ typically falls in the range $[0.1, 0.2]$. Adding ϵU ensures that the resulting chain is regular. The stationary distribution of P, denoted π, is called the PageRank vector, and the PageRank of an individual page (vertex) i is its i'th coordinate π_i.

2.3 Clusters in Markov Chains

Throughout, we let V denote the state space of the chain. Working from the above definitions, we define the concept of a cluster in a Markov chain as follows:

Fix a subset $C \subseteq V$ which will be our cluster for the rest of the discussion. Assume without loss of generality that the vertices of C correspond to the first $|C|$ rows and columns of P.

Let $\pi^{real} \in \mathbf{R}^{|C|}$ denote the projection of π onto C, normalized so that $\|\pi^{real}\|_1 = 1$ (from now on we let $\|\cdot\|$ denote the ℓ_1 norm). In other words,

$$\pi_i^{real} = \frac{\pi_i}{\pi(C)} \ .$$

The projected vector π^{real} is the (normalized) exact PageRank information restricted to the cluster C. (Presently we will define a small, related, Markov chain called P^{real}; it too will have stationary distribution π^{real}.) We are interested in efficient ways to approximate π^{real}.

Let P' be the submatrix of P corresponding to the rows and columns indexed by C. This is a *substochastic* matrix (the sum of the entries in each row is bounded by, but not necessarily equal to 1). We confine our attention to the case in which P' is regular (ensured for the PageRank Markov chain by the ϵ-reset). In this case, by the Perron-Frobenius Theorem, P' has a unique non-negative left principal eigenvector, which we denote by π^{est}, corresponding to an eigenvalue $0 < \lambda \leq 1$. We assume it is normalized ($\|\pi^{est}\| = 1$). In order to obtain a stochastic matrix with the same principal eigenvector π^{est}, we add a nonnegative multiple of π^{est} to each row of P' so that the resulting matrix, which we denote P^{est}, is stochastic. More precisely, if we denote the entries of P^{est} as p_{ij}^{est}, then

$$p_{ij}^{est} = p_{ij} + \pi_j^{est}\left(1 - \sum_{k \in C} p_{ik}\right) \ .$$

Thus, we redistribute the probability of escaping from C according to π^{est}.

It is not hard to verify that π^{est} is also the principal left eigenvector of P^{est} and that P^{est} is stochastic. Let ${\Phi_C}^{est}$ denote the conductance (within the small state space C) of P^{est}.

Definition 2. *The subset of states $C \subseteq V$ in a Markov chain is an (α, β)-*cluster *if $\pi(C) \leq 1/2$ and*

$${\Phi_C}^{est} \geq \alpha \quad \textit{and} \quad \Phi_V(C) \leq \beta \ .$$

If α is large and β is small, then C will be a cluster in the intuitive sense–a set of states within which it is easy to move, but from which it is difficult to escape. Also, note that the assertion ${\Phi_C}^{est} \geq \alpha$ depends on the substochastic matrix P' alone, and not on the entire matrix P.

Assume that C is an (α, β) cluster with relatively few states compared to the number of states in V. Clearly, π^{est} is easy to compute (because the state space is small), and π^{real} difficult. How good an estimate for π^{real} is π^{est}? The next section addresses this question, bounding $||\pi^{est} - \pi^{real}||$ in terms of α and β.

3 Bounding the Error $||\pi^{est} - \pi^{real}||$

To bound the error, we will consider two similar Markov chains, P^{real} and P^{est}, for which π^{real} and π^{est} are the respective stationary distributions, and apply previously known techniques to obtain a bound on $||\pi^{real} - \pi^{est}||$. The Markov chain P^{est} was defined above. We now show how to construct P^{real}.

Define the probability distribution

$$\tau_j^{real} = \frac{\sum_{k \notin C} \pi_k p_{kj}}{Q(V \backslash C, C)} \ .$$

This is the probability (at stationarity) that the chain moves to state j, given that it moves from $V \backslash C$ to C in one step. P^{real} is now defined by adding a nonnegative multiple of τ^{real} to each row of P' so that the resulting matrix is stochastic. Denoting the entries of P^{real} by p_{ij}^{real}, this means that $p_{ij}^{real} = p_{ij} + \tau_j^{real}\left(1 - \sum_{k \in C} p_{ik}\right)$. It is an easy exercise to verify that indeed the left principal eigenvector of P^{real} is π^{real}.

Thus, in both the construction of P^{real} and of P^{est}, we redirect probability drained from the substochastic matrix P' back into the system, according to τ^{real} and π^{est} respectively. For ease of notation in what follows, we let $\tau^{est} = \pi^{est}$, so that the τ's always refer to redirected probability mass.

To prove Theorem 4, we must bound $||\pi^{real} - \pi^{est}||$. Naturally, we will use Schweitzer's Theorem. We will see that the worst case occurs when τ^{real} is a point distribution, namely, concentrated at some pessimal choice of a page.

Theorem 4. *If C is an (α, β)-cluster, then the ℓ_1 difference between π^{real} and π^{est} is bounded by*

$$||\pi^{est} - \pi^{real}|| \leq c \frac{\beta}{\alpha^2} \log \frac{1}{\pi_{min}^{est}}$$

for some global constant $c > 0$, where $\pi_{min}^{est} = \min\{\pi_i^{est} \,|\, i \in C\}$.

Proof. By Schweitzer's Theorem, $\pi^{real} - \pi^{est} = \pi^{real} E Z^{est}$, where $E = P^{real} - P^{est}$. By choice of P^{real} and P^{est} we have $e_{ij} = (1 - \sum_{k \in C} p_{ik})(\tau_j^{real} - \tau_j^{est})$. and so $(\pi^{real} E)_j$ has the form

$$(\pi^{real} E)_j = \left(\sum_{i \in C} \pi_i^{real} \left(1 - \sum_{k \in C} p_{ik} \right) \right) (\tau_j^{real} - \tau_j^{est}) \ .$$

Note that the expression $\sum_{i \in C} \pi_i^{real} (1 - \sum_{k \in C} p_{ik})$ is exactly $\Phi_V(C)$, so $(\pi^{real} E)_j = \Phi(C)(\tau_j^{real} - \tau_j^{est})$.

We can therefore write

$$\pi^{real} E Z^{est} = \Phi_V(C)(\tau^{real} - \tau^{est}) Z^{est} \ .$$

Thus, using that $\tau^{est} = \pi^{est}$ (by definition) and $\pi^{est} Z^{est} = \pi^{est}$ (Fact 1), we get a bound in terms of the discrepancy of P^{est}:

$$\begin{aligned} \|\pi^{real} E Z^{est}\| &= \Phi_V(C) \sum_j |(\tau^{real} Z^{est})_j - (\tau^{real} Z^{est})_j| \\ &= \Phi_V(C) \sum_j |(\tau^{real} Z^{est})_j - \pi_j^{est}| \\ &\le \max_i |z_{ij}^{est} - \pi_j^{est}| = \mathcal{Z}^{est} \ . \end{aligned}$$

The last step follows from a convexity argument: $\sum_j |(\tau^{real} Z^{est})_j - \pi_j^{est}|$ is a convex function of τ^{real}, maximized at a point distribution.

Applying Fact 2, we have

$$\|\pi^{real} - \pi^{est}\| \le \Phi_V(C) \mathcal{Z}^{est} \le 4\mathcal{H}^{est} \le \frac{64}{(\Phi_C{}^{est})^2} \log \frac{1}{\pi_{min}^{est}},$$

From the assumption that C is an (α, β)-cluster, we may conclude the proof:

$$\|\pi^{real} - \pi^{est}\| \le 64 \frac{\beta}{\alpha^2} \log \frac{1}{\pi_{min}^{est}} .$$

4 A More Elementary Proof

The proof in this section is almost from first principles; in particular, it does not go through Schweitzer's Theorem. We begin by establishing some notation and making an observation. We then give some intuition for the approach.

We decompose the transition matrix P as follows $P = \begin{pmatrix} P' & \text{Out} \\ \text{In} & R \end{pmatrix}$, where $P' \in \mathbf{R}^{|C| \times |C|}$ is, as above, the transition matrix restricted to the rows and columns corresponding to our (α, β)-cluster $C \subseteq V$.

Consider the step evolution of P on its stationary distribution $\pi \in \mathbf{R}^n$ which we decompose as $\pi = (\pi(C)\pi^{real} \,|\, \epsilon)$ where $\epsilon \in \mathbf{R}^{n-|C|}, \epsilon \ge 0$. We get, $\pi = \pi P = (\pi(C)\pi^{real} P' + \epsilon \text{In} \,|\, \epsilon)$. Continuing inductively, we get that for all $t \ge 0$,

$$\pi = \pi P^t = \left(\pi(C)\pi^{real} P'^t + \sum_{i=1}^{t} \epsilon \mathrm{In} P'^{(t-i)} \middle| \epsilon \right).$$

The intuition behind this decomposition is as follows: the term $\pi(C)\pi^{real}P'^t$ corresponds to the original distribution π^{real} circulating in C. Some of it may drain out due to sub-stochasticity of P', but the direction of this vector converges to that of π^{est}. The part that will eventually contribute to $\|\pi^{real} - \pi^{est}\|$ is the *noise* term $\sum_{i=1}^{t} \epsilon \mathrm{In} P'^{t-i}$ entering C from $V \setminus C$, which we will bound using the fact that C is an (α, β) cluster.

Now, $\pi(C)\pi^{real} = \pi(C)\pi^{real}P'^t + \sum_{i=1}^{t} \epsilon \mathrm{In} P'^{(t-i)}$. Also, $\|\pi(C)\pi^{real}\| = \pi(C)$ $= \|\pi(C)\pi^{real}P'^t\| + \sum_{i=1}^{t} \|\epsilon \mathrm{In} P'^{(t-i)}\|$. Therefore,

$$\pi(C)\pi^{real} - \pi(C)\pi^{est} = \pi(C)\pi^{real}P'^t + \sum_{i=1}^{t} \epsilon \mathrm{In} P'^{(t-i)} - \left(\|\pi(C)\pi^{real}P'^t\| + \sum_{i=1}^{t} \|\epsilon \mathrm{In} P'^{(t-i)}\| \right) \pi^{est}$$

Taking norms, rearranging terms and applying the triangle inequality,

$$\|\pi(C)\pi^{real} - \pi(C)\pi^{est}\| \le \left\|\pi(C)\pi^{real}P'^t - \|\pi(C)\pi^{real}P'^t\|\pi^{est}\right\| + \sum_{i=1}^{t} \left\| \epsilon \mathrm{In} P'^{(t-i)} - \|\epsilon \mathrm{In} P'^{(t-i)}\|\pi^{est} \right\|. \quad (1)$$

Lemma 1. *For any lazy*[5] *Markov chain P and subset C there exist constants $\gamma > 0$ and $0 < \mu < 1$ such that for all positive vectors $w \in \mathbf{R}^{|C|}$*

$$\left\| wP'^t - \|wP'^t\|\pi^{est} \right\| \le \gamma(1-\mu)^t \left\| w - \|w\|\pi^{est} \right\| \le 2\gamma(1-\mu)^t \|w\| \quad (2)$$

holds with $\gamma = \sqrt{1/\pi^{est}_{min}}$ and $\mu = ({\Phi_C}^{est})^2/4$.

(The proof is sketched in Section 4.1.) In other words, the vector wP'^t tends to $\|wP'^t\|\pi^{est}$ exponentially fast with rate $(1-\mu)$. Note that $\|wP'^t - (\|wP'^t\|)\pi^{est}\|$ cannot exceed $2\|wP'^t\| \le 2\|w\|$, therefore, by Equation (2),

$$\left\| wP'^t - \|wP'^t\|\pi^{est} \right\| \le 2\|w\| \min\left\{ \gamma(1-\mu)^t, 1 \right\}. \quad (3)$$

We will use this to prove the theorem for lazy Markov chains and will argue at the end of the section that the restriction to lazy chains is irrelevant.

Plugging (3) into (1), we get that for all $t \ge 0$,

$$\|\pi(C)\pi^{real} - \pi(C)\pi^{est}\| \le 2\gamma(1-\mu)^t \|\pi(C)\pi^{real}P'^t\| + 2\sum_{i=1}^{t} \|\epsilon \mathrm{In}\| \min\{\gamma(1-\mu)^{(t-i)}, 1\}.$$

[5] Every state has transition probability of at least 1/2 to itself. Also commonly known as a strongly aperiodic chain.

Now we notice that by definition, $\|\epsilon \text{In}\| = Q(V\backslash C, C) = Q(C, V\backslash C)$. Dividing both sides by $\pi(C)$, recalling that $Q(C, V\backslash C)/\pi(C) = \Phi_V(C) \leq \beta$ by assumption, and taking the limit as $t \to \infty$, we obtain $\|\pi^{real} - \pi^{est}\| \leq 2\beta \sum_{i=0}^{\infty} \min\{\gamma(1 - \mu)^i, 1\}$. Let a be the minimal integer such that $\gamma(1-\mu)^a \leq 1$. Then $a \leq \lceil \frac{-\log \gamma}{\log(1-\mu)} \rceil$, and

$$\begin{aligned}\sum_{i=0}^{\infty} \min\{\gamma(1-\mu)^i, 1\} &= \sum_{i=0}^{a-1} 1 + \sum_{i=a}^{\infty} \gamma(1-\mu)^i \leq a + \sum_{i=0}^{\infty} (1-\mu)^i \\ &\leq \left\lceil \frac{-\log \gamma}{\log(1-\mu)} \right\rceil + \frac{1}{\mu} \leq \frac{\log \gamma}{\log(1+\mu)} + 1 + \frac{1}{\mu} \leq \frac{\log \gamma/\log 2 + 2}{\mu}\end{aligned} \tag{4}$$

(we used the fact that $\mu \log 2 \leq \log(1+\mu)$ for $0 < \mu < 1$). Therefore, $\|\pi^{real} - \pi^{est}\| \leq 2\beta(\log \gamma/\log 2 + 2)/\mu$. Taking $\gamma = \sqrt{1/\pi^{est}_{min}}$ and $\mu = ({\Phi_C}^{est})^2/4$ as in Lemma 1, and recalling that ${\Phi_C}^{est} \geq \alpha$ by the (α, β)-cluster assumption, we conclude that for lazy Markov chains, $\|\pi^{real} - \pi^{est}\| \leq 8\beta(\frac{1}{2}\log(1/\pi^{est}_{min})/\log 2 + 2)/\alpha^2$.

If $|C| > 1$, then $1/\pi^{est}_{min} \geq 2$, thus $\|\pi^{real} - \pi^{est}\| \leq (20/\log 2)\beta\alpha^{-2} \log \frac{1}{\pi^{est}_{min}}$. If $|C| = 1$ then $\|\pi^{real} - \pi^{est}\| = 0$. This proves Theorem 4 for lazy Markov chains. We now claim that the laziness requirement is non-restrictive. Indeed, we could replace P with a lazy Markov chain $\frac{1}{2}P + \frac{1}{2}I$. The vectors π^{real} and π^{est} for this chain are the same as for the original one. However, $\Phi_V(C)$ and ${\Phi_C}^{est}$ are decreased by a factor of 2. We conclude that for any Markov chain and (α, β)-cluster C,

$$\|\pi^{real} - \pi^{est}\| \leq \frac{(40/\log 2)\beta}{\alpha^2} \log \frac{1}{\pi^{est}_{min}} . \tag{5}$$

(Note that the constant 40 is conservative in the sense that we only assumed that $|C| \geq 2$, but it can replaced with $8 + \delta$ for any small $\delta > 0$ assuming $|C|$ is sufficiently large). It remains to prove Lemma 1.

4.1 A Bound on μ, γ Using Conductance

Recall that the matrix P^{est} is obtained by adding a nonnegative multiple of π^{est} to each row of P' such that the resulting matrix is stochastic. The vector π^{est} is the principle left eigenvector of P^{est} corresponding to the eigenvalue 1. Write this as $P^{est} = P' + T$, where the i'th row of T is $\pi^{est}\left(1 - \sum_{j \in C} p_{ij}\right)$. Now, for any vector $w > 0$ and integer $t \geq 0$,

$$wP'^t - \|wP'^t\|\pi^{est} = w(P^{est})^t - w\left((P^{est})^t - P'^t\right) - \|wP'^t\|\pi^{est} \tag{6}$$

Further manipulation shows:

$$wP'^t - \|wP'^t\|\pi^{est} = w(P^{est})^t - \|w(P^{est})^t\|\pi^{est} . \tag{7}$$

Bounds on the right hand expression have been extensively studied. It follows from [8] (Sect. 3) that if P^{est} is lazy then we get the required conclusion:

$$\left\| w(P^{est})^t - \|w(P^{est})^t\| \pi^{est} \right\| \le \sqrt{\frac{1}{\pi^{est}_{min}}} \left(1 - (\Phi_C{}^{est})^2/4\right)^t \left\| w - \|w\| \pi^{est} \right\| .$$

5 Experimental Results

As one of our main goals was to study large real-world Markov chains, we conducted a series of experiments on both synthetic and web graphs to evaluate how applicable our theoretical results might be in practice. We present two sets of results – first, in an idealized situation where we have planted clusters in a set of randomly generated graphs, and second, in a large crawl (over 90 million pages and 2.4 billion edges) of the actual web. We describe both of these below.

5.1 Synthetic Graphs

As a simple initial test, we constructed random graphs with planted clusters. In this $G_{n,p,m,q}$ model, we first generate a graph on n vertices according to $G_{n,p}$, where each directed edge appears with probability p. We then replace the subgraph on m of these vertices with a graph generated according to $G_{m,q}$, with $q \ge p$; these m vertices will form our planted cluster. In these small toy experiments, we held n at 1000 and m at 100.

We define a (non-reversible) Markov chain on this graph in the natural way – each vertex is a state, whose outgoing transition probability is divided among its out-neighbors. We expect the isolated stationary distribution of the planted cluster (π^{est} in the notation of Section 3) to be close to that induced by the stationary distribution of the large chain (π^{real}) when q is large and p is small.

We generated a series of random graphs fixing q at 0.3, while varying p from 0.01 to 0.03. Here the value of β should increase as p increases, leading to an increase in our error $\|\pi^{real} - \pi^{est}\|$. We generated 10 graphs for each value of p, and show the mean error in Figure 1.

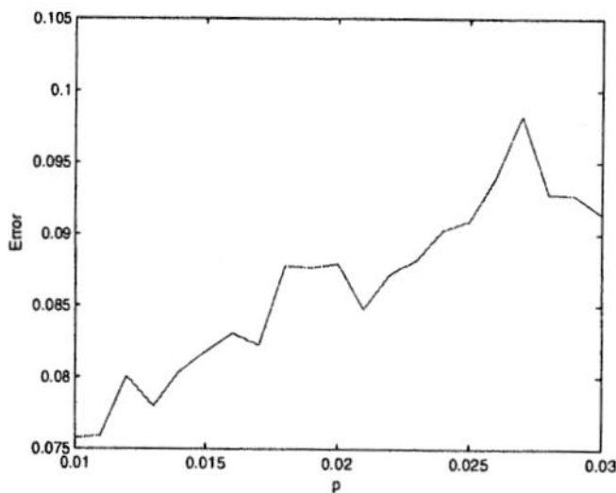

Fig. 1. Error for $G_{1000,p,100,0.3}$

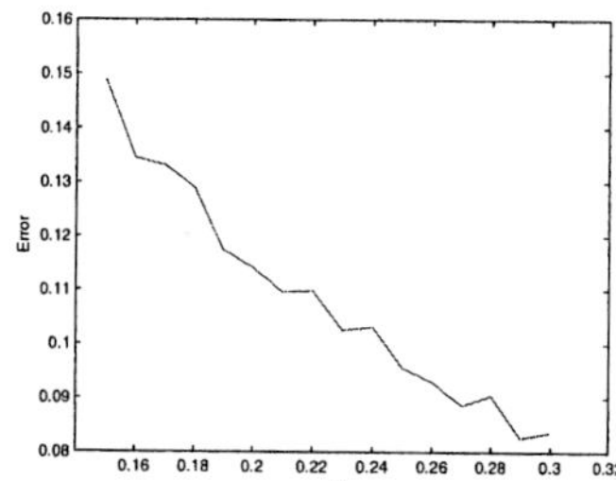

Fig. 2. Error for $G_{1000,0.015,100,q}$

For Figure 2, we fix p at 0.015, while varying q from 0.15 to 0.3. The value of α should increase as q increases, leading to a decrease in approximation error. (Note the vertical scales do not start at zero in these figures.)

The main observation here is that the approximations π^{est} are relatively accurate, even though the values of β for these clusters can actually be quite large. We are helped here not only by the clusters themselves, but by the global structure of the graph. Since in this random graph model the induced stationary distributions both inside and outside the cluster should not be too far from uniform, we are in the favorable situation where the incoming probability provided by the larger chain to the cluster (τ^{real}) is not far from that provided by the cluster in isolation ($\tau^{est} = \pi^{est}$).

5.2 Web Graphs

We then conducted large-scale experiments on a crawl of the web consisting of 90,560,988 web pages and 2,419,954,245 links. On this graph, we analyzed the classic vanilla PageRank Markov chain with an ϵ-reset of 0.1. Without the planted clusters in the synthetic graphs, the question of how to find clusters in the web graph becomes important. We imagine that clusters will be determined based on external information (either through textual analysis – e.g. all pages containing "Manchester United" or all pages belonging to a specific domain). In our experiments, we used the latter approach – selecting all pages that belong to `yahoo.com` (2,179,242 pages), `microsoft.com` (42,511 pages), `princeton.edu` (28,486 pages), and `stanford.edu` (72,970 pages) as four candidate clusters. These candidate clusters appear very promising at first, as at stationarity in the large PageRank chain, very little probability mass is carried out of the cluster via natural links (as opposed to the ϵ-reset). As the following results show, however, the differences between π^{real} and π^{est} for each of these clusters is very large:

	`yahoo.com`	`microsoft.com`	`princeton.edu`	`stanford.edu`
Error	0.624	1.761	0.700	0.739

Upon inspection, the reason for this becomes clear. Even though the actual links in the web graph keep almost all of the probability mass of each cluster within it, PageRank's ϵ-reset guarantees that β will be at least $\epsilon(1-\eta)$, where η is the fraction of pages in the web graph that belong to the cluster (and thus almost negligible). Also important is that unlike in the synthetic graphs above, the incoming probability to the cluster from the larger web (τ^{real}) is very different from the cluster's own stationary distribution, which we use for τ^{est}. This problem will likely be present in any proposed cluster in the web.

However, our earlier analysis indicates that we should still be able to accurately approximate a domain's induced PageRank if we have some estimate of τ^{real}. One natural idea is to run the global PageRank Markov chain P for a small number of iterations t, and then use the resulting probability distribution x^t to obtain an estimate τ^{est} for τ^{real}. Specifically, for any page j in the cluster C, τ_j^{est} will be proportional to $\sum_{i \notin C} x_i^t p_{ij}$.

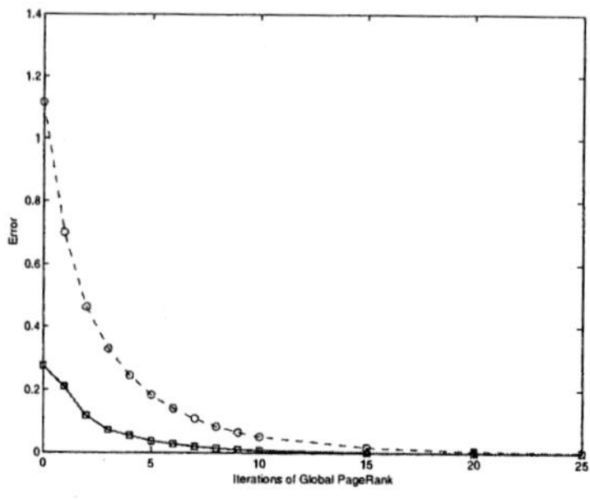

Fig. 3. Error for `princeton.edu`

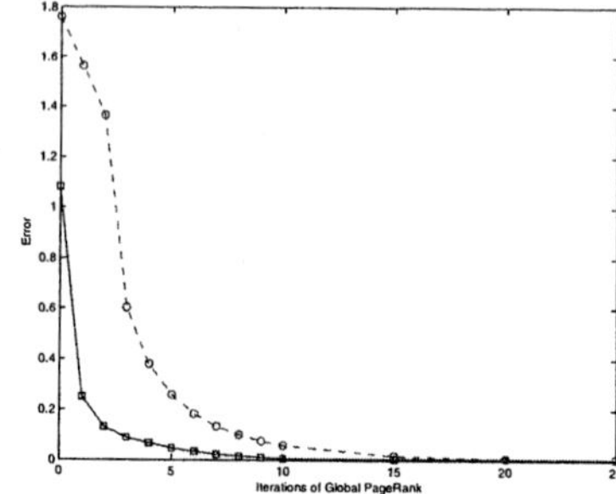

Fig. 4. Error for `microsoft.com`

The results of this approach are shown in Figures 3 and 4 for `princeton.edu` and `microsoft.com`. In both figures, the lower solid curve indicates our approximation error $\|\pi^{real} - \pi^{est}\|$, as a function of t, when we use x^t to generate τ^{est}. By comparison, the higher dashed curve indicates the error if we were to simply use the distribution on C induced by x^t; i.e. if we use $p_i^t/p^t(C)$ for each page i. We see that our error is small even for very low values of t (around 0.1 for $t = 2$ and 0.05 for $t = 5$), and much better than if we had used only the global PageRank vector x^t. (In fact, for `princeton.edu`, the error at $t = 0$ might be reasonable, and can be obtained without any computation on the crawl.) Thus, we can take advantage of clusters in the web even in this more challenging setting.

Acknowledgements. We thank Láci Lovász, Frank McSherry, and Alistair Sinclair for numerous helpful discussions.

References

[1] S. Brin and L. Page, "The anatomy of a large-scale hypertextual web search engine", *Computer Networks* **30(1-7)** pp. 101-117, 1998.

[2] G. Flake, S. Lawrence, and C. Giles, "Efficient Identification of Web Communities", *Proceedings of the Sixth International Conference on Knowledge Discovery and Data Mining (ACM SIGKDD-2000)*, pp.150–160, 2000.

[3] R. Kannan and S. Vempala and A. Vetta, "On clusterings: Good, bad and spectral", *Journal of the ACM*, **51(4)**, pp.540–556, 2004.

[4] J.G. Kemeny and J.L. Snell, *Finite Markov Chains*, New York: D. Van Nostrand Co, Inc., 1960

[5] L. Lovász and P. Winkler, "Mixing times", *Microsurveys in Discrete Probability* (ed. D. Aldous and J. Propp), DIMACS Series in Discrete Mathematics and Theoretical Computer Science, AMS, 1998, pp. 85–133.

[6] F. Chen and L. Lovász and I. Pak, "Lifting Markov chains to speed up mixing", *Proceedings of STOC*, 1995.

[7] N. Madras and D. Randall, "Markov Chain Decomposition for Convergence Rate Analysis", *The Annals of Applied Probability, 12*(2), pp.581–606, 2002.

[8] M. MIHAIL "Conductance and Convergence of Markov Chains - A Combinatorial Treatment of Expanders", *Proceedings of STOC*, 1989.

[9] R. MONTENEGRO, "Edge and vertex expansion bounds on eigenvalues of reversible Markov kernels", preprint.

[10] P.J. SCHWEITZER, "Perturbation Theory and Finite Markov Chains", *J. Applied Probability* 5(3), pp.401–404, 1968

An Architecture for Provably Secure Computation

Miklós Ajtai[1], Cynthia Dwork[2], and Larry Stockmeyer

[1] IBM Almaden Research Center
ajtai@almaden.ibm.com
[2] Microsoft Research
dwork@microsoft.com

Abstract. We describe an architecture requiring very few changes to any standard von Neumann machine that provably withstands coalitions between a malicious operating system and other users, in the sense that:

1. If the operating system permits a program to run, then the program produces the same outputs as it would produce if it were running on an ideal, single-user machine; moreover, even if the operating system behaves according to expectations only most of the time, the programs get executed.
2. The only information leaked by a program to the malicious coalition is the time and space requirements of the program.
3. If the malicious operating system is dynamically replaced by a good operating system, then the latter can quickly and correctly determine what memory resources are available for future programs, as well as how much time is left for each of the currently executing programs, and can distribute these resources without any restrictions. This can be accomplished without restarting currently executing programs.

To our knowledge, ours is the first attempt to provide provable guarantees along these lines, and the first treatment of any kind, provable or otherwise, for the third property.

1 Introduction

The problem of correctness of programs is central to many fields in computer science. In the area of formal methods, there has been extensive research on program specification and verification; in the theory community, work on checking, self-checking, and self-correcting programs, as well as checking memories, has been quite influential. These efforts typically assume a well-behaved operating system that does not, for example, tamper with the actual programs being run, and that cannot be exploited by a malicious program, such as a virus or a Trojan horse, to tamper with other programs or their results.

The trustworthiness of the operating system may have nothing to do with the intentions of the designer of the software. An operating system is so complex that many regard the task of proving correct such a large program to be infeasible in

J.R. Correa, A. Hevia, and M. Kiwi (Eds.): LATIN 2006, LNCS 3887, pp. 56–67, 2006.

practice (if not in theory). Such quasi-inherent vulnerabilities can have a chilling effect on software development.

There have been some efforts toward addressing these concerns; in particular, trusted platforms such as the Palladium chip (or NGSCB), the Trusted Mobile Platform [4], and the fine research on XOM (Execute Only Memory) [1–3][1]. These previous works are very holistic: they attempt to provide full programming functionality, with all the capabilities of current machines and languages. Perhaps for this reason, there is no full specification of the threat model, and no full description of what is provided in the face of such (an unspecified) threat. Some steps in this direction are taken in [3], which employs model checking to assist in verifying certain aspects of XOM[2].

We focus on what can *provably* be achieved using a slightly altered instruction set for a von Neumann machine. Although the scope of our results is more modest – we consider only a simple model, in which each user can submit a single program, there is no communication and no external memory – we believe that a theoretically sound approach is warranted, as in general it is hopeless to prove things about extremely complex systems. The proof of concept described here is an important step toward a completely general result.

We note that many natural modifications, for example, modifying the model to allow programs to arrive dynamically, may be achieved without substantial changes in the architecture or the notions of correctness and privacy. For simplicity, in this extended abstract we restrict our attention to the more basic setup.

We first define an ideal machine, which hosts a single user, by describing an instruction set for the machine. There is nothing unusual about the ideal machine. We then define *conforming programs*. These are programs that assume the memory to which they have access is arranged in a linked list, or *chain*, of blocks of a given size, and that produce the same outputs independent of which physical blocks are organized into this chain. This is a very rich class of programs, as any program can be written in a conforming fashion. Correctness for conforming programs is defined by their behavior on the ideal machine. That is, we only ensure correctness relative to behavior on the ideal machine, with no outside notion of specification etc for the programs.

Next we define an architecture for a real-life, multi-user machine. The multi-user machine is a von-Neumann machine with a few extra instructions and a few compound instructions, that is, sets of a small number of instructions that must be executed atomically (either all instructions in the set are executed or none of them is). The heart of our work is the construction and dismantling of the chain of memory blocks used for each program. Our programs begin with a declaration of their time and space requirements. The architecture ensures that execution of a program cannot begin until all its requested resources have been allocated to it; that memory allocated to a program cannot be accessed by any

[1] In these cases one of the principal applications is content protection; this is not our goal, and our work has no bearing on that topic.

[2] As the authors note, model checking does not provide a rigorous proof of correctness.

other program – including the operating system; and that resources cannot be taken away from the program before it has been allowed to run for its declared time bounds. The architecture also ensures that dismantling completes; in other words, the system ensures protection against memory leaks.

Actual allocation of resources – which programs are given memory to run in, and which programs may take steps when – is under control of the operating system. Thus, programs may be denied resources, and they may fail to make progress. Intuitively, however, we ensure that any progress is good progress: running for some number of steps on the multi-user machine yields the same outputs as running for the same number of steps on the ideal machine.

Finally, we give a complete characterization of the adversary and a rigorous proof that for any such adversary, our architecture protects against the following three adversarial goals. Here P is the program of an uncompromised user.

Causing Incorrect Outputs: Program P outputs a value different from what would be produced were P running on the ideal machine and given the same inputs.

Compromising Program Secrecy: Intuitively, the adversary "learns something about" the program P. This is formalized by distinguishing between two programs, P and P', having the same declared time and space requirements.

Poisoning the Well: A faulty operating system (possibly colluding with a subset of the users) destroys the data structures used for managing memory and for keeping track of which programs are currently executing. The adversary succeeds if, when the faulty operating system is replaced by a good operating system, the latter cannot in constant time find and correctly allocate new resources, assuming they are available, or de-allocate existing resources, or if it cannot permit currently executing programs to complete their execution.

To our knowledge, ours is the first work to even articulate protection against the third adversarial goal, on which we now elaborate. We may think of the operating system as "moody" in that sometimes it is operating (intuitively) correctly, and sometimes not. There can be many reasons for moody behavior. For example, being large and complex, certain parts of the system may be correctly written, while others are flawed, and a change in "mood" might correspond to a jump to a different part of the operating system. Alternatively, except in some unusual combination of circumstances, or in the absence of other malicious programs, the operating system might work correctly, but an unfortunate combination of events may occur, or a malicious program might manage to exploit a vulnerability in the system, that, for example, causes it to fail to correctly allocate resources. Finally, an operating system in a bad state might be re-started (without restarting the other, concurrently running, programs), which might restore it to a good state.

We would like that when the operating system is in a bad mood it cannot "poison the well," that is, make things impossible or even very difficult for a future, correct operating system. Such "poisoning" could be achieved by maliciously modifying the data structures describing available resources, for example by failing to return freed memory to a free list. Such a situation can frequently be corrected by rebooting the entire machine, but this could cause all progress to be lost on currently executing programs, which we view as unacceptable.

We prevent this situation from arising. Essentially our architecture specifies an interface, i.e., a set of data structures and methods (functions) for manipulating them. The architecture ensures that the operating system's ability to manipulate the data structures is restricted to the given set of methods. The methods in turn ensure that a certain set of simple programs can efficiently carry out various operating system tasks, such as accurately determining which memory blocks are available, allocating memory, allowing programs to be loaded into the machine and started, de-allocating memory, and enabling currently executing programs to take steps. We can write these simple programs, given the interface. Decisions about which ones to invoke, and when, are up to the operating system, but a bad operating system cannot make the efficient and correct execution of these procedures impossible; it cannot sabotage its own data structures. Thus we also protect against memory leaks, and ensure that the operating system can always be restarted without harming executing programs.

We close this section with some remarks about computational and physical assumptions. Our results do not rely on any computational assumptions. We use no cryptography; compartmentalization is ensured by the architecture. We assume that memory cannot be tampered with or otherwise accessed except by the CPU. We further assume that a user is connected to the machine via a secure channel (e.g. a terminal), and that there is a known upper bound n on the number of users. We treat the most basic case, in which each user i wishes to run a single program P_i. We assume programs from non-malicious users are conforming. No privacy or correctness guarantees are made for non-conforming programs, although naturally a non-conforming program will not be able to compromise the privacy or correctness of any conforming program. Finally, we do *not* impose an artificial restriction on the distribution of the resources between the various users; that is, we do not solve the problem by simply partitioning the memory once and for all into disjoint blocks so that each user can access only a single, predetermined, block. Such a solution would be inefficient, since under this arrangement the unused memory of one user cannot be made available to another user. We will describe a general condition on the machine which excludes these types of inefficiencies. Intuitively, programs are allocated only the resources they request; if a new program arrives and the resources it requests are available, in particular, if sufficient memory is available, and the operating system is in a good mood, then the operating system should be able to allocate the desired amount of memory to the new program.

In the next several sections we give a bit more detail about the basic components discussed above. To achieve the necessary degree of rigor to substantiate our claims of provability requires extremely detailed definitions and arguments. These are given in the full paper[3].

[3] To be made available on the Electronic Colloquium on Computational Complexity, `http://www.eccc.uni-trier.de/eccc`.

2 The Ideal Machine and Conforming Programs

The goal of a user is to execute a program. We assume that the programs were originally written for a von Neumann type random access machine with a single input/output channel. To make this assumption more explicit we fix a machine $\mathcal{N}_{q,m}$ of this type, consisting of m registers each containing q bits. Although we give (in the full paper) a complete description of the instruction set of this machine, the particular choice of the instructions is not important; our results could be formulated with any instruction set which includes input/output instructions for the single channel. The only exception is the instruction `INPUT` defined below. This is a special input instruction which will be able to cause a program to begin execution. (There is another input instruction `input` with the usual meaning.)

Several of the instructions for $\mathcal{N}_{q,m}$ have parameters. We assume that if the instruction requires k parameters and the instruction is in location u, then the arguments are in locations $u+1, \ldots, u+k$.

A program for $\mathcal{N}_{q,m}$ is a sequence of integers followed by an end-of-program delimiter. The special single-parameter input instruction `INPUT` x treats x as the starting location into which the program should be loaded (from the unique input/output channel). The effect is a loop during which a program instruction (integer in the sequence) is read and, if it is not the delimiter, it is stored in location x and x is incremented to $x+1$. When the delimiter is reached control is transferred to the initial instruction of the newly loaded program. More precisely, if the `INPUT` instruction is in location L of $\mathcal{N}_{q,m}$ then control is transferred to location $L+2$. We therefore have the following convention for invoking the `INPUT` x instruction: if the `INPUT` instruction is in location L of $\mathcal{N}_{q,m}$ then, on invocation, x should have value $L+2$ (since `INPUT` is a one-parameter instruction, with parameter x, we have that x itself is in location $L+1$).

In real machines programs are allocated memory in blocks (pages), and the actual memory used need not occupy a single contiguous region. The same will be true on our multi-user machine. The blocks will be organized into a linked list, and each block will have some fixed size, or number of registers, ξ. We will therefore make the simplifying assumption that the program originally written for the ideal machine $\mathcal{N}_{q,m}$ is already designed to tolerate storage allocations of this type. (We may think of the actual set of blocks allocated as a set of additional inputs provided to the program.) Such a program is called a *conforming* program. A conforming program has the property that its outcome depends only on the program itself (including its input), and not on the choice of the blocks allocated to it: the output of P is uniquely determined by P. Note that a conforming program never tries to access a register not allocated to it.

When a conforming program is run on the multi-user machine our notion of correctness will be with respect to its execution on $\mathcal{N}_{q,m}$. If allowed to run to completion, the outputs should be the same as those produced when the program is run on $\mathcal{N}_{q,m}$ (because the program is conforming, it is agnostic with regard to which blocks of $\mathcal{N}_{q,m}$ it is actually assigned.) More generally, since the program may not be run to completion, we require, intuitively, that at any point in the execution on the multi-user machine, the output sequence produced is an initial segment of the output sequence produced when run to completion on $\mathcal{N}_{q,m}$. It is important that in speaking about correctness, when we speak of output of a conforming program we mean only a sequence of output bits (or integers), without their timing.

We will discuss the privacy requirement after giving further details about the multi-user machine.

3 The Multi-user Machine and the Adversary

The intuition behind our secure machine, called $\mathcal{M}$, is that the operating system never exactly "does" anything, it just enables certain pre-specified sets of state transitions to be (atomically) executed. That is, the architecture specifies an interface – data structures and methods, or functions, for manipulating them – and the operating system's ability to manipulate these structures is restricted to the given set of methods. Given the interface it is easy to write a set of simple programs to efficiently carry out the various operating system tasks, such as allocating memory (if and only if available) and permitting a user's program to take a step. Decisions about which tasks to schedule, and when, are up to the operating system.

3.1 The Machine $\mathcal{M}$

In our simple model of multi-user computation, a number n of users share the resources of a single machine. Each user has its own dedicated input/output channel for communicating with the machine. We treat the operating system as a special user, who sends instructions to the machine through a special *control* channel.

We assume that the machine works in discrete time units. At each time the operating system sends an instruction, through the control channel, and as a result the machine changes the contents of a constant number of its registers. A *compound* instruction is a small number of simple instructions, to be executed atomically; that is, either all the simple instructions are executed or none are. We assume that the architecture supports atomic execution of compound instructions. These compound instructions are the methods, or functions, mentioned earlier. A simple instruction can be executed only as a part of a compound instruction. This restriction will ensure that, unlike in a machine with von Neumann type architecture, the operating system cannot read or change the contents of the registers in an arbitrary way. We will guarantee the desired properties of the machine through the right choice of the set of compound instructions.

We will require three types of instructions for the operating system. One type, roughly speaking, is needed for interacting with the user; for example, obtaining resource requests. Instructions of this type are not at all new. A second will be needed to enable the user's program to take steps. These are slightly unusual, and must ensure that the operating system does not learn the instructions executed by the user's program. Also, they have "side effects" involving bookkeeping, as the operating system monitors the number of steps taken by each program.

The third type of instruction is used in memory allocation and de-allocation. Memory management is the key to everything we do. Proper "compartmentalization" of the memory is key to privacy (no program, including the operating system, may learn the contents of memory allocated to another program), and to correctness (no program, including the operating system, may tamper with the contents of memory allocated to another program). Proper manipulation of memory, including the bookkeeping data structures, is essential to efficient allocation of resources (the bookkeeping data structures must accurately reflect the allocation of memory at all times).

$\mathcal{M}$ will have registers of the same size as $\mathcal{N}_{q,m}$, that is, q bits. Registers $0, 1, ..., m-1$ will have the same role as in $\mathcal{N}_{q,m}$. Apart from that, for each block of size ξ there will be a constant number of registers reserved for bookkeeping information regarding the block. These will be called the block registers. The information stored in them includes the user u, whose program P_u can use the registers in the corresponding block, and some information about the sequence of blocks $B_0, ..., B_i$ used by P_u. (We will call such a sequence a chain.) There will be a constant number of registers reserved for each user u as well. These registers will be called the user registers. These will contain information of the following types: the declared amount of necessary memory and time for program P_u, how much time has been used up already by program P_u, and which are the first and last blocks in the chain used by P_u. In the user registers there will be also some information about the running program P_u, which cannot be stored in the blocks while other programs are executed (e.g. the contents of the accumulator and the instruction pointer; the operating system will have no instruction permitting it to access these two registers).

We will consider the set of chains as a directed graph F, whose vertices are the blocks. If $B_0, ..., B_i$ is a chain $\langle B_0, B_1\rangle, ..., \langle B_{i-1}, B_i\rangle$ are edges of F. Since for each user there may be a chain, the graph F consists of several pairwise disjoint paths. Some of the nodes may be outside all of the paths; we will call these isolated nodes. We color the nodes in the chains with the integers $\{1, ..., n\}$, which represent the n users. The nodes of the chain which has been built for user u are colored by u. The isolated nodes have color 0. The vertices are always color coded by the contents of block registers in the corresponding blocks. One of the instructions enables making a chain longer, that is, to attach an isolated block at the end of a chain and color it with the appropriate color.

The `attach` instruction can only be issued by the operating system. Instruction `attach`(x, y, z) first checks that x and y are distinct block indices and that $z \in [n]$ (that is, z is a user name). If not, and if any of the following set of conditions holds, then the instruction becomes a no-op: Program z has started executing, block x or block y has already been allocated to another program or both have already been allocated to Program z, the chain is non-empty and block x is not the head of the chain. Otherwise, y is added to the chain, and becomes the new head. The instruction is either carried out completely or not at all. Therefore we may think of the operating system as being able to give instructions for graph operations. When the chain allocated to z is empty and both x and y are isolated, this starts a chain of color z, with x as the tail and y as the head. Note that the above conditions ensure that the chain never contains a loop.

Another graph operation makes it possible to dismantle a chain, by cutting off its last node. However it does not work while the program of the corresponding user is running. Whether this is the situation is determined by the content of one of the user registers. If the instruction which cuts down the last node of a chain colored u is applied at a time when the program P_u already used up its allocated time, then the contents of all of the registers are erased in the corresponding block.

The remaining type of instruction always relates to a specific user u. There is one instruction for each of the following tasks:

1. Ask for the first input from user u, which is the declared amount of needed time;
2. Ask for second input from user u, which is the declared amount of needed memory;
3. Check whether the number of blocks in the chain colored u is the same as the declared amount of needed memory, and if the answer is yes start the program P_u;
4. Check whether the user u has consumed its declared amount of time, and if not then execute the next step in program P_u.

Instructions of Type 3 can be executed quickly since one of the block registers corresponding to the last block of a chain contains the length of the chain, and as we have told already one of the user registers of user u contains the address of this last block.

Recall that running a program on the multi-user machine should be "just like" running the program on the ideal machine $\mathcal{N}_{q,m}$. The ideal machine has an accumulator and an instruction pointer. The secure machine will have special registers for each user that play the role of the accumulator and the instruction pointer (these are the first two registers in the chain assigned to the user). These registers are colored with the color corresponding to the user, and so remain private.

A special instruction `start`(u) is used to start the program of user u, as described in Type 3 above. Let L_0 be the address of the first register in the chain assigned to u. Once the bookkeeping has been verified (checking that the required amount of memory has been allocated and execution has not yet begun) the value L_0 is loaded into the instruction counter, the instruction `INPUT` is loaded into location L_0 (so the first instruction executed by user u will be the `INPUT` instruction, which will read in the program), location $L_0 + 1$ gets the value $L_0 + 2$ (so that the program will begin loading into location $L_0 + 2$). As we discussed in the previous section, once the program completes loading, the instruction pointer will be set to 2 + the location of the `INPUT` instruction, that is, location $L_0 + 2$, which is where the program was loaded.

Instructions of Type 4 are of the form exec(u). After testing that u is a user (a number in $[n]$) and doing some bookkeeping (e.g., making sure P_u has not exceeded its declared time), this instruction permits program P_u to take a step. Formally, the registers allocated to P_u describe a state of $\mathcal{N}_{q,m}$; the instruction indicated by the program counter for u causes a state transition to a new state of $\mathcal{N}_{q,m}$, and hence describes what should be the new configuration of the registers allocated to Program P_u. The effect of instruction exec(u) is to modify the registers allocated to Program P_u accordingly – unless some register involved is not allocated to u, in which case the instruction becomes a no-op. Note that, since each step on the ideal machine $\mathcal{N}_{q,m}$ involves only a small number of registers, each step of the multiuser machine also involves only a small number of registers, so the updates that must be performed atomically in the exec(u) instruction are not numerous (remain below some constant).

There is also a bookkeeping side effect of exec(u), which is to increase the contents of the register, owned by the operating system, that keeps track of how many steps Program x has executed.

The formal definition of each instruction describes exactly which registers are involved (may be modified by the instruction) and defines how the contents are modified. This will be important for making rigorous claims about the behavior of $\mathcal{M}$.

3.2 The Adversary

Users and the operating system are not trusted, and may collude arbitrarily against other users. Colluding parties may communicate out of band. We therefore think in terms of a single adversary, denoted $\mathcal{A}$. The adversary is assumed to have access to the declared time and space requirements of all programs, as well as knowledge of which blocks of storage they have been assigned, if any, and how many steps they have taken. Indeed, without loss of generality all information known to the operating system is assumed to be known to $\mathcal{A}$, even if the operating system is not faulty[4]. The adversary may additionally subvert users and the operating system, in an adaptive fashion. Any information known to a subverted user, in particular, the contents of all registers allocated to the user, become known to $\mathcal{A}$. We may also assume that the adversary learns any inputs the subverted user's program has received. In addition, if the operating system is subverted then the adversary controls whether or not resources are allocated to future programs, which memory locations will be allocated to which programs, whether or not allocated memory will be de-allocated, and whether allocations and de-allocations in progress when the operating system is first subverted will be completed; and the interleaving of steps between user programs.

Our computational model is sufficiently general that it allows for the possibility that the operating system may be restarted or reloaded (although we do not define such an event in the model).

---

[4] Even if the operating system is nonfaulty, a coalition of all users but x can gain information about the time and memory usage of P_x.

It is possible that the adversary subverts only a few users, leaving multiple users not subverted. Since we have no communication between users, we state all of our goals (apart from efficiency) from the point of view of a single, but arbitrary, user. Therefore, when we consider the correctness or secrecy of this user, we may assume that all of the other users are subverted and so the adversary gets all of the information available to all of the other users.

We say an architecture is secure if it prevents an adversary from achieving any of the three goals mentioned in the Introduction: causing incorrect output for a program P of an uncompromised user, compromising the privacy of program P, or poisoning the well. Intuitively, a secure architecture limits the adversary to temporarily mounting a denial of service attack. Service is restored as soon as the operating system returns to good behavior, so if $\mathcal{A}$ does not subvert the operating system then the programs of non-subverted users may run.

3.3 Weak Efficiency

Our efficiency requirement (see discussion of Poisoning the Well) may be stronger than necessary. Roughly speaking, it requires not only that the operating system, in a constant amount of time, be able to add a new register to the memory collected for User u, provided that there is still available free memory, but it also requires the operating system to know at all times where such an unused register can be found. That is, the operating system has to maintain a data structure where an unused register can be found in constant time.

Given a well-behaved operating system this can easily be done using known techniques, and the fact that the operating system has to maintain such a data structure is not an unreasonable requirement since something like this has to be maintained for the efficient use of the memory. However we require that the operating system has to do it in a secure way. That is, this data structure cannot be destroyed even by the operating system We sketch a technique for achieving exactly this in the next section.

On the other hand, for practical purposes, it may be sufficient that the operating system maintain such a data structure in the traditional unreliable way outside the machine $\mathcal{M}$, or inside but in an "insecure" way. This would mean that sometimes this data structure will be lost, but the operating system can always rebuild it easily. On the average such a solution may be less expensive in terms of resources than maintaining a secure data structure. Motivated by this, we define a weaker version of efficiency, in which we only require that the operating system can allocate an unused block of memory to a program in constant time, provided that such an unused block exists and the operating system knows its location. However weak efficiency does not guarantee that the operating system is always able to find this location in constant time.

Theorem 1 in the full paper states that when a conforming program is run on the machine $\mathcal{M}$ then both the correctness and privacy conditions are guaranteed. Moreover, the machine is *weakly efficient.* The proof is a conceptionally simple but very detailed induction on the states of the data structures. In the next section we sketch a modification of $\mathcal{M}$ that, in addition, is (strongly) efficient.

4 Achieving Strong Efficiency

We briefly sketch changes to $\mathcal{M}$ that will permit strong efficiency. We first describe the changes from an operational standpoint, and then remark on the conceptual elements of the proof.

At all times the free blocks will be connected via a doubly-linked cycle, represented by 2 additional registers in each set of block registers, called `forward`() and `backward`(). We assume that if the machine is rebooted all registers are zeroed out, so we want that the implicit meaning of 0 as a forward, respectively backward, pointer for block i is $i + 1 \bmod \chi$, respectively $i - 1 \bmod \chi$; here χ is the total number of memory blocks in the machine. We model the available blocks with a graph which contains at all time a single cycle. At time t the graph, F_t is defined as follows: for all blocks $1 \leq i, j \leq \chi$, available (that is, colored 0) at time t, there is an edge from block i to block j if and only if the forward pointer for block i contains the value $j - (i+1) \bmod \chi$ and the backward pointer for j contains $i - (j - 1) \bmod \chi$.

In addition, we define a new register, which can be read by the operating system, called `pick`. At all times $t \geq 0$ this will contain an (arbitrary) element of the graph F_t, that is, the index of some available block.

The `attach`(x, y, z) instruction now also causes y (and possibly x, if it has color 0) to be deleted from the cycle F_t, and `pick` to be updated. Similarly, the `detach`(x, z) instruction causes x to be added to the cycle. Note that the updates to the registers `forward`(), `backward`(), and `pick` are defined by the `attach` and `detach` operations, and are therefore only indirectly under the control of the operating system. This interface is key to ensuring that the operating system cannot poison the well.

We remark that a block allocated to a program that has not yet begun executing is still considered to be available, as the operating system may change its mind and de-allocate the block, possibly giving it instead to another user. Thus, a block corresponding to the head of a chain under preparation for a user whose program has not yet begun technically should also be part of the cycle F_t. Also, to ensure availability of blocks on chain allocated to a program P_u that has completed, we modify the `exec`(u) operation to take appropriate action when `timecount`(u) first exceeds $2 +$ `decltime`(u).

Note that our informal description of the modified attach and detach operations involved discussion of a graph F_t that depends on the entire history of the execution of the machine. The formal specification for the strongly efficient $\mathcal{M}_1$ involves extending the instruction set of $\mathcal{M}$ to incorporate histories (see Theorem 2 and its proof, in the full paper), but operationally there is no such complication.

5 Extensions

So far we have only considered the situation when each user has a single program to be executed. We briefly mention some other possible situations which can be handled with a similar but somewhat more complex set of instructions than the one that we have provided for the solution of the basic problem. Solutions to these problems are outlined in the full paper; however, there is much room for future research.

1. Each user has a fixed sequence of programs to be executed. The difficulty here is that the user may have prepared a sequence of inputs but the operating system may refuse an early program in the sequence (perhaps because insufficient memory is available), and the user must adapt the input sequence accordingly. Thus, the model must be modified to incorporate additional interaction between the machine and the user.

2. The amount of time, respecitvely, space, needed for the program is not known in advance; that is, the program, depending on the partial results, may ask for additional time or space.

In the two generalized problems mentioned above the notion of information protection changes in the sense that we must consider all of the requests for additional resources and their timing as public information that is not protected.

3. An interesting and important area is the question of communication among users. (A related topic is the handling of interrupts, such as the firing of a timer or the movement of a mouse.) We see many ways to address this, and several interesting questions arise concerning the appropriate changes to the definitions of security and efficiency. For example, in addition to information protection, one may also want protection from wasting time on unwanted communication initiated by others. We intend to return to these questions in another paper; see the full paper for some specific suggestions.

4. Many programs are long-lived. Such programs are essentially virtual users. There is no reason for these programs if they cannot communicate with other programs. Hence, the exact implementation of this concept depends very much on interprogram communication and signaling.

References

1. D. Lie, C. Thekkath, M. Mitchell, P. Lincoln, D. Boneh, J. Mitchell, and M. Horowitz, Architectural Support for Copy and Tamper Resistant Software. *Proceedings of the 9th International Conference on Architectural Support for Programming Languages and Operating Systems (ASPLOS-IX)*, pp. 169–177, 2000.
2. D. Lie, C. Thekkath and M. Horowitz, Implementing an Untrusted Operating System on Trusted Hardware. *Proceedings of the 19th ACM Symposium on Operating Systems Principles (SOSP)*, pp. 178–192, 2003
3. D. Lie, J. Mitchell, C. Thekkath and M. Horowitz, Specifying an Verifying Hardware for Tamper-Resistant Software, *Proceedings of the 2003 IEEE Symposium on Security and Privacy*, 2003.
4. Trusted Mobile Platform, http://www.trusted-mobile.org/

Scoring Matrices That Induce Metrics on Sequences*

Elói Araújo[1] and José Soares[2]

[1] FACET, Universidade Metodista de São Paulo, Brasil
[2] Instituto de Matemática e Estatística, Universidade de São Paulo, Brasil

Abstract. Scoring matrices are widely used in sequence comparisons. A scoring matrix γ is indexed by symbols of an alphabet. The entry in γ in row $\mathtt{a}$ and column $\mathtt{b}$ measures the cost of the *edit operation* of replacing symbol $\mathtt{a}$ by symbol $\mathtt{b}$.

For a given scoring matrix and sequences s and t, we consider two kinds of induced scoring functions. The first function, known as *weighted edit distance*, is defined as the sum of costs of the edit operations required to transform s into t. The second, known as *normalized edit distance*, is defined as the minimum quotient between the sum of costs of edit operations to transform s into t and the number of the corresponding edit operations.

In this work we characterize the class of scoring matrices for which the induced weighted edit distance is actually a metric. We do the same for the normalized edit distance.

Keywords: edit distance, normalized edit distance, metric.

1 Introduction

Comparison of sequences is an important problem in computer science which has several applications: computational biology [4], text processing [1], pattern recognition [7], pronunciation modeling [9], etc.

It is common to measure the distance between two sequences s and t by computing the minimum cost of transforming s into t through a sequence of weighted *edit operations*. These operations are: *insertion*, *deletion*, and *substitution* of symbols.

Let Σ be an alphabet and $\Sigma_\nabla = \Sigma \cup \{\nabla\}$, where $\nabla \notin \Sigma$. The symbol ∇ is used to represent insertions and deletions. A *scoring matrix* γ for Σ is a matrix whose elements are real numbers. The matrix γ has rows and columns indexed by symbols in Σ_∇. For $\mathtt{a}, \mathtt{b} \in \Sigma_\nabla$, we denote by $\gamma_{\mathtt{a}\to\mathtt{b}}$ the entry of γ in row $\mathtt{a}$ and column $\mathtt{b}$ and it represents the cost of the substitution of $\mathtt{a}$ for $\mathtt{b}$.

A simple weighted edit distance is known as Levenshtein distance [6]. The corresponding scoring matrix is such that $\gamma_{\mathtt{a}\to\mathtt{b}} = 0$ if $\mathtt{a} = \mathtt{b}$ and $\gamma_{\mathtt{a}\to\mathtt{b}} = 1$

* This research was partially supported by FAPESP/CNPq (PRONEX 2003/09925-5) and CNPq (Universal 478329/2004-0).

J.R. Correa, A. Hevia, and M. Kiwi (Eds.): LATIN 2006, LNCS 3887, pp. 68–79, 2006.

otherwise. However, not every scoring matrix induces a scoring function that can be properly called a weighted edit *distance*, in the sense that the scoring function might not be a metric. Sellers [10] shows that a sufficient condition for the scoring function to be a metric on Σ^* is that γ is a metric on Σ_∇. We show in this work that this condition is not necessary. We characterize the class of scoring matrices that induces a proper weighted edit distance. For example, it follows from Theorem 2 that the matrix

	a	b	c	∇
a	0	1	3	1
b	1	0	1	1
c	4	1	0	1
∇	1	1	1	0

induces a metric on Σ^*, even though $\gamma_{\mathsf{a}\to\mathsf{c}} \neq \gamma_{\mathsf{c}\to\mathsf{a}}$, $\gamma_{\mathsf{a}\to\mathsf{c}} \not\leq \gamma_{\mathsf{a}\to\nabla} + \gamma_{\nabla\to\mathsf{c}}$ and $\gamma_{\mathsf{c}\to\mathsf{a}} \not\leq \gamma_{\mathsf{c}\to\nabla} + \gamma_{\nabla\to\mathsf{a}}$.

Marzal and Vidal [7] defined another criterion to score alignments that depends not only on the edit operations involved but also on the number of such operations. This criterion is known as *normalized edit distance*. Similar to what happens with the conventional weighted edit distance, not every scoring matrix induces a proper normalized edit *distance*. We also characterize the class of scoring matrices that induces a proper normalized edit distance.

This paper is organized as follows. Sections 2 provides a brief description of the concepts, and we characterize the classes of matrices that induce, respectively, normalized edit distance and weighted edit distance. In Section 3 we prove the main result of this paper and we finalize in Section 4 with some remarks.

2 Preliminaries

We denote a sequence s over Σ by $s = s(1)s(2) \ldots s(n)$, where $s(i) \in \Sigma$. We say that the *length* of s, denoted by $|s|$, is n. We denote by ϵ the empty sequence. The sequence a^n is the sequence with length n consisting of the concatenation of n characters a. The sequence st represents the concatenation of the sequences s and t. The set of all sequences over Σ is denoted by Σ^*.

Let $\Sigma_\nabla = \Sigma \cup \{\nabla\}$, where $\nabla \notin \Sigma$. We call *space* the symbol ∇, which is used to represent an insertion or a deletion. An *alignment* of (s,t) is a pair of sequences (s',t') obtained by inserting spaces in the sequences s and t, in such a way that $|s'| = |t'|$ and there is no i such that $s'(i) = t'(i) = \nabla$. We say that $s'(i)$ and $t'(i)$ *are aligned* in (s',t') and that $|(s',t')| = |s'|$ is the *length* of the alignment (s',t'). We denote by $\mathcal{A}_{(s,t)}$ the set of all alignments of (s,t).

An alignment can be visualized placing s' above t', as showed in the following examples.

a	c	∇	c	b	∇	b	b	b	∇
c	∇	a	∇	a	c	∇	∇	c	b

∇	∇	∇	∇	a	c	c	b	b	b	b
c	a	a	c	c	b	∇	∇	∇	∇	∇

The figures above represent two different alignments of (accbbbb, caaccb). The left figure represents the alignment (ac∇cb∇bbb∇, c∇a∇ac∇∇cb), while the right figure represents the alignment (∇∇∇∇accbbbb, caaccb∇∇∇∇∇).

Given a scoring matrix γ, we define the functions v_γ and vN_γ that associate the following values for each alignment (s', t'):

$$v_\gamma(s', t') = \sum_{i=1}^{|(s',t')|} \gamma_{s'(i) \to t'(i)}$$

and

$$v\mathrm{N}_\gamma(s', t') = \begin{cases} 0 \text{ if } |s| = |t| = 0, \\ \sum_{i=1}^{|(s',t')|} \frac{\gamma_{s'(i) \to t'(i)}}{|(s',t')|} \text{ otherwise} . \end{cases}$$

We call $v_\gamma(A)$ the *score* of the alignment A, and $v\mathrm{N}_\gamma(A)$ the *normalized score* of the alignment A.

We also define functions opt_γ and $opt\mathrm{N}_\gamma$ as

$$opt_\gamma(s,t) = \min_{A \in \mathcal{A}_{(s,t)}} \{v_\gamma(A)\} \qquad \text{and} \qquad opt\mathrm{N}_\gamma(s,t) = \min_{A \in \mathcal{A}_{(s,t)}} \{v\mathrm{N}_\gamma(A)\} .$$

If $v_\gamma(s', t') = opt_\gamma(s,t)$ or $v\mathrm{N}_\gamma(s', t') = opt\mathrm{N}_\gamma(s,t)$ we say that the alignment (s', t') of (s,t) is *optimal* or *N-optimal*, respectively.

For a given set S, we say that a function *dist* is a *metric* on S if *dist* satisfies the following conditions. For each $s, t, u \in S$,

1. $dist(s,t) > 0$ if $s \neq t$, and $dist(s,t) = 0$ if $s = t$;
2. $dist(s,t) = dist(t,s)$;
3. $dist(s,t) \leq dist(s,u) + dist(u,t)$.

If opt_γ or $opt\mathrm{N}_\gamma$ is a metric on Σ^* we say that the scoring matrix γ *induces* a weighted edit distance or a normalized edit distance, respectively.

The most common class of scoring matrices that induces weighted edit distances is defined below. Let $\mathbb{M}^{\mathtt{C}}$ be the class of scoring matrices for Σ that have the following properties. For each $\mathsf{a}, \mathsf{b}, \mathsf{c} \in \Sigma_\nabla$,

1. $\gamma_{\mathsf{a} \to \mathsf{b}} > 0$ if $\mathsf{a} \neq \mathsf{b}$, and $\gamma_{\mathsf{a} \to \mathsf{b}} = 0$ if $\mathsf{a} = \mathsf{b}$;
2. $\gamma_{\mathsf{a} \to \mathsf{b}} = \gamma_{\mathsf{b} \to \mathsf{a}}$;
3. $\gamma_{\mathsf{a} \to \mathsf{c}} \leq \gamma_{\mathsf{a} \to \mathsf{b}} + \gamma_{\mathsf{b} \to \mathsf{c}}$.

In other words, γ is a metric on Σ_∇. Sellers [10] showed that scoring matrices in this class induce weighted edit distances.

However, the class $\mathbb{M}^{\mathtt{W}}$ defined below contains $\mathbb{M}^{\mathtt{C}}$ and, as mentioned below, it consists of all scoring matrices that induce weighted edit distance. Let $\mathbb{M}^{\mathtt{W}}$ be the class of scoring matrices for Σ that have the following properties. For each $\mathsf{a}, \mathsf{b}, \mathsf{c} \in \Sigma$,

1. $\gamma_{\mathsf{a} \to \nabla} = \gamma_{\nabla \to \mathsf{a}} > 0$;
2. $\gamma_{\mathsf{a} \to \mathsf{b}} > 0$ if $\mathsf{a} \neq \mathsf{b}$, and $\gamma_{\mathsf{a} \to \mathsf{b}} = 0$ if $\mathsf{a} = \mathsf{b}$;

3. if $\gamma_{a \to b} < \gamma_{a \to \nabla} + \gamma_{\nabla \to b}$, then $\gamma_{a \to b} = \gamma_{b \to a}$;
4. $\gamma_{a \to \nabla} \leq \gamma_{a \to b} + \gamma_{b \to \nabla}$;
5. $\min\{\gamma_{a \to c}, \gamma_{a \to \nabla} + \gamma_{\nabla \to c}\} \leq \gamma_{a \to b} + \gamma_{b \to c}$.

We also define the class $\mathbb{M}^{\mathtt{N}}$ of scoring matrices for Σ that have the following properties. For each $a, b, c \in \Sigma$,

1. $\gamma_{a \to \nabla} = \gamma_{\nabla \to a} > 0$;
2. $\gamma_{a \to b} > 0$ if $a \neq b$, and $\gamma_{a \to b} = 0$ if $a = b$;
3. if $\gamma_{a \to b} < \gamma_{a \to \nabla} + \gamma_{\nabla \to b}$, then $\gamma_{a \to b} = \gamma_{b \to a}$;
4. $\gamma_{a \to \nabla} \leq \gamma_{a \to b} + \gamma_{b \to \nabla}$;
5. $\min\{\gamma_{a \to c}, \gamma_{a \to \nabla} + \gamma_{\nabla \to c}\} \leq \gamma_{a \to b} + \gamma_{b \to c}$;
6. $\gamma_{a \to \nabla} \leq 2\gamma_{\nabla \to b}$.

The following theorem, proved in Section 3, states that $\mathbb{M}^{\mathtt{N}}$ consists of all scoring matrices that induce normalized edit distances.

Theorem 1. Let Σ be an alphabet and γ be a scoring matrix. Then $opt\mathrm{N}_\gamma$ is a metric on Σ^* if and only if $\gamma \in \mathbb{M}^{\mathtt{N}}$.

The similar theorem below states that $\mathbb{M}^{\mathtt{W}}$ consists of all scoring matrices that induce weighted edit distances. Its proof, omitted here, is an adaptation of the proof of Theorem 1.

Theorem 2. Let Σ be an alphabet and γ be a scoring matrix. Then opt_γ is a metric on Σ^* if and only if $\gamma \in \mathbb{M}^{\mathtt{W}}$.

It follows from the definitions that $\mathbb{M}^{\mathtt{C}} \subseteq \mathbb{M}^{\mathtt{W}}$, $\mathbb{M}^{\mathtt{N}} \subseteq \mathbb{M}^{\mathtt{W}}$, $\mathbb{M}^{\mathtt{C}} \not\subseteq \mathbb{M}^{\mathtt{N}}$, and $\mathbb{M}^{\mathtt{N}} \not\subseteq \mathbb{M}^{\mathtt{C}}$, as pictured in Figure 1.

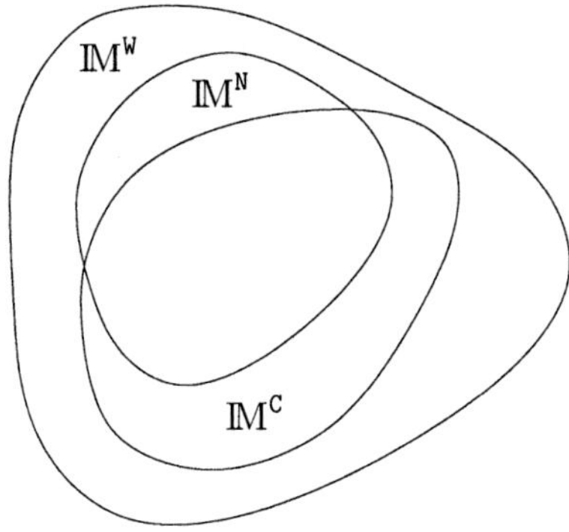

Fig. 1. Relationship among the classes $\mathbb{M}^{\mathtt{W}}$, $\mathbb{M}^{\mathtt{N}}$, and $\mathbb{M}^{\mathtt{C}}$

This work was motivated by the following remark made by Marzal and Vidal [7], who depicted the scoring matrix γ below:

	a	b	∇
a	0	5	5
b	5	0	1
∇	5	1	0

They commented that opt_γ is a metric on Σ_∇, but $opt\mathrm{N}_\gamma$ is not a metric on Σ_∇ since

$$3 = opt\mathrm{N}_\gamma(\mathtt{a}, \mathtt{b}) \not\leq opt\mathrm{N}_\gamma(\mathtt{a}, \mathtt{ab}) + opt_\gamma(\mathtt{ab}, \mathtt{b}) = \frac{1}{2} + \frac{7}{3} \,.$$

It is worth noting that $\gamma \in \mathbb{M}^{\mathrm{C}}$ but $\gamma \notin \mathbb{M}^{\mathrm{N}}$.

3 Scoring Matrices That Induce Normalized Edit Distances

This rather technical section is dedicated to the proof of Theorem 1, which states that the class $\mathbb{M}^{\mathrm{N}}$ consists of all scoring matrices that induce normalized edit distance.

Lemmas 3, 4, 7, 8, 10, and 11 are used to prove one implication of Theorem 1, namely, we prove that if $opt\mathrm{N}_\gamma$ is a metric on Σ^*, then $\gamma \in \mathbb{M}^{\mathrm{N}}$. More precisely, considering that $opt\mathrm{N}_\gamma$ is a metric, each one of these six lemmas shows that γ must obey one of the six properties stated in the definition of $\mathbb{M}^{\mathrm{N}}$.

Next we complete the proof of Theorem 1 by showing the other implication, namely, that if $\gamma \in \mathbb{M}^{\mathrm{N}}$, then $opt\mathrm{N}_\gamma$ satisfies the properties of reflexivity and strict positiveness (Lemma 14), symmetry (Lemma 15) and triangle inequality (Lemma 17).

Throughout this section we consider a fixed alphabet Σ and a fixed scoring matrix γ for Σ.

Lemma 3. Suppose that $opt\mathrm{N}_\gamma$ is a metric and that $\mathtt{a}$ is a symbol in Σ. Then $\gamma_{\mathtt{a}\to\nabla} = \gamma_{\nabla\to\mathtt{a}} > 0$.

Proof. Observe that the score of the alignment $(\mathtt{a}, \nabla)$ is $\gamma_{\mathtt{a}\to\nabla}$ and that $(\mathtt{a}, \nabla)$ is the unique alignment of $(\mathtt{a}, \epsilon)$. So, $\gamma_{\mathtt{a}\to\nabla} = v\mathrm{N}_\gamma\,(\mathtt{a}, \nabla) = opt\mathrm{N}_\gamma(\mathtt{a}, \epsilon)$.

Similarly, we have that $\gamma_{\nabla\to\mathtt{a}} = opt\mathrm{N}_\gamma(\epsilon, \mathtt{a})$.

If $opt\mathrm{N}_\gamma$ is a metric, then $opt\mathrm{N}_\gamma(\mathtt{a}, \epsilon) > 0$ and $opt\mathrm{N}_\gamma(\mathtt{a}, \epsilon) = opt\mathrm{N}_\gamma(\epsilon, \mathtt{a})$.

Therefore, $\gamma_{\mathtt{a}\to\nabla} = opt\mathrm{N}_\gamma(\mathtt{a}, \epsilon) = opt\mathrm{N}_\gamma(\epsilon, \mathtt{a}) = \gamma_{\nabla\to\mathtt{a}} > 0$. □

Lemma 4. Suppose that $opt\mathrm{N}_\gamma$ is a metric and that $\mathtt{a}$ and $\mathtt{b}$ are symbols in Σ. Then $\gamma_{\mathtt{a}\to\mathtt{b}} > 0$ if $\mathtt{a} \neq \mathtt{b}$, and $\gamma_{\mathtt{a}\to\mathtt{b}} = 0$ if $\mathtt{a} = \mathtt{b}$.

Proof. Suppose that $\mathtt{a} \neq \mathtt{b}$. Since *(i)* $opt\mathrm{N}_\gamma$ is a metric, *(ii)* the score of the alignment $(\mathtt{a}, \mathtt{b})$ is $\gamma_{\mathtt{a}\to\mathtt{b}}$, and *(iii)* $v\mathrm{N}_\gamma\,(\mathtt{a}, \mathtt{b}) \geq opt\mathrm{N}_\gamma(\mathtt{a}, \mathtt{b})$, it follows that $\gamma_{\mathtt{a}\to\mathtt{b}} = v\mathrm{N}_\gamma\,(\mathtt{a}, \mathtt{b}) \geq opt\mathrm{N}_\gamma(\mathtt{a}, \mathtt{b}) > 0$.

Now suppose that $\mathsf{a} = \mathsf{b}$. Observe that $\mathcal{A}_{(\mathsf{a},\mathsf{b})} = \{(\mathsf{a},\mathsf{b}), (\mathsf{a}\nabla, \nabla\mathsf{b}), (\nabla\mathsf{a}, \mathsf{b}\nabla)\}$. It follows from Lemma 3 that $v\mathrm{N}_\gamma(\mathsf{a}\nabla, \nabla\mathsf{b}) > 0$ and that $v\mathrm{N}_\gamma(\nabla\mathsf{a}, \mathsf{b}\nabla) > 0$. Since $opt\mathrm{N}_\gamma$ is a metric, it is true that $opt\mathrm{N}_\gamma(\mathsf{a},\mathsf{b}) = 0$. So, the only possible case is that $opt\mathrm{N}_\gamma(\mathsf{a},\mathsf{b}) = v\mathrm{N}_\gamma(\mathsf{a},\mathsf{b}) = \gamma_{\mathsf{a}\to\mathsf{b}} = 0$. □

We need the following notation to be used in the next lemmas. We denote by G the value of $\max_{\mathsf{a}\in\Sigma}\{\gamma_{\mathsf{a}\to\nabla}, \gamma_{\nabla\to\mathsf{a}}\}$ and by g a character of Σ such that $\max\{\gamma_{g\to\nabla}, \gamma_{\nabla\to g}\} = G$.

Lemma 5. Suppose that $opt\mathrm{N}_\gamma$ is a metric and that a and b are symbols in Σ. Then, for each $n \geq 0$, we have that

$$opt\mathrm{N}_\gamma(g^n\mathsf{a}, g^n\mathsf{b}) = \min\left\{\begin{array}{l} v\mathrm{N}_\gamma(g^n\mathsf{a}, g^n\mathsf{b}) = \gamma_{\mathsf{a}\to\mathsf{b}}/(n+1), \\ v\mathrm{N}_\gamma(g^n\mathsf{a}\nabla, g^n\nabla\mathsf{b}) = (\gamma_{\mathsf{a}\to\nabla} + \gamma_{\nabla\to\mathsf{b}})/(n+2) \end{array}\right\}.$$

Proof. (sketch) We first comment that

$$v\mathrm{N}_\gamma(g^n\mathsf{a}, g^n\mathsf{b}) = \frac{\gamma_{\mathsf{a}\to\mathsf{b}}}{n+1} \quad \text{and} \quad v\mathrm{N}_\gamma(g^n\mathsf{a}\nabla, g^n\nabla\mathsf{b}) = \frac{\gamma_{\mathsf{a}\to\nabla} + \gamma_{\nabla\to\mathsf{b}}}{n+2}.$$

So, to prove the lemma, we show that any alignment (s', t') of $(g^n\mathsf{a}, g^n\mathsf{b})$ is such that

$$v\mathrm{N}_\gamma(s', t') \geq \min\{\gamma_{\mathsf{a}\to\mathsf{b}}/(n+1), (\gamma_{\mathsf{a}\to\nabla} + \gamma_{\nabla\to\mathsf{b}})/(n+2)\}.$$

We examine four cases, covering all possible alignments of $(g^n\mathsf{a}, g^n\mathsf{b})$. The cases are: (i) a is aligned with b in (s', t'); (ii) a is aligned with g in (s', t'); (iii) g is aligned with b in (s', t'); and (iv) a and b are aligned with spaces in (s', t'). For each one of these cases, we show that value of the alignment is as desired. □

We need the following property in the next lemma.

Fact 6. Let a, b, and c be numbers, with $a < b + c$ and $n_0 = a/(b+c-a) - 1$. If $n > n_0$, then $a/(n+1) < (b+c)/(n+2)$.

Lemma 7. Suppose that $opt\mathrm{N}_\gamma$ is a metric and that a and b are symbols in Σ such that $\gamma_{\mathsf{a}\to\mathsf{b}} < \gamma_{\mathsf{a}\to\nabla} + \gamma_{\nabla\to\mathsf{b}}$. Then $\gamma_{\mathsf{a}\to\mathsf{b}} = \gamma_{\mathsf{b}\to\mathsf{a}}$.

Proof. Let n be any integer such that $n > \gamma_{\mathsf{a}\to\mathsf{b}}/(\gamma_{\mathsf{a}\to\nabla} + \gamma_{\nabla\to\mathsf{b}} - \gamma_{\mathsf{a}\to\mathsf{b}}) - 1$.

As consequence of Lemma 5 and Fact 6, we have that

$$opt\mathrm{N}_\gamma(g^n\mathsf{a}, g^n\mathsf{b}) = \frac{\gamma_{\mathsf{a}\to\mathsf{b}}}{n+1} < \frac{\gamma_{\mathsf{a}\to\nabla} + \gamma_{\nabla\to\mathsf{b}}}{n+2}.$$

Also, from Lemma 5, we have that

$$opt\mathrm{N}_\gamma(g^n\mathsf{b}, g^n\mathsf{a}) = \min\left\{\frac{\gamma_{\mathsf{b}\to\mathsf{a}}}{n+1}, \frac{\gamma_{\mathsf{b}\to\nabla} + \gamma_{\nabla\to\mathsf{a}}}{n+2}\right\}.$$

Suppose that $opt\mathrm{N}_\gamma(g^n\mathsf{b}, g^n\mathsf{a}) = (\gamma_{\mathsf{b}\to\nabla} + \gamma_{\nabla\to\mathsf{a}})/(n+2)$. As a consequence of Lemma 3 we have that

$$opt\mathrm{N}_\gamma(g^n\mathsf{a}, g^n\mathsf{b}) < \frac{\gamma_{\mathsf{a}\to\nabla} + \gamma_{\nabla\to\mathsf{b}}}{n+2} = \frac{\gamma_{\mathsf{b}\to\nabla} + \gamma_{\nabla\to\mathsf{a}}}{n+2} = opt\mathrm{N}_\gamma(g^n\mathsf{b}, g^n\mathsf{a}),$$

which is a contradiction, since $optN_\gamma$ is a metric. Therefore, $optN_\gamma(g^n\mathsf{b}, g^n\mathsf{a}) = (\gamma_{\mathsf{b}\to\mathsf{a}})/(n+1)$. It follows that

$$\frac{\gamma_{\mathsf{a}\to\mathsf{b}}}{n+1} = optN_\gamma(g^n\mathsf{a}, g^n\mathsf{b}) = optN_\gamma(g^n\mathsf{b}, g^n\mathsf{a}) = \frac{\gamma_{\mathsf{b}\to\mathsf{a}}}{n+1},$$

which implies that $\gamma_{\mathsf{a}\to\mathsf{b}} = \gamma_{\mathsf{b}\to\mathsf{a}}$. □

Lemma 8. Suppose that $optN_\gamma$ is a metric and that a and b are symbols in Σ. Then $\gamma_{\mathsf{a}\to\nabla} \le \gamma_{\mathsf{a}\to\mathsf{b}} + \gamma_{\mathsf{b}\to\nabla}$.

Proof. Since (a, ∇) is the unique alignment of (a, ϵ), we have that $optN_\gamma(\mathsf{a}, \epsilon) = vN_\gamma(\mathsf{a}, \nabla)$. It is also true that $optN_\gamma(\mathsf{a}, \epsilon) \le optN_\gamma(\mathsf{a}, \mathsf{b}) + optN_\gamma(\mathsf{b}, \epsilon)$, because $optN_\gamma$ is a metric. From the remarks above it follows that

$$\begin{aligned}\gamma_{\mathsf{a}\to\nabla} = vN_\gamma\left(\boxed{\begin{matrix}\mathsf{a}\\ \nabla\end{matrix}}\right) &= optN_\gamma(\mathsf{a}, \epsilon) \le optN_\gamma(\mathsf{a}, \mathsf{b}) + optN_\gamma(\mathsf{b}, \epsilon)\\ &\le vN_\gamma\left(\boxed{\begin{matrix}\mathsf{a}\\ \mathsf{b}\end{matrix}}\right) + vN_\gamma\left(\boxed{\begin{matrix}\mathsf{b}\\ \nabla\end{matrix}}\right) = \gamma_{\mathsf{a}\to\mathsf{b}} + \gamma_{\mathsf{b}\to\nabla}\,.\end{aligned}$$ □

We use the following fact in the next lemma.

Fact 9. Let $x \ne y$ be real numbers and n be such that $n > \max\left\{0, \frac{2y-x}{x-y}\right\}$. If $\frac{x}{n+2} \le \frac{y}{n+1}$ then $x < y$.

Lemma 10. Suppose that $optN_\gamma$ is a metric and that a, b, and c are symbols in Σ. Then

$$\min\{\gamma_{\mathsf{a}\to\mathsf{c}}, \gamma_{\mathsf{a}\to\nabla} + \gamma_{\nabla\to\mathsf{c}}\} \le \gamma_{\mathsf{a}\to\mathsf{b}} + \gamma_{\mathsf{b}\to\mathsf{c}}\,.$$

Proof. If $\gamma_{\mathsf{a}\to\nabla} + \gamma_{\nabla\to\mathsf{c}} = \gamma_{\mathsf{a}\to\mathsf{b}} + \gamma_{\mathsf{b}\to\mathsf{c}}$ then the lemma is proved. So, we may assume that $\gamma_{\mathsf{a}\to\nabla} + \gamma_{\nabla\to\mathsf{c}} \ne \gamma_{\mathsf{a}\to\mathsf{b}} + \gamma_{\mathsf{b}\to\mathsf{c}}$.

Let n be a positive integer such that

$$n > \frac{2(\gamma_{\mathsf{a}\to\mathsf{b}} + \gamma_{\mathsf{b}\to\mathsf{c}}) - (\gamma_{\mathsf{a}\to\nabla} + \gamma_{\nabla\to\mathsf{c}})}{(\gamma_{\mathsf{a}\to\nabla} + \gamma_{\nabla\to\mathsf{c}}) - (\gamma_{\mathsf{a}\to\mathsf{b}} + \gamma_{\mathsf{b}\to\mathsf{c}})}\,.$$

As consequence of Lemma 5, we have that

$$optN_\gamma(g^n\mathsf{a}, g^n\mathsf{b}) \le \frac{\gamma_{\mathsf{a}\to\mathsf{b}}}{n+1} \quad\text{and}\quad optN_\gamma(g^n\mathsf{b}, g^n\mathsf{c}) \le \frac{\gamma_{\mathsf{b}\to\mathsf{c}}}{n+1}\,. \tag{1}$$

Using Lemma 5, we inspect the two possible values of $optN_\gamma(g^n\mathsf{a}, g^n\mathsf{c})$.

If $optN_\gamma(g^n\mathsf{a}, g^n\mathsf{c}) = \gamma_{\mathsf{a}\to\mathsf{c}}/(n+1)$, then, since $optN_\gamma$ is a metric, and using (1), we have that

$$\frac{\gamma_{\mathsf{a}\to\mathsf{c}}}{n+1} = opt\mathrm{N}_\gamma(g^n\mathsf{a}, g^n\mathsf{c}) \le opt\mathrm{N}_\gamma(g^n\mathsf{a}, g^n\mathsf{b}) + opt\mathrm{N}_\gamma(g^n\mathsf{b}, g^n\mathsf{c}) \le \frac{\gamma_{\mathsf{a}\to\mathsf{b}}}{n+1} + \frac{\gamma_{\mathsf{b}\to\mathsf{c}}}{n+1},$$

which implies, since $n \ge 0$, that $\gamma_{\mathsf{a}\to\mathsf{c}} \le \gamma_{\mathsf{a}\to\mathsf{b}} + \gamma_{\mathsf{b}\to\mathsf{c}}$.

In a similar way, if $opt\mathrm{N}_\gamma(g^n\mathsf{a}, g^n\mathsf{c}) = (\gamma_{\mathsf{a}\to\nabla} + \gamma_{\nabla\to\mathsf{c}})/(n+2)$, then

$$\frac{\gamma_{\mathsf{a}\to\nabla} + \gamma_{\nabla\to\mathsf{c}}}{n+2} \le \frac{\gamma_{\mathsf{a}\to\mathsf{b}}}{n+1} + \frac{\gamma_{\mathsf{b}\to\mathsf{c}}}{n+1} = \frac{\gamma_{\mathsf{a}\to\mathsf{b}} + \gamma_{\mathsf{b}\to\mathsf{c}}}{n+1},$$

which implies, from Fact 9 and by the choice of n, that $\gamma_{\mathsf{a}\to\nabla} + \gamma_{\nabla\to\mathsf{c}} < \gamma_{\mathsf{a}\to\mathsf{b}} + \gamma_{\mathsf{b}\to\mathsf{c}}$. □

Lemma 11. Suppose that $opt\mathrm{N}_\gamma$ is a metric and that a and b are symbols in Σ. Then $\gamma_{\mathsf{a}\to\nabla} \le 2\gamma_{\nabla\to\mathsf{b}}$.

Proof. (sketch)

By contradiction, we assume that $\gamma_{\mathsf{a}\to\nabla} > 2\gamma_{\nabla\to\mathsf{b}}$. We choose k such that

$$k > \frac{2\gamma_{\nabla\to\mathsf{b}}}{\gamma_{\mathsf{a}\to\nabla} - 2\gamma_{\nabla\to\mathsf{b}}},$$

and consider the sequences a^k, b, and $\mathsf{a}^k\mathsf{b}$.

We can prove that $opt\mathrm{N}_\gamma(\mathsf{a}^k, \mathsf{b}) > opt\mathrm{N}_\gamma(\mathsf{a}^k, \mathsf{a}^k\mathsf{b}) + opt\mathrm{N}_\gamma(\mathsf{a}^k\mathsf{b}, \mathsf{b})$, showing that $opt\mathrm{N}_\gamma$ does not have the triangular inequality property. This contradicts the assumption that $opt\mathrm{N}_\gamma$ is a metric. □

Lemma 12. Let s and t be sequences in Σ^*. Then $opt\mathrm{N}_\gamma(s,t) \le G$.

Proof. Let (s', t') be the alignment of (s,t) such that $s' = s\nabla^{|t|}$ and $t' = \nabla^{|s|}t$. Then

$$opt\mathrm{N}_\gamma(s,t) \le v\mathrm{N}_\gamma(s',t') = \frac{|s|\gamma_{s(i)\to\nabla} + |t|\gamma_{\nabla\to t(j)}}{|(s',t')|} \le \frac{|(s',t')|G}{|(s',t')|} = G,$$

where the last inequality follows from the definition of G. □

Lemma 13. If $\mathsf{a}, \mathsf{b} \in \Sigma$ are aligned in an N-optimal alignment of (s,t), then

$$\gamma_{\mathsf{a}\to\mathsf{b}} < \gamma_{\mathsf{a}\to\nabla} + \gamma_{\nabla\to\mathsf{b}}.$$

Proof. Let (s', t') be an N-optimal alignment of (s,t) in which a and b are aligned. By contradiction, assume that $\gamma_{\mathsf{a}\to\mathsf{b}} \ge \gamma_{\mathsf{a}\to\nabla} + \gamma_{\nabla\to\mathsf{b}}$. Let j be an integer such that $s'(j) = \mathsf{a}$ and $t'(j) = \mathsf{b}$.

Then

$$v\mathrm{N}_\gamma\left(\boxed{\begin{array}{l} s'(1)\ldots s'(j-1)\; s'(j) \;\; \nabla \;\; s'(j+1)\ldots s'(|s'|) \\ t'(1)\ldots t'(j-1) \;\; \nabla \;\; t'(j)\; t'(j+1)\ldots t'(|t'|) \end{array}}\right) = \frac{v_\gamma(s',t') + \gamma_{\mathsf{a}\to\nabla} + \gamma_{\nabla\to\mathsf{b}} - \gamma_{\mathsf{a}\to\mathsf{b}}}{|(s',t')|+1} \le \frac{v_\gamma(s',t')}{|(s',t')|+1} < v\mathrm{N}_\gamma(s',t'),$$

contradicting the optimality of (s', t'). □

Lemma 14. Suppose $\gamma \in \mathbb{M}^{\mathrm{N}}$ and that s and t are sequences in Σ^*. Then $opt\mathrm{N}_\gamma(s,t) > 0$ if $s \neq t$, and $opt\mathrm{N}_\gamma(s,t) = 0$ if $s = t$.

Proof. Let (s',t') be an N-optimal alignment of (s,t). Since $\gamma \in \mathbb{M}^{\mathrm{N}}$, we have that, for each j, $\gamma_{s'(j)\to t'(j)} \geq 0$. Then

$$opt\mathrm{N}_\gamma(s,t) = v\mathrm{N}_\gamma\,(s',t') = \frac{\sum_j \gamma_{s'(j)\to t'(j)}}{|(s',t')|} \geq 0\,.$$

If $s = t$, then (s,t) is an alignment of (s,t) with $v\mathrm{N}_\gamma\,(s,t) = 0$. It follows from the inequality above that $opt\mathrm{N}_\gamma(s,t) = 0$.

If $s \neq t$, then for any N-optimal alignment (s',t') of (s,t), there exists i such that $s'(i) \neq t'(i)$, which implies, since $\gamma \in \mathbb{M}^{\mathrm{N}}$, that $\gamma_{s'(i)\to t'(i)} > 0$. Given that $\gamma \in \mathbb{M}^{\mathrm{N}}$, it holds that for any $j \neq i$, $\gamma_{s'(j)\to t'(j)} \geq 0$. We conclude that

$$opt\mathrm{N}_\gamma(s,t) = v\mathrm{N}_\gamma\,(s',t') = \frac{\gamma_{s'(i)\to t'(i)} + \sum_{j\neq i} \gamma_{s'(j)\to t'(j)}}{|(s',t')|} \geq \frac{\gamma_{s'(i)\to t'(i)}}{|(s',t')|} > 0\,.$$

□

Lemma 15. Suppose $\gamma \in \mathbb{M}^{\mathrm{N}}$ and that s and t are sequences in Σ^*. Then

$$opt\mathrm{N}_\gamma(s,t) = opt\mathrm{N}_\gamma(t,s)\,.$$

Proof. Let (s',t') be an N-optimal alignment of (s,t). For each j such that either $s'(j) = \nabla$ or $t'(j) = \nabla$, it holds that $\gamma_{s'(j)\to t'(j)} = \gamma_{t'(j)\to s'(j)}$ by the definition of $\mathbb{M}^{\mathrm{N}}$, item 1. For each j such that $s(j) \neq \nabla$ and $t(j) \neq \nabla$, it follows from Lemma 13 that $\gamma_{s'(j)\to t'(j)} < \gamma_{s'(j)\to\nabla} + \gamma_{\nabla\to t'(j)}$, which in turn implies that $\gamma_{s'(j)\to t'(j)} = \gamma_{t'(j)\to s'(j)}$ by the definition of $\mathbb{M}^{\mathrm{N}}$, item 3.

Using the remarks above we have that

$$\begin{aligned} opt\mathrm{N}_\gamma(s,t) = v\mathrm{N}_\gamma\,(s',t') &= \frac{\sum_j \gamma_{s'(j)\to t'(j)}}{|(s',t')|} \\ &= \frac{\sum_j \gamma_{t'(j)\to s'(j)}}{|(t',s')|} = v\mathrm{N}_\gamma\,(t',s') \geq opt\mathrm{N}_\gamma(t,s)\,. \end{aligned}$$

By analogous reasoning, we have that $opt\mathrm{N}_\gamma(t,s) \geq opt\mathrm{N}_\gamma(s,t)$, which allow us to conclude that $opt\mathrm{N}_\gamma(t,s) = opt\mathrm{N}_\gamma(s,t)$. □

We need this remark to be used in the next lemma.

Fact 16. Let x, y, z, w be real numbers, with $y, w > 0$. If $z/w \geq x/y$, then $(x+z)/(y+w) \geq x/y$.

Lemma 17. Suppose that $\gamma \in \mathbb{M}^{\mathrm{N}}$ and that s, t, and u are sequences in Σ^*. Then

$$opt\mathrm{N}_\gamma(s,u) \leq opt\mathrm{N}_\gamma(s,t) + opt\mathrm{N}_\gamma(t,u)\,.$$

Proof. Let A e B be N-optimal alignments of (s,t) and (t,u), respectively. Define

$$C_1 = \{i \mid s(i) \text{ is aligned with } \nabla \text{ in } A\},$$
$$C_2 = \{k \mid u(k) \text{ is aligned with } \nabla \text{ in } B\},$$
$$C_3 = \{j \mid t(j) \text{ is aligned with } \nabla \text{ in } A \text{ and } t(j) \text{ is aligned with } \nabla \text{ in } B\},$$
$$C_4 = \{(i,j) \mid s(i) \text{ is aligned with } t(j) \text{ in } A \text{ and } t(j) \text{ is aligned with } \nabla \text{ in } B\},$$
$$C_5 = \{(j,k) \mid t(j) \text{ is aligned with } \nabla \text{ in } A \text{ and } t(j) \text{ is aligned with } u(k) \text{ in } B\},$$
$$C_6 = \left\{ \begin{array}{l} (i,j,k) \mid s(i) \text{ is aligned with } t(j) \text{ in } A \text{ and } t(j) \text{ is aligned with} \\ u(k) \text{ in } B \text{ and } \gamma_{s(i)\to u(k)} \leq \gamma_{s(i)\to t(j)} + \gamma_{t(j)\to u(k)} \end{array} \right\},$$
$$C_7 = \left\{ \begin{array}{l} (i,j,k) \mid s(i) \text{ is aligned with } t(j) \text{ in } A \text{ and } t(j) \text{ is aligned with} \\ u(k) \text{ in } B \text{ and } \gamma_{s(i)\to u(k)} > \gamma_{s(i)\to t(j)} + \gamma_{t(j)\to u(k)} \end{array} \right\}.$$

So,

$$v_\gamma(A) = \sum_{i\in C_1} \gamma_{s(i)\to\nabla} + \sum_{j\in C_3} \gamma_{\nabla\to t(j)} + \sum_{(i,j)\in C_4} \gamma_{s(i)\to t(j)} + \sum_{(j,k)\in C_5} \gamma_{\nabla\to t(j)} + \sum_{(i,j,k)\in C_6} \gamma_{s(i)\to t(j)} + \sum_{(i,j,k)\in C_7} \gamma_{s(i)\to t(j)}$$

and

$$v_\gamma(B) = \sum_{k\in C_2} \gamma_{\nabla\to u(k)} + \sum_{j\in C_3} \gamma_{t(j)\to\nabla} + \sum_{(i,j)\in C_4} \gamma_{t(j)\to\nabla} + \sum_{(j,k)\in C_5} \gamma_{t(j)\to u(k)} + \sum_{(i,j,k)\in C_6} \gamma_{t(j)\to u(k)} + \sum_{(i,j,k)\in C_7} \gamma_{t(j)\to u(k)}\,.$$

We now define an alignment C of (s,u) according to the following three rules. For each $(i,j,k) \in C_6$, align $s(i)$ with $u(k)$. For each remaining $s(i)$ not yet aligned, align $s(i)$ with ∇. For each remaining $u(k)$ not yet aligned, align ∇ with $u(k)$. The score of such alignment is

$$v_\gamma(C) = \sum_{i\in C_1} \gamma_{s(i)\to\nabla} + \sum_{k\in C_2} \gamma_{\nabla\to u(k)} + \sum_{(i,j)\in C_4} \gamma_{s(i)\to\nabla} + \sum_{(j,k)\in C_5} \gamma_{\nabla\to u(k)} + \sum_{(i,j,k)\in C_6} \gamma_{s(i)\to u(k)} + \sum_{(i,j,k)\in C_7} \left(\gamma_{s(i)\to\nabla} + \gamma_{\nabla\to u(k)}\right).$$

If $(i,j) \in C_4$ then, by definition of $\mathbb{M}^{\mathrm{N}}$, item 4, we have that $\gamma_{s(i)\to t(j)} + \gamma_{t(j)\to\nabla} \geq \gamma_{s(i)\to\nabla}$. Thus

$$\sum_{(i,j)\in C_4} \gamma_{s(i)\to t(j)} + \sum_{(i,j)\in C_4} \gamma_{t(j)\to\nabla} \geq \sum_{(i,j)\in C_4} \gamma_{s(i)\to\nabla}\,. \tag{2}$$

Note also that if $(j,k) \in C_5$ then, by Lemma 13, it is true that $\gamma_{t(j)\to u(k)} < \gamma_{t(j)\to\nabla} + \gamma_{\nabla\to u(k)}$. The definition of $\mathbb{M}^{\mathrm{N}}$, item 3, implies that $\gamma_{t(j)\to u(k)} = \gamma_{u(k)\to t(j)}$. Thus,

$$\gamma_{\nabla\to t(j)} + \gamma_{t(j)\to u(k)} = \gamma_{\nabla\to t(j)} + \gamma_{u(k)\to t(j)} = \gamma_{t(j)\to\nabla} + \gamma_{u(k)\to t(j)}\,,$$

where the last equality follows from the definition of $\mathbb{M}^{\mathrm{N}}$, item 1. Using again the definition of $\mathbb{M}^{\mathrm{N}}$, item 4, we have that

$$\sum_{(j,k)\in C_5} \gamma_{\nabla\to t(j)} + \sum_{(j,k)\in C_5} \gamma_{t(j)\to u(k)} \geq \sum_{(j,k)\in C_5} \gamma_{\nabla\to u(k)} \,. \tag{3}$$

Using inequalities (2) and (3), we obtain that

$$\begin{aligned} v_\gamma(A) + v_\gamma(B) &\geq v_\gamma(C) + \sum_{j\in C_3} \gamma_{t(j)\to\nabla} + \sum_{j\in C_3} \gamma_{\nabla\to t(j)} \\ &= v_\gamma(C) + 2\sum_{j\in C_3} \gamma_{\nabla\to t(j)} \\ &\geq v_\gamma(C) + 2\sum_{j\in C_3} \frac{G}{2} = v_\gamma(C) + |C_3|G \,. \end{aligned} \tag{4}$$

The inequality (4) follows from $G \leq 2\gamma_{\nabla\to t(j)}$, which in turn follows from the definition of $\mathbb{M}^{\mathrm{N}}$, item 6.

Next we estimate the length of the alignments.

$$|A| = |C_1| + |C_3| + |C_4| + |C_5| + |C_6| + |C_7| \leq \sum_i |C_i| ,$$

$$|B| = |C_2| + |C_3| + |C_4| + |C_5| + |C_6| + |C_7| \leq \sum_i |C_i| ,$$

$$|C| = |C_1| + |C_2| + |C_4| + |C_5| + |C_6| + 2|C_7| \geq \left(\sum_i |C_i|\right) - |C_3| \,.$$

It follows that

$$\begin{aligned} opt\mathrm{N}_\gamma(s,t) + opt_\gamma(t,s) = v\mathrm{N}_\gamma(A) + v\mathrm{N}_\gamma(B) &= \frac{v_\gamma(A)}{|A|} + \frac{v_\gamma(B)}{|B|} \\ &\geq \frac{v_\gamma(A)}{\sum_i |C_i|} + \frac{v_\gamma(B)}{\sum_i |C_i|} \geq \frac{v_\gamma(C) + |C_3|G}{\sum_i |C_i|} \\ &\geq \frac{v_\gamma(C) + |C_3|G}{|C| + |C_3|} \,. \end{aligned}$$

Thus, to prove the lemma we show that $(v_\gamma(C) + |C_3|G)/(|C| + |C_3|) \geq opt\mathrm{N}_\gamma(s,u)$. If $|C_3| = 0$, the prove is done. So, we may assume that $|C_3| > 0$. We consider two cases.

If $v_\gamma(C)/|C| \geq |C_3|G/|C_3|$, then, by Fact 16 and Lemma 12, we have that

$$\frac{v_\gamma(C) + |C_3|G}{|C| + |C_3|} \geq G \geq opt\mathrm{N}_\gamma(s,u) \,.$$

If $v_\gamma(C)/|C| < |C_3|G/|C_3|$, then, by Fact 16 and from the observation that $v_\gamma(C)/|C|$ is an upper bound on $opt\mathrm{N}_\gamma(s,u)$, it follows that

$$\frac{v_\gamma(C) + |C_3|G}{|C| + |C_3|} \geq \frac{v_\gamma(C)}{|C|} \geq opt\mathrm{N}_\gamma(s,u) \,. \qquad \square$$

4 Final Remarks

Given a rational scoring matrix γ, the problem of deciding whether γ belongs to $\mathbb{M}^{\mathrm{W}}$ (or to $\mathbb{M}^{\mathrm{N}}$) can be easily solved by an $O(|\Sigma|^3)$-time algorithm. Items 1 and 2 of the definitions can be checked in time $O(|\Sigma|^2)$, while items 3, 4, 5 and 6 can be checked in time $O(|\Sigma|^3)$.

Gusfield [5], Pevzner [8], and Bafna, Lawler and Pevzner [2] developed approximation algorithms for the *multiple sequence alignment* problem. Such algorithms are based on scoring matrices in $\mathbb{M}^{\mathrm{C}}$ and they do no guarantee approximation bounds for scoring matrices in $\mathbb{M}^{\mathrm{W}}$. It would be interesting to design approximation algorithms to work for matrices in $\mathbb{M}^{\mathrm{W}}$.

Metric indexing algorithms [3], which require a metric between strings, are used for *proximity searching*. As pointed out by an anonymous referee, our characterization of scoring matrices allows to decide whether such algorithms can be used with a given scoring matrix.

References

1. Apostolico, A., Galil, Z. (eds.): Pattern matching algorithms. Oxford University Press, Oxford, 1997
2. Bafna, V., Lawler, E., Pevzner, P.A.: Approximation algorithms for multiple sequence alignment. Theoretical Computer Science **182** (1997) 233–244
3. Chávez, E., Navarro, G, Marroquin, J.L.: Searching in metric spaces. ACM Computing Surveys **33** (2001), no. 3, 273–321
4. Gusfield, D.: Algorithms on strings, trees, and sequences: computer science and computational biology. Cambridge University Press, New York, 1997
5. Gusfield, D.: Efficient methods for multiple sequence alignment with guaranteed error bounds. Bulletin of Mathematical Biology **55** (1993), no. 1, 141–154
6. Levenshtein, V.I.: Binary codes capable of correcting deletions, insertions and reversals. Soviet Physics Doklady **10** (1965), no. 8, 707–710
7. Marzal, A., Vidal, E.: Computation of normalized edit distance and applications. IEEE Transactions on Pattern Analysis and Machine Intelligence **15** (1993), no. 9, 926–932
8. Pevzner, P.A.: Multiple alignment, communication cost, and graph matching. SIAM Journal on Applied Mathematics **52** (1992), no. 6, 1763–1779
9. Ristad, E.S., Yianilos, P.N.: Learning string-edit distance. IEEE Transactions on Pattern Analysis and Machine Intelligence **20** (1998), no. 5, 522–532
10. Sellers, P.H.: On the theory and computation of evolutionary distances. SIAM Journal on Applied Mathematics **26** (1974), no. 4, 787–793

Data Structures for Halfplane Proximity Queries and Incremental Voronoi Diagrams

Boris Aronov[1,*], Prosenjit Bose[2,**], Erik D. Demaine[3,***], Joachim Gudmundsson[4], John Iacono[1,***], Stefan Langerman[5,†], and Michiel Smid[2]

[1] Department of CIS, Polytechnic University, Brooklyn, NY, USA
[2] School of Computer Science, Carleton University, Ottawa, ON, Canada
[3] Computer Science and Artificial Intelligence Lab, MIT, Cambridge, MA, USA
[4] National ICT Australia, Sydney, Australia
[5] Départment d' Informatique, Université Libre de Bruxelles, Brussels, Belgium

Abstract. We consider preprocessing a set S of n points in the plane that are in convex position into a data structure supporting queries of the following form: given a point q and a directed line ℓ in the plane, report the point of S that is farthest from (or, alternatively, nearest to) the point q subject to being to the left of line ℓ. We present two data structures for this problem. The first data structure uses $O(n^{1+\varepsilon})$ space and preprocessing time, and answers queries in $O(2^{1/\varepsilon} \log n)$ time. The second data structure uses $O(n \log^3 n)$ space and polynomial preprocessing time, and answers queries in $O(\log n)$ time. These are the first solutions to the problem with $O(\log n)$ query time and $o(n^2)$ space.

In the process of developing the second data structure, we develop a new representation of nearest-point and farthest-point Voronoi diagrams of points in convex position. This representation supports insertion of new points in counterclockwise order using only $O(\log n)$ amortized pointer changes, subject to supporting $O(\log n)$-time point-location queries, even though every such update may make $\Theta(n)$ combinatorial changes to the Voronoi diagram. This data structure is the first demonstration that deterministically and incrementally constructed Voronoi diagrams can be maintained in $o(n)$ pointer changes per operation while keeping $O(\log n)$-time point-location queries.

1 Introduction

Line simplification is an important problem in the area of digital cartography [6, 9, 13]. Given a polygonal chain P, the goal is to compute a simpler polygonal chain Q that provides a good approximation to P. Many variants of this

* Research supported in part by NSF grant ITR-0081964 and by a grant from US-Israel Binational Science Foundation.

** Research supported in part by NSERC.

*** Research supported in part by NSF grants CCF-0430849 and OISE-0334653.

† Chercheur qualifié du FNRS.

J.R. Correa, A. Hevia, and M. Kiwi (Eds.): LATIN 2006, LNCS 3887, pp. 80–92, 2006.

problem arise depending on how one defines *simpler* and how one defines *good approximation.* Almost all of the known methods of approximation compute distances between P and Q. Therefore, preprocessing P in order to quickly answer distance queries is a common subproblem to most line simplification algorithms.

Of particular relevance to our work is a line simplification algorithm proposed by Daescu et al. [7]. Given a polygonal chain $P = (p_1, p_2, \ldots, p_n)$, they show how to compute a subchain $P' = (p_{i_1}, p_{i_2}, \ldots, p_{i_m})$, with $i_1 = 1$ and $i_m = n$, such that each segment $[p_{i_j} p_{i_{j+1}}]$ of P' is a good approximation of the subchain of P from p_{i_j} to $p_{i_{j+1}}$. The amount of error is determined by the point of the subchain that is farthest from the line segment $[p_{i_j} p_{i_{j+1}}]$. To compute this approximation efficiently, the key subproblem they solve is the following:

Problem 1 (Halfplane Farthest-Point Queries). *Preprocess n points $p_1, p_2, \ldots, p_n$ in convex position in the plane into a data structure supporting the following query: given a point q and a directed line ℓ in the plane, report the point p_i that is farthest from q subject to being to the left of line ℓ.*

Daescu et al. [7] show that, with $O(n \log n)$ preprocessing time and space, these queries can be answered in $O(\log^2 n)$ time. On the other hand, a naïve approach achieves $O(\log n)$ query time by using $O(n^3)$ preprocessing time and $O(n^3)$ space. The open question they posed is whether $O(\log n)$ query time can be obtained with a data structure using subcubic and ideally subquadratic space.

In this paper, we solve this problem with two data structures. The first, relatively simple data structure uses $O(n^{1+\varepsilon})$ preprocessing time and space, and answers queries in $O(2^{1/\varepsilon} \log n)$ time. The second, more sophisticated data structure uses $O(n \log^3 n)$ space and polynomial preprocessing time, and answers queries in $O(\log n)$ time. Both of our data structures apply equally well to halfplane farthest-point queries, described above, as well as the opposite problem of halfplane nearest-point queries—together, *halfplane proximity queries.*

Dynamic Voronoi diagrams. An independent contribution of the second data structure is that it provides a new efficient representation for maintaining the nearest-point or farthest-point Voronoi diagram of a dynamic set of points. So far, point location in dynamic planar Voronoi diagrams has proved difficult because the complexity of the changes to the Voronoi diagram or Delaunay triangulation for an insertion can be linear at any one step. The randomized incremental construction avoids this worst-case behavior through randomization. However, for the deterministic insertion of points, the linear worst-case behavior cannot be avoided, even if the points being incrementally added are in convex position, and are added in order (say, counterclockwise). For this specific case, we give a representation of a (nearest-point or farthest-point) Voronoi diagram that supports $O(\log n)$-time point location in the diagram while requiring only $O(\log n)$ amortized pointer changes in the structure for each update. So as not to oversell this result, we note that we do not have an efficient method of determining which pointers to change (it takes $\Theta(n)$ time per change), so the significance of this representation is that it serves as a proof of the existence of an encoding of Voronoi diagrams that can be modified with few changes to the encodings while still supporting point location queries. However, we believe that our combinatorial observations about Voronoi diagrams will help lead to efficient dynamic Voronoi diagrams with fast queries.

Currently, the best incremental data structure supporting nearest-neighbor queries (one interpretation of "dynamic Voronoi diagrams") supports queries and insertions in $O(\log^2 n/\log\log n)$. This result uses techniques for decomposable search problems described by Overmars [14]; see [5]. Recently, Chan [4] developed a randomized data structure supporting nearest-neighbor queries in $O(\log^2 n)$ time, insertions in $O(\log^3 n)$ expected amortized time, and deletions in $O(\log^6 n)$ expected amortized time.

2 A Simple Data Structure

Theorem 2. *There is a data structure for halfplane proximity queries on a static set of n points in convex position that achieves $O(2^{1/\varepsilon}\log n)$ query time using $O(n^{1+\varepsilon})$ space and preprocessing.*

Our proof is based on starting from the naïve $O(n^3)$-space data structure mentioned in the introduction, and then repeatedly apply a space-reducing transformation. We assume that either all queries are halfplane farthest-point queries or all queries are halfplane nearest-point queries; otherwise, we can simply build two data structures, one for each type of query.

Both the starting data structure and the reduction use Voronoi diagrams as their basic primitive. More precisely, we use the farthest-site Voronoi diagram for the case of halfplane farthest-point queries, and the nearest-site Voronoi diagram for the case of halfplane nearest-point queries. When the points are in convex position and given in counterclockwise order, Aggarwal et al. [1] showed that either Voronoi diagram can be constructed in linear time. Answering point-location queries in either Voronoi diagram of points in convex position can be done in $O(\log n)$ time using $O(n)$ preprocessing and space [11].

The proof of this and other results can be found in the full paper [2]:

Lemma 3. *There is a static data structure for halfplane proximity queries on a static set of n points in convex position, called* Okey, *that achieves $O(\log n)$ query time using $O(n^3)$ space and preprocessing.*

Transform 4. *Given any static data structure $\mathcal{D}$ for halfplane proximity queries on a static set of n points in convex position that achieves $Q(n)$ query time using $M(n)$ space and preprocessing, and for any parameter $m \leq n$, there is a static data structure for halfplane proximity queries on a static set of n points in convex position, called $\mathcal{D}$-Dokey, that achieves $2\,Q(n) + O(\log n)$ query time using $\lceil n/m \rceil\, M(m) + O(n^2/m)$ space and preprocessing.*

By starting with the data structure Okey of Lemma 3, and repeatedly applying the Dokey transformation of Transformation 4, we obtain the structure Okey-Dokey-Dokey-Dokey-..., or Okey-Dokey^k, which leads to the following:

Corollary 5. *For every integer $k \geq 1$, Okey-Dokey^{k-1} is a data structure for halfplane proximity queries on a static set of n points in convex position that achieves $O(2^k \log n)$ query time using $O(n^{(2k+1)/(2k-1)})$ space and preprocessing.*

3 Grappa Trees

Our faster data structure for halfplane proximity queries requires the manipulation of binary trees with a fixed topology determined by a Voronoi diagram. To support efficient manipulation of such trees, we introduce a data structure called *grappa trees.* This data structure is a modification of Sleator and Tarjan's link-cut trees [16] that supports some unusual additional operations.

Definition 6. *Grappa trees solve the following data-structural problem: maintain a forest of rooted binary trees with specified topology subject to*

$T =$ Make-Tree(v): *Create a new tree T with vertex v (not in another tree).*

$T =$ Link(v, w, d, m_ℓ, m_r): *Given a vertex v in some tree T_v and the root w of a different tree T_w, add an edge (v, w) to make w a child of v, merging T_v and T_w into a new tree T. The value $d \in \{\ell, r\}$ specifies whether w becomes a left or a right child of v; such a child should not have existed previously. The new edge (v, w) is assigned a* left mark *of m_ℓ and a* right mark *of m_r.*

$(T_1, T_2) =$ Cut(v, w): *Delete the existing edge (v, w), causing the tree T containing it to split into two trees, T_1 and T_2. Here one endpoint of (v, w) becomes the root of the tree T_i that does not contain the root of T.*

Mark-Right-Spine(T, m): *Set the right mark of every edge on the right spine of tree T (i.e., the edge from the root of T to its right child, and recursively such edges in the right subtree of T) to the new mark m, overwriting the previous right marks of these edges.*

$(e, m_\ell^*, m_r^*) =$ Oracle-Search(T, O_e): *Search for the edge e in tree T. The data structure can find e only via* oracle queries*: given two incident edges (u, v) and (v, w) in T, the oracle $O_e(u, v, w, m_\ell, m_r, m'_\ell, m'_r)$ determines in constant time which of the subtrees of $T - v$ contains x.*[1] *(Note that edges (u, v) and (v, w) are considered to exist in $T - v$, even though one of their endpoints has been removed.) The data structure provides the oracle with the left mark m_ℓ and the right mark m_r of (u, v), as well as the left mark m'_ℓ and the right mark m'_r of (v, w), and at the end, it returns the left mark m_ℓ^* and the right mark m_r^* of the found edge e.*

Theorem 7. *There exists an $O(n)$-space constant-in-degree pointer-machine data structure that maintains a forest of grappa trees and supports each operation in $O(\log n)$ worst-case time per operation, where n is the total size of the trees affected by the operation.*

4 Rightification of a Tree: Flarbs

The fixed-topology binary search tree maintained by our faster data structure for halfplane proximity queries changes in a particular way as we add sites to a Voronoi diagram. We delay the specific connection for now, and instead define

[1] Given the number of arguments, it is tempting to refer to the oracle as $O(A, B, D, G, I, L, S)$, but we will resist that temptation.

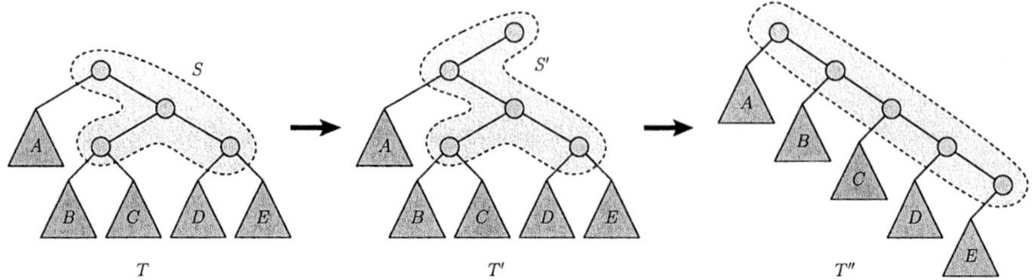

Fig. 1. An example of a flarb. The anchored subtree is highlighted.

the way in which the tree changes: a tree restructuring operation called a "flarb". Then we bound the work required to implement a sequence of n flarbs by showing that the total number of pointers changes (i.e., the total number of parent/left-child and parent/right-child relationships that change) is $O(n \log n)$. Thus, for the remainder of this section, we use the term *cost* to refer to (a constant factor times) the number of pointer changes required to implement a tree-restructuring operation, not the actual running time of the implementation. This bound on cost will enable us to implement a sequence of n flarbs via $O(n \log n)$ link and cut operations, for a total of $O(n \log^2 n)$ time.

The flarb operation is parameterized by an "anchored subtree" which it transforms into a "rightmost path". An *anchored subtree* S of a binary search tree T is a connected subgraph S of T that includes the root of T. A *right-leaning path* in a binary search tree T is a path monotonically descending through the tree levels, always proceeding from a node to its right child. A *rightmost path* in T is a right-leaning path that starts at the root of T.

The *flarb* operation[2] of an anchored subtree S of a binary search tree T is a transformation of T defined as follows; refer to Figure 1. First, we create a new root node r with no right child and whose left child subtree is the previous instance of T; call the resulting binary search tree T'. We extend the anchored subtree S of T to an anchored subtree S' of T' by adding r to S. Now we rearrange S' into a rightmost path on the same set of nodes, while maintaining the binary search tree order (in-order traversal) of all nodes. The resulting binary search tree T'' is the result of flarbing S in T.

Theorem 8. *A sequence of n flarb operations, starting from an empty tree, can be implemented at a cost of $O(\log n)$ amortized pointer changes per flarb.*

Proof. We use the potential method of amortized analysis, with a potential function inspired by the analysis of splay trees [17]. For any node x in a tree T, let $w(x)$ be the *modified weight* of the subtree rooted at x, which is the number of nodes in the subtree plus the number of null pointers in the tree. In

[2] "Flarb" is a clever abbreviation of a long technical term whose meaning we cannot reveal for reasons we cannot comment on at the moment, perhaps simply due to lack of space or of the aforementioned purported meaning. Note that this notion of flarb is different from that of [3].

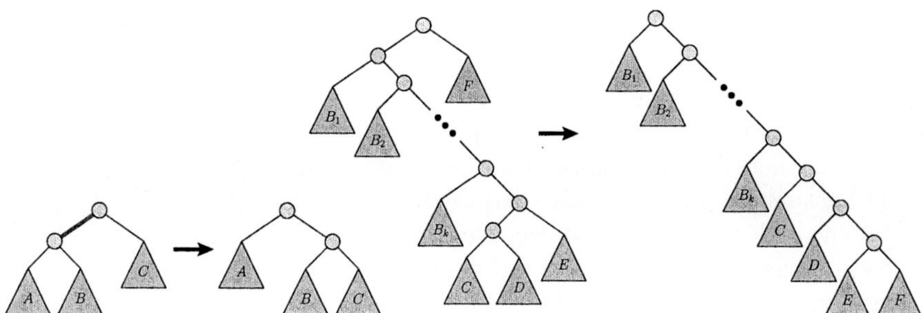

Fig. 2. A zig: The thick edge belongs to the anchored subtree S' and is light

Fig. 3. A zag

other words, we add dummy nodes as leaves in place of each null pointer in T, for the purpose of computing subtree size. Define $\varphi(x) = \lg \frac{w(left(x))}{w(right(x))}$. Clearly $|\varphi(x)| \leq \lg(2n-1)$, because the smallest possible subtree contains no real nodes and one dummy node, and the largest possible subtree contains $n-1$ real nodes and n dummy nodes. The potential of a tree T with n nodes is $\Phi(T) = \sum_x \varphi(x)$, with the sum taken over the (real) nodes x in T. Therefore, $|\Phi(T)| = O(n \log n)$.

For the purposes of the analysis, we use the following heavy-path decomposition of the tree. The *heavy path* from a node continues recursively to its child with the largest subtree, and the *heavy-path decomposition* is the natural decomposition of the tree into maximal heavy paths. Edges on heavy paths are called *heavy edges*, while all other edges are called *light edges*.

To analyze a flarb in a binary search tree T, we decompose the transformation into a sequence of several steps, and analyze each step separately.

First, the addition of the new root node r can be performed by changing a constant number of pointers in the tree. To implement rightification, we first execute several simplifying steps of two types, called "zig" and "zag",[3] in no particular order. A *zig* is executed whenever a light left edge is part of the anchored subtree S'; see Figure 2. The zig operation simply involves a right rotation on the edge in question. A *zag* is performed whenever there exists, within the anchored subtree S', a path that goes left one edge, right zero or more edges, and then left again one edge; see Figure 3. The zag operation performs a constant number of pointer changes to re-arrange the path in question into a right-leaning path. The full paper [2] shows that each zig or zag has zero amortized cost.

After all possible zigs and zags have been exhausted, we claim that the anchored subtree S' must have the form shown in Figure 4. Indeed, any tree that has no light left edge and no right-leaning path delimited by two left edges must have this form. In particular, because the rightmost path in this tree must be light, its length is at most $\lg(2n+1)$.

[3] Unlike most terminology in this paper, these terms are used for no particular reason. Cf. footnote 2.

The final *stretch* operation, which completes the flarb, simply converts this tree into a rightmost path by effectively concatenating the subsidiary right-leaning paths, incorporating them into the main path. Only $O(\log n)$ actual pointer changes are required. The potential does not increase because left subtrees of every node shrink and right subtrees grow, if they change at all. Thus, the amortized cost of the stretch is $O(\log n)$. □

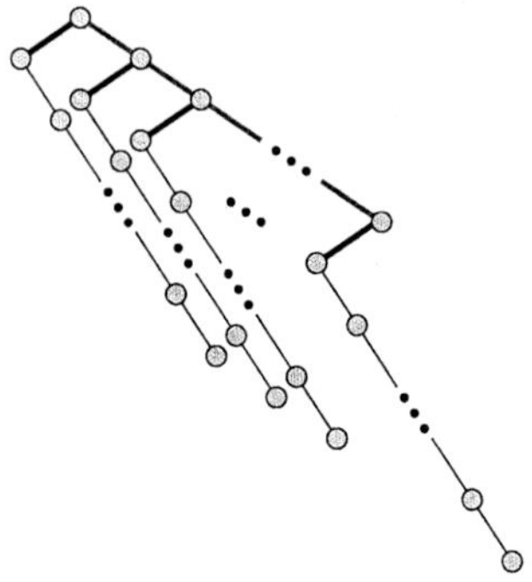

Fig. 4. S' before the final stretch. Thick light edges are light, and thick black edges are heavy.

5 Transformations

We focus on the farthest-point case, but the proofs apply to nearest-point too.

Transform 9. *Given a grappa tree data structure supporting each operation in $O(\log n)$ worst-case time, and given a data structure to incrementally maintain a tree created by n flarbs with $O(\log n)$ amortized pointer changes per flarb, we can construct an $O(n \log^2 n)$-space data structure that supports $O(\log n)$-time farthest-point queries on any prefix of a sequence of points in convex position in counterclockwise order.*

Proof. We construct an incremental data structure that supports $O(\log n)$-time farthest-point queries on the current sequence of points, $\langle p_1, p_2, \ldots, p_n \rangle$, and supports appending a new point p_{n+1} to the sequence provided that this change maintains the invariant that the vertices remain in convex position and in counterclockwise order. Thus the insertion order equals the index order and equals the counterclockwise traversal order of a convex polygon. The data structure runs on a pointer machine in which each node has bounded in-degree. Thus we can apply the partial-persistence transform of [10] and obtain the ability to support farthest-point queries on any prefix of the inserted points in $O(\log n)$ time. The space is proportional to the number of pointer changes during insertions.

We consider the ordered tree T formed by the finite segments of the farthest-point Voronoi diagram, ignoring their precise geometry; see Figure 5. More precisely, the *farthest-point Voronoi diagram* [15, Section 6.3] divides the plane into n cells by classifying each point q in the plane according to which of $p_1, p_2, \ldots, p_n$ is the farthest from q. The *farthest-point Delaunay triangulation* [12] is the dual of the farthest-point Voronoi diagram, i.e., it triangulates the convex polygon with vertices $p_1, p_2, \ldots, p_n$ by connecting two vertices whenever the corresponding Voronoi cells share an edge. We consider the dual tree T of this farthest-point Delaunay triangulation of the convex polygon, i.e., the dual graph excluding the infinite region exterior to the convex polygon. Each edge in this tree corresponds to (a nongeometric representation of) a finite edge of the farthest-point Voronoi diagram, which is the bisector of two of the points p_i and p_j that are adjacent in the Delaunay triangulation. Each node in the tree represents a vertex in the

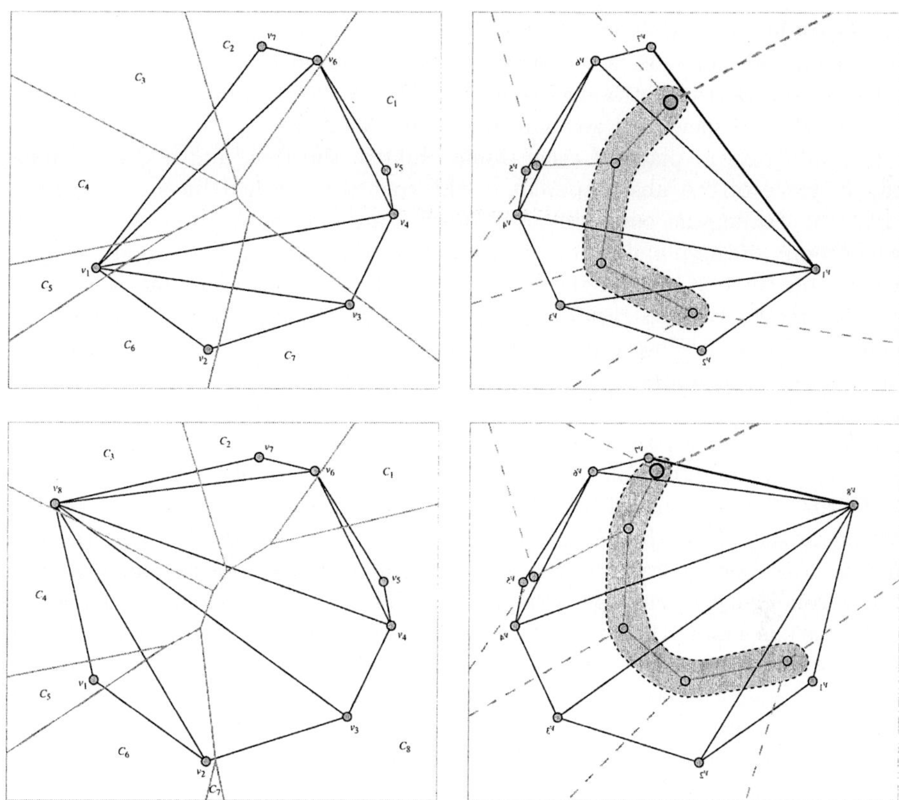

Fig. 5. Adding vertex v_8 in counterclockwise order. Top: Before. Bottom: After. Left: Farthest-point Voronoi diagram and its dual, the Delaunay triangulation. Right: Delaunay triangulation and its dual, the tree T with attached infinite rays drawn as dashed lines, drawn in mirror image so that geometric left versus right matches the order in the Voronoi diagram. The root vertex of T and its parent edge are emboldened.

farthest-point Voronoi diagram, or equivalently a triangle in the farthest-point Delaunay triangulation, and therefore has degree $d \leq 3$, where any degree deficit corresponds to $3 - d$ infinite rays in the farthest-point Voronoi diagram not represented in the tree T.

We can view the tree T as a binary search tree as follows. First, we root the tree at the node corresponding to the unique triangle in the Delaunay triangulation bounded by the edge connecting the first inserted point p_1 and the most recently inserted point p_n. We view the infinite ray emanating from the Voronoi vertex as the "parent edge" of this root node, defining the notion of *left child* versus *right child* of a node according to the counterclockwise order around the Voronoi vertex. (Note that this order is the opposite of the order defined by the triangulation, so in Figure 5 (right), we draw T in mirror image so that its geometric notions of left and right match that of the Voronoi diagram.) Second, we assign keys to nodes consistent with the in-order traversal. For each tree node corresponding to a Delaunay triangle with vertices p_i, p_j, p_k, where $i < j < k$, we assign a key of j. In other words, we assign the median of the three vertex labels of the Delaunay triangle to be the key of the corresponding tree node.

One way to view this key assignment is as follows. If we imagine adding an infinite rays in place of each absent child in the tree, and add an infinite ray in place of the absent parent of the root (the dashed lines in Figure 5, right), matching the counterclockwise order around the Voronoi vertex, then we decompose the plane into regions corresponding to Voronoi regions, each of which corresponds to a single point p_i. All of the nodes bounding p_i's region correspond to triangles incident to p_i. We assign the key i to the unique such node in T that is closest to the root of T, or equivalently the least common ancestor of such nodes, which is the inflection point between two descendant paths that bound the region. Two exceptions are $i = 1$ and $i = n$: the vertices incident to p_1 are those on the left spine of T, and the vertices incident to p_n are those on the right spine of T.

In this view, we also define the *left mark* of an edge to be the label of the region to the left of the edge, and similarly for the *right mark*. Thus, the two marks of an edge define the two points p_i and p_j whose bisector line contains the Voronoi edge. If an edge is the left edge of its parent node, then the edge's right mark is simply the key of that parent, because the right edge of the parent creates an inflection point at the parent. Similarly, if an edge is the right edge of its parent node, then the edge's left mark is the key of that parent. Intuitively, in either case, if we walk up from the edge on its "underside", then we immediately find a local maximum in the region. On the other hand, in either case, the other mark of the edge is the key of the parent node of the deepest ancestor edge that has the opposite orientation (left versus right): this bending point is the first inflection point we encounter as we walk up the tree on the "top side" of the edge. We use a grappa tree to represent T and the left and right marks of edges.

Next we consider the effect of inserting a new point p_{n+1}. As in the standard incremental algorithm for Delaunay construction [8, Section 9.3], we view the changes to the farthest-point Delaunay triangulation as first adding a triangle p_1, p_n, p_{n+1} and then flipping a sequence of edges to restore the farthest-point Delaunay property. The key property of the edge-flipping process is that all flipped edges end up incident to the newly inserted point p_{n+1}. Therefore these changes can be interpreted in the tree as adding a new root node, whose left child is the previous root, and then choosing a collection of nodes to move to the right path of the new root. This collection of nodes induces a connected subtree because the triangles involved in the flips form a connected set. (In particular, the flipping algorithm considers the neighbors of a triangle for flipping only if the triangle was already involved in a flip.) Thus, the changes correspond exactly to a flarb, with the flexibility of the flarb operation encompassing the various possibilities of which edges get flipped to maintain the farthest-point Delaunay property. Another way to view the addition of p_{n+1} is directly in the Voronoi diagram. The point p_{n+1} will capture the region R_{n+1} for which p_{n+1} is the farthest neighbor. The region R_{n+1} is a convex polygon. Outside R_{n+1}, the Voronoi diagram is unchanged, so all edges of the new Voronoi diagram are either bisectors of the same two points as before, or are edges of R_{n+1}. In T after the flarb, R_{n+1} corresponds to the right spine.

Each pointer change during a flarb operation can be implemented with one cut and one link operation. Therefore the grappa tree implements the $O(n \log n)$ total pointer updates from flarb operations in $O(n \log^2 n)$ total pointer updates. It remains to update the marks on the edges. By the incremental Voronoi/Delaunay view above, the only edges for which these marks might change are the edges incident to the new region R_{n+1}, i.e., the edges on the right spine. We update the right marks on all of these edges by calling Mark-Right-Spine($T, n+1$). The left mark of each edge on the right spine is simply the key of the parent node of the edge. During the execution of the flarb, various right paths were cut and pasted together with cuts and links to form the final right spine. The edges on the final right spine that were originally part of a right path in T already had a left mark equal to the key of their parent node. Any other edges on the final right spine were just added via links, so their left marks can be set accordingly by specifying the right m_ℓ argument to Link. Thus, the total number of pointer updates remains $O(n \log^2 n)$. This concludes the space bound of the data structure.

To support farthest-point queries, it suffices to build an oracle for the grappa tree's Oracle-Search. Specifically, given two incident edges (u, v) and (v, w), the oracle must determine which subtree of $T - v$ has the answer to the farthest-point query. Using the two marks on the two edges, two of which must be identical, we can determine the three vertices p_i, p_j, and p_k of the Delaunay triangle corresponding to vertex v in T. The vertex of the Voronoi diagram corresponding to v lies at the intersection of the three perpendicular bisectors between these three vertices of the Delaunay triangle. We draw three rays from this Voronoi vertex to each of the three corners of the Delaunay triangle. These three rays divide the plane into three sectors, and the Voronoi regions corresponding to the nodes in each subtree of $T - v$ lie entirely in one of these sectors, with exactly one subtree per sector. In constant time, we can decide which of the three sectors contains the query point q. The farthest-point Voronoi region containing the query point q is guaranteed to be incident to the corresponding subtree, and therefore we obtain a suitable answer for the oracle query. At the end, Oracle-Search will narrow the search to a specific edge of T, meaning that the query point q is in one of the two Voronoi regions incident to the corresponding Voronoi edge. In constant time, using the two labels on that edge of the tree, we can determine which side of the bisector contains q, and therefore which Voronoi region contains q, i.e., which point p_i is farthest from q. □

Transform 10. *Given an $O(n \log^2 n)$-space data structure that supports $O(\log n)$-time farthest-point queries on any prefix of a sequence of n points ordered in convex position in counterclockwise order, we can construct an $O(n \log^3 n)$-space data structure that supports $O(\log n)$-time farthest-point-left-of-line queries on n points in convex position.*

Combining Theorems 7 and 8 with Transforms 9 and 10, we obtain:

Corollary 11. *There is an $O(n \log^3 n)$-space data structure that supports $O(\log n)$-time halfplane proximity queries on n points in convex position.*

Corollary 12. *There is an $O(n)$-space data structure for maintaining a nearest-point or farthest-point Voronoi diagram of a sequence of points in convex position in counterclockwise order. The data structure supports inserting a new point at the end of the sequence, subject to preserving the invariants of convex position and counterclockwise order, in $O(\log n)$ amortized pointer changes per insertion; and supports point-location queries in $O(\log n)$ worst-case time.*

6 Open Problems and Conjectures

Several intriguing open problems remain open. One obvious question is whether the $O(n \log^3 n)$ space of our second data structure can be improved while keeping the optimal $O(\log n)$ query time. One specific conjecture in this direction is this:

Conjecture 13. *A sequence of n flarb operations, starting from an empty tree, can be implemented at a cost of $O(1)$ amortized pointer changes per flarb.*

We have no reason to believe that our $O(\log n)$ amortized bound is tight. Reducing the bound to $O(1)$ amortized would shave off a $O(\log n)$ factor from our space and preprocessing time. More importantly, it would increase our understanding of dynamic Voronoi diagrams, reducing the $O(\log n)$ amortized update time in Corollary 12 to $O(1)$ amortized. The potential function we use is inherently logarithmic; a completely new idea is needed here for further progress.

On the issue of improving our understanding of dynamic Voronoi diagrams, we pose the following problem:

Open Problem 14. *Is there a data structure for maintaining a Voronoi diagram of a set of points in convex position that allows point to be inserted in $\log^{O(1)} n$ time while supporting $O(\log n)$ point location queries?*

Here we relax the condition that the points be inserted in counterclockwise order, but maintain the restriction that they be in convex position. Although our potential function does not give the result, it is possible that a slight variation of it does.

Finally, it would be interesting to improve the construction time in our second data structure, in particular so that it completely subsumes the first data structure:

Open Problem 15. *Can the pointer changes caused by a flarb be found and implemented in $o(n)$ time, preferably $\log^{O(1)} n$ time?*

We have not been able to fully transform our combinatorial observations about the number of pointer changes into an efficient algorithm, because we lack efficient methods for finding which pointers change. Solving this question would improve our construction time by almost a linear factor, and would provide a reasonably efficient dynamic Voronoi data structure for inserting points in convex position in counterclockwise order.

Acknowledgments. This work was initiated at the Schloss Dagstuhl Seminar 04091 on Data Structures, organized by Susanne Albers, Robert Sedgewick, and Dorothea Wagner, and held February 22–27, 2004 in Germany. This work continued at the Korean Workshop on Computational Geometry and Geometric Networks, organized by Hee-Kap Ahn, Christian Knauer, Chan-Su Shin, Alexander Wolff, and René van Oostrum, and held July 25–30, 2004 at Schloss Dagstuhl in Germany; and at the 2nd Bertinoro Workshop on Algorithms and Data Structures, organized by Andrew Goldberg and Giuseppe Italiano, and held May 29–June 4, 2005 in Italy. We thank the organizers and institutions hosting both workshops for providing a productive research atmosphere. We also thank Alexander Wolff for introducing the problem to us.

References

1. A. Aggarwal, L. J. Guibas, J. B. Saxe, and P. W. Shor. A linear-time algorithm for computing the Voronoi diagram of a convex polygon. *Discrete Comput. Geom.*, 4(6):591–604, 1989.
2. B. Aronov, P. Bose, E. D. Demaine, J. Gudmundsson, J. Iacono, S. Langerman, and M. Smid. Data structures for halfplane proximity queries and incremental voronoi diagrams. arXiv:cs.CG/0512093, `http://arXiv.org/abs/cs.CG/0512093`
3. T. Calling. The adventures of Flarb Demingo! `http://www.thecalling.co.za/flarb_pictures.htm`, 2005. See also the fan site, `http://www2.fanscape.com/thecalling/streetteam/flarb.html`.
4. T. M. Chan. A dynamic data structure for 3-d convex hulls and 2-d nearest neighbor queries. In *Proc. 17th ACM-SIAM Sympos. Discrete Algorithms*, 2006.
5. Y.-J. Chiang and R. Tamassia. Dynamic algorithms in computational geometry. *Proc. IEEE*, 80(9):1412–1434, 1992.
6. R. G. Cromley. *Digital Cartography.* Prentice Hall, August 1991.
7. O. Daescu, N. Mi, C.-S. Shin, and A. Wolff. Farthest-point queries with geometric and combinatorial constraints. *Computat. Geom. Theory Appl.*, 2006. To appear.
8. M. de Berg, M. van Kreveld, M. Overmars, and O. Schwarzkopf. *Computat. Geom. Theory Appl.*. Springer, second edition, 1999.
9. B. D. Dent. *Cartography: Thematic Map Design.* William C Brown Pub, fifth edition, July 1998.
10. J. R. Driscoll, N. Sarnak, D. D. Sleator, and R. E. Tarjan. Making data structures persistent. *Journal of Computer and System Sciences*, 38(1):86–124, 1989.
11. H. Edelsbrunner, L. Guibas, and J. Stolfi. Optimal point location in a monotone subdivision. *SIAM Journal on Computing*, 15(2):317–340, 1986.
12. D. Eppstein. The farthest point Delaunay triangulation minimizes angles. *Computat. Geom. Theory Appl.*, 1(3):143–148, March 1992.
13. R. B. McMaster and K. S. Shea. *Generalization in Digital Cartography.* Association of American Cartographers, Washington D.C., 1992.

14. M. H. Overmars. *The Design of Dynamic Data Structures*, LNCS 156, 1983.
15. F. P. Preparata and M. I. Shamos. *Computational Geometry: An Introduction.* Springer, 1993.
16. D. D. Sleator and R. E. Tarjan. A data structure for dynamic trees. *Journal of Computer and System Sciences*, 26(3):362–391, June 1983.
17. D. D. Sleator and R. E. Tarjan. Self-adjusting binary search trees. *Journal of the ACM*, 32(3):652–686, July 1985.

The Complexity of Diffuse Reflections in a Simple Polygon

Boris Aronov*, Alan R. Davis**,
John Iacono***, and Albert Siu Cheong Yu†

Department of Computer and Information Science,
Polytecnic University, 5 MetroTech Center,
Brooklyn, NY 11201, USA
profdavisa@erols.com, siupaper@gmail.com
http://cis.poly.edu/~aronov, http://john.poly.edu

Abstract. The complexity of the visibility region formed by a point light source after k diffuse reflections in a simple n-sided polygon is $O(n^9)$, which is the first result polynomial in n, with no dependence on k. This bound is an exponential improvement over the previous bound of $O(n^{2\lceil (k+1)/2\rceil+1})$ due to Prasad et al. [8].

1 Introduction

Visibility problems in computational and combinatorial geometry have been studied extensively (see [3, 6, 9] and references therein). We confine our attention to results in the plane, more specifically those referring to visibility inside a simple polygon P with n vertices. Two points are *visible* to each other if the segment connecting them is contained in the polygon. The region visible from a point in P is a star-shaped polygon with at most n edges. The set of points of P visible from at least one point of a segment in P (the so-called "weak visibility polygon" from a segment) is a simple polygon with $O(n)$ edges and can be computed in linear time [5].

Aronov et al. [2, 1] and Davis [4] initiated the study of complexity of the region lit up by a single source of light in a simple polygon if reflection is allowed. Two models are considered. In both of them, any light incident upon a polygon corner is absorbed rather than reflected. In the *specular* reflection model, a light ray incident on a point in the interior of a polygon edge is reflected, as in geometric optics, with the angle of reflection equaling the angle of incidence. In the *diffuse* model which we consider in this paper, the light ray incident upon an interior point of an edge reflects in all possible interior directions. Aronov et al. [2] argue that for both diffuse and specular reflection the maximum complexity of the

* Research supported in part by NSF grant ITR-0081964 and by a grant from US-Israel Binational Science Foundation.

** Retired.

*** Research supported in part by NSF grants CCF-0430849 and OISE-0334653.

† Research supported in part by NSF grant CCF-0430849.

J.R. Correa, A. Hevia, and M. Kiwi (Eds.): LATIN 2006, LNCS 3887, pp. 93–104, 2006.

region lit up by a point light source with one reflection allowed is $\Theta(n^2)$. The results were generalized in [1] to any number k of reflections (where for simplicity we assume k is a constant and n can be arbitrarily large) and it was shown that for specular visibility this complexity is $O(n^{2k})$ and tight (at least for constant k). The case of multiple diffuse reflection is discussed by Prasad et al. [8], where they gave a bound of $O(n^{2\lceil (k+1)/2 \rceil + 1})$ on the complexity of the region lit up by a point with at most k diffuse reflections. Surprisingly, even though this bound is exponential in k (for arbitrarily large n), no constructions were known for diffuse reflection with complexity $\omega(n^2)$, irrespective of the number of reflections used. This gave rise to the conjecture in [8] that this in fact is the correct answer, for $k \geq 1$ reflections. As the analysis in [2], among other things, proves that the region visible from a point with one diffuse reflection is always simply connected, it has been suggested that this remains true when more diffuse reflections are allowed. However, Pal [7] gives an example when this conjecture fails already when two reflections are allowed.

In this paper, we partially settle the former conjecture on multiple diffuse reflections, namely we argue that the complexity of the region visible from a point with at most k diffuse reflections is $O(n^9)$, for any value of k.

2 Main Result

We obtain the main result, Theorem 1, in the old-fashioned way by presenting a sequence of lemmas that slowly lead to the theorem. A fixed simple polygon P with n edges is implicit in all notation.

Definition 1 (Time). *By time k, we mean the state of the visible region after exactly k diffuse reflections.*

Definition 2 (Edge). *We use the term edge to refer exclusively to an entire edge of the polygon P. The letter e, and its sub-and-superscripted variants, always refers to an edge.*

Definition 3 (Initial visibility region). *Initially, one specified point light source p is illuminated in P. At time 0, point q is illuminated if the interior of the segment pq is interior to the polygon P.*

Definition 4 (Illumination by diffuse reflection). *If point p on an edge is illuminated at time k, then point q is illuminated at time $k+1$ if the interior of the segment pq is interior to the polygon P. Points can only be illuminated in the manner described in the previous and current definitions.*

Definition 5 (Maximal illuminated segment). *We say segment x is maximally illuminated iff there does not exist an illuminated segment y such that $x \subset y$.*

Definition 6 (Triple). *We say (x, e, k) is illuminated if the maximally illuminated segment x on e is illuminated at time k.*

Definition 7 (Fundamental triple). *We say* (x, e, k) *is a fundamental illumination, or* $(x, e, k)^F$, *if* (x, e, k) *is illuminated with the restriction that* x *is the first illuminated segment from either end of* e *at time* k.

Definition 8 (Pentuple). *We say* (x, y, z, e, k) *is illuminated iff* x, y, *and* z *are maximally illuminated, adjacent, and disjoint segments of* e *illuminated at time* k. *To avoid symmetry problems, we assume there is a clockwise total ordering on disjoint line segments on the polygon* P *from an arbitrary vertex on* P, *and that* $x < y < z$ *with regards to this ordering.*

Definition 9 (Interior triple). *We say* (y, e, k) *is a interior triple, or* $(y, e, k)^I$, *if* (x, y, z, e, k) *is a pentuple.*

Lemma 1. *The complexity of the illuminated regions on the boundary of the polygon at time* k *is at most the number of fundamentally illuminated regions* $(x, e, k)^F$ *plus the number of the interiorly illuminated regions* $(y, e, k)^I$.

Proof. This is true since every triple must be either a fundamental triple or an interior triple, and there is only one illuminated region in either triple. Conversely, every segment that is lit has an associated triple.

2.1 Illuminations

Definition 10 (Illumination of triples, "$\rightarrow$" relation, defining light). *We say* (x, e, k) *illuminates* $(x', e', k+1)$, *or* $(x, e, k) \rightarrow (x', e', k+1)$ *for short, if:*

- $e \neq e'$.
- (x, e, k) *and* $(x', e', k+1)$ *are illuminated.*
- *If* x *was the only thing illuminated at time* k, *then at time* $k+1$ *there would be a segment illuminated on edge* e', *call it* x'', *and* $x'' \subseteq x'$. *We call the light from* x *to* x'' *the defining light of* $(x, e, k) \rightarrow (x', e', k+1)$. *(Verbal description:* x *illuminates either the whole maximally illuminated segment* x' *or one part of* x'.*)*

Definition 11 (Interior illumination, "$\xrightarrow{I}$" relation). *(See Figure 1.) We say that* $(y, e, k)^I$ *interior illuminates* $(y', e', k+1)^I$, *or* $(y, e, k) \xrightarrow{I} (y', e', k+1)$ *for short, iff*

- $(y, e, k) \rightarrow (y', e', k+1)$
- $\exists_{x,z,x',z'}(x, y, z, e, k)$ *and* $(x', y', z', e', k+1)$ *are pentuples. These are the defining pentuples of the interior illumination. (Note: This also implies* (y, e, k) *and* $(y', e', k+1)$ *are interior triples.)*
- $(x, e, k) \rightarrow (x'', e', k+1)$ *where* $x'' \neq y'$ *and* x'' *may or may not be the same as* x'
- $(z, e, k) \rightarrow (z'', e', k+1)$ *where* $z'' \neq y'$ *and* z'' *may or may not be the same as* z'

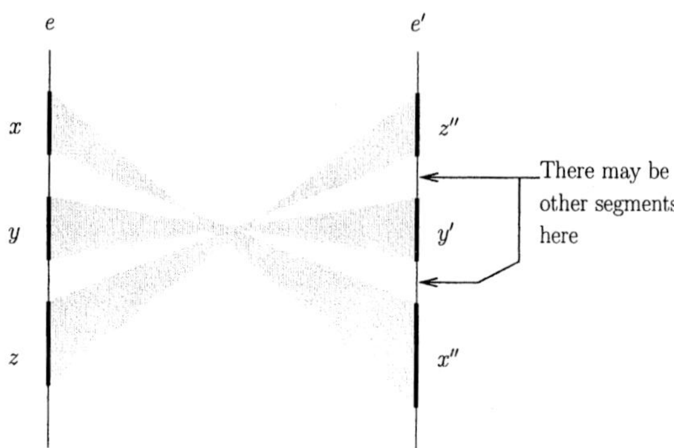

Fig. 1. An example of an interior illumination. The defining light of the illumination is the pink shaded region.

- *Verbal description: An interior segment y on edge e illuminates another interior segment y' on edge e' at time $k+1$. An adjacent segment of y, x, must also illuminate an segment x' on edge e', but it is not necessary that the segment x' is an adjacent segment of y'. Another adjacent segment of y, z, must also illuminate an segment z' on edge e', but it is not necessary that the segment z' is an adjacent segment of y'.*

The defining light of an interior illumination is the union of the defining lights of the three illuminations used in the definition.

Definition 12 ("$\overset{I}{\rightsquigarrow}$" relation). *We use $\overset{I}{\rightsquigarrow}$ to represent the transitive closure of the $\overset{I}{\rightarrow}$ relation on interior triple. Thus, $a \overset{I}{\rightsquigarrow} b$ if there is a directed path from a to b in G^I.*

Lemma 2. $(x, e, k) \overset{I}{\not\rightsquigarrow} (x', e, k')$ *for all $k' \geq k+2$. (Note: in this statement, and many others to follow, all variables are universally quantified unless otherwise noted.)*

Proof. This is true because for every edge e there is always one point not on e that can see all of e. Such a point can be found by extending a ray from the line at a suitably small angle.

Lemma 3 (Complete illumination). *At time $k = n$ the entire polygon is illuminated.*

Proof. This would be trivial, if the corners of the polygon could be illuminated from an incident edge in one step. However, from the definition of illumination, this is not the case. It is trivial that if $k = 2n$ the entire polygon is illuminated, since there is always one point that is visible from two points on adjacent edges. We omit the proof that the polygon is illuminated at time $k = n$ since it is more involved, and using the trivial $2n$ would not change any of the asymptotics of our results.

Note 1 (The end of time). From this point on we assume the time $k \leq n$, since beyond this time there is no additional complexity.

2.2 Graphs

Definition 13 (Graphs G, G^F, G^I). *Define a directed graph G with vertices consisting of the union of the fundamental and interior triples, and edges defined by the "$\rightarrow$" relation as defined in definition 10. We also define the graphs G^F and G^I which is the subgraph of G induced by the nodes representing fundamental triples and interior triples, respectively. Edges in G^I are defined by definition 11. That is, a node $(x, e, k)^I$ has an outgoing edge to $(x', e', k+1)^I$ in G^I iff $(x, e, k) \xrightarrow{I} (x', e', k+1)$.*

General idea: We first give an upper bound for the number of illuminated segments over all time and then we use this result to get an upper bound for the complexity of the visibility region. The total number of illuminated segments can be bounded by counting the number of nodes in G^F and G^I.

2.3 Bounding the Number of Fundamental Segments

Lemma 4. *There are at most $2n^2$ nodes in G^F.*

Proof. For a given e and k, there are at most 2 different segments x such that (x, e, k) is illuminated: Only the first segments from each end of e are fundamental triples. Since there are only n possible choices for e and the $k \leq n$ restriction of Note 1, this gives the result.

2.4 Bounding the Number of Interior Segments

Lemma 5 (Each interior segment can only illuminate n others). *Each illuminated segment can only illuminate n other segments. That is, for a given (x, e, k) there are only n segments y such that $(x, e, k) \rightarrow (y, e', k+1)$.*

Proof. This follows from the observation that the complexity of the region illuminated by an edge is at most linear, with at most one segment of each polygon edge appearing on its boundary [5].

Definition 14 (Source node). *A source node in G^I is defined to be an interior triple, $(x, e, k)^I$, with in-degree 0 in G^I.*

Lemma 6 (Bounding the number of source nodes). *There are only $4n^3$ source nodes in G^I.*

Proof. In graph G, the parent of a source node (x, e, k) in G^I is either (1) a fundamental triple or (2) an interior triple, $(x', e', k-1)^I$, with the restriction that $(x', e', k-1) \stackrel{I}{\not\rightarrow} (x, e, k)$ in G^I.

There are at most $2n^3$ source nodes whose parent is a fundamental triple. This is true because there are at most $2n^2$ nodes in G^F (Lemma 4), each of which can illuminate n segments (Lemma 5). Each illuminated segment can appear in at most 1 interior triple of G^I.

On the other hand, there are at most $2(n-1)(n^2)$ source nodes whose parent belongs to category 2 on all edges at all time. The harder observation is that all the illuminated segments on one edge can illuminate at most 2 source nodes whose parent belongs to category 2 on each edge at each time. The proof proceeds by contradiction.

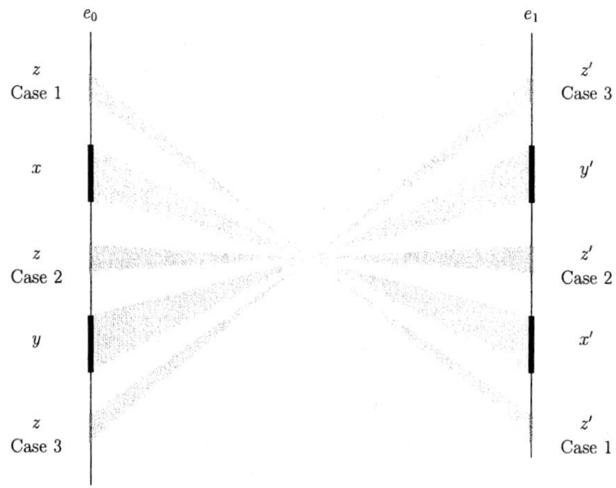

Fig. 2. No matter what the position of z is, if there are three illuminations between two edges, one must be interior

Let (x, e_0, k) illuminate the source node $(x', e_1, k+1)^I$ and (y, e_0, k) illuminate the source node $(y', e_1, k+1)^I$. Assume there exists the third segment (z, e_0, k) which illuminates the source node $(z', e_1, k+1)^I$. See Figure 2. If $x < z < y$, $(z', e-1, k+1)^I$ cannot be a source node in G^I by definitions 14 and 11. If $z < x < y$, $(x', e-1, k+1)^I$ cannot be a source node in G^I by definitions 14 and 11. If $x < y < z$, $(y', e-1, k+1)^I$ cannot be a source node in G^I by definitions 14 and 11. Therefore, all the illuminated segments on an edge can illuminate at most 2 source nodes whose parent belongs to category 2 on each edge at each time. Thus, all the illuminated segments on an edge can illuminate at most $2(n-1)$ source nodes whose parent belongs to category 2 on all edges at each time. Since there are n edges, at most $2(n-1)(n)$ source nodes have a parent whose belongs to category 2 at each time. This implies there are at most $2(n-1)(n^2)$ source nodes whose parent belongs to category 2 on all edges at all time.

Therefore, there are $2n^3$ source nodes whose parent belongs to category 1 and $2(n-1)(n^2)$ source nodes whose parent belongs to category 2. Totally there are $2n^3 + 2(n-1)(n^2)$ source nodes in G^I. We simply say there are at most $4n^3$ sources nodes in G^I for simplicity.

Definition 15 (Chord c of an illumination: $\xrightarrow{I}_c$). *We use the notion $(x, e_0, k) \rightarrow_c (x', e_1, k+1)$ to indicate $(x, e_0, k) \rightarrow (x', e_1, k+1)$ and that the defining light of this illumination passes through a chord c of the polygon.*

Lemma 7 (Good chord c of an interior illumination: $\xrightarrow{I\ g}_c$). *For any $(y, e_0, k) \xrightarrow{I}_c (y', e_1, k+1)$ there is a chord c inside the polygon such that the endpoints of c are vertices of the polygon; one endpoint lies on the blue dashed line and another endpoint lies on the red dotted line. See Figure 3.*

We call such a chord good, and use the notation $\xrightarrow{I\ g}_c$ to indicate that c is good. To avoid symmetry problems, if there is more than 1 vertex on the dotted blue or dashed red line, we select the vertex which is closest to the illuminated segment, y, to be the endpoint of the good chord. This implies a good chord is uniquely defined for every interior illumination.

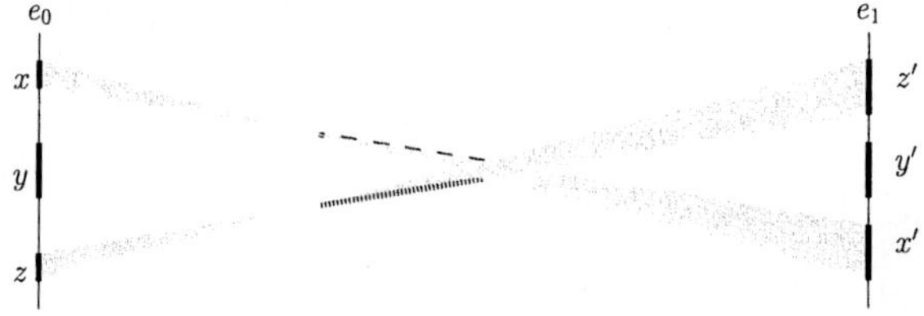

Fig. 3. If there are no polygon vertices on the boundary of the pink shaded region, z' or x' will be larger

Proof. If no vertex lies on the blue line, the lower endpoint of the illuminated segment, z', will have a different position. Similarly, if no vertex lies on the red line, the upper endpoint of the illuminated segment, x', will have a different position. Therefore, a good chord must always exist for any $(y, e_0, k) \xrightarrow{I}_c (y', e_1, k+1)$.

What is the purpose of a good chord? Through Lemma 8 to Lemma 11, We will prove that if a light passes through a good chord, it cannot go through the good chord in the opposite direction again in G^I. Therefore, a good chord will divide a simple polygon into two isolated regions.

Lemma 8. *If $(y, e_0, k) \xrightarrow{I\ g}_c (y', e_1, k+1)$ then for all e_2 $(y', e_1, k+1) \not\xrightarrow{I}_c (y'', e_2, k+2)$.*

Proof. We start by noting that there must be a good chord c by Lemma 7. In Figure 4, the right endpoint of the green line is the lower endpoint of the illuminated segment, y'. It passes through the highest point on the red line. Based on the basic geometry concept, the left endpoint of the green line must be located below the illuminated segment, x. Since the endpoint of the good chord c is on the red line, y' cannot illuminate anything through c above the green line. Therefore, the illuminated segment, y", cannot be above the illuminated segment, x. This implies that e_2 cannot be above e_0. By symmetry, we can conclude, e_2 must be the same as e_0. However, e_0 is totally illuminated at time k+2 by Lemma 2. Therefore, for all e_2, $(y', e_1, k+1) \not\xrightarrow{I}_c (y'', e_2, k+2)$.

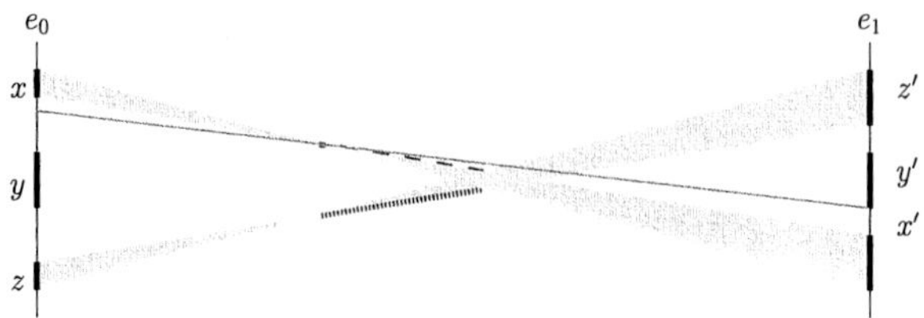

Fig. 4. The green shaded line goes from the bottom of y' to the left of the red dashed line. Its intersection with e_0 represents the highest point to the left of any chord c connecting the red dashed and blue dotted lines that can be illuminated by y'. Since the green shaded line is protected from the edges of the polygon by the pink shaded defining lights, it can never go to any edge above e_0.

Lemma 9 (On the intersection of chords). *For any two chords c and c' of the polygon P, if $(y, e, k) \overset{I\ g}{\rightarrow}_{c'} (y', e', k+1)$ and if c and c' intersect each other, then $(y, e, k) \overset{I}{\rightarrow}_{c} (y', e', k+1)$.*

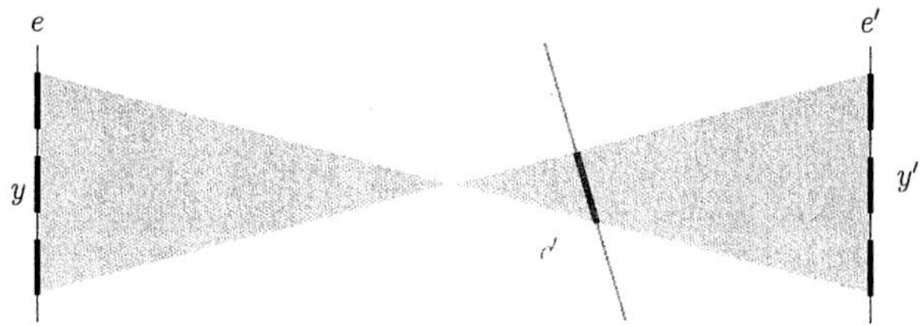

Fig. 5. If a chord c intersects c', the defining light from y to y' must go through c. This is because no endpoint of c is allowed inside the pink shaded region.

Proof. Refer to Figure 5. The endpoints of c' must be on the the border of the defining light of $(y, e, k) \overset{I\ g}{\rightarrow}_{c'} (y', e', k+1)$ by definition. The pink shaded region(not including the boundary) must not contain any vertex. Since c intersects with c', if one endpoint of c is on the left side of the grey line above the pink shaded region, another point of c must be on the right side of the grey line below the pink shaded region. By symmetry, if one endpoint of c is on the left side of the grey line below the pink shaded region, another point of c must be on the right side of the grey line above the pink shaded region. In either case, $(y, e, k) \overset{I}{\rightarrow}_{c} (y', e', k+1)$.

Lemma 10. *For all $t \geq 1$, if $(y, e_0, k) \overset{I\ g}{\rightarrow}_{c} (y', e_1, k+1)$ and if for all e_2 $(y', e_1, k+1) \overset{I}{\not\rightarrow}_{c} (y'', e_2, k+t)$ then for all e_2 $(y', e_1, k) \overset{I}{\not\rightarrow}_{c} (y''', e_2, k+t+1)$.*

Proof. Proof by contradiction. See Figure 6. Assume that for all e_2, $(y', e_1, k+1) \overset{I}{\not\rightarrow}_{c} (y'', e_2, k+t)$ and $(y', e_1, k+1) \overset{I}{\rightsquigarrow} (y^5, e_5, k+t)$ and $(y^5, e_5, k+t) \overset{I}{\rightarrow}_{c} (y^7, e_7, k+t+1)$. Let $(y', e_1, k+1) \overset{I}{\rightsquigarrow} (y^4, e_4, k+t-1)$ and $(y^4, e_4, k+t-1) \overset{I\ g}{\rightarrow}_{c'} (y^5, e_5, k+t)$. The good chord c' must exist by Lemma 7. If c' intersects with

c, then $(y^4, e_4, k+t-1) \xrightarrow{I}_c (y^5, e_5, k+t)$ by Lemma 9. This is a contradiction to the assumption that $(y', e_1, k+1) \not\xrightarrow{I}_c (y'', e_2, k+t)$ for all e_2. If c' does not intersect with c, then e_0 and e_4 are on one side of the good chord c' and e_5 is on another side of c'. By Lemma 8, for all e_6 $(y^5, e_5, k+t) \not\xrightarrow{I}_{c'} (y^6, e_6, k+t+1)$. It implies $(y^5, e_5, k+t) \not\xrightarrow{I}_c (y^7, e_7, k+t+1)$. This is a contradiction to the assumption.

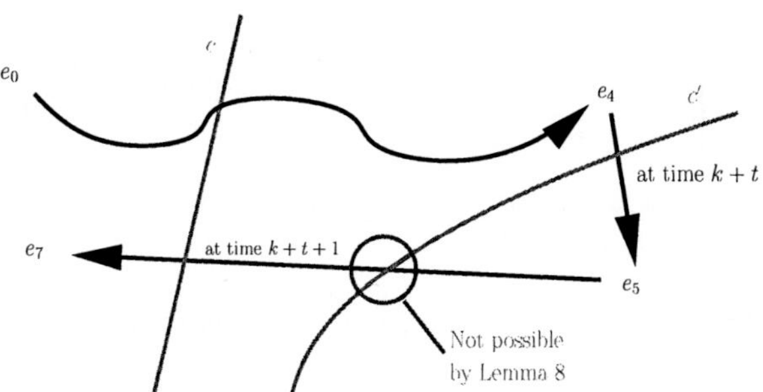

Fig. 6. The edge e_5 is on opposite sides of both c and c' from e_7. Since c and c' do not cross, no edge can be interior illuminated from e_5 through chord c without going through c' also. Since e_5 was just illuminated through c', the next illumination can not go though c' because of Lemma 8.

Lemma 11 (light cannot eventually go through the same chord twice). *If there exists a good chord such that* $(y, e_0, k) \xrightarrow{I\ g}_c (y', e_1, k+1)$ *then for all* e_2*,* y'' *and* $k' > k+1$ $(y', e_1, k) \not\xrightarrow{I}_c (y'', e_2, k')$.

Proof. The proof is by induction on k', using the previous three lemmas.

The following lemma will prove that all good chords of interior illuminations from the same node in G^I are disjoint.

Lemma 12. *If* $(y, e, k) \xrightarrow{I\ g}_{c'} (y', e', k+1)$ *and* $(y, e, k) \xrightarrow{I\ g}_c (y'', e'', k+1)$ *then* $(y, e, k) \not\xrightarrow{I}_c (y', e', k+1)$.

Proof. Assume $(y, e, k) \xrightarrow{I}_c (y', e', k+1)$. Refer to Figure 7. The pink shaded region (not including the boundary) must not contain any vertex. Suppose both endpoints of c are on the boundary of the pink shaded region. This implies there are two good chords for the defining light of $(y, e, k) \xrightarrow{I} (y', e', k+1)$. This can never happen by definition 7. Therefore, one endpoint of c must be above or below the pink shaded region. If that endpoint is on the right side of the grey line, it is not visible by (y, e, k) and thus, it cannot be an endpoint of the good chord c' by definition 7. This is a contradiction. If that endpoint is on the left side of the grey line, without loss of generality, we assume that endpoint is above the pink shaded region. Since the defining light of $(y, e, k) \xrightarrow{I} (y', e', k+1)$ must pass through chord c, another endpoint of c must be on the lower boundary of the pink shaded region or below the pink shaded region. This implies at least one endpoint of c' will block the defining light of $(y, e, k) \xrightarrow{I}{}^{g}_{c} (y'', e'', k+1)$. This is also a contradiction.

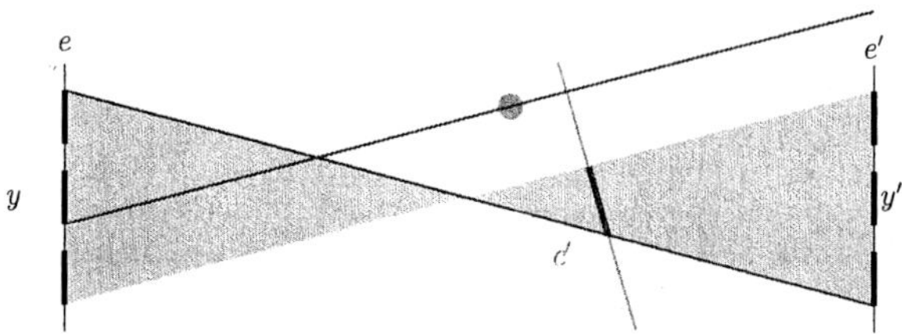

Fig. 7. If the grey point is an endpoint of c, then no light can originate on y, and pass through both the grey point and c'

Lemma 13. *No two nodes reachable from the same source node in G^I have the same e value.*

Proof. By Definition 7, every edge in G^I must pass through a good chord. By Lemma 12, all good chords of interior illuminations from the same node in G^I are disjoint and the light that goes through one good chord does not go through the others. By Lemma 11, the light can never go through a good chord twice. Thus, a good chord divides the polygon into two isolated regions, and this process recurses (Figure 8(a)). There is no path between nodes in different isolated regions (Figure 8(b)). Therefore, for each source node $(y_0, e_0, k_0)^I$, there are no directed paths such that $(y_0, e_0, k_0) \xrightarrow{I} (y_3, e_3, k_3)$ and $(y_0, e_0, k_0) \overset{I}{\rightsquigarrow} (y_4, e_3, k_4)$, when $y_3 \neq y_4$.

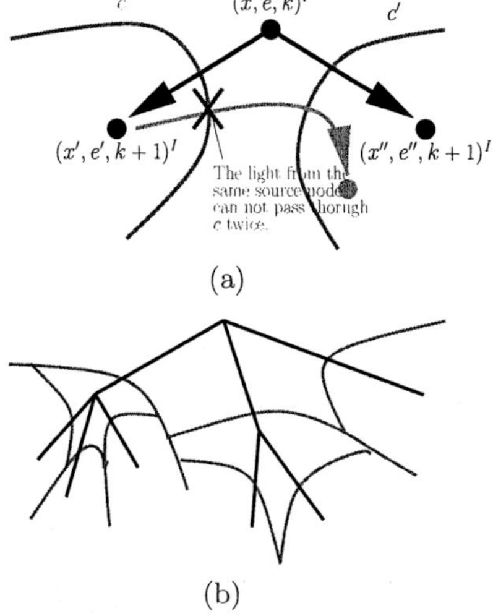

Fig. 8. Illustration of the proof of Lemma 13

Lemma 14. *The graph G^I has at most $4n^4$ nodes.*

Proof. In Lemma 6 there are at most $4n^3$ sources in G^I. By Lemma 13, there are at most $n-1$ nodes in G^I reachable from each of these sources.

2.5 The Complexity of the Visibility Region

Lemma 15. *The total complexity of the illuminated edges over all time is at most $4n^4 + 2n^2$.*

Proof. Follows from Lemmas 1, 4, and 14.

Lemma 16. *If x segments are illuminated in a polygon with n edges at time k, then the complexity of the illuminated region of the polygon (including the interior) at time $k+1$ is $O(nx^2)$.*

Proof. Since the region visible from one segment has complexity at most n [5], the intersection of x such regions is trivially $(xn)^2$. By observing that any region visible from one segment will intersect any segment not exterior to the polygon in exactly one place, this can be reduced to $O(nx^2)$.

Theorem 1. *The total complexity of the illuminated region at time k is $O(n^9)$.*

Proof. Lemma 15 gives a bound of $O(n^4)$ for the number of edges illuminated at time $k-1$. By applying Lemma 16, the $O(n^9)$ bound is obtained.

Theorem 2 ([8]). *The total complexity of the illuminated region at time k is $\Omega(n^2)$.*

Conjecture 1 ([8]). The total complexity of the illuminated region at time k is $\Theta(n^2)$.

We still believe this conjecture holds. Our proof over-counts in myriad ways and surely is not tight.

References

1. B. Aronov, A. R. Davis, T. K. Dey, S. P. Pal, and D. C. Prasad. Visibility with multiple reflections. *Discrete Comput. Geom.*, 20:61–78, 1998.
2. B. Aronov, A. R. Davis, T. K. Dey, S. P. Pal, and D. C. Prasad. Visibility with one reflection. *Discrete Comput. Geom.*, 19:553–574, 1998.
3. Tetsuo Asano, Subis K. Ghosh, and Thomas C. Shermer. Visibility in the plane. In Jörg-Rüdiger Sack and Jorge Urrutia, editors, *Handbook of Computational Geometry*, pages 829–876. Elsevier Science Publishers B.V. North-Holland, Amsterdam, 2000.
4. Alan R. Davis. *Visibility with reflection in triangulated surfaces.* PhD thesis, Polytechnic University, 1998.
5. H. ElGindy. An efficient algorithm for computing the weak visibility polygon from an edge in simple polygons. Manuscript, School Comput. Sci., McGill Univ., Montreal, PQ, 1984.

6. J. O'Rourke. Visibility. In Jacob E. Goodman and Joseph O'Rourke, editors, *Handbook of Discrete and Computational Geometry*, chapter 25, pages 467–480. CRC Press LLC, Boca Raton, FL, 1997.
7. Sudebkumar Prasant Pal and Dilip Sarkar. On multiple connectedness of regions visible due to multiple diffuse reflections. arXiv/cs.CG/0306010.
8. D. Chithra Prasad, Sudebkumar Prasant Pal, and Tamal K. Dey. Visibility with multiple diffuse reflections. *Computational Geometry*, 10(3):187–196, 1998.
9. Jorge Urrutia. Art gallery and illumination problems. In Jörg-Rüdiger Sack and Jorge Urrutia, editors, *Handbook of Computational Geometry*, pages 973–1027. North-Holland, 2000.

Counting Proportions of Sets: Expressive Power with Almost Order

Argimiro Arratia[1,⋆] and Carlos E. Ortiz[2,⋆⋆]

[1] Dpto. de Matemática Aplicada, Facultad de Ciencias, Universidad de Valladolid, Valladolid 47005, Spain
arratia@mac.cie.uva.es

[2] Department of Mathematics and Computer Science, Arcadia University, 450 S. Easton Road, Glenside, PA 19038-3295, U.S.A
ortiz@arcadia.edu

Abstract. We present a second order logic of proportional quantifiers, $\mathcal{SOLP}$, which is essentially a first order language extended with quantifiers that act upon second order variables of a given arity r, and count the fraction of elements in a subset of r–tuples of a model that satisfy a formula. Our logic is capable of expressing proportional versions of different problems of complexity up to **NP**-hard, and fragments within our logic capture complexity classes as **NL** and **P**, with auxiliary ordering relation. When restricted to monadic second order variables our logic of proportional quantifiers admits a semantic approximation based on almost linear orders, which is not as weak as other known logics with counting quantifiers, for it does not has the *bounded number of degrees property*. Moreover, we show in this almost ordered setting the existence of an infinite hierarchy inside our monadic language. We extend our inexpressibility result to an almost ordered (not necessarily monadic) fragment of $\mathcal{SOLP}$, which in the presence of full order captures **P**. To obtain all our inexpressibility results we developed combinatorial games appropriate for these logics.

Keywords: Proportional quantifiers, almost order, expressiveness, computational complexity, **P**, **NL**.

1 Introduction

An important open problem in Descriptive Complexity is to establish the existence of a logic, with recursive syntax and semantic, for describing all polynomial time computable problems, that is, for capturing the class **P**. The bottom line is that a solution to this problem should lead to a better understanding of the role of ordering in computations.

⋆ Supported by MEC, Projects MOISES (TIC2002-04019-C03-03), and SINGACOM (MTM-00958).

⋆⋆ Supported by a Faculty Award Grant from the Christian R. & Mary F. Lindback Foundation, and a Visiting Research Fellowship from Universidad de Valladolid, Spain.

J.R. Correa, A. Hevia, and M. Kiwi (Eds.): LATIN 2006, LNCS 3887, pp. 105–117, 2006.

As of today, all known logics that capture **P** need a built–in linear order as an extra symbol, so that the capturing may take place. The main issue is that a pre–defined ordering relation added to a logic and with its interpretation invariant through the models, makes the syntax of such logic non recursive (a consequence of Trahtenbrot's Theorem [2]); and thus this logic hardly classifies as "good" programming paradigm. On the other hand the presence of a built–in linear order, as part of the structures representing instances of computational problems, makes it very difficult for inexpressibility techniques from Model Theory, such as Ehrenfeucht-Fraïssé games, to succeed in showing meaningful computational lower bounds (e.g. see [5–§ 6.6]). To overcome this difficulty, and mindful of finding a logic in the aforesaid terms for **P**, various order–free extensions of first order logic (FO) have been proposed, most notably by the addition of some form of counting. However the demonstrated insufficient power of expressiveness of counting operators alone has led to the exploration (and exploitation) of some forms of pre–defined weak order and of the local nature of first order logic. The hope is that the logics with built-in weak form of order may have non-trivial expressive power, may be easier to separate, and eventually may shed light into the problem of separation of the corresponding logics with built-in order. In this context, the paper by Libkin and Wong [6] suggests that the above mentioned program may not be feasible because it shows an inherent expressive limitations of counting logics in the presence of auxiliary relations, which they call *preorders*, and their associated *almost–linear orders*. The main result of [6] is that a very powerful extension of FO with counting, denoted $\mathcal{L}^*_{\infty\omega}(C)$, which subsumes all known "pure" counting extensions of FO (meaning that fixpoint operators are not considered), in the presence of almost–linear orders, has the *bounded number of degrees property* (BNDP). The BNDP is a semantic property that limits the expressive power of logics that have it; such logics cannot express, for example, the transitive closure of a binary relation. (We will review all concepts in italics later in this paper.)

The purpose of this paper is to introduce a second order counting logic with built-in order that contains fragments whose expressive power is meaningful for Complexity Theory, and where the replacement of the built-in order by almost order does not yield logics with trivial expressive power, and where it should not be hard to obtain separation results. Our proposal consists of enhancing FO with quantifiers of the form $(P(X) \geq r)$ and $(P(X) \leq r)$ for rational $r \in (0, 1)$ and second order variable X of, say, arity $k > 0$, and whose meaning is that the cardinality of the set X is greater than or equal to (or less than or equal to) r times the cardinality of the set of k–tuples in the model. The logic obtained by adding these quantifiers, denoted by $\mathcal{SOLP}$ for *Second Order Logic of Proportions* (or *proportional quantifiers*), extends its first order counterpart $\mathcal{LP}$, which was introduced and studied by us in [1]. The intuition driving the definition of this logic is that by counting proportions as opposed to counting exact numbers of elements, the proportional quantifiers should be less susceptible to perturbations by the change of semantics from linear orders to almost-orders than the standard counting quantifiers.

Due to the proceedings' strict page limitations we must omit most of the proofs. The reader interested in learning all the details may request the extended version from the first author.

2 Second Order Logic of Proportional Quantifiers

Throughout this paper we use standard notation and concepts of Finite Model Theory as presented in the books [2] and [5]. Our vocabularies are finite and consists of relation symbols and constant symbols. Our structures are all finite, and if $\mathcal{A}$ is a structure over vocabulary τ, or τ–structure, and A is its universe, we either use $|\mathcal{A}|$ or $|A|$ to denote its size, that is, the number of elements in A.

In [1] we studied extensions of first order logic with quantifiers that count fractions of elements in a model that satisfy a given formula, and defined approximations to their semantics by giving interpretations of the formulae on finite structures where all predicates are restricted to act subject to an integer modulo. A natural extension is to have the proportional quantifiers act upon second order variables. This as we shall see gives more expressive power.

Definition 1. *The Second Order Logic of Proportional quantifiers, $\mathcal{SOLP}$, is the set of formulas of the form*

$$Q_1 \cdots Q_u \theta(x_1, \ldots, x_s, X_1, \ldots, X_r) \tag{1}$$

where $\theta(x_1, \ldots, x_s, X_1, \ldots, X_r)$ is a first order formula over some vocabulary τ with first order variables $x_1, \ldots, x_s$ and second order variables, $X_1, \ldots, X_r$; each Q_j $(j \leq u)$ is either $(P(X_i) \geq t_i)$ or $(P(X_i) \leq t_i)$, where t_i is a rational in $(0,1)$, for $i \leq r$. Whenever we want to make the underlying vocabulary τ explicit we will write $\mathcal{SOLP}(\tau)$.

We also define $\mathcal{SOLP}(\tau)[r_1, \ldots, r_k]$, for a given vocabulary τ and sequence $r_1, r_2, \ldots, r_k$ of distinct natural numbers, as the sublogic of $\mathcal{SOLP}(\tau)$ where the proportional quantifiers can only be of the form $(P(X) \leq q/r_i)$ or $(P(X) \geq q/r_i)$, for $i = 1, \ldots, k$ and q a natural number such that $0 < q < r_i$. Another fragment of $\mathcal{SOLP}$ which will be of interest for us is the Second Order Monadic Logic of Proportional quantifiers, *denoted $\mathcal{SOMLP}$, which is $\mathcal{SOLP}$ with the arity of the second order variables in* (1) *being all equal to 1.*

The interpretation for the proportional quantifiers is the natural one: Let X be a second order variable of arity k, $\overline{Y}$ a vector of second order variables, $\overline{x} = x_1, \ldots, x_m$ first order variables and $\phi(\overline{x}, \overline{Y}, X)$ a formula in $\mathcal{SOLP}(\tau)$ over some (finite) vocabulary τ, which does not contains X or any of the variables in $\overline{Y}$ as a relation symbol. Let r be a rational in $(0,1)$. Then

$$(P(X) \geq r)\phi(\overline{x}, \overline{Y}, X) \quad \text{and} \quad (P(X) \leq r)\phi(\overline{x}, \overline{Y}, X)$$

have the following semantics. For appropriate finite τ–structure $\mathcal{A}$, elements $\overline{a} = (a_1, \ldots, a_m)$ in A and vector of relations $\overline{B}$ over A, we have

$$\mathcal{A} \models (P(X) \geq r)\phi(\overline{a}, \overline{B}, X) \iff \text{there exists } S \subseteq A^k \text{ such that } \mathcal{A} \models \phi(\overline{a}, \overline{B}, S) \text{ and } |S| \geq r \cdot |A|^k$$

Similarly for $(P(X) \leq r)\phi(\overline{x}, \overline{Y}, X)$, substituting in the definition $\geq$ for $\leq$.

Example 1. Let $\tau = \{R, s, t\}$ where R is a ternary relation symbol, and s and t are constant symbols. Let r be a rational with $0 < r < 1$. We define

NOT-IN-CLOS$_{\leq r}$:= $\{\mathcal{A} = \langle A, R, s, t\rangle$: A has a set containing s but not t, closed under R, and of size at most a fraction r of $|A|$ $\}$.

$$\text{Let } \beta_{nclos}(X) := \forall x \forall u \forall v \ [X(s) \wedge \neg X(t) \wedge (X(u) \wedge X(v) \wedge R(u, v, x) \longrightarrow X(x))]$$

Then

$$\mathcal{A} \in \text{NOT-IN-CLOS}_{\leq r} \iff \mathcal{A} \models (P(X) \leq r)\beta_{nclos}(X)$$

We shall see in Section 3 that for $r = 1/2$ this problem is **P**–complete under first order reductions. (This result can be generalised to $r = 1/n$.) □

For **NP** we have the following problem.

Example 2. Let $\tau = \{E\}$, let r be a rational with $0 < r < 1$. We define

CLIQUE$_{\geq r}$:= $\{\mathcal{A} = \langle A, E\rangle : \langle A, E\rangle$ is a graph and at least a fraction r of the vertices form a complete graph $\}$

This problem can be defined by the sentence $(P(X) \geq r)\alpha_{cliq}(X)$, where

$$\alpha_{cliq}(X) := \forall x \forall y (X(x) \wedge X(y) \wedge x \neq y \longrightarrow E(x, y))$$

One can show that, for any rational $r \in (0, 1)$, CLIQUE$_{\geq r}$ is **NP**-complete via *logspace reducibilities.*

The following remark shows that $\mathcal{SOLP}$ extends the (classical) logic ∃SO.

Remark 1. Any formula in ∃SO is equivalent to a formula in $\mathcal{SOLP}[k]$, for any $k > 1$. Indeed, consider a formula of the form $\exists X \phi(X)$, where $\phi(X)$ is a first order formula with free second order variable X of arity $r > 0$. This can be expressed in $\mathcal{SOLP}[k]$ by the formula:

$$\left(P(X_1) \leq \frac{k-1}{k}\right)\left(P(X_2) \geq \frac{k-1}{k}\right) \phi(X_1) \vee \phi(X_2)$$

where X_1 and X_2 are variables of arity r.

3 Expressiveness of $\mathcal{SOLP}$ in the Presence of Order

By Remark 1, $\mathcal{SOLP}$ subsumes ∃SO. However, it adds extra information to the description of complexity classes, provided by the computing of bounds in the cardinality of sets in instances of problems. This we shall see in this section, where we impose constraints to the syntax of $\mathcal{SOLP}$ similar to Grädel's constraints for ∃SO in [4], and capture the classes **P** and **NL**, but as an extra information we have that **P** (and **NL**) $\subseteq \mathcal{SOLP}[2]$ and the first order part of the sentences describing this class is *universal Horn* (for **NL** it will be *universal Krom*). Furthermore, observe that all our examples of computational problems are definable in $\mathcal{SOMLP}$, the monadic fragment of $\mathcal{SOLP}$, some of them with not known expression (or non expressible) in monadic ∃SO.

Definition 2. *Let $\tau = \{R_1, \ldots, R_m, C_1, \ldots, C_s\}$ be some vocabulary with relation symbols R_1, …, R_m, and constant symbols C_1, …, C_s, and let X_1, …, X_r be second order variables of arity k_1, …, k_r, respectively. A first order formula α over $\tau \cup \{X_1, \ldots, X_r\}$, and extra binary relation symbol $=$ (equality) and the constant $\bot$ (standing for false), is a* universal Horn formula, *if α is a universally quantified conjunction of formulas over $\tau \cup \{X_1, \ldots, X_r\}$ of the form $\psi_1 \wedge \psi_2 \wedge \ldots \wedge \psi_s \longrightarrow \varphi$, where φ is either $X_i(\overline{u}_i)$ (where $\overline{u}_i$ denotes a k_i-tuple of first order terms, $i = 1, \ldots, r$) or $\bot$, and ψ_1, …, ψ_s are atomic or negation of atomic $(\tau \cup \{X_1, \ldots, X_r\})$-formulas except that any occurrence of the variables X_i must be positive (there are no restrictions on the predicates in τ or $=$). The logic $\mathcal{SOLP}Horn$ is the set of formulas of the form*

$$(P(X_1) \leq t_1) \cdots (P(X_r) \leq t_r)\alpha$$

where each t_i is a rational in $(0,1)$, and α is a universal Horn formula over some vocabulary τ and second order variables X_1, …, X_r.

By Example 1, the problem NOT-IN-CLOS$_{\leq r}$ is definable in $\mathcal{SOLP}Horn$. We can show that to test membership for a problem definable in $\mathcal{SOLP}Horn$ can be done deterministically in polynomial time.

Lemma 1. *The set of finite structures that satisfy a sentence θ in $\mathcal{SOLP}Horn$ is in* **P**. □

Thus, according to this lemma, our problem NOT-IN-CLOS$_{\leq r}$ is in **P**. We can prove that, for $r = 1/2$, it is complete for **P** via first order reductions. The idea is to define a reduction from the problem *Path System Accessibility* to NOT-IN-CLOS$_{\leq 1/2}$ using quantifier free first order formulae. An instance of the Path System Accessibility problem, which we abbreviate from now on as PS, is a finite structure $\mathcal{A} = \langle A, R, s, t\rangle$ or a *path system*, where the universe A consists of, say, n vertices, a relation $R \subseteq A \times A \times A$ (the *rules* of the system), a *source* $s \in A$, and a *target* $t \in A$ such that $s \neq t$. A positive instance of PS is a path system $\mathcal{A}$ where the target is *accessible* from the source, where a vertex v is accessible if it is the source s or if $R(x, y, v)$ holds for some accessible vertices x and y, possibly equal. In [7] Stewart shows that PS is complete for **P** via quantifier free first order reductions that include built-in order; in fact, via *projections* (see [7] for definitions and also [5–§ 11.2]). We get the following result.

Lemma 2. *The problem* NOT-IN-CLOS$_{\leq 1/2}$ *is complete for* **P** *via quantifier free projections (qfp's), that include the use of built-in successor.* □

Corollary 1. *Over finite structures, ordered with a built-in successor, the logic $\mathcal{SOLP}Horn$ captures* **P**. □

For logarithmic space bounded classes we have the following examples.

Example 3. Let $\tau = \{E, s\}$ where E is a binary relation symbol and s is a constant symbol. We think of τ-structures as graphs with a specify vertex s (the source). Let r be a rational with $0 < r < 1$. We define

$\mathrm{NCON}_{\geq r}$:= $\{\mathcal{A} = \langle A, E, s\rangle : \langle A, E\rangle$ is a graph and at least a fraction r of the vertices are **not** connected to $s\}$

Let $\alpha_{ncon}(Y)$ be the following formula

$$\alpha_{ncon}(Y) := \neg Y(s) \wedge \forall x \forall y (E(x,y) \wedge Y(x) \longrightarrow Y(y))$$

Then $\mathcal{A} \in \mathrm{NCON}_{\geq r} \iff \mathcal{A} \models (P(Y) \geq r)\alpha_{ncon}(Y)$.

Again, inspired on work by Grädel [4] we define:

Definition 3. *Let τ and $X_1, \ldots, X_r$ be as in Definition 2. A first order formula α over $\tau \cup \{X_1, \ldots, X_r\} \cup \{=, \bot\}$ is a* universal Krom formula, *if α is a universally quantified conjunction of clauses, where each clause is a disjunction of literals with at most two occurrences (positive or not) of the predicates $X_1, \ldots, X_r$, i.e. α is a 2-CNF formula with respect to the variables $X_1, \ldots, X_r$. The logic $\mathcal{SOLP}Krom$ is the set of formulas of the form*

$$(P(X_1) \geq t_1) \cdots (P(X_r) \geq t_r)\alpha$$

where each t_i is a rational in $(0,1)$, and α is a universal Krom formula over some vocabulary τ and second order variables $X_1, \ldots, X_r$.

The sentence defining $\mathrm{NCON}_{\geq r}$ is in $\mathcal{SOLP}Krom$. We can show that $\mathrm{NCON}_{\geq r}$ is in **NL**, the class of problems decidable by non deterministic logarithmic space bounded Turing machines; and, furthermore, that for $r = 1/2$ the problem $\mathrm{NCON}_{\geq r}$ is hard for **NL** via qfp's. Then with an argument similar to the one given for $\mathcal{SOLP}Horn$ one can show that satisfiability of sentences from $\mathcal{SOLP}Krom$ can be decided in **NL**, and conclude that over finite structures, ordered with built–in successor, $\mathcal{SOLP}Krom$ captures **NL**.

Remark 2. We can say more about the capturing of the class **P** by the logic $\mathcal{SOLP}$. The problem NOT-IN-CLOS$_{\leq 1/2}$ is complete via qfp's with order, and expressible in $\mathcal{SOLP}Horn[2]$; hence by reducing every problem K in **P** to NOT-IN-CLOS$_{\leq 1/2}$ with a quantifier free first order expressible reduction (which may include a successor relation), we get a sentence in $\mathcal{SOLP}Horn[2]$ defining K. Thus, $\mathbf{P} = \mathcal{SOLP}Horn[2]$ and obviously

$$\mathbf{P} \subseteq \mathcal{SOLP}[2] \subseteq \mathcal{SOLP}[2,3] \subseteq \mathbf{PSPACE} \tag{2}$$

The chain (2) motivate us to study the possibility of establishing a hierarchy in $\mathcal{SOLP}[2] \subseteq \mathcal{SOLP}[2,3] \subseteq \mathcal{SOLP}[2,3,5] \subseteq \ldots$, etc. We present in this paper the separation of fragments of these logics when a weak form of order is present, namely an almost linear order.

4 $\mathcal{SOLP}$ Restricted to Almost Orders

We begin with two preliminary definitions. The first is a slight modification of the notion of almost linear order from [6]; for it we remind the reader that a function $g : \mathbb{N} \to \mathbb{N}$ is sublinear if, for all $n \in \mathbb{N}$, $g(n) < n$.

Definition 4. *For a fixed positive integer k, a k-preorder over a set A is a binary, reflexive and transitive relation P in which every induced equivalence class of $P \cap P^{-1}$ has size at most k. An almost linear order over A, determined by a sublinear function $g : \mathbb{N} \to \mathbb{N}$, is a binary relation $\leq_g$ over A with a partition of the universe A into two sets B, C, such that B has cardinality $n - g(n)$ and $\leq_g$ restricted to B is a linear order, $\leq_g$ restricted to C is a 2-preorder, and for every $x \in C$ and every $y \in B$, $x \leq_g y$.*

Note that for any function $g : \mathbb{N} \to \mathbb{N}$, the almost linear order $\leq_g$ over a set A induces an equivalence relation $\sim_g$ in A defined by $a \sim_g b$ iff $a \leq_g b$ and $b \leq_g a$.

Definition 5. *Fix a sublinear $g : \mathbb{N} \to \mathbb{N}$ and let R be an n-ary relation on a set A. Let $\leq_g$ be an almost-order determined by g in A. We say that R is consistent with $\leq_g$ if for every pair of vectors $(a_1, \ldots, a_n)$ and $(b_1, \ldots, b_n)$ of elements in A with $a_i \sim_g b_i$ for every $i \leq n$, we have that*

$$R(a_1, \ldots, a_n) \text{ holds if and only if } R(b_1, \ldots, b_n) \text{ holds.}$$

Let $\mathcal{A} = \langle A, R_1^{\mathcal{A}}, \ldots, R_k^{\mathcal{A}}, C_1^{\mathcal{A}}, \ldots, C_s^{\mathcal{A}} \rangle$ be a τ-structure. We say that $\mathcal{A}$ is consistent with $\leq_g$ if and only if for every $i \leq k$, $R_i^{\mathcal{A}}$ is consistent with $\leq_g$.

By $\mathcal{SOLP}(\tau)_{\leq_g}$, for an almost order $\leq_g$, we understand the logic $\mathcal{SOLP}(\tau)$ with the almost order $\leq_g$ as additional built-in relation, and where we only consider models $\mathcal{A}$ that are consistent with $\leq_g$. Furthermore, for the formulas of the form $(P(X) \geq r)\phi(\overline{x}, \overline{Y}, X)$ and $(P(X) \leq r)\phi(\overline{x}, \overline{Y}, X)$, we require the following modification of the semantics: For an appropriate finite τ–model $\mathcal{A}$ consistent with $\leq_g$, for elements $\overline{a} = (a_1, \ldots, a_m)$ in A and an appropriate vector of relations $\overline{B}$, consistent with $\leq_g$, we should have

$$\mathcal{A} \models (P(X) \geq r)\phi(\overline{a}, \overline{B}, X) \iff \text{there exists } S \subseteq A^k, \textbf{ consistent with } \leq_g, \text{ such that } \mathcal{A} \models \phi(\overline{a}, \overline{B}, S) \text{ and} |S| \geq r \cdot |A|^k$$

Similarly for $(P(X) \leq r)\phi(\overline{x}, \overline{Y}, X)$, substituting in the condition $\geq$ for $\leq$.

The property of being consistent for $\leq_g$ holds in fact for all the formulas in $\mathcal{SOLP}(\tau)_{\leq_g}$. The proof is an easy induction in formulas.

Lemma 3. *Let $\mathcal{A}$ be a τ-structure which is consistent with $\leq_g$. Then, for every formula $\psi(\overline{x})$ in $\mathcal{SOLP}(\tau)_{\leq_g}$, the set $\psi^{\mathcal{A}} := \{\overline{a} \in \mathcal{A} : \mathcal{A} \models \psi(\overline{a})\}$ is consistent with $\leq_g$.* □

Definition 6. *We will use the expression "***almost** *second order proportional quantifier logic", and denote this by A–$\mathcal{SOLP}$, to refer to the collection of languages $\mathcal{SOLP}_{\leq_g}$ for every almost order $\leq_g$ given by a sublinear function g.*

Likewise, we denote A–$\mathcal{SOLP}[r_1,\ldots,r_k]$ the collection of all the languages $\mathcal{SOLP}_{\leq_g}[r_1,\ldots,r_k]$, for naturals $r_1,\ldots,r_k$, and A–$\mathcal{SOMLP}$, A–$\mathcal{SOMLP}[r_1,\ldots,r_k]$ for the corresponding monadic fragments.

For an illustration of the expressive power of the almost second order proportional quantifier logic, we shall give below a definition in A–$\mathcal{SOMLP}[2]$ of the set of models with almost order and with universe of even cardinality.

Example 4. Fix an almost order $\leq_g$, and consider the sentence

$$\Theta_2 := \left(P(B) \geq \frac{1}{2}\right)\left(P(C) \geq \frac{1}{2}\right)[\forall x(B(x) \vee C(x)) \wedge \forall y(B(y) \longrightarrow \neg C(y))]$$

Then for every structure $\mathcal{A}$, consistent with $\leq_g$,

$$\mathcal{A} \models \Theta_2 \text{ iff } |\mathcal{A}| := m \text{ is even}$$

The direction from left to right is clear: Θ expresses that B and C constitute a partition of $\mathcal{A}$. For the opposite direction, suppose m is even. There are $r \leq g(m)/2$ classes with two elements, say $\{a_1, b_1\}, \ldots, \{a_r, b_r\}$, and $l = m - 2r$ with one element, say there are $\{c_1\}, \ldots, \{c_l\}$. Hence, $m = 2r + l$ and since m is even, l must be even. We proceed to construct our disjoint sets C and B. Observe that for each $i = 1, \ldots, r$, both elements a_i and b_i must go into either $B^{\mathcal{A}}$ or $C^{\mathcal{A}}$, because $\mathcal{A}$ is consistent with $\leq_g$. With this in mind we do the following: If r is even then we can construct our even partition of same cardinality without much effort. If r is odd, then $r - 1 = 2k$ for some k, and so we put k classes (of two elements each) into $B^{\mathcal{A}}$, and the remaining $k + 1$ many 2-elements classes into $C^{\mathcal{A}}$. To compensate we put classes $\{c_1\}$ and $\{c_2\}$ in $B^{\mathcal{A}}$, and the remaining $l - 2$ 1-element classes are split evenly into $B^{\mathcal{A}}$ and $C^{\mathcal{A}}$. These sets $B^{\mathcal{A}}$ and $C^{\mathcal{A}}$ verify the formula $\alpha(B, C) := \forall x(B(x) \vee C(x)) \wedge \forall y(B(y) \longrightarrow \neg C(y))$ in $\mathcal{A}$ and have same cardinality. □

In a similar way, one can prove that for every natural $d > 2$, there exists a formula Θ_d, in the almost monadic second order proportional quantifier logic, with quantifiers of the form $P(X) \geq 1/d$ and $P(X) \geq (d-1)/d$ (i.e., contained in A–$\mathcal{SOMLP}[d]$), such that for structure $\mathcal{A}$, consistent with almost order $\leq_g$, $\mathcal{A} \models \Theta_d$ iff $|\mathcal{A}|$ is a multiple of d.

It was shown in [6] that a very powerful counting logic, $\mathcal{L}^*_{\infty\omega}(C)$, when restricted to almost orders, has the BNDP; hence, it has a very limited expressive power. The next example shows that this is not the case for A–$\mathcal{SOMLP}$.

Example 5. A–$\mathcal{SOMLP}$ **does not have the BNDP:** For a graph G, its degree set, $deg.set(G)$, is the set of all possible in- and out-degrees that are realised in G. A formula $\psi(x, y)$ on graphs has the Bounded Number of Degrees Property (BNDP) if there is a function $f : \mathbb{N} \to \mathbb{N}$ such that for any graph G with $deg.set(G) \subseteq \{0, \ldots, k\}$, $|deg.set(\psi[G])| \leq f(k)$, where $\psi[G]$ is the graph with same universe as G and edge relation given by ψ^G. These notions generalise to arbitrary τ-structures,and it is shown in [6] that every formula in $\mathcal{L}^*_{\infty\omega}(C)$, in the presence of almost-linear orders, has the BNDP and thus "*exhibits the very tame behaviour tipical for FO queries over unordered structures*" [6]. We shall see later that $A\text{-}\mathcal{SOMLP}$ presents a tame behaviour too since we can easily show separation results; however it differs from the counting logics considered by Libkin and Wong in [6] in that it does not have the BNDP.

Consider the quantifier free formula $path(x, y, U)$ in $A\text{-}\mathcal{SOMLP}(\{E\})$ that states that:

- $x \neq y$, $x \in U$ and $y \in U$;
- There is no element w of U such that $E(w, x)$ and there is no element w of U such that $E(y, w)$;
- $\exists w_1, w_2 \in U$ such that $E(x, w_1)$ and $E(w_2, y)$;
- For any element z in U different from x and y there exists unique $a, b \in U$ such that $E(a, z)$ and $E(z, b)$.

And let

$$\psi(x, y) := \left(P(U) \geq \frac{1}{2}\right) path(x, y, U)$$

This formula does not have the BNDP property for most sublinear functions g; for if we look at the models $\mathcal{A}$ consistent with $\leq_g$ and of cardinality $2n$, whose graph $E(x, y)$ is just the natural successor relation induced by $\leq_g$, i.e.

$$\underbrace{\begin{matrix}\bullet & & \bullet & & & \bullet \\ \uparrow\downarrow & \rightarrow & \uparrow\downarrow & \rightarrow & \ldots & \uparrow\downarrow \\ \bullet & & \bullet & & & \bullet\end{matrix}}_{g(2n)} \underbrace{\rightarrow \bullet \rightarrow \bullet \ldots \rightarrow \bullet}_{2n-g(2n)}$$

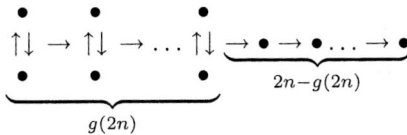

we see that E is consistent with $\leq_g$ and that $deg.set(\mathcal{A}) \subseteq \{1, 2, 3, 4\}$. However, the structure $\psi[\mathcal{A}]$ represents, for any n, the "transitive closure of length bigger or equal to half the size of the model $\mathcal{A}$", and thus $\lfloor n/2 \rfloor$, $\lfloor n/2 \rfloor + 1$, $\ldots \in deg.set(\psi[\mathcal{A}])$ for every g sublinear. □

5 Playing Games in $\mathcal{SOMLP}$

Definition 7. *Let τ be a vocabulary and $\mathcal{A}$ and $\mathcal{B}$ be two τ−structures, with $|B| = |A| + 1$. Let k and t be two positive integers. By $\mathcal{A} \prec_{(k,t)} \mathcal{B}$ we abbreviate the following statement:*

> *For every formula $\varphi(X_1, \ldots, X_t)$ of $FO(\tau \cup \{X_1, \ldots, X_t\})$ of (first order) quantifier rank $\leq k$ and unary second order variables $X_1, \ldots, X_t$, for all subsets $C_1, \ldots, C_t$ of A, there exist subsets $D_1, \ldots, D_t$ of B, such that*
>
> - *$|C_i| \leq |D_i| \leq |C_i| + 1$, for $i = 1, \ldots, t$, and*
> - *$\mathcal{A} \models \varphi(C_1, \ldots, C_t)$ implies $\mathcal{B} \models \varphi(D_1, \ldots, D_t)$*

The property $\mathcal{A} \prec_{(k,t)} \mathcal{B}$ basically states a first order elementary equivalence among the extended structures $\langle \mathcal{A}, C_1, \ldots, C_t \rangle$ and $\langle \mathcal{B}, D_1, \ldots, D_t \rangle$ with respect to first order formulas of the form $\varphi(X_1, \ldots, X_t)$, viewing $X_1, \ldots, X_t$ as extra unary relation symbols. This condition is sufficient for extending elementary equivalence to $\mathcal{A}$ and $\mathcal{B}$ with respect to sentences in $\mathcal{SOMLP}$.

Theorem 1. *Let $r_1, \ldots, r_s$ be distinct non zero natural numbers. Let τ be a vocabulary and $\mathcal{A}$ and $\mathcal{B}$ be two τ–structures, with $|A| = m$, $|B| = m+1$, $m+1 > r_i$ and $m \equiv_{r_i} -1$ for $i = 1, \ldots, s$. If $\mathcal{A} \prec_{(k,t)} \mathcal{B}$ then, for all sentence φ of $\mathcal{SOMLP}(\tau)[r_1, \ldots, r_s]$, of first order quantifier rank $\leq k$ and at most t unary second order variables (free or not), we have*

$$\mathcal{A} \models \varphi \textit{ implies } \mathcal{B} \models \varphi.$$

Our next goal is to characterise $\mathcal{A} \prec_{(k,t)} \mathcal{B}$ in terms of winning strategies for a Ehrenfeucht–Fraïssé type of games. Recall that, for a positive integer k, a k *rounds first order Ehrenfeucht–Fraïssé game* is played by two players, commonly known as *Spoiler* and *Duplicator*, and the game board consists of two structures $\mathcal{D}$ and $\mathcal{E}$ of the same vocabulary. The players alternatively select elements in the structures, doing so in the opposite structure as the one selected by his opponent and through k rounds, being Spoiler the first one to move in each round. Let d_1, $\ldots$, d_k be the elements selected in $\mathcal{D}$, and $e_1, \ldots, e_k$ the elements selected in $\mathcal{E}$. Duplicator wins if the substructure of $\mathcal{D}$ induced by $(d_1, \ldots, d_k)$ is isomorphic to the substructure of $\mathcal{E}$ induced by $(e_1, \ldots, e_k)$, under the function that maps d_i onto e_i, for $i = 1, \ldots, k$. The fundamental link between first order elementary equivalence and the k rounds first order Ehrenfeucht–Fraïssé game is given by the following theorem (cf. [2–§1.2] and [5–§6.1]).

Theorem 2 (Ehrenfeucht–Fraïssé). *For two structures $\mathcal{A}$ and $\mathcal{B}$ over the same vocabulary, and integer $k > 0$, the following two statements are equivalent:*

(i) $\mathcal{A} \equiv_k \mathcal{B}$ (i.e., every first order sentence of quantifier rank $\leq k$ that is true in $\mathcal{A}$ is also true in $\mathcal{B}$, and vice versa).
(ii) Duplicator has a winning strategy in the k rounds first order Ehrenfeucht–Fraïssé game played on $\mathcal{A}$ and $\mathcal{B}$. □

Our combinatorial game below is the classical game for monadic existential second order logic, to which we add strong restrictions on the possible cardinalities of both the structures upon the game is played and on the sets that the players choose as witnesses for second order variables (see [3] for definitions and a thorough analysis of games for monadic second order logic).

Definition 8. *Let τ be a relational vocabulary, s and k positive integers. Let $\mathcal{A}$ and $\mathcal{B}$ be two τ–structures such that $|B| = |A| + 1$. The proportional sets $(\mathcal{A}, \mathcal{B}, s, k)$–game (or simply the $(\mathcal{A}, \mathcal{B}, s, k)$–game) is played by Duplicator and Spoiler on $\mathcal{A}$ and $\mathcal{B}$ as follows:*

1. *Spoiler selects s subsets $S_1, \ldots, S_s$ of A.*
2. *Duplicator selects s subsets $T_1, \ldots, T_s$ of B, with $|S_i| \leq |T_i| \leq |S_i| + 1$, for $i = 1, \ldots, s$.*
3. *Both players play a k rounds first order Ehrenfeucht–Fraïssé game on the extended structures $\langle \mathcal{A}, S_1, \ldots, S_s\rangle$ and $\langle \mathcal{B}, T_1, \ldots, T_s\rangle$.*

Theorem 3. *Fix $k, s \in \mathbb{N}$, τ a vocabulary, $\mathcal{A}$ and $\mathcal{B}$ τ-structures with $|\mathcal{B}| = |\mathcal{A}| + 1$. $\mathcal{A} \prec_{(k,s)} \mathcal{B}$ if and only if Duplicator has a winning strategy in the $(\mathcal{A}, \mathcal{B}, s, k)$–game.* □

Now the tool for establishing non definability in $\mathcal{SOMLP}$ reads as follows.

Theorem 4. *Let $r_1, \ldots, r_n$ be distinct non zero natural numbers. Let τ be a relational vocabulary and K be a class of τ–structures. If for all positive integers k and s, there exists τ-structures $\mathcal{A}$ and $\mathcal{B}$ (that depend on k and s) such that $\mathcal{A} \in K$ and $\mathcal{B} \notin K$, $|\mathcal{B}| = |\mathcal{A}| + 1$, $|\mathcal{A}| \equiv_{r_i} -1$, for each $i = 1, \ldots, n$, and Duplicator has a winning strategy in the $(\mathcal{A}, \mathcal{B}, s, k)$–game, then K is not definable in $\mathcal{SOMLP}[r_1, \ldots, r_n]$.* □

5.1 Limitations in Expressive Power for A–$\mathcal{SOMLP}$

Recall that for a function g, the almost order $\leq_g$ on a universe A of a τ-structure $\mathcal{A}$, induces an equivalence relation $\sim_g$ on A. Let $[a]_g$ denote the $\sim_g$–equivalence class of $a \in A$, and $[A]_g := \{[a]_g : a \in A\}$. If, in addition, we ask of $\mathcal{A}$ to be consistent with $\leq_g$, then it makes sense to define the *quotient structure* $\mathcal{A}/_{\sim_g}$, as a τ-structure consisting of $[A]_g$ as its universe, and for a k-ary relation $R \in \tau$,

$$R^{\mathcal{A}/\sim_g} := \{([a_1]_g, \ldots, [a_k]_g) : (a_1, \ldots, a_k) \in R^{\mathcal{A}}\}$$

Furthermore, for a subset $B \subseteq A$ we define its $\leq_g$-*contraction* as $[B]_g := \{[b]_g : b \in B\}$; and for a subset $B \subseteq [A]_g$, its $\leq_g$-*expansion* is $(B)^g := \{a \in A : a \in [b]_g \text{ for some } [b]_g \in B\}$.

Definition 9. *Fix a sublinear function g and the almost order $\leq_g$. A $\leq_g$–cluster of models $\mathbf{C}$ is a collection of finite structures over same vocabulary τ, each consistent with $\leq_g$, and for each pair of τ-structures $\mathcal{A}$ and $\mathcal{B}$ in $\mathbf{C}$, their quotient under the equivalence relation $\sim_g$ are isomorphic, that is, $\mathcal{A}/_{\sim_g} \cong \mathcal{B}/_{\sim_g}$.*

Given $\mathcal{A}$ and $\mathcal{B}$ in the $\leq_g$-cluster $\mathbf{C}$, let F be an isomorphism from $\mathcal{A}/_{\sim_g}$ to $\mathcal{B}/_{\sim_g}$. Then, for $a \in A$ and $b \in B$, we write $a \equiv_{\mathbf{C}} b$ to indicate that $F([a]_g) = [b]_g$. Furthermore, for a subset $S \subseteq A$, the $\leq_g$-closure of S in $\mathcal{B}$ *is $cl_g(S, \mathcal{B}) := (F([S]_g))^g$ where $F([S]_g) := \{[b]_g \in [B]_g : F^{-1}([b]_g) \in [S]_g\}$.*

The following example gives an infinite family of sublinear functions that define almost orders.

Example 6. Fix $k \in \mathbb{N}$. Then $h_k(n) = 2r$, where $r \equiv_k n$, is a sublinear function. E.g., take $k = 3$, then $h_3(7) = 2$ and $h_3(8) = 4$. If $\mathcal{A}_7$ and $\mathcal{A}_8$ are sets of size 7 and 8 respectively, then $\mathcal{A}_7/_{\sim_{h_3}} \cong \mathcal{A}_8/_{\sim_{h_3}}$, and hence, they belong to the same $\leq_{h_3}$–cluster. □

The following lemma shows that pairs of structures, $\mathcal{A}$ and $\mathcal{B}$, that are in the same cluster and differ in one element, have the $\mathcal{A} \prec_{(k,s)} \mathcal{B}$ property.

Lemma 4. *Let g be a sublinear function and* $\mathbf{C}$ *an $\leq_g$-cluster of τ-models. Fix $\mathcal{A}$ and $\mathcal{B}$ in* $\mathbf{C}$*, with $|\mathcal{A}| = m$ and $|\mathcal{B}| = m+1$, and say $F : A/_{\sim_g} \to B/_{\sim_g}$ is the isomorphism among the quotient structures. Then:*

(i) For every first order formula $\phi(x_1, \ldots, x_s, \overline{Y})$ in $\mathcal{SOMLP}(\tau)$, for every a_1, $\ldots$, a_s in A, for every b_1, $\ldots$, b_s in B such that $a_i \equiv_{\mathbf{C}} b_i$, and for every sequence of subsets $S_1, \ldots, S_t$ of A, consistent with $\leq_g$, $\mathcal{A} \models \phi(a_1, \ldots, a_s, S_1, \ldots, S_t)$ iff $\mathcal{B} \models \phi(b_1, \ldots, b_s, cl_g(S_1, \mathcal{B}), \ldots, cl_g(S_t, \mathcal{B}))$;

(ii) If $S \subseteq A$ then $|S| \leq |cl_g(S, \mathcal{B})| \leq |S| + 1$.

Corollary 2. *Let g be a sublinear function and* $\mathbf{C}$ *an $\leq_g$-cluster of τ-models. For $\mathcal{A}, \mathcal{B} \in \mathbf{C}$, with $|\mathcal{A}| = m$, $|\mathcal{B}| = m+1$, and $k, s \in \mathbb{N}$, we have $\mathcal{A} \prec_{(k,s)} \mathcal{B}$.* □

Combining the previous corollary with Theorem 1 we get

Corollary 3. *Let r_1, $\ldots$, r_k be distinct non zero natural numbers. Let g be a sublinear function, $\leq_g$ an almost order and* $\mathbf{C}$ *an $\leq_g$-cluster of τ–structures. For every pair of structures $\mathcal{A}$, $\mathcal{B}$ in* $\mathbf{C}$*, such that $|\mathcal{A}| = m$, $|\mathcal{B}| = m+1$, $m+1 > r_i$ and $m \equiv_{r_i} -1$, for every $i \leq k$, we have that, $\mathcal{A} \models \varphi$ implies $\mathcal{B} \models \varphi$, for all sentences φ of $\mathcal{SOMLP}(\tau)[r_1, \ldots, r_k]$* □

Theorem 5. *Let $r, r_1, \ldots, r_k$ be distinct non zero natural numbers, pairwise relatively prime. Then A–$\mathcal{SOMLP}[r_1, \ldots, r_k] \subsetneq A$–$\mathcal{SOMLP}[r_1, \ldots, r_k, r]$.* □

Corollary 4. A-$\mathcal{SOMLP}[2] \subsetneq A$-$\mathcal{SOMLP}[2,3] \subsetneq A$-$\mathcal{SOMLP}[2,3,5] \subsetneq \ldots$

5.2 Limitations in Expressive Power for A–$\mathcal{SOLP}$

In this section we partially extend the separation result stated in Corollary 4 to second order variables of unbounded arity, that is, to A–$\mathcal{SOLP}$. It is a partial extension because we need to restrict our proportional quantifiers to be only of the form $(P(X) \leq 1/2)$, with X of arbitrary arity $r > 0$. Nonetheless, the result is interesting because it is precisely this type of quantifiers that defines $\mathcal{SOLP}Horn[2]$, which in the presence of order, captures **P**. Our main tool is a reshaping of Theorem 1 in the context of $\mathcal{SOLP}Horn[2]$.

Theorem 6. *Let τ be a vocabulary and $\mathcal{A}$ and $\mathcal{B}$ be two τ–structures, with $|A| = m$, $|B| = m+1$, $m+1 > 2$ and $m \equiv_2 -1$. If $\mathcal{A} \prec_{(k,t)} \mathcal{B}$ then, for all sentence φ of $\mathcal{SOLP}Horn(\tau)[2]$, of first order quantifier rank $\leq k$ and at most t second order variables (free or not), we have $\mathcal{A} \models \varphi$ implies $\mathcal{B} \models \varphi$* □

Theorem 7. *Let τ be a relational vocabulary and K be a class of τ–structures. If for all positive integers k and s, there exists τ-structures $\mathcal{A}$ and $\mathcal{B}$ (that depend on k and s) such that: $\mathcal{A} \in K$ and $\mathcal{B} \notin K$, $|\mathcal{B}| = |\mathcal{A}| + 1$, $|\mathcal{A}| \equiv_2 -1$, and Duplicator has a winning strategy in the $(\mathcal{A}, \mathcal{B}, s, k)$–game. Then K is not definable in $\mathcal{SOLP}Horn(\tau)[2]$.* □

Using as benchmark query: "the size of the model is a multiple of 3", which is definable in $A\text{-}\mathcal{SOLP}[2,3]$, we obtain

Corollary 5. $A\text{-}\mathcal{SOLP}Horn[2] \subsetneq A\text{-}\mathcal{SOLP}[2,3]$. □

References

1. Arratia, A., and Ortiz, C. (2004) Approximating the Expressive Power of Logics in Finite Models, *in: LATIN'04*, Lecture Notes in Comp. Sci. 2976 (Springer), 540–556.
2. Ebbinghaus, H.D., and Flum, J., *Finite Model Theory* (Springer-Verlag, 1995).
3. Fagin, R. (1997), Comparing the power of games on graphs, *Mathematical Logic Quarterly* **43**, 431–455.
4. Grädel, E. (1992), Capturing complexity classes by fragments of second order logic, *Theoretical Comp. Sci.* **101**, 35–57.
5. Immerman, N., *Descriptive Complexity* (Springer, 1998).
6. Libkin, L., and Wong, L., (2002) Lower bounds for invariant queries in logics with counting. *Theoretical Comp. Sci.* **288**, 153-180.
7. Stewart, I. (1994), Logical description of monotone NP problems, *J. Logic Computat.* **4** (4), 337-357.

Efficient Approximate Dictionary Look-Up for Long Words over Small Alphabets

Abdullah N. Arslan

University of Vermont, Department of Computer Science,
Burlington, VT 05405, USA
aarslan@cs.uvm.edu

Abstract. Given a dictionary $\mathcal{W}$ consisting of n binary strings of length m each, a d-query asks if there exists a string in $\mathcal{W}$ within Hamming distance d of a given binary query string q. The problem was posed by Minsky and Papert in 1969 as a challenge to data structure design. There is a tradeoff between time and space in solving the problem of answering a d-query. Recently developed time-efficient methods for text indexing with errors can be used to answer a d-query in $O(m)$ time. However, these methods use $O(n \log^d n)$ (or more) additional space which is not practical for large databases. We present a method for the problem assuming the standard RAM model of computation. We process the dictionary to construct an edge-labelled tree with distinct labels to siblings, and with bounded branching factor and height. Storing the resulting tree does not require asymptotically more space than the size of an ordinary trie that stores the given dictionary. We present an algorithm for the d-query problem that takes $O(m(3\log_{4/3} n - 1)^d (\log_2 n)^{d+1})$ time, and uses only $O(m)$ additional space. We also generalize the results for the case of the problem when a larger alphabet, or edit distance are used. We achieve $O(m(2|\Sigma| - 1)^d (\log_{2|\Sigma|/(2|\Sigma|-1)} n - 1)^d (\log_2 n)^{d+1})$ time complexity for the problem when Hamming distance is used. The time complexity increases by a factor of $O(d(2|\Sigma| - 1)^d (\log_2 n)^d)$ when we use edit distance. The algorithms are efficient when the approximate dictionary look-up involves long words defined over small alphabets. The algorithm can be modified such that it allows for words of different lengths as well as different lengths of query strings.

Keywords: d-query, approximate dictionary look-up, suffix tree, preprocessing, Hamming distance, edit distance, space efficient algorithm.

1 Introduction

Consider a dictionary $\mathcal{W}$ consisting of n binary strings of length m each. A *d-query* asks if there exists a string in $\mathcal{W}$ within Hamming distance d of a given binary query string q. Hamming distance between two strings is the number of positions at which the strings differ. The problem was originally posed by Minsky and Papert in 1969 [12] in which they asked if there is a data structure that supports fast d-queries. Algorithms for answering d-queries and its variations have

J.R. Correa, A. Hevia, and M. Kiwi (Eds.): LATIN 2006, LNCS 3887, pp. 118–129, 2006.

been a topic of interest in the literature [1–3, 5, 6, 11, 16]. *Approximate dictionary look-up* is a problem of dictionary look-up within distance d to a given query string q. It is essentially a d-query problem over a larger but finite alphabet, and it allows for various notions of proximity. *Approximate dictionary query* problem asks for not only one but all words that are close to the query string q.

A naive method for answering a d-query is to generate all possible strings differing from q in at most d positions, and perform $O(m^{d+1})$ exact queries using $O(m)$ additional space. If we use $O(m^d n)$ additional space to store all possible words within difference d of words in $\mathcal{W}$ we can answer a d-query in $O(m)$ time by performing one exact query. Therefore, there is a tradeoff between time and space. We are interested in finding a solution that does not require unreasonable space or time.

There are efficient algorithms for the 1-query problem (the d-query with $d = 1$) [11, 2, 16, 3]. They do not generalize to the d-query problem when $d > 1$.

Arslan and Eğecioğlu [1] study the approximate dictionary look-up problem in the standard RAM model, and they take into account all computations in the complexity analysis. They assume a trie representation for the dictionary $\mathcal{W}$. For the approximate dictionary look-up problem they present algorithms that use hybrid tree/dynamic programming approach [8, 13, 14] that combines tree traversal with partial computation of distances. Their method allows for the use of *simple edit distance* as well as the Hamming distance. The simple edit distance between two strings is the minimum number of edit operations (insert,delete, and substitute) required to transform one string into the other. The algorithm of Arslan and Eğecioğlu [1] answers a d-query in time $O(m^{d+1})$ using additional space $O(m)$.

Recently (during the development of this paper) several results for text indexing with errors have been published [4, 10]. These results improve the complexity of answering d-query. Results shown by Maaß [9] imply that when Hamming distance is used, and the dictionary is stored in a trie, the average time of trie-search to answer a d-query is $O(\log^{d+1}(nm))$. The method presented by Cole et al. [4] can be used to answer a d-query (where d can be the edit distance) in time $O(m + \log^d(nm) \log\log(nm))$, and it requires additional space $O(nm \log^d(nm))$ for indexing. Maaß and Nowak [10] have shown two results for text indexing with errors. Their results imply that when edit distance is used the d-query can be answered: 1) in $O(m)$ time using on average $O(n \log^d n)$ additional space for indexing. 2) in $O(m)$ average time using $O(n \log^d n)$ additional space for indexing. Although these methods are time-efficient, they are not practical for answering d-queries in very large databases.

In this paper we assume the standard RAM model of computation. We preprocess $\mathcal{W}$ to create an edge-labelled tree whose branching factor, and height are bounded from above by functions logarithmic in the number of words in $\mathcal{W}$, and the labels to siblings are distinct. The resulting tree does not require more space asymptotically than that required by a trie representation of $\mathcal{W}$. We use the hybrid tree/dynamic programming technique for approximate dictionary look-up. We assume that the alphabet Σ can be larger than a binary alphabet.

We study the Hamming distance, and edit distance cases separately for a given alphabet Σ. We develop algorithms similar to those presented by Arslan and Eğecioğlu [1]. We achieve $O(m(2|\Sigma|-1)^d(\log_{2|\Sigma|/(2|\Sigma|-1)} n-1)^d(\log_2 n)^{d+1})$ time complexity when Hamming distance is used by our first algorithm. In the second algorithm, the time complexity increases by a factor of $O(d(2|\Sigma|-1)^d(\log_2 n)^d)$ when edit distance is used. Our algorithms are efficient when the problem involves long words defined over a small alphabet. When we apply our algorithm for the Hamming distance case, the algorithm answers a d-query in time $O(m(3\log_{4/3} n-1)^d(\log_2 n)^{d+1})$ using only $O(m)$ space in run-time.

The outline of this paper is as follows: in Section 2 we describe how we preprocess dictionary $\mathcal{W}$ to create the tree that we use in our algorithms. We present our algorithms for the approximate dictionary look-up problem in Section 3. We first present the algorithm for the Hamming distance, and then the one for the simple edit distance. We summarize our results in Section 4.

2 Preprocessing

For simplicity we assume that dictionary $\mathcal{W}$ is stored as a trie $\mathcal{T}_W$ (or a Patricia tree, which is a trie in which the children with no siblings are merged with their parents). Otherwise, we can always create $\mathcal{T}_W$ for $\mathcal{W}$. $\mathcal{W}$ has words of lengths m each over an alphabet Σ where $|\Sigma| \geq 2$.

For any node v in a given tree $\mathcal{T}$ we denote by n_v the number of leaves rooted at subtree v. We first establish the following lemma about n_v.

Lemma 1. *Let $\mathcal{T}$ be a tree of height h, and branching factor b. There exists a node v in $\mathcal{T}$ such that $\frac{n}{2b} \leq n_v \leq \frac{n}{2}$.*

Proof. We construct an algorithm that finds node v such that the number of leaves n_v in the subtree rooted at v satisfies the inequalities in the lemma. The algorithm starts at the root, and throughout the entire search selects the node with highest leaf-counts among its siblings. By the pigeon-hole principle one child c of the root is a subtree with at least $\frac{n}{b}$ leaves because the branching factor of the tree is b, i.e. $n_c \geq \frac{n}{b}$. If the leaf-count n_c is also $\leq \frac{n}{2}$ then the search stops at node c since c satisfies the inequalities in the lemma, i.e. $v = c$. Otherwise, the algorithm continues at the subtree rooted at c with the largest leaf-count. The search will continue as long as the leaf-count for the current node is larger than $\frac{n}{2}$. When the leaf-count for the current node c finally is less than or equal to $\frac{n}{2}$ then by the pigeon-hole principle the leaf-count n_c is at least $\frac{n}{2b}$ for c that has the largest leaf-count among its siblings. The algorithm stops at node c and the leaf-count n_c is between $\frac{n}{2b}$ and $\frac{n}{2}$, i.e. $v = c$.

Corollary 1. *There exists a node v in $\mathcal{T}_W$ such that $\frac{n}{2|\Sigma|} \leq n_v \leq \frac{n}{2}$.*

For a given node $v \in \mathcal{T}_W$ we denote by $p_{r,v}$ the concatenation of labels of the edges on the path from the root r to node v. Note that $p_{r,v}$ is the common prefix of the words appearing in subtrie rooted at r. Another interpretation of

Corollary 1 is that there exists a prefix $p_{r,v}$ that is common to n_v words in $\mathcal{W}$ where $\frac{n}{2|\Sigma|} \leq n_v \leq \frac{n}{2}$.

We preprocess $\mathcal{T}_W$ to create an edge-labelled tree $\mathcal{S}$ with bounded branching factor and height, and in $\mathcal{S}$ from any node to its distinct children no label is a prefix of another. Function $Convert(r, n_r)$ in Figure 2 creates $\mathcal{S}$ shown in Figure 3 and returns its root r'. The function takes as a parameter node r of $\mathcal{T}_W$, and n_r. It creates a node r' in $\mathcal{S}$. If r is a leaf then the function returns r'. Otherwise, the algorithm reorganizes the tree rooted at r in $\mathcal{T}_W$ into a tree rooted at r' in $\mathcal{S}$. This is done by first determining the children of r', and recursively creating the subtrees in $\mathcal{S}$ rooted at these children from subtrees in $\mathcal{T}_W$. The algorithm performs this in a few main steps. First, it collects into a list nodes that are candidate to be children of r'. Second, these nodes are examined and the list is revised so that in the list of labels of arcs from r' to these nodes, no label is a prefix of another.

The algorithm uses a set L to keep track of nodes that are candidate for being children of r' in $\mathcal{S}$. It initializes L to be the empty set. At each iteration, the algorithm finds in $\mathcal{T}_W$ a node v of Lemma 1. The algorithm given in the proof of the lemma shows that there exists a node v such that $\frac{n}{2|\Sigma|} \leq n_v \leq \frac{n}{2}$. We may use any search algorithm which returns a node v with the leaf-count n_v satisfying these inequalities. We modify this algorithm such that it ignores any node and its subtries when the node is marked as "deleted". Finding a node with the largest leaf-count less than or equal to $\frac{n}{2}$ is advantageous because it yields to $\mathcal{S}$ with smaller height and branching factor. Once we find vertex v we delete the subtrie rooted at v logically. This involves marking it as deleted in $\mathcal{T}_W$, and the leaf-counts for all its ancestors in $\mathcal{T}_W$ are updated by subtracting n_v from each. This can easily be done if there are backward arcs. We can traverse $\mathcal{T}_W$ before the preprocessing, and add backward arcs. These arcs can be removed after the preprocessing is completed. Next, we iteratively find new vertices satisfying

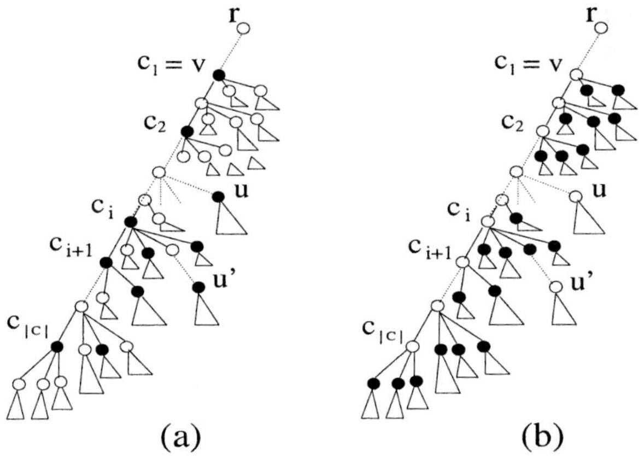

Fig. 1. (a) A sequence $C = c_1, c_2, \ldots, c_k$ of nodes obtained from L for v. Filled nodes are in L. (b) Filled nodes are added to F. Nodes u and u' are removed from L along with all nodes in C.

```
Algorithm Convert(r, n_r)
   create a new node r' in S
   if r is a leaf then return r'
   L := ∅
   while (n_r > 1) do {
     find node v of Lemma 1
     L := L ∪ {v}
     mark vertex v as ''deleted'' in T_W
     for every vertex w on the path from the root to v in T_W do
       update the leaf-count n_w := n_w − n_v
   }
   add to L the only (remaining) child v of r
   sort L into itself in ascending label lengths
   F := ∅
   while (|L| > 0) do {
     pick the next (shortest) node v in sorted list L; i := 1; c_i := v
     for every node w appearing after v ∈ L do {
       if p_{r,c_i} is a prefix of p_{r,w} then { i := i + 1; c_i := w}
     }
     remove all nodes in C from sorted list L
     for every node w ∈ L where c_1 is a prefix of p_{r,w} do {
       remove w from L
       clear the ''deleted'' mark on vertex w in T_W
       for every vertex u on the path from the root to w in T_W do
         update the leaf-count n_u := n_u + n_w
     }
     add to F all children of all nodes in c_1, c_2, ..., c_{|C|-1}
     add to F all siblings of all nodes in c_2, ..., c_{|C|}
     if c_{|C|} is a leaf then add it to F
       else add to F all children of node c_{|C|}
   }
   for every v ∈ F do {
     v' := Convert(v, n_v)
     make r' point to v' on label prefix p_{r,v}
   }
   return r'
```

Fig. 2. Function $Convert(r, n_r)$

Lemma 1 in remaining trees, and collect them in L. The iterations continue as long as there remain more than one leaves in $\mathcal{T}_W$, and finally the last leaf is also added to L. We note that at this stage the following are true: 1) $|L| \leq \log_{2|\Sigma|/(2|\Sigma|-1)} n$, and 2) For every vertex $v \in L$, $n_v \leq \frac{n_r}{2}$. Next, we create a list F from L such that F unlike L does not contain any two distinct nodes u, w where $p_{r,u}$ is a prefix of $p_{r,w}$. We sort L into itself in ascending order of label-lengths, and then initialize $F := \emptyset$. We iteratively pick the next node v in sorted list L, and for each v we remove nodes from L, and add nodes to F. Figure 1 illustrates possible cases on an example. We create a sequence C of nodes where the first element $c_1 = v$ by visiting every node in sorted list L appearing after v, and whenever we find a node w such that the label to the last node in the sequence p_{r,c_i} is a prefix of $p_{r,w}$ we set $c_i := w$ after incrementing i. In the resulting sequence $C = c_1(= v), c_2, \ldots, c_{|C|}$, c_i is a prefix of c_{i+1} for all $i, 1 \leq i < |C|$. We remove all the nodes in C from L. Then in new L we find all nodes w where $p_{r,v}$ is a prefix of $p_{r,w}$. We logically reattach the subtries rooted at these nodes to $\mathcal{T}_W$ by clearing the "deleted" mark, and updating for each ancestor u, $n_u := n_u + n_w$. We continue this process iteratively until no node remains in L. After the iterations end, for every node w in C we add to F all children of w (or only w if w is a leaf), and all siblings of w. We continue this process by picking the first node v with the shortest label in new L as long as $|L| > 0$. We note that when the iterations end, the following are true: 1) $|F| \leq (2|\Sigma| - 1)\log_{2|\Sigma|/(2|\Sigma|-1)} n$ because for each node initially in L there are at most $2|\Sigma| - 1$ nodes in F, 2) For every node $v \in F$, $n_v \leq \frac{n_r}{2}$. To see this consider a node in F. If v was also in L then the claim is immediately true. Otherwise v is a sibling of some node w in L, $n_w \leq \frac{n_r}{2}$, and because we always select a sibling with the largest leaf-count to place in L, $n_v \leq n_w$. 3) There are no two distinct nodes v and w in F such that $p_{r,v}$ is a prefix of $p_{r,w}$. Following the construction of F, the function creates a subtree for each node v in F by performing a recursive call $Convert(v, n_v)$ which creates a subtree for $\mathcal{S}$ from the subtrie of $\mathcal{T}_W$ rooted at v, and returns the root v' of the resulting subtree. The function makes v' a child of r', and sets prefix $p_{r,v}$ as the label of the arc connecting r' to v'.

Figure 3 illustrates the resulting tree $\mathcal{S}$. The following are true for $\mathcal{S}$:

- labels from any parent to its distinct children are distinct,
- the height h is $\leq \log_2 n$,
- the branching factor b is $\leq (2|\Sigma| - 1)\log_{2|\Sigma|/(2|\Sigma|-1)} n$. We expect that on average in practice we find a subtree with number of leaves close to half of the total number of leaves, and as a result the branching factor is much smaller in practice.

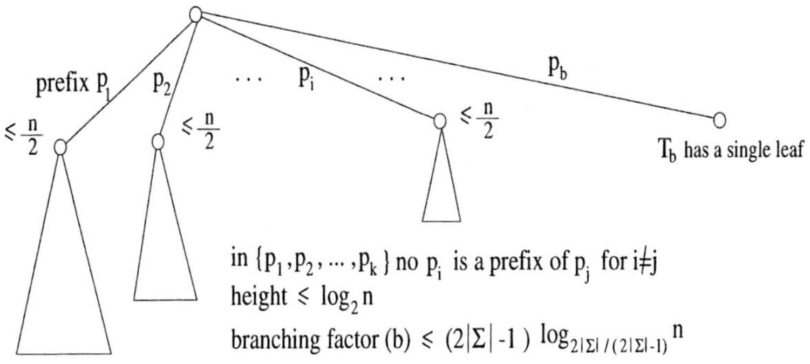

Fig. 3. The resulting tree, and the properties it satisfies

The preprocessing takes time $O(nmb)$ since each of $O(n)$ words gives rise to a leaf f in the resulting tree after the preprocessing spends $O(mb)$ time for following the edges, and for each ancestor u comparing u with its siblings to determine prefix relation, and placing u in sets L, and F. The space requirement is the same as that of $\mathcal{W}$ because in the resulting tree there are same number of leaves, and the preprocessing does not increase the depth of the leaves in the resulting tree compared to their depths in the original tree $\mathcal{T}_W$. Therefore, the number of nodes in both trees are asymptotically the same. The total length of the arc-labels are also the same asymptotically if we represent each label in S by using an index to a member, and start and end positions (or pointers) in $\mathcal{T}_W$.

We call a function F an *ordinary node-weight* function for a given tree T if F assigns non-negative integral weights to nodes such that for any node p at most one of its children has the same weight as p, and the weights of the other children are strictly larger than that of p.

The following is a generalization of a lemma in [1].

Lemma 2. *Let T be a tree whose height is h, and whose branching factor is b. Let the nodes of T be assigned weights by an ordinary node-weight function. Then the number of nodes N in T with weight $\leq d$ is $O(\ (b-1)^d\ h^{d+1})$.*

Proof. To find an upper bound for N we consider the complete tree C with branching factor b, and height h. Since the weights are assigned by an ordinary node-weight function, N is maximized when each parent node whose weight is w has exactly one child with weight w, and each of its other children has weight $w+1$. Let $L(l,w)$ denote the number of nodes with weight w at level l in C. Then $L(l,w)$ satisfies the recursion $L(l+1,w) = L(l,w) + (b-1)L(l,w-1)$ with $l \geq w$ and $L(l,0) = 1$. The solution of this recursion is $L(l,w) = \binom{l}{w}(b-1)^w$. Therefore the total number of nodes with weight w in C is $(b-1)^w \sum_{l=w}^{h} \binom{l}{w} = (b-1)^w \binom{h+1}{w+1}$, and therefore $N = \sum_{w=0}^{d} (b-1)^w \binom{h+1}{w+1} = O((b-1)^d h^{d+1})$.

3 Algorithms

Let $s[i..j]$ represent the substring $s_i s_{i+1} \ldots s_j$ of any given string $s = s_1 s_2 \ldots s_k$ with length k. With respect to a given query string q let function f assign a weight to each node v in the tree rooted at node r,

$$f(v) = H(p_{r,v}, q[1..|p_{r,v}|]) \tag{1}$$

where H denotes the Hamming distance. We note that f is an ordinary node-weight function for $\mathcal{S}$. Consider any parent node u, and its child w. The weight $f(w) = H(p_{r,w}, q[1..|p_{r,w}|]) = f(u) + H(p_{u,w}, q[(|p_{r,u}|+1)..(|p_{r,u}| + |p_{u,w}|)])$. Clearly $f(w) \geq f(u)$. If $f(w) = f(u)$ then for any other child of u the weight is larger than $f(u)$ because over all the arc labels from u only $p_{u,w}$ exactly matches $q[(|p_{r,u}|+1)..(|p_{r,u}| + |p_{u,w}|)]$, and among these arc labels no label is a prefix of another. We reach the following Corollary from Lemma 2:

Corollary 2. *Let N be the number of nodes in $\mathcal{S}$ with weight $\leq d$ as defined in (1). Then* $N = O((2|\Sigma|-1)^d(\log_{2|\Sigma|/(2|\Sigma|-1)} n - 1)^d(\log_2 n)^{d+1})$.

In particular when Σ is a binary alphabet $N = O((3\log_{4/3} n - 1)^d(\log_2 n)^{d+1})$.

We develop an algorithm shown in Figure 4 for the approximate dictionary look-up within Hamming distance d. Algorithm *DFT-LOOK-UP*$_H(r,q,d)$ checks if the tree rooted at r has a leaf whose Hamming distance from q is $\leq d$. The algorithm searches for a member in a depth-first manner. If $d < 0$ then it returns false since there is no such member. If r is a leaf and q is an empty string then the algorithm returns true since a member is found. Otherwise for each child the algorithm calculates a weight $d' = H(p_{r,c}, q[1..|p_{r,c}|])$, and recursively checks if the subtree rooted at each child contains a member within an updated distance $d - d'$. If any of these searches returns true then the algorithm returns true, otherwise, it returns false.

Since f in (1) is a ordinary node-weight function for $\mathcal{S}$, by Corollary 2 the algorithm in $\mathcal{S}$ visits $O((2|\Sigma|-1)^d(\log_{2|\Sigma|/(2|\Sigma|-1)} n - 1)^d(\log_2 n)^{d+1})$ nodes, and at each leaf spends time $O(m)$. The time complexity of the algorithm is, therefore, $O(m(2|\Sigma|-1)^d(\log_{2|\Sigma|/(2|\Sigma|-1)} n - 1)^d(\log_2 n)^{d+1})$. We can modify the algorithm such that the words in $\mathcal{W}$ as well as the query string can be of different lengths.

Next, we describe how we develop a similar algorithm for the problem when simple edit distances are used. Given two strings $X = x_1 \ldots x_m$ and $Y = y_1 \ldots y_m$, the simple edit distance $ed(X,Y)$ is the minimum number of edit operations which transform X into Y using three types of operations: insert, delete, and substitute. A common framework for computing an edit distance is the *edit graph* (see [8]) for definition), and it has a simple dynamic programming formulation [8]:

$$D_{i,j} = \min\{\ D_{i-1,j} + 1,\ D_{i-1,j-1} + H(x_i, y_j),\ D_{i,j-1} + 1\} \tag{2}$$

for all i, j, $0 \leq i, j \leq m$ with boundary conditions $D_{i,0} = i$, $D_{0,j} = j$.

With respect to a given query string q, let e be a function that assigns a weight to a given node v in $\mathcal{S}$ rooted at r as described in the following:

$$e(v) = \min\{ed(p_{r,v}, t) \mid t \text{ is a prefix of } q\} \tag{3}$$

```
Algorithm DFT-LOOK-UP_H(r, q, d)
   If d < 0 return FALSE
   If r is a leaf, and q is an empty string then return TRUE
   for each child c of r in S do {
      d' := H(p_{r,c}, q[1..|p_{r,c}|])
      if DFT-LOOK-UP_H(c, q[(|p_{r,c}| + 1)..|q|], d - d') then return TRUE
   }
   return FALSE
```

Fig. 4. Algorithm *DFT-LOOK-UP*$_H$ for approximate dictionary look-up within Hamming distance d

where *ed* denotes the simple edit distance. Note that in this definition $p_{r,v}$ and the prefix t of q can be of different lengths. We can see that e is not an ordinary node-weight function because it is possible that more than one children of a node can have the same weight as the parent node due to insert, and delete operations that can be performed on the labels to ancestor nodes.

Lemma 3. *Let N be the number of nodes in $\mathcal{S}$ with weight $\leq d$ as defined in (3). Then* $N = O((2|\Sigma| - 1)^{2d}(\log_{2|\Sigma|/(2|\Sigma|-1)} n - 1)^{2d}(\log_2 n)^{2d+1})$.

Proof. We imagine that we traverse the tree in breath-first manner starting at root at level 0, and consider the minimum possible weight for each node. Clearly for any node u, the weight $e(u)$ in (3) is less than or equal to $f(u)$ in (1). Suppose that initially for every node u in $\mathcal{S}$, $e(u) = f(u)$. If v is a parent node of u then $e(u) \geq e(v)$, i.e. e is non-decreasing.Due to possible delete, and insert operations on the label of the arc $p_{v,u}$ from v to u there may be nodes w in the subtree rooted at u such that w has more than one children sharing the same weight as w. When we studied an upper bound in Lemma 2 we considered that every node v has exactly one child with the same weight as v. This time, being overly pessimistic, we assume that all of v's children have the same weight as v if it is given that there are insertions, or deletions on $p_{v,u}$. We consider possibility of insertions, and deletions on all labels on arcs each connecting a node at level $i-1$ to a node at level i for a given i. This increases the number of nodes at level i with the same weight as their parents (and all $\leq d$) by a factor of $\leq b-1$, where b is the branching factor in $\mathcal{S}$. Since there are at most d insertions, or deletions, for each permutation of the levels they can occur, the number of nodes with the same weight as their parents (and all $\leq d$) is increased by a factor of $\leq (b-1)^d$. The number of levels where an insertion, or a deletion can occur is the same as the height of the tree, h. Since there can be at most d such operations, we need to consider $\binom{h}{d}$ possibilities. Putting all together, the product of $\binom{h}{d}(b-1)^d$ and the upper bound in Lemma 2 gives the upper bound in this lemma.

Next we propose Algorithm *DFT-LOOK-UP*$_{ed}$ for the d-query problem when edit distance is used. The steps of the algorithm are shown in Figure 5. The algorithm is based on depth-first traversal (DFT) of $\mathcal{S}$ during which the entries of the dynamic programming matrix are partially computed. To determine if two strings are within edit distance d it is sufficient to consider a diagonal band of the edit graph [15]. Algorithm *DFT-LOOK-UP*$_{ed}$ uses this observation (see Figure 7).

For a given node v in $\mathcal{S}$ rooted at r, we define $D_{v,j}$ where $\max\{0, i - \lfloor d/2 \rfloor\} \leq j \leq \min\{m, i + \lfloor d/2 \rfloor\}$, and $i = |p_{r,v}|$ (see Figure 7) as follows:

$$D_{v,j} = ed(p_{r,v}, q[1..j])$$

That is, $D_{v,j}$ is the minimum simple edit distance between $p_{r,v}$ and $q[1..j]$, and the weight of node v defined in (3) is

$$e(v) = \min\{\ D_{v,j} \mid \max\{0, i - \lfloor d/2 \rfloor\} \leq j \leq \min\{m, i + \lfloor d/2 \rfloor\} \text{ where } i = |p_{r,v}|\ \}.$$

```
Algorithm DFT-LOOK-UP_ed(d)
   D_{0,j} := j for all j,  1 ≤ j ≤ ⌊d/2⌋
   D_{i,0} := 0 for all i,  1 ≤ i ≤ m
   D_{0,0} := 0
   for each child v of the root r in S do
     if DFT-COMPUTE-D_ed(r, v) ≤ d then return YES
   return NO
```

Fig. 5. Algorithm *DFT-LOOK-UP*$_{ed}$ for dictionary look-up within edit distance d

```
Function DFT-COMPUTE-D_ed(v, u)
   i_start := |p_{r,v}|; i_end := |p_{r,v}| + |p_{v,u}|
   for i := i_start to i_end do
     for j := max{0, i − ⌊d/2⌋} to min{m, i + ⌊d/2⌋} do
       D_{i,j} := min{D_{i−1,j} + 1, D_{i−1,j−1} + H(p_{v,u}[i − i_start], q_j), D_{i,j−1} + 1}
   weight := min{D_{i_end,j} | max{0, i − ⌊d/2⌋} ≤ j ≤ min{m, i + ⌊d/2⌋}}
   if u is a leaf or weight > d then return weight
   if weight= d then {
    for j := max{0, i_end − ⌊d/2⌋} to min{m, i_end + ⌊d/2⌋} do {
     if D_{i_end,j} = d and there is a path from u to a leaf in S
         on q[(j + 1)..q_m] then return d }
    return d + 1
   }
   return min{DFT-COMPUTE-D_ed(u, w) | w is a child of u}
```

Fig. 6. Function *DFT-COMPUTE-D*$_{ed}$ for computing the minimum edit distance achieved in subtree rooted at u whose parent is v

If we process $\mathcal{S}$ in depth-first manner, we can compute $D_{v,j}$ for all nodes using a single matrix $D_{i,j}$ where $0 \leq i \leq m$.

Algorithm *DFT-LOOK-UP*$_{ed}$ starts with the initialization of scores for the first row, and it invokes Function *DFT-COMPUTE-D*$_{ed}$ for each child v of the root r. If any of these invocations returns a value $\leq d$ then the algorithm returns YES; otherwise it returns NO.

Given a parent node v, a child node u, Function *DFT-COMPUTE-D*$_{ed}(v, u)$ computes the shaded region of the edit graph shown in Figure 7 using $p_{v,u}$, and starting with the values in the row of parent node v. The minimum of the values in the row of u is set as the weight $e(u)$ of u . If this value is equal to d then the function examines every position j in the row of u where d is achieved. These are the only starting positions for a suffix of the query string q with which weight d is preserved in a subtree rooted at u . That is, these are the only positions which potentially lead to a leaf with weight d . Therefore the algorithm checks if starting from u at each such position j if there is a path to a leaf on $q[(j+1)..m]$. If the answer is yes then the algorithm returns d, otherwise it returns $d+1$ which is a number sufficiently large to yield a no answer when we only consider the subtree rooted at u . If the weight of u is smaller than d then the function traverses recursively the subtree rooted at u in depth-first manner, computes

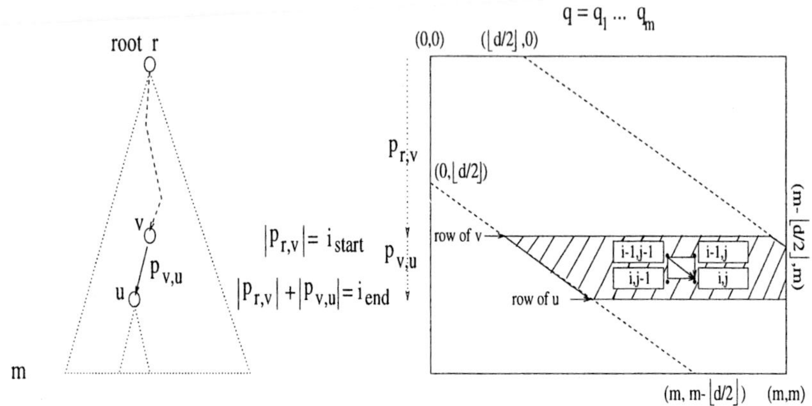

Fig. 7. The hybrid tree/dynamic programming approach used by Function *DFT-COMPUTE-D_{ed}*

and returns the minimum edit distance (leaf-weight) achievable in this subtree. Note that if the final value returned is $\leq d$ then it must be the weight of a leaf.

Since processing at each node takes time $O(dm)$, the algorithm's time complexity is $O(dm(2|\Sigma| - 1)^{2d}(\log_{2|\Sigma|/(2|\Sigma|-1)} n - 1)^{2d}(\log_2 n)^{2d+1})$ by Lemma 3, and it requires additional space $O(dm)$.

4 Conclusion

We present a method to preprocess a dictionary to create an edge-labelled tree with bounded branching factor and height, and with the property that from any node to its distinct children no label is a prefix of another. Size of the resulting tree is asymptotically the same as the space requirement of an ordinary trie that stores the dictionary. For approximate dictionary look-up we develop space-efficient algorithms which are also time-efficient when the alphabet-size is small. The main ideas in these algorithms can also be used for developing methods for text indexing with errors.

Acknowledgement

The author thanks Johannes Nowak for pointing out in an earlier version of the paper a flaw in preprocessing step. The author corrected it in this version.

References

1. A. N. Arslan and Ö. Eğecioğlu. Dictionary look-up within small edit distance, *Inter. J. of Found. of Comp. Sci.*, Vol. 15, No 1, pp. 57-71, February 2004.
2. G. S. Brodal and L. Gasieniec. Approximate dictionary queries, *In Proc. 7th Combinatorial Pattern Matching, LNCS, Vol. 1075, Springer, Berlin*, 65–74, 1996.

3. G. S. Brodal and S. Velkatesh. Improved bounds for dictionary look-up with one error. *IPL*, 75, 57–59, 2000.
4. R. Cole, L.-A. Gottlieb, and N. Lewenstein. Dictionary matching and indexing with errors and don't cares. *Proc. The 36th ACM STOC*, pp. 91-100, 2004.
5. D. Dolev, Y. Harari, N. Linial, N. Nisan and M. Parnas. Neighborhood preserving hashing and approximate queries. *Proc. The Fifth ACM SODA*, 1994.
6. D. Dolev, Y. Harari and M. Parnas. Finding the neighborhood of a query in a dictionary. *Proc. The Second Israel Symp. on Theory of Comp. and Sys.*, 1993.
7. P. Elias. Efficient storage and retrieval by content and address of static files. *J. ACM*, 21:246–260, 1974.
8. D. Gusfield. Algorithms on strings, trees, and sequences: computer science and computational biology. *Cambridge University Press*, 1997.
9. M. G. Maaß. Average-case analysis of approximate trie search. *CPM 2004, LNCS 3109*, pp. 472-484, 2004.
10. M. G. Maaß and J. Nowak. Text indexing with errors. *CPM 2005, LNCS 3537*, pp. 21-32, 2005.
11. U. Manber and S. Wu. An algorithm for approximate membership checking with applications to password security. *IPL*, 50, 191–197, 1994.
12. M. Minsky and S. Papert. Perceptrons. *MIT Press, Cambridge, MA*, 1969.
13. G. Navarro, R. Baeza-Yates, E. Sutinen and J. Tarhio. Indexing Methods for Approximate String Matching. *IEEE Data Engineering Bulletin*, 24(4):19-27, 2001.
14. H. Shang and T. H. Merrett. Tries for approximate string matching. *IEEE Trans. Knowl. Data Eng.*, 8(4):540-547, 1996.
15. E. Ukkonen. Algorithms for Approximate String Matching. *Information and Control*, 64, 100-118, 1985.
16. A. C. Yao and F. F. Yao. Dictionary look-up with one error. *J. of Algorithms*, 25(1), 194–202, 1997.

Relations Among Notions of Security for Identity Based Encryption Schemes

Nuttapong Attrapadung[1], Yang Cui[1], David Galindo[2], Goichiro Hanaoka[3], Ichiro Hasuo[2], Hideki Imai[1,3], Kanta Matsuura[1], Peng Yang[1], and Rui Zhang[1,⋆]

[1] Institute of Industrial Science, University of Tokyo
{nuts, cuiyang, zhang}@imailab.iis.u-tokyo.ac.jp,
{imai, kanta, pengyang}@iis.u-tokyo.ac.jp

[2] Institute for Computing and Information Sciences, Radboud University Nijmegen
{d.galindo, ichiro}@cs.ru.nl

[3] Research Center for Information Security, National Institute of Advanced Industrial Science and Technology
hanaoka-goichiro@aist.go.jp

Abstract. This paper shows that the standard security notion for identity based encryption schemes ($\mathcal{IBE}$), that is IND-ID-CCA2, captures the essence of security for all $\mathcal{IBE}$ schemes. To achieve this intention, we first describe formal definitions of the notions of security for $\mathcal{IBE}$, and then present the relations among OW, IND, SS and NM in $\mathcal{IBE}$, along with rigorous proofs. With the aim of comprehensiveness, notions of security for $\mathcal{IBE}$ in the context of encryption of multiple messages and/or to multiple receivers are finally presented. All of these results are proposed with the consideration of the particular attack in $\mathcal{IBE}$, namely the *adaptive chosen identity attack*.

1 Introduction

Identity based encryption ($\mathcal{IBE}$) is a public key encryption mechanism where an arbitrary string, such as the recipient's identity, can serve as a public key. This convenience yields the avoidance of the need to distribute public key certificates. On the other hand, in conventional public key encryption ($\mathcal{PKE}$) schemes, it is unavoidable to access the online public key directory in order to obtain the public keys. $\mathcal{IBE}$ schemes are largely motivated by many applications such as to encrypt emails with the recipient's email address.

Although the basic concept of $\mathcal{IBE}$ was proposed by Shamir [13] more than two decades ago, it is only very recent that the first fully functional scheme was proposed [6]. In 2001, Boneh and Franklin defined a security model and gave the first fully functional solution provably secure in the random oracle model. The notions of security proposed in their work are natural extensions to the standard ones for $\mathcal{PKE}$, namely indistinguishability-based ones.

⋆ Rui Zhang is supported by a JSPS Fellowship.

J.R. Correa, A. Hevia, and M. Kiwi (Eds.): LATIN 2006, LNCS 3887, pp. 130–141, 2006.

1.1 Motivation

So far in the literature, IND-ID-CCA2, is widely considered to be the "right" security notion which captures the essence of security for $\mathcal{IBE}$ [6,4,5,15]. However, such an issue has not been investigated rigorously, *yet.* This work aims to establish such an affirmative justification. Before discussing about how to define the "right" security notion for $\mathcal{IBE}$, we first glance back to the case of $\mathcal{PKE}$.

NOTIONS OF SECURITY FOR $\mathcal{PKE}$. A convenient way to formalize notions of security for cryptographic schemes is considering combinations of the various *security goals* and possible *attack models.* Four essential security goals being considered in the case of $\mathcal{PKE}$ are *one-wayness* (OW), *indistinguishability* (IND), *semantic security* (SS) [9, 11], and *non-malleability* (NM) [7], i.e. $\mathsf{G}_i \in$ {OW,IND,SS,NM}. The attack models are *chosen plaintext attack* (CPA) [11], *non-adaptive chosen ciphertext attack* (CCA1) [7] and *adaptive chosen ciphertext attack* (CCA2) [12], i.e. $\mathsf{A}_j \in$ {CPA,CCA1,CCA2}. Their combinations give nine security notions for $\mathcal{PKE}$, e.g. IND-CCA2.

SS is widely accepted as the natural goal of encryption scheme because it formalizes an adversary's inability to obtain any information about the plaintext from a given ciphertext. The equivalence between SS-CPA and IND-CPA has been given [11]; and the equivalences between SS-CCA1,2 and IND-CCA1,2 are given only recently [10,14]. On the other hand, NM formalizes an adversary's inability, given a challenge ciphertext y^*, to output a different ciphertext y' in such a way that the plaintexts x, x', underlying these two ciphertexts, are meaningfully related, e.g. $x' = x + 1$. The implications from IND-CCA2 to NM under any attack have been proved [3]. For these reasons, along with the convenience of proving security in sense of IND, in almost all concrete schemes, IND-CCA2 is considered to be the "right" standard security notion for $\mathcal{PKE}$.

TOWARDS DEFINING NOTIONS OF SECURITY FOR $\mathcal{IBE}$. Due to the particular mechanism, the adversaries are granted more power in $\mathcal{IBE}$ than in $\mathcal{PKE}$. Essentially, the adversaries have access to the *key extraction oracle*, which answers the private key of any queried public key (identity). Including this particular *adaptive chosen identity attack,* [1] we formalize the security notions for $\mathcal{IBE}$, e.g. IND-ID-CCA2, in such a way: G_i-ID-A_j, where $\mathsf{G}_i \in$ {OW,IND,SS,NM}, ID denotes the particular attack mentioned above, and $\mathsf{A}_j \in$ {CPA,CCA1,CCA2}. Boneh and Franklin are the first to define the security notion for $\mathcal{IBE}$, by naturally extending IND-CCA2 to IND-ID-CCA2.

Let us rigorously investigate whether IND-ID-CCA2 could be considered as the "right" notion for $\mathcal{IBE}$, besides the intuitive reason that it is analogous to IND-CCA2. The natural approach to justify such an appropriateness for $\mathcal{IBE}$ is, analogously to the case of $\mathcal{PKE}$, to (i) first define SS and NM based security notions for $\mathcal{IBE}$, (ii) and then establish the relations among the above security

[1] Actually in $\mathcal{IBE}$ there exists the other attack against identity, named *selective chosen identity attack.* In this paper we omit presenting the formal definitions of the security notions in this selective-ID secure sense, but it is easy to see that the implications and separations shown here also hold in the selective-ID case.

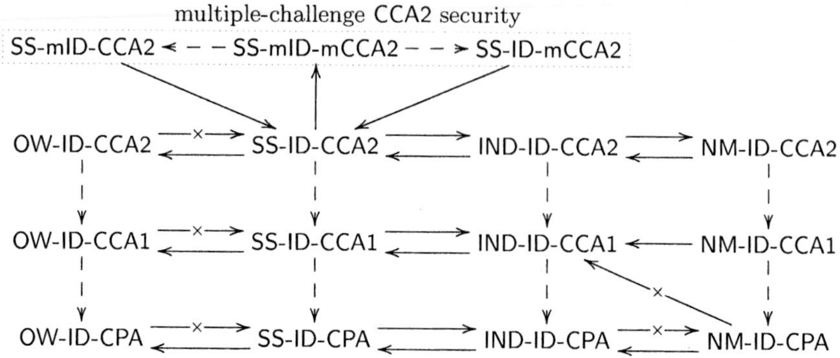

Fig. 1. Relations among the notions of security for $\mathcal{IBE}$

notions: to be more specific, the implications from IND-ID-CCA2 to all the other notions, i.e. IND-ID-CCA2 is the *strongest* notion of security for $\mathcal{IBE}$.

At the first place the intuition tells us that task (i) seems to be simply achievable by considering the analogy to the case of shifting IND-CCA to IND-ID-CCA as done in [6], and task (ii) could immediately follow from the relations among the notions as the case of $\mathcal{PKE}$, since we shift all the notions with the same additional attack power (namely, the accessibility to key extraction oracle). However, we emphasize that it will not follow simply and immediately until rigorous definitions for task (i) and rigorous proofs for task (ii) are presented. We managed to accomplish both tasks in this paper.

1.2 Our Contributions

Our contributions are three-fold.

First, we formally presented the definitions of the notions of security for $\mathcal{IBE}$ schemes. The overall definitions are built upon historical works [3, 6, 10].

Secondly, we rigorously proved the relations among these notions and achieved our conclusion that, IND-ID-CCA2 is the "right" notion of security for $\mathcal{IBE}$. It turns out that our intuition about those relations were right: the implication G_1-ID-$\mathsf{A}_1 \Rightarrow (\not\Rightarrow)\mathsf{G}_2$-ID-$\mathsf{A}_2$ will hold in $\mathcal{IBE}$ if and only if G_1-$\mathsf{A}_1 \Rightarrow (\not\Rightarrow)\mathsf{G}_2$-$\mathsf{A}_2$ holds in $\mathcal{PKE}$, respectively, where the corresponding security goals G_i and attack models A_j are mentioned above. The results of our second contribution are illustrated in Figure 1. An arrow is an implication, and there is path between **A** and **B** if and only if the security notion **A** implies the security notion **B**. A hatched arrow represents a separation which is proved in this paper. Dotted arrows refer to trivial implications. For each pair of notions we obtain an implication or a separation, which is either explicitly found in the diagram or deduced from it.

In the last place, we study the robustness of IND-ID-CCA2 secure schemes in the context of encryption of multiple messages and/or to multiple receivers. Concretely, inspired by [10], we propose several new attack models for the case of active adversaries: multiple-identity (mID-CCA2) attacks [2] (the adversary can adaptively query for encryptions of the same plaintext under different identities); multiple-plaintext (ID-mCCA2) attacks (the adversary chooses one fixed identity, and can adaptively query encryption of different plaintexts under that identity) and multiple-identity-plaintext attacks (mID-mCCA) (the adversary can adaptively query encryption of different plaintexts under different identities). It is shown that any IND-ID-CCA2 scheme also meets those stronger security levels.

Our results could be considered as having the same flavor as some historical results, to name just one, the equivalence between IND-CCA2 and SS-CCA2 for $\mathcal{PKE}$. There, although IND-CPA and SS-CPA were defined and proved equivalent in the year 1984 [11], the equivalence between IND-CCA2 and SS-CCA2 had not been proved rigorously until the year 2003 [14]. During this long period of time, people just *simply* believed that shifting the attack power from CPA to CCA2 will not affect the equivalence.

This paper is merged from two parallel works [1,8].

1.3 Organization

The rest of the paper is organized as follows: in Section 2 we review the formal definition of $\mathcal{IBE}$ schemes and several other basic terms. In Section 3 we define the formal definitions of notions of security for $\mathcal{IBE}$ schemes. In Section 4 we prove important relations among these notions, rigorously. In Section 5 we study the multi-challenge cases.

2 Preliminary

2.1 Identity Based Encryption

Formally, an identity based encryption scheme consists of four algorithms, i.e. $\mathcal{IBE} = (\mathcal{S}, \mathcal{X}, \mathcal{E}, \mathcal{D})$, where

- $\mathcal{S}$, the setup algorithm, takes a security parameter k and outputs system parameters $param$ and master-key mk. The system parameters include the message space $\mathcal{M}$, and the ciphertext space $\mathcal{C}$.
- $\mathcal{X}$, the extract algorithm, takes triple inputs as $param, mk$, and an arbitrary $id \in \{0,1\}^*$, and outputs a private key $sk = \mathcal{E}(param, mk, id)$. Here id is arbitrary.
- $\mathcal{E}$, the encrypt algorithm, takes triple inputs as $param$, $id \in \{0,1\}^*$ and a plaintext $x \in \mathcal{M}$. It outputs the corresponding ciphertext $y \in \mathcal{C}$.
- $\mathcal{D}$, the decrypt algorithm, takes triple inputs as $param$, $y \in \mathcal{C}$, and the corresponding private key sk. It outputs $x \in \mathcal{M}$.

The four algorithms must satisfy the standard consistency constraint, i.e. if and only if sk is the private key generated by the extract algorithm with the given id as the public key, then,

$$\forall x \in \mathcal{M} : \mathcal{D}(param, sk, y) = x, \text{ where } y = \mathcal{E}(param, id, x).$$

[2] This security definition has been previously considered in [2], but no proof of equivalence to IND-ID-CCA2 was given. Moreover, the attack we consider is stronger since it gives more power to the adversary.

2.2 Conventions

Notations. $\vec{x} \leftarrow \mathcal{D}(param, sk, \vec{y})$ denotes that the vector $\vec{x}$ is made up of the plaintexts corresponding to every ciphertext in the vector $\vec{y}$. $\hat{\mathcal{M}}$ denotes a subset of message space $\mathcal{M}$, where the elements of $\hat{\mathcal{M}}$ are distributed according to the distribution designated by some algorithm. Function $h : \hat{\mathcal{M}} \rightarrow \{0,1\}^*$ denotes the a-priori partial information about the plaintext and function $f : \hat{\mathcal{M}} \rightarrow \{0,1\}^*$ denotes the a-posteriori partial information.

Negligible Function. We say a function $\epsilon : \mathbb{N} \rightarrow \mathbb{R}$ is *negligible* if for every constant $c \geq 0$ there exits an integer k_c such that $\epsilon(k) < k^{-c}$ for all $k > k_c$.

R-related Relation. We consider R-related relation of arity t where t will be polynomial in the security parameter k. Rather than writing $R(x_1, x_2, \ldots, x_t)$ we write $R(x, \vec{x})$, denoting the first argument is special and the rest are bunched into a vector $\vec{x}$ where $|\vec{x}| = t-1$, and for every $x_i \in \vec{x}$, $R(x, x_i)$ holds.

Experiments. Let A be a probabilistic algorithm, and let $A(x_1, \ldots, x_n; r)$ be the result of running A on inputs $(x_1, \ldots, x_n)$ and coins r. Let $y \leftarrow A(x_1, \ldots, x_n)$ denote the experiment of picking r at random and let y be $A(x_1, \ldots, x_n; r)$. If S is a finite set then let $x \leftarrow S$ denote the operation of picking an element at random and uniformly from S. And sometimes we use $x \overset{R}{\leftarrow} S$ in order to stress this randomness. If α is neither an algorithm nor a set then let $x \leftarrow \alpha$ denote a simple assignment statement. We say that y can be output by $A(x_1, \ldots, x_n)$ if there is some r such that $A(x_1, \ldots, x_n; r) = y$.

3 Definitions of Security Notions for $\mathcal{IBE}$ Schemes

Let $A = (A_1, A_2)$ be an adversary, and we say A is polynomial time if both probabilistic algorithm A_1 and probabilistic algorithm A_2 are polynomial time. At the first stage, given the system parameters, the adversary computes and outputs a challenge template τ. A_1 can output some state information s which will be transferred to A_2. At the second stage the adversary is issued a challenge ciphertext y^* generated from τ by a probabilistic function, in a manner depending on the goal. We say the adversary A successfully breaks the scheme if she achieves her goal.

We consider four security goals, OW, IND, SS and NM. And we consider three attack models, ID-CPA,ID-CCA1,ID-CCA2, in order of increasing strength. The difference among the models is whether or not A_1 or A_2 is granted accesses to decryption oracles.[3]

[3] With regards to the adaptive chosen identity and selective chosen identity attacks, we only discuss in details the former case (full-ID security), while the results can be extended to the latter case (selective-ID security), since the strategies are similar. Roughly speaking, the target public key *id* should be decided by the adversary in advance, before the challenger runs the setup algorithm. The restriction is that the extraction query on *id* is prohibited.

Table 1. Oracle Sets $\mathcal{O}_1$ and $\mathcal{O}_2$ in the Definitions of the Notions for $\mathcal{IBE}$

	$\mathcal{O}_1 = \{\mathcal{XO}_1, \mathcal{EO}_1, \mathcal{DO}_1\}$
ID-CPA	$\{\mathcal{X}(param, mk, \cdot), \mathcal{E}(param, id, \cdot), \varepsilon\}$
ID-CCA1	$\{\mathcal{X}(param, mk, \cdot), \mathcal{E}(param, id, \cdot), \mathcal{D}(param, sk, \cdot)\}$
ID-CCA2	$\{\mathcal{X}(param, mk, \cdot), \mathcal{E}(param, id, \cdot), \mathcal{D}(param, sk, \cdot)\}$
	$\mathcal{O}_2 = \{\mathcal{XO}_2, \mathcal{EO}_2, \mathcal{DO}_2\}$
ID-CPA	$\{\mathcal{X}(param, mk, \cdot), \mathcal{E}(param, id, \cdot), \varepsilon\}$
ID-CCA1	$\{\mathcal{X}(param, mk, \cdot), \mathcal{E}(param, id, \cdot), \varepsilon\}$
ID-CCA2	$\{\mathcal{X}(param, mk, \cdot), \mathcal{E}(param, id, \cdot), \mathcal{D}(param, sk, \cdot)\}$

In Table 1, we describe the ability with which the adversary in different attack models accesses the *Extraction Oracle* $\mathcal{X}(param, mk, \cdot)$, the *Encryption Oracle* $\mathcal{E}(param, id, \cdot)$ and the *Decryption Oracle* $\mathcal{D}(param, sk, \cdot)$. When we say $\mathcal{O}_i = \{\mathcal{XO}_i, \mathcal{EO}_i, \mathcal{DO}_i\} = \{\mathcal{X}(param, mk, \cdot), \mathcal{E}(param, id, \cdot), \varepsilon\}$, where $i \in \{1, 2\}$, we mean $\mathcal{DO}_i$ is a function that returns an empty string ε on any input.

Remark 1. To have meaningful definitions, we insist that the target public key id should not be previously queried on, i.e. it is completely meaningless if the adversary has already known the corresponding private key of id.

3.1 One-Wayness

As far as we know, only one-wayness against full-identity chosen-plaintext attacks (referred to as OW-ID-CPA in the following definition) has been previously considered in the literature. Here we define one-wayness through a two-stage experiment. A_1 is run on the system parameters $param$ as input. At the end of A_1's execution she outputs (s, id), such that s is state information (possibly including $param$) which she wants to preserve, and id is the public key which she wants to attack. One plaintext x^* is *randomly* selected from the message space $\mathcal{M}$ beyond adversary's view. A challenge y^* is computed by encrypting x^* with the public key id. A_2 tries to computer what x^* was.

Definition 1 (OW-ID-CPA, OW-ID-CCA1, OW-ID-CCA2)

Let $\mathcal{IBE} = (\mathcal{S}, \mathcal{X}, \mathcal{E}, \mathcal{D})$ be an identity based encryption scheme and let $A = (A_1, A_2)$ be an adversary. For $\texttt{atk} \in \{\texttt{id-cpa}, \texttt{id-cca1}, \texttt{id-cca2}\}$ and $k \in \mathbb{N}$ let,

$$\mathbf{Adv}^{\texttt{ow-atk}}_{\mathcal{IBE},A}(k) = \Pr[\mathbf{Exp}^{\texttt{ow-atk}}_{\mathcal{IBE},A}(k) = 1] \tag{1}$$

where, for $b, d \in \{0, 1\}$,

Experiment $\mathbf{Exp}^{\texttt{ow-atk-b}}_{\mathcal{IBE},A}(k)$
  $(param, mk) \leftarrow \mathcal{S}(k)$; $(s, id) \leftarrow A_1^{\mathcal{O}_1}(param)$;
  $x^* \leftarrow \mathcal{M}$; $y^* \leftarrow \mathcal{E}(param, id, x^*)$; $x' \leftarrow A_2^{\mathcal{O}_2}(s, y^*, id)$;
  if $x' = x^*$ then $d \leftarrow 1$ else $d \leftarrow 0$;
  return d

We say that $\mathcal{IBE}$ is secure in the sense of OW-ATK, if $\mathbf{Adv}^{\texttt{ow-atk}}_{\mathcal{IBE},A}(k)$ is negligible for any A.

3.2 Indistinguishability

In this scenario A_1 is run on *param*, and outputs (x_0, x_1, s, id), such that x_0 and x_1 are plaintexts with the same length. One of x_0 and x_1 is *randomly* selected, say x_b, beyond adversary's view. A challenge y^* is computed by encrypting x_b with id. A_2 tries to distinguish whether y^* was the encryption of x_0 or x_1.

Definition 2 (IND-ID-CPA, IND-ID-CCA1, IND-ID-CCA2)

Let $\mathcal{IBE} = (\mathcal{S}, \mathcal{X}, \mathcal{E}, \mathcal{D})$ be an identity based encryption scheme and let $A = (A_1, A_2)$ be an adversary. For $\texttt{atk} \in \{\texttt{id-cpa}, \texttt{id-cca1}, \texttt{id-cca2}\}$ and $k \in \mathbb{N}$ let,

$$\mathbf{Adv}^{\texttt{ind-atk}}_{\mathcal{IBE},A}(k) = \Pr[\mathbf{Exp}^{\texttt{ind-atk-1}}_{\mathcal{IBE},A}(k) = 1] - \Pr[\mathbf{Exp}^{\texttt{ind-atk-0}}_{\mathcal{IBE},A}(k) = 1] \qquad (2)$$

where, for $b, d \in \{0, 1\}$ and $|x_0| = |x_1|$,

$$\begin{array}{l} \texttt{Experiment } \mathbf{Exp}^{\texttt{ind-atk-b}}_{\mathcal{IBE},A}(k) \\ \quad (param, mk) \leftarrow \mathcal{S}(k); \quad (x_0, x_1, s, id) \leftarrow A_1^{\mathcal{O}_1}(param); \\ \quad y^* \leftarrow \mathcal{E}(param, id, x_b); \quad d \leftarrow A_2^{\mathcal{O}_2}(x_0, x_1, s, y^*, id); \\ \quad \texttt{return } d \end{array}$$

We say that $\mathcal{IBE}$ is secure in the sense of IND-ATK, if $\mathbf{Adv}^{\texttt{ind-atk}}_{\mathcal{IBE},A}(k)$ is negligible for any A.

3.3 Semantic Security

In this scenario, A_1 is given *param*, and outputs $(\hat{\mathcal{M}}, h, f, s, id)$. Here the distribution of $\hat{\mathcal{M}}$ is designated by A_1, and $(\hat{\mathcal{M}}, h, f)$ is the challenge template τ. A_2 receives an encryption y^* of a random message x^* drawn from $\hat{\mathcal{M}}$. The adversary then outputs a value v. She hopes that $v = f(x^*)$. The adversary is successful if she can do this with a probability significantly more than any *simulator* does. The simulator tries to do as well as the adversary without knowing the challenge ciphertext y^* nor accessing any oracle.

Definition 3 (SS-ID-CPA, SS-ID-CCA1, SS-ID-CCA2)

Let $\mathcal{IBE} = (\mathcal{S}, \mathcal{X}, \mathcal{E}, \mathcal{D})$ be an identity based encryption scheme, let $A = (A_1, A_2)$ be an adversary, and let $A' = (A'_1, A'_2)$ be the simulator. For $\texttt{atk} \in \{\texttt{id-cpa}, \texttt{id-cca1}, \texttt{id-cca2}\}$ and $k \in \mathbb{N}$ let,

$$\mathbf{Adv}^{\texttt{ss-atk}}_{\mathcal{IBE},A,A'}(k) = \Pr[\mathbf{Exp}^{\texttt{ss-atk}}_{\mathcal{IBE},A}(k) = 1] - \Pr[\mathbf{Exp}^{\texttt{ss-atk}}_{\mathcal{IBE},A'}(k) = 1] \qquad (3)$$

where, for $b \in \{0, 1\}$,

Experiment $\mathbf{Exp}^{\texttt{ss-atk}}_{\mathcal{IBE},A}(k)$
$(param, mk) \leftarrow \mathcal{S}(k)$;
$(\hat{\mathcal{M}}, h, f, s, id) \leftarrow A_1^{\mathcal{O}_1}(param)$;
$x^* \stackrel{R}{\leftarrow} \hat{\mathcal{M}}$; $y^* \leftarrow \mathcal{E}(param, id, x^*)$;
$v \leftarrow A_2^{\mathcal{O}_2}(s, y^*, h(x^*), id)$;
`if` $v = f(x^*)$
`then` $d \leftarrow 1$ `else` $d \leftarrow 0$;
`return` d

Experiment $\mathbf{Exp}^{\texttt{ss-atk}}_{\mathcal{IBE},A'}(k)$
$(\hat{\mathcal{M}}, h, f, s, id) \leftarrow A_1'(k)$;
$x^* \stackrel{R}{\leftarrow} \hat{\mathcal{M}}$;
$v \leftarrow A_2'(s, |x^*|, h(x^*), id)$;
`if` $v = f(x^*)$
`then` $d \leftarrow 1$ `else` $d \leftarrow 0$;
`return` d

We say that $\mathcal{IBE}$ is secure in the sense of SS-ATK, if for any adversary A there exists a simulator such that $\mathbf{Adv}^{\texttt{ss-atk}}_{\mathcal{IBE},A}(k)$ is negligible .

We comment here that it is necessary to require in both cases τ is distributed identically, since both A and A' generate target public key id by themselves, i.e. τ is output by A and A' themselves.

3.4 Non-malleability

In this scenario, A_1 is given $param$, and outputs a triple $(\hat{\mathcal{M}}, s, id)$. A_2 receives an encryption y^* of a random message x_1. The adversary then outputs a description of a relation R and a vector $\vec{\boldsymbol{y}}$ of ciphertexts. We insist that $y \notin \vec{\boldsymbol{y}}$.[4] The adversary hopes that $R(x_1, \vec{\boldsymbol{x}})$ holds. We say she is successful if, she can do this with a probability significantly more than that, with which $R(x_0, \vec{\boldsymbol{x}})$ holds. Here x_0 is also a plaintext chosen uniformly from $\hat{\mathcal{M}}$, independently of x_1.

Definition 4 (NM-ID-CPA, NM-ID-CCA1, NM-ID-CCA2)

Let $\mathcal{IBE} = (\mathcal{S}, \mathcal{X}, \mathcal{E}, \mathcal{D})$ be an identity based encryption scheme and let $A = (A_1, A_2)$ be an adversary. For $\texttt{atk} \in \{\texttt{id-cpa}, \texttt{id-cca1}, \texttt{id-cca2}\}$ and $k \in \mathbb{N}$ let,

$$\mathbf{Adv}^{\texttt{nm-atk}}_{\mathcal{IBE},A}(k) = \Pr[\mathbf{Exp}^{\texttt{nm-atk-1}}_{\mathcal{IBE},A}(k) = 1] - \Pr[\mathbf{Exp}^{\texttt{nm-atk-0}}_{\mathcal{IBE},A}(k) = 1] \qquad (4)$$

where, for $b \in \{0, 1\}$ and $|x_0| = |x_1|$,

Experiment $\mathbf{Exp}^{\texttt{nm-atk-b}}_{\mathcal{IBE},A}(k)$
$(param, mk) \leftarrow \mathcal{S}(k)$; $(\hat{\mathcal{M}}, s, id) \leftarrow A_1^{\mathcal{O}_1}(param)$;
$x_0, x_1 \stackrel{R}{\leftarrow} \hat{\mathcal{M}}$; $y^* \leftarrow \mathcal{E}(param, id, x_1)$;
$(R, \vec{\boldsymbol{y}}) \leftarrow A_2^{\mathcal{O}_2}(s, y^*, id)$; $\vec{\boldsymbol{x}} \leftarrow \mathcal{D}(param, id, \vec{\boldsymbol{y}})$;
`if` $y \notin \vec{\boldsymbol{y}} \wedge \perp \notin \vec{\boldsymbol{x}} \wedge R(x_b, \vec{\boldsymbol{x}})$ `then` $d \leftarrow 1$ `else` $d \leftarrow 0$;
`return` d

We say that $\mathcal{IBE}$ is secure in the sense of NM-ATK, if $\mathbf{Adv}^{\texttt{nm-atk}}_{\mathcal{IBE},A}(k)$ is negligible for any A.

[4] The adversary is prohibited from performing copying the challenge ciphertext y^*. Otherwise, she could output the equality relation R, where $R(a, b)$ holds if and only if $a = b$, and output $\vec{\boldsymbol{y}} = \{y^*\}$, and be successful, *always*.

4 Relations Among the Notions of Security for $\mathcal{IBE}$ Schemes

In this section, we show that security proved in the sense of IND-ID-CCA2 is validly sufficient for implying security in any other sense in $\mathcal{IBE}$. We first extend the relation (equivalence) between IND-ATK and SS-ATK into $\mathcal{IBE}$ environment, and then extend the relation between IND-ATK and NM-ATK into $\mathcal{IBE}$ environment. At last we study the separation between IND-ATK and OW-ATK.

We demonstrate the relations among the notions of security for $\mathcal{IBE}$ as follows, where ATK $\in$ {ID-CPA,ID-CCA1,ID-CCA2},

4.1 Equivalence Between IND and SS

Theorem 1 (IND-ATK $\Leftrightarrow$ SS-ATK). *A scheme $\mathcal{IBE}$ is secure in the sense of* IND-ATK *if and only if $\mathcal{IBE}$ is secure in the sense of* SS-ATK.

Lemma 2 (IND-ATK $\Rightarrow$ SS-ATK). *If a scheme $\mathcal{IBE}$ is secure in the sense of* IND-ATK *then $\mathcal{IBE}$ is secure in the sense of* SS-ATK.

Proof. See Lemma 2 in [1] or Theorem 7 in [8]. □

Lemma 3 (SS-ATK $\Rightarrow$ IND-ATK). *If a scheme $\mathcal{IBE}$ is secure in the sense of* SS-ATK *then $\mathcal{IBE}$ is secure in the sense of* IND-ATK.

Proof. See Lemma 3 in [1] or Theorem 8 in [8]. □

Proof of Theorem 1. From Lemma 2 and 3, Theorem 1 follows immediately. ■

4.2 Relations Between IND and NM

Theorem 4 (IND-ID-CCA2 $\Rightarrow$ NM-ID-CCA2). *If a scheme $\mathcal{IBE}$ is secure in the sense of* IND-ID-CCA2 *then $\mathcal{IBE}$ is secure in the sense of* NM-ID-CCA2.

Proof. See Theorem 4 in [1] or Theorem 10 in [8]. □

Theorem 5 (NM-ATK $\Rightarrow$ IND-ATK). *If a scheme $\mathcal{IBE}$ is secure in the sense of* NM-ATK *then $\mathcal{IBE}$ is secure in the sense of* IND-ATK.

Proof. See Theorem 5 in [1] or Theorem 9 in [8]. □

Theorem 6 (IND-ID-CPA $\not\Rightarrow$ NM-ID-CPA). *If there is a scheme $\mathcal{IBE}$ secure in the sense of* IND-ID-CPA *then there also exists a scheme $\mathcal{IBE}'$ which is secure in the sense of* IND-ID-CPA, *but not secure in the sense of* NM-ID-CPA.

Proof. See Theorem 11 in [8]. □

Theorem 7 (IND-ID-CCA1 $\not\Leftarrow$ NM-ID-CPA). *If there is a scheme $\mathcal{IBE}$ secure in the sense of* NM-ID-CPA *then there also exists a scheme $\mathcal{IBE}'$ which is secure in the sense of* NM-ID-CPA, *but not secure in the sense of* IND-ID-CCA1.

Proof. See Theorem 12 in [8]. □

4.3 Separation Between IND and OW

Theorem 8 (IND-ATK $\not\Leftarrow$ OW-ATK). *If there is a scheme $\mathcal{IBE}$ secure in the sense of* OW-ATK *then there also exists a scheme $\mathcal{IBE}'$ which is secure in the sense of* OW-ATK, *but not secure in the sense of* IND-ATK.

Proof. See Theorem 6 in [8]. □

5 Semantical Security of $\mathcal{IBE}$ Schemes Under Multiple-Challenge CCA2

We present three notions of SS under multiple-challenge CCA2, following the conventional public-key version [10]. Here an adversary is allowed to make polynomially many challenge templates. Moreover each template is answered with a challenge ciphertext *immediately* (not after making all the templates), and the next challenge template can be generated according to the preceding templates and their answers. After this stage of asking many challenge templates *adaptively* and *in a related manner*, the adversary tries to guess information about the unrevealed plaintexts used in answering challenge templates.

We shall introduce three different types of multiple-challenge CCA2 attacks: mID-CCA2, ID-mCCA2, and mID-mCCA2.

In the definition of SS-ID-CCA2 an adversary consists of two algorithms A_1 and A_2, in such a way that A_1 outputs a challenge template, the challenger chooses a plaintext and presents its encryption, and then A_2 tries to guess information about the plaintext. In the multiple-challenge case this interaction is modelled by providing the adversary with a "*tester*" algorithm $T_{r,param}$ or T_r as its oracle. Here $T_{r,param}$ is given to an actual adversary (which obtains a ciphertext in addition to information leak), while T_r is given to its benign simulator (which only sees information leak). A challenge template[5] is then sent to one of these oracles as a query (called "*challenge query*").

Algorithm $T_{r,param}(P, id, h)$	Algorithm $T_r(P, h)$
return $(\mathcal{E}(param, id, P(r)), h(r))$	return $h(r)$

Intuitively the parameter r of a tester is understood as the multiple-challenge version of the coin tosses that the challenger uses to select plaintexts. It is a sufficiently long sequence of coin tosses $(r^1, r^2, \ldots, r^t)$ which is unrevealed to the adversary. Given the i-th challenge template (P^i, id^i, h^i) (or (P^i, h^i) from a simulator), the challenger chooses a plaintext by $P^i(r^1, r^2, \ldots, r^i)$ using the first

[5] In the previous sections a challenge template includes a distribution $\hat{\mathcal{M}}$ from which a challenge plaintext is picked. However, in this section we prefer to work with a deterministic "plain-text circuit" P which, given an input from $U_{poly(k)}$ kept secret for the adversary, outputs a challenge plaintext. The reason for doing so is some technical ease in the proofs. Here and in the following $U_{poly(k)}$ denotes the uniform distribution on $\{0,1\}^{p(k)}$ for some polynomial p.

i coin tosses in r. Note that now h leaks information on coin tosses r rather than plaintexts $P^i(r^1, r^2, \dots, r^i)$.[6]

As discussed in [10, 8], the only restriction here is that, extraction queries on challenge identities cannot be made.

Definition 5 (Semantic security under multiple-challenge CCA2)

Let $\mathcal{IBE} = (\mathcal{S}, \mathcal{X}, \mathcal{E}, \mathcal{D})$ be an identity based encryption scheme. $\mathcal{IBE}$ is secure in the sense of SS-mID-mCCA2 if the following holds. For every oracle PPT A ("SS-mID-mCCA2 adversary") with the following restriction on oracle queries: in any execution of $A^{\mathcal{X}_{mk}, \mathcal{D}_{mk}, T_{r,param}}(param)$, for each challenge query $(c, b) \leftarrow T_{r,param}(P, id, h)$ by A, A is prohibited to make (1) the extraction query $\mathcal{X}_{mk}(id)$ regardless of before or after the challenge query, or, (2) the decryption query $\mathcal{D}_{mk}(id, c)$ after the challenge query, there exists a PPT algorithm A' ("benign simulator of A") which is equally successful as A, in the following sense.

1. The difference between the advantage of the actual adversary A and that of the benign simulator A', namely
$$\Pr\left[v = f(r) \,\middle|\, \begin{array}{l} (param, mk) \leftarrow \mathcal{S}(k); \qquad r \leftarrow U_{poly(k)}; \\ (f, v) \leftarrow A^{\mathcal{X}_{mk}, \mathcal{D}_{mk}, T_{r,param}}(param) \end{array}\right]$$
$$- \Pr\left[v = f(r) \,\middle|\, r \leftarrow U_{poly(k)}; \quad (f, v) \leftarrow A'^{T_r}(1^k)\right]$$
is negligible as a function over k.
2. The two ensemples over $k \in \mathbb{Z}^+$:
$$\left[(t, f) \,\middle|\, \begin{array}{c} (param, mk) \leftarrow \mathcal{S}(k); \qquad r \leftarrow U_{poly(k)}; \\ (f, v) \leftarrow A^{\mathcal{X}_{mk}, \mathcal{D}_{mk}, T_{r,param}}(param) \text{ with trace } t \end{array}\right] \quad \text{and}$$
$$\left[(t, f) \,\middle|\, \begin{array}{c} r \leftarrow U_{poly(k)}; \\ (f, v) \leftarrow A'^{T_r}(1^k) \text{ with trace } t \end{array}\right]$$
are computationally indistinguishable. Here the *trace* of an execution of the actual adversary A is the sequence of (P, h)-part of the challenge queries (P, id, h) made by A. The *trace* of an execution of the simulator A' is simply the sequence of challenge queries A' makes.

SS-mID-CCA2 is defined analogously except that an adversary A is restricted to have the same plaintext circuit P and the same information leakage circuit h in all the challenge queries in the trace of an execution of A (the challenge identity id can vary).

SS-ID-mCCA2 is analogous to SS-mID-mCCA2 except that an adversary A must have the same challenge identity id in all the challenge queries in the trace of an execution of A (P and h can vary).

6 As is shown in Definition 5, the same goes to the information to guess: it is about the coin tosses r (i.e. $f(r)$ to guess) rather than plaintexts (i.e. $f(P(r))$ to guess).

Remark 2. Note that our mID-CCA2 attack is stronger than the attack consider in [2], since in the latter case the adversary has to commit at once to the identities on which it wants to be challenged, while in the present case the i-th identity can be chosen depending on the challenges received so far.

Theorem 9. *The three security notions under multiple-challenge* CCA2 *in Definition 5 are all equivalent to the single-challenge security* SS-ID-CCA2.

Proof. See Theorem 14 in [8]. □

References

1. N. Attrapadung, Y. Cui, G. Hanaoka, H. Imai, K. Matsuura, P. Yang, and R. Zhang. Relations among notions of security for identity based encryption schemes. Cryptology ePrint Archive, Report 2005/258, 2005. `http://eprint.iacr.org/2005/258`.
2. J. Baek, R. Safavi-Naini, and W. Susilo. Efficient multi-receiver identity-based encryption and its application to broadcast encryption. In *PKC 2005*, volume 3386 of *LNCS*, pages 380–397, 2005.
3. M. Bellare, A. Desai, D. Pointcheval, and P. Rogaway. Relations among notions of security for public-key encryption schemes. In *CRYPTO '98*, volume 1462 of *LNCS*, pages 26–45, 1998.
4. D. Boneh and X. Boyen. Secure identity based encryption without random oracles. In *CRYPTO '04*, volume 3152 of *LNCS*, pages 443–459, 2004.
5. D. Boneh, X. Boyen, and E. Goh. Hierarchical identity based encryption with constant size ciphertext. In *EUROCRYPT '05*, volume 3494 of *LNCS*, pages 440–456, 2005.
6. D. Boneh and M. Franklin. Identity-based encryption from the Weil pairing. In *CRYPTO '01*, volume 2139 of *LNCS*, pages 213–229, 2001.
7. D. Dolev, C. Dwork, and M. Naor. Non-malleable cryptography (extended abstract). In *STOC '91*, pages 542–552, 1991.
8. D. Galindo and I. Hasuo. Security notions for identity based encryption. Cryptology ePrint Archive, Report 2005/253, 2005. `http://eprint.iacr.org/2005/253`.
9. O. Goldreich. *Foundations of cryptography, Volumn II (revised, posted version Nr. 4.2)*. 2003. `http://www.wisdom.weizmann.ac.il/~oded/`.
10. O. Goldreich, Y. Lustig, and M. Naor. On chosen ciphertext security of multiple encryptions. Cryptology ePrint Archive, Report 2002/089, 2002. `http://eprint.iacr.org/`.
11. S. Goldwasser and S. Micali. Probabilistic encryption. *Journal of Computer and System Sciences*, 28:270–299, 1984.
12. C. Rackoff and D.R. Simon. Non-interactive zero-knowledge proof of knowledge and chosen ciphertext attack. In *CRYPTO '91*, volume 576 of *LNCS*, pages 433–444, 1991.
13. A. Shamir. Identity-based cryptosystems and signature schemes. In *CRYPTO '84*, volume 196 of *LNCS*, pages 47–53, 1985.
14. Y. Watanabe, J. Shikata, and H. Imai. Equivalence between semantic security and indistinguishability against chosen ciphertext attacks. In *PKC '03*, volume 2567 of *LNCS*, pages 71–84, 2003.
15. B. Waters. Efficient identity-based encryption without random oracles. In *EUROCRYPT '05*, volume 3494 of *LNCS*, pages 114–127, 2005.

Optimally Adaptive Integration of Univariate Lipschitz Functions

Ilya Baran[1], Erik D. Demaine[1], and Dmitriy A. Katz[2]

[1] MIT Computer Science and Artificial Intelligence Laboratory, 32 Vassar Street, Cambridge, MA 02139, USA
{ibaran, edemaine}@mit.edu

[2] Sloan School of Management, Massachusetts Institute of Technology, 50 Memorial Drive, Cambridge, MA 02142, USA
dimdim@mit.edu

Abstract. We consider the problem of approximately integrating a Lipschitz function f (with a known Lipschitz constant) over an interval. The goal is to achieve an error of at most ϵ using as few samples of f as possible. We use the adaptive framework: on all problem instances an adaptive algorithm should perform almost as well as the best possible algorithm tuned for the particular problem instance. We distinguish between DOPT and ROPT, the performances of the best possible deterministic and randomized algorithms, respectively. We give a deterministic algorithm that uses $O(\mathrm{DOPT}(f,\epsilon)\cdot\log(\epsilon^{-1}/\mathrm{DOPT}(f,\epsilon)))$ samples and show that an asymptotically better algorithm is impossible. However, any deterministic algorithm requires $\Omega(\mathrm{ROPT}(f,\epsilon)^2)$ samples on some problem instance. By combining a deterministic adaptive algorithm and Monte Carlo sampling with variance reduction, we give an algorithm that uses at most $O(\mathrm{ROPT}(f,\epsilon)^{4/3}+\mathrm{ROPT}(f,\epsilon)\cdot\log(1/\epsilon))$ samples. We also show that any algorithm requires $\Omega(\mathrm{ROPT}(f,\epsilon)^{4/3}+\mathrm{ROPT}(f,\epsilon)\cdot\log(1/\epsilon))$ samples in expectation on some problem instance (f,ϵ), which proves that our algorithm is optimal.

1 Introduction

We consider the problem of approximating a definite integral of a univariate Lipschitz function (with known Lipschitz constant) to within ϵ using the fewest possible samples. The function is given as a black box: sampling it at a parameter value is the only allowed operation. It is easy to show that $\Theta(\epsilon^{-1})$ samples are necessary and sufficient for a deterministic algorithm in the worst case (see, e.g., [1]). The results in [2] imply a Monte-Carlo method that requires only $\Theta(\epsilon^{-2/3})$ samples in the worst case.

The Adaptive Framework. The univariate Lipschitz integration problem becomes more interesting in the adaptive setting. The motivation is that, for a given ϵ, some problem instances have much lower complexity than others. For example, if $f(x) = Lx$, where L is the Lipschitz constant, then evaluating f at the endpoints of the interval over which the integral is taken is sufficient to solve the problem for any ϵ. Thus, it is desirable to have an algorithm that is guaranteed to use fewer samples on easier problem instances. Such an algorithm is called *adaptive*. We formalize this notion by defining

J.R. Correa, A. Hevia, and M. Kiwi (Eds.): LATIN 2006, LNCS 3887, pp. 142–153, 2006.

the difficulty of a problem as the performance of the best possible algorithm on that problem:

Definition 1. *Let $\mathcal{P}$ be a class of problem instances. Let $\mathcal{A}$ be the set of all correct algorithms for $\mathcal{P}$ (among some reasonable class of algorithms). Let* $\mathrm{COST}(A, P)$ *be the performance of algorithm $A \in \mathcal{A}$ on problem instance $P \in \mathcal{P}$. Define* $\mathrm{OPT}(P) = \min_{A \in \mathcal{A}} \mathrm{COST}(A, P)$. *We use* DOPT *when $\mathcal{A}$ is the set of deterministic algorithms and* ROPT *when $\mathcal{A}$ is the set of randomized algorithms that are correct on each $P \in \mathcal{P}$ with probability at least* $2/3$.

By definition, for every problem instance P, there is an algorithm whose cost on P is $\mathrm{OPT}(P)$. A good adaptive algorithm is a single algorithm whose cost is not much greater than $\mathrm{OPT}(P)$ for *every* problem instance P. Therefore, an adaptive guarantee is in general much stronger than a worst-case guarantee.

The ultimate goal of investigating a problem in the adaptive framework is to design an "optimally adaptive" algorithm. Suppose $\mathcal{P}$ is the set of problem instances and each problem instance $P \in \mathcal{P}$ has certain natural parameters, $v_1(P), \ldots, v_k(P)$, with the first parameter $v_1(P) = \mathrm{OPT}(P)$. An algorithm is *optimally adaptive* if its performance on every problem instance $P \in \mathcal{P}$ is within a constant factor of every algorithm's worst-case performance on the family of instances with the same values for the parameters: $\{P' \in \mathcal{P} \mid v_i(P') = v_i(P) \text{ for all } i\}$. Note that this definition depends on the choice of parameters, so in addition to OPT, we need to choose reasonable parameters, such as ϵ, the desired output accuracy.

Related Work. While approximate definite integration is well-studied both in numerical analysis (see, e.g., [3]) and in information-based complexity [4], those algorithms do not have provable guarantees about adaptivity. In that literature, the term "adaptive" typically refers to an algorithm that is allowed to pick samples based on previous sample values, which is quite different from our meaning.

For other problems, optimally adaptive algorithms have been previously designed in the context of set operations [5], aggregate ranking [6], and independent set discovery in [7]. Lipschitz functions also lend themselves well to adaptive algorithms. It is shown in [8] that Piyavskii's algorithm [9] for minimizing a univariate Lipschitz function performs $O(\mathrm{OPT})$ samples. [10] gives an adaptive algorithm for minimizing the distance from a point to a Lipschitz curve that is within a logarithmic factor of OPT. [11] gives adaptive algorithms for several problems on Lipschitz functions.

Our Results. In this paper we give a deterministic algorithm that makes at most $O(\mathrm{DOPT} \cdot \log(\epsilon^{-1}/\mathrm{DOPT}))$ samples. We also prove a matching lower bound on deterministic algorithms. When comparing to ROPT, however, we show that any deterministic adaptive algorithm uses $\Omega(\mathrm{ROPT}^2)$ samples on some problem instance. We present a randomized adaptive algorithm, LIPSCHITZ-MC-INTEGRATE, that always uses $O(\mathrm{ROPT}^{4/3} + \mathrm{ROPT} \cdot \log(\epsilon^{-1}))$ samples and prove a matching lower bound.

We therefore give optimally adaptive algorithms for the Lipschitz integration problem in the deterministic and randomized settings. Although the algorithms are simple, in both cases analyzing their adaptive performance is nontrivial. To our knowledge, LIPSCHITZ-MC-INTEGRATE is the first randomized optimally adaptive algorithm. Also, a simple corollary of the randomized lower bound is that the non-adaptive algorithm based on the results in [2] is optimal in the worst case.

Some of the results in this paper, primarily in Sections 3 and 4, are based on the first author's master's thesis [11]. Many of the proofs are omitted from this extended abstract.[1]

2 Problem Basics

We start by giving a precise formulation of the problem we consider:
Problem LIPSCHITZ-INTEGRATION:

Given: (f, a, b, L, ϵ)

Such that: $f\colon [a, b] \to \mathbb{R}$
and for $x_1, x_2 \in [a, b]$, $|f(x_2) - f(x_1)| \le L|x_2 - x_1|$

Compute: $I \in \mathbb{R}$ such that $\left| I - \int_a^b f(x)\, dx \right| \le \epsilon$

A randomized algorithm needs to be correct with probability at least $2/3$.

Some input parameters can be eliminated without loss of generality. The problem instance (f, a, b, L, ϵ) is equivalent to the problem instance $(\hat{f}, 0, 1, 1, \epsilon/L(b-a)^2)$ where $\hat{f}(x) = f\left(\frac{x-a}{b-a}\right)/L(b-a)$, so we can assume without loss of generality that $a = 0$, $b = 1$, and $L = 1$.

We now develop some basic tools we will need for discussing and analyzing the algorithms. Essentially, we show how to make use of the Lipschitz condition to bound the error of our estimates.

The Lipschitz condition allows an algorithm that has sampled f at two points to bound the value of the integral of f on the interval between them. We call the quality of this bound *area looseness*, and it depends on both the length of the interval and the values of f at the sampled points. A greater difference between values of f (a steeper function) results in a smaller area looseness. We define area looseness as follows (see Figure 1):

Definition 2. *Given a Lipschitz function f on $[0, 1]$, define the* area looseness *of a subinterval $[x_1, x_2]$ of $[0, 1]$ as $AL_f(x_1, x_2) = ((x_2 - x_1)^2 - (f(x_1) - f(x_2))^2)/2$. When it is clear which f we are talking about, we simply write $AL(x_1, x_2)$.*

Our analysis relies on area looseness being well behaved. The following proposition shows that it has the properties one would expect a bound on integration error to have and that an additional sample in the middle of the interval decreases total area looseness quickly.

[1] The full version of this paper is available at `http://www.mit.edu/~ibaran/papers/intfull.{pdf,ps}`

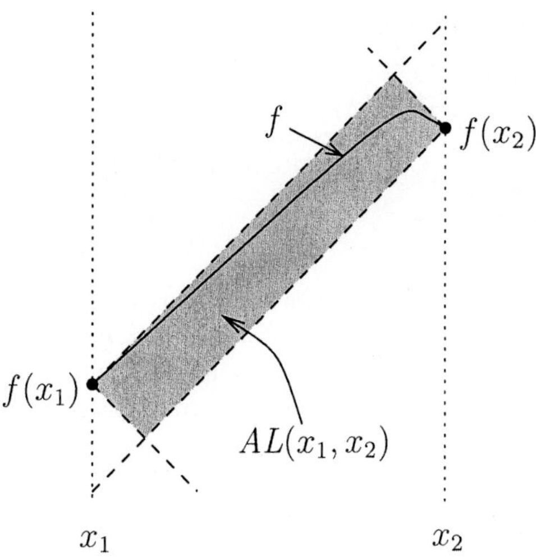

Fig. 1. Illustration of area looseness. Lipschitz bounds are dashed.

Proposition 1. *Area-looseness has the following properties:*

(1) $0 \leq AL(x_1, x_2) \leq (x_2 - x_1)^2/2$.
(2) If $x'_1 \leq x_1 < x_2 \leq x'_2$ *then* $AL(x_1, x_2) \leq AL(x'_1, x'_2)$.
(3) If $x \in [x_1, x_2]$, *then* $AL(x_1, x) + AL(x, x_2) \leq AL(x_1, x_2)$.
(4) $AL\big(x_1, \frac{x_1+x_2}{2}\big) + AL\big(\frac{x_1+x_2}{2}, x_2\big) \leq AL(x_1, x_2)/2$.

For the lower bounds, both on OPT and on adaptive algorithms, we need "extremal" Lipschitz functions, whose integral is either maximal or minimal, given the samples. We call these functions *HI* and *LO*. We also define *looseness*, the maximum difference between *HI* and *LO* over an interval.

Definition 3. *Given a Lipschitz function* f, *and* $0 \leq a < b \leq 1$, *define the Lipschitz functions* HI_a^b *and* LO_a^b *on* $[a, b]$ *as:* $HI_a^b(x) = \min(f(a) + x - a, f(b) + b - x)$ *and* $LO_a^b(x) = \max(f(a) - x + a, f(b) - b + x)$. *Also define* L_f *as* $L_f(a, b) = b - a - |f(b) - f(a)|$.

Proposition 2. *Given a Lipschitz function* f, *the functions* HI_a^b *and* LO_a^b *have the following properties:*

(1) If g *is Lipschitz,* $g(a) = f(a)$, *and* $g(b) = f(b)$, *then for* $x \in [a, b]$, $HI_a^b(x) \geq g(x) \geq LO_a^b(x)$.
(2) $AL(a, b)/(b - a) \leq \max_{x \in [a,b]} (HI_a^b(x) - LO_a^b(x)) = L(a, b) \leq 2AL(a, b)/(b - a)$.
(3) $\int_a^b HI_a^b(x)\, dx = (b - a)\frac{f(a)+f(b)}{2} + AL(a, b)/2$ *and* $\int_a^b LO_a^b(x)\, dx = (b - a)\frac{f(a)+f(b)}{2} - AL(a, b)/2$.

Proposition 3. *Given a Lipschitz function f, looseness has the following properties:*

(1) $0 \leq L(a,b) \leq b-a$.
(2) If $a' \leq a \leq b \leq b'$, *then* $L(a,b) \leq L(a',b')$.
(3) If $x_1 \leq x_2 \leq \cdots \leq x_n$, *then* $\sum_{i=1}^{n-1} L(x_i, x_{i+1}) \leq L(x_1, x_n)$.

3 Proof Sets

In order to compare the running time of an algorithm on a problem instance to DOPT, we define the concept of a proof set for a problem instance. A set P of points in $[0,1]$ is a *proof set* for problem instance (f,ϵ) and output x if for every f' that is equal to f on P, x is a correct output on (f',ϵ). In other words, sampling f at a proof set proves the correctness of the output. We say that a set of samples is a proof set for a particular problem instance without specifying the output if some output exists for which it is a proof set.

It is clear from the definition that sampling a proof set is the only way a deterministic algorithm can guarantee correctness: if an algorithm doesn't sample a proof set for some problem instance, we can feed it a problem instance that has the same value on the sampled points, but for which the output of the algorithm is incorrect. Conversely an algorithm can terminate as soon as it has sampled a proof set and always be correct. Thus, DOPT is equal to the size of a smallest proof set.

In order to analyze the deterministic algorithm, we will compare the number of samples it makes to the size of a proof set P. We will need some tools for doing this.

Let P be a nonempty finite set of points in $[0,1]$. Consider the execution of an algorithm which samples a function at points on the interval $[0,1)$ (if it samples at 1, ignore that sample). Let $s_1, s_2, \ldots, s_n$ be the sequence of samples that the algorithm performs in the order that it performs them. Let I_t be the set of unsampled intervals after sample s_t, i.e., the connected components of $[0,1) - \{s_1, \ldots, s_t\}$, except make each element of I_t half-open by adding its left endpoint, so that the union of all the elements of I_t is $[0,1)$. Let $[l_t, r_t)$ be the element of I_{t-1} that contains s_t.

Then sample s_t is a:

split	if	$[l_t, s_t) \cap P \neq \emptyset$ and $[s_t, r_t) \cap P \neq \emptyset$
squeeze	if	$[l_t, s_t) \cap P \neq \emptyset$ or $[s_t, r_t) \cap P \neq \emptyset$, but not both
fizzle	if	$[l_t, r_t) \cap P = \emptyset$.

These definitions are, of course, relative to P. See Figure 2. We can now bound the number of samples of different types:

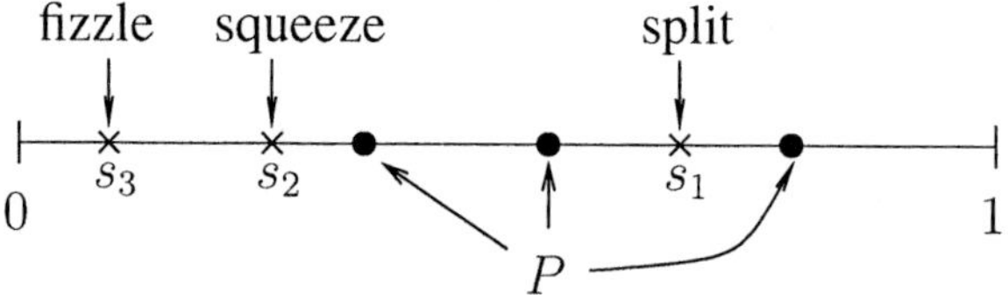

Fig. 2. Different types of samples

Proposition 4. *The number of splits is at most* $|P| - 1$.

Proposition 5. *Suppose that for all* i *and* j *with* $i \neq j$, $|s_i - s_j| > \epsilon$ *and that for all* t, $s_t = (l_t + r_t)/2$. *Then if* $|P| \leq \epsilon^{-1}/2$, *the number of squeezes is at most* $|P| \log_2(\epsilon^{-1}/|P|)$.

We now characterize proof sets for LIPSCHITZ-INTEGRATION.

Proposition 6. *Let* $P = \{x_1, x_2, \ldots, x_n\}$ *such that* $0 \leq x_1 < x_2 < \cdots < x_n \leq 1$. *Then* P *is a proof set for problem instance* (f, ϵ) *if and only if* $x_1^2 + (1 - x_n)^2 + \sum_{i=1}^{n-1} AL(x_i, x_{i+1}) \leq 2\epsilon$.

4 Deterministic Algorithm and Analysis

Proposition 6, together with Proposition 1 immediately shows the correctness of a trivial algorithm. Let $n = \lceil \epsilon^{-1}/4 \rceil$ and let the algorithm make n samples, at $\frac{1}{2n}, \frac{3}{2n}, \ldots, \frac{2n-1}{2n}$ and output the integral M as in the proof of Proposition 6. It is correct because the area-looseness of every interval is at most $(1/n)^2/2$. Because there are $n - 1$ intervals, the total area-looseness of all of them is at most $(n - 1)/(2n^2)$. Also, $x_1^2 = (1 - x_n)^2 = 1/(2n)^2$, so $x_1^2 + (1 - x_n)^2 + \sum_{i=1}^{n-1} AL(x_i, x_{i+1}) = n/(2n^2) \leq 2\epsilon$. Therefore, $\Theta(\epsilon^{-1})$ samples are always sufficient (and if, for instance, f is a constant, necessary).

We now give a deterministic adaptive algorithm. The algorithm maintains the total area-looseness of the current unsampled intervals, the unsampled intervals themselves in a linked list, and uses a priority queue to choose the unsampled interval with the largest area-looseness at every step and sample in the middle of it.

Let L be a linked list of (PARAMETER, VALUE) pairs and let Q be a priority queue of (AL, ELEM) pairs where the first element is a real number (and defines the order of Q) and the second element is a pointer into an element of L. The algorithm follows:

Algorithm. LIPSCHITZ-INTEGRATE

1. Add $(0, f(0))$ and $(1, f(1))$ to L and insert $(AL(0, 1), (0, f(0)))$ into Q
2. A-LOOSENESS $\leftarrow AL(0, 1)$.
3. Do while A-LOOSENESS $> 2\epsilon$:
  4. (AL, P_1) $\leftarrow$ EXTRACT-MAX[Q]
  5. $P_2 \leftarrow$ NEXT[L, P_1]
  6. $x \leftarrow$ (PARAMETER[P_1] + PARAMETER[P_2])/2
  7. AL$_1$ $\leftarrow AL$(PARAMETER[P_1], x), AL$_2$ $\leftarrow AL(x$, PARAMETER[P_2])
  8. Insert $(x, f(x))$ into L after P_1 and insert (AL$_1$, P_1) and (AL$_2$, $(x, f(x))$) into Q
  9. A-LOOSENESS $\leftarrow$ A-LOOSENESS $-$ AL $+$ AL$_1$ $+$ AL$_2$
10. Compute and output M using the values stored in L as described in Proposition 6.

The correctness of the algorithm is clear from Proposition 6: the algorithm stops precisely when the total area-looseness of the unsampled intervals is no more than 2ϵ. We need to analyze the algorithm's performance.

Theorem 1. *Algorithm* LIPSCHITZ-INTEGRATE *makes* $O(\text{DOPT} \cdot \log(\epsilon^{-1}/\text{DOPT}))$ *samples on problem instance* (f, ϵ).

Proof: We will actually compare the number of samples to $\mathrm{DOPT}(f, \epsilon/2)$ rather than to $\mathrm{DOPT}(f, \epsilon)$. We can do this because if we take a proof set for $\mathrm{DOPT}(f, \epsilon)$ and sample in the middle of every unsampled interval, then by Proposition 1 (4), we will obtain a proof set for $\mathrm{DOPT}(f, \epsilon/2)$. Thus, $\mathrm{DOPT}(f, \epsilon/2) \le 2 \cdot \mathrm{DOPT}(f, \epsilon) + 1$. So let P be a proof set for $(f, \epsilon/2)$ of size $\mathrm{DOPT}(f, \epsilon/2)$.

First, we argue that no interval of length smaller than 4ϵ is ever subdivided. Suppose for contradiction that among n intervals $I_1, \ldots, I_n$ of lengths $a_1, \ldots, a_n$, interval I_k with $a_k < 4\epsilon$ is chosen for subdivision. By Proposition 1 (1), $AL(I_i) \le a_i^2/2$, so $\sqrt{AL(I_k)} \le 2\epsilon$. On the other hand, $\sum a_i = 1$, so $\sum \sqrt{AL(I_i)} \le 1$. Multiplying the inequalities, we get $\sum AL(I_i) \le \sum \sqrt{AL(I_i) AL(I_k)} \le 2\epsilon$. But this implies that the algorithm should have terminated, which is a contradiction.

Now, we count the number of samples relative to P. The number of splits is $O(|P|)$ by Proposition 4. The above paragraph shows that we can use Proposition 5 to conclude that there are $O(|P| \log(\epsilon^{-1}/|P|))$ squeezes. We now show that there are $O(|P|)$ fizzles and so prove the theorem.

A fizzle occurs when an interval not containing a point of P is chosen for subdivision. Consider the situation after n points have been sampled. Let the sampled points be $0 = x_1 \le x_2 \le \cdots \le x_n = 1$. Because the total area-looseness of intervals between points of P is at most ϵ, by repeated application of Proposition 1 (2,3), we have $\sum_{[x_i, x_{i+1}) \cap P = \emptyset} AL(x_i, x_{i+1}) \le \epsilon$. The algorithm has not terminated, so the total area-looseness must be more than 2ϵ, which implies that $\sum_{[x_i, x_{i+1}) \cap P \ne \emptyset} AL(x_i, x_{i+1}) > \epsilon$. Because there are at most $|P|$ elements in the sum on the left hand side, the largest element must be greater than $\epsilon/|P|$. Therefore, there exists a k such that $[x_k, x_{k+1})$ contains a point of P and $AL(x_k, x_{k+1}) > \epsilon/|P|$. So if a fizzle occurs, the area-looseness of the chosen interval must be at least $\epsilon/|P|$.

Now let S_t be the set of samples made by the algorithm after time t. Define A_t as follows: let $\{y_1, y_2, \ldots, y_n\} = S_t \cup P$ with $0 = y_1 \le y_2 \le \cdots \le y_n$ and let $A_t = \sum_{i=1}^{n-1} AL(y_i, y_{i+1})$. Clearly, $A_t \ge 0$, $A_t \ge A_{t+1}$ (by Proposition 1 (3)), and therefore, $A_t \le A_0 \le 2\epsilon$. Every fizzle splits an interval between adjacent y's into two. Because the area-looseness of the interval before the split was at least $\epsilon/|P|$, by Proposition 1 (4), A_t decreases by at least $\epsilon/(2|P|)$ as a result of every fizzle. Therefore, there can be at most $4|P|$ fizzles during an execution. □

We prove a matching lower bound, showing that the logarithmic factor is necessary and that LIPSCHITZ-INTEGRATE is optimally adaptive:

Theorem 2. *For any deterministic algorithm and for any $\epsilon > 0$ and any integer k such that $0 < k < \epsilon^{-1}/2$, there exists a problem instance (f, ϵ) of* LIPSCHITZ-INTEGRATION *with $\mathrm{DOPT}(f, \epsilon) = O(k)$ on which that algorithm performs $\Omega(k \log(\epsilon^{-1}/k))$ samples.*

5 Algorithm LIPSCHITZ-MC-INTEGRATE

A standard strategy in a Monte Carlo integration algorithm is to sample at a point picked uniformly at random from an interval. The expected value of such a sample, scaled by the length of the interval, is precisely the value of the integral over the interval, so the goal is to minimize the variance. When the function is Lipschitz, the variance of the integral estimate based on such a sample can be as high as a constant times the fourth power of the length of the interval. However, if we use the fact that when the area looseness of an interval is low, we approximately know the function, we can adjust the sample to get an unbiased estimator of the integral over that interval whose variance is the square of the area looseness in the worst case. Procedure MC-SAMPLE shows how to do this.

Procedure MC-SAMPLE(x_1, x_2):

1. Let x be a random number, uniformly chosen from $[x_1, x_2]$
2. If $f(x_1) \leq f(x_2)$, then SAMPLE $\leftarrow \left(f(x) - x + \frac{x_1+x_2}{2}\right)$
3. Else SAMPLE $\leftarrow \left(f(x) + x - \frac{x_1+x_2}{2}\right)$
4. Return SAMPLE $\cdot\, (x_2 - x_1)$

Proposition 7. MC-SAMPLE(x_1, x_2) *returns an unbiased estimator of* $\int_{x_1}^{x_2} f(x)\,dx$ *that has variance at most* $AL^2(x_1, x_2)$.

In order to compute the integral over $[0, 1]$, we would like an estimator for that integral with low variance. If we split $[0, 1]$ into intervals whose total AL^2 is small and run MC-SAMPLE on each interval, we will get such an estimator, as shown in the following corollary.

Corollary 1. *Let* $0 = x_1 < x_2 < \cdots < x_n = 1$ *and suppose* $\sum_{i=1}^{n-1} AL^2(x_i, x_{i+1}) \leq \epsilon^2/3$. *Let* $\hat{I} = \sum_{i=1}^{n-1}$ MC-SAMPLE(x_i, x_{i+1}). *Let* $I = \int_0^1 f(x)\,dx$. *Then* $\Pr[|\hat{I} - I| \geq \epsilon] \leq 1/3$.

The remaining difficulty is to find a small number of intervals whose total AL^2 is smaller than $\epsilon^2/3$. Note that the deterministic adaptive algorithm in Section 4 finds a small number of intervals whose total AL is smaller than ϵ. We show that we can use the same idea here. Thus, to obtain a randomized adaptive algorithm, we use a deterministic adaptive algorithm to get a rough idea of the function and then use Monte Carlo sampling with variance reduction (MC-SAMPLE) to improve our estimate of the integral.

Let L be a linked list of (PARAMETER, VALUE) pairs and let Q be a priority queue of (AL, ELEM) pairs where the first element is a real number (and defines the order of Q) and the second element is a pointer into an element of L. The algorithm is as follows:

Algorithm. LIPSCHITZ-MC-INTEGRATE:

1. Add $(0, f(0))$ and $(1, f(1))$ to L and insert $(AL^2(0, 1), (0, f(0)))$ into Q
2. ALSQ $\leftarrow AL^2(0, 1)$.
3. Do while ALSQ $> \epsilon^2/3$:
  4. (AL, P_1) $\leftarrow$ EXTRACT-MAX$[Q]$
  5. $P_2 \leftarrow$ NEXT$[P_1]$
  6. $x \leftarrow$ (PARAMETER$[P_1]$ + PARAMETER$[P_2]$)$/2$
  7. AL$_1 \leftarrow AL^2$(PARAMETER$[P_1], x$), AL$_2 \leftarrow AL^2(x,$ PARAMETER$[P_2]$)
  8. Insert $(x, f(x))$ into L after P_1 and insert (AL$_1$, P_1) and (AL$_2$, $(x, f(x))$) into Q
  9. ALSQ $\leftarrow$ ALSQ $-$ AL $+$ AL$_1$ $+$ AL$_2$

10. $\hat{I} \leftarrow 0$.
11. For each element P of L except the last:
  12. $\hat{I} \leftarrow \hat{I} + \text{MC-SAMPLE}(\text{PARAMETER}[P], \text{PARAMETER}[\text{NEXT}[P]])$
13. Output $\hat{I}$

Correctness is guaranteed by Corollary 1 because the algorithm exits the loop in lines 3-9 only when the total AL^2 of intervals between points in L is no more than $\epsilon^2/3$.

6 Performance Analysis

For the analysis of the algorithm, let f be the Lipschitz function input to LIPSCHITZ-MC-INTEGRATE.

Lemma 1. *Given f, there exists a set of points $0 = x_1 < x_2 < \cdots < x_n = 1$ such that for $1 \leq i \leq n-2$, $AL(x_i, x_{i+1}) = 3\epsilon$, and $AL(x_{n-1}, x_n) \leq 3\epsilon$. Furthermore,* $\text{ROPT}(f, \epsilon) \geq (n-2)/3$.

Proof: We begin by constructing a set of points that satisfies the conditions. Obviously, x_1 should be 0. Suppose we have constructed the first k points and $x_k \neq 1$. If $AL(x_k, 1) \leq 3\epsilon$, set $x_{k+1} = 1$ and we are done. Otherwise, notice that f is continuous, so AL is also continuous. By Proposition 1 (1), $AL(x_k, x_k) = 0$. Therefore, by the intermediate value theorem, there is an $x \in [x_k, 1]$ such that $AL(x_k, x) = 3\epsilon$ and we set x_{k+1} to be that x.

Consider an algorithm A that is correct with probability at least $2/3$ on all inputs and consider its executions on f. Let e_i for $1 \leq i \leq n-2$ be the expected number of samples A performs in (x_i, x_{i+1}). We claim that in order for A to be correct, it must have $e_i \geq 1/3$ for all i and therefore, the total expected number of samples is $\sum_{i=1}^{n-2} e_i \geq (n-2)/3$.

Suppose for contradiction, that $e_i < 1/3$ for some i. Then, by Markov's inequality, the probability that A samples in (x_i, x_{i+1}) is less than $1/3$. Now consider two functions defined as follows: $\hat{f}_1(x) = \hat{f}_2(x) = f(x)$ everywhere except (x_i, x_{i+1}) and $\hat{f}_1(x) = LO_{x_i}^{x_{i+1}}(x)$ and $\hat{f}_2(x) = HI_{x_i}^{x_{i+1}}(x)$ on (x_i, x_{i+1}). By Proposition 2 (3), $\int_0^1 \hat{f}_2(x)dx - \int_0^1 \hat{f}_1(x) = AL(x_i, x_{i+1}) = 3\epsilon$, so no output is correct for both $\hat{f}_1$ and $\hat{f}_2$. Suppose, that we feed $\hat{f}_1$ and $\hat{f}_2$ with probability $1/2$ each as input to A. Conditioned on A not sampling in (x_i, x_{i+1}), the output of A is independent of which function was input. Therefore, conditioned on A not sampling in (x_i, x_{i+1}), the probability of error is at least $1/2$. Because $\hat{f}_1 = \hat{f}_2 = f$ not on (x_i, x_{i+1}), the probability of A not sampling on (x_i, x_{i+1}) is greater than $2/3$, so the probability of error is greater than $1/3$, which implies that A is invalid. □

Because the number of samples in steps 11–13 is smaller (by 1) than the number of samples in steps 1–9, we only focus on the samples in steps 1-9. For the analysis, we split the execution of the algorithm into two phases. The algorithm is in Phase 1 while there is a pair of adjacent elements x_i and x_{i+1} in L for which $AL(x_i, x_{i+1}) > 3\epsilon$. When all pairs of adjacent elements have AL at most 3ϵ, the algorithm is in Phase 2. Note that by Proposition 1 (2), area looseness between adjacent points in L never increases as the algorithm executes, so once it enters Phase 2, it never goes back to Phase 1. We now bound the number of samples made in steps 1–9 in the phases.

Lemma 2. *In Phase 1,* LIPSCHITZ-MC-INTEGRATE *makes* $O(\mathrm{ROPT}(f,\epsilon)\log(1/\epsilon))$ *samples on problem instance* (f,ϵ).

Proof: Let X be the set of x_i's constructed as in Lemma 1. We count the samples made by LIPSCHITZ-MC-INTEGRATE relative to X. By Proposition 4, there are at most $O(|X|)$ splits. We now need a lower bound on the size of intervals in Phase 1 to count the number of squeezes. We note that an interval whose length is smaller than $\sqrt{6\epsilon}$ has area looseness at most 3ϵ (by Proposition 1 (1)) and will therefore never be chosen for subdivision in Phase 1. Therefore, in Phase 1, every interval has length at least $\sqrt{6\epsilon}/2$. So by Proposition 5, there are at most $|X|\log((\sqrt{6\epsilon}/2)^{-1}/|X|) = O(|X|\log(1/\epsilon))$ squeezes. There are no fizzles because any interval whose area looseness is greater than 3ϵ must have a point of X (by Proposition 1 (2) and by construction of X). By Lemma 1, $|X| = O(\mathrm{ROPT}(f,\epsilon))$, so we have the claimed bound. □

Lemma 3. *In Phase 2,* LIPSCHITZ-MC-INTEGRATE *uses at most* $O(\mathrm{ROPT}(f,\epsilon)^{4/3} + \mathrm{ROPT}(f,\epsilon)\log(1/\epsilon))$ *samples on problem instance* (f,ϵ).

Proof: After Phase 1 is complete, L consists of points such that the area looseness between adjacent pairs is at most 3ϵ. Let $0 = y_1 < y_2 < \cdots < y_m = 1$ be the smallest subset of points in L (including 0 and 1) such that $AL(y_i, y_{i+1}) \leq 3\epsilon$ for all y. We claim that $m \leq 6 \cdot \mathrm{ROPT}(f,\epsilon)$. Consider the set of x_i's constructed as in Lemma 1. If y_i's are a minimal set of points with area looseness no greater than 3ϵ between adjacent ones, then every interval of the form $[x_i, x_{i+1}]$ has at most two y_i's (if there are three, the middle one is unnecessary). Therefore there are at most twice as many y_i's as x_i's.

Now assume the algorithm makes more samples in Phase 2 than in Phase 1 because otherwise, it makes $O(\mathrm{ROPT}(f,\epsilon)\log(1/\epsilon))$ samples and we are done. We apply Propostion 8 to prove this lemma. Let Y be the set of y_i's, let $Z^{(0)}$ be the set of points in L at the end of Phase 1 and let $t_0 = 550 \cdot \mathrm{ROPT}^{4/3}$. We have $A = \sum_{i=1}^{m-1} AL(y_i, y_{i+1}) \leq 18 \cdot \mathrm{ROPT} \cdot \epsilon$. By Proposition 8, after t_0 samples, the total AL^2 will be at most $\frac{4608 \cdot (6 \cdot \mathrm{ROPT})^2 \cdot (18 \cdot \mathrm{ROPT})^2 \epsilon^2}{550^3 \mathrm{ROPT}^4} \leq \epsilon^2/3$ so the algorithm will stop after t_0 steps. □

The following proposition shows that as our algorithm samples, the total squared area looseness declines as the cube of the number of samples. We prove it by associating a number with each interval that is an upper bound on its area looseness. We then show that these numbers are within a factor of four of each other and use this to show that that the sum of their squares decreases as the cube of the number of samples.

Proposition 8. *Let* $Y = \{y_1, \ldots, y_m\}$ *with* $0 = y_1 < \cdots < y_m = 1$*, and let* $A = \sum_{i=1}^{m-1} AL(y_i, y_{i+1})$*. Consider the sequence* $Z^{(0)}, Z^{(1)}, Z^{(2)}, \ldots$ *of sets of samples where* $Z^{(0)} \supseteq Y$ *is an arbitrary superset of* Y *and, for each* $t \geq 1$*,* $Z^{(t)} = Z^{(t-1)} \cup \{z^{(t)}\}$ *where* $z^{(t)}$ *is the midpoint* $(x^{(t)} + y^{(t)})/2$ *of the interval* $(x^{(t)}, y^{(t)})$ *of* $Z^{(t-1)}$ *with the largest area looseness* $AL(x^{(t)}, y^{(t)})$*. Then, for any* $t_0 \geq |Z_0|$*,* $\sum_{(x,y)\in\mathcal{I}(Z^{(t)})} AL^2(x,y) \leq (4608 m^2 A)/t_0^3$.

The upper bound follows immediately from the two lemmas we have shown.

Theorem 3. *On problem instance* (f, ϵ) *algorithm* LIPSCHITZ-MC-INTEGRATE *performs* $O(\mathrm{ROPT}^{4/3}(f, \epsilon) + \mathrm{ROPT}(f, \epsilon)\log(1/\epsilon))$ *samples.*

7 Randomized Lower Bounds

We first show that Lemma 1 is actually a tight (to within a constant factor) lower bound on ROPT by proving the following upper bound.

Lemma 4. *Given a Lipschitz function* f*, there is a set of points* $0 = x_1 < x_2 < \cdots < x_k = 1$ *such that for* $1 \le i \le k-2$*,* $AL(x_i, x_{i+1}) = \epsilon/4$*, and* $AL(x_{k-1}, x_k) \le \epsilon/4$*. Furthermore,* $\mathrm{ROPT}(f, \epsilon) \le 2k - 1$*.*

The above lemma implies that deterministic algorithms are not very powerful relative to ROPT. For instance, if $f(x) = 0$ for all x, $\mathrm{ROPT}(f, \epsilon) = O(\epsilon^{-1/2})$ by Lemma 4, but DOPT is $\Theta(\epsilon^{-1})$. Therefore every deterministic algorithm requires $\Omega(\mathrm{ROPT}^2)$ samples on some instances.

Theorem 4. *Given an* $\epsilon > 0$ *and an integer* k *such that* $0 < k < \epsilon^{-1}/2$*, there is a family of problem instances such that* $\mathrm{ROPT} = O(k)$ *on every member on the family, but any algorithm requires* $\Omega(k^{4/3} + k\log(1/\epsilon))$ *samples in expectation on some member of that family.*

A simple corollary shows that the nonadaptive method in [2] is optimal.

Corollary 2. *Any algorithm requires* $\Omega(\epsilon^{-2/3})$ *samples on some problem instance.*

8 Conclusion

We gave optimally adaptive deterministic and randomized algorithms for LIPSCHITZ-INTEGRATION. To simplify the analysis, we have been lax with constant factors in the randomized algorithm and the related proofs. Thus, it is possible to improve both the algorithm's performance and its analysis by constant factors.

A more interesting open problem is to design adaptive algorithms for definite integration over two or higher-dimensional domains or to prove that good adaptive algorithms do not exist. Although simple Monte Carlo methods readily extend to higher dimensions, designing and analyzing adaptive algorithms seems difficult.

References

1. Werschulz, A.G.: An overview of information-based complexity. Technical Report CUCS-022-02, Computer Science Department, Columbia University (2002)
2. Barabesi, L., Marcheselli, M.: A modified monte carlo integration. International Mathematical Journal **3**(5) (2003) 555–565
3. Davis, P.J., Rabinowitz, P.: Methods of Numerical Integration. second edn. Academic Press, San Diego (1984)

4. Traub, J., Wasilkowski, G., Woźniakowski, H.: Information-Based Complexity. Academic Press, New York (1988)
5. Demaine, E.D., López-Ortiz, A., Munro, J.I.: Adaptive set intersections, unions, and differences. In: Proceedings of the 11th Annual ACM-SIAM Symposium on Discrete Algorithms, San Francisco, California (2000) 743–752
6. Fagin, R., Lotem, A., Naor, M.: Optimal aggregation algorithms for middleware. Journal of Computer and System Sciences **66**(4) (2003) 614–656
7. Biedl, T., Brejova, B., Demaine, E.D., Hamel, A.M., López-Ortiz, A., Vinař, T.: Finding hidden independent sets in interval graphs. Theoretical Computer Science **310**(1–3) (2004) 287–307
8. Hansen, P., Jaumard, B., Lu, S.H.: On the number of iterations of piyavskii's global optimization algorithm. Mathematics of Operations Research **16**(2) (1991) 334–350
9. Piyavskii, S.: An algorithm for finding the absolute extremum of a function. USSR Computational Mathematics and Mathematical Physics **12** (1972) 57–67
10. Baran, I., Demaine, E.D.: Optimal adaptive algorithms for finding the nearest and farthest point on a parametric black-box curve. In: Proceedings of the 20th Annual ACM Symposium on Computational Geometry, Brooklyn, NY (2004) To appear.
11. Baran, I.: Adaptive algorithms for problems involving black-box lipschitz functions. Master's thesis, Massachusetts Institute of Technology, Cambridge, Massachusetts (2004) At `http://www.mit.edu/~ibaran/papers/mthesis.{pdf,ps}`.

Classical Computability and Fuzzy Turing Machines

Benjamín René Callejas Bedregal[1] and Santiago Figueira[2,*]

[1] Federal University of Rio Grande do Norte, Department of Informatics and Applied Mathematics, Laboratory of Logic and Computational Intelligence, Natal-RN, Brazil

[2] University of Buenos Aires, Department of Computer Science, FCEyN, Buenos Aires, Argentina

bedregal@dimap.ufrn.br, sfigueir@dc.uba.ar

Abstract. We work with fuzzy Turing machines (FTMs) and we study the relationship between this computational model and classical recursion concepts such as computable functions, r.e. sets and universality. FTMs are first regarded as *acceptors*. It has recently been shown in [23] that *these machines have more computational power than classical Turing machines.* Still, the context in which this formulation is valid has an unnatural implicit assumption. We settle necessary and sufficient conditions for a language to be r.e., by embedding it in a fuzzy language recognized by a FTM and we do the same thing for difference r.e. sets, a class of "harder" sets in terms of computability. It is also shown that there is no universal FTM. We also argue for a definition of computable fuzzy function, when FTMs are understood as *transducers.* It is shown that, in this case, our notion of computable fuzzy function coincides with the classical one.

1 Introduction

Classical computability admits several but equivalent models. Still, the fuzzification of these models may imply different and nonequivalent concepts of fuzzy computability. Even the same model can be fuzzified in several ways. These facts turn this subject very complex and interesting. A precursor of fuzzy computability was the proper founder of fuzzy set theory, Lotfi Zadeh, who in [24] defines the notion of fuzzy algorithm based on a fuzzification of Turing machines and Markov algorithms. However, that work was not deep enough in the recursion theoretical aspects of the mentioned models. Lately, Lee and Zadeh in [12] follow the same setting and Santos in [17, 18] proves that these two fuzzy models are equivalent. Unfortunately the research in this subject was not continued for more than a decade, revisited only in the works of Harkleroad [9] (for other works related to this topic, see for example [3, 2, 14, 7, 15]). More recently, with

* Partially supported by a grant of Fundación YPF.

J.R. Correa, A. Hevia, and M. Kiwi (Eds.): LATIN 2006, LNCS 3887, pp. 154–165, 2006.

the increasing interest in extrapolating Church-Turing thesis considering other aspects (for example interactions [8], real values [21], quantum universe [5], etc.), the research on fuzzy computability has gain new strength, mainly because it was shown by Wiedermann [22, 23] that it is possible to solve the halting problem (more precisely, it is possible to accept r.e. sets and co-r.e. sets) in a class of fuzzy Turing machines.

Section 2 are preliminaries and section 3 is devoted to present nondeterministic Turing machines, and fix notation to be extended later to the fuzzy context.

In section 4.1 we work with fuzzy Turing machines, when regarded as acceptors. We analyze carefully Wiedermann's statement mentioned above about the computational power of fuzzy Turing machines. We state it in a more rigorous manner and in Theorem 2 we impose necessary and sufficient conditions for a set to be r.e. in terms of associated fuzzy languages recognizable by fuzzy Turing machines. We also show that Wiedermann's statement is not completely correct since there are fuzzy Turing machines which could also "recognize" (in the sense used by Wiedermann) difference r.e. sets (and it is well known that these sets may be more complex than the r.e. or co-r.e. ones). In Theorem 3 we characterize the class of difference r.e. sets in terms of associated fuzzy languages recognized by fuzzy Turing machines.

In section 4.2 we deal with the recursive theoretical notion of universality. Theorem 4 shows that there is no universal fuzzy machine for the class of all fuzzy Turing machines. Some other narrower classes of fuzzy machines are considered for which there are fuzzy universality.

In section 4.3, we change the optic and we regard fuzzy Turing machines as transducers, that is as fuzzy devices computing functions, instead of just recognizing languages. We argue for a definition of fuzzy computable function, when this optic is taken, and in Theorem 5 we show that our proposed notion coincides with the classical one.

2 Elements of Fuzzy Theory

Let I be the unitary closed interval, i.e. $[0,1]$. A fuzzy set A in an universe U_A (a classical set) is characterized by its membership degree function

$$\mu_A : U_A \rightarrow I \ .$$

Thus, for each $x \in U_A$, $\mu_A(x)$ provides the belonging degree of the element x in the fuzzy set A. For each fuzzy set A, we define their support set as

$$S(A) = \{a \in U_A : \mu_A(a) > 0\}$$

and their crisp set as

$$C(A) = \{a \in U_A : \mu_A(a) = 1\} \ .$$

2.1 t-Norms

Triangular norms, or simply *t-norms*, were introduced by Schweizer and Skłar [19] with the intention of modelling the distance of probabilistic metric spaces. Moreover, Alsina, Trillas and Valverde [1] showed that this notion is adequate to model the conjunction in fuzzy logics or equivalently the intersection of fuzzy sets. A t-norm on I is any commutative and associative mapping $T : I \times I \to I$ such that 1 is the neutral element and is monotonic w.r.t. the natural order on I. Sometimes t-norms will be used in infix notation instead of the functional form. In this case, we will usually write the symbol $*$. Classical examples of t-norms are the following: $G(x,y) = \min\{x,y\}$ (Gödel t-norm), $P(x,y) = xy$ (product t-norm) and $L(x,y) = \max\{x+y-1, 0\}$ (Lukasiewicz t-norm).

An element $z \in (0,1)$ is said a *zero divisor* of a t-norm $*$ if there exists $y \in (0,1)$ such that $y * z = 0$. For example, each $z \in (0,1)$ is a zero divisor of L.

2.2 Fuzzy Functions

Zimmerman [25] considers several ways of fuzzifying the notion of function. Some other notions of fuzzy functions can also be found in [4, 15, 16].

In this article we propose the following one: Let A and B by fuzzy sets. A classical partial function $f : U_A \to U_B$ is a *fuzzy partial function* from A into B, if

$$\forall x \in U_A, f(x) \uparrow \text{ or } \mu_B(f(x)) \leq \mu_A(x). \tag{1}$$

This definition of fuzzy function differs from the one of Dubois and Prade ([25], Definition 7-1), which is based on the extension principle –we use $\leq$ in (1) whenever Dubois and Prade use $\geq$. Moreover, we consider partial functions instead of total functions. Our choice will be fully understood when we define the fuzzy function computed by a fuzzy Turing machine, in section 4.3.

Notice that Dubois and Prade's fuzzy function allows us to map an element with degree 0 –and therefore fully out of the set–, to an element with degree 1 –hence completely inside the set. According to our definition, whenever the input has degree 0, the output will also have degree 0. However, when the input has a significant degree (i.e. a degree greater than 0), then the output will not necessarily have a significant degree.

Let f be a fuzzy partial function. We define $S(f) : S(A) \to S(B)$ as the *support* of f, and $C(f) : C(A) \to C(B)$ as the *crisp* of f in the following way:

$$S(f)(x) = \begin{cases} f(x) & \text{if } \mu_B(f(x)) > 0; \\ \uparrow & \text{otherwise.} \end{cases} \qquad C(f)(x) = \begin{cases} f(x) & \text{if } \mu_B(f(x)) = 1; \\ \uparrow & \text{otherwise.} \end{cases}$$

3 Nondeterministic Turing Machines

In the literature, there are diverse definitions of nondeterministic Turing machines, NTM for short, and all of them are equivalent (see for example [10, 11, 13]). We use the following definition: A NTM is a septuple $\mathcal{T} = \langle Q, \Sigma, \Gamma, \delta, q_0, \Box, F \rangle$

where Q is a set of states, Σ is the input alphabet, Γ is the tape alphabet, $q_0 \in Q$ is the starting state, $\Box \in \Gamma$ is the blank symbol, $F \subseteq Q$ is the set of final states and $\delta \subseteq Q \times \Gamma \times Q \times \Gamma \times \{R, L\}$ is the set of instructions, i.e. the set of "next move" relation.

We will use the following string functions: $head(w)$ returns the leftmost symbol of w, $head^R(w)$ returns the rightmost symbol of w, $tail(w)$ returns the string w without its leftmost symbol and $tail^R(w)$ returns the string w without its rightmost symbol.

An *instantaneous description* of a NTM, ID for short, is a triple (u, q, v) meaning that the tape content is the string uv, the current state is q and the head is pointing at the leftmost symbol of v. For notational simplicity we will omit the parentheses and comma of IDs. A *valid move* from an ID uqv into an ID $u'pv'$ in the NTM $\mathcal{T}$, denoted by $uqv \vdash_{\mathcal{T}} u'pv'$, occurs whenever

$$\exists(q, head(v), p, b, R) \in \delta \text{ such that } u' = u \circ b \text{ and } v' = tail(v), \text{ or}$$

$$\exists(q, head(v), p, b, L) \in \delta \text{ such that } u = u' \circ head(v') \text{ and } v' = head^R(u) \circ b \circ tail(v).$$

As usual, an ID $u'pv'$ is reached from an ID uqv, denoted by $uqv \vdash^*_{\mathcal{T}} u'pv'$, if $uqv = u'pv'$ or there exists an ID $u''rv''$ such that

$$uqv \vdash_{\mathcal{T}} u''rv'' \text{ and } u''rv'' \vdash^*_{\mathcal{T}} u'pv' .$$

When a NTM $\mathcal{T}$ is regarded as an *acceptor*, we say that the string $w \in \Sigma^*$ is accepted by $\mathcal{T}$ if $q_0 w \vdash^*_{\mathcal{T}} uq_f v$ for some $u, v \in \Gamma^*$ and $q_f \in F$. As usual the *language accepted* by a NTM $\mathcal{T}$, denoted by $L(\mathcal{T})$, is the set of all strings accepted by $\mathcal{T}$.

When NTMs are understood as *transducers*, things change a little, so it is worth making a short digression in this point. For the same input, a NTM can give more of one output, hence it is natural to ask which one is the function computed by them. Some authors (for example [6]) consider that a NTM computes a function from Σ^* (the set of possible inputs) into $\mathcal{P}(\Gamma^*)$ (the powerset of possible outputs). Following this point of view, we would have a computability notion for functions with countable domain and uncountable rang, which go beyond Church-Turing thesis.

Other alternatives also have some problems. Therefore, we agree with Linz when he says in [13]: "Since it is not clear what role nondeterminism plays in computing functions, nondeterministic automata are usually viewed as acceptors." Hence, we believe that NTMs must only be considered as acceptors.

4 Fuzzy Turing Machines

Zadeh [24], Lee [12] and Santos [17] introduced the model of Fuzzy Turing machines and the languages accepted by this kind of machines, i.e. a class of fuzzy languages. Classical languages are linked to fuzzy languages through the support and crisp part of a fuzzy set. It turns out that this fuzzy machine model is computationally too powerful: in [23], Wiedermann claims that, in fact, its nondeterministic version accepts non recursively enumerable languages and that they can solve undecidable problems (these assertions will be fully analysed in section 4.1). On the other hand, the model is too restrictive from a fuzzy logic point of view, since it only considers the Gödel t-norm. The idea of this fuzzy Turing machine is to establish an *uncertainty degree* for the acceptance of a given string or, analogously, the *membership degree* of the string to the language. In order to compute this degree from individual degrees, a composition on the t-norm evaluation is used. Wiedermann [22, 23] introduced the class of fuzzy Turing machines as a fuzzy extension of the nondeterministic Turing machines, where each transition has a membership degree associated to it. In this case, he worked with arbitrary t-norms for the evaluation. We consider this same kind of fuzzy Turing machines:

Definition 1. *A* fuzzy Turing machine, FTM *for short, is a triple* $\mathcal{F} = \langle \mathcal{T}, *, \mu \rangle$ *where* $\mathcal{T} = \langle Q, \Sigma, \Gamma, \delta, q_0, \Box, F \rangle$ *is a* NTM, $*$ *is a t-norm and* μ *is a map which assigns a membership degree to each tuple in the "next move" relation* δ, *i.e.* $\mu : \delta \to I$.

An *instantaneous description* (ID) of a FTM $\mathcal{F}$ is a pair (uqv, d) where uqv is a classical ID for a Turing machine, i.e. uv is the string in the tape, the head is pointing to the leftmost symbol of v, the current state is q and d is the membership degree accumulated up to this moment.

A *valid move* from an ID (uqv, d) into and ID $(u'pv', d')$, denoted by $(uqv, d) \vdash_{\mathcal{F}} (u'pv', d')$, occurs whenever $uqv \vdash_{\mathcal{T}} u'pv'$ and

$$d' = \begin{cases} d * \mu(q, head(v), p, head^R(u'), R) & \text{if } tail^R(u') = u; \\ d * \mu(q, head(v), p, head(tail(u')), L) & \text{if } tail^R(u) = u'. \end{cases}$$

As with the NTM case, an ID $(u'pv', d')$ is reached from an ID (uqv, d), denoted by $(uqv, d) \vdash^*_{\mathcal{F}} (u'pv', d')$, if $(uqv, d) = (u'pv', d')$ or there exists an ID $(u''rv'', d'')$ such that $(uqv, d) \vdash_{\mathcal{F}} (u''rv'', d'')$ and $(u''rv'', d'') \vdash^*_{\mathcal{F}} (u'pv', d')$.

4.1 Fuzzy Turing Machines as Acceptors

The *degree of acceptance* in a FTM $\mathcal{F}$ of a string w is

$$\deg_{\mathcal{F}}(w, k) = \max\{d \in I : (q_0 w, k) \vdash^*_{\mathcal{F}} (uq_f v, d) \text{ for some } q_f \in F\} .$$

and $\deg_{\mathcal{F}}(w, k)$ becomes undefined when there is no accepting path of $\mathcal{F}(w)$. When $k = 1$ we will omit it and we will write $\deg_{\mathcal{F}}(w)$.

Since a language is just a set of strings, a natural definition for fuzzy language is "a fuzzy set of strings". Thus, the fuzzy language *accepted* by a FTM $\mathcal{F}$ is

$$L(\mathcal{F}) = \{(w, \deg_{\mathcal{F}}(w)) : w \in \Sigma^* \wedge (q_0 w, k) \vdash^*_{\mathcal{F}} (uq_f v, d) \text{ for some } q_f \in F\} .$$

In [22, 23], Wiedermann claims that fuzzy Turing machines can solve undecidable problems and that the languages accepted by these machines (when we consider a computable t-norm) are exactly the union of r.e. sets and co-r.e. sets.

Evidently there is some abuse in this terminology, since r.e. sets are *ordinary* languages and the languages accepted by fuzzy Turing machines are *fuzzy* languages. Hence, there is some kind of implicit *fuzzification* when he says that fuzzy Turing machines accept nonrecursive r.e. sets. This fuzzification is some kind of codifying the membership of an element to a set, by exploiting the degree of acceptance.

To explain what is the exact assertion of Wiedermann, let us first define a special way of fuzzifying ordinary sets into fuzzy sets. For any language A and for rationals a and b ($a, b \in I$) we define the following fuzzification of the set A:

$$F_A(a,b) = \{(w,a)\colon w \in A\} \cup \{(w,b)\colon w \notin A\} .$$

What Wiedermann actually does in the proof of Theorem 3.1 [23] is to show that for any r.e. set A, there is a FTM $\mathcal{F}$ which accepts the fuzzy language $F_A(1,b)$, where b is any rational such that $0 \leq b < 1$. In fact, it is not difficult to see that there is a FTM which accepts $F_A(a,b)$ for any fixed a and b with $0 \leq b < a \leq 1$. Even more, we can prove the following strongest result:

Theorem 1. *Let $A \subseteq \Sigma^*$ be any set and let a, b be rationals such that $0 \leq b < a \leq 1$. A is r.e. iff there is a* FTM *which accepts the fuzzy language $F_A(a,b)$.*

Proof. ($\Rightarrow$) Let A_s be the recursive approximation of A, i.e. $A_s(w) \in \{0,1\}$ and $A_0(w) = 0$ for any $s \in \mathbb{N}$ and $w \in \Sigma^*$. Besides, $A_s(w) \leq A_{s+1}(w)$, so that $A_s(w)$ changes at most one time –from 0 to 1– when we increase s, and $w \in A$ iff $\exists s\, A_s(w) = 1$. Let $\mathcal{F}$ be the FTM which on input w, it has a nondeterministic branch starting from state q_0:

– $\mathcal{F}$ passes from q_0 to the final state q_f via a transition with degree b, and
– $\mathcal{F}$ passes from q_0 to a procedure which scans $A_0(w), A_1(w), \ldots$ until it finds some t such that $A_t(w) = 1$ (all this procedure is carried on with transitions of degree 1). If this ever happens then $\mathcal{F}$ goes to the final state q_f via a transition with degree a and otherwise it keeps on searching (so it never reaches the final state).

Now, if $w \in A$ then there is a least s such that $A_s(w) = 1$, so there will be two accepting paths in $\mathcal{F}$: the one coming from the first nondeterministic branch, with accepting degree b, and the one coming from the second nondeterministic branch, with accepting degree a. Since $a > b$ then $(w,a) \in L(\mathcal{F})$. On the other hand, if $w \notin A$ then there is only one accepting path in the execution of $\mathcal{F}$ –the one coming from the first nondeterministic branch–, and hence $(w,b) \in L(\mathcal{F})$.

($\Leftarrow$) Suppose $\mathcal{F}$ is a FTM which accepts $F_A(a,b)$. The following procedure gives A_s, an r.e. approximation of A: search all the execution paths of $\mathcal{F}(w)$. If by stage s we find that $\mathcal{F}(w)$ arrives to a final state with accepting degree a then we let $A_s(w) = 1$. □

Here, the fuzzification used to interpret an ordinary language into a fuzzy language consists in defining w in the accepted language of $\mathcal{F}$ with membership degree a, for every $w \in A$; and w with membership degree b, for every $w \notin A$. It is worth noting that this result only applies when this particular way of fuzzifying r.e. sets of strings is used –that is, when working with $F_A(a,b)$. Although one could intuitively think that if there is a FTM which accepts $F_A(a,b)$, then there should be another FTM which accepts a "simple" transformation of $F_A(a,b)$, such as $F_A(b,a)$, the following proposition shows that this is not the case.

Proposition 1. *Let $A \subseteq \Sigma^*$ be a nonrecursive r.e. set and let a,b be rationals such that $0 \leq b < a \leq 1$, then the language $F_A(b,a)$ is not accepted by any* FTM.

Proof. Suppose A is as in the hypothesis and assume that there is a FTM $\mathcal{F}$ which accepts $F_A(b,a) = \{(w,b)\colon w \in A\} \cup \{(w,a)\colon w \notin A\}$. Then there would be an effective decision procedure for testing the membership of any string w to the set A, contradicting the assumption that A is nonrecursive. Here is the procedure: In parallel, run the enumeration of A (which exists by hypothesis) and simulate all the execution paths of $\mathcal{F}(w)$. Eventually we will find that either $w \in A$, or we find an accepting path of $\mathcal{F}(w)$ with membership degree a. Since $a > b$, then the path that we have found has maximum degree, and hence $w \notin A$. □

The above proposition shows that the fuzzification used by Wiedermann is intrinsically linked to the fact that A is r.e.; the result is not independent of the fuzzification used. Indeed, when Wiedermann [23] considers co-r.e. sets A, he changes the fuzzification, and in this case, he shows that there is a FTM which accepts $F_A(b,1)$, for any fixed rational $b \in [0,1)$. Hence, one has to be careful when saying that "languages *accepted* by FTM with computable t-norm coincide with the class of r.e. sets union co-r.e. sets": the notion of *acceptance* here involves a particular fuzzification, which differs in the r.e. case and the co-r.e. case.

We obtain the following corollaries from Theorem 1 and Proposition 1. Both follow immediately from the observation that $F_{\overline{A}}(b,a) = F_A(a,b)$.

Corollary 1. *Let $A \subseteq \Sigma^*$ be a set and let a,b be rationals such that $0 \leq b < a \leq 1$. A is co-r.e. iff there is a* FTM *which accepts the fuzzy language $F_A(b,a)$.*

Thus, A is recursive if and only if there are FTMs accepting the languages $F_A(a,b)$ and $F_A(b,a)$, respectively.

Corollary 2. *Let $A \subseteq \Sigma^*$ be a nonrecursive co-r.e. set and let a,b be rationals such that $0 \leq b < a \leq 1$, then the language $F_A(a,b)$ is not accepted by any* FTM.

It is not necessary to fix the values of the rationals a and b in the above results. In fact, using the same strategy than in Theorem 1, it is not difficult to prove:

Theorem 2. *A is r.e. if and only if there is some rational $r \in (0,1)$ and some* FTM *$\mathcal{F}$ such that $\deg_{\mathcal{F}}(w) > r$ iff $w \in A$.*

Proof. ($\Rightarrow$) Follows directly from Theorem 1.

($\Leftarrow$) Observe that we can simulate all the execution paths of $\mathcal{F}(w)$ in parallel. Whenever we see that $\mathcal{F}$ reaches a final state via an execution path with acceptance degree $d > r$, then $\deg_{\mathcal{F}}(w) \geq d > r$ and hence it is safe to assert $w \in A$. This procedure informally describes an effective r.e. approximation of A. □

So far we have been working with special fuzzifications of r.e. sets (and symmetrically, with co-r.e. sets). What about other sets which are more complex in terms of computability theory?

A set A is *difference r.e. (d.r.e.)* if $A = B \setminus C$, for some r.e. sets B and C. If A is d.r.e., then there is a recursive approximation of A, call it A_s, such that $\#\{s\colon A_s(w) \neq A_{s+1}(w)\} \leq 2$, $\lim_{s\to\infty} A_s(w) = A(w)$, and $A_0(w) = 0$ for all w. In other words, $A_s(w)$ starts in 0, it can only change to 1 and maybe go back to 0, when increasing s. This follows trivially from the definition of d.r.e. For more details, see [20].

It is well-known that there are d.r.e. sets which are neither r.e. nor co-r.e. Thus, we know that we cannot make a fuzzification of every d.r.e. set in the same way that we did it before. However, we can fuzzificate them in another way.

Theorem 3. *A is d.r.e. if and only if for any two rationals a and b, $0 \leq b < a \leq 1$, there is some* FTM *$\mathcal{F}$ such that $b < \deg_{\mathcal{F}}(w) < a$ iff $w \in A$.*

Proof. ($\Rightarrow$) Suppose A_s is a recursive approximation of A, i.e. $A_s(w)$ changes at most two times when $s \to \infty$. Imagine the FTM $\mathcal{F}$ which on input w, it starts from the initial state q_0 and makes the following three nondeterministic branches:

- With degree 0, $\mathcal{F}(w)$ goes to the accepting state q_f.
- With degree $\frac{a+b}{2}$, $\mathcal{F}(w)$ goes to a procedure which searches the least stage s such that $A_s(w) = 1$. Once this happens it passes to the accepting state q_f. If that never happens, it continues searching and it gets undefined.
- With degree 1, $\mathcal{F}(w)$ goes to a procedure which searches least s and t such that $s < t$ and $A_s(w) = 1$ and $A_t(w) = 0$. Once this happens it passes to the accepting state q_f. If that never happens, it continues searching and it gets undefined.

Now, suppose $w \in A$. Then there is a least s such that $A_s(w) = 1$. By the properties of A_s, we have that $\forall t \geq s\ A_t(w) = 1$. Then there is no accepting path via the third branch. The only two accepting paths transit via the first one, with accepting degree 0, and the second one, with accepting degree $\frac{a+b}{2} > 0$. Hence $(w, \frac{a+b}{2}) \in L(\mathcal{F})$. On the other hand, suppose that $w \notin A$. There are two possibilities: either $A_s(w)$ does not change or it changes two times. In the former case, the only accepting path goes via the first nondeterministic branch and hence $(w, 0) \in L(\mathcal{F})$; in the latter, the three are accepting paths, but the one with maximum degree is the third one, so $(w, 1) \in L(\mathcal{F})$.

($\Leftarrow$) Suppose a, b and $\mathcal{F}$ satisfy the conditions of this theorem. We simulate $\mathcal{F}(w)$ in stages: define $A_0(w) = 0$ and

$$A_{s+1}(w) = \begin{cases} 0 & \text{if by stage } s\text{, all accepting paths of } \mathcal{F}(w) \text{ have degree} \leq b; \\ 1 & \begin{array}{l}\text{if by stage } s\text{, there is an accepting path of } \mathcal{F}(w) \text{ with} \\ \text{degree} \in (b, a) \text{ and no accepting path with degree} \geq a;\end{array} \\ 0 & \begin{array}{l}\text{if by stage } s\text{, there is an accepting path of } \mathcal{F}(w) \text{ with} \\ \text{degree} \geq a.\end{array} \end{cases}$$

Clearly, the approximation $A_s(w)$ is recursive and changes at most two times, when $s \to \infty$ and hence A is d.r.e. Indeed, if $w \in A$ then eventually, at some stage s, we will find an accepting path of $\mathcal{F}$ with accepting degree $\in (b, a)$ and there cannot be any accepting path of $\geq a$. Then $\forall t \geq s\ A_t(w) = 1$. Otherwise, if $w \notin A$ then either all the accepting paths of $\mathcal{F}(w)$ have degree $\leq b$ or there is some accepting path of degree $\geq a$: in both cases we will have that there is some s such that $\forall t \geq s\ A_t(w) = 0$. □

4.2 Universal Fuzzy Turing Machines

In classical recursion theory, we have the notion of *universal machine*: in short a machine capable to simulate the behavior of every other machine. If $(\mathcal{M}_i)_{i \in \mathbb{N}}$ is an enumeration of all deterministic Turing machines (when seen as transducers), then $\mathcal{U}$ is said *universal* when $\mathcal{M}_i(w) \downarrow$ iff $\mathcal{U}(\langle w, i \rangle) \downarrow$ and if $\mathcal{M}_i(w) \downarrow$ then $\mathcal{M}_i(w) = \mathcal{U}(\langle w, i \rangle)$ (here $\langle \cdot, \cdot \rangle \colon \mathbb{N} \times \Sigma^* \to \Sigma^*$ is the usual pairing function). We also have a universal machine, when thinking of acceptors. In this case, $(\mathcal{M}_i)_{i \in \mathbb{N}}$ would correspond to an enumeration of all r.e. sets (identifying the domain of $\mathcal{M}_i$ with the i-th r.e. set) and $\mathcal{U}$ is said universal when $\mathcal{M}_i(w) \downarrow$ iff $\mathcal{U}(\langle w, i \rangle) \downarrow$.

Let $\mathcal{C}$ be the class of all FTMs with rational (or finitely representable, or even computable) degree membership and computable t-norm, i.e. fuzzy machines $\mathcal{F} = \langle \mathcal{T}, *, \mu \rangle$ where μ is computable and the range of μ is $\mathbb{Q} \cap I$ (or a set of finitely representable numbers in I). Since all the elements of each FTM are finitely representable, we can assign Gödel numbers to each FTM, and obtain $(\mathcal{F}_i)_{i \in \mathbb{N}}$, an enumeration of $\mathcal{C}$.

Following the notion of universality for classical computability, a *fuzzy universal machine* (regarded as an acceptor) $\mathcal{U}_F$ for the class $\mathcal{C}$ would be a special fuzzy machine with the ability to simulate the behavior of any other fuzzy machine in $\mathcal{C}$, that is $\mathcal{U}_F(\langle i, w \rangle) = \mathcal{F}_i(w)$. This means that for each $i \in \mathbb{N}$ and $w \in \Sigma^*$:

1. $\mathcal{F}_i(w) \downarrow$ iff $\mathcal{U}_F(\langle i, w \rangle) \downarrow$, and
2. if $\mathcal{F}_i(w) \downarrow$ then $\deg_{\mathcal{F}_i}(w) = \deg_{\mathcal{U}_F}(\langle i, w \rangle)$.

Although one could think that, as in the classical scenario, there should be such $\mathcal{U}_F$, the following result refutes the idea:

Theorem 4. *There is no universal fuzzy machine for the class $\mathcal{C}$.*

Proof. Suppose $\mathcal{U}_F = \langle \mathcal{T}_U, *, \mu \rangle$ where $\mathcal{T}_U = \langle Q, \Sigma, \Gamma, \delta, q_0, \square, F \rangle$ is a FTM as described above. Obviously, any computational path $t_1, \ldots, t_n$ of $\mathcal{U}_F$ ($t_i \in \delta$) will have degree $\mu(t_1) * \ldots * \mu(t_n) \leq 1$. Let

$$d = \max\{\mu(t) \colon w \in \Sigma^* \wedge t \in \delta \wedge \mu(t) < 1\} \cup \{0\} .$$

Any accepting path containing some $t \in \delta$ with $\mu(t) \leq d$ will have degree $\leq d$, hence $\mathcal{U}_F$ has no computational path with degree $\tilde{d} \in \mathbb{Q}$ such that $d < \tilde{d} < 1$. Now, let $\mathcal{F}$ be a FTM with Gödel number e such that $L(\mathcal{F}) = \{(w, \tilde{d}) \colon w \in \Sigma^*\}$. Clearly, $\mathcal{U}_F(\langle w, e \rangle) = \mathcal{F}(w)$, so $\mathcal{U}_F$ must accept $\langle w, e \rangle$ with membership degree $\tilde{d}$, and this is impossible. □

However, when we restrict ourselves to a smaller class, we still may have universality. Let $\mathcal{D}$ be a class of FTMs. We say that $\mathcal{U}$ is an universal FTM for the class $\mathcal{D}$, when $\mathcal{U}$ is able to simulate any other machine in $\mathcal{D}$, and $\mathcal{U} \in \mathcal{D}$.

For example, let $B \subset \mathbb{Q}$ and let $\mathcal{D}_B$ be the class of FTMs $\mathcal{F} = \langle \mathcal{T}, *, \mu \rangle$, where $\mathcal{T} = \langle Q, \Sigma, \Gamma, \delta, q_0, \Box, F \rangle$ is such that $\forall t \in \delta, \mu(t) \in B$. It is not difficult to see that if B is finite, there is a universal fuzzy machine for the class $\mathcal{D}_B$. Informally, if $B = \{b_1, \ldots, b_k\}$, this universal machine would have k special transitions $t_1, \ldots, t_k$ with $\mu(t_i) = b_i$, and will use them to actually pursue the degree of the simulated machine and input.

It is also interesting to observe that a class of FTMs such as $\mathcal{D}_B$, with finite B, is not the only situation where universality is admitted. For example, consider the product t-norm $P(x, y) = xy$ and $B' = \{2^{-i} : i \in \mathbb{N}^+\}$. We can see that there is a universal machine for the class $\mathcal{D}_{B'}$: A universal machine could have a unique special transition t with $\mu(t) = 1/2$ to actually obtain any number of B' by successive applications of the t-norm P.

Hence it is an interesting open question to characterize the class of FTMs which admit a universal machine.

4.3 Fuzzy Turing Machines as Transducers

We know that Turing machines have two roles: as a language acceptor machine and as a function computer (transducer). Hence, we can think of a FTM a as function computer, but with an additional membership degree. That is, it computes a fuzzy function from Σ^* into Γ^*, where the input as well as the output have a membership degree. Still, as mentioned at the end of section 3, NTMs as transducer, do not seem to be a reasonable approach, and therefore in this section we consider only deterministic FTM, denoted DFTM for short. Without loss of generality, we can assume that a deterministic Turing machine, DTM for short, has just a unique final state under which the machine halts when reached.

Let $\mathcal{F} = \langle \mathcal{T}, *, \mu \rangle$ be a DFTM. A fuzzy partial function $f : \Sigma^* \to \Gamma^*$ from the fuzzy set A into the fuzzy set B (i.e. Σ^* and Γ^* are the universes of A and B, respectively) is *computed* by $\mathcal{F}$ if f (when seen as a classical partial function) is computed by the DTM $\mathcal{T}$ and for each w, if $f(w)\downarrow$, then

$$\mu_B(f(w)) = \mu_A(w) * \mu(t_1) * \cdots * \mu(t_n) \tag{2}$$

where $t_1, \ldots, t_n$ is the computational path for $q_0 w \vdash_{\mathcal{T}}^* u q_f v$ with $uv = f(w)$ and q_f is the final state of $\mathcal{T}$. Clearly, a DFTM computes a fuzzy partial function for each fuzzification of Σ^*.

We say that a DFTM $\mathcal{F}$ S_**-computes* a partial function $f : \Sigma^* \to \Gamma^*$ if there exists a fuzzy partial function $\widetilde{f}$ computed by $\mathcal{F}$ such that $S(\widetilde{f}) = f$. Analogously, we say that a DFTM $\mathcal{F}$ C*-computes* a partial function $f : \Sigma^* \to \Gamma^*$ if there exists a fuzzy partial function $\widetilde{f}$ computed by $\mathcal{F}$ such that $C(\widetilde{f}) = f$.

Notice that the function S_*-computed by a DFTM $\mathcal{F}$ could change in case another t-norm is used, whereas the function C-computed by $\mathcal{F}$ is the same independently of the t-norm chosen.

Theorem 5. *Let $*$ be a t-norm without zero divisors and let $f : \Sigma^* \to \Gamma^*$ be a partial function. The following conditions are equivalent:*

1. *f is S_*-computable*
2. *f is C-computable*
3. *f is computable in the classical sense*

Proof. (1 ⇒ 2) Let $\mathcal{F} = \langle \mathcal{T}, *, \mu \rangle$ be a DFTM which S_*-computes f. Then, the DFTM $\mathcal{F}' = \langle \mathcal{T}, *, \mu' \rangle$ where for each $t \in \delta$,

$$\mu'(t) = \begin{cases} 1 & \text{if } \mu(t) > 0; \\ 0 & \text{otherwise.} \end{cases}$$

$\mathcal{F}'$ C-computes f, thanks to the non-existence of zero divisors of $*$.

(2 ⇒ 3) Let $\mathcal{F} = \langle \mathcal{T}, *, \mu \rangle$ be a DFTM which C-computes f, and let $\mathcal{T}' = \langle Q, \Sigma, \Gamma, \delta', q_0, \Box, F \rangle$ be the DTM obtained from $\mathcal{T}$ changing the transition relation by: $t \in \delta'$ iff $t \in \delta$ and $\mu(t) = 1$. Clearly, the function computed by $\mathcal{T}'$ is f.

(3 ⇒ 1) Let $\mathcal{T}$ be a DTM which computes f. Then, the DFTM $\mathcal{F} = \langle \mathcal{T}, *, \mu \rangle$, where

$$\mu(t) = \begin{cases} 1 & \text{if } t \in \delta; \\ 0 & \text{otherwise.} \end{cases}$$

S_*-computes (and also C-computes) f. □

Thus, in terms of classical computability, for t-norms without zero divisors, S_*-computability and C-computability are equivalent. Clearly, the same is valid for languages.

5 Final Remarks

The main goal of this paper is not to criticize Wiedermann's work, but rather to clear the context in which his result is valid. In this sense, we prove that considering the same kind of fuzzification the principal result of Wiedermann (Theorem 3.1 in [23]) is not valid. Other contributions are:

- To provide some results on the acceptation of d.r.e. languages via FTM. These sets might be more complex in terms of computability theory than r.e. and co-r.e. sets. In spite of this fact, FTMs can also embed this kind of sets in a fuzzy language (in the same way that Wiedermann embedded r.e. sets).
- To prove that it is not possible to achieve an universal fuzzy Turing machine. The difficulty comes when we try to simulate the degree of acceptance. It is important to notice that we are not trying to calculate the accepting degree as a written output. Instead, a universal fuzzy machine should genuinely copy the accepting degree of the simulated FTM, by using its own transitions.
- To provide some considerations on the notion of computability of functions by DFTMs and to prove that DFTMs have the same computational power than classical Turing machines (considering two ways of relating these concepts).

As further work, we pretend to establish a relationship between our results and the ones of Gerla in [7], who provides fuzzifications of several concepts of recursion theory –though some fuzzy notions do not coincide exactly with ours.

References

1. Alsina, C., Trillas E., Valverde, L.: On non-distributive logical connectives for fuzzy set theory. Busefal **3** (1980) 18–29
2. Biacino, L., Gerla, G.: Fuzzy subsets: A constructive approach. Fuzzy Sets and Systems **45** (1992) 161–168
3. Clares, B., Delgado, M.: Introduction of the concept of recursiveness of fuzzy functions. Fuzzy Sets and Systems **21** (1987) 301–310
4. Demirci, M.: Fuzzy functions and their applications. Journal of Mathematical Analysis and Appliactions **252(1)** (2000) 495–517
5. Deutsch, D.: Quantum theory, the Church-Turing principle and the universal quantum computer. Proc. Roy. Soc. London Ser. **A 400** (1985) 97–117
6. Fenner, S.A., Fortnow, L., Naik, A.V., Rogers, J.D.: Inverting onto functions. Information and Computation **186(1)** (2003) 90–103
7. Gerla, G.: Fuzzy Logic: Mathematical Tools for Approximate Reasoning. Springer-Verlag, Berlin Heidelberg New York (2001)
8. Goldin, D.Q., Smolka, S.A., Attie, P.C., Sonderegger, E.L.: Turing machines, transition systems and interaction. Information and Computation **194** (2004) 101–128
9. Harkleroad, L.: Fuzzy recursion, ret's and isols. Zeitschrift Fur Math. Logik und Grundlagen der Mathematik **30** (1984) 425–436
10. Harrison, M.A.: Introduction to formal language theory. Addison-Wesley publishing, Reading, Massachusetts (1978)
11. Hopcroft, J.E., Ullman, J.D.: Introduction to automata theory, languages and computation. Addison-Wesley publishing, Reading, Massachusetts (1979)
12. Lee, E.T., Zadeh, L.A.: Note on fuzzy languages. Information Sciences **1(4)** (1969) 421–434
13. Linz, P.: An Introduction to Formal Language and Automata. Jones and Bartlett Publisher 2001
14. Moraga, C.: Towards a Fuzzy Computability?. Mathware & Soft Computing **6** (1999) 163–172
15. Morales-Bueno, R., Conejo, R., Prez-de-la-Cruz, J.L., Triguero-Ruiz, F.: On a class of fuzzy computable functions. Fuzzy Sets and Systems **121** (2001) 505–522
16. Perfilieva, I.: Fuzzy function as an approximate solution to a system of fuzzy relation equations. Fuzzy Sets and Systems **147** (2004) 363–383
17. Santos, E.: Fuzzy algorithms. Information and Control **17** (1970) 326–339
18. Santos, E.: Fuzzy and probabilistic programs. Information Sciences **10** (1976) 331–335
19. Schweizer, B., Sklar, A.: Associative functions and abstract semigroups. Publ. Math. Debrecen **10** (1963) 69–81
20. Soare, R.: Recursively enumerable sets and degrees. Springer-Verlag, Berlin Heidelberg New York (1987)
21. Weihrauch, K.: Computable Analysis – An introduction. Springer Verlag, Berlin Heildelberg New York (2000)
22. Wiedermann, J.: Fuzzy Turing machines revised. Computating and Informatics **21(3)** (2002) 1–13
23. Wiedermann, J.: Characterizing the super-Turing computing power and efficiency of classical fuzzy Turing machines. Theoretical Computer Science **317** (2004) 61–69
24. Zadeh, L.A.: Fuzzy Algorithms. Information and Control **2** (1968) 94–102
25. Zimmermann, H.J.: Fuzzy Set Theory and its Applications. 4th edition. Kluwer Academic Publishers, Dorbrecht Boston London (2001)

An Optimal Algorithm for the Continuous/Discrete Weighted 2-Center Problem in Trees

Boaz Ben-Moshe, Binay Bhattacharya*, and Qiaosheng Shi

School of Computing Science, Simon Fraser University,
Burnaby B.C., V5A 1S6, Canada
{benmoshe, binay, qshi1}@cs.sfu.ca

Abstract. In this paper, an optimal algorithm to solve the continuous/discrete weighted 2-center problem is proposed. The method generalizes the "trimming" technique of Megiddo [5] in a nontrivial way. This result allows an improved $O(n \log n)$ time algorithm for the weighted 3-center and 4-center problems.

1 Introduction

The *p-center problem* is defined on a weighted undirected graph $G = (V, E)$, where $v \in V$ is associated with a non-negative weight w_v and $e \in E$ is associated with a non-negative length l_e. Let $A(G)$ denote the continuum set of points on the edges of G. $P_{x,y}$ denotes the *shortest path* in G from x to y, $x, y \in A(G)$, and $d(x, y)$ denotes the length of $P_{x,y}$. Let $S(X, G')$ denote the *maximum weighted distance from a set* $X : \{\alpha_1, \ldots, \alpha_p\}$ *to a subgraph* G', that is,

$$S(X, G') = \max_{v \in V(G')} \{w_v \cdot d(X, v)\}, \text{ where } d(X, v) = \min_{j=1,\ldots,p} d(\alpha_j, v).$$

The p-center problem is to determine a set X of p points in $A(G)$ so as to minimize $S(X, G)$. When all the weights w_v are equal to 1, we call it the *unweighted p-center problem*. When the p centers are restricted to be vertices of G, we call it *discrete p-center problem*. This continuous/discrete problem has been shown to be NP-hard on general graphs [4, 7].

Our study in this paper is restricted to tree graphs. Megiddo and Tamir [7] provided an $O(n \log^2 n \log \log n)$ procedure to solve the weighted p-center problem in tree graphs, which was improved to $O(n \log^2 n)$ by implementing the results by R. Cole [1]. For the discrete weighted p-center problem, it is also solvable in $O(n \log^2 n)$ [6]. In unweighted models, Frederickson [2] presented an $O(n)$ algorithm, where p can be variable.

In the special case of a path graph, $O(n)$ algorithms for the weighted models have already known. In fact, stronger results hold for this case. Suppose that

* Research of the second author was partially supported by MITACS and NSERC.

J.R. Correa, A. Hevia, and M. Kiwi (Eds.): LATIN 2006, LNCS 3887, pp. 166–177, 2006.

the nodes of the path are identified as points on the real line. The path topology then provides the ordering of these n points. However, even without this path topology, $O(n)$ algorithms are known for the continuous weighted 2 and 3 center problems on the real line. Moreover, by using the Helly property and implementing generalized linear programming (GLP), or LP-type approaches, randomized linear time algorithms can be obtained for the continuous weighted p-center problem and the discrete unweighted p-center problem, on the real line, for any fixed p [3].

The main result of this paper is a significant improvement of the upper bound of the continuous/discrete weighted p-center problem on a tree when $p = 2, 3$ and 4. We have proposed a linear-time algorithm for the weighted 2-center problem. Megiddo [5] used a "trimming" technique to solve the weighted 1-center problem in linear time. The problem of generalizing the trimming approach of Megiddo [5] to solve the p-center problem for $p > 1$ was open for a long time. In this paper we have used the interactions between the two centers to guide us in trimming the tree. As we will see that this generalization is non trivial. The improvement of the 2-center problem can then be utilized to provide better bounds for the 3-center and 4-center problems.

The paper is organized as follows. In Sect. 2, the properties of the weighted 2-center of a tree are established. These properties immediately give rise to an $O(n \log n)$ algorithm. Section 3 provides the main result of this paper - a linear-time algorithm to solve the continuous/discrete weighted 2-center problem in a tree. Section 4 briefly describes the improved upper bounds for the weighted p-center problem, $p = 3$ and 4 along with the conclusions.

2 An $O(n \log n)$ Algorithm

Let $T(V')$ be the induced subtree with vertex set $V' \subseteq V$. For a subtree T' of T, let $V(T'), E(T'), A(T')$ be the vertex set, the edge set and the continuum set of points on the edges of T', respectively. $\delta_{T'}(v)$ denotes the degree of v in T'.

Let $V_v(u)$ denote the set of vertices v' such that the vertex v lies on the simple path from the vertex u to v' ($v \neq u$). Let $T_v(u)$ denote the induced subtree rooted at v with the vertex set $V_v(u)$. See Fig. 1(a). A subtree T' is called a *real subtree* of T if the component $T \setminus T'$ is connected. We denote by ${}^-T'$ the subtree $T \setminus T'$. The vertex of a real subtree T' closest to ${}^-T'$ is called *the root* of T' and the edge linking T' and ${}^-T'$ is called *the root edge* of T'. For example, $T_1, \ldots, T_7$ in

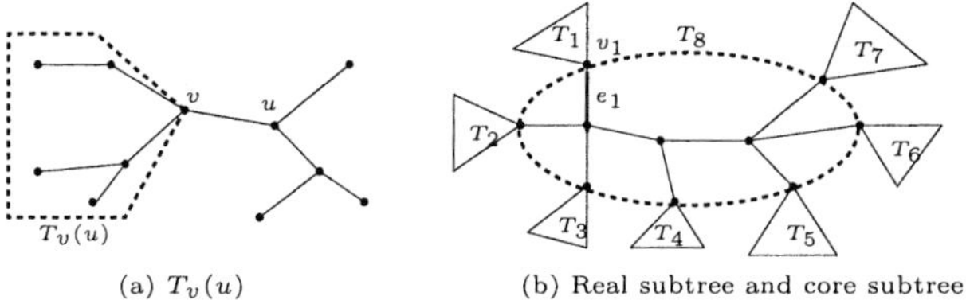

Fig. 1. Examples for $T_v(u)$, real subtree, and core subtree

Fig. 1(b) are real subtrees. v_1 is the root of T_1 and e_1 is the root edge of T_1. A subtree T' is called a *core subtree* of T if for $v \in V(T')$, either $\delta_{T'}(v) = 1$ or $\delta_{T'} = \delta_T(v)$. In Fig. 1(b), T_8 is a core subtree.

The *centroid* of a subtree T', which can be found in linear time, is a vertex $u \in V(T')$ *s.t.* each subtree with removal of u has the size at most $|V(T')|/2$.

An Overview of the Weighted 1-Center. Let r_T denote the *weighted-radius* of T, that is, $r_T = \min_{x \in A(T)} S(x, T)$. The service cost function $S(x, T)$ is convex on every simple path in T [4]. Based on this property, Kariv and Hakimi [4] designed an $O(n \log n)$ algorithm to locate the 1-center in a tree. Later, Megiddo [5] showed that it can be solved in linear time with a clever "trimming" technique. It is carried out in two phases. The first phase is to locate the component adjacent to the centroid o of current tree where the 1-center, say α, lies. The second phase answers the following *key question*: whether or not α lies within distance t to o. Once the answer to the key question is known, approximately 1/8 of the vertices in the current tree are discarded. The algorithm performs $O(\log n)$ such iterations. Each iteration takes linear time, linear in the size of the current tree.

Split-Edge. Let $C = \{\alpha_1, \ldots, \alpha_p\} \subset A(T)$ be a set of p centers in T. Let $V_i \subseteq V$ be the set of vertices closest to a particular center $\alpha_i \in C$. The edges whose endpoints belong to different subgraphs $G(V_i)$ are called *split-edges*. Thus, locating p centers in a tree is equivalent to finding a set of split-edges whose removal defines p connected components such that the maximum service cost of the 1-center of these components is equal to the optimal p-center cost of the entire tree.

It's trivial that the number of split edges is $p - 1$ for the p-center problem in a tree. In our problem, we need to locate one split-edge. An edge $e^* : u^*v^*$ is called *an optimal split-edge for the weighted 2-center problem* in T if it satisfies

$$\max\{r_{T_{u^*}(v^*)}, r_{T_{v^*}(u^*)}\} = \min_{e:uv \in E}\{\max\{r_{T_u(v)}, r_{T_v(u)}\}\}.$$

The weighted 2-center problem in T can be reformulated as a problem of finding a split-edge $e \in E(T)$ that minimizes $\phi(e : uv) = \max\{r_{T_u(v)}, r_{T_v(u)}\}$, called *service cost function of T for split-edge e*. It's easy to see that $\phi(e)$ is convex on every simple path of T. If a constant-size subtree contains an optimal split-edge, then the weighted 2-center can be computed in extra linear time by testing each edge in this subtree as a split-edge. Thus, the process will be terminated when we find that there exists an optimal split-edge in some constant-size subtree.

We call discarding one vertex *safe operation for an edge $e : uv$* if $r_{T_u(v)}$ and $r_{T_v(u)}$ stay unchanged before and after this operation. Discarding one vertex is a safe operation for a subtree T' if it is a safe operation for each edge in T'. Suppose that T' contains an optimal split-edge. After safely discarding some vertices for T', the local optimal solution of the new reduced tree with some split-edge in T' is an optimal solution of T. Let T_{cur} denote current tree. Let E_{opt} denote the set of edges containing an optimal split-edge. We always maintain the following invariant. The component composed of all the edges in E_{opt}, denoted by T_{opt}, is *a path* or *a core subtree*. If an optimal split-edge lies in a subtree T' of T_{opt}, clearly, all the safe operations for T_{opt} done so far are also safe operations for T'.

Lemma 1. *Suppose that T'_1, T'_2 are subtrees of T_{opt} and $E(T'_1 \cap T'_2) \neq \emptyset$. If T'_1 and T'_2 both contain optimal split-edges, then $T'_1 \cap T'_2$ contains an optimal split-edge.*

Lemma 2. *(Refer to Fig. 2(a).) Given an edge $uv \in E_{opt}$, if $r_{T_u(v)} \geq r_{T_v(u)}$, then an optimal split-edge lies in $\{uv, E(T_u(v))\} \cap E_{opt}$.*

Lemma 3. *Given an internal vertex v of the core subtree T_{opt}, suppose $T_1, \ldots, T_k$ $(k \geq 2)$ are the subtrees adjacent to v. Let T_1 and T_2 be the two components such that $S(v, T_1) \geq S(v, T_2)$ and $S(v, T_i) \leq S(v, T_2), 3 \leq i \leq k$. There exists an optimal split-edge in $\{vv_1, vv_2, E(T_1), E(T_2)\} \cap E_{opt}$.*

Proof. See Fig. 2(b). First, all the edges $vv_i \in E_{opt}, i = 1, \ldots, k$ since v is an internal vertex of the core subtree T_{opt}. We can see that the service cost $\phi(vv_1)$ is no more than the service cost with any split-edge in $E_{opt} \setminus \{vv_1, vv_2, E(T_1), E(T_2)\}$. Hence, an optimal split-edge lies in $\{vv_1, vv_2, E(T_1), E(T_2)\} \cap E_{opt}$. □

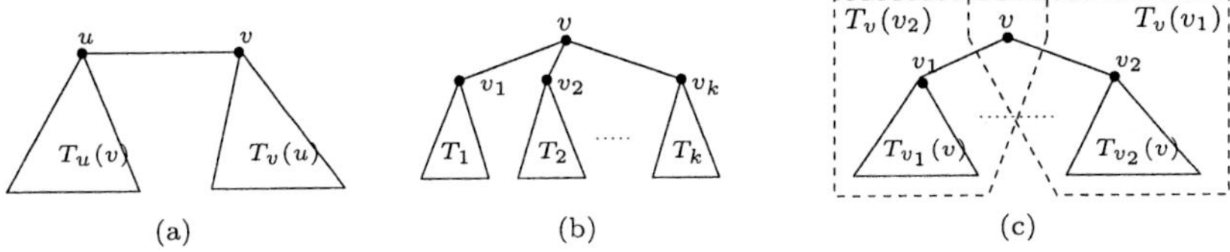

Fig. 2. Locate the component containing an optimal split-edge

The next lemma supports a binary-search technique for our problem.

Lemma 4. *(Refer to Fig. 2(c).) Given a vertex v in T_{cur}, we can find in linear time an optimal split-edge incident to v or find a vertex v' adjacent to v such that there is an optimal split-edge in $\{vv', E(T_{v'}(v))\} \cap E_{opt}$.*

Proof. It's trivial when $v \notin V(T_{opt})$ or v is a leaf of T_{opt}. Suppose v is an internal vertex of T_{opt}. We can find two vertices v_1, v_2 adjacent to v such that there is an optimal split-edge in $\{vv_1, vv_2, E(T_{v_1}(v)), E(T_{v_2}(v))\} \cap E_{opt}$ (by Lemma 3 if T_{opt} is a core subtree). Apply Lemma 2 on vv_1, vv_2. We have the following cases:

- If $r_{T_{v_1}(v)} \geq r_{T_v(v_1)}$ & $r_{T_{v_2}(v)} \geq r_{T_v(v_2)}$, vv_1, vv_2 both are optimal split-edges.
- If $r_{T_{v_1}(v)} \geq r_{T_v(v_1)}$ & $r_{T_{v_2}(v)} < r_{T_v(v_2)}$, there is an optimal split-edge in $\{vv_1, E(T_{v_1}(v))\} \cap E_{opt}$.
- Similarly, if $r_{T_{v_1}(v)} < r_{T_v(v_1)}$ & $r_{T_{v_2}(v)} \geq r_{T_v(v_2)}$, then there is an optimal split-edge in $\{vv_2, E(T_{v_2}(v))\} \cap E_{opt}$.
- Otherwise, at least one of vv_1, vv_2 is an optimal split-edge. An optimal split-edge is selected by evaluating the value of $\phi(vv_1)$ and $\phi(vv_2)$. □

Note that when T_{opt} is updated via Lemma 2 or Lemma 4, the new T_{opt} is still a path or a core subtree. Lemma 4 implies an $O(n \log n)$-time algorithm for the weighted 2-center problem in a tree, described as follows. Given that there is an optimal split-edge in T_{opt}, we test the centroid u of T_{opt} to find a subtree T' of T_{opt} adjacent to u that contains an optimal split-edge. Since the size of T' is at most half the size of T_{opt}, the process terminates within $O(\log n)$ examinations. The total cost is therefore $O(n \log n)$ time.

3 A Linear-Time Algorithm

Although it's hard to find an optimal split-edge quickly, we're able to obtain a subtree T_{opt}, with size no more than half-size of T_{cur}, in which an optimal split-edge lies. Given this reduced subtree T_{opt}, each connected component, with the removal of T_{opt}, must be served by one center. We'll see that there always exists a big component (at least half the size of T_{cur}) among them. Our objective is to eliminate a fraction of the vertices in this big component. More precisely, at least 1/16 of the vertices in T_{cur} are eliminated. The algorithm terminates after $O(\log n)$ iterations. The total running time is, therefore, linear.

Let o denote the centroid of T_{cur}, and let $v_1, \ldots, v_m$ be the vertices adjacent to o. If we find a real subtree $T_{v_s}(o)$ adjacent to o ($1 \leq s \leq m$) such that there is an optimal split-edge in $\{ov_s, E(T_{v_s}(o))\}$ (the solid bold part in Fig. 3(a)), then $T_o(v_s)$ is the big component served by one center in the optimal solution determined by some optimal split-edge in $\{ov_s, E(T_{v_s}(o))\}$. In the rest of this section, the 1-center serving the big component $T_o(v_s)$ is denoted by α_1, the other 1-center is denoted by α_2. As pointed out above, our goal is to safely discard a fraction of the vertices in $T_o(v_s)$. Let o' denote the centroid of $T_o(v_s)$. Like in Megiddo's method [5], the pruning stage is carried out in two phases. The first phase is to locate the component adjacent to o' where α_1 lies. In the second phase, the following key question is answered: does α_1 lie within the distance t to o'? The significance of t and how to determine t will be described later. It is very similar to the approach used in [5]. The main algorithm is sketched below.

Algorithm 1. Main algorithm for the weighted 2-center problem in T

```
T_opt = T, E_opt = E, T_cur = T.
repeat
    Get the centroid o of T_cur. Find a vertex v_s adjacent to o such that there is an
    optimal split-edge in {ov_s, E(T_{v_s}(o))} ∩ E_opt. Update E_opt, T_opt accordingly.
    Get the centroid o' of the subtree T_o(v_s). {The optimal split-edge is in T_{v_s}(0)}
    Find the component adjacent to o' that contains the center α_1 that serves T_o(v_s).
    Compute the value of t and answer the key question.
    Safely discard approximately 1/8 of the vertices in V(T_o(v_s)). Update T_cur.
until |E_opt| ≤ c (c is a predefined number)
Evaluate the service cost with each split-edge e in E_opt.
```

Analysis: We show in Sect. 3.1 and Sect. 3.2 that the steps in line 5 and line 6 can be implemented in linear time. Therefore,

Theorem 1. *The weighted 2-center in a tree can be solved in linear time.*

3.1 Phase 1: Locate the Component Adjacent to o' Where α_1 Lies

Refer to Fig. 3(a). With removal of the centroid o' of $T_o(v_s)$, T_{cur} is split into subtrees. Let $T_{v'_0}(o')$ denote the subtree among them that contains the vertex o. All the other subtrees are $T_{v'_1}(o'), \ldots, T_{v'_k}(o')$. Consider o' as the root of T_{cur}. Suppose that $S(o', T_{v'_1}(o')) \geq S(o', T_{v'_i}(o')), 2 \leq i \leq k$. Then, α_1 must lie in the component $\overline{o'v'_1} + T_{v'_1}(o')$ or in the component $\overline{o'v'_0} + T_{v'_0}(o')$. It can be decided in linear time by Lemma 5. Lemma 6 provides a more general result.

Lemma 5. *(Refer to Fig. 3(b).) Let T' be a real subtree of T_{cur} served by α_1. Let v denote the root of T' and x be a point on the root edge of T'. Whether α_1 lies in $\overline{vx} + T'$ can be decided in linear time.*

Lemma 6. *(Refer to Fig. 3(b).) Let T' be a real subtree of T_{cur} served by one center (either α_1 or α_2). Let v denote the root of T' and x be a point in the root edge of T'. Whether α_1 lies in $\overline{vx} + T'$ can be decided in linear time.*

Proof of Lemma 5. Let $U_x(S(x, T'))$ denote the set of vertices in T_{cur} with larger weighted distance to x than $S(x, T')$. It's trivial to see that α_1 lies in $\overline{vx}+T'$ if $U_x(S(x, T')) = \emptyset$. Suppose that $U_x(S(x, T')) \neq \emptyset$. Clearly, $U_x(S(x, T')) \subseteq V(^-T')$. Consider x as the root of T_{cur}. Let T'' be the smallest connected real subtree containing all the vertices in $U_x(S(x, T'))$. T'' must be a subtree of $^-T'$. Let u be the root of T''. Observe that α_1 lies in $\overline{vx} + T'$ if and only if T'' is served by α_2. First, let us see a useful lemma. Refer to Fig. 4(a). T_1 and T_2 are two real subtrees rooted at u', v' respectively, and b is a point in the root edge of T_1. T_2 contains all the vertices with larger weighted distance to b than $S(b, T_1)$.

Lemma 7. *If $E_{opt} \cap E(T_1) = E_{opt} \cap E(T_2) = \emptyset$, then in linear time we can find an optimal split-edge, or find a subpath of $P_{u',v'}$ containing an optimal split-edge.*

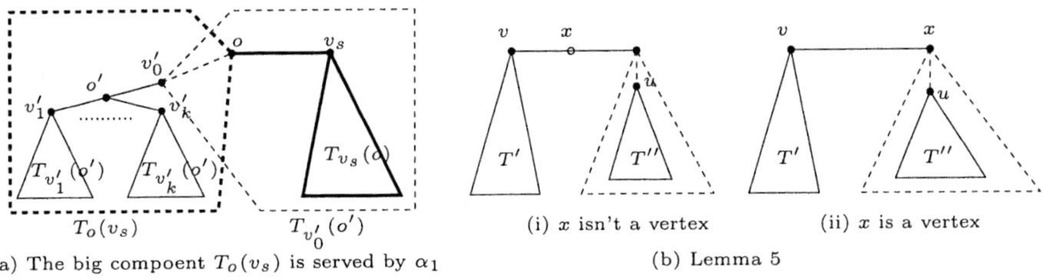

(a) The big compoent $T_o(v_s)$ is served by α_1 (b) Lemma 5

Fig. 3. The big component $T_o(v_s)$ served by α_1 and Lemma 5

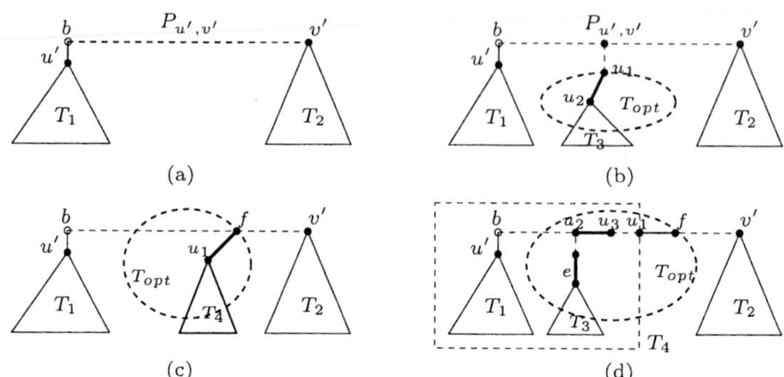

Fig. 4. Lemma 7

Proof. We have the following two cases.

- $E_{opt} \cap E(P_{u',v'}) = \emptyset$. See Fig. 4(b). Let u_1 be the vertex closest to $P_{u',v'}$ in T_{opt}. Examine u_1 using Lemma 4. In linear time, we can find an optimal split-edge incident to u_1 or find a vertex u_2 adjacent to u_1 such that there is an optimal split-edge in $\{u_1u_2, E(T_3)\} \cap E_{opt}$. In the latter case, u_1u_2 is an optimal split-edge since $r_{-T_3} \geq r_{T_3}$.
- $E_{opt} \cap E(P_{u',v'}) \neq \emptyset$. See Fig. 4(c)(d). Let f be the vertex closest to v' in T_{opt}. f must be in $P_{u',v'}$. Check f using Lemma 4. In linear time, we find a vertex u_1 adjacent to f *s.t.* there is an optimal split-edge in $\{fu_1, E(T_4)\} \cap E_{opt}$. If u_1 is not in $P_{u',v'}$, fu_1 is an optimal split-edge since $r_{-T_4} \geq r_{T_4}$ (Fig. 4(c)). Otherwise, u_1 is in $P_{u,v}$ (Fig. 4(d)). For any split-edge $e \in E_{opt}$ but $e \notin E(P_{u',v'})$, the service cost $\phi(e) = \max\{r_{T_3}, r_{-T_3}\} \geq S(b, T_1)$. Let u_2 be the vertex closest to e in $P_{u',v'}$. Let u_3 be the vertex adjacent to u_2 with $d(u_3, v') < d(u_2, v')$. It's trivial that $u_2u_3 \in E_{opt}$ since f is a leaf of T_{opt}. Since $\phi(u_2u_3) \leq \phi(e)$, there is an optimal split-edge in $E_{opt} \cap E(P_{u',v'})$. □

The linear-time checking process is briefly described as follows. Test vertex u with Lemma 4 and update E_{opt}, T_{opt} accordingly. If T_{opt} is a subtree of T'', then α_1 is in $\overline{vx} + T'$. Otherwise, by Lemma 7, either an optimal split-edge is found or we find that α_1 is in $\overline{vx} + T'$. This completes the proof of Lemma 5.

Proof of Lemma 6. Suppose that there is an edge $e \in E_{opt}$ in $P_{o',v}$ (otherwise, use Lemma 5). Let $U_x(S(x, T'))$ denote the set of vertices in T_{cur} with the larger weighted distance to x than $S(x, T')$. Consider two cases.

$U_x(S(x, T')) = \emptyset$: In this case, $\overline{vx} + T'$ contains α_1 if and only if there doesn't exist an optimal split-edge in $P_{o',v}$. Let u' be the closest vertex in T_{opt} to o' and let v' be the closest vertex in T_{opt} to v. u', v' should be on the path $P_{o',v}$ and $u' \neq v'$. $P_{u',v'}$ is the common subtree of $P_{o',v}, T_{opt}$.

- T_{opt} is a path. See Fig. 5(a). By Lemma 4 on u and v, we have two cases.
  - $P_{u',v'}$ contains an optimal split-edge. Then $\overline{vx} + T'$ doesn't contain α_1.
  - An optimal split-edge lies outside $P_{u',v'}$. Therefore $\overline{vx} + T'$ contains α_1.

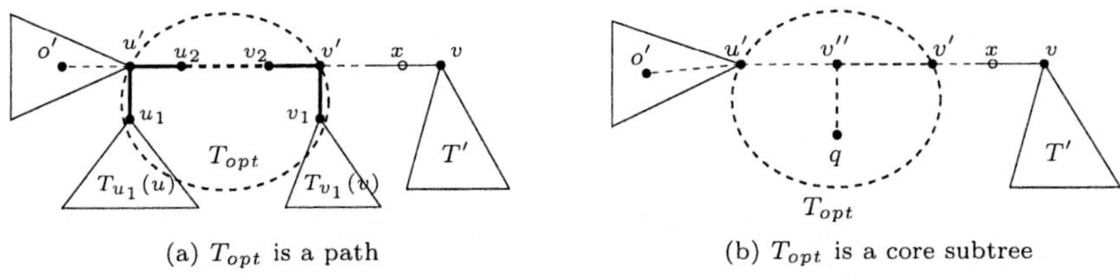

(a) T_{opt} is a path (b) T_{opt} is a core subtree

Fig. 5. Lemma 6: $U_x(S(x, T')) = \emptyset$

- T_{opt} is a core subtree. In this case, u', v' must be leaf vertices of T_{opt}. See Fig. 5(b). Let V'' denote the set of vertices z such that the simple path $P_{v,z}$ contains some edge in E_{opt}. Let q be a vertex in V'' with $w_q d(x, q) = \max_{z \in V''} w_z d(x, z)$. Then there exists an optimal split-edge on $P_{q,v}$ (otherwise, the service cost is larger than $w_q d(x, q)$). Update T_{opt} accordingly. Now, T_{opt} is a path. we can get the result by similar process.

$U_x(S(x, T')) \neq \emptyset$: Consider x as the root of T_{cur}. Let T'' be the smallest real subtree containing all the vertices in $U_x(S(x, T'))$. u is the root of T''. Similar to the proof of Lemma 5, we have two cases by testing u with Lemma 4:

- If updated T_{opt} is a subtree of T'', then $\overline{vx} + T'$ can't contain α_1.
- Otherwise, $E_{opt} \cap E(T'') = \emptyset$. Assume that we only find a subpath of $P_{u,v}$ containing an optimal split-edge by Lemma 7. If $T_o(v_s)$ is the subtree of T'', see Fig. 6(a), then α_1 can't lie in $\overline{vx} + T'$. Otherwise, $T_o(v_s)$ and T'' are disjoint (Fig. 6(b)). Let u' be the least common ancestor of u and o. Check u' by Lemma 4 and update E_{opt}, T_{opt} accordingly. If T_{opt} is the subpath of $P_{u,u'}$, then α_1 lies in $\overline{vx} + T'$. Otherwise, $\overline{vx} + T'$ doesn't contain α_1.

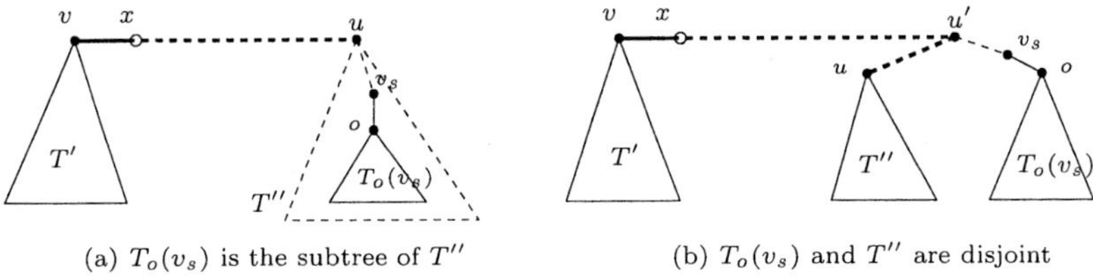

(a) $T_o(v_s)$ is the subtree of T'' (b) $T_o(v_s)$ and T'' are disjoint

Fig. 6. Lemma 6: $U_x(S(x, T')) \neq \emptyset$ and $E_{opt} \subseteq E(P_{v,v})$

3.2 Phase 2: Answer the Key Question

Having found the component where α_1 lies: $\overline{o'v'_1} + T_{v'_1}(o')$ (Case 1) or $\overline{o'v'_0} + T_{v'_0}(o')$ (Case 2), see Fig. 3(a), we need to check if α_1 lies within distance t to o'. The significance of t and the computation of t is described as follows [5]. Arbitrarily pair the vertices in $V(T_o(v_s)) \setminus V(T_{v'_1}(o'))$ for Case 1 and pair the vertices in $V(T_o(v_s)) \setminus V(T_{v'_0}(o')))$ for Case 2. Let $(u_1, u'_1), (u_2, u'_2), \ldots, (u_l, u'_l)$ be the pairs where $w_{u_i} > w_{u'_i}$. The case when $w_{u_i} = w_{u'_i}$ is easy to handle. Note that there will be at least $\lfloor n/8 \rfloor$ pairs since o' is the centroid of $T_o(v_s)$. For every such pair

$(u_i, u'_i), 1 \leq i \leq l$ let $t_i = (w_{u_i} d(u_i, o') - w_{u'_i} d(u'_i, o'))/(w_{u_i} - w_{u'_i})$. t is taken to be the median of these values. If $t > t_i$ then u'_i is dominated by u_i, and therefore is discarded; otherwise, u_i is discarded. In this way, we can eliminate at least $\lfloor n/16 \rfloor$ the vertices in T_{cur} after the key question is answered.

Case 1. We find all the points $y_1, \ldots, y_l$ in $T_{v'_1}(o')$ such that $d(o', y_i) = t, i = 1, \ldots, l$. Let $T_{y_1}(o'), \ldots, T_{y_l}(o')$ be the subtrees rooted at $y_1, \ldots, y_l$. Assume that $S(y_1, T_{y_1}(o')) = \max_{1 \leq i \leq l} S(y_i, T_{y_i}(o'))$. See Fig. 7(a). We evaluate the point y_1 and the real subtree $T_{y_1}(o')$ by Lemma 5. If α_1 lies in $T_{y_1}(o')$, then the answer to the key question is "NO"; otherwise, the answer to the key question is "YES"

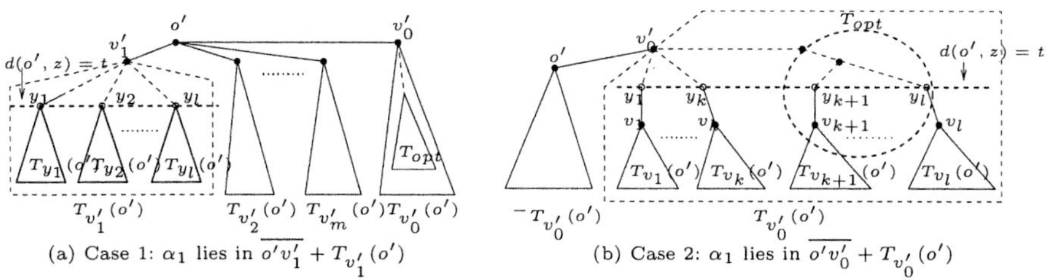

Fig. 7. Answer the key question: whether α_1 lies within the distance t to o'

since α_1 can't lie in $T_{y_2}(o'), \ldots, T_{y_l}(o')$. Therefore, for Case 1 we can check in linear time whether or not α_1 lies within distance t to o'.

Case 2. The vertices served by α_2 are contained in $T_{v'_0}(o')$. As in Case 1, we first find all the points $y_1, \ldots, y_l$ in $T_{v'_0}(o')$ such that $d(o', y_i) = t, i = 1, \ldots, l$. See Fig. 7(b). Let v_i be the vertex closest to y_i such that $d(o', v_i) \geq t$. The subtrees rooted at the vertices $v_1, \ldots, v_l$ are denoted by $T_{v_1}(o'), \ldots, T_{v_l}(o')$. The subtrees rooted at the points $y_1, \ldots, y_l$ are denoted by $T_{y_1}(o'), \ldots, T_{y_l}(o')$, which contain all points z in $T_{v'_0}(o')$ with $d(o', z) \geq t$. Let $\Gamma = \{T_{y_1}(o'), \ldots, T_{y_l}(o')\}$ and $\Delta = \{T_{v_1}(o'), \ldots, T_{v_l}(o')\}$. Without loss of generality, suppose that $T_{v_1}(o'), \ldots, T_{v_k}(o')$ are the subtrees that do not contain any edge of E_{opt} and also the path $P_{v_i, v'_0}, 1 \leq i \leq k$ do not contain any edge of E_{opt}. *So all the vertices in* $T_{v_i}(o'), 1 \leq i \leq k$ *are served by* α_1. Two things make the problem in Case 2 harder:

- First situation: There may exist an optimal split-edge in some subtree in Δ. Let Φ denote the set of subtrees in Δ containing some edges in E_{opt}.
- Second situation: There may exist an optimal split-edge on the path between o' and roots of some subtrees in Δ. Then these subtrees are served by α_2.

First situation. We compute $r_{T_{v_i}(o')}$ for $T_{v_i}(o') \in \Phi$. Since these subtrees are pairwise disjoint, all these values can be computed in linear time. Let $R = \max\{r_{T_{v_i}(o')}, T_{v_i}(o') \in \Phi\}$. If $r_{T_{v_i}(o')} < R, T_{v_i}(o') \in \Phi$ there exists an optimal split-edge outside subtree $T_{v_i}(o')$. If $r_{T_{v_{i_1}}(o')} = r_{T_{v_{i_2}}(o')} = R$ for some $i_1 \neq i_2, T_{v_{i_1}}(o'), T_{v_{i_2}}(o') \in \Phi$, then there exists an optimal split-edge outside of all the subtrees in Φ. The remaining case is when there is a unique subtree $T_{v_i^*}(o') \in \Phi$

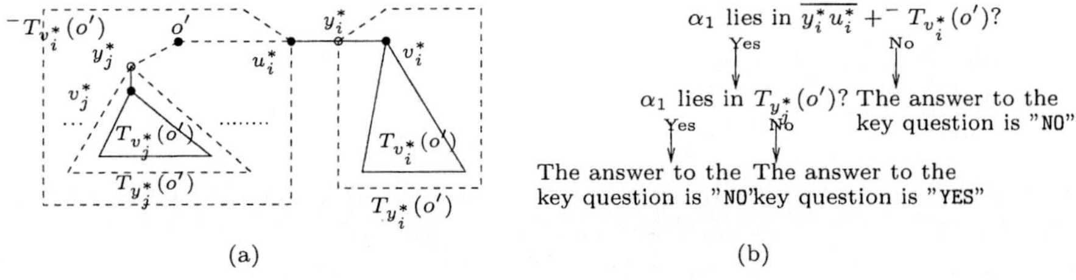

Fig. 8. $T_{v_i^*}$ contains an optimal split-edge

such that $r_{T_{v_i^*}(o')} = R$. Apply Lemma 4 on v_i^*. If $T_{v_i^*}(o') \in \Delta$ contains an optimal split-edge, then $^{-}T_{v_i^*}(o')$ is served by α_1. See Fig. 8(a). Let $S(y_j^*, T_{v_j^*}(o')) = \max_{T_{v_s}(o') \in \Delta \setminus \{T_{v_i^*}(o')\}} S(y_s, T_{v_s}(o'))$. The possibility of having the "NO" answer to the key question is that α_1 lies in either $T_{y_i^*}(o')$ or $T_{y_j^*}(o')$. The result can be obtained from Lemma 5, i.e., apply Lemma 5 on y_i^* and $^{-}T_{v_i^*}(o')$ served by α_1, and apply Lemma 5 on y_j^* and $T_{v_j^*}(o')$. The decision tree is shown in Fig. 8(b). Otherwise, there is an optimal split-edge outside all the subtrees in Δ. In this case, we encounter another problem, as described in the second situation.

Second situation. In this case, there is an optimal split-edge outside all the subtrees in Δ. That is, every subtree in Δ is served by 1-center (either α_1 or α_2). By Lemma 6, we can decide whether or not α_1 lies in $T_{y_i}(o') \in \Gamma, i = 1, \ldots, l$. However, the total cost is $O(n^2)$. We have an efficient method to achieve it. Let $T_{v_{i_1}^*}(o'), T_{v_{i_2}^*}(o')$ be the subtrees in Δ *s.t.* $S(y_{i_1}^*, T_{v_{i_1}^*}(o')) \geq S(y_{i_2^*}, T_{v_{i_2}^*}(o'))$ and $S(y_i, T_{v_i}(o')) \leq S(y_{i_2^*}, T_{v_{i_2}^*}(o')), T_{v_i}(o') \in \Delta \setminus \{T_{v_{i_1}}(o'), T_{v_{i_2}}(o')\}$. If α_1 lies in $T_{y_{i_1}^*}(o')$ or $T_{y_{i_2}^*}(o')$ (by Lemma 6), the answer to the key question is "YES". Suppose that α_1 does not lie in $T_{y_{i_1}^*}(o'), T_{y_{i_2}^*}(o')$. We need to determine whether α_1 lies in some subtree $T_{y_j}(o')$ in $\Gamma \setminus \{T_{y_{i_1}^*}(o'), T_{y_{i_2}^*}(o')\}$. Two necessary conditions for α_1 to lie in $T_{y_j}(o') \in \Gamma \setminus \{T_{y_{i_1}^*}(o'), T_{y_{i_2}^*}(o')\}$ are listed below.

- $T_{y_{i_1}^*}(o'), T_{y_{i_2}^*}(o')$ are served by α_2. Let θ be the lowest common ancestor of $v_{i_1}^*, v_{i_2}^*$ (o' is the root). There is an optimal split-edge in $E_{opt} \cap E(P_{\theta,o'})$.
- For each vertex v served by α_1, $w_v d(\alpha_1, v) \leq S(y_j, T_{v_j}(o'))$.

Let E_s be the set of edges $e \in E_{opt} \cap E(P_{\theta,o'})$ such that α_1 lies in some subtree in $\Gamma \setminus \{T_{y_{i_1}^*}(o'), T_{y_{i_2}^*}(o')\}$ with split-edge e. Observe that E_s contains an optimal split-edge if a subtree in $\Gamma \setminus \{T_{y_{i_1}^*}(o'), T_{y_{i_2}^*}(o')\}$ contains α_1 (It follows easily from two necessary conditions). If $E_s = \emptyset$ then α_1 can't lie in $\Gamma \setminus \{T_{y_{i_1}^*}(o'), T_{y_{i_2}^*}(o')\}$. Suppose that $E_s \neq \emptyset$. Let e_s be the edge closest to θ in E_s. See Fig. 9(a).

Lemma 8. *If E_s contains an optimal split-edge, then e_s is an optimal split-edge.*

Proof. For any edge $e' \in E_s$ ($e' : u_i u_{i+1}, e_s : u_j u_{j+1}$), the service cost $\phi(e_s) = \max\{r_{T_{u_j}(u_{j+1})}, r_{T_{u_{j+1}}(u_j)}\}$ is no more than $\phi(e') = \max\{r_{T_{u_i}(u_{i+1})}, r_{T_{u_{i+1}}(u_i)}\}$. Therefore, e_s is an optimal split-edge if E_s contains an optimal split-edge. □

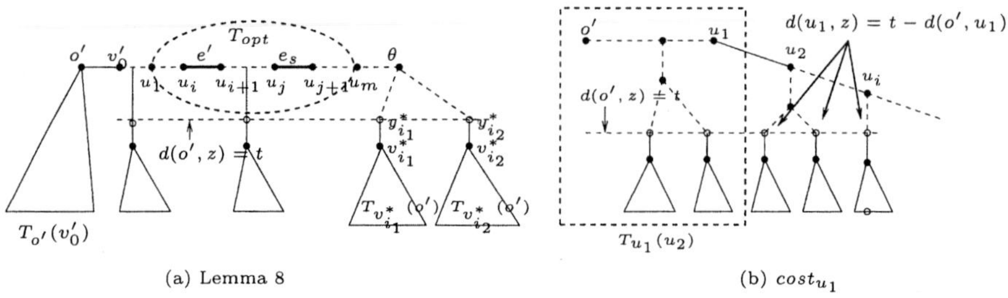

(a) Lemma 8 (b) $cost_{u_1}$

Fig. 9. Lemma 8 and compute $cost_{u_1}$

Therefore, it's enough to check if e_s is an optimal split-edge. Since $r_{T_{u_j}(u_{j+1})} \leq r_{T_{u_{j+1}}(u_j)}$, there is an optimal split-edge in $\{u_j u_{j+1}, E(T_{u_{j+1}}(u_j))\} \cap E_{opt}$. By Lemma 4, we can either find e_s is an optimal split-edge or find that there is an optimal split-edge outside E_s. In the latter case, all the subtrees in $\Gamma \setminus \{T_{y^*_{i_1}}(o'), T_{y^*_{i_2}}(o')\}$ don't contain α_1 in the optimal solution determined by any optimal split-edge in E_{opt}. Hence, the answer to the key question is "YES".

The last issue is to find such e_s in linear time if it exists. Refer to Fig. 9(b). The path $P_{o',\theta} \cap T_{opt}$ is denoted by $u_1, u_2, \ldots, u_m$. Let $cost_{u_i}$ denote the cost needed to cover all the vertices in subtree $T_{u_i}(u_{i+1})$ for a point outside subtree $T_{u_i}(u_{i+1})$ with the distance $t - d(o', u_i)$ to u_i (= with the distance t to o'), $i = 1, \ldots, m-1$. Obviously, all the values of $cost_{u_i}, i = 1, \ldots, m-1$ can be computed in linear time for a given t. Given a split-edge $u_i u_{i+1}, 1 \leq i < m$, let Γ_i denote the set of subtrees served by α_1 in Γ and $\sigma_i = \max\{S(y_k, T_{v_k}(o')), T_{y_k}(o') \in \Gamma_i\}$. α_1 can't lie in any subtree $T_{y_j}(o')$ with $S(y_j, T_{v_j}(o')) < \sigma_i$ for split-edge $u_i u_{i+1}$ and, if there are two subtrees $T_{y_j}(o'), T_{y_k}(o')$ in Γ_i with $S(y_j, T_{v_j}(o')) = S(y_k, T_{v_k}(o')) = \sigma_i$, then α_1 can't lie in any subtree in Γ_i. Let $T^i_{y_k}(o')$ be the unique subtree in Γ_i such that $S(y_k, T^i_{v_k}(o')) = \sigma_i$. For the edge $u_1 u_2$, $u_1 u_2 \in E_s$ if and only if $S(y^1_k, T_{u_1}(u_2)) = \sigma_1$. We can check if $u_i u_{i+1} \in E_s, i = 2, \ldots, m-1$ as follows:

- If $T^i_{y_k}(o') \in \Gamma_{i-1}$, that is, $T^i_{y_k}(o') = T^{i-1}_{y_k}(o')$, then
  - $u_{i-1} u_i \in E_s$: All the vertices in $T_{u_{i-1}}(u_i)$ with the weighted distance to y_k no more than $S(y_k, T^i_{v_k}(o'))$. We compute the cost needed to serve all the vertices in $T_{u_i}(u_{i+1}) \setminus T_{u_{i-1}}(u_i)$ by y_k. If the cost is greater than σ_i then $u_i u_{i+1} \notin E_s$. Otherwise, $E_s = E_s \cup \{u_i u_{i+1}\}$ and $e_s = u_i u_{i+1}$.
  - $u_{i-1} u_i \notin E_s$: There is at least one vertex in $T_{u_{i-1}}(u_i)$ that can't be covered by y_k within σ_{i-1}. Since $\sigma_i = \sigma_{i-1}$, α_1 can't lie in $T^i_{y_k}(o')$. Therefore, $u_i u_{i+1} \notin E_s$.
- Otherwise, $T^i_{y_k}(o') \in \Gamma_i \setminus \Gamma_{i-1}$. The cost needed to serve all the vertices in $T_{u_{i-1}}(u_i)$ for y_k is $cost_{u_{i-1}}$. We compute the cost needed to serve all the vertices in $T_{u_i}(u_{i+1}) \setminus T_{u_{i-1}}(u_i)$ for y_k. We then compute the maximum of this cost and $cost_{u_{i-1}}$. If the maximum cost is greater than σ_i then $u_i u_{i+1} \notin E_s$, otherwise, $E_s = E_s \cup \{u_i u_{i+1}\}$ and $e_s = u_i u_{i+1}$.

In i-th step, suppose that such subtree $T^i_{y_k}(o')$ is unique. We only need $|T_{u_i}(u_{i+1}) \setminus T_{u_{i-1}}(u_i)|$ time to compute the cost needed to serve all the vertices in $T_{u_i}(u_{i+1})$ by y_k. Thus, the running time of the algorithm is linear.

Lemma 9. *If $E_s \neq \emptyset$, e_s can be computed in linear time.*

Putting everything together, we establish Theorem 1. Adapting the algorithm for the discrete case is straightforward.

4 Conclusion

In this paper, an algorithm is given which finds weighted 2-center in trees in linear time. The proposed method is a nontrivial generalization of the "trimming" method of Megiddo [5]. With the linear-time algorithms for the weighted 1,2-center problems, the upper bound of the weighted 3-center and 4-center problems can be improved to $O(n \log n)$. It's not hard to get a method based on binary-search technique.

Theorem 2. *The weighted 3,4-center problems can be solved in $O(n \log n)$ time and linear space.*

One challenging work is to generalize this result for the weighted p-center problems in a tree graph (any fixed p). Currently, we have proved that one big component, defined as one connected component served by 1-center that contains a fraction of vertices in current tree, can be found in linear time. Also, lemmas similar to Lemma 5 and Lemma 6 are discovered.

References

1. R. Cole, "Slowing down sorting networks to obtain faster sorting algorithms", J. ACM 34 (1987) 200-208.
2. G.N. Frederickson, " Parametric search and locating supply centers in trees", WADS (1991) 299-319.
3. N. Halman, "Discrete and lexicographic helly theorems and their relations to LP-type problems", Ph.D. Thesis, Tel Aviv Univ., 2004.
4. O. Kariv and S.L. Hakimi, "An algorithmic approach to network location problems, Part I. The p-centers", SIAM J. Appl. Math., 37 (1979) 513-538.
5. N. Megiddo, "Linear-time algorithms for linear programming in R^3 and related problems", SIAM J. Comput. 12:4 (1983) 759-776.
6. N. Megiddo, A. Tamir, E. Zemel, and R. Chandrasekaran, "An $O(n \log^2 n)$ algorithm for the kth longest path in a tree with applications to location problems", SIAM J. Comput. 10:2 (1981) 328-337.
7. N. Megiddo, A. Tamir, "New results on the complexity of p-center problems", SIAM J. Comput. 12:4 (1983) 751-758.

An Algorithm for a Generalized Maximum Subsequence Problem

Thorsten Bernholt* and Thomas Hofmeister

Informatik 2, Universität Dortmund, 44221 Dortmund, Germany
Informatik 5, Universität Dortmund, 44221 Dortmund, Germany

Abstract. We consider a generalization of the maximum subsequence problem. Given an array $a_1, \ldots, a_n$ of real numbers, the generalized problem consists in finding an interval $[i, j]$ such that the length and the sum of the subsequence $a_i, \ldots, a_j$ maximize a given quasiconvex function f. Problems of this type occur, e.g., in bioinformatics. We show that the generalized problem can be solved in time $O(n \log n)$. As an example, we show how the so-called multiresolution criteria problem can be solved in time $O(n \log n)$.

1 Introduction and Preliminaries

The maximum subsequence problem is often used to show that different algorithmic approaches can lead to algorithms of varying efficiency. (See, e.g., Column 7 in [B].) Input to the problem is an array $a_1, \ldots, a_n$ of real numbers. For an interval $[i, j]$ of array elements, the sum of the interval is defined as $a_i + a_{i+1} + \cdots + a_j$. The maximum subsequence problem asks for an interval which has the maximum sum among all intervals. It is well-known that the problem can be solved by a dynamic programming approach in time $O(n)$.

In practice, there are other problems defined on array intervals which have to be solved. Examples from bioinformatics are, e.g., the longest biased interval, the longest non-negative sum interval [A], the maximum-sum segment [FLLTWY], the length-constrained heaviest segment [LJC], the range maximum-sum segment [CC] and DNA copy number data analysis [LABLY]. An example from statistics is the multiresolution criteria problem [DK]. All of these problems have in common that they assign a value $f(\ell, s)$ to an interval that depends on the length ℓ and the sum s of the interval only. To some of those problems, our algorithm can be applied. We first describe the class of functions f that is allowed in our problem.

Definition 1. *Let $D \subseteq \mathbb{R}^2$ be a nonempty convex set and let $f : D \to \mathbb{R}$. The function f is said to be quasiconvex if and only if for all points $s_1, s_2 \in D$ and all $\lambda \in [0, 1]$, we have $f(\lambda \cdot s_1 + (1 - \lambda) \cdot s_2) \leq \max\{f(s_1), f(s_2)\}$.*

---

* The financial support of the Deutsche Forschungsgemeinschaft (SFB 475, Reduction of complexity in multivariate data structures) is gratefully acknowledged.

J.R. Correa, A. Hevia, and M. Kiwi (Eds.): LATIN 2006, LNCS 3887, pp. 178–189, 2006.

Thus, quasiconvex functions assume their maximum value on a line segment in an endpoint of the segment. It is clear that for a quasiconvex function $f : D \to \mathbb{R}$, one can use induction to show the following: For all $\lambda_1, \ldots, \lambda_r \in [0,1]$ with $\sum_{i=1}^{r} \lambda_i = 1$, and all $s_1, \ldots, s_r \in D$, we have

$$f(\lambda \cdot s_1 + \cdots + \lambda_r \cdot s_r) \leq \max\{f(s_1), \ldots, f(s_r)\}.$$

Thus, on a convex set, a quasiconvex function assumes its maximum on an extremal point of the convex set. In Section 2, we will give examples of quasiconvex functions. We are now ready to formulate the generalized maximum subsequence problem which will be considered in this paper.

Definition 2. *For an interval $[i, j]$ (with $i \leq j$), we define its sum as* $\text{sum}(i, j) := a_i + a_{i+1} + \cdots + a_j$ *and its length as $\ell(i, j) := j - i + 1$. The generalized maximum subsequence problem can be described as follows:*

Input: *An array $a_1, \ldots, a_n$ and a quasiconvex function f.*
Output: *An interval $[i, j]$ such that its value*

$$w(i, j) := f(\ell(i, j), \text{sum}(i, j))$$

is maximal among all intervals. Alternatively, we are interested in the value $w(i, j)$ of such an interval.

Some remarks are in place: First, note that the maximum subsequence problem is the special case where $f(\ell, s) = s$. Second, we restrict ourselves to the task of computing the maximum value $w(i, j)$ instead of a corresponding interval itself. It will be obvious how such an interval can be computed as a side information in the algorithm. Third, when analyzing running times, we will assume that the evaluation of $f(\ell, s)$ can be done in time $O(1)$ (which is obviously the case for functions like, e.g., $|s|/\sqrt{\ell}$).

A trivial solution of the generalized maximum subsequence problem would be to enumerate all $\Theta(n^2)$ intervals $[i, j]$, evaluate $w(i, j)$ for each of them and output the maximum value. If implemented right, this can be done in time $\Theta(n^2)$. We will show in this paper that for every quasiconvex function f, the generalized maximum subsequence problem can be solved in time $O(n \log n)$.

2 Motivation

Our original motivation for investigating the generalized maximum subsequence problem came from a problem in statistics, more precisely, data analysis. Here, the so-called multiresolution criteria problem [DK] is useful for deciding whether residuals consist of white noise. The problem is based on a parameter which for an interval of the data is defined as $f(\ell, s) := |s|/\sqrt{\ell}$. As before, ℓ is the length and s is the sum of the interval. One then seeks for the interval with the largest parameter (or, the largest parameter itself). It turned out that our methods for tackling this problem could be generalized to the larger class of quasiconvex functions.

Let us first show that the function $f(\ell, s) := |s|/\sqrt{\ell}$ is quasiconvex. In case some of the used notions here should be unknown to the reader, we refer to Section 3. The following theorem is well-known in the literature and helpful for showing quasiconvexity (see [BV], p. 95/98):

Theorem 1. *Let $D \subseteq \mathbb{R}^2$ be a nonempty convex set. A function $f : D \to \mathbb{R}$ is quasiconvex if and only if the so-called sublevel sets $D_\alpha := \{x \in D \mid f(x) \leq \alpha\}$ are convex for all $\alpha \in \mathbb{R}$.*

We obtain the following:

Lemma 1. *Let $g : \mathbb{R}_+ \to \mathbb{R}_+$ be a concave function. Then $f(\ell, s) := |s|/g(\ell)$ is quasiconvex.*

Proof: Consider a sublevel set $D_\alpha = \{(\ell, s) \mid |s| \leq \alpha \cdot g(\ell)\}$. D_α is convex: For $\alpha < 0$, this is trivial since then, the sublevel set is empty. For $\alpha = 0$, we obtain a (convex) straight line and for $\alpha > 0$, we have that

$$D_\alpha = \{(\ell, s) \mid s \geq 0 \text{ and } s \leq \alpha \cdot g(\ell)\} \cup \{(\ell, s) \mid s \leq 0 \text{ and } s \geq -\alpha \cdot g(\ell)\}.$$

Since g is concave and $\alpha > 0$, it follows that $\alpha \cdot g$ is also concave and since g is mapping inputs to $\mathbb{R}_+$, it follows that D_α is convex. □

Since $\sqrt{\ell}$ is a concave function, we obtain that $|s|/\sqrt{\ell}$ is a quasiconvex function. Thus, our algorithm for the generalized maximum subsequence problem can be used to obtain the maximum parameter value for the multiresolution criteria problem in time $O(n \log n)$.

A similar function occurs in bioinformatics. For DNA copy number data analysis [LABLY], one might wish to find an interval $[i, j]$ which maximizes the value $\frac{1}{\sqrt{j-i+1}} \cdot \sum_{k=i}^{j} a_k$. This corresponds to choosing the function $f(\ell, s) := s/\sqrt{\ell}$.

Inspection of the paper [LABLY] shows that it is likely that the authors are in fact interested in the maximization of the function $|s|/\sqrt{\ell}$ instead of $s/\sqrt{\ell}$ (which would be the same function that we previously treated).

Nevertheless, one might also consider the function $s/\sqrt{\ell}$ which is no longer quasiconvex. Here, one can use a simple trick to make our algorithm applicable in "most" cases.

When in the input, there is at least one $a_i \geq 0$, then the maximal value OPT of the function $s/\sqrt{\ell}$ is also at least 0. (Choose the 1-element interval $[i, i]$ which has the value $a_i \geq 0$.) We can then consider the function

$$f'(\ell, s) := \begin{cases} f(\ell, s), & \text{if} s \geq 0. \\ 0, & \text{otherwise.} \end{cases}$$

The maximal value for this function f' is the same as for the function f and it can easily be shown that f' is quasiconvex (the proof is similar to the one for the function $|s|/\sqrt{\ell}$).

The assumption that not all values $a_1, \ldots, a_n$ in the input are negative is very likely to hold in the applications, since there, the a_i basically are the deviations of a random variable from its mean.

[LABLY] provides a linear time *approximation scheme* for the DNA copy number data analysis and provides an exact algorithm that is good for typical inputs but which could not be shown to have a worst-case running time better than $O(n^2)$. Thus, our approach improves upon the known worst-case bound for this slightly restricted version of the problem.

3 Basic Definitions

In this section, we recall a few definitions that are helpful for our purposes. The reader is also referred to, e.g., [M] or [BV].

Definition 3. *Given a set* $P = \{p_1, \ldots, p_m\} \subseteq \mathbb{R}^2$, a convex linear combination *of* P *is any point of the form* $\lambda_1 \cdot p_1 + \cdots + \lambda_m \cdot p_m$, *where* $\lambda_i \geq 0$ *for all* $1 \leq i \leq m$ *and* $\sum_{i=1}^{m} \lambda_i = 1$.

A set $S \subseteq \mathbb{R}^2$ *is convex if for each pair of points* $s_1, s_2 \in S$, *it holds that every convex linear combination of* $\{s_1, s_2\}$ *is also in* S.

Given a finite set M of points, we define the convex hull of M as the smallest subset C of M such that the convex polytope defined by the points in C contains all points of M. The convex hull can be split into its upper and lower part, the upper convex hull and the lower convex hull. The following figures show a point set M and its upper and lower convex hulls, respectively (marked by crosses).

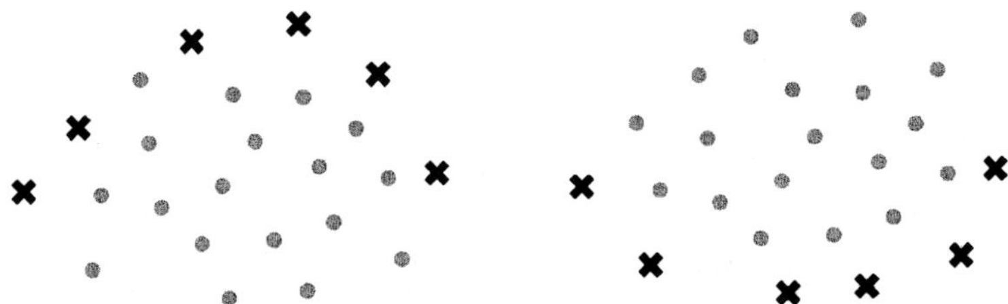

Definition 4. *Let* S *be a nonempty convex set. A function* f *in* n *variables is called convex on* S *if for all* $s, s' \in S$ *and all* $\lambda \in [0, 1]$ *it holds that*

$$f(\lambda \cdot s + (1 - \lambda) \cdot s') \leq \lambda \cdot f(s) + (1 - \lambda) \cdot f(s').$$

A function g *is called concave if* $-g$ *is convex.*

We remark that every convex function is also quasiconvex.

Definition 5. *Let* $p = (p_x, p_y)$ *be a point and* A *be a set of points. We say that* p *is upper-dominated by* A *if there is a point* p_{up} *such that*

i) $p_{up} = (p_x, p_y + c)$ *with* $c \geq 0$, *and*
ii) p_{up} *can be written as a convex linear combination of the points in* A.

(For a more informal view of upper-domination, see below).

In a completely analogous fashion, it is possible to define lower-domination of a point p by a set of points A, where in the definition, we replace p_{up} by a point $p_{below} = (p_x, p_y - c)$.

For a set A of points, the upper convex hull of A upper-dominates every point of A and, in fact, is the smallest such set of points.

We need a few simple properties of upper-domination (and lower-domination) which we prove for the sake of completeness. First, we show that upper-domination satisfies a certain "transitivity" property, more precisely:

Lemma 2. *If p is upper-dominated by the set A, and there is a point $q \in A$ which is upper-dominated by the set B, then p is upper-dominated by $(A \setminus \{q\}) \cup B$.*

Proof: Let $p = (p_x, p_y)$. By assumption, there is a point $p_{up} = (p_x, p_y + c)$ such that $c \geq 0$ and

$$p_{up} = \lambda_q \cdot q + \sum_{a \in A \setminus \{q\}} \lambda_a \cdot a$$

and there is a point $q_{up} = (q_x, q_y + c')$ such that $c' \geq 0$ and $q_{up} = \sum_{b \in B} \lambda'_b \cdot b$.

Consider now the point

$$p^* := \lambda_q \cdot q_{up} + \sum_{a \in A \setminus \{q\}} \lambda_a \cdot a.$$

It holds that $p^* - p_{up} = \lambda_q \cdot (q_{up} - q) = \lambda_q \cdot (0, c')$. This means that p^* agrees in the $x-$coordinate with p_{up} and p, and the $y-$coordinate of p^* is at least as large as p_y. Writing $p^* = \lambda_q \cdot (\sum_{b \in B} \lambda'_b \cdot b) + \sum_{a \in A \setminus \{q\}} \lambda_a \cdot a$ shows that it is a convex linear combination of the points in $(A \setminus \{q\}) \cup B$. □

Definition 6. *For two points $p = (p_x, p_y)$ and $q = (q_x, q_y)$ with $p_x < q_x$, we define the slope between p and q as*

$$inc(p, q) := \frac{q_y - p_y}{q_x - p_x}.$$

Definition 7. *A sequence $p_1 = (x_1, y_1), \ldots, p_n = (x_n, y_n)$ with $x_1 < \cdots < x_n$ is called concave if and only if $inc(p_1, p_2) \geq \cdots \geq inc(p_{n-1}, p_n)$, i.e., the sequence of slopes is monotone decreasing.*

Definition 8. *Given a concave sequence of points $p_1 = (x_1, y_1), \ldots, p_n = (x_n, y_n)$,* the graph f of the sequence *is defined as a (continous) function f defined on $[x_1, x_n]$. The function f is defined by setting $f(x) := y_i + \frac{y_{i+1} - y_i}{x_{i+1} - x_i} \cdot (x - x_i)$ if $x_i \leq x \leq x_{i+1}$.*

The figure below shows the graph of the upper convex hull and the graph of the lower convex hull of our example point set M.

Informally, one can say that a point p is upper-dominated by a set A if it lies below (or on) the graph of the upper convex hull of A.

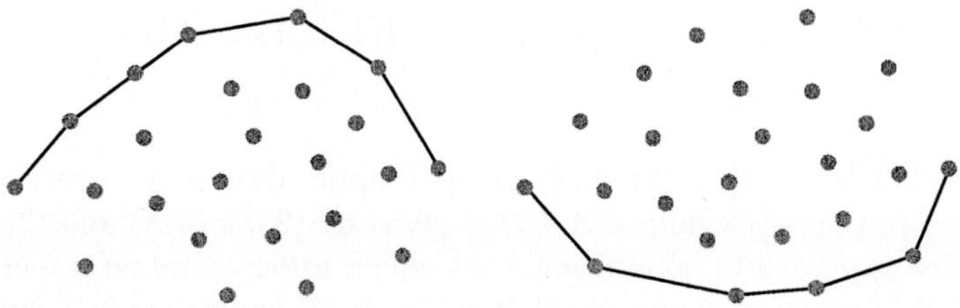

The following lemma should be rather obvious which is why we just give an informal sketch of the proof:

Lemma 3. *Let* $t_1, \ldots, t_n$ *be a concave sequence of points. Define the sequence* $\mathrm{diff}_1, \ldots, \mathrm{diff}_{n-1}$ *by* $\mathrm{diff}_i := t_{i+1} - t_i$ *and let* $\mathrm{diff}_{\ell(1)}, \ldots, \mathrm{diff}_{\ell(r)}$ *(with* $\ell(1) < \cdots < \ell(r)$*) be a subsequence of* $\mathrm{diff}_1, \ldots, \mathrm{diff}_{n-1}$*. Then the following holds:*

Every point of the sequence $s_1, \ldots, s_{r+1}$*, defined by* $s_1 := t_1$ *and* $s_{i+1} := s_i + \mathrm{diff}_{\ell(i)}$ *is upper-dominated by* $\{t_1, \ldots, t_n\}$*.*

Proof: (Informally): Let f be the graph of $t_1, \ldots, t_n$ and g be the graph of $s_1, \ldots, s_{r+1}$. Then $f(x) - g(x) \geq 0$ for all x in the domain of g, since in every point x, the increase of f in x is larger than the increase of g in x. □

The following figure provides a visualization of Lemma 3.

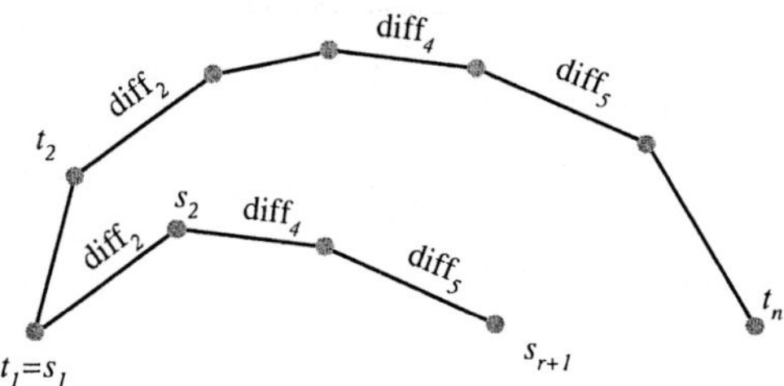

4 Joining Two Concave Sequences

In this section, we describe the main operation that our algorithm is based on.

For two point sets P and Q, we define the set $P + Q$ of points by

$$P + Q := \{p + q \mid p \in P, q \in Q\}.$$

This addition operation is also known under the name "Minkowski sum". (See, e.g., [dBvKOS]).

$P + Q$ may contain up to $|P| \cdot |Q|$ points. Given two point sets A and B, by *joining* A and B, we mean computing a concave sequence $t_1, \ldots, t_m$ of points from $A + B$ such that $\{t_1, \ldots, t_m\}$ upper-dominates every point of $A + B$. Note that this notion is different from the notion of "merging two (upper) convex hulls". Joining the sequences is basically the same as computing the upper convex hull of $A + B$ (which upper-dominates every point of $A + B$). The only (rather unimportant) difference is that the set $\{t_1, \ldots, t_m\}$ is not required to be minimal.

Example: Let $A = \{(1,1),(2,3),(3,4)\}$, $B = \{(1,1),(3,5),(4,6)\}$, then

$$A + B = \{(2,2),(4,6),(5,7),(3,4),(5,8),(6,9),(4,5),(6,9),(7,10)\}$$

and the set $\{(2,2),(3,4),(5,8),(6,9),(7,10)\}$ upper-dominates every point in $A+B$. The upper convex hull of $A+B$ is given by $(2,2),(5,8)$ and $(7,10)$.

The trivial approach to computing a set which upper-dominates every point of $A+B$ would be to first compute all points in $A+B$ and to apply a convex hull algorithm to the resulting set. This would take time at least $|A|\cdot|B|$ which would be too slow for our purposes. Instead, we apply an algorithm for computing the Minkowski sum of two convex polygons (see [dBvKOS], Theorem 13.11) to our setting. As a consequence, we can join A and B in time $O(|A|+|B|)$, if upper-dominating sets for A and B are already given. For the sake of completeness, we describe the algorithm for joining A and B below.

Let $p_1,\ldots,p_r$ and $q_1,\ldots,q_s$ be two concave sequences. We show how these two sequences can be joined in time $O(r+s)$. Define the two slope sequences $\Delta_1,\ldots,\Delta_{r-1}$ and $\Delta'_1,\ldots,\Delta'_{s-1}$ by

$$\Delta_i := inc(p_i,p_{i+1}) \text{ and } \Delta'_i := inc(q_i,q_{i+1}).$$

The two sequences are monotone decreasing. Thus, we can merge them in time $O(r+s)$ into one sequence $\Delta''_1,\ldots,\Delta''_{r+s-2}$ which is also monotone decreasing. For this purpose, the well-known merge step from the mergesort algorithm can be used. We can define the sequence t by the following algorithm. The algorithm works as follows: The points $t_1,\ldots,t_{r+s-1}$ are chosen one by one. When $t_\ell = p_i+q_j$ is already chosen, then it is checked whether the slope from t_ℓ to the point $p_{i+1}+q_j$ or to the point p_i+q_{j+1} is larger and the corresponding point is chosen as the next point $t_{\ell+1}$. Here is the procedure in algorithmic notation:

```
JOINING(p_1,...,p_r,q_1,...,q_s)  # Output is the sequence t_1,...,t_{r+s-1}.
i := 1; j := 1;  t_1 := p_1 + q_1;

while i + j ≤ r + s − 1 do
begin # The next run through the loop will define t_{i+j}.
   if i = r then j := j + 1; goto (*)
   if j = s then i := i + 1; goto (*)
   # Now the test whether Δ_i ≥ Δ'_j:
   if inc(p_i + q_j, p_{i+1} + q_j) ≥ inc(p_i + q_j, p_i + q_{j+1})
   then i := i + 1 else j := j + 1;
   (*) t_{i+j−1} := p_i + q_j.
end;
```

Theorem 2. *Every point in* $\{p_1,\ldots,p_r\}+\{q_1,\ldots,q_s\}$ *is upper-dominated by* $\{t_1,\ldots,t_{r+s-1}\}$. *The running time of the algorithm is* $O(r+s)$.

Proof: The statement on the running time is obvious. For the proof of the upper-domination, let us first note a few trivial properties of the t-sequence:

I) $t_1 = p_1 + q_1$ and $t_{r+s-1} = p_r + q_s$.
II) If $t_i = p_k + q_\ell$, then either $t_{i+1} = p_{k+1} + q_\ell$ or $t_{i+1} = p_k + q_{\ell+1}$.
III) The t-sequence is concave.

We claim that for every $1 \leq i \leq r$ and $1 \leq j \leq s$, the point $p_i + q_j$ is upper-dominated by $\{t_1, \dots, t_{r+s-1}\}$. To show this, choose $t_\ell = p_{i'} + q_{j'}$ in such a way that ℓ is the smallest index where $i' = i$ or $j' = j$. I.e., t_ℓ is the first point in the sequence which is of the form $p_i + q_{\dots}$ or $p_{\dots} + q_j$. W.l.o.g., we can assume that $t_\ell = p_i + q_{j'}$, where (due to property II)), $j' \leq j$. Consider the sequence of points

$$s_{j'} := t_\ell = p_i + q_{j'}, \quad s_{j'+1} := p_i + q_{j'+1}, \dots, \quad s_j := p_i + q_j.$$

The sequence of differences $(s_{j'+1} - s_{j'}, \dots, s_j - s_{j-1}) = (q_{j'+1} - q_{j'}, \dots, q_j - q_{j-1})$ is a subsequence of $t_{\ell+1} - t_\ell, \dots, t_{r+s-1} - t_{r+s-2}$: This is due to properties I) and II) and the fact that $t_\ell = p_i + q_{j'}$.

Thus, we can apply Lemma 3 to obtain that the points $s_{j'}, \dots, s_j$ are upper-dominated by $\{t_\ell, \dots, t_{r+s-1}\}$ and thus upper-dominated by $\{t_1, \dots, t_{r+s-1}\}$. Since $s_j = p_i + q_j$, the claim follows. □

It should also be clear that an analogous joining operation for lower-domination and convex sequences can be defined and that an analogous algorithm for implementing this joining operation exists.

5 The Algorithm for Solving the Generalized Maximum Subsequence Problem

The basic idea of our algorithm is the following: There are $\Theta(n^2)$ intervals of the form $[i, j]$ with $i \leq j$.

Each of them can be mapped to a point $p_{i,j} := (\ell(i,j), \mathrm{sum}(i,j))$ in $\mathbb{R}^2$. For the divide-and-conquer step that we use, it is important to realize that $p_{i,j} = p_{i,k} + p_{k+1,j}$ for every k with $i \leq k \leq j-1$.

Let $M := \{p_{i,j} \mid 1 \leq i \leq j \leq n\}$ be the set of all these points. The generalized maximum subsequence problem asks for the maximum of $f(m)$ where $m \in M$.

Since f is a quasiconvex function, it assumes its maximum value on M on a point in the convex hull of M. Since $|M| = \Omega(n^2)$ is possible, it is prohibitive to compute M first and then compute the convex hull of M by a standard approach.

Our algorithm (in its current form) does not explicitly construct the convex hull of M, but it evaluates f on a subset of M which is a superset of the convex hull of M. Thus, it finds a point where f assumes its maximum value. We obtain an overall running time of $O(n \log n)$.

Note that with the help of extra merge steps, one could also use our algorithm for computing the convex hull of M in time $O(n \log n)$. Although $|M| = \Omega(n^2)$ is possible, this does not contradict the known lower bound for convex hull algorithms as the set M is given implicitly.

Before we describe our algorithm, let us state another two simple properties of upper- and lower-domination:

Lemma 4

i) If a is upper-dominated by A and b is upper-dominated by B, then $a+b$ is upper-dominated by $A+B$.
ii) If p is upper-dominated by A and lower-dominated by B, then p is a convex linear combination of $A \cup B$.

The proof of *i)* is an easy exercise which uses that a point a which is a convex linear combination of A and a point b which is a convex linear combination of B satisfy that $a+b$ is a convex linear combination of $A+B$.

The proof of *ii)* uses the fact that p_{up} is a convex linear combination of A, p_{below} is a convex linear combination of B and that p is a point on the line between p_{up} and p_{below}.

Our algorithm uses a divide-and-conquer approach. If the array length n is equal to 1, it is easy to compute the maximum.

For $n > 1$, we divide the array into two halves of lengths $\lfloor n/2 \rfloor$ and $\lceil n/2 \rceil$ each, then solve the problem recursively in both halves. These two recursive calls take care of all intervals that are completely contained in the left half of the array and those that are completely contained in the right half of the array. Let OPT_{left} be the maximal function value in the left half and $\text{OPT}_{\text{right}}$ be the maximal function value in the right half.

It remains to compute $\text{OPT}_{\text{crossing}}$, the maximal function value of intervals that contain at least one element from the first half and at least one element from the second half. The final result is then the maximum of those three values, i.e., $\max\{\text{OPT}_{\text{left}}, \text{OPT}_{\text{right}}, \text{OPT}_{\text{crossing}}\}$.

Computing $\text{OPT}_{\text{crossing}}$ is the part where the joining procedure from Section 4 comes into play.

Figure 1 on the next page provides the algorithm ALGO(i,j) in algorithmic notation. In order to avoid notational mess caused by rounding, it is assumed in the description that the length of the input array is a power of two.

Theorem 3. *ALGO$(1,n)$ solves the generalized maximum subsequence problem in time $O(n \log n)$.*

Proof: By induction on the length of the array. The induction base is trivial. In the induction step, we know that OPT_{left} yields the maximal f-value for all intervals that lie completely in the subarray $a_i, \ldots, a_{\text{middle}}$ and that $\text{OPT}_{\text{right}}$ yields the maximal f-value for all intervals that lie completely in the subarray $a_{\text{middle}+1}, \ldots, a_j$. We call an interval *crossing* if it starts in the first half and ends in the second half. The set of points that crossing intervals are mapped to, is given by

$$M := \{(\ell(g,h), \text{sum}(g,h)) \mid i \le g \le \text{middle and middle} + 1 \le h \le j\}.$$

By definition, we have $\text{OPT}_{\text{crossing}} = \max\{f(m) \mid m \in M\}$. Since the algorithm computes $\max\{f(m) \mid m \in T\}$ in its last three lines, it remains to show that

$$\max\{f(m) \mid m \in T\} = \max\{f(m) \mid m \in M\}.$$

```
ALGO(i, j) # find the maximal value of f on the subarray a_i, ..., a_j
begin
   if i = j then return f(1, a_i); # Trivial case. Length is 1 and sum is a_i.
   middle := (j + i - 1)/2;
   length := middle - i + 1; # Length of a half.
   OPT_left := ALGO(i, middle);
   OPT_right := ALGO(middle + 1, j);

   # Now treat the crossing intervals. First treat the left half.
   b_1 := (1, a_middle); for k = 2 to length, set b_k = b_{k-1} + (1, a_{middle-k+1}).

   # Thus, b_k = (k, sum(middle - k + 1, middle)),
   # i.e., the first coordinate of b_k is k and its
   # second coordinate is the sum of the interval that ends
   # with array index middle and has length k.
   # Now the right half.

   In a similar fashion, compute b'_k = (k, sum(middle + 1, middle + k)),
   for all k = 1, ..., length.
   # I.e., the second coordinate of b'_k is the sum of the interval that
   # begins with array index middle + 1 and has length k.
(*)
   Compute the upper convex hull p_1, ..., p_r of the points b_1, ..., b_length.
   Compute the upper convex hull q_1, ..., q_s of the points b'_1, ..., b'_length.

   Join the two (concave) sequences p_1, ..., p_r and q_1, ..., q_s, using
   the joining algorithm from Section 4.
   The joining algorithm outputs a sequence t_1, ..., t_{r+s-1}.
(**)
   Repeat steps (*) to (**), with the lower instead of the upper convex hull and
   the analogous joining algorithm. Call the output sequence t'_1, ..., t'_{r'+s'-1}.
   Set T := {t_1, ..., t_{r+s-1}} ∪ {t'_1, ..., t'_{r'+s'-1}}.
   Evaluate f on all points in T and set OPT_crossing := max{f(x) | x ∈ T}.
   return max{OPT_left, OPT_right, OPT_crossing}
end
```

Fig. 1. The algorithm for the generalized maximum subsequence problem

We first show that "$\leq$" holds by showing that $T \subseteq M$:

Each crossing interval can be divided into two intervals, the first of which ends with position middle and the second of which starts with position middle + 1. Hence, it is clear that M is exactly the set $\{b_1, \ldots, b_{\text{length}}\} + \{b'_1, \ldots, b'_{\text{length}}\}$.

The points $p_1, \ldots, p_r$ are elements from $\{b_1, \ldots, b_{\text{length}}\}$, the points $q_1, \ldots, q_s$ are elements from $\{b'_1, \ldots, b'_{\text{length}}\}$, thus $\{t_1, \ldots, t_{r+s-1}\} \subseteq M$. In an analogous fashion, $\{t'_1, \ldots, t'_{r'+s'-1}\} \subseteq M$ can be shown. It follows that $T \subseteq M$.

We now show that "$\geq$" holds. Consider a point m from M, i.e., a point of the form $m = b_u + b'_v$ for some u and v.

Assume that $b_u + b'_v$ is upper-dominated by T and lower-dominated by T. We will show in a moment that this does indeed hold.

From Lemma 4, we then obtain that $b_u + b'_v$ is a convex linear combination of T and by the property of quasiconvex functions mentioned before Definition 2, it follows that $f(m) = f(b_u + b'_v) \leq \max\{f(x) \mid x \in T\}$.

We now show that $b_u + b'_v$ is upper-dominated by T: $\{p_1, \ldots, p_r\}$ is an upper convex hull for $b_1, \ldots, b_{\text{length}}$, i.e., b_u is upper-dominated by $\{p_1, \ldots, p_r\}$. Likewise, b'_v is upper-dominated by $\{q_1, \ldots, q_s\}$.

By Lemma 4, $b_u + b'_v$ is upper-dominated by $\{p_1, \ldots, p_r\} + \{q_1, \ldots, q_s\}$. By Theorem 2, every point in $\{p_1, \ldots, p_r\} + \{q_1, \ldots, q_s\}$ is upper-dominated by the computed set $\{t_1, \ldots, t_{r+s-1}\}$. By "transitivity", every point in $b_u + b'_v$ is thus upper-dominated by $\{t_1, \ldots, t_{r+s-1}\}$.

In a similar fashion, one can show that every $b_u + b'_v$ is lower-dominated by $\{t'_1, \ldots, t'_{r'+s'-1}\}$. Thus, every $b_u + b'_v$ is upper-dominated and lower-dominated by T, and we are done.

Now for the running time. In the algorithm, we compute the lower and upper convex hulls of the points $b_1, \ldots, b_{\text{length}}$ and $b'_1, \ldots, b'_{\text{length}}$. This can be achieved in linear time $O(n)$, since the x-coordinates of the sequences are already in sorted order.

Define $V(n)$ as the time our algorithm takes on inputs of length n. We then have the recursive inequality

$$V(n) \leq V(\lfloor n/2 \rfloor) + V(\lceil n/2 \rceil) + c \cdot n,$$

for some constant $c > 0$. The first two terms stem from the recursive calls, the term $c \cdot n$ estimates the time spent in computing the upper and lower hulls, joining the hulls and evaluating f on the candidate set. Here, we use the fact that the candidate set contains at most n points. It is well-known that the above inequality can be estimated by $V(n) = O(n \log n)$. □

6 Open Problems

It is a natural question whether the generalized maximum subsequence problem can also be solved in linear time $O(n)$ or whether it is so general that one can show a lower bound of $\Omega(n \log n)$ for at least one quasiconvex function f.

Acknowledgments

Our thanks go to Laurie Davies for bringing our attention to the multiresolution criteria problem. We also thank Friedrich Eisenbrand for pointing us to the notion of Minkowski sums.

References

[A] L. Allison, *Longest biased interval and longest non-negative sum interval,* Bioinformatics 19(10), p. 1294-1295, 2003.

[B] J. Bentley, *Programming Pearls,* Addison-Wesley, 1984.

[BV] S. Boyd and L. Vandenberghe, Convex Optimization, Cambridge University Press, 2004.

[CC] K.-Y. Chen and K.-M. Chao, *On the Range Maximum-Sum Segment Query Problem,* 15th International Symposium on Algorithms and Computation (ISAAC), p. 294-305, 2004.

[dBvKOS] M. de Berg, M. van Kreveld, M. Overmars, O. Schwarzkopf, *Computational Geometry – Algorithms and Applications,* Second Edition, Springer, 2000.

[DK] P. L. Davies and A. Kovac, *Local Extremes, Runs, Strings and Multiresolution (with discussion),* Annals of Statistics 29, p. 1-65, 2001.

[FLLTWY] T.-H. Fan, S. Lee, H.-I Lu, T.S. Tsou, T.-C. Wang, and A. Yao, *An Optimal Algorithm for Maximum-Sum Segment and Its Application in Bioinformatics,* Eighth Conference on Implementation and Application of Automata (CIAA), LNCS 2759, p. 251-257, 2003.

[LJC] Y.-L. Lin, T. Jiang, K.-M. Chao, *Efficient algorithms for locating the length-constrained heaviest segments, with applications to biomolecular sequence analysis,* Journal of Computer and System Sciences 65 (2002), p. 570-586.

[LABLY] D. Lipson, Y. Aumann, A. Ben-Dor, N. Linial and Z. Yakhini, *Efficient Calculation of Interval Scores for DNA Copy Number Data Analysis,* Ninth Annual International Conference on Research in Computational Molecular Biology (RECOMB), 2005.

[M] K. Mulmuley, *Computational geometry: An introduction through randomized algorithms,* Prentice-Hall, 1994.

Random Bichromatic Matchings

Nayantara Bhatnagar*,***, Dana Randall*,
Vijay V. Vazirani**, and Eric Vigoda***

College of Computing, Georgia Institute of Technology, Atlanta, GA 30332
{nand, randall, vazirani, vigoda}@cc.gatech.edu

Abstract. Given a graph with edges colored RED and BLUE, we wish to sample and approximately count the number of perfect matchings with exactly k RED edges. We study a Markov chain on the space of all matchings of a graph that favors matchings with k RED edges. We show that it is rapidly mixing using non-traditional canonical paths that can backtrack, based on an algorithm for a simple combinatorial problem. We show that this chain can be used to sample dimer configurations on a 2-dimensional toroidal region with k RED edges.

1 Introduction

Counting the number of matchings in a graph is a well-studied problem in combinatorics and computer science. Counting the number of perfect matchings in a bipartite graph is equivalent to computing the permanent of a matrix with 0, 1 entries. This problem is also of interest in statistical physics in the context of understanding the thermodynamic properties of a dimer system [3, 4]. Motivated by this application, Kastelyn showed that for planar graphs the number of perfect matchings can be computed exactly [9]. Recently Jerrum, Sinclair and Vigoda [6] gave an *fpras* (fully polynomial approximation scheme) approximating the number of perfect matchings in any bipartite graph, which is based on an *fpaus* (fully polynomial almost uniform sampler) for generating random perfect matchings.

A natural generalization of the matching problem is when the edges of the graph are colored RED or BLUE:

Problem: *Given a graph $G(V, E)$, a partition $E = R \cup B$, and $k \leq \frac{|V|}{2}$, count the number of perfect matchings in G with exactly k edges in R.*

The decision version of this problem is to find a matching with exactly k RED edges. These problems have been studied in combinatorial optimization [12] as well as statistical physics [2]. There are several open questions regarding both the decision and the counting versions of this problem. For the decision version of this problem, known as *exact matchings*, Mulmuley, Vazirani and Vazirani [11] give a randomized algorithm for general graphs. A deterministic algorithm is known only when the graph is complete or complete bipartite [8, 14].

* Supported in part by NSF grants CCR-0515105 and DMS-0505505.
** Supported in part by NSF grants 0311541, 0220343 and CCR-051586.
*** Supported in part by NSF grant CCR-0455666.

J.R. Correa, A. Hevia, and M. Kiwi (Eds.): LATIN 2006, LNCS 3887, pp. 190–201, 2006.

A special case of the counting problem, of interest in statistical physics, is where G is the $\sqrt{n} \times \sqrt{n}$ 2-dimensional lattice and the horizontal edges are RED, while the vertical edges are BLUE. We wish to count the number of dimer coverings with exactly k horizontal edges, as well as solve the sampling problem. Fisher [2] gave a closed form solution for the limiting distribution (as the size of the lattice tends to infinity) of configurations in terms of the activities λ and μ of horizontal and vertical dimers, where the weight of a configuration with k horizontal edges and k' vertical edges is given by $\lambda^k \mu^{k'}$. To our knowledge, ours is the first work to address the sampling/counting problem for general graphs.

We make progress on this problem for general graphs and solve the problem in some natural special cases. Our results for general graphs are best viewed in terms of the partition function for matchings. Throughout, let $\mathcal{M}$ denote the set of all matchings of an input graph G, and $\mathcal{P}$ denote the set of perfect matchings. The standard partition function on matchings

$$Z(\lambda) = \sum_{M \in \mathcal{M}} \lambda^{|M|}$$

can be approximated for all λ by the algorithm of Jerrum and Sinclair [5]. We show that we can approximate a modified partition function which puts most weight on *(k, ℓ)-matchings*, i.e. matchings of size ℓ with exactly k RED edges.

Theorem 1. *For any $G(V, E)$ with a partition of the edges $E = R \cup B$, activities $\lambda, \mu \leq 1$, any $\ell \leq |V|/2$ and $k \leq \ell$, there is an fpras for estimating the following partition function over weighted matchings:*

$$Z_{k,\ell}(\lambda, \mu) = \sum_{M \in \mathcal{M}} \lambda^{||M \cap R| - k|} \mu^{||M| - \ell|}. \tag{1}$$

An n-vertex graph is dense if it has minimum degree $d_{min} > n/2$. A bipartite graph with each partition of size n is dense if it has $d_{min} > n/4$.

Theorem 2. *For any dense graph $G(V, E)$, activity $\lambda \leq 1$, and $k \leq |V|/2$, there is an fpras for estimating the following partition function:*

$$\widehat{Z}_k(\lambda) = \sum_{P \in \mathcal{P}} \lambda^{||P \cap R| - k|}. \tag{2}$$

We approximate the partition functions within a factor $(1 \pm \varepsilon)$ w.p. $\geq 1 - \delta$. The running time in each case is polynomial in $1/\lambda, 1/\mu, 1/\varepsilon, \log(1/\delta)$ and the size of the graph.

We demonstrate the significance of these results on the 2-dimensional torus $\mathbb{Z}_{m_1} \times \mathbb{Z}_{m_2}$ for even m_1, m_2, taking the horizontal edges to be RED and the vertical edges to be BLUE. In particular, we present a polynomial time algorithm for approximately sampling and counting the set of perfect matchings (or *dimer coverings*) with exactly k RED edges. We note that there are algorithms to exactly count the number of perfect matchings on the 2-d torus [9] which can be extended to bichromatic matchings. However, our proof can be extended to the *monomer-dimer* model in which we approximately sample and count (k, ℓ)-matchings of the 2-d torus, giving the first solution to this problem.

Theorem 3. *Given any torus $\mathbb{Z}_{m_1} \times \mathbb{Z}_{m_2}$ with m_1 and m_2 even, any non-negative integer $k \leq m_1 m_2/2$ and any $\ell \geq k$, there is an fpaus for generating a random (k,ℓ)-matching of the torus and an fpras for estimating the number of such matchings that run in time polynomial in m_1 and m_2.*

Theorem 1 uses a Markov chain defined on the set of all matchings of the graph which puts most weight on (k,ℓ)-matchings. We use the *canonical paths technique* to bound the convergence rate of the Markov chain. Here, these paths are non-trivial to define, in contrast to the usual matching problem where the analysis of the path congestion was the harder task.

The combinatorial fact that enables us to define our paths is as follows. Consider a graph with edges colored RED and BLUE. For any k and for all perfect matchings P, P' with exactly k RED edges, there is a polynomial length path between P and P' along almost perfect matchings, with successive matchings differing by only a few edges, such that each contains close to k RED edges. We can reduce the problem of finding such a path to a combinatorial problem about moving two points along a two-dimensional landscape in a co-ordinated manner so that the sum of their heights stays constant. The canonical path from P to P' defined in [5] starts at the matching P and alternately deletes an edge of P' and adds an edge of P' along an alternating cycle. An interesting aspect of our canonical paths is that they may backtrack along portions of the alternating cycle, for instance we might delete edges of P' that were previously added.

Our second technical contribution is proving combinatorial inequalities that allow us to approximate the number of (k,ℓ)-matchings on the torus. Kenyon, Randall and Sinclair [10] showed that the number of near perfect matchings in the d-dimensional torus is polynomially related to the number of perfect matchings, thereby yielding polynomial time algorithms for approximately counting and uniformly sampling perfect matchings. In this paper, we generalize their result to show that, on the 2-d torus, this relationship holds even when we restrict to sets of matchings with exactly k RED edges. Our result builds on ideas of Temperley [13] and Burton and Pemantle [1] for constructing augmenting paths where every horizontal and vertical segment has even length.

2 Approximately Counting Bichromatic Matchings

We outline the proof of Theorem 1 in this section; similar ideas are used to prove Theorem 2. By a standard reduction, approximating the partition function $Z_{k,\ell}(\lambda,\mu)$ can be reduced to approximate sampling [7], so we concentrate on the sampling problem and defer the details of the fpras to the full version.

To solve the sampling problem we define a Markov chain on the set of matchings $\mathcal{M}$ which puts most weight on (k,ℓ)-matchings. The same chain was used by Jerrum and Sinclair [5], with the transition probabilities defined so that the stationary distribution was uniform over all matchings.

The Markov Chain $\mathcal{T}$. The state space is $\mathcal{M}$, the set of all matchings of G. Let $\ell \leq |V|/2$, $0 \leq k \leq \ell$ and $0 < \lambda, \mu \leq 1$. Define the weight of a matching M, as $w(M) = \lambda^{|k-|M \cap R||}\mu^{|\ell-|M||}$. The transitions $M_t \to M_{t+1}$ of $\mathcal{T}$ are defined as follows.

From a matching M_t, choose a random edge $e = (u, v) \in E$.

1) If $e \in M_t$ set $M' = M_t \setminus \{e\}$.
2) If $M \in \mathcal{N}(u, v)$, (i.e. u, v are unmatched), set $M' = M_t \cup \{e\}$.
3) If for $z \neq v$, $M_t \in \mathcal{N}(u, z)$ and $(w, v) \in M_t$, set $M' = (M_t \cup \{e\}) \setminus (w, v)$. Set $M_{t+1} = M'$ with probability $\frac{1}{2}\min(1, w(M')/w(M))$, else set $M_{t+1} = M_t$.

It is straightforward to verify that the Markov chain is connected, aperiodic and reversible and has stationary distribution proportional to $w(M)$.

Intuition for the Canonical Paths

In the canonical path method for bounding the mixing time of a Markov chain, for each pair of matchings I, F, we define a path from I to F along transitions of the chain. We need to bound the *congestion* of these paths through every transition to show that the Markov chain converges quickly.

The approach of Jerrum and Sinclair [5] to obtain this bound is to focus on a specific transition T. For each pair (I, F) whose path uses the transition T, we define an "encoding" E, which is also a matching; T and E let us recover (I, F), so E can be viewed as an injective map. Then the number of (I, F) pairs whose path uses T is at most the number of matchings, which is $|\Omega|$. This is sufficient to bound the congestion for unweighted matchings. For weighted matchings, we also need to show that $w(I)w(F) \leq w(T)w(E)poly(n)$. The encoding is defined as $E = (I \cup F) \setminus (M \cup M')$ where $T = M \to M'$, so E can be viewed as the complementary matching of T with respect to (I, F).

Suppose that $\ell = |V|/2$ so that we favor perfect matchings. If I and F are perfect matchings with k RED edges, they have maximum weight. The weight of transitions and encodings along the canonical path from I to F must be comparable to the weight of I and F. Hence, both T and E need to contain close to k RED edges, and simultaneously be close to a perfect matching (i.e., have only a constant number of unmatched vertices or "holes").

Consider the perfect matchings I, F, and suppose $I \oplus F$ (the symmetric difference of I and F) consists of a single alternating cycle. The transitions of the chain allow us to easily "unwind" this alternating cycle: remove one of the edges of I on the cycle, then perform a series of shifts (moves of type 3), and then add the final edge of the cycle of F.

To see the difficulty, suppose, as in Figure 1, this cycle alternates RED on I and BLUE on F on one half of the cycle, and BLUE on I and RED on F on

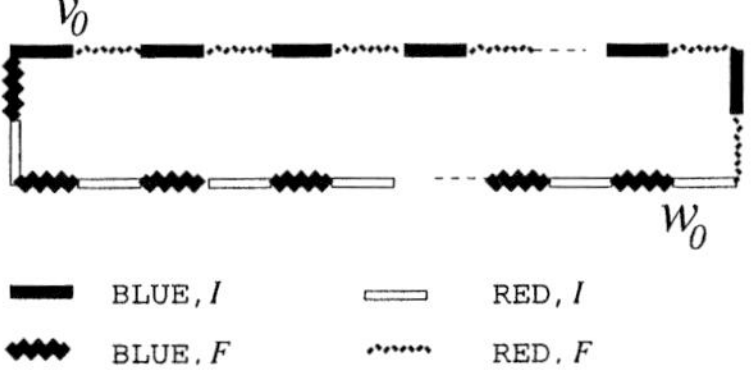

Fig. 1. An alternating cycle in $I \oplus F$

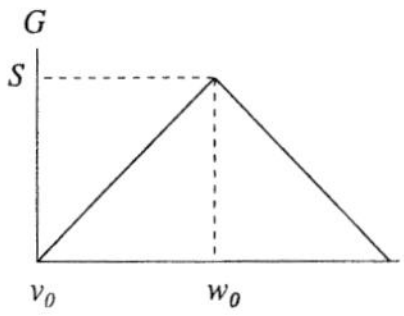

Fig. 2. Landscape for cycle

the other. Then, no matter where we start the unwinding there will be some intermediate matching with far more (or far less) RED edges than the intended k. Notice that in this example there are *two* vertices v_0, w_0 so that if we unwind from these two points *simultaneously* then we can ensure that the number of RED edges differs from k by at most a constant. It turns out that we can always choose two such positions to begin the unwinding of the cycle. To define the unwinding, it is helpful to look at the alternating cycle together with a function representing the number of RED edges gained along the cycle.

However, the protocol for unwinding is not straightforward and we may need need to backtrack (switch edges back from F to I) from one position to continue unwinding at the other. Hence, it is not obvious whether our paths can always make progress. Additionally, the picture is more complicated when $I \oplus F$ consists of multiple cycles and paths with varying lengths and numbers of RED edges. We focus on formalizing the problem of unwinding a single alternating cycle and defer the general case to the full version.

Paired Mountain Climbing

Consider the case that $I \oplus F$ is a single alternating cycle and I and F both contain exactly k RED edges. We would like to transform the cycle from I to F so that all the intermediate matchings have close to k RED edges.

For every other vertex v on the alternating cycle, assign $-1, 0$ or $+1$ to denote the change in the number of RED edges. Thus, for $e = (u, v) \in I, e' = (v, w) \in F$, $f(v) = 1_{e' \in R} - 1_{e \in R}$, where 1 is the indicator function. Fix a start vertex on the cycle, say v_0, and a direction for unwinding the cycle. For every vertex $v_{2\ell+1}$ on the cycle, let $G(v_{2\ell+1}) = \sum_{i=0}^{\ell} f(v_{2i+1})$, where $v_0 \to v_1 \to \cdots \to v_\ell$ is the alternating path from v_0 to v_ℓ. The function G defines a "landscape", as shown in Figure 2.

It can be shown that if $|I \cap R| = |F \cap R| = k$, then there always exists a vertex v_0 so that $G(v_0) = 0$, $G(v_\ell) \geq 0$ for all ℓ, and 0 again at the last vertex. We choose a companion start vertex for v_0 which is a (global) maximum, denote this vertex as w_0. Let $S = G(v_0) + G(w_0)$. We break the alternating cycle into a pair of alternating paths, $P = \{v_0, v_1, \ldots, v_n\}$ and $Q = \{w_0, \ldots, w_m\}$, where v_n is the vertex before w_0 and w_m is the vertex before v_0.

We now start unwinding the cycle at the vertices v_0 and w_0. If unwinding from one of the positions adds a RED edge, then from the other position we need to remove a RED edge by moving forward or backwards as necessary. Thus, if at some intermediate step we are at vertices v_i and w_j, we need that $(G(v_i) - G(v_0)) + (G(w_j) - G(w_0)) = 0$, i.e. $G(v_i) + G(w_j) = S$. The mountain climbing problem is to determine a (short) trajectory from (v_0, w_0) to (v_n, w_m) so that at each intermediate step (v_i, w_j) we have $G(v_i) + G(w_j) = S$. We may need to move backwards on one path in order to move forward on the other path, and this corresponds to rewinding the cycle.

We defer the details of the canonical paths for general I, F to the full version and focus instead on the algorithm for the mountain climbing problem which has all the ideas necessary to solve the problem in general.

The Algorithm for Mountain Climbing
A *landscape* is a function $P : [n] \to \mathbb{Z}_{\geq 0}$ such that for $1 \leq i \leq n-1$, $|P(i+1) - P(i)| = 1$ (see Figure 3). For $n, m > 1$, given landscapes $P : [n] \to \mathbb{Z}_{\geq 0}$ and $Q =: [m] \to \mathbb{Z}_{\geq 0}$, we say P and Q are *S-matched* if there is an integer S s.t.

i) $P(1) + Q(m) = P(n) + Q(1) = S$
ii) $P(1) = \min_i\{P(i)\}$, $P(n) = \max_i\{P(i)\}$, $Q(1) = \max_j\{Q(j)\}$, $Q(m) = \min_j\{Q(j)\}$.

A *traversal* of S-matched landscapes P, Q is a sequence $(i_1, j_1), \cdots, (i_\ell, j_\ell)$, s.t.

i) $i_1 = 1$, $j_1 = 1$, $i_\ell = n$, $j_\ell = m$
ii) For $1 \leq k \leq \ell - 1$, $|i_{k+1} - i_k| = 1$, $|j_{k+1} - j_k| = 1$ and $P(i_k) + Q(j_k) = S$.

Lemma 1. *Let P and Q be S-matched landscapes on $[n]$ and $[m]$ respectively. Then, there exists a traversal of P and Q of length at most $O(nm)$ and it can be found in time $O(nm)$.*

Proof. The proof is by induction on $n+m$. Let $S = min + max$, where $f_1 = g_m = min$ and $f_n = g_1 = max$. Assume that the $min < max$, otherwise, the problem is trivial. Also, we use "$(1, n, 1, m)$" as shorthand for the problem of determining a traversal for P, Q. We start by showing the inductive step and conclude with the base cases.

Case I: P has a maximum or minimum at i where $1 < i < n$.

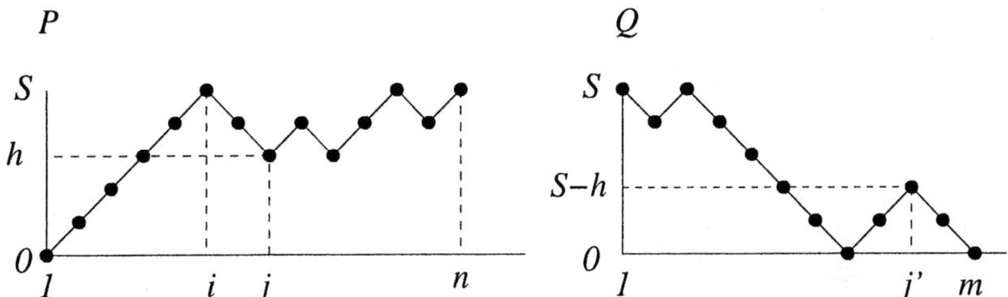

Fig. 3. Case Ia

Case Ia: Suppose that the first such point i is a maximum (Figure 3). Let h be the lowest value taken by P from i to n. Let j be the first point between i and n such that $P(j) = h$. Since both i and n are maxima of P, $i < j < n$. Let j' be the first point going from m to 1 (the direction here is important) such that $Q(j') = S - P(j)$. Note that it may be that $j' = 1$, but since m is a minimum of Q, $j' < m$. To find a traversal of P, Q, it is enough to concatenate the traversals for the following subproblems, in the given order: $(1, i, 1, m)$, (i, j, m, j'), (j, n, j', m). The functions on the shorter intervals take their values from P and Q. It can be verified that in each case, we obtain a problem of finding a traversal for smaller S-matched landscapes.

Case Ib: The first such point i is a minimum (Figure 4). Let h be the maximum value taken by P from 1 to i. Let j be the first point between 1 and i such that

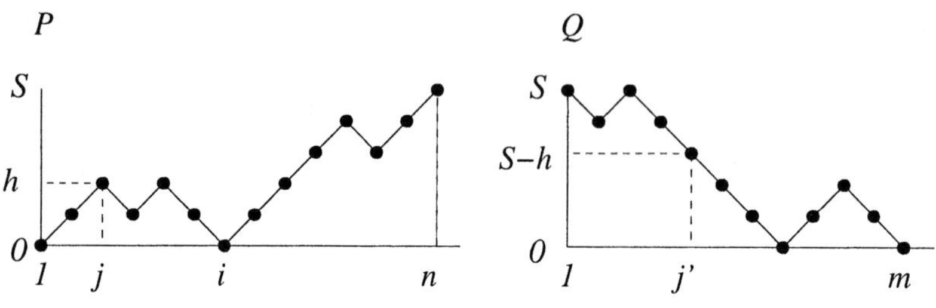

Fig. 4. Case Ib

$P(j) = h$. Since both 1 and i are minima of P, $1 < j < i$. Let j' be the first point after 1 where $Q(j') = S - h$. Since j is not a minimum of P, $j' > 1$. In this case, concatenate the traversals for the following subproblems in the given order to obtain a traversal of P, Q: $(1, j, 1, j')$, $(j, i, j', 1)$, $(i, n, 1, m)$.

Case II: Q has a maximum or minimum at i where $1 < i < m$. This case follows by symmetry from Case I.

Case III: The last case is when there is a unique maximum and minimum on P and Q. We concatenate the traversals for the subproblems $(1, 2, 1, 2)$ and $(2, n, 2, m)$, both of which are smaller problems than $(1, n, 1, m)$. It can be verified that in both cases we are reduced to the problem of finding a traversal for S-matched landscapes. Note that to show this, it is crucial to use the fact that P and Q have a unique minimum and maximum.

For the base case, let $n = 2$. Then, $m = 2$ since we may assume the paths have unique maximum and minimum, otherwise we go by induction. Since the paths are S-matched, the only possibility (upto a reversal of direction) is that P is a landscape going 'up', and Q is a landscape going 'down'. The traversal is the obvious one.

Finally, we show by induction that there is a traversal of P, Q of length at most $O(nm)$ and it can be found in time $O(nm)$. If $n = 2$, the traversal is obvious and is of length $O(m)$. If $n, m > 2$, in each of the three cases above, the traversal restricted to P is obtained by traversing edge-disjoint 'sublandscapes'. Hence, the length of the traversal is at most $O(nm)$ by induction. The proof above gives an $O(nm)$ algorithm. □

Our solution to the mountain climbing problem allows us to define the canonical paths for matchings I, F. The canonical paths are defined so that every pair of successive matchings along the path is a transition of the Markov chain and the size of an intermediate matching lies between the sizes of I and F and consists of $[I_R - 5, F_R + 1]$ RED edges, where $I_R = |I \cap R|, F_R = |F \cap R|$ and $I_R \leq F_R$. Essentially, we think of the concatenation of all the paths and cycles of $I \oplus F$ as one long landscape, and apply Lemma 1 without ever unwinding more than constantly many cycles or paths at any time. By the previous argument, the paths are at most of polynomial length. With standard machinery it is now straightforward to show that the Markov chain mixes in polynomial time. The details can be found in the full version. This completes the outline of the proof of Theorem 1.

3 Bichromatic Matchings on the Torus

Let G_{m_1,m_2} be the torus $\mathbb{Z}_{m_1} \times \mathbb{Z}_{m_2}$ with horizontal edges colored RED and vertical edges BLUE. It is known that the total number of near perfect matchings is polynomially related to the number of perfect matchings [10]. We first generalize this result to relate near-perfect matchings restricted to k RED edges and perfect matchings on close to k RED edges. Our goal will be to show that the number of matchings with k red edges does not vary much as we change the size of the matching. We will show this by defining maps from one set of matchings to the other that are invertible with a small amount of additional information. This will allow us to define an *fpras* for counting the number of (k,ℓ)-matchings of the torus. Let $\mathcal{N}_k^i$ be the set of $(k, m_1m_2/2-i)$-matchings of G_{m_1,m_2}. Let $\mathcal{P}_k = \mathcal{N}_k^0$. Let $\mathcal{N}_k(u,v)$ be the set of $(k, m_1m_2/2-1)$ matchings with holes at u and v.

Let m_1, m_2 be even. Let V_0 (white vertices) and V_1 (black vertices) be the even and odd sublattices of G_{m_1,m_2}. Further refining these sets, let V_{00} be the set of vertices both of whose co-ordinates are even: the sets V_{01}, V_{10} and V_{11} are defined analogously (Figure 5) . Note that if u, v are the holes of any near-perfect matching of G_{m_1,m_2}, then one of them is white while the other is black. Also, if m_1, m_2 are even, the number of RED edges in any perfect matching of G_{m_1,m_2} is even.

Theorem 4. *Let $m_1, m_2 \in \mathbb{Z}$ be even and $N = m_1m_2/2$. For $0 \le i \le N-1$ and $1 \le k \le N-1-i$, there is a map $f_i : \mathcal{N}_k^{i+1} \to \mathcal{N}_k^i \cup \mathcal{N}_{k+1}^i$ such that no matching of $\mathcal{N}_k^i \cup \mathcal{N}_{k+1}^i$ is mapped to by more than $O(N^3)$ matchings of $\mathcal{N}_k^{i+1}$.*

Proof. We first prove the theorem for $i = 0$. Let $N \in \mathcal{N}_k(u,v)$ and assume wlog that $u = (u_1, u_2)$ is in V_{00}. Define an alternating path L_1^u in N as follows: start at $z_0 = u$, and follow the unmatched RED edge to the vertex $z_1 = (u_1, u_2+1)$. Now, iteratively, if at an odd vertex z_{2i-1}, follow the unique matched edge to z_{2i} (see figure 6). From each even vertex z_{2i} along the path, take the unmatched edge in the *same direction* as the edge (z_{2i-1}, z_{2i}), so each segment (in the horizontal or vertical direction) of the path after the first step has even length. Continue in

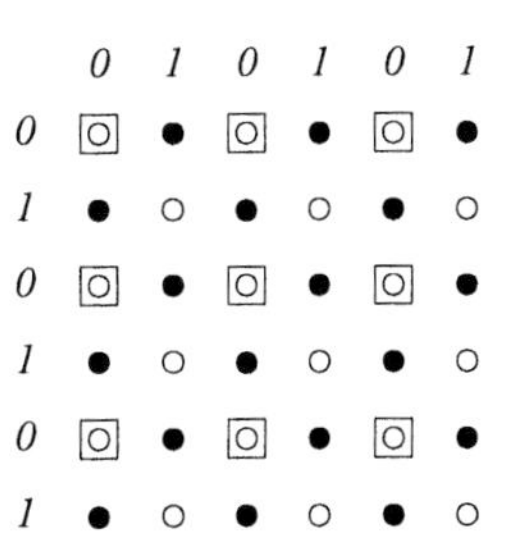

Fig. 5. The sublattice V_{00}

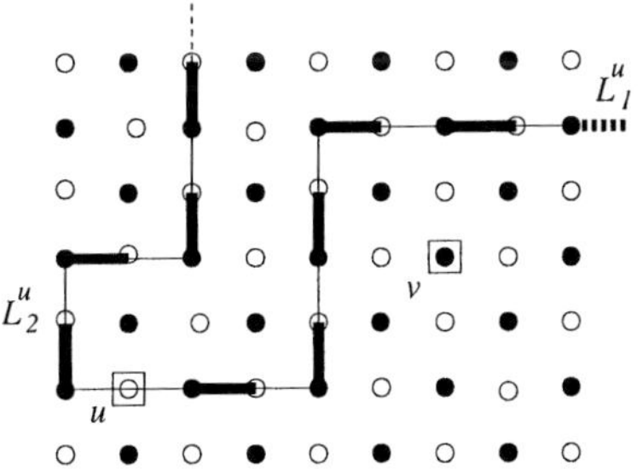

Fig. 6. Alternating paths L_1^u, L_2^u

this way until reaching v or revisiting a vertex, thus forming a cycle. The vertex set is finite, so one of these will occur. Define an alternating path L_2^u similarly, except, start with the edge from u to $(u_1, u_2 - 1)$. Note that every black vertex on these paths is in V_{01}, while the white vertices along the horizontal segments are in V_{00}, and those on the vertical segments are in V_{11}. Finally, define the paths K_1^u, K_2^u similarly, so that the first edges are to the vertices $(u_1 \pm 1, u_2)$. In this case, the black vertices on the path are in V_{10}, the white vertices on the vertical segments are in V_{00} and those on the horizontal segments are in V_{11}.

Let $v = (v_1, v_2) \in V_{01}$. We use these four paths to define an alternating path from u to v where the number of RED unmatched edges on the path is one more than the number of RED matched edges. Inverting along this path gives a perfect matching in $\mathcal{P}_{k+1}$. Given a perfect matching obtained in this way, we will be able to recover the near perfect matching with polynomial amount of information. We define the alternating path from u to v, by considering these cases.

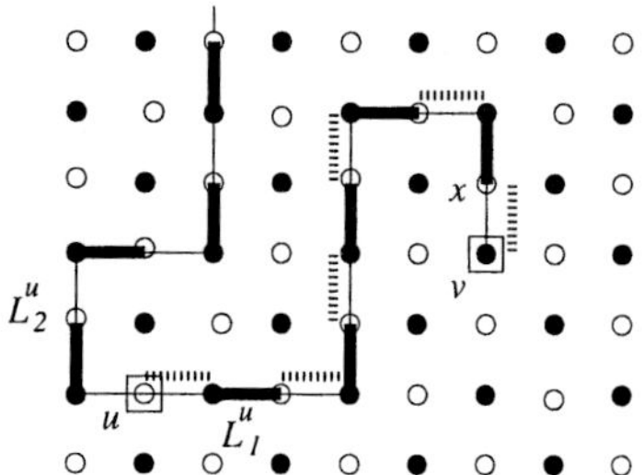

Fig. 7. L_1^u meets v

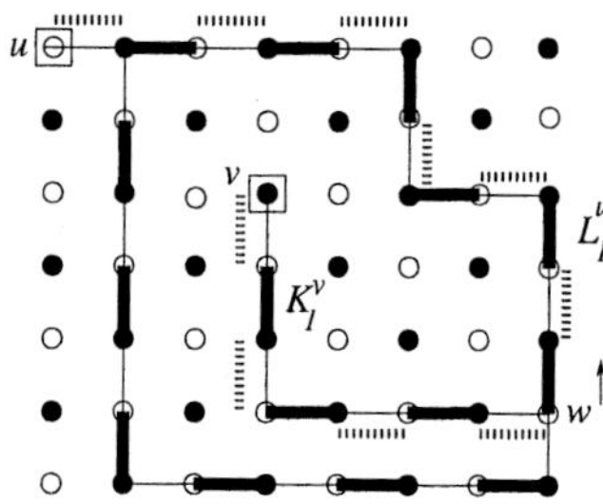

Fig. 8. The path K_1^v meets C_1

1. If one of the paths L_1^u or L_2^u reaches the vertex v before it cycles (Figure 7), then this is the alternating path. Say the path L_1^u reaches v. By construction, the number of unmatched RED edges along L_1^u is exactly one more than the number of matched RED edges, hence inverting along the path gives $P \in \mathcal{P}_{k+1}$. To invert the map, given $P \in \mathcal{P}_{k+1}$, start at v, if v is matched by a BLUE (RED) edge to the vertex x, the next unmatched edge along the path is taken to be the other BLUE (RED) edge incident with x. Continue in this way until u is reached.

2. If both paths L_1^u, L_2^u cycle without reaching v, we consider the following cases based on whether these cycles are contractible.

(a) At least one of the paths, say L_1^u , ends with a contractible cycle C_1 on the surface of the torus. It is easy to show that the interior of C_1 contains an odd number of vertices, and the number of black vertices exceeds the white vertices by 1. Hence, the interior of C_1 must contain an odd number of unmatched vertices and the unmatched vertex in the interior cannot be white: in particular, it cannot be u itself. So, v must lie in the interior of C_1. Consider the path K_1^v (see Figure 8). Since K_1^v cannot cycle in the interior of C_1 or end at an unmatched vertex, it must meet C_1. By construction, the white vertices of K_1^v are in V_{11}, hence the path meets the cycle on a vertical segment, say at the vertex w. The alternating path from u to v is defined by taking the subpath of L_1^u from u to w, and the subpath of K_1^v from v to w. The number of unmatched RED edges in the alternating path is one more than the number of matched RED edges, and so inverting along the path gives $P \in \mathcal{P}_{k+1}$ as required. Moreover, given P, and the edge incident to w in the original matching, the alternating path can be reconstructed.

(b) Both paths L_1^u, L_2^u end in non-contractible cycles on the surface of the torus. There are two possibilities, and we give a sketch of the arguments.

i) The cycles C_1 and C_2 are disjoint. This implies that the paths L_1^u and L_2^u are disjoint except at u. When a torus is cut along an incontractible simple cycle, we are left with a cylinder. If we cut along the cycles C_1 and C_2, we are left with 2 cylinders, one of which contains u and the paths L_1^u, L_2^u. The other cylinder can be shown to have an even number of vertices. Since the union of the two paths L_1^u, L_2^u is odd, the cylinder containing the paths has an odd number of vertices, and hence contains the vertex v. As before, K_1^v must hit one of the paths L_1^u or L_2^u since it cannot cycle on the cylinder.

ii) The cycles C_1, C_2 are not disjoint. In this case it can be shown that there exists a contractible cycle on the surface of the torus which can be cut out by starting at u along L_1^u, and ending at u along L_2^u (some edges may be used twice, once from above and once from below). As before, the interior contains an odd number of vertices which must be matched with each other, and hence must contain the vertex v. Since the cycle containing v is contractible, the path K_1^v must hit one of L_1^u, L_2^u.

In each case K_1^v hits the path from u on a vertical segment at a white vertex in V_{11} Given a matching in $\mathcal{P}_{k+1}, u, v$ and the vertex at which the paths from u and v meet, we can invert the map as described before.

In the case that $v \in V_{10}$, the same arguments can be made, except that we consider the paths $K_1^u, K_2^u, L_1^v, L_2^v$ instead of $L_1^u, L_2^u, K_1^v, K_2^v$ respectively. The difference is that the alternating paths constructed have one unmatched BLUE edge more than the number of matched BLUE edges along the path, so inverting edges with non-edges along the alternating path from u to v gives a matching in $\mathcal{P}_k$. This completes the proof for $i = 0$.

In the case that $i \neq 0$, suppose that $N \in \mathcal{N}_k^{i+1}$. Let u be the lexicographically first unmatched white vertex of N, and assume that $u \in V_{00}$. If one of L_1^u, L_2^u meets a black unmatched vertex v, then switching edges along the path from u to v gives a matching in $\mathcal{N}_{k+1}^i$. If not, then both L_1^u, L_2^u cycle.

Suppose L_1^u ends in a contractible cycle C_1. The interior of C_1 contains an odd number of vertices, including the vertices possibly on a segment of L_1^u starting at u. Hence, the interior contains an odd number of unmatched vertices. Since black vertices outnumber white vertices by one, the number of black unmatched vertices outnumber white unmatched vertices by one. In particular, the interior contains at least one black unmatched vertex, call it v.

Consider the paths K_1^v, K_2^v. If either one reaches a white unmatched vertex in the interior (including u), then switching edges along that path gives a matching in $\mathcal{N}_k^i$. Otherwise, if either one hits L_1^u, say at a vertex w, then we can switch edges along L_1^u from u to w, and then along K_1^v from v to w to obtain a matching either in $\mathcal{N}_{k+1}^i$ or $\mathcal{N}_k^i$ depending on whether v is in V_{01} or V_{10}. If the paths K_1^u, K_2^v do not hit a white unmatched vertex or L_1^u, they must cycle in a contractible cycle in the interior. Consider one of the paths, say K_1^v. Repeat the same argument as before, except now we consider white vertices u' in the interior of the cycle, and consider the paths $L_1^{u'}, L_2^{u'}$. Depending on whether u' is in V_{00} or V_{11} and the sublattice of v, alternating along the paths as before gives a matching either in $\mathcal{N}_k^i$ or $\mathcal{N}_{k+1}^i$. We can repeat this argument until we obtain an alternating path between a black and a white vertex, or, the interior of some cycle created by a vertex contains only one unmatched vertex. Since the single unmatched vertex cannot be the same as the vertex from which the cycle was created, this case can be solved in the same manner as the case when $i = 0$.

The remaining case, when L_1^u, L_2^u end in incontractible cycles, is similar to Case 2 above. □

Corollary 1. *Let m_1, m_2 be even, $N = m_1 m_2/2$. There is an algorithm to estimate the partition function $\widehat{Z}_k$ given in Equation (2) for every $\lambda \leq 1$ and k to within $(1 \pm \varepsilon)$ w.p. $\geq 1 - \delta$ in time polynomial in $N, \lambda, 1/\varepsilon$ and $\log(1/\delta)$.*

We can use similar arguments to relate the number of perfect matchings with k or $k+2$ Red edges.

Theorem 5. *Let m, n be even, $N = mn/2$. For every $0 \leq k \leq N-2$ even, $|\mathcal{P}_{k+2}|/p(N) \leq |\mathcal{P}_k| \leq p(N)|\mathcal{P}_{k+2}|$, where p is a polynomial.*

Proof. It suffices to show the upper bound for all k since the lower bound follows by switching the colors.

We construct a map from $\mathcal{P}_k$ to $\mathcal{P}_{k+2}$ as follows. Let $P \in \mathcal{P}_k$. Delete any vertical edge (u, v). Since $k \leq N-2$, there must be such an edge. Consider the paths L_1^u, L_2^u in $P \setminus (u, v)$. By parity, neither can reach v, and hence they must cycle on the surface of the torus. Since u is adjacent to v on the torus, neither path can end in a contractible cycle containing v in the interior. Hence both L_1^u, L_2^u end in incontractible cycles. By the arguments of Case 2 of the previous theorem, the path L_1^v must hit one of the paths L_1^u, L_2^u at a white vertex $w \in V_{11}$, i.e., on a vertical segment. Then, switching along the alternating path from u to v through w as before, we gain two Red edges, giving a matching in $\mathcal{P}_{k+2}$. The mapping is invertible given the vertex w and the vertices u, v, hence $|\mathcal{P}_k| \leq O(N^3)|\mathcal{P}_{k+2}|$. □

Using this Theorem and the estimator given by Corollary 1, we obtain an estimator for the set of perfect matchings of the torus with exactly k Red edges. The proof follows from standard arguments.

Theorem 6. *There is an algorithm to estimate $|\mathcal{P}_k|$ to within $1 \pm \varepsilon$ for every $0 < \varepsilon < 1$ with probability $\geq 1 - \delta$ in time polynomial in $N, 1/\varepsilon$ and $\log(1/\delta)$.*

These results can be generalized to approximating the size of the set of (k, ℓ)-matchings for any ℓ. By Theorem 1, we can approximate the partition function $\hat{Z}_{k,\ell}$ given in Equation (1) for every $\lambda, \mu \leq 1$ and $0 \leq k \leq \ell \leq n$. This estimator, together with the relations among sets of restricted matchings of arbitrary size (stated below) and the theorem of Kenyon, Randall and Sinclair [10] that the sizes of the sets $\mathcal{N}^i$ and $\mathcal{N}^{i+1}$ are polynomially related, gives an approximate counter for sets of restricted matchings of any size.

Theorem 7. *Let* m_1, m_2 *be even,* $N = m_1 m_2/2$. *For every* $1 \leq i \leq N-1$, *and* $0 \leq k \leq N-i-1$, *then for some polynomial* p, $|\mathcal{N}^i_{k+1}|/p(N) \leq |\mathcal{N}^i_k| \leq p(N)|\mathcal{N}^i_{k+1}|$.

The proof follows by constructing alternating paths as in Theorem 4.

Corollary 2. *There is an algorithm to estimate* $|\mathcal{N}^i_k|$ *to within* $1 \pm \varepsilon$ *for every* $0 < \varepsilon < 1$ *with probability* $\geq 1 - \delta$ *in time polynomial in* $N, 1/\varepsilon$ *and* $\log(1/\delta)$.

Acknowledgements. Thanks to a referee for suggesting a simplification of the statement of Lemma 1.

References

1. R. Burton and R. Pemantle. Local characteristics, entropy and limit theorems for spanning trees and domino tilings via transfer-impedances. *Annals of Probability* **21(3)**(1993), pp. 1329-1371.
2. M.E. Fisher. Statistical Mechanics of Dimers on a Plane Lattice. *Phys. Rev.* **124** (1961), pp.1664-1672.
3. R.H. Fowler and G.S. Rushbrooke. Statistical theory of perfect solutions. *Transactions of the Faraday Society* **33** (1937), pp. 1272-1294.
4. O.J. Heilmann and E.H.Leib. Theory of monomer-dimer systems. *Communications in Mathematical Physics* **25** (1972), pp 190-232.
5. M. Jerrum and A. Sinclair. Approximating the Permanent. *SIAM Journal of Computing* **18** (1989), pp. 1149-1178.
6. M. Jerrum, A. Sinclair, and E. Vigoda. A Polynomial-Time Approximation Algorithm for the permanent of a matrix with non-negative entries. *Journal of the ACM* **51(4)** 2004, pp. 671-697.
7. M. Jerrum, L. Valiant, and V. Vazirani. Random generation of combinatorial structures from a uniform distribution. *Theoretical Computer Science* **43** 1986, pp.169-188.
8. A.V. Karzanov. Maximum Matching of given weight in complete and complete bipartite graphs. *Kibernetica: 7-11, English translation in CYBNAW* **23(1)** (1987) pp. 8-13.
9. P. W. Kasteleyn. The statistics of dimers on a lattice,I. The number of dimer arrangements on a quadratic lattice. *Physica* **27** (1961) pp. 1209-1225.
10. C. Kenyon, D. Randall, and A. Sinclair. Approximating the number of monomer-dimer coverings of a lattice. *Journal of Statistical Physics* **83** (1996) pp. 637-659.
11. K. Mulmuley, U.V. Vazirani, and V. V. Vazirani. Matching is as easy as matrix inversion. *Combinatorica* **7(1)** (1987), pp.105-113.
12. C. Papadimitriou. Polytopes and Complexity. *Progress in Combinatorial Optimization* (W.R. Pulleyblank ed., Academic Press Canada, Don Mills, 1984) pp. 295-305.
13. H. N. V. Temperley. In Combinatorics, Proceedings of the British Combinatorial Conference 1973, London Mathematical Society Lecture Notes Series No.13. 1974, pp. 202-204.
14. T. Yi, K. G. Murty, C. Spera. Matchings in colored bipartite networks. *Discrete Applied Mathematics* **121(1-3)** (2002), pp. 261-277.

Eliminating Cycles in the Discrete Torus

Béla Bollobás[1], Guy Kindler[2], Imre Leader[3], and Ryan O'Donnell[2]

[1] Dept. of Mathematical Sciences, University of Memphis
[2] Theory Group, Microsoft Research
[3] Centre for Mathematical Sciences, University of Cambridge
bollobas@msci.memphis.edu, gkindler@microsoft.com,
leader@dpmms.cam.ac.uk, odonnell@microsoft.com

Abstract. In this paper we consider the following question: how many vertices of the discrete torus must be deleted so that no topologically nontrivial cycles remain?

We look at two different edge structures for the discrete torus. For $(\mathbb{Z}_m^d)_1$, where two vertices in $\mathbb{Z}_m$ are connected if their ℓ_1 distance is 1, we show a nontrivial upper bound of $d^{\log_2(3/2)}m^{d-1} \approx d^{0.6}m^{d-1}$ on the number of vertices that must be deleted. For $(\mathbb{Z}_m^d)_\infty$, where two vertices are connected if their ℓ_∞ distance is 1, Saks, Samorodnitsky and Zosin [8] already gave a nearly tight lower bound of $d(m-1)^{d-1}$ using arguments involving linear algebra. We give a more elementary proof which improves the bound to $m^d - (m-1)^d$, which is precisely tight.

1 Introduction

In this paper we consider a "vertex multicut" problem on discrete torus graphs. Let us begin by defining the two graphs of interest to us.

Definition 1. *The* ℓ_1 discrete torus of width m and dimension d, *denoted* $(\mathbb{Z}_m^d)_1$, *is the undirected graph on vertex set* $\mathbb{Z}_m^d$ *in which two vertices are connected if their* ℓ_1 *distance is 1.*

The ℓ_∞ discrete torus of width m and dimension d, *denoted* $(\mathbb{Z}_m^d)_\infty$, *is the undirected graph on vertex set* $\mathbb{Z}_m^d$ *in which two vertices are connected if their* ℓ_∞ *distance is 1.*

We will also write $(\mathbb{Z}^d)_1$ *and* $(\mathbb{Z}^d)_\infty$ *for the similarly defined infinite graphs on vertex set* $\mathbb{Z}^d$.

In each of these tori we are interested in the set of cycles that "wrap around" the torus in at least one dimension. Let us define this notion formally.

Definition 2. *A cycle in* $(\mathbb{Z}_m^d)_1$ *(respectively,* $(\mathbb{Z}_m^d)_\infty$*) is said to be noncontractible if, when regarded as a loop inside the solid torus, it is homotopically nontrivial.*

Main Problem. In this paper we want to study the minimal number of vertices in either discrete torus that must be deleted so that every noncontractible cycle is broken. In other words, we consider the problem of finding the set of vertices of minimal size that intersects every noncontractible cycle in $(\mathbb{Z}_m^d)_1$ or $(\mathbb{Z}_m^d)_\infty$.

J.R. Correa, A. Hevia, and M. Kiwi (Eds.): LATIN 2006, LNCS 3887, pp. 202–210, 2006.

We denote the minimal number for $(\mathbb{Z}_m^d)_1$ by $S_1(m,d)$ and the minimal number for $(\mathbb{Z}_m^d)_\infty$ by $S_\infty(m,d)$.

We note that there are some obvious bounds that hold for both $S_1(m,d)$ and $S_\infty(m,d)$. A lower bound of m^{d-1} follows by considering, in either graph, the m^{d-1} noncontractible cycles which are parallel to the first axis. These cycles are vertex-disjoint, so at least one vertex must be deleted from each of them. An obvious upper bound of $m^d - (m-1)^d$ is obtained by deleting the union of d "walls", one in each dimension; by a "wall" we mean a set of the form $\{x : x_i = a\}$ for some $i \in [d]$, $a \in \mathbb{Z}_m$.

1.1 History and Motivation

The problem discussed in this paper is a natural one in the context of the combinatorics of the discrete torus (see e.g. [2, 1, 3]), but it has other motivations as well.

Discrete Foams. Our problem is related to the isoperimetry of periodic tilings of space. The connection is apparent from the following formulation of our problem. We say that a finite set S in $\mathbb{Z}^d$ *generates a discrete foam for* $(\mathbb{Z}^d)_1$ *with periodicity* $m \cdot \mathbb{Z}^d$ if the set

$$\mathbb{Z}^d \setminus \{S + v\}_{v \in m \cdot \mathbb{Z}^d}$$

contains no paths in $(\mathbb{Z}^d)_1$ of infinite length. (We can give a similar definition for $(\mathbb{Z}^d)_\infty$.) It can easily be verified that our problem is identical to that of finding the minimal size of a set generating a discrete foam with periodicity $m \cdot \mathbb{Z}^d$.

This problem can be essentially regarded as that of finding a tiling of $\mathbb{Z}^d$ with periodicity $m \cdot \mathbb{Z}^d$ that has minimal vertex boundary; this is a discrete version of the problem of finding a (continuous) closed foam in $\mathbb{R}^d$ with periodicity $\mathbb{Z}^d$ and minimal surface area. Although there has been a lot of work on soap bubble and foam problems in $\mathbb{R}^d$ and even on the flat torus — see e.g. [7] — very little is known. We hope that discrete versions of the problem may prove to be a useful source for new observations regarding foams.

Directed Minimum Multicut. Another area in which our problem arises is in theoretical computer science, as was noted in a paper of Saks, Samorodnitsky and Zosin [8]. This paper studied the integrality gap of the natural linear programming formulation of the "directed minimum multicut" problem. This is the problem in which one is given a directed graph and d "source-sink" pairs of vertices $(s_1, t_1), \ldots, (s_d, t_d)$, and one is required to delete as few edges as possible so that there is no longer any s_i-to-t_i path. To obtain their integrality gap bound, Saks et al. translated the directed minimum multicut problem on a certain graph to an undirected vertex-deletion problem. Specifically, they looked at the graph $([m]^d)_\infty$ — i.e., the d-dimensional, width m *grid* with ℓ_∞ edges — and studied the following quantity:

Definition 3. *$S'_\infty(m,d)$ is the minimum number of vertices in $([m]^d)_\infty$ that need to be deleted to disconnect all d pairs of opposing walls.*

Clearly $S'_\infty(m,d) \geq S_\infty(m,d)$. Saks et al. proved a lower bound of $d(m-1)^{d-1}$ on $S'_\infty(m,d)$, but their proof immediately gives the same lower bound for $S_\infty(m,d)$. This result yielded an integrality gap arbitrarily close to d (which is the best possible) for the directed multicut problem. In this paper we improve the lower bound for $S_\infty(m,d)$ (and thus for $S'_\infty(m,d)$) to $m^d - (m-1)^d$, which exactly matches the upper bound mentioned earlier.

Parallel Repetition on Odd Cycles. Our original motivation came from a problem in the study of parallel repetition of two-prover one-round games [4,6], and in particular a question due to Feige [5] about how the max-cut problem on odd cycles behaves under parallel repetition.

The details of this problem are beyond the scope of this paper; suffice it to say that it can be reduced to a problem very similar to that of eliminating cycles in $(\mathbb{Z}_m^d)_\infty$ (we give more details in Section 3). However, it seems that solving that problem requires a proof of a lower bound on $S_\infty(m,d)$ that is "robust", in the sense that it should imply a nontrivial bound even under a certain relaxed hypothesis. The lower bound of Saks et al. relies on a linear algebraic argument, and this seems too fragile to give anything once hypotheses are relaxed. Our lower bound, on the other hand, is proven using more elementary methods; hence it seems to have more of a chance to be generalizable.

1.2 Our Results

We have two main results. Our first result is an improved upper bound on $S_1(m,d)$.

Theorem 1. $S_1(m,d) \leq d^{\log_2(3/2)} m^{d-1}$.

As far as we know, no nontrivial upper bound on $S_1(m,d)$ was previously known.

Our second result is a lower bound on $S_\infty(m,d)$ that precisely matches the obvious upper bound already discussed. This result improves on the lower bound of Saks, Samorodnitsky and Zosin [8] and eliminates their use of linear algebra.

Theorem 2. $S_\infty(m,d) \geq m^d - (m-1)^d$, *and hence* $S_\infty(m,d) = m^d - (m-1)^d$.

2 The Upper Bound on $S_1(m,d)$

Our main goal in this section is to prove Theorem 1, showing an upper bound for $S_1(m,d)$. Before doing this, we will motivate our bound by giving a tight construction in two dimensions which has size about $(3/2)m$.

2.1 A Tight Bound for $(\mathbb{Z}_m^2)_1$

It is easy to see that the following set of size at most $(3/2)m$ blocks all noncontractible cycles in $(\mathbb{Z}_m^2)_1$:

$$S = \{(x,x) : x \in \mathbb{Z}_m\} \cup \{(x,-x) : 0 \leq x \leq k/2\}.$$

Let us sketch a proof of this fact. It is well-known that in two dimensions, $(\mathbb{Z}_m^2)_\infty$ is dual to $(\mathbb{Z}_m^2)_1$. The set S contains a cycle in $(\mathbb{Z}_m^2)_\infty$ that winds once in the first dimension and no times in the second dimension — call such a cycle a $(1,0)$-cycle. This blocks all cycles in $(\mathbb{Z}_m^2)_1$ except those of type $(c,0)$. But S also contains a $(0,1)$-cycle in $(\mathbb{Z}_m^2)_\infty$, thus blocking all $(c,0)$-cycles in $(\mathbb{Z}_m^2)_1$, $c \neq 0$.

If we count precisely, we see that S actually has size $(3/2)m - 1$ when m is even and size $(3/2)m - 1/2$ when m is odd. We will now show these upper bounds are optimal by showing that $(3/2)m - 1$ is a lower bound.

So suppose $S \subset \mathbb{Z}_m^2$ blocks all noncontractible cycles. To block all $(1,0)$-cycles S must contain some (a,b)-cycle, C, in $(\mathbb{Z}_m^2)_\infty$ with $b \neq 0$. If either $|a|$ or $|b|$ is at least 2 then C contains at least $2m$ points. So we may assume that C is of type either $(0,1)$ or $(1,1)$. But now to block all cycles in $(\mathbb{Z}_m^2)_1$ that are parallel to C (i.e., have the same type as C), S must contain some other nontrivial cycle C' in $(\mathbb{Z}_m^2)_\infty$ not parallel to C. Hence we can conclude without loss of generality that one of the following three cases occurs in $(\mathbb{Z}_m^2)_\infty$: (i) S has a $(1,0)$-cycle and a $(0,1)$-cycle; (ii) S has a $(1,0)$-cycle and a $(1,1)$-cycle; or, (iii) S has a $(1,1)$-cycle and a $(1,-1)$-cycle.

For case (i), let C be the $(1,0)$-cycle and C' the $(0,1)$-cycle. Suppose that C contains t steps with vertical displacement of 1. Then it must also contain exactly t steps with vertical displacement -1, because its type is $(1,0)$. Thus C has length at least $\max(m, 2t)$. Also, C is contained in the union of $t+1$ horizontal lines, so it follows that C' must have at least $m - t - 1$ points not in C, since it has type $(0,1)$. Thus S has size at least $\max(m, 2t) + m - t - 1$, which is at least $(3/2)m - 1$, as claimed.

The argument for case (ii) is identical. For case (iii) things are even easier. In this case let C be the $(1,1)$-cycle, and note that C travels up at least m steps and right at least m steps. If C is to have fewer than $(3/2)m$ points by itself, then at least $m/2$ of these steps must be shared; i.e., C must have at least $m/2$ $(1,1)$-steps. Now let C' be the $(1,-1)$-cycle. Then C' needs to take at least m steps that are either horizontal, vertical, or $(1,-1)$-steps. Since none of these are the $m/2$ $(1,1)$-steps of C, we conclude that C and C' together have at least $(3/2)m$ vertices, as claimed.

2.2 Proof of Theorem 1

Having analyzed the case of $d = 2$, we will prove Theorem 1 by generalizing the example from the previous subsection to higher dimensions. Our proof uses the foam perspective described in Section 1. That is, we show a set that generates a discrete foam with periodicity $m \cdot \mathbb{Z}^d$ and has the size claimed in the theorem. To define the discrete foam boundary, it will help to first define a continuous foam.

We define inductively a set $B(r)$ in Euclidean space $\mathbb{R}^d$, where $d = 2^r$. The set $B(0)$ will be the set of all $x_1 \in \mathbb{R}^1$ satisfying

$$0 \leq x_1 < m.$$

In other words, $B(0) = [0, m)$. The inductive definition of $B(r) \subset \mathbb{R}^d$ is

$$(x_1, \ldots, x_d) \in B(r) \Leftrightarrow (x_1 + x_2, \ldots, x_{d-1} + x_d) \in B(r-1) \quad \text{and}$$
$$\left(\tfrac{x_1 - x_2}{2}, \quad \ldots, \tfrac{x_{d-1} - x_d}{2}\right) \quad \in B(r-1).$$

Thus we have that $B(1) \subset \mathbb{R}^2$ is the set of points (x_1, x_2) satisfying

$$0 \leq x_1 + x_2 < m$$

$$0 \leq x_1 - x_2 < 2m,$$

and $B(2) \subset \mathbb{R}^4$ is the the set of points (x_1, x_2, x_3, x_4) satisfying

$$0 \leq x_1 + x_2 + x_3 + x_4 < m$$

$$0 \leq x_1 - x_2 + x_3 - x_4 < 2m$$
$$0 \leq x_1 + x_2 - x_3 - x_4 < 2m$$
$$0 \leq x_1 - x_2 - x_3 + x_4 < 4m,$$

and it can easily be checked that $B(r)$ is the set of points $x \in \mathbb{R}^d$ satisfying $0 \leq H_r x < m u_r$, where H_r denotes the standard $2^r \times 2^r$ Hadamard matrix and u_r denotes the rth tensor power of the vector $(1, 2)$.

Let us also introduce the following notation: Let $L(r)$ denote the "lower boundary" of $B(r)$, containing all the points in $B(r)$ for which one of the inequalities hold as an equality; and, let $\overline{B}(r)$ be the closure of $B(r)$, which can also be obtained by replacing all strict inequalities by non-strict inequalities.

Since the Hadamard matrix is orthogonal, it is easy to see that $\overline{B}(r)$ is a closed rectangular box in $\mathbb{R}^d$ (although it is not axis-parallel). We will show that $B(r)$ tiles $\mathbb{R}^d$ with periodicity $m \cdot \mathbb{Z}^d$. This is a consequence of the following two propositions:

Proposition 1. *No two points of $B(r)$ are the same modulo $m \cdot \mathbb{Z}^d$.*

Proposition 2. *The volume of $B(r)$ is m^d.*

Proof. (Proposition 1.) The proof is by induction; the statement is clearly true for $r = 0$. For larger r, suppose x is in $B(r)$ and $x + m \cdot (a_1, \ldots, a_d)$ is also in $B(r)$, where the a_i's are integers. We wish to show that all a_i's equal 0. By definition, we know that

$$(x_1 + x_2, \ldots, x_{d-1} + x_d) \in B(r-1),$$

$$(x_1 + x_2 + m \cdot (a_1 + a_2), \ldots, x_{d-1} + x_d + m \cdot (a_{d-1} + a_d)) \in B(r-1).$$

By induction, then, we get

$$a_1 + a_2 = \cdots = a_{d-1} + a_d = 0. \tag{1}$$

It follows that $a_1 - a_2, \ldots, a_{d-1} - a_d$ are all even and thus $(a_1 - a_2)/2, \ldots,$ $(a_{d-1} - a_d)/2$ are all integers. But by definition we also know that

$$(\tfrac{x_1-x_2}{2},\ldots,\tfrac{x_{d-1}-x_d}{2}) \in B(r-1),$$

$$(\tfrac{x_1-x_2}{2}+m\tfrac{a_1-a_2}{2},\ldots,\tfrac{x_{d-1}-x_d}{2}+m\tfrac{a_{d-1}-a_d}{2}) \in B(r-1),$$

so by induction,

$$(a_1-a_2)/2 = \cdots = (a_{d-1}-a_d)/2 = 0. \tag{2}$$

Combining (1) and (2) we get that all a_i's are 0. This completes the induction.

Proof. (Proposition 2.) As mentioned, $B(r)$ is a rectangular box, so its volume is simply the product of its side lengths. The normal vectors to its sides are the rows of the Hadamard matrix H_r, which have length $\sqrt{d}$. Thus $B(r)$'s sides have length $(m/\sqrt{d})\cdot(u_r)_1,\ldots,(m/\sqrt{d})\cdot(u_r)_d$, where we recall the vector u_r is the rth tensor product of $(1,2)$. So to complete the proof it suffices to show that $\prod_{i=1}^{d}(u_r)_i = d^{d/2}$. This follows by induction since it is easy to see we have the recurrence $u_0 = 1$, $\prod_{i=1}^{d}(u_r)_i = 2^{d/2}(\prod_{i=1}^{d/2}(u_{r-1})_i)^2$.

We have now shown that $B(r)$ tiles $\mathbb{R}^d$ with periodicity $m\cdot\mathbb{Z}^d$. It follows easily that $L(r)$ generates a continious closed foam in $\mathbb{R}^d$ with preriodicity $m\cdot\mathbb{Z}^d$.

Let us now return to the discrete problem in which we are interested. A natural approach would be to show that $L(r)\cap\mathbb{Z}^d$ generates a discrete foam in $(\mathbb{Z}^d)_1$ with periodicity $m\cdot\mathbb{Z}^d$, which it indeed does, and to upper-bound $S_1(m,d)$ by counting the lattice points on $L(r)$. However, to avoid the need to approximate the number of lattice points on $L(r)$, we take a slightly different tack.

Let $L'(r)$ denote a thickening of $L(r)$ to width $1/\sqrt{d}$; in other words, $L'(r) = \{x\in B(r) : \operatorname{dist}(x,L(r)) \le 1/\sqrt{d}\}$. Note that $L(r)+v$ generates a continuous foam in $\mathbb{R}^d$ with periodicity $m\cdot\mathbb{Z}^d$ for any vector $v\in\mathbb{R}^d$. From this it's easy to see that $(L'(r)+v)\cap\mathbb{Z}^d$ generates a discrete foam in $(\mathbb{Z}^d)_1$ with periodicity $m\cdot\mathbb{Z}^d$; the reason is that the normals to the faces of $L(r)$ are of the form $(\pm1,\pm1,\ldots,\pm1)$, and so every edge of $(\mathbb{Z}^d)_1$ travels length at most $1/\sqrt{d}$ perpendicular to $L(r)$'s faces. Thus any infinite path in $(\mathbb{Z}^d)_1$ would have to pass through $L'(r)$.

We can now upper-bound $S_1(m,d)$ by counting the number of points in $(L'(r)+v)\cap\mathbb{Z}^d$ for any particular v. By volume considerations, it is clear that there exists a vector v such that

$$\#((L'(r)+v)\cap\mathbb{Z}^d) \le \operatorname{vol}(L'(r)) \le \operatorname{area}(L(r))/\sqrt{d}.$$

Thus to prove Theorem 1 it suffices to show that the surface area of $L(r)$ is at most $d^{\log_2(3/2)}m^{d-1}\cdot\sqrt{d}$. Since $B(r)$ is a rectangular box, the surface area of $L(r)$ is equal to the sum of the reciprocals of $B(r)$'s side lengths times its volume (i.e., m^d, by Proposition 2). $B(r)$'s side lengths equal $(m/\sqrt{d})\cdot(u_r)_i$, as was mentioned in the proof of Proposition 2, where the vector u_r is the rth tensor power of $(1,2)$. Thus to complete the proof we need to show that $\sum_{i=1}^{d} 1/(u_r)_i = d^{\log_2(3/2)} = (3/2)^r$. This can be proven by induction, as one can easily derive the recurrence $1/u_0 = 1$, $\sum_{i=1}^{d} 1/(u_r)_i = (3/2)\sum_{i=1}^{d/2} 1/(u_{r-1})_i$.

3 The Lower Bound on $S_\infty(m, d)$

In this section we prove Theorem 2. Our proof begins with the same strategy used by Saks et al. in [8], which involves *sections* and *tubes.*

Definition 4 (sections and tubes). *Given a direction $i \in [d]$ and a point $x \in \mathbb{Z}_m^d$, we define the* section based at x and perpendicular to direction i *to be the $(d-1)$-dimensional hypercube containing the points*

$$\{x + f : f \in \{0,1\}^{i-1} \times \{0\} \times \{0,1\}^{d-i}\}.$$

A tube in direction i *is the union of a section perpendicular to direction i with all of its translates by multiples of the vector $e_i = (0, \ldots, 0, 1, 0, \ldots, 0)$. A tube is therefore a union of m parallel sections.*

The lower bound of Saks et al., as well as our tight lower bound, is based on the following observation:

Observation 3. *If S is any set of vertices in $(\mathbb{Z}_m^d)_\infty$ that touches all noncontractible cycles, then S must contain at least one complete section from* every *tube.*

The proof of this observation is clear: if there were some tube for which every section had a vertex missed by S, then these vertices would form a noncontractible cycle, since all pairs of consecutive sections are completely mutually connected in $(\mathbb{Z}_m^d)_\infty$.

Given the observation above, we will now prove a lower bound of $m^d - (m-1)^d$ on the size of any subset S that contains a full section in every tube. In fact it suffices to forget about the tubes which "wrap around" the torus and think instead of the graph $([m]^d)_\infty$, which only contains the $d(m-1)^{d-1}$ tubes that are inside the grid. We prove the lower bound for any $S \subseteq [m]^d$ which contains a complete section from each one of these tubes.

The proof of Saks et al. showed that any $S \subseteq [m]^d$ containing at least one full section in each of these tubes contains at least $d(m-1)^{d-1}$ points. Their proof used a linear algebraic argument; it considered the dimension of the space spanned by indicators of the sections contained in S. We provide a more elementary argument, which gives a tight lower bound and seems to have more potential for generalizations. In particular, we would like to generalize the lower bound to the case where S is only known to contain a fixed fraction of the points of one section per tube. A good lower bound in this regime would translate to an advancement in the parallel repetition problem discussed briefly in Section 1.

Our proof goes by induction, where the key is to take a stronger induction statement. For this purpose, we define a *cube* to be a set of the form

$$\{x + f : f \in \{0,1\}^d\} \subseteq [m]^d;$$

in other words, a cube is the union of two consecutive sections. Theorem 2 follows immediately from the following:

Theorem 3. *Let S be a subset of the vertices of $[m]^d$ containing at least one complete section per tube and also containing at least c cubes. Then the cardinality of S is at least $m^d - (m-1)^d + c$.*

Proof. Let us first argue about the case $d = 2$ and $c = 0$. In this case we are considering the two-dimensional grid $[m]^2$. Tubes can be thought of as the $m-1$ vertical columns between the vertices and the $m-1$ horizontal rows between the vertices; sections can be thought of as horizontal edges and vertical edges (more accurately, as the pair of vertices making up these edges). Suppose S contains at least one horizontal edge per column and one vertical edge per row. When taken together, its clear that these $2m-2$ edges cannot form any cycle since they never have two edges "one above the other" (or "one to the left of the other"). Since an acyclic graph with $2m-2$ has exactly $2m-1 = m^2 - (m-1)^2$ vertices, the proof of the $d = 2$, $c = 0$ case is complete.

We next consider the $d = 2$ case for general c. In this case, we know that S contains at least $m-1$ vertical edges (sections) and it is clear that it must contain at least c more vertical edges because of the presence of c cubes (cubes are squares, in two dimensions). We have so far identified $m-1+c$ vertical edges contained in S. Now consider adding the $m-1$ horizontal sections that S must contain. The resulting set of $2m-2+c$ edges must still be acyclic since it has no two horizontal edges in the same tube. Thus it contains $2m-1+c = m^2-(m-1)^2+c$ vertices as required by the induction.

With the case $d = 2$ completely proven, we move to the induction on the dimension d. So suppose S is a subset of $[m]^d$ with at least one section per tube and also at least c cubes. Consider the set of sections perpendicular to the dth direction. We know that there are at least $(m-1)^{d-1}$ of these which are contained in S — one per tube going the dth direction. There must also be at least c tubes in the dth direction where S contains an additional section, because of the c cubes it contains. Let us stratify these sections according to what level $1, \ldots, m$ they are on in the dth direction. Specifically, say we have c_i of them on level i, where $c_1 + \cdots + c_m \geq (m-1)^{d-1} + c$.

We now view the ith level as an inductive instance in dimension $d-1$. Because S has at least one section per tube in $[m]^d$, it is easy to see that it also has at least one (lower-dimensional) section per (lower-dimensional) tube in $[m]^{d-1}$. It also has at least c_i cubes. So by induction, S has at least $m^{d-1}-(m-1)^{d-1}+c_i$ vertices on the ith level of $[m]^d$. Summing this over i yields at least

$$m(m^{d-1} - (m-1)^{d-1}) + (m-1)^{d-1} + c = m^d - (m-1)^d + c$$

as a lower bound for the number of points in S.

Theorem 2 follows from Theorem 3 by taking $c = 0$.

References

1. M. C. Azizoğlu and Ö. Eğecioğlu. The isoperimetric number of d-dimensional k-ary arrays. *International Journal of Foundations of Computer Science*, 10(3):289–300, 1999.
2. B. Bollobás and I. Leader. An isoperimetric inequality on the discrete torus. *SIAM J. Discret. Math.*, 3(1):32–37, 1990.
3. F. R. K. Chung and P. Tetali. Isoperimetric inequalities for cartesian products of graphs. *Combinatorics, Probability & Computing*, 7(2):141–148, 1998.

4. U. Feige. Error reduction by parallel repetition — the state of the art. Technical report, Weizmann Institute of Science, 1995.
5. U. Feige. Personal communication. 2005.
6. R. Raz. A parallel repetition theorem. *SIAM Journal on Computing*, 27(3):763–803, June 1998.
7. A. Rós. The isoperimetric problem. Lecture notes. MSRI summer school, 2001. `http://www.ugr.es/~aros/isoper.htm`.
8. M. E. Saks, A. Samorodnitsky, and L. Zosin. A lower bound on the integrality gap for minimum multicut in directed networks. *Combinatorica*, 24(3):525–530, 2004.

On Behalf of the Seller and Society: Bicriteria Mechanisms for Unit-Demand Auctions*

Claudson Bornstein[2], Eduardo Sany Laber[1], and Marcelo Mas[1]

[1] Departamento de Informatica, PUC-Rio, Rio de Janeiro, RJ, Brasil
laber@inf.puc-rio.br

[2] Instituto de Matematica and COPPE/Sistemas, UFRJ, Rio de Janeiro, RJ, Brasil

Abstract. This work obtains truthful mechanisms that aim at maximizing both the revenue and the economic efficiency (social welfare) of unit-demand auctions. In a unit-demand auction a set of k items is auctioned to a set of n consumers, and although each consumer bids on all items, no consumer can purchase more than one item.

We present a framework for devising polynomial-time randomized truthful mechanisms that are based on a new variant of the Vickrey-Clarke-Groves (VCG) mechanism. Instead of using reserve prices, this variant of VCG uses the number of objects that we wish to sell as a parameter. Our mechanisms differ in their selection of the number of items to be sold, and allow an interesting trade-off between revenue and economic efficiency, while improving upon the state-of-the-art results for the Unit-Demand Auctions problem (Guruswami et. al.[SODA 2005]).

Our probabilistic results depend on what we call the competitiveness of the auction, i.e., the minimum number of items that need to be sold in order to obtain a certain fraction of the maximum efficiency. We denote by $\mathcal{T}$ the optimal efficiency achieved by the VCG mechanism. Our efficiency-oriented mechanism achieves $\Omega(\mathcal{T})$ efficiency and $\Omega(\mathcal{T}/\ln(\min\{k,n\})$ revenue with probability that grows with the competitiveness of the auction. We also show that no truthful mechanism can obtain an $\omega(\mathcal{T}/\ln(\min\{k,n\})$ expected revenue on every set of bids. In fact, the revenue-oriented mechanism we present achieves $\Omega(\mathcal{T}/\ln(\min\{k,n\})$ efficiency and $\Omega(\mathcal{T}/\ln(\min\{k,n\})$ revenue, but the revenue can actually be much higher, even as large as $\Omega(\mathcal{T})$ for some bid distributions.

1 Introduction

Auction mechanism design has long been a field of interest in the Economics and Game Theory communities [13]. In recent years, with the rise in electronic commerce and high-profile auctions such as the Google IPO and FCC spectrum

* This work was partially supported by CNPq under grants 304093/2002-5, 300968/2003-5 and 476323/2004-5, and by FAPERJ under grant E-26/151.494/2005.

J.R. Correa, A. Hevia, and M. Kiwi (Eds.): LATIN 2006, LNCS 3887, pp. 211–223, 2006.

auctions, the field of mechanism design for auctions has drawn a lot of attention from theoretical Computer Science researchers [8, 6, 11, 1, 4].

This paper deals with Unit-Demand Auctions (UDA) where k distinct items are sold to a group of n consumers but at most one item can be sold to each consumer. This can reflect consumer preferences, say, in a real estate market, where consumers want to buy a single house to live in. A perhaps more realistic setting for UDA is a government license auction, in which the government imposes regulatory quotas on the outcome of the auction, so as to foster a competitive market.

We assume that no previous knowledge of the bid distribution is known, so that in fact the traditional Bayesian approach that relies on prior knowledge is not applicable.

1.1 The Model

Let $C = \{1, 2, \ldots, n\}$ be a set of consumers and let I be a set of k distinct items[1]. The auctions considered in this paper are sealed-bid auctions where each consumer submits a bid for each item in I. An *auction mechanism* is a function that maps any possible set of bids into a pair $(A, \mathbf{p})$, where A is an *allocation* that defines which item is sold to each consumer and $\mathbf{p}$ is the vector of prices determining the sale price of each allocated item. A consumer i can only be allocated to item j if his (her) bid for j is not smaller than j's price. We assume that the mechanism employed by the auctioneer is publicly known and the following assumptions are made about the consumers.

- Each consumer has $|I|$ private valuations, one for each item in I. The valuation of consumer i for item j, indicated by $v_{i,j} \geq 0$, is the maximum price for which i would be willing to buy item j.
- If consumer i buys item j, then his profit(utility) is $u_i = v_{i,j} - p_j$.
- Consumers are rational and will submit bids that try to maximize their utilities.
- Consumers do not collude.
- The consumers in the auction are indistinguishable from the perspective of the auctioneer.

Depending on the mechanism used by the auctioneer, the consumers might be able to increase their utility by presenting bids that misrepresent their valuations. An auction mechanism $\mathcal{A}$ is *truthful* if the best strategy for each consumer is to submit his own valuations regardless of his beliefs on the bidding strategies employed by the other consumers. Consumers cannot benefit from price speculation in a truthful auction and indeed rational consumers will bid their valuations in any auction that employs a truthful mechanism. By avoiding pricing games between the consumers and receiving the true valuations of the consumers as its

[1] We disregard the situation in which a number of copies of each item is available, but it can easily be modeled.

input, truthful auction mechanisms are in a much better position to optimize the outcome of the auction.

The revenue attained by an auction mechanism for a given set of bids is the sum of all prices paid as a result of the auction. The *economic efficiency* attained by an auction mechanism for a given set of bids is defined as the sum over all consumers of the valuations that each consumer attributes to the item he acquires. It relates to the social value of the auction and often enough, say, on the FCC spectrum auctions, maximizing the efficiency is or probably should be more important than maximizing the revenue even from the auctioneer's perspective.

A *randomized auction mechanism* is a probability distributions over deterministic auctions mechanisms. Following [6], we adopt a notion of randomized truthfulness in which a randomized truthful auction mechanism is a probability distribution over the set of deterministic truthful auction mechanisms.

The concepts presented so far can also be understood in terms of graphs. The matrix valuation $\mathbf{v} = (v_{i,j})_{i=1,\ldots,n}^{j=1,\ldots,k}$, which is the input of a truthful mechanism, can be viewed in terms of a weighted complete bipartite graph G among consumers and items, where the cost $c(e_{i,j})$ of edge $e_{i,j}$ associating the i-th consumer with the j-th item is $v_{i,j}$. Thus, a truthful deterministic auction mechanism $\mathcal{A}$ is a function that maps each weighted bipartite graph G onto a pair $(M, \mathbf{p})$, where $M = \cup_{i=1}^{|M|} e_i$ is a matching of G and $\mathbf{p} = (p_1, \ldots, p_{|M|})$ is a vector defining the sale price of every item allocated by M, that is, p_i is the sale price of the item touched by edge e_i. We must have $p_i \leq c(e_i)$, for $i = 1, \ldots, |M|$. The revenue and the efficiency of $\mathcal{A}$, for input G, are the sum of the prices assigned to the items of M and the sum of the costs of the edges of M, respectively. Clearly, for a fixed graph, the revenue cannot exceed the efficiency.

1.2 Our Results

Unlike most recent work on auction mechanisms, we design randomized truthful mechanisms that simultaneously concern with maximizing the revenue and the economic efficiency. The approach employed by our mechanisms consists of randomly dividing consumers in two groups, and using one group's bids to estimate a suitable number of items to sell to the consumers of the other group. It then uses a novel variant of the generalized VCG [15, 5, 9] mechanism that takes this limited number of items as a parameter to decide both the allocation and the sale prices. By adjusting our estimate of how many items should be sold we either obtain a efficiency-oriented mechanism or a revenue-oriented mechanism.

In order to quantify our results we introduce some definitions. For a valuation matrix $\mathbf{v}$, let $\mathcal{T}(\mathbf{v})$ be the maximum possible efficiency attained by a truthful mechanism. $\mathcal{T}(\mathbf{v})$ is exactly the cost of the maximum cost matching in the graph associated with $\mathbf{v}$ and, clearly, is an upper bound on both the revenue and the efficiency achieved by any truthful mechanism for input $\mathbf{v}$. In addition, let $\mathcal{F}(\mathbf{v})$ be the maximum possible revenue obtained by an 'omniscient' auctioneer, under the constraint that a single price must be used to sell all items. The probability that the mechanisms proposed in this paper attain a certain efficiency or revenue depends on the notion of δ-*competitiveness* of the valuation matrix $\mathbf{v}$, which is defined below.

Definition 1. *The* δ-competitiveness *of a valuation matrix* $\mathbf{v}$ *is the minimum number of items, or equivalently consumers, that can generate revenue at least* $\mathcal{T}(\mathbf{v})/\delta$*. In terms of the graph* G *associated with* $\mathbf{v}$*, the* δ-competitiveness *is the size of the smallest cardinality matching in* G *with cost at least* $\mathcal{T}(\mathbf{v})/\delta$*.*

This notion captures how the auction is dominated by a certain group of consumers. The higher the δ-competitiveness of $\mathbf{v}$ is, the higher is the number of consumers needed to dominate the auction, that is, to generate a $1/\delta$ fraction of the maximum possible efficiency. We remark that there is no connection between this measure and the notion of competitive ratio employed to analyze online algorithms [3].

Let s denotes $\min\{n, k\}$. Our efficiency-oriented mechanism simultaneously achieves $\Omega(\mathcal{T}(\mathbf{v})/\ln s)$ revenue and $\Omega(\mathcal{T}(\mathbf{v}))$ efficiency with failure probability that exponentially decreases with the growing of the 8-competitiveness of $\mathbf{v}$. In addition, we show that for every randomized truthful auction mechanism $\mathcal{A}$ there exists a valuation matrix $\mathbf{v}$, for which $\mathcal{A}$ attains expected revenue $O(\mathcal{T}(\mathbf{v}))/\ln s)$, which is matched by our mechanism.

On the other hand, our revenue-oriented mechanism simultaneously achieves revenue and efficiency $\Omega(\mathcal{F}(\mathbf{v}))$ with failure probability that exponentially decreases with the growing of the $(\ln s)$-competitiveness of $\mathbf{v}$. We note that proving an $\Omega(\mathcal{F}(\mathbf{v}))$ bound is stronger than proving an $\Omega(\mathcal{T}(\mathbf{v})/\ln s)$ bound, since the inequality $\mathcal{F}(\mathbf{v}) \geq \mathcal{T}(\mathbf{v})/\ln s$ always holds and, in fact, we may even have $\mathcal{F}(\mathbf{v}) = \Omega(\mathcal{T}(\mathbf{v}))$.

By combining this last mechanism with the VCG mechanism for UDA and with a mechanism that only sells one item, we obtain a *mixed* auction mechanism that achieves $\Omega(\mathcal{F}(\mathbf{v}))$ expected revenue and $\Omega(\mathcal{T}(\mathbf{v}))$ expected efficiency. However, in this case we do not have high concentration around the mean.

For a completely arbitrary valuation matrix $\mathbf{v}$ where, say, a single valuation is much higher than all the others, the maximum attainable revenue by truthful mechanisms can be arbitrarily far from both $\mathcal{T}(\mathbf{v})$ and $\mathcal{F}(\mathbf{v})$. This is a well known fact for single item auctions but it also applies to UDA . Some conditions on the valuation matrix are usually imposed in order to obtain any meaningful results. That's the same rationale behind the conditions that we impose on the competitiveness of the valuation matrix $\mathbf{v}$ so as to be able to prove bounds on the revenue achieved by truthful mechanisms (as a function of $\mathcal{T}(\mathbf{v})$ and $\mathcal{F}(\mathbf{v})$).

Finally, we shall mention that all mechanisms proposed in this paper run in polynomial time and, in addition, our results extend to bounded demand combinatorial auctions where every consumer may purchase a bundle with at most d items, where d is a constant which does not depend on n or k. In this case, however, our mechanisms require exponential time. A discussion on bounded demand combinatorial auctions is deferred to an extended version of this paper.

1.3 Related Work

If maximizing the efficiency is the unique goal in UDA then the generalized Vickrey-Clarke-Groves(VCG) mechanism [15, 5, 9] is the right choice. It attains the optimal efficiency $\mathcal{T}(\mathbf{v})$, for every input valuation $\mathbf{v}$, while giving incentive for truth-telling. However, this mechanism falls short in that it can have very poor revenue.

The problem of maximizing the revenue in UDA is mentioned as an open problem in [7]. In [10], Guruswami et. al. propose an interesting mechanism for maximizing the revenue in UDA. Their mechanism relies on previous knowledge on the range of bids and achieves $\Omega(\mathcal{T}(\mathbf{v})/\log h))$ revenue, where h is the ratio between the largest and lowest bid values. Basically, it consists of randomly selecting reserve prices for a VCG mechanism.

All our mechanisms compare favorably with the one proposed in [10]. Our efficiency-oriented mechanism assures an $\Omega(\mathcal{T}(\mathbf{v})/\ln s)$ lower bound on the revenue and simultaneously guarantees efficiency which is a constant factor of the optimal one. On the other hand, our revenue-oriented mechanism simultaneously assures $\Omega(\mathcal{F}(\mathbf{v}))$ revenue and efficiency. As we have already mentioned the inequality $\mathcal{F}(\mathbf{v}) \geq \mathcal{T}(\mathbf{v})/\ln s$ always holds and, in fact, for some auctions we may even have $\mathcal{F}(\mathbf{v}) = \Omega(\mathcal{T}(\mathbf{v}))$. In addition, as opposed to the one proposed in [10], both these mechanisms do not rely on previous knowledge about the range of the bids and, most importantly, they guarantee high concentration around the mean. In all fairness, we should be mention that the mechanism of [10] produces envy-free allocations whereas ours do not. The table below summarizes how our results compare to the one presented in [10].

Method	**Expected Revenue**	**Expected Efficiency**	**High Concentration**	**Envy-free allocations**
Efficiency-oriented	$\Omega(\mathcal{T}(\mathbf{v})/\ln s)$	$\Omega(\mathcal{T}(\mathbf{v}))$	yes	no
Revenue-oriented	$\Omega(\mathcal{F}(\mathbf{v}))$	$\Omega(\mathcal{F}(\mathbf{v}))$	yes	no
Guruswami et.al. [10]	$\Omega(\mathcal{T}(\mathbf{v})/\ln h)$	$\Omega(\mathcal{T}(\mathbf{v})/\ln h)$	no	yes

The economic efficiency and revenue are traded-off in [12] in the auctioning of multiple units of the same object. The resulting auction maximizes the expected economic efficiency while ensuring a minimum level of revenue in the auction. We obtain a similar trade-off for the UDA.

Finally, we shall mention that UDA can also be viewed as a combinatorial auction where only bundles (set of items) of size one can be sold. Some of the papers in combinatorial auctions that focus on maximizing the efficiency also discuss revenue issues [11, 1, 2, 4]. What is usually done is to compare the revenue achieved by the proposed auction mechanisms with that achieved by the VCG mechanism. However, the revenue achieved by VCG can be rather low, and even 0, thus making it a less desirable benchmark.

2 Graph Theoretical Results

In this section, we present some graph theoretical results that are important for the design and the analysis of the mechanisms proposed in this paper.

Let G be a weighted bipartite graph. For a consumer i, we use G_{-i} to denote the subgraph induced in G by the removal of i from its set of vertices. Let e' be the lowest weight edge in a matching M of G. We define $\mathcal{F}(M) = |M| \times c(e')$ and $\mathcal{T}(M) = \sum_{e \in M} c(e)$. We use $M_G^{\mathcal{T}}$ ($M_G^{\mathcal{F}}$) to denote the largest matching of G that maximizes $\mathcal{T}(\cdot)$ ($\mathcal{F}(\cdot)$). We define $\mathcal{F}_G = \mathcal{F}(M_G^{\mathcal{F}})$ and $\mathcal{T}_G = \mathcal{T}(M_G^{\mathcal{T}})$. Thus, the δ-competitiveness of G is the size of the smallest matching M that satisfies $\mathcal{T}(M) \geq \mathcal{T}_G/\delta$. The following propositions relate the metrics $\mathcal{T}(\cdot)$, $\mathcal{F}(\cdot)$ and the competitiveness of a graph. A similar result to the next proposition appears in [10]. Its proof, and that of Proposition 2 are omited here.

Proposition 1. *For every graph G, we have $\mathcal{F}_G \geq \mathcal{T}_G/\ln s$.*

Proposition 2. *For every graph G, the $(\ln s)$-competitiveness of G is at most $|M_G^{\mathcal{F}}|$.*

The next proposition shows that there exists a single matching M that has 'high' values for both $\mathcal{T}(\cdot)$ and $\mathcal{F}(\cdot)$. The existence of such a matching is key for our efficiency-oriented mechanism.

Proposition 3. *For every graph G, there is a matching M in G such that $\mathcal{F}(M) \geq \mathcal{T}_G/(2\ln s)$ and $\mathcal{T}(M) \geq \mathcal{T}_G/2$.*

Proof. Let $e_1, \ldots, e_s$ be the edges of $M_G^{\mathcal{T}}$ sorted in non-increasing order of weights and let i^* be the largest number such that $i^* \times c(e_{i^*}) \geq \mathcal{T}_G/2\ln s$. The existence of such an i^* is ensured in the proof of Proposition 1. Define M as $\cup_{j=1}^{i^*} e_j$. Clearly, $\mathcal{F}(M) \geq \mathcal{T}_G/2\ln s$.

For $j > i^*$, we have that $c(e_j) < \frac{\mathcal{T}_G}{2j \times \ln s}$. By adding these inequalities we obtain that

$$\sum_{j=i^*+1}^{s} c(e_j) \leq \frac{\mathcal{T}_G \times (\ln s - \ln i^*)}{2\ln s} \leq \frac{\mathcal{T}_G}{2}$$

Thus, $\mathcal{T}(M) = \sum_{j=1}^{i^*} c(e_j) \geq \mathcal{T}_G/2$ □

2.1 Approximation Matchings

Next, we introduce the concept of an approximation matching for a sequence of matchings. This is used in Section 3.2 as a technical tool bounding the probability of our efficiency-oriented mechanism. Roughly speaking, given a sequence $\mathcal{S}$ of matchings in a graph G, the approximation matching A for $\mathcal{S}$ has the property that for every matching S of $\mathcal{S}$ there is a sub-matching A' of A whose size is within a constant factor of the size of S and, moreover, $\mathcal{F}(A') \geq \min_{S \in \mathcal{S}}\{\mathcal{F}(S)\}/2$ and $\mathcal{T}(A') \geq \min_{S \in \mathcal{S}}\{\mathcal{T}(S)\}$.

For an increasing sequence of integers J, let $min(J) = \min\{j | j \in J\}$, $max(J) = \max\{j | j \in J\}$ and $pred(j)$ be the largest integer of J smaller than j, for $j \in J \setminus min(J)$.

Definition 2. *Let $(M_j)_{j \in J}$ be a sequence of matchings in G, where $|M_j| = j$, for every $j \in J$. We define the sequence $(A_j)_{j \in J}$ as follows:*

$$A_j = \begin{cases} M_j \textit{ if } j = min(J) \\ A_{pred(j)} \cup \{e|e \in M_j \textit{ and } A_{pred(j)} \cup e \textit{ is a matching}\}, \textit{ otherwise} \end{cases}$$

We call $A_{max(J)}$ the approximation matching for the sequence $(M_j)_{j \in J}$.

Example 1. Let $G = (V_1 \cup V_2, E)$ be a complete bipartite graph where $V_1 = \{1,3,5,7,9\}$ and $V_2 = \{2,4,6,8,10\}$. Let us consider the sequence of matchings M_2, M_3, M_5, where $M_2 = \{(1,2),(3,6)\}$, $M_3 = \{(1,2),(3,8),(5,10)\}$ and $M_5 = \{(1,2),(3,4),(5,6),(7,8),(9,10)\}$.

Then, we have $A_2 = \{(1,2),(3,6)\}$, $A_3 = \{(1,2),(3,6),(5,10)\}$ and A_5, the approximation matching, is $\{(1,2),(3,6),(5,10),(7,8)\}$.

We ommit the proof of the next lemma which states crucial properties regarding the approximation matching.

Lemma 1. *Let $(M_j)_{j \in J}$ be a sequence of matchings in G, where $|M_j| = j$, for every $j \in J$. Furthermore, let A be the approximation matching of $(M_j)_{j \in J}$. Then, for every $j \in J$, there is a sub-matching A' of A such that: (i)* $\max\{\min(J), j/2\} \leq |A'| \leq 2j$ *; (ii) $\mathcal{T}(A') \geq \mathcal{T}(M_{min(J)})$; (iii) $\mathcal{F}(A') \geq \min\{\mathcal{F}(M_i)|i \in J\}/2$ and (iv) If e' is the edge of lowest cost in A', then $c(e') \geq c(e)$, for every edge e that belongs to the matching $A \setminus A'$.*

3 Truthful Mechanisms for Unit-Demand Auctions

In this section we introduce a family of mechanisms for UDA that we denote by UDAF (Unit Demand Auctions Family). The mechanism $\mathcal{A}_l$ presented below is employed by all mechanisms of UDAF. $\mathcal{A}_l$ is a variation of the VCG auction mechanism where the parameter l determines the number of items that can be sold. The VCG mechanism for UDA coincides with mechanism $\mathcal{A}_l$, when $l = s = \min\{n, k\}$.

Mechanism $\mathcal{A}_l$(H: Graph)

1. Compute a matching M of H that maximizes $\mathcal{T}(\cdot)$ among all the matchings in H of size l, and assign consumers to items according to M.
2. If M assigns the consumer i to the item j, then the sale price of j is $p_j = c(e_{i,j}) - \mathcal{T}(M) + \mathcal{T}(M_{-i})$, where M_{-i} is the matching that maximizes $\mathcal{T}(\cdot)$ among all the matchings in H_{-i} of cardinality l.

The next lemma allows us to bound the revenue achieved by $\mathcal{A}_l$.

Lemma 2. *Let H be a weighted complete bipartite graph in which there is a matching M' with exactly $2l$ edges, all of them with weights at least y. Then the revenue of $\mathcal{A}_l$ for input H is at least $l \times y$.*

Proof. Let M be the matching determined by $\mathcal{A}_l(H)$. It suffices to argue that $p_j = c(e_{i,j}) - \mathcal{T}(M) + \mathcal{T}(M_{-i}) \geq y$, for every consumer i touched by M.

Since $|M'| \geq 2l$, it follows that there is $e \in M'$ such that $M \cup e - e_{ij}$ is a matching in H_{-i}. Thus, $\mathcal{T}(M_{-i}) \geq \mathcal{T}(M \cup e - e_{ij}) \geq \mathcal{T}(M) - c(e_{i,j}) + y$ and, as a consequence, $p_j = c(e_{i,j}) - \mathcal{T}(M) + \mathcal{T}(M_{-i}) \geq y$ □

Lemma 2 implies that for a suitable choice of l the revenue of $\mathcal{A}_l$ is $\Omega(\mathcal{F}_G)$. This lemma, along with Proposition 3, also guarantees the existence of a value l for which $\mathcal{A}_l$ has $\Omega(\mathcal{T}_G/\ln s)$ revenue and $\Omega(\mathcal{T}_G)$ efficiency. Unfortunately we do not know how to compute the optimum l without losing truthfulness.

Instead, our UDAF mechanisms first randomly splits the consumers into two groups and then uses one group to estimate a suitable (depending on the pursued goal) value of l. Finally, $\mathcal{A}_l$ is run for the consumers of the other group. What distinguishes one mechanism in UDAF family from the other is the function f employed to determine the number l of items to be sold. The definition of f will determine the economic efficiency, the revenue and the time complexity attained by the resulting mechanism.

Mechanism $UDAF_f$

1. Flip a fair coin n times to split the consumers into two groups, say, L (left) and R (right). Let G_L (G_R) be the bipartite graph induced by the consumers of L (R) and the set of all items.
2. Run $\mathcal{A}_{f(G_L)}$ on the graph G_R.

Lemma 3. *$UDAF_f$ is truthful for every choice of f.*

We note that the idea of randomly selecting a group of consumers to determine the prices of the items to be sold for the consumers in the remaining group has appeared before in the context of unlimited-supply auctions [7]. While this is a relatively simple concept, its successful application to UDA and the corresponding analysis are not as simple as one would assume at first glance. As an example, we devised the concept of approximation matchings (Secion 2.1) to help us deal with the technical aspects of this.

3.1 A Revenue-Oriented Mechanism

First, we investigate Rev, a definition for f that favors revenue. Rev estimates the size of the matching in G that maximizes $\mathcal{F}(\cdot)$.

Rev(G_L:graph)

1. Let M_{rev} be the largest (w.r.t. the number of edges) matching in G_L such that $\mathcal{F}(M_{rev}) \geq \mathcal{F}_{G_L}/3$.
2. Return $\lfloor |M_{rev}|/6 \rfloor$.

The next theorem gives a bound on the revenue attained by $UDAF_{Rev}$. The assumption in the theorem about the $O(\ln s)$ competitiveness of G being at least 500 is only used to assure that $|M_G^{\mathcal{F}}| \geq 500$. In this case, the proof Theorem 1 ensures that $UDAF_{Rev}$ achieves an expected revenue of $\Omega(\mathcal{F}_G)$.

Theorem 1. *Let G be a graph that has* $(\ln s)$*-competitiveness larger than* 500*. Then, $UDAF_{Rev}$ simultaneously attains revenue $\Omega(\mathcal{F}_G)$ and efficiency $\Omega(\mathcal{F}_G)$ with probability at least* $1 - \frac{74}{e^{cp/108}}$*, where cp is the* $(\ln s)$*-competitiveness of* G*.*

Proof. For every j, let M_j be a matching of size j in G which maximizes $\mathcal{F}(\cdot)$. The matching M_j is said to be *good* if $j \geq |M_G^{\mathcal{F}}|/3$ and $\mathcal{F}(M_j) \geq \mathcal{F}_G/9$. Let J be the set of integers defined as $J = \{j | M_j \text{ is a good matching }\}$.

For every $j \in J$, let C_j be the set of consumers of matching M_j. With respect to Step 1 of the UDAF mechanism, we define the event $\mathcal{E}_j$ as the event in which the number of consumers of C_j that lie in the left group is at least $j/3$ and at most $2j/3$. Furthermore, let $\mathcal{E} = \bigcup_{j \in J} \mathcal{E}_j$.

In what follows we make some observations under the assumption that $\mathcal{E}$ occurs. Recall that we use $M_G^{\mathcal{F}}$ to denote the largest matching of G that maximizes $\mathcal{F}(\cdot)$. Let M' be the sub-matching of $M_G^{\mathcal{F}}$ induced by the consumers of $M_G^{\mathcal{F}}$ that lie in G_L. Then, M' has at least $|M_G^{\mathcal{F}}|/3$ edges and $\mathcal{F}(M') \geq \mathcal{F}_G/3 \geq \mathcal{F}_{G_L}/3$. This implies that $\mathcal{F}_{G_L} \geq \mathcal{F}_G/3$ and $|M_{rev}| \geq |M_G^{\mathcal{F}}|/3$.

Since $\mathcal{F}(M_{rev}) \geq \mathcal{F}_{G_L}/3$ it follows that $\mathcal{F}(M_{rev}) \geq \mathcal{F}_G/9$ and, as a consequence, $|M_{rev}| \in J$. Therefore, there are at least $\lceil |M_{rev}|/3 \rceil$ consumers of $C_{|M_{rev}|}$ in the right group, which implies on the existence of a matching in G_R, say M^2, of size $\lceil |M_{rev}|/3 \rceil$, where every edge costs at least $\mathcal{F}_G/(9|M_{rev}|)$. Thus, it follows from Lemma 2 that the revenue of $\mathcal{A}_{\lfloor |M_{rev}|/6 \rfloor}$, for input G_R, is at least

$$\lfloor |M_{rev}|/6 \rfloor \times \mathcal{F}_G/(9|M_{rev}|) = \Omega(\mathcal{F}_G).$$

Since Proposition 1 guarantees that $\mathcal{F}_G \geq \mathcal{T}_G/\ln s$, it follows that the efficiency is $\Omega(\mathcal{T}_G/\ln s)$.

Now, we obtain a bound on the probability of event $\mathcal{E}$ happening. A direct application of the Chernoff Bound [14] ensures that the probability of event $\mathcal{E}_j$ not happening is at most $2e^{-j/36}$. Applying the union bound we get that the probability of failure of $\mathcal{E}$ is at most $\sum_{j \in J} 2e^{-j/36}$.

Now, we use the condition on the competitiveness of G. Since the $(\ln s)$-competitiveness of G is cp it follows from Proposition 2 that $|M_G^{\mathcal{F}}| \geq cp$. This implies that the minimum integer in J is at least $\lceil cp/3 \rceil$. Thus,

$$\sum_{j \in J} 2e^{-j/36} \leq \sum_{j=\lceil cp/3 \rceil}^{\infty} 2e^{-j/36} \leq \frac{2e^{-cp/108}}{1 - e^{-1/36}} \leq \frac{74}{e^{cp/108}} \qquad \square$$

With respect to the previous theorem, we note that the more competitive G is, the higher the probability of attaining the bounds for the revenue and for the efficiency.

If we do not concern ourselves with proving bounds on the probability of attaining a certain revenue and a certain efficiency we can obtain a simple mechanism that attains $\Omega(\mathcal{F}_G)$ *expected* revenue and $\Omega(\mathcal{T}_G)$ *expected* efficiency for every graph G with $\ln s$-competitiveness larger than 1.

Theorem 2. *Let Mixed be the auction mechanism that executes one of the following mechanisms with uniform probability: $UDAF_{Rev}$, $\mathcal{A}_1$ and VCG. Then,* *for every graph G with* $\ln s$*-competitiveness larger than 1, Mixed achieves expected revenue* $\Omega(\mathcal{F}_G)$ *and expected efficiency* $\Omega(\mathcal{T}_G)$*.*

Proof. Omitted. □

3.2 Favoring the Efficiency

Now, we investigate Eff, a definition of f that favors the efficiency. Eff estimates the size of the matching in G that satisfies the conditions in Proposition 3.

Eff(G_L: graph)

1. Let $e_1, \ldots, e_{|M^{\mathcal{T}}_{G_L}|}$ be the edges of $M^{\mathcal{T}}_{G_L}$ listed by non-increasing order of costs. Let s^* be the largest integer such that $s^* \times c(e_{s^*}) \geq \mathcal{T}_{G_L}/(2\ln|M^{\mathcal{T}}_{G_L}|)$.
2. Return $\lfloor s^*/12 \rfloor$.

The main result of this section is the following theorem.

Theorem 3. *Let* K' *be the 8−competitiveness of a graph G. Then, for input G,* $UDAF_{Eff}$ *simultaneously attains revenue* $\Omega(\mathcal{T}_G/\ln s)$ *and efficiency* $\Omega(\mathcal{T}_G)$ *with probability at least* $1 - \frac{148}{e^{K'/36}}$.

The proof consists of showing that with the probability stated above there is a matching M^* in G_R such that: (i) $s^*/6 \leq |M^*| \leq (4 \cdot s^*)/3$; (ii) $\mathcal{F}(M^*) = \Omega(\mathcal{T}_G/\ln s)$ and (iii) $\mathcal{T}(M^*) = \Omega(\mathcal{T}_G)$.

The next lemma shows that the existence of such a matching indeed ensures that the mechanism performs as desired.

Lemma 4. *If there is a matching* M^* *in* G_R *that satisfies properties (i)-(iii) above, then* $\mathcal{A}_{Eff(G_L)}$ *simultaneously attains efficiency* $\Omega(\mathcal{T}_G)$ *and revenue* $\Omega(\mathcal{T}_G/\ln s)$ *on* G_R*.*

Proof. Let M^2 be the matching of size $\lfloor s^*/12 \rfloor$ computed by $\mathcal{A}_{Eff(G_L)}$ on G_R. Since M^* has at most $(4 \cdot s^*)/3$ edges, the sum of the costs of the $\lfloor s^*/12 \rfloor$ most expensive edges of M^* is at least $\lfloor s^*/12 \rfloor \times 3\mathcal{T}(M^*)/(4 \cdot s^*)$. It follows that $\mathcal{T}(M^2) = \Omega(\mathcal{T}_G)$.

On the other hand, since $\mathcal{F}(M^*) = \Omega(\mathcal{T}_G/\ln s)$, then all the edges of M^* cost at least $K\mathcal{T}_G/(|M^*|\ln s)$, where K is the constant hidden in the asymptotic notation. Since $2|M^2| = 2\lfloor s^*/12 \rfloor \leq s^*/6 \leq |M^*|$, it follows from Lemma 2 that the revenue is at least $\lfloor s^*/12 \rfloor \times K\mathcal{T}_G/(|M^*|\ln s)$. By using the fact that $|M^*| \leq (4 \cdot s^*)/3$, we conclude that the revenue is $\Omega(\mathcal{T}_G/\ln s)$ □

Thus, it suffices to bound the probability that such a matching exists. The following definition is useful in our proofs.

Definition 3. *Given a matching M in G, let* C_j *be the set of consumers associated with the j most expensive edges of M. Let* $\mathcal{E}_j$ *be the event where at least*

$j/3$ consumers of C_j lie in the left group and at least $j/3$ lie in the right one. Finally, let $\mathcal{E}_M = \bigcup_{j=K'}^{|M|} \mathcal{E}_j$, where K' is the 8-competitiveness of G.

Two properties of $\mathcal{E}_M$ are useful for our analysis: the failure probability of $\mathcal{E}_M$ decreases exponentially with the increase of K' and if $\mathcal{E}_M$ occurs then the edges of M are "evenly" distributed between G_L and G_R in the sense that the sub-matching of M induced by the consumers that lie in G_L has approximately the same cost (w.r.t. $\mathcal{F}$ and $\mathcal{T}$) as the subgraph induced by the consumers that lie in G_R. The following two propositions formalize these observations.

Proposition 4. *The probability that $\mathcal{E}_M$ does not occur is at most $74e^{-K'/36}$.*

Proposition 5. *Let M be a matching in G, with $|M| > K'$. If $\mathcal{E}_M$ occurs then the sub-matching M' of M induced by the consumers of M that lie in G_L (G_R) satisfies the following properties: $\mathcal{F}(M') \geq \mathcal{F}(M)/3$ and $\mathcal{T}(M') \geq \mathcal{T}(M)/3 - \mathcal{T}_G/24$.*

Proposition 4 follows from a direct application of Chernoff bounds [14]. The proof of Proposition 5 is not as immediate but we defer it to an extended version of this paper.

We say that a matching M of G is *efficiency-good* if $\mathcal{F}(M) \geq \mathcal{T}_G/(7 \ln s)$ and $\mathcal{T}(M) \geq \mathcal{T}_G/7$. The existence of at least one efficiency-good matching is guaranteed by Proposition 3. Let J be an increasing sequence of integers such that $j \in J$ if and only if there is an efficiency-good matching in G of cardinality j. For every $j \in J$, let M_j be an arbitrary efficiency-good matching of size j. Note that the definition of efficiency-good matchings and the assumption over K' in Theorem 3 imply that $min(J) > K'$.

Proposition 6. *Let A be the approximation matching of $(M_j)_{j \in J}$. If the event $\mathcal{E}_{M_G^{\mathcal{T}}} \cup \mathcal{E}_A$ happens then there is a matching in G_R that meets the conditions (i)-(iii).*

Proof. First, we show that if $\mathcal{E}_{M_G^{\mathcal{T}}}$ occurs then there is an efficiency-good matching of size s^* in G. Let M be the matching formed by the s^* largest-weight edges of $M_{G_L}^{\mathcal{T}}$. Let M' be the sub-matching of $M_G^{\mathcal{T}}$ induced by the consumers of $M_G^{\mathcal{T}}$ that lie in G_L. It follows from Proposition 5 that $\mathcal{T}(M') \geq \mathcal{T}_G/3 - \mathcal{T}_G/24 \geq 7 \times \mathcal{T}_G/24$. Thus, $\mathcal{T}(M_{G_L}^{\mathcal{T}}) \geq \mathcal{T}(M') \geq 7 \times \mathcal{T}_G/24$. The definition of M and Proposition 3 imply that $\mathcal{T}(M) \geq 7 \times \mathcal{T}_G/48$ and $\mathcal{F}(M) \geq 7 \times \mathcal{T}_G/(48 \ln s)$. Thus, M is an efficiency-good matching in G.

Since there is a efficiency-good matching in G of size s^*, it follows from Lemma 1 that there is a sub-matching A' of the approximation matching A such that $\max\{min(J), s^*/2\} \leq |A'| \leq 2 \cdot s^*$, $\mathcal{T}(A') \geq \mathcal{T}_G/7$ and $\mathcal{F}(A') \geq \mathcal{T}_G/14 \ln s$.

Since $\mathcal{E}_A$ occurs, $|A'| \geq min(J) > K'$, and A' contains the $|A'|$ largest-weight edges of A then $\mathcal{E}_{A'}$ also happens. Let A'' be the sub-matching of A' induced by the consumers of A' that lie in G_R. It follows from Proposition 5 and the previous observation about A' that $s^*/6 \leq |A''| \leq (4 \cdot s^*)/3$, $\mathcal{T}(A'') \geq \mathcal{T}_G/168$ and $\mathcal{F}(A'') \geq \mathcal{T}_G/42 \ln s$. Thus, A'' meets the conditions (i)-(iii), which establishes our result. □

Proof of Theorem 3. If the event $\mathcal{E}_{M_G^{\mathcal{T}}} \cup \mathcal{E}_A$ happens, it follows from Proposition 6 and from Lemma 4 that, for input G_R, $\mathcal{A}_{Eff(G_L)}$ simultaneously achieves $\Omega(\mathcal{T}_G)$ efficiency and $\Omega(\mathcal{T}_G/\ln s)$ revenue.

On the other hand, it follows from Proposition 4 that $\mathcal{E}_{M_G^{\mathcal{T}}} \cup \mathcal{E}_A$ fails with probability at most $148e^{-K\prime/36}$. Thus, $\mathcal{E}_{M_G^{\mathcal{T}}} \cup \mathcal{E}_A$ happens with probability at least $1 - 148e^{-K\prime/36}$. □

4 Final Remarks

The mechanisms described here can be efficiently implemented. We also mention without proof a relatively straightforward upper bound for truthful Unit-Demand auctions.

Theorem 4. *For every randomized truthful mechanism $\mathcal{A}$ there is a graph $G_{\mathcal{A}}$ such that the expected revenue achieved by $\mathcal{A}$ on $G_{\mathcal{A}}$ is $O(\mathcal{T}_{G_{\mathcal{A}}}/\ln s)$.*

References

1. Aaron Archer, Christos Papadimitriou, Kunal Talwar, and Éva Tardos. An approximate truthful mechanism for combinatorial auctions with single parameter agents. In *Proceedings of the fourteenth Annual ACM-SIAM Symposium on Discrete Algorithms (SODA-03)*, pages 205–214, New York, January 12–14 2003. ACM Press.
2. Y. Bartal, R. Gonen, and N. Nisan. Incentive compatible multi-unit combinatorial auctions. In *Proceedings of Theoretical Aspects of Rationality and Knowledge*, 2003.
3. Allan Borodin and Ran El-Yaniv. *Online computation and competitive analysis.* Cambridge University Press, Cambridge, 1998.
4. Patrick Briest, Piotr Krysta, and Berthold Voecking. Approximation techniques for utilitarian mechanism design. In *Proceedings of the ACM Symposium on Theory of Computing*, 2005.
5. E. H. Clarke. Multipart pricing of public goods. *Public Choice*, 11:17–33, 1971.
6. A. Goldberg, J. Hartline, A. Karlin, A. Wright, and M. Saks. Competitive auctions. submitted, 2002.
7. Andrew V. Goldberg and Jason D. Hartline. Competitive auctions for multiple digital goods. In *ESA: European Symposium on Algorithms*, 2001.
8. Andrew V. Goldberg, Jason D. Hartline, and Andrew Wright. Competitive auctions and digital goods. In *Proceedings of the Twelfth Annual ACM-SIAM Symposium on Discrete Algorithms (SODA-01)*, pages 735–744, New York, January 7–9 2001. ACM Press.
9. T. Groves. Incentives in teams. *Econometrica*, 41:617–631, 1973.
10. Venkatesan Guruswami, Jason D. Hartline, Anna Karlin, David Kempe, Claire Kenyon, and Frank McSherry. On profit-maximizing envy-free pricing. In *Proceedings of the Sixteenth Annual ACM-SIAM Symposium on Discrete Algorithms (SODA-05)*, 2005.
11. Daniel Lehmann, Liadan Ita O'Callaghan, and Yoav Shoham. Truth revelation in approximately efficient combinatorial auctions. *Journal of the ACM*, 49(5):577–602, September 2002.

12. Anton Likhodedov and Tuomas Sandholm. Auction mechanism for optimally trading off revenue and efficiency. In *Proceedings of the 4th ACM Conference on Electronic Commerce (EC-03)*, pages 212–213, New York, June 9–12 2003. ACM Press.
13. Paul Milgrom. *Putting Auction Theory to Work.* Cambridge University Press, 2004.
14. Rajeev Motwani and Prabhakar Raghavan. *Randomized Algorithms.* Cambridge University Press, 1997.
15. W. Vickrey. Counterspeculation, auctions, and competive sealed tenders. *Journal of Finance*, 16:8–37, 1961.

Pattern Matching Statistics on Correlated Sources

Jérémie Bourdon[1] and Brigitte Vallée[2]

[1] LINA, CNRS and Université de Nantes, France
Jeremie.Bourdon@univ-nantes.fr
[2] GREYC, CNRS and Université de Caen, France
Brigitte.Vallee@info.unicaen.fr

Abstract. In pattern matching algorithms, two characteristic parameters play an important rôle : the number of occurrences of a given pattern, and the number of positions where a pattern occurrence ends. Since there may exist many occurrences which end at the same position, these two parameters may differ in a significant way. Here, we consider a general framework where the text is produced by a probabilistic source, which can be built by a dynamical system. Such "dynamical sources" encompass the classical sources –memoryless sources, and Markov chains–, and may possess a high degree of correlations. We are mainly interested in two situations : the pattern is a general word of a regular expression, and we study the number of occurrence positions – the pattern is a finite set of strings, and we study the number of occurrences. In both cases, we determine the mean and the variance of the parameter, and prove that its distribution is asymptotically Gaussian. In this way, we extend methods and results which have been already obtained for classical sources [for instance in [9] and in [6]] to this general "dynamical" framework. Our methods use various techniques: formal languages, and generating functions, as in previous works. However, in this correlated model, it is not possible to use a direct transfer into generating functions, and we mainly deal with generating operators which generate... generating functions.

1 Introduction

The problem of searching for a particular pattern in a text is an important problem in information theory. It is crucial to study precisely the number of *occurrences* of a given pattern in a typical text. Here, "typical" essentially means that the text is a random text produced by a probabilistic model that follows as far as possible the real complexity of the studied sequences. It is also very interesting to consider *positions of occurrence*, i.e., positions (in a text) where an occurrence of the pattern can terminate.

The two parameters – the number of occurrences, denoted in the following by Ω, and the number of occurrence positions, denoted by C – may differ in a significant way, since the number of occurrence positions is always bounded by

J.R. Correa, A. Hevia, and M. Kiwi (Eds.): LATIN 2006, LNCS 3887, pp. 224–237, 2006.

the text length, whereas this is not true for the number of occurrences. [There may exist many occurrences which end at the same occurrence position].

With a precise probabilistic study of these two parameters, one obtains sharp statistical heuristics (like Z-scores) which permit to describe the related algorithms, and perhaps improve them.

Various pattern matching problems. There are also different pattern matching problems, which differ according to the nature of the pattern.

String matching. This is the basic pattern matching problem. Here, a string w is a block of (consecutive) symbols $w = w_1 w_2 \dots w_s$ (of length s).

Set of strings. Previously, the string w should appear exactly in the text, while, in the approximate case, a few mismatches are considered acceptable. The *approximate string matching* is then expressed as a matching against *a set $\mathcal{L}$ of words* which contains all the valid approximations of the string.

Sequence of patterns. Here, the symbols no longer need to be consecutive in the text: we are interested in occurrences of the string w as a subsequence of the text T. The problem is different, and it is called the hidden word problem.

Regular expressions. Searching words from a regular language is surely the most general pattern matching problem, since all the three previous pattern matching problems all consist in finding words of a given regular language.

Motivations. *Molecular biology* [12, 17, 18] provides an important source of applications. As a rule, there, one searches for subsequences, not consecutive strings. There are plenty of examples: split genes where exons are interrupted by introns, starting and stopping signal in genes, etc.... In general, for gene searching [8], regular expressions are used as a general pattern model (such as the PROSITE format used to scan in protein databases).

In this general context, it is of obvious interest to discern what constitutes meaningful information from what is statistically unavoidable phenomenon. This leads to a probabilistic study. In information theory context, a source is a mechanism which emits symbols from an alphabet Σ. A text of length n is just an element of Σ^n, and the various models of sources are related to the choice of a probabilistic model on Σ^n. When the probabilistic model has been chosen, the main variables of interest — the number of occurrences Ω, and the number of occurrence positions C— become random variables, and it is crucial to study their distribution, in order to set *thresholds* from which appearance of a pattern becomes meaningful.

Previous results. The two classical models of sources are the memoryless sources (where each symbol m is always emitted with the same probability, and independently of the previous history) and Markov chains (where the probability of emitting m only depends on the unique symbol emitted before m). In both cases, these sources have a "bounded" memory and only provide idealized models, while real-life sources are often complex objects. Most of the results are obtained only for such idealized sources.

Number of occurrences Ω. The number of string occurrences in a random text has been intensively studied over the last two decades. Guibas and Odlyzko have revealed in 1981 the fundamental rôle played by autocorrelation. Régnier and Szpankowski [10, 11] established that the number of occurrences of a string is asymptotically normal under a diversity of models that include Markov chains. The number of occurrences of finite sets of (finite) strings also obeys the "Guibas and Odlyzko" principle, which now deals with correlation matrices.

In the case of the hidden word problems, Flajolet, Szpankowski and Vallée show that the distribution of Ω is asymptotically Gaussian for memoryless sources [6].

Number of occurrence positions C. Nicodème, Salvy, and Flajolet [9] showed that, for a simple[1] regular expression $\mathcal{E}$, the variable $C_n(\mathcal{E})$ is asymptotically normally distributed, both for memoryless sources and Markov chains.

Our results. We use here a general framework of sources related to dynamical systems theory which goes beyond the cases of memoryless and Markov sources [16, 4]. This model can describe non-Markovian processes, where the dependency on past history is unbounded, and as such, they attain a high level of generality. A probabilistic dynamical source is defined by two objects: a symbolic mechanism and a density. The mechanism, related to symbolic dynamics, associates an infinite word $M(x)$ to a real number $x \in [0, 1]$, and generalizes numeration systems. Once the mechanism has been fixed, the density f on the $[0, 1]$ interval can vary. This induces different probabilistic behaviors for sources of words.

In this context, string matching problems have been already considered: In [1], the authors study the parameter $\Omega(\mathcal{L})$ when $\mathcal{L}$ is a particular regular expression (namely, a generalized pattern), which provides a generalization for the hidden word problem. The mean and the variance of Ω_n are shown to be polynomial in n, and the exponent r depends on the number of freedom degrees of $\mathcal{L}$. However, the asymptotic distribution – expected to be Gaussian– is not obtained.

Here, we obtain two new results in this correlated model of dynamical sources. We prove here that many variables R defined on some set $\mathcal{R}$ follow asymptotically a gaussian law. We first provide a precise definition:

Definition [Asymptotic gaussian law]. *Consider a cost R defined on a set $\mathcal{R}$ and its restriction R_n to the subset $\mathcal{R}_n$ of size n. The cost R asymptotically follows a gaussian law if there exist three sequences a_n, b_n, r_n, with $r_n \to 0$, for which*

$$\Pr\left[(u,v) \in \mathcal{R}_n \;\middle|\; \frac{R_n(u,v) - a_n}{\sqrt{b_n}} \leq y\right] = \frac{1}{\sqrt{2\pi}} \int_{-\infty}^{y} e^{-t^2/2}\, dt + O(r_n)\,.$$

The sequence r_n defines the speed of convergence, denoted also by $r[R_n]$. The expectation $\mathbb{E}[R_n]$ and the variance $\mathbb{V}[R_n]$ satisfy $\mathbb{E}[R_n] \sim a_n$, $\mathbb{V}[R_n] \sim b_n$. The triple $(\mathbb{E}[R_n], \mathbb{V}[R_n], r_n)$ is a characteristic triple for the gaussian law of R.

[1] See Section 2.4 for a definition.

We now state our main result:[2]

Theorem. *Let $\mathcal{S}$ be a nice dynamical source.*

(i) Consider a simple regular expression $\mathcal{E}$ whose useful part of the automaton is primitive. The number of occurrence positions of $\mathcal{E}$ in a word of length n built by $\mathcal{S}$, denoted by $C_n(\mathcal{E})$, follows an asymptotic gaussian law with a characteristic triple given by $r[C_n(\mathcal{E})] = O(1/\sqrt{n})$,

$$\mathbb{E}[C_n(\mathcal{E})] = \gamma_{\mathcal{E}} \cdot n + \gamma'_{\mathcal{E}} + O(\mu_{\mathcal{E}}^n), \quad \mathbb{V}[C_n(\mathcal{E})] = \nu_{\mathcal{E}} \cdot n + \nu'_{\mathcal{E}} + O(\mu_{\mathcal{E}}^n),$$

The constants $\gamma_{\mathcal{E}}$ and $\nu_{\mathcal{E}}$ are expressible with the pression $\Lambda(t)$ of the operator $\mathbb{R}(e^t)$ defined in (8), namely $\gamma_{\mathcal{E}} = \Lambda'(0)$, $\nu_{\mathcal{E}} = \Lambda''(0)$, while $\mu_{\mathcal{E}} < 1$ is any real number strictly larger than the subdominant eigenvalue of $\mathbb{R}$.

(ii) Consider a finite set of words $\mathcal{W} \subset \Sigma^\star$. The number of occurrences of $\mathcal{W}$ in a text of length n built by $\mathcal{S}$, denoted by $\Omega_n(\mathcal{W})$, follows an asymptotic gaussian law with a characteristic triple given by $r[\Omega_n(\mathcal{W})] = O(1/\sqrt{n})$,

$$\mathbb{E}[\Omega_n(\mathcal{W})] = \alpha_{\mathcal{W}} \cdot n + \alpha'_{\mathcal{W}} + O(\eta_{\mathcal{W}}^n), \quad \mathbb{V}[\Omega_n(\mathcal{W})] = \beta_{\mathcal{W}} \cdot n + \beta'_{\mathcal{W}} + O(\eta_{\mathcal{W}}^n).$$

The constants $\alpha_{\mathcal{W}}$ et $\beta_{\mathcal{W}}$ are expressible with the pression $\Lambda(t)$ of the operator $\mathbb{B}(e^t)$ defined in (9), namely $\alpha_{\mathcal{W}} = \Lambda'(0)$, $\beta_{\mathcal{W}} = \Lambda''(0)$, while $\eta_{\mathcal{W}} < 1$ is any real number strictly larger than the subdominant eigenvalue of $\mathbb{B}$.

Methodology. For studying the parameter $C(\mathcal{E})$, Nicodème, Salvy and Flajolet describe in [9] a general method which directly translates a regular expression into rational generating functions. They use, as a main tool, the transition matrix of the automaton which recognizes the regular language $\Sigma^\star \cdot \mathcal{E}$, and the occurrence positions are related to the final states of the automaton. In [6], the authors also use similar methods, namely the de Bruijn graph, to study the parameter $\Omega(\mathcal{W})$. These two previous works, based on the "generating function methodology", as in the main books of the area [14, 13], operate a systematic translation of each language into its generating function. Due to correlations of a dynamical source, such a direct approach is no longer possible here. Instead, we perform what we call a "dynamical analysis" and we first operate a systematic translation into *generating operators*. In dynamical systems theory, an important tool is the *density transformer*; here, we give it the role of a "generating operator". Now, there are many instances of this methodology, applied in two main areas: text algorithms as in [2, 5, 16], or arithmetical algorithms as in [15]. Here, we deal with a mixed structure, where we insert generating operators inside the transition matrix of the automaton. We obtain an operator matrix which takes into account both the complexity of the source and the algebraic structure of the problem (namely an automaton).

[2] The word "nice" is defined in Def. 4, Section 3.3, the words "simple" and "useful" are defined in Def. 1, Section 2.4, the word "primitive" in Section 3.3.

2 Various Tools

We first introduce the languages and the related generating functions that intervene in the analysis of the characteristic parameters C and Ω. Next, we precise the probabilistic model. We define dynamical sources and introduce the generating operators that are a basic ingredient associated to our correlated sources.

2.1 Probabilistic Model and Generating Functions

As regards the probabilistic model, we consider a source that creates the text by emitting symbols from a finite alphabet Σ. For a given length n, a random text, denoted by T_n is an element of Σ^n which is drawn according to the induced probability on Σ^n, and, for any word w of length n, we denote by p_w the probability that the source emits a prefix equal to w. A language $\mathcal{L}$ is then a set of words. For any language, we denote by $\mathcal{L}_n$ the language formed with all the words w of $\mathcal{L}$ with length n. We aim at studying the random variables $Y = C$ (the number of occurrence positions) and $Y = \Omega$ (the number of occurrences). In both cases, we consider the restriction of Y to Σ^n, denoted by Y_n, and analyze its probabilistic behavior for $n \to \infty$. Our main tool is the moment generating function of Y_n, defined as

$$\mathbb{E}[\exp(tY_n)] := \sum_{w \in \Sigma^n} p_w \cdot \exp[tY(w)], \tag{1}$$

and the main challenge is to show that it behaves as a "quasi-power". Then, it will be possible to obtain an asymptotic Gaussian law:

Theorem 0. [Hwang] *Let Y_n be a sequence of variables whose moment generating functions satisfies $\mathbb{E}[\exp(tY_n)] = [\exp(nU(t)+V(t))] \cdot [1+O(W_n)]$, $W_n \to \infty$, with a uniform error term on the complex closed disk $|t| \le t_0$, $t_0 > 0$. Suppose that $U(t)$ and $V(t)$ are analytic in $|t| \le t_0$ and $U(t)$ satisfies $U''(0) \neq 0$. Then, Y_n follows an asymptotic gaussian law, with a characteristic triple given by*

$$\mathbb{E}[Y_n] = U'(0) \cdot n + V'(0) + O(W_n), \quad \mathbb{V}[Y_n] = U''(0) \cdot n + V''(0) + O(W_n),$$

$$r[Y_n] = O\left(\max(1/\sqrt{n}, W_n)\right).$$

2.2 Bivariate Generating Functions

The so-called probability generating function $F_Y(z,u)$ relative to parameter Y is defined as

$$F_Y(z,u) = \sum_{w \in \Sigma^\star} p_w \cdot u^{Y(w)} \cdot z^{|w|},$$

where $|w|$ denotes the length of w, the variables z and u respectively mark the length of the word and the parameter $Y(w)$. Remark that the moment generating function of parameter Y_n is closely related to $F_Y(z,u)$ via the relation

$$\mathbb{E}[\exp(tY_n)] = [z^n]F_Y(z, e^t) \tag{2}$$

where the notation $[z^n]G(z)$ denotes the coefficient of z^n in $G(z)$. Previous works, which deal with non correlated sources, directly work with the generating functions. Here, we cannot operate a direct translation from the problem into generating functions, and we mainly use generating operators.

2.3 Language vs Automaton

Let us first recall that an automaton is defined by $(\Sigma, \mathcal{Q}, \mathcal{F}, s, \delta)$, where Σ is an alphabet, $\mathcal{Q}$ is the (finite) set of states, $\mathcal{F} \subset \mathcal{Q}$ corresponds to the final states, $s \in \mathcal{Q}$ is the initial state and $\delta : \mathcal{Q} \times \Sigma \to \mathcal{Q}$ is the transition function of the automaton. In the following, the set $\mathcal{Q}$ will be always $\{0, \ldots, r-1\}$, and the state 0 will be the initial state.

The automaton recognizes a language $\mathcal{L}$ if, for all word $w := m_1 \ldots m_n$ of $\mathcal{L}$, there exists a path $q_1, q_2, \ldots, q_{n-1}$ of states and a final state f such that

$$\delta(s, m_1) = q_1, \quad \delta(q_i, m_{i+1}) = q_{i+1}, \quad [\text{for } 1 \le i \le n-2], \quad \delta(q_{n-1}, m_n) = f.$$

In this case, the language $\mathcal{L}$ is said to be a regular language. Every regular language can be described by a regular expression, composed of singletons and a finite number of unions, Cartesian products and star operations on those singletons. Conversely, it is possible to operate a direct translation from a regular expression to a deterministic finite automaton.

The transition matrix $\mathcal{T} := (\mathcal{T}_{i,j})$ is the $r \times r$ matrix whose element of index (i, j) is the set of symbols $m \in \Sigma$ for which there exists an edge from state i to state j labeled by m, namely $\mathcal{T}_{i,j} := \{m \in \Sigma; \quad \delta(i, m) = j\}$.

This matrix plays a fundamental rôle in the sequel. Thus, the component (i, j) of the matrix $\mathcal{T}^n$ is the language formed by all the words which allow to reach state j from state i in n steps. And, the component (i, j) of the matrix $\mathcal{T}^\star$ is the language formed by all the words which allow to reach state j from state i in an arbitrary number of steps. Finally, $\mathcal{L}_n = S \cdot \mathcal{T}^n \cdot F, \quad \mathcal{L} = S \cdot \mathcal{T}^\star \cdot F,$ where $F :=^t (f_1, \ldots, f_r)$ is a $\{0, 1\}$ column vector such that f_i equals 1 iff $i \in \mathcal{F}$, called the final vector and S is a row vector equal to $(1 \quad 0 \quad \cdots \quad 0)$.

2.4 Automata of Interest

To each parameter $[C(\mathcal{E})$ or $\Omega(\mathcal{W})]$, we associate an automaton which will be central in the study of this parameter.

Case $C(\mathcal{E})$- Automaton for the language $\mathcal{L} = \Sigma^\star \cdot \mathcal{E}$ associated to a regular expression $\mathcal{E}$. We consider the minimal automaton $\mathcal{A}$ which recognizes $\Sigma^\star \cdot \mathcal{E}$, and its decomposition into the acyclic graph of its strongly connected components (SCC):

Definition 1. *The expression $\mathcal{E}$ is simple if the minimal automaton $\mathcal{A}$ which recognizes $\Sigma^\star \cdot \mathcal{E}$ possesses a unique SCC which contains all the final states.*

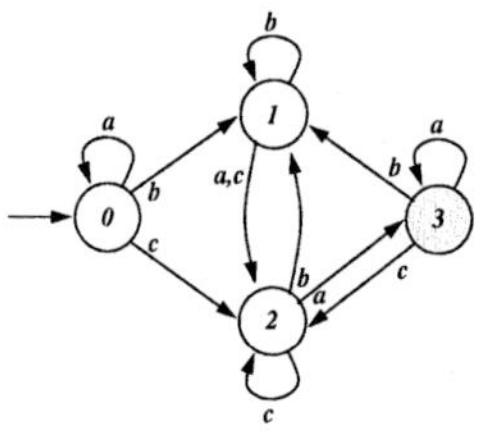

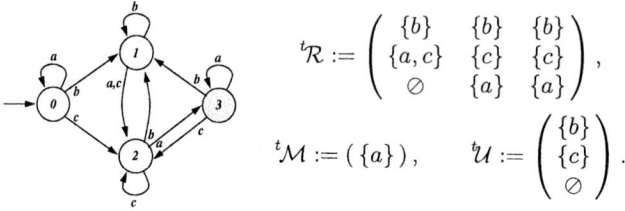

$$^t\mathcal{R} := \begin{pmatrix} \{b\} & \{b\} & \{b\} \\ \{a,c\} & \{c\} & \{c\} \\ \oslash & \{a\} & \{a\} \end{pmatrix}, \qquad ^t\mathcal{M} := (\,\{a\}\,), \qquad ^t\mathcal{U} := \begin{pmatrix} \{b\} \\ \{c\} \\ \oslash \end{pmatrix}.$$

$$\mathbb{M} := (\,\mathbf{G}[a]\,), \quad \mathbb{R}(u) := \begin{pmatrix} \mathbf{G}_{[b]} & \mathbf{G}_{[b]} & \mathbf{G}_{[b]} \\ \mathbf{G}_{[a]} + \mathbf{G}_{[c]} & \mathbf{G}_{[c]} & \mathbf{G}_{[c]} \\ 0 & u \cdot \mathbf{G}_{[a]} & u \cdot \mathbf{G}_{[a]} \end{pmatrix}, \quad \mathbb{U}(u) := \begin{pmatrix} \mathbf{G}_{[b]} \\ \mathbf{G}_{[c]} \\ 0 \end{pmatrix}$$

Fig. 1. The automaton relative to $\Sigma^\star.\mathcal{E}$ with $\mathcal{E} = (ba|c)^+a^+$, the transition matrices, and the marked operators

Generally speaking, it is possible that all final states do not belong to the same SCC. Here, we mainly consider the case when $\mathcal{E}$ is simple[3]. However, we explain (in the conclusion) how our method extends to the general case.

Proposition 1. *Let $\mathcal{E}$ be a simple regular expression and $\mathcal{A}$ be the minimal automaton which recognizes $\Sigma^\star \cdot \mathcal{E}$. Then, there exists a partition of its set of states $\mathcal{Q}$ into two sets $\mathcal{X}$ and $\mathcal{Y}$ for which the transition matrix $\mathcal{T}$ of this automaton can be written as*

$$\mathcal{T} = \begin{pmatrix} \mathcal{M} & \mathcal{U} \\ 0 & \mathcal{R} \end{pmatrix};$$

Here, $\mathcal{M}$ is the matrix restricted to $\mathcal{X}$, $\mathcal{R}$ is the matrix restricted to $\mathcal{Y}$, and $\mathcal{U}$ is the matrix from $\mathcal{X}$ to $\mathcal{Y}$. If $\mathcal{X}$ is non empty, it contains the initial state, while the graph $(\mathcal{Y}, \mathcal{R})$ is the SCC of the automaton, which contains all the final states and is called the useful part of the automaton.

Remarks. Then, the language $\mathcal{L}$ decomposes as $\mathcal{L} = S_\mathcal{X} \cdot \mathcal{M}^\star \cdot \mathcal{U} \cdot \mathcal{R}^\star \cdot F_\mathcal{Y}$, where $S_\mathcal{X}$ is the initial row vector restricted to $\mathcal{X}$, and $F_\mathcal{Y}$ is the final column vector restricted to $\mathcal{Y}$. Note that the language $\mathcal{L}_+$ of the words which contain at least one occurrence of the regular expression $\mathcal{E}$ satisfies $\mathcal{L}_+ \subset S_\mathcal{X} \cdot \mathcal{M}^\star \cdot \mathcal{U} \cdot \mathcal{R}^\star \cdot \mathbf{1}_\mathcal{Y}$, where $\mathbf{1}_\mathcal{Y}$ is a column vector, indexed with $\mathcal{Y}$, whose all components equal 1.

Example. See Figure 1 (at the end) for $\mathcal{E} := (ba|c)^+a^+$.

Case $\Omega(\mathcal{W})$- The de Bruijn automaton relative to an alphabet Σ and a length ℓ. In the sequel, ℓ will be the maximum length of a word of $\mathcal{W}$, minus 1. We consider a "sliding window" of length ℓ that scans a text of $\Sigma^\star$ and, at each stage, keeps in its (finite) memory the last ℓ letters read from the text. Formally, the de Bruijn graph is a finite automaton with state space $\mathcal{Q} = \Sigma^\ell$; when the symbol m is read, in a state $b \in \Sigma^\ell$, one erases the left symbol of b, which provides a word denoted by $\tau(b)$, and m is added on the right of $\tau(b)$, so that the new state is $\delta(b, m) = \tau(b) \cdot m$. A text of length $n \geq \ell$ is then associated to a

[3] This is also the only case which is considered in [9].

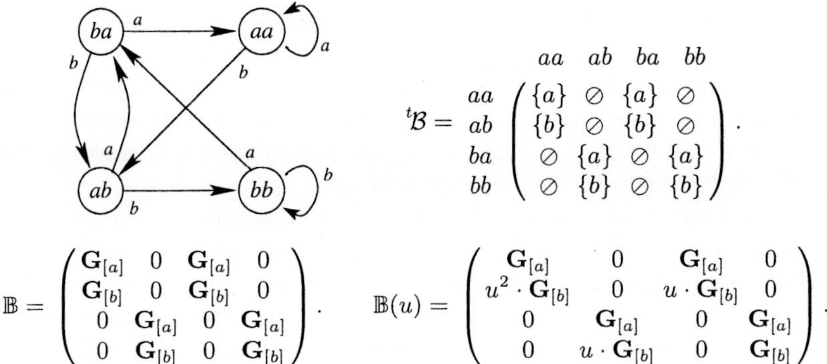

Fig. 2. The De Bruijn automaton, with its transition matrix, the operator, with its use for $\mathcal{W} = \{ab, aab, aba\}$ and the marked operator

path of length $n - \ell$ that begins at the state b formed with the first ℓ symbols of the text. This transition matrix is denoted by $\mathcal{B}$. Let us define the initial vector S as a row vector whose components are all the words of Σ^ℓ, and the final vector as a column vector whose components are all equal to 1. Then

$$\Sigma^n = S \cdot \mathcal{B}^{n-\ell} \cdot F, \qquad \Sigma^{\geq \ell} = S \cdot \mathcal{B}^\star \cdot F.$$

Example. See Fig. 2 (at the end) for the de Bruijn graph with $\Sigma := \{a, b\}, \ell = 2$.

We now present the probabilistic model for symbol generation. This model is based on dynamical systems. Here, probabilities are "generated" by operators, and the main generating functions of interest can be generated themselves by operators. Furthermore, unions and Cartesian products of sets translate into sums and compositions of the associated operators. This allows us to define a matrix generating operator related to a regular language.

2.5 Dynamical Sources

We first recall the definition of a dynamical system (of the interval). We refer to [16, 4] for more details. See Fig. 3 for an example.

Definition 2. *A dynamical system* $(\mathcal{I}, \mathcal{S})$ *is defined by four elements:*

(a) a finite alphabet Σ,
(b) a topological partition of $\mathcal{I} :=]0, 1[$ *with disjoint open intervals* $\mathcal{I}_m, m \in \Sigma$,
(c) an encoding mapping σ *which is constant and equal to* m *on each* $\mathcal{I}_m$,
(d) a shift mapping S *whose restriction to* $\mathcal{I}_m$ *is a bijection of class* $\mathcal{C}^2$ *from* $\mathcal{I}_m$ *to* $\mathcal{J}_m := S(\mathcal{I}_m)$. *The local inverse of* $S|_{\mathcal{I}_m}$ *is denoted by* h_m.

Such a dynamical system can be viewed as a "dynamical source", since, on an input x of $\mathcal{I}$, it outputs the word $M(x)$ formed with the sequence of symbols $\sigma S^j(x)$, i.e., $M(x) := (\sigma x, \sigma S x, \sigma S^2 x, \ldots)$.

The branches of S^k, and also its inverse branches, are then indexed by Σ^k, and, for any $w = m_1 \dots m_k \in \Sigma^k$, the mapping $h_w := h_{m_1} \circ h_{m_2} \circ \dots \circ h_{m_k}$ is a $\mathcal{C}^2$ bijection from $\mathcal{J}_w$ onto $\mathcal{I}_w$. It is possible that the word w cannot be produced by the source: this means that $\mathcal{J}_w$ is empty, and the inverse branch h_w does not exist. All the words that begin with the same prefix w correspond to real numbers x that belong to the same interval $\mathcal{I}_w$.

Such sources may possess a high degree of correlations, due to the *geometry* of the branches [i.e., the respective positions of intervals $\mathcal{I}_m$ and $\mathcal{J}_\ell := S(\mathcal{I}_\ell)$] and also to the *shape* of branches. [See [4] for more details]. For instance, the classical sources correspond to dynamical systems with affine branches, for which the derivatives are constant. Generally speaking, the probability of emitting a symbol m is closely related to the shape of branches, as we now see.

2.6 Probabilities and Generating Operators

When the interval $\mathcal{I}$ is endowed with some density g, this induces a probabilistic model on $\Sigma^{\mathbb{N}}$, and the probability p_w that a word begins with prefix w is the measure of the interval $\mathcal{I}_w$. Such a probability p_w is easily generated by an operator $\mathbf{G}_{[w]}$, defined as

$$\mathbf{G}_{[w]}[f](x) = |h'_w(x)| \; f \circ h_w(x) \mathbb{I}_{\mathcal{J}_w}(x), \tag{3}$$

since one has $p_w = \displaystyle\int_{\mathcal{I}_w} g(x)dx = \int_{\mathcal{J}_w} |h'_w(x)| g \circ h_w(x)dx = \int_0^1 \mathbf{G}_{[w]}[g](x)dx.$

Then, the operator $\mathbf{G}_{[w]}$ is called the generating operator of the prefix w. The generating operator $\mathbf{L}$ relative to a collection $\mathcal{L}$ of words is defined as the sum of all the generating operators relative to the words of $\mathcal{L}$, namely $\mathbf{L} := \sum_{w \in \mathcal{L}} \mathbf{G}_{[w]}$, and the generating operator $\mathbf{G}$ of the alphabet Σ

$$\mathbf{G} := \sum_{m \in \Sigma} \mathbf{G}_{[m]}. \tag{4}$$

plays a fundamental rôle here, since it is the density transformer of the dynamical system; it describes the evolution of densities on $\mathcal{I}$ under iterations of S: if X is a random variable with density g, then SX has density $\mathbf{G}[g]$.

For two prefixes w, w', the relation $p_{w.w'} = p_w p_{w'}$ is no longer true when the source has some memory, and is replaced by the following composition property

$$\mathbf{G}_{[w.w']} = \mathbf{G}_{[w']} \circ \mathbf{G}_{[w]}, \tag{5}$$

so that unions and Cartesian products of collections of words translate into sums and compositions of the associated generating operators. Remark just that, due to (5), the generating operator of $\mathcal{L} \times \mathcal{M}$ is $\mathbf{M} \circ \mathbf{L}$.

2.7 Matrix Generating Operators

Here, we transform the transition matrix of an automaton into a matrix generating operator that combines both information from the dynamical source and

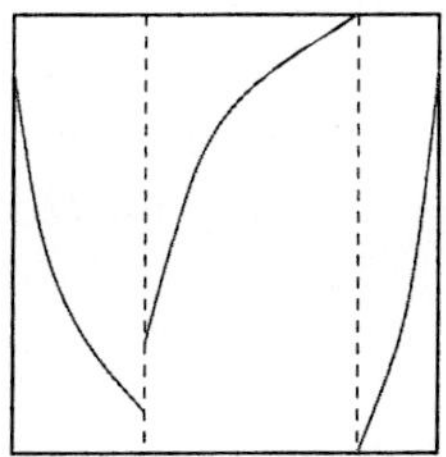

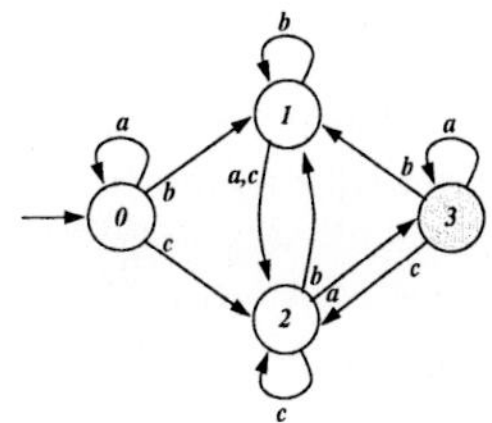

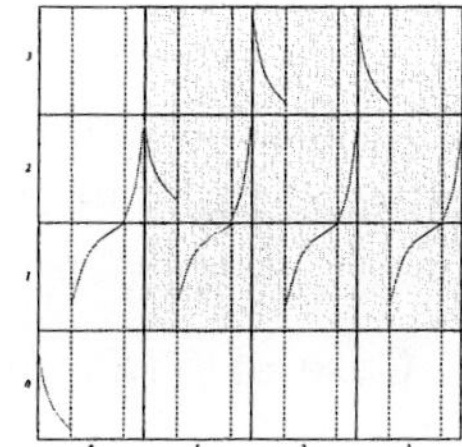

Fig. 3. The original source $\mathcal{S}$, the automaton $\mathcal{A}$ relative to $\Sigma^\star\mathcal{E}$, and the mixed sources $\mathcal{S}_\mathcal{A}$, $\mathcal{S}_\mathcal{R}$ when the states are restricted to be in $\mathcal{Y} := \{1, 2, 3\}$

the automaton. We associate to each element $\mathcal{T}_{j,i}$ of the matrix T, its generating operator $\mathbb{T}_{i,j}$

$$\mathbb{T}_{i,j} := \sum_{w \in \mathcal{T}_{j,i}} \mathbf{G}_{[w]}. \tag{6}$$

Then, $\mathbb{T}$ is a matrix generating operator which is related to ${}^t\mathcal{T}$, due to (5).

Examples. In the case when $\mathcal{L}$ is $\Sigma^\star \cdot \mathcal{E}$, there are three matrix operators, $\mathbb{M}, \mathbb{U}, \mathbb{R}$, respectively associated to matrices $\mathcal{M}, \mathcal{U}, \mathcal{R}$ [See Prop. 1]. For the de Bruijn graph, the generating operator is denoted by $\mathbb{B}$. See Figures 1 and 2 for examples.

2.8 The Mixed Source

We now build a source $\mathcal{S}_\mathcal{T}$ that combines both a transition matrix $\mathcal{T}$ of an automaton $\mathcal{A}$, and the original source $\mathcal{S}$. The set of states of $\mathcal{A}$ is $\mathcal{Q}$ and the matrix $\mathcal{T}$ has order r. The initial source $\mathcal{S}$ is defined by an interval $\mathcal{I}$, an alphabet Σ, a topological partition $(\mathcal{I}_m)_{m\in\Sigma}$ and a shift S whose each local inverse $h_m := (S|_{\mathcal{I}_m})^{-1}$ maps $\mathcal{J}_m :=]c_m, d_m[$ on $\mathcal{I}_m :=]a_m, b_m[$. The source $\mathcal{S}_\mathcal{T}$ [see Fig. 3 (at the end) for an example] is defined with the interval $\mathcal{I}^{[r]} = [0, r]$, the alphabet $\Gamma := \Sigma \times \mathcal{Q}$, a topological partition $(\mathcal{I}_{m,i})_{(m,i)\in\Gamma}$ and a shift function that maps $\mathcal{I}^{[r]}$ on $\mathcal{I}^{[r]}$. Each local inverse $h_{m,i}$ maps $\mathcal{J}_{m,i}$ on $\mathcal{I}_{m,i}$. More precisely, $\mathcal{I}_{m,i} = \mathcal{I}_m + i :=]a_m + i, b_m + i[$, $\mathcal{J}_{m,i} = \mathcal{J}_m + \delta(i,m) :=]c_m + \delta(i,m), d_m + \delta(i,m)[$, and $h_{m,i}(x) = h_m(x - \delta(i,m)) + i$. The density transformer $\mathfrak{G}$ of the source $\mathcal{S}_\mathcal{T}$ defined, as in (4), by

$$\mathfrak{G}[f](x) := \sum_{(m,i)\in\Sigma\times\mathcal{Q}} |h'_{m,i}(x)| \cdot f \circ h_{m,i}(x) \cdot \mathbb{1}_{\mathcal{J}_{m,i}}(x), \tag{7}$$

is conjugated to the matrix operator $\mathbb{T}$ defined in (6) via a mapping Ψ [namely $\mathfrak{G} = \Psi^{-1} \circ \mathbb{T} \circ \Psi$] which associates to g (defined on $\mathcal{I}^{[r]}$) the vector ${}^t[g_1, \ldots, g_r]$ where each g_i is defined on $\mathcal{I}$ by $g_i(x) := g|_{[i-1,1]}(x + i)$.

3 Probabilistic Behavior of Parameters C and Ω

Now, we come back to the two situations of interest. The next step consists in weighting operator matrices $\mathbb{T}$ in order to study our parameters $C(\mathcal{E})$ and $\Omega(\mathcal{W})$.

3.1 Case of $C(\mathcal{E})$

We consider here the language $\mathcal{L} := \Sigma^\star\mathcal{E}$ and the three matrix operators $\mathbb{M}, \mathbb{U}, \mathbb{R}$. We now mark the transitions which arrive at final states and define three new operators $\mathbb{R}(u)$, $\mathbb{U}(u), \mathbb{X}(u)$ by the relations.

$$\mathbb{R}(u)_{j,i} = u^{[\![j\in\mathcal{F}]\!]} \cdot \mathbb{R}_{j,i}, \quad \mathbb{U}(u)_{j,i} = u^{[\![j\in\mathcal{F}]\!]} \cdot \mathbb{U}_{j,i}, \quad ([\![\cdot]\!]\text{is Iverson's bracket}) \tag{8}$$

$$\mathbb{X}(z,u) := z \cdot \mathbb{U}(u) \circ (I - z\mathbb{M})^{-1}.S_{\mathcal{X}},$$

where the vector $S_{\mathcal{X}}$ is a column vector (of length $|\mathcal{X}|$) equal to ${}^t(1,0,\ldots,0)$.

Example. Figure 1 describes the marked matrix operators for $\mathcal{E} = (ba|c)^+a^+$.

3.2 Case of $\Omega(\mathcal{W})$

We consider here a set of finite words $\mathcal{W}$, and we choose the length ℓ of the de Bruijn graph to be equal to the maximal length of a word of $\mathcal{W}$, minus 1. this de Bruijn automaton is weighted with a counter that gets incremented each time a transition is effected, so that the value of the counter will contain at the end of the text the number $\Omega(\mathcal{W})$. A transition of the automaton, of the form $c = \delta(b,m)$ requires $b \cdot m \in \Sigma \cdot c$. When this transition is effected, one can "cash in" all the "new" occurrences of $\mathcal{W}$ which arise when reading the last letter m, i.e., all the occurrences of the pattern that *end* at the letter m. Precisely, for a transition $c = \delta(b,m)$ of the automaton, the number of occurrences of the pattern $\mathcal{W}$ contained in $b \cdot m$ and *ending* at the letter m is determined by either the pair (b,m) or the pair (b,c); we denote this number by $\phi(b,m)$ or $\psi(b,c)$, depending on context, so that $\phi(b,m) = \psi(b,c)$ whenever $c = \delta(b,m)$. Since the length of word $b \cdot m$ exactly equals $\ell + 1$ that is the maximum length of a word of $\mathcal{W}$, all the occurrences of $\mathcal{W}$ that end at m are contained in a text of the form $b \cdot m$ with $b \in \Sigma^\ell$ so that the relation $\phi(b,m) = \Omega(b \cdot m) - \Omega(b)$ holds. We build a operator matrix $\mathbb{B}(u)$ indexed by $\mathcal{Q} \times \mathcal{Q}$ as follows

$$\mathbb{B}(u)_{c,b} := u^{\phi(b,m)} \cdot [\![\, bm \in \Sigma c \,]\!] \cdot \mathbf{G}_{[m]} = u^{\Omega(bm)-\Omega(b)} \cdot [\![\, bm \in \Sigma c \,]\!] \cdot \mathbf{G}_{[m]}, \tag{9}$$

and the initial vector $\mathbb{X}(z,u)$ is a column vector defined by

$$(\mathbb{X}(z,u))_b = z^\ell \cdot u^{\Omega(b)} \cdot \mathbf{G}_{[b]} \,. \tag{10}$$

Example. Figure 2 describes the matrix $\mathbb{B}(u)$ relative to $\mathcal{W} := \{ab, aab, aba\}$.

In both cases, the operator $\mathbb{F}_Y(z,u) := (I - z\mathbb{T}(u))^{-1} \circ \mathbb{X}(z,u)$, with $\mathbb{T}(u) = \mathbb{R}(u)$ or $\mathbb{T}(u) = \mathbb{B}(u)$ itself generates, with (3), the generating function $F_Y(z,u)$, and we obtain, with (1):

Proposition 2. *The probability generating functions of parameters $Y = C$ and $Y = \Omega$ are expressible with the quasi-inverse of a matrix operator $\mathbb{T}(u)$,*

$$\mathbb{E}[u^{Y_n}] = [z^n] \cdot \left(\int_0^1 \left(\mathbf{1}_{\mathcal{X}} \cdot (I - z\mathbb{T}(u))^{-1} \circ \mathbb{X}(z,u)\right) [g](t) dt \right).$$

In the case $Y = C$, the operator $\mathbb{T}(u)$ equals $\mathbb{R}(u)$, and $\mathbb{R}(u), \mathbb{X}(z,u)$ involve the decomposition of Proposition 1 [see (8)]. In case $Y = \Omega$, the operator $\mathbb{T}(u)$ equals $\mathbb{B}(u)$ and $\mathbb{B}(u), \mathbb{X}(z,u)$ involve the de Bruijn graph [see (9,10)].

In the sequel, we prove that, provided that the source $\mathcal{S}$ and the transition matrix $\mathcal{T}$ possesses good properties, it is the same for the source $\mathcal{S}_{\mathcal{T}}$.

3.3 Nice Sources and Convenient Sources

Under quite general hypotheses, and on a convenient functional space, the density transformer admits $\lambda = 1$ as an eigenvalue of largest modulus. But, generally speaking, this is not a unique dominant eigenvalue isolated from the remainder of the spectrum.

Definition 3. *A dynamical source is said to be decomposable if, when acting on a convenient Banach space $\mathcal{F}$, the density transformer* **G** *[defined in (4)] possesses a unique dominant eigenvalue (equal to 1) separated from the remainder of the spectrum by a spectral gap, i.e., $\rho := \sup\{|\lambda| \ ; \quad \lambda \in \mathrm{Sp}\, \mathbf{G}, \lambda \neq 1\} < 1$.*

Remarks. Let us explain the terminology: Consider the dominant eigenfunction φ which is an invariant function for $\mathbf{G}$. Under the normalization condition $\int_0^1 \varphi(t)dt = 1$, this last object is unique too, and it is also the (unique) stationary density. Due to the existence of the spectral gap, the operator $\mathbf{G}$ decomposes into two parts, namely $\mathbf{G} = \lambda\mathbf{P} + \mathbf{N}$, where $\mathbf{P}$ is the projection of $\mathbf{G}$ onto the dominant eigenspace generated by φ, and $\mathbf{N}$, relative to the remainder of the spectrum, has a spectral radius equal to ρ, which is strictly less than 1. The operator $\mathbf{N}$ describes the correlations of the source. A decomposable dynamical source is ergodic and mixing with an exponential rate equal to ρ.

Most of the classical sources – memoryless sources, or primitive Markov chains – are easily proven to be decomposable. We now present sufficient conditions under which a general dynamical source will be proven to be decomposable, together with all its associated mixed sources $\mathcal{S}_{\mathcal{T}}$ [the proofs are omitted here].

Definition 4. *A dynamical source (on a finite alphabet) is said to be "nice" if it satisfies the two conditions*

(i) [Expansiveness] *There exist two constants C, D with $D > 1$ for which one has, for any $m \in \Sigma$, for any $x \in \mathcal{I}_m$, $D < |S'(x)| < C$.*

(ii) [Topologically mixing] *For any pair of two nonempty open sets* (V, W), *there exists* $n_0 \geq 1$ *such that* $S^{-n}V \cap W \neq \emptyset$ *for all* $n \geq n_0$.

Proposition 3. *A nice dynamical system is decomposable, with respect to the space* $BV(\mathcal{I})$ *of functions with bounded variation, endowed with the norm* $||f|| := \sup|f| + V(f)$ *[Here,* $V(f)$ *is the total variation of* f *on* $\mathcal{I}$*].*

We consider now the mixed source $\mathcal{S}_{\mathcal{T}}$. Recall that a transition matrix $\mathcal{T}$ is primitive if there exists a power of the matrix $\mathcal{T}$ whose coefficients are never the empty language. A strongly connected graph gives rise to a matrix $\mathcal{T}$ which is primitive if and only if the gcd of the lengths of its cycles equals 1. If it is not primitive, the gcd d of its cycle lengths is called the period, and $\mathcal{T}^d$ is primitive.

Proposition 4. *If* $\mathcal{S}$ *is a nice dynamical source, then the following holds:*

(i) the mixed source $\mathcal{S}_{\mathcal{T}}$ *relative to any primitive graph* $\mathcal{T}$ *is nice too.*
(ii) The mixed source $\mathcal{S}_{\mathcal{B}}$ *relative to a de Bruijn graph* $\mathcal{B}$ *is always nice.*
(iii) Define the period of a regular expression $\mathcal{E}$ *to be equal to the period of the useful part of* $\mathcal{R}$ *of its automaton. Then, for any regular language* $\mathcal{E}$ *of period* d*, the source* $\mathcal{S}^d_{\mathcal{R}}$ *(whose shift equals* $T^d_{\mathcal{R}}$*) is nice.*

3.4 Our Main Result

We are now ready for the proof of our main result.

Proof. We consider two graphs of interest: (i) in the case when we study $Y = C(\mathcal{E})$, the useful part $\mathcal{R}$ of the automaton $\mathcal{A}$ which recognizes the language $\Sigma^\star \cdot \mathcal{E}$ – (ii) in the case when we study $Y = \Omega(\mathcal{W})$, where $\ell + 1$ is the maximal length of the words of $\mathcal{W}$, the de Bruijn graph $\mathcal{B}$ of length ℓ.

With hypotheses of the present theorem, Propositions 3 and 4, and Definition 3, the density transformer $\mathfrak{G}$ has dominant spectral properties, and, by conjugation and perturbation theory, this transmits to the quasi-inverses of marked operators $\mathbb{R}(u)$ or $\mathbb{B}(u)$, when u is near 1, which admit a spectral decomposition too. Then, with Proposition 2, the moment generating functions of cost Y_n behave as approximate n-th powers. We end with Theorem 0 [7] (See 2.1). □

4 Conclusions

In this paper, as in [9], we restrict ourselves to the case when the expression $\mathcal{E}$ is simple. In the case when there does not exist a unique FSCC [see Section 2.4], all these FSCC's may play a rôle in the asymptotics, via their dominant eigenvalues. Our theorem extends to the general case by dealing with the super-dominant eigenvalues (which dominate the others).

Acknowledgements. We wish to thank Julien Clément, Maxime Crochemore, Philippe Flajolet, and Véronique Terrier for their help.

References

1. J. Bourdon, B. Vallée, Generalized pattern matching statistics. In Birkhauser, T.i.M., ed.: Mathematics and Computer Science II. (2002) 249–265
2. J. Bourdon, Size and path length of Patricia tries: dynamical sources context. Random Structures Algorithms **19** (2001) 289–315
3. E. Bender, F. Kochman, The distribution of subword counts is usually normal. European Journal of Combinatorics **14** (1993) 265–275
4. F. Chazal, V. Maume-Deschamps, B. Vallée, Systèmes dynamiques et algorithmique. In INRIA Research Report 5003. (2003) 121–150
5. J. Clément, P. Flajolet, B. Vallée, Dynamical sources in information theory: a general analysis of trie structures. Algorithmica **29** (2001) 307–369
6. P. Flajolet, W. Szpankowski, B. Vallée, Hidden word statistics. to appear in Journal de l'ACM (2005)
7. H.K. Hwang, Théorèmes limites pour les structures combinatoires et les fonctions arithmétiques. PhD thesis, Ecole Polytechnique, Palaiseau, France (1994)
8. P. Nicodème, T. Doerks, M. Vingron, Proteome analysis based on motif statistics. Bioinformatics **18** (2002) 161–171
9. P. Nicodème, B. Salvy, P. Flajolet, Motif statistics. Theoretical Computer Science **287** (2002) 593–617
10. M. Régnier, W. Szpankowski, On the approximate pattern occurrences in a text. In: Proc. SEQUENCE'97, IEEE Computer Society. (1997) 253–264
11. M. Régnier, W. Szpankowski, On pattern frequency occurrences in a Markovian sequence. Algorithmica **22** (1998) 631–649
12. I. Rigoutsos, A. Floratos, L. Parida, Y. Gao, D. Platt, The emergence of pattern discovery techniques in computational biology. J. of Met. Eng. **2** (2000) 159–177
13. R. Sedgewick, P. Flajolet, An introduction to the analysis of algorithms. Foreword by D. E. Knuth. Amsterdam: Addison-Wesley. xv, 492 p. (1996)
14. W. Szpankowski, Average case analysis of algorithms on sequences. Wiley-Interscience Series in Discrete Mathematics and Optimization (2001)
15. B. Vallée, Euclidean Dynamics to appear in Discrete and Continuous Dynamical Systems, 2005, web page: `www.info.unicaen.fr/~brigitte`
16. B. Vallée, Dynamical sources in information theory: fundamental intervals and word prefixes. Algorithmica **29** (2001) 262–306
17. A. Vanet, L. Marsan, M.F. Sagot, Promoter sequences and algorithmical methods for identifying them. Research in Microbiology **150** (1999) 779–799
18. M. S. Waterman, Introduction to Computational Biology. Chapman & Hall (1995)

Robust Model-Checking of Linear-Time Properties in Timed Automata*

Patricia Bouyer, Nicolas Markey, and Pierre-Alain Reynier

Lab. Spécification & Vérification,
CNRS & ENS de Cachan – France
{bouyer, markey, reynier}@lsv.ens-cachan.fr

Abstract. Formal verification of timed systems is well understood, but their *implementation* is still challenging. Raskin *et al.* have recently brought out a model of parameterized timed automata in which the transitions might be slightly delayed or expedited. This model is used to prove that a timed system is *implementable* with respect to a safety property, by proving that the parameterized model *robustly* satisfies the safety property. We extend here the notion of implementability to the broader class of linear-time properties, and provide PSPACE algorithms for the robust model-checking of Büchi-like and LTL properties. We also show how those algorithms can be adapted in order to verify bounded-response-time properties.

Keywords: Implementability, robust verification, timed systems.

1 Introduction

Verification and control of real-time systems. In the last thirty years, formal verification of systems has become a very active field of research in computer science, with numerous success stories. Formal verification aims at checking that (the model of) a system satisfies (a formula expressing) its specifications. The importance of taking real-time constraints into account in verification has quickly been understood, and the model of timed automata (defined by Alur & Dill [2]) has become one of the most established models for real-time systems, with well studied underlying theory and development of mature model-checking tools, such as UPPAAL [13] and KRONOS [7].

Implementation of real-time systems. Implementing mathematical models on physical machines is an important step for applying theoretical results on practical examples. This step is well understood for many untimed models that have been studied (*e.g.* finite automata, pushdown automata). In the timed setting, while timed automata are widely-accepted as a framework for modelling the real-time aspects of timed systems, it is known that they cannot be faithfully

* Work supported by ACI "Sécurité Informatique" CORTOS (Control and Observation of Real-Time Open Systems), a program of the French Ministry of research.

J.R. Correa, A. Hevia, and M. Kiwi (Eds.): LATIN 2006, LNCS 3887, pp. 238–249, 2006.

implemented on finite-speed CPUs (*e.g,.* the authors of [8] provide an example of a timed automaton that performs transitions exactly at dates n and $n+1/n$).

Studying the "implementability" of timed automata is thus a challenging question of obvious theoretical and practical interest.

A semantical point of view. In [10], a new semantics, called the AASAP-semantics (AASAP stands for "Almost ASAP"), has been introduced for timed automata. It takes into account the inherent digital aspect of hardware, the non-instancy of hardware communication, and several characteristics of a real processor. The point is then to decide whether a given classical controller correctly supervises the system under the AASAP-semantics. In [10], solving this problem is reduced to that of checking whether there exists a delay reaction Δ for the controller to supervise the system: given a system Sys and a controller Cont, their interaction is denoted $[\![\mathsf{Sys} \parallel \mathsf{Cont}]\!]_\Delta$ where Δ is the parameter representing the reaction delay of the controller (and in practice this is the classical parallel composition where clock constraints are enlarged by Δ).

The problem is then to decide, given a property $\mathcal{P}$ to be satisfied, whether there exists some $\Delta \in \mathbb{Q}_{\geq 0}$ s.t. $[\![\mathsf{Sys} \parallel \mathsf{Cont}]\!]_\Delta$ satisfies $\mathcal{P}$. It is thus a problem of *robust model-checking.* The special case of safety properties (stating that a set of bad configurations cannot be reached) has been solved in [9] through a region-based algorithm.

Our contribution. In this paper, we solve the robust model-checking problem for more general specifications like Büchi and LTL properties (*e.g.*, that "something occurs infinitely often", or that "a request is eventually granted"). The algorithm we propose is based on an extension of the classical region automaton construction which roughly captures all behaviors of the system, even those which may deviate due to constraint enlargement. Our algorithm is in PSPACE, which appears to be optimal. We also develop a PSPACE algorithm for verifying simple timed properties (namely, the bounded-response-time and bounded-invariance properties). Our algorithm is *ad hoc*, but it is a first step towards the verification of more general timed specifications.

Related work. Our approach contrasts with another modeling-based solution [1], where the behavior of the platform is modeled as a timed automaton. This framework is very expressive, but suffers from not verifying the "faster-is-better" property ("if an automaton can be implemented on some hardware, it can also be implemented on faster hardwares"). A notion of robust timed automata has been proposed and studied in [11, 14], where not all traces are accepted, but only those belonging to an accepting tube. This approach is topological, and is not related to ours (in fact, it drops some behaviors of the system while we add some), though this is also a semantical approach to robustness. Finally, in [16, 4, 9], a small perturbation on slopes of clocks is allowed. In the case of safety properties and under some natural assumptions, this approach is equivalent to constraint enlargement, as proved in [9].

Outline of the paper. In Section 2, we introduce basic definitions, we define the problem of robust model-checking, and we make clear the link between our work and the results of [10, 9]. Then, we provide in Section 3 our model-checking algorithm for co-Büchi properties, and in Section 4 its application to LTL properties. Finally, we present in Section 5 our first results for timed properties, and conclude with a landscape on possible future works.

Only sketches of the proofs are done in this paper. The complete proofs are available in the associated technical report [6].

2 Definitions

2.1 Timed Automata

Timed automata. Let $\mathcal{C}$ be a finite set of variables, named *clocks*. We denote by $\mathcal{G}$ the set of *clock contraints* generated by the following grammar:

$$\mathcal{G} \ni g ::= g \wedge g \mid c \sim n$$

where c ranges over $\mathcal{C}$, n ranges over $\mathbb{N}$ and [1] $\sim \in \{\leq, \geq\}$.

A *timed automaton* is a tuple $\mathcal{A} = (L, \ell_0, \mathcal{C}, \Sigma, \delta)$ where L is a finite set of *locations*, $\ell_0 \in L$ is the initial location, $\mathcal{C}$ is a finite set of *clocks*, Σ is a finite set of *actions*, and $\delta \subseteq L \times \mathcal{G} \times \Sigma \times 2^{\mathcal{C}} \times L$ is the set of transitions. We assume w.l.o.g. that transitions are labeled by their name, and we identify Σ with δ.

We define a parameterized semantics for $\mathcal{A}$ which we denote by $[\![\mathcal{A}]\!]_\Delta$. Notice that, in the definitions below, the standard semantics of timed automata can be recovered by letting $\Delta = 0$. In that case, we omit the subscript Δ.

Given a parameter $\Delta \in \mathbb{Q}_{\geq 0}$, whether a *clock valuation* $v \colon \mathcal{C} \to \mathbb{R}^+$ satisfies a constraint g within Δ, written $v \models_\Delta g$, is defined inductively as follows:

$$\begin{cases} v \models_\Delta c \leq n & \text{iff } v(c) \leq n + \Delta \\ v \models_\Delta c \geq n & \text{iff } v(c) \geq n - \Delta \\ v \models_\Delta g_1 \wedge g_2 & \text{iff } v \models_\Delta g_1 \text{ and } v \models_\Delta g_2 \end{cases}$$

A *state* of $[\![\mathcal{A}]\!]_\Delta$ is a pair (ℓ, v) where $\ell \in L$ and $v \colon \mathcal{C} \to \mathbb{R}^+$ assigns to each clock its current value. Intuitively, in a given position (ℓ, v), there are two possible behaviors for $[\![\mathcal{A}]\!]_\Delta$:

- it can either perform an *action transition*, namely a transition of δ. This requires that there exists $(\ell, g, \sigma, r, \ell') \in \delta$ s.t. $v \models_\Delta g$. In that case, the automaton ends up in state $(\ell', v[r \leftarrow 0])$, where $v[r \leftarrow 0]$ is the valuation mapping clocks in r to 0 and the other clocks to their valuation given by v;
- or it can perform a *delay transition*, *i.e.* let a certain amount of time t elapse. In that case, the automaton ends up in state $(\ell, v+t)$ where $v+t$ represents the valuation $c \mapsto v(c) + t$ for all $c \in \mathcal{C}$.

In the first case we write $(\ell, v) \xrightarrow{\sigma}_\Delta (\ell', v[r \leftarrow 0])$, whereas we write $(\ell, v) \xrightarrow{t}_\Delta (\ell, v+t)$ in the second case. The graph $[\![\mathcal{A}]\!]_\Delta$ is thus an infinite transition system.

[1] We simplify the notations by assuming that all inequalities are non-strict. As argued in [9], this does not change the expressive power of the model under the enlarged semantics.

Paths in timed automata. A *trace* in a timed automaton $\mathcal{A} = (L, \ell_0, \mathcal{C}, \Sigma, \delta)$ is a (finite or infinite) sequence of consecutive transitions $(\delta_i)_{i \in I}$.

A *path* of $[\![\mathcal{A}]\!]_\Delta$ over a trace $(\delta_i)_{i \in I}$ is a sequence $(\ell_0, v_0) \xrightarrow{d_0}_\Delta (\ell_0, v_0 + d_0) \xrightarrow{\delta_0}_\Delta (\ell_1, v_1) \xrightarrow{d_1}_\Delta (\ell_1, v_1 + d_1) \ldots$ where for each $i \in I$, $d_i \in \mathbb{R}^+$. The (unique) trace corresponding to a path π is referred to as $\mathsf{trace}(\pi)$.

Let $T = (\delta_i)_{i \in I}$ be a trace of $\mathcal{A}$. A state (ℓ', v') is said to be *reachable* from a set of states S following T in $[\![\mathcal{A}]\!]_\Delta$ if there exists a path over T in $[\![\mathcal{A}]\!]_\Delta$ starting in some $(\ell, v) \in S$ and containing (ℓ', v'). We write $\mathsf{Reach}^T_\Delta(S)$ for the set of states that are reachable from S following trace T. We note $\mathsf{Reach}_\Delta(S)$ for the union over all possible traces T of $\mathsf{Reach}^T_\Delta(S)$. This set represents all states that are reachable in $[\![\mathcal{A}]\!]_\Delta$ from S.

Region automaton. In order to symbolically reason about the infinite state space of timed automata, [2] defines an equivalence relation (of finite index) as follows. Let $\mathcal{A}$ be a timed automaton, and M be the largest integer occuring in $\mathcal{A}$. Two valuations v and v' are equivalent iff the following conditions hold on valuations v and v':[2]

- for all $c \in \mathcal{C}$, either $v(c)$ and $v'(c)$ are greater than M, or $\lfloor v(c) \rfloor = \lfloor v'(c) \rfloor$;
- for all $c, c' \in \mathcal{C}$, if both $v(c)$ and $v(c')$ are lower than M, then
  - $\langle v(c) \rangle \leq \langle v(c') \rangle$ iff $\langle v'(c) \rangle \leq \langle v'(c') \rangle$;
  - $\langle v(c) \rangle = 0$ iff $\langle v'(c) \rangle = 0$.

This defines an equivalence relation, whose equivalence classes are referred to as *regions*. We write $[v]$ for the region containing v, and $\overline{r}$ for the topological closure of the region r. The set of regions is finite and exponential in the size of the timed automaton. We define the *region automaton* as the finite automaton $\mathcal{R}(\mathcal{A}) = (\Gamma, \gamma_0, \rightarrow)$ where

- Γ is the set $\{(\ell, r) \mid \ell \in L,\ r \text{ region}\}$,
- γ_0 is the initial state (ℓ_0, r_0) where r_0 is the region which contains the valuation v_0 with $v_0(c) = 0$ for every $c \in \mathcal{C}$,
- $\rightarrow \subseteq \Gamma \times (\Sigma \cup \{\tau\}) \times \Gamma$ and $((\ell, r), \sigma, (\ell', r')) \in \rightarrow$ iff $(\ell, r) \neq (\ell', r')$ and
  - either $\sigma \in \Sigma$ and $(\ell, v) \xrightarrow{\sigma} (\ell', v')$ is a transition of $[\![\mathcal{A}]\!]$ for some $v \in r$ and $v' \in r'$,
  - or σ is the symbol τ, and there exists $t \in \mathbb{R}^+$ s.t. $(\ell, v) \xrightarrow{t} (\ell', v')$ is a transition of $[\![\mathcal{A}]\!]$ for some $v \in r$ and $v' \in r'$.

The notions of path in the region automaton, trace of a path, ... are defined in the usual way. It is well known that this automaton is *time-abstract bisimilar* to the original timed automaton, which implies that, under the standard semantics, all reachability and Büchi-like properties can be checked equivalently on the original timed automaton or on the region automaton. We assume that classical properties of region automata are known, and refer to [2] for more details.

---

[2] $\lfloor v(c) \rfloor$ represents the integer part of $v(c)$ and $\langle v(c) \rangle$ represents its fractional part.

2.2 Robust Verification of Linear-Time Properties

In this section, after several remarks on the implementability of timed systems, we present the problem of robust verification for linear-time properties.

Implementability of timed systems. Controllers of programs built using a classical synthesis algorithm may be seen as idealized controllers which are difficult to implement. We should be able to guarantee that a controller built for satisfying some property $\mathcal{P}$ can be implemented in such a way that an implementation of the controller also satisfies the property $\mathcal{P}$. In [10], a simplified model of hardware is given, with specifications (the frequency of the clock and the speed of the CPU) given as characteristic parameters of the platform on which the controller will be implemented. Two important properties are then proved: 1) first, "faster is better", which means that if a program behaves correctly (w.r.t. the property $\mathcal{P}$) on a given hardware, then it will also behave correctly on a faster hardware, 2) for a program $\mathcal{A}$ to be correctly implemented on a platform as the one described above, it is sufficient to prove its correctness on $[\![\mathcal{A}]\!]_\Delta$ for some $\Delta > 0$. This naturally leads to the definition of robust satisfaction below.

Robust model-checking. We assume that we are given a property $\mathcal{P}$ for paths of timed automata, and we note $\models$ the classical satisfaction relation for $\mathcal{P}$. Given a timed automaton $\mathcal{A}$, with initial state (ℓ_0, v_0), we define the *robust satisfaction relation* $\models\!\!\!\equiv$ as follows:

$$\mathcal{A} \models\!\!\!\equiv \mathcal{P} \stackrel{\text{def}}{\iff} \exists \Delta > 0 \text{ s.t. for all paths } \pi \text{ of } [\![\mathcal{A}]\!]_\Delta \text{ starting in } (\ell_0, v_0),\ \pi \models \mathcal{P}.$$

Intuitively, if the property $\mathcal{P}$ holds *robustly*, then it is possible to find a sufficiently fast hardware (somehow given by the parameter Δ) to implement the automaton $\mathcal{A}$ correctly w.r.t. $\mathcal{P}$, because, as explained above and proved in [10],

$$\mathcal{A} \models\!\!\!\equiv \mathcal{P} \implies \mathcal{A} \text{ implementable w.r.t. } \mathcal{P}.$$

This result holds for properties quantifying universally over paths, and thus holds for LTL properties, but not for CTL properties.

In the sequel we address the *robust model-checking problem*: "given a timed automaton $\mathcal{A}$ and a path property $\mathcal{P}$, can we decide whether $\mathcal{A} \models\!\!\!\equiv \mathcal{P}$?" This problem has been solved in [9] for basic safety properties of the type "avoid bad states", with several restrictions on timed automata.

Restrictions on timed automata. A *progress cycle* in the region automaton of $\mathcal{A}$ is a cyclic path along which all the clocks are reset, and that does not only contain the initial region (*i.e.* the region where all the clocks are set to 0). We do the following hypotheses on timed automata:

Restriction 1. *We assume timed automata $\mathcal{A}$ satisfy the following requirements:*

- *clocks are bounded by some constant M,*
- *all the cycles in the region automaton $\mathcal{R}(\mathcal{A})$ are* progress cycles.

The first hypothesis is not really restrictive since bounded timed automata are as expressive as standard timed automata (see for example [5]). Note that it entails that any time-divergent path contains infinitely many action transitions. In the following we will only consider such infinite time-divergent paths. The second point is a classical restriction [16], and in the framework of bounded timed automata, it is less restrictive than classical strong non-*Zenoness* assumptions.

Robust model-checking of safety properties. The following result has then been proved in [9]: let $\mathcal{A}$ be a timed automaton (satisfying Restriction 1) with initial state (ℓ_0, v_0), let Bad be a set of bad locations of $\mathcal{A}$, and define the set $\mathsf{Reach}^*(S) = \bigcap_{\Delta>0} \mathsf{Reach}_\Delta(S)$, where S denotes a set of states, then:

1. checking whether $\exists \Delta > 0$ s.t. $\mathsf{Reach}_\Delta(\ell_0, v_0) \cap \mathsf{Bad} = \emptyset$ is equivalent to check whether $\mathsf{Reach}^*(\ell_0, v_0) \cap \mathsf{Bad} = \emptyset$,
2. checking whether $\mathsf{Reach}^*(\ell_0, v_0) \cap \mathsf{Bad} = \emptyset$ is decidable, and PSPACE-complete.

These results rely on the classical region automaton construction where a strongly connected component (SCC for short) of the region automaton is added to the set $\mathsf{Reach}^*(\ell_0, v_0)$ as soon as it can be reached: indeed, if an SCC can be partly reached, then by iterating the SCC, all points of the SCC can also be reached.

Example 1 ([16, 9]). Consider the automaton depicted on Figure 1. For this automaton, it is possible to compute the sets $\mathsf{Reach}(\ell_0, v_0)$ and $\mathsf{Reach}^*(\ell_0, v_0)$. We obtain, for locations ℓ_1 and ℓ_2, the two sets described on Figure 2. The difference is due to the iteration of the cycle around ℓ_1 and ℓ_2.

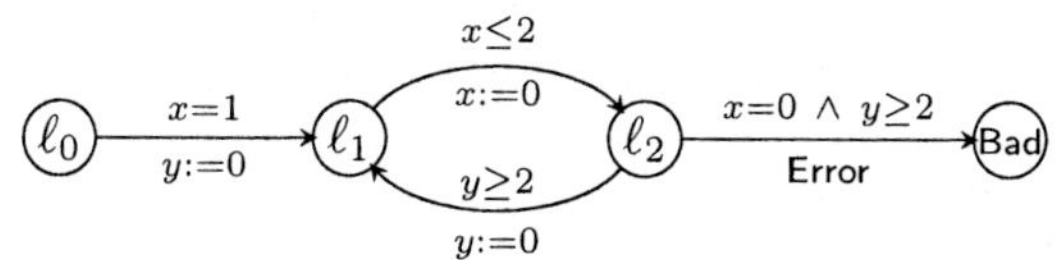

Fig. 1. A timed automaton $\mathcal{A}$

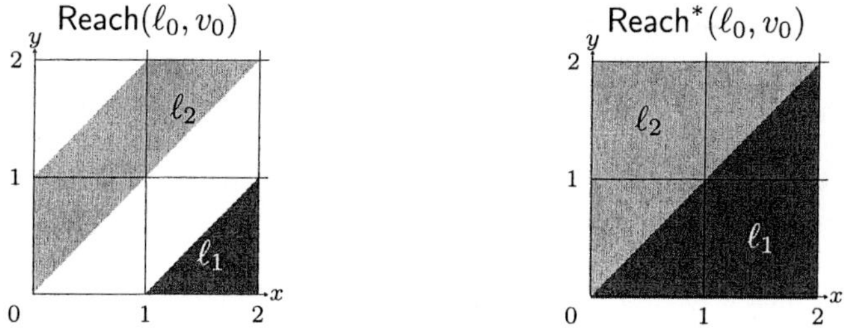

Fig. 2. Differences between $\mathsf{Reach}(\ell_0, v_0)$ and $\mathsf{Reach}^*(\ell_0, v_0)$

In the next sections, we solve the robust model-checking for co-Büchi, LTL, and bounded-response-time properties.

3 Robust Model-Checking of Co-Büchi Conditions

In this section, we are interested in co-Büchi conditions: given a set B of locations in the timed automaton, a path π satisfies co-Büchi(B) iff its trace contains finitely many transitions entering a location in B. Following Section 2.2, this immediately defines the notion of robust satisfaction for a co-Büchi condition in a timed automaton. We also recall the notion of satisfying a co-Büchi condition for the region automaton: it satisfies a co-Büchi condition B iff every path starting in γ_0 (the initial state) only runs in states of B finitely often.

Extended region automaton $\mathcal{R}^*$. We build an extension of the region automaton that takes into account the possible "deviations" of the underlying timed automaton. Let $\mathcal{A}$ be a timed automaton, and $\mathcal{R}(\mathcal{A})$ be its corresponding region automaton. We define the extended region automaton $\mathcal{R}^*(\mathcal{A})$ as follows:

- states of $\mathcal{R}^*(\mathcal{A})$ are states of $\mathcal{R}(\mathcal{A})$, *i.e.* pairs (ℓ, r) where ℓ is a location of $\mathcal{A}$ and r is a region for automaton $\mathcal{A}$
- transitions of $\mathcal{R}^*(\mathcal{A})$ are transitions of $\mathcal{R}(\mathcal{A})$ (we assume labels of transitions are names of transitions in $\mathcal{A}$), and transitions $(\ell, r) \xrightarrow{\gamma} (\ell, r')$ when $\overline{r} \cap \overline{r'} \neq \emptyset$ and (ℓ, r') is in an SCC of $\mathcal{R}(\mathcal{A})$.

The γ-transitions which are added to the classical region automaton indicate that an SCC can be reached and iterated, and then, as already written in Subsection 2.2, all configurations along the SCC can be reached.

Decidability of the robust model-checking for co-Büchi conditions. The following result is the main result of this paper. The extension from simple reachability to repeated reachability is not trivial since the method used in [9], based on the distance between new reachable states, is not sufficient in our context. Instead, we prove that the extended region automaton roughly recognizes all paths of the system, even those which deviate from standard semantics.

Theorem 2. *Let $\mathcal{A}$ be a timed automaton and B a set of locations of $\mathcal{A}$. Then*

$$\mathcal{A} \models\!\!\!\equiv \text{co-Büchi}(B) \iff \mathcal{R}^*(\mathcal{A}) \models \text{co-Büchi}(B)$$

Proof (Sketch). We first prove the left-to-right implication by contradiction. Assume that $\mathcal{A} \models\!\!\!\equiv \text{co-Büchi}(B)$, and that $\mathcal{R}^*(\mathcal{A}) \not\models \text{co-Büchi}(B)$. We can thus pick some $\Delta > 0$ s.t. every path of $[\![\mathcal{A}]\!]_\Delta$ starting in (ℓ_0, v_0) satisfies the co-Büchi condition, and pick a path π in $\mathcal{R}^*(\mathcal{A})$ not satisfying the co-Büchi condition. We will build from π a path in $[\![\mathcal{A}]\!]_\Delta$ not satisfying the co-Büchi condition, and thus obtain a contradiction. To this aim we state the following Lemma:

Lemma 3. *Let π be a path in $\mathcal{R}(\mathcal{A})$ labelled by T, starting in (ℓ, r), and ending in (ℓ', r') such that there is a transition $(\ell', r') \xrightarrow{\gamma} (\ell', r'')$ in $\mathcal{R}^*(\mathcal{A})$ (due to a cyclic path over some trace τ). Then, for every $\Delta > 0$,*

1. *for every valuation $v' \in \overline{r'}$, there exists a valuation $v \in \overline{r}$ and a path in $[\![\mathcal{A}]\!]_\Delta$ from (ℓ, v) to (ℓ', v') over trace T;*
2. *for every valuation $v' \in \overline{r'} \cap \overline{r''}$, for every valuation $v'' \in \overline{r''}$, there exists a path in $[\![\mathcal{A}]\!]_\Delta$ over trace τ^k (for some $k \geq 0$) from (ℓ', v') to (ℓ', v'').*

Splitting the path π into subpaths not containing γ-transitions, we can apply the first point of the above lemma to each subpath. We thus obtain real paths in $[\![\mathcal{A}]\!]_\Delta$, which we can glue together using the second point.

Conversely, assume that $\mathcal{A} \not\models$ co-Büchi(B). This entails that, for any positive Δ, there is a path in $[\![\mathcal{A}]\!]_\Delta$ entering infinitely many times a state in B. Since B is finite, there exists a location $f \in B$ that witnesses the Büchi condition for paths π_Δ for arbitrarily small Δ. We will build a path of $\mathcal{R}^*(\mathcal{A})$ satisfying the Büchi condition $\{f\}$, using the following lemma:

Lemma 4. *Given a timed automaton $\mathcal{A}$, there exists a positive value Δ s.t. for any (finite) path ρ in $[\![\mathcal{A}]\!]_\Delta$, there exists a path in $\mathcal{R}^*(\mathcal{A})$ whose trace, when removing γ-transitions, is the same as the trace of ρ.*

The proof is done by induction on the length of ρ. Using this lemma, we can fix such a value of Δ and apply it to any prefix of the corresponding path π, which satisfies the Büchi condition $\{f\}$. We can take a prefix of π containing $k+1$ times the discrete state f, which leads to a path of $\mathcal{R}^*(\mathcal{A})$ satisfying the Büchi condition $\{f\}$. □

As a corollary, and using the PSPACE-hardness of the robust model-checking of safety properties [9], we get:

Corollary 5. *The robust model-checking for co-Büchi acceptance conditions is* PSPACE*-complete.*

Remark 1. We prove Theorem 2 for co-Büchi conditions, because we need those conditions for verifying LTL properties (see Section 4). However, we could have adapted our construction to Büchi conditions (this would require to unfold once the SCCs in $\mathcal{R}^*(\mathcal{A})$), or other standard acceptance conditions on infinite runs.

4 Robust Model-Checking of LTL

We now show how our results on robust model-checking of co-Büchi conditions can be used to robustly model-check LTL properties on timed automata. We use the classical construction of Büchi automata which recognize exactly the models of LTL formulae, and then apply the results of the previous section.

Definition 6 (Logic LTL). *The logic* LTL *over finite set of actions Σ is defined by the following grammar: (a ranges over the set of actions Σ)*

$$\mathsf{LTL} \ni \varphi ::= a \mid \varphi \vee \varphi \mid \neg\varphi \mid \mathbf{X}\,\varphi \mid \varphi\,\mathbf{U}\,\varphi$$

***Semantics of* LTL.** We define the semantics of LTL over traces of timed automata, which naturally induces a semantics over paths of timed automata: a path π will satisfy an LTL formula if and only if its trace $\mathsf{trace}(\pi)$ satisfies this formula. We thus assume we are given an infinite path π and denote by $T = (\delta_i)_{i\in\mathbb{N}} \in \Sigma^\omega$ its trace. Given a natural number $j \in \mathbb{N}$, we denote by T^j the trace $(\delta_i)_{i\geq j}$. The satisfaction relation for LTL over traces is denoted $\models$ and is defined inductively as follows (we omit the semantics of standard boolean operators):

$$\begin{aligned} T \models a &\iff \delta_0 = a \\ T \models \mathbf{X}\,\varphi &\iff T^1 \models \varphi \\ T \models \varphi_1 \,\mathbf{U}\, \varphi_2 &\iff \exists i \geq 0 \text{ s.t. } T^i \models \varphi_2 \text{ and } \forall\, 0 \leq j < i, T^j \models \varphi_1 \end{aligned}$$

In the following, we equivalently write $\pi \models \varphi$ for $\mathsf{trace}(\pi) \models \varphi$ and use classical shortcuts like $\mathbf{F}\,\varphi$ (which holds for $\top\,\mathbf{U}\,\varphi$ where $\top$ denotes the "true" formula) or $\mathbf{G}\,\varphi$ (which holds for $\neg(\mathbf{F}\,(\neg\varphi))$).

Remark 2. It is worth noticing that the semantics we consider is the so-called *pointwise* semantics where formulae are interpreted only when an action occurs, which is quite different from the *interval-based* semantics where formulae can be interpreted at any time (see [17, 15] for a discussion on these semantics).

***Robust model-checking of* LTL.** The *robust satisfaction relation* for LTL is thus derived from the general definition given in Section 2.2:

$$\mathcal{A} \mathrel{|\!\!\!\equiv} \varphi \iff \exists \Delta > 0 \text{ s.t. } \forall \pi \text{ path of } [\![\mathcal{A}]\!]_\Delta \text{ starting in } (\ell_0, v_0),\ \pi \models \varphi.$$

We recall the following classical result on LTL:

Proposition 7 ([18]). *Given an* LTL *formula* φ*, we can build a Büchi automaton* $\mathcal{B}_\varphi$ *(with initial state* q_φ *and repeated states* Q_φ*) which accepts the set* $\{T \in \Sigma^\omega \mid T \models \varphi\}$.

We now state that the robust model-checking of LTL is decidable.

Theorem 8. *Given a timed automaton* $\mathcal{A}$*, and an* LTL *formula* φ*, we denote by* $\mathcal{C} = \mathcal{A} \times \mathcal{B}_{\neg\varphi}$ *the timed Büchi automaton obtained by a strong synchronization over actions of automata* $\mathcal{A}$ *and* $\mathcal{B}_{\neg\varphi}$*. We then have the following equivalence:*

$$\mathcal{A} \mathrel{|\!\!\!\equiv} \varphi \iff \mathcal{C} \mathrel{|\!\!\!\equiv} \mathsf{co\text{-}B\ddot{u}chi}(L \times Q_{\neg\varphi}).$$

It remains to notice that the timed Büchi automaton $\mathcal{A} \times \mathcal{B}_{\neg\varphi}$ satisfies all Restrictions 1 (bounded clocks and only progress cycles) as soon as $\mathcal{A}$ does. Since we have shown in Section 3 how to robustly model-check co-Büchi properties, we get the following result:

Corollary 9. *The robust model-checking of* LTL *over timed automata is decidable and* PSPACE*-complete.*

Classically, the verification of LTL over finite structures is PSPACE-complete, but the complexity is only NLOGSPACE in the size of the system we analyze. In the case of timed automata, both standard and robust model-checking problems for LTL are PSPACE-complete, but they are PSPACE in both the size of the structure and the size of the formula.

5 Towards Robust Model-Checking of Timed Properties

The logic MTL [12, 3] extends the logic LTL with time restrictions on "until" modalities. We present here a first positive step towards the robust model-checking of MTL formulae. We consider the following *bounded-response-time property* $\varphi = \mathbf{G}\,(a \rightarrow \mathbf{F}_{\leq c}\, b)$. where a and b denote actions (elements of Σ), c belongs to $\mathbb{Q}^+$, and $\rightarrow$ denotes the classical "imply" operator. This formula expresses that event a is always followed in less than c time units by a b. This property thus constrains the reaction delays of the system. The robust satisfaction of such a property (defined as in Subsection 2.2) ensures that the system, even under small perturbations, will satisfy this quantitative property given by the bounded delay c.

To formally define the satisfiability of φ over a path, we need timing informations about the path. We thus define the *time length* of a path between two actions as follows. Let consider an infinite path π:

$$(\ell_0, v_0) \xrightarrow{d_0} (\ell_0, v_0 + d_0) \xrightarrow{\delta_0} (\ell_1, v_1) \cdots (\ell_k, v_k) \xrightarrow{d_k} (\ell_k, v_k + d_k) \xrightarrow{\delta_k} \cdots$$

Given two indices $i_1 < i_2$, we define the *time length* of π between actions δ_{i_1} and δ_{i_2}, denoted by $\mathsf{time}(\delta_{i_1}, \delta_{i_2})$ by the value $\sum_{j=i_1+1}^{i_2} d_j$. We then say that path π satisfies the formula φ, denoted by $\pi \models \varphi$, whenever:

$$\forall\, i \geq 0, \text{ if } \delta_i = a, \text{ then } \exists\, j > i \text{ s.t. } \delta_j = b \text{ and } \mathsf{time}(\delta_i, \delta_j) \leq c.$$

In particular, if π satisfies φ then π also satisfies the LTL property $\mathbf{G}\,(a \rightarrow \mathbf{F}\, b)$.

We now state the following result:

Theorem 10. *The robust model-checking of bounded-response-time properties is decidable in* PSPACE *over timed automata.*

Proof (Sketch). Let $\varphi = \mathbf{G}\,(a \rightarrow \mathbf{F}_{\leq c}\, b)$. We assume $\mathcal{A}$ is a timed automaton which satisfies the untimed property $\mathbf{G}\,(a \rightarrow \mathbf{F}\, b)$. The proof is based on the following equivalence:

$\mathcal{A} \not\models \varphi \iff$ there is a state α in $\mathsf{Reach}^*(\ell_0, v_0)$ s.t. there is a finite path in $[\![\mathcal{A}]\!]$ from α starting with an a, ending after the first b such that the time elapsed between these two actions is greater than c.

The right hand-side of the above equivalence is decidable, one solution is to use *corner-points* because paths with maximal time length always run through corner-points [5]. Such an algorithm has a PSPACE complexity. □

Remark 3. The above proof is somehow *ad-hoc*, as it is very specific to the formula which is considered. However it can for example be adapted to bounded-invariance properties like $\mathbf{G}\,(a \to \mathbf{G}_{\leq c}\neg b)$.

6 Conclusion

In this paper, we have extended the results of [9] in order to decide a sufficient condition for the implementability of a timed automaton. To that aim, we have defined a notion of *robust satisfaction* for linear-time properties and provided PSPACE algorithms for the robust model-checking of Büchi-like and LTL properties (these algorithms are b.t.w. optimal). We have also made a first step towards the robust model-checking of MTL formulae, through the verification of bounded-response-time property.

It is worth noticing that our results may extend easily to another case of perturbations: in [16], Puri considers drifts in the rates of clocks, instead of enlarging guards. In fact, both extensions happen to have the same impact on the set of reachable states [16, 9], and it seems quite natural to think that our proofs may be adapted to the case of drifts on clocks. Furthermore, the case of bounded-response-time properties is encouraging and we are trying to extend it to more general timed properties. Another direction to be studied is that of semantics: indeed we have pointed out that we consider in this paper a pointwise semantics for LTL. It could be interesting to study whether our results extend to the more involved interval-based semantics. Finally, it could also be a great challenge to extend this approach to branching-time properties. This requires to adapt the robust semantics, and also to bring new keys to make the link with implementability. This may lead to robust model-checking of logics like CTL, or even TCTL.

References

1. K. Altisen and S. Tripakis. Implementation of timed automata: An issue of semantics or modeling? In *Proc. 3rd Int. Work. Formal Modeling and Analysis of Timed Systems (FORMATS'05)*, volume 3829 of *LNCS*, pages 273–288. Springer, 2005.
2. R. Alur and D. Dill. A theory of timed automata. *Theoretical Computer Science*, 126(2):183–235, 1994.
3. R. Alur and T. A. Henzinger. Real-time logics: Complexity and expressiveness. *Information and Computation*, 104(1):35–77, 1993.
4. R. Alur, S. La Torre, and P. Madhusudan. Perturbed timed automata. In *Proc. 8th Int. Work. Hybrid Systems: Computation and Control (HSCC'05)*, volume 3414 of *LNCS*, pages 70–85. Springer, 2005.
5. G. Behrmann, A. Fehnker, T. Hune, K. G. Larsen, P. Pettersson, J. Romijn, and F. Vaandrager. Minimum-cost reachability for priced timed automata. In *Proc. 4th Int. Work. Hybrid Systems: Computation and Control (HSCC'01)*, volume 2034 of *LNCS*, pages 147–161. Springer, 2001.
6. P. Bouyer, N. Markey, and P.-A. Reynier. Robust model-checking of timed automata. Tech. Report LSV-05-08, LSV, ENS Cachan, France, 2005.

7. M. Bozga, C. Daws, O. Maler, A. Olivero, S. Tripakis, and S. Yovine. KRONOS: a model-checking tool for real-time systems. In *Proc. 10th Int. Conf. Computer Aided Verification (CAV'98)*, volume 1427 of *LNCS*, pages 546–550. Springer, 1998.
8. F. Cassez, T. A. Henzinger, and J.-F. Raskin. A comparison of control problems for timed and hybrid systems. In *Proc. 5th Int. Work. Hybrid Systems: Computation and Control (HSCC'02)*, volume 2289 of *LNCS*, pages 134–148. Springer, 2002.
9. M. De Wulf, L. Doyen, N. Markey, and J.-F. Raskin. Robustness and implementability of timed automata. Tech. Report 2004.30, Centre Fédéré en Vérification, Belgium, Dec. 2005. Revised version.
10. M. De Wulf, L. Doyen, and J.-F. Raskin. Almost ASAP semantics: From timed models to timed implementations. *Formal Aspects of Computing*, 17(3):319–341, 2005.
11. V. Gupta, T. A. Henzinger, and R. Jagadeesan. Robust timed automata. In *Proc. Int. Work. Hybrid and Real-Time Systems (HART'97)*, volume 1201 of *LNCS*, pages 331–345. Springer, 1997.
12. R. Koymans. Specifying real-time properties with metric temporal logic. *Real-Time Systems*, 2(4):255–299, 1990.
13. K. G. Larsen, P. Pettersson, and W. Yi. UPPAAL in a nutshell. *Journal of Software Tools for Technology Transfer*, 1(1–2):134–152, 1997.
14. J. Ouaknine and J. B. Worrell. Revisiting digitization, robustness and decidability for timed automata. In *Proc. 18th Ann. Symp. Logic in Computer Science (LICS'03)*, pages 198–207. IEEE Comp. Soc. Press, 2003.
15. J. Ouaknine and J. B. Worrell. On the decidability of metric temporal logic. In *Proc. 19th Ann. Symp. Logic in Computer Science (LICS'05)*, pages 188–197. IEEE Comp. Soc. Press, 2005.
16. A. Puri. Dynamical properties of timed automata. In *Proc. 5th Int. Symp. Formal techniques in Real-Time and Fault-Tolerant Systems (FTRTFT'98)*, volume 1486 of *LNCS*, pages 210–227. Springer, 1998.
17. J.-F. Raskin. *Logics, Automata and Classical Theories for Deciding Real-Time.* PhD thesis, University of Namur, Namur, Belgium, 1999.
18. P. Wolper, M. Y. Vardi, and A. P. Sistla. Reasoning about infinite computation paths. In *Proc. 24th Ann. Symp. Foundations of Computer Science (FOCS'83)*, pages 185–194. IEEE Comp. Soc. Press, 1983.

The Computational Complexity of the Parallel Knock-Out Problem

Hajo Broersma, Matthew Johnson, Daniël Paulusma, and Iain A. Stewart

Department of Computer Science, Durham University,
Science Laboratories, South Road, Durham DH1 3LE, UK
{hajo.broersma, matthew.johnson2, daniel.paulusma,
i.a.stewart}@durham.ac.uk

Abstract. We consider computational complexity questions related to parallel knock-out schemes for graphs. In such schemes, in each round, each remaining vertex of a given graph eliminates exactly one of its neighbours. We show that the problem of whether, for a given graph, such a scheme can be found that eliminates every vertex is NP-complete. Moreover, we show that, for all fixed positive integers $k \geq 2$, the problem of whether a given graph admits a scheme in which all vertices are eliminated in at most k rounds is NP-complete. For graphs with bounded tree-width, however, both of these problems are shown to be solvable in polynomial time.

Keywords: Parallel knock-out; graphs; computational complexity.

1 Introduction

In this paper, we consider *parallel knock-out schemes* for finite undirected simple graphs. These were introduced by Lampert and Slater [5]. Such a scheme proceeds in rounds: in the first round each vertex in the graph selects exactly one of its neighbours, and then all the selected vertices are eliminated simultaneously. In subsequent rounds this procedure is repeated in the subgraph induced by those vertices not yet eliminated. The scheme continues until there are no vertices left, or until an isolated vertex is obtained (since an isolated vertex will never be eliminated).

A graph is *reducible* if there exists a parallel knock-out scheme that eliminates the whole graph. The *parallel knock-out number* of a graph G, denoted by $\text{pko}(G)$, is the minimum number of rounds in a parallel knock-out scheme that eliminates every vertex of G. If G is not reducible, then $\text{pko}(G) = \infty$. Consider the following decision problem.

PARALLEL KNOCK-OUT (PKO)
Instance: A graph G.
Question: Is G reducible?

In [5], it was claimed that PKO is NP-complete even when restricted to the class of bipartite graphs. No proof was given; the reader was referred to a paper

J.R. Correa, A. Hevia, and M. Kiwi (Eds.): LATIN 2006, LNCS 3887, pp. 250–261, 2006.

that was in preparation. Our attempts to obtain and verify this proof have been unsuccessful. We shall obtain the result as a corollary to a stronger theorem (Theorem 1 below) by considering a related problem, which is defined for each positive integer k.

PARALLEL KNOCK-OUT (k) (PKO(k))
Instance: A graph G.
Question: Is $\text{pko}(G) \leq k$?

That there is a polynomial algorithm to decide PKO(1) follows easily from a piece of graph theory folklore (see [1] for details). Our first result classifies the complexity of PKO(k), $k \geq 2$.

Theorem 1. *For $k \geq 2$, PKO(k) is NP-complete even if instances are restricted to the class of bipartite graphs.*

In [1], it was shown, using a dynamic programming approach, that the parallel knock-out number for trees can be computed in polynomial time. It was asked whether this result could be extended to graphs with bounded tree-width. In our second result, we give an affirmative answer.

Theorem 2. *The problem PKO(k) can be solved in linear time on graphs with bounded tree-width.*

We will also show that PKO can be solved in polynomial time on graphs with bounded tree-width.

The paper is organised as follows. In the next two sections we introduce a number of definitions and simple results. In Section 4 and Section 5 are the proofs and corollaries of Theorems 1 and 2 respectively.

2 Preliminaries

Graphs in this paper are denoted by $G = (V, E)$. An edge joining vertices u and v is denoted uv. In the *null graph*, $V = E = \emptyset$. For graph terminology not defined below, refer to [2].

For a vertex $u \in V$ we denote its *neighbourhood*, that is, the set of adjacent vertices, by $N(u) = \{v \mid uv \in E\}$. The *degree* of a vertex is the number of edges incident with it, or, equivalently, the size of its neighbourhood.

For a graph G, a *KO-selection* is a function $f : V \to V$ with $f(v) \in N(v)$ for all $v \in V$. If $f(v) = u$, we say that vertex v *fires at* vertex u, or that vertex u *is knocked out* by vertex v.

For a KO-selection f, we define the corresponding *KO-successor* of G as the subgraph of G that is induced by the vertices in $V \setminus f(V)$; if H is the KO-successor of G we write $G \rightsquigarrow H$. Note that every graph without isolated vertices has at least one KO-successor. A graph G is called *KO-reducible*, if there exists a finite sequence

$$G \rightsquigarrow G_1 \rightsquigarrow G_2 \rightsquigarrow \cdots \rightsquigarrow G_r,$$

where G_r is the null graph. If no such sequence exists, then $\text{pko}(G) = \infty$. Otherwise, the parallel knock-out number $\text{pko}(G)$ of G is the smallest number r for which such a sequence exists. A sequence of KO-selections that transform G into the null graph is called a *KO-reduction scheme*. A single step in this sequence is called a *round* of the KO-reduction scheme. A subset of V is *knocked out* in a certain round if every vertex in the subset is knocked out in that round.

We make some simple observations that we will use later on.

Observation 1. *Let G be a graph on at least three vertices. If G contains two vertices of degree 1 that share the same neighbour, then G is not KO-reducible.*

Observation 2. *Let u_1, u_2, u_3, u_4 be four vertices of a KO-reducible graph G such that $N(u_2) = \{u_1, u_3\}$, $N(u_3) = \{u_2, u_4\}$ and $N(u_4) = \{u_3\}$. If u_1 is knocked out in the first round of a KO-reduction scheme, then u_1 fires at u_2 in the first round.*

An odd path $u_1u_2 \ldots u_{2k+1}$ is called a *centred path* of G with *centrevertex* u_{k+1} if $G - \{u_{k+1}\}$ contains as components the path $u_1u_2 \ldots u_k$ and the path $u_{k+2}u_{k+3} \ldots u_{2k+1}$.

Observation 3. *Let $P = u_1u_2 \ldots u_7$ be a centred path of a KO-reducible graph G. In the first round of any KO-reduction scheme u_1 and u_2 fire at each other, u_3 fires at u_2, u_6 and u_7 fire at each other, u_5 fires at u_6, u_4 fires at u_3 or u_5, and u_4 will not be knocked out. In the second round of any KO-reduction scheme u_4 and its remaining neighbour in P fire at each other.*

3 NP-Complete Problems

In this section, we consider two NP-complete problems that we will use in the proof of Theorem 1. We refer to [4] and [6] for further details.

DOMINATING SET (DS)
Instance: A graph $G = (V, E)$ and a positive integer p.
Question: Does G have a *dominating set* of size at most p, that is, is there a subset $V' \subseteq V$ such that $|V'| \leq p$ and every vertex of G is in V' or adjacent to a vertex in V'?

A *hypergraph* $J = (Q, \mathcal{S})$ is a pair of sets where $Q = \{q_1, \ldots, q_m\}$ is the vertex set and $\mathcal{S} = \{S_1, \ldots, S_n\}$ is the set of *hyperedges*. Each member S_j of $\mathcal{S}$ is a subset of Q.

HYPERGRAPH 2-COLOURABILITY (H2C)
Instance: A hypergraph $J = (Q, \mathcal{S})$.
Question: Is there a *2-colouring* of $J = (Q, \mathcal{S})$, that is, a partition of Q into sets B and W such that, for each $S \in \mathcal{S}$, $B \cap S \neq \emptyset$ and $W \cap S \neq \emptyset$.

The *incidence graph* I of a hypergraph $J = (Q, \mathcal{S})$ is a bipartite graph with vertex set $Q \cup \mathcal{S}$ where (q, S) forms an edge if and only if $q \in S$.

With a hypergraph $J = (Q, \mathcal{S})$ we can associate another hypergraph $J' = (X, \mathcal{Z})$ called the *triple* of J; triples of hypergraphs will play a crucial role in our NP-completeness proofs in the next section. It requires a little effort to define the vertices X and hyperedges $\mathcal{Z}$ of the triple of J.

Recall that $Q = \{q_1, \dots, q_m\}$ and $\mathcal{S} = \{S_1, \dots, S_n\}$. For $1 \le i \le m$, let $\ell(i)$ be the number of hyperedges in $\mathcal{S}$ that contain q_i, let $Q_i = \{q_i^1, \dots, q_i^{\ell(i)}\}$ and let $U_i = \{u_i^1, \dots, u_i^{\ell(i)}\}$. The union of all such sets is the vertex set of J', that is

$$X = \bigcup_{i=1}^{m} (Q_i \cup U_i).$$

Now the hyperedges:

- for $1 \le i \le m$, for $1 \le k \le \ell(i)$, let $P_i^k = \{q_i^k, u_i^k\}$,
- for $1 \le i \le m$, for $1 \le k \le \ell(i) - 1$, let $R_i^k = T_i^k = \{u_i^k, q_i^{k+1}\}$, and
- for $1 \le i \le m$, let $R_i^{\ell(i)} = T_i^{\ell(i)} = \{u_i^{\ell(i)}, q_i^1\}$.

Let $\mathcal{P}_i = \{P_i^1, \dots, P_i^{\ell(i)}\}$, $\mathcal{R}_i = \{R_i^1, \dots, R_i^{\ell(i)}\}$, and $\mathcal{T}_i = \{T_i^1, \dots, T_i^{\ell(i)}\}$, and let

$$\mathcal{P} = \bigcup_{i=1}^{m} \mathcal{P}_i, \quad \mathcal{R} = \bigcup_{i=1}^{m} \mathcal{R}_i, \quad \mathcal{T} = \bigcup_{i=1}^{m} \mathcal{T}_i.$$

For $1 \le j \le n$, there is also a hyperedge S'_j. If in J, S_j contains q_i, then in J', S'_j contains a vertex of Q_i. In particular, if S_j is the kth hyperedge that contains q_i in J, then S'_j contains q_i^k. For example, if q_1 is in S_1, S_4 and S_7 in J, then $\ell(1) = 3$ and in J' there are vertices q_1^1, q_1^2, q_1^3 with $q_1^1 \in S'_1$, $q_1^2 \in S'_4$, and $q_1^3 \in S'_7$.

Let $\mathcal{S}' = \{S'_1, \dots, S'_n\}$. The set of hyperedges for J' is

$$\mathcal{Z} = \mathcal{S}' \cup \mathcal{P} \cup \mathcal{R} \cup \mathcal{T}.$$

We denote the incidence graph of the triple J' by I'. See Figure 1 for an example that illustrates the case where q_1 belongs to S_1, S_4 and S_7.

Proposition 1. *$J = (Q, \mathcal{S})$ has a 2-colouring $B \cup W$ if and only if $J' = (X, \mathcal{Z})$ has a 2-colouring $B' \cup W'$ such that for each $1 \le i \le m$ either $Q_i \subseteq B'$ and $U_i \subseteq W'$, or $Q_i \subseteq W'$ and $U_i \subseteq B'$.*

Proof. Suppose $B \cup W$ is a 2-colouring of J. Define a partition $B' \cup W'$ of X as follows. If q_i is in B, then each q_i^k is in B' and each u_i^k is in W'. If q_i is in W, then each q_i^k is in W' and each u_i^k is in B'. Obviously, $B' \cup W'$ is a 2-colouring of J' with the desired property.

Suppose we have a 2-colouring $B' \cup W'$ of J' such that for each $1 \le i \le m$ either $Q_i \subseteq B'$ and $U_i \subseteq W'$, or $Q_i \subseteq W'$ and $U_i \subseteq B'$. Then let $q_i \in B$ if and only if $Q_i \subseteq B'$, and let $W = Q \setminus B$. Clearly, if S_j contains only elements from B (respectively W), then S'_j would contain only elements from B' (respectively W'). Hence $B \cup W$ is a 2-colouring of J. □

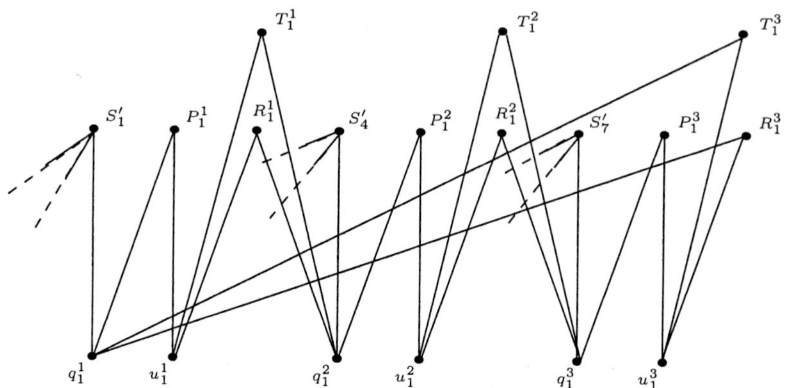

Fig. 1. Part of the incidence graph of the triple of a hypergraph

4 Complexity Classification

Theorem 1. *For $k \geq 2$, PKO(k) is* NP*-complete even if instances are restricted to the class of bipartite graphs.*

Proof. The proof is in three cases.

Case 1. $k = 2$. We use reduction from DS. Given $G = (V, E)$ and a positive integer $p \leq |V|$, we shall construct a bipartite graph B such that $\text{pko}(B) = 2$ if and only if G has a dominating set D where $|D| \leq p$.

Let the vertex set of B be the disjoint union of $V = \{v_1, \ldots, v_n\}$, $V' = \{v'_1, \ldots, v'_n\}$ and $W = \{w_1, \ldots, w_{n-p}\}$. Let the edge set of B contain

- $v_i v'_i$, $1 \leq i \leq n$,
- $v_i v'_j$ and $v'_i v_j$, for each edge $v_i v_j \in E$, and
- $v_i w_h$, $1 \leq i \leq n$, $1 \leq h \leq n - p$.

Suppose that G has a dominating set $D = \{v_1, \ldots, v_d\}$ where $d \leq p$. Note that every vertex in V' is adjacent to a vertex of D in B. We shall describe a 2-round KO-reduction scheme for B. In round 1

- for $1 \leq i \leq n$, v_i fires at v'_i,
- for $1 \leq j \leq p$, v'_j fires at v_j,
- for $p + 1 \leq j \leq n$, v'_j fires at a vertex in D, and
- for $1 \leq h \leq n - p$, w_h fires at a vertex in D.

Thus each vertex in $\{v_1, \ldots, v_p\}$ and V' is eliminated, and each vertex in $V \setminus \{v_1, \ldots, v_p\}$ and W survives to round 2. As the surviving vertices induce the balanced complete bipartite graph $K_{n-p,n-p}$ in B, it is clear that every surviving vertex can be eliminated in one further round.

Now suppose that B has a 2-round KO-reduction scheme. Let D be the subset of V containing vertices that are fired at in round 1. As every vertex in V' fires at — and so is adjacent to — a vertex in D, D is a dominating set in G (since each vertex in V' is joined only to copies of itself and its neighbours). We must show that $|D| \leq p$. Let $V_S = V \setminus D$ and $V'_S \subset V' \cup W$ be the sets of vertices that survive round 1. As round 2 is the final round,

$$|V_S| = |V'_S|. \tag{1}$$

As $|V' \cup W| = 2n - p$ and at most n vertices in $V' \cup W$ are fired at in round 1, $|V'_S| \geq n - p$. Thus, by (1), $|V_S| \geq n - p$. Therefore

$$\begin{aligned} |D| &= |V| - |V_S| \\ &\leq n - (n - p) \\ &= p. \end{aligned}$$

Case 2. $k = 3$. Let $J = (Q, \mathcal{S})$ be an instance of H2C. Let I' be the incidence graph of its triple $J' = (X, \mathcal{Z})$. Recall that $\mathcal{Z} = \mathcal{S}' \cup \mathcal{P} \cup \mathcal{R} \cup \mathcal{T}$. From I', we obtain a further bipartite graph G by connecting each vertex with a path as follows:

- For each vertex x in X, w add a path $H^x = y_1^x y_2^x y_3^x$ and join x to y_1^x.
- For each vertex R in $\mathcal{R}$, add a path $H^R = y_1^R \dots y_4^R$ and join R to y_1^R.
- For each vertex T in $\mathcal{T}$, add a path $H^T = y_1^T \dots y_4^T$ and join T to y_1^T.
- For each vertex P in $\mathcal{P}$, add a path $H^P = y_1^P \dots y_7^P$ and join P to the centrevertex y_4^P.
- For each vertex S' in $\mathcal{S}'$, add a path $H^{S'} = y_1^{S'} \dots y_7^{S'}$ and join S' to the centrevertex $y_4^{S'}$.

Figure 2 illustrates G. We shall prove that J is 2-colourable if and only if $\text{pko}(G) \leq 3$. Throughout the proof, G_1 and G_2 denote the graphs induced by the surviving vertices after, respectively, 1 and 2 rounds of a KO-reduction scheme.

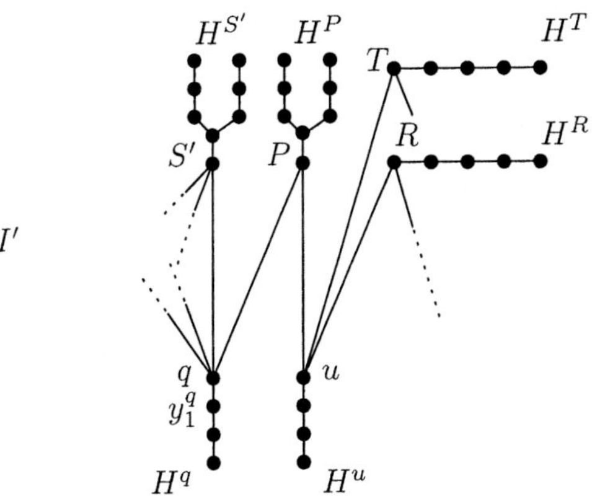

Fig. 2. The graph G in Case 2

Suppose $B \cup W$ is a 2-colouring of J. By Proposition 1, J' has a 2-colouring $B' \cup W'$. We define a three-round KO-reduction scheme for G.

Round 1. Vertices of degree 1 and their neighbours fire at each other. Each H^P with $P \in \mathcal{P}$ and each $H^{S'}$ with $S' \in \mathcal{S}'$ is a centred path of G, and the vertices fire as in Observation 3. For each $z \in \mathcal{R} \cup \mathcal{T}$, vertex y_1^z fires at y_2^z and y_2^z fires at y_3^z. Each vertex in $\mathcal{Z}$ fires at one of its neighbours in B'. Each vertex x in X fires at its neighbour y_1^x in H^x. Each y_1^x with $x \in B'$ fires at x. Each y_1^x with $x \in W'$ fires at y_2^x.

Thus every vertex in W' and no vertex in B' survives. Also every vertex in $\mathcal{Z}$ survives. Each vertex $z \in \mathcal{R} \cup \mathcal{T}$ is adjacent to a vertex y_1^z of degree 1, and each vertex $z \in \mathcal{S}' \cup \mathcal{P}$ is adjacent to a vertex y_4^z whose only other neighbour is a vertex y_3^z of degree 1.

Round 2. Because $B' \cup W'$ is a 2-colouring of $J = (X, \mathcal{Z})$, every vertex in $\mathcal{Z}$ has a neighbour in W' in G_1. For each $S_j' \in \mathcal{S}'$ we choose one neighbour in W' and let W'' be the set of selected vertices. Since no two vertices in $\mathcal{S}'$ have a common neighbour in X, $|W''| = n$. The vertices in G_1 fire as follows. Vertices of degree 1 and their neighbours fire at each other. Each vertex $P \in \mathcal{P}$ with a neighbour in $W' \backslash W''$ fires at this neighbour. Otherwise P fires at y_4^P. Each $x \in X$ fires at its neighbour in $\mathcal{P}$. Each $S' \in \mathcal{S}'$ fires at $y_4^{S'}$.

Thus the vertex set of G_2 is $W'' \cup \mathcal{S}'$.

Round 3. Each $S' \in \mathcal{S}'$ and its unique neighbour in W'' fire at each other, which leaves us with the null graph.

Now we suppose that $\mathrm{pko}(G) \leq 3$. We assume that a particular KO-reduction scheme for G is given and prove that J has a 2-colouring.

Claim 1. If a vertex in a set Q_i is knocked out in the first round, then all vertices in Q_i are knocked out in the first round.

Suppose that vertex $q_i^k \in Q_i$ is knocked out in the first round. We show that q_i^{k+1} (with $q_i^{\ell(i)+1} = q_i^1$) is also knocked out in the first round.

If $q_i^k \in Q_i$ is knocked out in the first round, then, by Observation 2, q_i^k fires at $y_1^{q_i^k}$. Suppose q_i^{k+1} is *not* knocked out in the first round. Observation 3 implies that P_i^{k+1} must fire at u_i^{k+1} and P_i^k must fire at either q_i^k or u_i^k. If P_i^k fires at u_i^k, then by Observation 2 u_i^k fires at $y_1^{q_i^k}$. Since vertices in $H^{P_i^k}$ must fire as in Observation 3, this means that G_1 contains a component isomorphic to a path on three vertices. By Observation 1 G_1 is not KO-reducible. Hence, P_i^k fires at q_i^k.

For the same reason R_i^{k+1} or T_i^{k+1} cannot fire at u_i^k, and consequently, fire at $y_1^{R_i^{k+1}}$ and $y_1^{T_i^{k+1}}$ respectively. Due to Observation 2 this implies that $y_1^{R_i^{k+1}}$ fires at $y_2^{R_i^{k+1}}$, and $y_1^{T_i^{k+1}}$ fires at $y_2^{T_i^{k+1}}$.

In G_1 both T_i^k and R_i^k have exactly the same neighbours, namely u_i^k and q_i^{k+1}. If T_i^k and R_i^k fire at a different neighbour in the second round, then due to Observation 2 both will be isolated vertices in G_2. Suppose T_i^k and R_i^k fire at the same neighbour. Then in all possible schemes G_2 will contain two vertices of degree 1 having the same neighbour. Observation 1 implies that G_2 is not KO-reducible. We conclude that q_i^{k+1} must be knocked out in the first round as well, and this proves the claim.

Claim 2. If a vertex in a set U_i is knocked out in the first round, then all vertices in U_i are knocked out in the first round.

This claim is proven by using the same arguments as in Claim 1.

By Claim 1 and Claim 2 we may define a set $B' \subseteq X$ as follows. All vertices of a set Q_i or U_i are in B' if and only if the set is knocked out in the first round. Let $W' = X \backslash B'$.

Claim 3. For all $1 \leq i \leq m$, either $Q_i \subseteq B'$ and $U_i \subseteq W'$, or $Q_i \subseteq W'$ and $U_i \subseteq B'$.

Let $1 \leq i \leq m$. By Observation 3, each vertex $P_i^k \in \mathcal{P}_i$ must fire at either q_i^k or u_i^k in the first round. The previous two claims imply that Q_i or U_i is knocked out in the first round. Suppose both sets are knocked out in the first round. Then, by Observation 2, u_i^1 fires at $y_1^{u_i^1}$ and q_i^1 fires at $y_1^{q_i^1}$. Then, by Observation 3, P_i^1 will not be knocked out in any round. The claim is proved.

By Claim 3, all vertices in $\mathcal{Z} \backslash \mathcal{S}'$ have one neighbour in B' and one neighbour in W'. Let S_j' be a vertex in $\mathcal{S}$. By Observation 3, S_j' fires at a neighbour in $\bigcup_{i=1}^m Q_i$. By definition, this neighbour is in B'. By both Observation 2 and Observation 3, S_j' is knocked out by a neighbour in $\bigcup_{i=1}^m Q_i$ that is not knocked out in the first round. By definition, this neighbour is in W'. It is now clear that $B' \cup W'$ is a 2-colouring of J' such that for each $1 \leq i \leq m$ either $Q_i \subseteq B'$ and $U_i \subseteq W'$, or $Q_i \subseteq W'$ and $U_i \subseteq B'$. Hence, by Proposition 1, J also has a 2-colouring.

Case 3. $k \geq 4$. We use reduction from H2C. From an instance $J = (Q, \mathcal{S})$ we construct the graph G as in the previous case. We claim that J is 2-colourable if and only if $\text{pko}(G) \leq k$.

Suppose that J is 2-colourable. As we have seen in the previous case this implies that $\text{pko}(G) \leq 3 \leq k$.

Suppose that $\text{pko}(G) \leq k$. Then G is KO-reducible. Note that in the proof of the previous case we only assume that G is KO-reducible. Hence we can copy the proof of the previous case. This completes the proof of Theorem 1. □

Corollary 1. *The* PKO *problem is* NP*-complete, even if instances are restricted to the class of bipartite graphs.*

Proof. We use reduction from H2C. From an instance $J = (Q, \mathcal{S})$ we construct the graph G as in the proof of Theorem 1. We claim that J is 2-colourable if and only if G is KO-reducible.

Suppose that J is 2-colourable. As we have seen in the proof of Theorem 1 this implies that $\text{pko}(G) \leq 3$. Hence G is KO-reducible.

Suppose that G is KO-reducible. We copy the proof of Case 2 of Theorem 1. □

Exact Parallel Knock-Out (k) (EPKO(k))
Instance: A graph G.
Question: Is $\text{pko}(G) = k$?

Corollary 2. *The* EPKO(k) *problem is polynomially solvable for $k = 1$ and is* NP*-complete for $k \geq 2$, even if instances are restricted to the class of bipartite graphs.*

Proof. For the case $k = 1$ we only have to exclude the null graph. Let $k \geq 2$. In [1] a family of trees Y_ℓ is constructed with $\text{pko}(Y_\ell) = \ell$ for $\ell \geq 1$. For the case $k = 2$ we only have to add a disjoint copy of the tree Y_2 (a path on 7 vertices) to the graph B in the proof of Case 1 in Theorem 1. For $k \geq 3$ it suffices to add a disjoint copy of the tree Y_k to the graph G constructed in the proof of Case 2 in Theorem 1. Note that the size of a tree Y_k only depends on k and not on the size of our input graph G (so we do not need the exact description of this family). □

5 Bounded Tree-Width

In this section we use *monadic second-order logic*; that is, that fragment of second-order logic where quantified relation symbols must have arity 1. For example, the following sentence, which expresses that a graph (whose edges are given by the binary relation E) can be 3-coloured, is a sentence of monadic second-order logic:

$$\exists R \exists W \exists B \Bigg\{ \forall x \Bigg(\big(R(x) \vee W(x) \vee B(x)\big) \wedge \neg(R(x) \wedge W(x)) \wedge \neg(R(x) \wedge B(x)) \wedge \neg(W(x) \wedge B(x)) \Bigg) \wedge \forall x \forall y \Bigg(E(x,y) \Rightarrow (\neg(R(x) \wedge R(y)) \wedge \neg(W(x) \wedge W(y)) \wedge \neg(B(x) \wedge B(y))) \Bigg) \Bigg\}$$

(the quantified unary relation symbols are R, W and B, and should be read as sets of 'red', 'white' and 'blue' vertices, respectively). Thus, in particular, there exist **NP**-complete problems that can be defined in monadic second-order logic.

A seminal result of Courcelle [3] is that on any class of graphs of bounded tree-width, every problem definable in monadic second-order logic can be solved in time linear in the number of vertices of the graph. Moreover, Courcelle's result holds not just when graphs are given in terms of their edge relation, as in the example above, but also when the domain of a structure encoding a graph G consists of the disjoint union of the set of vertices and the set of edges, as well as unary relations V and E to distinguish the vertices and the edges, respectively, and also a binary incidence relation I which denotes when a particular vertex is incident with a particular edge (thus, $I \subseteq V \times E$). The reader is referred to [3] for more details and also for the definition of tree-width which is not required here. To prove Theorem 2, we need only prove the following proposition.

Proposition 2. *For $k \geq 1$, PKO(k) can be defined in monadic second order logic.*

Proof. Recall that a parallel knock-out scheme for a graph $G = (V, E)$ is a sequence of graphs

$$G \rightsquigarrow G_1 \rightsquigarrow G_2 \rightsquigarrow \cdots \rightsquigarrow G_r,$$

where G_r is the null graph. Let $W_0 = V$ and, for $1 \leq i \leq r$, let W_i be the vertex set of G_i. If we can write a formula $\Phi(W_i, W_{i+1})$ of monadic second-order logic that says

> *there exists a KO-selection f_i on W_i such that the vertex set of the KO-successor is W_{i+1},*

then we could prove the proposition with the following sentence Ω_k which is satisfied if and only if G is in PKO(k):

$$\begin{aligned}&\exists W_0 \exists W_1 \cdots \exists W_k (\forall v (W_0(v) \Leftrightarrow V(v))\\ &\wedge \Phi(W_0, W_1) \wedge \Phi(W_1, W_2) \wedge \cdots \wedge \Phi(W_{k-1}, W_k)\\ &\wedge (\forall v (\neg W_k(v) \Leftrightarrow V(v))).\end{aligned}$$

(Here and elsewhere we have presupposed that each W_i is a set of vertices; we could easily include additional clauses to check this explicitly.)

The following claim will help us write $\Phi(W_i, W_{i+1})$.

Claim 4. There is a KO-selection f_i on W_i such that W_{i+1} is the vertex set of the KO-successor if and only if there is a partition V_1, V_2, V_3 of W_i and subsets E_1, E_2, E_3 of E such that

(a) for $j = 1, 2, 3$, each vertex in V_j is incident with exactly one edge of E_j, this edge joins it to a vertex in $W_i \setminus V_j$, and this accounts for every edge in E_j (so $|V_j| = |E_j|$).
(b) $W_{i+1} \subset W_i$ and, for $j = 1, 2, 3$, $W_{i+1} \cap V_j$ is the set of vertices in V_j not incident with edges in $E_{j'}$ for any $j' \neq j$.

We will prove the claim later. First we use it to write $\Phi(W_i, W_{i+1})$.

The following formula $\psi(V_1, E_1, V_2, E_2, V_3, E_3, W_i)$ checks that the sets V_1, V_2 and V_3 partition W_i, that the sets E_1, E_2, E_3 are edges in the graph, and that (a) is satisfied.

$$\begin{aligned}&\forall v((V_1(v) \vee V_2(v) \vee V_3(v)) \Leftrightarrow W_i(v)) \wedge \forall v(\neg(V_1(v) \wedge V_2(v))\\ &\wedge \neg(V_1(v) \wedge V_3(v)) \wedge \neg(V_2(v) \wedge V_3(v)))\\ &\wedge \forall x((E_1(x) \vee E_2(x) \vee E_3(x)) \Rightarrow E(x))\\ &\wedge \forall x(E_1(x) \Rightarrow \exists u \exists v(V_1(u) \wedge (V_2(v) \vee V_3(v)) \wedge I(u,x) \wedge I(v,x)))\\ &\wedge \forall x(E_2(x) \Rightarrow \exists u \exists v(V_2(u) \wedge (V_1(v) \vee V_3(v)) \wedge I(u,x) \wedge I(v,x)))\\ &\wedge \forall x(E_3(x) \Rightarrow \exists u \exists v(V_3(u) \wedge (V_1(v) \vee V_2(v)) \wedge I(u,x) \wedge I(v,x)))\\ &\wedge \forall v(V_1(v) \Rightarrow \exists! x(I(v,x) \wedge E_1(x)))\\ &\wedge \forall v(V_2(v) \Rightarrow \exists! x(I(v,x) \wedge E_2(x)))\\ &\wedge \forall v(V_3(v) \Rightarrow \exists! x(I(v,x) \wedge E_3(x)))\end{aligned}$$

(The semantics of $\exists!$ is 'there exists exactly one'; clearly, this abbreviates a more complex though routine first-order formula.) The following formula checks that (b) is satisfied and is denoted $\chi(V_1, E_1, V_2, E_2, V_3, E_3, W_i, W_{i+1})$.

$$\begin{aligned}\forall v(W_{i+1}(v) \Leftrightarrow (W_i(v) \wedge ((V_1(v) \wedge \neg\exists x((E_2(x) \vee E_3(x)) \wedge I(v,x)))\\ \vee(V_2(v) \wedge \neg\exists x((E_1(x) \vee E_3(x)) \wedge I(v,x)))\\ \vee(V_3(v) \wedge \neg\exists x((E_1(x) \vee E_2(x)) \wedge I(v,x)))).\end{aligned}$$

And now we can write $\Phi(W_i, W_{i+1})$:

$$\begin{aligned}\exists V_1 \exists E_1 \exists V_2 \exists E_2 \exists V_3 \exists E_3 (\psi(V_1, E_1, V_2, E_2, V_3, E_3, W_i)\\ \wedge\chi(V_1, E_1, V_2, E_2, V_3, E_3, W_i, W_{i+1})).\end{aligned}$$

It only remains to prove Claim 4. Suppose that we have sets V_1, V_2, V_3, E_1, E_2 and E_3 that satisfy the conditions of the claim. Then to define the KO-selection f_i, for $j = 1, 2, 3$, for each vertex $v \in V_j$, let v fire at the unique neighbour joined to v by an edge in E_j. It is easy to check that W_{i+1} is the vertex set of the KO-successor.

Now suppose that we have a KO-selection f_i. Let H_i be the spanning subgraph of G_i with edge set $\{vf_i(v) \mid v \in W_i\}$. The firing can be represented as an orientation of H: orient each edge from v to $f_i(v)$ (some edges may be oriented in both directions). As each vertex has exactly one edge oriented away from it, each component of the oriented graph contains one directed cycle, of length at least 2, with a pendant in-tree attached to each vertex of the cycle; see Figure 3.

We find the sets $V_1, V_2, V_3, E_1, E_2, E_3$; the edge sets contain only edges of H_i. We may assume that H_i is connected (else we can find the sets componentwise). Let the vertices of the unique cycle in the orientation be $v_1, \ldots, v_c$ where the edges are $v_l v_{l+1}$, $1 \leq l \leq c-1$, and $v_c v_1$. So H_i contains vertices $v_1, \ldots, v_c$ with a pendant tree (possibly trivial) attached to each.

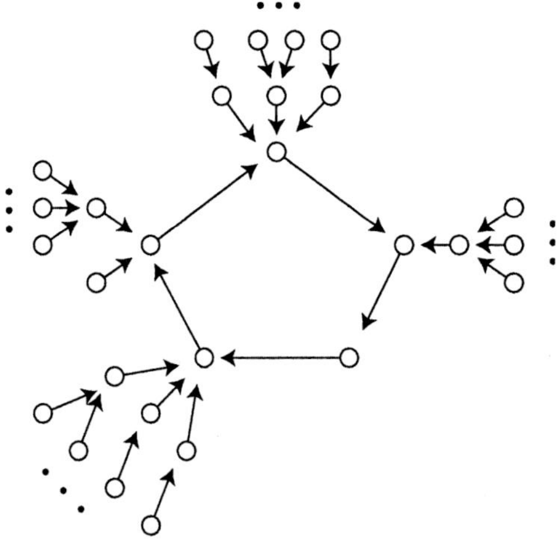

Fig. 3. A representation of vertices firing

For $1 \leq l \leq c$, let U_e^l be the set of vertices in the pendant tree attached to v_l whose distance from v_l is even (but not zero), and let U_o^l be the vertices in the tree at odd distance from v_l. Let

$$V_1 = \bigcup_{l \text{ odd}} U_o^l \cup \bigcup_{l \text{ even}} U_e^l \cup \{v_l : l \text{ is even}, l \neq c\},$$
$$V_2 = \bigcup_{l \text{ odd}} U_e^l \cup \bigcup_{l \text{ even}} U_o^l \cup \{v_l : l \text{ is odd}, l \neq c\}, \text{ and}$$
$$V_3 = \{v_c\},$$

and, for $i = 1, 2, 3$, let E_i contain $vf_i(v)$ for each $v \in V_i$. It is clear that the sets we have chosen satisfy the conditions of the claim.

This completes the proof of the claim and of the proposition. □

Theorem 2 follows from the proposition. And, noting that EPKO(k) is defined by the monadic second-order sentence $\Omega_k \wedge \neg\Omega_{k-1}$, we have the following result.

Corollary 3. *For $k \geq 1$, EPKO(k) is solvable in linear time on any class of graphs with bounded tree-width.*

Finally, we note that to check whether a graph G is reducible it is sufficient to check whether pko(G) $= k$, for $1 \leq k \leq \Delta$, where Δ is the maximum degree of G. Thus G is reducible if and only if the sentence $\Omega_\Delta \vee \Omega_{\Delta-1} \vee \cdots \vee \Omega_1$ is satisfied. This gives us our last result.

Corollary 4. *On any class of graphs with bounded tree-width,* PKO *can be solved in polynomial time.*

References

1. H. Broersma, F.V. Fomin, R.Královič, and G.J. Woeginger. Eliminating graphs by means of parallel knock-out schemes, to appear in *Discrete Mathematics.*
2. J.A. Bondy, and U.S.R.Murty (1976). *Graph Theory with Applications.* Macmillan, London and Elsevier, New York.
3. B. Courcelle (1990). The monadic second-order logic of graphs. I. Recognizable sets of finite graphs, *Information and Computation 85*, 12–75.
4. M.R. Garey and D.S. Johnson (1979). *Computers and Intractability: A Guide to the Theory of NP-Completeness.* Freeman, San Francisco.
5. D.E. Lampert and P.J. Slater (1998). Parallel knockouts in the complete graph. *American Mathematical Monthly 105*, 556–558.
6. L. Lovász (1973). Covering and coloring of hypergraphs, *Proceedings of the 4th Southeastern Conference on Combinatorics, Graph Theory, and Computing,* Utilitas Mathematica, 3–12.

Reconfigurations in Graphs and Grids

Gruia Calinescu[1], Adrian Dumitrescu[2,*], and János Pach[3,**]

[1] Department of Computer Science, Illinois Institute of Technology, Chicago, IL 60616, USA
calinesc@iit.edu

[2] Department of Computer Science, University of Wisconsin–Milwaukee, 3200 N. Cramer Street, Milwaukee, WI 53211, USA
ad@cs.uwm.edu

[3] Courant Institute of Mathematical Sciences, 251 Mercer Street, New York, NY 10012-1185, USA
pach@cims.nyu.edu

Abstract. Let G be a connected graph, and let V and V' two n-element subsets of its vertex set $V(G)$. Imagine that we place a chip at each element of V and we want to move them into the positions of V' (V and V' may have common elements). A move is defined as shifting a chip from v_1 to v_2 ($v_1, v_2 \in V(G)$) on a path formed by edges of G so that no intermediate vertices are occupied. We give upper and lower bounds on the number of moves that are necessary, and analyze the computational complexity of this problem under various assumptions: labeled versus unlabeled chips, arbitrary graphs versus the case when the graph is the rectangular (infinite) planar grid, etc. We provide hardness and inapproximability results for several variants of the problem. We also give a linear-time algorithm which performs an optimal (minimum) number of moves for the unlabeled version in a tree, and a constant-ratio approximation algorithm for the unlabeled version in a graph. The graph algorithm uses the tree algorithm as a subroutine.

1 Introduction

Consider a set system (set) of n pairwise disjoint objects in the Euclidean space that need to be brought from a given start (initial) configuration S into a desired goal (target) configuration T. In many cases, the problem admits the following abstraction: we have an underlying finite or infinite graph, the start configuration is represented by a set of n chips at n start vertices and the target configuration by another set of n target vertices. A vertex can be both a start and target position. The case when the chips are labeled or unlabeled give two different variants of the problem. In one move a chip can follow an arbitrary path in the graph and end up at another vertex, provided the path (including the end

* Research partially supported by NSF CAREER grant 0444188.

** Research partially supported by grants from NSF, NSA, OTKA, and from the US-Israeli Binational Research Foundation.

J.R. Correa, A. Hevia, and M. Kiwi (Eds.): LATIN 2006, LNCS 3887, pp. 262–273, 2006.

vertex) is free of other chips. The *motion planning* problem for such a system is that of computing a sequence of object motions (schedule) that achieves this task. If such a sequence of motions exists, we say that the problem is *feasible* and we say that it is *infeasible* otherwise. To avoid trivial questions, we always assume the graph is connected.

In certain applications, objects are indistinguishable, therefore the chips are unlabeled; for instance, a modular robotic system consists of a number of identical modules (robots), each of which having identical capabilities [9, 10]. In another application, the chips are indivisible packets (copies) of the same data that need to be moved from one site to another of a wide-area communication network without ever exceeding the capacities of the communication buffers at each site [5, 14].

In this variant with unlabeled chips, the problem is easier and always feasible; therefore we are interested in minimizing the number of moves. For the variant with labeled chips it may be the case that the problem is infeasible: it is known for instance that the 15-puzzle on a 4×4 grid — introduced by Sam Loyd in 1878 — has a solution if and only if the start permutation is an even permutation [13, 18] (see [3] for a recent approach).

Other reconfiguration rules (models) for systems of disks in the plane have been examined recently: [1, 8, 7]. These models do not fall in the graph reconfiguration framework in this paper, because a disk may partially overlap several target positions. A model that fits in the graph reconfiguration framework has been analyzed in [9]: it deals with reconfiguration of modular systems acting in a grid-like environment, where moves must maintain connectivity of the whole system, and the motion rules are very local: a chip can only move to an adjacent position in one move.

The general form of the reconfiguration problem we consider is to find a reconfiguration sequence with a minimum number of moves. Depending on whether we refer to the graph or grid version, or to the labeled or unlabeled version, we call the problem U-GRAPH-RP, L-GRAPH-RP, U-GRID-RP or L-GRID-RP.

Consider for example the reconfiguration problem in the infinite grid with unlabeled (or labeled) chips. The following simple algorithm does $2n$ moves for reconfiguration of n objects (chips). In the first step (n moves), move in a suitable order all the chips away in the free grid space. In the second step (n moves), bring the chips "back" to target positions. We will show that minimizing the number of moves is intractable in both (labeled and unlabeled) variants. A move is a *target move* if it moves a chip to a final target position. Otherwise it is a *non-target* move. Our lower bounds use the the following argument: if no target chip coincides with a start chip (so each chip must move), a schedule with x non-target moves consists of at least $n + x$ moves.

Previous related work. Most of the work done so far concerns labeled versions of the reconfiguration problem, and we give here only a very brief survey.

For the generalization of the 15-puzzle on an arbitrary graph (with $k = v - 1$ labeled chips in a graph on v vertices), Wilson gave an efficiently checkable characterization of the solvable instances of the problem, depending on whether the graph is bipartite or not [19]. Kornhauser *et al.* have extended his result to any $k \leq v - 1$ and provided bounds on the number of moves for solving any solvable instance [14].

Ratner and Warmuth have shown that finding a solution with minimum number of moves for the $(N \times N)$-extension of the 15-puzzle is intractable [17], so the reconfiguration problem in graphs with labeled chips is NP-hard.

Auletta *et al.* gave a linear time algorithm for the *pebble motion on a tree* [5]. This problem is the labeled variant of the same reconfiguration problem we study here, however each move is along one edge only.

Papadimitriou *et al.* studied a problem of *motion planning on a graph* in which there is a mobile robot at one of the vertices s, that has to reach to a designated vertex t using the smallest number of moves, in the presence of obstacles (pebbles) at some of the other vertices [15]. Robot and obstacle moves are done along edges, and obstacles have no destination assigned and may end up in any vertex of the graph. The problem has been shown to be NP-complete even for planar graphs, and a ratio $O(\sqrt{n})$ polynomial time approximation algorithm was given in [15].

Dumitrescu *et al.* have addressed several basic questions in the analysis of modular metamorphic systems [10]. In particular the next two questions have been shown to be decidable: (*i*) whether a given set of motion rules maintains connectivity; (*ii*) whether a goal configuration is reachable from a given initial configuration (at specified locations) using a given set of motion rules. Other seemingly similar questions have been shown to be undecidable.

Our results are:

(1) The reconfiguration problem in graphs with unlabeled chips U-GRAPH-RP is NP-hard, and even APX-hard.
(2) The reconfiguration problem in graphs with labeled chips L-GRAPH-RP is APX-hard.
(3) For the infinite planar rectangular grid, both the labeled and unlabeled variants L-GRID-RP and U-GRID-RP are NP-hard.
(4) There exists a ratio 3 approximation algorithm for the unlabeled version in graphs U-GRAPH-RP. Thereby we get a ratio 3 approximation algorithm for the unlabeled version U-GRID-RP in the (infinite) rectangular grid.
(5) We show that n moves are always enough (and sometimes necessary) for the reconfiguration of n unlabeled chips in graphs. For the case of trees, we present a linear time algorithm which performs an optimal (minimum) number of moves.
(6) We show that $7n/4$ moves are always enough, and $3n/2$ are sometimes necessary, for the reconfiguration of n labeled chips in the infinite planar rectangular grid (L-GRID-RP).

2 Unlabeled Chips in Graphs and Trees

Let G be a connected graph, and let V and V' two n-element subsets of its vertex set $V(G)$. Imagine that we place a chip at each element of V and we want to move them into the positions of V' (V and V' may have common elements). A move is defined as shifting a chip from v_1 to v_2 ($v_1, v_2 \in V(G)$) along a path in G so that no intermediate vertices are occupied.

Theorem 1. *In G one can get from any n-element initial configuration V to any n-element final configuration V' using at most n moves, so that no chip moves twice. Moreover, for the case of a tree T with r vertices, there is a $O(r)$-time algorithm which performs the optimal (minimum) number of moves.*

Proof. It is sufficient to prove the theorem for trees. We argue by induction on the number of chips. Take the smallest tree T containing V and V', and consider an arbitrary leaf l of T. Assume first that the leaf l belongs to V: say $l = v$. If v also belongs to V', the result trivially follows by induction, so assume that this is not the case. Choose a path P in T, connecting v to an element v' of V' such that no internal point of P belongs to V'. Apply the induction hypothesis to $V \setminus \{v\}$ and $V' \setminus \{v'\}$ to obtain a sequence of at most $n-1$ moves, and add a final (unobstructed) move from v to v'.

The remaining case when the leaf l belongs to V' is symmetric: say $l = v'$; choose a path P in T, connecting v' to an element v of V such that no internal point of P belongs to V. Move first v to v' and append the sequence of at most $n-1$ moves obtained from the induction hypothesis applied to $V\setminus\{v\}$ and $V'\setminus\{v'\}$.

We further refine this algorithm so as to minimize the number of moves. We call a vertex that is both a start and target position an *obstacle*. We have four types of vertices: free vertices, chip-only vertices, target-only vertices, and obstacles. Denote by c (resp. t) the number of chip-only (resp. target-only) vertices, and by o the number of obstacles. We have $c + o = o + t = n$. We call a tree *balanced* if it contains an equal number of chip-only and target-only vertices. Clearly, the initial tree T is balanced. If there exists an obstacle whose removal from T breaks T into balanced subtrees, we keep this obstacle fixed and proceed recursively (by induction) on the subtrees. If no obstacle removal breaks T into balanced subtrees, then all obstacles must move (each at least once), hence the number of moves necessary is at least $o + c = n$, and the algorithm in the first part of our proof can be used to obtain an optimal schedule.

The above observation together with postorder traversal keeping additional information for every node, form the basis of the linear time reconfiguration algorithm (omitted due to lack of space). □

Remark. Theorem 1 implies that in the infinite rectangular grid, we can get from any starting position to any ending position of the same size n in at most n moves. It is perhaps interesting to compare this to the problem of sliding congruent unlabeled disks in the plane: here one can come up with "cage-like" constructions that require at least about $\frac{16n}{15}$ moves [7].

3 Hardness Results for the Variants on Graphs

Theorem 2. *The unlabeled version in graphs* U-GRAPH-RP *is NP-complete. Moreover, assuming $P \neq NP$, there is an absolute constant $\epsilon_1 > 0$ such that no*

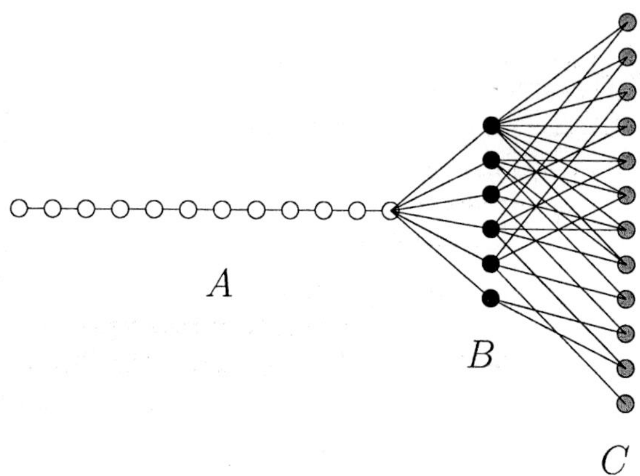

Fig. 1. The "broom" graph G corresponding to a set cover instance with $|U| = 12$, and $|\mathcal{F}| = 6$. The vertices occupied only by chips are white, those occupied by both chips and targets are black, and those occupied only by targets are shaded. An optimal reconfiguration takes 15 moves (an optimal set cover has size 3).

polynomial-time algorithm has approximation guarantee at most $1 + \epsilon_1$. *That is,* U-GRAPH-RP *is APX-hard.*

Proof. The decision version of U-GRAPH-RP is clearly in NP, so we only have to prove its NP-hardness. We reduce the *set cover* problem SC to U-GRAPH-RP. An instance of the set cover problem consists of a family $\mathcal{F}$ of subsets of a finite set U. The problem is to decide whether there is a set cover of size k for $\mathcal{F}$, i.e., a subset $\mathcal{F}' \subseteq \mathcal{F}$, with $|\mathcal{F}'| \leq k$, such that every element in U belongs to at least one member of $\mathcal{F}'$. SC is known to be NP-complete [11].

Consider an instance of SC represented by a bipartite graph $(B \cup C, E)$, where $U = C$, $\mathcal{F} = B$, and edges describe the membership relation. Construct the undirected graph G shown in Fig. 1, with $|A| = |C|$. The chips are $S = A \cup B$ and the targets are $T = B \cup C$. Clearly, G can be constructed in polynomial time. The reduction is complete once we establish the following claim (proof omitted).

Claim. There is a set cover consisting of at most q sets if and only if reconfiguration in G can be done using at most $|A| + q$ moves.

To prove the approximation hardness we use the same reduction, and the fact that 3-SC, the set cover problem in which the size of each set in $\mathcal{F}$ is bounded from above by 3 is APX-hard [16, 2]. We omit the details. □

Remark. A similar reduction can be made for the labeled version. The chips in A have targets in C, labeled as in Fig. 2 (here $|A| = |C| = m$).

The obstacle chips in B coincide with their targets. Each vertex in B is adjacent to a "twin" free vertex. The reduction follows from the next claim (proof omitted).

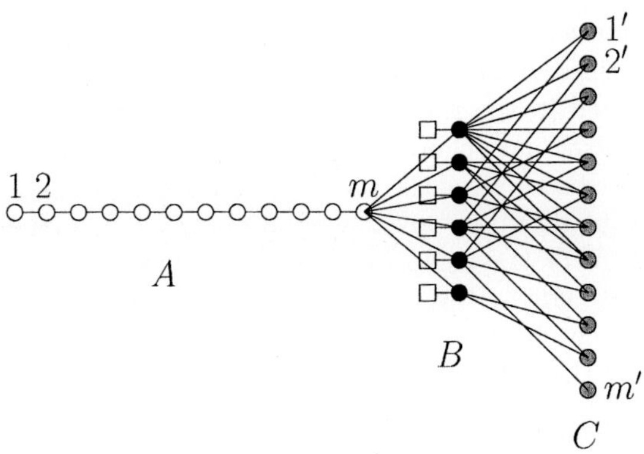

Fig. 2. The graph constructed in the reduction for labeled chips. Free vertices are drawn as squares. An optimal reconfiguration takes 18 moves (an optimal set cover has size 3).

Claim. There is a set cover consisting of at most q sets if and only if reconfiguration in G can be done using at most $|A| + 2q$ moves.

We thus obtain:

Theorem 3. *The labeled version in graphs* L-GRAPH-RP *is APX-hard.*

4 An Approximation Algorithm for the Unlabeled Version in Graphs

Theorem 4. *There exists a* 3*-approximation algorithm for* U-GRAPH-RP.

Proof. The algorithm is obtained by applying the local ratio method of Bar-Yehuda [6] to a graph H whose construction we describe below.

The vertex set of the input graph G is partitioned into four sets:

$C = V \setminus V'$, the chip-only vertices
$A = V' \setminus V$, the target-only vertices
$B = V \cap V'$, the obstacles
$F = V(G) \setminus (V \cup V')$, the free vertices

Then $V(H) = A \cup B \cup C$. For every pair of vertices u and v of H we put in $E(H)$ the edge uv if $uv \in E(G)$ or there is a path in G from u to v having all the internal vertices from F.

In H, we use the local ratio method to find a set of edges Q such that every connected component D of $(V(H), Q)$ has an equal number of chip-only and target-only vertices. We call this the U-STEINER problem. U-STEINER is a network design problem given by a 0-1 *proper function* [12], problem for which both the primal-dual schema and the local ratio method give a 2-approximation. The function, defined over all sets of vertices, is one if the set is unbalanced and zero otherwise. Recall that a set is *balanced* if it contains an equal number of chip-only and target-only vertices. It can be shown that this is a proper function, however this fact is not needed in the ratio 3 approximation algorithm, whose proof is self-contained. From the next claim, one can easily obtain a 4-approximation for U-GRAPH-RP, as shown after the proof of claim.

One more piece of notation: given a solution Q for U-GRAPH-RP in G, consider the edges of G traversed by the moving chips during the reconfiguration process. These edges, together with their endpoints (including the free vertices through which chips pass through) form a number $k \geq 1$ of connected components. We say that Q has k connected components. Write $c = |C| = |A|$.

Claim 1. Given a feasible solution Q for U-GRAPH-RP in G with m moves and k connected components, there is a feasible solution for U-STEINER in H with at most $m + c - k$ edges. Conversely, given a feasible solution Q for U-STEINER in H with e edges, there is a feasible solution for U-GRAPH-RP in G with at most $e - c + k$ moves, where k is the number of connected components of Q which intersect A (and C).

Proof of Claim 1. For the first part, let S be the set of vertices of G involved in the moves of Q, and let S_i, for $i = 1, 2, \ldots, k$, be the connected components of S. Then the number of moves inside S_i is at least $|S_i \cap C| + |S_i \cap B|$. Let $S'_i = S_i \cap V(H)$ and note that S'_i is also connected. In each S'_i pick a tree T_i; the union of the trees T_i is the feasible solution of U-STEINER. In each T_i, the number of edges is $|V(T_i)| - 1 = |V(T_i) \cap C| + |V(T_i) \cap B| + |V(T_i) \cap A| - 1$. Summing over i gives the needed equality.

For the second part, let T_i, for $i = 1, \ldots, k$, be one such connected component with e_i edges. Then $|A \cap V(T_i)| = |C \cap V(T_i)|$ and using Theorem 1 one can move all the chips of $V(T_i)$, including those sitting on obstacles, to all the targets of $V(T_i)$ using $|V(T_i)| - |C \cap T_i|$ moves: the chips from $(C \cup B) \cap V(T_i)$ move along the edges of T_i, passing if necessary through vertices of F (in G). Since $|V(T_i)| = |E(T_i)| + 1$, this second part of the claim follows by adding up over the components, since no move puts a chip on a vertex of F. (As a side remark, if Q is an optimal solution in H, each T_i is a tree intersecting A and C). □

Here is a short account for the ratio 4 approximation algorithm: By the first part of Claim 1 applied to an optimal solution for U-GRAPH-RP in G with m_{OPT} moves and k_{OPT} components, the number of edges e_{OPT} in an optimal solution for U-STEINER in H satisfies

$$e_{OPT} \leq m_{OPT} + c - k_{OPT} \leq m_{OPT} + c \leq 2m_{OPT}.$$

Therefore, by the second part of Claim 1, the number of moves m in the solution for U-GRAPH-RP in G returned by the algorithm satisfies (since $k_1 \leq c$, where k_1 is the number of components in the solution for U-STEINER in H)

$$m \leq e - c + k_1 \leq e \leq 2e_{OPT} \leq 4m_{OPT}.$$

We now present the ratio 3 approximation algorithm. We have to enter the details of the local ratio method to get this ratio (instead of 4). The local ratio algorithm approximately solves U-STEINER instances with non-negative weight δ on the edges. The algorithm below, given in [6], is recursive. Each recursive call is given a set S of already selected edges and a non-negative edge weight function, and it returns a set of edges (in step 8). Given a set of edges $F \subseteq E(H)$, call a connected component X of $(V(H), F)$ *balanced* if $|V(X) \cap A| = |V(X) \cap C|$ and *unbalanced* otherwise. For the first call of the algorithm, $S = \emptyset$, and $\delta = 1$ on all edges. By feasible solution we mean feasible solution for the original U-STEINER instance.

1. The input parameters are the set of edges $S \subseteq E(H)$ and non-negative weight δ on $E(H)$.
2. If all the connected components of $(V(H), S)$ are balanced, return $\emptyset$.
3. Define weight function δ_1 on edges of $E(H) \backslash S$ as follows: edges going between two unbalanced components of $(V(H), S)$ get weight 1, edges going between one unbalanced component of $(V(H), S)$ and one balanced component of $(V(H), S)$ get weight 1/2, and all the other edges get weight 0.
4. Compute a positive real number α such that the weight function δ_2 on edges of H given by $\delta_2 = \delta - \alpha \cdot \delta_1$ is non-negative and for at least one edge e, we have $\delta_2(e) = 0 < \delta(e)$.
5. Let $M = \{e \mid \delta_2(e) = 0\}$.
6. recursively solve the instance with parameters $S \cup M$ and weight δ_2 on $E(H)$, producing set of edges L such that $S \cup M \cup L$ is a feasible solution.
7. Obtain minimal $M' \subseteq M$ such that $S \cup M' \cup L$ is a feasible solution.
8. Return $L \cup M'$.

We are guaranteed that $M \neq \emptyset$ and thus the algorithm terminates. For the approximation ratio, we need the following claim.

Claim 2. During the execution of a recursive call of the algorithm, let k be the number of unbalanced components of $(V(H), S)$. Then $\delta_1(P) \geq k/2$ for any feasible solution P. The set of edges Q returned by the recursive call of the algorithm satisfies $\delta_1(Q) \leq k - q$, where q is the number of connected components of $(V(H), Q \cup S)$.

Proof of Claim 2. Consider a feasible solution P. If an edge of P goes between two unbalanced components of $(V(H), S)$ we assign 1/2 to each such component, and if it goes from one unbalanced component of $(V(H), S)$ to one balanced component of $(V(H), S)$, we assign 1/2 to the unbalanced component. Each unbalanced component of $(V(H), S)$ must have at least one edge of P going to some other component, and thus it is assigned at least 1/2. Therefore $\delta_1(P) \geq k/2$.

Consider now the edges Q selected (returned) by one recursive call of the algorithm and let Q_i, for $i = 1, \ldots, q$ be the connected components of $(V(H), Q \cup S)$. Fix one component Q_i. Inside Q_i, contract to a single vertex the vertices from the same component of $(V(H), S)$, obtaining $\bar{Q}_i$. The minimal property of Q as ensured by Step 7 of the algorithm ensures that $\bar{Q}_i$ is acyclic (and thus it is a tree) and every leaf of $\bar{Q}_i$ is an unbalanced component of S. Note that all the edges of Q_i with positive δ_1-weight are in $\bar{Q}_i$. Root $\bar{Q}_i$ at an arbitrary vertex v given by an unbalanced component of S.

For every $u \in (V(\bar{Q}_i) \setminus \{v\})$ given by an unbalanced component of S, charge to u either one or two edges, of total δ_1-weight 1, as follows: consider the path from u to v in $\bar{Q}_i$ and let u' be the next vertex on this path given by an unbalanced component of S. If the path has only one edge, charge this edge to u. If this path has more than one edge, charge to u the first and last edge of the path. It is easy to check that every edge of positive δ_1-weight of $\bar{Q}_i$ is charged at least once. Thus $\delta_1(Q_i) = \delta_1(\bar{Q}_i) \le s_i - 1$, where s_i is the number of unbalanced component of S included in Q_i; here we used that v is not being charged.

Summing over i yields the second part of the claim. □

We continue with the proof of Theorem 4. First we note that, by using Claim 2 and induction, the local ratio algorithm ensures that its output LR satisfies, for any feasible solution P, the following:

$$\delta(LR) = \alpha\delta_1(LR) + \delta_2(LR) \le \alpha 2\delta_1(P) + 2\delta_2(P) = 2\delta(P). \tag{1}$$

Let OPT be a solution with a minimum number of moves, $m(OPT)$ be the number of moves of OPT, and OP be the set of edges of the U-STEINER feasible solution obtained from OPT in Claim 1. Let LR be the set of edges selected by the local ratio approximation algorithm when applied to the U-STEINER instance, $k(LR)$ be the number of connected components of LR which intersect A, and $m(LR)$ be the number of moves of the solution obtained in Claim 1 from LR. The weight functions δ, δ_1, and δ_2 refer to the first call of the local ratio algorithm, as does the real α, which we note is at least 1. We have:

$$\begin{aligned}
m(LR) &\le |LR| + k(LR) - c \quad \text{by Claim 1} \\
&= \delta(LR) + k(LR) - c \\
&= \alpha\delta_1(LR) + \delta_2(LR) + k(LR) - c \\
&\le \alpha(2\delta_1(OP) - k(LR)) + \delta_2(LR) + k(LR) - c \quad \text{by Claim 2} \\
&\le 2\alpha\delta_1(OP) - k(LR) + \delta_2(LR) + k(LR) - c \quad \text{since } \alpha \ge 1 \\
&\le 2\alpha\delta_1(OP) + 2\delta_2(OP) - c \quad \text{by Equation 1} \\
&\le 2\delta(OP) - c \\
&= 2|OP| - c \\
&\le 2(m(OPT) + c) - c \quad \text{by Claim 1} \\
&= 2m(OPT) + c \\
&\le 3m(OPT), \quad \text{since } m(OPT) \ge c.
\end{aligned}$$

This concludes the proof of Theorem 4. □

Remark. In the graph version with unlabeled chips, if we count as a move every edge traversed by a chip, minimizing the number of moves can be solved in polynomial time, as described below. Construct a complete weighted bipartite graph $B = (V \cup V', F)$ with bipartition V: the vertices containing chips and V': the vertices containing targets (with obstacles in both sides of the bipartition). The weight of an edge in F is equal to the length of the shortest path connecting the endpoints of the edge in G. Now apply an algorithm for Minimum Weight Perfect Matching in B, and move accordingly: if the path a chip c_1 would take to reach its destination has another chip c_2 on it, have the two chips switch destinations and continue moving c_2. One can check that the number of moves does not exceed the weight of the perfect matching. On the other side, the optimum solution must move chips to targets and cannot do better than the total length of the shortest paths in a minimum matching.

5 Chips in Grids

In this section we analyze the reconfiguration problem with labeled, respectively unlabeled chips, in an infinite grid. However a finite section of the grid clearly suffices for this purpose. Due to lack of space, we omit the proofs of the following two theorems. The reductions are from Rectilinear Steiner Tree.

Theorem 5. *The unlabeled version in the grid* U-GRID-RP *is NP-complete.*

Theorem 6. *The labeled version in the grid* L-GRID-RP *is NP-complete.*

5.1 Labeled Chips: Bounds on the Number of Moves

Theorem 7. *Given a pair of start and target configurations S and T, each with n labeled chips, one can move chips from S to T using at most $7n/4$ moves. On the other hand, $3n/2$ moves are sometimes necessary (for n even).*

Proof. The lower bound is trivial (however it does not appear to be trivial to improve on it!): take a pair of chips labeled 1 and 2, say next to each other, and have the target positions switch them; that is $t_1 = s_2$ and $t_2 = s_1$. Clearly three moves are needed to rearrange this group of two, and by repeating it (with different labels), one gets a pair of configurations which require $3n/2$ moves.

We now describe a reconfiguration algorithm which executes no more than $7n/4$ moves (as mentioned in the introduction, the problem can be solved trivially in $2n$ moves).

Let S and T be the start and target configurations. Consider the directed graph G with n edges (loops allowed) given by $S \to T$. Vertices are grid points of $S \cup T$ (the number of vertices is between n and $2n$). Each edge originates at a start chip and ends at some (free or occupied) target cell. Note that each in-degree and out-degree is at most one, so G can be partioned into a collection of disjoint paths and cycles (and loops).

Consider the rows of S numbered from top to bottom: $1, 2, \ldots, r$. Let D (resp. E) be the set of odd (resp. even) rows; we can assume without loss of generality that $|D| \leq |E|$, thus $|D| \leq n/2$. Let A be the set of elements of E whose target lie in E, and let B (resp. C) be the the set of elements of E whose target lie in rows of D congruent to 1 (resp. to 3) modulo 4. Write $a = |A|$, $b = |B|$, $c = |C|$. We can assume without loss of generality that $c \leq b$.

1. Move (far) away the elements of D (row by row, and for each row, move elements one by one, say from left to right) to form a set of corridors.
2. Move away the elements of C (elements of the even rows are adjacent to corridors, therefore any subset of chips of an even row can be moved away).
3. Select and move away an element from each cycle of the directed graph G remaining among the elements of A (not from the loops).
4. Fill (say, from left to right) the odd rows congruent to 1 modulo 4 with the elements of B and elements far away as follows: note that each even row is adjacent to an odd row congruent to 3 modulo 4; take out an element of B from the even row through the empty corridor congruent to 3 modulo 4, and then back in the target odd row congruent to 1 modulo 4.
5. Fill the even rows using the adjacent empty corridors (congruent to 3 modulo 4), with elements from A and elements far away. The elements of A move directly to their destination and such a move is possible as long as some elements of A still need to move, since we moved away one element from each cycle of the directed graph G contained in A.
6. Fill (say, from left to right) the odd rows congruent to 3 modulo 4 (the corridors) with elements far away.

The number of non-target moves is at most

$$n - (a + b + c) + \frac{a}{2} + c \leq n - \frac{a}{2} - b \leq \frac{3n}{4},$$

since $a + 2b \geq a + b + c \geq n/2$. Therefore the total number of moves is not more than $n + 3n/4 = 7n/4$. □

Remark. The above lower bound clearly holds even in the stronger *lifting model*, when chips can be lifted and placed back in the plane (see [7, 8] for related aspects of disk reconfiguration problems).

Acknowledgement. The authors thank Sergey Bereg and Marius Zimand for several conversations on the topic.

References

1. M. Abellanas, S. Bereg, F. Hurtado, A. G. Olaverri, D. Rappaport, and J. Tejel, Moving coins, *Computational Geometry: Theory and Applications*, to appear.
2. P. Alimonti and V. Kann, Hardness of approximating problems on cubic graphs, *Proceedings of the 3rd Italian Conference on Algorithms and Complexity*, LNCS 1203, Springer-Verlag (1997), 288–298.
3. A. Archer, A modern treatment of the 15 puzzle, *American Mathematical Monthly*, **106** (1999), 793–799.
4. S. Arora, Nearly linear time approximation schemes for Euclidean TSP and other geometric problems, *J. of the ACM*, **45**(5):1–30, 1998.
5. V. Auletta, A. Monti, M. Parente, and P. Persiano, A linear-time algorithm for the feasibility of pebble motion in trees, *Algorithmica*, **23** (1999), 223–245.

6. R. Bar-Yehuda, One for the Price of Two: A Unified approach for approximating covering problems, *Algorithmica*, **27** (2000), 131–144.
7. S. Bereg, A. Dumitrescu, and J. Pach, Sliding disks in the plane, *Japan Conference on Discrete and Computational Geometry 2004* (J. Akiyama, M. Kano and X. Tan, editors). To appear in LNCS Proceedings.
8. S. Bereg and A. Dumitrescu, The lifting model for reconfiguration, *Discrete & Computational Geometry*, accepted. A preliminary version in *Proceedings of the 21st Annual Symposium on Computational Geometry*, (SOCG '05), Pisa, Italy, June 2005, 55–62.
9. A. Dumitrescu and J. Pach, Pushing squares around, *Graphs and Combinatorics*, to appear. A preliminary version in *Proceedings of the 20-th Annual Symposium on Computational Geometry*, (SOCG '04), NY, June 2004, 116–123.
10. A. Dumitrescu, I. Suzuki and M. Yamashita, Motion planning for metamorphic systems: feasibility, decidability and distributed reconfiguration, *IEEE Transactions on Robotics and Automation*, **20(3)**, 2004, 409–418.
11. M. Garey and D. Johnson, *Computers and Intractability: A Guide to the Theory of NP-Completeness.*, W. H. Freeman and Company, 1979.
12. M. X. Goemans and D. P. Williamson, The primal-dual method for approximation algorithms and its application to network design problems, in *Approximation Algorithms for NP-Hard Problems*, edited by D. S. Hochbaum, PWS Publishing Co., 1995.
13. W. W. Johnson, Notes on the 15 puzzle. I., *American Journal of Mathematics*, **2** (1879), 397–399.
14. D. Kornhauser, G. Miller, and P. Spirakis, Coordinating pebble motion on graphs, the diameter of permutation groups, and applications, *Proceedings of the 25-th Symposium on Foundations of Computer Science*, (FOCS '84), 241–250.
15. C. Papadimitriou, P. Raghavan, M. Sudan, and H. Tamaki, Motion planning on a graph, *Proceedings of the 35-th Symposium on Foundations of Computer Science*, (FOCS '94), 511–520.
16. C. Papadimitriou and M. Yannakakis, Optimization, approximation, and complexity classes, *Journal of Computer and System Sciences*, **43** (1991), 425–440.
17. D. Ratner and M. Warmuth, Finding a shortest solution for the $(N \times N)$-extension of the 15-puzzle is intractable, *Journal of Symbolic Computation*, **10** (1990), 111–137.
18. W. E. Story, Notes on the 15 puzzle. II., *American Journal of Mathematics*, **2** (1879), 399–404.
19. R. M. Wilson, Graph puzzles, homotopy, and the alternating group, *Journal of Combinatorial Theory, Series B*, **16** (1974), 86–96.

$\mathcal{C}$-Varieties, Actions and Wreath Product

Laura Chaubard

LIAFA, Université Paris VII and CNRS, Case 7014,
2 Place Jussieu, 75251 Paris Cedex 05, France
Laura.Chaubard@liafa.jussieu.fr

Abstract. Motivated by open problems in language theory, logic and circuit complexity, Straubing generalized Eilenberg's variety theory, introducing the $\mathcal{C}$-varieties. As a further contribution to this theory, this paper first studies a new $\mathcal{C}$-variety of languages, lying somewhere between star-free and regular languages. Then, continuing the early works of Esik-Ito, we extend the wreath product to $\mathcal{C}$-varieties and generalize the wreath product principle, a powerful tool originally designed by Straubing for varieties. We use it to derive a characterization of the operations $L \to LaA^*$ and $L \to La$ on languages. Finally, we investigate the decidability of the operation $\mathbf{V} \to \mathbf{V} * \mathbf{LI}$ (the wreath product by locally trivial semigroups) and solve it explicitely in several non-trivial cases.

1 Introduction

Algebraic methods in automata theory were introduced soon after Kleene's seminal paper on automata, but became fully recognized after the pioneering works of Schützenberger, Eilenberg, Simon, Brzozowski-Simon and McNaughton (see [2, 10]). In particular, Eilenberg's theory of varieties gives an appealing framework to study classes of recognizable languages closed under Boolean operations, quotients, and inverse morphisms. The success of this algebraic approach goes far beyond automata and formal languages. It sheds a new light on an increasing number of research fields, including model theory and logic, circuits, communication complexity, discrete dynamic systems, etc. Naturally, along with this enlargment of the scope, comes a strong need for theoretical developments.

Motivated by open problems in language theory as well as in logic and circuit complexity [16], Straubing [17] recently introduced the notion of $\mathcal{C}$-varieties. A similar notion was introduced independently by Esik and Ito [4]. The formal definition of a $\mathcal{C}$-variety of languages is quite similar to Eilenberg's except that it only requires closure under inverse of morphisms belonging to some natural class $\mathcal{C}$. Typically, this class $\mathcal{C}$ can be the class of all length-preserving morphisms, of all length-multiplying morphisms, of all non-erasing morphisms, etc. The main advantage of this latter approach is to cover families of languages that could not be studied in Eilenberg's theory. Examples include languages occurring in circuit complexity, temporal logic [3, 4, 5], and languages of generalized star height $\leq n$ for a given n. On the algebraic side, Eilenberg considered finite semigroups or

J.R. Correa, A. Hevia, and M. Kiwi (Eds.): LATIN 2006, LNCS 3887, pp. 274–285, 2006.

monoids, while Straubing considers *stamps*, which are surjective morphisms from a free monoid onto a finite monoid.

Although this new framework offers a theoretical correspondence between stamps and languages, only few examples have been so far studied on both aspects. Our first result is a new example of $\mathcal{C}$-variety, lying between star-free and rational languages, which admits a natural algebraic description.

The new-born theory of $\mathcal{C}$-varieties also badly needs algebraic tools. Early papers by Kunc [7], Pin-Straubing [11], and Esik-Ito [4] have already shown the way by introducing the equational theory for $\mathcal{C}$-varieties, the Mal'cev product and the cascade product. This paper is a further contribution to the theory. Elaborating on [4], we propose a more general definition of the wreath product and an extended version of the Wreath Product Principle (WPP), a powerful tool originally designed by Straubing [14, 12] for Eilenberg varieties.

The classical WPP provides a description of the languages recognized by a wreath product of monoids. This looks like a rather technical result but it has far-reaching consequences. To give a few examples, it can be used to characterize the languages expressible by first order formulas of Büchi's sequential calculus or by various fragments of temporal logic [17]. Here, we use our extended version to derive a characterization of the operations $L \to LaA^*$ and $L \to La$ on languages, extending a result of [10]. These operations play an important role in the study of the so-called polynomial operations, but are also used in linear temporal logic.

Next, we consider a particular instance of wreath product. Let us first describe its algebraic background. Recall that in a semigroup S, an element e is *idempotent* if $e^2 = e$. Idempotents play a crucial role in semigroup theory. In particular, the set $eSe = \{ese \mid s \in S\}$ is a subsemigroup of S called the *local semigroup at* e. We are interested in the class called **LI** (which stands for "*locally trivial*") of finite semigroups whose local semigroups are all trivial (that is, $eSe = \{e\}$ for each idempotent e). The operation $\mathbf{V} \to \mathbf{V} * \mathbf{LI}$ (where $*$ denotes the wreath product) is essential in Eilenberg's theory since several major results in language theory [1, 10, 12, 18] boil down to particular instances of this operation. It is somewhat connected with Hanf's locality lemma in model theory [6] and has also motivated pure semigroup developments, culminating in the proof of the Delay Theorem [15, 19]. A crucial consequence of this latter result is that if $\mathbf{V}$ is a decidable variety of finite semigroups then $\mathbf{V} * \mathbf{LI}$ is also decidable. Extending the Delay Theorem to $\mathcal{C}$-varieties appears to be a challenging problem. In this paper, we offer a positive solution in a few nontrivial particular cases.

Our approach to the WPP led us to modify the ground algebraic objects and to consider *actions*, which are very close to automata, rather than stamps. It can be shown that the two approaches are essentially equivalent [4]. However, the use of actions makes the definition of the wreath product far more transparent.

Our paper is organised as follows. Definitions and standard results on varieties and $\mathcal{C}$-varieties are presented in Section 2. Section 3 characterizes thoroughly the lp-variety defined by the single identity $a^\omega = a^{\omega+1}$. Section 4.1 introduces sequential and wreath products, whereas the wreath product principle itself is stated in Section 4.2. Section 5 gives a characterization of the operations

$L \to LaA^*$ and $L \to La$ on languages. Section 6 describes languages in $\mathbf{V} * \mathbf{LI}$ and finally Section 7 gathers examples of $\mathcal{C}$-varieties of the form $\mathbf{V} * \mathbf{LI}$ where $\mathbf{V}$ is defined by a single identity.

2 Varieties

2.1 Eilenberg's Variety Theory

Varieties of Finite Monoids. A (finite) monoid is a finite set equipped with a binary associative operation and a unit element. Given two monoids M and N, a *monoid morphism* is a map $\varphi : M \to N$ satisfying $\varphi(1) = 1$ and $\varphi(uv) = \varphi(u)\varphi(v)$ for all u, v in M. A monoid M is a *submonoid* of a monoid T if there exists an injective morphism $\varphi : M \to T$. A monoid N is a *quotient* of a monoid M if there exists an onto morphism $\varphi : M \to N$. A monoid N *divides* a monoid T if N is a quotient of a submonoid of T. The *product* of two monoids M_1 and M_2 is the set $M_1 \times M_2$ equipped with the product $(x_1, x_2)(y_1, y_2) = (x_1y_1, x_2y_2)$. A *variety of (finite) monoids* is a class of finite monoids closed under division and finite product. Throughout the paper, all monoids are either finite or free.

Varieties of Languages. Monoids can be seen as language recognizers in the following way. Given a finite alphabet A, a language $L \subseteq A^*$ is *recognized by* a monoid M if there exist a subset $F \subseteq M$ and a morphism $\varphi : A^* \to M$ such that $L = \varphi^{-1}(F)$. The *syntactic monoid* of a language L is the quotient $A^*/{\sim_L}$ where the congruence $\sim_L$ is defined on A^* by $u \sim_L v$ iff for all x, y in $A^*, xuy \in L \Leftrightarrow xvy \in L$. It is the smallest monoid that recognizes L. A language is said to be *recognizable* if it is recognized by some finite monoid, or equivalently, if its syntactic monoid is finite. The famous Kleene theorem asserts that recognizable and regular languages coincide. A *Boolean algebra* is a class of languages that is closed under finite union, finite intersection, and complement. A *class of recognizable languages* $\mathcal{V}$ assigns to each finite alphabet A a set $\mathcal{V}(A^*)$ of recognizable languages of A^*. A *variety of languages* is a class of recognizable languages $\mathcal{V}$ such that

(1) for every alphabet A, $\mathcal{V}(A^*)$ is a Boolean algebra,
(2) if $L \in \mathcal{V}(A^*)$ and $a \in A$ then $a^{-1}L$ and La^{-1} are in $\mathcal{V}(A^*)$,
(3) if $\varphi\colon A^* \to B^*$ is a morphism, $L \in \mathcal{V}(B^*)$ implies $\varphi^{-1}(L) \in \mathcal{V}(A^*)$.

Eilenberg's Theorem. Given a variety of finite monoids $\mathbf{V}$, the class $\mathcal{V}$ of all languages recognized by a monoid in $\mathbf{V}$ is a variety of languages. Eilenberg's theorem asserts that the correspondence $\mathbf{V} \to \mathcal{V}$ is one-to-one and onto [2].

Eilenberg developed a similar theory for varieties of finite semigroups. In this case, the corresponding languages never contain the empty word and condition (3) is restricted to non-erasing morphisms. The theory of $\mathcal{C}$-varieties enables to unify these two notions.

Example 1. Commutative monoids in which all elements are idempotent form the variety $\mathbf{J_1}$. The corresponding variety of languages $\mathcal{J}_1$ is such that for each alphabet A, $\mathcal{J}_1(A^*)$ is the Boolean algebra generated by the languages B^*, where $B \subseteq A$.

Example 2. **R** is the variety of all monoids in which right division is a partial order. In other words, a monoid M is in **R** iff for all x, u, v, y in M, $x = yu$ and $y = xv$ imply $x = y$. For each alphabet A, $\mathcal{R}(A^*)$ is the smallest Boolean algebra closed under the operation $L \to LaA^*$, for each a in A.

Example 3. **LI** is the variety of *locally trivial* semigroups. In other words, S is in **LI** iff $eSe = \{e\}$ for each idempotent e. The associated class of languages are the *prefix-suffix testable languages*, that is, finite unions of finite languages or languages of the form $pA^* \cap A^*s$ with p, s in A^*.

2.2 $\mathcal{C}$-Varieties

Actions. $\mathcal{C}$-varieties were originally defined with stamps, but *actions* are more appropriate to the purpose of this paper. Since both theories are essentially equivalent [4], we adopt the point of view of actions here and refer the reader to [17, 11] for a presentation of stamps. Let P be a finite set and let A be a finite alphabet. A (*right*) *action* from A on P is a map $P \times A \to P$, denoted $(p, a) \mapsto p \cdot a$. An action (P, A) can therefore be viewed as the transition function of a deterministic finite automaton (DFA) with set of states P on the alphabet A. The action of A on P can be recursively extended into an action of A^* on P by setting $p \cdot 1 = p$ and $p \cdot ua = (p \cdot u) \cdot a$, for all $p \in P$, $u \in A^*$ and $a \in A$. Let $\mu : A^* \to P^P$ be the function which maps the word u onto the transformation $p \mapsto p \cdot u$. The set $\mu(A^*)$ (resp. $\mu(A^+)$) is called the *transformation monoid* (resp. *transformation semigroup*) of the action (P, A). A *constant action* on the alphabet A is an action (P, A) such that $p \cdot a = p$, for each $p \in P$ and $a \in A$.

$\mathcal{C}$-Varieties of Actions. Straubing [17] generalized the variety theory in order to study classes of recognizable languages that are not varieties in Eilenberg's sense. His definition involves classes of morphisms between free monoids, closed under composition and containing all length-preserving morphisms. Examples include the classes of all *length-preserving* morphisms (the image of each letter is a letter), of all *length-multiplying* morphisms (for some integer k, the length of the image of a word is k times the length of the word), all *non-erasing* morphisms (the image of each letter is a nonempty word), all *length-decreasing* morphisms (the image of each letter has length at most 1) and all morphisms.

Let $\mathcal{C}$ be a fixed class of morphisms satisfying the above conditions. An action (P, A) $\mathcal{C}$-*divides* an action (Q, B) if there is a pair (η, f) where $f \colon A^* \to B^*$ is in $\mathcal{C}$ and $\eta : Q \to P$ is a surjective partial function such that for each $q \in \mathrm{Dom}(\eta)$ and each $a \in A$, $\eta(q) \cdot a = \eta(q \cdot f(a))$. Such a pair (η, f) is called a $\mathcal{C}$-*division*. The *(restricted) product* of two actions (P_1, A) and (P_2, A) is the action $(P_1 \times P_2, A)$ defined by $(p_1, p_2) \cdot a = (p_1 \cdot a, p_2 \cdot a)$. A $\mathcal{C}$-*variety of actions* is a class of actions containing all constant actions and closed under $\mathcal{C}$-division and product. When $\mathcal{C}$ is the class of all (resp. length-preserving, length-multiplying, non-erasing) morphisms, we use the term *all-variety* (resp. *lp-variety*, *lm-variety*, *ne-variety*). Notice that if $\mathcal{C}' \subseteq \mathcal{C}$, each $\mathcal{C}$-variety of actions is also a $\mathcal{C}'$-variety of actions.

$\mathcal{C}$-Varieties of Languages. A $\mathcal{C}$-variety of languages shares the same defining properties as a variety of languages except that Condition (3) is weakened to:

(3′) If $\varphi\colon A^* \to B^*$ is a morphism of $\mathcal{C}$, $L \in \mathcal{V}(B^*)$ implies $\varphi^{-1}(L) \in \mathcal{V}(A^*)$.

A language $L \subseteq A^*$ is *recognized by an action* (Q, A) if there exist an *initial state* q_0 in Q and a set of *final states* $F \subseteq Q$ such that $L = \{u \in A^* \mid q_0 \cdot u \in F\}$.

Straubing's Theorem. Given a $\mathcal{C}$-variety of actions $\mathbf{W}$, we denote by $\mathcal{W}(A^*)$ the set of languages recognized by some action (P, A) in $\mathbf{W}$. Straubing's theorem states that the class $\mathcal{W}$ is a $\mathcal{C}$-variety of languages and the correspondence $\mathbf{W} \to \mathcal{W}$ is one-to-one and onto [17].

As was mentioned before, Eilenberg's varieties can actually be considered as a particular case of $\mathcal{C}$-varieties. Indeed, given a variety of monoids $\mathbf{V}$, the class of actions whose transformation monoid is in $\mathbf{V}$ is an *all*-variety of actions [17]. In the sequel, both varieties will be denoted by $\mathbf{V}$. Likewise, given a variety of semigroups $\mathbf{V}$, the class of actions whose transformation semigroup is in $\mathbf{V}$ is an *ne*-variety of actions.

Example 4. The class of all actions (Q, A) such that for each $a \in A$ and $q \in Q$, $q \cdot a = q \cdot a^2$ is an *lp*-variety of actions. The corresponding languages are the *stutter invariant* languages and appear in connection with temporal logic [8]. Given such a language L, a word uav is in L if and only if $ua^+v \subseteq L$.

Example 5. Let **MOD** be the class of all actions (Q, A) whose transition monoid is a cyclic group and such that for $a, b \in A$ and $q \in Q$, $q \cdot a = q \cdot b$. Then **MOD** is an *lm*-variety. The corresponding *lm*-variety of languages $\mathcal{M}od$ satisfies that for each alphabet A, $\mathcal{M}od(A^*)$ is the Boolean algebra generated by the languages of the form $A^k(A^n)^*$, for $0 \leq k < n$.

2.3 Identities

Both varieties of finite monoids and $\mathcal{C}$-varieties of actions have an equational characterization [13, 7, 11]. The formal definition of identities requires the introduction of profinite topologies. For the sake of simplicity, we shall only consider a weaker notion on a few basic examples, which will be sufficient for our purpose.

An *ω-term* is built from letters of an alphabet using the usual concatenation product and a unary operator ω. Thus abc, a^ω and $(ab^\omega c)^\omega ab$ are examples of ω-terms. Given an ω-term t on the alphabet B, an action (Q, A) and a morphism $f\colon B^* \to A^*$, the action of $f(t)$ on Q is defined recursively as follows. Let $q \in Q$. If t is a letter, then $f(t)$ is a word, and $q \cdot f(t)$ has its usual value. If t and t' are ω-terms, $q \cdot f(tt') = (q \cdot f(t)) \cdot f(t')$. If $t = u^\omega$, then $q \cdot f(t) = q \cdot f(u)^n$, where n is the least integer such that $q \cdot f(u)^n = q \cdot f(u)^{2n}$. In other words, the map from Q to Q defined by u^ω is the unique idempotent power of the map from Q to Q defined by u.

Let u, v be ω-terms on a finite alphabet B. An action (Q, A) is said to *satisfy the identity $u = v$ with respect to $\mathcal{C}$* if, for every morphism $f\colon B^* \to A^*$ in $\mathcal{C}$ and each q in Q, $q \cdot f(u) = q \cdot f(v)$. A $\mathcal{C}$-variety $\mathbf{V}$ satisfies a given identity if every action in $\mathbf{V}$ satisfies this identity with respect to $\mathcal{C}$. The class of all actions satisfying the identity $u = v$ (wrt $\mathcal{C}$) is a $\mathcal{C}$-variety denoted by $[u = v]$. A language $L \subseteq A^*$ *satisfies the identity* $u = v$ wrt $\mathcal{C}$ if for every $\mathcal{C}$-morphism $f\colon B^* \to A^*$, $f(u) \sim_L f(v)$. Note that the interpretation of the symbols occurring in the identities depends on the choice of $\mathcal{C}$. For instance, if $\mathcal{C}$ is the class of all morphisms, symbols should be interpreted as words, as in the classical Eilenberg's setting. For *lp*-varieties, on the contrary, they should be interpreted as letters, while for *lm*-varieties they stand for words of the same length.

Example 6. We can now describe all examples given before by identities. Indeed, $\mathbf{J_1}$ is the *all*-variety satisfying $x = x^2$ and $xy = yx$. Assuming $\mathcal{C} = lp$, the stutter-invariant variety is defined by $a = a^2$, whereas **MOD** is defined by the identity $a^{\omega-1}b = 1$.

Example 7. For an integer k, the *ne*-variety of actions $\mathbf{LI}_k$ (respectively $\mathbf{D}_k$) is defined by the identity $x_1 \cdots x_k y x_1 \cdots x_k = x_1 \cdots x_k$ (respectively $y x_1 \cdots x_k = x_1 \cdots x_k$). Thus $\mathbf{D}_k \subset \mathbf{LI}_k$. The languages corresponding to $\mathbf{LI}_k$ are the *prefix-suffix k-testable* languages, which are finite unions of languages of the form $\{u\}$ with $|u| < k$, or $pA^* \cap A^*s$ with $|p| = |s| = k$.

3 The *lp*-Variety $[a^\omega = a^{\omega+1}]$

In this section we describe the *lp*-variety of languages $\mathcal{U}$ associated to the *lp*-variety $[a^\omega = a^{\omega+1}]$.

Given a deterministic automaton $\mathcal{A} = (Q, A, E, i, F)$, and a letter $a \in A$, a path $(q_0, a, q_1)(q_1, a, q_2) \cdots (q_n, a, q_0)$ is a *letter-counter* if $n > 0$ and for $0 < i \leq n$, $q_i \neq q_0$. An automaton is said to be *letter-counter free* if it does not contain any letter-counter. The following lemma is straightforward.

Lemma 1. *A language is in $\mathcal{U}(A^*)$ if and only if its minimal (deterministic) automaton is letter-counter free.*

Recall that a language $X \subseteq A^*$ is a *code* if X^* is free. The language X is a *prefix code* if, for all words u, v in A^*, u and uv in X^* implies v in X^*. In accordance with the terminology *pure codes* [9], a code will be said to be *letter-pure* if it contains no word of the form a^n with a in A and $n > 1$. The identity $a^\omega = a^{\omega+1}$ may be reminiscent of the identity defining aperiodic semigroups, but our description of the languages is rather similar to Kleene theorem.

Theorem 1. *The family $\mathcal{U}(A^*)$ is the smallest family of languages containing the singletons and closed under finite union, product and star operation restricted to letter-pure prefix codes.*

4 Wreath Product Principle

4.1 Sequential Products and Wreath Products

The sequential product of actions corresponds to the notion of cascade product of automata in the work of Esik and Ito [4]. It mimics the composition on actions.

The *sequential product* of two actions $(P, Q \times A)$ and (Q, A), denoted by $(P, Q \times A) \circ (Q, A)$, is the action $(P \times Q, A)$ defined by $(p, q) \cdot a = (p \cdot (q, a), q \cdot a)$. Observe that for $u = a_1 \cdots a_n$, $(p, q) \cdot u = \big(p \cdot (q, a_1)(q \cdot a_1, a_2) \cdots (q \cdot a_1 \cdots a_{n-1}, a_n), q \cdot u\big)$.

The sequential product can be extended to $\mathcal{C}$-varieties of actions. In accordance with the traditional terminology on varieties, we call it the *wreath product*. Let $\mathcal{C}$ be a class of morphisms and let $\mathbf{V}$ and $\mathbf{W}$ be two $\mathcal{C}$-varieties. A $(\mathbf{V}, \mathbf{W})$-*sequential product* is an action of the form $(P, Q \times A) \circ (Q, A)$ with $(P, Q \times A)$ in $\mathbf{V}$ and (Q, A) in $\mathbf{W}$. We define $\mathbf{V} * \mathbf{W}$ to be the class of all actions that $\mathcal{C}$-divide a $(\mathbf{V}, \mathbf{W})$-sequential product.

For technical reasons, we restrict ourselves to some specific classes of morphims, that nevertheless include all classical examples.

Definition 1. *A class $\mathcal{C}$ of morphisms between finitely generated free monoids is called* convenient *if it is closed under composition, contains all length-preserving morphisms, and satisfies that membership of a morphism $f : A^* \to B^*$ in $\mathcal{C}$ depends only on the set of integers $\{|f(a)| \mid a \in A\}$.*

We did not find any natural example of nonconvenient classes. The simplest, though already quite artificial, example of a nonconvenient class we can think of is the class of all morphisms $\varphi : A^* \to B^*$ such that, for each letter a in A, there exists b in B such that $f(a) \in b^+$.

Although the definition of the wreath product depends on the class $\mathcal{C}$, the following proposition shows that one can write $\mathbf{V} * \mathbf{W}$ without referring to $\mathcal{C}$, as soon as $\mathcal{C}$ is convenient. Indeed, in this case $\mathbf{V} * \mathbf{W}$ appears to be the class of all actions that *lp*-divide a $(\mathbf{V}, \mathbf{W})$-sequential product.

Proposition 1. *Let $\mathcal{C}$ be a convenient class of morphisms and let $\mathbf{V}$ and $\mathbf{W}$ be two $\mathcal{C}$-varieties. An action (P, A) is in $\mathbf{V} * \mathbf{W}$ if and only if there exist a $(\mathbf{V}, \mathbf{W})$-sequential product $(T, Q \times A) \circ (Q, A)$ and a lp-division (η, Id_{A^*}) from (P, A) into $(T, Q \times A) \circ (Q, A)$.*

Proof. Let (P, A) be an action in $\mathbf{V} * \mathbf{W}$. By definition, there exist a $(\mathbf{V}, \mathbf{W})$-sequential product $(T, Q \times B) \circ (Q, B)$ and a $\mathcal{C}$-division (η, f) from (P, A) into $(T, Q \times B) \circ (Q, B)$. Define a morphism $g \colon (Q \times A)^* \to (Q \times B)^*$ by $g(q, a) = (q, b_1)(q \cdot b_1, b_2) \cdots (q \cdot b_1 \cdots b_{k-1}, b_k)$, where $b_1 \cdots b_k = f(a)$. Define an action $(T, Q \times A)$ by $t \cdot (q, a) = t \cdot g(q, a)$. The map g is in $\mathcal{C}$ because f is in $\mathcal{C}$ and $\mathcal{C}$ is convenient. The pair (Id_T, g) is a $\mathcal{C}$-division from $(T, Q \times A)$ into $(T, Q \times B)$ and thus $(T, Q \times A)$ is in $\mathbf{V}$. In the same way, define an action (Q, A) by setting $q \cdot a = q \cdot f(a)$. This action is in $\mathbf{W}$ since (Id_Q, f) is a $\mathcal{C}$-division from (Q, A) into (Q, B). Consider now the $(\mathbf{V}, \mathbf{W})$-sequential product $S = (T, Q \times A) \circ (Q, A)$. One verifies easily that (η, Id_{A^*}) is a $\mathcal{C}$-division from (P, A) into S. □

Assuming $\mathcal{C}$ convenient leads to the following "convenient" properties of $\mathbf{V} * \mathbf{W}$.

Theorem 2. *Let $\mathcal{C}$ be a convenient class of morphisms and let $\mathbf{V}, \mathbf{W}$ be two $\mathcal{C}$-varieties of actions. Then, the class $\mathbf{V} * \mathbf{W}$ is a $\mathcal{C}$-variety of actions that contains $\mathbf{W}$. Further, the wreath product is an associative operation on $\mathcal{C}$-varieties of actions which extends the classical wreath product on Eilenberg's varieties.*

4.2 Wreath Product Principle

The wreath product principle gives a description of languages recognized by an action of $\mathbf{V} * \mathbf{W}$. Proposition 1 enables us to readily extend the results of [4] on lp-varieties to all $\mathcal{C}$-varieties, for $\mathcal{C}$ convenient. Therefore, from now on we assume that $\mathcal{C}$ is convenient.

Recall that a *transducer* is a 7-tuple $\mathcal{T} = (Q, A, B, q_0, \cdot, *, F)$ where $\mathcal{A} = (Q, A, q_0, \cdot, F)$ is a DFA, B is a finite alphabet called the *output alphabet*, and $(q, a) \mapsto q * a \in B^*$ is called the *output function.* The function *realized* by the transducer $\mathcal{T}$ is the partial function $\sigma : A^* \to B^*$ defined by $\sigma(u) = q_0 * u$, for each word u accepted by $\mathcal{A}$. The transducer $\mathcal{T}$ is a $\mathcal{C}$*-transducer* if the output morphism $(q, a) \mapsto q{*}a$ belongs to $\mathcal{C}$. A $\mathcal{C}$*-sequential function* is a partial function that can be realized by a $\mathcal{C}$-transducer. Notice that if $\mathcal{C} = lp$, an lp-transducer is just a Mealy automaton. The following proposition illustrates the natural links between sequential products and $\mathcal{C}$-sequential functions.

Proposition 2. *Let $\mathbf{V}, \mathbf{W}$ be $\mathcal{C}$-varieties of actions. Let $\mathcal{V}$ (resp. $\mathcal{U}$) be the $\mathcal{C}$-variety of languages associated to $\mathbf{V}$ (resp. $\mathbf{V} * \mathbf{W}$). Then, if L is a language of $\mathcal{V}(B^*)$ and $\sigma\colon A^* \to B^*$ is a sequential function realized by a transducer $\mathcal{T}$ (denoted as above) whose input function $(q, a) \mapsto q \cdot a$ is an action of $\mathbf{W}$, then $\sigma^{-1}(L)$ is in $\mathcal{U}(A^*)$.*

We now focus on specific lp-sequential functions in order to state the WPP. Given an action (Q, A) and $q_0 \in Q$, define the function $\sigma_{q_0} : A^* \to (Q \times A)^*$ by

$$\sigma_{q_0}(a_1 \cdots a_n) = (q_0, a_1)(q_0 \cdot a_1, a_2) \cdots (q_0 \cdot a_1 \cdots a_{n-1}, a_n) \qquad (*)$$

The function σ_{q_0} is realized by a Mealy automaton with initial state q_0, input action (Q, A), output function defined by $q * a = (q, a)$ and all states final. Now, a sequential function σ is said to be *associated with* (Q, A) if $\sigma = \sigma_q$ for some q in Q. Recall that a *positive Boolean algebra* on A^* is a set of languages of A^* that is closed under finite intersection and finite union.

Proposition 3. (WPP) Let $\mathcal{U}$ be the $\mathcal{C}$-variety of languages associated with $\mathbf{V} * \mathbf{W}$.

(1) For each alphabet A, $\mathcal{U}(A^*)$ is the smallest positive Boolean algebra containing $\mathcal{W}(A^*)$ and the languages of the form $\sigma^{-1}(V)$, where σ is associated with an action (Q, A) in $\mathbf{W}$ and V is in $\mathcal{V}\big((Q \times A)^*\big)$.

(2) Each language in $\mathcal{U}(A^*)$ is a finite union of languages of the form $W \cap \sigma^{-1}(V)$ where W is in $\mathcal{W}(A^*)$, σ is associated with an action (Q, A) in $\mathbf{W}$ and V is in $\mathcal{V}\big((Q \times A)^*\big)$.

We now specialize the WPP.

5 Operations on Languages

5.1 The Operation $L \to LaA^*$

We extend here a standard result of [10]. Consider the *all*-variety $\mathbf{J}_1$ of actions whose transition monoid is idempotent and commutative.

Proposition 4. *Let* $\mathbf{V}$ *be a* $\mathcal{C}$*-variety of actions, and let* $\mathcal{V}$ *(resp.* $\mathcal{U}$*) be the* $\mathcal{C}$*-variety of languages corresponding to* $\mathbf{V}$ *(resp.* $\mathbf{J}_1 * \mathbf{V}$*). Then,* $\mathcal{U}(A^*)$ *is the Boolean algebra generated by the languages* L *and* LaA^* *with* L *in* $\mathcal{V}(A^*)$*,* $a \in A$*.*

Now, since $\mathbf{R}$ is the closure of $\mathbf{J}_1$ under wreath product, the following holds.

Corollary 1. *Let* $\mathbf{V}$ *be a* $\mathcal{C}$*-variety of actions, and let* $\mathcal{V}$ *(resp.* $\mathcal{U}$*) be the* $\mathcal{C}$*-variety of languages corresponding to* $\mathbf{V}$ *(resp.* $\mathbf{R} * \mathbf{V}$*). Then,* $\mathcal{U}(A^*)$ *is the smallest Boolean algebra containing* $\mathcal{V}(A^*)$ *and closed under the operation* $L \mapsto LaA^*$*.*

5.2 The Operation $L \to La$

Recall that $\mathbf{D}_1$ is an *ne*-variety of actions defined by the identity $xy = y$. We give here a characterization of languages corresponding to the $\mathcal{C}$-variety $\mathbf{D}_1 * \mathbf{V}$, extending a result of [15].

Proposition 5. *Let* $\mathbf{V}$ *be a* $\mathcal{C}$*-variety of actions, where* $\mathcal{C} \subseteq ne$*. Let* $\mathcal{V}$ *(resp.* $\mathcal{U}$*) be the* $\mathcal{C}$*-variety of languages corresponding to* $\mathbf{V}$ *(resp.* $\mathbf{D}_1 * \mathbf{V}$*). Then, for each alphabet* A*,* $\mathcal{U}(A^*)$ *is the Boolean algebra generated by the languages* L *and* La *with* L *in* $\mathcal{V}(A^*)$ *and* a *in* A*.*

Since $\mathbf{D} = \bigcup_k \mathbf{D}_k$ is the closure of $\mathbf{D}_1$ under wreath product, the following holds.

Corollary 2. *Let* $\mathbf{V}$ *be a* $\mathcal{C}$*-variety of actions, where* $\mathcal{C} \subseteq ne$*. Let* $\mathcal{V}$ *(resp.* $\mathcal{U}$*) be the* $\mathcal{C}$*-variety of languages corresponding to* $\mathbf{V}$ *(resp.* $\mathbf{D} * \mathbf{V}$*). Then, for any alphabet* A*,* $\mathcal{U}(A^*)$ *is the smallest Boolean algebra containing* $\mathcal{V}(A^*)$ *and closed under the operation* $L \mapsto Lu$*, for each word* u *in* A^**.*

6 Wreath Products of the Form $\mathbf{V} * \mathbf{LI}_k$

This section points out differences between $\mathcal{C}$-varieties and varieties of monoids. As an example, given a $\mathcal{C}$-variety $\mathbf{V}$, it is not true in general that $\mathbf{V} * \mathbf{LI}_k = \mathbf{V} * \mathbf{D}_k$, whereas it holds for non-trivial varieties of semigroups. For instance, if $\mathcal{C} = lp$, $ba_1 \cdots a_k = ca_1 \cdots a_k$ is an identity of $[a = b] * \mathbf{D}_k$ but not of $\mathbf{LI}_k$. Besides, the standard characterization of $\mathbf{V} * \mathbf{LI}_k$ is still a necessary condition but there is no evidence that it should also remain sufficient, because the "spelling function" used is *not* non-erasing, preventing us from using Proposition 2 . This explains some of the obstacles that arise when exploring wreath products of the form $\mathbf{V} * \mathbf{LI}_k$.

We now introduce the "spelling function" δ_k. Let A be an alphabet and for each $k > 0$, let $C_k = A^k$. The function $\delta_k \colon A^+ \to C_k^*$ is defined on $A^{k-1}A^*$ by:

$$\delta_k(a_1 \cdots a_n) = \begin{cases} 1 & \text{if } n = k-1 \\ [a_1 \cdots a_k][a_2 \cdots a_{k+1}] \cdots [a_{n-k+1} \cdots a_n] & \text{if } n \geq k \end{cases}$$

Theorem 3. *Let* $\mathbf{V}$ *be a* $\mathcal{C}$*-variety of actions, where* $\mathcal{C} \subseteq ne$ *and let* A *an alphabet. Denote by* $\mathcal{V}$ *(resp.* $\mathcal{U}$*) the* $\mathcal{C}$*-variety of languages corresponding to* $\mathbf{V}$ *(resp.* $\mathbf{V} * \mathbf{LI}_k$*). Then, any language in* $\mathcal{U}(A^*)$ *is a finite union of languages of the form* $\{u\}$ *with* $|u| < k$ *or* $pA^* \cap \delta_{k+1}^{-1}(K) \cap A^*s$*, where* $p, s \in A^k$ *and* K *is in* $\mathcal{V}(C_{k+1}^*)$*.*

Sketch of the proof. By Proposition 3, a language L in $\mathcal{U}(A^*)$ is a finite union of languages of the form $W \cap \sigma^{-1}(V)$, where W is prefix-suffix k-testable, σ is associated with an action (Q, A) in $\mathbf{LI}_k$, and V is in $\mathcal{V}((Q \times A)^*)$. Let $q \in Q$ and denote by σ the lp-sequential function asociated with q and the action (Q, A) (see Formula $(*)$). Define a length-preserving morphism α_q from C^*_{k+1} into $(Q \times A)^*$ by setting $\alpha_q([a_1 \cdots a_{k+1}]) = (q \cdot a_1 \cdots a_k, a_{k+1})$. Let $u = a_1 \cdots a_n$ be a word of length $n \geq k$ and set $p = a_1 \cdots a_k$. Since (Q, A) is in $\mathbf{LI}_k$, for each $i \geq k$ we have $q \cdot a_1 \cdots a_i = q \cdot pa_{i-k+1} \cdots a_i$. Thus, for each $i \geq k$, $(q \cdot a_1 \cdots a_i, a_{i+1}) = \alpha_{q \cdot p}([a_{i-k+1} \cdots a_{i+1}])$ and finally, $\sigma(u) = \sigma(p)\alpha_{q \cdot p}(\delta_{k+1}(u))$. Therefore, $u \in \sigma^{-1}(V) \cap pA^* \Leftrightarrow u \in \delta^{-1}_{k+1}(K_p) \cap pA^*$, with $K_p = \alpha^{-1}_{q \cdot p}(\sigma(p)^{-1}V)$. One then verifies that K_p is in $\mathcal{V}(C^*_{k+1})$. Finally, denoting by F the finite set of words of length $< k$ in $\sigma^{-1}(V)$, the following suffices to conclude:

$$\sigma^{-1}(V) = F \cup \Big(\bigcup_{p \in A^k} \sigma^{-1}(V) \cap pA^* \Big) = F \cup \Big(\bigcup_{p \in A^k} \delta^{-1}_{k+1}(K_p) \cap pA^* \Big). \qquad \square$$

7 Wreath Products of the Form V ∗ LI

As in the case of semigroups, $\mathbf{V} * \mathbf{LI} = \bigcup_{k>0} \mathbf{V} * \mathbf{LI}_k$. Thus, Theorem 3 can be extended as follows:

Proposition 6. *Let $\mathcal{U}$ be the $\mathcal{C}$-variety of languages corresponding to $\mathbf{V} * \mathbf{LI}$. Then, $\mathcal{U}(A^*)$ is contained in the smallest Boolean algebra containing the prefix-suffix testable languages and those of the form $\delta_k^{-1}(K)$ for $k > 0$ and $K \in \mathcal{V}(C^*_k)$.*

From now on, assume that $\mathcal{C} = lp$. We will use small capital letters "$\textsc{a}, \textsc{b}, \textsc{u}, \textsc{v}, \textsc{x}$" to denote letters of C_k and words of C^*_k. Let $\textsc{a} = [a_1 \cdots a_k]$, $\textsc{b} = [b_1 \cdots b_k]$ be two letters in C_k. The transition $\textsc{ab}$ is said to be *correct* if and only if $a_2 \cdots a_k = b_1 \cdots b_{k-1}$. It is *incorrect* otherwise. Finally, let P_k be the set $\{[a^k] \mid a \in A\}$.

7.1 The lp-Variety $[a = a^2] * \mathbf{LI}$

Theorem 4. $[a = a^2] * \mathbf{LI} = [a^\omega = a^{\omega+1}]$.

Sketch of the proof. Let $\mathcal{V}$ (resp. $\mathcal{U}$, $\mathcal{W}$) be the lp-variety of languages associated with $[a = a^2]$ (resp. $[a = a^2] * \mathbf{LI}$, $[a^\omega = a^{\omega+1}]$) and let A be an alphabet. If $A = \{a\}$, the result is trivial, so assume that $|A| \geq 2$. By Proposition 6, $\mathcal{U}(A^*)$ is included in the smallest Boolean algebra $\mathcal{B}$ containing the prefix-suffix testable languages and those of the form $\delta_k^{-1}(K)$ where $K \in \mathcal{V}(C^*_k)$ and $k > 0$. One verifies that $\mathcal{W}$ contains $\mathcal{B}$, which yields $\mathcal{U}(A^*) \subseteq \mathcal{B} \subseteq \mathcal{W}(A^*)$. Conversely, let L in $\mathcal{W}(A^*)$. Since L satisfies $a^\omega = a^{\omega+1}$, there exists an integer $k \geq 2$ such that L satisfies $a^k = a^{k+1}$. Let p be a word of A^k. We define an lp-sequential function $\sigma_p \colon A^* \to C^*_{k+1}$ by $\sigma_p(u) = \delta_{k+1}(pu)$. Note that σ_p is injective and that its range consists of words with no incorrect transition. Further, σ_p can be realized by a Mealy automaton whose transition semigroup is in $\mathbf{D}_k$. Now, denote by K_p the language $\sigma_p(L)$ and let $\tau \colon C_{k+1} \to C^*_{k+1}$ be the rational substitution defined, for each letter $\textsc{b}$ in C_{k+1}, by $\tau(\textsc{b}) = \textsc{b}^+$. The rest of the proof relies on the two following lemmas, whose technical proofs we omit.

Lemma 2. *Let $p = p_1 \cdots p_k$ be a word in A^k such that $p_{k-1} \neq p_k$. Then the language $\tau(K_p)$ satisfies the identity $a = a^2$.*

Lemma 3. *Let $u = u_1 \cdots u_n$ be a word in A^* and let $p = p_1 \cdots p_k$ be a word in A^k such that $p_k \neq u_1$. If $\sigma_p(u)$ is in $\tau(K_p)$, then u is in L.*

Let $\mathcal{F} = \{p_1 \cdots p_k \in A^k \mid p_{k-1} \neq p_k\}$. Elaborating on Lemma 3, one can show that $L = \bigcap_{p \in \mathcal{F}} \sigma_p^{-1}(\tau(K_p))$. Then, by Proposition 2 and Lemma 2, the language $\sigma_p^{-1}(\tau(K_p))$ is in $\mathcal{U}(A^*)$ for each $p \in \mathcal{F}$, and thus L is also in $\mathcal{U}(A^*)$. □

Using the same techniques as in Theorem 4, one can prove additional results.

Proposition 7

(1) *Let n, ℓ be positive integers. Then $[a^n = a^{n+\ell}] * \mathbf{LI} = [a^\omega = a^{\omega+\ell}]$.*

(2) *The lp-variety $[ab^\omega = b^\omega a] * \mathbf{LI}$ is the lp-variety of all actions.*

7.2 Wreath Products of the Form $[u = v] * \mathbf{LI}$

The following theorem enables to treat simple particular cases.

Theorem 5. *Let u and v be words in $\{a, b\}^*$ of respective length n and $n + \ell$. Assume that each of the words u and v has at least three distinct factors of length 2. Then $[u = v] * \mathbf{LI} = [a^\omega = a^{\omega+\ell}]$. In particular, if u and v have the same length, $[u = v] * \mathbf{LI}$ is the lp-variety of all actions.*

Sketch of the proof. Let $B = \{a, b\}$ and let A be a finite alphabet. For a word x in A^* we denote by $F_2(x)$ the set of its factors of length 2.

Lemma 4. *Let k be an integer ≥ 2 and let x be a word in B^* such that $|F_2(x)| \geq 3$. Then for each lp-morphism $\lambda \colon B^* \to C_k^*$, $\lambda(x)$ has no incorrect transition if and only if $\lambda(a) = \lambda(b) \in P_k$.*

Let $\mathcal{V}$ (resp. $\mathcal{U}$, $\mathcal{W}$) be the *lp*-variety of languages associated with $[u = v]$ (resp. $[u = v] * \mathbf{LI}$, $[a^\omega = a^{\omega+\ell}]$). Let $L \in \mathcal{W}(A^*)$ and choose an integer $k \geq 2$ such that for each $c \in A$, $c^k \sim_L c^{k+\ell}$. Let $p = p_1 \cdots p_k$ be a word in A^k such that $p_{k-1} \neq p_k$. As before, define $\sigma_p \colon A^* \to C_{k+1}^*$ by $\sigma_p(x) = \delta_{k+1}(px)$, and set $K_p = \sigma_p(L) = \delta_{k+1}(pL)$. Thanks to Lemma 4, we obtain that for each *lp*-morphism $\lambda \colon B^* \to C_{k+1}^*$, $\lambda(u) \sim_{K_p} \lambda(v)$, which is equivalent to saying that K_p is in $\mathcal{V}(C_{k+1}^*)$. Finally, since σ_p is injective, we have $L = \sigma_p^{-1}(K_p)$, and Proposition 2 enables us to conclude that L is in $\mathcal{U}(A^*)$. The reverse inclusion is trivial since $[u = v] \subseteq [a^n = a^{n+\ell}]$. □

8 Conclusion

This paper first studies thoroughly a new $\mathcal{C}$-variety that admits a natural characterization both algebraically and in terms of languages. It extends the notion of wreath product as well as the characterization of languages associated with a wreath product, and then deals with non-trivial examples of wreath products of the form $[u = v] * \mathbf{LI}$. It would be no surprise if these varieties could be characterized thanks to combinatorial arguments similar to the one used in Theorem 5. Further, a challenging prospect would be to adapt the Delay Theorem to $\mathcal{C}$-varieties. Unfortunately, the classical proof [15] does not carry over to $\mathcal{C}$-varieties because of the restriction made on the class $\mathcal{C}$.

Aknowledgments. I would like to thank Howard Straubing for giving me access to his most recent results on $\mathcal{C}$-varieties. A very special thank to Jean-Éric Pin for his kind advice, valuable ideas and constant support!

References

1. J. A. Brzozowski and I. Simon. Characterizations of locally testable events. *Discrete Math.*, 4:243–271, 1973.
2. S. Eilenberg. *Automata, languages, and machines. Vol. B.* Academic Press, 1976.
3. Z. Ésik. Extended temporal logic on finite words and wreath products of monoids with distinguished generators. *LNCS*, 2450:43–58, 2003.
4. Z. Ésik and M. Ito. Temporal logic with cyclic counting and the degree of aperiodicity of finite automata. *Acta Cybernetica*, 16:1–28, 2003.
5. Z. Ésik and K. G. Larsen. Regular languages defined by Lindström quantifiers. *Theoret. Informatics Appl.*, 37:179–242, 2003.
6. W. Hanf. Model-theoretic methods in the study of elementary logic. In *Theory of Models*, pages 132–145. North-Holland, 1965.
7. M. Kunc. Equational description of pseudovarieties of homomorphisms. *Theoret. Informatics Appl.*, 37:243–254, 2003.
8. D. Peled and T. Wilke. Stutter-invariant temporal properties are expressible without the next-time operator. *Inf. Process. Lett.*, 63(5):243–246, 1997.
9. J.-É. Pin. *Varieties of formal languages.* North Oxford, London, 1986.
10. J.-É. Pin. Syntactic semigroups. In G. Rozenberg and A. Salomaa, editors, *Handbook of formal languages*, volume 1, chapter 10. Springer Verlag, 1997.
11. J.-É. Pin and H. Straubing. Some results on $\mathcal{C}$-varieties. *Theoret. Informatics Appl.*, 39:239–262, 2005.
12. J.-É. Pin and P. Weil. The wreath product principle for ordered semigroups. *Communications in Algebra*, 30:5677–5713, 2002.
13. J. Reiterman. The Birkhoff theorem for finite algebras. *Alg. Univ.*, 14(1):1–10, 1982.
14. H. Straubing. Families of recognizable sets corresponding to certain varieties of finite monoids. *J. Pure Appl. Algebra*, 15(3):305–318, 1979.
15. H. Straubing. Finite semigroup varieties of the form $\mathbf{V} * \mathbf{D}$. *J. Pure Appl. Algebra*, 36(1):53–94, 1985.
16. H. Straubing. *Finite automata, formal logic, and circuit complexity.* Birkhäuser Boston Inc., 1994.
17. H. Straubing. On logical descriptions of regular languages. In *LATIN 2002*, number 2286 in Lect. Notes Comp. Sci., pages 528–538, Berlin, 2002. Springer.
18. D. Thérien and A. Weiss. Graph congruences and wreath products. *J. Pure Appl. Algebra*, 36, 1985.
19. B. Tilson. Categories as algebra: an essential ingredient in the theory of monoids. *J. Pure Appl. Algebra*, 48(1-2):83–198, 1987.

Local Construction of Planar Spanners in Unit Disk Graphs with Irregular Transmission Ranges

Edgar Chávez[1], Stefan Dobrev[2,*], Evangelos Kranakis[3,**], Jaroslav Opatrny[4,*], Ladislav Stacho[5,*], and Jorge Urrutia[6]

[1] Escuela de Ciencias Fisico-Matemáticas de la Universidad Michoacana de San Nicolás de Hidalgo, México
[2] School of Information Technology and Engineering (SITE), University of Ottawa, 800 King Eduard, Ottawa, Ontario, Canada, K1N 6N5
[3] School of Computer Science, Carleton University, 1125 Colonel By Drive, Ottawa, Ontario, Canada K1S 5B6
[4] Department of Computer Science, Concordia University, 1455 de Maisonneuve Blvd West, Montréal, Québec, Canada, H3G 1M8
[5] Department of Mathematics, Simon Fraser University, 8888 University Drive, Burnaby, British Columbia, Canada, V5A 1S6
[6] Instituto de Matemáticas, Universidad Nacional Autónoma de México, Área de la investigación cientifica, Circuito Exterior, Ciudad Universitaria, Coyoacán 04510, México, D.F. México

Abstract. We give an algorithm for constructing a connected spanning subgraphs(panner) of a wireless network modelled as a unit disk graph with nodes of irregular transmission ranges, whereby for some parameter $0 < r \leq 1$ the transmission range of a node includes the entire disk around the node of radius at least r and it does not include any node at distance more than one. The construction of a spanner is distributed and local in the sense that nodes use only information at their vicinity, moreover for a given integer $k \geq 2$ each node needs only consider all the nodes at distance at most k hops from it. The resulting spanner has maximum degree at most $3 + \frac{6}{\pi r} + \frac{r+1}{r^2}$, when $0 < r < 1$ (and at most five, when $r = 1$). Furthermore it is shown that the spanner is planar provided that the distance between any two nodes is at least $\sqrt{1 - r^2}$. If the spanner is planar then for $k \geq 2$ the sum of the Euclidean lengths of the edges of the spanner is at most $\frac{kr+1}{kr-1}$ times the sum of the Euclidean lengths of the edges of a minimum weight Euclidean spanning tree.

1 Introduction

The problem of constructing connected spanning subgraphs(spanners) (e.g., minimum cost spanning trees, triangulated spanners, planar spanners) for "various

* Research supported in part by NSERC (Natural Science and Engineering Research Council of Canada) grant.

** Research supported in part by NSERC (Natural Science and Engineering Research Council of Canada) and MITACS (Mathematics of Information Technology and Complex Systems) grants.

J.R. Correa, A. Hevia, and M. Kiwi (Eds.): LATIN 2006, LNCS 3887, pp. 286–297, 2006.

types" of geometric graphs has been considered extensively in the current literature due to its many applications ranging from VLSI design, to efficient communication in networks and medical imaging (see Eppstein [8]). A variety of optimization results have been derived that considered tradeoffs among weight, diameter, dilation, and max degree between the original graph and the resulting spanner. Nevertheless the majority of these results (e.g., Eppstein [8], Arya, Das, Mount, Salowe, and Smid [1], Arya and Smid [2], Narasimhan and Smid [18], Bose, Gudmundsson and Smid [4]) consider only centralized, non-distributed algorithms that do not take into account the dynamic changes taking place in a communication network.

In recent years, the problem of producing efficiently a planar spanner has been given new research impetus in communication networks due to its applicability in more dynamically changing environments consisting of wireless interconnected nodes. In this case, in addition to considering the previously mentioned parameters of weight, diameter, dilation, and max degree, a new condition of *locality of communication* becomes important: nodes should take into account information by consulting only other nodes within their "close" geographic vicinity. In fact, locality in wireless networking is a necessity imposed by the geographic limitations of the networking environment.

Moreover, there are two important issues in wireless networking. The first one is to be able to perform locally and efficiently important communication tasks, like routing. Ultimately, this is easily resolved if the underlying graph is planar using face routing (see Kranakis, Singh, and Urrutia [11]). The second one is a "local" construction of a "simple" planar spanner from the given wireless network. In fact, Bose, Morin, Stojmenovic, and Urrutia [5] address this problem for wireless networks corresponding to unit disk graphs by constructing a planar spanner in a local and distributed manner using the Gabriel test (see Gabriel and Sokal [9]).

In addition to the Gabriel test, there are known algorithms for constructing locally and distributively a planar subgraph of bounded degree and constant stretch factor for unit disk graphs. However the resulting degree is rather high (more than 25), the constructions are relatively complicated, and the cost of such graph can be much higher than the cost of a Minimum cost Spanning Tree (MST) (e.g., see Li, Calinescu, and Wan [15], Wang and Li [20], Li and Wang [16]). In a recent paper Li, Wang, and Song [17] give an algorithm for constructing a spanner from the relative neighborhood graph [19] of a unit disk graph. This spanner has maximum degree at most six, and its total weight is a constant multiple of the total weight of the MST, where the weight of a graph is defined as the sum of Euclidean lengths of the edges. In this paper we consider the problem of constructing a spanner of networks which are more general than those represented by a unit disk graph.

A unit disk graph is a representation of a wireless network in which all nodes have the same circular transmission range. Clearly, this is an idealized representation and it does not need to correspond to actual situations. Typically, the nodes in a network are not exactly identical and some obstacles in the terrain containing the nodes may result in the transmission ranges of nodes to be irregular. In this paper we are taking into consideration the fact that the transmission range of each node of a network could be irregular to "some degree" (see Figure 1). We assume that in a given network there is an additional parameter r, a positive real number less than or equal 1. The transmission range of a node in the network is assumed to be a region contained within the unit disk around the node, but this region contains all points at distance less than r. Thus any two nodes at distance at most r can communicate directly, but no nodes at distance more than 1 can communicate directly. Two nodes at distance more than r and at most 1 may or may not communicate directly. An example of a transmission range of node u is shown in Figure 1 as the darker area. We shall consider the static case in which the irregularity of each node is fixed and does not change with time. We call the geometric representation of such a network a *unit disk graph with irregularity* r. This class of unit disk graphs with irregular transmission ranges was first introduced by Barrière, Fraigniaud, Narayanan, and Opatrny [3] in order to propose robust position-based routing. The problem of constructing a spanner for unit disk graphs with irregular transmission ranges is more complex, for example the usual planarization algorithms like the Gabriel test or the relative neighborhood graph algorithm do not work for them.

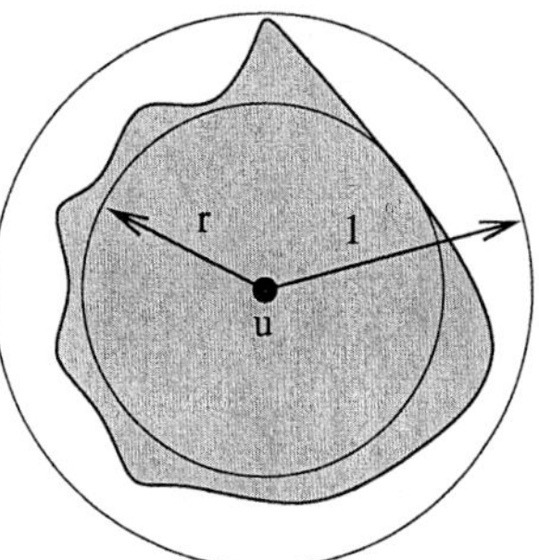

Fig. 1. The irregular transmission area of node u

1.1 Results and Outline of the Paper

We give an algorithm for constructing a spanner of a connected unit disk graph with irregularity r. The construction is *local* in the sense that nodes use only information at their vicinity: given $k \geq 2$, each node needs only to consider all the nodes at distance at most k hops from it, i.e., nodes joined to it by paths of length at most k. The resulting spanner has maximum degree at most $3 + \frac{6}{\pi r} + \frac{r+1}{r^2}$, when $0 < r < 1$ (and at most five, when $r = 1$). Moreover, it is shown that the spanner is *planar* provided that the distance between any two nodes is at least $\sqrt{1 - r^2}$. For $k \geq 2$ the sum of the euclidean lengths of the edges of the spanner is at most $\frac{kr+1}{kr-1}$ times the sum of the euclidean lengths of the edges of a minimum weight euclidean spanning tree if the spanner is planar. The class of graphs whereby the distance between any two nodes is at least λ (in our graphs $\lambda = \sqrt{1 - r^2}$) were first called *civilized* by Doyle and Snell [7][page 136] and have also been referred to as λ-*precision* by Hunt, Marathe, Radhakrishnan, Ravi, Rosenkrantz, and Stearns [10], and $\Omega(1)$-*constant* by Kuhn, Wattenhofer, and Zollinger [14] (see also Kuhn, Wattenhofer, Zhang, and Zollinger [13]).

Our results extend work of Li, Wang, and Song [17] mentioned above from the case $r = 1$ to arbitrary irregularity factor r. Note that even in the special case $r = 1$, we obtain explicit bounds on degree and cost of the spanner rather that asymptotic bounds. Also, our proofs use only elementary techniques and do not rely on [6].

An outline of the paper is as follows. Section 2 gives definitions needed for the algorithm, the main one being a definition of a linear order of the edges of the graph, while Section 3 gives the main result on constructing a spanner and proves the correctness of our algorithm.

2 Preliminaries

A graph G is *geometric* if it is embedded into the Euclidean plane and the edges are straight line segments between the nodes. The edge selection in our algorithms will depend on a linear order on edges of the input geometric graph G.

2.1 Linear Order on Edges

Let $|u, v|$ denote the Euclidean distance between nodes u and v. Intuitively, we can define a linear order on the edges of G

- by first considering the Euclidean length,
- if two edges have the same length, the one with rightmost, topmost endnode is larger, and finally
- if two edges of same length share their rightmost, topmost node, then their second endnode is considered; the edge with the right most, top most second endnode is defined as larger.

Formally, we have the following definition.

Definition 1 (Compatible Linear Order). *Each edge $\{u, v\}$ is assigned a 5-tuple $(|u, v|, x_1, y_1, x_2, y_2)$, where x_1, y_1 and x_2, y_2 are the coordinates of the endnodes of the edge with either $x_1 > x_2$ or $x_1 = x_2$ and $y_1 > y_2$. Clearly this gives a unique 5-tuple to any edge, and 5-tuples assigned to any two edges are distinct. The linear order $\prec$ is defined by using the lexicographical ordering of the assigned 5-tuples.*

Notice that in the order $\prec$, we first consider the Euclidean length of edges and the coordinates are used for ordering edges of the same length. The input graph G may have many minimum cost spanning trees (MSTs) when the Euclidean length of edges is the cost function. However, if we break the ties by the linear order $\prec$, then G has a unique MST $T^{\prec}$ which can be computed for example by Kruskal's algorithm.

Definition 2. *For a given geometric graph H, define cost(H) as the sum of Euclidean lengths of the edges of H.*

Definition 3. *Given a graph G and a vertex v of G, we denote by $N_k[v]$ the distance k closed neighborhood of v, i.e. the nodes of G reachable from v by a path with at most k edges. Note that $v \in N_k[v]$. Sometimes, the graph induced by vertices in $N_k[v]$ will be denoted by the same symbol $N_k[v]$.*

3 Constructing a Spanner

This section is the core of our paper. Subsection 3.1 gives the main algorithm for constructing spanners directly from a unit disk graph, while Subsection 3.2 states the main theorem (Theorem 1) and its complete proof.

3.1 Spanner Algorithm

Consider algorithm **LocalMST**$_k$, for $k \geq 2$, which was presented by Li, Wang, and Song in [17].

Algorithm. LocalMST$_k$
Input. A connected geometric graph G with the linear order $\prec$;
Output. Graph $G_k^{\prec}$

Run the following algorithm at each node v of G:

1. Learn your distance k neighborhood $N_k[v]$.
2. Construct locally the unique MST $T_k(v)$ of $N_k[v]$.
3. Broadcast in $N_1[v]$ the edges of $N_1[v]$ which have been retained in $T_k(v)$ (i.e. $N_1[v] \cap T_k(v)$).
4. The output graph $G_k^{\prec}$ is defined as follows: an edge is selected into $G_k^{\prec}$ if and only if it was retained by both of its incident nodes.

Clearly, this is a distributed algorithm. To learn its distance k neighborhood, v first broadcasts its coordinates to all its neighbors. After having learnt its distance k neighborhood it broadcasts it to all its neighbours. It can then construct the unique MST $T_k(v)$ (which is selected using Kruskal's algorithm [12] and the linear order $\prec$) of $N_k[v]$ and broadcasts edges in $N_1[v] \cap T(v)$ to all nodes in $N_1[v]$. The parameter k determines the desired locality of our algorithm, and thus the resulting graph $G_k^{\prec}$ is constructed "locally", each node v uses only knowledge of $N_k[v]$ and the results of its neighbours.

3.2 Main Result and Proof of Correctness

Let G be a unit geometric graph with irregularity r and $k \geq 2$. We show that the graph $G_k^{\prec}$ constructed by the above algorithm has interesting properties summarized in the following theorem.

Theorem 1. *If G is a connected geometric unit disk graph with irregularity r and $k \geq 2$, then the graph $G_k^{\prec}$ has the following properties.*

a) $G_k^{\prec}$ is connected;
b) if the distance between any two nodes of the graph is at least $\sqrt{1-r^2}$, then the graph $G_k^{\prec}$ is planar;
c) $\Delta(G_k^{\prec}) \leq \begin{cases} 5 & \text{if } r = 1, \\ 3 + \frac{6}{\pi r} + \frac{r+1}{r^2} & \text{if } 0 < r < 1; \end{cases}$
d) If $G_k^{\prec}$ is planar and $kr > 1$, then $cost(G_k^{\prec}) \leq \frac{kr+1}{kr-1} \times cost(T^{\prec})$.

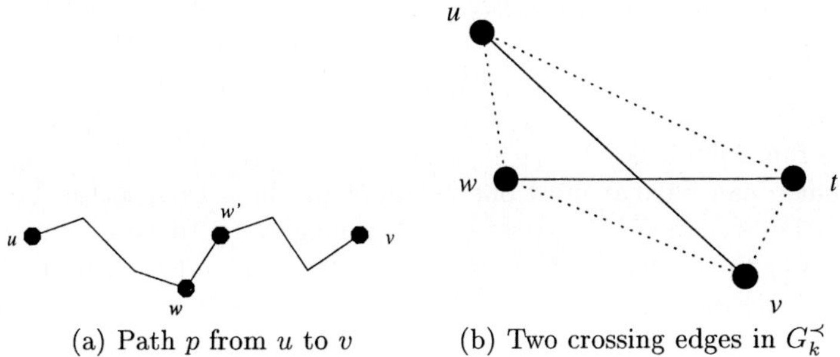

(a) Path p from u to v (b) Two crossing edges in $G_k^{\prec}$

Fig. 2.

Proof. The proof of part a) follows from the following claim.

Claim. $T^{\prec} \subseteq G_k^{\prec}$.

Proof. We argue by contradiction. Let the edge $\{u, v\}$ be retained in T, but rejected in $G_k^{\prec}$. Without loss of generality we may assume it was rejected in $T_k(v)$. Since $\{u, v\}$ was retained in T, there is no other path in T joining u and v. Since $\{u, v\}$ was rejected by $T_k(v)$, there exists a path, say p, in $T_k(v)$ joining u and v and using only edges smaller than $\{u, v\}$.

Let $\{w, w'\}$ be an edge in p such that $\{w, w'\} \notin T$ (see Figure 2(a)). It follows that there is a path in T joining w and w' and using only edges smaller than the edge $\{w, w'\}$. As this argument applies to each such edge of p, there must be a path in T joining u and v using only edges smaller then $\{u, v\}$. This contradicts the fact that the edge $\{u, v\}$ was retained in T.

To prove part b), assume by way of contradiction, that $G_k^{\prec}$ is not planar and let $\{u, v\}$ and $\{w, t\}$ be two crossing edges in $G_k^{\prec}$. Without loss of generality we may assume that the angle $\angle uwv$ is the largest angle in the quadrilateral $uwvt$ (see Figure 3.2). Clearly, this angle is at least $\pi/2$.

Since $|u, v| \leq 1$ we have $|u, w|^2 + |w, v|^2 \leq 1$. Thus $|u, w|^2 \leq 1 - |w, v|^2 \leq r^2$ since $|w, v| \geq \sqrt{1 - r^2}$ by our assumption. Therefore, $\{u, w\}$ is an edge in G. The same argument shows that $\{w, v\}$ is an edge in G.

We will show that the diagonal $\{u, v\}$ will not be selected into $G_k^{\prec}$ by u. Assume u computes $T_k(u)$ using Kruskal's algorithm. Either $\{u, w\}$ is retained in $T_k(u)$, or there already exists a path in $T_k(u)$ consisting of smaller edges connecting u and w. Analogously, the same is true for $\{w, v\}$. This means that at the moment $\{u, v\}$ is considered by u for inclusion into $T_k(u)$, there already exists a path in $T_k(u)$ connecting u and v and hence $\{u, v\}$ will be rejected by u, which contradicts the fact that edge $\{u, v\}$ is in $G_k^{\prec}$. Note that from our assumption on distance between vertices of G and the assumption that G is connected, we have $\sqrt{1 - r^2} \leq r$ and thus $r \geq \sqrt{\frac{1}{2}}$.

To prove part c), let u be any vertex of G. Partition the unit circle around u into six equal size sectors each with angle at u equal to $\pi/3$. Figure 3 depicts such a sector by the dark shaded area. Since G is finite, we may assume that the edges of these sectors do not pass through any neighbor of u. Hence, for any two neighbors v and w of u inside any fixed sector, the angle $\angle wuv$ is less that $\pi/3$. Then $|v,w| < \max\{|u,v|, |u,w|\}$. If $|v,w| \le r$, then one of $\{u,v\}$, $\{u,w\}$ would have been replaced in $G_k^{\prec}$ by $\{v,w\}$. Thus, we conclude $|v,w| > r$. If $r = 1$, it follows that u can have at most one neighbor inside of each sector. So u has at most six neighbors in $G_k^{\prec}$. Suppose u has six neighbors. However, this may only occur if u is in the center of a perfect hexagon formed by its neighbors. However, in this case only two incident edges will be retained, as four of the incident edges will be deleted as the largest edges of an incident equilateral triangle. Hence u has at most five neighbors as claimed.

Suppose now, $r < 1$. Consider a fixed sector S defined above. Draw a circle of radius $r/2$ around every neighbor of u in this sector. It follows that these circles are disjoint and all are inside the region determined by the union of the circle of radius $r/2$ centred at u, the sector of radius $1 + r/2$ centred at u and containing the sector S, and two rectangles with sides $1 + r/2$ and $r/2$ (see Figure 3).

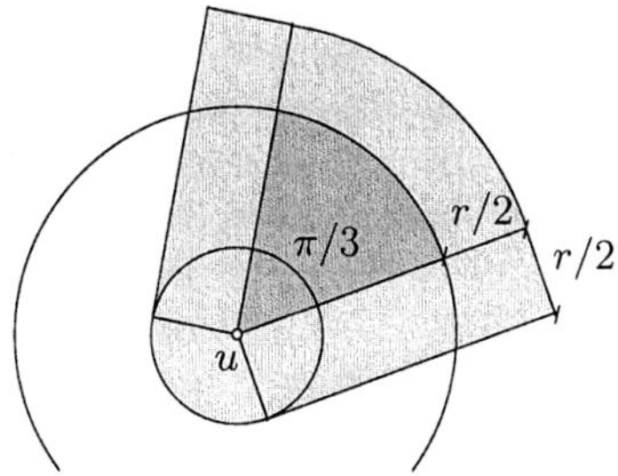

Fig. 3. The light shaded area contains all disjoint circles of radius $r/2$ around all neighbors of u inside the dark shaded area

Hence an upper bound on the number of neighbors of u in the sector S is the number of circles of radius $r/2$ that can be packed into this area. This number is at most

$$\frac{\pi(r/2)^2 + \frac{\pi(1+r/2)^2}{6} + 2(1+r/2)r/2}{\pi r^2} < \frac{1}{2} + \frac{1}{\pi r} + \frac{r+1}{6r^2}.$$

Summing up through all six sectors, we obtain that u has at most $3 + \frac{6}{\pi r} + \frac{r+1}{r^2}$ neighbors.

The following claim captures a crucial property of the graph $G_k^{\prec}$ that helps to prove part d).

Claim. Every cycle C in $G_k^{\prec}$ has the Euclidean length greater than $\max\{(k+1)r, kr + l\}$, where l is the length of a longest edge in C.

Proof. Let C be a cycle in $G_k^{\prec}$ and l be the Euclidean length of the largest edge, say $\{u,v\}$, of this cycle. Without loss of generality we may assume that

v in counterclockwise from u (see Figure 5(b)). Since C was retained in $G_k^{\prec}$, the edge $\{u,v\}$ must have been retained in $T_k(u)$. However, this means that there exists a node $z \in C$ such that $z \notin N_k[u]$, otherwise u would have seen the whole cycle C and therefore rejected edge $\{u,v\}$ as the largest edge of C. Since $z \notin N_k[u]$, the path from u to z clockwise around C contains a vertex of $N_i[u] \setminus N_{i-1}[u]$ for all $1 \leq i \leq k$. Let w_i be the furthest such a vertex in $N_i[u] \setminus N_{i-1}[u]$. Similarly, also the path from u to z counter clockwise around C contains a vertex of $N_i[u] \setminus N_{i-1}[u]$ for all $1 \leq i \leq k$, and let x_i be the furthest such a vertex in $N_i[u] \setminus N_{i-1}[u]$. It follows that $w_1, w_2, \ldots, w_k$ are in clockwise order around C while $x_1, x_2, \ldots, x_k$ are is counter clockwise order.

By definition, if $k \geq 2$ then the Euclidean distances $|u, w_2|$, $|u, x_2|$, $|z, w_{k-1}|$, $|z, x_{k-1}|$, and for all $1 \leq i \leq k-2$ $|w_i, w_{i+2}|$ and $|x_i, x_{i+2}|$ are all greater than r. From the triangle inequality, we have.

If k is odd then the Euclidean length

$$\begin{aligned}|C| &\geq |u, w_{k-1}| + |w_{k-1}, z| + |u, x_{k-1}| + |x_{k-1}, z| \\ &> \frac{k-1}{2}r + r + \frac{k-1}{2}r + r \\ &= (k+1)r,\end{aligned}$$

or similarly

$$\begin{aligned}|C| &\geq |u, w_2| + |w_2, w_{k-1}| + |w_{k-1}, z| + |u, x_{k-1}| + |x_{k-1}, z| \\ &> l + \frac{k-3}{2}r + r + \frac{k-1}{2}r + r \\ &= kr + l,\end{aligned}$$

If k is even then the Euclidean length

$$\begin{aligned}|C| &\geq |w_1, x_1|_C + |w_1, w_{k-1}| + |w_{k-1}, z| + |x_1, x_{k-1}| + |x_{k-1}, z| \\ &> |w_1, x_1| + \frac{k-2}{2}r + r + \frac{k-2}{2}r + r + r \\ &= |w_1, x_1|_C + kr \\ &\geq kr + l,\end{aligned}$$

where $|w_1, x_1|_C$ denotes the Euclidean length of the counterclockwise path from w_1 to x_1 on the cycle C. To complete the proof it remains to show that $|w_1, x_1|_C > r$. Let w_1' be the clockwise successor of w_1 along C. By definition of w_1 and the fact that $z \notin N_k[u]$, the vertex $w_1' \notin N_1[u]$. Let $|w_1, w_1'| = l_1$. We have $|w_1, u|_C \geq r - \epsilon_1$ and $|u, x_1|_C \geq r - \epsilon_2$ for some $\epsilon_1 > 0$ and $\epsilon_2 > 0$.

From the triangle inequality and the fact that $w_1' \notin N_1[u]$, we get $r - \epsilon_1 + l_1 > r$. From this and the fact that l is the largest edge in C, we get $l \geq l_1 > \epsilon_1$. Since v is reachable from u, we know that $r - \epsilon_2 \geq l$. Combining with $l > \epsilon_1$ we get $\epsilon_1 + \epsilon_2 < r$ and thus $|w_1, x_1|_C \geq r - \epsilon_1 + r - \epsilon_2 > r$.

Finally we prove that if $G_k^{\prec}$ is planar, then $\text{cost}(G_k^{\prec}) \leq \frac{kr+1}{kr-1} \times \text{cost}(T^{\prec})$. Let $C_1, C_2, \ldots, C_f$ be the faces in $G_k^{\prec}$. First note that the sum of the Euclidean lengths of the faces is equal to twice the sum of the Euclidean lengths of all edges. This implies that $\text{cost}(G_k^{\prec})$ is equal to half the sum of the Euclidean lengths of the faces, which by Claim 3.2 is bounded from below by $(krf + \sum_{i=1}^{f} l_i)/2$ where l_i is the longest edge in C_i.

Since, $T^{\prec} \subseteq G_k^{\prec}$, it follows from the well-known Euler's formula that the spanning tree $T^{\prec}$ can be obtained by deleting some $f - 1$ edges $e_1, e_2, \ldots, e_{f-1}$ from $G_k^{\prec}$. Therefore we obtain that $\text{cost}(G_k^{\prec}) \leq \text{cost}(T^{\prec}) + \sum_{j=1}^{f-1} |e_j|$. We want to upper bound the last sum by $\sum_{i=1}^{f} l_i$. To do this, we need to assign each edge e_j to a unique face C_i so that $e_j \in C_i$. For this, consider the bipartite graph H with partite sets $X = \{e_1, e_2, \ldots, e_{f-1}\}$ and $Y = \{C_1, C_2, \ldots, C_f\}$ in which a vertex in X is joined by an edge to the two faces it is incident on. Consider a subset $X' \subseteq X$. We claim that $|N(X')| > |X'|$. Indeed, if for some X', the edges in X' are incident only to $|X'|$ faces, then after removal of these edges we obtain a new planar graph which will have the same number of nodes, will have $|X'|$ less edges and $|X'| - 1$ less faces that $G_k^{\prec}$, which is a contradiction with Euler's formula. It follows from the well-known Hall's matching theorem that H has a matching saturating X. Now, assign the edge e_j to the face determined by the matching. We may assume (after appropriate relabelling) that e_j is assigned to C_j for $j = 1, \ldots, f - 1$. Since the length of a longest edge in C_j is l_j, we have $\sum_{j=1}^{f-1} |e_j| \leq \sum_{j=1}^{f-1} l_j$, and hence $\text{cost}(G_k^{\prec}) \leq \text{cost}(T^{\prec}) + \sum_{j=1}^{f-1} l_j$. This implies that

$$\text{cost}(T^{\prec}) \geq krf/2 - \sum_{i=1}^{f-1} l_i/2 \geq (krf - f + 1)/2).$$

Notice that by the assumption, $kr > 1$ and hence the last expression is positive. Consequently, $\frac{\text{cost}(G_k^{\prec})}{\text{cost}(T^{\prec})}$ is at most

$$\frac{\text{cost}(T^{\prec}) + \sum_{j=1}^{f-1} l_j}{\text{cost}(T^{\prec})} \leq 1 + \frac{f-1}{\text{cost}(T^{\prec})} \leq 1 + \frac{f}{(krf - f + 1)/2} \leq \frac{kr+1}{kr-1}$$

This completes the proof of the theorem.

To see that $G_k^{\prec}$ is not necessarily planar, consider the example of a graph G on Figure 4 for which any connected spanner must retain all edges. As a corollary we also obtain the following result.

Corollary 1. *If G is a connected geometric unit disk graph and $k \geq 2$, then the graph $G_k^{\prec}$ has the following properties.*

a) $G_k^{\prec}$ is connected;
b) $G_k^{\prec}$ is planar;
c) $\Delta(G_k^{\prec}) \leq 5$;
d) cost$(G_k^{\prec}) \leq \frac{k+1}{k-1} \times$ cost$(T^{\prec})$.

This corollary gives identical results as those of Li, Wang, and Song [17] as far as the connectedness and the planarity is concerned, but it improves the maximum degree to 5 and gives an explicit value of the cost factor of the spanner.

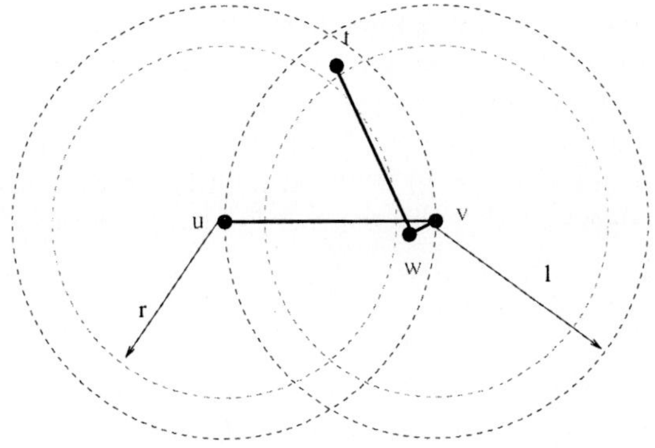

Fig. 4. Configuration of vertices that gives a non-planar spanner

Observe that without the ordering on the edges of the geometric graph, the algorithm to obtain G does not work, because it could produce a disconnected graph.

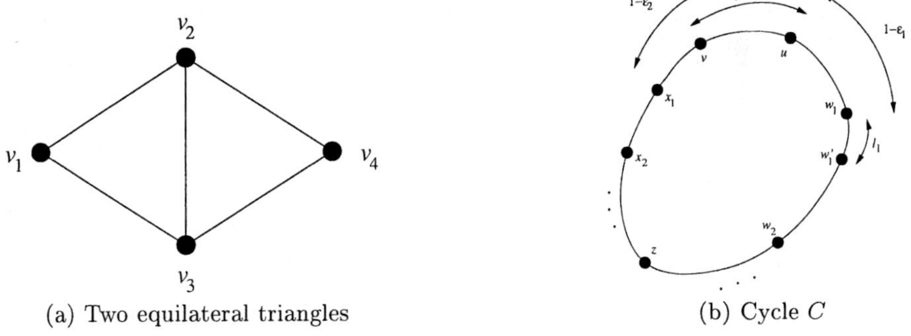

(a) Two equilateral triangles

(b) Cycle C

Fig. 5.

A simple counterexample with four nodes is depicted in Figure 5(a). It consists of four nodes v_1, v_2, v_3, v_4 such that the distance between any pair of them, but one (say v_1 and v_4) is equal to 1. Our nodes are the vertices of two equilateral triangles with disjoint interiors that share an edge (in this case $\{v_2, v_3\}$). Without a total ordering induced on the edges of this graph, we can get a disconnected graph.

4 Conclusions

In this paper, we gave a new local, distributed algorithm for constructing a planar spanner of a connected unit disk graph with nodes having irregular transmission ranges, give bounds on the degree of the spanner, and a sufficient condition on the graph to obtain a planar spanner. When the spanner is planar, we give an explicit bound on the the cost factor of the spanner. Examples that compare our result to the well-known RNG (Relative Neighborhood [19]) and GG (Gabriel [9]) graphs can be found in the full paper. It would be interesting to derive a cost factor in case when the spanner is not planar. Another interesting problem is to see whether our techniques can be extended to obtain a distributed algorithm that constructs a low cost spanner of a given geometric unit graph (possibly with irregular transmission range) which in addition guarantee a low geometric stretch factor of edges.

References

1. S. Arya, G. Das, D. M. Mount, J. S. Salowe, and M. Smid. Euclidean spanners: Short, thin, and lanky. In *Proceedings of the Twenty-Seventh Annual ACM Symposium on the Theory of Computing*, pages 489–498, Las Vegas, Nevada, 29 May–1 June 1995.
2. S. Arya and M. Smid. Efficient construction of a bounded degree spanner with low weight. *Algorithmica*, 17:33–54, 1997.
3. L. Barrière, P. Fraigniaud, L. Narayanan, and J. Opatrny. Robust position-based routing in wireless ad-hoc networks with irregular transmission ranges. *Wireless Communications and Mobile Computing Journal*, 2003.
4. P. Bose, J. Gudmundsson, and M. Smid. Constructing plane spanners of bounded degree and low weight. *Algorithmica*, to appear.
5. P. Bose, P. Morin, I. Stojmenovic, and J. Urrutia. Routing with guaranteed delivery in ad hoc wireless networks. *Wireless Networks*, 7:609–616, 2001.
6. G. Das and G. N. J. Salowe. A new way to weigh malnourished euclidean graphs. In *Proceedings of the 6th Annual ACM-SIAM Symposium on Discrete Algorithms, SODA'95 (San Francisco, California, January 22-24, 1995)*, pages 215–222, Philadelphia, PA, 1995. ACM SIGACT, SIAM, Society for Industrial and Applied Mathematics.
7. P. Doyle and J. Snell. *Random Walks and Electric Networks.* The Carus Mathematical Monographs. The Mathematical Association of America, 1984.
8. D. Eppstein. Spanning trees and spanners. In J.-R. Sack and J. Urrutia, editors, *Handbook of Computational Geometry.* Elsevier, 2000.
9. K. R. Gabriel and R. R. Sokal. A new statistical approach to geographic variation analysis. *Systemic Zoology*, 18:259–278, 1972.
10. H. H. III, M. Marathe, V. Radhakrishnan, S. Ravi, D. Rosenkrantz, and R. Stearns. Nc-approximation schemes for np- and pspace-hard problems for geometric graphs. *J. Algorithms*, 26(2):238ŋ–274, 1998.
11. E. Kranakis, H. Singh, and J. Urrutia. Compass routing on geometric networks. In *Proc. of 11th Canadian Conference on Computational Geometry*, pages 51–54, August 1999.
12. J. B. Kruskal, Jr. On the shortest spanning subtree of a graph and the traveling salesman problem. *Proc. Amer. Math. Soc.*, 7:48–50, 1956.
13. F. Kuhn, R. Wattenhofer, Y. Zhang, and A. Zollinger. Geometric ad-hoc routing: of theory and practice. In *Proceedings of the Twenty-Second ACM Symposium on Principles of Distributed Computing (PODC 2003), July 13-16, 2003, Boston, Massachusetts, USA. ACM*, pages 63–72. ACM Press, 2003.
14. F. Kuhn, R. Wattenhofer, and A. Zollinger. Asymptotically optimal geometric mobile ad-hoc routing. In *Proceedings of 6th International Workshop on Discrete Algorithms and Methods for Mobile Computing and Communications (Dial-M)*, pages 24ŋ–33. ACM Press, 2002.

15. X.-Y. Li, G. Calinescu, and P.-J. Wan. Distributed construction of planar spanner and routing for ad hoc wireless networks. In *Proceedings of the 21st Annual Joint Conference of the IEEE Computer and Communications Society (INFOCOM-02)*, pages 1268–1277, Piscataway, NJ, USA, June 23–27 2002.
16. X.-Y. Li and Y. Wang. Efficient construction of low weight bounded degree planar spanner. In *COCOON: Annual International Conference on Computing and Combinatorics*, 2003.
17. X.-Y. Li, Y. Wang, and W.-Z. Song. Applications of k-local mst for topology control and broadcasting in wireless ad hoc networks. In *Proceedings of the 23rd Annual Joint Conference of the IEEE Computer and Communications Society (INFOCOM-04)*, Piscataway, NJ, USA, Mar. 7–11 2004.
18. G. Narasimhan and M. Smid. Approximating the stretch factor of Euclidean graphs. *SIAM Journal on Computing*, 30(3):978–989, June 2001.
19. G. T. Toussaint. The relative neighborhood graph of a finite set. *Pattern Recognition*, 12:261–268, 1980.
20. Y. Wang and X.-Y. Li. Localized construction of bounded degree and planar spanner for wireless ad hoc networks. In *DialM: Proceedings of the Discrete Algorithms and Methods for Mobile Computing & Communications*, 2003.

An Efficient Approximation Algorithm for Point Pattern Matching Under Noise

Vicky Choi[1] and Navin Goyal[2]

[1] Department of Computer Science, Virginia Tech, USA
vchoi@cs.vt.edu

[2] Department of Computer Science, McGill University, Canada
navin@cs.mcgill.ca

Abstract. Point pattern matching problems are of fundamental importance in various areas including computer vision and structural bioinformatics. In this paper, we study one of the more general problems, known as LCP (largest common point set problem): Let P and Q be two point sets in $\mathbb{R}^3$, and let $\epsilon \geq 0$ be a tolerance parameter, the problem is to find a rigid motion μ that maximizes the cardinality of subset I of Q, such that the Hausdorff distance $\mathsf{dist}(P, \mu(I)) \leq \epsilon$. We denote the size of the optimal solution to the above problem by $\mathsf{LCP}(P, Q)$. The problem is called exact-LCP for $\epsilon = 0$, and tolerant-LCP when $\epsilon > 0$ and the minimum interpoint distance is greater than ϵ. A β-distance-approximation algorithm for tolerant-LCP finds a subset $I \subseteq Q$ such that $|I| \geq \mathsf{LCP}(P, Q)$ and $\mathsf{dist}(P, \mu(I)) \leq \beta\epsilon$ for some $\beta \geq 1$.

This paper has three main contributions. (1) We introduce a new algorithm, called *T-hashing*, which gives the fastest known deterministic 4-distance-approximation algorithm for tolerant-LCP. (2) For the exact-LCP, when the matched set is required to be large, we give a simple sampling strategy that improves the running times of all known deterministic algorithms, yielding the fastest known deterministic algorithm for this problem. (3) We use expander graphs to speed-up the T-hashing algorithm for tolerant-LCP when the size of the matched set is required to be large, at the expense of approximation in the matched set size. Our algorithms also work when the transformation μ is allowed to be scaling transformation.

1 Introduction

The general problem of finding large similar common substructures in two point sets arises in many areas ranging from computer vision to structural bioinformatics. In this paper, we study one of the more general problems, known as the *largest common point set problem* (LCP), which has several variants to be discussed below.

Problem Statement. Given two point sets in $\mathbb{R}^3$, $P = \{p_1, \ldots, p_m\}$ and $Q = \{q_1, \ldots, q_n\}$, and an error parameter $\epsilon \geq 0$, we want to find a rigid motion μ that maximizes the cardinality of subset $I \subseteq Q$, such that $\mathsf{dist}(P, \mu(I)) \leq \epsilon$. For an optimal set I, denote $|I|$ by $\mathsf{LCP}(P, Q)$. There are two commonly used distance measures between point sets: *Hausdorff distance* and *bottleneck distance*. The Hausdorff distance $\mathsf{dist}(P, Q)$ between two point sets P and Q is given by $\max_{q \in Q} \min_{p \in P} ||pq||$.

J.R. Correa, A. Hevia, and M. Kiwi (Eds.): LATIN 2006, LNCS 3887, pp. 298–310, 2006.

The bottleneck distance $\mathsf{dist}(P, Q)$ between two point sets P and Q is given by $\min_f \max_{q \in Q} ||f(q) - q||$, where $f : Q \to P$ is an injection. Thus we get two versions of the LCP depending on which distance is used.

Another distinction that is made is between the *exact*-LCP and the *threshold*-LCP. In the former we have $\epsilon = 0$ and in the latter we have $\epsilon > 0$. The exact-LCP is computationally easier than the threshold-LCP; however, it is not useful when the data suffers from round-off and sampling errors, and when we wish to measure the resemblance between two point sets and do not expect exact matches. These problems are better modelled by the threshold-LCP, which turns out to be harder, and various kinds of approximation algorithms have been considered for it in the literature (see below). A special kind of threshold-LCP in which one assumes that the minimum interpoint distance is greater than the error parameter 2ϵ is called *tolerant*-LCP. Tolerant-LCP is nicer to deal with than threshold-LCP and at the same time it appears to captures many problems arising in practice. Notice that for the tolerant-LCP, the Hausdorff and bottleneck distances are the same and thus there is no need to specify which distance is in use. In practice, it is often the case that the size of the solution set I to the LCP is required to be at least a certain fraction of the minimum of the sizes of the two point sets: $|I| \geq \frac{1}{\alpha} \min(|P|, |Q|)$, where α is a positive constant. This version of the LCP is known as the α-LCP. A special case of the LCP which requires matching the entire set Q is called *Pattern Matching* (PM) problem. Again, we have exact-PM, threshold-PM, and tolerant-PM versions.

In this paper, we focus on approximation algorithms for tolerant-LCP and tolerant-α-LCP. There are two natural notions of approximation. (1) *Distance approximation:* The algorithm may find a transformation that brings a set $I \subseteq Q$ of size at least $\mathsf{LCP}(P, Q)$ within distance ϵ' for some constant $\epsilon' > \epsilon$. (2) *Size-approximation:* The algorithm guarantees that $|I| \geq (1 - \delta)\mathsf{LCP}(P, Q)$, for constant $\delta \in [0, 1)$.

Previous work. The LCP has been extensively investigated in computer vision (e.g. [27]), computational geometry (e.g. [7]), and also finds applications in computational structural biology (e.g. [29]). For the exact-LCP problem, there are four simple and popular algorithms: *alignment* (e.g. [22, 4]), *pose clustering* (e.g. [27]), *geometric hashing* (e.g. [26]) and *generalized Hough transform* (GHT) (e.g. [19]). These algorithms are often confused with one another in the literature. A brief description of these algorithms can be found in the full version of this paper. Among these four algorithms, the most efficient algorithm is GHT.

As we mentioned above, the tolerant-LCP (or more generally, threshold-LCP) is a better model of many situations that arise in practice. However, it turns out that it is considerably more difficult to solve the tolerant-LCP than the exact-LCP. Intuitively, a fundamental difference between the two problems lies in the fact that for the exact-LCP the set of rigid motions, that may potentially correspond to the solution, is *discrete* and can be easily enumerated. Indeed, the algorithms for the exact-LCP are all based on the (explicit or implicit) enumeration of rigid motions that can be obtained by matching triplets to triplets. On the other hand, for the tolerant-LCP this set is *continuous*, and hence the direct enumeration strategies do not work. Nevertheless, the optimal rigid motions can be characterized by a set of high degree polynomial equations as in [8]. A similar characterization was made by Alt and Guibas in [6] for the 2D tolerant-PM problem and by the authors in [13] for the 3D tolerant-PM. All known algorithms for the threshold-LCP use these characterizations and involve solving systems of high degree equations which leads to "numerical instability problem" [6]. Note that exact-LCP and the exact solution for tolerant-LCP are two distinct problems. (Readers are cautioned not to confuse these two problems as in Gavrilov et al. [15].) Ambühl et al. [8] gave an algorithm for tolerant-LCP with running time $O(m^{16}n^{16}\sqrt{m+n})$. The algorithm in [13] for threshold-PM can be adapted to solve the tolerant-LCP in $O(m^6n^6(m+n)^{2.5})$ time. Both algorithms are for bottleneck distances. These algorithms can be modified to solve threshold-LCP under Hausdorff distance with a better running time by replacing the maximum bipartite graph matching algorithm which runs in $O(n^{2.5})$ with the $O(n \log n)$ time algorithm for nearest neighbor search. Both of these algorithms are for the general threshold-LCP, but to the best of our knowledge, these algorithms are the only known exact algorithms for the tolerant-LCP also.

Goodrich et al [16] showed that there is a small discrete set of rigid motions which contains a rigid motion approximating (in distance) the optimal rigid motion for the threshold-PM problem, and thus the threshold-PM problem can be solved approximately by an enumeration strategy. Based on this idea and the alignment approach of enumerating all possible such discrete rigid motions, Akutsu [3], and Biswas and Chakraborty [10, 9] gave distance-approximation algorithms with running time $O(m^4n^4\sqrt{m+n})$ for the threshold-LCP under bottleneck distance, which can be modified to give $O(m^3n^4 \log m)$ time algorithm for the tolerant-LCP.

Similar to the tolerant-LCP, the exact algorithm for tolerant-PM is difficult, even in 2D (see [6]). For this reason, Heffernan and Schirra [20] introduced approximate *decision* algorithms to approximate the minimum Hausdorff distance between two point sets. Given $\epsilon > 0$, their algorithm answers correctly (YES/NO) if ϵ is not too close to the optimal value ϵ^* (which is the minimum Hausdorff distance between the two point sets) and DON'T KNOW if the answer is too close to the optimal value. Notice that this approximation framework can not be directly adopted to the LCP problem because in the LCP case there are two parameters – size and distance – to be optimized. This appears to be mistaken by Indyk et al. in [21, 15] where their approximation algorithm for tolerant-LCP is not well defined. Cardoze and Schulman [11] gave an approximation algorithm (with possible false positives) but the transformations are restricted to translations for the LCP problem. Given α, let $\epsilon_{min}(\alpha)$ denote the smallest ϵ for which α-LCP exists; given ϵ, let $\alpha_{min}(\epsilon)$ denote the smallest α for which α-LCP exists. Biswas and Chakraborty [10, 9] combined the idea from Heffernan and Schirra and the algorithm of Akutsu [3] to give a size-approximation algorithm which returns $\alpha_u > \alpha_l$ such that $\min\{\alpha : \epsilon > 8\epsilon(\alpha)\} \geq \alpha_u \geq \alpha_{min}(\epsilon)$ and $\alpha_{min}(\epsilon) > \alpha_l \geq \max\{\alpha : \epsilon < \frac{1}{8}\epsilon_{min}(\alpha)\}$. However, all these approximation algorithms still take high running time of $\tilde{O}(m^3n^4)$ (the notation $\tilde{O}$ hides poly log factors in m and n).

In practice, the tolerant-LCP is solved heuristically by using the geometric hashing and GHT algorithms for which rigorous analyses are only known for the exact-LCP. For example, the algorithms in [14, 27] are for tolerant-LCP but the analyses are for exact-LCP only. Because of its practical performance, the exact version of GHT was carefully analyzed by Akutsu et al. [4], and a randomized version of the exact version of geometric hashing in 2D was given by Irani and Raghavan [22]. The tolerant version of GHT (and geometric hashing) is based on the corresponding exact version by replacing the exact matching with the approximate matching which requires a distance measure to compare the keys. We can no longer identify the optimal rigid motion by the maximum votes as in the exact case. Instead, the tolerant version of GHT *clusters* the rigid motions (which are points in a six-dimensional space) and heuristically approximates the optimal rigid motion by a rigid motion in the largest cluster. Thus besides not giving any guarantees about the solution, this heuristic requires clustering in six dimensions, which is computationally expensive.

Other Related Work. There is some closely related work that aims at computing the *minimum* Hausdorff distance for PM (see, e.g., [12] and references therein). Also, the problems we are considering can be thought of as the point pattern matching problem under *uniform distortion*. Recently, there has been some work on point pattern matching under *non-uniform distortion* [24, 5].

Our results. There are three results in this paper. First, we introduce a new distance-approximation algorithm for tolerant-LCP algorithm, called *T-hashing*.

Theorem 1. *T-hashing (see Algorithm 1.) finds a rigid motion μ such that there is a subset I of Q that is at least as large as* $\mathsf{LCP}(P, Q)$ *with error parameter ϵ and each point of $\mu(I)$ is within distance 4ϵ from some point of P.*

T-hashing is simple and more efficient than the known distance-approximation algorithms (which are alignment-based) for tolerant-LCP. The running time of T-hashing is $O(m^3 n^3 \log m)$ in the worst case. For general input, we expect the algorithm to be much faster because it is simpler and more efficient than the previous heuristics that are known to be fast in practice. This is because our clustering step is simple (sorting linearly ordered data) while the clustering step in those heuristics requires clustering high-dimensional data.

Second, based on a combinatorial observation, we improve the algorithms for exact-α-LCP by a linear factor for pose clustering or GHT and a quadratic factor for alignment or geometric hashing. This also corrects a mistake by Irani and Raghavan [22].

Finally, we achieve a similar speed-up for T-hashing using a sampling approach based on expander graphs at the expense of approximation in the matched set size. We remark that this result is mainly of theoretical interest because of the large constant factor involved. Expander graphs have been used before in geometric optimization for fast deterministic algorithms [1, 23]; however, the way we use these graphs appears to be new. Our results also hold when we extend the set of transformations to scaling; for simplicity we restrict ourselves to rigid motions in this paper.

Terminology and Notation. For a transformation μ, denote by I_μ the set of points in $\mu(Q)$ that are within distance ϵ of some point in P. We call I_μ the matched set of μ and say that μ is an $|I_\mu|$-matching. We call the transformation μ that maximizes $|I_\mu|$ the *maximum matching* transformation. A *basis* is a minimal (for containment relation) ordered tuple of points which is required to uniquely define a rigid motion. For example, in 2D every ordered pair is a basis; while in 3D, every non-collinear triplet is a basis. In Figure 1, a rigid motion in 3D is specified by mapping a basis (q_1, q_2, q_3) to another basis (p_1, p_2, p_3). We call a key used to represent an ordered tuple S a *rigid motion*

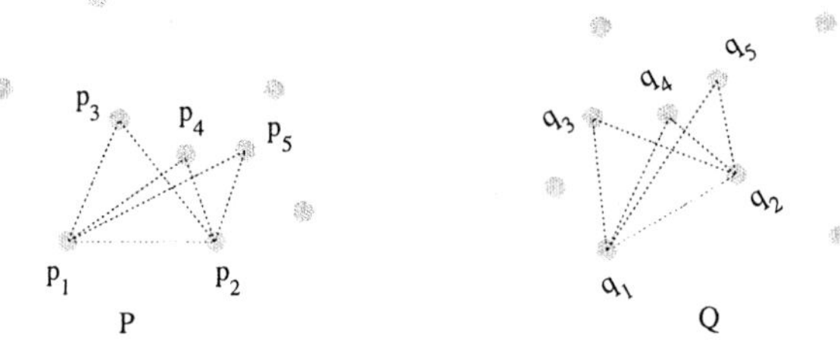

Fig. 1. In this example, the rigid motion is obtained by matching $C_Q = \{q_1, q_2, \ldots, q_5\}$ in Q to $C_P = \{p_1, p_2, \ldots, p_5\}$ in P. We have $\mathsf{LCP}(P, Q) = |C_P| = |C_Q| = 5$. The corresponding 5-matching transformation μ can be discovered by matching (q_1, q_2) to (p_1, p_2), the rigid motions μ_i that transform (q_1, q_2, q_i) to (p_1, p_2, p_i) for $i = 3, 4, 5$ are all the same and thus $\mu = \mu_3 = \mu_4 = \mu_5$ will get 3 votes, which is the maximum.

invariant key if it satisfies the following: (1) the key remains the same for all $\mu(S)$ where μ is any rigid motion, and (2) for any two ordered tuples S and S' with the same rigid motion invariant key there is a unique rigid motion μ such that $\mu(S) = S'$. For example, as rigid motion preserves orientation and distances among points, given a non-degenerate triangle Δ, the 3 side lengths of Δ together with the orientation (the sign of the determinant of the ordered triplet) form a rigid motion invariant key for Δ in $\mathbb{R}^3$. Henceforth, for simplicity of exposition, in the description of our algorithms we will omit the orientation part of the key.

Outline. The paper is organized as follows. In Section 2, we introduce our new distance-approximation algorithm for tolerant-LCP, called *T-hashing*. In Section 3, we show how a simple deterministic sampling strategy based on the pigeonhole principle yields speed-ups for the exact-α-LCP algorithms. In Section 4, we show how to use expander graphs to further speed up the T-hashing algorithm for tolerant-α-LCP at the expense of approximation in the matched set size.

2 T-Hashing

In this section, we introduce a new distance-approximation algorithm, called *T-hashing*, for tolerant-LCP. The algorithm is based on a simple geometric observation. It can be seen as an improvement of a known GHT-based heuristic such that the output has theoretical guarantees.

2.1 Review of GHT

First, we review the idea of the pair-based version of GHT for exact-LCP. See the full version of this paper for more details. For each congruent pair, say (p_1, p_2) in P and (q_1, q_2) in Q, and for each of the remaining points $p \in P$ and $q \in Q$, if (q_1, q_2, q) is congruent to (p_1, p_2, p), compute the rigid motion μ that matches (q_1, q_2, q) to (p_1, p_2, p). We then cast one vote for μ. The rigid motion that receives the maximum number of votes corresponds to the maximum matching transformation sought. See Figure 1 for an example.

2.2 Comparable Rigid Motions by Dihedral Angles

For the exact-LCP, one only needs to compare rigid motions by equality (for voting). For the tolerant-LCP, one needs to measure how close two rigid motions are. In $\mathbb{R}^3$, each rigid motion can be described by 6 parameters (3 for translations and 3 for rotations). How to define a distance measure between rigid motions? We will show below that the rigid motions considered in our algorithm are related to each other in a simple way that enables a natural notion of distance between the rigid motions.

Observation. In the pair-based version of GHT as described above, the rigid motions to be compared have a special property: the rigid motions transform a common pair — they all match (q_1, q_2) to (p_1, p_2) in Figure 1. Two such transformations no longer differ in all 6 parameters but differ in only one parameter. To see this, we first recall that a dihedral angle is the angle between two intersecting planes; see Figure 2 for an example. In general, we can decompose the rigid motion for matching (q_1, q_2, q_3) to (p_1, p_2, p_3) into two parts: first, we transform (q_1, q_2) to (p_1, p_2) by a transformation ϕ_1; then we rotate the point $\phi_1(q_3)$ about $\overrightarrow{p_1 p_2}$ by an angle θ, where θ is the dihedral angle between the planes (p_1, p_2, p_3) and $(\phi_1(q_1), \phi_1(q_2), \phi_1(q_3))$. This will bring q_3 to coincide with p_3. Thus, we have the following lemma:

Lemma 1. *Let (p_1, p_2, p_3) and (q_1, q_2, q_3) be two congruent non-collinear triplets, and let ϕ_1 be a rigid motion that takes q_i to p_i for $i = 1, 2$. Let ϕ_2 be the rotation about $\overrightarrow{p_1 p_2}$ by an angle θ, where θ is the dihedral angle between the planes (p_1, p_2, p_3) and $(\phi_1(q_1), \phi_1(q_2), \phi_1(q_3))$. Then the unique rigid motion that takes (p_1, p_2, p_3) to (q_1, q_2, q_3) is equal to $\phi_2 \circ \phi_1$.*

We now state another lemma that will be useful in the description and proof of correctness of T-hashing. Let (p_1, p_2, p) and q be four points as shown in Figure 2. Consider the rotations about $\overrightarrow{p_1 p_2}$ that take q to within ϵ of p. The rotation angles of these transformations form a subinterval of $[0, 2\pi)$. This is because a circle C (corresponding to the trajectory of p) intersects with the sphere B (around p with radius ϵ) at at most two points (corresponding to a subinterval of $[0, 2\pi)$), as shown in Figure 2. That is, we have the following lemma:

Lemma 2. *Let $p_1, p_2, p, q \in \mathbb{R}^3$ be four points (not necessarily non-collinear), then the rotation angles of transformations that rotate q about $\overrightarrow{p_1 p_2}$ to within ϵ of p form a subinterval of $[0, 2\pi)$.*

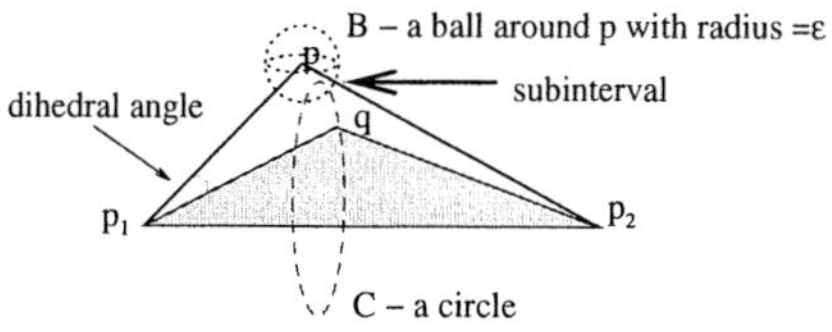

Fig. 2. The dihedral angle is the angle between planes formed by (p_1, p_2, q) and (p_1, p_2, p). The rotation angles of transformations that rotate q about $\overrightarrow{p_1 p_2}$ to within ϵ of p form a subinterval of $[0, 2\pi)$.

2.3 Approximating the Optimal Rigid Motion by the "Diametric" Rigid Motion

For a point set $S \subset \mathbb{R}^3$, we call a pair of points $\{p, q\} \in S^2$ *diameter-pair* if $||p - q|| =$ diameter(S). A rigid motion of Q that takes q_1 to p_1 and q_2 on the line p_1p_2 and closest possible to p_2 is called a (p_1, p_2, q_1, q_2)-rigid motion. Based on an idea similar to the one behind Lemma 2.4 in Goodrich et al. [16], we have the following lemma:

Lemma 3. *Let μ be a rigid motion such that each point of $\mu(S)$, where $S \subseteq Q$, is within distance ϵ of a point in P. Let $\{q_1, q_2\}$ be a diameter-pair of S. Let $p_i \in P$ be the closest point to $\mu(q_i)$ for $i = 1, 2$. Then we have a (p_1, p_2, q_1, q_2)-rigid motion μ' of Q such that each point of $\mu'(S)$ is within 4ϵ of a point in P.*

Proof Sketch. Translate $\mu(q_1)$ to p_1; this translation shifts each point by at most ϵ. Next, rotate about p_1 such that $\mu(q_2)$ is closest to p_2 (which implies $\mu'(q_1)$, $\mu'(q_2)$ and p_2 are collinear). Since $\{q_1, q_2\}$ is a diameter-pair, this rotation moves each point by at most 2ϵ. Thus, each point is at most $\epsilon + \epsilon + 2\epsilon = 4\epsilon$ from its matched point. □

2.4 Approximation Algorithm for Tolerant-LCP

We first describe the idea of our algorithm T-hashing. Input is two point sets in $\mathbb{R}^3$, $P = \{p_1, \ldots, p_m\}$ and $Q = \{q_1, \ldots, q_n\}$ with $m \geq n$, and $\epsilon \geq 0$. Suppose that the optimal rigid motion μ_0 was achieved by matching a set $I_{\mu_0} = \{q_1, q_2, \ldots, q_k\} \subseteq Q$ to $J_{\mu_0} = \{p_1, p_2, \ldots, p_k\} \subseteq P$. WLOG, assume that $\{q_1, q_2\}$ is the diameter pair of I_{μ_0}. Then by Lemma 3, there exists a (p_1, p_2, q_1, q_2)-rigid motion μ of Q such that $\mu(I_{\mu_0})$ is within 4ϵ of a point in P. Since we do not know the matched set, we do not know a diameter-pair for the matched set either. Therefore, we exhaustively go through each possible pair. Namely, for each pair $(q_1, q_2) \in Q$ and each pair $(p_1, p_2) \in P$, if they are approximately congruent then we find a (p_1, p_2, q_1, q_2)-rigid motion μ of Q that matches as many remaining points as possible. Note that (p_1, p_2, q_1, q_2)-rigid motions are determined up to a rotation about the line p_1p_2. By Lemma 2, the rotation angles that bring $\mu(q_i)$ to within 4ϵ of p_i form a subinterval of $[0, 2\pi)$. And the number of non-empty intersection subintervals corresponds to the size of the matched set. Thus, to find μ, for each pair $(p, q) \in P \setminus \{p_1, p_2\} \times Q \setminus \{q_1, q_2\}$, we compute the dihedral angle interval according to Lemma 2. The rigid motion μ sought corresponds to an angle ϕ that lies in the maximum number of dihedral intervals. The details of the algorithm are described in Algorithm 1.

Time Complexity. For each triplet in Q, using kd-tree for range query, it takes $O(m^{3\cdot(1-\frac{1}{3})} + m^3 + m^3 \log m^2) = O(m^3 \log m)$ for lines 11–20. For each pair (q_1, q_2) and (p_1, p_2), we spend time $O(mn)$ to find the subintervals for the dihedral angles, and time $O(mn \log m)$ to sort these subintervals and do the scan to find an angle that lies in the maximum number of subintervals. Thus the total time is $O(m^3 n^3 \log m)$.

3 Improvement by Pigeonhole Principle

In this section we show how a simple deterministic sampling strategy based on the pigeonhole principle yields speed-ups for the four basic algorithms for exact-α-LCP.

Algorithm 1. T-hashing

procedure PREPROCESSING
  for each pair (p_1, p_2) of P **do**
    Compute and insert the key of $||p_1p_2||$ into a dictionary $\mathcal{D}_1$;
  end for
  for each triplet (p_1, p_2, p_3) of P **do**
    Compute and insert the *rigid motion invariant* key for (p_1, p_2, p_3) into a dictionary $\mathcal{D}_2$;
  end for
end procedure
procedure RECOGNITION
  for each pair (q_1, q_2) of $\binom{Q}{2}$ **do** ▷ This can be reduced by the edge set of an expander of Q.
    if $[||q_1q_2|| - 2\epsilon, ||q_1q_2|| + 2\epsilon]$ exists in $\mathcal{D}_1$ **then**
      Initialize an empty dictionary $\mathcal{D}_3$ of pairs;
      for each remaining point $q \in Q$ **do**
        Compute and search the range $[||q_1q_2|| - 2\epsilon, ||q_1q_2|| + 2\epsilon] \times [||qq_1|| - 2\epsilon, ||qq_1|| + 2\epsilon] \times [||qq_2|| - 2\epsilon, ||qq_2|| + 2\epsilon]$ of (q_1, q_2, q) in $\mathcal{D}_2$; ▷ e.g. using a kd-tree.
        for each entry (p_1, p_2, p) found **do**
          If (p_1, p_2) exists in $\mathcal{D}_3$, increase its vote; otherwise insert (p_1, p_2) into $\mathcal{D}_3$ with one vote;
          Append the matched pair (q, p) to the list associated with (p_1, p_2);
        end for
      end for
    end if ▷ Compute the maximum transformation that matches (q_1, q_2) to (p_1, p_2).
    for each pair (p_1, p_2) in the dictionary $\mathcal{D}_3$ **do**
      Compute a transformation ϕ that brings q_1 to p_1 and q_2 closest to p_2;
      For each matched pair (q, p) of the associated list of (p_1, p_2), compute an interval of dihedral angles such that $\phi(q)$ is within 4ϵ of p;
      Sort all the intervals of dihedral angles; and find a dihedral angle ψ that occurs in the largest number V of intervals;
      Compute the transformation μ by the composition of ϕ and the rotation about p_1p_2 by angle ψ; ▷ μ brings $V + 2$ points of Q to within 4ϵ of some matched points in P.
      Keep the maximum matched set size and the corresponding transformation;
    end for
  end for
end procedure

Specifically, we get a linear speed-up for pose clustering and GHT, and quadratic speed-up for alignment and geometric hashing. It appears to have been erroneously concluded previously that no such improvements were possible deterministically [22].

In pose clustering or GHT, suppose we know a pair (q_1, q_2) in Q that is in the sought matched set, then the transformation sought will be the one receiving the maximum number of votes among the transformations computed for (q_1, q_2). Thus if we have chosen a pair (q_1, q_2) that lies in the matched set, then the maximum matching transformation will be found. We are interested in the question "can we find a pair in the matched set without exhaustive enumeration"? The answer is yes: we only need to try a linear number of pairs (q_1, q_2) to find the maximum matching transformation or conclude that there is none that matches at least $\frac{n}{\alpha}$ points.

We are given a set $Q = \{q_1, \ldots, q_n\}$, and let $I \subseteq Q$ be an unknown set of size $\geq \frac{n}{\alpha}$ for some constant $\alpha > 1$. We need to discover a pair (p, q) with $p, q \in I$ by using queries of the following type. A query consist of a pair (a, b) with $a, b \in Q$. If we have $a, b \in I$, the answer to the query is YES, otherwise the answer is NO. Thus our goal is to devise a deterministic query scheme such that as few queries are needed as possible in the worst case (over the choice of I) before a query is answered YES. Similarly, one can ask the question about querying triplets to discover a triplet entirely in I.

Theorem 2. *For an unknown set $I \subseteq Q$ with $|I| \geq \frac{n}{\alpha}$ and $|Q| = n$ using queries as described above,*

(1) it suffices to query $O(\alpha n)$ pairs to discover a pair in I;
(2) it suffices to query $O(\alpha^2 n)$ triplets to discover a triplet in I.

The simple proof, which is based on the pigeonhole principle, can be found in the full version of this paper. Consequently, pose clustering or GHT can be sped up by a linear factor; alignment or geometric hashing can be sped up by a quadratic factor.

4 Expander-Based Sampling

While for the exact-α-LCP the simple pigeonhole sampling served us well, for the tolerant-α-LCP we do not know any such simple scheme for choosing pairs. The reason is that now we not only need to guarantee that each large set contain some sampled pairs, but also that each large set contain a sampled pair with large length (diameter-pair) as needed for the application of Lemma 3 in the T-hashing algorithm. Our approach is based on expander graphs (see, e.g., [2]). Informally, expander graphs have linear number of edges but the edges are "well-spread" in the sense that there is an edge between any two sufficiently large disjoint subsets of vertices. Let G be an expander graph with Q as its vertex set. We show that for each $S \subseteq Q$, if $|S|$ is not too small, then there is an edge (u, v) in G such that $(u, v) \in S^2$ and $||uv||$ approximates the diameter of S.

By choosing the pairs for the T-hashing algorithm from the edge set of G (the rest of the algorithm is same as before), we obtain a bicriteria – distance and size – approximation algorithm as stated in Theorem 4 below. We first give a few definitions and recall a result about expander graphs that we will need to prove the correctness of our algorithm.

Definition 1. *Let S be a finite set of points of $\mathbb{R}^r$ for $r \geq 1$, and let $0 \leq k \leq n$. Define* $\mathsf{diameter}(S, k) = \min_{T:|T|=k} \mathsf{diameter}(S \setminus T)$.

That is, $\mathsf{diameter}(S, k)$ is the *minimum* of the diameter of the sets obtained by deleting k points from S. Clearly, $\mathsf{diameter}(S, 0) = \mathsf{diameter}(S)$.

Let U and V be two disjoint subsets of vertices of a graph G. Denote by $e(U, V)$ the set of edges in G with one end in U and the other in V. We will make use of the following well-known theorem about the eigenvalues of graphs (see, e.g. [25], for the proof and related background).

Theorem 3. *Let G be a d-regular graph on n vertices. Let $d = \lambda_1 \geq \lambda_2 \geq \ldots \geq \lambda_n$ be the eigenvalues of the adjacency matrix of G. Denote $\lambda = \max_{2 \leq i \leq n} |\lambda_i|$. Then for every two disjoint subsets $U, W \subset V$,*

$$\left| |e(U, W)| - \frac{d|U||W|}{n} \right| \leq \lambda \sqrt{|U||W|}. \tag{1}$$

Corollary 1. *Let $U, W \subset V$ be two disjoint sets with $|U| = |W| > \frac{\lambda n}{d}$. Then G has an edge in $U \times W$.*

Proof. It follows from (1) that if $\frac{d|U||W|}{n} > \lambda\sqrt{|U||W|}$ then $|e(U, W)| > 0$, and since $|e(U, W)|$ is integral, $|e(U, W)| \geq 1$. But the above condition is clearly true if we take U and W as in the statement of Corollary 1. □

There are efficient constructions of graph families known with $\lambda < 2\sqrt{d}$ (see, e.g., [2]). Let us call such graphs *good expander graphs.* We can now state our main result for this section.

Theorem 4. *For an α-LCP instance (P, Q) with $\mathsf{LCP}(P, Q) > \frac{n}{\alpha}$, the T-hashing algorithm with expander-based sampling using a good expander graph of degree $d > 2500\alpha^2$ finds a rigid motion μ in time $O(m^3 n^2 \log m)$ such that there is a subset I satisfying the following criteria:*

(1) size-approximation criterion: $|I| \geq \mathsf{LCP}(P, Q) - \frac{50}{\sqrt{d}}n$;

(2) distance-approximation criterion: each point of $\mu(I)$ is within distance 6ϵ from a point in P.

Thus by choosing d large enough we can get as good size-approximation as desired. The constants in the above theorem have been chosen for simplicity of the proof and can be improved slightly.

For the proof we first need a lemma showing that choosing the query pairs from a graph with small $\lambda(G)$ (the second largest eigenvalue of G) gives a long (in a well-defined sense) edge in every not too small subset of vertices.

Lemma 4. *Let G be a d-regular graph with vertex set $Q \subset \mathbb{R}^3$, and $|Q| = n$. Let $S \subseteq Q$ be such that $|S| > \frac{25\lambda(G)n}{d}$. Then there is an edge $\{s_1, s_2\} \in E(G) \cap S^2$ such that $||s_1 s_2|| \geq \frac{\mathsf{diameter}(S, \frac{25\lambda(G)}{d}n)}{2}$.*

Proof. For a positive constant c to be chosen later, remove cn pairs from S as follows. First remove a diameter pair, then from the remaining points remove a diameter pair, and so on. Let T be the set of points in the removed pairs and T^p the set of removed pairs. The remaining set $S \setminus T$ has diameter $\geq \mathsf{diameter}(S, 2cn)$ by the definition of $\mathsf{diameter}(S, 2cn)$, and hence each of the removed pairs has length $\geq \mathsf{diameter}(S, 2cn)$. For $B, C \subset S$ let $||B, C|| = \min_{b \in B, c \in C} ||bc||$.

Claim. The set T defined above can be partitioned into three sets B, C, E, such that $|B|, |C| \geq \frac{cn}{6}$, and $||B, C|| \geq \frac{\text{diameter}(S,2cn)}{2}$.

Proof. Fix a Cartesian coordinate system and consider the projections of the pairs in T^p on the x-, y- ,and z-axes. It is easy to see that for at least one of these axes, at least $\frac{cn}{3}$ pairs have projections of length $\geq \frac{\text{diameter}(S,2cn)}{\sqrt{3}}$. Suppose without loss of generality that this is the case for the x-axis, and denote the set of projections of pairs on the x-axis with length $\geq \frac{\text{diameter}(S,2cn)}{\sqrt{3}}$ by T_x^p, and the set of points in the pairs in T_x^p by T_x. We have $|T_x| \geq 2cn/3$. Now consider a *sliding window* W on the x-axis of length $\frac{\text{diameter}(S,2cn)}{2}$, initially at $-\infty$, and slide it to $+\infty$. At any position of W, each pair in T_x^p has at most 1 point in W, as the length of any pair is more than the length of W. Thus at any position, W contains $\leq |T_x^p| = |T_x|/2$ points. It is now easy to see by a standard continuity argument that there is a position of W, call it $\bar{W}$, where there are $\geq \frac{|T_x|}{4} \geq \frac{cn}{6}$ points of T_x both to the left and to the right of $\bar{W}$.

Now, B is defined to be the set of points in T whose projection is in T_x and is to the left of $\bar{W}$; similarly C is the set of points in T whose projection is in T_x and is to the right of $\bar{W}$. Clearly any two points, one from B and the other from C, are $\frac{\text{diameter}(S,2cn)}{2}$-apart. □

Coming back to the proof of Lemma 4, the property that we need from the query-graph is that for any two disjoint sets $B, C \subset S$ of size $\delta|S|$, where δ is a small positive constant, the query-graph should have an edge in $B \times C$.

By Corollary 1 if $|B| \geq \frac{cn}{6} > \frac{\lambda n}{d}$, and $|C| \geq \frac{cn}{6} > \frac{\lambda n}{d}$, that is, if $c > \frac{6\lambda}{d}$, then G has an edge in $B \times C$. Taking $c = \frac{12.5\lambda}{d}$ completes the proof of Lemma 4. □

Proof of Theorem 4. If we take G to be a good expander graph then Lemma 4 gives that G has an edge of length $\geq \frac{\text{diameter}(S,\frac{50}{\sqrt{d}}n)}{2}$. Let S also be a solution to tolerant-LCP for input (P, Q) with error parameter $\epsilon > 0$. We have that one of the sampled pairs has length at least $\frac{\text{diameter}(S,\frac{50}{\sqrt{d}}n)}{2}$. Thus applying an appropriate variant (replacing the diameter pair by the sampled pair with large length as guaranteed by Lemma 4) of Lemma 3, we get a rigid motion μ such that there is a subset I satisfying the following:

(1) $|I| \geq |S| - \frac{50}{\sqrt{d}}n$ for any $d > 2500\alpha^2$;
(2) Each point of I is within $6\epsilon (= \epsilon + \epsilon + 4\epsilon)$ of a point in M. □

Acknowledgment. We thank S. Muthukrishnan and Ali Shokoufandeh for the helpful comments and advice.

References

1. Ajtai, M., Megiddo, N.: A Deterministic Poly(log log N)-Time N-Processor Algorithm for Linear Programming in Fixed Dimension. Proc. 24th ACM Symp. on Theory of Computing (1992) 327–338
2. Alon, N., Spencer, J.: The Probabilistic Method. Wiley-Interscience 2000
3. Akutsu, T.: Protein Structure Alignment Using Dynamic Programming and Iterative Improvement. IEICE Trans. on Information and Systems, **12** (1996) 1629–1636

4. Akutsu, T., Tamaki, H., Tokuyama, T.: Distribution of Distances and Triangles in a Point Set and Algorithms for Computing the Largest Common Point Sets. Discrete & Computational Geometry, **20**(3) (1998) 307–331
5. Akutsu, T., Kanaya, K., Ohyama A., Fujiyama, A.: Point matching under non-uniform distortions. Discrete Applied Mathematics, **127**(1) (2003) 5–21
6. Alt, H., Guibas, L.J.: Discrete Geometric Shapes: Matching, Interpolation, and Approximation. In: Sack, J.-R., Urrutia J. (eds): Handbook of Computational Geometry, Elsevier Science Publishers B.V. North-Holland Amsterdam (1999) 121–153
7. Alt, H., Mehlhorn, K., Wagener, H., Welzl, E.: Congruence, Similarity, and Symmetries of Geometric Objects. Discrete & Computational Geometry, **3** (1988) 237–256
8. Ambühl, C., Chakraborty, S., Gärtner, B.: Computing Largest Common Point Sets under Approximate Congruence. ESA 2000, Lecture Notes in Computer Science Vol. 1879. Springer-Verlag, (2000) 52–63
9. Biswas, S., Chakraborty, S.: Fast Algorithms for Determining Protein Structure Similarity. Workshop on Bioinformatics and Computational Biology, Hyderabad, India, (2001)
10. Chakraborty, S., Biswas, S.: Approximation Algorithms for 3-D Commom Substructure Identification in Drug and Protein Molecules. Workshop on Algorithms and Data Structures (WADS) 1999. Lecture Notes in Computer Science, Vol. 1663. Springer-Verlag (1999)
11. Cardoze, D., Schulman, L.: Pattern Matching for Spatial Point Sets. Proc. 39th Annual IEEE Symp. on Foundations of Computer Science (1998) 156–165
12. Chew, L.P., Dor, D., Efrat, A., Kedem, K.: Geometric Pattern Matching in d-Dimensional Space, Discrete and Computational Geometry, **21** (1999) 257–274
13. Choi, V., Goyal, N.: A Combinatorial Shape Matching Algorithm for Rigid Protein Docking. Combinatorial Pattern Matching (CPM) 2004, Lecture Notes in Computer Science, Vol. 3109, Springer (2004) 285–296
14. Finn, P., Kavraki, L., Latombe, J-C., Motwani, R., Shelton, C., Venkatasubramanian, S., Yao, A.: RAPID: Randomized Pharmacophore Identification in Drug Design The 13th Symposium on Computational Geometry, 1997. Computational Geometry: Theory and Applications **10** (4), (1998)
15. Gavrilov, M., Indyk, P., Motwani, R., Venkatasubramanian, S.: Combinatorial and Experimental Methods for Approximate Point Pattern Matching. Algorithmica **38**(1) (2003) 59–90
16. Goodrich, M.T., Mitchell, J.S.B., Orletsky, M.W.: Approximate Geometric Pattern Matching Under Rigid Motions. IEEE Trans. Pattern Anal. Mach. Intell. **21**(4) (1999) 371–379
17. Grimson, W.E.L, Huttenlocher, D.P.: On the Sensitivity of Geometric Hashing. Proc. of the 3rd Int'l Conference on Computer Vision, (1990) 334–338
18. Grimson, W.E.L., Huttenlocher, D.P.: On the Sensitivity of Hough Transform for Object Recognition. IEEE Trans. on Pattern Analysis and Machine Intell. **12**(3) (1990)
19. Hecker, Y., Bolle, R.: On Geometric Hashing and the Generalized Hough Transform. IEEE Trans. on Systems, Man and Cybernetics, **24** (1994) 1328–1338
20. Heffernan, P.J., Schirra, S.: Approximate Decision Algorithms for Point Set Congruence. Comput. Geom. **4** (1994) 137–156
21. Indyk, P., Motwani, R., Venkatasubramanian, S.: Geometric Matching under Noise: Combinatorial Bounds and Algorithms. The 10th ACM-SIAM Symp. on Discrete Algorithms (1999)
22. Irani, S., Raghavan, P.: Combinatorial and Experimental Results for Randomized Point Matching Algorithms. The 12th Symposium on Computational Geometry, 1996. Comput. Geom. **12**(1-2) (1999) 17–31
23. Katz, M., Sharir, M.: An Expander-based Approach to Geometric Optimization. SIAM J. Comput., **26**(5) 1997 1384–1408
24. Kenyon, C., Rabani, Y., Sinclair, A.: Low Distortion Maps Between Point Sets. Proc. of the 36th Annual ACM Symp. on Theory of Computing (2004)

25. Krivelevich, M., Sudakov, B.: Pseudo-random Graphs. Preprint. Available at `http://www.math.tau.ac.il/~krivelev/papers.html`
26. Lamdan, Y., Wolfson, H.J.: Geometric Hashing: A General and Efficient Model-based Recognition Scheme. The 2nd Int'l Conference on Computer Vision (1988) 238–249
27. Olson, C.F.: Efficient Pose Clustering Using a Randomized Algorithm. Int'l J. of Computer Vision, **23**(2) (1997) 131–147
28. Wolfson, H.J., Rigoutsos, I.: Geometric Hashing: an Overview. IEEE Comput. Science and Engineering, **4** (1997) 10–21
29. Nussinov, R., Wolfson, H.J.: Efficient Detection of Three - Dimensional Motifs In Biological Macromolecules by Computer Vision Techniques, Proc. of the Nat'l Academy of Sciences, **88** (1991) 10495 – 10499

Oblivious Medians Via Online Bidding
(Extended Abstract)

Marek Chrobak[1,*], Claire Kenyon[2], John Noga[3], and Neal E. Young[1]

[1] Department of Computer Science, University of California, Riverside
[2] Computer Science Department, Brown University
[3] Department of Computer Science, California State University, Northridge

Abstract. Following Mettu and Plaxton [22, 21], we study oblivious algorithms for the k-medians problem. Such an algorithm produces an incremental sequence of facility sets. We give improved algorithms, including a $(24 + \epsilon)$-competitive deterministic polynomial algorithm and a $2e \approx 5.44$-competitive randomized non-polynomial algorithm. Our approach is similar to that of [18], which was done independently.

We then consider the competitive ratio with respect to *size*. An algorithm is *s-size-competitive* if, for each k, the cost of F_k is at most the minimum cost of any set of k facilities, while the size of F_k is at most sk. We present optimally competitive algorithms for this problem.

Our proofs reduce oblivious medians to the following *online bidding* problem: faced with some unknown threshold $T \in \mathbb{R}^+$, an algorithm must submit "bids" $b \in \mathbb{R}^+$ until it submits a bid $b \geq T$, paying the sum of its bids. We describe optimally competitive algorithms for online bidding.

Some of these results extend to approximately metric distance functions, oblivious fractional medians, and oblivious bicriteria approximation.

When the number of medians takes only two possible values k or l, for $k < l$, we show that the optimal cost-competitive ratio is $2 - 1/l$.

1 Introduction and Summary of Results

An instance of the k-median problem is specified by a finite set $\mathcal{C}$ of customers, a finite set $\mathcal{F}$ of facilities, and, for each customer u and facility f, a distance $d_{uf} \geq 0$ from u to f representing the cost of serving u from f. The cost of a set of facilities $X \subseteq \mathcal{F}$ is $cost(X) = \sum_{u \in \mathcal{C}} d_{uX}$, where $d_{uX} = \min_{f \in X} d_{uf}$. For a given k, the *(offline) k-median problem* is to compute a *k-median*, that is, a set $X \subseteq \mathcal{F}$ of cardinality k for which $cost(X) = opt_k$ is minimum (among all sets of cardinality k). *Metric* k-median refers to the case where the distance function is metric (the shortest u-to-f path has length d_{uf} for each u and f).

The k-median problem is a well-known NP-hard facility location problem. Substantial work has been done on efficient approximation algorithms that, given k, find a set F_k of k medians of approximately minimum cost [2, 1, 6, 5, 13, 12, 24]. In particular, for the metric version Arya et al. show that, for any $\epsilon > 0$, a set F_k of cost at most $(3 + \epsilon)opt_k$ can be found in polynomial time [2].

* Research supported by NSF Grant CCR-0208856.

J.R. Correa, A. Hevia, and M. Kiwi (Eds.): LATIN 2006, LNCS 3887, pp. 311–322, 2006.

problem:	**cost-competitive metric**		**size-competitive**		**bidding**
time:	polynomial	non-polynomial	polynomial	non-polynomial	polynomial
deterministic	$24 + \epsilon$	8	$\mathbf{O(\log n)}$	**4**	**4**
randomized	$6e + \epsilon < 16.31$	$2e < 5.44$	$\mathbf{O(\log n)}$	$\mathbf{e < 2.72}$	$\mathbf{e < 2.72}$

Fig. 1. Competitive ratios shown for oblivious medians and online bidding. Ratios in bold are optimal.

Oblivious medians is an online version of the k-median problem where k is not specified in advance [22, 21]. Instead, authorizations for additional facilities arrive over time. A (possibly randomized) *oblivious algorithm* produces a sequence $\bar{F} = (F_1, F_2, \ldots, F_n)$ of facility sets which must satisfy the oblivious constraint $F_1 \subseteq F_2 \subseteq \cdots \subseteq F_n \subseteq \mathcal{F}$. In general, in an oblivious solution, the F_k's cannot all simultaneously have minimum cost. The algorithm is said to be *c-cost-competitive*, or to have *cost-competitive ratio* of c, if it produces a (possibly random) sequence $\bar{F}$ of sets which is c-cost-competitive, that is, such that for each k, the set F_k has size at most k and (expected) cost at most $c \cdot opt_k$. For offline solutions we use the term "approximate" instead of "competitive".

Mettu and Plaxton [22, 21] give a c-cost-competitive linear time oblivious algorithm with $c \approx 30$. Our first contribution is to improve this ratio. The problem is difficult both because (1) the solution must be oblivious, and (2) even the offline problem is NP-hard. To study separately the effects of the two difficulties, we consider both polynomial and non-polynomial algorithms.

Theorem 1. **(a)** *Oblivious metric medians has non-polynomial deterministic and randomized algorithms that are* 8*-cost-competitive and* $2e$*-cost-competitive, respectively.* **(b)** *If metric* k*-median has a polynomial* c*-cost-approximation algorithm, then the oblivious problem has polynomial deterministic and randomized algorithms that are* $8c$*-cost-competitive and* $2ec$*-cost-competitive, respectively.*

As it is known that there is a polynomial $(3 + \epsilon)$-cost-approximation algorithm for the offline metric medians [2], Theorem 1 implies the cost-competitive ratios shown in Fig. 1. Theorem 1 was recently and independently discovered by Lin, Nagarajan, Rajaraman and Williamson [18]. For polynomial algorithms, they improved the result further using a Lagrangian-multiplier-preserving approximation algorithm for facility location; they obtained 16-cost-competitive and randomized $4e$-competitive polynomial algorithms for metric medians.

We also consider here oblivious algorithms that are *s-size-competitive*: they are allowed to use extra medians, but must achieve the optimal cost for each k. An algorithm is *s-size-competitive* if it produces a sequence $\bar{F}$ such that each set F_k has cost at most opt_k and size at most sk. (If the algorithm is randomized, it must produce a random sequence such that each set F_k costs at most opt_k and has *expected* size at most sk.)

To our knowledge, size-competitive algorithms for oblivious medians have not been studied, although other online problems have been analyzed in an analogous setting of *resource augmentation* (e.g. [14, 7, 17]). We completely characterize the optimal size-competitive ratios for oblivious medians:

Theorem 2. **(a)** *Oblivious medians has non-polynomial deterministic and randomized oblivious algorithms that are* 4*-size-competitive and* e*-size-competitive, respectively.* **(b)** *No deterministic or randomized oblivious algorithm is less than* 4*-size-competitive or* e*-size-competitive, respectively.* **(c)** *If offline k-median has a polynomial c-size-competitive algorithm, then the oblivious problem has polynomial deterministic and randomized algorithms that are* $4c$*-size-competitive and* ec*-size-competitive, respectively.*

The upper and lower bounds in Theorem 2 hold for both the metric and non-metric problems. Part (c) on polynomial algorithms is included for completeness, as is the following result for offline k-medians (proof omitted):

Theorem 3. *Offline k-medians has a polynomial $O(\log(n))$-size-approximation algorithm.*

This improves the best previous result — a bicriteria approximation algorithm that finds a facility set of size $\ln(n+n/\epsilon)k$ and cost $(1+\epsilon)opt_k$ [24]. Our algorithm finds a true (not bicriteria) approximate solution: a facility set of size $O(\log k)$ and cost at most opt_k.

Theorems 2 and 3 imply the size-competitive ratios shown in Fig. 1. Note also that no polynomial algorithm (oblivious or offline) is $o(\log n)$-size-competitive unless P=NP, even for the metric case.

To analyze oblivious medians, we reduce the size- and cost-competitive oblivious problems to the following folklore "*online bidding problem*": An algorithm repeatedly submits "bids" $b \in \mathbb{R}^+$, until it submits a bid b that is at least as large as some unknown threshold $T \in \mathbb{R}^+$. Its cost is the total of the submitted bids. The algorithm is β-competitive if, for any $T \in \mathbb{R}^+$, its cost is at most βT (or, if the algorithm is randomized, its expected cost is at most βT). More generally, the algorithm may be given in advance a closed universe $\mathcal{U} \subseteq \mathbb{R}^+$, with a guarantee that the threshold T is in $\mathcal{U}$ and a requirement that all bids be in $\mathcal{U}$.

For $\mathcal{U} = \mathbb{R}^+$, it is known that an optimal deterministic strategy bids increasing powers of 2, and that there is a better randomized strategy which bids (randomly translated) powers of e. We complete this characterization by proving that the randomized strategy is optimal.

Theorem 4. **(a)** *Online bidding has deterministic and randomized algorithms that are* 4*-competitive and* e*-competitive, respectively. Furthermore, if $\mathcal{U}$ is finite, the algorithms run in time polynomial in $|\mathcal{U}|$.* **(b)** *No deterministic or randomized algorithm is less than* 4*-competitive or* e*-competitive, respectively, even when restricted to instances of the form $\mathcal{U} = \{1, 2, ..., n\}$ for some integer n.*

Weighted medians. All of our results extend to the weighted version, where we allow the facilities and the customers to have non-negative weights w. In this case, for a facility set X, one constrains the total weight $\sum_{f\in X} w(f)$ to be at most k, and one takes $cost(X) = \sum_{u\in C} w(u)d_{uX}$.

Approximate triangle inequality. Mettu and Plaxton show that their oblivious median algorithm also works in "λ-approximate" metric spaces, achieving cost-competitive-ratio $O(\lambda^4)$ [22, 21]. We reduce this ratio to $O(\lambda^2)$. We say that the cost function d is a *λ-relaxed metric* if $d_{fy} \le \lambda(d_{fx} + d_{xg} + d_{gy})$ for any $f, g \in \mathcal{F}$ and $x, y \in \mathcal{C}$. (This condition is somewhat less restrictive than the one in [22, 21]. A related concept was studied in [10].) Theorem 1 generalizes as follows (proofs omitted):

Theorem 5. (a) *Oblivious λ-relaxed metric medians has (non-polynomial) deterministic and randomized algorithms that are $8\lambda^2$-cost-competitive and $2e\lambda^2$-cost-competitive, respectively.* **(b)** *If offline λ-relaxed metric k-median has a polynomial c-cost-approximation algorithm, then the oblivious problem has deterministic and randomized polynomial algorithms that are $8\lambda^2 c$-cost-competitive and $2e\lambda^2 c$-cost-competitive, respectively.*

The kl-medians problem. A natural question to ask is whether better competitive ratios are possible if the number of medians can take only some limited number of values. As shown in [22, 21], no algorithm can be better than 2-competitive even when there are only two possible numbers of medians, either 1 or k, for some large k. Here, we solve the deterministic kl-median problem (where the number of medians is either k or $l > k$).

Theorem 6. *For any $k < l$, there is a deterministic oblivious algorithm for kl-medians with competitive ratio $2 - 1/l$, and no better ratio is possible.*

Oblivious fractional medians. A *fractional k-median* is a solution to the linear program which is the relaxation of the standard integer program for the k-median problem. The natural *oblivious* version of this fractional problem is to find a $c \ge 1$ and, for every integer $k \in [n]$ simultaneously, a pair $(x_{if}^{(k)}), (y_f^{(k)})$ meeting the constraints of the linear program, as well as $y_f^{(k)} \le y_f^{(k+1)}$ (for all f) and $\sum_u \sum_f x_{uf} d_{uf} \le c \cdot opt_k$ (where opt_k is the minimum cost of any fractional k-median). The goal is to minimize the competitive ratio c.

The proof of the theorem below (omitted) extends the proof of Theorem 1, along with the observation that the randomized algorithm for the fractional problem can be derandomized without increasing the competitive ratio.

Theorem 7. *Oblivious fractional metric medians has a deterministic polynomial algorithm that is $2e$-cost-competitive.*

Bicriteria approximations. Combining Theorem 2, Theorem 8, and offline bicriteria results from [2, 19, 20, 16], we can obtain oblivious, polynomial algorithms with the following bicriteria (c, s)-competitiveness guarantees for oblivious metric medians. The first quantity c is the cost-competitive ratio and the second quantity s is the size-competitive ratio: (a) $(3+\epsilon, 4)$, for any $\epsilon > 0$, (b) $(2+\epsilon, 4(1+2\epsilon^{-1}))$, for any $\epsilon > 0$, (c) $(1+\epsilon, 4(3+5\epsilon^{-1}))$, for any $\epsilon > 0$.

Notation. Throughout we use the following terminology for online bidding. Given the universe $\mathcal{U}$, the algorithm outputs a *bid set* $\mathcal{B} \subseteq \mathcal{U}$. Against a particular threshold T, the algorithm pays for the bids $\{b \in \mathcal{B} : b \leq T^+\}$, where $T^+ = \min\{b \in \mathcal{B} : b \geq T\}$. The bid set $\mathcal{B}$ is *β-competitive* if, for any $T \in \mathcal{U}$, this payment is at most βT. Also, $\mathbb{R}^+$ denotes the set of non-negative reals, $\mathbb{Z}$ the set of integers, and $\mathbb{N}^+$ the set of positive integers. For $n \in \mathbb{N}^+$, let $[n] = \{1, 2, \ldots, n\}$.

Plan of the paper. We prove our upper bounds on competitive algorithms for oblivious medians (Theorem 1 for cost-competitive algorithms and Theorem 2(a) for size-competitive algorithms) by reducing oblivious medians to online bidding (Theorem 8, below) and then proving the upper bounds for online bidding (Theorem 4). We prove our lower bounds on size-competitive algorithms for oblivious medians (Theorem 2(b)) by reducing online bidding to size-competitive medians (Theorem 9, below) and then proving the lower bounds for online bidding in Theorem 4. We prove the reductions in Section 2 and analyze online bidding in Section 3. In Section 4 we prove Theorem 6.

2 Oblivious Medians and Online Bidding

We start by showing that oblivious medians can be reduced to online bidding. We show that (a) $2c\beta$-cost-competitive oblivious metric medians reduces (in polynomial time) to β-competitive online bidding and c-cost-approximate offline medians, and (b) $s\beta$-size-competitive oblivious medians reduces (in polynomial time) to β-competitive online bidding and s-size-approximate offline medians.

Note that part (b) holds even for non-metric medians. Also, if allowing non-polynomial time, one can take F_k^* to be the optimal k-median in Theorem 8, which is both 1-cost-approximate and 1-size-approximate; then the oblivious solution $\bar{F}$ is (a) 2β-cost-competitive or (b) β-size-competitive.

Theorem 8. *Let $\beta \geq 1$ and assume that there exists a polynomial β-competitive algorithm for online bidding. Fix an instance of k-median.*

(a) *In the metric case, suppose that for each $i \in [n]$ we have a set of facilities F_i^* with $|F_i^*| \leq i$ and cost$(F_i^*) \leq c \cdot opt_i$. Then in polynomial time we can compute an* oblivious *solution $(F_i)_i$ where $|F_i| \leq i$ and cost$(F_i) \leq 2c\beta \cdot opt_i$.*

(b) *Suppose that for each $i \in [n]$, we have a set of facilities F_i^* with $|F_i^*| \leq s \cdot i$ and cost$(F_i^*) \leq opt_i$. Then in polynomial time we can compute an* oblivious *solution $(F_i)_i$ where $|F_i| \leq s\beta \cdot i$ and cost$(F_i) \leq opt_i$.*

If the algorithm for online bidding is randomized, then the computations in (a) and (b) are also randomized.

Proof. We first prove part (a) of Theorem 8 in the deterministic case. The proof in the randomized setting is similar and we omit it.

For convenience, we introduce distances between facilities: given two $f, g \in \mathcal{F}$, let $d'_{fg} = \min_{x \in \mathcal{C}}(d_{fx} + d_{xg})$. This extension satisfies the triangle inequality. By assumption, each F_k^* is c-cost-approximate: $|F_k^*| \le k$ and $cost(F_k^*) \le c \cdot opt_k$. Assume without loss of generality that $cost(F_k^*) \le cost(F_{k+1}^*)$ for all k.

The algorithm constructs the oblivious solution $(F_i)_i$ from $(F_i^*)_i$ in several steps. First, fix some index set $\mathcal{K} \subseteq [n]$, with $1 \in \mathcal{K}$, by a method to be described later, and let $\kappa(1), \kappa(2), \ldots, \kappa(m)$ denote the indices in $\mathcal{K}$ in increasing order. Next, compute F_k just for $k \in \mathcal{K}$. Start by defining $F_{\kappa(m)} = F_{\kappa(m)}^*$. Then, working backwards, inductively define $F_{\kappa(i)}$ to contain the facilities within $F_{\kappa(i+1)}$ that are "closest" to $F_{\kappa(i)}^*$.

More precisely, given two subsets A, B of $\mathcal{F}$, let $\Gamma(A, B)$ denote a subset Γ of B, minimal with respect to inclusion, and such that $d'_{\mu\Gamma} = d'_{\mu B}$ for all $\mu \in A$ (breaking ties arbitrarily). Obviously, $|\Gamma(A,B)| \le |A|$, and $\Gamma(A,B)$ can be computed in polynomial time given A and B. Then $F_{\kappa(i)} = \Gamma(F_{\kappa(i)}^*, F_{\kappa(i+1)})$.

Finally, define F_k for $k \in [n] \setminus \mathcal{K}$ as follows. Let $k^- = \max\{i \in \mathcal{K} : i \le k\}$ (it is well defined, since $1 \in \mathcal{K}$.) Define $F_k = F_{k^-}$. To complete the construction, it remains to describe how to compute $\mathcal{K}$, which we momentarily defer.

To analyze the size, note that $|F_k| \le k$, because for $k \in \mathcal{K}$, by definition of Γ we have $|F_k| \le |F_k^*| \le k$, while for $k \notin \mathcal{K}$, we have $|F_k| = |F_{k^-}| \le k^- < k$.

To analyze the cost, we use the following lemma. (The proof can be found in [8] and is also implicit in [13].)

Lemma 1. *Assume that the distance function is metric. Consider two sets $A, B \subseteq \mathcal{F}$ and let $\Gamma = \Gamma(A, B)$. Then for every $x \in X$ we have $c_{x\Gamma} \le 2c_{xA} + c_{xB}$.*

We now claim that

$$cost(F_k) \;\le\; 2 \sum_{\ell \ge k^-, \ell \in \mathcal{K}} cost(F_\ell^*). \tag{1}$$

Indeed, for indices $k \in \mathcal{K}$, we have $k = k^-$, and (1) follows from Lemma 1 summed over all x and from the construction of F_k (for $k = \kappa(m), \ldots, \kappa(1)$). For $k \notin \mathcal{K}$, inequality (1) holds as well, simply because $F_k = F_{k^-}$, the bound holds for $k = k^-$, and $(k^-)^- = k^-$.

Since $cost(F_k^*) \le c\, opt_k$, to make F $2c\beta$-cost-competitive we will choose $\mathcal{K}$ so that, for all k,

$$\sum_{\ell \ge k^-, \ell \in \mathcal{K}} cost(F_\ell^*) \;\le\; \beta\, cost(F_k^*). \tag{2}$$

To compute the set $\mathcal{K}$, let $\mathcal{U} = \{cost(F_n^*), cost(F_{n-1}^*), \ldots, cost(F_1^*)\}$ and take $\mathcal{B}$ to be any β-competitive bid set for universe $\mathcal{U}$. Define $\mathcal{K} = \{\kappa(m), \kappa(m-1), \ldots, \kappa(1)\}$ to be a minimal set (containing 1) such that the bid set is $\mathcal{B} = \{cost(F_{\kappa(m)}^*), cost(F_{\kappa(m-1)}^*), \ldots, cost(F_{\kappa(1)}^*)\}$. Then the left-hand side of (2) is exactly the sum of the bids paid from the bid set for threshold $T = cost(F_k^*)$. Since the bid set is β-competitive, this is at most $\beta\, cost(F_k^*)$, so (2) holds. This completes the proof of part (a).

We now prove part (b) of Theorem 8. By assumption each F_k^* is s-size-approximate, that is, $|F_k^*| \le sk$ and $cost(F_k^*) \le opt_k$.

Fix some β-competitive bid set $\mathcal{B}$. Let $\mathcal{B}_k$ be the set of bids in $\mathcal{B}$ paid against threshold $T = k$ with $\mathcal{U} = [n]$. Define $F_k = \bigcup_{b \in \mathcal{B}_k} F_b^*$. Then $\bar{F} = (F_1, F_2, ..., F_n)$

is an oblivious solution because $\mathcal{B}_k \subseteq \mathcal{B}_\ell$ for $\ell \geq k$. Further, $cost(F_k) \leq opt_k$ because F_k contains F_b^* for some $b \geq k$, so $cost(F_k) \leq cost(F_b^*) \leq opt_b \leq opt_k$. Since $\mathcal{B}$ is β-competitive, we have $|F_k| \leq \sum_{b \in \mathcal{B}_k} |F_b^*| \leq \sum_{b \in \mathcal{B}_k} sb \leq s\beta k$.

Our next reduction shows that competitive online bidding reduces to size-competitive oblivious medians. Note that, together with Theorem 8(b), this implies that online bidding and size-competitive oblivious medians are equivalent.

Theorem 9. *Let $s \geq 1$ and assume that, for oblivious medians (metric or not), there is a (possibly randomized) s-size-competitive algorithm. Then, for any integer n, there is a (randomized) s-competitive algorithm for online bidding with $\mathcal{U} = [n]$.*

Proof. We give the proof in the deterministic setting. (The proof in the randomized setting is similar and we omit it.) For any arbitrarily large m, we construct sets $\mathcal{C}$ of customers and $\mathcal{F}$ of facilities, a metric distance function d_{uf}, for $u \in \mathcal{C}$ and $f \in \mathcal{F}$. The facility set $\mathcal{F}$ will be partitioned into sets $M_1, M_2, \ldots, M_m$, where $|M_k| = k$ for each k, with the following properties: (i) For all k, $cost(M_k) > cost(M_{k+1})$, and (ii) For all k, and for every set F of facilities, if $cost(F) \leq cost(M_k)$ then there exists $\ell \geq k$ such that M_ℓ is contained in F. These conditions imply that each M_k is the unique optimum k-median.

Assume for the moment that there exists such a metric space, and consider an s-size-competitive oblivious median $\bar{F}$ for it. Let $\mathcal{B} = \{k : M_k \subseteq F_k\}$. We show that $\mathcal{B}$ is an s-competitive bid set for universe $\mathcal{U} = [m]$. Against any threshold $T \in [m]$, the total of the bids paid will be

$$X = \sum \{k : k < T, M_k \subseteq F_k\} \ + \ \min\{\ell : \ell \geq T, M_\ell \subseteq F_\ell\} \tag{3}$$

Now, $\sum\{k : k < T, M_k \subseteq F_k\} \leq \sum\{k : k < T, M_k \subseteq F_T\}$ since $\bar{F}$ is a nested sequence. Similarly, we have

$$\min\{\ell : \ell \geq T, M_\ell \subseteq F_\ell\} \leq \min\{\ell : \ell \geq T, M_\ell \subseteq F_T\}$$

(By (ii), $M_\ell \subseteq F_T$ for some $\ell \geq T$, so the minimum on the right is well-defined for $T \in [m]$.) Thus:

$$\begin{aligned}
X &\leq \sum\{k : k < T, M_k \subseteq F_T\} + \min\{\ell : \ell \geq T, M_\ell \subseteq F_T\} \\
&= \sum\{|M_k| : k < T, M_k \subseteq F_T\} + \min\{|M_\ell| : \ell \geq T, M_\ell \subseteq F_T\} \ \text{ since } |M_k| = k \\
&\leq \sum\{|M_k| : M_k \subseteq F_T\} \\
&\leq |F_T| \quad \text{since the } M_k\text{'s are disjoint} \\
&\leq sT \quad \text{since } \bar{F} \text{ is } s\text{-size-competitive.}
\end{aligned}$$

Thus, the bid set $\mathcal{B}$ is s-competitive for universe $\mathcal{U} = [m]$.

We now present the construction of the metric space satisfying conditions (i) and (ii). Let $\mathcal{C}$ be the set of integer vectors $\bar{u} = (u_1, u_2, \ldots, u_m)$ where $u_\ell \in [1, \ell]$ for all $\ell = 1, 2, \ldots, m$. For each $\ell \in [1, m]$, introduce a set $M_\ell = \{\mu_{\ell,1}, \mu_{\ell,2}, \ldots, \mu_{\ell,\ell}\}$, and for each node $\bar{u}$ in $\mathcal{C}$, connect $\bar{u}$ to μ_{ℓ,u_ℓ} with an edge of length $\delta_\ell = 1 + (m!)^{-\ell}$. The set of facilities is $\mathcal{F} = \bigcup_\ell M_\ell$. All distances between points in $\mathcal{C} \cup \mathcal{F}$ other than those specified above are determined by shortest-path distances. The resulting distance function satisfies the triangle inequality.

We have $cost(M_j) = m!\delta_j$ for each $j \in [1, m]$, so (i) holds. We prove (ii) by contradiction. Fix some index j and consider a set $F \subseteq \mathcal{F}$ that does not contain M_ℓ for any $\ell \geq j$: for each $\ell \geq j$ there is $i_\ell \leq \ell$ such that $\mu_{\ell,i_\ell} \notin F$. Define a customer $\bar{u}$ as follows: $u_i = 1$ for $\ell = 1, \ldots, j-1$ and $u_i = i_\ell$ for $\ell = j, \ldots, m$. Then the facility $\mu_{\ell,i} \in F$ serving this $\bar{u}$ must have $\ell < j$ or $i \neq i_\ell$. Either way, it is at distance at least δ_{j-1} from $\bar{u}$. Since each other customers pays strictly more than 1, we get $cost(F) > m! - 1 + \delta_{j-1} = m!\delta_j = cost(M_j)$ – a contradiction.

3 Online Bidding

In this section we prove Theorem 4. For completeness, we give proofs of the (folklore) deterministic and randomized upper bounds and deterministic lower bound. The upper bound uses a doubling algorithm that has been used in several papers, first in [15, 23] and later in [11, 3, 4, 9]. Our main new contribution in this section is a new randomized lower bound that matches the upper bound. (The proof of Lemma 3 was communicated to us by Yossi Azar.)

Lemma 2. *For online bidding, there is a deterministic 4-competitive algorithm. If $\mathcal{U}$ is finite, the algorithm runs in time polynomial in $|\mathcal{U}|$.*

Proof. First consider the case $\mathcal{U} = \mathbb{R}^+$. Define the algorithm to produce the set of bids $\{0\} \cup \{2^j : j \in \mathbb{N}\}$. Let $i = \lceil \log_2 T \rceil$, where $T > 0$ is the threshold: the algorithm pays $\sum_{j \leq i} 2^j = 2^{i+1} \leq 4T$, hence is 4-competitive.

Next, we reduce the general case to the case $\mathcal{U} = \mathbb{R}^+$. Knowing that $T \in \mathcal{U}$, the algorithm, when it would have bid $b \notin \mathcal{U}$, will instead bid the next smaller bid in $\mathcal{U}$ (if there is one, and otherwise the bid is skipped). This only decreases the cost the algorithm pays against any threshold $T \in \mathcal{U}$. Note that the modified algorithm can be implemented in time polynomial in $|\mathcal{U}|$ if $\mathcal{U}$ is finite.

Lemma 3. *For online bidding, no deterministic algorithm can be better than 4-competitive, even for $\mathcal{U} = \mathbb{N}^+$.*

Proof. let x_n be the nth bid, $s_n = \sum_1^n x_i$ and $y_n = s_{n+1}/s_n$. Suppose, for a contradiction, that there exists $a < 4$ such that $s_{n+1}/x_n < a$ for all n. Rewriting, we get $y_{n+1} \leq (1 - 1/y_n)a$. Since $1 - 1/z < z/4$, this implies $y_{n+1} < (y_n/4)a$; thus $y_n < (a/4)^n y_0$, and so eventually $s_{n+1} < s_n$, which is a contradiction.

Lemma 4. *For online bidding, there is a randomized e-competitive algorithm. If $\mathcal{U}$ is finite, then the algorithm runs in time polynomial in $|\mathcal{U}|$.*

Proof. First we consider the case $\mathcal{U} = \mathbb{R}^+$. Pick a real number $\xi \in (0, 1]$ uniformly at random, then choose the set of bids $\mathcal{B} = \{0\} \cup \{e^{i+\xi} : i \in \mathbb{N}\}$.

For the analysis, let random variable b be the largest bid paid by the algorithm against threshold $T > 0$. The total paid by the algorithm is less than $\sum_{i=0}^{\infty} be^{-i} = be/(e-1)$. Since b/T is distributed like e^{ξ} where ξ is distributed uniformly in $[0,1)$, the expectation of b is $T\int_0^1 e^z\,dz = T(e-1)$. Thus, the expected total payment is eT, and the algorithm is e-competitive.

The general case reduces to the case $\mathcal{U} = \mathbb{R}^+$ just as in the proof of Lemma 2.

Lemma 5. *Fix any $n \in \mathbb{N}^+$. Suppose $\mu : [n] \to \mathbb{R}^+$ and $\pi : [n] \to \mathbb{R}^+$ satisfy*

$$\sum_{T=t}^{n} \frac{1}{T}\pi(T) \;\geq\; \frac{1}{b}\sum_{T=t}^{b} \mu(T) \quad (\forall b, t : 1 \leq t \leq b \leq n). \tag{4}$$

For online bidding with $\mathcal{U} = [n]$, there is no randomized algorithm with competitive ratio better than $\sum_{T=1}^{n} \mu(T) \;/\; \sum_{T=1}^{n} \pi(T)$.

Proof. Consider a random set $\mathcal{B}$ of bids generated by any β-competitive randomized algorithm when $\mathcal{U} = [n]$. Without loss of generality, the maximum bid in $\mathcal{B}$ is n.

Let $\mathcal{B} = \{b_1, b_2, \ldots, b_m = n\}$ be the ordered sequence of bids in $\mathcal{B}$. Consider the sequence of intervals $([1, b_1], [b_1+1, b_2], [b_2+1, b_3], \ldots, [b_{m-1}+1, b_m])$, which exactly covers the points $1, 2, \ldots, n$. Let $x(t, b)$ denote the probability (over all random $\mathcal{B}$) that $[t, b]$ is one of these intervals. The algorithm pays bid b against threshold T if and only if, for some integer $t \leq T$, $[t, b]$ is one of these intervals. Thus, for any threshold T and bid b, $\sum_{t=1}^{T} x(t, b)$ is the probability that bid b is made against threshold T. (We will use this below.)

We claim that β, x form a feasible solution to the following linear program (LP):

$$\text{minimize}_{\beta,x}\ \beta \ \text{ subject to } \begin{cases} \beta - \displaystyle\sum_{b=1}^{n} \frac{b}{T} \sum_{t=1}^{T} x(t,b) \geq 0 \ (\forall T \in [n]) \\ \displaystyle\sum_{b=T}^{n} \sum_{t=1}^{T} x(t,b) \geq 1 \ (\forall T \in [n]) \\ x(t,b) \geq 0 \ (\forall t, b \in [n]). \end{cases}$$

The first constraint is met because, for any threshold T, $\sum_{t \leq T; b} b\, x(t, b)$ is the expected sum of the bids made by the algorithm if T is the threshold. This is at most βT because the algorithm has competitive ratio β. The second constraint is met because for any threshold T, the algorithm must have at least one bid above the threshold, hence at least one $[t, b]$ with $t \leq T \leq b$.

Thus, the value of this linear program (LP) is a lower bound on the optimal competitive ratio of the randomized algorithm. To get a lower bound on the value of (LP), we use the dual (DLP) (where the dual variables $\mu(T)$ correspond to the first set of constraints and $\pi(T)$ to the second set of constraints):

$$\text{maximize}_{\mu,\pi} \sum_{T=1}^{n} \mu(T) \quad \text{subject to} \quad \begin{cases} \displaystyle\sum_{T=1}^{n} \pi(T) \leq 1 \\ \displaystyle\sum_{T=t}^{b} \mu(T) - \sum_{T=t}^{n} \frac{b}{T}\pi(T) \leq 0 \ (\forall t, b \in [n]) \\ \mu(T), \pi(T) \geq 0 \ (\forall T \in [n]). \end{cases}$$

Now, given any μ and π meeting the condition of the lemma, if we scale μ and π by dividing by $\sum_T \pi(T)$, we get a feasible dual solution whose value is $\sum_T \mu(T) \,/ \sum_T \pi(T)$. Since the value of any feasible dual solution is a lower bound on the value of any feasible solution to the primal, it follows that the competitive ratio β of the randomized algorithm is at least $\sum_T \mu(T) \,/ \sum_T \pi(T)$.

Lemma 6. *There exists $\mu : [n] \to \mathbb{R}^+$ and $\pi : [n] \to \mathbb{R}^+$ satisfying Condition (4) of Lemma 5 and such that $\sum_T \mu(T)/\sum_T \pi(T) \geq (1 - o(1))e$.*

Proof. Fix U arbitrarily large and let $n = \lceil U^2 \log U \rceil$. Let $\alpha > 0$ be a constant to be determined later: We will choose α so that Condition (4) holds, and then show that the corresponding lower bound is $e(1 - o(1))$ as $U \to \infty$. Define

$$\mu(T) = \begin{cases} \alpha/T & \text{if } U \leq T \leq U^2 \\ 0 & \text{otherwise} \end{cases} \quad \text{and} \quad \pi(T) = \begin{cases} 1/T & \text{if } U \leq T \leq U^2 \log U \\ 0 & \text{otherwise.} \end{cases}.$$

If $T \geq U^2$, then the right-hand side of Condition (4) has value 0, so the condition holds trivially. On the other hand, since $\pi(T)$ and $\mu(T)$ are zero for $T < U$, if the condition holds for $T = U$, then it also holds for $T < U$. So, we need only verify the condition for T in the range $U \leq T \leq U^2$. The expression on the left-hand side of (4) then has value

$$\sum_{T=t}^{U^2 \log U} \frac{1}{T^2} \geq \int_t^{1+U^2 \log U} \frac{1}{T^2}\, dT = \frac{1}{t} - \frac{1}{1 + U^2 \log U} \geq \frac{1}{t}(1 - o(1)).$$

In comparison, the expression on the right-hand side has value at most

$$\max_{b \geq t} \frac{1}{b} \sum_{T=t}^{b} \frac{\alpha}{T} \leq \alpha \max_{b \geq t} \frac{1}{b} \int_{t-1}^{b} \frac{1}{T}\, dT = \alpha \max_{b \geq t} \frac{1}{b} \ln \frac{b}{t-1} = \frac{\alpha}{e\, t(1 - o(1))}.$$

(The second equation follows by calculus, for the maximum occurs when $b = e(t-1)$.) Thus, Condition (4) is met for $\alpha = (1 - o(1))e$. Then, Lemma 5 gives a lower bound on the competitive ratio of

$$\frac{\sum_T \mu(T)}{\sum_T \pi(T)} = \frac{\sum_{T=U}^{U^2} \alpha/T}{\sum_{T=U}^{U^2 \log U} 1/T} = (1 - o(1))\alpha \frac{\ln(U^2/U)}{\ln((U^2 \log U)/U)} = (1 - o(1))e.$$

Theorem 4 follows directly from Lemmas 2, 3, 4, 5, and 6.

4 Oblivious Algorithms for *kl*-Medians

In this section we sketch the proof of Theorem 6. Formally, in the kl-median problem we need to compute two sets $F_k \subseteq F_l$ with $|F_k| = k$ and $|F_l| = l$, minimizing the competitive ratio $R = \max\{cost(F_k)/opt_k, cost(F_l)/opt_l\}$.

The lower bound is a slight refinement of the one in [22, 21]. The metric space contains l customers, where customers j is connected to facility g_j by an edge of length $\delta = 1/l$. All customers are also connected to a facility f with edges of length 1.

Let $G = \{g_1, \ldots, g_l\}$. Then G is the optimal l-median. We have $cost(f) = l$, $cost(G) = l\delta$, $cost(g_i) = \delta + (l-1)(2+\delta)$, and $cost(G - g_i + f) = (l-1)\delta + 1$. So for $\delta = 1/l$, we get:

$$\frac{cost(g_i)}{cost(f)} = 2 - 1/l \quad \text{and} \quad \frac{cost(G - g_i + f)}{cost(G)} = 2 - 1/l.$$

The upper bound is achieved as follows. Let F and G denote, respectively, the optimum k-median and the optimum l-median. The algorithm choosese the better of two options: either (a) $F_k = F$ and $F_l = F \cup G - X$, where $X \subseteq G$ is a set of cardinality k that minimizes $cost(F \cup G - X)$, or (b) $F_k = Y$, where $Y \subseteq G$ is a set of cardinality k that minimizes $cost(Y)$, and $F_l = G$.

The competitive analysis of this algorithm is based on a probabilistic argument and will appear in the full version of this paper.

Acknowledgments. We are grateful to anonymous referees for suggestions to improve the presentation. We also wish to thank Yossi Azar for pointing out references to previous work on online bidding and simplifying the proof of Lemma 3.

References

1. A. Archer, R. Rajagopalan, and D.B. Shmoys. Lagrangian relaxation for the k-median problem: new insights and continuity properties. In *Proc. 11th European Symp. on Algorithms (ESA)*, pages 31–42, 2003.
2. V. Arya, N. Garg, R. Khandekar, K. Munagala, and V. Pandit. Local search heuristic for k-median and facility location problems. In *Proc. 33rd Symp. Theory of Computing (STOC)*, pages 21–29. ACM, 2001.
3. S. Chakrabarti, C.A. Phillips, A.S. Schulz, D.B. Shmoys, C. Stein, and J. Wein. Improved scheduling algorithms for minsum criteria. In *Automata, Languages and Programming*, pages 646–657, 1996.
4. M. Charikar, C. Chekuri, T. Feder, and R. Motwani. Incremental clustering and dynamic information retrieval. In *Proc. 29th Symp. Theory of Computing (STOC)*, pages 626–635. ACM, 1997.
5. M. Charikar and S. Guha. Improved combinatorial algorithms for the facility location and k-median problems. In *Proc. 40th Symp. Foundations of Computer Science (FOCS)*, pages 378–388. IEEE, 1999.
6. M. Charikar, S. Guha, E. Tardos, and D.B. Shmoys. A constant-factor approximation algorithm for the k-median problem. In *Proc. 31st Symp. Theory of Computing (STOC)*, pages 1–10. ACM, 1999.

7. C. Chekuri, A. Goel, S. Khanna, and A. Kumar. Multi-processor scheduling to minimize flow time with ϵ-resource augmentation. In *Proc. 36th Symp. Theory of Computing (STOC)*, pages 363–372. ACM, 2004.
8. M. Chrobak, C. Kenyon, and N.E. Young. The reverse greedy algorithm for the k-median problem. *Information Processing Letters*, 97:68–72, 2006.
9. S. Dasgupta and P.M. Long. Performance guarantees for hierarchical clustering. *Journal of Computer and System Sciences*, 70(4):555–569, 2005.
10. R. Fagin and L. Stockmeyer. Relaxing the triangle inequality in pattern matching. *IJCV: International Journal of Computer Vision*, 30:219–231, 1998.
11. M. Goemans and J. Kleinberg. An improved approximation ratio for the minimum latency problem. In *Proc. 7th Symp. on Discrete Algorithms (SODA)*, pages 152 – 158. ACM/SIAM, 1996.
12. K. Jain, M. Mahdian, and A. Saberi. A new greedy approach for facility location problems. In *Proc. 34th Symp. Theory of Computing (STOC)*, pages 731–740. ACM, 2002.
13. K. Jain and V.V. Vazirani. Approximation algorithms for metric facility location and k-median problems using the primal-dual schema and lagrangian relaxation. *Journal of ACM*, 48:274–296, 2001.
14. B. Kalyanasundaram and K. Pruhs. Speed is as powerful as clairvoyance. *J. ACM*, 47:214–221, 2000.
15. M. Kao, J.H. Reif, and S. Tate. Searching in an unknown environment: An optimal randomized algorithm for the cow-path problem. *Information and Computation*, 131(1):63–80, 1996. Preliminary version appeared in the Proceedings of the Symp. on Discrete Algorithms, Austin, TX, Jan 1993.
16. M.R. Korupolu, C.G. Plaxton, and R. Rajaraman. Analysis of a local search heuristic for facility location problems. *Journal of Algorithms*, 37:146–188, 2000.
17. E. Koutsoupias. Weak adversaries for the k-server problem. In *Proc. 40th Symp. Foundations of Computer Science (FOCS)*, pages 444–449. IEEE, 1999.
18. G. Lin, C. Nagarajan, R. Rajamaran, and D.P. Williamson. A general approach for incremental approximation and hierarchical clustering. In *Proc. 17th Symp. on Discrete Algorithms (SODA)*. ACM/SIAM, 2006.
19. J-H. Lin and J.S. Vitter. Approximation algorithms for geometric median problems. *Information Processing Letters*, 44:245–249, 1992.
20. J-H. Lin and J.S. Vitter. ϵ-approximations with minimum packing constraint violation (extended abstract). In *Proc. 24th Symp. Theory of Computing (STOC)*, pages 771–782. ACM, 1992.
21. R. Mettu and C.G. Plaxton. The online median problem. *SIAM Journal on Computing*, 32:816–832, 2003.
22. R.R. Mettu and C.G. Plaxton. The online median problem. In *Proc. 41st Symp. Foundations of Computer Science (FOCS)*, pages 339–348. IEEE, 2000.
23. R. Motwani, S. Phillips, and E. Torng. Nonclairvoyant scheduling. *Theoretical Computer Science*, 130(1):17–47, 1994.
24. N.E. Young. K-medians, facility location, and the Chernoff-Wald bound. In *Proc. 11th Symp. on Discrete Algorithms (SODA)*, pages 86–95. ACM/SIAM, January 2000.

Efficient Computation of the Relative Entropy of Probabilistic Automata

Corinna Cortes[1], Mehryar Mohri[1,2,⋆], Ashish Rastogi[2], and Michael D. Riley[1]

[1] Google Research, New York, NY, USA
[2] Courant Institute of Mathematical Sciences, New York University, New York, NY, USA

Abstract. The problem of the efficient computation of the relative entropy of two distributions represented by deterministic weighted automata arises in several machine learning problems. We show that this problem can be naturally formulated as a *shortest-distance problem* over an intersection automaton defined on an appropriate semiring. We describe simple and efficient novel algorithms for its computation and report the results of experiments demonstrating the practicality of our algorithms for very large weighted automata. Our algorithms apply to *unambiguous* weighted automata, a class of weighted automata that strictly includes *deterministic* weighted automata. These are also the first algorithms extending the computation of entropy or of relative entropy beyond the class of deterministic weighted automata.

1 Introduction

The relative entropy, or Kullback-Leibler divergence, is used in a variety of contexts as a measure of the discrepancy of two distributions p and q [5]. It is an asymmetric difference that, from the point of view of coding theory, measures the number of additional bits needed to encode p, when using an optimal code for q in place of an optimal code for p.

The problem of the efficient computation of the relative entropy of two distributions represented by weighted automata arises in several machine learning problems. Weighted automata are used extensively in text and speech processing to model different aspects of language such as morphology, phonology, or syntax [12]. The output of a large-vocabulary speech recognition system or that of a complex information extraction system is typically represented as a weighted automaton compactly representing a large set of alternative sequences [17]. Weighted automata are also used in other applications such as image processing [6].

When a weighted automaton is obtained as a result of training on a large data set, the quality of the learning algorithm can be measured by computing the relative entropy of the automaton inferred and that of the target automaton.

⋆ This work was partially funded by the New York State Office of Science Technology and Academic Research (NYSTAR).

J.R. Correa, A. Hevia, and M. Kiwi (Eds.): LATIN 2006, LNCS 3887, pp. 323–336, 2006.

Similarly, in some grammar inference applications, the convergence of an iterative algorithm relies on the magnitude of the relative entropy of two consecutive weighted automata. The relative entropy is also often used for clustering large sets of automata, such as those output by a speech recognition or information extraction system.

This motivates the design of efficient algorithms for the computation of the relative entropy of two weighted automata. One approximate solution would consist of sampling sequences from the distributions represented by each of the automata and of using those to compute the KL-divergence by simply summing their contributions. But, sample sizes guaranteeing a small approximation error could be very large, which would significantly increase the computation, while still providing only an approximate solution.

We present a detailed analysis of the problem of the computation of the relative entropy of weighted automata in the case where they are *deterministic* or, more generally, *unambiguous*, i.e., no two successful paths are labeled with the same string. We show that the problem can be formulated naturally as a *single-source shortest-distance problem* over an intersection automaton defined on an appropriate semiring that we will refer to as the *entropy semiring*. We describe simple and efficient algorithms for the computation of relative entropy and report the results of experiments demonstrating the practicality of our algorithms for very large weighted automata.

A procedure for the approximate computation of the relative entropy was given by [3]. The procedure applies to deterministic weighted automata and cannot be generalized to the case of unambiguous weighted automata because of the specific sum decomposition it is based on (the partitioning assumed in [3] [eq. 15, page 6] does not hold for unambiguous automata). Our algorithms apply to the larger class of unambiguous weighted automata. For some unambiguous weighted automata, the size of any equivalent deterministic weighted automaton is exponentially larger. Since the size of the machine directly affects the complexity of the computation, it is important to be able to compute the entropy directly from the unambiguous automaton. We give the first *exact* algorithms for the computation of the relative entropy. We also describe approximate algorithms that are conceptually simpler than the procedure of [3] and have a better time and space complexity.

The paper is organized as follows. Section 2 introduces the preliminary semiring and automata definitions used in the remaining of the paper. Section 3 introduces the entropy semiring and formulates the computation of the relative entropy in terms of shortest-distances over that semiring. Section 4 describes both an exact and a fast approximate algorithm for the computation of the relative entropy. Section 5 briefly reports the results of experiments demonstrating the practicality of our algorithms for very large weighted automata.

2 Preliminaries

Weighted automata are automata in which each transition carries some weight in addition to the usual alphabet symbol [7, 18, 1]. For various operations to be well-defined, the weight set must have the algebraic structure of a semiring [10]. A semiring is a ring that may lack negation.

Definition 1. *A* semiring *is a system* $(\mathbb{K}, \oplus, \otimes, \overline{0}, \overline{1})$ *such that:* $(\mathbb{K}, \oplus, \overline{0})$ *is a commutative monoid with* $\overline{0}$ *as the identity element for* $\oplus$*;* $(\mathbb{K}, \otimes, \overline{1})$ *is a monoid with* $\overline{1}$ *as the identity element for* $\otimes$*;* $\otimes$ *distributes over* $\oplus$*: for all* a, b, c *in* $\mathbb{K}$*:* $(a \oplus b) \otimes c = (a \otimes c) \oplus (b \otimes c)$ *and* $c \otimes (a \oplus b) = (c \otimes a) \oplus (c \otimes b)$*, and* $\overline{0}$ *is an annihilator for* $\otimes$*:* $\forall a \in \mathbb{K}, a \otimes \overline{0} = \overline{0} \otimes a = \overline{0}$*.*

Some familiar semirings are the Boolean semiring $(\{0, 1\}, \vee, \wedge, 0, 1)$ or the tropical semiring $(\mathbb{R}_+ \cup \{\infty\}, \min, +, \infty, 0)$ related to classical shortest-paths problems and algorithms. A semiring is idempotent if for all $a \in \mathbb{K}$, $a \oplus a = a$. It is *commutative* when $\otimes$ is commutative.

Definition 2. *A* weighted automaton $A = (\Sigma, Q, I, F, E, \lambda, \rho)$ *over a semiring* $(\mathbb{K}, \oplus, \otimes, \overline{0}, \overline{1})$ *is a 7-tuple where:* Σ *is the finite alphabet of the automaton,* Q *is a finite set of states,* $I \subseteq Q$ *the set of initial states,* $F \subseteq Q$ *the set of final states,* $E \subseteq Q \times \Sigma \cup \{\epsilon\} \times \mathbb{K} \times Q$ *a finite set of transitions,* $\lambda : I \to \mathbb{K}$ *the initial weight function mapping* I *to* $\mathbb{K}$*, and* $\rho : F \to \mathbb{K}$ *the final weight function mapping* F *to* $\mathbb{K}$*.*

The weighted automata considered in this paper are assumed not to contain ϵ-transitions. A pre-processing ϵ-removal algorithm can be used to remove such transitions for the automata considered here [14]. Furthermore, it is assumed that the automata are *trim*, i.e. all states in the automata are both accessible and co-accessible.

We denote by $|A| = |E| + |Q|$ the size of an automaton $A = (\Sigma, Q, I, F, E, \lambda, \rho)$, that is the sum of the number of states and transitions of A. Given a transition $e \in E$, we denote by $i[e]$ its input label, $p[e]$ its origin or previous state and $n[e]$ its destination state or next state, $w[e]$ its weight (weighted automata case). Given a state $q \in Q$, we denote by $E[q]$ the set of transitions leaving q.

A *path* $\pi = e_1 \cdots e_k$ in A is an element of E^* with consecutive transitions: $n[e_{i-1}] = p[e_i]$, $i = 2, \ldots, k$. We extend n and p to paths by setting: $n[\pi] = n[e_k]$ and $p[\pi] = p[e_1]$. We denote by $P(q, q')$ the set of paths from q to q' and by $P(q, x, q')$ the set of paths from q to q' with input label $x \in \Sigma^*$. The labeling functions i and the weight function w can also be extended to paths by defining the label of a path as the concatenation of the labels of its constituent transitions, and the weight of a path as the $\otimes$-product of the weights of its constituent transitions: $i[\pi] = i[e_1] \cdots i[e_k]$, $w[\pi] = w[e_1] \otimes \cdots \otimes w[e_k]$.

The output weight associated by an automaton A to an input string $x \in \Sigma^*$ is defined by:

$$[\![A]\!](x) = \bigoplus_{\pi \in P(I, x, F)} \lambda[p[\pi]] \otimes w[\pi] \otimes \rho[n[\pi]]. \tag{1}$$

Our algorithms for the computation of the entropy of a weighted automata or the computation of the relative entropy of two automata apply to *unambiguous weighted automata*. A weighted automaton is said to be *unambiguous* if for any

$x \in \Sigma^*$ it admits at most one accepting path labeled with x. Thus, the class of unambiguous weighted automata includes *deterministic* weighted automata. A weighted automaton A is said to be *deterministic* or *subsequential* if it has a deterministic input, that is if it has a unique initial state and if no two transitions leaving the same state share the same input label.

Fig. 1 (a) shows an unambiguous weighted automaton that does not admit an equivalent deterministic weighted automaton (the proof will be included in a future journal version). Previous work on the computation of the relative entropy [3] was limited to deterministic finite automata. We present the first algorithms for the computation of the relative entropy of unambiguous weighted automata.

Let $s[A]$ denote the $\oplus$-sum of the weights of all successful paths of A when it is defined and in $\mathbb{K}$. $s[A]$ can be viewed as the *shortest-distance* from the initial states to the final states. When the sum of the weights of all paths from any state p to any state q is well-defined and in $\mathbb{K}$, we can define the *shortest distance* from $p \in Q$ to $q \in Q$ as:

$$d[p, q] = \bigoplus_{\pi \in P(p,q)} w[\pi], \tag{2}$$

where the summation is defined to be $\overline{0}$ when $P(p, q) = \emptyset$. Let A be a weighted automaton defined over the probability semiring $(\mathbb{R}_+, +, \times, 0, 1)$. We will say that A is *probabilistic* if for any state $q \in Q$, $\bigoplus_{\pi \in P(q,q)} w[\pi]$, the sum of the weights of all cycles at q, is well-defined and in $\mathbb{K}$ and $\sum_{x \in \Sigma^*} [\![A]\!](x) = 1$. *Stochastic automata* are probabilistic automata such that at each state the weights of the outgoing transitions and the final weight sum to one.

Let A_1 and A_2 be two weighted automata with $A_i = (\Sigma, Q_i, I_i, F_i, E_i, \lambda_i, \rho_i)$ for $i = 1, 2$. The intersection A of A_1 and A_2 is denoted by $A = A_1 \cap A_2$. It is a weighted automaton accepting the language $L(A_1) \cap L(A_2)$ and defined by the tuple $A = (\Sigma, Q_1 \times Q_2, I_1 \times I_2, F_1 \times F_2, E, (\lambda_1, \lambda_2), (\rho_1, \rho_2))$, where the transitions E are defined according to the following rule:

$$(q_1, a, w_1, q_2) \in E_1 \text{ and } (q_1', a, w_1', q_2') \in E_2 \Rightarrow ((q_1, q_1'), a, (w_1 \otimes w_1'), (q_2, q_2')) \in E.$$

There exists a general algorithm for the computation of the intersection over an arbitrary semiring, even in presence of ϵ-transitions [16]. The time complexity of the algorithm is quadratic $O(|A_1||A_2|)$ since in the worst case the outgoing transitions of each state of A_1 match all those of each state of A_2.

3 Formulation of the Problem

The problem that we are interested in is that of computing $D(A\|B)$, the relative entropy of two unambiguous probabilistic automata A and B.

3.1 Relative Entropy

The entropy $H(p)$ of a probability mass function p defined over a discrete set $\mathcal{X}$ is defined as [5]:

$$H(p) = -\sum_{x \in \mathcal{X}} p(x) \log p(x), \tag{3}$$

where by convention $0 \log 0 = 0$. The relative entropy, or Kullback-Leibler divergence of two probability mass functions defined over a discrete set $\mathcal{X}$ is defined as:

$$D(p\|q) = \sum_{x \in \mathcal{X}} p(x) \log \frac{p(x)}{q(x)} = \mathrm{E}_p[\log \frac{p(X)}{q(X)}], \tag{4}$$

where we use the standard conventions: $0 \log \frac{0}{q} = 0$ and $p \log \frac{p}{0} = \infty$. It is easy to show using Jensen's inequality and the concavity of the log function that the relative entropy is a non-negative number and that $D(p\|q) = 0$ if and only if $p = q$. Note that $D(p\|q)$ is not a metric because it is not symmetric and does not satisfy the triangle inequality.

These definitions can be naturally extended to probabilistic automata which define distributions over sets of strings. The relative entropy of A and B can be written as the sum of two terms:[1]

$$D(A\|B) = \sum_x [\![A]\!](x) \log [\![A]\!](x) - \sum_x [\![A]\!](x) \log [\![B]\!](x). \tag{5}$$

The next section introduces a semiring, the *entropy semiring*, showing that each term can be viewed as a single-source shortest-distance for an automaton defined over that semiring.

3.2 Entropy Semiring

Let $\mathbb{K}$ denote $(\mathbb{R} \cup \{+\infty, -\infty\}) \times (\mathbb{R} \cup \{+\infty, -\infty\})$. For pairs (x_1, y_1) and (x_2, y_2) in $\mathbb{K}$, define the following :

$$(x_1, y_1) \oplus (x_2, y_2) = (x_1 + x_2, y_1 + y_2) \tag{6}$$

$$(x_1, y_1) \otimes (x_2, y_2) = (x_1 x_2, x_1 y_2 + x_2 y_1) \tag{7}$$

Lemma 1. *The system* $(\mathbb{K}, \oplus, \otimes, (0,0), (1,0))$ *defines a commutative semiring.*

Proof. The proof is rather straightforward and will be included in the journal version. □

We call the semiring just defined the *entropy semiring* due to its relevance in the computation of the entropy and the relative entropy. This semiring arises in other contexts and can be defined in terms of an S-module [2, 8].

3.3 Semiring Formulation

The unambiguous weighted automata A and B are not necessarily *complete*: at some states, there may be no outgoing transition labeled with a given element of the alphabet $a \in \Sigma$. We can however make them complete in a way similar to the standard construction in the unweighted case. We introduce a new state q_0 with final weight 0, add self-loops with weight 0 at that state labeled with all

[1] The first term is simply $-H(A)$, where $H(A)$ is the entropy of A.

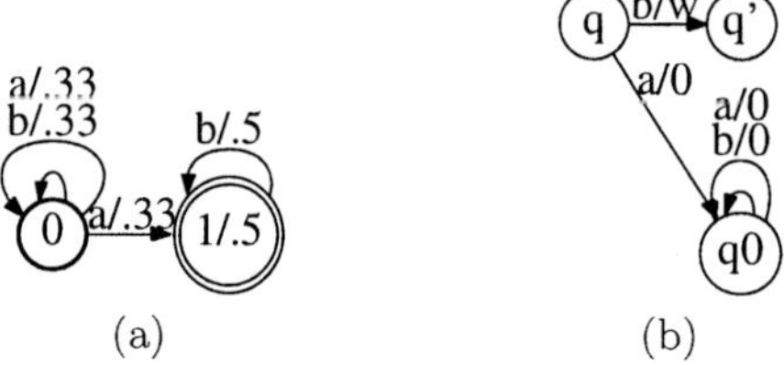

(a) (b)

Fig. 1. (a) An unambiguous weighted finite automaton that cannot be determinized. 0 is the initial state and 1 the final state. The automaton accepts the set of strings $(a^*b^*)^*ab^*$. (b) Illustration of the completion operation.

elements of the alphabet, and for any $a \in \Sigma$ and $q \in Q$, add a transition from state q to q_0 labeled with a with weight 0 when q does not have an outgoing transition labeled with a (see Figure 1 (b)). This construction leads to a complete and unambiguous weighted automaton equivalent to the original one since the transitions added have all weight 0. The completion operation is only applied to handle the boundary case when there exists a string $x \in \Sigma^*$ such that $[\![B]\!](x) = 0$ and $[\![A]\!](x) \neq 0$. In this case, the completion operation ensures that the future computation of the relative entropy would correctly lead to ∞. Note that the completion operation can be done on-demand. States and transitions can be created only when necessary for the application of other operations. We can thus assume that A and B are unambiguous and complete. At the cost of introducing a super-initial and a super-final state, we can also assume in the following, without loss of generality, that the initial weight λ and the final weights $\rho(q)$ are all equal to 1 in A and B.

Let $\log A$ denote the weighted automaton derived from A by replacing each weight $w \in \mathbb{R}_+$ by $\log w$ and let $\Phi_1(A)$ ($\Phi_2(A)$) denote the weighted automaton over the entropy semiring derived from A by replacing each weight w by the pair $(w, 0)$ (resp. $(1, w)$). The construction of $\log A$, $\Phi_1(A)$, or $\Phi_2(A)$ from A is straightforward and can be done in linear time.

Proposition 1. *The relative entropy of A and B satisfies the following identity in the entropy semiring:*

$$(0, D(A\|B)) = s[\Phi_1(A) \cap \Phi_2(\log A)] - s[\Phi_1(A) \cap \Phi_2(\log B)]. \tag{8}$$

Thus, the relative entropy is expressed in terms of single-source shortest-distance computations over the entropy semiring.

Proof. Since A is unambiguous and complete, both $\Phi_1(A)$ and $\Phi_2(\log A)$ are also unambiguous and complete. Thus, for a given string x, there is at most one accepting path in $\Phi_1(A)$ or $\Phi_2(\log A)$ labeled with x. Then, by definition of intersection, the weight associated by $\Phi_1(A) \cap \Phi_2(\log A)$ to a string x is

$$([\![A]\!](x), 0) \otimes (1, \log[\![A]\!](x)) = ([\![A]\!](x), [\![A]\!](x) \log[\![A]\!](x)). \tag{9}$$

Thus, the shortest-distance from the initial states to the final states in $\Phi_1(A) \cap \Phi_2(\log A)$ is

$$s[\Phi_1(A) \cap \Phi_2(\log A)] = \bigoplus_x (\llbracket A \rrbracket(x), \llbracket A \rrbracket(x) \log \llbracket A \rrbracket(x)) \tag{10}$$

$$= (\sum_x \llbracket A \rrbracket(x), \sum_x \llbracket A \rrbracket(x) \log \llbracket A \rrbracket(x)) \tag{11}$$

$$= (1, \sum_x \llbracket A \rrbracket(x) \log \llbracket A \rrbracket(x)). \tag{12}$$

Similarly, we can show that

$$s[\Phi_1(A) \cap \Phi_2(\log B)] = (1, \sum_x \llbracket A \rrbracket(x) \log \llbracket B \rrbracket(x)). \tag{13}$$

The statement of the proposition follows directly from the identities 12 and 13 and Equation 5. □

Thus, the computation of the relative entropy is reduced to two single-source shortest-distance computations over the entropy semiring. The next section discusses two general algorithms for computing these distances.

4 Algorithms

This section describes two algorithms for computing a single-source shortest distance over the entropy semiring, an exact algorithm, and a more efficient and more practical approximate algorithm.

4.1 Exact Solution

A generalization of the classical Floyd-Warshall algorithm can be used to compute all-pairs shortest distances $d[p, q]$ $(p, q \in Q)$ over a *closed semiring* not necessarily idempotent [13, 15]. This algorithm can thus also be used to compute $s[A]$ for a weighted automaton A over a non-idempotent semiring, which is needed for our purpose.

In what follows, we assume a definition of closed semirings [11] that is more general than the classical one used by Cormen *et al.* [4] in that it does not assume idempotence. This is because idempotence is not necessary for the proof of the correctness of the generic all-pairs shortest-distance algorithms of Floyd-Warshall and Gauss-Jordan [13, 15]. More generally, given a graph or automaton A, we introduce the following definition.

Definition 3. *A semiring is* closed for A *if the infinite sum (closure) is defined for any cycle weight c of A and if associativity, commutativity, and distributivity apply to countable sums of cycle weights.*

Clearly, the generic Floyd-Warshall algorithm can also be applied to any automaton A for which the semiring considered is closed. The following lemma shows that the entropy semiring has the desired property.

Lemma 2. *Let A be a weighted automaton over the entropy semiring such that for any cycle weight $w = (x, y)$, $0 \leq x < 1$. Then, the entropy semiring is closed for A.*

Proof. For any $(x, y) \in \mathbb{K}$ and $k \geq 0$, define R_k as:

$$R_k = \overbrace{(x, y) \otimes \ldots \otimes (x, y)}^{k \text{ times}}. \tag{14}$$

with $R_0 = (1, 0)$. We can show by induction that $R_k = (x^k, kyx^{k-1})$. The base case is readily established for $k = 0$. Assume that the hypothesis holds for all $i < k$. Then

$$\begin{aligned} R_k &= R_{k-1} \otimes (x, y) \\ &= (x^{k-1}, (k-1)yx^{k-2}) \otimes (x, y) \\ &= (x^k, kyx^{k-1}). \end{aligned} \tag{15}$$

For $N \geq 0$, define S_N by: $S_N = \bigoplus_{i=0}^{N} R_i$. It is easy to prove by induction as above that S_N verifies

$$S_N = \Big(\frac{1 - x^{N+1}}{1 - x}, y \cdot \big[\frac{1 - x^N}{(1 - x)^2} - \frac{Nx^N}{1 - x}\big]\Big). \tag{16}$$

Thus, for $0 \leq x < 1$, the closure of (x, y) is well-defined and in $\mathbb{K}$:[2]

$$(x, y)^* = \lim_{N \to \infty} S_N = (\frac{1}{1 - x}, \frac{y}{(1 - x)^2}). \tag{17}$$

The associativity, commutativity, and distributivity properties follow the associativity, commutativity, and distributivity of the sums S_N with other elements of the entropy semiring and the corresponding properties of their pointwise limits. □

Let A be a probabilistic automaton, then the weight u of a cycle must verify $0 \leq u < 1$, otherwise the automaton is not closed. The weight of a cycle of $\Phi_1(A) \cap \Phi_2(\log A)$ is $(u, u \log u)$ (see Equation 9), where u is the weight of a cycle of A, and similarly, the weight of a cycle of $\Phi_1(A) \cap \Phi_2(\log B)$ is of the form $(u, u \log v)$, where v is the weight of a matching cycle in B.

Thus, the entropy semiring is closed both for $\Phi_1(A) \cap \Phi_2(\log B)$ and $\Phi_1(A) \cap \Phi_2(\log A)$ and the generic Floyd-Warshall algorithm can be applied to compute the shortest-distances $s[\Phi_1(A) \cap \Phi_2(\log B)]$ and $s[\Phi_1(A) \cap \Phi_2(\log A)]$.

The generic Floyd-Warshall admits an in-place implementation [13]; the following gives the corresponding pseudocode.

[2] The right-hand side can be written as: $(x^*, y(x^*)^2)$, if we denote by $x^* = \sum_{n=0}^{\infty} x^n$.

```
for i ← 1 to |Q|
    do for j ← 1 to |Q|
          do d[i, j] ← ⊕_{e∈P(i,j)} w[e]
for k ← 1 to |Q|
    do for i ← 1 to |Q|
          do for j ← 1 to |Q|
                do d[i, j] ← d[i, j] ⊕ (d[i, k] ⊗ d[k, k]* ⊗ d[k, j])
return d
```

The $\oplus$- and $\otimes$-operations of the entropy semiring can be performed in constant time. For (x, y) with $0 \leq x < 1$, the closure $(x, y)^* = (\frac{1}{1-x}, \frac{y}{(1-x)^2})$ can also be computed in constant time. Thus, the running time complexity of the algorithm is $\Theta(|E| + |Q|^3)$ and its space complexity is $\Omega(|Q|^2)$ when applied to a weighted automaton $A = (Q, I, F, \Sigma, \delta, \sigma, \lambda, \rho)$ over the tropical semiring.

The intersection $\Phi_1(A) \cap \Phi_2(\log A)$ can be computed in linear time $O(|A|)$ but the worst cost computation of $\Phi_1(A) \cap \Phi_2(\log B)$ is quadratic, $O(|A||B|)$. The total time complexity of the computation of the relative entropy is thus in $\Theta(|A \cap B|^3)$. Its space complexity is in $\Theta(|A \cap B|^2)$.

This provides an exact algorithm for the computation of the relative entropy. The cubic time complexity of the algorithm with respect to the size of the intersection automaton makes it rather slow for large automata.

Its quadratic lower bound complexity with respect to the size of the intersection machine makes it prohibitive for use in many applications. In text and speech processing applications, a weighted automaton may have several hundred million states and transitions. Even, if A has only about 100,000 states and $A \cap B$ has about the same number of states, the algorithm requires maintaining a matrix d with 10 billion entries.

The next section presents an algorithm that exploits the sparseness of the graph and does not impose these space requirements.

4.2 Approximate Solution

A generic single-source shortest-distance algorithm was presented for directed graphs defined over a *k-closed* semiring in [15]. The algorithm can be viewed as a generalization to these semirings of classical shortest-paths algorithms. This generalization is not trivial and does not require the semiring to be idempotent. The algorithm is also generic in the sense that it works with any queue discipline.

Definition 4. *Let $k \geq 0$ be an integer. A semiring $(\mathbb{K}, \oplus, \otimes, \overline{0}, \overline{1})$ is* k-closed *if:*

$$\forall a \in \mathbb{K}, \bigoplus_{n=0}^{k+1} a^n = \bigoplus_{n=0}^{k} a^n. \tag{18}$$

More generally, we will say that $\mathbb{K}$ is *k-closed for a graph G or automaton A*, if Equation 18 holds for all cycle weights $a \in \mathbb{K}$.

By definition, the entropy semiring is k-closed for any acyclic automaton A and thus the generic single-source shortest distance can be used to compute the relative entropy exactly in such cases. But, in general, the entropy semiring is not k-closed for a non-acyclic automaton A since by definition of S_N,

$$\forall k > 0, S_{k+1} - S_k = R_{k+1} = (x^{k+1}, (k+1)yx^k). \tag{19}$$

But, given a weighted automaton A over the entropy semiring such that all cycle weights $w = (x, y)$ verify $0 \leq x < 1$, there exists K_A sufficiently large such that for all $k \geq K_A$, $||S_{k+1} - S_k||_\infty \leq \epsilon$. Indeed, let X denote the maximum value of x for all cycles and Y the maximum $|y|$. Then, for $k \geq \frac{\log(Y/\epsilon)}{\log(1/X)}$, $||S_{k+1} - S_k||_\infty \leq \epsilon$ for all (x, y). This leads us to consider an approximate version of the generic single-source shortest distance algorithm in non-acyclic cases, where the equality test is replaced by an ϵ-equality: $u =_\epsilon v$ if $||u - v||_\infty \leq \epsilon$. The following gives the pseudocode of the modified algorithm.

```
1  for i ← 1 to |Q|
2      do d[i] ← r[i] ← 0̄
3  d[s] ← r[s] ← 1̄
4  S ← {s}
5  while S ≠ ∅
6          do q ← head(S)
7             DEQUEUE(S)
8             r′ ← r[q]
9             r[q] ← 0̄
10            for each e ∈ E[q]
11            do if d[n[e]] ≠_ε d[n[e]] ⊕ (r′ ⊗ w[e])
12                  then d[n[e]] ← d[n[e]] ⊕ (r′ ⊗ w[e])
13                       r[n[e]] ← r[n[e]] ⊕ (r′ ⊗ w[e])
14                       if n[e] ∉ S
15                         then ENQUEUE(S, n[e])
```

$d[q]$ denotes the tentative shortest distance from the source s to q. $r[q]$ keeps track of the sum of the weights added to $d[q]$ since the last queue extraction of q. The attribute r is needed for the shortest-distance algorithm to work in non-idempotent cases. The algorithm uses a queue S to store the set of states to consider for the relaxation steps of lines 11-15 [15]. Any queue discipline, e.g., FIFO, shortest-first, topological (in the acyclic case), can be used. The test of line 11 is based on an ϵ-equality.

Different queue disciplines yield different running times for our algorithm. The choice of the best queue discipline to use can be based on the structure of the two automata, which can be exploited to obtain a more efficient algorithm to compute the relative entropy. More specifically, let Q, E denote (respectively) the set of states and edges in the intersection automata. Further, let $N(q)$ denote the number of times a state q is inserted in the queue. Then, using the Fibonacci heap with a shortest first queue discipline (as in Dijkstra's algorithm), the complexity of the algorithm is given by:

$$O(|Q| + |E| \max_{q \in Q} N(q) + \log |Q| \sum_{q \in Q} N(q)). \tag{20}$$

If the underlying automata are acyclic, then using the queue discipline corresponding to the topological order yields the best time complexity, and the problem can be solved in linear time: $O(|Q| + |E|)$.

Using a breadth-first queue discipline (as in the Bellman-Ford shortest distance algorithm), updates to the shortest distance estimates in iteration k can be formulated as $D^k = MD^{k-1}$, where M is the *matrix associated to the automaton*, that is the matrix representing the weighted graph defined by the automaton. Note that the matrix multiplication here is over the $\oplus$ and $\otimes$ operations of the semiring, so that $D^k[i] = \oplus_{j=1}^{|Q|} M[i,j] \otimes D^{k-1}[j]$.

We now analyze the convergence rate of the approximate algorithm with the breadth-first queue discipline. Let us focus only on the first component of the distance pair. Let M_1 be the matrix obtained by taking the first part of each element of M. Assume that the matrix M is a stochastic matrix.

By the Perron-Frobenius theorem, we know that the largest eigenvalue is 1 and has a multiplicity of 1. Furthermore, all other eigenvalues λ are such that $|\lambda| < 1$. Using the Jordan canonical form of M, it is not hard to show that the matrix multiplication operation converges in $O(|\lambda_2|^k)$, where λ_2 is the second largest eigenvalue of M (see [9] for a similar analysis). Thus, the updates in the kth iteration are proportional to λ_2^k, hence, $k = \frac{\log(1/\epsilon)}{\log(1/|\lambda_2|)}$. Plugging in this expression for $N(q)$, the overall complexity of the approximate algorithm is:

$$O(|Q| + (|E| + |Q|)\frac{\log(1/\epsilon)}{\log(1/|\lambda_2|)}). \tag{21}$$

For ϵ exponentially smaller than $|\lambda_2|$ ($\epsilon = |\lambda_2|^d$), the cost in complexity is only linear: $O(|Q| + d(|E| + |Q|))$.

It is possible to use different queue disciplines in different parts of the graph and improve the running time of the algorithm. For example, for a large graph with several strongly connected components, one can use a topological order on the component graph, with shortest-first queue discipline in each strongly connected component [15]. If there are k strongly connected components, with the ith component having n_i vertices, then the running time is given by $O(|Q| + |E| \max_{q \in Q} N(q) + \log|\max_i n_i| \sum_{q \in Q} N(q))$. If the largest component has $O(n/k)$ vertices, then this improves the general complexity by an additive factor of $\sum_{q \in Q} N(q) \log k$. Our experience with such computations for very large graphs of several million states shows that the generic topological order with the shortest-fist queue discipline within each strongly connected component often leads to the most efficient results in practice.

4.3 Comparison with Previous Work

In [3], the author describes a *procedure* for an approximate computation of the relative entropy of two deterministic stochastic automata. The procedure is based on an iterative method (which can be viewed as approximating the inverse of a matrix) for computing, for a stochastic automaton A, the probability of each state q, that is the sum of the weights of all paths going through q. The convergence is claimed but not proved and no bound is indicated on the maximum number of iterations.

The author reports no complexity result for the procedure described, which makes it difficult to compare with our algorithm. Our most favorable estimate of its complexity is $\Omega(|A|^2|B|^2(T+|\Sigma|))$, where T denotes the maximum number of iterations executed. This is because the procedure requires using a matrix of size $|A|^2|B|^2$. The complexity of the procedure also depends on the size of the alphabet, which, in some applications such as natural language processing applications, may be very large. Furthermore, the lower bound space complexity of this procedure is $\Omega(|A|^2|B|^2)$. This makes it unsuitable for computing the relative entropy of large weighted automata. Note that the experiments reported by the author were carried out with very small grammars of about 30 rules. Nevertheless, the procedure bears some resemblance with our approximate algorithm. It can be viewed as an alphabet-dependent non-sparse implementation of that algorithm for the particular case of a FIFO queue discipline.

5 Experiments

We implemented both the generic Floyd-Warshall algorithm and the approximate algorithm for the computation of the relative entropy of unambiguous probabilistic automata.

To avoid the numerical instability issues related to the multiplications of probabilities, we used instead negative log probabilities. This corresponds to taking the image of the entropy semiring by the semiring morphism $\log \times I$ where I is the identity over the second element of the weights.

To evaluate the efficiency of our approximate algorithm for computing the relative entropy we created two n-gram statistical models trained on a large corpus – one a bigram model ($n = 2$) and one a trigram model ($n = 3$). The minimal deterministic weighted automaton representing the bigram model had about 200,000 transitions, that of the trigram model about 400,000 transitions. It took about 3s on a single 2GHz Intel processor with 128MB of RAM to compute the relative entropy of these large weighted automata using a FIFO queue discipline. With a shortest-first queue discipline, the time was reduced to 2s.

6 Conclusion

We described several algorithms for the computation of the relative entropy of two deterministic weighted automata or the entropy of a single deterministic weighted automaton by formulating the problem as a shortest-distance computation over the entropy semiring. We presented both an exact algorithm and an approximate algorithm that was shown to be very efficient even for very large automata of several hundred thousand transitions. The results demonstrate the benefit of a semiring-theory formulation of the problem. Our algorithms can be used similarly to compute the so-called unnormalized relative entropy of two weighted automata, which is defined by:

$$D(A\|B) = \sum_x [\![A]\!](x) \log \frac{[\![A]\!](x)}{[\![B]\!](x)} - [\![A]\!](x) + [\![B]\!](x) \tag{22}$$

simply by replacing Φ_1 and Φ_2 by Φ'_1 and Φ'_2, where $\Phi'_1(A)$ ($\Phi'_2(A)$) is the weighted automaton over the entropy semiring derived from A by replacing each weight w with the pair $(w, 1)$ (resp (w, w)). The entropy semiring can also be used to give a conceptually simple formulation of the computation of the relative entropy of tree automata and to derive similar computation algorithms.

References

1. Jean Berstel and Christophe Reutenauer. *Rational Series and Their Languages.* Springer-Verlag: Berlin-New York, 1988.
2. Stephen Bloom and Zoltan Ésik. *Iteration Theories.* Springer-Verlag, Berlin, 1991.
3. Rafael C. Carrasco. Accurate computation of the relative entropy between stochastic regular grammars. *Informatique Théorique et Applications (ITA)*, 31(5):437–444, 1997.
4. Thomas H. Cormen, Charles E. Leiserson, and Ronald L. Rivest. *Introduction to Algorithms.* The MIT Press: Cambridge, MA, 1992.
5. Thomas M. Cover and Joy A. Thomas. *Elements of Information Theory.* John Wiley & Sons, Inc., New York, 1991.
6. Karel Culik II and Jarkko Kari. Digital Images and Formal Languages. In Grzegorz Rozenberg and Arto Salomaa, editors, *Handbook of Formal Languages*, volume 3, pages 599–616. Springer, 1997.
7. Samuel Eilenberg. *Automata, Languages and Machines*, volume A–B. Academic Press, 1974–1976.
8. Jason Eisner. Expectation Semirings: Flexible EM for Finite-State Transducers. In *Proceedings of the ESSLLI Workshop on Finite-State Methods in NLP*, 2001.
9. G. H. Golub and C. F. V. Loan. *Matrix Computations.* The Johns Hopkins University Press, Baltimore, 1996.
10. Werner Kuich and Arto Salomaa. *Semirings, Automata, Languages.* Number 5 in EATCS Monographs on Theoretical Computer Science. Springer-Verlag, Berlin, Germany, 1986.
11. Daniel J. Lehmann. Algebraic Structures for Transitive Closures. *Theoretical Computer Science*, 4:59–76, 1977.
12. Mehryar Mohri. Finite-State Transducers in Language and Speech Processing. *Computational Linguistics*, 23(2), 1997.
13. Mehryar Mohri. General Algebraic Frameworks and Algorithms for Shortest-Distance Problems. Technical Memorandum 981210-10TM, AT&T Labs - Research, 62 pages, 1998.
14. Mehryar Mohri. Generic Epsilon-Removal and Input Epsilon-Normalization Algorithms for Weighted Transducers. *International Journal of Foundations of Computer Science*, 13(1):129–143, 2002.
15. Mehryar Mohri. Semiring Frameworks and Algorithms for Shortest-Distance Problems. *Journal of Automata, Languages and Combinatorics*, 7(3):321–350, 2002.

16. Mehryar Mohri, Fernando C. N. Pereira, and Michael Riley. Weighted Automata in Text and Speech Processing. In *Proceedings of the 12th biennial European Conference on Artificial Intelligence (ECAI-96), Workshop on Extended finite state models of language, Budapest, Hungary*. John Wiley and Sons, Chichester, 1996.
17. Mehryar Mohri, Fernando C. N. Pereira, and Michael Riley. Weighted Finite-State Transducers in Speech Recognition. *Computer Speech and Language*, 16(1):69–88, 2002.
18. Arto Salomaa and Matti Soittola. *Automata-Theoretic Aspects of Formal Power Series*. Springer-Verlag, 1978.

A Parallel Algorithm for Finding All Successive Minimal Maximum Subsequences

Ho-Kwok Dai[1] and Hung-Chi Su[2]

[1] Computer Science Department, Oklahoma State University, Stillwater, Oklahoma 74078, USA
[2] Department of Computer Science, Arkansas State University, State University, Arkansas 72467, USA
dai@cs.okstate.edu, suh@csm.astate.edu

Abstract. Efficient algorithms for finding multiple contiguous subsequences of a real-valued sequence having large cumulative sums, in addition to its combinatorial appeal, have widely varying applications such as in textual information retrieval and bioinformatics. A maximum contiguous subsequence of a real-valued sequence is a contiguous subsequence with the maximum cumulative sum. A minimal maximum contiguous subsequence is a minimal contiguous subsequence (with respect to subsequential containment) among all maximum ones of the sequence. We present a logarithmic-time and optimal linear-work parallel algorithm on the parallel random access machine model that finds all successive minimal maximum subsequences of a real-valued sequence.

1 Introduction

Combinatorial and algorithmic problems in sequences and trees arise in widely varying applications such as in textual information retrieval and bioinformatics. The area of large-scale (sub)sequence comparison, alignment, and analysis is central in computational biology. Efficient algorithms for finding multiple contiguous subsequences of a real-valued sequence having large cumulative sums, in addition to its combinatorial appeal, can help identify multiple statistically significant subsequences (with respect to a scoring scheme) in biomolecular sequences.

Given a sequence $X = (x_\eta)_{\eta=1}^{n}$ of n real-valued terms, the cumulative sum of a non-empty contiguous subsequence $(x_\eta)_{\eta=i}^{j}$, where i and j are in the index range $[1, n]$ and $i \leq j$, is $\sum_{\eta=i}^{j} x_\eta$ (and that of the empty sequence is 0). All subsequences addressed in our study are contiguous in real-valued sequences; the terms "subsequence" and "supersequence" will hereafter abbreviate "contiguous subsequence" and "contiguous supersequence", respectively. A maximum subsequence of X is one with the maximum cumulative sum. A minimal maximum subsequence of X is a minimal subsequence (with respect to subsequential containment) among all maximum subsequences of X. Note that: (1) X is non-positively valued if and only if the empty subsequence is the

J.R. Correa, A. Hevia, and M. Kiwi (Eds.): LATIN 2006, LNCS 3887, pp. 337–348, 2006.

unique minimal maximum subsequence of X, and (2) the minimality constraint on the maximum cumulative sum translates into that all non-empty prefixes and suffixes of a minimal maximum subsequence of X have positive cumulative sums.

Very often in practical applications it is not sufficient to find just a single or even all maximum subsequences of a sequence X, what is rather required is to find many or all pairwise disjoint subsequences having cumulative sums above a prescribed threshold. Observe that subsequences having major overlap with a maximum subsequence tend to have good cumulative sums. Intuitively, we define the sequence of all successive minimal maximum subsequences $(S_1, S_2, \ldots)$ of X inductively as follows:

1. The sequence S_1 is a (non-empty) minimal maximum subsequence of X, and
2. Assume that the sequence $(S_1, S_2, \ldots, S_i)$ of non-empty subsequences of X, where $i \geq 1$, has been constructed, the subsequence S_{i+1} is a (non-empty) minimal subsequence (with respect to subsequential containment) among all non-empty maximum subsequences (with respect to cumulative sum) that are disjoint from each of $\{S_1, S_2, \ldots, S_i\}$.

As in the definition of minimal maximum subsequence, the minimality constraint on the maximum cumulative sums of $S_1, S_2, \ldots$ is equivalent to the non-existence of non-empty prefixes nor suffixes with non-positive cumulative sums in each of $\{S_1, S_2, \ldots\}$.

Efficient algorithms for computing the sequence of all successive minimal maximum subsequences of a given sequence are essential for statistical inference in large-scale biological sequence analysis. In biomolecular sequences, high (sub)sequence similarity usually implies significant structural or functional similarity (the first fact of biological sequence analysis in [5]). When incorporating good scoring schemes, this provides a powerful statistical paradigm for identifying biologically significant functional regions in biomolecular sequences (see [7], [10], [8], and [12]), such as transmembrane regions [3] and deoxyribonucleic acid-binding domains [9] in protein analyses.

A common approach is to employ an application-dependent scoring scheme that assigns a score to each single constituent of an examined biomolecular sequence, and then find all successive minimal maximum subsequences of the underlying score sequence having large cumulative sums above a prescribed threshold. A theory of logarithmic odds ratios, developed in [7], yields an effective logarithmic likelihood-ratio scoring function in this context. The non-positivity of the expected score of a random single constituent tends to delimit unrealistic long runs of contiguous positive scores.

We present a logarithmic-time and optimal linear-work parallel algorithm on the parallel random access machine (PRAM) model that finds all successive minimal maximum subsequences of a real-valued sequence. Our study is motivated by the linear-time sequential algorithm for this computation problem [12], which introduces an equivalent non-procedural characterization of the sequence of all successive minimal maximum subsequences.

2 Related Work and Structural Characterization of Successive Minimal Maximum Subsequences

For computing a single (minimal) maximum subsequence of a length-n real-valued sequence of X, a simple sequential algorithm solves this problem in $O(n)$ optimal time. The algorithm maintains/updates two subsequences (their index ranges) and their corresponding cumulative sums of each successive prefix of X: a maximum subsequence and the maximum suffix of the prefix in an inductive and on-line fashion.

A parallel algorithm [1] on the PRAM model solves the problem in $O(\log n)$ parallel time using a total of $O(n)$ operations (work-optimal). The algorithm computes the delimiting indices $\alpha, \beta \in [1, n]$ (index range $[\alpha, \beta]$) of a maximum subsequence of X by using prefix sums as follows. For the input sequence $X = (x_\eta)_{\eta=1}^n$, let $(s_\eta)_{\eta=1}^n$ denote the sequence of prefix sums of X (that is, $s_i = \sum_{\eta=1}^i x_\eta$ for $i \in [1, n]$) and $(sm_\eta)_{\eta=1}^n$ denote the sequence of suffix maxima: for $i \in [1, n]$, $sm_i = \max\{s_\eta \mid \eta \in [i, n]\}$ and $\beta(i) = \min \arg\max\{s_\eta \mid \eta \in [i, n]\}$ (the least index at which the maximum s_η is attained). The knowledge of $(sm_\eta)_{\eta=1}^n$ and $(\beta(\eta))_{\eta=1}^n$ allows us to compute the maximum cumulative sum, denoted by m_i, of a subsequence with starting index i (and ending index $\beta(i)$): for $i \in [1, n]$, $m_i = sm_i - s_i + x_i$. Hence the (maximum) cumulative sum m of a maximum subsequence of X is given by $m = \max\{m_\eta \mid \eta \in [1, n]\}$ with delimiting indices $\alpha = \max \arg\max\{m_\eta \mid \eta \in [1, n]\}$ (the greatest index at which the maximum m_η is attained) and $\beta = \beta(\alpha)$.

The prefix-sums algorithm can be adapted to compute the statistics: $(s_\eta)_{\eta=1}^n$, $((sm_\eta, \beta(\eta)))_{\eta=1}^n$, $(m_\eta)_{\eta=1}^n$, and (m, α, β) on the exclusive-read exclusive-write PRAM model in the stated time- and work-bounds. The algorithm is also implemented on the PRAM model augmented with broadcasting with selective reduction [1], which is an additional form of concurrent access to shared memory in three phases: concurrent broadcasting to all memory locations, and concurrent selection and reduction of selected data into a single datum in all memory locations. With appropriate broadcast instruction, selection rule, and reduction operator, the algorithm on the augmented PRAM model runs in $O(1)$ parallel time using a total of $O(n)$ operations (work-optimal).

For the problem of finding the sequence of all successive minimal maximum subsequences of a length-n real-valued sequence X, a recursive divide-and-conquer strategy can apply the linear-time sequential algorithm above to compute a minimal maximum subsequence of X whose deletion results in a prefix and a suffix for recursion. The algorithm has a (worst-case) time complexity of $\Theta(n^2)$. Empirical analyses of the algorithm [12] on synthetic data sets (sequences of independent and identically distributed uniform random terms with negative mean) and score sequences of genomic data indicate that the running time grows at $\Theta(n \log n)$.

In order to circumvent the iterative dependency in computing the sequence of all successive minimal maximum subsequences, Ruzzo and Tompa [12] prove a structural characterization of the sequence as follows. Denote by $\mathrm{MAX}(X)$ the set of all successive minimal maximum subsequences of a real-valued sequence X.

Theorem 1. *[12] For a non-empty real-valued sequence X, a non-empty subsequence S of X is in* MAX(X) *if and only if:*

1. *[Monotonicity; see Lemma 3 below] The subsequence S is monotone: every proper subsequence of S has its cumulative sum less than that of S, and*
2. *[Maximality of Monotonicity] The subsequence S is maximal in X with respect to monotonicity, that is, every proper supersequence of S contained in X is not monotone.*

Hence, we also term MAX(X) as the set of all maximal monotone subsequences of X.

The following lemma gives a structural decomposition of X into MAX(X).

Lemma 2. *[12] Let X be a non-empty real-valued sequence. Then:*

1. *Every non-empty monotone subsequence of X is contained in a maximal monotone subsequence in* MAX(X)*; in particular, every positive term of X is contained in a maximal monotone subsequence in* MAX(X)*, and*
2. *The set* MAX(X) *is a pairwise disjoint collection of all maximal monotone subsequences of X.*

Based on the structural characterization of MAX(X), they present a sequential algorithm that computes MAX(X) in $O(n)$ optimal sequential time and $O(n)$ space (worst case). The algorithm generalizes the one above for computing a single maximum subsequence in a similar inductive and on-line fashion. For each successive prefix $P_i = (x_\eta)_{\eta=1}^{i}$ of X for $i \in [1, n-1]$, the algorithm maintains the prefix sum of X ending at index i and a complete list L_i of $k(i)$ pairwise disjoint subsequences of P_i in MAX(P_i): $S_1, S_2, \ldots, S_{k(i)}$. The sufficient statistics for each $S \in$ MAX(P_i) are its index range $[\alpha(S), \beta(S)]$ and its starting and ending prefix sums (prefix sums of X ending at indices $\alpha(S) - 1$ and $\beta(S)$, respectively). For a positive term x_{i+1}, the augmented list $L_{i+1} = (S_1, S_2, \ldots, S_{k(i)}, (x_{i+1}))$ does not constitute the desired MAX(P_{i+1}) in general — due to the violation of the maximality of monotonicity from integrating the monotone subsequence (x_{i+1}) into L_i. The algorithm restores the maximality of monotonicity by a sequence of merges, each merging backward as many trailing terms of L_{i+1} as possible into a single monotone trailing term of L_{i+1} using the sufficient statistics above.

3 Parallel Algorithm on PRAM Model Computing MAX

We now devise a parallel algorithm on the PRAM model that computes MAX(X) for a real-valued sequence X in logarithmic parallel time and optimal linear work.

3.1 Characterization of Monotonicity Via Prefix Sums

For a real-valued sequence $X = (x_\eta)_{\eta=1}^{n}$, denote by $s_i(X)$ the i-th prefix sum $\sum_{\eta=1}^{i} x_\eta$ of X for $i \in [1, n]$, and $s_0(X) = 0$. For a subsequence Y of X, denote by $\sigma(Y)$ the cumulative sum of Y ($\sigma(Y) = 0$ if Y is empty). The following characterization of monotonicity leads to an effective computation of the index range of a non-trivial monotone subsequence containing a given term of X.

Lemma 3. *Let X be a non-empty real-valued sequence and Y be a non-empty subsequence of X with index range $[\alpha, \beta]$. The following statements are equivalent:*

1. *Y is monotone.*
2. *The starting prefix sum $s_{\alpha-1}(X)$ of Y is the unique minimum and the ending prefix sum $s_\beta(X)$ of Y is the unique maximum of all $s_i(X)$ for all $i \in [\alpha-1, \beta]$.*
3. *All non-empty prefixes and non-empty suffixes of Y have positive cumulative prefix sums.*

Proof. We prove that statement 1 implies statement 2 by contrapositive. If $s_{\alpha-1}(X)$ is not the unique minimum of all $s_i(X)$ for all $i \in [\alpha-1, \beta]$, that is, there exists $i \in [\alpha, \beta]$ such that $s_{\alpha-1}(X) \geq s_i(X)$, we consider the proper subsequence Y' of Y with index range $[i+1, \beta]$. The cumulative sum of Y' is $\sigma(Y') = s_\beta(X) - s_i(X)$ (note that Y' is empty with zero cumulative sum if $i = \beta$). Now, $\sigma(Y') = s_\beta(X) - s_i(X) \geq s_\beta(X) - s_{\alpha-1}(X) = \sigma(Y)$. The existence of such Y' gives that Y is not monotone. If $s_\beta(X)$ is not the unique maximum of all $s_i(X)$ for all $i \in [\alpha-1, \beta]$, then an analogous argument leads also to the non-monotonicity of Y.

To prove that statement 2 implies statement 1, we assume that the starting prefix sum $s_{\alpha-1}(X)$ and the ending prefix sum $s_\beta(X)$ are the unique minimum and unique maximum, respectively, of all $s_i(X)$ for all $i \in [\alpha-1, \beta]$. Note that the assumption implicitly gives that $s_{\alpha-1}(X) < s_\beta(X)$, or equivalently $\sigma(Y) > 0$. Let Y' be an arbitrary proper subsequence of Y with index range $[\alpha', \beta']$. If Y' is empty, then $\sigma(Y') = 0 < \sigma(Y)$, as desired. Therefore, we may assume that $\alpha \leq \alpha' \leq \beta' \leq \beta$, and $\alpha < \alpha'$ or $\beta' < \beta$. Observe that:

$$\begin{aligned}\sigma(Y) &= s_\beta(X) - s_{\alpha-1}(X) \\ &= (s_\beta(X) - s_{\beta'}(X)) + (s_{\beta'}(X) - s_{\alpha'-1}(X)) + (s_{\alpha'-1}(X) - s_{\alpha-1}(X)) \\ &= (s_\beta(X) - s_{\beta'}(X)) + \sigma(Y') + (s_{\alpha'-1}(X) - s_{\alpha-1}(X)) \ .\end{aligned}$$

The terms $s_\beta(X) - s_{\beta'}(X)$ and $s_{\alpha'-1}(X) - s_{\alpha-1}(X)$ represent the cumulative sums of the suffix and the prefix of Y that immediately follows and precedes Y', respectively. The unique maximality of $s_\beta(X)$ and the unique minimality of $s_{\alpha-1}(X)$ of all $s_i(X)$ for all $i \in [\alpha-1, \beta]$, together with $\alpha < \alpha'$ or $\beta' < \beta$, give that both terms are non-negative but at least one of them is positive. Hence $\sigma(Y) > \sigma(Y')$, which establishes the monotonicity of Y.

The equivalence of statements 2 and 3 follows from that of the uniqueness of minimality of $s_{\alpha-1}(X)$ (maximality of $s_\beta(X)$) and the absence of non-empty prefixes (non-empty suffixes, respectively) of Y with non-positive cumulative sums. □

The monotonicity constraint applies only to the delimiting indices of a monotone subsequence, but not to the entire index range.

3.2 Computation of a Refinement of MAX

The parallel algorithm above for finding a single maximum subsequence focuses on computing the primary statistics for locating concurrently the delimiting indices of a maximum subsequence with each possible starting index, but it disregards the monotonicity condition of all such subsequences. More work is required to process the secondary statistics in order to reveal the structural decomposition of X into MAX(X).

The key to a parallel implementation of finding MAX(X) for a length-n sequence $X = (x_\eta)_{\eta=1}^{n}$ lies in the concurrent computation of the ending index of the maximal monotone subsequence constrained with the starting index $i \in [1, n]$. Lemma 3 suggests to consider only positive terms x_i of X for the desired computation. We abbreviate the prefix sums $s_\eta(X)$ to s_η for all $\eta \in [0, n]$.

Let $\epsilon : \{i \in [1, n] \mid x_i > 0\} \to [1, n]$ be the function that $\epsilon(i)$ denotes the ending index of the maximal monotone subsequence of X constrained with the starting index i. We show that the function ϵ is composed of the following two functions:

1. The function $\epsilon' : \{i \in [1, n] \mid x_i > 0\} \to [2, n + 1]$ defined by:

$$\epsilon'(i) = \begin{cases} \min\{\eta \in [i + 1, n] \mid s_{i-1} \geq s_\eta\} & \text{if the minimum exists,} \\ n + 1 & \text{otherwise,} \end{cases}$$

   locates the least index $\eta \in [i, n]$ (closest to i), effectively $\eta \in [i + 1, n]$ since $x_i > 0$, such that $s_{i-1} \geq s_\eta$ if it exists.
2. The function $\epsilon'' : \{(i, j) \in [1, n]^2 \mid i \leq j\} \to [1, n]$ defined by:

$$\epsilon''(i, j) = \min \arg\max\{s_\eta \mid \eta \in [i, j]\} ,$$

   locates the least index $\eta \in [i, j]$ (closest to i) such that s_η is the maximum prefix sum of those of X over the index range $[i, j]$.

Figure 1 illustrates the computation of the two functions ϵ' and ϵ'' and their composition to yield ϵ, as stated in the following lemma.

Lemma 4. *For the functions ϵ, ϵ', and ϵ'' defined above, $\epsilon(i) = \epsilon''(i, \epsilon'(i) - 1)$ for all $i \in [1, n]$.*

Proof. We first show that the subsequence of X with the index range $[i, \epsilon''(i, \epsilon'(i) - 1)]$ is monotone. Since $x_i > 0$, we have $s_{i-1} < s_i$ $(= s_{i-1} + x_i)$. By the definition of ϵ', the prefix sum s_{i-1} of X is the unique minimum of all s_η for all $\eta \in [i - 1, \epsilon'(i) - 1]$. Observe that $\epsilon''(i, \epsilon'(i) - 1) \in [i, \epsilon'(i) - 1] \subseteq [i - 1, \epsilon'(i) - 1]$, thus s_{i-1} is the unique minimum of all s_η for all $\eta \in [i - 1, \epsilon''(i, \epsilon'(i) - 1)]$. As $s_{i-1} < s_i$ and by the definition of ϵ'', $s_{\epsilon''(i, \epsilon'(i)-1)}$ is the unique maximum of all s_η for all $\eta \in [i - 1, \epsilon'(i) - 1]$. This establishes the monotonicity of the subsequence over the index range $[i, \epsilon''(i, \epsilon'(i) - 1)]$. Hence $\epsilon(i) \geq \epsilon''(i, \epsilon'(i) - 1)$ by the definition of ϵ.

To prove that $\epsilon(i) \leq \epsilon''(i, \epsilon'(i) - 1)$, we notice that: (1) the maximal monotone subsequence of X constrained with the starting index i ends at the index $\epsilon(i)$, and

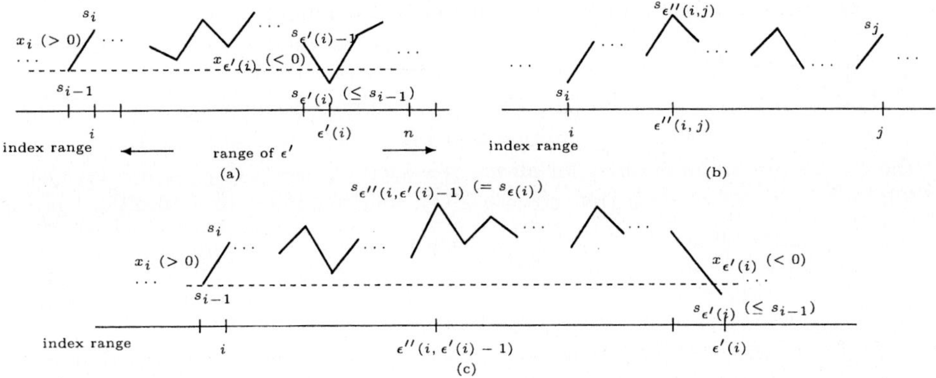

Fig. 1. (a) For each index $i \in [1, n]$ with $x_i > 0$, ϵ' computes the least index $\eta \in [i+1, n]$ such that $s_{i-1} \geq s_\eta$ if it exists; (b) For all indices $i, j \in [1, n]$ with $i \leq j$, ϵ'' computes the least index $\eta \in [i, j]$ such that s_η is the maximum prefix sum of those of X over the index range $[i, j]$; (c) For each index $i \in [1, n]$ with $x_i > 0$, ϵ computes the ending index of the maximal monotone subsequence constrained with the starting index i ($\epsilon(i) = \epsilon''(i, \epsilon'(i) - 1)$)

(2) Lemma 3 says that s_{i-1} is the unique minimum of all s_η for all $\eta \in [i-1, \epsilon(i)]$. The definition of ϵ' gives that $[i-1, \epsilon(i)] \subseteq [i-1, \epsilon'(i)-1]$, that is, $\epsilon(i) \leq \epsilon'(i)-1$. Moreover, this results in $\max\{s_\eta \mid \eta \in [i-1, \epsilon(i)]\} \leq \max\{s_\eta \mid \eta \in [i-1, \epsilon'(i) - 1]\}$, which implies that $\min \arg\max\{s_\eta \mid \eta \in [i-1, \epsilon(i)]\} \leq \min \arg\max\{s_\eta \mid \eta \in [i-1, \epsilon'(i) - 1]\}$. Lemma 3 indicates that $s_{\epsilon(i)}$ is the unique maximum of all s_η for all $\eta \in [i-1, \epsilon(i)]$, which gives that $\epsilon(i) = \min \arg\max\{s_\eta \mid \eta \in [i-1, \epsilon(i)]\} \leq \min \arg\max\{s_\eta \mid \eta \in [i-1, \epsilon'(i) - 1]\} = \epsilon''(i, \epsilon'(i) - 1)$, and this completes the proof. □

The concurrent computation of ϵ, when applied to all the positive terms x_i in X, generates the statistics $\mathrm{MON}(X) = \{[i, \epsilon(i)] \mid i \in [1, n] \text{ with } x_i > 0\}$ for the set of all index ranges of all maximal monotone subsequences of X constrained with given positive starting terms. The following lemma reveals the structural decomposition of X into MON(X), which refines MAX(X) and provides a basis for a parallel computation of MAX(X) from MON(X).

Lemma 5. *For a real-valued sequence X,* MON(X) *respects the following parenthesis structure:*

1. *Every positive term of X has its index as the starting index of a unique index range in* MON(X),
2. *For every pair of index ranges in* MON(X)*, either they are disjoint or one is contained in another, and*
3. *For every maximal monotone subsequence of X in* MAX(X)*, its index range is in* MON(X).

Proof. Parts 1 and 3 follow from the definition of ϵ. For part 2, consider two index ranges $[i, \epsilon(i)], [j, \epsilon(j)] \in \mathrm{MON}(X)$ such that $i < j \leq \epsilon(i)$. We show that $[j, \epsilon(j)] \subseteq [i, \epsilon(i)]$.

Since $i < j \le \epsilon(i)$ and s_{i-1} is the unique minimum of all s_η for all $\eta \in [i-1, \epsilon(i)]$, we have $s_{i-1} < s_{j-1}$. Also, s_{j-1} is the unique minimum of all s_η for all $\eta \in [j-1, \epsilon'(j)]$. (Note that if $\epsilon(i) > \epsilon'(j)$, then $\epsilon(i) > \epsilon'(j) > \epsilon(j)$ by Lemma 4 — as desired in part (2). Hence we may assume that $\epsilon(i) \le \epsilon'(j)$.) Combining these with the assumption that $i < j \le \epsilon(i) \le \epsilon'(j)$ yields that s_{i-1} is the unique minimum of all s_η for all $\eta \in [i-1, \epsilon(i)] \cup [j-1, \epsilon'(j)] = [i-1, \epsilon'(j)]$. The definition of ϵ' gives that $\epsilon'(i) \ge \epsilon'(j)$, which implies that $\max\{s_\eta \mid \eta \in [i, \epsilon'(i)-1]\} \ge \max\{s_\eta \mid \eta \in [j, \epsilon'(j)-1]\}$.

Consider the candidate index position for $\min \arg\max\{s_\eta \mid \eta \in [i, \epsilon'(i)-1]\}$ ($= \epsilon(i)$) over the index range $[i, \epsilon'(i)-1]$ when partitioned into three disjoint subranges $[i, j) \cup [j, \epsilon'(j)-1] \cup (\epsilon'(j)-1, \epsilon'(i)-1]$.

Case when $\epsilon(i) \in [i, j)$: This can not occur due to the assumption that $i < j \le \epsilon(i)$.

Case when $\epsilon(i) \in [j, \epsilon'(j)-1]$: When both $\min \arg\max\{s_\eta \mid \eta \in [i, \epsilon'(i)-1]\}$ ($= \epsilon(i)$) and $\min \arg\max\{s_\eta \mid \eta \in [j, \epsilon'(j)-1]\}$ ($= \epsilon(j)$) are in $[j, \epsilon'(j)-1]$, we have $\epsilon(i) = \epsilon(j)$.

Case when $\epsilon(i) \in (\epsilon'(j)-1, \epsilon'(i)-1]$: Clearly $\epsilon(j) \le \epsilon(i)$.

Combining the three cases gives that $\epsilon(j) \le \epsilon(i)$, as desired. This completes the proof of part 2. □

3.3 Computation of MAX from MON

For a real-valued sequence X, we compute MAX(X) based on the statistics of MON(X) via an appropriately defined function on the parenthesis structure of MON(X). Let $\mathcal{I}$ be a set of (some) index ranges of $[1, n]$ such that for every pair of index ranges of $\mathcal{I}$, either they are disjoint or one is contained in another. We define the function (range-composition) $\circ : \mathcal{I}^2 \to \mathcal{I}$ that selects the rightmost or outmost index range from its two operands in $\mathcal{I}$: for all $[i_1, j_1], [i_2, j_2] \in \mathcal{I}$,

$$[i_1, j_1] \circ [i_2, j_2] = \begin{cases} [i_2, j_2] & \text{if } j_1 < i_2 \ ([i_1, j_1] \cap [i_2, j_2] = \emptyset) \ , \\ [i_1, j_1] & \text{if } j_2 < i_1 \ ([i_1, j_1] \cap [i_2, j_2] = \emptyset) \ , \\ [i_2, j_2] & \text{if } [i_1, j_1] \subseteq [i_2, j_2] \ , \\ [i_1, j_1] & \text{if } [i_1, j_1] \supseteq [i_2, j_2] \ . \end{cases}$$

The function $\circ$ is associative since, for all $I_1, I_2, I_3 \in \mathcal{I}$, both $(I_1 \circ I_2) \circ I_3$ and $I_1 \circ (I_2 \circ I_3)$ compute the same index range in $\{I_1, I_2, I_3\}$: the outmost one if I_1, I_2, and I_3 are nested, or the rightmost and then outmost one otherwise.

Lemma 6. *Let X be a length-n real-valued sequence $(x_\eta)_{\eta=1}^n$ with m positive terms, and $\mathcal{I} = (I_\eta)_{\eta=1}^m$ be the length-m sequence that enumerates all m index ranges of* MON(X), *$I_\eta = [i_\eta, \epsilon(i_\eta)]$ (with $x_{i_\eta} > 0$) for all $\eta \in [1, m]$, according to the starting index i_η ($i_1 < i_2 < \cdots < i_m$). Denote by T_k the k-th prefix sum of $\mathcal{I}$ (with respect to the range-composition $\circ$), that is, $T_k = I_1 \circ I_2 \circ \cdots \circ I_k$ for $k \in [1, m]$. Then, for every $k \in [1, m]$, T_k gives the index range of the unique maximal monotone subsequence in* MAX(X) *that contains the positive i_k-th term x_{i_k} of X.*

Proof. We prove the statement in the lemma by an induction on k. For the basis of the induction that $k = 1$, we notice that $x_{i_k} = x_{i_1}$ is the first positive term in X, and $x_j \leq 0$ for all $j \in [1, i_1)$. Thus, according to Lemma 3, a monotone subsequence of X containing x_{i_1} can not have a starting index in $[1, i_1)$. The maximal monotone subsequence of X constrained with the starting index i_1 (with index range $[i_1, \epsilon(i_1)]$ $(= I_1 = T_1)$) must be the (unconstrained) maximal monotone subsequence containing the i_1-th term x_{i_1} of X. This proves the basis.

For the induction step, assume that the statement is true for all $l \leq k$, where $k \geq 1$. We show that T_{k+1} is the index range of the (unconstrained) maximal monotone subsequence I in MAX(X) that contains the positive term $x_{i_{k+1}}$ of X. Denote by $[\alpha, \beta]$ the index range of I. By Lemma 3, x_α is a positive term of X, which gives that $\alpha \in \{i_1, i_2, \ldots, i_k, i_{k+1}\}$. Let $\alpha = i_j$ for some $j \in [1, k+1]$.

We claim that the j-th prefix sum T_j $(= I_1 \circ I_2 \circ \cdots \circ I_{j-1} \circ I_j) = I_j$. The case when $j = 1$ is immediate. To prove the claim when $j > 1$, we first notice that, similar to the basis, the maximal monotone subsequence of X constrained with the starting index i_j (with index range $I_j = [i_j, \epsilon(i_j)]$ is the (unconstrained) maximal monotone subsequence containing the i_j-th term x_{i_j} of X. This gives that $[\alpha, \beta] = [i_j, \epsilon(i_j)] = I_j$. Applying the induction hypothesis (in the case of $l = j-1$ with $1 \leq l \leq k$), the $(j-1)$-th prefix sum $T_{j-1} = I_1 \circ I_2 \circ \cdots \circ I_{j-1}$ computes the index range of the (unconstrained) maximal monotone subsequence I' in MAX(X) that contains the i_{j-1}-th term $x_{i_{j-1}}$ of X. By the partitioning structure of X into MAX(X) stated in Lemma 2, the two (unconstrained) maximal monotone subsequences I' and I (with index ranges T_j and $[\alpha, \beta]$ $(= [i_j, \epsilon(i_j)] = I_j)$, respectively) are disjoint, and $[\alpha, \beta]$ is the rightmost index range of the two. Therefore we have $T_j = T_{j-1} \circ I_j = I_j$, as desired in the claim.

We now have the $(k+1)$-th prefix sum $T_{k+1} = T_j \circ (I_{j+1} \circ I_{j+2} \circ \cdots \circ I_{k+1}) = I_j \circ (I_{j+1} \circ I_{j+2} \circ \cdots \circ I_{k+1})$. Notice that all the indices $i_j, i_{j+1}, \ldots, i_{k+1} \in [\alpha, \beta] = I_j$, the parenthesis structure of MON(X) stated in Lemma 5 gives that $I_\eta \subseteq I_j$ for all $\eta \in [j+1, k+1]$. Hence $T_{k+1} = I_j \circ (I_{j+1} \circ I_{j+2} \circ \cdots \circ I_{k+1}) = I_j$ since I_j is the outmost index range of all I_η for all $\eta \in [j, k+1]$; that is, $T_{k+1} = [\alpha, \beta]$, as desired. This completes the induction step. By induction, the lemma is proved. □

3.4 PRAM Algorithm for Computing MAX Within the Work-Time Framework

We present the overall algorithm within the work-time framework, and provide algorithmic details and analysis. Improvements to the algorithm are also addressed.

algorithm Compute_MAX
input: A length-n real-valued sequence X in an array $x[1..n]$.
output: The sequence of all successive minimal maximum subsequences (that is, all maximal monotone subsequences) of X occupying the low-order subarray of an array $M[1..\lceil \frac{n}{2} \rceil]$ $(|\text{MAX}(X)| \leq \lceil \frac{n}{2} \rceil)$.
begin
1. Compute the prefix sums of X in an array $s[1..n]$ such that $s[i] = \sum_{\eta=1}^{i} x[\eta]$ for all $i \in [1, n]$;

2. Compute the function ϵ in an array $\epsilon[1..n]$ such that $\epsilon[i]$ denotes the ending index of the maximal monotone subsequence of X constrained with the starting index i as follows:
   2.1. Compute the function ϵ' in an array $\epsilon'[1..n]$, in which $\epsilon'[i]$ is the least index $\eta \in [i+1, n]$ such that $s[i-1] \geq s[\eta]$ if it exists, and $n+1$ otherwise;
   2.2. Compute the function ϵ'' in an array $\epsilon''[1..n, 1..n]$, in which $\epsilon''[i, j]$ is the least index $\eta \in [i, j]$ such that $s[\eta] = \max\{s[k] \mid k \in [i, j]\}$;
   2.3. Compose ϵ from ϵ' and ϵ'' as follows:
   for all $i \in [1, n]$ in parallel do
   $\epsilon[i] := \epsilon''[i, \epsilon'[i] - 1]$;
   end for;
3. Compute $\text{MON}(X) = \{[i, \epsilon[i]] \mid i \in [1, n] \text{ with } x[i] > 0\}$ (ordered according to the starting index) and pack all the index ranges of $\text{MON}(X)$ in the low-order subarray of the array $M[1..\lceil \frac{n}{2} \rceil]$ ($|\text{MON}(X)| \leq \lceil \frac{n}{2} \rceil$);
4. Compute $\text{MAX}(X)$ in the array $M[1..\lceil \frac{n}{2} \rceil]$ as follows:
   4.1. Compute the prefix sums of the (non-trivial) low-order subarray of $M[1..\lceil \frac{n}{2} \rceil]$ using the range-composition function $\circ$ for the prefix computation;
   4.2. Pack all the distinct elements (pairwise disjoint index ranges) in the (non-trivial) low-order subarray of $M[1..\lceil \frac{n}{2} \rceil]$ in place, while maintaining their relative order (according to the starting index);

end Compute_MAX;

Step 1 is implemented by the prefix-sums algorithm [11] that runs in $O(\log n)$ time, using $O(n)$ operations on the exclusive-read exclusive-write (EREW) PRAM.

As for step 2, the computation of ϵ' (in step 2.1) is reduced to (within the resource bounds of $O(\log n)$ time and $O(n)$ operations) the problem of all nearest smaller values of the sequence $(s_{\eta-1} \mid x_\eta > 0)_{\eta=1}^{n}$, which can be solved by an algorithm (see [2] and [4]) that runs in $O(\log n)$ time, using $O(n)$ operations on the EREW PRAM. The computation of ϵ'' (in step 2.2) is reduced to (within the same resource bounds) the problem of range-minima, which can be solved by an algorithm (see [6]) that runs in $O(\log n)$ time, using $O(n)$ operations on the EREW PRAM. The composition of ϵ (in step 2.3) is executed in $O(1)$ time, using $O(n)$ operations on the EREW PRAM.

Step 3 is reduced to the problem of array packing, which can be solved by the prefix-sums algorithm.

As for step 4, step 4.1 is a direct application of the prefix-sums algorithm, and step 4.2 is reduced to array packing as in step 3.

Theorem 7. *For a length-n real-valued sequence X, the algorithm Compute_MAX computes the set $\text{MAX}(X)$ of all successive minimal maximum subsequences (that is, all maximal monotone subsequences) of X in $O(\log n)$ time using $O(n)$ operations (work-optimal) on the EREW PRAM model.*

A few potential performance improvements can be applied to the algorithm without any asymptotic increase in the complexity bounds. A pre-processing step can compress the sequence X into an alternating one X° by summing up contiguous positive terms and contiguous non-positive terms into single terms, and maintains the correspondence between the compressed index ranges and single indices. The two structures MAX(X) and MAX(X°) are essentially identical via the correspondence due to Lemma 3. The work can be implemented by the prefix-sums algorithm within the resource bounds.

A generalization to the computation of ϵ is its bidirectional extension to $\overline{\epsilon} : \{i \in [1,n] \mid x_i > 0\} \to [1,n]^2$ such that $\overline{\epsilon}(i) = (\epsilon^-(i), \epsilon^+(i))$, where $\epsilon^+(i)$ is $\epsilon(i)$ and $\epsilon^-(i)$ denotes the starting index of the maximal monotone subsequence of X constrained with the ending index i. The computation of ϵ^- is analogous to that of ϵ^+, and $\overline{\epsilon}$ yields a coarser partition of MAX(X).

The task of computing MAX(X) can be partitioned into independent ones based on the existence of positive terms $x_i > 0$, where $i \in [1,n]$, which satisfy at least one of the following conditions that mirror the computation of ϵ':

	condition	term discarded	partition (X', X'')
c_1 :	non-existence of $\max\{\eta \in [0, i-2] \mid s_{i-1} > s_\eta\}$	x_{i-1}	$[1, i-2], [i, n]$
c_2 :	non-existence of $\max\{\eta \in [0, i-2] \mid s_i > s_\eta\}$	x_{i-1}	$[1, i-2], [i, n]$
c_3 :	non-existence of $\min\{\eta \in [i+1, n] \mid s_i < s_\eta\}$	x_{i+1}	$[1, i], [i+2, n]$
c_4 :	non-existence of $\min\{\eta \in [i+1, n] \mid s_{i-1} < s_\eta\}$	x_{i+1}	$[1, i], [i+2, n]$

Since c_2 implying c_1 and c_4 implying c_3, the condition $c_1 \vee c_2 \vee c_3 \vee c_4$ is logically equivalent to $c_1 \vee c_3$. If either c_1 or c_3 is met, then Lemma 3 gives that no monotone subsequence of X can overlap with both X' and X'' non-trivially, which establishes that MAX(X) is the disjoint union of MAX(X') and MAX(X''). The detection of the condition c_1 (c_3) can be reduced to the problem of all nearest smaller values (all nearest larger values, respectively) of the sequence $(s_{\eta-1} \mid x_\eta > 0)_{\eta=1}^n$ ($(s_\eta \mid x_\eta > 0)_{\eta=1}^n$, respectively).

4 Conclusion

The problem of computing the set of all successive minimal maximum subsequences of a real-valued sequence has wide applications such as in textual information retrieval and bioinformatics. The MAX-computation has real practical importance as it appears as a subroutine in biological sequence analysis. Hence there is a natural need for computing MAX in parallel and its implementation on practical parallel systems. Our parallel algorithm computes MAX in logarithmic parallel time and optimal linear work on the EREW PRAM model. The MAX-computation has linear sequential complexity, and is solved very efficiently by an optimal linear-time sequential algorithm, hence achieving good speed-ups on a practical parallel system is a challenge. For biomolecular sequence analysis, alternative definitions of maximum subsequences and more complex scoring schemes exist in the literature. Another direction for research is to adapt or modify the optimal (sequential and parallel) algorithms accordingly.

References

1. S. G. Akl and G. R. Guenther. Applications of broadcasting with selective reduction to the maximal sum subsegment problem. *International Journal of High Speed Computing*, 3(2):107–119, 1991.
2. O. Berkman, D. Breslauer, Z. Galil, B. Schieber, and U. Vishkin. Highly-parallelizable problems. In *Proceedings of the 21st Annual ACM Symposium on Theory of Computing*, pages 309–319. Association for Computing Machinery, 1989.
3. V. Brendel, P. Bucher, I. R. Nourbakhsh, B. E. Blaisdell, and S. Karlin. Methods and algorithms for statistical analysis of protein sequences. In *Proceedings of the National Academy of Sciences U. S. A., 89(6):2002-2006*, 1992.
4. D. Z. Chen. Efficient geometric algorithms on the EREW PRAM. *IEEE Transactions on Parallel and Distributed Systems*, 6(1):41–47, 1995.
5. D. Gusfield. *Algorithms on Strings, Trees, and Sequences: Computer Science and Computational Biology.* Cambridge University Press, New York, 1997.
6. J. JáJá. *An Introduction to Parallel Algorithms.* Addison-Wesley, 1992.
7. S. Karlin and S. F. Altschul. Methods for assessing the statistical significance of molecular sequence features by using general scoring schemes. In *Proceedings of the National Academy of Sciences U. S. A., 87(6):2264-2268*, 1990.
8. S. Karlin and S. F. Altschul. Applications and statistics for multiple high-scoring segments in molecular sequences. In *Proceedings of the National Academy of Sciences U. S. A., 90(12):5873-5877*, 1993.
9. S. Karlin and V. Brendel. Chance and statistical significance in protein and DNA sequence analysis. *Science*, 257(5066):39–49, 1992.
10. S. Karlin and A. Dembo. Limit distributions of maximal segmental score among Markov-dependent partial sums. *Advances in Applied Probability*, 24:113–140, 1992.
11. R. E. Ladner and M. J. Fischer. Parallel prefix computation. *Journal of the Association for Computing Machinery*, 27(4):831–838, 1980.
12. W. L. Ruzzo and M. Tompa. A linear time algorithm for finding all maximal scoring subsequences. In *Proceedings of the Seventh International Conference on Intelligent Systems for Molecular Biology*, pages 234–241. International Society for Computational Biology, 1999.

De Dictionariis Dynamicis Pauco Spatio Utentibus
(*lat.* On Dynamic Dictionaries Using Little Space)*

Erik D. Demaine[1], Friedhelm Meyer auf der Heide[2], Rasmus Pagh[3], and Mihai Pătraşcu[1]

[1] Massachusetts Institute of Technology
edemaine, mip@mit.edu
[2] University of Paderborn
fmadh@upb.de
[3] IT University of Copenhagen
pagh@itu.dk

Abstract. We develop dynamic dictionaries on the word RAM that use asymptotically optimal space, up to constant factors, subject to insertions and deletions, and subject to supporting perfect-hashing queries and/or membership queries, each operation in constant time with high probability. When supporting only membership queries, we attain the optimal space bound of $\Theta(n \lg \frac{u}{n})$ bits, where n and u are the sizes of the dictionary and the universe, respectively. Previous dictionaries either did not achieve this space bound or had time bounds that were only expected and amortized. When supporting perfect-hashing queries, the optimal space bound depends on the range $\{1, 2, \ldots, n+t\}$ of hashcodes allowed as output. We prove that the optimal space bound is $\Theta(n \lg\lg \frac{u}{n} + n \lg \frac{n}{t+1})$ bits when supporting only perfect-hashing queries, and it is $\Theta(n \lg \frac{u}{n} + n \lg \frac{n}{t+1})$ bits when also supporting membership queries. All upper bounds are new, as is the $\Omega(n \lg \frac{n}{t+1})$ lower bound.

1 Introduction

The dictionary is one of the most fundamental data-structural problems in computer science. In its basic form, a dictionary allows some form of "lookup" on a set S of n objects, and in a dynamic dictionary, elements can be inserted into and deleted from the set S. However, being such a well-studied problem, there are many variations in the details of what exactly is required of a dictionary, particularly the lookup operation, and these variations greatly affect the best possible data structures. To enable a systematic study, we introduce a unified view consisting of three possible types of queries that, in various combinations, capture the most common types of dictionaries considered in the literature:

Membership: Given an element x, is it in the set S?

* A full version of this paper is available as arXiv:cs.DS/0512081.

J.R. Correa, A. Hevia, and M. Kiwi (Eds.): LATIN 2006, LNCS 3887, pp. 349–361, 2006.

Retrieval: Given an element x in the set S, retrieve r bits of data associated with x. (The outcome is undefined if x is not in S.) The associated data can be set upon insertion or with another update operation. We state constant time bounds for these operations, which ignore the $\Theta(r)$ divided by word size required to read or write r bits of data.

Perfect hashing: Given an element x in the set S, return the *hashcode* of x. The data structure assigns to each element x in S a unique hashcode in $[n+t]$,[1] for a specified parameter t (e.g., $t = 0$ or $t = n$). Hashcodes are *stable*: the hashcode of x must remain fixed for the duration that x is in S. (Again the outcome is undefined if x is not in S.)

Standard hash tables generally support membership and retrieval. Some hash tables with open addressing (no chaining) also support perfect hashing, but the expected running time is superconstant unless $t = \Omega(n)$. However, standard hash tables are not particularly space efficient if n is close to u: they use $O(n)$ words, which is $O(n \lg u)$ bits for a universe of size u, whereas only $\log_2 \binom{u}{n} = \Theta(n \lg \frac{u}{n})$ bits (assuming $n \leq u/2$) are required to represent the set S.[2]

Any dictionary supporting membership needs at least $\log_2 \binom{u}{n}$ bits of space. But while such dictionaries are versatile, they are large, and membership is not always required. For example, Chazelle et al. [3] explore the idea of a static dictionary supporting only retrieval, with several applications related to Bloom filters. For other data-structural problems, such as range reporting in one dimension [10, 1], the only known way to get optimal space bounds is to use a dictionary that supports retrieval but not membership. The retrieval operation requires storing the r-bit data associated with each element, for a total of at least rn bits. If r is asymptotically less than $\lg \frac{u}{n}$, then we would like to avoid actually representing the set S. However, as we shall see, we still need more than rn bits even in a retrieval-only dictionary.

Perfect hashing is stronger than retrieval, up to constant factors in space, because we can simply store an array mapping hashcodes to the r-bit data for each element. Therefore we focus on developing dictionaries supporting perfect hashing, and obtain retrieval for free. Conversely, lower bounds on retrieval apply to perfect hashing as well. Because hashcodes are stable, this approach has the additional property that the associated data never moves, which can be useful, e.g. when the data is large or is stored on disk.

Despite substantial work on dictionaries and perfect hashing (see Section 1.2), no dynamic dictionary data structure supporting any of the three types of queries simultaneously achieves (1) constant time bounds with high probability and (2) *compactness* in the sense that the space is within a constant factor of optimal.

1.1 Our Results

We characterize the optimal space bound, up to constant factors, for a dynamic dictionary supporting any subset of the three operations, designing data

[1] The notation $[k]$ represents the set $\{0, 1, \ldots, k-1\}$.

[2] Throughout this paper, $\lg x$ denotes $\log_2(2+x)$, which is positive for all $x \geq 0$.

structures to achieve these bounds and in some cases improving the lower bound. To set our results in context, we first state the two known lower bounds on the space required by a dictionary data structure. First, as mentioned above, any dictionary supporting membership (even static) requires $\Omega(n \lg \frac{u}{n})$ bits of space, assuming that $n \leq u/2$. Second, any dictionary supporting retrieval must satisfy the following recent and strictly weaker lower bound:

Theorem 1. [10] *Any dynamic dictionary supporting retrieval (and therefore any dynamic dictionary supporting perfect hashing) requires $\Omega(n \lg \lg \frac{u}{n})$ bits of space in expectation, even when the associated data is just $r = 1$ bit.*

Surprisingly, for dynamic dictionaries supporting perfect hashing, this lower bound is neither tight nor subsumed by a stronger lower bound. In Section 5, we prove our main lower-bound result, which complements Theorem 1 depending on the value of t:

Theorem 2. *Any dynamic dictionary supporting perfect hashing with hashcodes in $[n+t]$ must use $\Omega(n \lg \frac{n}{t+1})$ bits of space in expectation, regardless of the query and update times, assuming that $u \geq n + (1+\varepsilon)t$ for some constant $\varepsilon > 0$.*

Our main upper-bound result is a dynamic dictionary supporting perfect hashing that matches the sum of the two lower bounds given by Theorems 1 and 2. Specifically, Section 4 proves the following theorem:

Theorem 3. *There is a dynamic dictionary that supports updates and perfect hashing with hashcodes in $[n+t]$ (and therefore also retrieval queries) in constant time per operation, using $O(n \lg \lg \frac{u}{n} + n \lg \frac{n}{t+1})$ bits of space. The query and space complexities are worst-case, while updates are handled in constant time with high probability.*

To establish this upper bound, we find it necessary to also obtain optimal results for dynamic dictionaries supporting both membership and perfect hashing. In Section 3, we find that the best possible space bound is a sum of two lower bounds in this case as well:

Theorem 4. *There is a dynamic dictionary that supports updates, membership queries, and perfect hashing with hashcodes in $[n+t]$ (and therefore also retrieval queries) in constant time per operation, using $O(n \lg \frac{u}{n} + n \lg \frac{n}{t+1})$ bits of space. The query and space complexities are worst-case, while updates are handled in constant time with high probability.*

In the interest of Theorems 3 and 4, we develop a family of *quotient hash functions.* These hash functions are permutations of the universe; they and their inverses are computable in constant time given a small-space representation; and they have natural distributional properties when mapping elements into buckets. (In contrast, we do not know any hash functions with these properties and, say, 4-wise independence.) These hash functions may be of independent interest.

Table 1 summarizes our completed understanding of the optimal space bounds for dynamic dictionaries supporting updates and any combination of the three types of queries in constant time with high probability. All upper bounds are new, as are the lower bounds for perfect hashing with or without membership.

Table 1. Optimal space bounds for all types of dynamic dictionaries supporting operations in constant time with high probability. The upper bounds supporting retrieval without perfect hashing can be obtained by substituting $t = n$. The $\Theta(n \lg \frac{u}{n})$ bounds assume $n \leq u/2$; more precisely, they are $\Theta(\log_2 \binom{u}{n})$.

Dictionary queries supported	Optimal space	Reference
retrieval	$\Theta(n \lg\lg \frac{u}{n} + nr)$	O §4; Ω [10]
retrieval+perfect hashing	$\Theta(n \lg\lg \frac{u}{n} + n \lg \frac{n}{t+1} + nr)$	O §4; Ω §5, [10]
membership	$\Theta(n \lg \frac{u}{n})$	O §3; Ω std.
membership+retrieval	$\Theta(n \lg \frac{u}{n} + nr)$	O §3; Ω std.
membership+retrieval+perfect hashing	$\Theta(n \lg \frac{u}{n} + n \lg \frac{n}{t+1} + nr)$	O §3; Ω §5

1.2 Previous Work

There is a huge literature on various types of dictionaries, and we do not try to discuss it exhaustively. A milestone in the history of constant-time dictionaries is the realization that the space and query bounds can be made worst case (construction and updates are still randomized). This was achieved in the static case by Fredman, Komlós, and Szemerédi [7] with a dictionary that uses $O(n \lg u)$ bits. Starting with this work, research on the dictionary problem evolved in two orthogonal directions: creating dynamic dictionaries with good update bounds, and reducing the space.

In the dynamic case, the theoretical ideal is to make updates run in constant time per operation with high probability. After some work, this was finally achieved by the high-performance dictionaries of Dietzfelbinger and Meyer auf der Heide [4]. However, this desiderate is usually considered difficult to achieve, and most dictionary variants that have been developed since then fall short of it, by having amortized and/or expected time bounds (not with high probability).

As far as space is concerned, the goal was to get closer to the information theoretic lower bound of $\log_2 \binom{u}{n}$ bits for membership. Brodnik and Munro [2] were the first to solve static membership using $O(n \lg \frac{u}{n})$ bits, which they later improved to $(1+o(1)) \log_2 \binom{u}{n}$. Pagh [11] solves the static dictionary problem with space $\log_2 \binom{u}{n}$ plus the best lower-order term known to date. For the dynamic problem, the best known result is by Raman and Rao [13], achieving space $(1 + o(1)) \log_2 \binom{u}{n}$. Unfortunately, in this structure, updates take constant time amortized and in expectation (not with high probability). These shortcomings seem inherent to their technique.

Thus, none of the previous results simultaneously achieve good space and update bounds, a gap filled by our work. Another shortcoming of the previous results lies in the understanding of dynamic dictionaries supporting perfect hashing. The dynamic perfect hashing data structure of Dietzfelbinger et al. [5] supports membership and a weaker form of perfect hashing in which hashcodes are not stable, though only an amortized constant number of hashcodes change per update. This structure achieves a suboptimal space bound of $O(n \lg u)$ and updates take constant time amortized and in expectation. No other dictionaries can answer perfect hashing queries except by associating an explicit hashcode with each element, which requires $\Theta(n \lg n)$ additional bits. Our result for membership and perfect hashing is the first achieving $O(n \lg \frac{u}{n})$ space, even for weak update bounds. A more fundamental problem is that all dynamic data structures supporting perfect hashing use $\Omega(n \lg \frac{u}{n})$ space, even when we do not desire membership queries so the information theoretic lower bound does not apply.

Perfect hashing in the static case has been studied intensely, and with good success. There, it is possible to achieve good bounds with $t = 0$, and this has been the focus of attention. When membership is required, a data structure using space $(1 + o(1)) \lg \binom{u}{n}$ was finally developed by [12]. Without membership, the best known lower bound is $n \log_2 e + \lg \lg u + O(\lg n)$ bits [6], while the best known data structure uses $n \log_2 e + \lg \lg u + O(n \frac{(\lg \lg n)^2}{\lg n} + \lg \lg \lg u)$ bits [8]. Our lower bound depending on t shows that in the dynamic case, even $t = O(n^{1-\varepsilon})$ requires $\Omega(n \lg n)$ space, making the problem uninteresting. Thus, we identify an interesting hysteresis phenomenon, where the dynamic nature of the problem forces the data structure to remember more information and use more space.

Retrieval without membership was introduced as "Bloomier filters" by Chazelle et al. [3]. The terminology is by analogy with the Bloom filter, a static structure supporting approximate membership (a query we do not consider in this paper). Bloomier filters are static dictionaries supporting retrieval using $O(nr + \lg \lg u)$ bits of space. For dynamic retrieval of $r = 1$ bit without membership, Chazelle et al. [3] show that $\Omega(n \lg \lg u)$ bits of space can be necessary in the case $n^{3+\varepsilon} \le u \le 2^{n^{O(1)}}$. Their bound is improved in [10], giving Theorem 1. On the upper-bound side, the only previous result is that of [10]: dynamic perfect hashing for $t = \Theta(n/\lg u)$ using space $O(n \lg \lg u)$. Our result improves $\lg \lg u$ to $\lg \lg \frac{u}{n}$, and offers the full tradeoff depending on t.

1.3 Details of the Model

A few details of the model are implicit throughout this paper. The model of computation is the Random Access Machine with cells of $\lg u$ bits (the word RAM). Because we ignore constant factors, we assume without loss of generality that u, t, and b are all exact powers of 2.

In dynamic dictionaries supporting perfect hashing, n is not the current size of the set S, but rather n is a fixed upper bound on the size of S. Similarly, t is a fixed parameter. This assumption is necessary because of the problem statement: hashcodes must be stable and the hashcode space is defined in terms of n and t. This assumption is not necessary for retrieval queries, although we effectively assume it through our reduction to perfect hashing. Our results leave open whether a dynamic dictionary supporting only retrieval can achieve space bounds depending on the current size of the set S instead of an upper bound n; such a result would in some sense improve the first row of Table 1.

On the other hand, if we want a dynamic dictionary supporting membership but not perfect hashing (but still supporting retrieval), then we can rebuild the data structure whenever $|S|$ changes by a constant factor, and change the upper bounds n and t then. This global rebuilding can be deamortized at the cost of increasing space by a constant factor, using the standard tricks involving two copies of the data structure with different values of n and t.

Another issue of the model of memory allocation. We assume that the dynamic data structure lives in an infinite array of word-length cells. At any time, the space usage of the data structure is the length of the shortest prefix of the array containing all nonblank cells. This model charges appropriately for issues such as external fragmentation (unlike, say, assuming that the system provides memory-block allocation) and is easy to implement in practical systems. See [13] for a discussion of this issue.

Finally, we prove that our insertions work in constant time with high probability, that is, with probability $1 - 1/n^c$ for any desired constant $c > 0$. Thus, with polynomially small probability, the bounds might be violated. For a with-high-probability bound, the data structure could fail in this low-probability event. To obtain the bounds also in expectation and with zero error, we can freeze the high-performance data structure in this event and fall back to a simple data structure, e.g., a linked list of any further inserted elements. Any operations (queries or deletions) on the old elements are performed on the high-performance data structure, while any operations on new elements (e.g., insertions) are performed on the simple data structure. The bounds hold in expectation provided that the data structure is used for only a polynomial amount of time.

2 Quotient Hash Functions

We define a quotient hash function in terms of three parameters: the universe size u, the number of buckets b, and an upper bound n on the size of the sets of interest. A quotient hash function is simply a bijective function $h : [u] \to [b] \times [\frac{u}{b}]$. We interpret the first output as a bucket, and the second output as a "quotient" which, together with the bucket, uniquely identifies the element. We write $h(x)_1$ and $h(x)_2$ when we want to refer to individual outputs of h.

We are interested in sets of elements $S \subset [u]$ with $|S| \leq n$. For such a set S and an element x, define $\mathcal{B}_h(S, x) = \{y \in S \mid h(y)_1 = h(x)_1\}$, i.e. the set of elements mapped to the same bucket as x. For a threshold t, define $\mathcal{C}_h(S, t) = \{x \in S \mid \#\mathcal{B}_h(S, x) \geq t\}$, i.e. the set of elements which map to buckets containing at least t elements. These are elements that "collide" beyond the allowable threshold.

Theorem 5. *There is an absolute constant $\alpha < 1$ such that for any u, n and b, there exists a family of quotient hash functions $\mathcal{H} = \{h : [u] \to [b] \times [\frac{u}{b}]\}$ satisfying:*

- *an $h \in \mathcal{H}$ can be represented in $O(n^\alpha)$ space and sampled in $O(n^\alpha)$ time.*
- *h and h^{-1} can be evaluated in constant time on a RAM;*
- *for any fixed $S \subset [u], |S| \leq n$ and any $\delta < 1$, the following holds with high probability over the choice of h:*

$$\begin{cases} \text{if } b \geq n, & \#\mathcal{C}_h(S, 2) \leq 2\frac{n^2}{b} + n^\alpha \\ \text{if } b < n, & \#\mathcal{C}_h(S, (1+\delta)\frac{n}{b} + 1) \leq 2ne^{-\delta^2 n/(3b)} + n^\alpha \end{cases}$$

It is easy to get an intuitive understanding of these bounds. In the case $b \geq n$, the expected number of collisions generated by universal hashing (2-independent hashing) would be $\frac{n^2}{b}$. For $b < n$, we can compare against a highly independent hash function. Then, the expected number of elements that land in overflowing buckets is $ne^{-\delta^2 n/(3b)}$, by a simple Chernoff bound. Our family matches these two bounds, up to a constant factor and an additive error term of $O(n^\alpha)$, which are both negligible for our purposes. The advantage of our hash family is two-fold. First, it gives quotient hash functions, which is essential for our data structure. Second, the number of overflowing elements is guaranteed with high probability, not just in expectation.

Construction of the hash family. Due to space limitations, we only sketch the construction, without proofs. First, we reduce the universe to n^c, for some big enough c, by applying a random 2-independent permutation on the original universe. We keep only the first $c \lg n$ bits of the result, and make the rest part of the quotient.

We now interpret the universe as a two-dimensional table, with $n^{3/4}$ columns, and $\frac{u}{n^{3/4}}$ rows. The plan is to use this column structure as a means of generating independence. Imagine a hash function that generates few collisions in expectation, but not necessarily with high probability. However, we can apply a *different* random hash function inside each column. The expectation is unchanged, but now Chernoff bounds can be used to show that we are close to the expectation with high probability, because the behavior of each column is independent.

However, to put this plan into action, we need to guarantee that the elements of S are spread rather uniformly across columns. We do this by applying a random circular shift to each row: consider a highly independent hash function mapping row numbers to $[n^{3/4}]$; inside each row, apply a circular shift by the hash function of that row. Note that the number of rows can be pretty large (larger than n), so we cannot afford a truly random shift for each row. However, the number of rows is polynomial, and we can use Siegel's family of highly independent hash functions [14] to generate highly independent shifts, which turns out to be enough.

In the case $b \geq n$, our goal is to get close to the collisions generated by a 2-independent permutation, but with high probability. As explained above, we can achieve this effect through column independence: apply a random 2-universal permutation inside each column. To complete the construction, break each column into $\frac{b}{n^{3/4}}$ equal-sized buckets. The position within a bucket is part of the quotient. A classic Chernoff bound (using column independence), can show that imposing the bucket granularity does not generate too many collisions.

In the case $b < n$, the ideal size of each bucket is $\frac{n}{b}$ elements. We are interested in buckets of size exceeding $(1+\delta)\frac{n}{b}$, and want to bound the number of elements in such buckets close to the expected number for a highly independent permutation. As explained already, we do not know any family of highly independent permutations that can be represented with small space and evaluated efficiently. Instead, we will revert to the brute-force solution of representing truly random permutations. To use this idea and keep the space small, we need two tricks. The first trick is to generate and store fewer permutations than columns. It turns out that re-using the same permutation for multiple columns still gives enough independence.

Note, however, that we cannot even afford to store a random permutation inside a single column, because columns might have more than n elements. However, we can reduce columns to size $\sqrt{n}$ as follows. Use the construction from above for $b' = n^{5/4}$. This puts elements into $n^{5/4}$ *first-order buckets* ($\sqrt{n}$ buckets per column), with a negligible number of collisions. Thus, we can now work at the granularity of first-order buckets, and ignore the index within a bucket of an element. We now group columns into $n^{1/4}$ equal-sized groups. For each group, generate a random permutation on $\sqrt{n}$ positions, and apply it to the first-order buckets inside each column of the group.

In the full version, we also describe how our construction can be used to get good concentration bounds for dynamic sets S.

3 Solution for Membership and Perfect Hashing

There are two easy cases. First, if $u = \Omega(n^{1.5})$, then the space bound is $\Theta(n \lg u)$. In this case, a solution with hashcode range exactly $[n]$ can be obtained by using a high-performance dictionary [4]. We store an explicit hashcode as the data associated with each value, and maintain a list of free hashcodes. This takes $O(n \lg n + n \lg u) = O(n \lg u)$ bits. Second, if $t = O(n^{\alpha})$, for $\alpha < 1$, then the space bound is $\Theta(n \lg n)$. Because $u = O(n^{1.5})$, we can use the same brute-force solution. In the remaining cases, we can assume $t \leq \frac{n^2}{u}$ (we are always free to decrease t), so that the space bound is dominated by $\Theta(n \lg \frac{n}{t})$.

The data structure is composed of three levels. An element is inserted into the first level that can handle it. The first-level filter outputs hashcodes in the range $[n + \frac{t}{3}]$, and handles most elements of S: at most $c_1 t$ elements (for a constant $c_1 \leq \frac{1}{3}$ to be determined) are passed on to the second level, with high probability. The goal of the second-level filter is to handle all but $O(\frac{n}{\lg n})$ elements with high probability. If $c_1 t \leq \frac{n}{\lg n}$, this filter is not used. Otherwise, we use this filter, which outputs hashcodes in the range $[\frac{t}{3}]$. Finally, the third level is just a brute-force solution using a high-performance dictionary. Because it needs to handle only $\min\{O(\frac{n}{\lg n}), c_1 t\}$ elements, the output range can be $[\frac{t}{3}]$ and the space is $O(n)$ bits. This dictionary can always be made to work with high probability in n (e.g. by inserting dummy elements up to $\Omega(\sqrt{n})$ values).

A query tries to locate the element in all three levels. Because all levels can answer membership queries, we know when we've located an element, and we can just obtain a hashcode from the appropriate level. Similarly, deletion just removes the element from the appropriate level.

The first-level filter. Let $\mu = c_2 (\frac{n}{t})^3$, for a constant c_2 to be determined. We use a quotient hash function mapping the universe into $b = \frac{n}{\mu}$ buckets. Then, we expect μ elements per bucket, but we will allow for an additional $\mu^{2/3}$ elements. By Theorem 5, the number of elements that overflow is with high probability at most $ne^{-\Omega(\sqrt[3]{\mu})} + n^\alpha$. For big enough c_2, this is at most $\frac{c_1}{2}t$ (remember that we are in the case when n^α is negligible).

Now we describe how to handle the elements inside each bucket. For each bucket, we have a hashcode space of $[\mu + \mu^{2/3}]$. Then, the code space used by the first-level filter is $n + \frac{n}{\sqrt[3]{\mu}} \leq n + \frac{t}{3}$ for big enough c_2. We use a high-performance dictionary inside each bucket, which stores hashcodes as associated data. We also store a list of free hashcodes to facilitate insertions. To analyze the space, observe that a hashcode takes only $O(\lg \frac{n}{t})$ bits to represent. In addition, the high-performance dictionary need only store the *quotient* of an element. Indeed, the element is uniquely identified by the quotient and the bucket, so to distinguish between the elements in a bucket we only need a dictionary on the quotients. Thus, we need $O(\lg \frac{u\mu}{n}) = O(\lg \frac{u}{n} + \lg \frac{n}{t})$ bits per element.

The last detail we need to handle is what happens when an insertion in the bucket's dictionary fails. This happens with probability μ^{-c_3} for each insertion, where c_3 is any desired constant. We can handle a failed insertion by simply passing the element to the second level. The expected number of elements whose insertion at the first level failed is $n\mu^{-c_3} \leq \frac{c_1}{4}t$ for big enough c_3. Since we can assume $t = \Omega(n^{5/6})$, we have $\mu = O(\sqrt{n})$ and $b = \Omega(\sqrt{n})$. This means we have $\Omega(\sqrt{n})$ dictionaries, which use independent random coins. Thus, a Chernoff bound guarantees that we are not within twice this expectation with probability at most $e^{-\Omega(t/\sqrt{n})} = e^{-n^{\Omega(1)}}$ because $t = \Omega(n^\alpha)$. Thus, at most $c_1 t$ elements in total are passed to the second level with high probability.

The second-level filter. We first observe that this filter is used only when $\lg \frac{u}{n} = O(\sqrt[4]{\lg n})$. Indeed, $t \leq \frac{n^2}{u}$, so when $\lg \frac{u}{n} = \Omega(\sqrt[4]{\lg n})$, we have $t = o(\frac{n}{\lg n})$, and we can skip directly to the third level.

We use a quotient hash function mapping the universe to $b = \frac{c_1 t}{\sqrt[4]{\lg n}}$ buckets. We allow each bucket to contain up to $2\sqrt[4]{\lg n}$ elements; overflow elements are passed to the third level. By Theorem 5, at most $n/2^{\Omega(\sqrt[4]{\lg n})} = o(n/\lg n)$ elements are passed to the third level, with high probability. Because buckets contain $O(\sqrt[4]{\lg n})$ elements of $O(\sqrt[4]{\lg n})$ bits each, we can use word-packing tricks to handle buckets in constant time. However, the main challenge is space, not time. Observe that we can afford only $O(\lg \frac{u}{n})$ bits per element, which can be much smaller than $O(\sqrt[4]{\lg n})$. This means that we cannot even store a permutation of the elements inside a bucket. In particular, it is information-theoretically impossible even to store the elements of a bucket in an arbitrary order!

Coping with this challenge requires a rather complex solution: we employ $O(\lg \lg n)$ levels of filters and permutation hashing inside each bucket. Let us describe the level-i filter inside a bucket. First, we apply a random permutation to the bucket universe (the quotient of the elements inside the bucket). Then, the filter breaks the universe into $\frac{c_4 \sqrt[4]{\lg n}}{2^i}$ equal-sized *tiles*. The filter consists of an array with one position per tile. Such a position could either be empty, or it stores the index within the tile of an element mapped to that tile (which is a quotient induced by the permutation at this level). Observe that the size of the tiles doubles for each new level, so the number of entries in the filter array halves. In total, we use $h = \frac{1}{8} \lg\lg n$ filters, so that the number of tiles in any filter is $\Omega(\sqrt[8]{\lg n})$. Conceptually, an insertion traverses the filters sequentially starting with $i = 0$. It applies permutation i to the element, and checks whether the resulting tile is empty. If so, it stores the element in that tile; otherwise, it continues to the next level. Elements that cannot be mapped in any of the h levels are passed on to the third level of our big data structure. A deletion simply removes the element from the level where it is stored. A perfect-hash query returns the identifier of the tile where the element is stored. Because the number of tiles decreases geometrically, we use less than $2c_4 \sqrt[4]{\lg n}$ hashcodes per bucket. We have $\frac{c_1 t}{\sqrt[4]{\lg n}}$ buckets in total and we can make c_1 as small as we want, so the total number of hashcodes can be made at most $\frac{t}{3}$.

We now analyze the space needed by this construction. Observe that the size of the bucket universe is $v = u \cdot \frac{\sqrt[4]{\lg n}}{c_1 t}$. Thus, at the first level, the filter requires $\lg \frac{v}{c_4 \sqrt[4]{\lg n}}$ bits to store an index within each tile. At each consecutive level, the number of bits per tile increases by one (because tiles double in size), but the number of tiles halves. Thus, the total space is dominated by the first level, and it is $O(\lg \frac{u}{t}) = O(\lg \frac{u}{n} + \lg \frac{n}{t})$ bits per element.

The full version of the paper contains the proof that the number of unfiltered elements is small, as well as further implementation details.

4 Solution for Perfect Hashing

The data structure supporting perfect hashing but not membership consists of one quotient hash function, selected from the family of Theorem 5, and two instances of the data structure of Theorem 4 supporting perfect hashing and membership. The quotient hash function divides the universe into b buckets, and we set $b = \frac{cn^2}{t+1}$ for a constant $c \geq 1$ to be determined.

The first data structure supporting perfect hashing and membership stores the set B of buckets occupied by at least one element of S. An entry in B effectively represents an element of S that is mapped to that bucket. However, we have no way of knowing the exact element. The second data structure supporting perfect hashing and membership stores the additional elements of S, which at the time of insertion were mapped to a bucket already in B.

Insertions check whether the bucket containing the element is in B. If not, we insert it. Otherwise, we insert the element into the second data structure. Deletions proceed in the reverse order. First, we check whether the element is listed in the second data structure, in which case we delete it from there. Otherwise, we delete the bucket containing the element from the first data structure.

The range of the first perfect hash function should be $[n + \frac{t}{2}]$. For the second one, it should be $[\frac{t}{2}]$; we show below that this is sufficient with high probability. Thus, we use $[n+t]$ distinct hashcodes in total. To perform a query, we first check whether the element is listed in the second data structure. If it is, we return the label reported by that data structure (offset by $n + \frac{t}{2}$ to avoid the hashcodes from the first data structure). Otherwise, because we assume that the element is in S, it must be represented by the first data structure. Thus, we compute the bucket assigned to the element by the quotient hash function, look up that bucket in the first data structure, and return its label.

It remains to analyze the space requirement. We are always free to reduce t, so we can assume $t = O(n/\lg \frac{u}{n})$, simplifying our space bound to $O(n \lg \frac{n}{t+1})$. Because $|B| \leq n$, the first data structure needs space $O(\lg \binom{b}{n} + n \lg \frac{n}{t/2+1}) = O(n \lg \frac{b}{n} + n \lg \frac{n}{t+1}) = O(n \lg \frac{n}{t+1})$. Because $b \geq n$, our family of hash functions guarantees that, with high probability, the number of elements of S that were mapped to a nonempty bucket at the time of their insertion is at most $\frac{2n^2}{b} + n^\alpha = \frac{2(t+1)}{c} + n^\alpha$. If $n^\alpha < \frac{t}{8}$, this is at most $\frac{t}{4}$ for sufficiently large c. If $t = O(n^\alpha)$, we can use a brute-force solution: first, construct a perfect hashing structure with $t = n$ (this is possible through the previous case); then, relabel the used positions in the $[2n]$ range to a minimal range of $[n]$, using $O(n \lg n)$ memory bits. Given this bound on the number of elements in the second structure, note that the number of hashcodes allowed $(t/2)$ is double the number of elements. Thus the space required is $O(\lg \binom{u}{t/4} + t) = O(t \lg \frac{u}{t}) = O(\frac{n}{\lg(u/n)} \lg \frac{u}{n/\lg(u/n)}) = O(n)$.

5 Lower Bound for Perfect Hashing

This section proves Theorem 2 assuming $u \geq 2n$. We defer case of smaller u to the full version. Our lower bound considers the dynamic set S which is initially $\{n+1, \ldots, 2n\}$ and is transformed through insertions and deletions into $\{1, \ldots, n\}$. More precisely, we consider $\frac{n}{2t}$ stages. In stage i, we pick a random subset $D_i \subseteq S \cap \{n+1, \ldots, 2n\}$, of cardinality $2t$. Then, we delete the elements in D_i, and we insert elements $I_i = \{(i-1)2t + 1, \ldots, i \cdot 2t\}$. Note than, in the end, the set is $\{1, \ldots, n\}$. By the easy direction of Yao's minimax principle, we can fix the random bits of the data structure, such that it uses the same expected space over the input distribution.

Our strategy is to argue that the data structure needs to remember a lot of information about the history, i.e. there is large hysteresis in the output of the perfect hash function. Intuitively, the $2t$ elements inserted in each stage need to be mapped to only $3t$ positions in the range: the t positions free at the beginning of the stage, and the $2t$ positions freed by the recent deletes. These free positions are quite random, because we deleted random elements. Thus, this choice is very constrained, and the data structure needs to remember the constraints.

Let h be a function mapping each element in $[2n]$ to the hashcode it was assigned; this is well defined, because each element is assigned a hashcode exactly once (though for different intervals of time). We argue that the vector of sets $(h(I_1), \ldots, h(I_{n/2t}))$ has entropy $\Omega(n \lg \frac{n}{t})$. One can recover this vector by querying the final state of the data structure, so the space lower bound follows.

We first break up the entropy of the vector by: $H(h(I_1), \ldots, h(I_{n/2t})) = \sum_j H(h(I_j) \mid h(I_1), \ldots, h(I_{j-1}))$. Now observe that the only randomness up to stage j is in the choices of $D_1, \ldots, D_{j-1}$. In other words, $D_1, \ldots, D_{j-1}$ determine $h(I_1), \ldots, h(I_{j-1})$. Then, $H(h(I_1), \ldots, h(I_{n/2t})) \geq \sum_j H(h(I_j) \mid D_1, \ldots, D_{j-1})$. To alleviate notation, let $D_{<j}$ be the vector $(D_1, \ldots, D_{j-1})$.

Now we lower bound each term of the sum. Let F_j be the set of free positions in the range at the beginning of stage j. Because we made the data structure deterministic, F_j is fixed by conditioning on $D_{<j}$. Because I_j can be mapped to free positions only after D_j is deleted, we find that $h(I_j) \subset F_j \cup h(D_j)$. Note that $|h(I_j)| = 2t$, but $|F_j| = t$. Thus, $|h(D_j) \setminus h(I_j)| \le t$.

Now we argue that the entropy of $h(D_j)$ is large. Indeed, D_j is chosen randomly from $S \cap \{n+1, \ldots, 2n\}$, a set of cardinality $n - 2t(j-1)$. Conditioned on $D_{<j}$, the set $S \cap \{n+1, \ldots, 2n\}$ is fixed, so its image through h is fixed. Then, choosing D_j randomly is equivalent to choosing $h(D_j)$ randomly from a fixed set of cardinality $n - 2t(j-1)$. So $H(h(D_j) \mid D_{<j}) = \lg \binom{n-2t(j-1)}{2t}$. Now consider $h(D_j) \setminus h(I_j)$. This is a set of cardinality at most t from the same set of $n - 2t(j-1)$ positions. Thus, $H(h(D_j) \setminus h(I_j) \mid D_{<j}) \le \lg \binom{n-2t(j-1)}{t} + t$.

Using $H(a, b) \le H(a) + H(b)$, we have $H(h(D_j) \mid D_{<j}) \le H(h(D_j) \cap h(I_j) \mid D_{<j}) + H(h(D_j) \setminus h(I_j) \mid D_{<j})$. Of course, $H(h(I_j) \mid D_{<j}) \ge H(h(I_j) \cap h(D_j) \mid D_{<j})$. This implies $H(h(I_j) \mid D_{<j}) \ge H(h(D_j) \mid D_{<j}) - H(h(D_j) \setminus h(I_j) \mid D_{<j}) \ge \lg \binom{n-2t(j-1)}{2t} - \lg \binom{n-2t(j-1)}{t} - t$. Using $\binom{a}{b} / \binom{a}{c} = \binom{a-c}{b}$, we have $H(h(I_j) \mid D_{<j}) \ge \lg \binom{n-t(2j-1)}{2t} - t$. For $j \le \frac{n}{3t}$, we have $H(h(I_j) \mid D_{<j}) = \Omega(t \lg \frac{n}{t})$. We finally obtain $H(h(I_1), \ldots, h(I_j)) = \Omega(n \lg \frac{n}{t})$.

Acknowledgement. Part of this work was done at the Oberwolfach Meeting on Complexity Theory, July 2005.

References

1. Stephen Alstrup, Gerth Brodal, and Theis Rauhe. Optimal static range reporting in one dimension. In STOC'01, pages 476–482.
2. Andrej Brodnik and J. Ian Munro. Membership in constant time and almost-minimum space. *SIAM J. Computing*, 28(5):1627–1640, 1999. See also ESA'94.
3. Bernard Chazelle, Joe Kilian, Ronitt Rubinfeld, and Ayellet Tal. The Bloomier filter: an efficient data structure for static support lookup tables. In SODA'04.
4. Martin Dietzfelbinger and Friedhelm Meyer auf der Heide. A new universal class of hash functions and dynamic hashing in real time. In ICALP'90, pages 6–19.
5. Martin Dietzfelbinger, Anna Karlin, Kurt Mehlhorn, Friedhelm Meyer Auf Der Heide, Hans Rohnert, and Robert E. Tarjan. Dynamic perfect hashing: Upper and lower bounds. *SIAM J. Computing*, 23(4):738–761, 1994. See also FOCS'88.
6. M. L. Fredman and J. Komlós. On the size of separating systems and families of perfect hash functions. *SIAM J. Algebraic and Discrete Methods*, 5(1):61–68, 1984.
7. Michael L. Fredman, János Komlós, and Endre Szemerédi. Storing a sparse table with 0(1) worst case access time. JACM, 31(3):538–544, 1984. See also FOCS'82.
8. Torben Hagerup and Torsten Tholey. Efficient minimal perfect hashing in nearly minimal space. In STACS'01, pages 317–326.
9. Eyal Kaplan, Moni Naor, and Omer Reingold. Derandomized constructions of k-wise (almost) independent permutations. In RANDOM'05, pages 354–365.
10. Christian Worm Mortensen, Rasmus Pagh, and Mihai Pătraşcu. On dynamic range reporting in one dimension. In STOC'05, pages 104–111.
11. Rasmus Pagh. Low redundancy in static dictionaries with constant query time. *SIAM Journal on Computing*, 31(2):353–363, 2001. See also ICALP'99.

12. Rajeev Raman, Venkatesh Raman, and S. Srinivasa Rao. Succinct indexable dictionaries with applications to encoding k-ary trees and multisets. In SODA'02.
13. Rajeev Raman and S. Srinivasa Rao. Succinct dynamic dictionaries and trees. In ICALP'03, pages 357–368.
14. Alan Siegel. On universal classes of extremely random constant-time hash functions. *SIAM Journal on Computing*, 33(3):505–543, 2004.

Customized Newspaper Broadcast: Data Broadcast with Dependencies

Sandeep Dey and Nicolas Schabanel

École normale supérieure de Lyon, UMR CNRS ENS-LYON INRIA UCBL n°5668, 46 allée d'Italie, 69364 Lyon Cedex 07, France
dey.sandeep@gmail.com
http://perso.ens-lyon.fr/nicolas.schabanel

Abstract. Broadcasting has been proved to be an efficient means of disseminating data in wireless communication environments (such as Satellite, mobile phone networks; other typical broadcast networks are Videotext systems). Recent works provide strong evidence that correlation-based broadcast can significantly improve the average service time of broadcast systems. Most of the research on data broadcasting was done under the assumption that user requests are for a single item at a time and are independent of each other. However in many real world applications, such as web servers, dependencies exist among the data items, for instance: web pages on a server usually share a lot of items such as logos, style sheets, title-bar... and all these components have to be downloaded together when any individual page is requested. Such web server could take advantage of the correlations between the components of the pages, to speed up the broadcast of popular web pages. This paper presents a theoretical analysis of data dependencies and provides a polynomial time 4-approximation as well as theoretical proofs that our correlation-based approach can improve by an arbitrary factor the performances of the system. To our knowledge, our solutions are the first provably efficient algorithms to deal with dependencies involving more than two data items.

Topics: Approximation algorithms, Wireless and Push-based broadcast scheduling, Bluetooth and Satellite networks.

1 Introduction

Motivations. Broadcasting has been proved to be an efficient means of disseminating data in wireless communication environments (such as Satellite, mobile phone networks; other typical broadcast networks are Videotext systems). Recent works [15, 14, 10] provide strong evidence that correlation-based broadcast can significantly improve the average service time of broadcast systems. Most of the research on data broadcasting was done under the assumption that user requests are for a single item at a time and are independent of each other. However in many real world applications, such as web servers, dependencies exist among the data items, for instance: web pages on a server usually share a lot

J.R. Correa, A. Hevia, and M. Kiwi (Eds.): LATIN 2006, LNCS 3887, pp. 362–373, 2006.

of items such as logos, style sheets, titlebar..., and all these components have to be downloaded together when any individual page is requested. Such web server could take advantage of the correlations between the components of the pages, to speed up the broadcast of popular web pages.

This paper presents a theoretical analysis of data dependencies and provides theoretical proofs that a correlation-based approach can improve arbitrarily the performance of the system. To our knowledge, our solutions are the first provably efficient algorithms to deal with dependencies involving more than two data items. Our results rely essentially on the construction of a new lower bound, based on a non-linear convex minimization program, in combination with existing heuristics.

Background. Broadcast (or push-based) server usually maintains a profile of the typical users, *i.e.*, the popularity of each item (see [1]), and schedules the broadcasts of each item accordingly, obliviously of the effective requests made by the users, so as to minimize the average service time for the users. The users connect at random instants and monitor the broadcast channel until the information they are interested in is broadcast. The user profiles can easily be obtained by analyzing the log files of the server or by asking the user to list its interests at the subscription to the service (see [1]). Such profiles provide not only informations on the popularity of each item, but also informations on their *correlations*. Several heuristics [15, 14, 18] have been proposed to take advantage of these dependencies.

Very little is known theoretically on the performance of these algorithms. As far as we know, the only papers that addressed this question theoretically are [8, 6]. In [8], the authors give optimal polynomial time algorithms for broadcasting a set of $n = 2$ items given arbitrary user profiles. In [6], the authors design lower bounds and constant factor approximation algorithms to design cyclic schedules (*i.e.*, schedules where each item is broadcast exactly once per cycle) for the case where correlations are restricted to dependencies between pairs of items.

Earlier theoretical work, including NP-hardness results and approximation algorithms, on the databroadcast problem with independent requests can be found in [2, 5, 17, 19, 16, 7]. Related work on on-demand broadcast where request asks for a single item, can be found in [11, 4]. Current results in this last setting include intractability results as well as competitive algorithms with resource augmentation. As far as we know, introducing explicit dependencies is still an unexplored question in this setting.

The Customized Newspaper Broadcast Problem. We adopt the following setting. n unit length news items are made available on a broadcast server, *e.g.*: Weather, Sport, Stock exchange, International,... Each user is interested in a given subset of the news items with some probability, *e.g.*: {Weather, Stock exchange, Sport}, {Weather, International}, or {International, Stock exchange},... He then connects at a random time, monitors the broadcast channel until he is served, *i.e.*, until all the news items he is interested in have been broadcast, *e.g.*: Weather and International. The goal for the server is to find a schedule of the news items that minimizes the average service time of the user, given the request probability for each available *subset* of items. These probabilities are naturally obtained by asking the user to check its topics of interests in the list of available items when subscribing to the news system (which is now a widespread practice over the Internet). In the terms of [6] and [15, 14], our model of user requests is refered as "AND" and "unordered" requests, respectively. Note that the constraint that all news item have unit length is *not* restrictive, since any longer news item can be split into a set of unit length packets, which furthermore, fits the actual situation of a web server on Internet where all files are cut into equal size packets before been sent.

Our Contribution. Our main contribution consists in providing a new lower bound for the cost of an optimum schedule (Propositions 3 and 4), which is shown to be tight up to a constant factor. This lower bound is expressed as a non-linear convex minimization problem and is solved to obtain polynomial time 4-approximation algorithms (Theorems 2 and 3). These are, as far as we know, the first algorithms with bounded guarantee when dependencies involve more than two items. We also use this lower bound to show that correlation-based schedulers can indeed improve the quality of the previously known solutions by an arbitrary factor with respect to the optimum cost (Example 1). Our algorithms use previously known heuristics as subroutines. In particular, perfectly periodic schedules introduced in [7, 9] find here a new interesting application since their regularity can solve efficiently dependencies between items by forcing an order on them. Our algorithms yield significant improvements to system performances by managing correlations and are also not too complicated to implement (and would be in particular well adapted to time multiplexing environments such as Bluetooth networks). Interestingly enough, the classic randomized algorithm is shown to be inefficient in this setting (Theorem 1 and Example 2).

The next section gives a formal description of the problem and states its NP-hardness. Then, Section 3 presents our new lower bound, and shows that previous approaches ignoring correlations can lead to arbitrarily bad performances. Section 4 analyzes a classic randomized algorithm and shows that it achieves an approximation ratio of $2H_n$ exactly. Our deterministic 4-approximation algorithm is given in Section 5. Finally in Section 6, we extend our results to the setting where the broadcast of each item has a cost (*e.g.*, see [5]).

2 Notations and Preliminaries

The problem. The input consists of:

- n unit length news items $M_1, \ldots, M_n$,
- $\zeta = \{S_1, \ldots, S_k\}$ a set of k non-empty distinct sets of news items, $S_j \subseteq \{1, \ldots, n\}$, and
- positive request probabilities $(p_S)_{S\in\zeta}$ for each set S in ζ, such that $\sum_{S\in\zeta} p_S = 1$.

A *schedule* $\mathcal{S}$ is an infinite sequence $\mathcal{S} = \mathcal{S}(0)\mathcal{S}(1)\mathcal{S}(2)\mathcal{S}(3)\ldots$, where $\mathcal{S}(t) \in \{\perp, 1, \ldots, n\}$ for all t. If $\mathcal{S}(t) = i$, we say that *news item M_i is broadcast at time t in schedule* $\mathcal{S}$, *i.e.*, between time t and $t+1$; if $\mathcal{S}(t) = \perp$, no item is broadcast at time t. A schedule $\mathcal{S}$ is *periodic with period T* if $\mathcal{S}(t+T) = \mathcal{S}(t)$, for all time t. Such a periodic schedule is completely determined by its *cycle*, *i.e.*, the finite sequence $\mathcal{S}(0)\mathcal{S}(1)\ldots\mathcal{S}(T-1)$.

The *cost*, $\mathrm{COST}(\mathcal{S})$, of a schedule $\mathcal{S}$ is defined as the *average service time* to a random request, where the average is taken over the moments when requests occur *and* over the type S_j of news items subset requested. In our model, each client asks for news item set $S \in \zeta$ with probability p_S, connects at a random (integer) instant t according to some Poisson process, and is *served* when all the news items M_i, $i \in S$, have been broadcast. We denote by $\mathrm{ST}(\mathcal{S}, S, t)$ the *service time* of schedule $\mathcal{S}$ to a request for news items set S arriving at time t.

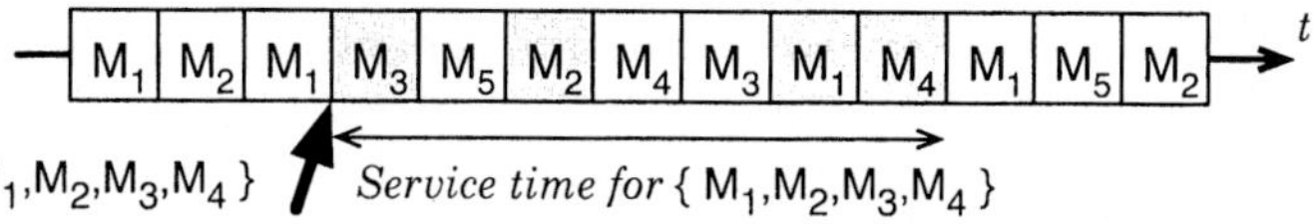

M1 M2 M1 M3 M5 M2 M4 M3 M1 M4 M1 M5 M2 → t

{ M1,M2,M3,M4 } *Service time for* { M1,M2,M3,M4 }

If $t_i \geqslant t$ is the first instant when news item M_i is broadcast on or after time t in $\mathcal{S}$, then $\mathrm{ST}(\mathcal{S}, S, t) = \max_{i \in S}(t_i + 1)$ (we consider that a request is served when the broadcast of the last requested item ends). Abusing the notation, $\mathrm{ST}(\mathcal{S}, M_i, t)$ will refer to $1 + t_i$. The average service time to a random request arriving at time t is then

$$\mathrm{AST}(\mathcal{S}, t) = \sum_{S \in \zeta} p_S \cdot \mathrm{ST}(\mathcal{S}, S, t).$$

Since the requests arrive according to some Poisson process, requests are uniformly distributed over any given time interval I (see [12]). Thus, the average service time $\mathrm{AST}(\mathcal{S}, I)$ of the schedule $\mathcal{S}$ during time interval $I = [t_1, t_2]$ is

$$\mathrm{AST}(\mathcal{S}, I) = \frac{1}{t_2 - t_1} \sum_{t=t_1}^{t_2-1} \mathrm{AST}(\mathcal{S}, t).$$

The cost of schedule $\mathcal{S}$ is then defined as the asymptotic value of this quantity as t goes to infinity:

$$\mathrm{COST}(\mathcal{S}) = \limsup_{t \to \infty} \mathrm{AST}(\mathcal{S}, [0, t]).$$

Note that if $\mathcal{S}$ is periodic with period T, its cost is simply defined as:

$$\mathrm{COST}(\mathcal{S}) = \frac{1}{T} \sum_{t=0}^{T-1} \mathrm{AST}(\mathcal{S}, t).$$

Our goal is to compute a schedule with minimum cost. We denote by $\mathrm{OPT} = \inf_{\mathcal{S}} \mathrm{COST}(\mathcal{S})$ the optimum cost of a schedule for a given instance.

The customized newspaper problem is a generalization of the preemptive data-broadcast setting which is shown to be strongly NP-hard in [19]. Thus,

Proposition 1 ([19]). *The customized newspaper problem is strongly NP-hard.*

Reduction to periodic schedules. General schedules are hard to handle since the frequency of each item may vary widely over time. The following lemma shows that we can restrict our study to periodic schedules which are much simpler to deal with. The following proposition follows the lines of [16].

Proposition 2. $\mathrm{OPT} = \inf_{\mathcal{S}\ periodic} \mathrm{COST}(\mathcal{S})$.

3 Lower Bounding Optimal Cost

We present now a new lower bound on which our approximation algorithms rely. This lower bound takes into account the correlations between the requests, *i.e.*, the fact that a given item might be requested by different types of requests for different sets of news items. As opposed to previous approaches (*e.g.*, [2, 5]), our lower bound cannot be solved by means of Lagrangian relaxation but can be expressed as a convex minimization program and solved by using the ellipsoid algorithm [13].

Section 3.2 shows that taking into account these correlations is indeed needed to estimate correctly the optimum cost, by showing that previously known methods (*e.g.*, [2, 5, 16, 19]), that only used the probability that a given item is requested (ignoring possible correlations), can construct schedules with cost as large as $\Omega(\sqrt{n})$ times the optimum value.

3.1 Lower Bound

According to Proposition 2, any lower bound on the cost of periodic schedules is a lower bound on the optimum cost. We now focus on periodic schedules.

Lemma 1. *Let $\mathcal{S}$ be a periodic schedule with period T, such that each news item M_i is broadcast exactly n_i times per cycle. Then*

$$\mathrm{COST}(\mathcal{S}) \geqslant \mathrm{LB}(\tau),$$

where $\mathrm{LB}(\tau) = \frac{1}{2} + \frac{1}{2}\sum_{S\in\zeta} p_S \max_{i\in S} \tau_i$, *with* $\tau_i = T/n_i$ *and* $1/\tau_1 + \cdots + 1/\tau_n \leqslant 1$.

Proof. Consider a request for a set $S \in \zeta$ arriving at time t. Since, this request waits for the broadcast of each news item M_i with $i \in S$, its service time is at least:

$$\mathrm{ST}(\mathcal{S}, S, t) \geqslant \mathrm{ST}(\mathcal{S}, M_i, t), \quad \text{for all } i \in S.$$

Then, by taking the average over t, the average service time to a request for set S is at least:

$$\mathrm{AST}(\mathcal{S}, S) \geqslant \mathrm{AST}(\mathcal{S}, M_i), \quad \text{for all } i \in S,$$

where $\mathrm{AST}(\mathcal{S}, M_i)$ denotes the average service time to a request that would ask for news item M_i alone. It is known from previous work (*e.g.*, [2, 5]) that if M_i is broadcast n_i times in a periodic schedule with period T, then $\mathrm{AST}(\mathcal{S}, M_i) \geqslant$

$\frac{1}{2} + \frac{T}{2n_i}$. Consider indeed $t_1, \ldots, t_{n_i}$ the time elapsed between the ends of each of the n_i broadcasts of M_i during a cycle of $\mathcal{S}$. With probability t_j/T, the request for M_i falls in an interval of length t_j, waits on average $\left(\frac{1}{t_j}\sum_{t=1}^{t_j}(t_j - t)\right) = \frac{t_j - 1}{2}$ time for the broadcast of M_i to begin, and is finally served one unit of time later, when the download of M_i is completed. Thus, $\mathrm{AST}(\mathcal{S}, M_i) = 1 + \sum_{j=1}^{n_i} \frac{t_j(t_j-1)}{2T}$. But $t_1 + \cdots + t_{n_i} = T$, then $\mathrm{AST}(\mathcal{S}, M_i) = \frac{1}{2} + \sum_{j=1}^{n_i} \frac{t_j^2}{2T}$. Furthermore, the sum of the square of the t_js is classically minimized when they are all equal to T/n_i, which finally yields:

$$\mathrm{AST}(\mathcal{S}, M_i) \geqslant \frac{1}{2} + \frac{T}{2n_i} = \frac{1}{2} + \frac{\tau_i}{2},$$

where $\tau_i = T/n_i$. Since, this holds for all $i \in S$,

$$\mathrm{AST}(\mathcal{S}, S) \geqslant \max_{i \in S} \mathrm{AST}(\mathcal{S}, M_i) \geqslant \frac{1}{2} + \frac{1}{2} \max_{i \in S} \tau_i.$$

Finally, summing over all $S \in \zeta$ gives

$$\mathrm{COST}(\mathcal{S}) = \sum_{S \in \mathcal{S}} p_S \, \mathrm{AST}(\mathcal{S}, S) \geqslant \frac{1}{2} + \frac{1}{2} \sum_{S \in \zeta} p_S \max_{i \in S} \tau_i = \mathrm{LB}(\tau).$$

As no more than T news items can be broadcast in a time interval of length T, we have $\sum_{i=1}^{n} n_i \leqslant T$, *i.e.*,

$$\sum_{i=1}^{n} \frac{1}{\tau_i} \leqslant 1.$$

We now state our lower bound, on which the algorithms presented in Sections 4 and 5 rely.

Proposition 3. *The following non-linear convex minimization problem* LB *is a lower bound on the optimum cost* OPT.

$$\mathrm{LB} = \begin{cases} \underset{\tau > 0}{\text{Minimize}} \; \frac{1}{2} + \frac{1}{2} \sum_{S \in \zeta} p_S \max_{i \in S} \tau_i \\ \\ \text{such that: } \frac{1}{\tau_1} + \cdots + \frac{1}{\tau_n} \leqslant 1 \end{cases} \quad (1)$$

There exists a unique solution τ^* to the minimization problem LB, *and one can compute in polynomial time a feasible solution τ' such that* $\mathrm{LB}(\tau') \leqslant \mathrm{LB} + \frac{1}{4}$.

Proof. By Proposition 2 and Lemma 1, clearly $\mathrm{LB} \leqslant \mathrm{OPT}$. Now, the objective function $\mathrm{LB}(\tau)$ is a continuous convex function, and is minimized over a strictly convex closed domain $D = \{\tau \in (\mathbb{R}_+^*)^n : 1/\tau_1 + \cdots + 1/\tau_n \leqslant 1\}$. Note that the Round Robin schedule that broadcasts cyclically M_1 to M_n, shows that $\mathrm{OPT} \leqslant n$ and then $\mathrm{LB} \leqslant n$. Since, $\mathrm{LB}(\tau) > n$ as soon as for some $S \in \zeta$ and some $i \in S$, $\tau_i > n/p_S$, the minimum of $\mathrm{LB}(\tau)$ is in fact obtained on the compact set $D' = D \cap (0, \frac{n}{\min_{S\in\zeta} p_S}]^n$. Since $\mathrm{LB}(\tau)$ is continuous, there exists $\tau^* \in D'$ such that $\mathrm{LB}(\tau^*) = \mathrm{LB}$. Assume that two such optimal solutions exist, say τ_1 and τ_2. Since the objective function and the domain are convex, $\tau' = \frac{\tau_1+\tau_2}{2}$ is also an optimal solution. But τ' lies in the interior of the domain, and then scaling it down by some factor $\lambda < 1$ allows to obtain a better feasible solution: $\mathrm{LB}(\lambda\tau') < \mathrm{LB}(\tau') = \mathrm{LB}$, contradiction. The optimal solution τ^* to LB is then unique. Furthermore, $1/\tau_1^* + \cdots + 1/\tau_n^* = 1$.

A feasible solution τ' within an additive error of $\frac{1}{4}$ of the optimum value can be computed in polynomial time using the ellipsoid method (*e.g.*, see [13]). The only ingredient needed is a separation oracle. Note that non-linear minimization problem LB can be restated as follows:

$$\mathrm{LB} = \begin{cases} \displaystyle\operatorname*{Minimize}_{\substack{\tau_1,\ldots,\tau_n > 0 \\ \sigma_{S_1},\ldots,\sigma_{S_k} > 0}} \quad \frac{1}{2} + \sum_{S\in\zeta} p_S \sigma_S & \\ \text{such that:} \qquad \tau_i \leqslant \sigma_S & \forall S, \forall i \in S \\ \qquad \displaystyle\frac{1}{\tau_1} + \cdots + \frac{1}{\tau_n} \leqslant 1 & \end{cases} \tag{2}$$

We just need to provide a separation oracle for each of these constraints. Only the last one is non-linear. But, for any solution $\tilde{\tau}$ violating this constraint, the tangent hyperplane to the differentiable convex surface $\partial D = \{\tau : \frac{1}{\tau_1} + \cdots + \frac{1}{\tau_n} = 1\}$ at the projection of $\tilde{\tau}$ on ∂D, provides in polynomial time a separation oracle for $\tilde{\tau}$.

3.2 Requests Correlations Mislead Previous Approaches

We show in this section that previously known algorithms (*e.g.*, [2, 3, 5, 17, 19, 7]) can generate schedules with cost *arbitrarily larger* than the optimum value when requests are correlated.

In previous approaches, only the requests probability π_i for *each individual item* M_i are used. Since item M_i is requested for each request for a set S containing i, the probability π_i that an individual item M_i is requested by some user, is proportional to $\sum_{S:i\in S} p_S$, *i.e.*,

$$\pi_i = \sum_{S:i\in S} \frac{p_S}{\langle\zeta\rangle}, \quad \text{where } \langle\zeta\rangle \text{ denotes the average size of a set } \langle\zeta\rangle = \sum_{S\in\zeta} |S| p_S.$$

All previous approaches (*e.g.*, [2, 5, 17, 19]) then construct a schedule such that each item M_i is broadcast every $\Theta(\vartheta_i)$ where ϑ is given by the "square-root rule" (*e.g.*, [2, 5]):

$$\vartheta_i = \frac{\sum_{j=1}^n \sqrt{\pi_j}}{\sqrt{\pi_i}}.$$

The example bellow shows that previously known solutions using these ϑ_is can generate schedules with cost arbitrarily large with respect to the optimum.

Example 1. Consider $2n$ news items $A_1, \ldots, A_n, B_1, \ldots, B_n$ and n sets:

$$\zeta = \{\{A_1, \ldots, A_n, B_1\}, \ldots, \{A_1, \ldots, A_n, B_n\}\},$$

where each set is requested with probability $\frac{1}{n}$. The request probability for each individual item is then $\pi_A = \frac{1}{n+1}$ for the A_is and $\pi_B = \frac{1}{n(n+1)}$ for the B_js. Thus, for all A_is, $\vartheta_A = \frac{n\sqrt{\pi_A}+n\sqrt{\pi_B}}{\sqrt{\pi_A}} = \Theta(n)$ and, for all B_js, $\vartheta_B = \frac{n\sqrt{\pi_A}+n\sqrt{\pi_B}}{\sqrt{\pi_B}} = \Theta(n\sqrt{n})$. Consider now a schedule that broadcasts each A_i every $\Theta(\vartheta_A)$ and each item B_j every $\Theta(\vartheta_B)$, as in previous approaches.

Assuming that the requests were independent, each item A_i and each item B_j are respectively requested with probabilities π_A and π_B, and the average service time of this schedule would be:

$$n \cdot \pi_A \cdot \Theta(\vartheta_A) + n \cdot \pi_B \cdot \Theta(\vartheta_B) = \Theta(n \cdot \frac{1}{n} \cdot n + n \cdot \frac{1}{n^2} \cdot n\sqrt{n}) = \Theta(n).$$

But, since the requests are not independent, the cost of any schedule that broadcasts each item every $\Theta(\vartheta)$, is at least, according to Lemma 1:

$$\mathrm{LB}(\Theta(\vartheta)) = \frac{1}{2} + \frac{1}{2} \cdot n \cdot \frac{1}{n} \cdot \max(\Theta(\vartheta_A), \Theta(\vartheta_B)) = \Omega(n\sqrt{n}).$$

In fact, the cost of the Round Robin schedule that broadcasts each item in turn cyclically, is only $\Theta(n)$ (every request is served after at most $2n$ time units).

We conclude that treating correlated requests individually can yield schedules with cost *as large as $\Omega(\sqrt{n})$ times the optimum*, where n is the number of items. □

As a consequence, considering the dependencies between requests for different items is essential to obtain good performances to data broadcast systems. The next sections show how we use our lower bound to compute efficient broadcast schedules.

4 Randomized Approximation

The lower bound in Proposition 3 suggests that each news item M_i should be broadcast every τ_i^*, which is approximated by τ_i' given by the ellipsoid method. We analyze here a classic randomized scheduler that chooses the next item M_i to be broadcast with probability $1/\tau_i'$. We show that it achieves a $2H_n$ factor approximation, and that our analysis is tight by providing a family of tight instances.

Lemma 2. *The expected cost of the random schedule $\mathcal{S}$ output by Algorithm 1 is at most*

$$\mathbb{E}[\mathrm{COST}(\mathcal{S})] \leqslant H_n \sum_{S \in \zeta} p_S \max_{i \in S} \tau_i',$$

where $H_n = 1 + \frac{1}{2} + \cdots + \frac{1}{n}$.

Algorithm 1. Randomized scheduler

Computes τ' given by Proposition 3.
for all $t \geqslant 0$ **do**
  Pick $i \in \{1, \ldots, n\}$ with probability $1/\tau'_i$.
  Broadcast M_i at time t.
end for

Proof. The proof relies on a classic coupon collector argument (*e.g.*, see [12]). Consider a request for a set S of size q. At any time t, each news item in S is broadcast with probability at least $\rho_S = \min_{i \in S} 1/\tau'_i$. Let t_j be the random variable for the time elapsed since the issue of the request until j distinct items of S have been downloaded. The expected service time to the request is clearly $\mathbb{E}[t_q]$. Let $T_j = t_j - t_{j-1}$ for $j = 1, \ldots, q$, with $t_0 = 0$. Clearly, $\mathbb{E}[t_q] = \sum_{j=1}^{q} \mathbb{E}[T_j]$. Since after time t_{j-1}, $j-1$ distinct items have been downloaded, the probability to get a new item from S in each time slot between t_{j-1} and t_j is at least $\rho_{S,j} = (q - j + 1)\rho_S$. Thus,

$$\mathbb{E}[T_j] \leqslant \sum_{t \geqslant 0} (t+1)(1 - \rho_{S,j})^t \rho_{S,j} = \frac{1}{(q-j+1)\rho_S} = \frac{\max_{i \in S} \tau'_i}{q-j+1}.$$

The expected service time to a request for S in the random schedule $\mathcal{S}$ is then at most:

$$\mathbb{E}[\mathrm{AST}(\mathcal{S}, S)] \leqslant \sum_{j=1}^{q} \frac{1}{q-j+1} \max_{i \in S} \tau'_i \leqslant H_n \max_{i \in S} \tau'_i.$$

Then summing over all the sets $S \in \zeta$ yields the result.

Theorem 1. *Algorithm 1 is a polynomial time randomized $2H_n$-approximation for the customized newspaper problem.*

The next example shows that our analysis of the randomized scheduler is tight.

Example 2. Consider k disjoint sets $S_1, \ldots, S_k$ of q news items each, $S_j = \{M_{1,j}, \ldots, M_{q,j}\}$ for $j = 1, \ldots, k$. Each set is requested with probability $1/k$. By symmetry of the instance, the unique solution to LB is obviously $\tau^*_{i,j} = kq$ for all $1 \leqslant i \leqslant q$ and $1 \leqslant j \leqslant k$. Since each item of each set S_j is broadcast independently in each time slot with probability $1/kq$, by the classic coupon collector argument (*e.g.*, see [12]), the expected cost of the random schedule $\mathcal{S}$ is then exactly:

$$\mathbb{E}[\mathrm{COST}(\mathcal{S})] = H_q \cdot k \cdot q.$$

Now consider the Round Robin schedule $\mathcal{R}$ that broadcasts each item of each set cyclically as follows: $M_{1,1} \ldots M_{q,1} M_{1,2} \ldots M_{q,2} \ldots M_{1,k} \ldots M_{q,k} M_{1,1} \ldots$. Taking $k = \ln m$ and $q = m$, we get for $m \to \infty$:

$$\mathbb{E}[\mathrm{COST}(\mathcal{S})] \sim H_m \cdot m \ln m \quad \text{and} \quad \mathrm{COST}(\mathcal{R}) \sim \frac{m \ln m}{2},$$

which implies that the optimum cost is at least at a factor $2H_n$ of the optimum value where $n = m \ln m$ is the number of news items. □

Algorithm 2. Determistic 4-approximation algorithm

Computes τ' given by Proposition 3.
for all $i \in \{1,\ldots,n\}$ **do**
  $\beta_i \leftarrow 2^{\lceil \log_2 \tau'_i \rceil}$
end for
Broadcast the news items (M_i) according to a perfectly periodic schedule built on the periods (β_i).

5 Deterministic 4-Approximation

The main issue with the randomized algorithm above is that items of a given set appear in a random order which is inefficient and introduces a H_n factor to the cost (due to a "coupon collector phenomenon"). A special type of schedules, known as *perfectly periodic schedules*, is of particular interest here. These schedules were implicitly introduced in [3] in a very simple setting and studied in details by [7, 9] in the context of bluetooth and sensor networks. In these networks, the clients access to a communication channel by means of time multiplexing. In these protocols, the ith client is given a period β_i and an offset o_i, and can emit only during time slots $o_i + k\beta_i$, with $k = 1, 2, \ldots$, each of these slots being different for each client. Such a schedule is said to be perfectly periodic since each client i gets access to the channel exactly every β_i time. These perfectly periodic schedules find an interesting new application here: their regularity solves efficiently the dependencies between items by forcing an order on them.

In [3, 7], the authors show the following lemma that states that if the requested periods β_is are power of 2 and satisfies the maximum bandwidth constraint $1/\beta_1 + \cdots + 1/\beta_n \leqslant 1$, then one can construct in polynomial time a perfectly periodic schedule for this set of periods.

Lemma 3 (Perfectly periodic schedules [3, 7]). *Given a set of periods (β_i) such that for all i, $\beta_i = 2^{j_i}$ for some integer j_i, and $1/\beta_1 + \cdots + 1/\beta_n \leqslant 1$, one can construct in polynomial time a perfectly periodic schedule that broadcasts each news item M_i exactly every β_i, for all i.*

Algorithm 2 follows the lines of the "power-of-two" heuristic given in [3]: first round up each period τ'_i (given by Proposition 3) to the closest power of 2, β_i, and then constructs a perfectly periodic schedules for the set of periods (β_i).

The next theorem shows that with our choice of τ' given by our lower bound, this algorithm achieves an approximation ratio of 4. This algorithm is, to our knowledge, the first to obtain a constant factor approximation for the data broadcast problem with dependencies involving more than two items.

Theorem 2. *Algorithm 2 is a polynomial time deterministic 4-approximation for the customized newspaper problem.*

Proof. Since the periods τ'_i are rounded up to the closest power of 2, β_i, we have $\tau'_i \leqslant \beta_i \leqslant 2\tau'_i$. Then $1/\beta_1 + \cdots + 1/\beta_n \leqslant 1$ and Lemma 3 gives in polynomial time a perfectly periodic schedule $\mathcal{S}$ that broadcasts each item M_i exactly every β_i. Every request for a set $S \in \zeta$ then waits at most β_i before downloading each M_i, with $i \in S$. The average service time to a request for S is then bounded by

$$\mathrm{AST}(\mathcal{S}, S) \leqslant 1 + \max_{i \in S} \beta_i \leqslant 1 + 2 \max_{i \in S} \tau_i'.$$

Summing over $S \in \zeta$ yields finally the following bound on the cost of $\mathcal{S}$:

$$\mathrm{COST}(\mathcal{S}) \leqslant 1 + 2 \sum_{S \in \zeta} p_S \max_{i \in S} \tau_i' \leqslant 1 + 4\left(\mathrm{LB}(\tau') - \frac{1}{2}\right) \leqslant 1 + 4\left(\mathrm{LB} - \frac{1}{4}\right) \leqslant 4\,\mathrm{OPT}.$$

6 Adding Broadcast Costs

A classic extension of data broadcast problem includes broadcast costs (*e.g.*, see [5]). An instance of the customized newspaper broadcast problem *with* broadcast costs associates a cost c_i to each news item M_i, which is applied every time M_i is broadcast. The goal is now to find a schedule $\mathcal{S}$ that minimizes to sum of two quantites: the average service time to a random request, $\mathrm{COST}(\mathcal{S})$, and the average broadcast cost, $\mathrm{BC}(\mathcal{S})$. The *average broadcast cost* of $\mathcal{S}$ over a time interval $I = [t_1, t_2]$ is defined as:

$$\mathrm{BC}(\mathcal{S}, I) = \frac{1}{t_2 - t_1} \sum_{t=t_1}^{t_2 - 1} c_{\mathcal{S}(t)}$$

The average broadcast cost of $\mathcal{S}$ is then defined as the asymptotic value of this quantity:

$$\mathrm{BC}(\mathcal{S}) = \limsup_{t \to \infty} \mathrm{BC}(\mathcal{S}, [0, t]).$$

The cost of a schedule is then defined as $\mathrm{COST}_{BC}(\mathcal{S}) = \mathrm{COST}(\mathcal{S}) + \mathrm{BC}(\mathcal{S})$. We denote by OPT_{BC} the optimum cost: $\mathrm{OPT}_{BC} = \inf_{\mathcal{S}} \mathrm{COST}_{BC}(\mathcal{S})$. As before, the following lower bound is obtained by bounding the cost of periodic schedules from which we derive a deterministic 4-approximation (proofs are omitted due to space constraints).

Proposition 4. *LB_{BC} is a lower bound on the optimum cost OPT_{BC}, where:*

$$\mathrm{LB}_{BC} = \begin{cases} \underset{\tau > 0}{\text{Minimize}}\ \dfrac{1}{2} + \dfrac{1}{2} \displaystyle\sum_{S \in \zeta} \left(p_S \max_{i \in S} \tau_i\right) + \sum_{i=1}^{n} \frac{c_i}{\tau_i} \\[2ex] \text{such that: } \dfrac{1}{\tau_1} + \cdots + \dfrac{1}{\tau_n} \leqslant 1 \end{cases}$$

There exists a unique solution τ_{BC}^* to the convex minimization problem LB_{BC}, and one can compute in polynomial time a feasible solution τ_{BC}' such that $\mathrm{LB}_{BC}(\tau_{BC}') \leqslant \mathrm{LB}_{BC} + \frac{1}{4}$.

Theorem 3. *Using periods τ_{BC}' instead of τ' in Algorithms 1 and 2 yields respectively a randomized $2H_n$-approximation and a deterministic 4-approximation for the customized newspaper problem with broadcast costs.*

References

1. S. Acharya. *Broadcast Disks: Dissemination-based Management for Assymmetric Communication Environments.* PhD thesis, Brown University, May 1998.
2. M. H. Ammar and J. W. Wong. The design of teletext broadcast cycles. In *Performance Evaluation*, volume 5(4), pages 235–242, 1985.
3. S. Anily, C. A. Glass, and R. Hassin. The scheduling of maintenance service. 1995.
4. N. Bansal, M. Charikar, S. Khanna, and J. S. Naor. Approximation the average response time in broadcast scheduling. In *Proc. of ACM-SIAM SODA*, 2005.
5. A. Bar-Noy, R. Bhatia, J. S. Naor, and B. Schieber. Minimizing service and operation costs of periodic scheduling. *Math. Of Op. Research*, 27(3):518–544, 2002.
6. A. Bar-Noy, J. S. Naor, and B. Schieber. Pushing dependent data in clients-providers-servers systems. *Wireless Networks*, 9:421–230, 2003.
7. A. Bar-Noy, A. Nisgav, and B. Patt-Shamir. Nearly optimal perfectly-periodic schedules. In *Proc. of ACM PODC'01*, pages 107–116, 2001.
8. A. Bar-Noy and Y. Shilo. Optimal broadcasting of two files over an asymmetric channel. *J. Parallel Distrib. Comput.*, 60(4):474–493, 2000.
9. Z. Brakerski, V. Dreizin, and B. Patt-Shamir. Dispatching in perfectly-periodic schedules. *J. Algorithms*, 49(2):219–239, 2003.
10. K. Cai, H. Lin, and C. Chen. Correlation-based data broadcasting in wireless networks. In *LNCS Proc. of 2nd BNCOD*, volume 3567, Jul. 2005.
11. J. Edmonds and K. Pruhs. Multicast pull scheduling: When fairness is fine. *Algorithmica*, 36(3):315–330, 2003.
12. W. Feller. *An Introduction to Probability Theory*, volume I. John Willey & Sons, 3rd edition, 1968.
13. M. Grötschel, L. Lovász, and A. Schrijver. *Geometric Algorithms and Combinatorial Optimization.* Springer, 1988.
14. J.-L. Huang and M.-S. Chen. Broadcast program generation for unordered queries with data replication. In *Proc. of the ACM SAC*, pages 866–870, Mar. 2003.
15. J.-L. Huang and M.-S. Chen. Dependent data broadcasting for unordered queries in a multiple channel mobile environment. *IEEE Trans. on Knowledge and Data Engineering*, 16(9):1143–1156, 2004.
16. C. Kenyon and N. Schabanel. The data broadcast problem with non-uniform transmission time. *Algorithmica*, 35:147–175, 2002.
17. C. Kenyon, N. Schabanel, and N. Young. Polynomial-time approximation scheme for data broadcast. In *Proc. of ACM STOC*, pages 659–666, 2000.
18. F. J. Ovalle-Martínez, J. S. González, and I. Stojmenović. A parallel hill climbing algorithm for pushing dependent data in clients-providers-servers systems. *Mobile Network and Applications*, 9:257–264, 2004.
19. N. Schabanel. The data broadcast problem with preemption. In *LNCS Proc. of the 17th STACS*, volume 1770, pages 181–192, 2000.

On Minimum k-Modal Partitions of Permutations

Gabriele Di Stefano[1], Stefan Krause[2], Marco E. Lübbecke[3], and Uwe T. Zimmermann[2]

[1] Dipartimento di Ingegneria Elettrica, Universita dell'Aquila, Monteluco di Roio, I-67040, L'Aquila
gabriele@ing.univaq.it
[2] Institut für Mathematische Optimierung, Technische Universität Braunschweig, Pockelsstraße 14, D-38106, Braunschweig
{stefan.krause, u.zimmermann}@tu-bs.de
[3] Technische Universität Berlin, Institut für Mathematik, Sekr. MA 6-1, Straße des 17. Juni 136, D-10623, Berlin
m.luebbecke@math.tu-berlin.de

Abstract. Partitioning a permutation into a minimum number of monotone subsequences is $\mathcal{NP}$-hard. We extend this complexity result to minimum partitioning into k-modal subsequences, that is, subsequences having at most k internal extrema. Based on a network flow interpretation we formulate both, the monotone and the k-modal version, as mixed integer programs. This is the first proposal to obtain provably optimal partitions of permutations. From these models we derive an LP rounding algorithm which is a 2-approximation for minimum monotone partitions and a $(k+1)$-approximation for minimum (upper) k-modal partitions in general; this is the first approximation algorithm for this problem. In computational experiments we see that the rounding algorithm performs even better in practice. For the associated online problem, in which the permutation becomes known to an algorithm sequentially, we derive a logarithmic lower bound on the competitive ratio for minimum monotone partitions, and we analyze two (bin packing) online algorithms. These findings immediately apply to online cocoloring of permutation graphs; they are the first results concerning online algorithms for this graph theoretical interpretation.

Keywords: Mixed integer program; approximation algorithm; LP rounding; online algorithm; $\mathcal{NP}$-hardness; monotone sequence; k-modal sequence; cocoloring.

MSC (2000): 90C11, 90C27, 05A05, 68Q25.

1 Introduction

Given a sequence S of distinct integers, we seek a partition into a minimum number of subsequences (not necessarily consecutive elements in S) with particular monotony properties. Research in this direction dates back to the famous

J.R. Correa, A. Hevia, and M. Kiwi (Eds.): LATIN 2006, LNCS 3887, pp. 374–385, 2006.

Erdős/Szekeres theorem of 1935 stating that every sequence of n distinct reals contains a monotone subsequence of length $\lceil\sqrt{n}\rceil$, see the review [11]. Greedily extracting longest monotone subsequences in an iterative way yields a partition into at most $2\lfloor\sqrt{n}\rfloor$ monotone subsequences in $O(n^{1.5})$, see [2]. However, finding a minimum size partition into monotone subsequences is $\mathcal{NP}$-hard [12]. For fixed k and l (not part of the input), a partition into exactly k increasing and l decreasing subsequences can be computed in $O(n^{k+l})$, see [4]. A minimum monotone partition can be approximated within a factor of 1.71 in $O(n^{2.5})$, see [8].

A natural generalization asks for partitions into k-modal subsequences; that are sequences having at most k internal local extrema. In particular for 1-modal, or *unimodal*, subsequences Chung [5] proves that any permutation of length n contains such a subsequence of length $\lceil\sqrt{3(n-1/4)}-1/2\rceil$. Chung also mentions the guaranteed length of $\lceil\sqrt{2n+1/4}-1/2\rceil$ for contained *upper unimodal* subsequences, i.e., subsequences with no internal minimum. She refers to a simple proof obtained by Steele and Chvátal (among others, unpublished, but see [6] for a proof). For the guaranteed length of contained k-modal subsequences, Chung [5] gives the upper bound $\sqrt{(2k+1)n}$. Steele [10] proves that the average length of k-modal subsequences of a permutation of size n asymptotically grows as $2\sqrt{(k+1)n}$. Based on these bounds, one can derive results on the size of the partitions generated by recursively extracting a respective longest subsequence. In particular, this greedy approach yields an upper unimodal partition of size $O(\sqrt{n})$ in $O(n^{2.5})$ time [6]. Even though a more general discussion is possible, we only consider k-modal sequences where the first internal extremum is a maximum, i.e., a generalization of *upper unimodal* sequences.

Our Contribution. We show that partitioning a permutation into a minimum number of k-modal (in particular: unimodal) subsequences is $\mathcal{NP}$-hard. On the positive side, we propose a linear programming (LP) rounding algorithm which is the first approximation algorithm for this problem: Its approximation factor is $k+1$ for upper k-modal partitions. In fact, an easy observation allows us to derive a $1.71(k+1)$-approximation first. Not only because of the practical motivation described below, we are interested in actually computing optimum partitions. To this end we introduce mixed integer programming (MIP) formulations which can be easily extended to respect a variety of practical side constraints. We further give the first negative and (weakly) positive results concerning online algorithms for minimum monotone partitions. These findings immediately apply to cocoloring of permutation graphs, for which no online algorithms were known either.

Motivation and Application. In railroad shunting yards incoming freight trains are split up and re-arranged according to their destinations. In stations and depots passenger trains and trams are parked overnight or during low traffic hours. In either case we are given an ordering of arriving *units*, and we have to decide for each unit on which track it will be stored [3, 6, 13]. Our choice is limited by the fixed number of available tracks and by the mode tracks may be accessed: Entrance and exit may be on one or on both ends. The parked units have to leave each track one by one without additional reordering. Our task is to choose a track for each unit, and the goal is to use as few tracks as possible.

The relation to our problems is that units on each track represent a subsequence of the incoming sequence of units. The different entry/exit combinations lead in particular to monotone and unimodal subsequences [6]. This relation may seem to be artificial, and we concede that the purpose of this paper primarily is to study the more theoretical background; however, the MIP models we propose can be tailored to fully capture the "real-world" situation, see our conclusions.

2 Preliminaries

Our results hold for any sequence $S = [s_1, s_2, \ldots, s_n]$ of n distinct reals, but we assume S to be a permutation of the first n integers. A *subsequence* σ of S is a sequence $\sigma = [s_{i_1}, s_{i_2}, \ldots, s_{i_m}]$ with $1 \leq i_j < i_h \leq n$ for all $j < h$. A sequence is called *increasing* if $s_i < s_j$ for $i < j$. It is called *decreasing* if $s_i > s_j$ for $i < j$. These two cases are also subsumed under *monotone*. An *internal extremum* of S is an index i with $2 \leq i \leq n-1$ and $s_{i-1} < s_i$, $s_{i+1} < s_i$ or $s_{i-1} > s_i$, $s_{i+1} > s_i$. A sequence is *k-modal* if is has at most k internal extrema; in particular in this paper, usually the first extremum should be a maximum, i.e., the first sequence is increasing (then we speak of *upper k-modal*). Particularly well known is the case of 1-modal (i.e., *unimodal*) sequences.

We use an intuitive set notation and language to work with sequences; e.g., when referring to all the elements contained in two sequences we speak of their union. A *partition* of S of size m is a collection P of m disjoint subsequences of S, the union of which is precisely S. For a given S we are interested in finding a partition P of minimum size. The type of subsequences allowed in P gives the name of the resulting minimization problem, that is, (monotone), (unimodal), or (upper k-modal). A *cover* of S is a collection of subsequences, the union of which contains each element in S *at least* once. Eliminating multiply covered elements, one can turn a cover into a partition without increasing the number of subsequences. This is why our problems are also known as covering a permutation [12].

Related Concepts. The easiest of our partitions are well studied in a graph theoretical context. The *permutation graph* $G = (S, E)$ associated with a permutation S has an edge (s_i, s_j) if and only if $s_i > s_j$ and $i < j$. An increasing subsequence in S corresponds to an independent set in G, and a decreasing subsequence in S corresponds to a clique in G.

A partition of the vertices of a graph into independent sets is called a *coloring*. A minimum partition of a permutation graph into *either* independent sets *or* cliques can be given in $O(n \log n)$ (see e.g., [9]). *Cocoloring* a graph asks for partitioning its vertex set into a minimum number of parts in which each part is either an independent set or a clique (so the partition may contain a mixture of both). Thus, in problem (monotone) we compute an optimal cocoloring of a permutation graph. Problem (k-modal) can be interpreted as a particular coloring problem on hypergraphs [6].

3 Complexity

In this extended abstract we present all statements for (upper k-modal), but for the sake of brevity proofs are given only for (upper unimodal). Restricting attention to this case essentially captures the necessary ideas needed for the generalization; all details are in the full paper.

Theorem 1. *Problem (k-modal) is strongly $\mathcal{NP}$-hard.*

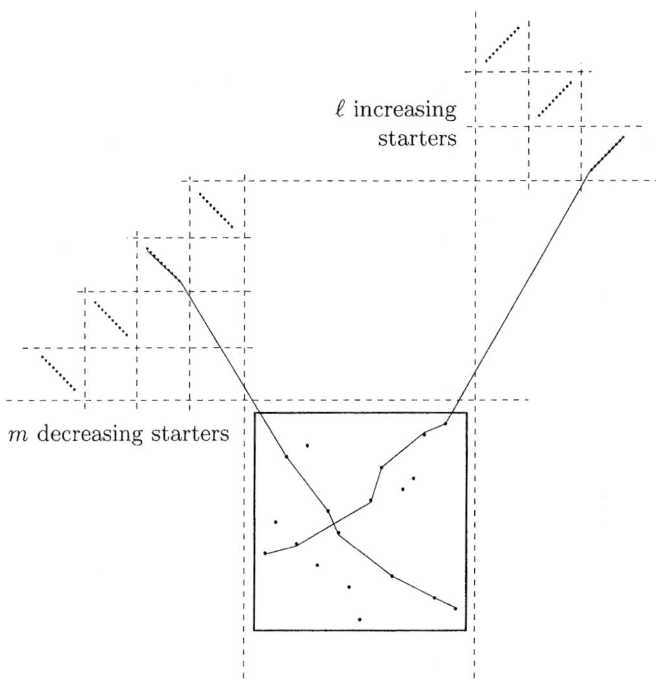

Fig. 1. Point map of the construction used in the proof of Theorem 1; $m = 4$, $\ell = 3$

Proof. Clearly, (upper unimodal) is in $\mathcal{NP}$; we will drop the attribute *upper* in the remainder. We use a reduction from (monotone) which is strongly $\mathcal{NP}$-hard [12]. In fact, one can solve (monotone) by solving a series of p restricted problems of partitioning S into at most $\ell = 1, \ldots, p$ increasing and at most $m = p - \ell$ decreasing subsequences. We reduce to this restricted version.

We represent elements and subsequences as points and lines. Having arranged the points corresponding to the elements of a given permutation S, we construct an extended arrangement of points which can be covered by p unimodal lines if and only if the original set of points can be covered by ℓ increasing and m decreasing lines. In fact, there always is an optimal solution to our construction which uses monotone lines only. We briefly use the notion *bounding rectangle* for an axis-parallel rectangle containing the points corresponding to the given permutation, and no other points of our construction.

Above and to the left of the bounding rectangle we introduce m sets of points called the *decreasing starters.* Each of them contains $2p$ points which form a strictly decreasing line. The decreasing starters themselves are arranged in a chain going upwards and rightwards such that their respective ranges of x- and y-coordinates are disjoint. Above them and to the right of the bounding rectangle we introduce ℓ sets of points called the *increasing starters.* Each of them contains $2p$ points which form a strictly increasing line. The increasing starters themselves are arranged in a chain going downwards and rightwards such that their respective ranges of x- and y-coordinates are disjoint. Since $p \leq n$, this construction is polynomial.

If there is a cover of the given permutation's points with m decreasing and ℓ increasing lines, then these lines can be extended to $p = \ell + m$ unimodal, in fact monotone, lines as indicated in the figure such that all starters are covered.

On the other hand, assume that we are given a cover of p unimodal lines for the extended point set. The decreasing starters have to be covered by m distinct decreasing lines, and the increasing starters have to be covered by ℓ distinct increasing lines. Actually, since the increasing starters are above the decreasing starters, the arrangement enforces that all of these $p = \ell + m$ lines have to be distinct. These can pass through the bounding rectangle, and we obtain the claimed solution to the original problem. □

4 Exact Approaches: Mixed Integer Programs

In this section we develop mixed integer programs (MIPs) for computing optimal partitions (see e.g., [9] for background on linear and integer programming). We first solve the problem of partitioning into increasing subsequences via a linear program (LP) which in fact is a minimum cost flow model. We embark on this expensive approach because we can extend this model to monotone and k-modal covers by means of additional binary variables. We describe the construction of the respective directed graphs from which the MIP models can be easily derived. When we speak of inserting a directed edge $e = (i, j)$, we imply inserting the *tail* node i of e, and the *head* node j of e, if they are not already present. Unless otherwise stated, there are no capacity bounds on edges except non-negativity. We denote the source of the respective graph by s and denote the sink by t.

A Network Flow Linear Program. We construct a directed graph as follows. Corresponding to element s_i, $i = 1, \ldots, n$, we introduce an edge e_i with a lower capacity bound of 1 and zero cost. We connect the source s to the tail of each e_i with unit cost edges. The head of each e_i is connected to the sink t with zero cost edges. Additionally, we insert a zero cost edge going from the head of e_i to the tail of e_j if and only if $i < j$ and $s_i < s_j$ (that is, we model increasing subsequences; the decreasing case is similar).

We seek a minimum cost flow from s to t. Since our graph is acyclic, an optimal flow can be decomposed into s-t-paths [1]. By construction, each of these paths uses exactly one edge incident to s, and the objective value is the number of paths. Each path uses a subset of the edges e_i. Our construction ensures that the sequence of the elements s_i corresponding to the edges e_i in each path is an increasing subsequence of S. Since the lower bound on the edges e_i is 1, all these edges must be contained in some s-t-path; the subsequences of S corresponding to the paths form a minimum partition into increasing subsequences.

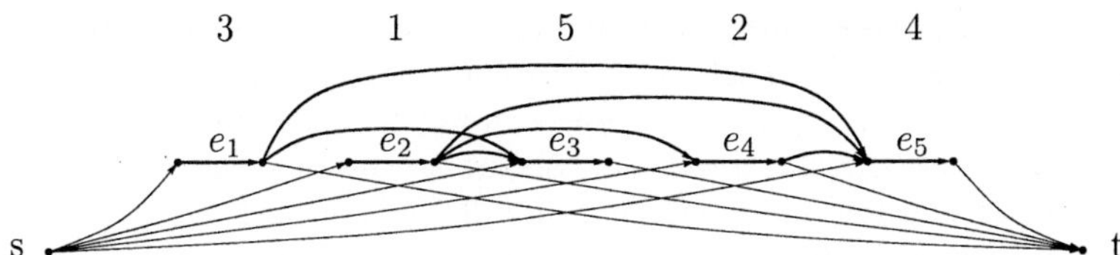

Fig. 2. The graph for the network flow model in the increasing case, $S = [3, 1, 5, 2, 4]$

A Flow Based MIP for Monotone Partitions. One can easily find a minimum monotone cover if we fix for each element whether it occurs in an increasing or in a decreasing subsequence: This results in two independent instances. We use this fact to model the monotone case. We use two complementary copies of the above network flow model, one part corresponding to increasing subsequences, and one complemented part for decreasing subsequences. For each e_i in the increasing part there is a corresponding copy e'_i in the decreasing part. The increasing part remains as before, and in the decreasing part there is an edge going from the head of e'_i to the tail of e'_j if and only if $i < j$ and $s_i > s_j$. The two parts share the source s and the sink t. We introduce binary variables x_i and x'_i and set the lower bound on the edges e_i to x_i and of e'_i to x'_i in the increasing and the decreasing part, respectively, where we require that $x_i + x'_i = 1$.

Again, an optimal flow decomposes into s-t-paths; these correspond to monotone subsequences of S, and the objective function value gives the number of paths. Since exactly one of e_i or e'_i has a lower bound of 1 these subsequences form a minimum monotone cover.

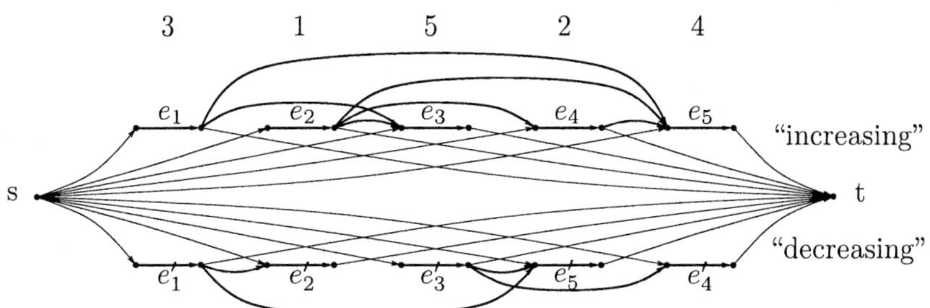

Fig. 3. The graph for the network flow based MIP model in the monotone case, $S = [3, 1, 5, 2, 4]$

A Flow Based MIP for (Upper) Unimodal Partitions. For the unimodal case, we start with the graph constructed for the monotone case. From the increasing part we omit the edges incident to t. From the decreasing part we omit the edges incident to s. For each i we add an edge connecting the head of e_i to the head of e'_i. Again, e_i and e'_i each get a lower bound of x_i and x'_i, respectively, where x_i and x'_i are binary variables with $x_i + x'_i = 1$.

In this graph an s-t-path uses exactly one edge incident to s, at least one edge e_i in the increasing part (corresponding to an increasing subsequence) and possibly some edges e'_i in the decreasing part (corresponding to a decreasing subsequence). Together, a path represents an upper unimodal subsequence. The variables x_i and x'_i control whether s_i occurs in the increasing part of such a sequence (including its maximum) or in its decreasing part. Note that also degenerate cases are considered, that is, monotone sequences are possible parts of a solution. The binary variables ensure that each s_i occurs in at least one unimodal sequence, therefore an optimal solution to this MIP gives a minimum unimodal cover. This construction generalizes to (upper k-modal) via the construction of an extended network of $k+1$ *layers*.

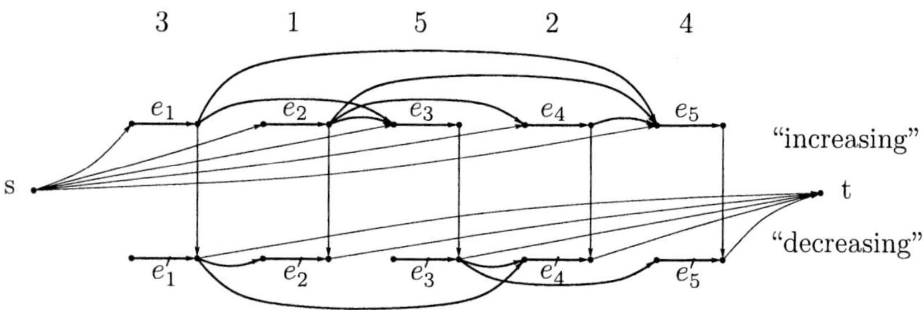

Fig. 4. The graph for the network flow based MIP model in the upper unimodal case, $S = [3, 1, 5, 2, 4]$, "upper unimodal" meaning—as always in this extended abstract—at most one internal maximum

5 Approximation Algorithms

Fomin, Kratsch, and Novelli [8] give a factor 1.71 approximation algorithm for finding a minimum partition of a partially ordered set into chains and antichains. In particular, this is a 1.71 approximation algorithm for the (monotone) problem. It is an open question whether there exists a polynomial time approximation scheme (PTAS). We derive a $1.71(k+1)$-approximation algorithm for (k-modal).

Lemma 1. *An α-approximate solution for (monotone) is a $(k+1)\alpha$-approximate solution for (k-modal). An α-approximate solution for (k-modal) can be converted to a $(k+1)\alpha$-approximate solution for (monotone).*

Proof. Denote by z^{α}_{mon} and by z^{α}_{k} the size of an α-approximate partition for (monotone) and for (k-modal), respectively. Since any k-modal sequence can be split into at most $k+1$ monotone subsequences, the optimal partition sizes z_{mon} and z_k relate as $z_{\text{mon}} \leq (k+1) \cdot z_k$. This gives

$$z_{\text{mon}}^{\alpha} \leq \alpha \cdot z_{\text{mon}} \leq (k+1) \cdot \alpha \cdot z_k,$$

proving the first part of the lemma. Any monotone sequence is k-modal, and therefore $z_k \leq z_{\text{mon}}$. Together with the above mentioned splitting of a k-modal sequence we immediately obtain

$$(k+1) \cdot z_k^{\alpha} \leq (k+1) \cdot \alpha \cdot z_k \leq (k+1) \cdot \alpha \cdot z_{\text{mon}},$$

which proves the second part. □

Using our network flow MIP models from the preceeding section, we are able to improve on this factor. We obtain a $(k+1)$-approximation algorithm for (upper k-modal). We state the result and the proof for (monotone) only.

Algorithm LP Rounding for (monotone)
Solve the LP relaxation of the MIP model for (monotone). For each element $i = 1, \ldots, n$, fix $x_i = 0$ if $x_i < 0.5$, and fix $x_i = 1$ if $x_i \geq 0.5$. Solve the resulting "fixed" LP again, and output the subsequences of S corresponding to the s-t-paths in an optimal solution.

Lemma 2. LP Rounding *is a 2-approximation algorithm for (monotone).*

Proof. For each $i = 1, \ldots, n$, if we fix $x_i = 1$ we increase the lower bound on e_i from at least 0.5 to 1.0. If we fix $x_i = 0$, this implies to fix $x_i' = 1$, and we increase the lower bound on e_i' from at least 0.5 to 1.0. The respective lower bound is at most doubled.

Denote by z the objective function value of an optimal solution x to the linear programming relaxation. Doubling the flow value of every s-t-flow in x gives a feasible solution to the fixed problem with objective function value at most $2z$. This is an upper bound for the optimal flow's objective function value in the fixed problem, yielding the claimed approximation factor.

This result generalizes to (upper k-modal) since we have $k+1$ variables per element, so at least one has fractional value at least $1/(k+1)$. Polynomial time solvability of linear programs follows from the ellipsoid method [9]. □

We note that the integrality gap of our MIP model for (monotone) is at least $\frac{3}{2}$ as is shown e.g., by the sequence $[6, 2, 1, 4, 3, 5]$: The optimal LP value is 2.0, the optimal integral objective is 3.0. From our computational experience we conjecture that the correct gap is smaller than 2, and that the analysis of the performance of LP Rounding can be improved.

6 Online Algorithms

Not only in view of our practical motivation it is natural to ask for the online version of our problems in which the permutation becomes known sequentially.

We have to assign elements to subsequences without looking at the remaining elements of the permutation, see e.g., [7] for background on online algorithms. For partitions into increasing subsequences the (optimal) greedy algorithm is in fact an online algorithm [6]. Already for (monotone) the situation is much worse.

Theorem 2. *There is no constant factor competitive online algorithm for (monotone).*

Proof. Consider any online algorithm $\mathcal{A}$. Depending on the decisions made by $\mathcal{A}$ we construct a sequence S with $n = 2^h - 1$ elements. We start with the range of numbers $a = 1$ to $b = n$. The first element of S is $(a+b)/2 = 2^{h-1}$, and $\mathcal{A}$ has to open a subsequence. We arbitrarily set $a = 2^{h-1}+1$ or $b = 2^{h-1}-1$, and serve $(a+b)/2$ as second element. In general, $\mathcal{A}$ has three options (of which in fact only two are actually possible). We describe this for the second iteration. First note that a decision to append to an existing subsequence decides upon whether that sequence is increasing or decreasing.

If $\mathcal{A}$ decides to append in an increasing way we set $b = 2^{h-1} - 1$. If $\mathcal{A}$ decides to append in a decreasing way we set $a = 2^{h-1} + 1$. In either case we have a connected range of $2^{h-1} - 1$ numbers none of which can be appended to an already existing subsequence. If a new subsequence is opened we adapt either a or b arbitrarily as above. We iterate with the new values of a and b, and it follows by induction that $\mathcal{A}$ generates at least $h/2$ subsequences for the first h elements of S (since each subsequence contains at most two elements).

Let $a_1, \ldots, a_h$ and $b_1, \ldots, b_h$ be the values of a and b throughout the first h iterations described above. The ith element of S is either $a_{i+1} - 1$ or $b_{i+1} + 1$. Since the sequences $a_i, \ldots, a_h$ and $b_1, \ldots, b_h$ are increasing and decreasing, respectively, the first h elements of S can be covered by an increasing subsequence of $a_1 - 1, \ldots, a_h - 1$ and by a decreasing subsequence of $b_1 + 1, \ldots, b_h + 1$.

If the remaining elements of S are arranged in an increasing way the optimal solution contains 3 subsequences. However, the solution determined by $\mathcal{A}$ contains at least h subsequences. Therefore, $\mathcal{A}$ is $\log_2(n+1)/6$-competitive at best. □

Since we are not aware of any previous results on online algorithms for cocoloring, it is interesting in its own right to restate this result in graph theoretical terms.

Restatement of Theorem 2. *The problem of cocoloring a permutation graph does not allow an online algorithm with constant competitive ratio.*

We next discuss the performance of two online algorithms for (monotone) and (unimodal). Both are reminiscent of simple bin packing online algorithms.

Online algorithm Next Fit
Keep adding elements to one and the same subsequence as long as monotony (unimodularity) is not violated. Then start a new subsequence and leave the previous ones unchanged.

Lemma 3. NEXT FIT *is $n/4$-competitive for (monotone) and (unimodal).*

Proof. Any two elements of the input sequence S form a monotone (unimodal) subsequence. Thus, we have $n/2$ as a trivial upper bound for the number of subsequences determined by NEXT FIT. If S itself is monotone (unimodal) the algorithm finds the optimal solution. Otherwise, the optimal solution consists of at least two subsequences giving a competitive ratio of $n/4$. To see that this bound is tight consider the sequence $S = [n, 1, n-1, 2, \ldots]$. In the monotone and the unimodal case NEXT FIT will determine a solution consisting of $n/2$ subsequences with two elements each. The optimal solution consists of two sequences in both cases. Therefore, NEXT FIT is exactly $n/4$-competitive. □

Next we make use of the fact that we know the set of pending elements, which are the numbers in $1, \ldots, n$ we have not yet seen in the input sequence. Interestingly, this does not help the competitive ratio.

Online algorithm Best Fit
We start with n increasing and n decreasing subsequences with an initial dummy element of 0 and $n+1$, respectively, that will be removed when the respective first element is added. An iteration is as follows. Let s be the current element of the input sequence and let t_i be the last element of the ith subsequence. Select an index i such that s can feasibly be added to the ith subsequence and such that the number of pending elements that are between s and t_i is minimum. Resolve ties arbitrarily but prefer already started subsequences. In the end, throw away all unused subsequences.

Lemma 4. BEST FIT *is $n/4$-competitive for (monotone) and (unimodal).*

Proof. If the input permutation is itself feasible, BEST FIT is optimal. Otherwise, by definition, it generates at most $n/2$ feasible subsequences and is thus at least $n/4$-competitive. To see that the upper bound is tight, we consider the permutation $S = [2, 1, 4, 3, \ldots, 2k, 2k-1, \ldots]$. The algorithm generates decreasing two-element subsequences $[2k, 2k-1]$ for all k, but the optimal partition contains only the two increasing subsequences $[2, 4, \ldots]$, $[1, 3, \ldots]$. □

7 Conclusions

We studied partitions of permutations into subsequences with particular monotony properties. The theoretical hardness legitimates applying computationally expensive algorithms like solving (probably large scale) mixed integer programs. These, in addition to their practical usefulness, yield (small) constant factor approximation algorithms via LP rounding.

In the full paper we computationally evaluate our proposals for random permutations. As a brief summary at this point, permutations of more than 100 elements can be partitioned optimally within a few seconds or minutes by solving our MIPs. The greedy algorithm, which iteratively extracts a longest subsequence of the requested type, runs in a split second and yields an acceptable solution quality on the average and also in the (empirical) worst case. The quality of solutions obtained with the LP ROUNDING algorithm significantly stays below the theoretically guaranteed approximation factor. However, the simple NEXT FIT online algorithm also empirically performs as poorly as predicted by the competitive analysis, whereas the BEST FIT online algorithm gives somewhat better results on average, as was to be expected.

There are several extensions motivated from practice which we did not explicitly consider in this more theoretical study, but which can be easily incorporated in our models. One such extension is a bounded track length, that is, subsequences must not contain more than a fixed number of elements. Solutions to our network flow based models become resource constrained shortest paths in this case which may be of independent theoretical interest. In particular, we have developed a set covering model which is most flexible in terms of (practical) extendibility. It is able to capture more "dirty" side constraints which do not directly fit into the context of this extended abstract.

There remain several open questions, spawned by our work:

- What is the exact approximability status of (monotone) and (k-modal), in particular, does there exist a PTAS? Can our LP techniques lead to an improvement over the 1.71 approximation for (monotone)? Such a result would be quite fascinating since the known algorithm [8] already elegantly exploits the combinatorial nature of the problem.
- Considering the competitiveness lower bound of Theorem 2 one would be interested in an online algorithm matching this bound. Which competitiveness ratio is possible when look-ahead is allowed?
- The crucial property we use in the construction of the graphs underlying our MIP models, and which ensures that paths correspond to increasing or decreasing subsequences, is the transitivity of the ordering of elements. We would have liked to generalize our positive results for permutations to partially ordered sets (corresponding to comparability graphs). However, in general, this property is lost for the complement of a comparability graph. Is there a network flow based model similar to ours which allows LP rounding, thus yielding a constant factor approximation?

Acknowledgment. We would like to thank Laura Heinrich-Litan for pointing us to the literature on cocoloring.

References

1. R.K. Ahuja, T.L. Magnanti, and J.B. Orlin. *Network Flows: Theory, Algorithms and Applications.* Prentice-Hall, Inc., Englewood Cliffs, New Jersey 07632, 1993.
2. R. Bar-Yehuda and S. Fogel. Partitioning a sequence into few monotone subsequences. *Acta Inform.*, 35(5):421–440, 1998.
3. U. Blasum, M.R. Bussieck, W. Hochstättler, C. Moll, H.-H. Scheel, and T. Winter. Scheduling trams in the morning. *Math. Methods Oper. Res.*, 49(1):137–148, 1999.

4. A. Brandstädt and D. Kratsch. On partitions of permutations into increasing and decreasing subsequences. *Elektron. Informationsverarb. Kybernet.*, 22(5/6):263–273, 1986.
5. F.R.K. Chung. On unimodal subsequences. *J. Combin. Theory Ser. A*, 29:267–279, 1980.
6. G. Di Stefano and M.L. Koči. A graph theoretical approach to the shunting problem. *Electr. Notes Theor. Comput. Sci.*, 92:16–33, 2004.
7. A. Fiat and G.J. Woeginger. *Online Algorithms—The State of the Art*, volume 1442 of *Lecture Notes in Computer Science*. Springer, 1998.
8. F.V. Fomin, D. Kratsch, and J.-C. Novelli. Approximating minimum cocolourings. *Inform. Process. Lett.*, 84(5):285–290, 2002.
9. A. Schrijver. *Combinatorial Optimization: Polyhedra and Efficiency*. Springer, Berlin, 2003.
10. J.M. Steele. Long unimodal subsequences: A problem of F.R.K. Chung. *Discrete Math.*, 33:223–225, 1981.
11. J.M. Steele. Variations on the monotone subsequence theme of Erdős and Szekeres. In D. Aldous, P. Diaconis, J. Spencer, and J.M. Steele, editors, *Discrete Probability and Algorithms*, pages 111–131. Springer-Verlag, New-York, 1995.
12. K. Wagner. Monotonic coverings of finite sets. *Elektron. Informationsverarb. Kybernet.*, 20(12):633–639, 1984.
13. T. Winter and U.T. Zimmermann. Real-time dispatch of trams in storage yards. *Ann. Oper. Res.*, 96:287–315, 2000.

Two Birds with One Stone: The Best of Branchwidth and Treewidth with One Algorithm

Frederic Dorn and Jan Arne Telle

Department of Informatics, University of Bergen,
Bergen, Norway

Abstract. In this paper we introduce semi-nice tree-decompositions and show that they combine the best of both branchwidth and treewidth. We first give simple algorithms to transform a given tree-decomposition or branch-decomposition into a semi-nice tree-decomposition. We then give two templates for dynamic programming along a semi-nice tree-decomposition, one for optimization problems over vertex subsets and another for optimization problems over edge subsets. We show that the resulting runtime will match or beat the runtimes achieved by doing dynamic programming directly on either a branch- or tree-decomposition. For example, given a graph G on n vertices with path-, tree- and branch-decompositions of width pw, tw and bw respectively, the Minimum Dominating Set problem on G is solved in time $O(n2^{\min\{1.58\,pw, 2\,tw, 2.38\,bw\}})$ by a single dynamic programming algorithm along a semi-nice tree-decomposition.

1 Introduction

The three graph parameters treewidth, branchwidth and pathwidth were all introduced by Robertson and Seymour as tools in their seminal proof of the Graph Minors Theorem. The treewidth $tw(G)$ and branchwidth $bw(G)$ of a graph G satisfy the relation $bw(G) \leq tw(G) + 1 \leq \frac{3}{2}\,bw(G)$ [16], and thus whenever one of these parameters is bounded by some fixed constant on a class of graphs, then so is the other. Tree-decompositions have traditionally been the choice when solving NP-hard graph problems by dynamic programming to give FPT algorithms when parameterized by treewidth, see e.g. [5, 15] for overviews. Of the various algorithmic templates suggested for this over the years the nice tree-decompositions [14] with binary Join and unary Introduce and Forget operations are preferred for their simplicity and have been widely used both for showing new results, for pedagogical purposes, and in implementations. Tree-decompositions are in fact moving into the computer science curriculum, *e.g.* twenty pages of a new textbook on Algorithm Design [13] is devoted to this topic.

Recently there have been several papers [10, 7, 6, 12, 11, 8] showing that for graphs of bounded genus the base of the exponent in the running time of these FPT algorithms could be improved by instead doing the dynamic programming along a branch-decomposition of optimal branchwidth. Dynamic programming along either a branch- or tree-decomposition of a graph both share the

J.R. Correa, A. Hevia, and M. Kiwi (Eds.): LATIN 2006, LNCS 3887, pp. 386–397, 2006.

property of traversing a tree bottom-up and combining solutions to problems on certain subgraphs that overlap in a bounded-size separator of the original graph. But there are also important differences, *e.g.* the subgraphs mentioned above are for tree-decompositions usually induced by subsets of vertices and for branch-decompositions by non-overlapping sets of edges. Various optimization tricks have been presented to speed up the algorithms, some of these come from the field of tree-decompositions [3, 2] and others from the field of branch-decompositions [10, 7]. As already mentioned it seems that for planar graphs the branchwidth parameter is the better choice, at least for worst-case runtime. There are other graph classes where treewidth is better. In most situations the input graphs contain some graphs where branchwidth is better and others where treewidth is better. If we already have implementations of heuristic algorithms for both branchwidth and treewidth, then the better choice for the dynamic programming stage will rely on the output of these heuristics for a given graph. Both from a theoretical and also applied viewpoint it therefore seems necessary, for each optimization problem, to design and possibly implement two separate dynamic programming algorithms, one for tree-decompositions and another for branch-decompositions. In this paper we show that a single dynamic programming algorithm will suffice to get the best of both treewidth and branchwidth.

For this purpose we introduce semi-nice tree-decompositions that maintain much of the simplicity of the nice tree-decompositions. However, the vertices of a Join are partitioned into 3 sets D,E,F and the binary Join operation treat vertices in each set differently in order to improve runtime. Symmetric Difference vertices D are those that appear in only one of the children, Forget vertices F are those for which all their neighbors have already been considered, and Expensive vertices E are the rest (the formal definitions follow later.) We first show how to transform a given branch-decomposition or tree-decomposition into a semi-nice tree-decomposition. We then give two templates for dynamic programming on semi-nice tree-decompositions, one for vertex subset problems and the other for edge subset problems.

For vertex subset problems we improve the runtime for the Join update operation during dynamic programming. Along the way we also simplify the proof of monotonicity of table entries for domination-type problems of [2] by a slight change in the definition of the vertex states used. Our results are also a step towards meeting the 'research challenge', first proposed in [3], of lowering to $O(n\lambda^k)$ the runtime of dynamic programming on treewidth k graphs for solving a problem having λ vertex-states. For edge subset problems the two subgraphs for whom solutions are combined in the Join operation are defined to not overlap at all in edges. Edges on vertices common to the two subgraphs are instead introduced in a later Forget operation. In their paper [6] on heuristics for TSP (travelling salesman problem) Cook and Seymour state that when carrying out dynamic programming to solve optimization problems that deal with edge sets branchwidth is a more natural framework than treewidth. We claim that our template shares this property of being a natural framework for edge set problems.

We employ this approach to various problems, such as dominating set problems, some of which had previously been solved for tree-decompositions in [17, 3] and for branch-decompositions in [10], to TSP solved for branch-decompositions in [6] and tree-decompositions in [4], and in the long version to this paper [9] to (k, r)-center solved for branch-decompositions in [7]. In each case we match or improve the running time of the algorithms given in those papers. We do this by combining and extending the various optimization tricks for branchwidth and treewidth used in those papers into our dynamic programming algorithm on semi-nice tree-decompositions. Table 1 gives the resulting worst-case runtime on various domination-type problems that are NP-hard for general graphs. For treewidth the previous best results [3] arise from treating all vertices in the Join as Expensive vertices, thus $tw = E$ in column Join of Table 1 instead of $tw = D+E+F$ as we have. For branchwidth the entry for Minimum Dominating set in the first row of Table 1 matches the previous best [10], while the results for each of the other problems are new. We emphasize that for any problem this is the first time that a single dynamic programming algorithm achieves the best of both treewidth and branchwidth.

Table 1. The number of vertex states and time for a Join operation with Expensive vertices E, Forgettable vertices F and Symmetric Difference vertices D. Worst-case runtime expressed also by treewidth tw and branchwidth bw of the input graph, and the cutoff point at which treewidth is the better choice. To not clutter the table, we leave out pathwidth pw, allthough for each problem there is a cutoff at which pathwidth would have been best.

	States	Join	Total time	tw faster
Min Dom set	3	$O(3^{D+F}4^{E})$	$O(n2^{\min\{2tw,2.38bw\}})$	$tw \leq 1.19bw$
Min/Max Ind Dom set	3	$O(3^{D+F}4^{E})$	$O(n2^{\min\{2tw,2.38bw\}})$	$tw \leq 1.19bw$
∃/Min/Max Perfect Code	3	$O(3^{D}4^{E+F})$	$O(n2^{\min\{2tw,2.58bw\}})$	$tw \leq 1.29bw$
Min Perfect Dom set	3	$O(3^{D}4^{E+F})$	$O(n2^{\min\{2tw,2.58bw\}})$	$tw \leq 1.29bw$
Max 2-Packing	3	$O(3^{D}4^{E+F})$	$O(n2^{\min\{2tw,2.58bw\}})$	$tw \leq 1.29bw$
Min Total Dom set	4	$O(4^{D+F}6^{E})$	$O(n2^{\min\{2.58tw,3bw\}})$	$tw \leq 1.16bw$
∃/Min/Max Perf Total Dom	4	$O(4^{D}5^{F}6^{E})$	$O(n2^{\min\{2.58tw,3.16bw\}})$	$tw \leq 1.22bw$

2 Semi-nice Tree-Decompositions

We use standard graph notation and terminology, e.g. for a subset $S \subseteq V(G)$ of the vertices of a graph G we let $N(S) = \{v \notin S : \exists u \in S \wedge uv \in E(G)\}$ be the set of vertices not in S that are adjacent to some vertex in S. For clarity we speak of nodes of a tree and vertices of a graph. To simplify expressions involving the cardinality of a set X, we write e.g. 2^X when we actually mean $2^{|X|}$.

A tree-decomposition $(T, \mathcal{X})$ of a graph G is an arrangement of the vertex subsets $\mathcal{X}$ of G, called bags, as nodes of the tree T such that for any two adjacent vertices in G there is some bag containing them both, and for each vertex of G the bags containing it induce a connected subtree. When we say bag we may refer both to the tree node and the associated vertex subset, sometimes even both at the same time, e.g. 'the intersection of two adjacent bags'. The width of the tree-decomposition $(T, \mathcal{X})$ is simply the size of the largest bag minus one.

A branch-decomposition (T, μ) of a graph G is a ternary tree T, i.e. with all inner nodes of degree three, together with a bijection μ from the edge-set of G to the leaf-set of T. For every edge e of T define a vertex subset of G called $mid(e)$ consisting of those vertices $v \in V(G)$ for which e lies on the path in T between two leaves whose mapped edges are incident to v (note that this is a non-standard but equivalent way of defining these so-called middle sets.) The width of (T, μ) is the size of the largest $mid(e)$ thus defined.

For a graph G its treewidth and branchwidth is the smallest width of any tree-decomposition and branch-decomposition of G, respectively, while its pathwidth is the smallest width of a tree-decomposition $(T, \mathcal{X})$ where T is a path.

We introduce semi-nice tree-decompositions and two lemmas on transforming a given branch- or tree-decomposition into a semi-nice tree-decomposition. A tree-decomposition $(T, \mathcal{X})$ is *semi-nice* if T is a rooted binary tree with each non-leaf of T being either a:

- **Introduce** node X with a single child C and $C \subset X$.
- **Forget** node X with a single child C and $X \subset C$.
- **Join** node X with two children B, C and $X = B \cup C$.

For an Introduce node we call $X \setminus C$ the 'introduced vertices' and for a Forget node $C \setminus X$ the 'forgotten vertices'. It follows by properties of a tree-decomposition that a vertex can be introduced in several nodes but is forgotten in at most one node. Note that the nice tree-decompositions [14] require that a Join node has $X = B = C$, Introduce has $|X| = |C| + 1$, and Forget has $|X| = |C| - 1$, but are otherwise identical to the semi-nice tree-decompositions.

For a Join node X with children B, C and parent A (the root node being its own parent) we define a partition of $X = B \cup C$ into 3 sets D, E, F:

- **Symmetric Difference** $D = (C \setminus B) \cup (B \setminus C)$
- **Expensive** $E = A \cap B \cap C$
- **Forgettable** $F = (B \cap C) \setminus A$

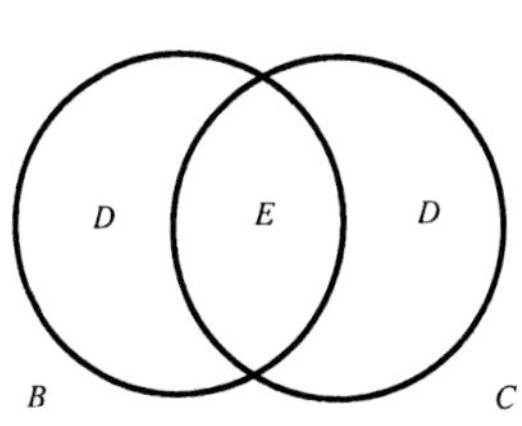

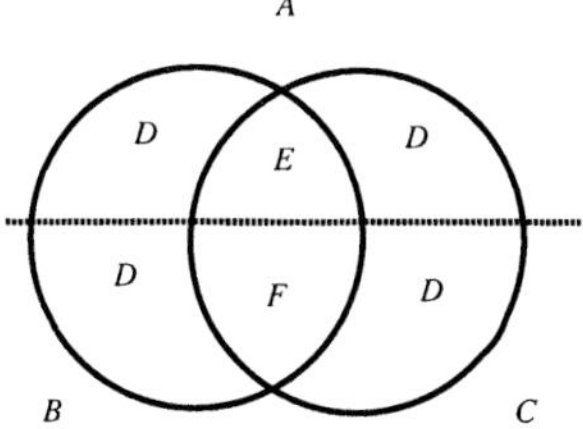

Fig. 1. Two Venn diagrams illustrating the children B,C of a Join node $X = B \cup C$ and its partition D, E, F. On the right the parent A is a Forget node represented by the part of $B \cup C$ above the dashed line. On the left the parent A is not a Forget node and we then have $B \cup C \subset A$ and $F = \emptyset$. In both cases what we call the *New* edges go between $B \setminus C$ and $C \setminus B$.

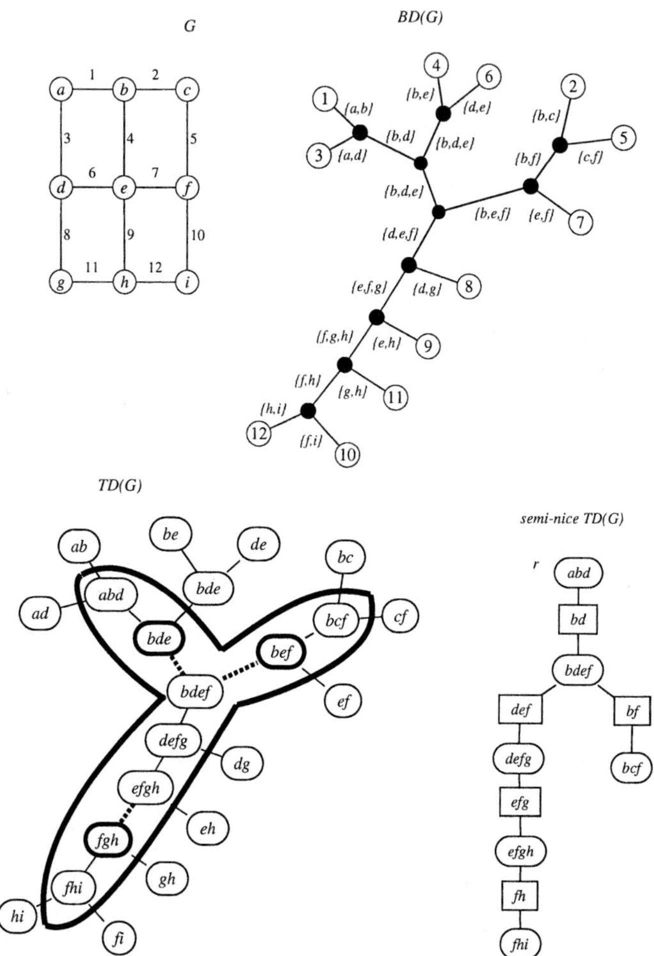

Fig. 2. *Algorithm:* 1) Transform branch-decomposition into a tree-decomposition on the same tree, 2) Transform tree-decomposition into a small tree-decomposition having $O(n)$ nodes, 3) Transform tree-decomposition into a sparse semi-nice tree-decomposition. We illustrate the algorithm with above figure. On the upper left a 3×3 grid graph G. On the upper right an optimal branch-decomposition with leaves labeled by edges of G as given by μ and the sets $mid(e)$. Step 1) is well-known (see e.g. [10] for a correctness proof): On the lower left a tree-decomposition formed with leaf-bags given by μ^{-1} and inner bags given by the union of adjacent $mid(e)$. In step 2) all nodes outside the bold line are then removed. The edges drawn in a dashed line are contracted. For step 3) we apply Lemma 1. On the lower right the resulting semi-nice tree-decomposition with new nodes emphasized rectangularly and arranged below arbitrary root node r.

D, E, F is a partition of X by definition. Note that if the parent A of $X = B \cup C$ is an Introduce or Join node then $B \cup C \subset A$ and we get $F = \emptyset$. See Figure 1. The *Forgettable* vertices are useful for any node whose parent is a Forget node, and their definition for an Introduce or leaf node X with parent

A is simply $F = X \setminus A$. We say that a neighbor u of a vertex $v \in X$ has been *considered* at node X of T if $u \in X$ or if $u \in X'$ for some descendant node X' of X. Clearly, if X is a Forget node forgetting v then all neighbors of v must have been considered at X. For fast dynamic programming we want sparse semi-nice tree-decompositions where vertices are forgotten as soon as possible.

Definition 1. *A semi-nice tree-decomposition is* sparse *if whenever a node X containing a vertex $v \in X$ has the property that all neighbors of v have been considered, then the parent of X is a Forget node forgetting v.*

Note that for a Join node with Forget parent A and children B, C of a sparse semi-nice tree-decomposition any vertex in $B \setminus A \cup C$ has a neighbor in $C \setminus A \cup B$ and vice-versa.

Lemma 1. *Given a tree-decomposition $(T, \mathcal{X})$ of width k of a graph G with n vertices, we can make it into a sparse semi-nice tree-decomposition $(T', \mathcal{X}')$ of width k in time $O(k^2 n)$ while keeping the E-sets in the partition of each Join node as small as the given tree-decomposition allows.*

For proofs see [9]. See Figure 2 for an illustration of the transformation from a given branch-decomposition to a semi-nice tree-decomposition described in the next lemma.

Lemma 2. *Given a branch-decomposition (T, μ) of a graph G with n vertices and m edges we can compute a sparse semi-nice tree-decomposition $(T', \mathcal{X})$ with $O(n)$ nodes in time $O(m)$ such that for any bag X of T' we have some $t \in V(T)$ with incident edges e, f, g such that $X \subseteq mid(e) \cup mid(f) \cup mid(g)$ and if X is a Join node with partition D, E, F then $E \subseteq mid(e) \cap mid(f) \cap mid(g)$ and $F \subseteq mid(f) \cap mid(g) \setminus mid(e)$ and $D \subseteq mid(e) \setminus mid(f) \cap mid(g)$.*

3 Dynamic Programming for Vertex Subset Problems

In this section we give the algorithmic template for doing fast dynamic programming on a semi-nice tree-decomposition $(T, \mathcal{X})$ of a graph G to solve an optimization problem related to vertex subsets on G. The runtime will in this section be given simply as a function of the D, E, F partition of the Join bags, and $X \setminus F, F$ partition of the other bags. In the final section we will then express the runtimes by pathwidth, branchwidth or treewidth of the graph. We introduce the template by giving a detailed study of the algorithm for Minimum Dominating sets, and then consider generalizations to various other vertex subset problems like Perfect Code, 2-Packings. We study these variants and the (k, r)-center problem in the long version of this paper where we also give all ommitted proofs [9].

As usual, we compute in a bottom-up manner along the rooted tree T a table of solutions for each node X of T. Let G_X denote the subgraph of G induced by vertices $\{v \in X' : \mathrm{X}' = X \text{ or } X' \text{ a descendant of } X \text{ in } T\}$. The table $Table_X$ at X will store solutions to the optimization problem on G_X indexed by certain equivalences classes of solutions. The solution to the problem on G is found by an optimization over the table at the root of T. To develop a specific algorithm one must define the tables involved and then show how to Initialize the table at a leaf node of T, how to compute the tables of Introduce, Forget and Join nodes given that their children tables are already computed, and finally how to do the Optimization at the root.

We use the Minimum Dominating Set problem as an example, whose tables are described by the use of three so-called vertex states:

- **Dom** (Dominating)
- **NbrD** (Neighbor is Dominating)
- **Free** (Temporary state)

Each index s of $Table_X$ at a node X represents an assignment of states to vertices in the bag X. For index $s : X \rightarrow \{Dom, NbrD, Free\}$ the vertex subset S of G_X is *legal* for s if:

- $V(G_X) \setminus X = (S \cup N(S)) \setminus X$
- $\{v \in X : s(v) = Dom\} = X \cap S$
- $\{v \in X : s(v) = NbrD\} \subseteq X \cap N(S)$

$Table_X(s)$ is defined as the cardinality of the smallest S legal for s, or we have $Table_X(s) = \infty$ if no S is legal for s.

Informally, the 3 constraints are that S is a dominating set of $G_X \setminus X$, that vertices with state Dom are exactly $X \cap S$, and that vertices with state NbrD have a neighbor in S. Note that vertices with state Free are simply constrained not to be in S. Since this is also a constraint on vertices with state NbrD a subset S which is legal for an index s would still be legal even if some vertex with state NbrD instead had state Free. This immediately implies the monotonicity property $Table_X(t) \leq Table_X(s)$ for pairs of indices t and s where $\forall v \in X$ either $t(v) = s(v)$ or $t(v) = Free$ and $s(v) = NbrD$.

Let us also remark that the $Table_X$ data structure should be an array. To simplify the update operations we should associate integers 0,1,2 with each vertex state so that an index is a 3-ary string of length $|X|$. Moreover, the ordering of vertices in the indices of $Table_X$ should respect the ordering in $Table_C$ for any child node C of X and in case C is the only child of X then all vertices in the larger bag should precede those in the smaller bag. We find this by computing a total order on $V(G)$ respecting the partial order given by the ancestor/descendant relationship of the Forget nodes forgetting vertices $v \in V(G)$.

The table $Table_X$ at a Forget node X will have 3^X indices, one for each of the possible assignments $s : X \rightarrow \{Dom, NbrD, Free\}$. We assume a machine model with words of length 3^X, to avoid complexity issues related to fast array accesses. Assume Forget node X has child C with $Table_C$ already computed. The correct value for $Table_X(s)$ is the minimum of $\{Table_C(s^+)\}$ over all indices s^+ where $s^+(v) = s(v)$ if $v \in X$ and $s^+(v) \in \{Dom, NbrD\}$ otherwise. For this reason we call the state Free a Temporary state. The Forget update operation takes time $O(3^X 2^{C \setminus X})$.

Note that the Forget update operation had no need for the indices of the table at the child where a forgotten vertex in $C \setminus X$ had state Free. This observation allows us to save some space and time for the Forgettable vertices of a bag having a Forget parent.

If X is a leaf node with Forgettable vertices F then $Table_X$ has only $3^{X \setminus F}2^F$ indices, in accordance with the above observation, and is computed in a brute-force manner. This takes time $O(X3^{X \setminus F}2^F)$, since for each index s we must check if $Table_X(s)$ should be equal to the number of vertices in state Dom, or if there is a vertex in state NbrD with no neighbor in state Dom in which case $Table_X(s) = \infty$.

If X is an Introduce node with Forgettable vertices F and child C then $Table_X$ has $3^{X \setminus F}2^F$ indices and the correct value at $Table_X(s)$ is:

- ∞ if $Table_C(s) = \infty$ or if $\exists x \in X \setminus C$ with $s(x) = NbrD$ but no neighbor of x in state Dom.
- $Table_C(s) + |\{v \in X \setminus C : s(v) = Dom\}|$ otherwise

The Introduce update operation thus takes time $O(X3^{X \setminus F}2^F)$.

The correct values for $Table_X$ at a Join node X with partition D, E, F and children B, C are computed in three steps, where the last three steps account for new adjacencies that have not been considered in any child table (we call these 'new edges'):

1. $\forall s : Table_X(s) = \min\{Table_B(s_b) + Table_C(s_c) - |B \cap C \cap \{v : s(v) = Dom\}|\}$ over (s_b, s_c) such that triple (s, s_b, s_c) is necessary (see below).
2. $New = \{uv \in E(G) : u \in B \setminus C \wedge v \in C \setminus B\}$
3. $\forall R \subseteq D : New(R) = \{u \in D \setminus R : \exists v \in R \wedge uv \in New\}$
4. $\forall s : Table_X(s) = Table_X(s')$ where $s'(v) = Free$ if $v \in D \wedge s(v) = NbrD \wedge v \in New(\{u : s(u) = Dom\})$ and otherwise $s'(v) = s(v)$.

We describe and count the necessary triples of indices (s, s_b, s_c) for the Join update using the method of [10], by first considering the number of *necessary vertex state triples* $(s(v), s_b(v), s_c(v))$ such that vertex state $s_b(v)$ and $s_c(v)$ in B and C respectively will yield the vertex state $s(v)$ in X:

- $v \in B \setminus C \subseteq D$: 3 triples (Dom,Dom,-), (NbrD,NbrD,-), (Free,Free,-)
- $v \in C \setminus B \subseteq D$: 3 triples (Dom,-,Dom), (NbrD,-,NbrD), (Free,-,Free)
- $v \in F$: 3 triples (Dom,Dom,Dom), (NbrD,Free,NbrD), (NbrD,NbrD,Free)
- $v \in E$: 4 triples (Dom,Dom,Dom), (NbrD,Free,NbrD), (NbrD,NbrD,Free), (Free,Free,Free)

Lemma 3. *The Join update just described for a node X with partition D, E, F is correct and takes time $O(3^{D+F}4^E)$.*

For a proof see [9]. Finally, at the root node R of T we compute the smallest dominating set of G by the minimum of $\{Table_R(s) : s(v) \in \{Dom, NbrD\} \forall v \in R\}$. This takes time $O(2^R)$. Correctness of the algorithm follows by induction on the tree-decomposition, in the standard way for such dynamic programming algorithms. For the timing we have the Join operation usually being the most expensive, although there are graphs, e.g. when pathwidth=treewidth, for which the leaf Initialization or Introduce operations are the most expensive. However, the Forget and Root optimization operations will never be the most expensive.

Theorem 1. *Given a semi-nice tree-decomposition $(T, \mathcal{X})$ of a graph G on n vertices we can solve in time $O(n(\max\{4^E 3^{D+F}\} + \max\{X 3^{X \setminus F} 2^F\}))$ the Min Dominating Set Problem on G with maximization over Join nodes of T with partition D, E, F and over Initialization and Introduce nodes with bag X and Forgettable set F, respectively.*

For problems over vertex subsets having other domination-type constraints we get slightly different runtimes. A general class of such constraints are parameterized by two subsets of natural numbers σ and ρ. A subset of vertices S is a (σ, ρ)-set if $\forall v \in S$ we have $|N(v) \cap S| \in \sigma$ and $\forall v \notin S$ we have $|N(v) \cap S| \in \rho$ [17]. Some well-studied and natural types of (σ, ρ)-sets are when σ is either all natural numbers $\mathbb{N}$, all positive numbers $\mathbb{N}^+$, or $\{0\}$, and with ρ being either all positive numbers, or $\{1\}$. The six resulting constraints are Dominating set ($\sigma = \mathbb{N}$, $\rho = \mathbb{N}^+$); Perfect Dominating Set ($\sigma = \mathbb{N}$,$\rho = \{1\}$); Independent Dominating set ($\sigma = \{0\}$,$\rho = \mathbb{N}^+$); Perfect Code ($\sigma = \{0\}$,$\rho = \{1\}$); Total Dominating set ($\sigma = \mathbb{N}^+$, $\rho = \mathbb{N}^+$); Total Perfect Dominating set ($\sigma = \mathbb{N}^+$,$\rho = \{1\}$). For Perfect Code and Total Perfect Dom set it is NP-complete simply to decide if a graph has any such set, for Ind Dom set it is NP-complete to find either a smallest or largest such set, while for the remaining three problems it is NP-complete to find a smallest set. The thesis [1] considers these six constraints, and give dynamic programming algorithms on nice tree-decompositions that take into account monotonicity properties to arrive at fast runtimes. See column Join in Table 1 for an overview of our results and [9] for exact calculations. The previous best results for these problems [1] correspond to our results when treating all vertices as Expensive, so we have moved closer to the goal of λ^{D+E+F} time for a problem with λ vertex states. These algorithms can of course be extended also to more general (σ, ρ)-sets. For example, if $\sigma = \{0, 1, ..., p\}$ and $\rho = \{0, 1, ..., q\}$ we are asking for a subset $S \subseteq V(G)$ such that S induces a subgraph of maximum degree at most p with each vertex in $V(G) \setminus S$ having at most q neighbors in S. For this case we would use $p+q+2$ vertex states and get runtime $O((p + q + 2)^D (s(p) + s(q))^{E+F})$, where $s(i)$ is the number of pairs of ordered non-negative integers summing to i. Thus, for the Maximum 2-Packing problem (also known as Max Strong Stable set), which is of this form with $p = 0$ and $q = 1$, we get an $O(3^D 4^{E+F})$ algorithm.

4 Dynamic Programming for Edge Subset Problems

Problems like Hamiltonian cycle and Travelling Salesman ask for a subset of edges of the input graph with a given property. An index of the table storing solutions to subproblems will likewise represent a class of edge subsets of the subgraph considered so far. Consider a Join node X with children B, C, and assume that B and C store solutions for the subgraphs G'_B and G'_C. For these edge subset problems the Join operation at X is simplified if we can assume that the two subgraphs G'_B and G'_C do not overlap in edges. To accomplish this we define the subgraph G'_X for edge subset problems to be the graph we get from taking the subgraph G_X as used for vertex subset problems and removing all edges having both endpoints in the set X.

Definition 2. *For edge subset problems the subgraph G'_X of G for which solutions are stored in a table at node X of the tree T is the graph on vertex set $V(G'_X) = \{v \in X' : X' = X \text{ or } X' \text{ a descendant of } X \text{ in } T\}$ and edge set $E(G'_X) = \{uv \in E(G) : \{u, v\} \subseteq V(G'_X) \text{ and at most one of } u \text{ and } v \text{ in } X\}$.*

The implication is that the Join update is simplified, since there is no overlap of edges in the two subgraphs. The Introduce operation becomes trivial, simply adding isolated vertices to the existing subgraph. Likewise, the Initialize-Table operation is trivial since it considers a subgraph without edges. On the other hand the Forget operation becomes more complicated. Let X be a Forget node with child B, thus with $B \setminus X$ the forgotten vertices. Note that an edge between a forgotten vertex $u \in B \setminus X$ and a vertex $v \in X$ has not been considered so far in the algorithm, since it does not belong to G'_B. However, such an edge does belong to G'_X and it will in fact be considered for the first time during the Forget operation at X. This consideration of new adjacencies performed by the Forget operation for edge problems is almost identical to what is performed by the Introduce operation for vertex problems. The Root-Optimization step at root node X becomes trivial since we simply ensure that $|X| = 1$, by a preceding Forget operation.

A comparison with the template given for vertex problems and the one just described shows that for edge problems the Forget-operation is more complicated but the other operations are less complicated. However, note that the gain we get in the runtime of the Join operation for vertex subset problems from the Forgettable vertices F is no longer easily achieved under the edge subset template, since the vertices in F have not had all their adjacencies considered at the time of the Join.

Cook and Seymour [6] give a heuristic algorithm for the Traveling Salesman Problem (TSP). Their paper contains a subroutine which for a subgraph of the input graph solves the TSP problem exactly by dynamic programming along a branch-decomposition. Their paper is not focused on runtime but we can estimate the running time of their dynamic programming algorithm for exact solution of TSP on a heuristically generated branch-decomposition of width k as $O(c^{1.5\,k\,logk}m)$ for some constant c. Their update operation on middle sets of the branch-decomposition is directly transferred as the update we need for our Join operation, as the subgraphs we are considering do not overlap in edges. When forgetting vertex v we have to consider all neighbors of v in X since these edges have not been accounted for earlier. In the Forget-operation we do this independently for every index of $Table_X$ and every forgotten vertex. Compared to their algorithm, the runtime for our more complicated Forget-operation gives only an additional polynomial factor in the size of the Forget node X. Without going into details in this extended abstract we claim that a dynamic programming algorithm solving TSP on a semi-nice tree-decomposition can in this way be developed exactly as in the paper [6] and with the same exponential runtime.

5 Runtime by Branchwidth, Treewidth or Pathwidth

In this section we assume that we are given a branch-decomposition of width bw or a tree-decomposition of width tw and first transform these into a semi-nice tree-decomposition by the algorithms of Section 2. We then run any of the algorithms described in Sections 3 or 4 to express the runtime to solve those problems as a function of bw or tw. This runtime will match or improve the best results achieved by dynamic programming directly on the branch- or tree-decompositions. For a proof see the long version [9].

Theorem 2. *We can solve Minimum Dominating set by dynamic programming on a semi-nice tree-decomposition in time:* $O(2^{3\log_4 3bw}n) = O(2^{2.38\,bw}n)$ *if given a branch-decomposition* (T,μ) *of width* bw*;* $O(2^{2\,tw}n)$ *if given a tree-decomposition of width* tw*;* $O(2^{1.58\,pw}n)$ *if given a path-decomposition of width* pw*; and* $O(2^{\min\{1.58\,pw, 2\,tw, 2.38\,bw\}})$ *if given all three. For other domination-type problems we get runtimes as in Table 1.*

For certain classes of graphs, e.g. grid graphs, pathwidth is indeed the best parameter. The runtime we get for Minimum Dominating set as a function of branchwidth bw is essentially the same as that achieved by the algorithm of [10] working directly on the branch-decomposition (the runtime there is expressed with multiplicative factor m instead of our n but for a graph with branchwidth bw we have $m = O(n\,bw)$.) See Table 1 for a summary of the results for each domination-type problem. For the TSP problem we have already argued in Section 4 that our algorithm matches the runtime of the algorithm of [6] that works directly on a branch-decomposition.

Acknowledgements. We would like to thank Jochen Alber and Rolf Niedermeier for suggesting the comparison of dynamic programming approaches on tree-decompositions and branch-decompositions.

References

1. J. Alber, *Exact algorithms for np-hard problems on networks: Design, analysis, and implementation*, PhD Thesis, Universität Tübingen, (2002).
2. J. Alber, H. Bodlaender, H. Fernau, T. Kloks, and R. Niedermeier, *Fixed parameter algorithms for Dominating Set and related problems on planar graphs*, Algorithmica, 33 (2002), pp. 461–493.
3. J. Alber and R. Niedermeier, *Improved tree decomposition based algorithms for domination-like problems*, in LATIN'02: Theoretical informatics (Cancun), vol. 2286 of Lecture Notes in Computer Science, Berlin, 2002, Springer, pp. 613–627.

4. S. ARNBORG AND A. PROSKUROWSKI, *Linear time algorithms for np-hard problems restricted to partial k-trees*, Discrete Applied Math, 23 (1989), pp. 11–24.
5. H. BODLAENDER, *Treewidth: Algorithmic techniques and results.*, in MFCS'97: Mathematical Foundations of Computer Science 1997, 22nd International Symposium (MFCS), vol. 1295 of Lecture Notes in Computer Science, Springer, 1997, pp. 19–36.
6. W. COOK AND P. SEYMOUR, *Tour merging via branch-decomposition*, INFORMS Journal on Computing, 15 (2003), pp. 233–248.
7. E. D. DEMAINE, F. V. FOMIN, M. T. HAJIAGHAYI, AND D. M. THILIKOS, *Fixed-parameter algorithms for the (k,r)-center in planar graphs and map graphs*, in ICALP'03: Automata, languages and programming, vol. 2719 of Lecture Notes in Comput. Sci., Berlin, 2003, Springer, pp. 829–844.
8. F. DORN, E. PENNINKX, H. L. BODLAENDER, AND F. V. FOMIN, *Efficient exact algorithms on planar graphs: Exploiting sphere cut branch decompositions*, in ESA'05: 13th Annual European Symposium on Algorithms, vol. 3669 of Lecture Notes in Comput. Sci., Berlin, 2005, Springer, pp. 95–106.
9. F. DORN AND J. TELLE, *Two birds with one stone: the best of branchwidth and treewidth with one algorithm*, long version, (2005). `http://www.ii.uib.no/~frederic/DT.pdf`.
10. F. V. FOMIN AND D. M. THILIKOS, *Dominating sets in planar graphs: branch-width and exponential speed-up*, in SODA'03: Proceedings of the Fourteenth Annual ACM-SIAM Symposium on Discrete Algorithms (Baltimore, MD, 2003), New York, 2003, ACM, pp. 168–177.
11. ——, *Fast parameterized algorithms for graphs on surfaces: Linear kernel and exponential speed-up*, in ICALP'04:Automata, Languages and Programming: 31st International Colloquium, vol. 3142 of Lecture Notes in Computer Science, Berlin, 2004, Springer, pp. 581–592.
12. ——, *A simple and fast approach for solving problems on planar graphs.*, in STACS'04: 22nd Ann. Symp. on Theoretical Aspect of Computer Science, vol. 2996 of Lecture Notes in Computer Science, Berlin, 2004, Springer, pp. 56–67.
13. J. KLEINBERG AND E. TARDOS, *Algorithm design*, Addison-Wesley, 2005.
14. T. KLOKS, *Treewidth*, vol. 842 of Lecture Notes in Computer Science, Springer-Verlag, Berlin, 1994. Computations and approximations.
15. B. REED, *Treewidth and tangles, a new measure of connectivity and some applications*, Surveys in Combinatorics, 1997.
16. N. ROBERTSON AND P. SEYMOUR, *Graph minors X. Obstructions to tree-decomposition.*, Journal of Combinatorial Theory Series B, 52 (1991), pp. 153–190.
17. J. A. TELLE AND A. PROSKUROWSKI, *Algorithms for vertex partitioning problems on partial k-trees*, SIAM J. Discrete Math, 10 (1997), pp. 529–550.

Maximizing Throughput in Queueing Networks with Limited Flexibility

Douglas G. Down* and George Karakostas**

Department of Computing and Software,
McMaster University,
Hamilton, ON, Canada
downd@mcmaster.ca, karakos@mcmaster.ca

Abstract. We study a queueing network where customers go through several stages of processing, with the *class* of a customer used to indicate the stage of processing. The customers are serviced by a set of *flexible* servers, i.e., a server may be capable of serving more than one class of customer and the sets of classes that the servers are capable of serving may overlap. We would like to choose an assignment of servers that achieves the maximal capacity of the given queueing network, where the maximal capacity is λ^* if the network can be stabilized for all arrival rates $\lambda < \lambda^*$ and cannot possibly be stabilized for all $\lambda > \lambda^*$. We examine the situation where there is a restriction on the number of servers that are able to serve a class, and reduce the maximal capacity objective to a maximum throughput allocation problem of independent interest: the TOTAL DISCRETE CAPACITY CONSTRAINED PROBLEM *(TDCCP)*. We prove that solving *TDCCP* is in general NP-complete, but we also give exact or approximation algorithms for several important special cases.

Keywords: Queueing networks, scheduling, approximation algorithms.

1 Introduction

Consider a system (which we will henceforth call a queueing network), in which discrete entities (or customers) progress through a series of operations (referred to as classes). At each class, a processing step must be performed that requires an amount of time that can be modelled as a random variable. There is infinite waiting room in a queue at each class. Processing at a class is performed by one or more servers. In a traditional queueing network, servers are dedicated to a class. If the queue at a class is empty, the dedicated server(s) there is forced to idle. In the operations research literature, there has been much recent interest in a generalization of this model, in which the servers are flexible, i.e. the customers progress through the network as before, but servers may be capable of

* Research supported by NSERC grant 239150-2001.

** Research supported by an NSERC Discovery grant.

J.R. Correa, A. Hevia, and M. Kiwi (Eds.): LATIN 2006, LNCS 3887, pp. 398–409, 2006.

performing processing at more than one class (and as such, a decision must be made at each point in time as to where a server is located). A typical example of this is a production system where the classes are machines performing manufacturing steps, customers are the parts being produced, and the flexible servers are workers cross-trained to operate multiple machines (see Hillier and So [12], for example). Such models also arise in areas such as power control for wireless networks (Armony and Bambos [4]) and parallel computer systems (Squillante et al. [17]). For an extensive overview of the literature, see Andradóttir et al. [3] and Hopp and van Oyen [13]. A precise mathematical definition of the model is given in the next section.

The design problem in which we are interested is to choose a (dynamic) assignment of servers to classes to address a particular performance objective. In this paper, we are interested in determining the maximal capacity of a given network, where we define the maximal capacity to be λ^* if the network can be stabilized for all arrival rates $\lambda < \lambda^*$ and cannot possibly be stabilized for any $\lambda > \lambda^*$. A number of authors have examined flexible server systems with throughput as a performance measure. In addition to [3] and [8], these include Tassiulas and Ephremides [19], Tassiulas and Bhattacharya [18], Andradóttir, Ayhan, and Down [2], Andradóttir and Ayhan [1], Bischak [6], Zavadlav, McClain, and Thomas [20].

The above references do not constrain the number of servers that may be at a class. This is not a realistic assumption for most settings, as for example, one may have budgetary constraints for training and as a result, one would like to restrict the amount of cross-training, but still achieve reasonable throughput (as compared to a system with no such constraints). In order to address this issue, we need a means to calculate the maximal capacity of a constrained network. We believe that this is the first attempt to address such a problem.

In the area of queueing networks, the use of fluid limits to characterize stability is a standard technique, originating in the work of Rybko and Stolyar [15] and Dai [7]. The central idea in this approach is that one can equate stability of a (stochastic) queueing network with that of a related deterministic model (the fluid model). We emphasize here that determining stability conditions for the original stochastic queueing network is typically extremely difficult. The fluid model approach provides a rigorous connection between the two models and one hopes that the deterministic model is easy (or at the very least easier) to analyze. For the flexible server setting, the fluid limit methodology has been used to break down the determination of maximal capacity to two steps (see [3] and Dai and Lin [8]).

1. Determine the maximal capacity of the fluid model and an optimal allocation of each server's effort.
2. Use the allocation to construct a scheduling policy for the original network.

The framework in [3] gives a standard means by which to perform the second step. Also, if there is no constraint on the number of servers that can be at a class at any one time, it is shown in [3] that the first step reduces to solving a linear programming problem, so as a result the entire problem has been reduced to one that is very tractable. For the more realistic constrained problem considered here, we find that the analysis of the deterministic fluid model is much more difficult. We shall call such a problem the TOTAL DISCRETE CAPACITY CONSTRAINED PROBLEM *(TDCCP)*. For this problem, the second step in the above procedure is unchanged. It is the first step in the procedure which sees the most significant change, and the resulting optimization problem is the focus of this paper.

We show that this problem is NP-complete even for special cases. Hence we look for approximation algorithms for such hard special cases and for the general problem. We achieve an approximation factor of 1/10 for the important case of service rates that depend only on the servers (and not on the classes)[1], and these approximation techniques extend also to the general case (but with a worse approximation factor.) Even more importantly, some of these techniques give exactly the same approximation factors for the *budgetary constraint* version of the problem. In this generalization, a *per service unit cost* of assigning a particular server to a particular class is given, as well as a budget that our total assignment cost should not exceed. Some of our approximation algorithms produce solutions that are within the previous approximation factors without violating the budget.

2 Queueing Network Model

The model we consider is a generalization of that in Andradóttir, Ayhan, and Down [3]. For completeness, we present the model in its entirety.

2.1 Network Topology

Consider a network where the location of a customer is given by its class k. We assume that there are K distinct classes, with a buffer of infinite size for each class. Arrivals to a class may occur from inside or outside of the network. Customers arriving from outside of the network do so according to an arrival process with independent and identically distributed (i.i.d.) interarrival times $\{\xi(n)\}$. The associated arrival rate is $\lambda = 1/E[\xi(1)]$. An arrival from outside of the network is routed to class k with probability $p_{0,k}$, with $\sum_{k=1}^{K} p_{0,k} = 1$. Within the network, customers circulate as follows. Upon completion of service at class i, a customer becomes one of class k with probability $p_{i,k}$. The customer leaves the network with probability $1 - \sum_{k=1}^{K} p_{i,k}$. We define the routing matrix P to have (i,k) entry $p_{i,k}$ for $i,k = 1,\ldots,K$ and I to be the $K \times K$ identity matrix. We assume that all customers eventually leave the network, which is equivalent to $(I - P')$ being invertible. (Note that the (i,k) entry of $(I - P')^{-1}$ is the expected number of future visits to class k of a class i customer.)

For technical reasons, we assume that the interarrival times are unbounded and spread out. For more details see [3].

[1] In Section 4.1 we show that the special case of service rates which depend only on the classes can be solved optimally.

2.2 Service Mechanism

The network is populated by M servers which service customers within a class according to First Come, First Served order. When switching from class i to class k for the nth time, server j incurs a (possibly zero) switching time of $\zeta^j_{i,k}(n)$. It is assumed that the sequence $\{\zeta^j_{i,k}(n)\}$ is i.i.d. for every $j = 1, \ldots, M$ and $i, k = 1, \ldots, K$. Further, we assume that $\{\zeta^j_{i,i}(n)\}$ is identically zero for all i and j.

Several servers may be simultaneously at a class, in which case they work in parallel. If server j is capable of working at class k, the service time of the nth customer served by server j at class k is given by $\eta_{j,k}(n)$, where the sequence $\{\eta_{j,k}(n)\}$ is assumed to be i.i.d. for each j and k. The associated mean service time for server j at class k is $m_{j,k} = E[\eta_{j,k}(1)]$, with associated service rate $\mu_{j,k} = 1/m_{j,k}$. If server j cannot work at class k, we set $\mu_{j,k} = 0$. We only consider nonpreemptive policies.

The difference between the model in [3] and that considered here is that we put an upper limit, $c_k \leq M$, on the number of servers that can be assigned to a class (a server is assigned to a class if it spends any time at class k).

3 Total Discrete Capacity Constrained Problem

We are first interested in computing the *capacity*. A network operating under a service policy π is said to have capacity λ^π if the system is stable for all values of the arrival rate $\lambda < \lambda^\pi$. We wish to calculate a tight upper bound on the capacity that a given system can achieve (called the *maximal capacity*). In the course of doing so, we identify a means to construct server assignment policies that have capacity that is arbitrarily close to the maximal capacity.

3.1 The Allocation Program

First, we solve the traffic equations for the network, which give the total arrival rate to class k, λ_k, if the network is stable. Here we have

$$\lambda_k = p_{0,k}\lambda + \sum_{i=1}^{K} p_{i,k}\lambda_i,$$

for $k = 1, \ldots, K$. This system of equations is known to have a unique solution if $(I - P')$ is invertible. If we let a_i, $1 \leq i \leq K$ be the unique solution with $\lambda = 1$, then $\lambda_k = \lambda a_k$ is the unique solution of the traffic equations for an arbitrary value of λ.

Let $\delta_{j,k}$ be the proportion of time that server j is working at class k. These proportions exist under the policies considered below. The resulting optimization problem (with variables $\delta_{j,k}$ and λ) that will give us the assignment of servers to classes is:

$$\max \lambda \tag{MP}$$

$$\text{s.t.} \sum_{j=1}^{M} \mu_{j,k}\delta_{j,k} \geq \lambda a_k, \ k = 1, \ldots, K \tag{1}$$

$$\sum_{k=1}^{K} \delta_{j,k} \leq 1, \ j = 1, \ldots, M, \tag{2}$$

$$\delta_{j,k} \geq 0, \ k = 1, \ldots, K, j = 1, \ldots, M, \tag{3}$$

$$\sum_{j=1}^{M} \chi\{\delta_{j,k} > 0\} \leq c_k, \ k = 1, \ldots, K, \tag{4}$$

where $\chi\{\cdot\}$ is the indicator function. The constraints in (MP) have the following interpretations. The first, (1), says that the service rate allocated to class k must be greater than the arrival rate. The second and third constraints, (2) and (3), prevent over allocations and negative allocations of a server, respectively. Finally, the constraint (4) limits the flexibility, by only allowing c_k servers to be assigned to work at class k. Let a solution of (MP) be given by λ^*, $\{\delta^*_{j,k}\}$. We will see that λ^* is the desired maximal capacity and $\{\delta^*_{j,k}\}$ is the set of proportional allocations of server j to classes k required to achieve λ^*.

Obviously, the difficulty in solving (MP) comes from the integral constraints (4). Note that in these constraints, although the allocation variables $\delta_{j,k}$ are fractional, the capacity each one is allocated is either 0 or 1 (depending on whether $\delta_{j,k}$ is 0 or not). To the best of our knowledge, we are not aware of other scheduling problems with such constraints. In Section 6 we show that even a simpler variant of the problem is NP-complete. First we consider special cases in Section 4: If the $\mu_{j,k}$'s are independent of j, i.e., $\mu_{j,k} = \mu_k$ for all j, then the problem can be solved in polynomial time. If the $\mu_{j,k}$'s are independent of k, i.e., $\mu_{j,k} = \mu_j$ for all k, then the problem is NP-complete, but can be approximated within a factor 1/10, or better under certain assumptions. For the general case, we show in Section 5 that in polynomial time one can find a solution within a factor $1/10w_{max}$, where $w_{max} := \max_j \max_{k_1,k_2,\mu_{j,k_2}\neq 0} \frac{\mu_{j,k_1}}{\mu_{j,k_2}}$. The bulk of the remainder of the paper is concerned with how one can solve (MP). Before doing this, we complete the connection between solving (MP) and the problem of finding the maximal capacity in the original queueing network.

For the original queueing network, we consider the set of generalized round robin policies. A generalized round robin policy π is given by a set of integers $\{\ell^\pi_{j,k}\}$ and an ordered list of classes V^π_j. Server j servers classes in V^π_j in cyclic order, with server j performing $\ell^\pi_{j,k}$ services at each class in V^π_j (unless server j idles, in which case the server moves to the next class in V^π_j). If the classes in V^π_j are all empty, the server idles at an arbitrary class in V^π_j. The details of how to construct a generalized round robin policy π given a set of required proportional allocations $\{\delta^*_{j,k}\}$ is given in Section 3.3 of [3]. As this can be used directly, we give no further discussion of the construction here.

Define $Q_k(t)$ to be the number of class k customers present at time t and $Q(t)$ be a vector with kth entry $Q_k(t)$. The following theorem gives the strong connection between maximizing capacity in the queueing network and the solution to (MP).

Theorem 1. **(i)** *Any capacity less than λ^* may be achieved. More specifically, for an arrival process with rate $\lambda < \lambda^*$, there exists a dynamic server assignment policy such that the distribution of the queue length process $\{Q(t)\}$ converges to a steady-state distribution φ as $t \to \infty$.*
(ii) *A capacity larger than λ^* cannot be achieved. More specifically, for an arrival process with rate $\lambda > \lambda^*$, as $t \to \infty$, $P(|Q(t)| \to \infty) = 1$.*

The proof of this theorem is a trivial extension of that of Theorem 1 in [3] and is thus omitted. The derivation of the additional constraint (4) is a straightforward exercise, the remainder of the proof is unchanged.

Theorem 1 says that the difficult stochastic optimization problem can be converted into a deterministic optimization problem. The mapping of the solution to the deterministic problem back to a solution to the original stochastic problem does not depend on the complexity of the deterministic problem (it simply uses the resulting solution). For the remainder of the paper, we thus focus on solving (MP). In [3], the deterministic problem is simply (MP), with the constraint (4) removed. This is easily seen to be a linear programming problem, and so there is the appealing result that a difficult stochastic problem becomes a simple deterministic problem. However, in our case, the resulting deterministic problem can also be difficult, as will be seen below. From this point, we refer to the required deterministic optimization problem as the TOTAL DISCRETE CAPACITY CONSTRAINED PROBLEM (TDCCP).

4 Solving TDCCP - Special Cases

It is instructive to first look at several special cases of TDCCP that give an idea of the inherent complexity.

4.1 The Case $\mu_{j,k} = \mu_k$ for all j

Suppose that the service rates are independent of the server and that each server is capable of working at every class, so $\mu_{j,k} = \mu_k$ for $j = 1, \ldots, M$. Here, (MP) can be rewritten as

$$\begin{aligned} &\max \lambda \text{ s.t.} \\ \textstyle\sum_{j=1}^{M} \delta_{j,k} &\geq \lambda a_k/\mu_k, && k = 1, \ldots, K \\ \textstyle\sum_{k=1}^{K} \delta_{j,k} &\leq 1, && j = 1, \ldots, M, \\ \delta_{j,k} &\geq 0, && k = 1, \ldots, K, j = 1, \ldots, M, \\ \textstyle\sum_{j=1}^{M} \chi\{\delta_{j,k} > 0\} &\leq c_k, && k = 1, \ldots, K, \end{aligned}$$

where $\chi\{\cdot\}$ is the indicator function.

Proposition 1. *If $\mu_{j,k} \equiv \mu_k$, the maximal capacity is*

$$\lambda^* = \min\left(\frac{M}{\sum_{k=1}^K a_k/\mu_k}, \min_{1\le k\le K} \frac{c_k\mu_k}{a_k}\right).$$

The proof of Proposition 1 appears in the full version.

4.2 The Case $\mu_{j,k} = \mu_j$ for all k

Suppose now that the service rates depend only on the server and that each server is capable of working at every class, so $\mu_{j,k} = \mu_j$ for $k = 1, \ldots, K$. In this case (MP) can be written as

$$\begin{aligned} &\max \lambda \text{ s.t.} \\ \sum_{j=1}^M x_{j,k} &\ge \lambda a_k, \; k = 1, \ldots, K \\ \sum_{k=1}^K x_{j,k} &\le \mu_j, \;\; j = 1, \ldots, M \\ x_{j,k} &\ge 0, \quad k = 1, \ldots, K, j = 1, \ldots, M \\ \sum_{j=1}^M \chi\{x_{j,k} > 0\} &\le c_k, \;\; k = 1, \ldots, K \end{aligned} \qquad \text{(MP}')$$

where we performed the substitution $x_{j,k} := \mu_j \delta_{j,k}, \; \forall j, k$. This case is already NP-complete, as is shown in Theorem 2 in Section 6.

(MP′) actually is an instance of the *maximum concurrent multicommodity k-splittable flow problem* which can be stated as follows: Let $G = (V, E)$ be a directed or undirected graph with integral edge capacities $u_e > 0$, for all $e \in E$. There are l source-sink pairs (s_i, t_i), $i = 1, \ldots, l$, one for each of l different commodities. For each commodity i there is also a demand d_i, and a bound k_i on the number of different paths allowed for this commodity. Then the maximum concurrent multicommodity k-splittable flow problem is asking for a flow assignment to paths in G that respects the edge capacities and the splittability bounds for the commodities, and routes the maximum possible fraction of all commodity demands *simultaneously.* This, together with several other versions of k-splittable problems, are studied in [5]. Also, when $k_i = 1$, $\forall i$, then these problems are called just *unsplittable* (instead of 1-splittable).

Problem (MP′) is a special case of the multicommodity k-splittable flow problem: the K classes can be seen as K commodities of demand a_k, $k = 1, \ldots, K$, each with a splittability upper bound of $0 < c_k \le M$. These commodities are routed on the network of Figure 1. All commodities have the same source s, but commodity i has its own sink t_i. Each of the vertices t_i, $i = 1, \ldots, K$ is connected to all vertices u_j, $j = 1, \ldots, M$, and s is connected to all vertices v_j, $j = 1, \ldots, M$. The edge (v_j, u_j) has capacity μ_j for all $j = 1, \ldots, M$, while the rest of the edges have infinite capacity. Note that a solution to the maximum concurrent multicommodity k-splittable flow problem on this instance will also give a solution to our original problem (MP′), since every flow path that carries flow f of commodity k through edge (v_j, u_j) corresponds to setting $\delta_{j,k} := f$. And vice versa, a solution to (MP′) gives us also a path flow assignment that achieves the same value for the minimum fraction of commodity demand that is satisfied in the maximum concurrent multicommodity k-splittable flow problem above.

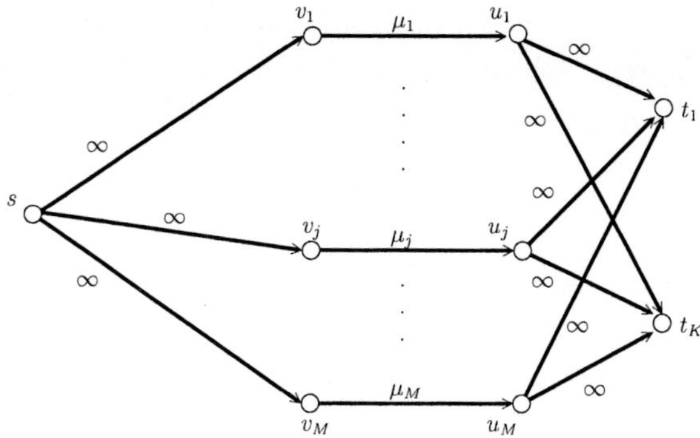

Fig. 1. The graph for our special k-splittable flow instance

Baier et al. [5] show that any ρ-approximation algorithm for the maximum concurrent unsplittable flow problem yields a $\rho/2$-approximation algorithm for the maximum concurrent k-splittable flow problem. Dinitz et al. [9] present an algorithm that achieves an approximation factor of $\rho = 1/5$ in running time $O(KM(K+M))$, using ideas by Kolliopoulos and Stein [14]. Therefore the solution we get for our problem has a guaranteed *worst-case* performance of at least $1/10$ of the optimum. Note that in our case, the usual *balancing* assumption

$$\text{max demand} \leq \text{min capacity}$$

does not hold, hence the somewhat worse approximation ratios achieved, as compared to the ratios achieved if this assumption holds.

A Different Approximation Algorithm. The previous algorithm cannot take advantage of the better approximation factor of 2 for congestion in the unsplittable flow problem, because the balancing assumption doesn't hold in our case. Here we follow a different path, in order to provide an approximation algorithm that under certain assumptions achieves a factor better than $1/10$ for the case $\mu_{j,k} = \mu_j$ for all k. We will reduce our problem to the generalized assignment problem, and then we will use the approximation algorithm by Shmoys and Tardos [16].

The first step of the new algorithm is the same as before: we transform the given problem into an exactly-k-splittable flow problem, with a loss of a factor of $1/2$. Hence commodity k is split into c_k commodities (k,i), $i = 1, \ldots, c_k$, each with demand a_k/c_k.

During the second step, we solve the following concurrent flow problem in the network defined above, which in turn is a relaxation of the concurrent unsplittable flow:

$$\begin{aligned} &\max \lambda \text{ s.t.} \\ \textstyle\sum_{j=1}^{M} x_{j,(k,i)} &\geq \lambda \tfrac{a_k}{c_k}, \ \forall k,i \\ \textstyle\sum_{(k,i)} x_{j,(k,i)} &\leq \mu_j, \quad \forall j \\ x_{j,(k,i)} &\geq 0, \quad \forall i,j,k \end{aligned} \qquad \text{(LP-NEW)}$$

If x^*, λ^* is the optimal solution for (LP-NEW), then define $\lambda_{(k,i)} := (\sum_{j=1}^M x^*_{j,(k,i)})/(a_k/c_k)$. Obviously $\lambda_{(k,i)} \geq \lambda^* > 0, \forall (k,i)$. Also, we define $y_{j,(k,i)} := \frac{c_k}{\lambda_{(k,i)} a_k} x^*_{j,(k,i)}$ and $p_{j,(k,i)} := \frac{a_k}{c_k \mu_j}, \forall i, j, k$. Then y satisfies the following system of inequalities:

$$\begin{aligned} \textstyle\sum_{j=1}^M y_{j,(k,i)} &= 1, \qquad \forall k, i \\ \textstyle\sum_{(k,i)} p_{j,(k,i)} y_{j,(k,i)} &\leq 1/\lambda^*, \forall j \\ y_{j,(k,i)} &\geq 0, \qquad \forall i, j, k \end{aligned}$$

This is exactly the relaxation of the problem (without costs) of scheduling unrelated parallel machines that [16] studies. We can think of the commodities (k,i) as jobs, the edges of capacities μ as machines, $p_{j,(k,i)}$ as the processing time of job (k,i) on machine j, $1/\lambda^*$ as the makespan, and y as a feasible (fractional) assignment of jobs to machines that achieves this makespan. Suppose that there is some $\rho > 0$ such that $p_{j,(k,i)} \leq \rho/\lambda^*, \forall i, j, k$. Then Theorem 2.1 of [16] implies that their algorithm produces an (integral) assignment of jobs to machines $\hat{y}$ with makespan at most $(1+\rho)/\lambda^*$. This algorithm is the third step of our algorithm.

Our solution assigns $\hat{x}_{j,(k,i)} := \frac{\lambda_{(k,i)} a_k}{c_k} \hat{y}_{j,(k,i)}$ (note that for every (k,i), these values are going to be 0 for all j except one.) It is easy to prove the following:

Lemma 1. *The solution produced by the algorithm above is within* $1/2(1+\rho)$ *of the optimum.*

Proof. The solution $\hat{x}$ satisfies the constraints of (LP-NEW) for $\lambda := \lambda^*/(1+\rho)$. Hence it approximates the maximum concurrent unsplittable flow within a factor of 1/2. Together with the approximation factor of 1/2 from the first step, this implies the lemma.

4.3 The Case $\mu = \alpha \cdot \beta^T$

This case can be generalized to any $M \times K$ matrix μ which is the product of an $M \times 1$ vector α and the transpose of a $K \times 1$ vector β, i.e., $\mu = \alpha \cdot \beta^T$ (in other words, the service rates satisfy $\mu_{j,k} = \alpha_j \beta_k$). Then it is easy to see that the initial problem (MP) is equivalent to

$$\begin{aligned} &\max \lambda \text{ s.t.} \\ &\textstyle\sum_{j=1}^M x_{j,k} \geq \lambda b_k, \; k = 1, \ldots, K \\ &\textstyle\sum_{k=1}^K x_{j,k} \leq \alpha_j, \; j = 1, \ldots, M \\ &x_{j,k} \geq 0, \quad k = 1, \ldots, K, j = 1, \ldots, M \\ &\textstyle\sum_{j=1}^M \chi\{x_{j,k} > 0\} \leq c_k, \quad k = 1, \ldots, K \end{aligned} \qquad \text{(MP}''\text{)}$$

where $x_{j,k} := \alpha_j \delta_{j,k}$, for all j, k, and $b_k := a_k/\beta_k$. (MP$''$) then falls into the case of Section 4.2.

4.4 Extension to TDCCP with Costs

We can extend TDCCP by introducing *costs* to the assignment of servers to classes. Let $c_{j,k}$ be the per unit cost of assigning server j to class k. Hence, if $\delta_{j,k}$

is the fraction of its effort dedicated by j to k, then the cost incurred is $c_{j,k}\delta_{j,k}$. For example, the assignment of a worker to a machine where he has no expertise may incur a bigger cost (because of training needs, damages because of deficient products he produces etc.) than the cost of an experienced worker to the same machine. Together with these costs $c_{j,k}$, we are also given a *budget* that cannot be exceeded by our final assignment. Hence we are asked for an assignment of servers to classes that respect the given budget and maximizes the throughput.

The algorithms of [5], [16] used in Section 4.2 are *cost preserving.* When in the first step we transform the budget-constrained k-splittable flow problem into a budget-constrained exactly-k-splittable flow problem, [5] proves that the optimal solution of the latter is not only an 1/2 approximation of the former, but it also respects the initial budget constraint. Also the algorithm of [16] we use in Section 4.2 produces an assignment that always respects the budgetary constraint (although it may not produce the optimal makespan).[2]

5 Solving TDCCP - General Case

For the general case, let

$$w_j := \frac{\mu_j^{max}}{\mu_j^{min}}, \; j = 1, 2, \ldots, M$$

where $\mu_j^{max} := \max_k\{\mu_{j,k}\}$ and $\mu_j^{min} := \min_k\{\mu_{j,k}\}$. Note that $\mu_{j,k} = 0$ implies that $\delta_{j,k} = 0$, so we will assume that $\mu_{j,k} > 0$ for all j, k. Also, let $w_{max} := \max_j\{w_j\}$, and let δ^*, λ^* be the optimal solution to (MP). Instead of the original problem (MP), we will try to solve (approximately) the following problem:

$$\begin{aligned} &\max \lambda \text{ s.t.} \\ \textstyle\sum_{j=1}^M \mu_{j,k}\delta_{j,k} &\geq \lambda a_k, \quad k = 1, \ldots, K \\ \textstyle\sum_{k=1}^K \mu_{j,k}\delta_{j,k} &\leq \mu_j^{max}, \; j = 1, \ldots, M \\ \delta_{j,k} &\geq 0, \qquad k = 1, \ldots, K, j = 1, \ldots, M \\ \textstyle\sum_{j=1}^M \chi\{\delta_{j,k} > 0\} &\leq c_k, \qquad k = 1, \ldots, K. \end{aligned} \tag{NEW MP}$$

It is clear that, as in Section 4.2, we can set $x_{j,k} := \mu_{j,k}\delta_{j,k}$ in (NEW MP) to get exactly the same formulation as (MP′). Hence we can apply the same techniques we applied in Section 4.2, to obtain an approximate solution $\hat{x}, \hat{\lambda}$, which is within 1/10 of the optimum solution (of (MP′)). Then we output the following solution to the original problem:

$$\delta_{j,k} := \frac{\hat{x}_{j,k}}{w_j \mu_{j,k}}, \; \forall j, k. \tag{5}$$

The following proposition is proven in the full version:

Proposition 2. *Solution (5) is a feasible solution for (MP), and achieves a λ of value at least $\lambda^*/10w_{max}$.*

[2] Obviously the costs in the budgetary constraint in each of the LP formulations above are scaled following the scaling of the assignment variables.

This result extends to the case of a budgetary constraint problem, i.e. the approximation factor can be achieved without violating the (given) budget (cf. Section 4.4.)

6 NP-Completeness

We reduce a slight variation of the classical PARTITION problem (see [SP12] in [10]) to the version of our problem that is studied in Section 4.2, which, by abusing the terminology a little bit, we will call problem (MP′):

MP′
Instance: We are given (MP′) and $\lambda^* \in \mathbb{R}$.
Question: Is the solution of (MP′) greater than or equal to λ^*?

Obviously this problem is in NP (given λ^* and a solution to MP′, one can easily check whether its objective is greater than or equal to λ^*). The PARTITION problem variation we reduce it to is the following:

PARTITION
Instance: Finite set A of even cardinality and a size $s(a) \in \mathbb{Z}^+$ for each item $a \in A$.
Question: Is there a subset $A' \subseteq A$ of cardinality $|A|/2$ and such that $\sum_{a \in A'} s(a) = \sum_{a \in A \setminus A'} s(a)$?

Given the PARTITION instance, we identify the elements of A with the numbers $1, 2, \ldots, |A|$. Let $S := \sum_{j=1}^{|A|} s(j)$ be the total size. We set $K := 2, M := |A|$ and $\mu_j := s(j),\ j = 1, \ldots, |A|$. We also set $c_1 = c_2 := |A|/2$, and $a_1 = a_2 := 1$. Finally we set $\lambda^* := S/2$. Therefore we get an instance of (MP′) in polynomial time. From now on, when we refer to (MP′), we actually refer to this specific instance we constructed. We can prove (proof in the full version) the following

Theorem 2. PARTITION *has a solution iff (MP′) achieves* $\lambda \geq \lambda^*$.

References

1. S. Andradóttir and H. Ayhan. Throughput maximization for tandem lines with two stations and flexible servers. *Operations Research*, 53:516–531, 2005.
2. S. Andradóttir, H. Ayhan, and D.G. Down. Server assignment policies for maximizing the steady-state throughput of finite queueing systems. *Management Science*, 47:1421-1439, 2001.
3. S. Andradóttir, H. Ayhan and D.G. Down. Dynamic server allocation for queueing networks with flexible servers. *Operations Research*, 51:952-968, 2003.
4. M. Armony and N. Bambos. Queueing networks with interacting service resources. *Proceedings of the 37th Annual Allerton Conference on Communications, Control, and Computing*, 42-51, 1999.
5. G. Baier, E. Köhler, and M. Skutella. On the k-splittable flow problem. *Proceedings of ESA'02.*
6. D.P. Bischak. Performance of a manufacturing module with moving workers. *IIE Transactions*, 28:723-733, 1996.

7. J.G. Dai. On positive Harris recurrence of multiclass queueing networks: A unified approach via fluid limit models. *Annals of Applied Probability*, 5:49-77, 1995.
8. J.G. Dai and W. Lin. Maximum pressure policies in stochastic processing networks. *Operations Research*, 53:197–218, 2005.
9. Y. Dinitz, N. Garg, and M. Goemans. On the single-source unsplittable flow problem. *Combinatorica*, 19 (1999), pp. 17–41.
10. M. R. Garey and D. S. Johnson. *Computers and Intractability: a guide to the theory of NP-Completeness.* W. H. Freeman and Co., 1979.
11. M. Grötschel, L. Lovász, and A. Schrijver. *Geometric algorithms and combinatorial optimization*, Chapter 6. Springer-Verlag, 1993.
12. F.S. Hillier and K.C. So. On the simultaneous optimization of server and work allocations in production line systems with variable processing times. *Operations Research*, 44:435-443, 1996.
13. W.J. Hopp and M.P. van Oyen. Agile workforce evaluation: A framework for cross-training and coordination. *IIE Transactions*, 36:919–940, 2004.
14. S. G. Kolliopoulos and C. Stein. Approximation algorithms for single-source unsplittable flow. *SIAM J. Computing* 31:919-946, 2002.
15. A.N. Rybko and A.L. Stolyar. Ergodicity of stochastic processes describing the operation of open queueing networks. *Problems of Information Transmission*, 28:199-220, 1992.
16. D. B. Shmoys and . Tardos. An approximation algorithm for the generalized assignment problem. *Mathematical Programming A*, Vol. 62(3), pp. 461 - 474, 1993.
17. M.S. Squillante, C.H. Xia, D.D. Yao, and L. Zhang. Threshold-based priority policies for parallel-server systems with affinity scheduling. *Proceedings of the 2001 American Control Conference*, 2992-2999, 2001.
18. L. Tassiulas and P.B. Bhattacharya. Allocation of independent resources for maximal throughput. *Stochastic Models*, 16:27-48, 2000.
19. L. Tassiulas and A. Ephrimedes. Stability properties of constrained queueing systems and scheduling policies for maximum throughput in multihop radio networks. *IEEE Transactions on Automatic Control*, 37:1936-1948, 1992.
20. E. Zavadlav, J.O. McClain, and L.J. Thomas. Self-buffering, self-balancing, self-flushing production lines. *Management Science*, 42:1151-1164, 1996.

Network Flow Spanners

Feodor F. Dragan and Chenyu Yan

Department of Computer Science, Kent State University,
Kent, OH 44242, USA
dragan@cs.kent.edu, cyan@cs.kent.edu

Abstract. In this paper, motivated by applications of ordinary (distance) spanners in communication networks and to address such issues as bandwidth constraints on network links, link failures, network survivability, etc., we introduce a new notion of *flow spanner*, where one seeks a spanning subgraph $H = (V, E')$ of a graph $G = (V, E)$ which provides a "good" approximation of the source-sink flows in G. We formulate few variants of this problem and investigate their complexities. A special attention is given to the version where H is required to be a tree.

1 Introduction

Given a graph $G = (V, E)$, a spanning subgraph $H = (V, E')$ of G is called a *spanner* if H provides a "good" approximation of the distances in G. More formally, for $t \geq 1$, H is called a t*–spanner* of G [5, 21, 20] if $d_H(u, v) \leq t \cdot d_G(u, v)$ for all $u, v \in V$, where $d_G(u, v)$ is the distance in G between u and v. Sparse spanners (where $|E'| = O(|V|)$) found a number of applications in various areas; especially, in distributed systems and communication networks. In [21], close relationships were established between the quality of spanners (in terms of stretch factor t and the number of spanner edges $|E'|$), and the time and communication complexities of any synchronizer for the network based on this spanner. Also sparse spanners are very useful in message routing in communication networks; in order to maintain succinct routing tables, efficient routing schemes can use only the edges of a sparse spanner [22]. It is well-known that the problem of determining, for a given graph G and two integers $t, m \geq 1$, whether G has a t-spanner with m or fewer edges, is NP-complete (see [20]).

The sparsest spanners are tree spanners. They occur in biology and can be used as models for broadcast operations. Tree t-spanners were considered in [3]. It was shown that, for a given graph G, the problem to decide whether G has a spanning tree T such that $d_T(u, v) \leq t \cdot d_G(u, v)$ for all $u, v \in V$ is NP–complete for any fixed $t \geq 4$ and is linearly solvable for $t = 1, 2$. For more information on spanners consult [1, 2, 3, 5, 6, 7, 18, 20, 21].

In this paper, motivated by applications of spanners in communication networks and to address such issues as bandwidth constraints on network links, link failures, network survivability, etc., we introduce a new notion of *flow spanner*, where one seeks a spanning subgraph $H = (V, E')$ of a graph G which provides a

J.R. Correa, A. Hevia, and M. Kiwi (Eds.): LATIN 2006, LNCS 3887, pp. 410–422, 2006.

"good" approximation of the source-sink flows in G. We formulate few variants of this problem and investigate their complexities. In this preliminary investigation, a special attention is given to the version where H is required to be a tree.

2 Problem Formulations and Results

A *network* is a 4-tuple $N = (V, E, c, p)$ where $G = (V, E)$ is a connected, finite, and simple graph, c(e) are nonnegative edge *capacities*, and $p(e)$ are nonnegative edge *prices*. We assume that graph G is undirected in this paper, although similar notions can be defined for directed graphs, too. In this case, $c(e)$ indicates the maximum amount of flow edge $e = (v, u)$ can carry (in either v to u direction or in u to v direction), $p(e)$ is the cost that the edge will incur if it carries a non-zero flow. Given a source s and a sink t in G, an *(s,t)-flow* is a function f defined over the edges that satisfies capacity constraints, for every edge, and conservation constraints, for every vertex, except the source and the sink. The net flow that enters the sink t is called the *(s,t)-flow*. Denote by $F_G(s,t)$ the *maximum (s,t)-flow* in G. Note that, since G is undirected, $f(v,u) = -f(u,v)$ for any edge $e = (v, u) \in E$ and $F_G(x,y) = F_G(y,x)$ for any two vertices (source and sink) x and y (by reversing the flow on each edge).

Let $H = (V, E')$ be a subgraph of G, where $E' \subseteq E$. For any two vertices $u, v \in V(G)$, define $flow\_stretch(u,v) = \frac{F_G(u,v)}{F_H(u,v)}$ to be the *flow–stretch factor* between u and v. Define the *flow–stretch factor* of H as $fs_H = \max\{flow\_stretch(u,v) : u, v \in V(G)\}$. When the context is clear, the subscript H will be omitted. Similarly, define the *average flow–stretch factor* of the subgraph H as follows $afs_H = \frac{2}{n(n-1)} \sum_{u,v \in V} \frac{F_G(u,v)}{F_H(u,v)}$.

The general problem, we are interested in, is to find a *light flow–spanner* H of G, that is a spanning subgraph H such that fs_H (or afs_H) is as small as possible and at the same time the total cost of the spanner, namely $\mathcal{P}(H) = \sum_{e \in E'} p(e)$, is as low as possible. The following is the decision version of this problem.

Problem: Light Flow–Spanner

Instance: An undirected graph $G = (V, E)$, non-negative edge capacities $c(e)$, non-negative edge costs $p(e)$, $e \in E(G)$, and two positive numbers t and B.

Output: A light flow–spanner $H = (V, E')$ of G with flow–stretch factor $fs_H \leq t$ and total cost $\mathcal{P}(H) \leq B$, or "there is no such spanner".

We distinguish also few special variants of this problem.

Problem: Sparse Flow–Spanner

Instance: An undirected graph $G = (V, E)$, non-negative edge capacities $c(e)$, unit edge costs $p(e) = 1$, $e \in E(G)$, and two positive numbers t and B.

Output: A sparse flow–spanner $H = (V, E')$ of G with flow–stretch factor $fs_H \leq t$ and $\mathcal{P}(H) = |E'| \leq B$, or "there is no such spanner".

Problem: Sparse Edge-Connectivity–Spanner

Instance: An undirected graph $G = (V, E)$, unit edge capacities $c(e) = 1$, unit edge costs $p(e) = 1$, $e \in E(G)$, and two positive numbers t and B.
Output: A sparse flow–spanner $H = (V, E')$ of G with flow–stretch factor $fs_H \leq t$ and $\mathcal{P}(H) = |E'| \leq B$, or "there is no such spanner".

Note that here the maximum (s,t)-flow in H is actually the maximum number of edge-disjoint (s,t)-paths in H, i.e., the edge-connectivity of s and t in H. Thus, this problem is named the *Sparse Edge-Connectivity–Spanner* problem. Spanning subgraph H provides a "good" approximation of the vertex-to-vertex edge-connectivities in G. The following is the version of this Edge-Connectivity Spanner problem with arbitrary costs on edges.

Problem: Light Edge-Connectivity–Spanner

Instance: An undirected graph $G = (V, E)$, unit edge capacities $c(e) = 1$, arbitrary non-negative edge costs $p(e)$, $e \in E(G)$, and two positive numbers t and B.
Output: A light flow–spanner $H = (V, E')$ of G with flow–stretch factor $fs_H \leq t$ and total cost $\mathcal{P}(H) \leq B$, or "there is no such spanner".

In Section 4, using a reduction from the 3-dimensional matching problem, we show that the Sparse Edge-Connectivity–Spanner problem is NP-complete, implying that all other three problems are NP-complete as well.

Replacing in all four formulations "$fs_H \leq t$" with "$afs_H \leq t$", we obtain four more variations of the problem: *Light Average Flow–Spanner*, *Sparse Average Flow–Spanner*, *Sparse Average Edge-Connectivity–Spanner* and *Light Average Edge-Connectivity–Spanner*, respectively. These four problems are topics of our current investigations.

In Section 5, we investigate two simpler variants of the problem: *Tree Flow–Spanner* and *Light Tree Flow–Spanner* problems.

Problem: Tree Flow–Spanner

Instance: An undirected graph $G = (V, E)$, non-negative edge capacities $c(e)$, $e \in E(G)$, and a positive number t.
Output: A *tree t-flow–spanner* $T = (V, E')$ of G, that is a spanning tree T of G with flow–stretch factor $fs_T \leq t$, or "there is no such tree spanner".

Problem: Light Tree Flow–Spanner

Instance: An undirected graph $G = (V, E)$, non-negative edge capacities $c(e)$, non-negative edge costs $p(e)$, $e \in E(G)$, and two positive numbers t and B.
Output: A *light tree t-flow–spanner* $T = (V, E')$ of G, that is a spanning tree T of G with flow–stretch factor $fs_T \leq t$ and total cost $\mathcal{P}(T) \leq B$, or "there is no such tree spanner".

In a similar way one can define also the *Tree Average Flow–Spanner* and *Light Tree Average Flow–Spanner* problems. Notice that our tree t-flow-spanners are different from the well-known *Gomory-Hu trees* [14]. Gomory-Hu trees represent the structure of all s-t maximum flows of undirected graphs in a compact way, but they are not necessarily spanning trees.

We show that the Tree Flow–Spanner problem has easy polynomial time solution while the Light Tree Flow–Spanner problem is NP-complete. In Section 6, we propose two approximation algorithms for the Light Tree Flow–Spanner problem.

3 Related Work

In [11], a network design problem, called *smallest k-ECSS problem* is considered, which is close to our Sparse Edge-Connectivity–Spanner problem. In that problem, given a graph G along with an integer k, one seeks a spanning subgraph H of G that is k-edge-connected and contains the fewest possible number of edges. The problem is known to be MAX SNP-hard [9], and the authors of [11] give a polynomial time algorithm with approximation ratio $1 + 2/k$ (see also [4] for an earlier approximation result). It is interesting to note that a sparse k-edge-connected spanning subgraph (with $O(k|V|)$ edges) of a k-edge-connected graph can be found in linear time [19]. In our Sparse Edge-Connectivity–Spanner problem, instead of trying to guarantee the k-edge-connectedness in H for all vertex pairs, we try to closely approximate by H the original (in G) levels of edge-connectivities.

Paper [12] deals with the *survivable network design problem (SNDP)* which can be considered as a generalization of our Light Edge-Connectivity–Spanner problem. In SNDP, we are given an undirected graph $G = (V, E)$, a non-negative cost $p(e)$ for every edge $e \in E$ and a non-negative connectivity requirement r_{ij} for every (unordered) pair of vertices i, j. One needs to find a minimum-cost subgraph in which each pair of vertices i, j is joined by at least r_{ij} edge-disjoint paths. The problem is NP-complete since the Steiner Tree Problem is a special case, and [13] gives an efficient approximate solution. If connectivity requirements are at most k (for some integer k), then a solution found is within a factor $2\mathcal{H}(k) = 2(1 + \frac{1}{2} + \frac{1}{3} + \ldots + \frac{1}{k})$ of optimal. See also [10, 12, 16, 24] for some earlier results. By setting $r_{ij} := \lceil F_G(i, j)/t \rceil$ for each pair of vertices i, j, our Light Edge-Connectivity–Spanner problem (with given flow–stretch factor t) can be reduced to SNDP.

Another related problem, which deals with the maximum flow, is investigated in [8, 17]. In that problem, called *MaxFlowFixedCost*, given a graph $G = (V, E)$ with non-negative capacities $c(e)$ and non-negative costs $p(e)$ for each edge $e \in E$, a source s and a sink t, and a positive number B, one must find an edge subset $E' \subseteq E$ of total cost $\sum_{e \in E'} p(e) \leq B$, such that in spanning graph $H = (V, E')$ of G the flow from s to t is maximized. Paper [8] shows that this problem, even with uniform edge-prices, does not admit a $2^{log^{1-\epsilon} n}$-ratio approximation for any constant $\epsilon > 0$ unless $NP \subseteq DTIME(n^{polylog\ n})$. In [17], a polynomial time F^*-approximation algorithm for the problem is presented, where F^* denotes the maximum total flow. In our Sparse Flow–Spanner problem we require from spanning subgraph H to approximate maximum flows for all vertex pairs simultaneously.

To the best of our knowledge our spanner-like all-pairs problem formulations are new.

4 Hardness of the Flow–Spanner Problems

This section is devoted to the proof of the NP-completeness of the Sparse Edge-Connectivity–Spanner problem and other Flow–Spanner problems.

Theorem 1. *Sparse Edge-Connectivity–Spanner problem is NP-complete.*

Proof. It is obvious that the problem is in NP. To prove its NP-hardness, we will reduce the 3-dimensional matching (3DM) problem to this one, by extending a reduction idea from [10].

Let $M \subseteq W \times X \times Y$ be an instance of 3DM, with $|M| = p$ and $W = \{w_i | i = 1, 2, \cdots, q\}, X = \{x_i | i = 1, \cdots, q\}$ and $Y = \{y_i | i = 1, \cdots, q\}$. One needs to check if M contains a matching, that is, a subset $M' \subseteq M$ such that $|M'| = q$ and no two triples of M' share a common element from $W \cup X \cup Y$.

Define $Deg(a)$ to be the number of triples in M that contain a, $a \in W \cup X \cup Y$. We construct a graph $G = (V, E)$ as follows (see Fig. 1). For each triple $(w_i, x_j, y_k) \in M$, there are four corresponding vertices $a_{ijk}, \overline{a}_{ijk}, d_{ijk}$ and $\overline{d}_{ijk}$ in V. d_{ijk} and $\overline{d}_{ijk}$ are called *dummy vertices.* Denote $D := \{d_{ijk} | (w_i, x_j, y_k) \in M\}$, $\overline{D} := \{\overline{d}_{ijk} | (w_i, x_j, y_k) \in M\}$, $A := \{a_{ijk} | (w_i, x_j, y_k) \in M\}$, $\overline{A} := \{\overline{a}_{ijk} | (w_i, x_j, y_k) \in M\}$. Additionally, for each $a \in X \cup Y$, we define a vertex a and $2Deg(a) - 1$ dummy vertices $d_1(a), \cdots, d_{2Deg(a)-1}(a)$ of a. For each $w_i \in W$, we define a vertex w_i and $4Deg(w_i) - 3$ dummy vertices $d_1(w_i), \cdots, d_{4Deg(w_i)-3}(w_i)$

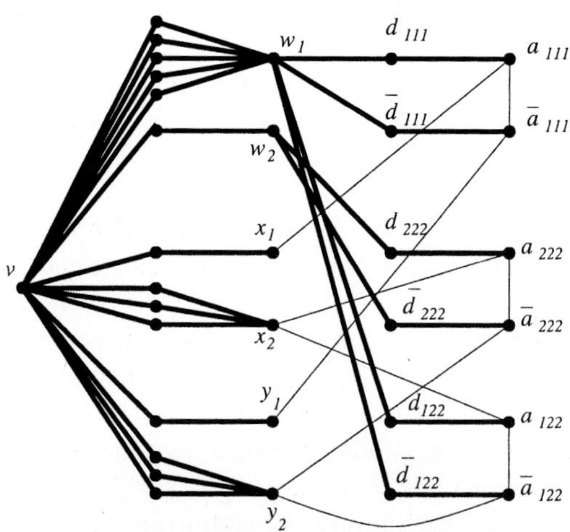

Fig. 1. Graph created for a 3DM instance: $M = \{(w_1, x_1, y_1), (w_2, x_2, y_2), (w_1, x_2, y_2)\}$, $W = (w_1, w_2), X = (x_1, x_2)$ and $Y = (y_1, y_2)$. The edges from E_d are shown in bold.

of w_i. There is an extra vertex v in V. Let N_d be the dummy vertices (note that $D, \overline{D} \subset N_d$). So, the vertex set V of G is $V = \{v\} \cup W \cup X \cup Y \cup A \cup \overline{A} \cup N_d$.

For each dummy vertex $d_i(a) \in N_d$ $(a \in W \cup X \cup Y)$ put $(a, d_i(a)), (v, d_i(a))$ into E_d. Also put $(w_i, d_{ijk}), (d_{ijk}, a_{ijk}), (w_i, \overline{d}_{ijk}), (\overline{d}_{ijk}, \overline{a}_{ijk})$ into E_d. Now, the edge set E of G is $E = E_d \cup \{(a_{ijk}, \overline{a}_{ijk}), (a_{ijk}, x_j), (\overline{a}_{ijk}, y_k) | (w_i, x_j, y_k) \in M\}$. This completes the description of $G = (V, E)$. Clearly, each dummy vertex has exactly two neighbors in G, and each vertex of $A \cup \overline{A}$ has exactly 3 neighbors in G. Also, each w_i has $4Deg(w_i) - 3 + 2Deg(w_i) = 6Deg(w_i) - 3$ neighbors and each $a \in X \cup Y$ has $2Deg(a) - 1 + Deg(a) = 3Deg(a) - 1$ neighbors in G.

Set $t = 3/2$ and $B = |E_d| + p + q$. We claim that M contains a matching M' if and only if G has a flow–spanner $H = (V, E')$ with flow–stretch factor $\leq t$ and with B edges. Proof of this claim is presented in the journal version. □

Corollary 1. *The Light Flow–Spanner, the Sparse Flow–Spanner and the Light Edge-Connectivity–Spanner problems are NP-complete.*

5 Tree Flow–Spanners

In this section, we show that the Light Tree Flow–Spanner problem is NP-complete while the Tree Flow–Spanner problem can be solved efficiently by any Maximum Spanning Tree algorithm.

Theorem 2. *The Light Tree Flow–Spanner problem is NP-complete.*

Proof. The problem is obviously in NP. One can non-determenistically choose a spanning tree and test in polynomial time whether it satisfies the cost and the flow–stretch bounds. To prove its NP-hardness, we will reduce the 3SAT problem to this one.

Let x_i be a variable in the 3SAT instance. Without loss of generality, assume that the 3SAT instance does not have clause of type $(x_i \vee \overline{x}_i \vee x_j)$ (note j may be equal to i). Since such a clause is always true no matter what value x_i gets, it can be eliminated without affecting the satisfiability.

From a 3SAT instance one can construct a graph $G = (V, E)$ as follows. Let $x_1, x_2, \cdots, x_n$ be the variables and $C_1, \cdots, C_q$ be the clauses of 3SAT. Let k_i be the number of clauses containing either literal x_i or literal $\overline{x}_i$. Create $2k_i$ vertices for each variable x_i in G. Denote those vertices by $V(x_i) = \{x_i^1, x_i^2, \cdots, x_i^{k_i}\}$ and $\overline{V}(x_i) = \{\overline{x}_i^1, \cdots, \overline{x}_i^{k_i}\}$. All these vertices are called *variable vertices.* Put an edge $(x_i^l, \overline{x}_i^l)$ into $E(G)$, for $1 \leq l \leq k_i$. Set $p(x_i^l, \overline{x}_i^l) = c(x_i^l, \overline{x}_i^l) = 1$. For each integer l, where $1 \leq l < k_i$, put (x_i^l, x_i^{l+1}) and $(\overline{x}_i^l, \overline{x}_i^{l+1})$ into $E(G)$ and set their prices and capacities to be 2.

For each clause C_j, create a *clause vertex* C_j in G. At the beginning, mark all the vertices corresponding to the variables as "free". Do the following for $j = 1, 2, \ldots, q$. If x_i (or $\overline{x}_i$) is in C_j, then find the smallest integer l such that x_i^l (or $\overline{x}_i^l$) is "free" and put (C_j, x_i^l) $((C_j, \overline{x}_i^l)$, respectively) into $E(G)$. Mark x_i^l and $\overline{x}_i^l$ as "busy". Set $c(C_j, x_i^l) = p(C_j, x_i^l) = 3$ (respectively, $c(C_j, \overline{x}_i^l) = p(C_j, \overline{x}_i^l) = 3$).

Graph G has also an extra vertex v. For each variable x_i, put edges (v, x_i^1) and $(v, \overline{x}_i^1)$ into $E(G)$. Set their prices and capacities to 2. This completes the description of G. Obviously, the transformation can be done in polynomial time.

For each variable x_i, let H_i be the subgraph of G induced by vertices $\{v, x_i^1, \cdots, x_i^{k_i}, \overline{x}_i^1, \cdots, \overline{x}_i^{k_i}\}$. Name all the edges with capacity 2 *assignment edges*, the edges with capacity 1 *connection edges* and the edges with capacity 3 *consistent edges.* The path $(v, x_i^1, x_i^2, \cdots, x_i^{k_i})$ is called *positive path* of H_i and the path $(v, \overline{x}_i^1, \cdots, \overline{x}_i^{k_i})$ is called *negative path* of H_i.

Let $N = k_1 + k_2 + \cdots + k_n$. Set $B = 3N + 3q$ and $fs_T = 8$. We need to show that the 3SAT is satisfiable if and only if the graph G has a tree flow–spanner with total cost less than or equal to B and flow–stretch factor at most 8. Here, we prove the "only if" direction. A proof for the "if" direction is presented in the journal version.

Let T be a tree flow–spanner of G such that $fs_T \leq 8$ and $\sum_{e \in E(T)} p(e) \leq B$. Obviously, T must have at least q consistent edges. Assume T has r assignment edges, s connection edges and $t+q$ consistent edges. Clearly, $r, s, t \geq 0$ and, since T has $2N + q$ edges (because G has $2N + q + 1$ vertices), $r + s + t = 2N$. From $\sum_{e \in E(T)} p(e) \leq B = 3N + 3q$ we conclude also that $2r + s + 3t \leq 3N$. Hence, $2r + s + 3t - 2(r + s + t) \leq -N$, i.e., $t \leq s - N$. If $s < N$, then $t < 0$, which is impossible. Therefore, T must include all N connection edges of G, implying $s = N$ and $r + t = N$, $2r + 3t \leq 2N$. From $2r + 3t - 2(r + t) \leq 0$ we conclude that $t \leq 0$. So, t must be 0, and therefore, T contains exactly q consistent edges, exactly N assignment edges and all N connection edges. This implies that, for

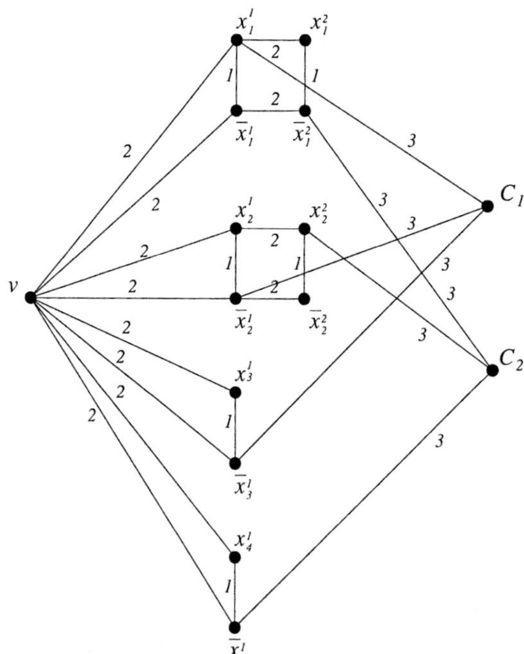

Fig. 2. Graph created from expression $(x_1 \vee \overline{x}_2 \vee \overline{x}_3) \wedge (\overline{x}_1 \vee x_2 \vee \overline{x}_4)$

every variable x_i, exactly one edge from $\{(x_i^1, v), (\overline{x}_i^1, v)\}$ is in $E(T)$. Since in T each clause vertex must be adjacent to at least one variable vertex and there are q consistent edges in T, each clause vertex is a pendant vertex of T (is adjacent in T to exactly one variable vertex). By construction of G, for each variable vertex x_i^l, any path between x_i^l and v in G either totally lies in H_i or has to use at least one clause vertex. Since all clause vertices are pendant in T, the path between x_i^l and v in T must totally lie in H_i. Similarly, the path between $\overline{x}_i^l$ and v in T must totally lie in H_i.

Now, we show how to assign true/false to the variables of the 3SAT instance to satisfy all its clauses. For each variable x_i, if $(x_i^1, v) \in E(T)$ then assign true to x_i, otherwise assign false to x_i. We claim that, if a clause vertex C_j is adjacent to a variable vertex x_i^l (or to a variable vertex $\overline{x}_i^l$) in T, then x_i is assigned true (false, respectively). The claim can be proved by contradiction. Assume x_i is assigned false, i.e., $(\overline{x}_i^1, v) \in E(T)$ and $(x_i^1, v) \notin E(T)$, but C_j is adjacent to a variable vertex x_i^l in T. As it was mentioned in the previous paragraph, the path $P_T(x_i^l, v)$ between x_i^l and v in T must totally lie in H_i. Since $(x_i^1, v) \notin E(T)$, edge (x_i^1, v) cannot be in $P_T(x_i^l, v)$. By construction of H_i, any path in H_i from x_i^l to v not using edge (x_i^1, v) must contain at least one connection edge. This means that the path $P_T(C_j, v)$ contains at least one connection edge, too. Since all connection edges have capacity 1, $F_T(C_j, v) = 1$. On the other hand, $F_G(C_j, v) = 9$. Hence, $flow\_\ stretch(C_j, v) = 9 > 8$, contradicting with $fs_T \leq 8$. This contradiction proofs the claim. Now, since every clause contains at least one true literal (note $(x_i^l, C_j) \in E(G)$ implies clause C_j contains x_i), the 3SAT instance is satisfiable.

This completes the proof of the theorem. □

Let $G = (V, E)$ be graph of an instance of the Light Tree Flow–Spanner problem. Let c^* be the maximum edge capacity of G and c_* be the minimum edge capacity of G. Note that, if $\frac{c^*}{c_*} = 1$, then the Light Tree Flow–Spanner problem can be solved in polynomial time by simply finding a minimum spanning tree T_p of G, where the weight of an edge $e \in E(G)$ is $p(e)$. From the proof of Theorem 2, one concludes that when $\frac{c^*}{c_*} \geq 3$, the Light Tree Flow–Spanner problem is NP-complete.

Now we turn to the Tree Flow-Spanner problem on a graph $G = (V, E)$ (recall that in this problem $p(e) = 1$ for any $e \in E$).

Lemma 1. *Let T_c be a maximum spanning tree of a graph G (with edge weights $c(\cdot)$) and T be an arbitrary spanning tree of G. Then, for any two vertices $u, v \in V(G)$, $F_{T_c}(u, v) \geq F_T(u, v)$.*

Lemma 1 implies that a maximum spanning tree T_c of a graph G, where the edge capacities are interpreted as edge weights, is an optimal tree flow–spanner of G. Hence, the following theorem holds.

Theorem 3. *Given an undirected graph $G = (V, E)$, with non-negative capacities on edges, and a number $t > 0$, whether G admits a tree flow–spanner with flow–stretch factor at most t can be determined in polynomial time (by any maximum spanning tree algorithm).*

6 Approximation Algorithms

In this section, we present some approximation algorithms for the Light Tree Flow–Spanner problem. Let $G = (V, E)$ be an undirected graph with non-negative edge capacities $c(e)$ and non-negative edge costs $p(e)$, $e \in E(G)$. For given two positive numbers t and B we want to check if a spanning tree T^* of G with flow–stretch factor $fs_{T^*} \leq t$ and total cost $\mathcal{P}(T^*) \leq B$ exists or not. If such a tree exists then we say that the Light Tree Flow–Spanner problem on G has a solution. We will say that a spanning tree T of a graph G gives an *(α, β)-approximate solution* to the Light Tree Flow–Spanner problem on G if the inequalities $fs_T \leq \alpha t$ and $\mathcal{P}(T) \leq \beta B$ hold for T. A polynomial time algorithm producing an (α, β)-approximate solution to any instance of the Light Tree Flow–Spanner problem admitting a solution is called an *(α, β)-approximation algorithm* for the Light Tree Flow–Spanner problem.

Lemma 2. *If $\frac{c^*}{c_*} \leq k$, where $c^* := max\{c(e) : e \in E\}$ and $c_* := min\{c(e) : e \in E\}$, then there is a $(k, 1)$-approximation algorithm for the Light Tree Flow–Spanner problem.*

This result will be used in our main approximation algorithm. Let $G = (V, E)$ be an undirected graph with non-negative edge capacities $c(e)$ and non-negative edge costs $p(e)$, $e \in E(G)$. Assume that G has a spanning tree T^* with $fs_{T^*} \leq t$ and $\mathcal{P}(T^*) \leq B$. In what follows, we describe a polynomial time algorithm which, given a parameter (any real number) r larger than 1 and smaller than t $(1 < r \leq t - 1)$, produces a spanning tree T of G such that $fs_T \leq r(t-1)t$ and $\mathcal{P}(T) \leq 1.55 \log_r(r(t-1))B$. Thus, it is an $(r(t-1), 1.55 \log_r(r(t-1)))$-approximation algorithm for the Light Tree Flow–Spanner problem.

Assume that the edges of G are ordered in a non-decreasing order of their capacities, i.e., we have an ordering $e_1, e_2, \cdots, e_m$ of the edges of G such that $c(e_1) \leq c(e_2) \cdots \leq c(e_m)$. Let $1 < r \leq t - 1$. If $c(e_m)/c(e_1) \leq r(t-1)$, then Lemma 2 suggests to construct a minimum spanning tree of G using $p(e)$s as the edge weights. This tree is an $(r(t-1), 1)$-approximate solution, and hence we are done. Assume now that $c(e_m)/c(e_1) > r(t-1)$. We cluster all the edges of G into groups as follows. First group consists of all the edges whose capacities are in the range $[l_1 = c(e_m)/r, h_1 = c(e_m)]$. Then, we find the largest capacity $c(e_i)$ such that $c(e_i) < c(e_m)/r$ and form the second group of edges. It consists of all edges whose capacities are in the range $[l_2 = c(e_i)/r, h_2 = c(e_i)]$. We continue this process until a group of edges whose capacities are in the range $[l_k, h_k]$ with $c(e_1) \geq l_k$ is formed.

Let $G_i = (V, E_i)$ be a subgraph of G formed by $E_i = \{e \in E(G) : l_i \leq c(e) \leq h_1\}$. Let $G_1^i, G_2^i, \cdots, G_{p_i}^i$ be those connected components of G_i which contain at least two vertices. Consider another subgraph $G_i' = (V, E_i')$ of G formed by $E_i' = \{e \in E(G) : h_i/(r(t-1)) \leq c(e) \leq h_1\}$. $G_1'^i, G_2'^i, \cdots, G_{q_i}'^i$ are used to denote those connected components of G_i' which contain at least two vertices.

Let $u, v \in V(G)$ be two arbitrary vertices. Choose the minimum i such that u and v are connected in G_i and let G_j^i be the connected component of G_i which contains u and v. Let $G'^{i}_{j'}$ be the connected component of G'_i such that $G^i_j \subseteq G'^{i}_{j'}$ (clearly, such a connected component exists). The following lemma holds (proof is presented in the journal version).

Lemma 3. *If G has a tree flow–spanner T^* with flow–stretch factor $\leq t$, then the path $P_{T^*}(u,v)$ connecting u and v in T^* must totally lie in $G'^{i}_{j'}$.*

From Lemma 3, our approximation algorithm for the Light Tree Flow–Spanner problem is obvious.

PROCEDURE 1. Construct a light tree flow–spanner for a graph G

Input: An undirected graph G with non-negative edge capacities $c(e)$ and non-negative edge costs $p(e)$, $e \in E(G)$; positive real numbers t and $1 < r \leq t-1$.
Output: A spanning tree T of G.

Method:
set $G_f := (V, E_f)$, where $E_f = \{e \in E(G) : p(e) = 0\}$;
for $i = 1$ **to** k **do**
let $G_i := (V, E_i)$ be a subgraph of G with $E_i := \{e \in E(G) : l_i \leq c(e) \leq h_1\}$;
let $G^i_1, \cdots, G^i_{p_i}$ be those conn. comp. of G_i which contain at least two vertices;
let $G'_i := (V, E'_i)$ be a subgraph of G with $E'_i := \{e \in E(G) : \frac{h_i}{r(t-1)} \leq c(e) \leq h_1\}$;
let $G'^{i}_1, \cdots, G'^{i}_{q_i}$ be those conn. comp. of G'_i which contain at least two vertices;
set $V_t := \bigcup_{1 \leq j \leq p_i} V(G^i_j)$;
in each conn. comp. G'^{i}_j $(1 \leq j \leq q_i)$, find an approximate minimum weight Steiner tree T'^{i}_j where terminals are $V(G'^{i}_j) \cap V_t$ and $p(e)$s are the edge weights;
set $E_f := E_f \bigcup \{\bigcup_{1 \leq j \leq q_i} \{e \in E(T'^{i}_j) : p(e) > 0\}\}$;
for each edge $e \in \bigcup_{1 \leq j \leq p_i} E(G^i_j)$, set $p(e) := 0$;
find a maximum spanning tree T of G_f using the capacities as the edge weights;
return T.

Below, the quality of the tree T constructed by above procedure is analyzed.

Lemma 4. *If G admits a tree t-flow–spanner, then $fs_T \leq r(t-1)t$.*

Proof. Let $u, v \in V(G)$ be two arbitrary vertices and T^* be a tree t-flow–spanner of G. Choose the smallest integer i such that u and v are connected in G_i. Let $P_G(u,v)$ be an arbitrary path between u and v in G and $e \in P_G(u,v)$ be an edge on the path with smallest capacity. By the choice of i, we have $c(e) \leq h_i$.

Without loss of generality, assume $u, v \in G^i_j$. According to Procedure 1, u and v will be connected by a path $P_{T'^{i}_j}(u,v)$ in T'^{i}_j. Let $e' \in P_{T'^{i}_j}(u,v)$ be an edge with minimum capacity in $P_{T'^{i}_j}(u,v)$. It is easy to see that $c(e') \geq h_i/(r(t-1))$.

We claim that after iteration i, there is a path $P_{G_f}(u,v)$ between u and v in G_f such that for any edge $e \in P_{G_f}(u,v)$, the inequality $c(e) \geq h_i/(r(t-1))$ holds. We prove this claim by induction on i. All edges of $P_{T'^{i}_j}(u,v)$ with current $p(e)$ greater than 0 are added to E_f. E_f contains also each edge for which original $p(e)$ was 0. Therefore, if G_f does not contain an edge $e = (a,b) \in E(P_{T'^{i}_j}(u,v))$, then current $p(e)$ of e was 0, and this implies $c(e) > h_i$. According to Procedure 1, a, b must be in a connected component of G_l where $1 \leq l < i$. Hence, by induction, at lth iteration, a and b must be connected by a path $P_{G_f}(a,b)$ such that, for each edge $e \in P_{G_f}(a,b)$, the inequality $c(e) \geq h_l/(r(t-1)) > h_i/(r(t-1))$ holds. By concatenating such paths and the edges put into G_f during ith iteration, one can find a path between u and v which satisfies the claim.

Since T is a maximum spanning tree of G_f (where the edge weights are their capacities), similarly to the proof of Lemma 1, one can show that for any edge $e \in P_T(u,v)$, $c(e) \geq h_i/(r(t-1))$ holds. This implies $F_{T^*}(u,v) \leq h_i \leq r(t-1)F_T(u,v)$. Since T^* has flow–stretch factor $\leq t$, we have $F_G(u,v) \leq tF_{T^*}(u,v)$, and therefore $\frac{F_G(u,v)}{F_T(u,v)} \leq r(t-1)t$. This concludes our proof. □

Lemma 5. *If G has a tree t-flow–spanner T^* with cost $\mathcal{P}(T^*)$, then $\mathcal{P}(T) \leq 1.55 \log_r(r(t-1))\mathcal{P}(T^*)$.*

Proof. By Lemma 3, one knows that for any two vertices u, v of G^i_j, $P_{T^*}(u,v)$ totally lies in $G'^i_{j'}$ where $G^i_j \subseteq G'^i_{j'}$. Hence, the smallest subtree of T^* spanning all vertices of $V_t \cap G'^i_{j'}$ is totally contained in $G'^i_{j'}$. We can use in Procedure 1 an 1.55-approximation algorithm of Robins and Zelikovsky [23] to construct an approximation to a minimum weight Steiner tree in $G'^i_{j'}$ spanning terminals $V_t \cap V(G'^i_{j'})$. It is easy to see that $\mathcal{P}_i(G_f) \leq 1.55\ \mathcal{P}_i(T^*)$, where $\mathcal{P}_i(G_f)$ is the total cost of the Steiner trees constructed by Procedure 1 on ith iteration and $\mathcal{P}_i(T^*)$ is the total cost of the edges from T^* which have capacities in the range $[h_i/(r(t-1)), h_i]$ and are used to connect vertices in V_t. Therefore, $\mathcal{P}(G_f) \leq \sum_{1\leq i\leq k} \mathcal{P}_i(G_f) \leq 1.55 \sum_{1\leq i\leq k} \mathcal{P}_i(T^*)$. We will prove that $\sum_{1\leq i\leq k} \mathcal{P}_i(T^*) \leq \log_r(r(t-1))\mathcal{P}(T^*)$. To see this, we show that each edge of T^* appears at most l times in $\sum_{1\leq i\leq k} \mathcal{P}_i(T^*)$, where $\frac{1}{r^l} \geq \frac{1}{r(t-1)}$. Then $l \leq \log_r(r(t-1))$ will follow.

Consider an edge $e \in G'_i$ with $p(e) \neq 0$. We have $h_i/(r(t-1)) \leq c(e) \leq h_i$. According to Procedure 1, after ith iteration, all the edges with capacity in $[h_i/r, h_i]$ have 0 cost. After $(i+1)$th iteration, all the edges with capacity in $[h_i/r^2, h_i]$ have 0 cost. After $(i+l-1)$th iteration, all the edges with capacity in $[h_i/r^l, h_i]$ have 0 cost. To have $p(e) > 0$, the inequality $h_i/r^l \geq h_i/(r(t-1))$ must hold. So, $l \leq \log_r(r(t-1))$ and therefore $\mathcal{P}(G_f) \leq 1.55\ \log_r(r(t-1))\ \mathcal{P}(T^*)$. Since T is a spanning tree of G_f, the lemma clearly follows. □

In the remaining part, we describe how to get a tree flow–spanner T of G with flow–stretch factor $\leq t$ and total cost at most $(n-1)\mathcal{P}(T^*)$, provided G has a tree t-flow–spanner T^*. The algorithm is as follows.

PROCEDURE 2. Construct a light tree t-flow–spanner for a graph G

Input: An undirected graph G with non-negative edge capacities $c(e)$ and non-negative edge costs $p(e)$, $e \in E(G)$; a positive real number t.
Output: A tree t-flow–spanner T of G.

Method:
set $G_f := (V_f, E_f)$, where $V_f = V, E_f = \emptyset$;
construct a complete graph $G' = (V, E')$;
for each $(u,v) \in E'$, let $w(u,v) := F_G(u,v)$ be the weight of the edge;
construct a maximum spanning tree T' of the weighted graph G';

for each edge $(u, v) \in E(T')$ **do**
  let $G_{w(u,v)}$ be a subgraph of G obtained from G by removing all edges e with $c(e) < w(u,v)/t$;
  find a connected component $G_{u,v}$ of $G_{w(u,v)}$ such that $u, v \in V(G_{u,v})$;
  if we cannot find such a connected component, then
    return "G does not have any flow tree t-spanner";
  find a shortest (w.r.t. the costs of the edges) path $P_{G_{u,v}}(u, v)$ between u and v;
  set $E_f := E_f \cup E(P_{G_{u,v}}(u, v))$;
find a maximum spanning tree T of G_f using the edge capacities as their weights;
return T.

The following lemma is true (proof is presented in the journal version).

Lemma 6. *$fs_T \leq t$ and $\mathcal{P}(T) \leq (n-1)\ \mathcal{P}(T^*)$.*

Summarizing the discussion of this section, we state

Theorem 4. *There exist $(r(t-1), 1.55 \log_r(r(t-1)))$-approximation and $(1, n-1)$-approximation algorithms for the Light Tree Flow–Spanner problem.*

References

1. I. Althöfer, G. Das, D. Dobkin, D. Joseph, and J. Soares, On sparse spanners of weighted graphs, *Discrete Comput. Geom.*, 9 (1993), 81–100.
2. S. Baswana and S. Sen, A simple linear time algorithm for computing a $(2k-1)$-spanner of $o(n^{1+1/k})$ size in weighted graphs, *ICALP'03*, LNCS 2719, pp. 384–396.
3. L. Cai and D.G. Corneil, Tree spanners, *SIAM J. Discr. Math.*, 8 (1995), 359–387.
4. J. Cheriyan and R. Thurimella, Approximating Minimum-Size k-Connected Spanning Subgraphs via Matching, *SIAM J. Comput.*, 30 (2000), 528-560.
5. L.P. Chew, There are planar graphs almost as good as the complete graph, *J. of Computer and System Sciences,* 39 (1989), 205–219.
6. M. Elkin and D. Peleg, $(1+\epsilon, \beta)$-spanner constructions for general graphs, *STOC'01*, pp. 173–182, 2001.
7. Y. Emek and D. Peleg, Approximating Minimum Max-Stretch spanning Trees on unweighted graphs, *SODA'04*, pp. 261-270, 2004.
8. G. Even, G. Kortsarz, and W. Slany, On network design problems: fixed cost flows and the Covering Steiner Problem, to appear in *Transactions on Algorithms.*
9. C.G. Fernandes, A Better Approximation Ratio for the Minimum Size k-Edge-Connected Spanning Subgraph Problem, *J. Algorithms,* 28 (1998), 105-124.
10. G.N. Frederickson and J. JáJá, Approximation algorithms for several graph augmentation problems, *SIAM Journal on Computing,* 10 (1981), 270-283.
11. H.N. Gabow, M.X. Goemans, E. Tardos, and D.P. Williamson, Approximating the smallest k-edge connected spanning subgraph by LP-rounding, *SODA 2005*, pp. 562–571, 2005.
12. H.N. Gabow, M.X. Goemans, D.P. Williamson, An efficient approximation algorithm for the survivable network design problem, *Math. Program.*, 82 (1998), 13-40.
13. M.X. Goemans, A.V. Goldberg, S.A. Plotkin, D.B. Shmoys, É. Tardos, D.P. Williamson, Improved Approximation Algorithms for Network Design Problems, *SODA 1994*, 223–232.

14. R.E. Gomory, T.C. Hu, Multi-terminal network flows, *J. SIAM*, 9 (1961), 551–570.
15. R. Hassin and A. Levin, Minimum restricted diameter spanning trees, *Proc.* 5th *Int. Workshop on Approx. Algorithms for Combinatorial Optimization,* LNCS 2462, 2002, 175-184.
16. S. Khuller and U. Vishkin, Biconnectivity Approximations and Graph Carvings, *STOC'92*, pp. 759-770, 1992.
17. S.O. Krumke, H. Noltemeier, S. Schwarz, H.-C. Wirth, and R. Ravi, Flow Improvement and Network Flows with Fixed Costs. *OR'98*, Springer, 1998. ftp://www.mathematik.uni-kl.de/pub/scripts/krumke/or98-flow.pdf
18. A.L. Liestman, T. Shermer, Additive graph spanners, *Networks,* 23(1993), 343-364.
19. H. Nagamochi, T. Ibaraki, A Linear-Time Algorithm for Finding a Sparse k-Connected Spanning Subgraph of a k-Connected Graph, *Algorithmica*, 7 (1992), 583–596.
20. D. Peleg and A.A. Schäffer, Graph Spanners, *J. Graph Theory*, 13 (1989), 99-116.
21. D. Peleg and J.D. Ullman, An optimal synchronizer for the hypercube, In *Proc.* 6th *ACM SPDC,* Vancouver, 1987, 77–85.
22. D.Peleg and E.Upfal, A tradeoff between space and efficiency for routing tables, *STOC'98,* 43-52, 1988.
23. G. Robins and A. Zelikovsky, Improved Steiner tree approximation in graphs, *SODA'00*, 770–779, 2000.
24. D.P Williamson, M.X. Goemans, M. Mihail, and V.V. Vazirani, A primal-dual approximation algorithm for generalized Steiner network problems, *STOC'93*, pp. 708–717, 1993.

Finding All Minimal Infrequent Multi-dimensional Intervals

Khaled M. Elbassioni

Max-Planck-Institut für Informatik, Saarbrücken, Germany
elbassio@mpi-sb.mpg.de

Abstract. Let $\mathcal{D}$ be a database of transactions on n attributes, where each attribute specifies a (possibly empty) real closed interval $I = [a, b] \subseteq \mathbb{R}$. Given an integer threshold t, a multi-dimensional interval $I = ([a_1, b_1], \ldots, [a_n, b_n])$ is called *t-frequent*, if (every component interval of) I is contained in (the corresponding component of) at least t transactions of $\mathcal{D}$ and otherwise, I is said to be *t-infrequent*. We consider the problem of generating all *minimal* t-infrequent multi-dimensional intervals, for a given database $\mathcal{D}$ and threshold t. This problem may arise, for instance, in the generation of association rules for a database of time-dependent transactions. We show that this problem can be solved in quasi-polynomial time. This is established by developing a quasi-polynomial time algorithm for generating maximal independent elements for a set of vectors in the product of lattices of intervals, a result which may be of independent interest. In contrast, the generation problem for *maximal* frequent intervals turns out to be NP-hard.

1 Introduction

Consider a database in which each transaction is associated with a time stamp indicating the start and end times of the transaction. For instance, [15] gives an example of a cellular phone company (or more generally a service provider) which records the time and length for each phone call made by each customer. Then it is useful, for the purpose of both improving the service and making more profit, to determine the intervals of time during which the number of calls exceeds a given threshold (*frequent intervals*), or the intervals of time during which the number of calls lies below some threshold (*infrequent intervals*). Clearly, the property of an interval being infrequent is *monotone*: if an interval I was occupied by less than t customers' phone calls, then the same is true for any interval containing I. Thus we may restrict our attention to *maximal* frequent and *minimal* infrequent intervals. In [15] an algorithm was proposed to enumerate all maximal frequent intervals from a given database.

More generally, one may consider a database of transactions, each of which describes an episode of events appearing over time. For instance, in the above example, we may store in the database the different calls made by each customer in different days. Then an interesting observation, that may be deduced from the database, can take the form "Fewer than 10% of the customers make calls on

J.R. Correa, A. Hevia, and M. Kiwi (Eds.): LATIN 2006, LNCS 3887, pp. 423–434, 2006.

Saturday between 1-2 AM, on Sunday between 1-2 AM, and on Monday between 9-10 AM", or "At least 60% of the customers who use the service between 5-9 PM in the first 5 days of the month tend also to use the service between 5-9 PM in the last five days". These examples illustrate the requirement for discovering correlation or *association rules* [1] between occurrences of events over time. As in the case of mining association rules between sets of items in a database (see e.g. [1, 2, 3]), a fundamental problem that arises in our case is the generation of frequent and infrequent multi-dimensional intervals (as opposed to frequent and infrequent sets in [1]). As was suggested in [12, 16, 17] for the case of frequent itemsets, it might be much more economical to represent the frequent and infrequent intervals by their boundary, defined as the union of *maximal* frequent and *minimal* infrequent intervals, since typically the number of intervals in such a boundary is much smaller. This motivates us to investigate the complexities of the problems of *jointly* and *separately* generating these two families. It turns out that they exhibit the same behavior as that, discovered in [5], for maximal frequent and minimal infrequent sets. More precisely, let $\mathcal{F}_{\mathcal{D},t}$ and $\mathcal{G}_{\mathcal{D},t}$ denote respectively the families of maximal frequent and minimal infrequent multi-dimensional intervals for a given database $\mathcal{D}$ and an integer threshold t. Then it will be shown that we can generate, in *incremental quasi-polynomial* time [1], the union $\mathcal{F}_{\mathcal{D},t} \cup \mathcal{G}_{\mathcal{D},t}$ (in some mixed way, and we do not control the order in which the elements of these two families are generated). It will be also illustrated that this result implies that the family of minimal infrequent intervals can also be generated in incremental quasi-polynomial time. Finally, we show also that the problem of incrementally generating the family $\mathcal{F}_{\mathcal{D},t}$ separately is NP-hard in general.

The paper is organized as follows. In the next section, we formally define the problems considered and state our results, and in Section 3, we briefly survey some related work. Following this, Section 4 explains how to view our problem as that of generating maximal frequent/minimal infrequent vectors in the product of lattices of intervals, constructed from the given database $\mathcal{D}$. In section 5, we reduce the problem of generating minimal infrequent intervals into the so called *dualization* problem in products of lattices of intervals. Finally, In Section 6 we show that this latter problem can be solved in quasi-polynomial time.

2 Problem Definition and Our Results

Let $\mathcal{D}$ be a database of records each of which has n attributes, where each attribute specifies a (possibly empty) real closed interval $I = [a, b] \subseteq \mathbb{R}$, $a, b \in \mathbb{R}$. Denote by $\mathbb{B}_n$ the set of all n-dimensional intervals (or boxes, or hyper-rectangles): $\mathbb{B}_n \stackrel{\text{def}}{=} \{(I_1, \ldots, I_n) \ : \ I_1, \ldots,\ I_n$ are closed intervals of $\mathbb{R}\}$. Henceforth, we shall refer to an n-dimensional interval simply as an interval when it is understood from the context that it has n dimensions. Let us denote

[1] i.e. given a partial list $\mathcal{X}$ of elements that have been already generated, generating a new element requires time $O(k^{\text{polylog}(k)})$, where $k = n + |\mathcal{D}| + |\mathcal{X}|$.

by "$\preceq$" the precedence relation of the partial order defined on $\mathbb{B}_n$, that is, given two intervals $I = (I_1, \ldots, I_n)$ and $(J_1, \ldots, J_n)$ in $\mathbb{B}_n$, let us say that $I \preceq J$ if and only if $I_i \subseteq J_i$ for all $i = 1, \ldots, n$. For $I \in \mathbb{B}_n$, let $S_{\mathcal{D}}(I)$ be the set of transactions of $\mathcal{D}$ that *support* I, i.e. $S_{\mathcal{D}}(I) \stackrel{\text{def}}{=} \{J \in \mathcal{D} \; : \; J \succeq I\}$. Given an integer threshold $0 \leq t \leq |\mathcal{D}|$, an interval I is said to be *t-frequent* if $|S_{\mathcal{D}}(I)| \geq t$ and *maximal* t-frequent if $|S_{\mathcal{D}}(J)| \leq t-1$ for all $J \succ I$. Similarly an interval I is called *t-infrequent* if $|S_{\mathcal{D}}(I)| \leq t-1$ and *minimal* t-infrequent if decreasing any interval component of I makes it t-frequent. Denote by $\mathcal{F}_{\mathcal{D},t}$ and $\mathcal{G}_{\mathcal{D},t}$ respectively the families of maximal frequent and minimal infrequent multi-dimensional intervals for a given database $\mathcal{D}$ and an integer threshold t, and by $\mathcal{F}^-_{\mathcal{D},t}$ and $\mathcal{G}^+_{\mathcal{D},t}$ the families of t-frequent and t-infrequent intervals. In this paper, we consider the following problem of incrementally generating all minimal infrequent intervals:

SEP-GEN-$(\mathcal{G}_{\mathcal{D},t}, \mathcal{X})$: *Given a sublist* $\mathcal{X} \subseteq \mathcal{G}_{\mathcal{D},t}$ *of minimal t-infrequent intervals, either find a new element in* $\mathcal{G}_{\mathcal{D},t} \setminus \mathcal{X}$ *or declare that the given sublist is complete:* $\mathcal{X} = \mathcal{G}_{\mathcal{D},t}$.

Similarly problem *SEP-GEN-*$(\mathcal{F}_{\mathcal{D},t}, \mathcal{X})$ of separately generating all maximal t-frequent intervals can be defined. We prove the following positive and negative results.

Theorem 1. *Problem SEP-GEN-*$(\mathcal{G}_{\mathcal{D},t}, \mathcal{Y})$ *can be solved in incremental quasi-polynomial time* $k^{O(\log^2 k)}$, *where* $k = n + |\mathcal{D}| + |\mathcal{Y}|$.

Proposition 1. *There exist instances of problem SEP-GEN-*$(\mathcal{F}_{\mathcal{D},t}, \mathcal{X})$ *which are NP-hard.*

On our way to proving Theorem 1, we also investigate the complexity of the *joint* generation of minimal infrequent and maximal frequent intervals:

JOINT-GEN$(\mathcal{D}, t, \mathcal{X}, \mathcal{Y})$: *Given two collections* $\mathcal{X} \subseteq \mathcal{F}_{\mathcal{D},t}$ *and* $\mathcal{Y} \subseteq \mathcal{G}_{\mathcal{D},t}$, *either find a new element in* $(\mathcal{F}_{\mathcal{D},t} \setminus \mathcal{X}) \cup (\mathcal{G}_{\mathcal{D},t} \setminus \mathcal{Y})$, *or declare that these collections are complete:* $(\mathcal{X}, \mathcal{Y}) = (\mathcal{F}_{\mathcal{D},t}, \mathcal{G}_{\mathcal{D},t})$.

Theorem 2. *Problem JOINT-GEN*$(\mathcal{D}, t, \mathcal{X}, \mathcal{Y})$ *can be solved in incremental quasi-polynomial time.*

Theorems 1 and 2 indicate that problems SEP-GEN-$(\mathcal{G}_{\mathcal{D},t}, \mathcal{Y})$ and JOINT-GEN$(\mathcal{D},\, t, \mathcal{X}, \mathcal{Y})$ are, most likely, not NP-hard, since no NP-complete problem is known to be solvable in quasi-polynomial time.

In contrast to these results, we can show that the separate generation problems SEP-GEN-$(\mathcal{F}^-_{\mathcal{D},t}, \mathcal{X})$ and SEP-GEN-$(\mathcal{G}^+_{\mathcal{D},t}, \mathcal{X})$ for t-frequent and t-infrequent intervals can be solved with (amortized) polynomial delay (i.e. the average time required to generate an element of $\mathcal{F}^-_{\mathcal{D},t}$ is bounded by a polynomial in n and $|\mathcal{D}|$). This follows, for instance, from a straightforward generalization of the well-known *Apriori* algorithm [3], applied to a product of lattices constructed from the database in a certain way. We omit the proof of the following theorem from this abstract.

Theorem 3. *Given a database $\mathcal{D}$ of transactions each of which is composed of n time intervals, and an integer t, all t-frequent intervals can be computed with amortized delay of $O(n^3|\mathcal{D}|\sum_{i=1}^{n}|\mathbb{P}_i|)$ per generated interval, and a total number of $O(\sum_{i=1}|\mathbb{P}_i|)$ scans of the database, where $\mathbb{P}_i$ is the set of distinct end-points appearing in the ith column of the database. All t-infrequent intervals can be also computed with the same amortized delay.*

We remark that we can also obtain a polynomial delay algorithm for generating $\mathcal{G}^+_{\mathcal{D},t}$ and $\mathcal{F}^-_{\mathcal{D},t}$, but at the cost of increasing the number of scans of the database.

3 Some Related Work

The problem of enumerating frequent sets arises in the context of mining association rules from binary data, see e.g. [1], mining correlations [6], episodes [18], and many other applications. In [3], an algorithm called *Apriori* was suggested to find all frequent sets from a binary database. Improvements on this algorithm as well as other methods were subsequently proposed, see e.g. [21, 22]. Further work had also considered non-binary databases, for example, databases where items belong to sets of *taxonomies* (or *is-a hierarchies*) [13, 14, 19], and databases with categorical or quantitative attributes [13, 20]. While the Apriori algorithm generates all frequent sets with amortized polynomial delay, it was shown in [5] that the generation of maximal frequent sets is NP-hard. It was also shown in the same paper that the generation of minimal infrequent sets can be solved in incremental quasi-polynomial time. In this paper, we establish similar results for the case of multi-dimensional intervals.

The problem of finding frequent 1-dimensional intervals, in a discrete domain, was considered in [23], where an Apriori-based algorithm was suggested. In [15], an algorithm for finding maximal frequent 1-dimensional intervals, in a continuous domain, was proposed. Another related problem is the generation of empty or sparse boxes in multi-dimensional data, considered in [4, 9]. In this problem, it is required to generate all inclusion-wise maximal hyper-rectangles that contain no point of the database in their interior. A polynomial-time algorithm was presented in [9] to solve the problem in 2-dimensions. This problem was shown to be solvable in quasi-polynomial time in [4] using a similar approach to the one used in this paper. The main difficulty that arises in dealing with frequent intervals is that they may contain some components representing empty intervals, a problem which did not appear in the case of maximal sparse boxes.

4 Embedding the Problem into the Products of Lattices of Intervals

4.1 The Lattice of Intervals

Let $\mathbb{I}_1, \ldots, \mathbb{I}_n \subseteq \mathbb{R}^n$ be n sets of real closed intervals. For $i = 1, \ldots, n$, let $\mathcal{L}_i$ be the *lattice of intervals* whose elements are all possible intersections and

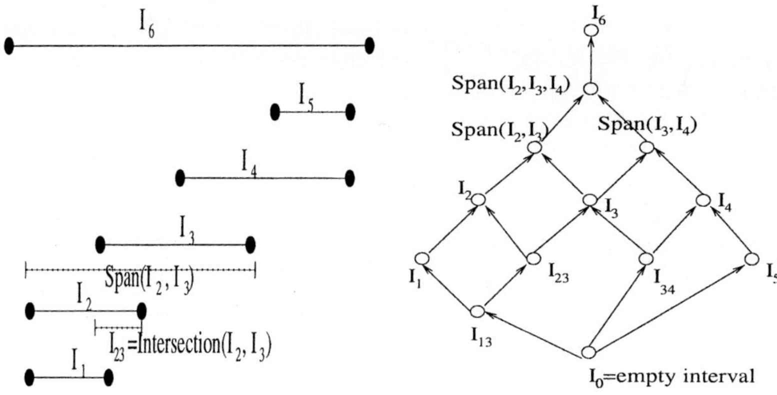

a: A set of intervals $\mathbb{I}_1$. b: The corresponding lattice of intervals $\mathcal{L}_1$.

Fig. 1. The lattice of intervals

spans defined by the intervals in $\mathbb{I}_i$, and ordered by containment: The meet of any two intervals in $\mathcal{L}_i$ is their *intersection*, and the join is their *span*, i.e., the minimum interval containing both of them (see Figure 1 for an example). Let $\mathcal{L} \stackrel{\text{def}}{=} \mathcal{L}_1 \times \cdots \times \mathcal{L}_n$ be the Cartesian product of these n lattices. Throughout we shall denote by $\succeq$ the precedence relation in $\mathcal{L}$ (and also in $\mathcal{L}_1, \ldots, \mathcal{L}_n$, i.e. if $p = (p_1, \ldots, p_n) \in \mathcal{L}$ and $q = (q_1, \ldots, q_n) \in \mathcal{L}$, then $p \preceq q$ in $\mathcal{L}$ if and only if $p_1 \preceq q_1$ in $\mathcal{L}_1$, ..., $p_n \preceq q_n$ in $\mathcal{L}_n$) and use $\vee$ and $\wedge$ to denote the join and meet operators over $\mathcal{L}$. We shall also denote by $l = (l_1, \ldots, l_n)$ and $u = (u_1, \ldots, u_n)$ the minimum and maximum elements of $\mathcal{L}$, respectively. For $x \in \mathcal{L}_i$, denote by $x^{\perp}$ the set of immediate predecessors of x, i.e.

$$x^{\perp} = \{y \in \mathcal{L}_i \mid y \prec x,\ (\nexists z \in \mathcal{L}_i : y \prec z \prec x)\}.$$

Similarly, denote by $x^{\top}$ the set of immediate successors of x. The following is a simple property satisfied by any lattice of intervals.

Proposition 2. *Let $\mathcal{L}_i$ be a lattice of intervals. Then (i) $|x^{\top}| \leq 2$ for all $x \neq l_i$ in $\mathcal{L}_i$, and (ii) $|x^{\perp}| \leq 2$ for all $x \in \mathcal{L}_i$.*

It is easy to see that $|\mathcal{L}_i| = O(|\mathbb{I}_i|^2)$ and that, if l_i represents the empty interval, then $|l_i^{\top}| \leq |\mathbb{I}_i|$. Clearly every element in $\mathcal{L}$ represents an n-dimensional interval in $\mathbb{B}_n$, and the precedence relation in $\mathcal{L}$ corresponds to that in $\mathbb{B}_n$, i.e. if $p \preceq q$ in $\mathcal{L}$, then the corresponding intervals $I, J \in \mathbb{B}_n$ satisfy $I \preceq J$. Although $\mathcal{L}$ is a proper subset of $\mathbb{B}_n$, for our purposes the elements of $\mathcal{L}$ represent the set of all possible *extremal* intervals that are of interest to us, as we shall see in the next subsection.

4.2 Lattices of Intervals Defined by the Database

Given a database of n-dimensional intervals $\mathcal{D}$, and $i \in [n]$, let $\mathbb{P}_i = \{p_i^1, p_i^2, \ldots, p_i^{k_i}\}$ be the set of end-points of intervals appearing in the ith column of $\mathcal{D}$.

Clearly $k_i \leq 2|\mathcal{D}|$, and assuming that $p_i^1 < p_i^2 < \ldots < p_i^{k_i}$, we obtain a set $\mathbb{I}_i = \{[p_i^1, p_i^2], [p_i^2, p_i^3], \ldots, [p_i^{k_i-1}, p_i^{k_i}]\}$ of at most $2|\mathcal{D}|$ intervals. Now we let $\mathcal{L}_i$ be the lattice of intervals defined by the set $\mathbb{I}_i$, for $i = 1, \ldots, n$, and let $\mathcal{L} = \mathcal{L}_1 \times \cdots \times \mathcal{L}_n$. Clearly, each record in $\mathcal{D}$ appears as an element in $\mathcal{L}$, i.e. $\mathcal{D} \subseteq \mathcal{L}$. For $x \in \mathcal{L}$, let $S(x) = \{y \in \mathcal{D} \mid y \succeq x\}$. Given an integer threshold t, let us say that an element $x \in \mathcal{L}$ is t-frequent (with respect to $\mathcal{D}$) if $|S(x)| \geq t$ and maximal t-frequent if $|S(y)| < t$ for all $y \succ x$. Similarly we define t-infrequent and minimal t-infrequent elements of $\mathcal{L}$.

Now, it is easy to see that the maximal t-frequent elements of $\mathcal{L}$ are in one-to-one correspondence with the maximal t-frequent intervals defined by $\mathcal{D}$, in the obvious way: if $x = (x_1, \ldots, x_n) \in \mathcal{L}$ is a maximal frequent element, then the corresponding interval $(I_1, \ldots, I_n)$ (where I_i corresponds to x_i, for $i = 1, \ldots, n$) is the corresponding maximal frequent interval. The situation with minimal infrequent intervals is just a bit more complicated: if $x = (x_1, \ldots, x_n) \in \mathcal{L}$ is a minimal infrequent element then the corresponding minimal infrequent interval $(I_1, \ldots, I_n)$ is computed as follows. For $i = 1, \ldots, n$, if $x_i = l_i$ is the minimum element of $\mathcal{L}_i$, then $I_i = \emptyset$. If x_i represents a point $p_i \in \mathbb{R}$ then $I_i = [p_i, p_i]$. Otherwise, let $[a_i, b_i]$ and $[c_i, d_i]$ be the two intervals corresponding to the two immediate predecessors of x_i, where we assume $a_i < c_i$ (note that $c_i \leq b_i$). If $a_i = b_i$ and $c_i = d_i$ then x_i corresponds to the interval $[a_i, c_i]$ and we have an infinite number of minimal infrequent intervals defined (uniquely) by I_i, namely $I_i = [p_i, p_i]$ for all points p_i in the open interval (a_i, c_i). Finally, if $a_i < b_i$ and $c_i < d_i$, then $I_i = [c_i - \epsilon, b_i + \epsilon]$ for an infinitesimal constant $\epsilon > 0$. Consequently, in all cases, our problems reduce to finding maximal t-frequent/minimal t-infrequent elements in the lattice product $\mathcal{L}$.

5 Enumerating Minimal Infrequent Intervals

5.1 Dualization Problem in Products of Lattices of Intervals

For a subset $\mathcal{A} \subseteq \mathcal{L}$ of n-dimensional intervals in $\mathcal{L}$, denote by $\mathcal{A}^+ = \{x \in \mathcal{L} \mid x \succeq a, \text{ for some } a \in \mathcal{A}\}$ and $\mathcal{A}^- = \{x \in \mathcal{L} \mid x \preceq a, \text{ for some } a \in \mathcal{A}\}$. Any element in $\mathcal{L} \setminus \mathcal{A}^+$ is called *independent of* $\mathcal{A}$. Let $\mathcal{I}(\mathcal{A})$ be the set of all maximal independent elements for $\mathcal{A}$ (also referred to as the *dual* of $\mathcal{A}$):

$$\mathcal{I}(\mathcal{A}) \stackrel{\text{def}}{=} \{p \in \mathcal{L} \mid p \notin \mathcal{A}^+ \text{ and } (q \in \mathcal{L}, q \succeq p, q \neq p \Rightarrow q \in \mathcal{A}^+)\}.$$

Given $\mathcal{A} \subseteq \mathcal{L}$, we consider the problem of incrementally generating $\mathcal{I}(\mathcal{A})$:

DUAL$(\mathcal{L}, \mathcal{A}, \mathcal{B})$: *Given subsets* $\mathcal{A} \subseteq \mathcal{L}$ *and* $\mathcal{B} \subseteq \mathcal{I}(\mathcal{A})$, *either find a new element* $x \in \mathcal{I}(\mathcal{A}) \setminus \mathcal{B}$, *or prove that* $\mathcal{A}$ *and* $\mathcal{B}$ *form a dual pair:* $\mathcal{B} = \mathcal{I}(\mathcal{A})$.

Clearly, the entire set $\mathcal{I}(\mathcal{A})$ can be generated by initializing $\mathcal{B} = \emptyset$ and iteratively solving the above problem $|\mathcal{I}(\mathcal{A})| + 1$ times. When each lattice $\mathcal{L}_i = \{0, 1\}$, the problem reduces to the well-known hypergraph transversal problem, for which the best-known algorithm [10] runs in time $k^{o(\log k)}$, where $k = |\mathcal{A}| + |\mathcal{B}|$. An extension of this algorithm, for solving the dualization problem for general lattices was given in [8], and runs in time $poly(n, \mu(\mathcal{L})) + m^{\gamma(W(\mathcal{L}))\cdot o(\log m)}$, where $m = |\mathcal{A}| + |\mathcal{B}|$, $\gamma(W) = O(W^2 \ln W)$, $\mu = \mu(\mathcal{L}) \stackrel{\text{def}}{=} \max\{|\mathcal{L}_i| \ : \ i \in [n]\}$, and $W = W(\mathcal{L}) \stackrel{\text{def}}{=} \max_{i\in[n]}\{W(\mathcal{L}_i)\}$ is the maximum width of the n lattices, i.e. the maximum size of an antichain in the n lattices. Note that for the lattice of intervals $\mathcal{L}_i$, defined by a set of intervals $\mathbb{I}_i$, we have $W(\mathcal{L}_i) = O(|\mathbb{I}_i|)$ and $|\mathcal{L}_i| = O(|\mathbb{I}_i|^2)$. Thus, for this special case, the result of [8] gives an *exponential* algorithm in the total number of intervals of $\sum_{i=1}^n |\mathbb{I}_i|$. Here, we shall strengthen this result, in the case of products of lattices of intervals, as follows:

Theorem 4. *Problem $DUAL(\mathcal{L}, \mathcal{A}, \mathcal{B})$ can be solved in $k^{O(\log^2 k)}$ time, if $\mathcal{L}$ is a product of interval lattices, where $k = |\mathcal{A}| + |\mathcal{B}| + \sum_{i=1}^n |\mathcal{L}_i|$.*

The proof of Theorem 4 will be given in Section 6. In the next section, we show how to use this result to prove Theorems 1 and 2.

5.2 Proof of Theorems 1 and 2

In this section, we argue that the generation problems JOINT-GEN$(\mathcal{D}, t, \mathcal{X}, \mathcal{Y})$ and SEP-GEN$(\mathcal{G}_{\mathcal{D},t}, \mathcal{X})$ reduce in polynomial time to dualization in products of lattices of intervals. For the former problem, the reduction follows from a straightforward generalization of a known result, relating the time complexity of joint generation to that of dualization:

Proposition 3 ([7, 11]). *Problem JOINT-GEN$(\mathcal{D}, t, \mathcal{X}, \mathcal{Y})$ can be solved in time* poly$(n, |\mathcal{D}|, |\mathcal{X}|, |\mathcal{Y}|)$ $+T_{dual}$ *where T_{dual} denotes the time required to solve problem $DUAL(\mathcal{L}, \mathcal{A}, \mathcal{B})$.*

For the latter problem, we use Proposition 3 together with a combinatorial Lemma from [5], to show that the family $\mathcal{G}_{\mathcal{D},t}$ is *uniformly dual-bounded* in the sense that

$$|\mathcal{I}(\mathcal{X}) \cap I(\mathcal{G}_{\mathcal{D},t})| \leq |\mathcal{D}||\mathcal{X}|, \tag{1}$$

for any non-empty $\mathcal{X} \subseteq \mathcal{G}_{\mathcal{D},t}$. Inequality (1) implies that, if we apply joint generation to problem SEP-GEN$(\mathcal{G}_{\mathcal{D},t}, \mathcal{X})$, we generate, in addition to the elements of the required family $\mathcal{G}_{\mathcal{D},t}$, only a polynomial number of unrequired elements belonging to the family $\mathcal{F}_{\mathcal{D},t} = \mathcal{I}(\mathcal{G}_{\mathcal{D},t})$. This proves Theorem 1. It remains to show (1), which follows from the following Lemma:

Lemma 1 ([5]). *Let $t \in \mathbb{R}_+$ be a given positive threshold, and $\mathcal{S} \neq \emptyset$ and $\mathcal{T}$ be two families of subsets of a finite set V such that (i) for all $X \in \mathcal{S}$ and $Y \in \mathcal{T}$, we have $|Y| \geq t > |X|$, (ii) for every $Y' \neq Y'' \in \mathcal{T}$ there exists an $X \in \mathcal{S}$ such that $X \supseteq Y' \cap Y''$. Then $|\mathcal{T}| \leq |V||\mathcal{S}|$.*

To apply the lemma to get (1), let $V = \mathcal{D}$, $\mathcal{S} = \{S(x) \ : \ x \in \mathcal{X}\}$ and $\mathcal{T} = \{S(y) \ : \ y \in \mathcal{I}(\mathcal{X}) \cap I(\mathcal{G}_{\mathcal{D},t})\}$, and observe that $|S(y)| \geq t > |S(x)|$ for all $x \in \mathcal{X}$ and all $y \in \mathcal{Y} \stackrel{\text{def}}{=} \mathcal{I}(\mathcal{X}) \cap I(\mathcal{G}_{\mathcal{D},t})$, since $\mathcal{X} \subseteq \mathcal{G}_{\mathcal{D},t}$ and $\mathcal{Y} \subseteq \mathcal{F}_{\mathcal{D},t}$. Furthermore, given two distinct elements $y', y'' \in \mathcal{Y}$, it follows by their maximality in $\mathcal{L} \setminus \mathcal{X}^+$ that $y' \vee y'' \succeq x$, for some $x \in \mathcal{X}$, and thus $S(y') \cap S(y'') = S(y' \vee y'') \subseteq S(x) \in \mathcal{S}$.

6 Dualization Algorithm

6.1 Preliminaries

Let $\mathcal{L} = \mathcal{L}_1 \times \cdots \times \mathcal{L}_n$ where each $\mathcal{L}_i$ is a lattice defined by a set of intervals $\mathbb{I}_i$. We denote respectively by l_i and u_i the minimum and maximum elements of $\mathcal{L}_i$. Given two subsets $\mathcal{A} \subseteq \mathcal{L}$, and $\mathcal{B} \subseteq \mathcal{I}(\mathcal{A})$, we say that $\mathcal{B}$ is *dual to* $\mathcal{A}$ if $\mathcal{B} = \mathcal{I}(\mathcal{A})$. Given any $\mathcal{Q} \subseteq \mathcal{L}$, let us denote by

$$\mathcal{A}(\mathcal{Q}) = \{a \in \mathcal{A} \mid a^+ \cap \mathcal{Q} \neq \emptyset\}, \qquad \mathcal{B}(\mathcal{Q}) = \{b \in \mathcal{B} \mid b^- \cap \mathcal{Q} \neq \emptyset\},$$

the subsets of $\mathcal{A}, \mathcal{B}$ whose *ideal* and *filter* respectively intersect $\mathcal{Q}$.

To solve problem DUAL($\mathcal{L}, \mathcal{A}, \mathcal{B}$), we decompose it into a number of smaller subproblems which are solved recursively. In each such subproblem, we start with a sub-lattice $\mathcal{Q} = \mathcal{Q}_1 \times \cdots \times \mathcal{Q}_n \subseteq \mathcal{L}$ (initially $\mathcal{Q} = \mathcal{L}$), and two subsets $\mathcal{A}(\mathcal{Q}) \subseteq \mathcal{A}$ and $\mathcal{B}(\mathcal{Q}) \subseteq \mathcal{B}$, and we want to check whether $\mathcal{A}(\mathcal{Q})$ and $\mathcal{B}(\mathcal{Q})$ are dual in $\mathcal{Q}$. To estimate the reduction in problem size from one level of the recursion to the next, we measure the change in the "volume" of the problem defined as $v = v(\mathcal{A}, \mathcal{B}, \mathcal{L}) \stackrel{\text{def}}{=} |\mathcal{A}||\mathcal{B}| \sum_{i=1}^{n} |\mathcal{L}_i|$. Since $\mathcal{B} \subseteq \mathcal{I}(\mathcal{A})$ is assumed, the following condition holds for the original problem and all subsequent subproblems:

$$a \not\preceq b, \quad \text{for all } a \in \mathcal{A}, b \in \mathcal{B}. \tag{2}$$

We stop decomposing a problem when one of the sets $\mathcal{A}$ or $\mathcal{B}$ becomes sufficiently small, in which case the problem is easily solvable in polynomial time.

Let us say that a coordinate $i \in [n]$ is *essential* for an element $a \in \mathcal{A}$ ($b \in \mathcal{B}$), if $a_i \succ l_i$ (respectively, $b_i \prec u_i$). Let us denote by $E(x)$ the set of essential coordinates of a element $x \in \mathcal{A} \cup \mathcal{B}$. The following lemma generalizes a known fact for dual Boolean functions [10].

Lemma 2. *If $\mathcal{A}, \mathcal{B} \subseteq \mathcal{L}$, then either (i) there exists an element $x \in \mathcal{A} \cup \mathcal{B}$ with few essential coordinates: $|E(x)| \leq \log m$, where $m = |\mathcal{A}| + |\mathcal{B}|$, or (ii) an element $x \in \mathcal{L} \setminus (\mathcal{A}^+ \cup \mathcal{B}^-)$ can be found in polynomial time.*

Lemma 3. *Let $\mathcal{A}, \mathcal{B}$ be a pair of dual subsets of $\mathcal{L}$ with $|\mathcal{A}||\mathcal{B}| \geq 1$. Then there exists a coordinate $i \in [n]$ and a element $z \in \mathcal{L}_i$, such that either:*

(i) $|\{a \in \mathcal{A} \mid a_i \succeq z\}| \geq 1$ and $|\{b \in \mathcal{B} \mid b_i \not\succeq z\}| \geq \frac{|\mathcal{B}|}{\log m}$, or

(ii) $|\{b \in \mathcal{B} \mid b_i \preceq z\}| \geq 1$ and $|\{a \in \mathcal{A} \mid a_i \not\preceq z\}| \geq \frac{|\mathcal{A}|}{\log m}$.

6.2 The Algorithm - Proof of Theorem 4

Given subsets $\mathcal{A}, \mathcal{B} \subseteq \mathcal{L}$ that satisfy (2), we proceed as follows:

Step 1. If $\max\{|\mathcal{A}|, |\mathcal{B}|\} \leq 1$, the problem can be solved in $poly(\sum_{i=1}^{n} |\mathcal{L}_i|)$ time.

Step 2. For each $k \in [n]$: if $a_k \notin \mathcal{L}_k$ for some $a \in \mathcal{A}$ ($b_k \notin \mathcal{L}_k$ for some $b \in \mathcal{B}$), set $a_k \leftarrow \bigwedge\{x \mid x \in a_k^+ \cap \mathcal{L}_k\}$ (respectively, set $b_k \leftarrow \bigvee\{x \mid x \in b_k^- \cap \mathcal{L}_k\}$).

Step 3. Check if there is an $x \in \mathcal{A} \cup \mathcal{B}$ with at most $\log m$ essential coordinates. If no such x can be found, a new element in $\mathcal{L} \setminus (\mathcal{A}^+ \cup \mathcal{B}^-)$ can be obtained as described in Lemma 2. Otherwise, we proceed to the next step.

Step 4. If $x = a^o \in \mathcal{A}$, find an $i \in E(a^o)$, and a $z = a_i^o \in \mathcal{L}_i$, for which condition (i) of Lemma 3 is satisfied. Assume without loss of generality that $i = 1$.

In the following steps, we shall decompose $\mathcal{L}_1$ into two (not necessarily disjoint) sub-lattices $\mathcal{L}_1'$ and $\mathcal{L}_1''$, and let $\mathcal{L}' = \mathcal{L}_1' \times \mathcal{L}_2 \times \cdots \times \mathcal{L}_n$, and $\mathcal{L}'' = \mathcal{L}_1'' \times \mathcal{L}_2 \times \cdots \times \mathcal{L}_n$ be the sub-lattices of $\mathcal{L}$ induced by this decomposition. It will follow then that $\mathcal{A}$ and $\mathcal{B}$ are dual in $\mathcal{L}$ *if and only if*

$$\mathcal{A}(\mathcal{L}'), \mathcal{B}(\mathcal{L}') \text{ are dual in } \mathcal{L}', \text{ and } \mathcal{A}(\mathcal{L}''), \mathcal{B}(\mathcal{L}'') \text{ are dual in } \mathcal{L}'', \qquad (3)$$

each of which is a dualization problem over the product of lattices of intervals. Note that $\mathcal{A}(\mathcal{L}') = \{a \in A \mid a_1^+ \cap \mathcal{L}_1' \neq \emptyset\}$ and $\mathcal{B}(\mathcal{L}') = \{b \in B \mid b_1^- \cap \mathcal{L}_1' \neq \emptyset\}$; $\mathcal{A}(\mathcal{L}'')$ and $\mathcal{B}(\mathcal{L}'')$ are defined similarly. Let $\epsilon = 1/\log m$.

Step 4.1. If $\mathcal{L}_1$ is a total order (chain), then use the following decomposition of $\mathcal{L}_1$: $\mathcal{L}_1' \leftarrow z^+ \cap \mathcal{L}_1$, $\mathcal{L}_1'' \leftarrow \mathcal{L}_1 \setminus \mathcal{L}_1'$. Then $|\mathcal{B}(\mathcal{L}')| \leq (1-\epsilon)|\mathcal{B}|$ by the selection of a^o, and $|\mathcal{A}(\mathcal{L}'')| \leq |\mathcal{A}| - 1$ since $(a^o)^+ \cap \mathcal{L}'' = \emptyset$. This reduces the original problem, of volume $v = |\mathcal{A}||\mathcal{B}|\sum_{i=1}^n |\mathcal{L}_i|$ into two subproblems (3) of volumes

$$v' \leq |\mathcal{A}||\mathcal{B}|(1-\epsilon)(\sum_{i=1}^n |\mathcal{L}_i| - 1) \leq (1-\epsilon)v,$$

$$v'' \leq (|\mathcal{A}| - 1|)|\mathcal{B}|(\sum_{i=1}^n |\mathcal{L}_i| - 1) \leq v - 1.$$

Step 4.2. Otherwise ($\mathcal{L}_1$ is not a chain), let w be the *largest* element, with respect to the precedence relation "$\preceq$" on the lattice $\mathcal{L}_1$, such that $|w^\perp| = 2$ (see Figure 2–a). Denote respectively by q and y the two immediate predecessors of w. Let $I_q = [a, b]$ and $I_y = [c, d]$ be the two intervals represented by q and y respectively, and assume that $a < c$ (and therefore $b < d$). It is not hard to see that q^- is a lattice of intervals and that $\mathcal{L}_1 \setminus q^-$ is a chain. Now we consider three cases:

(i) if $z \succ w$, we use the decomposition $\mathcal{L}_1' \leftarrow z^+ \cap \mathcal{L}_1$, $\mathcal{L}_1'' \leftarrow \mathcal{L}_1 \setminus \mathcal{L}_1'$. Otherwise, the choice of z implies that either cases (ii) or (iii) hold.

(ii) $|\{b \in \mathcal{B} \mid b_1 \in q^-\}| \geq \frac{\epsilon}{2}|\mathcal{B}|$: in this case, we decompose $\mathcal{L}_1$ as $\mathcal{L}_1' \leftarrow \mathcal{L}_1 \cap q^-$, $\mathcal{L}_1'' \leftarrow \mathcal{L}_1 \setminus q^-$.

(iii) $|\{b \in \mathcal{B} \mid b_1 \in y^-| \geq \frac{\epsilon}{2}|\mathcal{B}|$: in this case, we decompose $\mathcal{L}_1$ as $\mathcal{L}_1' \leftarrow \mathcal{L}_1 \cap y^-$, $\mathcal{L}_1'' \leftarrow \mathcal{L}_1 \setminus y^-$.

In case (i), we get again that $|\mathcal{B}(\mathcal{L}')| \leq (1-\epsilon)|\mathcal{B}|$ and $|\mathcal{A}(\mathcal{L}'')| \leq |\mathcal{A}| - 1$, and consequently, the resulting problems are of respective volumes $v' \leq (1-\epsilon)v$ and $v'' \leq v - 1$. In case (ii), we get $|\mathcal{B}(\mathcal{L}'')| \leq (1-\epsilon/2)|\mathcal{B}|$ and $|\mathcal{L}_1'| \leq |\mathcal{L}_1| - 1$, and therefore, the resulting two problems have volumes $v' \leq v-1$ and $v'' \leq (1-\epsilon/2)v$. Similarly, in case (iii), we get also that $v' \leq v - 1$ and $v'' \leq (1-\epsilon/2)v$.

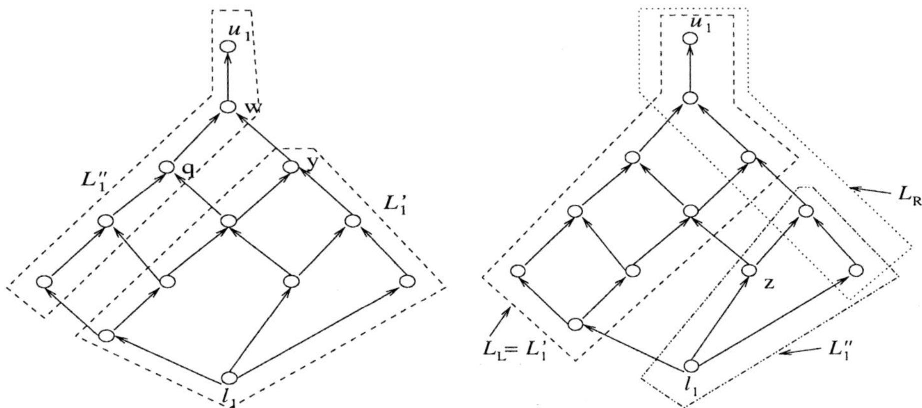

a: Decomposition rule used in Step 4.2. b: Decomposition rule used in Step 5.1.

Fig. 2. Decomposing the lattice $\mathcal{L}_1$

Step 5. Now assume that $x = b^o \in \mathcal{B}$, and find an $i \in E(b^o)$, and a $z = b_i^o \in \mathcal{L}_i$, for which condition (ii) of Lemma 3 is satisfied. Assume again, without loss of generality, that $i = 1$.

Step 5.1. If $\min(\mathcal{L}_1) = l_1$ does not represent the empty interval, or $z \succ l_1$, then let $I_z = [a, b]$ be the interval corresponding to z, and let $\mathcal{L}_L \subseteq \mathcal{L}_1$ be the lattice of intervals $I = [c, d]$ for which $c < a$, and likewise, $\mathcal{L}_R \subseteq \mathcal{L}_1$ be the lattice of intervals $I = [e, f]$ for which $f > b$ (see Figure 2–b). Note that these definitions imply that $(\mathcal{L}_L \cup \{l_1\}) \cap z^- = \{l_1\}$, $(\mathcal{L}_R \cup \{l_1\}) \cap z^- = \{l_1\}$, and $\mathcal{L}_L \cup z^- \cup \mathcal{L}_R = \mathcal{L}$. Note also that $\mathcal{L}_L \cup \mathcal{L}_R \neq \emptyset$ since $z \neq u_1 = \max(\mathcal{L}_1)$. By our selection of z, either

$$\text{(i) } |\{a \in \mathcal{A} \mid a_1 \in \mathcal{L}_L \setminus \{l_1\}\}| \geq \frac{\epsilon}{2}|\mathcal{A}|, \text{ or (ii) } |\{a \in \mathcal{A} \mid a_1 \in \mathcal{L}_R \setminus \{l_1\}\}| \geq \frac{\epsilon}{2}|\mathcal{A}|.$$

In case (i), we decompose $\mathcal{L}_1$ as follows: $\mathcal{L}_1' \leftarrow \mathcal{L}_L$, $\mathcal{L}_1'' \leftarrow (\mathcal{L}_1 \setminus \mathcal{L}_1') \cup \{l_1\}$. Note that both $\mathcal{L}_1'$ and $\mathcal{L}_1''$ are also lattices of intervals, that $|\mathcal{L}_1'| \leq |\mathcal{L}_1| - 1$ since $z \notin \mathcal{L}_1'$, and that $\mathcal{A}(\mathcal{L}'') \leq (1 - \epsilon/2)|\mathcal{A}|$, since $w \not\preceq y$ for all $w \in \mathcal{L}_1' \setminus \{l_1\}$ and $y \in \mathcal{L}_1'' \setminus \{l_1\}$ (indeed, if $I_w = [c, d]$ is the interval corresponding to $w \in \mathcal{L}_1' \setminus \{l_1\}$ and $I_y = [e, f]$ is the interval corresponding to $y \in \mathcal{L}_1'' \setminus \{l_1\}$, then $c < a$ while $e \geq a$ and thus $I_w \not\subseteq I_y$). Therefore, we get, in this case, two subproblems of volumes $v' \leq v - 1$ and $v'' \leq (1 - \epsilon/2)v$. In case (ii), we let similarly $\mathcal{L}_1' \leftarrow \mathcal{L}_R$ and $\mathcal{L}_1'' \leftarrow (\mathcal{L}_1 \setminus \mathcal{L}_1') \cup \{l_1\}$, and we decompose the original problem into two subproblems of volumes $v' \leq v - 1$ and $v'' \leq (1 - \epsilon/2)v$, respectively.

Step 5.2. If $z = l_1$ represents the empty interval, then we let z' be any immediate successor of z, and let $\mathcal{L}_L$ and $\mathcal{L}_R$ be the lattices of intervals as defined in Step 5.1, but with respect to $I_{z'} = [a, b]$ instead of I_z. Note in this case that any interval $[c, d]$ in $\mathcal{L}_L$ either must be strictly to the left of $I_{z'}$, i.e. with $d < a$, or must contain $I_{z'}$. Similarly, any interval $[e, f]$ in $\mathcal{L}_R$ either must be strictly to the right of $I_{z'}$, i.e. with $e > b$, or must contain $I_{z'}$. We consider four cases:

(i) No interval of $\mathbb{I}_1$, corresponding to an element of $\mathcal{L}_1$, lies strictly to the left or strictly to the right of $I_{z'}$: the choice of z, in this case, implies that $|\{a \in \mathcal{A} \mid a_1 \succeq z'\} \geq \epsilon|\mathcal{A}|$. Thus using the decomposition $\mathcal{L}_1' \leftarrow (z')^+$ and $\mathcal{L}_1'' \leftarrow \{z\}$ results in two subproblems of volumes $v' \leq v-1$ and $v'' \leq (1-\epsilon)v$.

(ii) No interval lies strictly on the right of $I_{z'}$, but there is at least one that lies strictly to its left: by our choice of z, one of the sets $\{a \in \mathcal{A} \;:\; a_1 \in \mathcal{L}_L\}$ or $\{a \in \mathcal{A} \mid a_1 \succeq z'\}$ have size at least $\frac{\epsilon}{2}|\mathcal{A}|$. In the former case we use the decomposition $\mathcal{L}_1' \leftarrow \mathcal{L}_L$, $\mathcal{L}_1'' \leftarrow \mathcal{L}_1 \setminus \mathcal{L}_1'$, and get two subproblems of volumes $v' \leq v-1$ and $v'' \leq (1-\epsilon)v$. In the latter case, we let $\mathcal{L}_1''$ be the lattice of intervals lying strictly to the left of $I_{z'}$ and $\mathcal{L}_1' \leftarrow (z')^+ \cup \{z\}$, and get two subproblems of volumes $v' \leq v-1$ and $v'' \leq (1-\epsilon)v$.

(iii) No interval lies strictly on the left of $I_{z'}$, but there is at least one that lies strictly to its right: we use a similar decomposition as in case (ii) above.

(iv) There is at least one interval that lies strictly to the left of $I_{z'}$, and at least one interval strictly to its right: in this case, we know that either $|\{a \in \mathcal{A} \mid a_1 \in \mathcal{L}_L \cup (z')^+\}| \geq \epsilon|\mathcal{A}|/2$ or $|\{a \in \mathcal{A} \mid a_1 \in \mathcal{L}_R \cup (z')^+\}| \geq \epsilon|\mathcal{A}|/2$. In the former case, we use the decomposition $\mathcal{L}_1' \leftarrow \mathcal{L}_L \cup (z')^+ \cup \{z\}$, $\mathcal{L}_1'' \leftarrow \mathcal{L}_1 \setminus \mathcal{L}_1' \cup \{z\}$, and in the latter case, we use the decomposition $\mathcal{L}_1' \leftarrow \mathcal{L}_R \cup (z')^+ \cup \{z\}$, $\mathcal{L}_1'' \leftarrow \mathcal{L}_1 \setminus \mathcal{L}_1' \cup \{z\}$. In both cases, we get two subproblems of volumes $v' \leq v-1$ and $v' \leq (1-\epsilon/2)v$.

Thus, in all cases, we apply the algorithm recursively to the resulting subproblems, and obtain the recurrence

$$C(v) \leq 1 + C((1-\epsilon/2)v) + C(v-1),$$

where $C(v)$ is the number of recursive calls required to solve a problem of volume v. Together with $C(v) = 1$, this recurrence evaluates to $C(v) \leq v^{2\log v/\epsilon}$. Since $v \leq m^2 n\mu$, we get that the running time of the algorithm is $O((m^2 n\mu)^{2\log m \log(m^2 n\mu)})$.

References

1. R. Agrawal, T. Imielinski and A. Swami, Mining association rules between sets of items in massive databases, in *Proc. the 1993 ACM-SIGMOD Int. Conf. Management of Data*, pp. 207-216.
2. R. Agrawal, H. Mannila, R. Srikant, H. Toivonen and A. I. Verkamo, Fast discovery of association rules, in *Advances in Knowledge Discovery and Data Mining* (U. M. Fayyad, G. Piatetsky-Shapiro, P. Smyth and R. Uthurusamy, eds.), pp. 307-328, AAAI Press, Menlo Park, California, 1996.
3. R. Agrawal and R. Srikant, Fast algorithms for mining association rules in large databases, in *Proc. 20th Int. Conf. Very Large Data Bases (VLDB'94)*, pp. 487–499, 1994.
4. E. Boros, K. Elbassioni, V. Gurvich, L. Khachiyan and K. Makino, An Intersection Inequality for Discrete Distributions and Related Generation Problems, in *Automata, Languages and Programming, 30-th International Colloquium, ICALP 2003, Lecture Notes in Computer Science (LNCS)* 2719 (2003) pp. 543–555.

5. E. Boros, V. Gurvich, L. Khachiyan and K. Makino, On the complexity of generating maximal frequent and minimal infrequent sets, in *19th Int. Symp. Theoretical Aspects of Computer Science, (STACS)*, March 2002, LNCS 2285, pp. 133–141.
6. S. Brin, R. Motwani, and C. Silverstein, Beyond market basket: Generalizing association rules to correlations, in *Proc. the 1997 ACM-SIGMOD Int. Conf. Management of Data*, pp. 265–276.
7. J. C. Bioch and T. Ibaraki, Complexity of identification and dualization of positive Boolean functions, *Information and Computation* 123 (1995) pp. 50–63.
8. K. Elbassioni, An algorithm for dualization in products of lattices and its applications, in *Proc. 10th Annual European Symposium on Algorithms (ESA 2002), LNCS* 2461, pp. 424–435, September 2002.
9. J. Edmonds, J. Gryz, D. Liang and R. J. Miller, Mining for empty rectangles in large data sets, in *Proc. 8th Int. Conf. Database Theory (ICDT)*, Jan. 2001, *LNCS* 1973, pp. 174–188.
10. M. L. Fredman and L. Khachiyan, On the complexity of dualization of monotone disjunctive normal forms, *Journal of Algorithms*, 21 (1996) pp. 618–628.
11. V. Gurvich and L. Khachiyan, On generating the irredundant conjunctive and disjunctive normal forms of monotone Boolean functions, *Discrete Applied Mathematics*, 96-97 (1999) pp. 363–373.
12. D. Gunopulos, R. Khardon, H. Mannila, and H. Toivonen, Data mining, hypergraph transversals and machine learning, in *Proc. 16th ACM PODS*, (1997) pp. 12–15.
13. J. Han, Y. Cai and N. Cercone, Data driven discovery of quantitative rules in relational databases, *IEEE Trans. Knowledge and Data Engineering*, Vol. 5 No. 1, (1993) pp. 29-40.
14. J. Han and Y. Fu, Discovery of multiple-level association rules from large databases. In *Proc. 21st Int. Conf. Very Large Data Bases (VLDB'95)*, pp. 420-431, 1995.
15. J.-L. Lin, Mining maximal frequent intervals, in *Proc. 18th Annual ACM Symp. Applied Computing*, pp. 426–431, Melbourne, FL, Sept. 2003.
16. H. Mannila and H. Toivonen, Multiple uses of frequent sets and condensed representations, in *Proc. 2nd Int. Conf. Knowledge Discovery and Data Mining* 1996, pp. 189–194.
17. H. Mannila and H. Toivonen, Levelwise search and borders of theories in knowledge discovery, *Data Mining and Knowledge Discovery*, 1(3), 1997, pp. 241–258.
18. H. Mannila, H. Toivonen, and A. I. Verkamo, Discovery of frequent episodes in event sequences. *Data Mining and Knowledge Discovery*, 1(3), 1997, pp. 259–289.
19. R. Srikant and R. Agrawal, Mining generalized association rules. In *Proc. 21st Int. Conf. Very Large Data Bases (VLDB'95)*, pp. 407–419, 1995.
20. R. Srikant and R. Agrawal, Mining quantitative association rules in large relational tables, in *Proc. the 1996 ACM-SIGMOD Int. Conf. Management of Data*, pp. 1–12, 1996.
21. A. Savasere, E. Omiecinski and S. Navathe, An efficient algorithm for mining association rules in large databases, in *Proc. 21st Int. Conf. Very Large Data Bases (VLDB'95)*, pp. 432–444, 1995.
22. H. Toivonen, Sampling large databases for association rules, in *Proc. 22nd Int. Conf. Very Large Data Bases (VLDB'96)*, pp. 134–145, 1996.
23. H.-C. Yu, Efficient data mining for frequent intervals, Master thesis, Department of Information Management, National Taiwan University, Taiwan, July 2002.

Cut Problems in Graphs with a Budget Constraint

Roee Engelberg[1], Jochen Könemann[2,⋆], Stefano Leonardi[3], and Joseph (Seffi) Naor[4,⋆⋆]

[1] Computer Science Department, Technion, Haifa 32000, Israel
roee@cs.technion.ac.il
[2] Department of Combinatorics and Optimization, University of Waterloo, 200 University Avenue West, Waterloo, ON N2L 3G1, Canada
jochen@math.uwaterloo.ca
[3] Dipartimento di Informatica e Sistemistica, Università di Roma "La Sapienza", Via Salaria 113, 00198 Roma, Italy
leon@dis.uniroma1.it
[4] Microsoft Research, One Microsoft Way, Redmond, WA 98052, USA
naor@cs.technion.ac.il

Abstract. We study budgeted variants of classical cut problems: the *Multiway Cut* problem, the *Multicut* problem, and the *k-Cut* problem, and provide approximation algorithms for these problems. Specifically, for the budgeted multiway cut and the k-cut problems we provide constant factor approximation algorithms. We show that the budgeted multicut problem is at least as hard to approximate as the sparsest cut problem, and we provide a bi-criteria approximation algorithm for it.

1 Introduction

Given an undirected graph $G = (V, E)$ with a positive cost function on the edges $c : E \to \mathbb{Z}^+$, and a subset of vertices $S \subseteq V$, called *terminals*, the well-known *multiway cut* problem is to find a minimum cost subset of edges whose removal disconnects the terminals from each other. The study of the multiway cut problem was initiated by Dahlhaus et al. [6], who proved that it is MAX-SNP-hard even when restricted to instances with 3 terminals and unit edge cost. They also gave a $(2-\frac{2}{k})$-approximation algorithm for the problem, where $|S| = k$.

In [4], Călinescu et al. introduced a $(1.5 - \frac{1}{k})$-approximation algorithm. They considered a linear programming relaxation for the multiway cut problem which embeds the given graph into the $(k-1)$-dimensional simplex. The algorithm of [4] rounds an optimal solution to the linear programming relaxation; its bound was later improved to ~ 1.3438 by [11].

In this paper we study two budgeted variants of the multiway cut problem that differ in their objective function. In the budgeted variants, given an instance of the

⋆ This work was done while being on leave at the Dipartimento di Informatica e Sistemistica at Università di Roma "La Sapienza", Italy.

⋆⋆ On leave from the Computer Science Department, Technion, Haifa 32000, Israel.

J.R. Correa, A. Hevia, and M. Kiwi (Eds.): LATIN 2006, LNCS 3887, pp. 435–446, 2006.

multiway cut problem together with an additional positive integer B, the *budget*, the problem is to find a subset of edges whose cost is within the given budget and whose removal maximizes the value of the given objective function.

We say that a pair of terminals (s_i, s_j) is *separated* if there is no path between s_i and s_j, and that a terminal s_i is *isolated* if there is no path between s_i and any other terminal. The number of *isolated* terminals is the objective function of the first *budgeted* variant of the multiway cut problem, referred to as the *budgeted isolating multiway cut* (*BIMC*) problem. In the second *budgeted* variant, referred to as the *budgeted separating multiway cut* (*BSMC*) problem, the objective function is the number of *separated* pairs of terminals. We also consider the *weighted* versions of both *BSMC* and *BIMC*.

An application of the weighted BSMC problem is network design against denial-of-service attacks in networks. In [3], Aura et al. suggest a formal framework for the study of the single-server inhibition attack, which is a common scenario for modelling a denial of service attack. One of the problems they consider is finding the best attack whose cost is within a given budget constraint. In this problem, every client has a non-zero weight denoting its importance. The cost of an attack is the total cost of the disconnected links in the network, and the *value* of the attack is the total weight of the clients separated from the given server. This problem can be considered as a weighted BSMC by setting the weight of every (server, client) pair to be the client's weight.

A well known generalization of the multiway cut problem is the *multicut* problem, which is the problem of finding a minimum cost cut separating a given set of source-sink pairs of vertices. Indeed, the multiway cut problem is a special case of the multicut problem in which the set of source-sink pairs consists of all the pairs of a given set of terminals. Consider the following budgeted variant of the multicut problem. Given is a set of source-sink pairs of vertices together with a budget. Let the source-sink pairs be associated with a non-negative weight. The goal is to find a cut whose cost is within the budget that separates a maximum weight set of source-sink pairs. Thus, this budgeted multicut problem is precisely the weighted version of the BSMC problem.

Finally, given an undirected graph, we consider the problem of finding a set of edges whose cost is within a given budget and whose removal partitions the graph into a maximum number of connected components. This problem, referred to as the *budgeted graph disconnection* (*BGD*) problem, can be thought of as the budgeted version of the *k-cut* problem. In the k-cut problem, an integer k is given and the goal is to find a minimum cost edge set whose removal partitions the graph into at least k connected components.

1.1 Our Results

The hardness of the multiway cut problem implies that both BIMC and BSMC cannot be efficiently solved unless $P = NP$. Although the problem definitions of BIMC and BSMC are closely related, they capture different aspects of the theory of cuts, and therefore differ in their level of hardness. Thus, we study each of the problems independently.

For the BIMC and weighted BIMC problems we give constant factor approximation algorithms that match some of the lower bounds we prove. Our algorithms basically use a greedy approach. In the weighted case we improve on the greedy approach by using an FPTAS for the *knapsack* problem.

We show that weighted BSMC/budgeted multicut is at least as hard to approximate as the *Sparsest Cut* problem is.[1] We show that a natural linear programming relaxation has an unbounded integrality gap. Nevertheless, based on this relaxation, we introduce a constant factor approximation algorithm for the weighted BSMC on *trees*, which implies a constant integrality gap of the relaxation for tree instances. We further note that a better constant factor approximation can be obtained for the weighted BSMC on *trees* through the work of Sviridenko [15]. We then consider the weighted BSMC problem on general graphs. We achieve a bi-criteria approximation of $(\frac{e}{e-1}, O(\log^2 n \log\log n))$ using a recent hierarchical decomposition of graphs by Räcke (see [13] and [9]).

Interestingly, we show that BSMC is related to the budgeted variant of the *Sparsest Cut* problem. Specifically, we prove that for certain weight functions, an approximation algorithm for BSMC can be used to derive an approximation algorithm for the budgeted sparsest cut problem, and vice versa.

Finally, we give a constant factor approximation algorithm for BGD by using the Gomory-Hu tree (see [8]). Our algorithm uses ideas similar to those of the algorithm of Saran and Vazirani [14] for the k-cut problem.

1.2 Related Work

To the best of our knowledge, all of the above mentioned *budgeted* cut problems are studied for the first time here. Nevertheless, there is a vast literature on budgeted optimization problems and we mention the following relevant works.

Vohra and Hall [16] considered a budgeted variant for the classical *set cover* problem, while Khuller et al. [12] studied its weighted variant. They gave a constant factor approximation algorithm for the problem that is based on the greedy approach, and showed that their result is tight under a (weak) assumption on the hardness of NP. Their result points out the possible gap between the hardness of a problem and the hardness of its budgeted variant, as the *set cover* problem cannot be approximated within a factor of $(1-\epsilon)\ln n$ for any $\epsilon > 0$ under the same assumption on the hardness of NP. By improving a former work by Wolsey [17], Sviridenko [15] generalized the result of Khuller et al. for the problem of maximizing any submodular function subject to a budget constraint. We note that this framework does not capture most of the problems we deal with in this paper, but it does capture the weighted BSMC on *trees*.

[1] For the sake of comparison, we note the recent series of results regarding the sparsest cut problem. In [1] an $O(\sqrt{\log n})$ approximation is presented for the uniform case. For the general sparsest cut problem, [5] gave an $O(\log^{\frac{3}{4}} k)$-approximation, which was improved to an $O(\sqrt{\log k} \log\log k)$-approximation by [2].

2 Preliminaries

In this section we formally define the problems considered in this paper. In all of these problems, we are given an undirected graph $G = (V, E)$ with a positive cost function on the edges $c : E \to \mathbb{Z}^+$, and a positive *budget* B.

Problem 1 (Budgeted Graph Disconnection (BGD)). Find a subset of edges $C \subseteq E$ of cost at most B whose removal partitions the graph into the maximum number of connected components.

In the following problems, we are additionally given a subset of vertices $S \subseteq V$ (let $k = |S|$), called *terminals*.

Definition 1 (*Separation and Isolation*). *Let $S \subseteq V$ be a set of terminals. Given a subset of edges $C \subseteq E$, we say that vertices s and s' $(s' \neq s)$ are* separated *by C, or, equivalently, that C is a* separating cut *of (s, s'), if every path between s and s' contains* at least *one edge from C. We say that a vertex $s \in S$ is* isolated *by C, or equivalently, that C is an* isolating cut *of s, if for every $s' \in S$, $s' \neq s$, s and s' are separated by C.*

Definition 2. *Given a weight function on the terminals, $w : S \to \mathbb{Z}^+$, the* isolation weight *of a given subset of edges $C \subseteq E$, is the sum of the weights of the terminals isolated by C. Given a weight function on the pairs of terminals, $w : S \times S \to \mathbb{Z}^+$, the* separation weight *of a given subset of edges $C \subseteq E$, is the sum of the weights of the pairs of terminals separated by C.*

Problem 2 (Weighted Budgeted Isolating Multiway Cut (weighted BIMC)). Given a weight function on the terminals, $w : S \to \mathbb{Z}^+$, find a subset of edges $C \subseteq E$ of cost at most B whose isolation weight is maximized.

Without loss of generality we assume that there exists $s \in S$ such that the cost of the minimum cost isolating cut of s is at most B. We denote by *BIMC* the special case of weighted BIMC where $w(s) = 1$ for every $s \in S$.

Problem 3 (Weighted Budgeted Separating Multiway Cut (weighted BSMC)). Given a weight function on the pairs of terminals, $w : S \times S \to \mathbb{Z}^+$, find a subset of edges $C \subseteq E$ of cost at most B whose separation weight is maximized.

Without loss of generality we assume that for every pair $s, s' \in S$, the cost of the minimum cost separating cut of s and s' is at most B. We denote by *BSMC* the special case of weighted BSMC where $w(s, s') = 1$ for every $s, s' \in S$.

With respect to the same input, we define the Sparsest Cut problem. Given a non-empty subset of vertices $U \subset V$, the cut *associated with* U, denoted by $(U, \overline{U})$, is $\{e = (u, v) \in E : u \in U, v \notin U\}$. The *Sparsity* of the cut $(U, \overline{U})$ is given by $\frac{c(U,\overline{U})}{w(U,\overline{U})}$, where $w(\cdot)$ is the separation weight.

Problem 4 (Sparsest Cut). Find a non-empty subset of vertices $U \subset V$ such that the sparsity of its associated cut is minimized.

Lastly, we say that an algorithm ALG is a *bi-criteria approximation with parameters* (α, β) for a given maximization budget problem Π, or simply an (α, β)-*approximation for* Π, if for every instance of Π with budget B, ALG outputs a solution whose value is at least $|OPT|/\alpha$ and whose cost is at most βB, where $|OPT|$ is the value of the optimal solution with respect to the *given budget* B.

3 The Budgeted Isolating Multiway Cut Problem

In this section, we study BIMC and weighted BIMC problems. First we show some hardness results, including integrality gaps of two possible linear relaxations. These integrality gaps suggest that an approximation algorithm which is based on them cannot outperform the constant factor approximation algorithm we give for BIMC. Finally, we give two approximation algorithms for weighted BIMC, the second of which matches one of the lower bounds we show.

3.1 Hardness Results

The proof of the next two propositions is given in the full version of this paper.

Proposition 1. *Unless* $P = NP$, *there is no* α-*approximation for the BIMC problem for all* $\alpha > 1/3$.

Proposition 2. *Unless* $P = NP$ *there is no* α-*approximation for the BIMC problem for every* $\alpha > 1 - 2/OPT$, *where* $OPT > 2$ *is the number of isolated terminals in an optimal solution. Moreover, there is no* α-*approximation for every* $\alpha > 1 - 2/k$, *when the number of terminals is a fixed* k.

Integrality Gap of Linear Programming Relaxations. We consider two linear programming relaxations for the BIMC problem, and show in the full version of this paper that their integrality gap is at least 2. Hence, we argue that using these relaxations, one cannot achieve an approximation factor for BIMC better than the constant factor approximation presented in the next subsection.

In what follows we assume that for every $s \in S$, the cost of the minimum cost isolating cut of s is at most B (if not, a slight modification can be made in the relaxations and the relevant claims still hold). The first relaxation is a straight forward formulation.

$$\begin{array}{lll}
\max \sum_{s \in S} x_s & & \text{(N-ISO-LP)} \\
\text{s.t.} & & \\
\quad x_s - \sum_{e \in P_{s,s'}} y_e \leq 0 & \text{for every } s, s' \in S \ (s \neq s') & \\
 & \text{and path } P_{s,s'} \text{ from } s \text{ to } s' & \\
\quad \sum_{e \in E} c(e) \cdot y_e \leq B & & \\
\quad 0 \leq x_s \leq 1 & \text{for every } s \in S & \\
\quad 0 \leq y_e & \text{for every } e \in E &
\end{array}$$

The second formulation we consider is derived from the linear programming relaxation of the multiway cut problem presented in [4]. We assume that

$S = \{s_1, \ldots, s_k\}$, and embed the given graph into the k-dimensional simplex. We "reserve" the 0-coordinate for the connected component that contains all the terminals *not* isolated by the solution, and the ith coordinate for the connected component that contains terminal s_i, if terminal s_i is isolated by the solution. Thus, we only allow terminal s_i to be mapped to either the "0" component, or the ith component.

$$\begin{aligned}
&\max \textstyle\sum_{s_i \in S} x^i_{s_i} && \text{(CKR-ISO-LP)}\\
&\text{s.t.}\\
&\quad x^i_{s_i} + x^0_{s_i} = 1 && \text{for } 1 \le i \le k\\
&\quad \textstyle\sum_{i=0}^{k} x^i_v = 1 && \text{for every } v \in V \setminus S\\
&\quad x^i_v \ge 0 && \text{for every } v \in V \text{ and } 0 \le i \le k\\
&\quad y_e = \tfrac{1}{2} \textstyle\sum_{i=0}^{k} |x^i_u - x^i_v| && \text{for every } e = (u,v) \in E\\
&\quad \textstyle\sum_{e \in E} c(e) \cdot y_e \le B
\end{aligned}$$

3.2 A Greedy Approximation Algorithm for BIMC

The following greedy algorithm for BIMC is a variant of the algorithm presented in [6] for the multiway cut problem. As [6] mentioned, note that a minimum cost isolating cut for $s_i \in S$ can be computed efficiently by merging the terminals in $S \setminus \{s_i\}$ into a single node r and computing a minimum cut separating r and s_i.

Algorithm *GR-ISO*: First, for each $s \in S$, find a minimum cost isolating cut for s, and denote it by C_s. Then, sort the cuts in a non-decreasing order of their cost. Output the maximal sequence of cuts, starting from the cheapest, whose total cost is at most B.

Lemma 1. *Let l denote the value of an optimal solution. Algorithm GR-ISO achieves an approximation factor of $\frac{1}{2}$ if l is even, and $\frac{1}{2} - \frac{1}{2l}$ if l is odd.*[2]

Proof. Let OPT be an optimal solution, and let I denote the set of terminals isolated by OPT. We assume without loss of generality that there is no edge in OPT that can be removed without changing the set of isolated terminals. Let $G' = (V, E \setminus OPT)$. For $s \in I$, let OPT_s be the edges in OPT that have an endpoint in the connected component of s in G'.

Consider the following charging scheme for the terminals in I. Charge the cost of every edge $e \in OPT$ as follows: if there exist two distinct terminals $s \in I$ and $s' \in I$ such that $e \in OPT_s$ and $e \in OPT_{s'}$, charge each of the terminals with $c(e)/2$; else, charge the terminal $s \in I$ such that $e \in OPT_s$ with $c(e)$. Denote by $c(s)$ the total cost charged to terminal s. Obviously, $\sum_{s \in I} c(s) = c(OPT) \le B$ (every edge in OPT is clearly paid for by the charging scheme) and $c(C_s) \le c(OPT_s) \le 2c(s)$ for every $s \in I$ (OPT_s is an isolating cut for s). Let A_l be the set of the first l terminals as sorted by the algorithm. Notice that $\sum_{s \in A_l} c(C_s) \le \sum_{s \in I} c(C_s) \le 2 \sum_{s \in I} c(s) \le 2B$. Thus, the cost of the first $\lfloor l/2 \rfloor$ terminals is $\le B$, and the lemma follows from the definition of *GR-ISO*. □

The above analysis is tight as we show in the full version of this paper.

[2] For the trivial case in which $l = 1$ the algorithm finds an optimal solution.

3.3 Approximation Algorithms for the Weighted BIMC Problem

We present two algorithms for the weighted BIMC problem. The first one is a generalization of algorithm *GR-ISO*.

Algorithm *GR-ISO*$_w$. First, for each $s \in S$, find a minimum cost isolating cut for s, and denote it by C_s. Then, sort the cuts with $c(C_s) \leq B$ in a non-decreasing order of the ratio between their cost and their terminal's weight ($c(C_s)/w(s)$). Let $\{C_i\}_{i=1}^k$ be the resulting sequence of cuts. Let $\{C_i\}_{i=1}^m$ be the maximal prefix of $\{C_i\}_{i=1}^k$ with a total cost of at most B. Output the heavier cut (with respect to isolation weight) between $\bigcup_{i=1}^m C_i$ and C_{m+1} (if $m = k$ then $\bigcup_{i=1}^m C_i$ is an optimal solution). In the full version of this paper we prove that Algorithm *GR-ISO*$_w$ achieves an approximation factor of $\frac{1}{4}$.

A $(\frac{1}{3} - \epsilon)$-Approximation. The analysis of algorithm *GR-ISO*$_w$ hints that improving the approximation factor requires an efficient use of the given budget. To this end, we use the FPTAS for the *Knapsack* problem [10], denoted by $A(\pi, \epsilon)$, where π is the Knapsack instance.

Algorithm *PACK*$_w(\epsilon)$. For each $s \in S$, find a minimum cost isolating cut for s, and denote it by C_s. Construct an instance of the *Knapsack* problem, π: treat each terminal $s \in S$ such that $c(C_s) \leq B$ as an item whose profit is $w(s)$ and whose size is $c(C_s)$, and let B be the "knapsack capacity". Run $A(\pi, \epsilon)$ and denote by P the resulting subset of terminals. Finally, Output $\bigcup_{s \in P} C_s$.

Let *OPT* be an optimal solution for the weighted BIMC instance. Since every terminal s with $c(C_s) > B$ cannot be isolated by either *OPT* or *PACK*$_w(\epsilon)$, we ignore such terminals in what follows. Let I denote the set of the terminals isolated by *OPT* and l be the isolation weight of *OPT*, i.e., the value of the optimal solution. Denote by $|OPT(\pi)|$ the value of the optimal solution for the *Knapsack* instance π.

Lemma 2. $|OPT(\pi)| \geq \frac{1}{3}l$.

Proof. Let $U = \{X \subseteq I \mid \sum_{s \in X} w(s) \geq \frac{1}{3}l\}$, i.e., U is the set of the subsets of I of profit $\geq \frac{1}{3}l$. Let Y be a set in U of minimum *size in* π (notice that there must exist such a subset). Assume to the contrary that $|OPT(\pi)| < \frac{1}{3}l$, and in particular that $\sum_{s \in Y} c(C_s) > B$. It follows from our assumption, that for every $s \in S$, $w(s) < \frac{1}{3}l$. Thus, there are at least two terminals in Y, and moreover, $\sum_{s \in Y} w(s) < \frac{2}{3}l$ (otherwise, by taking off a terminal from Y we get a contradiction for the minimality of Y in U with respect to size). By similar arguments to those used in the proof of Lemma 1, we get that $\sum_{s \in I} c(C_s) \leq 2B$. Thus, $\sum_{s \in I \setminus Y} c(C_s) < B$ and $\sum_{s \in I \setminus Y} w(s) > \frac{1}{3}l$ and thus $I \setminus Y$ is a feasible solution to π with the desired value. □

It follows from Lemma 2 and the FPTAS for *Knapsack* that Algorithm *PACK*$_w(\epsilon)$ achieves an approximation factor of $\frac{1}{3} - \epsilon$. We can show that Lemma 2 is tight for arbitrarily large values of k by constructing appropriate examples.

4 Weighted Budgeted Separating Multiway Cut

In this section, we study weighted BSMC. We show that approximating it is at least as hard as the sparsest cut problem. We present a natural linear programming relaxation for the problem and show that it has an unbounded integrality gap for general graphs. However, we give a constant factor approximation algorithm for weighted BSMC on *trees*, which is based on this relaxation. We further note that a better approximation is achieved by Sviridenko's [15] framework. Finally, we use a hierarchical decomposition of graphs by Räcke [13, 9] to obtain a bi-criteria approximation of $(\frac{e}{e-1}, O(\log^2 n \log\log n))$ for arbitrary graphs.

4.1 Hardness Results

Hardness with Respect to the Sparsest Cut Problem. We firstly prove a lemma and a corollary whose proofs are given in the full version of this paper.

Lemma 3. *Given a non-empty cut $C \subseteq E$ that partitions G into $r > 2$ connected components, there is an algorithm that finds a cut $C' \subset C$ such that $c(C')/w(C') \leq c(C)/w(C)$ and C' partitions G into $r-1$ connected components.*

Corollary 1. *Given a non-empty cut $C \subseteq E$, there is an algorithm that finds a non-empty subset of vertices $U \subseteq V$ such that the sparsity of the cut associated with U is at most $c(C)/w(C)$.*

The following theorem shows that the weighted BSMC problem is at least as hard to approximate as the sparsest cut problem is (up to a constant).

Theorem 1. *Let ALG be an (α, β)-approximation for weighted BSMC. Then, there exists a $(1+\epsilon)\alpha\beta$-approximation for Sparsest Cut, for every $\epsilon > 0$.*

Proof. Assume we are given an instance of the Sparsest Cut problem, denote it by π, and let OPT_π denote its optimal solution, and $|OPT_\pi| = \frac{c(OPT_\pi, \overline{OPT_\pi})}{w(OPT_\pi, \overline{OPT_\pi})}$ denote the optimal solution's sparsity. Denote by (π, B) the input for the weighted BSMC problem that consists of the instance π and the budget B, and let $OPT_{\pi,B}$ be a corresponding optimal solution. Then, since $(OPT_\pi, \overline{OPT_\pi})$ is a feasible solution for the weighted BSMC problem on (π, B) for every $B \geq c(OPT_\pi, \overline{OPT_\pi})$, then $w(OPT_\pi, \overline{OPT_\pi}) \leq w(OPT_{\pi,B})$ for every $B \geq c(OPT_\pi, \overline{OPT_\pi})$.

For $\lceil \log_{1+\epsilon} c(C_{min}) \rceil \leq i \leq \lceil \log_{1+\epsilon} c(E) \rceil$, where C_{min} is the minimum cost cut in G, let C_{B_i} be the cut returned by $ALG(\pi, B_i = (1+\epsilon)^i)$. Then, by applying Corollary 1 on each C_{B_i} we can obtain a non-empty subset of vertices $U_i \subseteq V$ where the sparsity of the cut associated with U_i is at most $\frac{c(C_{B_i})}{w(C_{B_i})} \leq \frac{\beta B_i}{w(OPT_{\pi,B_i})/\alpha} = \alpha\beta \frac{B_i}{w(OPT_{\pi,B_i})}$. Let $j = \lceil \log_{1+\epsilon} c(OPT_\pi, \overline{OPT_\pi}) \rceil$. Then, $\frac{B_j}{w(OPT_{\pi,B_j})} \leq (1+\epsilon) \frac{c(OPT_\pi, \overline{OPT_\pi})}{w(OPT_\pi, \overline{OPT_\pi})} = (1+\epsilon)|OPT_\pi|$. We conclude that the sparsity of $(U_j, \overline{U_j})$ is at most $(1+\epsilon)\alpha\beta|OPT_\pi|$, and the theorem follows by choosing the sparsest cut among $\{(U_i, \overline{U_i})\}_{\lceil \log_{1+\epsilon} c(C_{min}) \rceil \leq i \leq \lceil \log_{1+\epsilon} c(E) \rceil}$. □

Integrality Gap of a Linear Programming Relaxation. We present a natural linear programming relaxation for the weighted BSMC problem, and in the full version of the paper we show that its integrality gap of is $\Omega(n)$. This implies that an algorithm based on this linear relaxation would have poor performance. Nevertheless, in what follows we show an approximation algorithm for the special case of trees based on the same relaxation. In what follows we assume w.l.o.g. that $c(e) \leq B$ for every $e \in E$.

$$
\begin{aligned}
&\max \textstyle\sum_{s_i,s_j \in S} w(s_i,s_j) \cdot x_{ij} && \text{(SEP-LP)}\\
&\text{s.t.}\\
&\quad x_{ij} - \textstyle\sum_{e \in P_{i,j}} y_e \leq 0 && \text{for every } s_i, s_j \in S\\
&&& \text{and path } P_{i,j} \text{ from } s_i \text{ to } s_j\\
&\quad \textstyle\sum_{e \in E} c(e) \cdot y_e \leq B\\
&\quad 0 \leq x_{ij} \leq 1 && \text{for every } s_i, s_j \in S\\
&\quad 0 \leq y_e && \text{for every } e \in E
\end{aligned}
$$

4.2 Approximation Algorithms for Weighted BSMC in Trees

Let P_{ij} denote the unique path in the tree between s_i and s_j. The dual LP of SEP-LP is:

$$
\begin{aligned}
&\min B \cdot \gamma + \textstyle\sum_{s_i,s_j \in S} \beta_{ij} && \text{(SEP-DLP)}\\
&\text{s.t.}\\
&\quad c(e) \cdot \gamma - \textstyle\sum_{i,j: e \in P_{ij}} \alpha_{ij} \geq 0 && \text{for every } e \in E\\
&\quad \alpha_{ij} + \beta_{ij} \geq w(s_i,s_j) && \text{for every } s_i, s_j \in S\\
&\quad \gamma, \alpha_{ij}, \beta_{ij} \geq 0 && \text{for every } s_i, s_j \in S
\end{aligned}
$$

We define the *worthiness of an edge e with respect to C*, a feasible solution, as $\Gamma_C(e) = \frac{\sum_{i,j: e \in P_{ij}} w(s_i,s_j) \cdot (1 - x_{ij})}{c(e)}$, where x is the corresponding solution of SEP-LP. The following algorithm greedily updates the solution as long as the budget is not exceeded, while maintaining the corresponding solution of SEP-LP.

Algorithm *GR-SEP*. Initialize: $h = 0$, $C_0 = \emptyset$, $\forall\ s_i, s_j \in S$, $x_{ij} = 0$, and $\forall\ e \in E$, $y_e = 0$. While there exists an edge $e \in E \setminus C_h$, execute the following loop: Let e_h be a lowest cost edge among the edges with the maximum value of Γ_{C_h}. If $c(e_h) > B - c(C_h)$, output the better solution between $\{e_h\}$ and $C = C_h$. Otherwise, $C_{h+1} \leftarrow C_h \cup \{e_h\}$, set $y_{e_h} = 1$ and $x_{ij} = 1$ for all the pairs (s_i, s_j) separated by e_h, and let $h \leftarrow h + 1$. If $E \setminus C_h$ is empty, output $C = C_h$.

Observation 1. *By the definition of the worthiness of an edge, and the fact that edges are only added to the solution during the algorithm, for every* $e \in E$ *and* $0 < h \leq |C|$, $\Gamma_{C_h}(e) \leq \Gamma_{C_{h-1}}(e)$, *i.e. the worthiness of an edge can only decrease during the algorithm.*

Corollary 2. *If* $C \neq E$, *then* $c(e_{|C|}) \cdot \Gamma_C(e_{|C|}) \leq c(e_{|C|}) \cdot \Gamma_{C_0}(e_{|C|}) = w(\{e_{|C|}\})$, *i.e., adding the edge* $e_{|C|}$ *to* C *increases its separation weight by at most the separation weight of* $\{e_{|C|}\}$.

Theorem 2. *Algorithm GR-SEP achieves an approximation factor of $\frac{1}{3}$.*

Proof. If the algorithm outputs $C = E$, the solution is optimal. Otherwise, it is the case that $\sum_{h=0}^{|C|-1} c(e_h) + c(e_{|C|}) > B$. Denote by $w(\mathit{GR\text{-}SEP})$ the value of the solution output by *GR-SEP*. Consider the following dual solution: $\beta_{ij} = w(s_i, s_j) \cdot x_{ij}, \alpha_{ij} = w(s_i, s_j) \cdot (1 - x_{ij}), \gamma = \Gamma_C(e_{|C|})$. Since $\Gamma_C(e_{|C|}) \geq \Gamma_C(e)$ for every $e \notin C$, this is a feasible dual solution. Let z denote its value. Then,

$$
\begin{aligned}
z = B \cdot \gamma + \sum_{s_i, s_j \in S} \beta_{ij} &= B \cdot \Gamma_C(e_{|C|}) + \sum_{s_i, s_j \in S} w(s_i, s_j) \cdot x_{ij} && (1)\\
&< \left(\sum_{h=0}^{|C|-1} c(e_h) + c(e_{|C|}) \right) \cdot \Gamma_C(e_{|C|}) + w(C) && (2)\\
&\leq \sum_{h=0}^{|C|-1} c(e_h) \cdot \Gamma_C(e_{|C|}) + w(\{e_{|C|}\}) + w(C) && (3)\\
&\leq \sum_{h=0}^{|C|-1} c(e_h) \cdot \Gamma_{C_h}(e_h) + w(\{e_{|C|}\}) + w(C) && (4)\\
&= w(C) + w(\{e_{|C|}\}) + w(C) \leq 3w(\mathit{GR\text{-}SEP}), && (5)
\end{aligned}
$$

Where: Inequality (2) follows from the definition of the algorithm *GR-SEP*, (3) follows from Corollary 2, and (4) follows by noticing that for every $0 < h \leq |C|$, $\Gamma_{C_h}(e_h) \leq \Gamma_{C_{h-1}}(e_{h-1})$. Thus, the theorem follows by weak duality. □

Remark 1. In [15], Sviridenko introduces a greedy $\frac{e-1}{e}$-approximation algorithm for the problem of maximizing a submodular function subject to a budget constraint. We note that on tree instances (unlike general graphs), the separation weight is a submodular function and thus the weighted BSMC problem on trees can be solved using Sviridenko's algorithm. We note that Sviridenko's algorithm's running time is $\Omega(n^3)$, while the running time of *GR-SEP* is $O(n^2)$.

4.3 Weighted BSMC - General Graphs

In this subsection we introduce an $(\frac{e}{e-1}, O(\log^2 n \log\log n))$-approximation algorithm for weighted BSMC. Since the weighted budgeted variant of Multicut is equivalent to weighted BSMC, we conclude that a $(\frac{e}{e-1}, O(\log^2 n \log\log n))$-approximation exists for this problem as well.

In [13], Räcke describes a hierarchical decomposition of any undirected graph $G = (V, E)$ into a tree T_G, where there is a $1-1$ correspondence between V and the leaves of T_G. T_G has the property that any feasible multi-commodity flow function in T_G can be routed in G causing a congestion bounded by a function of G's parameters, denoted by β. By min-cut-max-flow theorems this implies a corresponding bounded ratio between the cost of cuts in G and the cost of cuts in T_G. In [9], Harrelson et al. give a polynomial-time construction of T_G with $\beta = O(\log^2 n \log\log n)$, which we use in the following algorithm.

Algorithm *SEP*. Let $B' = 2\beta B$. Construct a decomposition tree, T_G, of G. For every $e = (u, v) \in T_G$ with a cost $> B'$, merge the vertices u and v. Let T'_G be the resulted tree. Run Sviridenko's algorithm on T'_G with budget B', and output the associated cut in G.

Theorem 3. *Algorithm SEP is a* $(\frac{e}{e-1}, O(\log^2 n \log\log n))$*-approximation for the weighted BSMC problem.*

Proof. Let OPT be an optimal solution, and let I denote the set of pairs of terminals separated by OPT. Let OPT_{T_G} be a minimum cost cut separating I in T_G. By [7], $c(OPT_{T_G}) \leq 2MCF_I(T_G)$, where $MCF_I(T_G)$ is the value of the maximum multi-commodity flow in T_G between the pairs in I. By the construction of T_G and its property, $MCF_I(T_G) \leq \beta MCF_I(G)$. Since $MCF_I(G)$ lower bounds the cost of any cut separating I in G, $MCF_I(G) \leq c(OPT)$, and thus we get $c(OPT_{T_G}) \leq 2\beta c(OPT) \leq 2\beta B = B'$. Particularly, OPT_{T_G} does not contain any edge with cost more than B', and thus OPT_{T_G} is a feasible solution for the weighted BSMC problem on T'_G with budget B', with value $w(OPT_{T_G}) \geq w(OPT)$. From [15], running Sviridenko's algorithm will return a solution C whose cost is at most B' and whose value is at least $\frac{e-1}{e} w(OPT_{T_G})$. By the properties of the decomposition tree, the associated cut in G has a cost of at most B' and a separation weight of at least $w(C)$ and the theorem follows.

□

4.4 Further Discussion

Although it remains an open question whether it is possible to improve upon the bi-criteria approximation, or even achieve a uni-criteria approximation, in this subsection we review some related ideas and point out some possible directions towards solving the problem. First, consider the following budget problem, whose input is the same as the input for the weighted BSMC problem.

Problem 5 (Budgeted Sparsest Cut). Find a non-empty subset of vertices $U \subset V$ such that $c(U, \overline{U}) \leq B$ and the sparsity of $(U, \overline{U})$ is minimized.

In order to understand the relationship between weighted BSMC and Budgeted Sparsest Cut, we look for results similar to those of Subsection 4.1. Notice that the algorithm of Corollary 1 actually finds a cut whose cost is at most $c(C)$. Hence, Theorem 1 can be easily generalized to obtain the following.

Theorem 4. *Let ALG be an* (α, β)*-approximation for weighted BSMC. Then, there exists a* $((1+\epsilon)\alpha\beta, \beta)$*-approximation for the Budgeted Sparsest Cut problem for every* $\epsilon > 0$.

Specifically, notice that a uni-criteria approximation for weighted BSMC implies a uni-criteria approximation for the Budgeted Sparsest Cut problem. This result suggests that the budgeted sparsest cut is not harder than BSMC. Nevertheless, it seems that the budgeted sparsest cut is not much easier, as we argue in the full paper, and we prove for certain weight functions that this is indeed the case.

5 The Budgeted Graph Disconnection Problem

The following algorithm for BGD is a variant of the algorithm presented in [14] for the k-cut problem.

Algorithm GR-PAR. First, compute a Gomory-Hu tree T for G. Then, sort the edges of T in a non-decreasing order of their cost. Finally, choose the maximal sequence of edges starting from the cheapest whose cost is at most B, and output the union of the cuts in G corresponding to these edges. Let l be the value of an optimal solution. We can prove that Algorithm *GR-PAR* achieves an approximation factor of $\frac{1}{2} + \frac{1}{l}$ if l is even, and of $\frac{1}{2} + \frac{1}{2l}$ if l is odd.

References

[1] S. Arora, S. Rao, and U. Vazirani. Expander flows, geometric embeddings, and graph partitionings. In *Proc. of STOC'04*, pages 222–231, 2004.
[2] S. Arora, J. R. Lee and A. Naor. Euclidean distortion and the sparsest cut. In *Proc. of STOC'05*, pages 553–562, 2005.
[3] T. Aura, M. Bishop and D. Sniegowski. Analyzing single-server network inhibition. In *Proc. of CSFW'00*, p. 108–117, 2000.
[4] G. Călinescu, H. Karloff and Y. Rabani. An Improved Approximation Algorithm for Multiway Cut. *JCSS*, 60(3):564–574, 2000.
[5] S. Chawla, A. Gupta and H. Räcke. Embeddings of negative-type metrics and an improved approximation to generalized sparsest cut. In *Proc. of SODA'05*, p. 102–111, 2005.
[6] E. Dahlhaus, D. Johnson, C. Papadimitriou, P. Seymour and M. Yannakakis. The Complexity of Multiterminal Cuts. *SIAM J. on Computing*, 23:864–894, '94. Preliminary version appeared in *Proc. of the 24th ACM Symposium on Theory of Computing*, p. 241–251, 1992.
[7] N. Garg, V.V.Vazirani and M. Yannakakis. Primal-dual approximation algorithms for integral flow and multicut in trees. *Algorithmica*, 18:3–20, 1997.
[8] R. Gomory and T. Hu. Multiterminal network flows. *J. of SIAM*, 9:551–570, '61.
[9] C. Harrelson, K. Hidrum and S. Rao. A polynomial time tree decomposition to minimize congestion. In *Proc. of SPAA'03*, p. 34–43, 2003.
[10] O.H. Ibarra and C.E. Kim. Fast approximation algorithms for the knapsack and sum of subset problems. *J. of the ACM*, 22:463–468, 1975.
[11] D. Karger, P. Klein, C. Stein, M. Thorup and N. Young. Rounding algorithms for a geometric embedding of minimum multiway cut. *STOC'99*, p. 668–678, '99.
[12] S. Khuller, A. Moss and J. Naor. The Budgeted Maximum Coverage Problem. *Information Processing Letters*, 70(1):39–45, '99.
[13] H. Räcke. Minimizing congestion in general networks. *In FOCS'02*, p. 43–52, '02.
[14] H. Saran and V.V. Vazirani. Finding k-cuts within twice the optimal. *SIAM J. on Comp.*, 24:101–108, 1995.
[15] M. Sviridenko. A note on maximizing a submodular set function subject to knapsack constraint. *Operations Research Letters*, 32:41–43, 2004.
[16] R.V. Vohra and N.G. Hall. A probabilistic analysis of the maximal covering location problem. *Discrete Applied Mathematics*, 43(2):175–183, 1993.
[17] L. Wolsey. Maximizing real-valued submodular functions: primal and dual heuristics for location problems. *Mathematics of Operations Research*, 7:410–425, '82.

Lower Bounds for Clear Transmissions in Radio Networks *

Martín Farach-Colton, Rohan J. Fernandes, and Miguel A. Mosteiro

Department of Computer Science, Rutgers University,
Piscataway, NJ 08854, USA
{farach, rohanf, mosteiro}@cs.rutgers.edu

Abstract. We show new lower bounds for collision-free transmissions in Radio Networks. Our main result is a tight lower bound of $\Omega(\log n \log(1/\epsilon))$ on the time required by a uniform randomized protocol to achieve a clear transmission with success probability $1 - \epsilon$ in a one-hop setting. This result is extended to non-uniform protocols as well. A new lower bound is proved for the important multi-hop setting of nodes distributed as a connected Random Geometric Graph. Our main result is tight for a variety of problems.

Keywords: Ad-hoc network, Broadcast, Contention resolution, Dominating set, Leader election, Lower bound, Maximal independent set, Radio network, Random geometric graphs, Station selection, Sensor network, Wake up, Weak Sensor Model. ACM-class: F.2.2.

1 Introduction

Any network where transmissions may collide needs a protocol for *collision-free transmissions*. Different networks provide different information about collisions. For example, on some hardware, transmitters can distinguish amongst three states at each time step: no transmission, single transmission, and collision, whereas on other hardware, transmitters can not distinguish between no transmission and collisions. In some networks, transmitters know an upper bound on their number. Sometimes, transmitters may not *snoop*, i.e., listen to the channel when not transmitting; whereas at the other extreme, transmitters may only snoop, i.e., they get no information on the channel when they are transmitting. In some networks collisions are transitive. The properties of a shared channel have a profound impact on the protocols usable on such a channel.

Sensor networks are a heavily studied example of a shared-channel network. A sensor network consists of small devices with processing, sensing and communication capabilities. These *sensor nodes* are randomly deployed over an area in order to achieve sensing tasks after self-organizing as a wireless radio network. Sensor nodes have strong limitations and operate under harsh conditions. Some of the important limitations of

* This research was supported in part by DIMACS, Center for Discrete Mathematics & Theoretical Computer Science, grants numbered NSF CCR 00-87022, NSF EIA 02-05116 and Alfred P. Sloan Foundation 99-10-8.

J.R. Correa, A. Hevia, and M. Kiwi (Eds.): LATIN 2006, LNCS 3887, pp. 447–454, 2006.

sensor nodes include: lack of collision detection hardware, non-simultaneous transmission and reception, and one channel of communication. We call any such network a Radio Network. Additionally, nodes in sensor networks wake up at arbitrary times. Sensor networks are even more restricted in various ways that will not concern us here. The Radio Network restrictions, along with these further restrictions, are part of the Weak Sensor Model presented in [1].

Algorithms for achieving a *clear*, that is, uncolliding, transmission have been studied in several shared-channel contention settings[1]. Hayashi, Nakano and Olariu [3] presented the first $O(\log^2 n)$ algorithm for clear transmission with high probability in one-hop Radio Networks. A $O(\log n \log(1/\epsilon))$-time algorithm for a clear transmission with probability $1-\epsilon$ in a one-hop Radio Network was introduced in [2]. Strikingly, when $\epsilon = 1/n$ the same time bound can be obtained for the much more complicated problem of computing a Maximal Independent Set (MIS) in the multi-hop Weak Sensor Model [7]. Kushilevitz and Mansour [5] proved the first lower bound of $\Omega(\log n)$ on the expectation of the running time of any randomized algorithm for clear transmissions in radio networks. A lower bound of $\Omega(\log n \log(1/\epsilon)/(\log\log n + \log\log(1/\epsilon))$ for achieving a clear transmission with probability $1-\epsilon$ in a one-hop, globally-synchronized Radio Network was proved in [4]. The latter lower bound is tighter than the previous one if ϵ is $o(1/\log n)$.

Our Results: The gap between the lower bound for achieving something so simple as a clear transmission and upper bounds for more complicated problems such as MIS is tantalizingly narrow: respectively $\Omega(\log^2 n/\log\log n)$ and $O(\log^2 n)$, when ϵ is $\Theta(1/n^c)$. In this paper, we close this gap by proving a stronger lower bound: it takes time $\Omega(\log n \log(1/\epsilon))$ to solve the problem of achieving a clear transmission with probability $1-\epsilon$ in a one-hop setting, which implies, for example, the $\Omega(\log n)$ lower bound on the expectation of any randomized algorithm for clear transmission. Our lower bounds apply to any network with the following characteristics:

- *Shared channel of communication:* All nodes communicate with their neighbors using broadcasts that are transmitted on a shared channel.
- *Lack of a collision detection mechanism:* Nodes do not have the ability to distinguish between a collision on the channel or lack of a transmission.
- *Non-simultaneous transmission and reception:* Nodes cannot snoop on the channel while transmitting.
- *Local synchronization:* Time is assumed to be divided into slots and all nodes have the same clock frequency.
- *Adversarial wake-up schedule:* Nodes are woken up by an adversary.

Indeed, we will prove our lower bound with the following weak adversary: the adversary may chose an $i \in [1, \log n]$, and 2^i nodes wake up at time 0. Our techniques

[1] In a one-hop Radio Network, the clear transmission problem is equivalent to the so-called *wake-up* and *leader election* problems. These problems differ in multi-hop networks, although, a clear transmission is necessary to achieve wake-up and leader election since, indeed, a clear transmission is necessary to solve any problem on a Radio Network. Since we are interested in lower bounds, we will cite bounds for the clear transmission problem in previous papers, even when the bounds were originally stated for the other problems.

also give us a lower bound of $\Omega(\log\log n \log(1/\epsilon))$ on clear transmissions in the well-studied case of sensor nodes distributed uniformly at random with enough nodes to ensure connectivity, and thus for more complicated problems such as MIS. There was no non-trivial lower bound known for this problem, and the best upper bound known is $O(\log^2 n)$ with high probability, proved for the more complicated problem of sensor network initialization in the Weak Sensor Model [1].

1.1 Roadmap

In Section 2, we show the main lower bound, for uniform protocols in one-hop networks. In Section 3, we extend this result in two ways. We show a lower bound for nonuniform protocols, and a lower bound for nodes distributed geometrically.

2 Uniform Protocols in One-Hop Radio Networks

In this section, we prove a lower bound on randomized uniform protocols and extend the result to nonuniform protocols in Section 3.

We first define what the clear transmission problem is in the one-hop setting. The nodes are all connected to a common broadcast channel and each transmission is available for snooping to all non-transmitting nodes. The connectivity of the nodes can be modelled as a clique. In this case we assume that all nodes know an upper bound on the number of their neighbors. In this setting, a clear transmission is achieved if exactly one node transmits in a time slot.

As explained in Section 1, we prove our lower bounds under the assumption of the existence of a weak adversary that, at a given time, wakes up (i.e. turns on) some subset of nodes. We call them *active* nodes. Upon waking up, the active nodes start the execution of a protocol to achieve a clear transmission. All non-active nodes do not participate in the protocol.

We define a *randomized uniform protocol* for clear transmission to be a sequence $p_1, p_2, \ldots$ where each node transmits with probability p_ℓ in the ℓth time step after waking up. Given our adversary, this means that all active nodes transmits with same probability as each other in each time slot.

We seek a lower bound on the number of time-slots required to achieve a clear transmission with probability $(1-\epsilon)$. We simplify the analysis in two ways. First, we further weaken the adversary by requiring that the number of nodes participating can only be one of $\{2^i | 0 \leq i \leq \log_2 n\}$. Secondly, we assume that all $p_\ell \in \{2^{-j} | 1 \leq j \leq \log_2 n\}$. If this assumption is not true of a particular algorithm A, we can always produce an algorithm A' from A by replacing one attempt in A by a constant number of attempts in A' where the probabilities of transmission in A' have been rounded off to the closest power of $1/2$.

One of the principal benefits of our weak adversary is that, the probability P_ℓ of a clear transmission by time ℓ is the same for any permutation of $p_1, p_2, \ldots, p_\ell$. Therefore, we need not bother with what order the steps are taken in, but only how many times the protocol fires with each probability.

Let t_j be the number of time-slots that nodes are transmitting with probability 2^{-j}. Let p_{ij} denote the probability that 2^i nodes fail to clear when they all transmit with probability 2^{-j}. Thus we know that:

$$
\begin{aligned}
p_{ij} &= 1 - 2^i \frac{1}{2^j}\left(1 - \frac{1}{2^j}\right)^{2^i - 1} \\
&= 1 - 2^{i-j}(1 - 2^{-j})^{2^i - 1}
\end{aligned}
$$

The total probability of failure for any number of active nodes, 2^i, needs to be bounded by:

$$
\begin{aligned}
&\prod_j p_{ij}^{t_j} \le \epsilon \\
\iff &\sum_j t_j \ln(p_{ij}) \le \ln(\epsilon).
\end{aligned}
$$

A lower bound is achieved by minimizing the total number of time-slots needed to satisfy the previous constraints. This can be formulated as the following *primal* linear program:

$$
\begin{aligned}
&\text{Minimize } \mathbf{1}^T \mathbf{t}, \\
&\text{subject to:} \\
&\qquad \mathbf{P}\mathbf{t} \ge \boldsymbol{\epsilon} \\
&\qquad \mathbf{t} \ge \mathbf{0} \\
&\text{where:} \\
&\qquad \mathbf{t} \triangleq [t_j], \\
&\qquad \boldsymbol{\epsilon} \triangleq [-\ln(\epsilon)], \\
&\qquad \mathbf{P} \triangleq [-\ln(p_{ij})],
\end{aligned}
$$

which yields the following *dual*:

$$
\begin{aligned}
&\text{Maximize } \boldsymbol{\epsilon}^T \mathbf{u}, \\
&\text{subject to:} \\
&\qquad \mathbf{P}^T \mathbf{u} \le \mathbf{1} \\
&\qquad \mathbf{u} \ge \mathbf{0}.
\end{aligned}
$$

The primal linear program has a finite minimum solution, and hence its dual has a finite maximum solution. The value of the objective function for every feasible solution of the dual is a lower bound on the minimum value of the objective function for the primal. Thus any feasible solution for the dual will give a lower bound on the number of time-slots required to achieve a clear transmission with failure probability ϵ.

Suppose that the jth row, $\mathbf{P}_j^T$, of $\mathbf{P}^T$ has the maximum row sum, and let $r(\mathbf{P}^T) = \mathbf{P}_j^T \mathbf{1}$. Now we set $\mathbf{u} = [1/r(\mathbf{P}^T)]$. This value of $\mathbf{u}$ satisfies all constraints of the dual. The value of the objective function of the dual is simply $\boldsymbol{\epsilon}^T \mathbf{u}$. To obtain the value of the objective function of the dual we need to find the row of $\mathbf{P}^T$ with the largest row sum which is the same as the column of $\mathbf{P}$ with the largest column sum.

Lemma 1. *The trace of every column vector of the constraint matrix* $\mathbf{P}$ *of the primal is in* $O(1)$.

Proof. We begin by stating the following useful inequality [6, §2.68]:

$$e^{-x/(1-x)} \leq 1 - x \leq e^{-x}, 0 < x < 1. \tag{1}$$

The sum of the elements of a column j of $\mathbf{P}$ is:

$$\begin{aligned} S_j &\leq \sum_i -\ln(1 - 2^{i-j}(1-2^{-j})^{2^i-1}) \\ &\leq \sum_i -\ln\left(e^{-2^{i-j}(1-2^{-j})^{2^i-1}/(1-2^{i-j}(1-2^{-j})^{2^i-1})}\right) \text{ (By Inequality 1)} \\ &= \sum_i \frac{2^{i-j}(1-2^{-j})^{2^i-1}}{1-2^{i-j}(1-2^{-j})^{2^i-1}}. \end{aligned}$$

Let $y_{ij} \triangleq 2^{i-j}(1-2^{-j})^{2^i-1}$.

$$\begin{aligned} S_j &= \sum_i \frac{y_{ij}}{1-y_{ij}} \\ &\leq \sum_i \frac{y_{ij}}{1-y_{max}} \text{ (where } y_{max} = \max_{ij}\{y_{ij}\}). \end{aligned}$$

Now we derive an upper bound on y_{max}:

$$\begin{aligned} y_{max} &= \max_{ij} y_{ij} \\ &= \max_{ij} 2^{i-j}(1-2^{-j})^{2^i-1} \\ &\leq \max_{ij} 2^{i-j}e^{-2^{i-j}+2^{-j}} \text{ (By Inequality 1)} \\ &\leq \max_{ij} \sqrt{e}\frac{2^{i-j}}{e^{2^{i-j}}} \; (\because j \geq 1) \\ &\leq \frac{1}{\sqrt{e}} \text{ (The function is maximized, when } i = j). \end{aligned}$$

Therefore:

$$S_j \leq \frac{\sqrt{e}}{\sqrt{e}-1}\sum_i y_{ij}$$

We derive an upper bound on the right hand side sum.

$$\begin{aligned} \sum_i y_{ij} &= \sum_i 2^{i-j}(1-2^{-j})^{2^i-1} \\ &\leq \sum_i 2^{i-j}(e^{-2^{-j}})^{2^i-1} \text{ (By Inequality 1)} \\ &= \sum_i 2^{i-j}e^{-2^{i-j}+2^{-j}} \end{aligned}$$

$$\leq \sqrt{e}\left(\sum_{i\geq j} 2^{i-j}e^{-2^{i-j}} + \sum_{i<j} 2^{i-j}e^{-2^{i-j}}\right) (\because j \geq 1)$$

$$\leq \sqrt{e}\left(\sum_{k\geq 0} 2^{k}e^{-2^{k}} + \sum_{k\geq 1} 2^{-k}e^{-2^{-k}}\right)$$

$$\leq \sqrt{e}\left(\sum_{k\geq 0} 2^{k}e^{-2^{k}} + \sum_{k\geq 1} 2^{-k}\right)$$

$$\in O(1) \text{ (Because both the sums are bounded by a constant)}$$

$$\implies S_j \in O(1).$$

Theorem 1. *Every uniform randomized algorithm to achieve a clear transmission with probability* $1-\epsilon$ *in a one-hop Radio Network requires* $\Omega(\log n \log(1/\epsilon))$ *time-slots.*

Proof. From lemma 1, we know that $r(P^T) \in O(1)$, then $\boldsymbol{\epsilon}^T\mathbf{u} = \lceil -\ln(\epsilon)\rceil \cdot \lceil 1/\mathbf{P}^T_{max}\mathbf{1}\rceil \in O(\log n \log(1/\epsilon))$. From this we can conclude that the dual linear program has a feasible solution with objective function evaluating to $\Omega(\log n \log(1/\epsilon))$. Since we showed earlier that the solution to the primal linear program gives a lower bound on the number of time-slots required to achieve a clear transmission with probability $1-\epsilon$, the statement of the theorem holds.

3 Extensions

In this section, we show how to obtain lower bounds for nonuniform protocols and for geometric distributions of nodes.

3.1 Randomized Nonuniform Protocols in One Hop Radio Networks

In this section we prove our lower bound for the case in which processors may run different algorithms using their unique ID's to break symmetry. We call this a *nonuniform protocol*. Recall that we model a randomized protocol to achieve a clear transmission as a schedule, or temporal sequence, of probabilities of transmission such that, at time slot i an active node transmits with probability p_i. In the case of the randomized uniform protocols, we assume that nodes either have no ID or the protocol does not make use of it to break symmetry. Then, given that no information can be obtained from a shared-channel before a clear transmission, all active nodes transmit with the same probability in the same time slot. On the other hand, if nodes have unique ID's, they may use different schedules of probabilities of transmission and achieve a clear transmission faster. We prove in this section that in fact having unique ID's does not help.

As in [5], we prove our lower bound by showing a reduction from a nonuniform protocol to a uniform one. We first state our result formally.

Theorem 2. *Every randomized nonuniform protocol to achieve a clear transmission with probability* $1-\epsilon$ *in a one-hop Radio Network requires* $\Omega(\log n \log(1/\epsilon))$ *time slots.*

Proof. For the sake of contradiction, assume that there exists a randomized nonuniform protocol $\mathcal{A}$ that achieves a clear transmission with probability $1 - \epsilon$ in T time slots, where $T \in o(\log n \log(1/\epsilon))$. Then, we can define a randomized uniform protocol $\mathcal{A}'$ that achieves the same running time as follows.

For each node
Choose uniformly at random an integer $i \in [1, n^2/\epsilon]$. [2]
Simulate protocol $\mathcal{A}$ *using* i *as ID.*

Each node running the protocol $\mathcal{A}'$ obtains a unique ID with probability at least $1-\epsilon$. This is true because the probability that some pair of nodes chooses the same ID is ϵ/n^2 and there are $\binom{n}{2}$ possible pairs. Given that the random choice of the ID can be done in constant time, the protocol $\mathcal{A}'$ is a randomized uniform protocol that achieves a clear transmission with probability $1 - 2\epsilon$ in $o(\log n \log 1/\epsilon) = o(\log n \log 1/2\epsilon)$ time slots, which is a contradiction with Theorem 1.

3.2 Randomized Protocols for Geometrically Distributed Nodes

We begin with some preliminaries on geometrically distributed nodes, before getting to the lower bound. In the Random Geometric Graph Model $\mathcal{G}_{n,r,\ell}$, n nodes are distributed uniformly at random in $[0, \ell]^2$, and nodes are connected by an edge iff they are at Euclidean distance at most r, the *connectivity radius.* The node density depends on the relative values of n,r and ℓ. A specific instance of $\mathcal{G}_{n,r,\ell}$ is a *Random Geometric Graph (RGG)*, also referred to as $G(n, r, \ell)$. In a $G(n, r, \ell)$, the asymptotic behavior of route stretch is studied as $\ell \to \infty$ while maintaining sufficient density to preserve connectivity. In this paper we will assume that parameter conditions to ensure connectivity are always satisfied [8].

Here we consider the problem of achieving a clear transmission under the following conditions:

The nodes are connected by a broadcast channel to some subset of nodes and each transmission made by a node is available to its neighbors only, but it can interfere with all transmissions originating in a two-hop neighborhood. The specific case we will derive a lower bound for is the case of nodes consistent with the Weak Sensor Model distributed randomly in the plane with limited transmission range but adequate density to ensure connectivity. The connectivity of the nodes can be modelled as a *Random Geometric Graph (RGG).* In this case, we assume that nodes know an upper bound on the number of their neighbors with a probability given by the parameter conditions for connectivity.

In this setting, we say that a clear transmission occurred if exactly one node is transmitting and no other nodes within two hops of it are transmitting. Then, the clear transmission problem in a multi-hop setting is solved after every node either produces or receives a clear transmission.

In a $G(n, r, \ell)$ satisfying the connectivity conditions explained previously, the number of nodes contained in any circle of radius $\Theta(r)$ is $\Theta(\log n)$ with high probability.

[2] Under the assumptions of the Weak Sensor Model, nodes have only $O(\log n)$ bits of memory. Therefore, this lower bound applies also to sensor networks when $\epsilon \geq 1/n^\gamma$, for some constant $\gamma > 0$.

This can be proved by a simple application of the Chernoff-Hoeffding bounds. Then, we complete our lower bounds with the following corollary, which can be obtained as a simple application of Theorems 1 and 2 to this setting.

Corollary 1. *Every randomized protocol to solve the clear transmission problem with probability* $1 - \epsilon$ *in a Radio Network with geometrically distributed nodes requires* $\Omega(\log \log n \log(1/\epsilon))$ *time slots, where* $\epsilon \geq 1/n^\gamma$ *for some constant* $\gamma > 0$.

Proof. Replacing the appropriate density for any one-hop neighborhood in this setting, i.e. $\Theta(\log n)$ instead of n, in theorem 2 the corollary follows.

References

1. M. Farach-Colton, R. J. Fernandes, and M. A. Mosteiro. Bootstrapping a hop-optimal network in the weak sensor model. In *Proc. of the 13th Annual European Symposium on Algorithms*, 2005.
2. L. Gasieniec, A. Pelc, and D. Peleg. The wakeup problem in synchronous broadcast systems. In *Proceedings 19th Annual ACM Symposium on Principles of Distributed Computing*, 2000.
3. T. Hayashi, K. Nakano, and S. Olariu. Randomized initialization protocols packet radio networks. In *IPPS*, 1999.
4. T. Jurdziński and G. Stachowiak. Probabilistic algorithms for the wakeup problem in single-hop radio networks. In *ISAAC*, 2002.
5. E. Kushilevitz and Y. Mansour. An $\Omega(D \log(N/D))$ lower bound for broadcast in radio networks. *SIAM J. Comput.*, 27(3):702–712, 1998.
6. D. S. Mitrinović. *Elementary Inequalities*. P. Noordhoff Ltd. - Groningen, 1964.
7. T. Moscibroda and R. Wattenhofer. Maximal independent sets in radio networks. In *PODC*, 2005.
8. S. Muthukrishnan and G. Pandurangan. The bin-covering technique for thresholding random geometric graph properties. In *Proc. of ACM-SODA*, 2005.

Asynchronous Behavior of Double-Quiescent Elementary Cellular Automata

Nazim Fatès, Damien Regnault, Nicolas Schabanel, and Éric Thierry

LIP (UMR CNRS ÉNSL UCBL INRIA 5668),
46 allée d'Italie, 69364 Lyon Cedex 07, France
{nazim.fates, damien.regnault, nicolas.schabanel,
eric.thierry}@ens-lyon.fr

Abstract. In this paper we propose a probabilistic analysis of the relaxation time of elementary finite cellular automata (i.e., $\{0,1\}$ states, radius 1 and unidimensional) for which both states are quiescent (i.e., $(0,0,0) \mapsto 0$ and $(1,1,1) \mapsto 1$), under α-asynchronous dynamics (i.e., each cell is updated at each time step independently with probability $0 < \alpha \leqslant 1$). This work generalizes previous work in [1], in the sense that we study here a continuous range of asynchronism that goes from full asynchronism to full synchronism. We characterize formally the sensitivity to asynchronism of the relaxation times for 52 of the 64 considered automata. Our work relies on the design of probabilistic tools that enable to predict the global behaviour by counting local configuration patterns. These tools may be of independent interest since they provide a convenient framework to deal *exhaustively* with the tedious case analysis inherent to this kind of study. The remaining 12 automata (only 5 after symmetries) appear to exhibit interesting complex phenomena (such as polynomial/exponential/infinite phase transitions).

1 Introduction

The aim of this article is to analyze the asynchronous behavior of unbounded finite cellular automata. Cellular automata are widely used to model systems involving a huge number of interacting elements such as agents in economy, particles in physics, proteins in biology, distributed systems, etc. In most of these applications, in particular in many real system models, agents are not synchronous. Depending on the transition rules, the behaviour of the system may vary widely when asynchronism increases in the dynamics. More generally one can ask how much does asynchronous in real system perturbs computation. In spite of this lack of synchronism, real living systems are very resilient over time. One might then expect the cellular automata used to model these systems to be robust to asynchronism and to other kind of failure as well (such as misreading the states of the neighbors). It turns out that the resilience to asynchronism widely varies from one automata to another (e.g., [2, 3]). Only few theoretical studies exist on the influence of asynchronism. Most of them usually focus on one specific cellular

J.R. Correa, A. Hevia, and M. Kiwi (Eds.): LATIN 2006, LNCS 3887, pp. 455–466, 2006.

automata (e.g., [4, 5, 6]) and do not address the problem globally. Recently, Gács shows in [7] that it is undecidable to determining if in a given automota, the sequences of changes of states followed by a given cell is independent of the history of the updates. Related work on the existence of stationary distribution on infinite configurations for probabilistic automata can be found in [8].

One can see cellular automata as physical systems where cell states change according to local constraints (the transition rules). One typical example consists of a network where each cell have two states, e.g., "I have a token" and "I don't have a token", and where transitions from one state to the other depends on the states of the neighbours, e.g., "I get a token if both of my neighbors have one" or "I have a token if and only if my right neighbor has one", etc. One natural question for such systems, ask for the *relaxation time*, i.e. the time needed to reach a stable configuration (e.g., "everyone has a token" or "no one has a token"). As opposed to classic work in asynchronous distributed computing, where one tries to *design* efficient transitions rules that guarantees fast convergence to a stable configuration (e.g., [9]), we study here how asynchrony affects the global evolution of the system given an *arbitrary* set of local constraints, and in particular how does asynchronicity affects its relaxation time. In [1], the authors carried out a complete analysis of the class of one-dimensional double quiescent elementary cellular automata (DQECA), where each cell has two states 0 and 1 which are quiescent (i.e., where each cell for which every cell in its neighbourhood are in the same state, remains in the same state) and where each cell updates according to its state and the states of its two immediate neighbours. They study the behaviour of these automata under fully asynchronous dynamics, where only one random cell is updated at each time step. They show that one can classify the 64 DQECAs in six categories according to their relaxation times under full asynchronism (either constant, logarithmic, linear, quadratic, exponential or infinite) and furthermore that the relaxation time characterizes their behaviour, i.e., that all automata with equivalent relaxation times present the same kind of space-time diagrams.

The present paper extends this study to a continuous range of asynchyronism from fully asynchronous dynamics to fully synchronous dynamics: the α-*asynchronous dynamics*, with $0 < \alpha \leqslant 1$. In this setting, each cell is updated independently with probability α at each time step. When α varies from 1 down to 0, the α-asynchronous dynamics evolves from the fully synchronous regime to a more and more asynchronous regime. As α approaches 0, the probability that the updates involve at most one cell tends to 1, and the dynamics gets closer and closer to a kind of fully asynchronous dynamics up to a time rescaling by a factor $1/\alpha$. Abusing of the notation, we thus refer the fully asynchronous dynamics as the 0-asynchronous regime.

Figure 1 page 457 presents the space-time diagrams of the 24 representatives of the DQECAs as α increases (by steps of 0.25) starting from the same random configuration of length $n = 100$. The last column plots the density of black cells at time step $t = 1000/\alpha$ on one single random configuration. This class exhibits a rich variety of behaviours. Thirteen representatives of the DQECAs (ECAs 204 to 128, 198, and 142 on Fig. 1) appear to be marginally sensitive to asynchronism.

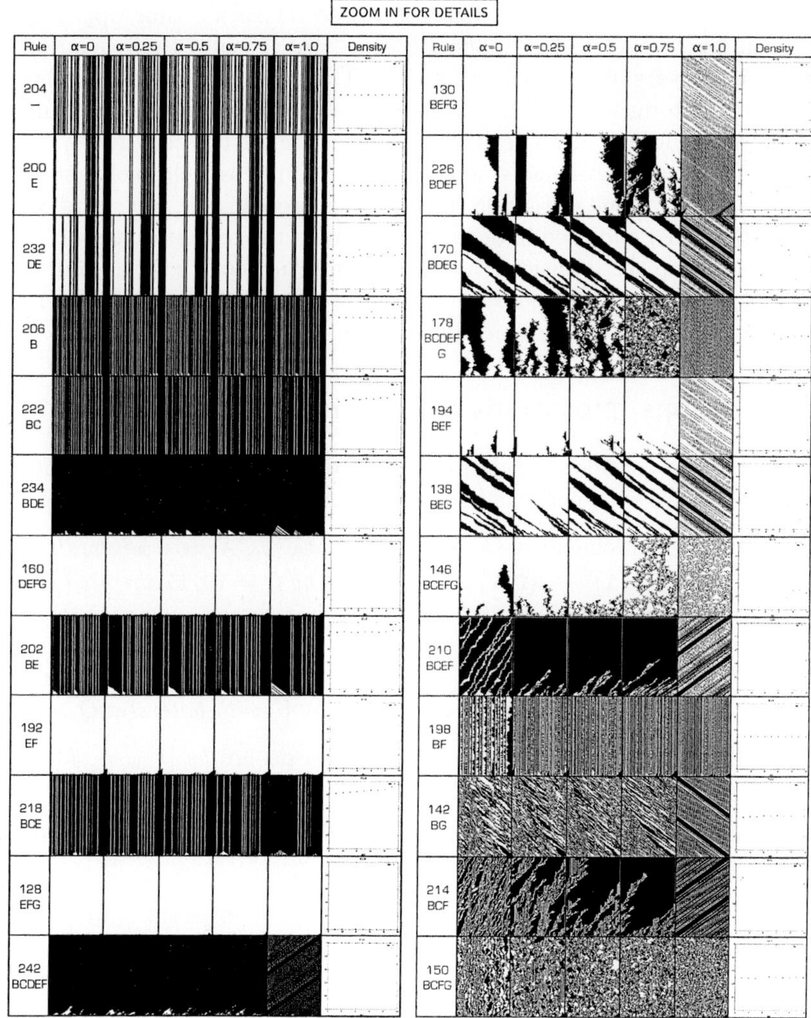

Fig. 1. Behaviour of DQECAs as a function of the synchronicity rate α

Six of them (ECAs 242 to 170, 194, and 138 on Fig. 1) present a brutal transition from the synchronous to asynchronous dynamics: they converge in polynomial time to an all-zero or all-one configuration as soon as (even a small amount of) asynchronism is introduced, while diverge under synchronous dynamics. One can observed that their space-time diagrams exhibit random walks like behaviour. The most interesting behaviour are observed on the remaining five representatives. The relaxation time of ECAs 210 and 214 are respectively exponential and infinite under fully asynchronous dynamics, and both infinite under synchronous dynamics, but appears to be polynomial under α-asynchronous dynamics. The relaxation time as well as the time-space diagrams of ECAs 178 and 146 evolve continuously as α increases, but seem to present an interesting phase transition at some α_c and α'_c, respectively, such that the relaxation time appears to be polynomial for $\alpha < \alpha'_c$, and exponential for $\alpha > \alpha'_c$. Finally, the relaxation time of ECA 150 appears to be exponential when $0 < \alpha < 1$, and is infinite otherwise.

Section 2 introduces the main definitions and presents our main result. Section 3 presents the key phenomena that differentiate the different dynamics: fully synchronous, α-asynchronous (studied here), and fully asynchronous (studied in [1]). These observations will guide the design of probabilistic tools that are presented in Section 4 and used in Section 5 to bound the relaxation times. Finally, Section 6 sums up the intuitions, hints and conjectures on the behaviours of the remaining automata that could not be treated theoretically here, leaving the determination of their relaxation times open.

2 Definitions, Notations and Main Results

In this paper, we consider the class of the two-state cellular automata on finite size configurations with periodic boundary conditions.

Definition 1. *An* Elementary Cellular Automata (ECA) *is given by its* transition function $\{\delta : \{0,1\}^3 \to \{0,1\}\}$. *We denote by* $Q = \{0,1\}$ *the set of* states. *A state* q *is* quiescent *if* $\delta(q,q,q) = q$. *An ECA is* double-quiescent (DQECA) *if both states* 0 *and* 1 *are quiescent.*

We denote by $U = \mathbb{Z}/n\mathbb{Z}$ *the set of* cells. *A* finite configuration *with periodic boundary conditions* $x \in Q^U$ *is a word indexed by* U *with letters in* Q.

Definition 2. *For a given pattern* $w \in Q^*$, *we denote by* $|x|_w = \#\{i \in U : x_{i+1} \dots x_{i+|w|} = w\}$ *the number of occurrences of* w *in configuration* x.

We will use the following labels introduced in [1] which will simplify the analysis of the probabilistic evolution of the ECAs.

Notation 1. We say that a transition is *active* if it changes the state of the cell where it is applied. Each ECA is fully determined by its active transitions. We label each active transition by a letter as follow:

label	A	B	C	D	E	F	G	H
$x\ y\ z$	000	001	100	101	010	011	110	111
$\delta(x,y,z)$	1	1	1	1	0	0	0	0

We label each ECA by the set of its active transitions. Note that with these notations, the DQECAs are exactly the ECAs having a label containing neither A nor H.

We consider three kinds of dynamics for ECAs: the *synchronous dynamics*, the α-*asynchronous dynamics* and the *fully asynchronous dynamics*. The synchronous dynamics is the classic dynamics of cellular automata, where the transition function is applied at each (discrete) time step on each cell simultaneously.

Definition 3 (Synchronous Dynamics). *The* synchronous dynamics $S_\delta : Q^U \to Q^U$ *of an ECA* δ, *associates deterministically to each configuration* x *the configuration* y, *such that for all* $i \in U$, $y_i = \delta(x_{i-1}, x_i, x_{i+1})$.

Definition 4 (Asynchronous Dynamics). *An* asynchronous dynamics AS_δ *of an ECA* δ *associates to each configuration* x *a random configuration* y*, such that* $y_i = x_i$ *for* $i \notin S$*, and* $y_i = \delta(x_{i-1}, x_i, x_{i+1})$ *for* $i \in S$*, where* S *is a random subset of* U *chosen by a daemon. We consider two types of asynchronous dynamics:*

- *in the* α-asynchronous dynamics, *the daemon selects at each time step each cell* i *in* S *independently with probability* α *where* $0 < \alpha \leqslant 1$*. The random function which associates the random configuration* y *to* x *according to this dynamics is denoted* AS_δ^α*.*
- *in the* fully asynchronous dynamics, *the daemon chooses a cell* i *uniformly at random and sets* $S = \{i\}$*. The random function which associates the random configuration* y *to* x *according to this dynamics is denoted* AS_δ^F*.*

For a given ECA δ*, we denote by* x^t *the random variable for the configuration obtained after* t *applications of the asynchronous dynamics function* AS_δ *on configuration* x*, i.e.,* $x^t = (AS_\delta)^t(x)$*. Note that* $(x^t)_{t\in\mathbb{N}}$ *is an homogeneous Markov chain on* Q^n*. Remark that* AS_δ *could equivalently be seen as a function with two arguments, the configuration* x *and the random subset* $S \subseteq U$ *chosen according to the processes listed above.*

Definition 5 (Fixed point). *We say that a configuration* x *is a* fixed point *for* δ *under asynchronous dynamics if* $AS_\delta(x) = x$ *whatever the choice of* S *is (the cells to be updated).* $\mathfrak{F}_\delta$ *denotes the set of fixed points for* δ*.*

The set of fixed points for the considered asynchronous dynamics is clearly identical to $\{x : S_\delta(x) = x\}$ the set of fixed points of the synchronous dynamics. The set of fixed points of an automaton can be easily deduce from its labeling as shown in [1]. Every DQECA admits two *trivial fixed points*, $\mathtt{0}^n$ and $\mathtt{1}^n$.

Definition 6 (Relaxation Time). *Given an ECA* δ *and a configuration* x*, we denote by* $T_\delta(x)$ *the random variable for the time elapsed until a fixed point is reached from configuration* x *under an asynchronous dynamics, i.e.,* $T_\delta(x) = \min\{t : x^t \in \mathfrak{F}_\delta\}$*. The* relaxation time *of ECA* δ *is* $\max_{x\in Q^U} \mathbb{E}[T_\delta(x)]$*.*

If $\alpha < 1$ the process $(x^t)_{t\in\mathbb{N}}$ converges to a stationary distribution, but we will abusively say that an ECA *diverges from an initial configuration* x if the probability to reach a fixed point from x is 0. We can now state our main theorem.

Theorem 1 (Main result). *Under* α*-asynchronous dynamics, among the sixty-four DQECAs, we can determine the behaviour of 52 of them:*

- *forty-eight converge almost surely to a random fixed point from any initial configuration, and the relaxation times of these forty-eight convergent DQECAs are* 0, $\Theta(\frac{\ln n}{\ln(1-\alpha)})$, $\Theta(\frac{n}{\alpha})$, $\Theta(\frac{n}{\alpha} + \frac{1}{\alpha(1-\alpha)})$, $O(\frac{n}{\alpha(1-\alpha)})$, $O(\frac{n}{\alpha^2(1-\alpha)})$, $\Theta(\frac{n^2}{\alpha(1-\alpha)})$.
- *two diverge from any initial configuration that is neither* $\mathtt{0}^n$*, nor* $\mathtt{1}^n$*, nor* $(\mathtt{01})^{n/2}$ *when* n *is even.*

Table 1. DQECAs under asynchronous and synchronous dynamics (see Section 2)

ECA (#)	Rule	01	10	010	101	Shift	Spawn	Fork	Annihil.	Full As.	α-Asynchr.	Synchr.
204 (1)	·	·	·	·	·	·	·	·	·	0	0	0
200 (2)	E	·	·	+	·	·	·	·	·	$\Theta(n \ln n)$	$\Theta(\frac{\ln n}{\ln(1-\alpha)})$	1
232 (1)	DE	·	·	+	+	·	·	·	·	$\Theta(n \ln n)$	$O(\frac{n}{\alpha} + \frac{1}{\alpha(1-\alpha)})$	∞
206 (4)	B	$\leftarrow$	·	·	·	·	·	·	·	$\Theta(n^2)$	$\Theta(\frac{n}{\alpha})$	$\Theta(n)$
222 (2)	BC	$\leftarrow$	$\rightarrow$	·	·	·	·	·	+	$\Theta(n^2)$	$\Theta(\frac{n}{\alpha})$	$\Theta(n)$
192 (4)	EF	$\rightarrow$	·	+	·	·	·	·	·	$\Theta(n^2)$	$\Theta(\frac{n}{\alpha})$	$\Theta(n)$
128 (2)	EFG	$\rightarrow$	$\leftarrow$	+	·	·	·	·	+	$\Theta(n^2)$	$\Theta(\frac{n}{\alpha})$	$\Theta(n)$
234 (4)	BDE	$\leftarrow$	·	+	+	+	·	·	·	$\Theta(n^2)$	$\Theta(\frac{n}{\alpha} + \frac{1}{\alpha(1-\alpha)})$	∞
202 (4)	BE	$\leftarrow$	·	+	·	+	·	·	·	$\Theta(n^2)$	$\Theta(\frac{n}{\alpha} + \frac{1}{\alpha(1-\alpha)})$	∞
250 (2)	BCDE	$\leftarrow$	$\rightarrow$	+	+	+	·	+	+	$\Theta(n^2)$	$O(\frac{n}{\alpha} + \frac{1}{\alpha(1-\alpha)})$	∞
218 (2)	BCE	$\leftarrow$	$\rightarrow$	+	·	+	·	+	+	$\Theta(n^2)$	$O(\frac{n}{\alpha} + \frac{1}{\alpha(1-\alpha)})$	∞
242 (4)	BCDEF	$\leftrightsquigarrow$	$\rightarrow$	+	+	+	+	+	+	$\Theta(n^2)$	$\Theta(\frac{n}{\alpha(1-\alpha)})$	∞
130 (4)	BEFG	$\leftrightsquigarrow$	$\leftarrow$	+	·	+	+	·	+	$\Theta(n^2)$	$\Theta(\frac{n}{\alpha(1-\alpha)})$	∞
170 (2)	BDEG	$\leftarrow$	$\leftarrow$	+	+	+	·	·	·	$\Theta(n^3)$	$\Theta(\frac{n^2}{\alpha(1-\alpha)})$	∞
138 (4)	BEG	$\leftarrow$	$\leftarrow$	+	·	+	·	·	·	$\Theta(n^3)$	$\Theta(\frac{n^2}{\alpha(1-\alpha)})$	∞
226 (2)	BDEF	$\leftrightsquigarrow$	·	+	+	+	+	·	·	$\Theta(n^3)$	$O(\frac{n^2}{\alpha(1-\alpha)})$	∞
194 (4)	BEF	$\leftrightsquigarrow$	·	+	·	+	+	·	·	$\Theta(n^3)$	$O(\frac{n}{\alpha^2(1-\alpha)})$	∞
178 (1)	BCDEFG	$\leftrightsquigarrow$	$\leftrightsquigarrow$	+	+	+	+	+	+	$\Theta(n^3)$	phase transition ? poly. for $\alpha < \alpha_c$? exp. for $\alpha > \alpha_c$?	∞
146 (2)	BCEFG	$\leftrightsquigarrow$	$\leftrightsquigarrow$	+	·	+	+	+	+	$\Theta(n^3)$	phase transition ? poly. for $\alpha < \alpha'_c$? exp. for $\alpha > \alpha'_c$?	∞
210 (4)	BCEF	$\leftrightsquigarrow$	$\rightarrow$	+	·	+	+	+	+	$\Theta(n2^n)$	polynomial ?	∞
214 (4)	BCF	$\leftrightsquigarrow$	$\rightarrow$	·	·	·	+	·	+	∞	polynomial ?	∞
150 (1)	BCFG	$\leftrightsquigarrow$	$\leftrightsquigarrow$	·	·	·	+	·	+	∞	exponential ?	∞
198 (2)	BF	$\leftrightsquigarrow$	·	·	·	·	+	·	·	∞	∞	∞
142 (2)	BG	$\leftarrow$	$\leftarrow$	·	·	·	·	·	·	∞	∞	∞

– *two converge with a small probability from few initial configurations when n is even and diverge otherwise.*

The twelves others (5 after symetries) have different behaviours that we cannot prove presently. Some seem to exhibit a phase transition but their mathematical analysis remains a challenging problem. All the results and the conjectures (with question marks) are summed up in table 1.

3 Key Observations

Due to 0/1 and reversal symmetries of configurations, we shall w.l.o.g. only consider the 24 DQECAs listed in Tab. 1 among the 64 DQECAs. For each of these 24 DQECAs, the number of the equivalent automata under symmetries is written within parentheses after their classic ECA code in the table.

From now on, we only consider the α-asynchronous dynamics; this will be implicit in all the following propositions. Our results rely on the study of the evolution of the *0-regions* and *1-regions* in the space-time diagram (i.e., of the intervals of consecutive 0s or consecutive 1s in configuration x^t). We will now enumerate the different ways the regions can be affected.

First we consider the cases where a cell updates and none of its two neighbours update:

- Transitions D and E are thus responsible for decreasing the number of regions in the space-time diagram: D "erases" the isolated 1s and E the isolated 0s.
- Transitions B and F act on patterns 01. Intuitively, transition B moves a pattern 01 to the left, and transition F moves it to the right. In particular, patterns 01 perform a kind of random walk for DQECA with both transitions B and F if no others phenomena occurs. The arrows in Tab. 1 represent the different behavior of the patterns: $\leftarrow$ or $\rightarrow$, for left or right moves of the patterns 01 or 10; $\leftrightsquigarrow$, for random walks of these patterns.
- Similarly, transitions C and G act on patterns 10. Transition C moves a pattern 10 to the right, and transition G moves it to the left.

One important observation made during the study of the fully asynchronous dynamics in [1] is that the number of regions can only decrease and each activation of D or E makes the number of regions decrease by one. This statement is not true anymore under the α-asynchronous dynamics, as we will see now. Here are the new phenomena when two or three neighboring cells update at the same time:

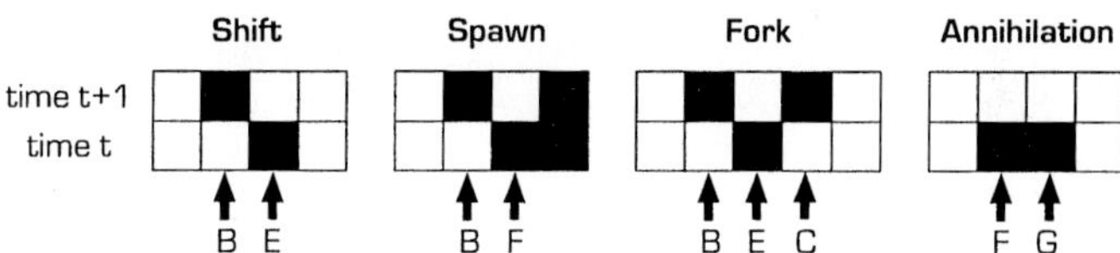

- **Shift** phenomenon occurs with the activation of rules B and E, or C and E, or F and D, or G and D together: in this case an isolated 0 or an isolated 1 is shifted. Here even if a transition D or E is activated, no regions is erased.
- **Spawn** phenomenon occurs with the activation of rules B and F, or C and G together: a pattern 0011 can create a new region. This is an important phenomenon because it increases the number of regions by one each time it occurs.
- **Fork** phenomenon occurs with the activation of rules B, C and E or F, G and D together: here three neighboring cellules update at the same time and an isolated point is duplicated. This phenomenon increases the number of regions by one each time it occurs.
- **Annihilation** phenomenon occurs with the activation of rules B and C or F and G together: the activation of these two rules erases a region of length 2. This is a very important new phenomenon because it is another way to decrease the number of region. In particular, it is the only way to decrease the number of regions in automaton where neither D nor neither E is activated.

The next section is devoted to the tools which will be used to prove our main theorem.

4 Lyapunov Functions Based on Local Neighbourhoods

Definition 7 (Mask). *A* mask $\dot{m}$ *is a word on* $\{0, 1, \dot{0}, \dot{1}\}$ *containing exactly one dotted letter in* $\{\dot{0}, \dot{1}\}$. *We say that the cell* i *in configuration* x *matches the mask* $\dot{m} = m_{-k} \dots m_{-1} \dot{m}_0 m_1 \dots m_l$ *if* $x_{i-k} \dots x_i \dots x_{i+l} = m_{-k} \dots m_0 \dots m_l$. *We denote by* m *the undotted word* $m_{-k} \dots m_0 \dots m_l$.

Fact 2. *The number of cells matching a given mask* $\dot{m}$ *in a configuration* x *is exactly* $|x|_m$, *the number of occurrences of the undotted word* m.

Definition 8 (Masks basis). *A* masks basis $\mathcal{B}$ *is a finite set of masks such that for any configuration* x *and any cell* i, *there exists an unique* $\dot{m} \in \mathcal{B}$ *that matches cell* i.

A masks basis $\mathcal{B}$ can be represented by a binary tree where the children of a node are labelled by adding 0 and 1 to the node label, on the right or the left (the children of the root receive $\dot{0}$ and $\dot{1}$), and where the masks of $\mathcal{B}$ are the labels of the leaves. Reciprocally, any binary tree observing these properties defines a unique masks basis by taking the labels of its leaves. Figure 2b page 463 illustrates the construction of the tree for the masks basis $\mathcal{B} = \{1\dot{1}, 00\dot{1}0, 00\dot{1}1, 010\dot{1}, 110\dot{1}, \dot{0}0, 0\dot{0}10, 0\dot{0}11, 01\dot{0}1, 11\dot{0}1\}$.

Masks bases will be used to define Lyapunov weight functions from local patterns. It provides an efficient tool to validate exhaustive case analysis.

Definition 9 (Local weight function). *A* local weight function f *is a function from a masks basis* $\mathcal{B}$ *to* $\mathbb{Z}$. *The local weight of the cell* i *in configuration* x *given by* f *is* $F(x, i) = f(\dot{m})$ *where* $\dot{m}$ *is the unique mask in* $\mathcal{B}$ *matching cell* i. *The* weight of a configuration x *given by* f *is defined as* $F(x) = \sum_i F(x, i)$.

Fact 3. *Given a local weight function* $f : \mathcal{B} \to \mathbb{Z}$, *the weight of configuration* x *is equivalently defined as:* $F(x) = \sum_{\dot{m} \in \mathcal{B}} f(\dot{m}) \cdot |x|_m$.

Notation 2. *For a given random sequence of configurations* $(x^t)_{t \in \mathbb{N}}$ *and a weight function* F *on the configurations, we denote by* $(\Delta F(x^t))_{t \in \mathbb{N}}$ *the random sequence* $\Delta F(x^t) = F(x^{t+1}) - F(x^t)$.

The next lemma provides upper bounds on stopping times for the markovian sequence of configurations $(x^t)_{t \in \mathbb{N}}$ subject to a weight function F decreasing or remaining constant on average (a *Lyapunov function*). Its proof can be found in [1].

Lemma 1. *Let* $m \in \mathbb{Z}_+$ *and* $\epsilon > 0$. *Consider* (x^t) *a random sequence of configurations, and* F *a weight function such that* $(\forall x)\ F(x) \in \{0, \dots, m\}$. *Assume that if* $F(x^t) > 0$, *then* $\mathbb{E}[\Delta F(x^t) | x^t] \leqslant -\epsilon$. *Let* $T = \min\{t : F(x^t) = 0\}$ *denote the random variable for the first time* t *where* $F(x^t) = 0$. *Then,* $\mathbb{E}[T] \leqslant \frac{m + F(x^0)}{\epsilon}$.

5 Relaxation Times

Due to space constraints, we only present the theorem for the relaxation time of the DQECA **BEF**. The results for Identity, **E**, **EF**, **EFG**, **DE**, **B**, **BC**, **BDE**, **BE**, **BCDE**, **BCE**, **BCDEF**, **BEFG**, **BDEG**, **BEG**, **BDEF**, **BF**, **BG** are given in Tab. 1 (check our websites for the full version of the paper).

5.1 Automaton BEF(194)

The fixed points of this automaton are 0^n and 1^n. Fixed point 1^n cannot be reached from any other configuration. Under fully asynchronous dynamics, the length of any 1-region follows a random walk, and thus converges in $O(n^3)$ in expectation. Here, the Spawn phenomenon (rule **B** and **F** applied together to cells $i-1$ and i) can transform the pattern $000\dot{1}11$ into the pattern 001011 with probability α^2. Even if the number of 0s and 1s are the same in these two patterns, in the pattern 001011 two 1s can become 0s at the next step (by applying rules **E** and **F**), while only one 0 can become a 1 at the next step (by applying rule **B**). So the creation of isolated 0s tends to decrease the number of 1s at the next step, leading to a speed up from a cubic relaxation time under fully asynchronous dynamics to a linear relaxation time in α-asynchronous dynamics with the respect of the size of the configuration. We consider the following variant. Let $a = -2c+2, b = -1, c = -\lfloor \frac{3}{\alpha} \rfloor - 1$. We use the masks basis and local weight function f given on page 463. We have: $F(x) = a|x^t|_1 + b|x^t|_{011} + c|x^t|_{101}$. For all configuration x, $F(x) \in \{0, \ldots, 2n(\lfloor \frac{3}{\alpha} \rfloor + 4)\}$ and $F(x) = 0$ if and only if $x = 0^n$.

Lemma 2. *For all non-fixed point configuration x^t,* $\mathbb{E}[\Delta F(x^t)] \leqslant -\alpha(1-\alpha)|x^t|_{01}$.

Proof. By linearity of expectation: $\mathbb{E}[\Delta F(x)] = \sum_{i=0}^{n-1} \mathbb{E}[\Delta F(x,i)]$. We evaluate the variation of $F(x,i)$ using the masks basis of Figure 2b.

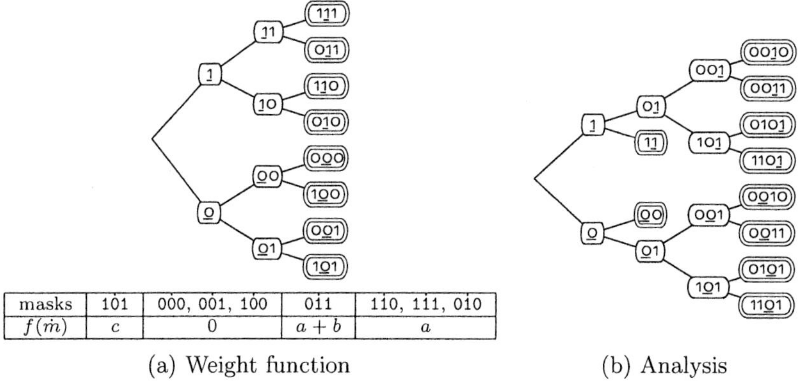

masks	$1\dot{0}1$	$0\dot{0}0$, $0\dot{0}1$, $1\dot{0}0$	$0\dot{1}1$	$1\dot{1}0$, $1\dot{1}1$, $0\dot{1}0$
$f(\dot{m})$	c	0	$a+b$	a

(a) Weight function (b) Analysis

Fig. 2. Mask basis for BEF

Consider that at step t, cell i matches:

- **mask** $1\dot{1}$: $F(x^t, i) = a$. With probability 1 at the step $t+1$, cell i matches mask $\dot{1}$. So $F(x^{t+1}, i) \in \{a, a+b\}$. Since $b < 0$, $F(x^{t+1}, i) \leqslant F(x^t, i)$. Thus, $\mathbb{E}[(\Delta F(x^t, i)] \leqslant 0$.
- **mask** $\dot{0}0$: $F(x^t, i) = 0$. With probability 1 at the step $t+1$, cell i matches mask $\dot{0}$. So $F(x^{t+1}, i) \in \{0, c\}$. Since $c < 0$, $F(x^{t+1}, i) \leqslant F(x^t, i)$. Thus, $\mathbb{E}[(\Delta F(x^t, i)] \leqslant 0$.
- **mask** $001\dot{0}$ **(and** $0\dot{0}10$ **together):**

With probability	$\alpha(1-\alpha)$	$\alpha(1-\alpha)$	$(1-\alpha)^2$	α^2
At the step $t+1$, cell i matches mask	$0\dot{0}$	$1\dot{1}$	$0\dot{1}$	$1\dot{0}$
and $\Delta F(x^t, i-1)$	$= 0$	$= a+b$	$= 0$	$= a$
and $\Delta F(x^t, i)$	$= -a$	$= 0$	$= 0$	$= -a$

Thus, $\mathbb{E}[\Delta F(x^t, i) + \Delta F(x^t, i-1)] = -a\alpha(1-\alpha) + (a+b)\alpha(1-\alpha) = b\alpha(1-\alpha) = -\alpha(1-\alpha)$.
- **mask** $001\dot{1}$ **(and** $0\dot{0}11$ **together):**

With probability	$\alpha(1-\alpha)$	$\alpha(1-\alpha)$	$(1-\alpha)^2$	α^2
at the step $t+1$, cell i matches mask	$0\dot{0}$	$1\dot{1}$	$0\dot{1}$	$1\dot{0}$
and $\Delta F(x^t, i-1)$	$= 0$	$= a+b$	$= 0$	$= c-a-b$
and $\Delta F(x^t, i)$	$= -a-b$	$= -b$	$= 0$	$= a$

Thus, $\mathbb{E}[\Delta F(x^t, i) + \Delta F(x^t, i-1)] = (-a-b)\alpha(1-\alpha) + a\alpha(1-\alpha) + (c-b)\alpha^2 \leqslant \alpha(1-\alpha) - 2\alpha \leqslant -\alpha(1-\alpha)$.
- **mask** $110\dot{1}$ **(and** $11\dot{0}1$ **together):**

With probability	α	$(1-\alpha)$
at the step $t+1$, the cell i matches mask	$0\dot{0}$	$0\dot{1}$
and $\Delta F(x, i-1)$	$= -c$	$= 0$
and $\Delta F(x, i)$	$= -a-b$	$= 0$

Thus, $\mathbb{E}[\Delta F(x^t, i) + \Delta F(x^t, i-1)] = (-a-b-c)\alpha(1-\alpha) \leqslant -\alpha(1-\alpha)$.
- **mask** $010\dot{1}$ **(and** $01\dot{0}1$ **together):**

With probability	α	$(1-\alpha)^2$	$\alpha(1-\alpha)$
at the step $t+1$, the cell i matches mask	$0\dot{0}$	$10\dot{1}$	$00\dot{1}$
and $\Delta F(x^t, i-1)$	$= -c$	$= 0$	$= -c$
and $\Delta F(x^t, i)$	$= -a-b$	$= 0$	$= 0$

Thus, $\mathbb{E}[\Delta F(x^t, i) + \Delta F(x^t, i-1)] = (-a-b-c)\alpha(1-\alpha) - c\alpha(1-\alpha) \leqslant -\alpha(1-\alpha)$.

Finally $\sum_{i=0}^{n-1} \mathbb{E}[\Delta F(x^t, i)] \leqslant -\alpha(1-\alpha)(|x^t|_{0010} + |x^t|_{0011} + |x^t|_{1011} + |x^t|_{0101}) \leqslant -\alpha(1-\alpha)|x^t|_{01}$. So, as long as x^t is not a fixed point, we have $\mathbb{E}[\Delta F(x^t)] \leqslant -\alpha(1-\alpha)|x^t|_{01} \leqslant -\alpha(1-\alpha)$.

Theorem 4. *Under α-asynchronous dynamics, DQECA* ***BEF*** *converges a.s. to a fixed point from any initial configuration. The relaxation time is $O(\frac{n}{\alpha^2(1-\alpha)})$.*

Proof. Using Lemma 1 and Lemma 2, automaton **BEF** converges a.s. from any intial configuration (except 1^n) to 0^n. The relaxation time is $O(\frac{n}{\alpha} \times \frac{1}{\alpha(1-\alpha)}) = O(\frac{n}{\alpha^2(1-\alpha)})$.

6 Conjectures

This section presents the remaining twelve DQECAs for which the mathematical analysis is not achieved yet. However by means of simulation and by the study of special patterns, we can give some insights of the phenomena which guide their dynamics and differentiate them from the other DQECAs.

Automaton ***BCDEFG****(178).* The fixed points of this automaton are exactly 0^n and 1^n. Simulations show a phase transition concerning the relaxation time, which can be also clearly observed on time-space diagrams and seems to occur at $\alpha = \alpha_c \approx 0,5$. If $\alpha < \alpha_c$, the overall behaviour of the automaton does not drastically change when α varies: 0- and 1-regions merge into larger regions reducing their number, and it seems to converge to 0^n or 1^n with an $O(n^2/\alpha)$ expected time. While if $\alpha > \alpha_c$, large 0- and 1-regions crumble quickly at their frontiers and patterns of $0101\cdots01$ fill the space between the regions. The closer α is to 1, the smaller is the probability of formation of large regions. In that case, we conjecture that the relaxation time is exponential in n.

Automaton ***BCEFG****(146).* The fixed points of this automaton are exactly 0^n and 1^n. This automaton shows a phase transition which seems to appear when $\alpha = \alpha'_c \approx 0,67$. When $\alpha < \alpha'_c$, 1-regions quickly disappear and the expected convergence time is conjectured to be polynomial in n. When α is close to 1, like the automaton **BCDEFG**, large 1-regions do not survive because they tend to crumble very quickly. On the other hand, isolated 1s are easily deleted and seem to multiply faster than they disappear. In that case, we conjecture that the relaxation time is exponential in n.

Automaton ***BCF****(214).* The fixed points of this automaton are 0^n, 1^n and $(01)^{n/2}$ (if n is even). When starting from another configuration, it is impossible to reach one of these fixed points in the fully asynchronous dynamics, since the number of regions remains constant. With the α-asynchronous dynamics, due to the Annihilation phenomenon, any configuration converges a.s. to a fixed point within a finite time. The sizes of large 0-regions decrease quickly. Only regions with two 0s may disappear, but 10011 patterns may evolve into 11111 or 10101 with the same probability. This could lead to an increase of small regions, tending to slow down the convergence. However a sequence of consecutive small 0-regions slows down the spawning phenomenon: in a 1001001 pattern, the first 00 region can not split. Thus the number of regions tends to decrease. We conjecture that the relaxation time is polynomial in n and contains an $O(\frac{1}{\alpha^2(1-\alpha)})$ term corresponding to the deletion of 00 regions.

Automaton ***BCFG****(150).* The fixed points of this automaton are 0^n, 1^n and $(01)^{n/2}$ (if n is even). In the fully asynchronous dynamics, this automaton does not converge to a fixed point since it is impossible to suppress a region. However in the α-asynchronous dynamics, due to the Annihilation phenomenon, this automaton converges a.s. to a fixed point within a finite time. Simulations suggest that the relaxation time is exponential in n.

Automaton ***BCEF****(210).* The fixed points of this automaton are exactly 0^n and 1^n. In the fully asynchrnous dynamics, this automaton converges to 0^n with a exponential expected time. In both fully asynchrounous and α-asynchronous, dynamics, the sizes of regions of 0 tend to decrease quickly. However in the fully asynchronous dynamics, they may only disappear by merging, and the size of the last 0-region will converge to 0 in exponential expected time. The α-asynchronous dynamics introduces the Annihilation phenomenon. On simulations, the convergence to fixed points seems to be polynomial. This case seems similar to the **BCF** automaton, but the analysis is a bit more complicated since 0-regions may merge, and this must be taken into account in the proof of bounds for the relaxation time.

References

1. Fatès, N., Morvan, M., Schabanel, N., Thierry, E.: Asynchronous behaviour of double-quiescent elementary cellular automata. In: LNCS 3618 Proc. of the 30th MFCS. (2005)
2. Bersini, H., Detours, V.: Asynchrony induces stability in cellular automata based models. In: Proc. of the 4th Artificial Life. (1994) 382–387
3. Fatès, N., Morvan, M.: An experimental study of robustness to asynchronism for elementary cellular automata. Submitted, `arxiv:nlin.CG/0402016` (2004)
4. Fukś, H.: Probabilistic cellular automata with conserved quantities. arxiv:nlin.CG/0305051 (2005)
5. Fukś, H.: Non-deterministic density classification with diffusive probabilistic cellular automata. Phys. Rev. E **66** (2002)
6. Schönfisch, B., de Roos, A.: Synchronous and asynchronous updating in cellular automata. BioSystems **51** (1999) 123–143
7. Gács, P.: Deterministic computations whose history is independent of the order of asynchronous updating. `http://arXiv.org/abs/cs/0101026` (2003)
8. Louis, P.Y.: Automates Cellulaires Probabilistes : mesures stationnaires, mesures de Gibbs associées et ergodicité. PhD thesis, Université de Lille I (2002)
9. Fribourg, L., Messika, S., Picaronny, C.: Coupling and self-stabilization. In: LNCS Proc. of 18th DISC. Volume 3274. (2004) 201–215
10. Grimmet, G., Stirzaker, D.: Probability and Random Process. 3rd edn. Oxford University Press (2001)

Lower Bounds for Geometric Diameter Problems

Hervé Fournier[1] and Antoine Vigneron[2]

[1] Laboratoire PRiSM, Université de Versailles St-Quentin
herve.fournier@prism.uvsq.fr
[2] Unité Mathématiques et Informatique Appliquées,
INRA, Domaine de Vilvert, F–78352 Jouy–en–Josas cedex
antoine.vigneron@polytechnique.org

Abstract. The diameter of a set P of n points in $\mathbb{R}^d$ is the maximum Euclidean distance between any two points in P. If P is the vertex set of a 3–dimensional convex polytope, and if the combinatorial structure of this polytope is given, we prove that, in the worst case, deciding whether the diameter of P is smaller than 1 requires $\Omega(n \log n)$ time in the algebraic computation tree model. It shows that the $O(n \log n)$ time algorithm of Ramos for computing the diameter of a point set in $\mathbb{R}^3$ is optimal for computing the diameter of a 3–polytope. We also give a linear time reduction from Hopcroft's problem of finding an incidence between points and lines in $\mathbb{R}^2$ to the diameter problem for a point set in $\mathbb{R}^7$.

Keywords: Computational geometry; Lower bound; Diameter; Convex polytope; Hopcroft's problem.

1 Introduction

The diameter problem for a set P of n points in $\mathbb{R}^d$ is to compute the largest distance between any two points in P. In other words, if we denote by $\mathrm{d}(\cdot,\cdot)$ the Euclidean distance in $\mathbb{R}^d$, it consists in finding $\mathrm{diam}(P) = \max\{\mathrm{d}(x,y) \mid x,y \in P\}$. It is a fundamental problem in computational geometry and has been studied extensively [3, 5, 12, 13, 16, 17]. If $P \subset \mathbb{R}^2$, then its diameter can be computed in $O(n \log n)$ time [16], which is optimal in the algebraic computation tree model [2, 4]. The three dimensional case remained open for a much longer time, but eventually Clarkson and Shor [8] designed an optimal $O(n \log n)$ time randomized algorithm to compute the diameter of a set of n points in $\mathbb{R}^3$, and Ramos [17] found a deterministic counterpart.

The $\Omega(n \log n)$ lower bound for computing the diameter of $P \subset \mathbb{R}^2$ can be broken if P is given as the sequence of the vertices of a convex polygon sorted along its boundary, in which case an $O(n)$ time algorithm is known [16]. Our main result (Theorem 1) is to show that the same speed–up cannot be achieved in $\mathbb{R}^3$, when P is the vertex set of a convex polytope, and the combinatorial structure of this polytope is given. In the worst case $\Omega(n \log n)$ time is required to compute the diameter of P. More precisely, we show that deciding whether the diameter of P is smaller than 1 requires an algebraic computation tree with

J.R. Correa, A. Hevia, and M. Kiwi (Eds.): LATIN 2006, LNCS 3887, pp. 467–478, 2006.

depth $\Omega(n \log n)$. We prove this result by applying Ben–Or's technique [2, 4, 16] to a suitable family of polytopes. Our lower bound implies that the algorithm by Ramos [17] is optimal for computing the diameter of a 3–polytope.

Similar problems of closing the gap between an $\Omega(n)$ lower bound and an $O(n \log n)$ upper bound have been studied recently. Chazelle et al. [6] mention that it is possible to compute the convex hull of two 3–polytopes in linear time, and it is not known whether the convex hull of a subset of the n vertices of a convex polytope can be computed in $O(n)$ time. On the other hand, given the Delaunay triangulation of a set P of n points in $\mathbb{R}^2$ (which is a special case of 3–dimensional convex hulls [9]), it is possible to compute the Delaunay triangulation of any subset of P in $O(n)$ time.

Hopcroft posed the following well known problem [11]. Given n lines and n points in $\mathbb{R}^2$, decide whether there is a point contained in a line. Matoušek [14] gave an $O\left(n^{4/3} 2^{O(\log^* n)}\right)$ time algorithm for this problem, but no $O(n^{4/3})$ time algorithm has been found so far. The only lower bound known for an algebraic computation tree is $\Omega(n \log n)$, and Erickson gave an $\Omega(n^{4/3})$ lower bound in a weaker model of computation [11]. Thus finding a reduction from Hopcroft's problem to any other problem suggests that this problem is difficult to solve in $o(n^{4/3})$ time. Erickson gave several such reductions to various geometric problems [10], for instance he showed that ray shooting in polyhedral terrains and halfspace emptiness checking in $\mathbb{R}^5$ are harder than Hopcroft's problem. In this paper, we show that the same is true for the diameter problem in $\mathbb{R}^7$. More precisely, we show that there is a linear time reduction from Hopcroft's problem to the diameter problem in $\mathbb{R}^7$ using a real random access machine [16] (real–RAM). We give a similar reduction to the red–blue diameter problem in $\mathbb{R}^6$. Our approach is based on a linearization argument. Using the lifting transformation and advanced data structures for ray shooting [15], the diameter of a set of n points in $\mathbb{R}^d$ can be computed in $O(n^{2-2/(\lceil d/2 \rceil+1)} \log^{O(1)} n)$ time, which is $O(n^{1.6} \log^{O(1)} n))$ for $d = 7$.

2 Notation and Preliminary

We work in fixed dimension d, so d is an integer such that $d = O(1)$. When $d = 3$, we use an orthonormal coordinate frame $Oxyz$ of $\mathbb{R}^3$. For all $a, b \in \mathbb{R}^d$, we denote by $\mathrm{d}(a, b)$ the Euclidean distance between a and b. For any set P of n points in $\mathbb{R}^d$, the diameter of P, that we denote by $\mathrm{diam}(P)$, is given by

$$\mathrm{diam}(P) = \max_{a,b \in P} \mathrm{d}(a, b).$$

Given two finite point sets $A, B \subset \mathbb{R}^d$, where the points in A are called the *red points* and the points in B are called the *blue points*, the *red–blue* diameter of (A, B) is

$$\mathrm{diam}(A, B) = \max_{a \in A, b \in B} \mathrm{d}(a, b).$$

The convex hull of P is denoted by $\mathrm{CH}(P)$. For any $a \in \mathbb{R}^d$ and $r > 0$, we denote by $\mathrm{B}(a, r)$ the open Euclidean ball with center a and radius r. If B is a non–empty subset of $\mathbb{R}^d$, we denote by $\mathrm{d}(a, B)$ the distance between a and B, that is

$$\mathrm{d}(a, B) = \inf_{b \in B} \mathrm{d}(a, b).$$

We denote by $\mathrm{m}(a, b)$ the midpoint of the line segment ab. We use the notation $\|\cdot\|$ for the L_2 norm. In other words, for all $a, b \in \mathbb{R}^d$, we have $\|a - b\| = \mathrm{d}(a, b)$. We denote by $\langle a, b \rangle$ the inner product of a and b. We use the notation $\bar{u} = (u_1, u_2, \ldots, u_m)$ to denote a sequence, and the concatenation of two sequences is written with a coma: $((1, 2), (3, 4)) = (1, 2, 3, 4)$.

A *3–polytope* is a 3–dimensional convex polytope. The *combinatorial structure* of a 3–polytope P is the set of all inclusion relations between its vertices, edges and facets. In our lower bound arguments, we assume that the combinatorial structure of P is given together with the following information: the coordinates of the vertices of P and, for each facet f of P, the edges of f are given as a sequence ordered along the boundary of f.

2.1 Models of Computation

The real–RAM model is the model of computation that is most commonly used to analyze geometric algorithms [16]. It is a random access machine that can store a real number in each register, and access it in constant time. It can perform comparisons and arithmetic operations $(+, -, \times, /)$ between real numbers at unit cost.

In order to prove lower bounds under the real–RAM model, we will prove lower bounds using an *algebraic computation tree* [4], which is a stronger model of computation. We will only use the algebraic computation tree model for decision problems, so following Ben–Or [2], we use the following definition where leaves are labeled by YES or NO. We denote by $x = (x_1, x_2, \ldots, x_n) \in \mathbb{R}^n$ the input to our problem. An algebraic computation tree T is a binary tree where each node is either a computation node (a degree one node, with one son), a branching node (a degree two node, with two sons), or a leaf. A computation node u is either labeled by an input number x_i, or it is associated with an arithmetic operation. Each such operation can either be taken in $\{+, -, \times, /, \sqrt{\cdot}\}$, or it can be a multiplication by a real constant; the operands are values obtained at computation nodes that are ancestors of u. At each branching node v, we compare with 0 the value obtained at a computation node that is an ancestor of v; each comparison can be taken in $\{>, \geqslant, =\}$. According to the result of this comparison, the program branches to one son of v or the other. So, according to the value of the input point x, the program follows a path in T that leads to leaf labeled YES or NO. We say that that T *decides* the set $W \subset \mathbb{R}^n$ if, for all $x \in W$, we reach a leaf labeled YES, and for all $x \notin W$, we reach a leaf labeled NO.

The depth of T gives an upper bound on the number of arithmetic and branching instructions it needs to decide whether a given $x \in \mathbb{R}^n$ is in W. On the other hand, a lower bound on the depth of all algebraic computation trees that decide W gives a lower bound on the worst case running time of any real–RAM that decides W.

3 Diameter of a 3-Polytope

In this section, we show that computing the diameter of a 3–polytope requires $\Omega(n \log n)$ time in the algebraic computation tree model. Our approach is the following. We first construct a family of 3–polytopes that have the same combinatorial structure, but do not all have the same diameter. Then we apply Ben–Or's technique [2, 4, 16].

We will use the inequalities

$$\forall \theta \in \left[-\frac{\pi}{2}, \frac{\pi}{2}\right], \ \frac{\theta^2}{4} \leqslant 1 - \cos\theta \leqslant \frac{\theta^2}{2}, \tag{1}$$

with strict inequalities if $\theta \neq 0$.

Let $n > 0$ be an integer. Let α and φ denote two real numbers such that $0 < \alpha \leqslant \frac{1}{4}$ and $0 < \varphi \leqslant \frac{1}{4}$. Both are to be thought of as small enough, to be chosen later. Then we define $\psi = \frac{\varphi}{n}$, $\gamma = \frac{\alpha}{n}$, $t = \left(1 - \cos\left(\frac{1}{2}\psi\right)\right) / \left(1 + \cos\left(\frac{1}{2}\psi\right)\right)$ and $r = 1 - t$. The length r has the following property (see Fig. 1): if e, f, g and h are four points such that $ef = eg = r$, $\angle feg = \angle feh = \frac{1}{2}\psi$ and $\angle efh = \frac{\pi}{2}$, then the midpoint $\mathrm{m}(g, h)$ is at distance 1 from e.

Now we define three sets of points in $\mathbb{R}^3$ (see Fig. 2). For all $i \in \{-n, -n+1, \ldots, n\}$, we define

$$a_i = \begin{pmatrix} \frac{1}{2}(1 - \cos(i\gamma)) \\ 0 \\ \frac{1}{2}\sin(i\gamma) \end{pmatrix}$$

and we denote $A = \{a_i \mid -n \leqslant i \leqslant n\}$. For all $i \in \{-n, -n+1, \ldots, n-1\}$ and $s \in \{-1, 1\}$, let

$$c_i^s = \begin{pmatrix} r\cos\left(\left(i + \frac{1}{2}\right)\psi\right) \\ r\sin\left(\left(i + \frac{1}{2}\right)\psi\right) \\ \frac{1}{2}s\alpha \end{pmatrix}$$

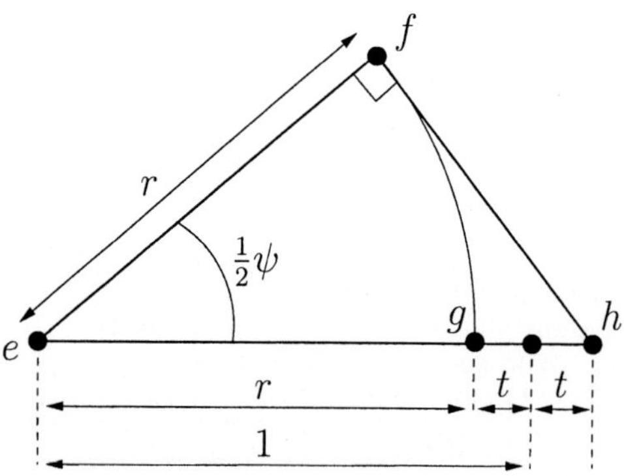

Fig. 1. Geometric interpretation of r and t

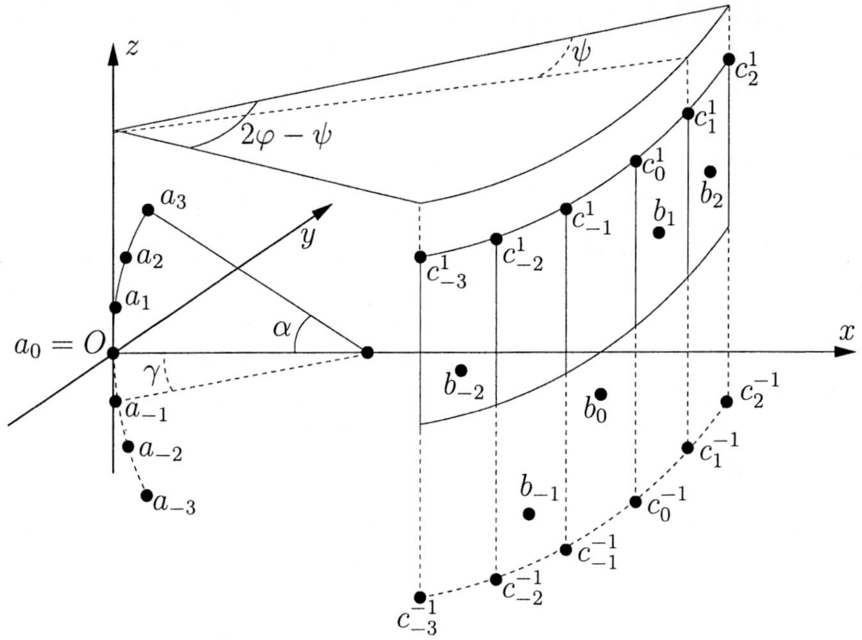

Fig. 2. The sets $A, B(\bar{\beta})$ and C when $n = 3$ and $\bar{\beta} \in [-\alpha, \alpha]^{2n-1}$

and $C = \{c_i^s \mid -n \leqslant i < n,\ s \in \{-1, 1\}\}$. Now for a parameter $\beta \in \mathbb{R}$ and for all $j \in \{-n+1, -n+2, \dots, n-1\}$, we define

$$b_j(\beta) = \begin{pmatrix} \cos(j\psi) - \frac{1}{2}(1 - \cos\beta) \\ \sin(j\psi) \\ \frac{1}{2}\sin(\beta) \end{pmatrix}$$

For all $\bar{\beta} = (\beta_{-n+1}, \beta_{-n+2}, \dots, \beta_{n-1}) \in \mathbb{R}^{2n-1}$, we define $B(\bar{\beta}) = \{b_j(\beta_j) \mid -n+1 \leqslant j \leqslant n-1\}$.

The following lemma shows that, for α small enough and $\bar{\beta} \in [-\alpha, \alpha]^{2n-1}$, the graph of $\mathrm{CH}(A \cup B(\bar{\beta}) \cup C)$ does not depend on the angle sequence $\bar{\beta}$.

Lemma 1. *Assume that $\alpha < 2t\cos(\frac{1}{2}\psi)$ and $\bar{\beta} \in [-\alpha, \alpha]^{2n-1}$. Then the graph of $\mathrm{CH}(A \cup B(\bar{\beta}) \cup C)$ is the the union of the graph of $\mathrm{CH}(A \cup C)$ and the set of the edges connecting each $b_j(\beta_j)$ to the points c_{j-1}^1, c_{j-1}^{-1}, c_j^1 and c_j^{-1} (see Fig. 3).*

Proof. Let H_j be the vertical plane containing $\{c_j^1, c_j^{-1}\}$ and normal to $(O, m(c_j^1, c_j^{-1}))$ (see Fig. 4). Let H'_j be the vertical plane containing the points $c_{j-1}^1, c_{j-1}^{-1}, c_j^1$ and c_j^{-1}. Let H^+ (resp. H^-) be the horizontal plane with equation $z = \frac{1}{2}\alpha$ (resp. $z = -\frac{1}{2}\alpha$). Let Δ_j be the interior of the polytope defined by the planes H_{j-1}, H_j, H'_j, H^+ and H^-. By elementary trigonometry (see also Fig. 1),

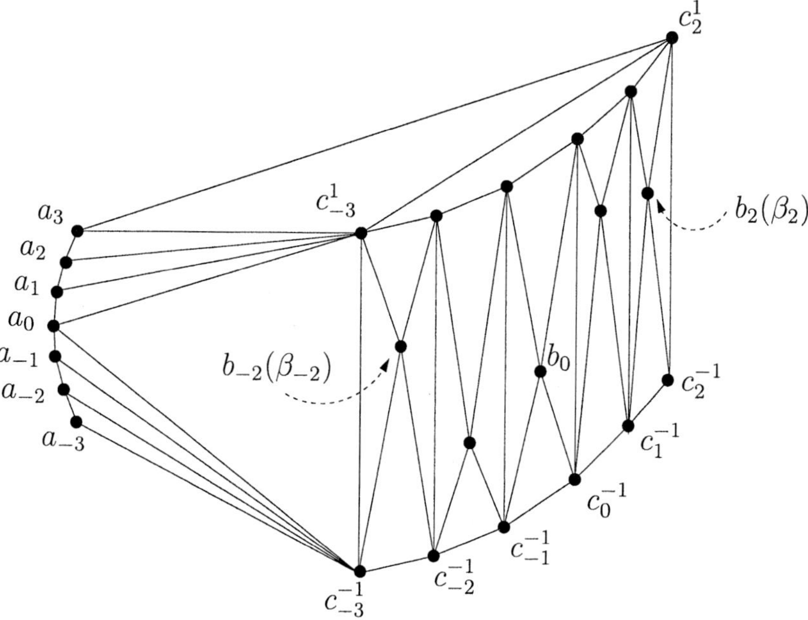

Fig. 3. This figure shows $\mathrm{CH}(A \cup B(\bar{\beta}) \cup C)$. The graph of $\mathrm{CH}\,(A \cup C)$ is obtained by removing all the vertices in $B(\bar{\beta})$ and the adjacent edges.

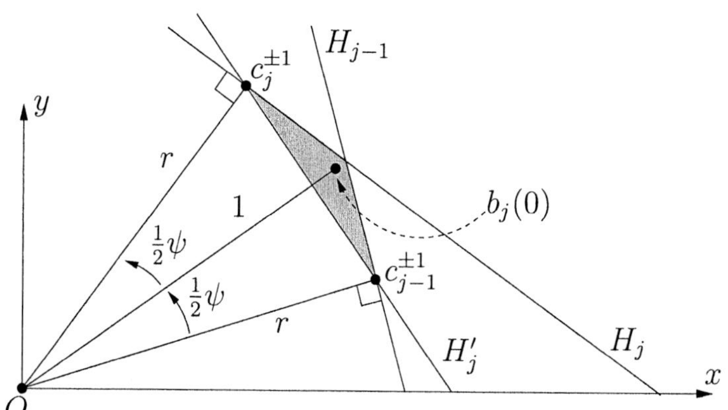

Fig. 4. The shaded area is Δ_j, seen from above

we can show that

$$\begin{cases} b_j(0) \in \Delta_j, \\ d(b_j(0), H^+) = d(b_j(0), H^-) = \frac{1}{2}\alpha, \\ d(b_j(0), H'_j) > t, \text{ and} \\ d(b_j(0), H_{j-1}) = d(b_j(0), H_j) = t\cos(\frac{1}{2}\psi). \end{cases}$$

Our assumption that $\frac{1}{2}\alpha < t\cos(\frac{1}{2}\psi)$ implies that $\mathrm{B}(b_j(0), \frac{1}{2}\alpha) \subset \Delta_j$. For all j, we have $\beta_j \in [-\alpha, \alpha]$, so $b_j(\beta_j) \in \mathrm{B}(b_j(0), \frac{1}{2}\alpha)$ and thus $b_j(\beta_j) \in \Delta_j$. Therefore, the only facet of $\mathrm{CH}(A \cup C)$ that is visible from $b_j(\beta_j)$ is the facet $c_{j-1}^{-1} c_{j-1}^{1} c_j^{1} c_j^{-1}$, and no point in $B(\bar{\beta}) \setminus \{b_j(\beta_j)\}$ is visible from $b_j(\beta_j)$; the result follows.

Lemma 2. *Assume that* $\varphi \leqslant \frac{1}{4n}$ *and* $j \in \{-n+1, -n+2, \ldots, n-1\}$. *Then the set* $\{b_j(\beta) \mid \beta \in [-\alpha, \alpha]$ *and* $\mathrm{diam}(A, \{b_j(\beta)\}) < 1\}$ *has at least* $2n$ *connected components.*

Proof. Let us first compute $\mathrm{d}^2(a_i, b_j(\beta))$. By developing the sum of squares and factoring, we obtain

$$\begin{aligned}
\mathrm{d}^2(a_i, b_j(\beta)) &= \frac{1}{4}\left(2 - (\cos(i\gamma) + \cos\beta) - 2\cos(j\psi)\right)^2 + \sin^2(j\psi) \\
&\quad + \frac{1}{4}(\sin(i\gamma) - \sin\beta)^2 \\
&= 1 + \frac{1}{4}(\cos(i\gamma) + \cos\beta)^2 + \cos^2(j\psi) - (\cos(i\gamma) + \cos\beta) \\
&\quad -2\cos(j\psi) + (\cos(i\gamma) + \cos\beta)\cos(j\psi) + \sin^2(j\psi) \\
&\quad + \frac{1}{4}(\sin(i\gamma) - \sin\beta)^2 \\
&= \frac{5}{2} + \frac{1}{2}(\cos(i\gamma)\cos\beta - \sin(i\gamma)\sin\beta) - (\cos(i\gamma) + \cos\beta) \\
&\quad -2\cos(j\psi) + (\cos(i\gamma) + \cos\beta)\cos(j\psi),
\end{aligned}$$

and thus

$$\mathrm{d}^2(a_i, b_j(\beta)) = 1 - \frac{1}{2}(1 - \cos(i\gamma + \beta)) + (1 - \cos(j\psi))(2 - \cos(i\gamma) - \cos\beta). \quad (2)$$

The result follows directly from the following two claims:

Claim 1. Let $i \in \{-n, -n+1, \ldots, n\}$ and $\beta = -i\gamma$. Then $\mathrm{d}(a_i, b_j(\beta)) \geqslant 1$.
This is obvious from equation (2) since the second term evaluates to 0.

Claim 2. Let $k \in \{-n+1, -n+2, \ldots, n\}$ and $\beta = (k - \frac{1}{2})\gamma$. Then diam $(A, \{b_j(\beta)\}) < 1$.
Let $i \in \{-n, -n+1, \ldots, n\}$. Let $\nu = \beta + i\gamma$. From equation 2 we get

$$\mathrm{d}^2(a_i, b_j(\beta)) = 1 - \frac{1}{2}(1 - \cos\nu) + (1 - \cos(j\psi))(2 - \cos(i\gamma) - \cos\beta).$$

Note that $|\nu| \leqslant 2\alpha \leqslant \frac{1}{2} < \frac{\pi}{2}$. Moreover, by the choice of β, we have $|\nu| \geqslant \frac{1}{2}\gamma$. Thus Equation (1) yields $1 - \cos\nu > \frac{1}{4}\nu^2 \geqslant \frac{1}{16}\gamma^2$. Besides we have $1 - \cos(j\psi) \leqslant 1 - \cos\varphi < \frac{1}{2}\varphi^2$ and $2 - \cos(i\gamma) - \cos(\beta) < \frac{1}{2}(i\gamma)^2 + \frac{1}{2}\beta^2 \leqslant \alpha^2$. These inequalities imply that $\mathrm{d}^2(a_i, b_j(\beta)) < 1 - \frac{1}{32}\gamma^2 + \frac{1}{2}\varphi^2\alpha^2$. Remember that $\alpha = n\gamma$, so we obtain $\mathrm{d}^2(a_i, b_j(\beta)) < 1 + \frac{1}{2}\alpha^2\left(\varphi^2 - \frac{1}{16n^2}\right)$. As $\varphi \leqslant \frac{1}{4n}$, we conclude that $\mathrm{d}(a_i, b_j(\beta)) < 1$.

Lemma 3. *Assume that $\alpha \leqslant \frac{1}{2}t$. Then for any $\bar{\beta} \in [-\alpha, \alpha]^{2n-1}$, we have*

$$\operatorname{diam}(A \cup B(\bar{\beta}) \cup C) = \operatorname{diam}(A, B(\bar{\beta})).$$

Proof. Clearly we have $d(a_i, a_j) \leqslant \alpha \leqslant \frac{1}{4}$ and $d(c_i^s, c_{i'}^{s'}) \leqslant 2r\varphi + \alpha \leqslant \frac{3}{4}$. In the same way, $d(b_j(\beta), b_{j'}(\beta')) \leqslant 2\varphi + \alpha \leqslant \frac{3}{4}$ and $d(a_i, c_j^s) \leqslant r + \alpha$. Similarly we have

$$d(b_j(\beta), c_i^s) \leqslant d(b_j(0), b_j(\beta)) + d(b_j(0), c_i^s) \leqslant \frac{\alpha}{2} + 2r\varphi + \frac{\alpha}{2} + t \leqslant 2r\varphi + \frac{3}{2}t.$$

By our assumption that $\varphi \leqslant \frac{1}{4}$ and by Equation (1), we have $t \leqslant \frac{1}{128}$ so $d(b_j(\beta), c_i^s) \leqslant \frac{3}{4}$. On the other hand, $d(a_i, b_j(\beta)) \geqslant d(a_0, b_j(0)) - d(a_0, a_i) - d(b_j(0), b_j(\beta)) > 1 - \frac{1}{2}\alpha - \frac{1}{2}\alpha = 1 - \alpha$. The result follows from the facts that $1 - \alpha \geqslant \frac{3}{4}$ and $1 - \alpha \geqslant r + \alpha$.

In order to be able to apply lemmas 1, 2 and 3, we need to find values of $\alpha \in (0, \frac{1}{4}]$ and $\varphi \in (0, \frac{1}{4}]$ such that the following three conditions hold simultaneously: $\alpha < 2t\cos(\frac{1}{2}\psi)$, $\varphi \leqslant \frac{1}{4n}$ and $\alpha \leqslant \frac{1}{2}t$. We achieve it as follows. We first choose $\varphi = \varphi_n = \frac{1}{4n}$, so as to satisfy the second condition. Notice that ψ and t are now fixed. Then, choosing $\alpha = \alpha_n > 0$ small enough[1] ensures that the other two conditions are met.

From now on, we assume that $\alpha = \alpha_n$ and $\varphi = \varphi_n$ have been chosen as above. We define the sequences $\bar{a} = (a_{-n}, a_{-n+1}, \ldots, a_n)$ and $\bar{c} = (c_{-n}, c_{-n+1}, \ldots, c_{n-1})$. For any $\bar{\beta} \in \mathbb{R}^{2n-1}$ we also define the sequence

$$\bar{b}(\bar{\beta}) = (b_{-n+1}(\beta_{-n+1}), b_{-n+2}(\beta_{-n+2}), \ldots, b_{n-1}(\beta_{n-1})).$$

We define the set of sequences

$$\mathcal{S}_n = \{(\bar{a}, \bar{b}(\bar{\beta}), \bar{c}) \mid \bar{\beta} \in [-\alpha, \alpha]^{2n-1}\} \subset \mathbb{R}^{24n}.$$

Thus, each element of $\mathcal{S}_n$ is a sequence of $8n$ points in $\mathbb{R}^3$.

Lemma 4. *The set $\mathcal{S}_n$ can be decided by an algebraic computation tree with depth $O(n)$.*

Proof. Given three sequences $\bar{u}$, $\bar{v}$ and $\bar{w}$ of respectively $2n+1$, $2n-1$ and $4n$ points in $\mathbb{R}^3$, we want to check in linear time if there exists $\bar{\beta} \in [-\alpha, \alpha]^{2n-1}$ such that $(\bar{u}, \bar{v}, \bar{w}) = (\bar{a}, \bar{b}(\bar{\beta}), \bar{c})$. As this computation tree is allowed to use real parameters, it is trivial to check that $\bar{u} = \bar{a}$ and $\bar{w} = \bar{c}$. Now it remains to check that there exists $\bar{\beta} \in [-\alpha, \alpha]^{2n-1}$ such that $\bar{v} = \bar{b}(\bar{\beta})$. We denote $\bar{v} = (v_{-n+1}, v_{-n+2}, \ldots, v_{n-1})$. For each integer $j \in \{-n+1, -n+2, \ldots, n-1\}$, we only need to check that v_j belongs to:

- the sphere of center $(\cos(j\psi) - \frac{1}{2}, \sin(j\psi), 0)$ and radius $\frac{1}{2}$,
- the plane $y = \sin(j\psi)$,
- and the halfspace $x \geqslant \cos(j\psi) - \frac{1}{2} + \cos(\alpha)$.

---

[1] Notice that α_n can be taken as large as $2^{-10}n^{-4}$: since $t = (1 - \cos(\frac{1}{2}\psi))/(1 + \cos(\frac{1}{2}\psi)) \geqslant \frac{1}{2}(\frac{1}{4}(\frac{1}{2}\psi^2)^2)$, it implies that $\alpha_n < \frac{1}{2}t$.

This can obviously be decided by a computation tree of linear depth.

Now we consider the following subset of $\mathcal{S}_n$:

$$\mathcal{E}_n = \{(\bar{a}, \bar{b}(\bar{\beta}), \bar{c}) \mid \bar{\beta} \in [-\alpha, \alpha]^{2n-1} \text{ and } \operatorname{diam}(A \cup B(\bar{\beta}) \cup C) < 1\}.$$

Lemma 5. *An algebraic computation tree that, given a sequence $\bar{s}$ of $8n$ points in $\mathbb{R}^3$, decides whether $\bar{s} \in \mathcal{E}_n$, has depth $\Omega(n \log n)$.*

Proof. By Lemma 3 we have

$$\mathcal{E}_n = \{(\bar{a}, \bar{b}(\bar{\beta}), \bar{c}) \mid \bar{\beta} \in [-\alpha, \alpha]^{2n-1} \text{ and } \operatorname{diam}(A, B(\bar{\beta})) < 1\},$$

and thus

$$\mathcal{E}_n = \{\bar{a}\} \times \prod_{j=-n+1}^{n-1} \{b_j(\beta) \mid \beta \in [-\alpha, \alpha] \text{ and } \operatorname{diam}(A, \{b_j(\beta)\}) < 1\} \times \{\bar{c}\}.$$

By Lemma 2 we know that $\mathcal{E}_n$ has at least $(2n)^{2n-1}$ connected components. We conclude by applying Ben–Or's bound [2, 4, 16] to $\mathcal{E}_n$.

The graph of a 3–polytope is planar, so a 3–polytope with n vertices has $O(n)$ edges and facets. Therefore, we can encode its combinatorial structure, the coordinates of its n vertices, and the ordering of the edges of each facet around its boundary, using $O(n)$ real numbers—for instance, using a doubly–connected edge list [9]. In the theorem below, we assume that the input is given using this encoding.

Theorem 1. *Assume that an algebraic computation tree T_n decides whether the diameter of a 3–polytope with n vertices is smaller than 1. Then T_n has depth $\Omega(n \log n)$.*

Proof. We denote by $(\bar{s}, \bar{g})$ the input of the tree T_{8n}, where $\bar{s} = (s_1, s_2, \ldots, s_{8n})$ denotes a sequence of $8n$ points in $\mathbb{R}^3$, and $\bar{g}$ encodes the graph of the convex hull of $S = \{s_1, s_2, \ldots, s_{8n}\}$. By Lemma 4, there is an algebraic computation tree U_n with depth $O(n)$ that decides whether $\bar{s} \in \mathcal{S}_n$. By plugging U_n to each accepting leaf of T_{8n}, we obtain an algebraic computation tree T'_{8n} that accepts 3–polytopes $(\bar{s}, \bar{g})$ such that $\bar{s} \in \mathcal{S}_n$ and $\operatorname{diam}(S) < 1$. In other words, T'_{8n} accepts 3–polytopes $(\bar{s}, \bar{g})$ such that $\bar{s} \in \mathcal{E}_n$. By Lemma 1, all 3-polytopes $(\bar{s}, \bar{g})$ accepted by T'_{8n} have the same graph $\bar{g} = \bar{g}_0$. Therefore, substituting the input part $\bar{g}$ with $\bar{g}_0$ in this tree gives an algebraic computation tree T''_{8n} that decides whether $\bar{s} \in \mathcal{E}_n$. If we denote by d_n the depth of T_n, then the depth of T''_{8n} is $d_{8n} + O(n)$. On the other hand, Lemma 5 tells us that T''_{8n} has depth $\Omega(n \log n)$. It follows that $d_n = \Omega(n \log n)$.

Ramos gave an $O(n \log n)$ upper bound on the complexity of computing the diameter of a 3-polytope in the real–RAM model [17]. So Theorem 1 implies that the complexity of deciding if the diameter of a 3–polytope is smaller than 1 is $\Theta(n \log n)$, and it implies that the complexity of computing the diameter of a 3–polytope is $\Theta(n \log n)$. This is true both in the (non-uniform) computation tree model and in the (uniform) real–RAM model.

4 Diameter is Harder Than Hopcroft's Problem

Hopcroft posed the following problem: given a set L of lines and a set P of points in $\mathbb{R}^2$, decide whether there is a line $\ell \in L$ and a point $p \in P$ such that $p \in \ell$. We will show that the diameter problem for a point set in $\mathbb{R}^7$ is harder than Hopcroft's problem. We first show a reduction to the red–blue diameter problem. In the following two propositions, we deal with a real–RAM that can use the constant $\sqrt{2}$. We will explain at the end of this section how we can avoid using this constant.

Proposition 1. *There is a linear-time reduction from Hopcroft's problem to the red–blue diameter problem in $\mathbb{R}^6$ using a real–*RAM *that uses the constant* $\sqrt{2}$.

Proof. Let $(a_1, \ldots, a_n, b_1, \ldots, b_p)$ be an instance of Hopcroft's problem. For all i, the point $a_i = (u_i, v_i, w_i)$ corresponds to the line with equation $u_i x + v_i y + w_i = 0$. Each point b_i is given by its coordinates $(x_i, y_i) \in \mathbb{R}^2$. We denote $c_i = (x_i, y_i, z_i = 1)$. So our instance of Hopcroft's problem has a positive answer if and only if $\langle a_i, c_j \rangle = 0$ for some i and j.

We denote $a'_i = a_i/\|a_i\|$ and $c'_i = c_i/\|c_i\|$. We define the function $\theta : \mathbb{R}^3 \to \mathbb{R}^6$ by

$$\theta(x, y, z) = \frac{1}{x^2 + y^2 + z^2}(x^2, y^2, z^2, \sqrt{2}xy, \sqrt{2}xz, \sqrt{2}yz).$$

Now let the points given by $f_i = \theta(a_i)$ be the red points, and let the points $g_i = \theta(c_i)$ be the blue points. Notice that $\|f_i\|^2 = \|a_i\|^4/\|a_i\|^4 = 1$, and $\|g_i\|^2 = 1$. It implies that $\|f_i - g_j\|^2 = \|f_i\|^2 + \|g_j\|^2 - 2\langle f_i, g_j \rangle = 2 - 2\|a_i\|^{-2}\|c_j\|^{-2}(u_i x_j + v_i y_j + w_i z_j)^2 = 2 - 2\langle a'_i, c'_j \rangle^2$. Thus, the red–blue diameter of the 6 dimensional point sets $\{f_i \mid 1 \leqslant i \leqslant n\}$ and $\{g_i \mid 1 \leqslant i \leqslant p\}$ is 2 if and only if our instance of Hopcroft's problem is positive.

A simple modification of the proof of Proposition 1 gives a reduction to the diameter problem in $\mathbb{R}^7$.

Proposition 2. *There is a linear-time reduction from Hopcroft's problem to the diameter problem in $\mathbb{R}^7$ using a real–*RAM *that uses the constant* $\sqrt{2}$.

Proof. With the notations from the previous proposition, we define $\hat{f}_i = (f_i, 1) \in \mathbb{R}^7$ and $\hat{g}_j = (g_j, -1) \in \mathbb{R}^7$. One have $\|\hat{f}_i - \hat{f}_j\|^2 = \|f_i - f_j\|^2 \leqslant (\|f_i\| + \|f_j\|)^2 \leqslant 4$, and $\|\hat{g}_i - \hat{g}_j\|^2 \leqslant 4$ in the same way. But $\|\hat{f}_i - \hat{g}_j\|^2 = \|f_i - g_j\|^2 + 4 \geqslant 4$. Thus, the diameter of $\{\hat{f}_1, \ldots, \hat{f}_n, \hat{g}_1, \ldots, \hat{g}_p\}$ is realized by a couple of points of the form $(\hat{f}_i, \hat{g}_j)$.

In propositions 1 and 2, we allowed the use of the constant $\sqrt{2}$ by the real–RAM machine. It can be avoided at the expense of increasing the dimension if we replace our function $\theta : \mathbb{R}^3 \to \mathbb{R}^6$ in the proof of Proposition 1 by the function $\theta' : \mathbb{R}^3 \to \mathbb{R}^9$ defined as follows:

$$\theta'(x, y, z) = \frac{1}{x^2 + y^2 + z^2}(x^2, y^2, z^2, xy, xy, xz, xz, yz, yz).$$

Thus we obtain the following result:

Proposition 3. *Using a real-*RAM *without constant, there are linear–time reductions from Hopcroft's problem to the red–blue diameter problem in* $\mathbb{R}^9$ *and to the diameter problem in* $\mathbb{R}^{10}$.

5 Concluding Remarks

Our lower bounds naturally apply to other computational geometry problems that the diameter problem reduces to, for instance the problems of computing the all–pairs farthest neighbors, the external farthest neighbors and the maximum Euclidean spanning tree [1]. Cheong, Shin and Vigneron [7] gave randomized algorithms to solve these three problems in near linear time when the input points are in $\mathbb{R}^3$ and are in convex position. Our results show that, even if the graph of the convex hull of the input points is known, these problems require $\Omega(n \log n)$ time in the algebraic computation tree model.

As we noted earlier, our lower bound for computing the diameter of a convex polytope leaves no room for improvement. Our results on the diameter for point sets in higher dimension, however, are not known to be optimal. First there is no lower bound other than $\Omega(n \log n)$ for Hopcroft's problem in the algebraic computation model. Second, even assuming that Hopcroft's problem cannot be solved in $o(n^{4/3})$ time, our result is not entirely satisfactory because the best known algorithm for the red–blue diameter problem [15] in $\mathbb{R}^6$ runs in $O(n^{1.5} \log^{O(1)} n)$ time. On the other hand, the red–blue diameter in $\mathbb{R}^4$ can be computed in $O(n^{4/3} \log^{O(1)} n)$ time, so it would be interesting to prove that this problem is harder than Hopcroft's problem. (Similarly, Erickson [10] asked whether the diameter in $\mathbb{R}^4$ is harder than halfspace emptiness checking in $\mathbb{R}^5$.)

Another intriguing question is the following. In propositions 1 and 2 we find reductions from Hopcroft's problem to diameter problems using a real–RAM that can use the constant $\sqrt{2}$. In proposition 3, we use a real–RAM without constant, and we obtain reductions to diameter problems in 3 dimensions higher. Is it possible to find such a reduction without increasing the dimension?

Acknowledgment

We would like thank Otfried Cheong and the anonymous referees for their helpful comments.

References

1. P. Agarwal, J. Matoušek, and S. Suri. Farthest neighbors, maximum spanning trees and related problems in higher dimensions. *Computational Geometry: Theory and Applications*, 1(4):189–201, 1992.
2. M. Ben-Or. Lower bounds for algebraic computation trees. In *Proceedings of the 15th Annual ACM Symposium on Theory of Computing*, pages 80–86, 1983.
3. S. Bespamyatnikh. An efficient algorithm for the three-dimensional diameter problem. *Discrete and Computational Geometry*, 25(2):235–255, 2000.

4. P. Bürgisser, M. Clausen, and M. Shokrollahi. *Algebraic Complexity Theory*. Springer, 1997.
5. T. Chan. Approximating the diameter, width, smallest enclosing cylinder, and minimum-width annulus. *International Journal of Computational Geometry and Applications*, 12(1–2):67–85, 2002.
6. B. Chazelle, O. Devillers, F. Hurtado, M. Mora, V. Sacristán, and M. Teillaud. Splitting a delaunay triangulation in linear time. *Algorithmica*, 34(1):39–46, 2002.
7. O. Cheong, C. Shin, and A. Vigneron. Computing farthest neighbors on a convex polytope. *Theoretical Computer Science*, 296(1):47–58, 2003.
8. K. Clarkson and P. Shor. Applications of random sampling in computational geometry, II. *Discrete and Computational Geometry*, 4:387–421, 1989.
9. M. de Berg, M. van Kreveld, M. Overmars, and O. Schwarzkopf. *Computational Geometry: Algorithms and Applications*. Springer-Verlag, Berlin, Germany, 2nd edition, 2000.
10. J. Erickson. On the relative complexities of some geometric problems. In *Proceedings of the 7th Canadian Conference on Computational Geometry*, pages 85–90, 1995.
11. J. Erickson. New lower bounds for Hopcroft's problem. *Discrete and Computational Geometry*, 16:389–418, 1996.
12. S. Har-Peled. A practical approach for computing the diameter of a point set. In *Proceedings of the Seventeenth Annual Symposium on Computational Geometry*, pages 177–186, New York, June 3–5 2001. ACM Press.
13. G. Malandain and J.-D. Boissonnat. Computing the diameter of a point set. *International Journal of Computational Geometry and Applications*, 12(6):489–510, 2002.
14. J. Matoušek. Range searching with efficient hierarchical cuttings. *Discrete and Computational Geometry*, 10(2):157–182, 1993.
15. J. Matoušek and O. Schwarzkopf. On ray shooting in convex polytopes. *Discrete and Computational Geometry*, 10(2):215–232, 1993.
16. F. Preparata and I. Shamos. *Computational geometry: An introduction*. Texts and Monographs in Computer Science. Springer-Verlag, New York, 2nd edition, 1985.
17. E. Ramos. An optimal deterministic algorithm for computing the diameter of a three-dimensional point set. *Discrete and Computational Geometry*, 26:233–244, 2001.

Connected Treewidth and Connected Graph Searching

Pierre Fraigniaud[1,*] and Nicolas Nisse[2,**]

[1] CNRS, Lab. de Recherche en Informatique, Université Paris-Sud, 91405 Orsay, France
[2] Lab. de Recherche en Informatique, Université Paris-Sud, 91405 Orsay, France
pierre@lri.fr, nisse@lri.fr

Abstract. We give a constructive proof of the equality between *treewidth* and *connected treewidth*. More precisely, we describe an $O(nk^3)$-time algorithm that, given any n-node width-k tree-decomposition of a connected graph G, returns a connected tree-decomposition of G of width $\leq k$. The equality between treewidth and connected treewidth finds applications in *graph searching* problems. First, using equality between treewidth and connected treewidth, we prove that the *connected* search number $\mathbf{cs}(G)$ of a connected graph G is at most $\log n + 1$ times larger than its search number. Second, using our constructive proof of equality between treewidth and connected treewidth, we design an $O(\log n\sqrt{\log OPT})$-approximation algorithm for connected search, running in time $O(t(n) + nk^3 \log^{3/2} k + m \log n)$ for n-node m-edge connected graphs of treewidth at most k, where $t(n)$ is the time-complexity of the fastest algorithm for approximating the treewidth, up to a factor $O(\sqrt{\log OPT})$.

1 Introduction

The *treewidth* of a graph is a central concept in the theory of Graph Minors developped by Robertson and Seymour. Roughly speaking, the treewidth, $\mathbf{tw}(G)$, of a graph G measures "how far" the graph G is from a tree. More formally, a *tree-decomposition* of graph G is a pair (T, X) where T is a tree, and $X = \{X_v, v \in V(T)\}$ is a collection of subsets of $V(G)$ satisfying the following three conditions:

- **C1:** $V(G) = \cup_{v \in V(T)} X_v$;
- **C2:** For any edge e of G, there is a set X_v such that both end-points of e are in X_v;
- **C3:** For any triple u, v, w of nodes in $V(T)$, if v is on the path from u to w in T, then $X_u \cap X_w \subseteq X_v$.

* Additional supports from the INRIA Project "Grand Large", and from the Project PairAPair of the ACI "Masse de Données".
** Additional supports from the Project Fragile of the ACI "Sécurité & Informatique".

J.R. Correa, A. Hevia, and M. Kiwi (Eds.): LATIN 2006, LNCS 3887, pp. 479–490, 2006.

Condition **C3** can be rephrased as: for any node x of G, $\{v \in V(T) \mid x \in X_v\}$ is a subtree of T. The sets X_v, $v \in V(T)$, are often called *bags*. The *width*, $\omega(T, X)$, of a tree-decomposition (T, X) is defined as $\max_{v \in V(T)} |X_v| - 1$, i.e., the width of (T, X) is roughly the maximum size of its bags. The treewidth $\mathbf{tw}(G)$ is defined as $\min \omega(T, X)$ where the minimum is taken over all tree-decompositions (T, X) of G. Hence the treewidth of any tree is 1, the treewidth of outerplanar graphs is ≤ 2, and the treewidth of an n-node complete graph is $n - 1$.

Treewidth is related to other types of graph-decompositions. In particular, Seymour and Thomas [15] introduced the concept of *carving*. For the sake of simplicity, we restrict ourself to edge-carving, i.e., *branch-decomposition* [14]. A branch-decomposition of a graph G is a pair (T, f) where T is a tree with all its internal nodes of degree 3, and f is a one-to-one mapping between the leaves of T and the edges of G. Given an edge e of T, removing e from T results in two trees $T_1^{(e)}$ and $T_2^{(e)}$. An *e-cut* of a branch-decomposition is defined as the pair $\{E_1^{(e)}, E_2^{(e)}\}$, where $E_i^{(e)} \subset E(G)$ is the set of leaves of $T_i^{(e)}$ for $i = 1, 2$. For any edge-set $E \subseteq E(G)$, let $\delta(E)$ denote the set of nodes of G with one incident edge in E and another in $E(G) \setminus E$. The *width* of a branch decomposition (T, f) is defined as $\omega(T, f) = \max_e |\delta(E_1^{(e)})|$ where the maximum is taken over all e-cuts in T. The *branchwidth*, $\mathbf{bw}(G)$, of G is then defined as $\min \omega(T, f)$ where the minimum is taken over all branch-decompositions (T, f) of G. It was proved in [14] that: $\mathbf{bw}(G) - 1 \leq \mathbf{tw}(G) \leq 3\,\mathbf{bw}(G)/2$.

Both tree-decomposition and branch-decomposition can be requested to be *connected*. An e-cut of a tree-decomposition (T, X) of a graph G is defined as the pair $\{X_1^{(e)}, X_2^{(e)}\}$, where $X_i^{(e)} \subseteq V(G)$ is the set of nodes of G in $\bigcup_{v \in V(T_i^{(e)})} X_v$ for $i = 1, 2$.

- A tree-decomposition is *connected* if, for any of its e-cuts, the two subgraphs of G, induced by $X_1^{(e)}$ and $X_2^{(e)}$ are connected. The connected treewidth, $\mathbf{ctw}(G)$, of a connected graph G, is defined as the minimum width of any connected tree-decomposition of G.
- A branch-decomposition is *connected* if, for any of its e-cut, the two subgraphs of G induced by $E_1^{(e)}$ and $E_2^{(e)}$ are connected. The connected branchwidth, $\mathbf{cbw}(G)$, of a connected graph G, is defined as the minimum width of any connected branch-decomposition of G.

A major result about branchwidth is that if a 2-edge-connected graph G has a branch-decomposition of width k, then it has a connected branch-decomposition of width $\leq k$ (see [15]). Therefore:

$$\text{For any 2-edge-connected graph } G,\ \mathbf{cbw}(G) = \mathbf{bw}(G). \tag{1}$$

The proof of (1) in [15] is non constructive, but it can be transformed into a constructive one (cf. [8]). The same result as (1) was proved for treewidth, by combining results in [9] and [11]. Indeed, on one hand, it was shown in [9] that a "clique tree" of a minimal triangulation H of a connected graph G is an optimal tree-decomposition of G (i.e., of width $\mathbf{tw}(G)$). On the other hand, [11] proved that the set Δ_H of the minimal separators of H is exactly the set of pairwise "parallel" minimal separators in G, and for any $S \in \Delta_H$, S induces the same connected components in H and G, which implies that the clique tree is in fact a connected tree-decomposition. As a consequence:

$$\text{For any connected graph } G,\ \mathbf{tw}(G) = \mathbf{ctw}(G). \tag{2}$$

Note that the equality $\mathbf{ctw} = \mathbf{tw}$ holds for any connected graph, whereas the equality $\mathbf{cbw} = \mathbf{bw}$ holds for 2-edge-connected graphs. The proof of (2) by combination of [9] and [11] is non constructive.

One of the contributions of this paper is the description of a constructive proof of (2). This result has an impact on the design of connected search strategies. Indeed, treewidth is related to several variants of the *graph searching* problem (see, e.g., [3, 6, 10, 12]). In graph searching, a fugitive is hidden in a graph G. A team of *searchers* is aiming at capturing this fugitive. These searchers can be placed at nodes, removed from nodes, and moved along the edges. The fugitive is assumed to be arbitrary fast, and permanently aware of the positions of the searchers. The graph searching problem asks for the design of search strategies using a minimum number of searchers. The search number of a graph G varies depending on the relative power of the fugitive and the searchers. If the searchers are permanently aware of the position of the fugitive, then the optimal size of the team is essentially $\mathbf{tw}(G)$ [3]. On the other hand, if the searchers are unaware of the position of the fugitive, then the optimal size of the team, $\mathbf{s}(G)$, is essentially the *pathwidth* $\mathbf{pw}(G)$ of G [3]. (A path-decomposition of G is a tree-decomposition (T, X) of G, where T is a path. The pathwidth is defined as $\min \omega(T, X)$ where the minimum is taken over all path-decompositions (T, X) of G.) For any graph G, we have [3]: $\mathbf{pw}(G) \leq \mathbf{s}(G) \leq \mathbf{pw}(G) + 2$. For instance, $\mathbf{s}(P_n) = 1$ for the n-node path P_n, $\mathbf{s}(C_n) = 2$ for the n-node cycle C_n, $\mathbf{s}(K_n) = n$ for the n-node clique K_n, and $\mathbf{s}(T) \leq \log_3(n-1) + 1$ for any n-node tree T [10].

It has been argued (cf., e.g., [1] and the references therein) that several practical applications (e.g., network security, speleological rescue [5], etc.) require the search strategy be connected, i.e., at any time of the search strategy, the portion of the searched graph is a connected subgraph. All searchers are initially placed at the same node, and clear the graph by moving along the edges from that initial node. In [1] is described a polynomial-time algorithm for computing the *connected* search number, $\mathbf{cs}$, of trees. There are n-node graphs G for which $\mathbf{cs}(G) > \mathbf{s}(G)$. (For instance, there are trees with connected search number $\lfloor \log_2 n \rfloor$). A major challenge regarding the connected search number is actually to bound the "cost of connectedness", that is to bound the ratio connected search number over search number. In [2], it is proved that the connected search number of a tree is at most twice its search number, and this bound is tight. Deriving bounds for arbitrary graphs is more complex, for at least two reasons. First, the set of graphs with connected search number at most k is not minor-closed, as opposed to the non-connected setting. Second, there are graphs for which no optimal connected search strategy is monotone [16], as opposed to the non-connected setting [4]. (Roughly, a search strategy is monotone if the fugitive cannot "recontaminate" a part of the graph that has been cleared). Nevertheless, using the concept of branchwidth, [8] shows that, for arbitrary connected m-edge graph G, the connected search number, $\mathbf{cs}(G)$, satisfies $\mathbf{cs}(G)/\mathbf{s}(G) \leq \lceil \log m \rceil + 1$. In [8] is also described an $O(t(n) + m^3)$-time $O(\log n \log OPT)$-approximation algorithm for connected search in arbitrary graphs, where $t(n)$ is the time complexity of the fastest algorithm for approximating the treewidth of an n-node graph, up to a factor $O(\log OPT)$.

Our Results

1. We give a constructive proof of the equality between the treewidth and the connected treewidth of connected graphs. This proof is obtained via the design of a polynomial-time algorithm transforming an n-node tree-decomposition of width k into a connected tree-decomposition of width $\leq k$, in time $O(nk^3)$.
2. We prove that $\mathbf{cs}(G)/\mathbf{s}(G) \leq \log n + 1$ via the design of a connected search strategy based on a connected tree-decomposition of the graph.
3. We combine this design with our algorithm for connected tree-decomposition, resulting in an $O(t(n) + nk^3 \log^{3/2} k + m \log n)$-time $O(\log n \sqrt{\log OPT})$-approximation algorithm for connected search, where $t(n)$ is the time complexity of the fastest algorithm for approximating the treewidth of an n-node graph, up to a factor $O(\sqrt{\log OPT})$, and k is the treewidth of the graph.

The two latter results improve [8].

2 Subconnected Tree-Decomposition

Given a tree-decomposition (T, X) of a graph G, and $u \in V(T)$, we denote by (T, X, u) the tree-decomposition (T, X) rooted at node u. For $v \in V(T)$, we denote by T_v the subtree of (T, X, u) rooted at v. The subgraph of G induced by the nodes in the bags of T_v is denoted by $G[T_v]$.

Definition 1. *A rooted tree-decomposition* (T, X, u) *is* subconnected *at* $v \in V(T)$ *if, for any* $w \in V(T_v)$, $G[T_w]$ *is a connected graph.* (T, X, u) *is* subconnected *if it is subconnected at* u.

Note that, alternatively, one can define the subconnectedness of (T, X, u) in v as: (1) $G[T_v]$ is connected, and (2) for any child w of v in T_v, (T, X, u) is subconnected at w.

We now describe an elementary procedure, called `split`, whose iterative application transforms a tree-decomposition into a subconnected tree-decomposition. The procedure `split` takes as input (1) a rooted tree-decomposition (T, X, u) of a connected graph G, and (2) a node $v \in V(T_u)$, $v \neq u$, such that, for every child w of v in T_u, (T, X, u) is subconnected at w. `split` returns a rooted tree-decomposition (T', X', u) of G that is equal to (T, X, u), except at node v.

Roughly speaking, node v in (T, X, u) is replaced by several nodes $v_1, \ldots, v_\ell$ in (T', X', u). The v_i's have the parent of v in (T, X, u) as parent in (T', X', u). The children of v in (T, X, u) are distributed among the v_i's. Procedure `split` satisfies that (T', X', u) is subconnected at every v_i, $i = 1, \ldots, \ell$. Hereafter, we describe formally this procedure:

Procedure `split`: Let (T, X, u) be a rooted tree-decomposition of a connected graph G, of width k. Let $v \in V(T_u)$, $v \neq u$, such that:

- (T, X, u) is not subconnected at v;
- for every child w of v, (T, X, u) is subconnected at w.

Since G is connected, and since (T, X, u) is not subconnected at v but subconnected at every child of v, the subgraph of G induced by the nodes in the bag X_v is not connected. Let Y_i, $i = 1, \ldots, r$ be the decomposition of X_v in connected components (i.e., the $G[Y_i]$'s are the connected components of $G[X_v]$). Let v' be the parent of v in T_u. Procedure `split` proceeds as follows:

Case 1: v is a leaf of T_u. `split` replaces v by r nodes $v_1, \ldots, v_r$, all connected to v', and every corresponding bag X_{v_i} is set to Y_i.

Case 2: v has s children $w_1, \ldots, w_s$, $s \geq 1$. Let Z_i, $i = 1, \ldots, t$, be the connected components of $G[T_v]$. As we will prove later, there is a partition $\{I_i, i = 1, \ldots, t\}$ of $\{1, \ldots, r\}$, and a partition $\{J_i, i = 1, \ldots, t\}$ of $\{1, \ldots, s\}$ such that, for every $i = 1, \ldots, t$:

$$Z_i = \Big(\cup_{j \in I_i} Y_j \Big) \cup \Big(\cup_{j \in J_i} G[T_{w_j}] \Big). \tag{3}$$

Procedure `split` replaces v by t nodes $v_1, \ldots, v_t$, all connected to v' (cf. Fig. 1). For every $i = 1, \ldots, t$, node v_i is the parent of w_j for all $j \in J_i$, and the bag X_{v_i} corresponding to v_i is set to $\cup_{j \in I_i} Y_j$.

Using the same notations as in the description of Procedure `split`, we have:

Lemma 1. *Procedure* `split` *applied to node v returns a rooted tree-decomposition (T', X', u) of G, of width $\leq k$. The tree-decomposition (T', X', u) differs from (T, X, u) only at v, which is replaced by some nodes $v_1, \ldots, v_\ell$. (T', X', u) is subconnected at v_i, $i = 1, \ldots, \ell$. Moreover, for every node z such that (T, X, u) is subconnected at z, (T', X', u) remains subconnected at z.*

Proof. The lemma clearly holds in Case 1. Hence we concentrate our attention to Case 2. In this case, $\ell = t$. Obviously, since the modification of T occurs at node v only, for every node z such that (T, X, u) is subconnected at z, (T', X', u) remains subconnected at z. Hence, we focus on the transformation of v into $v_1, \ldots, v_t$. First, let us show that Equation (3) holds. Let H be the bipartite graph whose one partition consists of r nodes $Y_1, \ldots, Y_r$, and the other partition consists of s nodes $w_1, \ldots, w_s$ (cf. Fig. 1). There is an edge between Y_i and w_j if and only if there is a node x of G that belongs to $Y_i \cap X_{w_j}$. By construction, there is a one-to-one correspondence between the connected components of

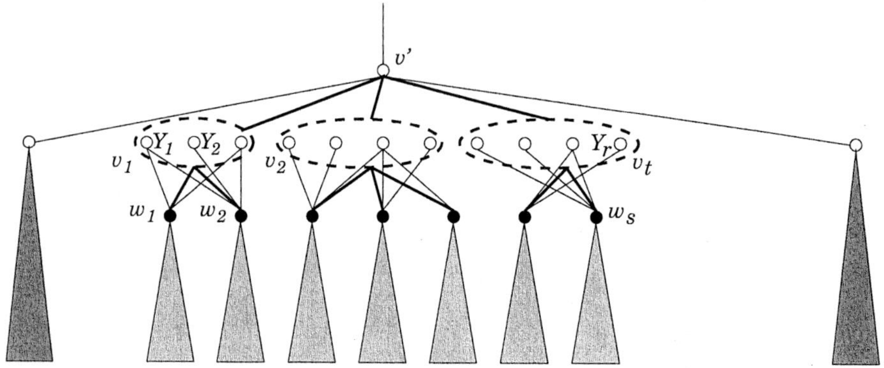

Fig. 1. The procedure `split` replaces node v by t nodes $v_1, \ldots, v_t$. The Y_i's form a partition of X_v in connected components. The w_i's are the children of v. Node v' is the parent of v.

$G[T_v]$ and the connected components of H. Equation (3) follows from this correspondence. From the construction of the v_i's, based on Equation (3), (T', X', u) is subconnected at v_i, $i = 1, \ldots, t$. Thus, it remains to prove that (T', X') is a tree-decomposition of G, of width $\leq k$.

Since $X'_{v_i} = \cup_{j \in I_i} Y_j$, and the I_i's form a partition of $\{1, \ldots, r\}$, we get that $\cup_{i=1,\ldots,r} X'_{v_i} = \cup_{j=1,\ldots,r} Y_j = X_v$.

Thus every node of G appears in at least one bag, i.e., **C1** holds. Every edge of G appears in at least one bag too, because the Y_i's are the connected components of X_v, and thus there is no edge between nodes that belong to two different Y_i's. I.e., **C2** holds.

Non surprisingly, Condition **C3** of tree-decomposition is the most tricky to check. Let x, y, z be three pairwise distinct nodes of T' with y on the path between x and z. Let us show that $X'_x \cap X'_z \subseteq X'_y$. Obviously, this claim holds if $v_i \notin \{x, y, z\}$ for all $i = 1, \ldots, r$, because, in this case, the considered bags of T' are exactly those of T. The claim also holds if the path P from x to z contains two v_i's, because, in this case, $X'_x \cap X'_z = \emptyset$. Thus, in the following, we consider the case where the path P contains exactly one v_i. There are two subcases, depending on whether the node of P that belongs to $\{v_1, \ldots, v_t\}$ is y, or one of the two end-points x or z.

Assume that $x = v_i$ for some $i \in \{1, \ldots, r\}$, and that $v_j \notin \{y, z\}$ for all $j \neq i$. Then, $X'_x \cap X'_z \subseteq X_v \cap X'_z = X_v \cap X_z \subseteq X_y = X'_y$, and thus Condition **C3** holds.

Assume that $y = v_i$ for some $i \in \{1, \ldots, r\}$, and that $v_j \notin \{x, z\}$ for all $j \neq i$. Assume, w.l.o.g., that $x \in V(T'_{v_i})$. Node z then belongs either to $V(T'_{v_i})$, or to the subtree of T containing v', obtained after removing the edge $\{v, v'\}$ from T. In both cases, $X'_x \cap X'_z = X_x \cap X_z \subseteq X_v$. Moreover, by construction of the bipartite graph H (cf. Fig. 1), $X'_x \cap X_v \subseteq \cup_{j \in I_i} Y_j$. Therefore, $X'_x \cap X'_z \subseteq \cup_{j \in I_i} Y_j = X'_{v_i}$, and thus Condition **C3** holds.

To complete the proof, observe that applying Procedure `split` can only decrease the width of the tree-decomposition since one bag is split into several smaller bags. Hence, the width of (T', X', u) is $\leq k$. □

(cf. [3]). A crusade $(E_0, E_1, \cdots, E_r)$ in a graph G is connected if E_i induces a connected subgraph of G for every $i = 1, \ldots, r$. It is easy to check that if k denotes the smallest number for which there is a connected k-crusade in G, then $\mathbf{cs}(G) \leq k+1$.

Claim 3. For any connected n-nodes graph G, and for any $e \in E(G)$, there exists a connected k-crusade $(E_0, E_1, \ldots, E_r)$ of G with $k \leq \mathbf{tw}(G) \log n$ and $E_1 = \{e\}$.

The proof of Claim 3 is by induction on n. If $n \in \{1, 2\}$, the result obviously holds. Let $n \geq 3$ and let us assume that, for any $n' < n$, Claim 3 holds. Let G be an n-node connected graph, and let $e \in E(G)$. Let (T, X) be a connected tree decomposition of G, of width $\mathbf{tw}(G)$. For a subtree T' of T, let us denote by $G[T']$ the subgraph of G induced by the nodes in the bags of T'. Theorem 2.5 in [13] specifies that, for any tree-decomposition:

- either there exists $u \in V(T)$ such that removing u from T results in subtrees $T_1, \cdots, T_q$ of T, with $|V(G[T_i])| \leq n/2$ for every $1 \leq i \leq q$,
- or there exists $\{u, u'\} \in E(T)$ such that removing $\{u, u'\}$ from T results in subtrees $T_1, \cdots, T_q$ of T, with $|V(G[T_i])| \leq n/2$ for every $1 \leq i \leq q$.

We consider the two cases ("node-centroid" and "edge-centroid") separatly.

Let us first consider the former case. Let $u \in V(T)$ such that for all $1 \leq i \leq q$, $|V(G[T_i])| \leq n/2$. By definition of connected tree decomposition, $G[T_i]$ is a connected subgraph of G. Assume, w.l.o.g., that there exists $v \in V(T_1)$ such that $e \in X_v$. By induction, let $(E_0^{(1)}, E_1^{(1)}, \ldots, E_{r_1}^{(1)})$ be a connected k-crusade of $G[T_1]$, with $k \leq \mathbf{tw}(G[T_1]) \log |G[T_1]|$, and $E_1^{(1)} = \{e\}$. We set $E_i = E_i^{(1)}$ for $i = 0, \ldots, r_1$. Since $\mathbf{tw}(G[T_1]) \leq \mathbf{tw}(G)$ and $|G[T_1]| \leq n/2$, we get $|\delta(E_i)| \leq \mathbf{tw}(G)(\log n - 1)$ for $i = 0, \ldots, r_1$. Since G is connected, there is $f \in E(G) \setminus E_{r_1}$ such that the subgraph induced by $E_{r_1} \cup \{f\}$ is connected.

- If $f \in X_u$, then let $E_{r_1+1} = E_{r_1} \cup \{f\}$. For computing $|\delta(E_{r_1+1})|$, consider any node $x \in \delta(E_{r_1+1})$. This node is incident to an edge in E_{r_1+1} and to an edge in $E(G) \setminus E_{r_1+1}$. Therefore $x \in X_u \cup (\cup_{v \in T_1} X_v)$ and $x \in \cup_{v \notin T_1} X_v$. As a consequence, by the third property of a tree-decomposition, $x \in X_u$ and thus $|\delta(E_{r_1+1})| \leq |X_u| \leq \mathbf{tw}(G)$.
- If $f \notin X_u$, then there exists $i \in \{2, \cdots, q\}$ and $v \in V(T_i)$ such that $f \in X_v$. Assume, w.l.o.g., that $i = 2$. Let $(E_0^{(2)}, E_1^{(2)}, \ldots, E_{r_2}^{(2)})$ be a connected k-crusade of $G[T_2]$, with $k \leq \mathbf{tw}(G[T_2]) \log |G[T_2]| \leq \mathbf{tw}(G)(\log n - 1)$ and $E_1^{(2)} = \{f\}$. For every $1 \leq i \leq r_2$, we set $E_{r_1+i} = E_{r_1} \cup E_i^{(2)}$. By construction, for any $1 \leq i \leq r_2$, $G[E_{r_1+i}]$ is a connected subgraph of G. Let $i \in \{1, \ldots, r_2\}$, and $x \in \delta(E_{r_1+i})$. We have $x \in \delta(E_i^{(2)}) \cup X_u$, and thus $|\delta(E_{r_1+i})| \leq \mathbf{tw}(G)(\log n - 1) + \mathbf{tw}(G) \leq \mathbf{tw}(G) \log n$.

We iterate the process until all edges of G are in the crusade, completing the analysis of the first case, i.e., where the "centroid" is a node. Due to lack of space, the second case, where the "centroid" is an edge, is omitted, but can be treated similarly as above. This completes the proof of Claim 3.

Item 1 of Theorem 2 is a direct consequence of Claim 3. Indeed, let k be the smallest integer such that there exists a connected k-crusade in G. We have seen that $\mathbf{cs}(G) \leq k+1$. Thus, by Claim 3, $\mathbf{cs}(G) \leq \mathbf{tw}(G)\log n + 1$. Since $\mathbf{tw}(G) \leq \mathbf{s}(G)$, we get $\mathbf{cs}(G) \leq \mathbf{s}(G)\log n + 1$ and thus $\mathbf{cs}(G)/\mathbf{s}(G) \leq \log n + 1$. Item 2 is obtained by combining claim 3 with the algorithm of Feige et al. [7]. □

References

1. L. Barrière, P. Flocchini, P. Fraigniaud, and N. Santoro. Capture of an intruder by mobile agents. In 14th ACM Symp. on Parallel Algorithms and Architectures (SPAA), pages 200-209, 2002.
2. L. Barrière, P. Fraigniaud, N. Santoro, and D. Thilikos. Connected and Internal Graph Searching. In 29th Workshop on Graph Theoretic Concepts in Computer Science (WG), Springer-Verlag, LNCS 2880, pages 34–45, 2003.
3. D. Bienstock. Graph searching, path-width, tree-width and related problems (a survey). DIMACS Series in Discrete Mathematics and Theoretical Computer Science 5, pages 33-49, 1991.
4. D. Bienstock and P. Seymour. Monotonicity in graph searching. Journal of Algorithms 12, pages 239-245, 1991.
5. R. Breisch. An intuitive approach to speleotopology. Southwestern Cavers VI(5), pages 72-78, 1967
6. J. A. Ellis, I.H. Sudborough, J.S. Turner. The Vertex Separation and Search Number of a Graph Information and computation 113, pages 50-79, 1994.
7. U. Feige, M. Hajiaghayi, and J. Lee. Improved approximation algorithms for minimum-weight vertex separators. In 37th ACM Symposium on Theory of Computing (STOC), 2005.
8. F. Fomin, P. Fraigniaud, D. Thilikos. The Price of Connectedness in Expansions. Technical Report LSI-04-28-R, UPC Barcelona, 2004.
9. M. C. Golumbic. Algorithmic graph theory and perfect graphs. Computer Science and Applied Mathematics, 1980.
10. N. Megiddo, S. Hakimi, M. Garey, D. Johnson and C. Papadimitriou. The complexity of searching a graph. Journal of the ACM 35(1), pages 18-44, 1988.
11. A. Parra and P. Scheffler. Characterizations and algorithmic applications of chordal graphe embeddings. Discrete Applied Mathematics 79, pages 171-188, 1997.
12. T. Parson. Pursuit-evasion in a graph. Theory and Applications of Graphs, Lecture Notes in Mathematics, Springer-Verlag, pages 426-441, 1976.
13. N. Robertson and P. D. Seymour. Graph minors II, Algorithmic Aspects of Tree-Width. Journal of Algorithms 7, pages 309-322, 1986.
14. N. Robertson and P. D. Seymour. Graph minors X, Obstructions to tree-decomposition. Journal Combin. Theory B, Vol. 52, pages 153-190, 1991.
15. P. Seymour and R. Thomas. Call routing and the rat-catcher. Combinatorica 14(2), pages 217–241, 1994.
16. B. Yang, D. Dyer, and B. Alspach. Sweeping Graphs with Large Clique Number. In 5th International Symposium on Algorithms and Computation (ISAAC), Springer, LNCS 3341, pages 908-920, 2004.

A Faster Algorithm for Finding Maximum Independent Sets in Sparse Graphs

Martin Fürer*

Pennsylvania State University, University Park, PA 16802, USA
furer@cse.psu.edu
http://www.cse.psu.edu/~furer

Abstract. An algorithm is presented for finding a maximum independent set in a connected graph with n vertices and m edges in time $O(\text{poly}(n)1.2365^{m-n})$. As a consequence, we find a maximum independent set in a graph of degree 3 in time $O(\text{poly}(n)1.1120^n)$, which improves the currently best results of $O(1.1254^n)$ of Chen, Kanj and Xia.

Keywords: Maximum independent set, exponential time algorithm, sparse graph, NP-hard.

1 Introduction

Pioneering more efficient algorithms for exponential time problems, Tarjan and Trojanowski [1] have shown that maximum independent sets can be found in time $O(2^{n/3}) = O(1.2599^n)$ for worst case graphs, a drastic improvement over the previous trivial upper bound of $O(\text{poly}(n)2^n)$. Actually, as they remark, their running time is slightly faster. If $\tau(s,t)$ is the positive root of

$$x^{-s} + x^{-t} = 1$$

then their running time is $O(\tau(1,7)^n) = O(1.255^n)$. After improvements by Jian [2] and Robson [3] for arbitrary graphs, Beigel [4] improved the running time for sparse graphs. For a graph with m edges his running time is $O(1.082^m)$, implying a running time of $O(1.126^n)$ for degree 3 graphs. Later, Robson [5] has improved the running time for arbitrary graphs to $O(1.1888^n)$. Finally, Chen, Kanj and Xia [6] have obtained a running time of $O(1.1254^n)$ for degree 3 graphs.

Our new algorithm runs in time $O(\text{poly}(n)1.2365^{m-n})$ and polynomial space, resulting in improved running times for very sparse graphs. In particular, our running time is $O(\text{poly}(n)1.1120^n)$ for finding a Maximum Independent Set in a graph with degree 3. Our improvement is obtained mainly by a novel handling of small separators, cutting off a constant size subgraph. We don't require the existence of such separators, but if we find them, we can use them as an asset rather than obstacle (as in previous algorithms).

* Research supported in part by NSF Grant CCR-0209099.

J.R. Correa, A. Hevia, and M. Kiwi (Eds.): LATIN 2006, LNCS 3887, pp. 491–501, 2006.

2 The Algorithm

Our algorithm is based on recursive decomposition. As usual, we view the computation as a tree, where at each branching node some vertex v is either included or not included in the independent set being chosen.

We also solve some simple cases not producing any branching. This allows us to assume without loss of generality that the input graph G has the following properties.

- G is connected.
- Every vertex of G has degree at least 3.

We now present a recursive description of our algorithm to find a maximum independent set on an input graph $G = (V, E)$. Whenever we say some set V' of vertices is *deleted*, we mean to continue with the subgraph of G induced by $V - V'$. Merging an independent set V' of vertices means replacing them by one new vertex v and replacing every edge to a vertex of V' by an edge to v. Often when we say, the algorithm does something with a vertex fulfilling some condition, it is possible that several vertices fulfill such a condition. In such a case, the algorithm may just pick anyone of them. The neighborhood $N(v)$ of a vertex v is the set of vertices u with an edge from v. We define $\bar{N}(v)$ as $N(v) \cup \{v\}$.

We start with a number of steps that are always done when possible.

2.1 Simple Non-branching Steps

- If the graph is not connected, do a recursive call for each connected component.
- If there is just one vertex v, return the independent set $I = \{v\}$.
- If there is a vertex v of degree 1, delete it and its neighbor. Make a recursive call and add v to the returned independent set.
- If there is a vertex v of degree 2 with two adjacent neighbors, then delete v and its neighbors. Make a recursive call and add v to the returned independent set.
- If there is a vertex v of degree 2 with two non-adjacent neighbors u and u', then delete v and merge its neighbors u and u' (replacing them by a new vertex w). Make a recursive call. If the returned independent set I contains w, then return $I \cup \{u, u'\} - \{w\}$, else return $I \cup \{v\}$.
- If u *dominates* v, i.e., $\bar{N}(u) \subseteq \bar{N}(v)$, then delete v. (In case of equality, delete either one.)

If u dominates v, then deletion of v is justified, because every solution containing v can be modified into an equally good solution with v replaced by u. The cases of v having degree 1, and v having exactly two neighbors, which are adjacent, are actually special cases of domination.

The handling of the cases with v having exactly 2 neighbors is justified by the fact that a single chosen neighbor of v can always be replaced by v. The procedure can also be described in 2 steps as follows. If the degree of any vertex

v of G is 2, we construct a new graph G' obtained by deleting v and merging the 2 neighbors of v into one vertex. If the two neighbors of v are adjacent, then the new vertex has a self-loop, meaning it cannot participate in any independent set and should be deleted. Every independent set in G' of size k can easily be modified to an independent set in G of size $k+1$ and vice versa.

2.2 Handling Separators of Size Two or Less

Our algorithm introduces a new way of handling small separators. These computation steps are not always done when possible, but only when their applicability is discovered (while pursuing another goal). Alternatively, we could always test for the applicability of these steps without doing any harm to the complexity of our algorithm.

Handling of separators includes a pretty obvious step for non-biconnected graphs and a novel extension to non-triconnected graphs. Doing these steps whenever possible would always be beneficial, but it would not improve the worst case time bound. Therefore, we only do these steps when they can be done in constant time, because the component in question is so small.

- If our algorithm discovers an articulation point v (a vertex whose deletion makes the remaining graph disconnected), then it has also found a small (size bounded by a constant) non-trivial (size at least 2) subset $V' \subseteq V$ containing v with the property that there are no edges between $V' - \{v\}$ and $V - V'$. Determine both, a maximum independent set I' in the subgraph induced by $V' - \{v\}$ and a maximum independent set I'' in the subgraph induced by V'. The sizes of I' and I'' differ by at most 1.

  If they differ by 1, then make a recursive call with the subgraph induced by $(V - V') \cup \{v\}$ (the remaining graph including v). After returning from this call with an independent set I, the algorithm returns $I \cup (I'' - \{v\})$. If the sizes of I and I' are the same, then make a recursive call with the subgraph induced by $V - V'$ (the remaining graph excluding v). After returning from this call with an independent set I, the algorithm returns $I \cup I'$.
- If there is a vertex separator $\{u, v\}$ of size 2 (a pair of vertices whose deletion makes the remaining graph disconnected), then it has also found a small (size bounded by a constant) nontrivial (size at least 3) subset $V' \subseteq V$ containing u and v and having the property that there are no edges between $V' - \{u, v\}$ and $V - V'$. Form the four graphs induced by the vertex sets $V' - \{u, v\}$, $V' - \{v\}$, $V' - \{u\}$, and V'. Make a recursive call for each to obtain independent sets I', I'', I''', and I''''. Now, there are several cases to consider. Let s be the size of I', the base size.

  Case A: $|I''''| = s+2$. In this case $|I''| = |I'''| = s+1$. Make a recursive call with the graph induced by $(V - V') \cup \{u, v\}$. When the independent set I is returned, let J be either I', I'', I''', or I'''', whichever agrees with I on $\{u, v\}$. Now the algorithm returns $I \cup J$.

  Case B: $|I''| = |I'''| = |I''''| = s + 1$. Make a recursive call with the graph induced by $(V - V') \cup \{u, v\}$ and an edge $\{u, v\}$ added. When the independent set I is returned, let J be either I', I'', or I''', whichever agrees with I on $\{u, v\}$. Now the algorithm returns $I \cup J$.

Case C: $|I''| = s$ and $|I'''| = |I''''| = s + 1$. Make a recursive call with the graph induced by $(V - V') \cup \{v\}$. When the independent set I is returned, let J be either I', or I''', whichever agrees with I on $\{v\}$. Now the algorithm returns $I \cup J$. (There is a symmetric case with u and v interchanged.)

Case D: $|I''| = |I'''| = s$ and $|I''''| = s + 1$. Form the graph induced by $(V - V') \cup \{u, v\}$ and merge the vertices u and v into a new vertex w. Make a recursive call with this graph. When the independent set I is returned, two cases are considered. If $w \in I$ then the algorithm returns $(I - \{w\}) \cup I''''$. In this case $\{u, v\} \subseteq I''''$. Otherwise, the algorithm returns $I \cup I'$.

Case E: $|I''| = |I'''| = |I''''| = s$. Make a recursive call with the graph induced by $V - V'$. When the independent set I is returned, the algorithm returns $I \cup I'$.

We have not yet specified what the algorithm does when no small separator is discovered.

Lemma 1. *If the algorithm is correct when no small separator is discovered, then it is always correct.*

Proof. The proof is by induction on the number of times a small separator is discovered. It is based on the obvious fact that by removing one vertex from a graph, the size of a maximum independent set can decrease by at most 1. □

The main part of the algorithm consists of picking some vertex of maximal degree, branching on including or not including this vertex, and doing recursive calls on the two resulting smaller graphs.

Our measure of progress is $m - n$, the difference between the number of edges and the number of vertices. This is also equal to half the number of vertices of degree 3 in a graph with degrees 2 and 3 only.

If branching causes the creation of two graphs with the measure $m - n$ decreased by s and t respectively, then we say the *branching number* is $\tau(s, t)$, where $\tau(s, t)$ is defined as the positive root of

$$x^{-s} + x^{-t} = 1$$

Standard arguments imply a running time of $O(\text{poly}(n)\tau(s, t)^{m-n})$, if $\tau(s, t)$ is the maximal branching number of a computation tree.

2.3 The Degree 3 Case

We use the notation $N(v)$ for the neighborhood of v, i.e., the set of vertices of distance 1 from v. We also use $N^k(v)$ to denote the set of vertices of distance k from v.

First we describe the algorithm for a degree 3 graph. We have already eliminated all vertices of degree less than 3. The elimination of degree 2 vertices can produce higher degree vertices. That case will be treated in Section 2.4. Here, we assume the degree is 3. Thus the graph is regular of degree 3, and we pick any vertex r as a candidate for branching. If there are 2 edges within $N(r)$, then one neighbor x of r has no connection to $N^2(r)$. In this simple case, we actually delete r as it is dominated by x (Figure 1). $m - n$ decreases without branching. If there is just one edge within $N(r)$, then we actually branch on the neighbor x of r not incident to this edge (Figure 2). On the branch with x not selected, we have the great advantage of selecting r by domination. The resulting branching number is at most $\tau(4, 4)$.

From now on, we assume there are no edges within $N(r)$. When handling a vertex r, the algorithm only looks at a bounded number of levels $N^k(r)$. Whenever the algorithm investigates any level $N^k(r)$, then we assume that there are still vertices at the next level $N^{k+1}(r)$. Otherwise the size of the graph is constant, immediately guaranteeing a good running time.

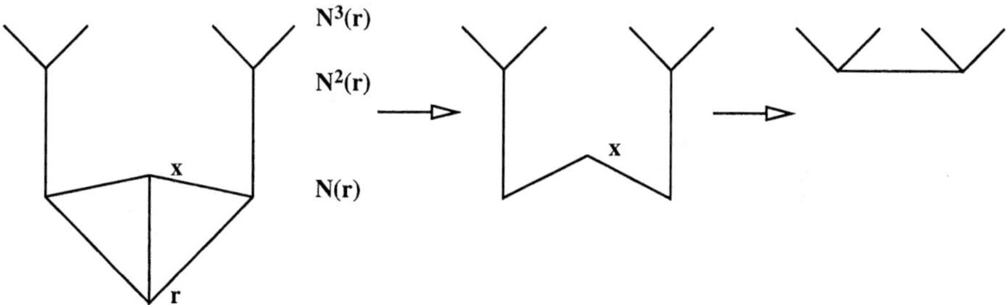

Fig. 1. Two edges within $N(r)$

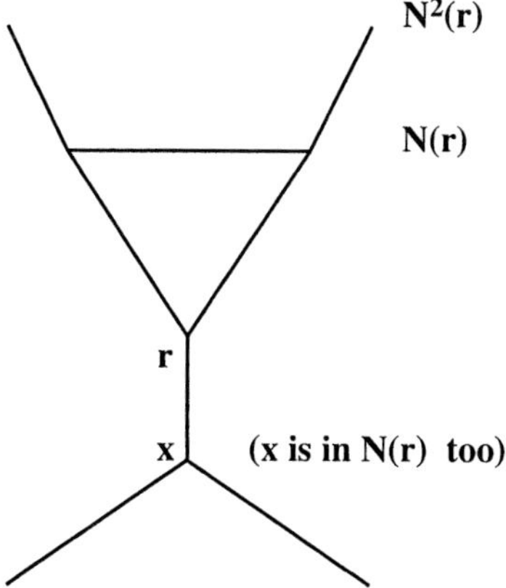

Fig. 2. Just one edge within $N(r)$

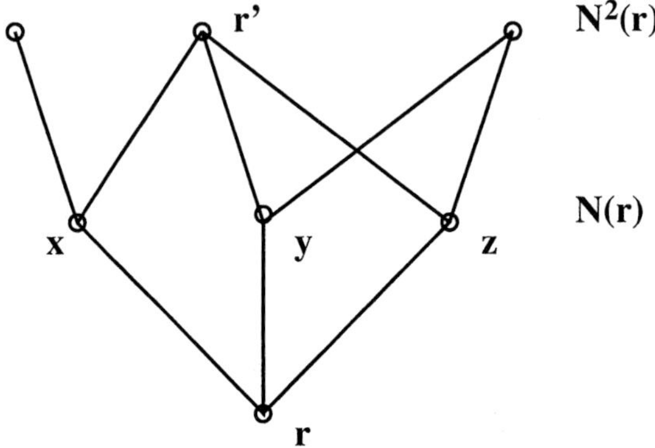

Fig. 3. Every vertex of $N^2(r) - \{r'\}$ is connected to $N^2(r)$

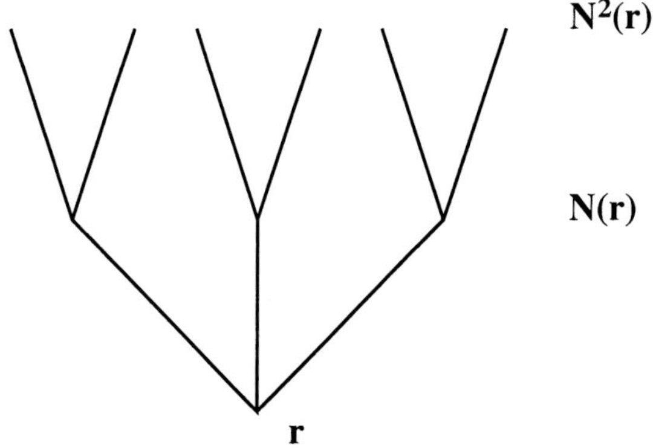

Fig. 4. The simplest case

Another special case occurs when the vertex r has a mirror image r' with the same neighbors x, y, z as r (Figure 3). The bipartite subgraph induced by $\{r, r', x, y, z\}$ has only two maximal independent sets, $\{r, r'\}$ and $\{x, y, z\}$. One of them is part of a maximum independent set. This implies a simple reduction of deleting $\{r, r'\}$ and merging $\{x, y, z\}$ into one vertex.

Now we come to the interesting cases. The simplest case occurs when all 3 of r's neighbors have 2 distinct neighbors in $N^2(r)$ (Figure 4). When r is not selected, then we delete 3 edges and 1 vertex, when r is selected, then we delete 9 edges and 4 vertices, resulting in a branching number of $\tau(2, 5)$. We want to argue now that we obtain the same branching number, not just in this simple case, but in all remaining cases.

We consider the process of selecting r, deleting $\bar{N}(r)$, and deleting any degree 1 vertices for as long as possible. The difficult part is to show that in this process the number of deleted edges is 5 more than the number of deleted vertices. For this purpose, we introduce the notion of a *live edge*, which is any edge that is deleted in this process. As we assume there are no edges within $N(r)$, we have at least the following live edges: All 3 edges from r to $N(r)$, and all 6 edges from

$N(r)$ to $N^2(r)$. Furthermore, the 4 vertices of $\bar{N}(r)$ are deleted resulting in the desired temporary balance of 5 for the difference of edge deletions and vertex deletions.

We think of the life edges being directed away from r to visualize the direction of further deletions. Whenever any vertex x outside $\bar{N}(r)$ is reached by only one live edge, we are happy. The degree of x is decreased from 3 to 2, but x is not deleted and the process stops there. If two live edges reach the same vertex x, they make x a singleton. Thus x and the third edge e incident on x are deleted, maintaining the desirable balance, but making e live. We direct e away from x (Figure 5). Finally, when 3 live edges point to the same vertex x, then x is

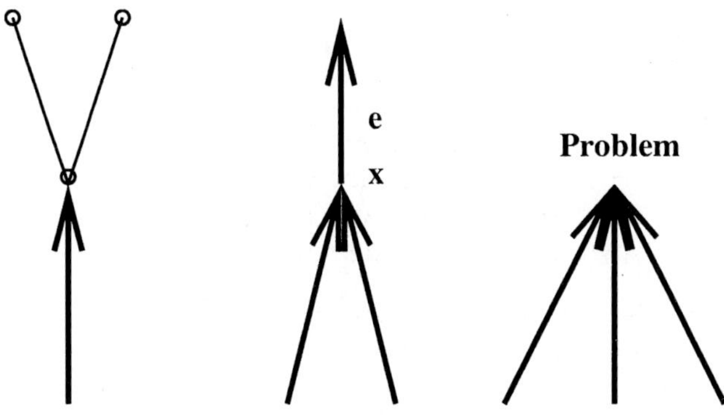

Fig. 5. Live edges

deleted and we have a problem. To avoid it, we actually don't branch on r if this would happen, as we will discuss it below.

Now we do the analysis in more detail. Because of previous handling of the other cases, we assume there are no edges within $N(r)$. Obviously, every vertex of $N^2(r)$ has at least one edge from a vertex of $N(r)$. Because of the degree bound 3 of G, the subgraph induced by $N^2(r)$ has a degree of at most 2, and is of size at most 6.

If there is any connected component within $N^2(r)$ not connected to $N^3(r)$, then it is a path of length $k-3$ or a cycle of length k with k connections from $N(r)$ for some k in the range $3 \leq k \leq 6$. The case of a cycle is fine, as all incoming live edges are stopped (Figure 6). The case of a path of length 0 has already been handled: The single vertex r' on the path is the mirror image of r. The remaining case of a longer path P is not good (3 live edges point to the same vertex) and requires special treatment. In this case, instead of branching at r, we handle the separator of size $6 - k \leq 2$ consisting of the complement of the path P in $N^2(r)$ (Figure 7). In this case, a small separator cuts off a constant size piece of the graph. We have described its handling in the previous subsection.

If a connected component in $N^2(r)$ is connected to $N^3(r)$, then this component is a path P connected at one or both of its endpoints to $N^3(r)$. If it is connected at both end points, then there is exactly 1 live edge entering from

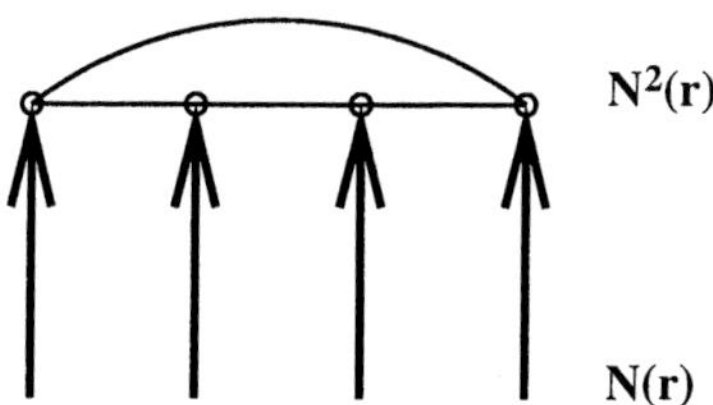

Fig. 6. A cycle in level 2 has no connection to $N^3(r)$

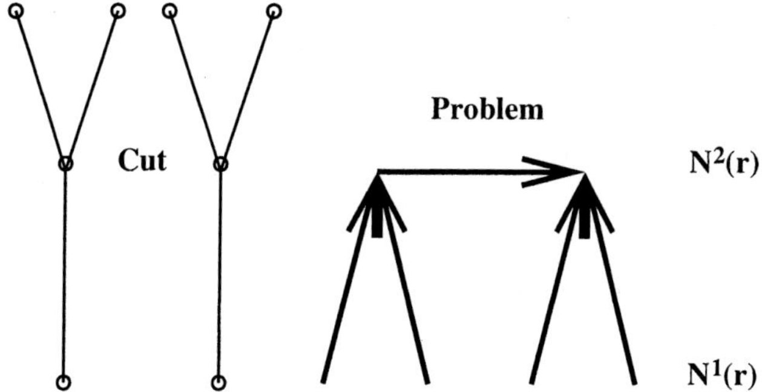

Fig. 7. One component is not connected to $N^3(r)$

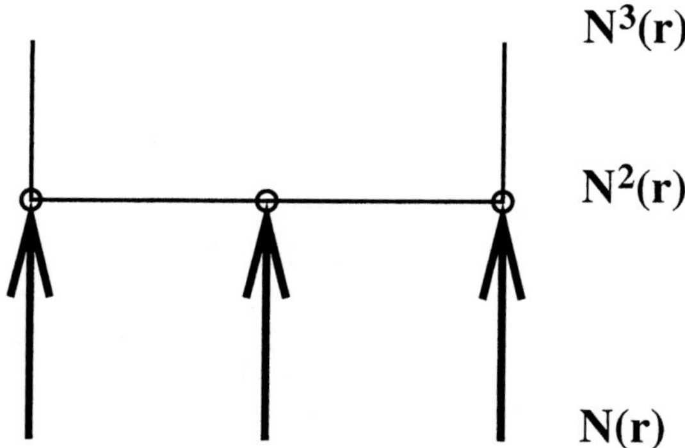

Fig. 8. Both ends are connected to $N^3(r)$

$N(r)$ into each vertex of this component. So each such edge decreases a degree from 3 to 2 in a vertex of $N^2(r)$ (Figure 8). If only 1 endpoint of P is connected to a vertex $x \in N^3(r)$, then the opposite end point has 2 edges from $N(r)$, while all other vertices of the path have one such edge. If the whole path has length 0, then its only vertex has two edges from $N(r)$ and one to x. Now the whole path consists of live edges directed towards x (Figure 9). There are at most 3 live edges reaching level 3. If 2 life edges e, e' meet in a vertex x of $N^3(r)$, then the other edge $\{x, y\}$ is live and directed towards y. All is fine if this is the only

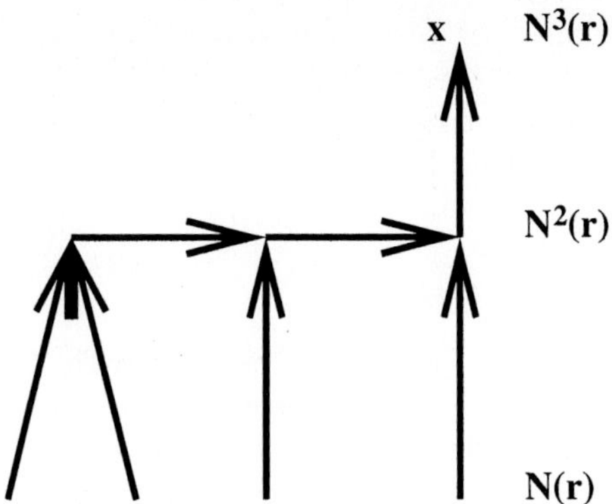

Fig. 9. One end is connected to $N^3(r)$

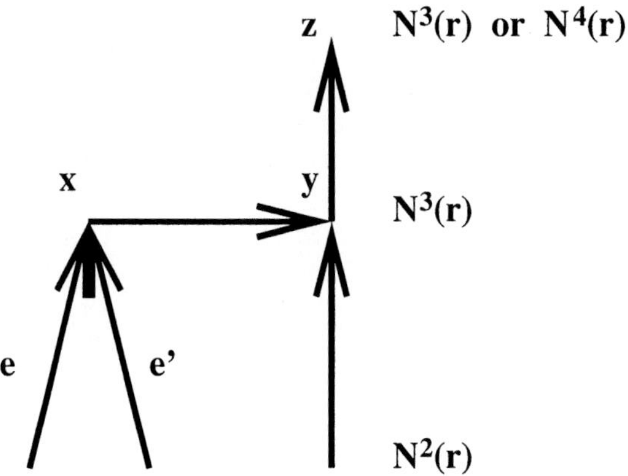

Fig. 10. Three live edges reaching $N^3(r)$

live edge reaching y. Otherwise, y is still on level 3, and the third live edge from level 2 reaches y too (Figure 10). Then the third edge $\{y, z\}$ of y is the only live edge reaching z. This leave just one trivial case: All 3 live edges from level 2 point to the same vertex x on level 3. In this case, x is the only vertex on level 3, and the whole graph is of size 8.

2.4 The Higher Degree Case

We assume now the degree is greater than 3. We also assume the degree of every vertex is at least 3. Pick any vertex r of maximum degree. If at most 3 vertices of $N^2(r)$ are connected to $N^3(r)$, then branch on one of them (which has w.l.o.g. at least 2 edges to $N^3(r)$) and use the remaining 2 (or fewer) vertices as a separator. The branching number is at most $\tau(6, 6)$.

In the remaining case, $N(r) \geq 4$, $N^2(r) \geq 4$, and every vertex has degree at least 3. Branching on r results in a branching number of at most $\tau(3, 5)$, which can be seen as follows.

Let e_{ij} be the number of edges between $N^i(r)$ and $N^j(r)$. Then $e_{01} = |N(r)| \geq 4$, $e_{12} \geq |N^2(r)| \geq 4$, and $2e_{11} + e_{12} \geq 2|N(r)| \geq 8$. When r is selected, we delete $\Delta v = |N(r)|+1$ vertices and $e_{01}+e_{11}+e_{12} = e_{01}+\frac{1}{2}e_{12}+\frac{1}{2}(2e_{11}+e_{12}) \geq |N(r)| + 2 + 4 = \Delta v + 5$ edges. When r is not selected, we delete $\Delta v = 1$ vertex and $|N(r)| \geq 4 = \Delta v + 3$ edges.

2.5 Putting Things Together

There remains one important problem to be dealt with. When we branch on a vertex r in a connected graph, either selecting or not selecting r, the remaining graph might actually be disconnected. For most complexity measures, these would just be lucky cases speeding up the running time, as each connected component can be handled separately.

With our $m - n$ measure, we have to be more careful. There is a potential problem, when some connected components are trees. The tree components can be handled fine. For example, by maintaining a list of degree one vertices, trees can be handled in time $O(m) = O(mc^{m-n})$ for any c, because $m - n = -1$. The problem is with the remaining components, because for some of them $m - n$ might be greater than in the whole graph.

When branching on a vertex r, not selecting r, then no tree can ever be cut off. Otherwise, instead of branching at r the algorithm would actually have proceeded with selecting vertices of degree 1.

When branching on a vertex r, selecting r, then some trees can actually be cut off. In this situation, we just delete such a tree, before we look at our measure $m - n$. To cut off a tree of size k, we need $k + 2$ edges from $N(r)$, because of our lower bound of 3 for all vertex degrees. Furthermore, the tree itself has $k - 1$ edges. We compare this situation with the one where pairs of edges $\{u_1, v_1\}$, $\{u_2, v_2\}$ with $u_1, u_2 \in N(r)$ and v_1, v_2 on the tree, are replaced by internal edges u_1, u_2 inside $N(r)$. Thus these edges to the tree only count 1/2 each. Therefore, a tree of size k (causing the deletion of k vertices) is also responsible for

$$\left\lfloor k - 1 + \frac{k+2}{2} \right\rfloor = \left\lfloor \frac{3k}{2} \right\rfloor$$

extra edge deletions, a surplus of $\lfloor \frac{k}{2} \rfloor \geq 0$ edge deletions over vertex deletions.

As a result, splitting off a tree is still a good case. The excess of vertices in a tree is not balanced by an excess of edges in another component, but compensated by an excess of edge deletions during the split off.

We have seen that the worst case branching number of the whole algorithm is $\tau(2, 5) = 1.237$ implying the following result on the running time.

Theorem 1. *The running time of our independent set algorithm is at most* $O(poly(n)\tau(2,5)^{m-n}) = O(poly(n)1.237^{m-n})$.

This result is of interest in very sparse graphs. It greatly improves the running time for degree 3 graphs (with $m \leq 1.5n$).

Corollary 1. *For degree 3 graphs, the running time of our independent set algorithm is* $O(poly(n)\sqrt{\tau(2,5)}^n) = O(poly(n)\tau(4,10)^n) = O(poly(n)1.112^n)$.

3 Final Remarks

In a separate paper, we will combine this algorithm with the ideas of [7] to obtain better running times for graphs of degree greater than three.

As the simplest case of an induced tree in $\bar{N}(r) \cup N^2(r)$ is also our worst case, it seems very hard to get any further improvement for degree 3. On the other hand, it seems that the degree 2 vertices produced in every step of the algorithm almost never group pairwise. At least, one can conjecture that by selecting the branching point r appropriately, one can always avoid such pairing. As a consequence, the elimination of the degree 2 vertices would then always produce some higher degree vertices implying a better running time.

References

1. Tarjan, R.E., Trojanowski, A.E.: Finding a maximum independent set. SIAM J. Comput. **6** (1977, September) 537–546
2. Jian, T.: An $O(2^{0.304n})$ algorithm for solving maximum independent set problems. IEEE Trans. Comput. **35** (1986) 847–851
3. Robson, J.M.: Algorithms for maximum independent sets. J. Algorithms **7** (1986) 425–440
4. Beigel, R.: Finding maximum independent sets in sparse and general graphs. In: SODA '99: Proceedings of the tenth annual ACM-SIAM symposium on Discrete algorithms, Society for Industrial and Applied Mathematics (1999) 856–857
5. Robson, J.M.: Finding a maximum independent set in time $O(2^{n/4})$? Technical report, LaBRI, Université Bordeaux I (2001)
6. Chen, J., Kanj, I.A., Xia, G.: Labeled search trees and amortized analysis: Improved upper bounds for NP-hard problems. In: ISAAC: 14th International Symposium on Algorithms and Computation. Volume 2906 of LNCS., Springer (2003) 148–157
7. Dahllöf, V., Jonsson, P., Wahlström, M.: Counting models for 2SAT and 3SAT formulae. Theoretical Computer Science **332** (2005) 265–291

The Committee Decision Problem*

Eli Gafni[1], Sergio Rajsbaum[2], Michel Raynal[3], and Corentin Travers[3]

[1] Department of Computer Science, UCLA, Los Angeles, CA 90095, USA
[2] Instituto de Matemáticas, UNAM, D. F. 04510, Mexico
[3] IRISA, Campus de Beaulieu, 35042 Rennes Cedex, France
eli@cs.ucla.edu, rajsbaum@math.unam.mx
{raynal|corentin.travers}@irisa.fr

Abstract. We introduce the (b,n)-*Committee Decision Problem* (CD) - a generalization of the consensus problem. While set agreement generalizes consensus in terms of the number of decisions allowed, the CD problem generalizes consensus in the sense of considering many instances of consensus and requiring a processor to decide in at least one instance. In more detail, in the CD problem each one of a set of n processes has a (possibly distinct) value to propose to each one of a set of b consensus problems, which we call *committees*. Yet a process has to decide a value for at least one of these committees, such that all processes deciding for the same committee decide the same value. We study the CD problem in the context of a wait-free distributed system and analyze it using a combination of distributed algorithmic and topological techniques, introducing a novel reduction technique.

We use the reduction technique to obtain the following results. We show that the $(2,3)$-CD problem is equivalent to the musical benches problem introduced by Gafni and Rajsbaum in [10], and both are equivalent to $(2,3)$-set agreement, closing an open question left there. Thus, all three problems are wait-free unsolvable in a read/write shared memory system, and they are all solvable if the system is enriched with objects capable of solving $(2,3)$-set agreement. While the previous proof of the impossibility of musical benches was based on the Borsuk-Ulam (BU) Theorem, it now relies on Sperner's Lemma, opening intriguing questions about the relation between BU and distributed computing tasks.

Keywords: Asynchronous distributed system, Wait-free computing, Shared memory, Consensus, Set Agreement, Musical benches.

1 Introduction

In a distributed asynchronous system of n processes where at most t of them can fail by stopping, the (k,n)-*set agreement* problem [7] abstracts away a basic coordination problem: processes have input values, and they must agree on at most k of these values. The problem has no solution if the shared-memory has

* This work has been supported by grants from LAFMI (Franco-Mexican Lab in Computer Science) and PAPIIT-UNAM.

J.R. Correa, A. Hevia, and M. Kiwi (Eds.): LATIN 2006, LNCS 3887, pp. 502–514, 2006.

only read/write registers when $k \leq t$ [4, 17, 21] but is solvable if either $k > t$ or else more powerful communication primitives are available in the system. Set agreement and *consensus*, when $k = 1$, have motivated a lot of research (e.g., [2, 18]) and helped to expand our understanding of distributed computing. The *wait-free* case of $t = n-1$ has been shown to be fundamental (e.g., [12, 13, 17]), because from this case results can be derived for any value of t [4, 6], and the wait-free techniques can be generalized to other synchronous and partially synchronous models (e.g., [15, 16]), and even models with stronger communication primitives (e.g., [14]). In this paper we concentrate on the wait-free model.

One of the important uses of consensus arises in a distributed state machine (e.g., [20]): the processes are executing a sequence of operations, and they need to agree on the result of each one of the operations, before they can execute the next one. This and other forms of long-lived versions of consensus (e.g., [3]) that we are aware of are sequential, in that processes propose values, then they agree on one of them, and only then they proceed to the next instance of consensus and propose another value. However, it is also very natural to consider concurrent versions of the problem, where a process p_i proposes a vector V_i of values, and each one of them is intended to one of b different consensus problems, called *committees.* We require that processes deciding on the same committee must decide the same value for that committee. Thus, if the processes participate concurrently in b different applications, we can guarantee wait-free progress in at least one application, without using strong communication objects.

We call this generalization of consensus the *committee decision problem* (CD). Notice that the usual termination requirement of consensus is weakened: a process has to decide a value v for only one of the committees, which it can choose; that is, if its decision is the pair (j, v), then all processes choosing to decide for the j-th committee decide the same value v. The decisions should satisfy the standard agreement and validity requirements of consensus: the value decided for a committee was proposed by some process to that committee, and every process deciding on the committee decides the same value. In addition to its possible applications, there seem to be various interesting generalizations that may motivate new research, such as:

- The number of different committees that are decided is at most k.
- At most k different values are decided for each committee.
- A process that decides must decide in at least k committees.

The CD problem cannot be solved when $n = 2$ and $b = 1$, since this is exactly equal to consensus for two processes, which has no solution [12]. On the other hand, it is easily solvable when $b \geq n$: p_i decides on its own proposal, for the i-th committee, $(i, V_i[i])$. In this paper we concentrate on the binary $(2, 3)$-CD problem, where the proposals are taken from the set $\mathcal{V} = \{0, 1\}$, and there are $b = 2$ committees, and $n = 3$ processes. We state our results for this fundamental case to simplify the presentation (avoiding more algebraic topology notation), and defer the most general phrasing to the full version. We prove that the $(2, 3)$-CD problem is equivalent to the *musical benches* problem of Gafni and Rajsbaum [10], and both are equivalent to $(2, 3)$-set agreement, closing an open question left there. Thus, all three problems are wait-free unsolvable in a read/write shared memory system, and they are all solvable if the system is enriched with objects capable of solving $(2, 3)$-set agreement (such as Test&Set).

Our paper is a follow up to [10], that introduced the musical benches problem, and showed the first connection between distributed computing and the Borsuk-Ulam theorem.[1] In the musical benches problem there are 3 processes, the first two, p_{-1}, p_1, wake up in the first bench (consensus instance), while a third one wakes up in the 2nd bench, either p_{-2} or p_2, but not both. In executions without conflict, namely when only one of p_{-1}, p_1 wakes up, each process decides its own index. Otherwise, the only requirement is that processes decide at most one index in $\{-1, 1\}$ and one index in $\{-2, 2\}$.

The musical benches problem tries to model a new distributed coordination difficulty: processes jump from bench to bench trying to find one in which they may be alone or not in conflict with one another. It resembles the consensus problem in the sense that at least two processes must agree on the value for one committee. However, it is not as clean a generalization as the CD is. Our first aim was to show that the two problems are equivalent, but while investigating the CD problem, we found that both are equivalent to $(2, 3)$-set agreement, while in [10] we only knew that musical benches is somewhere in between $(2, 3)$-set agreement and read-write memory in terms of difficulty. We believe these equivalences are interesting, because although the problems are equivalent in the sense that one can be reduced to any other, they are not the same, a situation reminiscent of NP-complete problems. Having an arsenal of problems that we know are not solvable in read-write memory allows us to judge other problems unsolvable through reductions [9], rather than only through direct topological arguments. Indeed, distributed computing theory development has been promoted by the identification of problems that capture essential coordination difficulties.

The results in this paper are obtained through a novel reduction technique that combines distributed algorithmic ideas with topological intuition. The reduction technique consists of taking a read/write shared memory wait-free protocol, A, and identifying one or more executions, at the end of which an object solving some problem B is invoked. If the resulting protocol solves a problem C (for *any* object that implements a solution to problem B), we have shown that a solution to B implies a solution to C. Although reducing one problem to another is an old idea, our version here has some novel features that stem from the topological perspective of papers such as [15, 16, 17, 21]. We first consider the set of executions of A as a geometric object, called a *complex*. In the case of $n = 3$, each execution is drawn as a triangle, or *simplex*, where its corner vertices are labeled with the views (local states) of each one of the processes at the end of the execution. We then identify the triangles (or sometimes edges corresponding to 2-process executions) on which we are going to invoke

[1] Although we do not use it in this paper, the reader may be interested to know that the theorem is "one of the most useful tools offered by by elementary algebraic topology to the outside world" [19]. It implies Sperner's lemma, but not the opposite.

the object B. Then we replace these triangles by the complex representing the set of possible responses of an implementation instance of B, and obtain the combined complex representing the protocol reduction. The goal is to obtain a protocol whose complex gives enough flexibility[2] to associate a decision function with each one of its vertices and solve the desired problem, C. See for example Figure 1, where we start with the simplex representing the inputs to the $(2, 3)$-set agreement problem, we then execute a wait-free protocol where we identify two triangles to be removed and replaced by the set of possible responses of an arbitrary musical benches implementation, and the vertices of the resulting complex (obtained by gluing in the later complex into the hole of the former), can be colored with decisions (placed in the figure by each one of the vertices) that map into the $(2, 3)$-set agreement outputs, represented by a hollow triangle. We have thus created a hole, which gives the desired flexibility to the final complex, and allows for an appropriate decision function to be designed. More details appear in Section 3.1, that includes more formal topology definitions and explanations about Figure 1. A good introduction to basic topology is [1].

The rest of the paper is organized as follows. Section 2 defines the problems of CD, set agreement and musical benches, and some additional preliminaries. Section 3 describes an algorithm to solve $(2, 3)$-set agreement using a musical benches object, and an algorithm to solve $(2, 3)$-set agreement using a CD object. Section 4 shows that the CD problem is wait-free solvable using a $(2, 3)$-set agreement object. Due to space limitation, proofs are omitted. Additional details and full proofs can be found in [11].

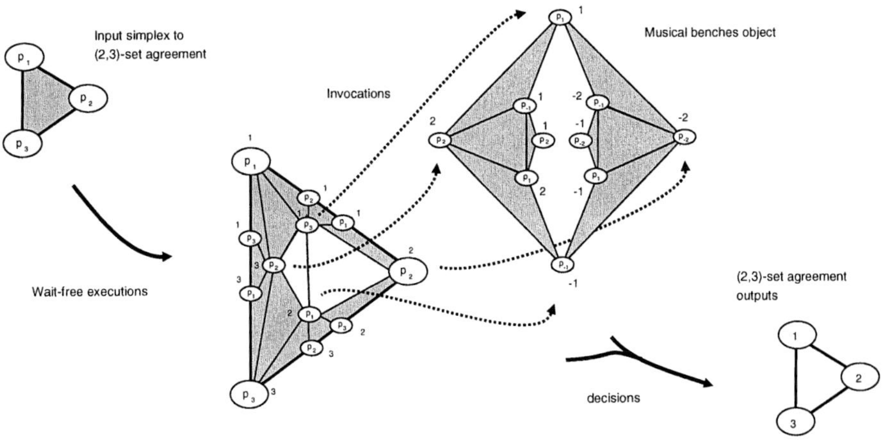

Fig. 1. Solving $(2, 3)$-set agreement using (one example of) a musical benches object

[2] The actual complex obtained depends on the actual solution to B used, but any such complex should exhibit that flexibility. Two features add flexibility: holes and more vertices.

2 Three Problems and Preliminaries

This paper considers the usual asynchronous shared memory model, composed of single-writer/multi-reader registers, and studies wait-free algorithms, where any number of processes can fail by crashing. A full description of these concepts can be found in textbooks such as [2, 18].

2.1 The Problems

The usual notion of *task* is a one-shot decision problem specified in terms of an input/output relation Δ. The processes start with private input values, and must eventually decide on output values, by writing to a write-once variable. An *input vector* I specifies in its i-th entry, $I[i]$, the input value of process p_i, and we say p_i *proposes* $I[i]$ in the execution; similarly, an *output vector* J specifies a decision value $J[i]$ for each process p_i. The task defines a set of legal input vectors, and for each one, Δ specifies a set of legal output vectors. Thus, given input vector I, the processes decide a vector J such that individually p_i decides $J[i]$. It is sometimes convenient to consider *inputless tasks,* where a process has only one possible input value, namely its own id.

Set Agreement. The *k-set agreement* problem is a generalization of consensus where processes must decide on at most k different values, out of the input values. The corresponding inputless version for three processes, p_1, p_2, p_3, and $k = 2$, denoted $(2, 3)$-set agreement, is illustrated in Figure 2 (ids associated to each output value are omitted for clarity). It is defined by the set of input vectors consisting of (p_1, p_2, p_3) and all its subvectors, and the relations: $\Delta(p_i) = \{(i)\}$, $\Delta(p_i, p_j) = \{(i, i), (j, j), (i, j), (j, i)\}$ and, $\Delta(p_i, p_j, p_k)$ equal to all vectors of i, j, k with at most two different values (this requirement is represented in the figure by the hole; the possible outputs have no triangle, only edges and vertices). Set agreement is not wait-free solvable [4, 17, 21], due to a generalization of the consensus impossibility connectivity argument to higher dimensions; wait-free executions induce a "flat structure" subdividing the input triangle, and in the figure one can see that a flat triangle is required to be mapped to a hollow one (preserving the boundary), which is impossible.

Committee Decision Problem. In the (b, n)*-committee decision* (CD) problem n processes are trying to solve b consensus instances, called *committees*, and

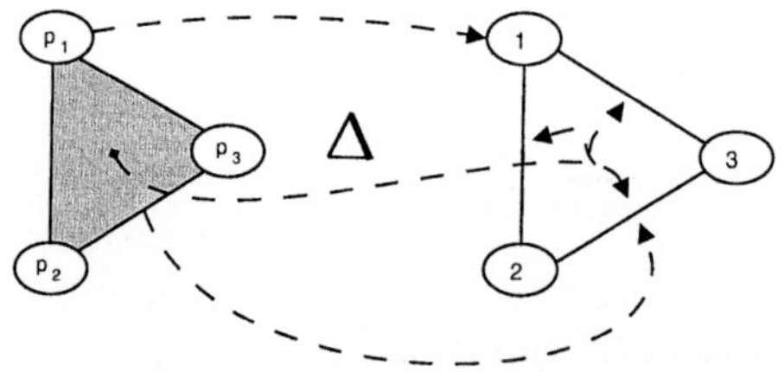

Fig. 2. The inputless $(2, 3)$-set agreement problem (some arrows of Δ omitted)

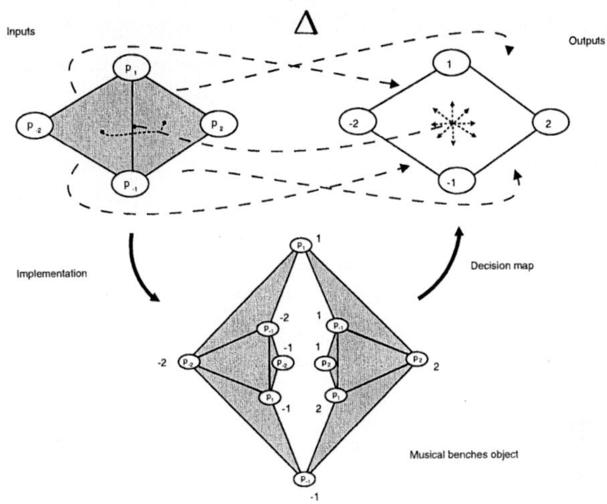

Fig. 3. Musical benches task with a musical benches object

each process is required to make a decision for at least one of them. More explicitly, in an execution, each process p_i proposes a vector V_i of b entries: $V_i[\ell]$ is the value proposed by p_i for committee ℓ. A process decides a pair (ℓ, v) where ℓ, $1 \leq \ell \leq b$ denotes a committee, and v a value proposed by a process for committee ℓ. The problem is defined by the three requirements:

- **Termination.** No process takes infinitely many steps without deciding.
- **Validity.** If a process decides (ℓ, v) then $\exists\ j$ such that $v = V_j[\ell]$.
- **Agreement.** Assume p_i, p_j decide (ℓ_i, v_i) and (ℓ_j, v_j) respectively. Then $\ell_i = \ell_j \Rightarrow v_i = v_j$.

We concentrate our attention on the binary $(2,3)$-CD problem, where $n = 3$, $b = 2$ and the proposed values are taken from $\mathcal{V} = \{0,1\}$. We refer to this version as the *CD problem.*

Musical Benches. We can think of 2-process binary consensus as a *bench* with two places, designated 1 and -1. Processes p_1 and p_{-1}, wake up at places 1 and -1, respectively. In a solo execution a process must return the place it wakes up in. Otherwise, in an execution where both participate, they return the same place. We add a second bench, with places $2, -2$, and wake up either process p_2 at slot 2, or p_{-2} at slot -2, but not both. In executions with no conflict, i.e., either p_{-1} or p_1 wake up but not both, the participating processes return the places they wake up in. Only if both p_{-1} and p_1 wake up, then any participating process can go to any seat. This is the *musical benches* problem of [10], shown there to have no wait-free solution.

The musical benches problem is illustrated in Figure 3, disregarding ids and omitting the dotted arrows of Δ for single vertices, to avoid cluttering the figure. In the figure there is also an example of an object implementing the musical benches problem. Each vertex is labeled on the inside with a process p_i, and on the outside with the value d returned from the object to p_i. The corner vertices correspond to executions where the process invokes the object alone, and therefore, a p_i vertex is labeled with value i. An edge joining two such vertices represents an execution where both processes invoke the object alone. Notice that there are two paths connecting the corners p_1, p_{-1}, with vertices labeled p_1 or p_{-1}, representing executions where only these processes invoke the object. For example, they are two edges incident to the p_1 corner, one representing an execution where the object returns 1 to p_{-1} and another where it returns -2 to p_{-1}. Executions where p_{-2} participates appear on the left side of the hole, while executions where p_2 participates appear on the right side of the hole. Notice also that no two vertices with the same id have the same value. One can check that this object indeed satisfies the musical benches specification given by Δ.

2.2 Participating Set Problem

Preparing for the next section we recall the k-participating set problem [10], a generalization of the one in [5] that can access a set agreement object. We present here the case of 3 levels, and either $k = 2$, that has access to $(2,3)$-set agreement, or $k = 3$, the original problem of [5] that has no access to set agreement. That is, we have our first simple example of a reduction, in this case from the 2-participating set problem to $(2,3)$-set agreement. The 3-participating set problem shows that read write shared memory complex can be flattened to a subdivided simplex, as in the left side of Figure 4. Using a $(2,3)$-set agreement implementation, as in the right side of the figure, the center triangle is removed and we can create a subdivided simplex with a hole. A process p_i computes a set of ids S_i, such that

1. $\forall i : i \in S_i$,
2. $\forall i, j : S_i \subseteq S_j \vee S_j \subseteq S_i$,
3. $\forall i, j : i \in S_j \Rightarrow S_i \subseteq S_j$,
4. $|\{j : |S_j| = 3\}| \leq k$.

The first three are the requirements of the participating set problem in [5]. Sets satisfying these properties correspond to the subdivided simplex in Figure 4.

For completeness a protocol solving the k-participating set appears in Figure 5. The 4-th property is achieved through the set agreement object, invoked by p_i with the operation setAg(i), when $k = 2$. Invoking the set agreement operation has the effect of removing the simplex in the center of the subdivision

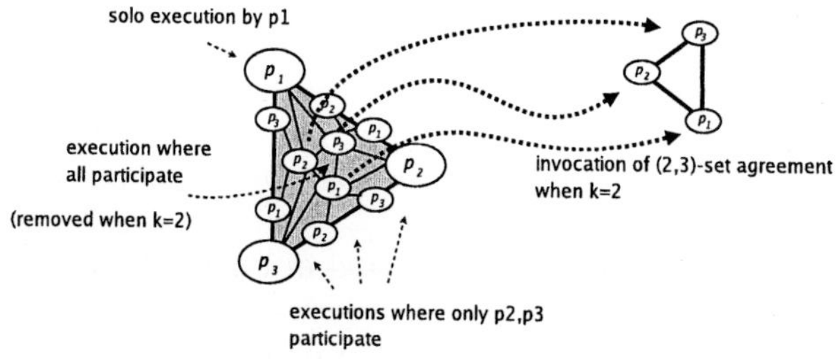

Fig. 4. The k-participating set views for $k = 3$; when $k = 2$ the center triangle is removed

```
Initially: Level[j] = 4 ∀j ∈ {1,2,3}; k = 2 or k = 3;
Function k-PARTICIPATINGSET(i)
Init OK_i ← false;
repeat Level[i] ← Level[i] − 1;
        for j = 1 to 3 do level_i[j] ← Level[j] enddo
        S_i ← {j : level_i[j] ≤ Level[i], j ∈ {1,2,3}};
        if |S_i| = 3 and k = 2 then ans_i ← (2,3)-SETAG(i);
                              if ans_i = i then OK_i ← true endif
        else OK_i ← true endif
until (|S_i| ≥ Level_i[i]) ∧ OK_i;
return(S_i)
```

Fig. 5. From $(2,3)$-set agreement to k-Participating set (code for p_i)

(impossible that the three processes produce sets of size 3), and leaving just its boundary (at most two processes may produce sets of size 3).

3 Solving $(2,3)$-Set Agreement

An algorithm to solve musical benches using $(2,3)$-set agreement is described in [10]. In Section 3.1 we describe an algorithm to solve $(2,3)$-set agreement using a musical benches object. Therefore, the musical benches problem is equivalent to $(2,3)$-set agreement. In Section 3.2 we describe an algorithm to solve $(2,3)$-set agreement using a CD object.

3.1 Solving $(2,3)$-Set Agreement with Musical Benches

Informally, the idea is very simple. In the musical benches one of two combinations of 3 processors start with 3 distinct inputs. They eventually halt with at most 2 distinct outputs. Thus the problem possess that "narrowing of choices" property that set agreement exhibit. The only problem we face is how to interface between the requirement of set agreement and those of musical benches. Resolving this is the crux of the paper: Employ read-write first and then glue the musical benches to replace two adjacent simplexes.

A protocol that solves $(2,3)$-set agreement using musical benches appears in Figure 6, and it is illustrated in Figure 1. Each process p_i starts by invoking the participating set protocol of Figure 5 with $k = 3$. Once it gets back a set S_i, it invokes a musical benches protocol with a parameter $h_{mb}(i, S_i)$ defined as follows:

$$h_{mb}(i, S_i) = \begin{cases} -1 \text{ if } i = 1 \text{ and } S_i = \{1,2,3\} \\ +1 \text{ if } i = 3 \text{ and } S_i = \{1,2,3\} \\ +2 \text{ if } i = 2 \text{ and } S_i = \{1,2,3\} \\ -2 \text{ if } i = 2 \text{ and } S_i = \{2\} \\ \perp \;\; \text{otherwise} \end{cases}$$

That is, the musical benches protocol is invoked only when $h_{mb}(i, S_i) \neq \perp$, and if so, each process p_i makes a decision, $f_{mb}(bench)$, that depends on the answer *bench* returned by the musical benches protocol, as follows

$$f_{mb}(bench) = \begin{cases} 1 \text{ if } bench = 1 \text{ or } bench = -2 \\ 2 \text{ if } bench = -1 \\ 3 \text{ if } bench = 2 \end{cases}$$

or if p_i did not invoke the musical benches protocol, then it returns $g(i, S_i)$. The only requirement is that $g(i, S_i)$ returns an id in S_i, to satisfy the *validity* requirement of the set agreement problem (a decision was proposed by somebody).

Each vertex on the left of Figure 1 is labeled in the inside with the corresponding process p_i, and on the outside with its decision. The boundary of the removed triangles fits the boundary of the musical benches object. We stress that the object in the figure is just an example of one possible implementation of the musical benches problem; the protocol works for any implementation. Each of the vertices of the musical benches object is labeled in the inside with the corresponding process p_i, and on the outside with the value returned by the object. Thus, if we consider a vertex on the boundary of the hole (left side of the figure), say the corner p_2, it corresponds to an execution where p_2 runs solo, gets $S_2 = \{2\}$ from the participating set object, invokes the musical benches with $h_{mb}(2, \{2\}) = -2$, and gets back -2 (the label by the corresponding vertex on the right side of the figure) and decides $f_{mb}(-2) = 2$ (the label by p_2's corner vertex on the left side of the figure). A p_i vertex of the left side of the figure where the musical benches object is not invoked is labeled with $g(i, S_i)$ (this particular g is just an example).

```
Function (2,3)-SetAg-from-Benches(i)
(01) S_i ← 3-ParticipatingSet(i);
(02) if h_mb(i, S_i) ≠ ⊥ then
(03)    bench_i ← MusicalBenches(h_mb(i, S_i));
(04)    return f_mb(bench_i)
(05) else return g(i, S_i) endif
```

Fig. 6. From Musical Benches to $(2, 3)$-Set Agreement (code for p_i)

Lemma 1. *The* $(2, 3)$-SetAg-from-Benches *protocol solves* $(2, 3)$*-set agreement using any musical benches implementation.*

3.2 Solving $(2, 3)$-Set Agreement with Committee Decision

The technique of Section 3.1 can be used to solve $(2, 3)$-set agreement with CD. The SetAg-from-CD protocol of Figure 8 is similar to the one in Figure 6, except that a CD object is invoked instead of invoking a musical benches object, and the the functions h_{mb}, f_{mb} and g change.

Each process p_i starts by invoking the participating set protocol of Figure 5 with $k = 3$. Once it gets back a set S_i, it checks if $h_{cd}(i, S_i) = \bot$. If so it decides according to the function $g_{cd}(i, S_i)$ (values by the vertices on the left side of Figure 7):

$$g_{cd}(i, S_i) = \begin{cases} i \text{ if } |S_i| = 1 \text{ else:} \\ 1 \text{ if } (i = 1 \text{ and } 2 \in S_i) \text{ or } (i = 2 \text{ and } 1 \in S_i) \text{ or } (i = 3 \text{ and } 1 \in S_i), \\ 2 \text{ if } i = 3 \text{ and } 2 \in S_i, \\ 3 \text{ otherwise.} \end{cases}$$

Else, $h_{cd}(i, S_i) \neq \bot$, and it invokes a CD protocol with the parameter $h_{cd}(i, S_i)$ defined as follows. This is illustrated in the right side of Figure 7, where an example of a CD object is presented (not all the object is depicted, only the values returned for the proposed input vectors).

$$h_{cd}(i, S_i) = \begin{cases} (-1, -2) \text{ if } i = 1 \text{ and } S_i = \{1, 2, 3\}, \\ (+1, +2) \text{ if } i = 3 \text{ and } S_i = \{1, 2, 3\}, \\ (-1, +2) \text{ if } i = 2 \text{ and } S_i = \{1, 2, 3\}, \\ (+1, -2) \text{ if } i = 2 \text{ and } S_i = \{2\}, \\ \bot \qquad \text{otherwise.} \end{cases}$$

Once the CD object returns a value the process p_i stores it in a local variable *bench*. In the right side of Figure 7, the vectors proposed to the CD are depicted only in the 4 corners for lack of space; every vertex is labeled with the value returned by the object. Notice that no two vertices with the same id and proposed vectors have the same returned value associated (this is why the boundary can be subdivided here, but not in a musical benches object). The process then computes a decision $f_{cd}(i, bench)$, defined as follows:

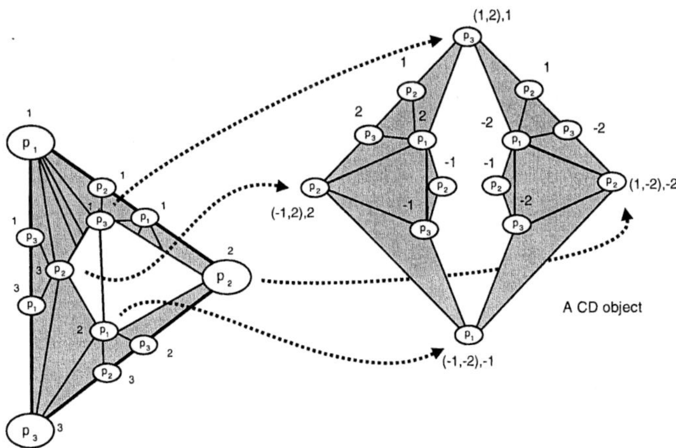

Fig. 7. To solve set agreement each process p_i invokes a CD object. On the left figure, decisions are the values by the vertices; on the right figure values by the vertices are returned by the object.

$$f_{cd}(i, bench) = \begin{cases} 1 \text{ if } i = 1 \text{ and } bench = 1 \text{ or } bench = 2\ , \\ 2 \text{ if } i = 1 \text{ and } bench = -2, \\ 3 \text{ if } i = 1 \text{ and } bench = -1, \\ 1 \text{ if } i = 3 \text{ and } bench = 1 \text{ or } bench = 2, \\ 2 \text{ if } i = 2 \text{ and } bench = 1 \text{ or } bench = -2, \\ 3 \text{ if } i = 2 \text{ and } bench = -1 \text{ or } bench = 2, \\ 2 \text{ if } i = 3 \text{ and } bench = -2. \\ 3 \text{ if } i = 3 \text{ and } bench = -1, \end{cases}$$

```
Function (2,3)-SetAg-from-CD(i)
S_i ← 3-ParticipatingSet(i);
if h_cd(i, S_i) ≠ ⊥ then
    bench_i ← CD(h_cd(i, S_i));
    return f_cd(bench_i)
else return g_cd(i, S_i) endif
```

Fig. 8. From CD to (2, 3)-Set Agreement (code for p_i)

Lemma 2. *The* (2, 3)-SETAG-FROM-CD *protocol solves* (2, 3)*-set agreement using any CD implementation.*

4 Solving Committee Decision with (2, 3)-Set Agreement

This section shows that the (2, 3)-CD problem is wait-free solvable using a (2, 3)-set agreement object. Since in Section 3.2 we showed the opposite reduction, we have that both problems are equivalent. The wait-free impossibility of solving (2, 3)-set agreement [4, 17, 21] implies that (2, 3)-CD is wait-free unsolvable. In [10] a protocol that solved the musical benches problem with access to a (2, 3)-set agreement object is described. This protocol can be adapted to solve the CD problem; the main difference is the decision function. The protocol works as follows. Each process p_i gets a vector V_i as input to the CD problem. It

```
Function (2,3)-CD-from-SetAg(V_i)
Init view_i ← ∅; id_i ← [⊥, ⊥, ⊥];
Prop[i] ← V_i;
S_i ← 2-ParticipatingSet(i);
if |S_i| = 3 then id[i] ← i;
                  for j = 1 to 3 do id_i[j] ← id[j] enddo
                  view_i ← {j : id_i[j] ≠ ⊥, j ∈ {1, 2, 3}}
endif
return f(S_i, view_i)
```

Fig. 9. From (2, 3)-set agreement to (2, 3)-CD (code for p_i)

first writes it to a shared array, *Prop*, in position *Prop*[*i*]. Then p_i invokes the 2-PARTICIPATINGSET(i) function of Figure 5, and gets back a set S_i of process ids, satisfying the 2-PARTICIPATINGSET properties (see section 2.2): Once p_i gets a set S_i back from the 2-PARTICIPATINGSET object, if $|S_i| = 3$ it executes lines (03)–(05) which have the effect of proposing its *id* to a read/write object, and gets back a set $view_i$ of ids, of processes that invoked the object. The aim is to subdivide the boudary of removed center triangle of the protocol complex. Finally, process p_i decides a value $f(S_i, view_i)$. Due to space limitation, the corresponding figure and the definition of the decision function are omitted. More details can be found in the technical report [11].

Lemma 3. *The* (2,3)-CD-FROM-SETAG *protocol solves* (2,3)-*CD using any* (2,3)-*set agreement object.*

As a consequence of Lemmas 1, 2, and 3 we have our main result.

Theorem 1. *Musical benches can be wait-free solved iff CD can be wait-free solved iff* (2,3)-*set agreement can be wait-free solved.*

References

1. Armstrong M.A., *Basic Topology,* Springer-Verlag, 251 pages, 1983.
2. Attiya H. and Welch J., *Distributed Computing: Fundamentals, Simulations and Advanced Topics,* McGraw–Hill, 451 pages, 1998.
3. Bar-Noy A., Deng X., Garay J., Kameda T., Optimal amortized distributed consensus. *Info. and Comp.,* 120(1):93-100, 1995.
4. Borowsky E. and Gafni E., Generalized FLP Impossibility Results for t-Resilient Asynchronous Computations. *Proc. 25th ACM Symposium on the Theory of Computing (STOC'93),* ACM Press, pp. 91-100, 1993.
5. Borowsky E. and Gafni E., Immediate Atomic Snapshots and Fast Renaming (Extended Abstract). *Proc. 12th ACM Symposium on Principles of Distributed Computing (PODC'93),* ACM Press, pp. 41-51, 1993.
6. Borowsky E., Gafni E., Lynch N. and Rajsbaum S., The BG Distributed Simulation Algorithm. *Distributed Computing,* 14(3):127–146, 2001.
7. Chaudhuri S., More *Choices* Allow More *Faults:* Set Consensus Problems in Totally Asynchronous Systems. *Information and Computation,* 105:132-158, 1993.
8. Fischer M.J., Lynch N.A. and Paterson M.S., Impossibility of Distributed Consensus with One Faulty Process. *Journal of the ACM*, 32(2):374-382, 1985.
9. Gafni E. DISC/GODEL presentation: R/W Reductions (DISC'04), 2004. http://www.cs.ucla.edu/~ eli/eli/godel.ppt
10. Gafni E. and Rajsbaum S., Musical Benches. *Proc. 19th Int. Symposium on Distributed Computing (DISC'05)*, Springer Verlag LNCS #3724, pp. 63–77, 2005.
11. Gafni E., Rajsbaum R., Raynal M. and Travers C., The Committee Decision Problem. *Tech Report #1745*, IRISA, University of Rennes 1 (France), 2005. http://www.irisa.fr/bibli/publi/pi/2005/1745/1745.html
12. Herlihy M.P., Wait-Free Synchronization. *ACM Transactions on programming Languages and Systems*, 11(1):124-149, 1991.

13. Herlihy M., Rajsbaum S., New Perspectives in Distributed Computing. *Proc. 24th International Symposium Mathematical Foundations of Computer Science (MFCS'99),* Springer Verlag LNCS #1672, pp. 170–186, 1999.
14. Herlihy H., Rajsbaum S., Algebraic spans. *Mathematical Structures in Computer Science*, 10(4): 549–573, 2000.
15. Herlihy, M. Rajsbaum, S. and Tuttle, M. Unifying Synchronous and Asynchronous Message-Passing Models. *Proc. 17th ACM Symposium on Principles of Distributed Computing (PODC'98),* pp. 133–142, 1998.
16. Herlihy, M. Rajsbaum, S. and Tuttle, M. An axiomatic approach to computing the connectivity of synchronous and asynchronous systems. *Proc. of the 6th workshop on Geometric and Topological Methods in Concurrency and Distributed Computing (GETCO'04),* 2004.
17. Herlihy M.P. and Shavit N., The Topological Structure of Asynchronous Computability. *Journal of the ACM*, 46(6):858-923, 1999.
18. Lynch N., Distributed Algorithms. *Morgan Kaufmann Pub.*, San Francisco (CA), 872 pages, 1996.
19. Matousek J., *Using the Borsuk-Ulam Theorem,* Lectures on Topological Methods in Combinatorics and Geometry, 2003, Springer.
20. Lamport L., The Part-Time Parliament. *ACM Transactions On Computer Systems*, 16(2):133-169, 1998.
21. Saks, M. and Zaharoglou, F., Wait-Free k-Set Agreement is Impossible: The Topology of Public Knowledge. *SIAM Journal on Computing*, 29(5):1449-1483, 2000.

Common Deadline Lazy Bureaucrat Scheduling Revisited*

Ling Gai and Guochuan Zhang

Department of Mathematics, Zhejiang University,
Hangzhou 310027, China
zgc@zju.edu.cn

Abstract. Lazy bureaucrat scheduling is a new class of scheduling problems introduced by Arkin et al. [1]. In this paper we focus on the case where all the jobs share a common deadline. Such a problem is denoted as CD-LBSP, which has been considered by Esfahbod et al. [2]. We first show that the worst-case ratio of the algorithm *SJF* (Shortest Job First) is two under the objective function [*min-time-spent*], and thus answer an open question posed in [2]. We further present two approximation schemes A_k and B_k both having worst-case ratio of $\frac{k+1}{k}$, for any given integer $k > 0$, under the objective function [*min-makespan*] and [*min-time-spent*] respectively. Finally, we prove that the problem CD-LBSP remains NP-hard under several objective functions, even if all jobs share the same release time.

1 Introduction

Scheduling has been studied extensively for decades. Generally in a scheduling problem the workers are required to stay in their peak level to minimize the total cost or maximize the throughput, etc. However, in some different situation, the worker (we call him/her a bureaucrat) may not want to do much work, which motivates a new kind of scheduling problems, called lazy bureaucrat scheduling, introduced by Arkin et al. [1]. In this problem, the bureaucrat wants to do things as little (or easy) as possible, which is in the reverse objective of the classical scheduling problem. Of course there is a basic assumption, called the *busy requirement*, that the bureaucrat must keep working as long as there are some executable jobs, otherwise the problem would become trivial and the optimal strategy for the bureaucrat would be just stay idle without doing anything. Arkin et al. [1] studied the problem extensively and many complexity and algorithmic results were obtained. An interesting example was given in [1], in which the "lazy bureaucrat" tries to minimize the amount of the jobs done so that he is able to go home earlier or have more free time. We refer to [1] for more details. Hepner and Stein [5] recently dealt with several problems in the preemptive case where the jobs can be paused and restarted later. In the literature one can also

* Research supported by NSFC grant No. 10231060 and No. 60573020.

J.R. Correa, A. Hevia, and M. Kiwi (Eds.): LATIN 2006, LNCS 3887, pp. 515–523, 2006.

find the general optimization problems in reverse, such as the maximum TSP [4], the maximum cut problem [3] and the longest path problem [6], which may lead to a better understanding of the structure and algorithmic complexity of the original problem.

In this paper we focus on the case where all the jobs share the same deadline. Such a lazy bureaucrat scheduling problem is denoted as CD-LBSP, which was first considered in [2]. More precisely, there is a set of jobs $J_1, \ldots, J_n$. Job J_i arrives at time a_i and has processing time of p_i, for $i = 1, 2, \ldots, n$. Each job is associated with a weight. There is a common deadline D for all jobs. Job J_i can be executed only if its starting time is in between a_i and $D - p_i$. In other words, a job can only be started at or after its release time, and once it is started, it cannot be interrupted and must be completed by the common deadline D. At any time a job is called *executable* if it can be executed. Throughout the paper we will alternatively use the terms machine and bureaucrat without causing any confusion.

Basically there are four objective functions [1, 2]:

1. [*min-time-spent*]: Minimize the total amount of time spent working.
2. [*min-weighted-sum*]: Minimize the weighted sum of completed jobs.
3. [*min-makespan*]: Minimize the makespan, the maximum completion time of the jobs.
4. [*min-number-jobs*]: Minimize the total number of completed jobs.

It is easy to see that objective functions 1 and 4 are special cases of 2. If all jobs arrive at the same time, objective functions 1 and 3 are the same.

Previous Results. Arkin et al. [1] proved that the general LBSP is NP-hard in the strong sense under all objective functions and is not approximable to within any fixed factor. Esfahbod et al. [2] considered the problem CD-LBSP where all deadlines are the same. They showed CD-LBSP is NP-hard under all objective functions. It was also proved that the objective function [*min-number-of-jobs*] (and thus [*min-weighted-sum*]) is not approximable within any fixed factor unless P=NP. For [*min-makespan*] they presented an approximation algorithm *SJF* (Shortest Job First) with worst-case ratio of 2.

Our Contribution. In this paper, we concentrate on the problem CD-LBSP. We first show that algorithm *SJF* has a worst-case ratio of 2 under the objective function [*min-time-spent*], and thus answer an open question in [2]. Then we devise a new algorithm A_k under objective function [*min-makespan*] and show that the worst-case ratio is at most $(k+1)/k$, for any given integer $k > 0$. This result implies the existence of a PTAS and significantly improves the previous result. Furthermore, we give another algorithm B_k which leads to a PTAS for objective function [*min-time-spent*] by employing algorithm A_k a polynomial number of times. Finally we show the special case that all jobs arrive at the same time is still NP-hard under several objective functions.

Organization of the Paper. The remainder of the paper is organized as follows. Section 2 gives the worst-case analysis of algorithm *SJF* for the objective function [*min-time-spent*]. The new algorithms A_k and B_k are shown in Section 3 and Section 4, respectively. Section 5 presents the NP-hardness proof for the special case where all jobs arrive at the same time.

2 Algorithm *SJF*

The algorithm *SJF* works as follows. Whenever the machine is idle and there are some executable jobs, schedule the one (among the executable jobs) with the shortest processing time (ties broken arbitrarily). Esfabhod et al. [2] showed that *SJF* is a 2-approximation algorithm under the objective function [*min-makespan*] and this bound is tight. In the same paper they left an open question which asks for the worst-case ratio of algorithm *SJF* under the objective function [*min-time-spent*]. We will answer this question by showing that *SJF* has the same worst-case ratio of 2. The instance in [2], devoted to the objective function [*min-makespan*], still applies to the objective function [*min-time-spent*]. For the sake of completeness we present a simple instance to show the lower bound of 2. Just take three jobs, two of processing time 1 and the other of $1+\varepsilon$, where $\varepsilon > 0$ is an arbitrarily small number. The common deadline is 2. The algorithm *SJF* will produce a schedule of time spent 2 while the optimum is just $1+\varepsilon$. It implies a lower bound of 2 for algorithm *SJF*. In the following we only need to show

$$\frac{T_{SJF}(I)}{T^*(I)} \leq 2 \tag{1}$$

holds for any instance I, where $T^*(I)$ and $T_{SJF}(I)$ are the total time spent in an optimal schedule and in the schedule generated by algorithm *SJF*, respectively. If it is not true, namely, if there exists some instance violating the above inequality, consider a minimum counterexample I_s in terms of the number of jobs. The instance I_s has two properties:

1. $T_{SJF}(I_s)/T^*(I_s) > 2$.
2. For any instance I where $|I| < |I_s|$, i.e., the number of jobs in I is smaller than that in I_s, we have $T_{SJF}(I)/T^*(I) \leq 2$.

Since $T_{SJF}(I_s) \leq D$, we have $T^*(I_s) < D/2$. Let σ^* be an optimal schedule and σ be the schedule generated by *SJF* on instance I_s, respectively. Consider the first time point t_1 at which the machine becomes idle in schedule σ^*. Clearly $t_1 < D/2$. Let t_2 be the first time point at which the machine becomes idle in schedule σ. We want to show that $t_1 = t_2$. If this does not hold, there could be two cases.

Case 1. $t_1 > t_2$. Consider those jobs scheduled in σ^*, that arrive before time t_2. Clearly the total processing time of them is greater than t_2 (otherwise $t_1 \leq t_2$). Thus there must be some job J that is started before time t_2 in σ^* but is not scheduled in σ. Let p be the processing time of job J. Then $p > D-t_2 > D-t_1 > t_1$, since J cannot fit in the schedule σ within the interval $[t_2, D]$. This conflicts with the fact that $p \leq t_1$. Thus this case cannot happen.

Case 2. $t_2 > t_1$. In this case there must be some job J of processing time p that is started before time t_1 in σ but is not scheduled in σ^*. Clearly $p + t_1 > D$. This shows that $p > D/2$ and J is the largest job scheduled in σ. Moreover there is some job $J' \in \sigma^* - \sigma$ (i.e., $J' \in \sigma^*$ but $J' \notin \sigma$). At the time t that J is started in σ, there are no small executable jobs (with processing time of at most $D/2$). Thus the release time of J' is larger than t, since otherwise J' should be scheduled instead of job J by the rule of algorithm *SJF*. It follows that the machine must become idle at time $t < t_1$ in schedule σ^*. This is a contradiction with the definition of t_1.

Therefore we have proved that $t_1 = t_2 < D/2$. Clearly in σ any job scheduled by t_1 has a processing time less than $D/2$, that must also be scheduled in σ^*, and vice versa. This implies that σ and σ^* schedule the same jobs by time t_1. Consider a new instance I by removing all jobs arriving before time t_1. Note that $T_{SJF}(I)/T^*(I) = (T_{SJF}(I_s) - t_1)/(T^*(I_s) - t_1) > 2$. It implies that I is a counterexample with less jobs than I_s. This conflicts with the assumption that I_s is a minimum counterexample. Hence the inequality (1) holds for any instances. We conclude that

Theorem 1. *The worst-case ratio of algorithm* SJF *is 2 for CD-LBSP under objective function [min-time-spent].* □

3 A PTAS for Minimizing Makespan

Esfahbod et al. [2] showed a 2-approximation algorithm under the objective [min-makespan]. In the following, we will show that for any fixed integer k, there is an approximation algorithm with a bound of $(k+1)/k$. We first investigate the following property for an optimal schedule.

Lemma 1. *There must exist an optimal schedule that obeys the first come first serve rule (*FCFS*) for minimizing makespan.*

Proof. It is easy to show with an exchanging strategy. We start with an optimal schedule that does not obey the *FCFS* rule. There must exist two adjacent jobs J_i and J_j with release times $a_i < a_j$, while job J_j is scheduled before job J_i in the optimal schedule. Swapping two such jobs does not disturb the remaining schedule and the optimality remains. Continue this process until any two adjacent jobs satisfy the property that the job arriving later is started later. Finally we get an optimal schedule obeying *FCFS*. □

For any given integer $k > 0$, distinguish the jobs as follows: A job is called *big* if the job processing time is larger than $D/(k+1)$; otherwise it is called *small*. Let S be the set of small jobs and L be the set of big jobs. Suppose that $|L| = m \leq n$. Clearly in any schedule at most k big jobs can be executed. Consider all subsets of L that consists of up to k big jobs. The number of such subsets is bounded from above by $N_k = \binom{m}{k} + \binom{m}{k-1} + \cdots + \binom{m}{1} + \binom{m}{0} = O(m^{k+1})$. Denote them by L_i, for $i = 1, 2, \ldots, N_k$.

Algorithm A_k

1. Sort the jobs in nondecreasing order of release times. Schedule the ordered jobs one by one as long as it is executable (i.e., apply *FCFS* rule to the jobs) and obtain a schedule σ_0.
2. For $i = 1, 2, \ldots, N_k$, schedule the jobs in $L_i \cup S$ with *FCFS*. Denote the schedule by σ_i. Discard schedule σ_i if
   - some job $\in L_i \cup S$ cannot be scheduled by the common deadline D with *FCFS*, or
   - at some point t where the machine is idle a job $J_j \notin L_i \cup S$ could be started, namely $p_j \leq D - t$.
3. Select the best one (with the least makespan) among all remaining schedules including σ_0.

The discarding strategy at Step 2 ensures that a schedule is kicked out if and only if it is not feasible (not satisfying the busy requirement).

Theorem 2. *For any given k, A_k is an approximation algorithm with worst-case ratio of at most $1 + 1/k$ under objective function [min-makespan].*

Proof. Consider an optimal schedule σ^* satisfying the *FCFS* rule, i.e., the executed jobs are in the order of their arrivals. If σ^* does not execute all small jobs, then the optimal makespan C^* is at least $D - D/(k+1) = kD/(k+1)$. Note that the makespan C_{A_k} given by A_k is at most D. Thus $C_{A_k}/C^* \leq 1 + 1/k$. Now assume that σ^* executes all small jobs together with $k_1 \leq k$ big jobs. More precisely, σ^* schedules those jobs with the *FCFS* rule. Obviously there must be a feasible schedule identical to σ^* in Step 2. In this case we get an optimal schedule with algorithm A_k.

The combination of the two cases implies that the worst-case ratio of algorithm A_k is $1 + 1/k$.

Finally we estimate the running time of algorithm A_k. It takes time $O(n \log n)$ to schedule jobs with *FCFS*. The running time of the algorithm is determined at Step 2. It takes time $O(n^{k+2} \log n)$. As k is a fixed integer (constant), algorithm A_k is polynomial. □

For any fixed small number $\varepsilon > 0$, let $k = \lceil 1/\varepsilon \rceil$, we get an approximation scheme with worst-case ratio of at most $1 + \varepsilon$. Thus we reach the following conclusion.

Corollary 1. *There is a PTAS under objective function [min-makespan].* □

4 A PTAS for Minimizing Time Spent

In this section, we will show that for any fixed integer k, there exists an approximation algorithm B_k with a bound of $1 + 1/k$ under objective function [*min-time-spent*].

Suppose that there are m distinct release times $0 = T_1 < T_2 < \cdots < T_m$. Obviously if $m = 1$ the two objective functions [*min-time-spent*] and [*min-makespan*] are equivalent. If $m > 1$, the two objective functions may make difference since some idle space may be introduced in a schedule. To deal with [*min-time-spent*] our main idea is to analyze a schedule phase by phase. We will recursively apply algorithm A_k in each phase and try to figure out the configuration of an optimal schedule.

Based on algorithm A_k we define a series of algorithms A_k^i, for $i = 1, 2, \ldots, m$. Algorithm A_k^i works exactly as algorithm A_k under the following two conditions: (1) Only the jobs arriving at or after T_i are scheduled by A_k, and (2) a job is *big* if its processing time is larger than $(D - T_i)/(k+1)$ and *small* otherwise.

Let $M(i,j)$ denote the makespan produced by algorithm A_k^i in scheduling all jobs arriving in the closed interval $[T_i, T_j]$, while the corresponding schedule is denoted by $S(i,j)$, for $0 \le i \le j \le m$. For simplicity denote $M(i,i)$ as $M(i)$. Let $P(j)$ be the time spent by applying algorithm B_k (to be defined recursively below) to the jobs arriving at or before T_j, while the corresponding schedule is denoted by σ_j, for $j = 1, \ldots, m$.

Algorithm B_k

1. $P(0) = 0$, $P(1) = M(1)$; Let σ_1 be the corresponding schedule.
2. For $j = 2, \ldots, m$ do:
   $P(j) = \min\{P(j-1) + M(j), P(j-2) + M(j-1,j)), \cdots, P(1) + M(2,j), M(1,j)\}$; Suppose that $P(j)$ is determined by $P(h) + M(h+1,j)$ for some $1 \le h \le j-1$. Let M_h be the makespan of schedule σ_h, i.e., the completion time of the jobs arriving before T_{h+1}. We construct σ_j as follows.
   - If $M_h \le T_{h+1}$, then σ_h has no intersection with $S(h+1,j)$, the schedule of jobs arriving in $[T_{h+1}, T_j]$. Combining the two schedules we get a feasible schedule σ_j.
   - If $M_h > T_{h+1}$, we keep schedule σ_h and delay the schedule $S(h+1,j)$ (without changing the job order and schedule the jobs as early as possible) to time M_h. We thus get a feasible schedule σ_j.
3. The final schedule $\sigma = \sigma_m$.

Before analyzing the above algorithm we need to show, if $M_h > T_{h+1}$ for some h, schedule σ_j is feasible. In other words, the jobs scheduled in $S(h+1,j)$ can be completed by time D after the delay. If it is not true, then $M_h + M(h+1,j) \ge M_h + \sum_{J_i \in S(h+1,j)} p_i > D$. Let T_l be the ending time of the last idle interval of schedule σ_h (if no idle time, $T_l = T_1$). Obviously $T_l + M(l,j) \le D$, thus $P(h) + M(h+1,j) \ge P(l-1) + (M_h - T_l) + M(h+1,j) > P(l-1) + D - T_l \ge P(l-1) + M(l,j)$. This conflicts with the assumption that $P(j)$ is determined by $P(h) + M(h+1,j)$. Thus σ_j is feasible.

Algorithm B_k employs algorithm A_k as a subroutine. To calculate $P(m)$ we need to run A_k for $O(m^2)$ times. The running time of B_k is $O(n^{k+4} \log n)$. It is polynomial for any fixed integer $k > 0$.

Theorem 3. *For any given k, the worst-case ratio of algorithm B_k is at most $1 + 1/k$ under objective function [min-time-spent].*

Proof. We prove this theorem by induction. If there is only one (common) release time T_1, i.e., $m = 1$, B_k works exactly the same as A_k and the makespan $M(1)$ is equal to the time spent $P(1)$. By Theorem 2 we get the bound of $1 + 1/k$.

Assume the theorem holds for l distinct release times, i.e., holds for $m = l$. We want to show it is still true for $m = l + 1$. Consider an optimal schedule σ^*. If there is no idle time before all its jobs are completed, we have already get the conclusion with the following two points:

- In σ^*, the time spent is equal to the makespan.
- The time spent by algorithm B_k is at most $M(1, l + 1)$ which is at most $1 + 1/k$ times the makespan of σ^*, by Theorem 2.

Now assume that there is indeed some idle time before all its jobs (of σ^*) are completed. Denote the last idle interval by $[x, y]$. Obviously y must be some job release time (otherwise the job that starts at time y would have been scheduled earlier due to the busy requirement). Let $y = T_h$, $h \leq l+1$. The optimal schedule σ^* is divided into two parts σ_1^* and σ_2^*, where σ_1^* is an optimal schedule for those jobs arriving before T_{l+1}, and σ_2^* is an optimal schedule for the jobs arriving at or after T_{l+1}. Denote by OPT_1 and OPT_2 the two objective values of σ_1^* and σ_2^*, respectively. Then $OPT = OPT_1 + OPT_2$ is the objective value of σ^*. By induction we have $P(h - 1) \leq (1 + 1/k)OPT_1$. Note that there is no idle time in σ_2^*. Thus $M(h, l + 1) \leq (1 + 1/k)OPT_2$ by Theorem 2. On the other hand, $P(l + 1) \leq P(h - 1) + M(h, l + 1)$. Therefore we have

$$P(l + 1) \leq (1 + 1/k)(OPT_1 + OPT_2) = (1 + 1/k)OPT.$$

The theorem is proved. □

Analogously as Corollary 1 we have

Corollary 2. *There is a PTAS under objective function [min-time-spent].* □

5 NP-Hardness for Common Release Time

It has been shown in the literature [1, 2] that the lazy bureaucrat scheduling problem is NP-hard if all jobs have the same release time or have the same deadline. In this section we improve the above result by considering a special case that jobs come at the same time and have to be done by the same deadline if scheduled.

Theorem 4. *Even when the jobs share the same release time, the CD-LBSP is still NP-hard under objective functions [*min-number-jobs*] and [*min-weighted-sum*].*

Proof. We only need to prove the NP-hardness for [*min-number-jobs*]. The reduction is from the *subset sum problem*: Given a set of integers $S = \{x_1, \ldots, x_n\}$ and a target integer k, does there exist a subset $S' \subseteq S$, such that $\sum_{x_i \in S'} x_i = k$? Without loss of generality, we assume that $k > 1$.

For any given instance of the subset sum problem we construct an instance of CD-LBSP containing $2n+1$ jobs, all having the same release time 0 and the common deadline $D = k$. Corresponding to each integer $x_i \in S$, define an element job m_i with processing time $p_i = x_i$, for $i = 1, 2, \ldots, n$. Moreover we introduce $n+1$ tiny jobs, each with a processing time of $1/(n+1)$. One can easily verify that the answer to the subset sum problem is "yes" if and only if the number of the executed jobs is at most n. □

As for the other two objective functions [*min-makespan*] and [*min-time-spent*] we give the following remark. Instead of the original scheduling problem we will show the NP-hardness for a slightly different problem, namely the CD-LBSP *with multiplicities of jobs* where some jobs are replicable.

Note that for the special problem in any schedule there will be no idle time before the schedule ends. Therefore, the two objective functions [*min-makespan*] and [*min-time-spent*] are equivalent. We consider [*min-makespan*] with a reduction from the *subset sum problem* (see above). For any given instance of the subset sum problem we construct an instance of CD-LBSP containing $n+k+1$ jobs, all having the same release time 0 and the common deadline $D = k + 1 - 2/k^2$. Corresponding to each integer $x_i \in S$, define an element job m_i with processing time $p_i = x_i$, for $i = 1, 2, \ldots, n$. Moreover we introduce $k+1$ tiny jobs, each with a processing time of $1 - 1/k^2$.

Although we create an exponential number of jobs ($n+k+1$ jobs) in the scheduling problem, the length of the input is still polynomial, since all of the $k+1$ tiny jobs have the same processing time. Thus the reduction can be done in polynomial time.

The bureaucrat aims at minimizing the makespan. We claim that the answer to the subset sum problem is "yes" if and only if there exists a feasible schedule in which the bureaucrat executes only element jobs and the makespan is exactly k.

If the answer to the subset sum problem is "yes", then there exists a subset of element jobs, the sum of whose processing times is k. Schedule these jobs one by one and finish them at time k. The length of the remaining time interval $[k, D]$ is smaller than any job processing time in the instance. Thus the schedule ends with a makespan of k.

If the answer to the subset sum problem is "no", we will show that the makespan of any feasible schedule of the scheduling problem is larger than k. Assume that the total processing time of the element jobs executed is smaller than k. In this case, the bureaucrat must execute some tiny jobs. Note that the total processing time of $k+1$ tiny jobs is $k + 1 - (k+1)/k^2 > k$ (since $k > 1$). There must be some tiny jobs unscheduled, since otherwise the makespan is already over k. Thus the makespan of any feasible schedule is larger than $D - (1 - 1/k^2) = k - 1/k^2$. Note that any job processing time in the instance can be divided by $1/k^2$. It means that the makespan can be divided by $1/k^2$, namely, the makespan is at least k. However, the total processing time of any number k_1 ($1 \le k_1 \le k+1$) of tiny jobs is not an integer. In other words the makespan is not an integer and thus it is larger than k.

Acknowledgement. The authors would like to thank Joe Mitchell, Valentin Polishchuk and the anonymous referee for their insightful comments which led to the improvement of this paper.

References

1. Arkin, E.M., Bender, M.A., Mitchell, J.S.B., and Skiena, S.S.: The lazy bureaucrat scheduling problem. Information and Computation **184** (2003) 129-146
2. Esfahbod, B. Ghodsi, M., and Sharifi, A.: Common-deadline lazy bureaucrat scheduling problems. In: Dehne, F.K.H.A. Sack,J.-R, Smid, M.H.M (ed.s): Algorithms and Data Structures. Lecture Notes in Computer Science, Vol. 2748. Springer-Verlag (2003) 59-66
3. Goemans, M.X. and Williamaon, D.P.: Improved approximation algorithms for maixmum cut and satisfiability problems using semidefinite programming. Journal of ACM **42** (1995) 1115-1145
4. Hassin, R. and Rubinstein, S.: An approximation algorithm for the maximum traveling salesman problem. Information Processing Letter **67** (1998) 125-130
5. Hepner, C. and Stein, C.: Minimizing makespan for the lazy bureaucrat problem. In: Penttonen, M, Schmidt, E.M. (ed.s): Algorithm Theory. Lecture Notes in Computer Science, Vol. 2368. Springer-Verlag (2002) 40-50
6. Karger, D., Motwani, R., Ramkumar, G.: On approximating the longest path in a graph. Algorithmica **18** (1997) 82-98

Approximate Sorting*

Joachim Giesen[1], Eva Schuberth[1,2], and Miloš Stojaković[1]

[1] Institute for Theoretical Computer Science,
ETH Zürich, CH-8092, Zürich
[2] Swiss Federal Laboratories for Materials Testing and Research,
CH-8600, Dübendorf

Abstract. We show that any comparison based, randomized algorithm to approximate any given ranking of n items within expected Spearman's footrule distance $n^2/\nu(n)$ needs at least $n(\min\{\log \nu(n), \log n\} - 6)$ comparisons in the worst case. This bound is tight up to a constant factor since there exists a deterministic algorithm that shows that $6n(\log \nu(n) + 1)$ comparisons are always sufficient.

Keywords: Sorting, Ranking, Spearman's footrule metric, Kendall's tau metric.

1 Introduction

Our motivation to study approximate sorting comes from the following market research application. We want to find out how a respondent ranks a set of products. In order to simulate real buying situations the respondent is presented pairs of products out of which he has to choose one that he prefers, i.e., he has to perform paired comparisons. The respondent's ranking is then reconstructed from the sequence of his choices. That is, a procedure that presents a sequence of product pairs to the respondent in order to obtain the product ranking is nothing else than a comparison based sorting algorithm. We can measure the efficiency of such an algorithm in terms of the number of (pairwise) comparisons needed in order to obtain the ranking. The information theoretic lower bound on sorting [7] states that there is no procedure that can determine a ranking by posing less than $n \log \frac{n}{e}$ paired comparison questions to the respondent, i.e., in general $\Omega(n \log n)$ comparisons are needed. Even for only moderately large n that easily is too much since respondents often get worn out after a certain number of questions and do not answer further questions faithfully anymore. On the other hand, it might be enough to know the respondent's ranking approximately. In this paper we pursue the question of how many comparisons are necessary and sufficient in order to approximately rank n products.

* Partly supported by the Swiss National Science Foundation under the grant "Robust Algorithms for Conjoint Analysis" and by the joint Berlin/Zurich graduate program Combinatorics, Geometry and Computation, financed by ETH Zurich and the German Science Foundation (DFG).

J.R. Correa, A. Hevia, and M. Kiwi (Eds.): LATIN 2006, LNCS 3887, pp. 524–531, 2006.

In order to give sense to the term "approximately" we need some metric to compare rankings. Assume that we are dealing with n products. Since a ranking is a permutation of the products, this means that we need a metric on the permutation group S_n. Not all of the metrics, e.g., the Hamming distance that counts how many products are ranked differently, are meaningful for our application. For example, if in the respondent's ranking one exchanges every second product with its predecessor, then the resulting ranking has maximal Hamming distance to the original one. Nevertheless, this ranking still tells a lot about the respondent's preferences. In marketing applications Kendall's tau metric [4] is frequently used since it seems to capture the intuitive notion of closeness of two rankings and also arises naturally in the statistics of certain random rankings [8].

Our results. Instead of working with Kendall's metric we use Spearman's footrule metric [4] which essentially is equivalent to Kendall's metric, since the two metrics are within a constant factor of each other [4]. The maximal distance between any two rankings of n products in Spearman's footrule metric is less than n^2. We show that in order to obtain a ranking at distance $n^2/\nu(n)$ to the respondent's ranking with any strategy, a respondent has in general to perform at least $n\,(\min\{\log \nu(n), \log n\} - 6)$ comparisons in the worst case, i.e., there is an instance for which any comparison based algorithm performs at least $n\,(\min\{\log \nu(n), \log n\} - 6)$ comparisons. Moreover, if we allow the strategy to be randomized such that the obtained ranking is at expected distance $n^2/\nu(n)$ to the respondent's ranking, we can show that the same bound on the minimum number of comparisons holds.

On the other hand, there is a deterministic strategy (algorithm), suggested in [2], that shows that $6n(\log \nu(n) + 1)$ comparisons are always sufficient.

Related work. At first glance our work seems related to work done on pre-sorting. In pre-sorting the goal is to pre-process the data such that fewer comparisons are needed afterwards to sort them. For example in [5] it is shown that with $O(1)$ pre-processing one can save $\Theta(n)$ comparisons for Quicksort on average. Pre-processing can be seen as computing a partial order on the data that helps for a given sorting algorithm to reduce the number of necessary comparisons. The structural quantity that determines how many comparisons are needed in general to find the ranking given a partial order is the number of linear extensions of the partial order, i.e., the number of rankings consistent with the partial order. Actually, the logarithm of this number is a lower bound on the number of comparisons needed in general [6]. Here we study another structural measure, namely, the maximum diameter in the Spearman's metric of the set of rankings consistent with a partial order. Our results shows that with $o(n \log n)$ comparisons one can make this diameter asymptotically smaller than the diameter of the set of all rankings. That is not the case for the number of linear extensions which stays in $\Theta(2^{n \log n})$.

Notation. The logarithm log in this paper is assumed to be binary, and by id we denote the identity (increasing) permutation of $[n]$.

2 Algorithm

The idea of the ASORT algorithm is to partition the products into a sorted sequence of equal-sized bins such that the elements in each bin have smaller rank than any element in subsequent bins. This approach was suggested by Chazelle [2] for near-sorting. The output of the algorithm is the sequence of bins. Note that we do not specify the ordering of elements inside each bin, but consider any ranking consistent with the ordering of the bins. We will show that any such ranking approximates the actual ranking of the elements in terms of Spearman's footrule metric

$$D(\pi, \mathrm{id}) = D(\pi) = \sum_{i=1}^{n} |i - \pi(i)|,$$

where $\pi(i)$ is the rank of the element of rank i in an approximate ranking, i.e., $|i - \pi(i)|$ measures deviation of the approximated rank from the actual rank. Note that for any ranking the distance in the Spearman's footrule metric to id is at most $\frac{n^2}{2}$.

Since for every i the value $|i - \pi(i)|$ is bounded by n divided by the number of bins, we see that the approximation quality depends on the number of bins.

The algorithm ASORT iteratively performs a number of median searches, each time placing the median into the right position in the ranking. Here the median of n elements is defined to be the element of rank $\lfloor \frac{n+1}{2} \rfloor$.

ASORT (B : set, m : int)
1 $B_{01} := B$ // B_{ij} is the j'th bin in the i'th round
2 **for** $i := 1$ **to** m **do**
3   **for** $j := 1$ **to** 2^{i-1} **do**
4   compute the *median* of $B_{(i-1)j}$
5     $B_{i(2j-1)} := \{x \in B_{(i-1)j} \mid x \leq \mathit{median}\}$
6     $B_{i(2j)} := \{x \in B_{(i-1)j} \mid x > \mathit{median}\}$
7   **end for**
8 **end for**
9 **return** $B_{m1}, \ldots, B_{m(2^m)}$

To compute the median in line 4 and to partition the elements in line 5 and 6 we use the deterministic algorithm by Blum et al. [1] that performs at most $5.73n$ comparisons in order to compute the median of n elements and to partition them according to the median. We note that in putting the algorithm ASORT to practice one may want to use a different median algorithm, like, e.g., RANDOMIZEDSELECT [3].

In the following we determine the number of comparisons the algorithm ASORT needs on input B with $|B| = n$ in order to guarantee a prescribed approximation error of the actual ranking for any ranking consistent with the ordering of the bins $B_{m1}, \ldots, B_{m(2^m)}$ computed by the algorithm.

Lemma 1. *For every $x \in B_{ij}$, where $0 \le i \le m$ and $1 \le j \le 2^i$, it holds*

$$\sum_{k=1}^{j-1} |B_{ik}| + 1 \le \text{rank}(x) \le \sum_{k=1}^{j} |B_{ik}|.$$

Proof. The lemma can be proven by induction on the number of rounds. By construction, the elements in B_{01} have rank at least 1 and at most $n = |B_{01}| = \sum_{k=1}^{1} |B_{0k}|$. The claim for $i = 0$ follows if we set the empty sum $\sum_{k=1}^{0} |B_{0k}|$ to 0.

Now assume that the statement holds after the $(i-1)$'th round. The algorithm partitions every bin $B_{(i-1)j}$ into two bins $B_{i(2j-1)}$ and $B_{i(2j)}$. Again by construction the elements in bin $B_{i(2j-1)}$ have rank at least

$$\sum_{k=1}^{j-1} |B_{(i-1)k}| + 1 = \sum_{k=1}^{j-1} (|B_{i(2k-1)}| + |B_{i(2k)}|) + 1 = \sum_{k=1}^{(2j-1)-1} |B_{ik}| + 1,$$

and at most

$$\sum_{k=1}^{(2j-1)-1} |B_{ik}| + |B_{i(2j-1)}| = \sum_{k=1}^{2j-1} |B_{ik}|.$$

Similarly, the elements in bin $B_{i(2j)}$ have rank at least $\sum_{k=1}^{2j-1} |B_{ik}| + 1$ and at most $\sum_{k=1}^{j} |B_{(i-1)k}| = \sum_{k=1}^{2j} |B_{ik}|$. □

Lemma 2. $\lfloor \frac{n}{2^i} \rfloor \le |B_{ij}| \le \lceil \frac{n}{2^i} \rceil$ *for $0 \le i \le m$ and $1 \le j \le 2^i$.*

Proof. We prove by induction that in any round i the sizes of any two bins differ by at most 1, i.e., $\big||B_{ij}| - |B_{ik}|\big| \le 1$ for $0 \le i \le m$ and $1 \le j, k \le 2^i$. The statement of the lemma then follows since by an averaging argument and the integrality of the bin sizes, the size of each bin must be of size either $\lceil \frac{n}{2^i} \rceil$ or $\lfloor \frac{n}{2^i} \rfloor$.

For $i = 0$ all n elements of B are in bin B_{01}. The claim for $i = 0$ follows since $\lfloor n \rfloor \le n \le \lceil n \rceil$.

Now assume that the statement holds for $i - 1$. Take two bins $B_{(i-1)j}$ and $B_{(i-1)k}$. We distinguish two cases.

Case 1. $B_{(i-1)j}$ and $B_{(i-1)k}$ have the same size c. If c is even, then both bins get split up into two bins each and the resulting four bins all have the same size. If c is odd, then each of the bins gets split up into two bins of sizes $\lfloor \frac{c}{2} \rfloor$ and $\lceil \frac{c}{2} \rceil$, respectively, which differ by 1.

Case 2. Without loss of generality, $|B_{(i-1)j}| = c$ and $|B_{(i-1)k}| = c + 1$. If c is even, then $B_{(i-1)j}$ gets split up into two bins both of size $\frac{c}{2}$ and $B_{(i-1)k}$ gets split up into two bins of size $\frac{c}{2}$ and $\frac{c}{2} + 1$, respectively. If c is odd, then $B_{(i-1)j}$ gets split up into two subsets of size $\frac{c+1}{2}$ and $\frac{c+1}{2} - 1$, respectively, and $B_{(i-1)k}$ gets split up into two bins of size $\frac{c+1}{2}$. In any case the bins differ in size by at most 1. □

Lemma 3. *In m rounds the algorithm* ASort *performs less than $6nm$ comparisons.*

Proof. The algorithm by Blum et al. [1] needs at most $5.73n$ comparisons to find the median of n elements and to partition the elements with respect to the median. In the i'th round ASort partitions the elements in every bin $B_{ij}, 1 \leq j \leq 2^i$ with respect to their median. Thus the i'th round needs at most

$$\sum_{j=1}^{2^i} 5.73|B_{ij}| = 5.73 \sum_{j=1}^{2^i} |B_{ij}| = 5.73n \leq 6n$$

comparisons. As the algorithm runs for m rounds the overall number of comparisons is less than $6nm$. □

Theorem 1. *Let $r = \frac{n^2}{\nu(n)}$. Any ranking consistent with the ordering of the bins computed by* ASort *in $\log \nu(n) + 1$ rounds, i.e., with less than $6n(\log \nu(n) + 1)$ comparisons, has a Spearman's footrule distance of at most r to the actual ranking of the elements from B.*

Proof. Using the definition of Spearman's footrule metric and Lemmas 1 and 2 we can conclude that the distance of the ranking of the elements in B to any ranking consistent with the ordering of the bins computed by ASort in m rounds can be bounded by

$$\begin{aligned} \sum_{j=1}^{2^m} \frac{|B_{mj}|^2}{2} &\leq 2^m \frac{(\lceil \frac{n}{2^m} \rceil)^2}{2} \\ &\leq 2^{m-1} \left(\frac{n}{2^m} + 1 \right)^2 \\ &\leq 2^{m-1} \left(\frac{2n}{2^m} \right)^2, \text{ since } 2^m \leq n \\ &= \frac{n^2}{2^{m-1}}. \end{aligned}$$

Plugging in $\log \nu(n) + 1$ for m gives a distance less than r as claimed in the statement of the theorem. The claim for the number of comparisons follows from Lemma 3. □

3 Lower Bound

For $r > 0$, by $B_D(\mathrm{id}, r)$ we denote the ball centered at id of radius r with respect to the Spearman's footrule metric, so

$$B_D(\mathrm{id}, r) := \{\pi \in S_n : D(\pi, \mathrm{id}) \leq r\}.$$

Next we estimate the number of permutations in a ball of radius r.

Lemma 4

$$\left(\frac{r}{en}\right)^n \le |B_D(\mathrm{id}, r)| \le \left(\frac{2e(r+n)}{n}\right)^n .$$

Proof. Every permutation $\pi \in S_n$ is uniquely determined by the sequence $\{\pi(i)-i\}_i$. Hence, for any sequence of non-negative integers d_i, $i = 1, \ldots, n$, there are at most 2^n permutations $\pi \in S_n$ satisfying $|\pi(i) - i| = d_i$.

If $d_D(\pi, \mathrm{id}) \le r$, then $\sum_i |\pi(i) - i| \le r$. Since the number of sequences of n non-negative integers whose sum is at most r is $\binom{r+n}{n}$, we have

$$|B_D(\mathrm{id}, r)| \le \binom{r+n}{n} 2^n \le \left(\frac{2e(r+n)}{n}\right)^n .$$

Next, we give a lower bound on the size of $B_D(\mathrm{id}, r)$. Let $s := \lceil \frac{n^2}{r} \rceil$, and let us first assume that n is divisible by s. We divide the index set $[n]$ into s blocks of size n/s, such that for every $i \in \{1, 2, \ldots, s\}$ the ith block consists of elements $(i-1)\frac{n}{s}+1, (i-1)\frac{n}{s}+2, \ldots, i\frac{n}{s}$. For every s permutations $\pi_1, \pi_2, \ldots, \pi_s \in S_{n/s}$ we define the permutation $\rho \in S_n$ to be the concatenation of the permutations applied to corresponding blocks, so $\rho := \pi_1(b_1)\pi_2(b_2)\ldots\pi_s(b_s)$. Note that the distance of ρ to id with respect to Spearman's footrule metric is at most $n \cdot n/s \le r$, since $|\rho(i) - i| \le n/s$, for every $i \in [n]$. Obviously, for every choice of $\pi_1, \pi_2, \ldots, \pi_s$ we get a different permutation ρ, which means that we have at least

$$\left(\left(\frac{n}{s}\right)!\right)^s \ge \left(\frac{r}{en}\right)^n$$

different permutations in $B_D(\mathrm{id}, r)$.

If n is not divisible by s, we divide $[n]$ into s blocks of size either $\lceil n/s \rceil$ or $\lfloor n/s \rfloor$, again apply an arbitrary permutation on each of them and we can obtain the same bound in an analogous fashion. □

Using the upper bound from the last lemma, we now give a lower bound for the worst case running time of any comparison based approximate sorting algorithm.

Theorem 2. *Let $\mathcal{A}$ be a randomized approximate sorting algorithm based on comparisons, let $\nu = \nu(n)$ be a function, and let $r = r(n) = \frac{n^2}{\nu(n)}$.*

If for every input permutation $\pi \in S_n$ the expected Spearman's footrule distance of the output to id *is at most r, then the algorithm performs at least $n(\min\{\log\nu, \log n\} - 6)$ comparisons in the worst case.*

Proof. Let k be the smallest integer such that $\mathcal{A}$ performs at most k comparisons for every input. For a contradiction, let us assume that

$$k < n\left(\min\{\log\nu, \log n\} - 6\right).$$

First, we are going to prove

$$\frac{1}{2}n! > 2^k \left(\frac{2e(2r+n)}{n}\right)^n . \tag{1}$$

Since $\log \nu - 6 > k/n$, we have $\frac{\nu}{2^6} > 2^{k/n}$ and since $\nu = \frac{n^2}{r}$ we get

$$\frac{n}{2e} > 2^{k/n} \frac{2e \cdot 2r}{n}. \tag{2}$$

On the other hand, from $\log n - 6 > k/n$ we get $\frac{n}{2^6} > 2^{k/n}$ implying

$$\frac{n}{2e} > 2^{k/n} \frac{2e \cdot n}{n}. \tag{3}$$

Putting (2) and (3) together, we obtain

$$\frac{n}{e} > 2^{k/n} \frac{2e(2r+n)}{n}.$$

Hence

$$\frac{1}{2} n! \geq \left(\frac{n}{e}\right)^n > 2^k \left(\frac{2e(2r+n)}{n}\right)^n,$$

proving (1).

We denote by R the source of random bits for $\mathcal{A}$. One can see R as the set of all infinite 0-1 sequences, and then the algorithm is given a random element of R along with the input. For a permutation $\pi \in S_n$ and $\alpha \in R$, we denote by $\mathcal{A}(\pi, \alpha)$ the output of the algorithm with input π and random bits α.

We fix $\widetilde{\alpha} \in R$ and run the algorithm for every permutation $\pi \in S_n$. Note that with the random bits fixed the algorithm is deterministic. For every comparison made by the algorithm there are two possible outcomes. We partition the set of all permutations S_n into classes such that all permutations in a class have the same outcomes of *all the comparisons* the algorithm makes. Since there is no randomness involved, we have that for every class C there exists a $\sigma \in S_n$ such that for every $\pi \in C$ we have $\mathcal{A}(\pi, \widetilde{\alpha}) = \sigma \circ \pi$, where $\circ$ is the multiplication in the permutation group S_n. In particular, this implies that the set $\{\mathcal{A}(\pi, \widetilde{\alpha}) : \pi \in C\}$ is of size $|C|$. On the other hand, since the algorithm in this setting is deterministic and the number of comparisons of the algorithm is at most k, there can be at most 2^k classes. Hence, each permutation in S_n is the output for at most 2^k different input permutations. From Lemma 4 we have $|B_D(\mathrm{id}, 2r)| \leq \left(\frac{2e(2r+n)}{n}\right)^n$, and this together with (1) implies that at least

$$n! - 2^k \left(\frac{2e(2r+n)}{n}\right)^n > \frac{1}{2} n!$$

input permutations have output at distance to id more than $2r$.

Now, if both the random bits $\alpha \in R$ and the input permutation $\pi \in S_n$ are chosen at random, the expected distance of the output $\mathcal{A}(\pi, \alpha)$ to id is more than r. Therefore, there exists a permutation π_0 such that for a randomly chosen $\alpha \in R$ the expected distance $d_D(\mathcal{A}(\pi_0, \alpha), \mathrm{id})$ is more than r. Contradiction. $\square$

4 Conclusion

Motivated by an application in market research we studied the problem to approximate a ranking of n items. The metric we use to compare rankings is Spearman's footrule metric, which is within a constant factor to Kendall's tau metric that is frequently used in marketing research. We showed that any comparison based, randomized algorithm in the worst case needs at least

$$n\,(\min\{\log \nu(n), \log n\} - 6)$$

comparisons to approximate a given ranking of n items within expected distance $n^2/\nu(n)$. This result is complemented by an algorithm that shows that $6n(\log \nu(n) + 1)$ comparisons are always sufficient.

In particular, this means that in some cases substantially less comparisons have to be performed than for sorting exactly, provided that a sufficiently large error is allowed. That is, as long as the desired expected error is of order $n^{2-\alpha}$ for constant α one needs $\Omega(n \log n)$ comparisons, which asymptotically is not better than sorting exactly. But to achieve expected error of order $n^{2-o(1)}$ only $o(n \log n)$ comparisons are needed.

Acknowledgments. We are indebted to Jiří Matoušek for comments and insights that made this paper possible.

References

1. M. Blum, R. W. Floyd, V. Pratt, R. L. Rivest, and R. E. Tarjan. Linear time bounds for median computations. In *STOC '72: Proceedings of the fourth annual ACM symposium on Theory of computing*, pages 119–124. ACM Press, 1972.
2. B. Chazelle. The soft heap: An approximate priority queue with optimal error rate. *Journal of the ACM*, 47(6):1012–1027, 2000.
3. T. H. Cormen, C. E. Leiserson, and R. L. Rivest. *Introduction to Algorithms.* The MIT Press/McGraw-Hill, 1990.
4. P. Diaconis and R. L. Graham. Spearman's footrule as a measure of disarray. *Journal of the Royal Statistical Society*, 39(2):262–268, 1977.
5. H. K. Hwang, B. Y. Yang, and Y. N. Yeh. Presorting algorithms: an average-case point of view. *Theoretical Computer Science*, 242(1-2):29–40, 2000.
6. J. Kahn and J. H. Kim. Entropy and sorting. In *STOC '92: Proceedings of the twenty-fourth annual ACM symposium on Theory of computing*, pages 178–187. ACM Press, 1992.
7. D. E. Knuth. *The Art of Computer Programming*, volume 3. Addison Wesley, 1973.
8. C. L. Mallows. Non-null ranking models. *Biometrica*, 44:114–130, 1957.

Stochastic Covering and Adaptivity*

Michel Goemans[1] and Jan Vondrák[2]

[1] Department of Mathematics, MIT, Cambridge, MA 02139, USA
goemans@math.mit.edu

[2] Microsoft Research, Redmond, WA 98052, USA
vondrak@microsoft.com

Abstract. We introduce a class of "stochastic covering" problems where the target set X to be covered is fixed, while the "items" used in the covering are characterized by probability distributions over subsets of X. This is a natural counterpart to the stochastic packing problems introduced in [5]. In analogy to [5], we study both adaptive and non-adaptive strategies to find a feasible solution, and in particular the *adaptivity gap*, introduced in [4].

It turns out that in contrast to deterministic covering problems, it makes a substantial difference whether items can be used repeatedly or not. In the model of Stochastic Covering with item multiplicity, we show that the worst case adaptivity gap is $\Theta(\log d)$, where d is the size of the target set to be covered, and this is also the best approximation factor we can achieve algorithmically. In the model without item multiplicity, we show that the adaptivity gap for Stochastic Set Cover can be $\Omega(d)$. On the other hand, we show that the adaptivity gap is bounded by $O(d^2)$, by exhibiting an $O(d^2)$-approximation non-adaptive algorithm.

1 Introduction

Stochastic optimization deals with problems involving uncertainty on the input. We consider a setting with multiple stages of building a feasible solution. Initially, only some information about the probability distribution of the input is available. At each stage, an "item" is chosen to be included in the solution and the precise properties of the item are revealed (or "instantiated") after we commit to selecting the item irrevocably. The goal is to optimize the expected value/cost of the solution. This model follows the framework of Stochastic Packing [4, 5] where the problem is to select a set of items with random sizes, satisfying given capacity contraints. We obtain profit only for those items that fit within the capacity; as soon as a capacity constraint is violated, the procedure terminates and we do not receive any further profit. It is an essential property of this model that once an item is selected, it cannot be removed from the solution. The objective is to maximize the expected profit obtained.

* Research Supported in part by NSF grants CCR-0098018 and ITR-0121495, and ONR grant N00014-05-1-0148.

J.R. Correa, A. Hevia, and M. Kiwi (Eds.): LATIN 2006, LNCS 3887, pp. 532–543, 2006.

In this paper, we study a class of problems in a sense dual to Stochastic Packing: Stochastic Covering. Here, items come with random sizes again but the goal is to select sufficiently many items so that given covering constraints are satisfied. An example is the Stochastic Set Cover problem where the ground set X is fixed, while the items are characterized by probability distributions over subsets of X. We select items knowing only these distributions. Each item turns out to be a certain subset of X and we repeat this process until the entire set X is covered. For each item used in the solution, we have to pay a certain cost. The objective is then to minimize the expected cost of our solution.

A central paradigm in this setting is the notion of *adaptivity*. Knowing the instantiated properties of items selected so far, we can make further decisions based on this information. We call such an approach an *adaptive policy*. Alternatively, we can consider a model where this information is not available and we must make all decisions beforehand. This means, we choose an ordering of items to be selected, until a feasible solution is obtained, only based on the known probability distributions. Such an approach is called a *non-adaptive policy*. A fundamental question is, what is the benefit of being adaptive? We measure this benefit quantitatively as the ratio of expected costs incurred by optimal adaptive vs. non-adaptive policies (*the adaptivity gap*). A further question is whether a good adaptive or non-adaptive policy can be found efficiently.

1.1 Definitions

Now we define the class of problems we are interested in. The input comes in the form of a collection of *items*. Item i has a scalar value v_i and a vector size $\boldsymbol{S}(i)$. Unless otherwise noted, we assume that $\boldsymbol{S}(i)$ is a random vector with nonnegative components, while v_i is a deterministic nonnegative number. The random size vectors of different items are assumed independent.

We start with the deterministic form of a general covering problem, which is known under the name of a *Covering Integer Program* (see [17]). The forefather of these problems is the well-known Set Cover.

Definition 1 (CIP). *Given a collection of sets $\mathcal{F} = \{S(1), S(2), \ldots, S(n)\}$, $\bigcup_{i=1}^{n} S(i) = X$, Set Cover is the problem of selecting as few sets as possible so that their union is equal to X.*

More generally, given a nonnegative matrix $A \in \mathbb{R}_{+}^{n \times d}$ and vectors $\boldsymbol{b} \in \mathbb{R}_{+}^{d}, \boldsymbol{v} \in \mathbb{R}_{+}^{n}$, a Covering Integer Program (CIP) is the problem of minimizing $\boldsymbol{v} \cdot \boldsymbol{x}$ subject to $A\boldsymbol{x} \geq \boldsymbol{b}$ and $\boldsymbol{x} \in \{0,1\}^{d}$. Set Cover corresponds to the case where A is a 0/1 matrix, with columns representing the sets $S(1), \ldots, S(n)$.

We define a Stochastic Covering problem as a stochastic variant of CIP where the columns of A, representing items sizes $\boldsymbol{S}(i)$, are random. The "demand vector" $\boldsymbol{b}$ is considered deterministic. By scaling, we can assume that $\boldsymbol{b} = \mathbf{1} = (1, 1, \ldots, 1)$.

Definition 2 (Stochastic Covering). *Stochastic Covering (SP) is a stochastic variant of a CIP where A is a random matrix whose columns are independent random nonnegative vectors, denoted $\boldsymbol{S}(1), \ldots \boldsymbol{S}(n)$. Stochastic Set Cover is a* *special case where the $\boldsymbol{S}(i)$ are random 0/1 vectors. A given instantiation of a set of items F is feasible if $\sum_{i\in F}\boldsymbol{S}(i) \geq \mathbf{1}$. The value of $\boldsymbol{S}(i)$ is instantiated and fixed once we include the item i in F. Once this decision is made, the item cannot be removed.*

When we refer to a stochastic optimization problem "with multiplicity", it means that each item on the input comes with an unlimited number of identical copies. This makes sense for deterministic CIP as well, where we could allow arbitrary integer vectors $\boldsymbol{x} \in \mathbb{Z}_+^n$. In the stochastic case, this means that the probability distributions for the copies of each item are identical; their instantiated sizes are still independent random variables.

We require a technical condition that the set of all items is feasible with probability 1. For Stochastic Covering with multiplicity, it is sufficient to require that the set of all items is feasible with positive probability.

For all variants of Stochastic Covering problems, we consider adaptive and non-adaptive policies.

Definition 3 (Adaptive and non-adaptive policies.). *A non-adaptive policy is a fixed ordering of items to be inserted.*

An adaptive policy is a function $\mathcal{P} : 2^{[n]} \times \mathbb{R}_+^d \rightarrow [n]$. The interpretation of $\mathcal{P}$ is that given a configuration $(\mathcal{I}, \boldsymbol{b})$ where $\mathcal{I}$ represents the items still available and $\boldsymbol{b}$ the remaining demand, $\mathcal{P}(\mathcal{I}, \boldsymbol{b})$ determines which item should be chosen next among the items in $\mathcal{I}$.

The cost incurred by a policy is the total value of the items used until a feasible covering is found. For an instance of a Stochastic Covering problem, we define

- *ADAPT = minimum expected cost incurred by an adaptive policy.*
- *NONADAPT = minimum expected cost incurred by a non-adaptive policy.*
- *$\alpha = NONADAPT/ADAPT$ is the adaptivity gap.*

For a class of Stochastic Covering problems, we define α^* as the maximum possible adaptivity gap.

1.2 Our Results

We present several results on Stochastic Covering problems. We develop an LP bound on the adaptive optimum, based on the notion of "mean truncated size". For Stochastic Covering with multiplicity, we show that the worst case adaptivity gap is $\Theta(\log d)$. We prove the upper bound by presenting an efficient non-adaptive $O(\log d)$-approximation algorithm, based on the LP bound.

For Stochastic Covering without multiplicity, we have results in the special case of Stochastic Set Cover. We show that the adaptivity gap in this case can be $\Omega(d)$ and it is bounded by $O(d^2)$. Again, the upper bound is constructive, by an efficient non-adaptive approximation algorithm. Also, we show an adaptive $O(d)$-approximation algorithm for Stochastic Set Cover. This, however, does not bound the worst-case adaptivity gap which could be anywhere between $\Omega(d)$ and $O(d^2)$. We leave this as an open question.

1.3 Previous Work

Stochastic Optimization with Recourse. Recently, stochastic optimization has come to the attention of the computer science community. An optimization model which has been mainly under scrutiny is the *two-stage stochastic optimization with recourse* [10, 8, 15]. In the first stage, only some information about the probability distribution of possible inputs is available. In the second stage, the precise input is known and the solution must be completed at any cost. The goal is to minimize the expected cost of the final solution. This model has been also extended to multiple stages [9, 16]. However, an essential difference between this model and ours is whether the randomness is in the properties of items forming a solution or in the demands to be satisfied. Let's illustrate this on the example of Set Cover: Shmoys and Swamy consider in [15] a Stochastic Set Cover problem where the sets to be chosen are deterministic and there is a random target set A to be covered. In contrast, we consider a Stochastic Set Cover problem where the target set is fixed but the covering sets are random. This yields a setting of a very different flavor.

Stochastic Knapsack. The first problem analyzed in our model of multi-stage optimization with adaptivity was the Stochastic Knapsack [4]. The motivation for this problem is in the area of *stochastic scheduling* where a sequence of jobs should be scheduled on a machine within a limited amount of time. The goal is to maximize the expected profit received for jobs completed before a given deadline. The jobs are processed one by one; after a job has been completed, its precise running time is revealed - but then it is too late to remove the job from the schedule. Hence the property of *irrevocable decisions*, which is essential in the definition of our stochastic model.

In [4, 6], we showed that adaptivity can provide a certain benefit which is, however, bounded by a constant factor. A non-adaptive solution which achieves expected value at least $1/4$ of the adaptive optimum is achieved by a greedy algorithm which runs in polynomial time. Thus the adaptivity gap is upper-bounded by 4. Concerning adaptive approximation, we showed that for any $\epsilon > 0$, there is a polynomial-time adaptive policy achieving at least $1/3 - \epsilon$ of the adaptive optimum.

Stochastic Packing. Stochastic Packing problems generalize the Stochastic Knapsack in the sense that we allow multidimensional packing contraints. This class contains many combinatorial problems: set packing, maximum matching, b-matching and general Packing Integer Programs (PIP, see e.g. [14]). In the stochastic variants of these problems we consider items with random vector sizes which are instantiated upon inclusion in the solution. Our motivation for this generalization is *scheduling with multiple resources.*

The analysis of Stochastic Packing in [5] reveals a curious pattern of results. Let us present it on the example of Stochastic Set Packing. Here, each item is defined by a value and a probability distribution over subsets $A \subseteq X$ where X is a ground set of cardinality $|X| = d$. A feasible solution is a collection of disjoint sets. It is known that for deterministic Set Packing, the greedy algorithm provides an $O(\sqrt{d})$-approximation, and there is a closely matching inapproximability result which states that for any fixed $\epsilon > 0$, a polynomial-time $d^{1/2-\epsilon}$-approximation algorithm would imply $NP = ZPP$ [2].

For Stochastic Set Packing, it turns out that the adaptivity gap can be as large as $\Theta(\sqrt{d})$. On the other hand, this is the worst case, since there is a polynomial-time non-adaptive policy which gives an $O(\sqrt{d})$-approximation of the *adaptive optimum*. Note that even with an adaptive policy, we cannot hope for a better approximation factor, due to the hardness result for deterministic Set Packing.

These results hint at a possible underlying connection between the quantities we are investigating: deterministic approximability, adaptivity gap and stochastic approximability. Note that there is no reference to computational efficiency in the notion of adaptivity gap, so a direct connection with the approximability factor would be quite surprising.

In this paper, we are investigating the question whether such phenomena appear in the case of covering problems as well.

Covering Integer Programs. Stochastic Covering problems can be seen as generalizations of Covering Integer Programs (CIP, see [17]). The forefather of Covering Integer Programs is the well-known Set Cover problem. For Set Cover, it was proved by Johnson [11] that the greedy algorithm gives an approximation factor of $\ln d$. This result was extended by Chvátal to the weighted case [3]. The same approximation guarantee can be obtained by a linear programming approach, as shown by Lovász [13]. Finally, it was proved by Uriel Feige [7] that these results are optimal, in the sense that a polynomial-time $(1-\epsilon)\ln d$ approximation algorithm for Set Cover would imply $NP \subset TIME(n^{O(\log \log n)})$.

Note. Usually the cardinality of the ground set is denoted by n, but to be consistent with Stochastic Packing problems, we view this parameter as "dimension" and denote it by d.

For general Covering Integer Problems, the optimal approximation has been found only recently [1, 12]. The approximation factor turns out to be again $O(\log d)$ but the approximation algorithm is more sophisticated. Also, the natural LP can have an arbitrarily large integrality gap.

2 Stochastic Covering with Multiplicity

Let's start with the class of Stochastic Covering problems where each item can be used arbitrarily many times.

Lemma 1. *There are instances of Stochastic Set Cover with multiplicity where the adaptivity gap is at least* $0.45 \ln d$.

Proof. Consider item types for $i = 1, 2, \ldots, d$ where $\boldsymbol{S}(i) = \mathbf{0}$ or $\boldsymbol{e}_i$ with probability $1/2$. All items have unit values. An adaptive policy inserts an expected number of 2 items of each type until the respective component is covered; $ADAPT \leq 2d$.

Assume that a nonadaptive policy at some point has inserted k_i items of type i, for each i. Denote the total size at that point by $\boldsymbol{S}$. We estimate the probability that this is a feasible solution:

$$\Pr[\boldsymbol{S} \geq \mathbf{1}] = \prod_{i=1}^{d} \Pr[S_i \geq 1] = \prod_{i=1}^{d} (1 - 2^{-k_i}) \leq \exp\left(-\sum_{i=1}^{d} 2^{-k_i}\right).$$

Assume that $k = \sum_{i=1}^{d} k_i = d \log_2 d$. By convexity, the probability of covering is maximized for a given d when $k_i = k/d = \log_2 d$, and then still $\Pr[\boldsymbol{S} \geq \mathbf{1}] \leq 1/e$. Thus whatever the non-adaptive policy does, there is probability $1 - 1/e$ that it needs more than $d \log_2 d$ items, which means $NONADAPT \geq (1-1/e) d \log_2 d > 0.9 d \ln d$.

Now we would like to prove that $O(\log d)$ is indeed the worst possible adaptivity gap, not only for Stochastic Set Cover with multiplicity but for all Stochastic Covering problems with multiplicity. First, we need a bound on the adaptive optimum. For this purpose, we define the *mean truncated size* of an item, in analogy to [4].

Definition 4. *We define the mean truncated size of an item with random size* $\boldsymbol{S}$ *by components as*

$$\mu_j = \mathbf{E}[\min\{S_j, 1\}].$$

For a set of items $\mathcal{A}$*, we let* $\mu_j(\mathcal{A}) = \sum_{i \in \mathcal{A}} \mu_j(i)$.

We prove that in expectation, the mean truncated size of the items inserted by any policy must be at least the demand required for each coordinate.

Lemma 2. *For a Stochastic Covering problem and any adaptive policy, let* $\mathcal{A}$ *denote the (random) set of items which the policy uses to achieve a feasible covering. Then for each component* j,

$$\mathbf{E}[\mu_j(\mathcal{A})] \geq 1.$$

Proof. Consider component j. Denote by $M_j(c)$ the minimum expected $\mu_j(\mathcal{A})$ for a set $\mathcal{A}$ that an adaptive policy needs to insert in order to satisfy remaining demand c in the j-th component. We prove, by induction on the number of available items, that $M_j(c) \geq c$. Suppose that an optimal adaptive policy, given remaining demand c, inserts item i. Denote by $cover(i, c)$ the indicator variable of the event that item i covers the remaining demand (i.e., $S_j(i) \geq c$, and in that case the policy terminates). We denote by $\tilde{s}_j(i)$ the truncated size $\tilde{s}_j(i) = \min\{S_j(i), 1\}$:

$$\begin{aligned} M_j(c) &= \mu_j(i) + \mathbf{E}[(1 - cover(i,c)) M_j(c - \tilde{s}_j(i))] \\ &= \mathbf{E}[\tilde{s}_j(i) + (1 - cover(i,c)) M_j(c - \tilde{s}_j(i))] \end{aligned}$$

and using the induction hypothesis,

$$M_j(c) \geq \mathbf{E}[\tilde{s}_j(i) + (1 - cover(i,c))(c - \tilde{s}_j(i))] = c - \mathbf{E}[cover(i,c)(c - \tilde{s}_j(i))] \geq c$$

since $cover(i,c) = 1$ only if $\tilde{s}_j(i) \geq c$.

Note that this lemma does not depend on whether multiplicity of items is allowed or not. In any case, having this bound, we can write an LP bounding the expected cost of the optimal adaptive policy. We introduce a variable x_i for every item i which can be interpreted as the probability that a policy ever inserts item i. For problems with item multiplicities, x_i represents the expected number of copies of item i inserted by a policy. Then the expected cost of the policy can be written as $\sum_i v_i x_i$. Due to Lemma 2, we know that $\mathbf{E}[\mu_j(\mathcal{A})] = \sum_i x_i \mu_j(i) \geq 1$ for any policy. So we get the following lower bound on the expected cost of any adaptive policy.

Theorem 1. *For an instance of Stochastic Covering with multiplicity, ADAPT $\geq$ LP where*

$$LP = \min \left\{ \sum_i x_i v_i : \begin{array}{c} \sum_i x_i \boldsymbol{\mu}(i) \geq \mathbf{1} \\ x_i \geq 0 \end{array} \right\}.$$

For a problem without multiplicity, $x_i \geq 0$ would be replaced by $x_i \in [0,1]$.

Now we are ready to prove an upper bound on the adaptivity gap, for $d \geq 2$. (The case $d = 1$ can be viewed as a special case of $d = 2$.)

Theorem 2. *For Stochastic Covering with multiplicity in dimension $d \geq 2$,*

$$NONADAPT \leq 12 \ln d \ ADAPT$$

and the corresponding non-adaptive policy can be found in polynomial time.

Proof. Consider the LP formulation of Stochastic Covering with multiplicity:

$$LP = \min \left\{ \sum_i x_i v_i : \sum_i x_i \boldsymbol{\mu}(i) \geq \mathbf{1},\ x_i \geq 0 \right\}.$$

We know from Theorem 1 that $ADAPT \geq LP$. Let x_i be an optimal solution. We inflate the solution by a factor of $c \ln d$ (hence the need to be able to repeat items) and we build a random multiset $\mathcal{F}$ where item i has an expected number of copies $y_i = x_i\ c \ln d$. This can be done for example by including $\lfloor y_i \rfloor$ copies of item i deterministically and another copy with probability $y_i - \lfloor y_i \rfloor$. Then the total size of set $\mathcal{F}$ in component j can be seen as a sum of independent random $[0,1]$ variables and the expected total size is

$$\mathbf{E}[S_j(\mathcal{F})] = \sum_i y_i \mathbf{E}[S_j(i)] \geq \sum_i y_i \mu_j(i) \geq c \ln d.$$

By Chernoff bound, with $\mu = \mathbf{E}[S_j(\mathcal{F})] \geq c \ln d$ and $\delta = 1 - 1/\mu$:

$$\Pr[S_j(\mathcal{F}) < 1] = \Pr[S_j(\mathcal{F}) < (1-\delta)\mu] < e^{-\mu\delta^2/2} \leq e^{-\mu/2+1} \leq \frac{e}{d^{c/2}}.$$

We choose $c = 9$ and then by the union bound

$$\Pr[\exists j; S_j(\mathcal{F}) < 1] < \frac{e}{d^{3.5}}.$$

For $d \geq 2$, $\mathcal{F}$ is a feasible solution with a constant nonzero probability at least $1 - e/2^{3.5}$. Its expected cost is

$$\mathbf{E}[v(\mathcal{F})] = \sum_i y_i v_i = 9 \ln d \; LP \leq 9 \ln d \; ADAPT.$$

If $\mathcal{F}$ is not a feasible solution, we repeat; the expected number of iterations is $1/(1 - e/2^{3.5}) < 4/3$. Therefore

$$NONADAPT \leq 12 \ln d \; ADAPT.$$

This randomized rounding algorithm can be derandomized using pessimistic estimators in the usual way.

3 General Stochastic Covering

Now we turn to the most general class of Stochastic Covering problems, where each item can be used only once (unless the input itself contains multiple copies of it) and the random item sizes are without any restrictions. Unfortunately, in this setting there is little that we are able to do. We can write a linear program bounding the adaptive optimum, analogous to Theorem 1:

$$LP = \min \left\{ \sum_i x_i v_i : \begin{array}{l} \sum_i x_i \boldsymbol{\mu}(i) \geq \mathbf{1} \\ x_i \in [0,1] \end{array} \right\}.$$

However, this LP can be far away from the actual adaptive optimum, even for $d = 1$. Consider one item of size $S(1) = 1 - \epsilon$ and an unlimited supply of items of size $S(2) = 1$ with probability ϵ and $S(2) = 0$ with probability $1 - \epsilon$. I.e., $\mu(1) = 1 - \epsilon$, $\mu(2) = \epsilon$. All items have unit values. We can set $x_1 = x_2 = 1$ and this gives a feasible solution with $LP = 2$. However, in the actual solution the item of size $1 - \epsilon$ does not help; we need to wait for an item of the second type to achieve size 1. This will take an expected number of $1/\epsilon$ items, therefore $ADAPT = 1/\epsilon$.

This example illustrates a more general issue with any approach using mean truncated sizes. As long as we do not use other information about the probability distribution, we would not distinguish between the above instance and one where the actual sizes of items are $\mu(1) = 1 - \epsilon$ and $\mu(2) = \epsilon$. Such an instance would indeed have a solution of cost 2. Thus using only mean truncated sizes, we cannot prove any approximation result in this case. It would be necessary to use a more complete description of the distributions of $S(i)$, but we leave this question outside the scope of this paper.

4 Stochastic Set Cover

Perhaps the circumstances are more benign in the case of Set Cover, i.e. 0/1 size vectors. However, the adaptivity gap is certainly not bounded by $O(\log d)$, when items cannot be used repeatedly.

Lemma 3. *For Stochastic Set Cover without multiplicity, the adaptivity gap can be $d/2$.*

Proof. Consider $\boldsymbol{S}(0) = \mathbf{1} - \boldsymbol{e}_k$, where $k \in \{1, 2, \ldots, d\}$ is uniformly random, $v_0 = 0$, and $\boldsymbol{S}(i) = \boldsymbol{e}_i$ deterministic, $v_i = 1$, for $i = 1, 2, \ldots, d$. An adaptive policy inserts item 0 first; assume its size is $\boldsymbol{S}(0) = \mathbf{1} - \boldsymbol{e}_k$. Then we insert item k which completes the covering for a cost equal to 1. An optimal non-adaptive policy still inserts item 0 first, but then, for any ordering of the remaining items that it chooses, the expected cost incurred before it hits the one which is needed to complete the covering is $d/2$.

The question is whether this is the worst possible example. First, let's consider the problem in dimension 1, where the size of each item is just a Bernoulli random variable. Thus the instance is completely characterized by the mean size values. In this case, a greedy algorithm yields the *optimal solution.*

Lemma 4. *For Stochastic Set Cover of one element ($d = 1$), assume the items are ordered, so that*

$$\frac{v_1}{\mu(1)} \leq \frac{v_2}{\mu(2)} \leq \frac{v_3}{\mu(3)} \leq \cdots \frac{v_n}{\mu(n)}$$

(call such an ordering "greedy"). Then inserting items in a greedy ordering yields a covering of minimum expected cost. The adaptivity gap in this case is equal to 1.

Proof. First, note that in this setting, adaptivity cannot bring any advantage. Until a feasible solution is obtained, we know that all items must have had size 0. An adaptive policy has no additional information and there is only one possible configuration for every subset of available items. Thus there is an optimal item to choose for each subset of available items and an optimal adaptive policy is in fact a fixed ordering of items.

For now, we assume that the items are not ordered and we consider any ordering (not necessarily the greedy ordering), say $(1, 2, 3, \ldots)$. The expected cost of a feasible solution found by inserting in this order is

$$C = \sum_{k=1}^{n} v_k \prod_{j=1}^{k-1} (1 - \mu(j)).$$

Let's analyze how switching two adjacent items affects C. Note that switching i and $i+1$ affects only the contributions of these two items - the terms corresponding to $k < i$ and $k > i + 1$ remain unchanged. The difference in expected cost will be

$$\begin{aligned}
\Delta C &= v_i \left(\prod_{j=1}^{i-1} (1-\mu(j)) \right) (1-\mu(i+1)) \; + \; v_{i+1} \left(\prod_{j=1}^{i-1} (1-\mu(j)) \right) \\
&\quad - v_i \left(\prod_{j=1}^{i-1} (1-\mu(j)) \right) \; - \; v_{i+1} \left(\prod_{j=1}^{i} (1-\mu(j)) \right) \\
&= \left(\mu(i)\mu(i+1) \prod_{j=1}^{i-1} (1-\mu(j)) \right) \left(\frac{v_{i+1}}{\mu(i+1)} - \frac{v_i}{\mu(i)} \right).
\end{aligned}$$

Therefore, we can switch any pair of elements such that $\frac{v_i}{\mu(i)} \geq \frac{v_{i+1}}{\mu(i+1)}$ and obtain an ordering whose expected cost has not increased.

Assume that $\mathcal{O}$ is an arbitrary greedy ordering and $\mathcal{O}^*$ is a (possibly different) optimal ordering. If $\mathcal{O} \neq \mathcal{O}^*$, there must be a pair of adjacent items in $\mathcal{O}^*$ which are swapped in $\mathcal{O}$. By switching these two items, we obtain another optimal ordering $\mathcal{O}'$. We repeat this procedure, until we obtain $\mathcal{O}$ which must be also optimal.

The Adaptive Greedy Algorithm. For Stochastic Set Cover in dimension d, we generalize the greedy algorithm in the following way: For each component $i \in [d]$, we find an optimal ordering restricted only to component i; we denote this by $\mathcal{O}_i$. Then our greedy adaptive algorithm chooses at any point a component j which has not been covered yet, and inserts the next available item from $\mathcal{O}_j$. Observe that this algorithm is *adaptive* as the decision is based on which components have not been covered yet.

Corollary 1. *For Stochastic Set Cover in dimension d, the greedy adaptive policy achieves expected cost*

$$GREEDY \leq d \cdot ADAPT.$$

Proof. When the policy chooses an item from $\mathcal{O}_j$, we charge its cost to a random variable X_j. Note that items from $\mathcal{O}_j$ can be also charged to other variables but an item which is charged to X_j can be inserted only after all items preceding it in $\mathcal{O}_j$ have been inserted already. Thus the value of X_j is at most the cost of covering component j, using the corresponding greedy ordering, and $\mathbf{E}[X_j] \leq ADAPT$. Consequently, $GREEDY = \sum_{i=1}^{d} \mathbf{E}[X_i] \leq d \cdot ADAPT$.

Thus we have a d-approximation, but this approximation algorithm is *adaptive* so it doesn't settle the adaptivity gap for Stochastic Set Packing. The final answer is unknown. The best upper bound we can prove is the following.

Theorem 3. *For Stochastic Set Cover,*

$$NONADAPT \leq d^2 \cdot ADAPT$$

and the corresponding non-adaptive policy can be found in polynomial time.

Proof. Consider the greedy ordering $\mathcal{O}_j$ for each component j. We interleave $\mathcal{O}_1, \mathcal{O}_2, \ldots, \mathcal{O}_d$ in the following way: We construct a single sequence $\mathcal{O}^* = (i(1), i(2), i(3), \ldots)$ where $i(t)$ is chosen as the next available item from $\mathcal{O}_{j(t)}$; $j(t)$ to be defined. We set $X_i(0) = 0$ for each $1 \leq i \leq d$, $X_{j(t)}(t) = X_{j(t)}(t-1) + v_{i(t)}$ and $X_k(t) = X_k(t-1)$ for $k \neq j(t)$. In other words, we charge the cost of $i(t)$ to $X_{j(t)}$ which denotes the "cumulative cost" of component $j(t)$. At each time t, we choose the index $j(t)$ in order to minimize $X_{j(t)}(t)$ among all possible choices of $j(t)$.

Consider a fixed component k and the time τ when component k is covered. This is not necessarily by an item chosen from $\mathcal{O}_k$, i.e. $j(\tau)$ doesn't have to be k. If $j(\tau) = k$, denote by q the item from $\mathcal{O}_k$ covering component k: $q = i(\tau)$. If $j(\tau) \neq k$, denote by q the next item to be chosen from $\mathcal{O}_k$ if component k had not been covered yet. We denote by $\mathcal{Q}_k$ the prefix of sequence $\mathcal{O}_k$ up to (and including) item q. We claim that for any j, $X_j(\tau)$ is at most the cost of $\mathcal{Q}_k$: If there is some $X_j(\tau) > v(\mathcal{Q}_k)$, the last item that we charged to X_j should not have been chosen; we should have chosen an item from $\mathcal{O}_k$ which would have kept X_k still bounded by $v(\mathcal{Q}_k)$ and thus smaller than $X_j(\tau)$. Therefore $X_j(\tau) \leq v(\mathcal{Q}_k)$. For the total cost Z_k spent up to time τ when component k is covered, we get

$$Z_k = \sum_{j=1}^{d} X_j(\tau) \leq d\ v(\mathcal{Q}_k).$$

Now consider the set of items $\mathcal{Q}_k$ which is a prefix of $\mathcal{O}_k$. The probability that $\mathcal{Q}_k$ has length at least l is at most the probability that an (optimal) policy covering component k using the ordering $\mathcal{O}_k$ needs to insert at least l items from $\mathcal{O}_k$; this is the probability that the first $l-1$ items in $\mathcal{O}_k$ attain size 0 in component k. If $ADAPT_k$ denotes the minimum expected cost of an adaptive policy covering component k, we get $\mathbf{E}[v(\mathcal{Q}_k)] \leq ADAPT_k \leq ADAPT$ and $\mathbf{E}[Z_k] \leq d \cdot ADAPT$.

Finally, the total cost spent by our policy is $Z = \max_k Z_k$, since we have to wait for the last component to be covered. Therefore,

$$\mathbf{E}[Z] = \mathbf{E}[\max_{1 \leq k \leq d} Z_k] \leq \sum_{k=1}^{d} \mathbf{E}[Z_k] \leq d^2\ ADAPT.$$

5 Concluding Remarks

We have seen that allowing or not allowing items to be used repeatedly makes a significant difference in Stochastic Covering. The case where items can be used repeatedly is basically solved, with the worst-case adaptivity gap and polynomial-time approximation factor being both on the order of $\Theta(\log d)$. This would support the conjecture that there is some connection between the adaptivity gap and the optimal approximation factor for the deterministic problem. However, the general class of Stochastic Covering problems without item multiplicity does not follow this pattern. The adaptivity gap for Set Cover can be $\Omega(d)$, while the optimal approximation in the deterministic case, as well as the integrality gap of the associated LP, is $O(\log d)$.

Our main open question is what is the worst possible adaptivity gap for Set Cover. We conjecture that it is $\Theta(d)$ but we are unable to prove this. Also, it remains to be seen what can be done for general Stochastic Covering when the complete probability distributions of item sizes are taken into account.

References

1. Robert D. Carr, Lisa K. Fleischer, Vitus J. Leung, and Cynthia A. Phillips. Strengthening integrality gaps for capacitated network design and covering problems. In *SODA*, pages 106–115, 2000.
2. C. Chekuri and S. Khanna. On multidimensional packing problems. *SIAM J. Computing*, 33(4):837–851, 2004.
3. V. Chvátal. A greedy heuristic for the set covering problem. *Math. Oper. Res*, 4:233–235, 1979.
4. B. Dean, M. X. Goemans, and J. Vondrák. Approximating the stochastic knapsack: the benefit of adaptivity. In *FOCS*, pages 208–217, 2004.
5. B. Dean, M. X. Goemans, and J. Vondrák. Adaptivity and approximation for stochastic packing problems. In *SODA*, pages 395–404, 2005.
6. B. Dean, M. X. Goemans, and J. Vondrák. Approximating the stochastic knapsack: the benefit of adaptivity. *Journal version, submitted*, 2005.
7. U. Feige. A threshold of ln n for approximating set cover. *JACM*, 45(4):634–652, 1998.
8. A. Gupta, M. Pál, R. Ravi, and A. Sinha. Boosted sampling: approximation algorithms for stochastic optimization. In *SODA*, pages 417–426, 2004.
9. A. Gupta, M. Pál, R. Ravi, and A. Sinha. What about wednesday? approximation algorithms for multistage stochastic optimization. In *APPROX*, pages 86–98, 2005.
10. N. Immorlica, D. Karger, M. Minkoff, and V. Mirrokni. On the costs and benefits of procrastination: Approximation algorithms for stochastic combinatorial optimization problems. In *SODA*, pages 184–693, 2004.
11. D. Johnson. Approximation algorithms for combinatorial problems. *J. Comp. Syst. Sci.*, 9:256–216, 1974.
12. S. Kolliopoulos and N. Young. Tight approximation results for general covering integer programs. In *FOCS*, pages 522–531, 2001.
13. L. Lovász. On the ratio of the optimal integral and fractional covers. *Disc. Math.*, 13:256–278, 1975.
14. P. Raghavan. Probabilistic construction of deterministic algorithms: approximating packing integer programs. *J. Comp. and System Sci.*, 37:130–143, 1988.
15. D. Shmoys and C. Swamy. Stochastic optimization is (almost) as easy as deterministic optimization. In *FOCS*, pages 228–237, 2004.
16. D. Shmoys and C. Swamy. Sampling-based approximation algorithms for multistage stochastic optimization. In *FOCS*, pages 357–366, 2005.
17. A. Srinivasan. Improved approximations of packing and covering problems. In *STOC*, pages 268–276, 1995.

Algorithms for Modular Counting of Roots of Multivariate Polynomials

Parikshit Gopalan[1,⋆], Venkatesan Guruswami[2,⋆⋆], and Richard J. Lipton[1,⋆]

[1] College of Computing,Georgia Tech,
Atlanta, GA, USA
{parik, rjl}@cc.gatech.edu

[2] Department of Computer Science & Engineering,
University of Washington, Seattle, WA, USA
venkat@cs.washington.edu

Abstract. Given a multivariate polynomial $P(X_1, \cdots, X_n)$ over a finite field $\mathbb{F}_q$, let $N(P)$ denote the number of roots over $\mathbb{F}_q^n$. The modular root counting problem is given a modulus r, to determine $N_r(P) = N(P) \bmod r$. We study the complexity of computing $N_r(P)$, when the polynomial is given as a sum of monomials. We give an efficient algorithm to compute $N_r(P)$ when the modulus r is a power of the characteristic of the field. We show that for all other moduli, the problem of computing $N_r(P)$ is NP-hard. We present some hardness results which imply that that our algorithm is essentially optimal for prime fields. We show an equivalence between maximum-likelihood decoding for Reed-Solomon codes and a root-finding problem for symmetric polynomials.

1 Introduction

Given a polynomial $P(X_1, \cdots, X_n)$ of degree d in n variables over a field $\mathbb{F}_q$ of characteristic p in sparse representation, i.e. written as a sum of m monomials, let $N(P)$ denote the number of solutions to $P(X_1, \cdots, X_n) = 0$ over $\mathbb{F}_q$. The problem of computing $N(P)$ exactly is known to be #P-complete. In this paper we study the complexity of the modular counting problem, which is given a modulus r, compute $N_r(P) = N(P) \bmod r$. We also study the related problem of deciding whether $N(P) > 0$ i.e. if the equation $P = 0$ is feasible over $\mathbb{F}_q$.

1.1 Problem History and Motivation

The problem of counting roots of a polynomial over a finite field is a fundamental and well studied problem in algebra with applications to several areas including coding theory and cryptography[1]. Ehrenfeucht and Karpinski showed that computing $N(P)$ is #P complete even when we restrict the degree to be three [2]. Hence one has to look for approximation algorithms, or algorithms that work for some special class of polynomials.

⋆ Supported by NSF grant CCR-3606B64.

⋆⋆ Supported in part by NSF grant CCF-0343672 and a Sloan Research Fellowship.

J.R. Correa, A. Hevia, and M. Kiwi (Eds.): LATIN 2006, LNCS 3887, pp. 544–555, 2006.

Randomized algorithms for computing $N(P)$ approximately were given by Karpinski and Luby for $\mathbb{F}_2$ [3] and Grigoriev and Karpinksi for $\mathbb{F}_q$ [4]. A more randomness efficient algorithm for $\mathbb{F}_2$ was given by Luby, Velikovic and Wigderson [5]. The problem has been extensively studied for equations in few variables. Schoof gives an exact algorithm to count the number of points on an elliptic curve over $\mathbb{F}_q$ [6]. The counting problem for plane curves has been well studied [7,8,9]. Von zur Gathen *et.al* show that the counting problem for sparsely represented curves is #P-complete [10]. Huang and Wong give efficient algorithms for both the feasibility and counting problems when the number of variables is a fixed constant [11]. The related problem of computing the Zeta-function of an algebraic variety is well studied (see [12] and the references therein).

The problem of computing $N_r(P)$ has been studied in the literature in many different contexts. A famous theorem due to Chevalley and Warning states that if P is a polynomial over a field $\mathbb{F}_q$ of characteristic p and $\deg(P) < n$, then $N_p(P) \equiv 0$ [1]. This was considerably strengthened by Ax who shows that if $k = \lceil \frac{n-d}{d} \rceil$ then $N_{q^k}(P) = 0$ (see [13]). This was extended to systems with many equations by Katz. Wan gives a simpler proof of the Ax-Katz theorem over $\mathbb{F}_p$ [13]. Moreno and Moreno observed that by reducing a system of equations over $\mathbb{F}_q$ to a system over $\mathbb{F}_p$ and then applying the Ax-Katz bound for prime fields, one can get a bound that often beats the Ax-Katz bound over $\mathbb{F}_q$. They introduced the notion of p-weight degree $w_p(P)$ of a polynomial which is upper bounded by $\deg(P)$ (see Section 3). They showed that if $q = p^h$, and if $k = \lceil h\frac{n-w_p(P)}{w_p(P)} \rceil$ then $N_{p^k}(P) = 0$. Schoof's algorithm for counting the number of points on an elliptic curve proceeds by computing $N_r(P)$ for several small primes r and using Chinese Remaindering [6]. Wan describes methods to compute the reduction of the zeta-function of a curve modulo p^k [12]. Thus all these results are related to the problem of computing $N_r(P)$ for various moduli r. Our work appears to be the first to address the complexity of computing $N_r(P)$.

1.2 Our Results

We give a simple algorithm for computing $N_{p^k}(P)$ given $P(X_1, \cdots, X_n)$ in sparse representation over a field $\mathbb{F}_q$ where $q = p^h$. The running time of our algorithm is $O(nm^{2qk})$ where m is the sparsity of the polynomial i.e. the number of monomials with non-zero coefficients. The algorithm proceeds in two steps. There is a lifting step, where we define an indicator polynomial for the zeroes of the polynomial over $\mathbb{F}_q^n$, and *lift* it to an indicator polynomial modulo p over a ring of characteristic 0. We then amplify this polynomial to get an indicator modulo p^k and sum each monomial modulo p^k over the lift of $\mathbb{F}_q^n$. This high level structure is similar to the proof of the Chevalley-Warning theorem [1] and Wan's proof of the Ax-Katz theorem over prime fields [13]. For a prime field, we lift the problem from $\mathbb{F}_p$ the integers. For non-prime fields, the lifting is from $\mathbb{F}_q$ to an appropriate ring of algebraic integers.

We also present a more naive algorithm to compute $N_{p^k}(P)$ for a polynomial over $\mathbb{F}_q$, which works by reducing the problem to the $\mathbb{F}_p$ case. While the running time of this algorithm is exponential in the degree of the polynomial, it is only singly exponential in the extension degree h of $\mathbb{F}_q$ over $\mathbb{F}_p$, as opposed to the previous algorithm which is doubly exponential in h. This suggests that there might be an algorithm over $\mathbb{F}_q$ with running time singly exponential in h and polynomial in the degree.

The amplification step of our algorithm uses constructions of low-degree modulus amplifying polynomials from complexity theory. Such polynomials were first constructed for the proof of Toda's theorem [14]. Subsequently, better constructions were given by Yao [15] and by Beigel and Tarui [16] to prove upper bounds on a circuit class called ACC. Ours appears to be the first work to make algorithmic use of these polynomials. The construction of Beigel *et.al* gives degree $2k-1$. We show a matching lower bound on the degree of any such polynomial using Mason's theorem.

On the hardness side, we show that over any field $\mathbb{F}_q$ of characteristic p, if r is not a power of p, the problem of computing $N_r(P)$ given the polynomial P in sparse representation is NP-hard under randomized reductions. More precisely, the problem of deciding whether $N_r(P)$ belongs to a particular congruence class modulo r is NP-hard. We study the related feasibility problem for sparse polynomials, which is to decide if $N(P) > 0$. While the problem is easy for constant size fields, we show that it becomes NP-complete, when either the characteristic p or the extension degree h becomes large. As consequence of this, we show that exponential dependence on p and h in our algorithms is unavoidable, since the corresponding counting problems are hard when these parameters are large. Also, when $k = n$, then $N_{p^k}(P) = N(P)$ hence having k in the exponent is also unavoidable. Thus our algorithm for $\mathbb{F}_p$ with running time is $O(nm^{2pk})$ is asymptotically optimal.

Finally we pose the problem of feasibility for symmetric polynomials over $\mathbb{F}_q$, which are sparsely represented over the basis of elementary symmetric polynomials. Our motivation for studying this problem comes from the maximum-likelihood decoding problem for Reed-Solomon codes. Building on work of Guruswami and Vardi [17], we show that this decoding problem is equivalent to a certain root-finding problem for symmetric multilinear polynomials over $\mathbb{F}_q$.

This paper is organized as follows: in Section 2 we discuss modulus amplifying polynomials. We present our algorithmic results in Section 3 and our hardness results in Section 4. We discuss maximum-likelihood decoding of Reed-Solomon codes in Section 5.

2 On Modulus Amplifying Polynomials

Definition 1. *A univariate integer polynomial $A_k(X)$ is k-modulus amplifying if for every integer r, the following condition holds:*

$$\begin{aligned} x \equiv 0 \bmod r \;&\Rightarrow\; A_k(x) \equiv 0 \bmod r^k \\ x \equiv 1 \bmod r \;&\Rightarrow\; A_k(x) \equiv 1 \bmod r^k \end{aligned} \tag{1}$$

We use the following Lemma by Beigel and Tarui.

Lemma 1. *[16] The polynomial $A_k(X) \in \mathbb{Z}[X]$ is k-modulus amplifying iff:*

$$A_k(X) \equiv \begin{cases} 0 & \bmod X^k \\ 1 & \bmod (X-1)^k \end{cases} \tag{2}$$

Beigel *et.al* derive the polynomial $A_k(X)$ by truncating the power series expansion of $(1-X)^{-k}$. We give an alternate derivation of their construction in the full version of this paper.

Lemma 2. *[16] The following polynomial is k-modulus amplifying:*

$$A_k(X) = X^k \sum_{i=0}^{k-1} \binom{2k-1}{k+i} X^i (1-X)^{k-1-i}$$

Since $A_k(X)$ must be divisible by X^k, it must have degree at least k. The running time of our algorithms depends exponentially on the degree of $A_k(X)$ so even a factor 2 saving in the degree would be significant. But we will show that the degree needs to be $2k-1$. The proof uses Mason's theorem which proves the ABC-conjecture for polynomials [18]. Let $z(P)$ denote the number of *distinct* roots of a polynomial over the complex numbers.

Mason's Theorem. [18] *Given polynomials $A(X), B(X), C(X) \in \mathbb{Z}[X]$ which are relatively prime such that $A(X) + B(X) = C(X)$,*

$$\max\{\deg(A), \deg(B), \deg(C)\} \leq z(ABC) - 1$$

Here $z(ABC)$ is the number of distinct complex roots of $A(X)B(X)C(X)$.

Lemma 3. *If $A_k(X)$ is k-modulus amplifying, then* $\deg(A_k) \geq 2k-1$.

Proof. Note that $A(X) = U(X)X^k = V(X)(X-1)^k + 1$ by Lemma 1. Hence

$$U(X)X^k - V(X)(X-1)^k = 1$$

Assume that $\deg(U) = d$. Since the leading term cancels out with the leading term of $V(X)(X-1)^k$, we have $\deg(V) = d$. We set

$$A(X) = U(X)X^k, \quad B(X) = V(X)(X-1)^k, \quad C(X) = 1$$

It is clear that these are relatively prime, so we can apply Mason's theorem. Note that the maximum degree is $d+k$. The product polynomial is

$$A(X)B(X)C(X) \;=\; U(X)V(X)X^k(X-1)^k$$

which can have at most $2+2d$ distinct roots over the complex numbers. Hence

$$d + k \leq 2d + 2 - 1 \;\Rightarrow\; k - 1 \leq d$$

This shows that the degree of $A_k(X) = U(X)X^k$ is at least $2k-1$. □

We note that modulus amplifying polynomials work for every modulus r. In our algorithms, it suffices that the polynomial is amplifying for a specific modulus p, the characteristic of $\mathbb{F}_q$. It is interesting to ask if the same lower bound holds asymptotically for polynomials that are amplifying only for the modulus p.

3 Algorithms for Counting Roots

We use the notation $\mathbf{X} = (X_1, \cdots, X_n)$ for a vector of variables and $\mathbf{x} = (x_1, \cdots, x_n)$ for a vector of constants. Given a vector $D = (d_1, \cdots, d_n)$ in $\mathbb{Z}^n$, we use $\mathbf{X}^D$ to denote the monomial $\prod_i X_i^{d_i}$.

3.1 Modular Counting over Prime Fields

We define a *lift* of $\mathbb{F}_p$ to $\mathbb{Z}$ which maps $i \in \mathbb{F}_p$ to the integer i. We use the same notation for $i \in \mathbb{F}_p$ and its lift in $\mathbb{Z}$, whether i belongs to $\mathbb{F}_p$ or $\mathbb{Z}$ will be clear from the context. We can similarly lift vectors (polynomials) over $\mathbb{F}_p$ to vectors (polynomials) over $\mathbb{Z}$.

The input to the algorithm is a polynomial $P(\mathbf{X}) = \sum_D c_D \mathbf{X}^D$ over $\mathbb{F}_p$. We first lift it to $\mathbb{Z}[\mathbf{X}]$ and then define a polynomial $Q(\mathbf{X}) \in \mathbb{Z}[\mathbf{X}]$ using:

$$Q(\mathbf{X}) = A_k(1 - P(\mathbf{X})^{p-1})$$

Let $Q(\mathbf{X}) = \sum_E c_E \mathbf{X}^E$ where the sum is over at most $2km^{(p-1)(2k-1)}$ monomials. $Q(\mathbf{X})$ satisfies the following relations for $\mathbf{x} \in \mathbb{F}_p^n$:

$$\begin{aligned} P(\mathbf{x}) = 0 \text{ over } \mathbb{F}_p &\Rightarrow Q(\mathbf{x}) \equiv 1 \bmod p^k \text{ over } \mathbb{Z} \\ P(\mathbf{x}) \neq 0 \text{ over } \mathbb{F}_p &\Rightarrow Q(\mathbf{x}) \equiv 0 \bmod p^k \text{ over } \mathbb{Z} \end{aligned}$$

Hence $N(P) \equiv \sum_{\mathbf{x} \in \mathbb{F}_p^n} Q(\mathbf{x}) \equiv \sum_{\mathbf{x} \in \mathbb{F}_p^n} \sum_E c_E \mathbf{x}^E \equiv \sum_E c_E \sum_{\mathbf{x} \in \mathbb{F}_p^n} \mathbf{x}^E \bmod p^k$

where the sum is over the lift of $\mathbb{F}_p^n$ to $\mathbb{Z}^n$. To sum each monomial, observe that

$$\sum_{\mathbf{x} \in \mathbb{F}_p^n} \mathbf{x}^E \equiv \prod_{i=1}^{n} \left(\sum_{x_i=0}^{p-1} x_i^{e_i} \right)$$

Each e_i is at most $2pk$. Note that we cannot use the substitution $X^p = X$ since this need not hold modulo p^k. Thus the time to compute the sum for each monomial is bounded by $O(np^2k^2)$. Hence we can compute $N_{p^k}(P)$ in time $O(2km^{(p-1)(2k-1)}np^2k^2) = O(m^{2pk}n)$. We summarize the algorithm below.

Computing $N_{p^k}(P)$ over $\mathbb{F}_p$.
Input: $P(\mathbf{X}) = \sum_D c_D \mathbf{X}^D$ over $\mathbb{F}_p$.
1. Compute the integer polynomial

$$Q(\mathbf{X}) = A_k(1 - P(\mathbf{X})^{p-1}) = \sum c_E \mathbf{X}^E$$

2. Compute $c_E \sum_{\mathbb{F}_p^n} \mathbf{X}^E \bmod p^k$ for each E and output the sum.

Theorem 1. *Given a polynomial $P(\mathbf{X}) \in \mathbb{F}_p[\mathbf{X}]$ in n variables with m monomials, there is an $O(m^{2pk}n)$ algorithm to compute $N_{p^k}(P)$. For fixed p and k, $N_{p^k}(P)$ can be computed in polynomial time.*

3.2 Modular Counting over Arbitrary Fields

Let $q = p^h$ and let $\mathbb{F}_q = \mathbb{F}_p(\alpha)$ be a degree h field extension of $\mathbb{F}_p$ generated by α. Let $H(X) \in \mathbb{F}_p[X]$ be the monic irreducible polynomial of degree h so that $P(\alpha) = 0$. We will assume that the $H(X)$ is given as input. We lift $H(X)$ to the integers, and then define the quotient $\mathbb{Z}(\alpha) = \mathbb{Z}[X]/(H(X))$ where α is a formal root of $H(X)$. In fact $H(X)$ is irreducible over $\mathbb{Z}$, but we will not use this fact.

Lemma 4. *There is an isomorphism between $\mathbb{Z}[\alpha]/(p)$ and $\mathbb{F}_q$.*

Proof. Note that $\mathbb{Z}[\alpha]/(p) = \mathbb{Z}[X]/(H(X), p) = \mathbb{F}_p[X]/(H(X))$ where in the last expression, $H(X)$ is taken to be a polynomial over $\mathbb{F}_p$. By our choice of $H(X)$, this quotient is precisely $\mathbb{F}_q = \mathbb{F}_p(\alpha)$. It is easy to check that mapping $\alpha \in \mathbb{Z}[\alpha]/(p)$ to $\alpha \in \mathbb{F}_q$ gives an isomorphism. $\square$

Note that this idea of first going modulo p is used to characterize primes in the ring of Gaussian integers [19]. We can lift $\mathbb{F}_q$ to $\mathbb{Z}(\alpha)$ by sending $\alpha \in \mathbb{F}_q$ to $\alpha \in \mathbb{Z}(\alpha)$ and sending $i \in \mathbb{F}_p$ to $i \in \mathbb{Z}$. We now describe the algorithm for computing $N_{p^k}(P)$ over $\mathbb{F}_q$. Given a polynomial $P(\mathbf{X}) = \sum_D c_D \mathbf{X}^D$ over $\mathbb{F}_q$, lift it to $\mathbb{Z}(\alpha)[\mathbf{X}]$ and then define a polynomial $Q(\mathbf{X}) \in \mathbb{Z}(\alpha)[\mathbf{X}]$ using:

$$Q(\mathbf{X}) = (1 - P(\mathbf{X})^{p-1})$$

Let $Q(\mathbf{X}) = \sum_E c_E \mathbf{X}^E$ where the sum is over at most $m^{(p-1)(2k-1)}$ monomials. $Q(\mathbf{X})$ satisfies the following conditions

$$\begin{aligned} P(\mathbf{x}) = 0 \text{ over } \mathbb{F}_q &\Rightarrow Q(\mathbf{x}) \equiv 0 \bmod p \text{ over } \mathbb{Z}(\alpha) \\ P(\mathbf{x}) \neq 0 \text{ over } \mathbb{F}_q &\Rightarrow Q(\mathbf{x}) \equiv 1 \bmod p \text{ over } \mathbb{Z}(\alpha) \end{aligned}$$

Finally define $R(\mathbf{X}) \in \mathbb{Z}(\alpha)[\mathbf{X}]$ as $R(\mathbf{X}) = A_k(Q(\mathbf{X}))$. It is easy to see that $A_k(X)$ is modulus amplifying even for $\mathbb{Z}(\alpha)$. Hence

$$\begin{aligned} P(\mathbf{x}) = 0 \text{ over } \mathbb{F}_q &\Rightarrow R(\mathbf{x}) \equiv 0 \bmod p^k \text{ over } \mathbb{Z}(\alpha) \\ P(\mathbf{x}) \neq 0 \text{ over } \mathbb{F}_q &\Rightarrow R(\mathbf{x}) \equiv 1 \bmod p^k \text{ over } \mathbb{Z}(\alpha) \\ \text{Hence } N(P) &\equiv \sum_{\mathbf{x} \in \mathbb{F}_q^n} R(\mathbf{x}) \bmod p^k \end{aligned}$$

We can compute this sum by writing $R(\mathbf{X}) = \sum_E c_E \mathbf{x}^E$ and summing each monomial individually over the lift of $\mathbb{F}_q$. It is easy to see that $R(\mathbf{X})$ has at most $2km^{(2k-1)(q-1)}$ monomials. So the running time is bounded by $O(nm^{2qk})$.

Computing $N_{p^k}(P)$ over $\mathbb{F}_q$.
Input: $\mathbb{F}_q = \mathbb{F}_p(\alpha)$ given by the irreducible polynomial $H(X)$ of α over $\mathbb{F}_p$. $P(\mathbf{X}) = \sum_D c_D \mathbf{X}^D$ over $\mathbb{F}_q$.

1. Let α satisfy $H(\alpha) = 0$ over $\mathbb{Z}$. Lift $P(\mathbf{X})$ to $\mathbb{Z}(\alpha)[\mathbf{X}]$.
2. Compute the polynomial $R(\mathbf{X}) \in \mathbb{Z}(\alpha)$ given by

$$R(\mathbf{X}) = A_k(1 - P(\mathbf{X})^{q-1}) = \sum c_E \mathbf{X}^E$$

3. Compute $c_E \sum_{\mathbb{F}_q^n} \mathbf{X}^E \bmod p^k$ for each E and output the sum.

Here the sum is over the lift of $\mathbb{F}_q$ to $\mathbb{Z}(\alpha)$. We treat α as a formal symbol satisfying $H(\alpha) = 0$ over $\mathbb{Z}$. All arithmetic operations are preformed modulo p^k.

Theorem 2. *Given a polynomial $P(\mathbf{X}) \in \mathbb{F}_q[\mathbf{X}]$ in n variables with m monomials, there is an $O(nm^{2qk})$ algorithm to compute $N_{p^k}(P)$. For fixed q, p and k, $N_{p^k}(P)$ can be computed in polynomial time.*

3.3 Reduction from $\mathbb{F}_q$ to $\mathbb{F}_p$

Let $P(\mathbf{X}) = \sum_D c_D \mathbf{X}^D$ where $c_D \in \mathbb{F}_q$ be the input. For each variable X_i we substitute $X_i = Y_{i,0} + Y_{i,1}\alpha \cdots Y_{i,h-1}\alpha^{h-1}$ where $Y_{i,j} \in \mathbb{Z}_p$. Thus we replace the monomial $\prod_i X_i^{d_1}$ of total degree d by $\prod_i (Y_{i,0} + Y_{i,1}\alpha \cdots Y_{i,h-1}\alpha^{h-1})^{d_i}$. Naively, this expression has sparsity h^d. We can improve this bound using the notion of p-weight degree due to Moreno and Moreno [20].

Definition 2. *Given an integer $d = d_0 + d_1 p \cdots + d_t p^t$, define its p-weight $\sigma(d) = \sum_j d_j$. The p-weight degree of a monomial $\mathbf{X}^D = \prod_i X_i^{d_i}$ is defined as $w_p(\mathbf{X}^D) = \sum_i \sigma(d_i)$. The p-weight degree $w_p(P)$ of a polynomial $P(\mathbf{X})$ is the maximum of the p-weight degree over all monomials.*

Note that $\sigma_p(d) \leq d$, hence the p-weight degree of a monomial is bounded by its degree. Returning to the monomial $\prod_i X_i^{d_i}$, let $d_i = \sum_t d_{it} p^t$. Then,

$$\begin{aligned}\prod_i X_i^{d_i} &= \prod_i \left(\sum_{j=0}^{h-1} Y_{i,j}\alpha^j\right)^{\sum_t d_{it}p^t} = \prod_{i,t} \left(\sum_{j=0}^{h-1} Y_{i,j}\alpha^j\right)^{d_{it}p^t} \\ &= \prod_{i,t} \left(\sum_{j=0}^{h-1} Y_{i,j}^{p^t}\alpha^{jp^t}\right)^{d_{it}} = \prod_{i,t} \left(\sum_{j=0}^{h-1} Y_{i,j}\alpha^{jp^t}\right)^{d_{it}}\end{aligned}$$

where we use $Y_{i,j} \in \mathbb{F}_p$ hence $Y_{i,j}^{p^t} = Y_{i,j}$. Let $c_D = \sum_u c_u \alpha^u$. Then

$$c_D \prod_i X_i^{d_i} = \sum_{u=0}^{h-1} c_u \alpha^u \prod_{i,t} \left(\sum_{j=0}^{h-1} Y_{i,j}\alpha^{jp^t}\right)^{d_{it}} = \sum_E c_E \mathbf{Y}^E \alpha^{f(E)} \tag{3}$$

where $c_E \in \mathbb{Z}_p$ and $f(E)$ is some function of E. This summation involves $h^{1+\sum d_{it}} = h^{w_p(\mathbf{X}^D)+1}$ monomials. Repeating this for every monomial, we get a sum over at most mh^w monomials, where $w = w_p(P) + 1$:

$$P(\mathbf{Y}) = \sum_E c_E \mathbf{Y}^E \alpha^{f(E)} \tag{4}$$

Since $\{\alpha^0, \cdots, \alpha^{h-1}\}$ is a basis for $\mathbb{F}_q$ over $\mathbb{Z}_p$, we can each $\alpha^{f(E)}$ as a linear combination over this basis. Grouping the various powers of α gives

$$P(\mathbf{Y}) = P_0(\mathbf{Y}) + P_1(\mathbf{Y})\alpha + \cdots + P_{h-1}(\mathbf{Y})\alpha^{h-1} \tag{5}$$

Each polynomial $P_\ell(\mathbf{Y})$ has sparsity at most mh^w, since each monomial from Equation (4) contributes at most one monomial to $P_\ell(\mathbf{Y})$. Since the powers of α are linearly independent over $\mathbb{F}_p$, this sum is 0 iff for $0 \leq \ell \leq h-1$ the coefficient of α^ℓ is 0 over $\mathbb{F}_p$. This implies that for each ℓ, we must have $P_\ell(\mathbf{Y}) = 0$ over $\mathbb{F}_p$. We can combine these into a single equation $Q(\mathbf{Y}) = 0$ over $\mathbb{F}_p$ where

$$Q(\mathbf{Y}) = 1 - \prod_{\ell=0}^{h-1} (1 - P_\ell(\mathbf{Y})^{p-1})$$

The roots of $Q(\mathbf{Y})$ over $\mathbb{F}_p$ are in one-to-one correspondence with the roots of $P(\mathbf{X})$ over $\mathbb{F}_q$, so we can use the $\mathbb{F}_p$ algorithm on $Q(\mathbf{Y})$. Since $Q(\mathbf{Y})$ only takes 0/1 values we can directly apply A_k to $1 - Q(\mathbf{Y})$. The total running time can be bounded by $O(n(mh^w)^{2hpk})$. In addition to p, k there is an exponential dependence on h and the (p-weight) degree.

Theorem 3. *Let $P(\mathbf{X})$ be a polynomial in n variables with m monomials over $\mathbb{F}_q$ where $q = p^h$. Let $w_p(P)$ be p-weight degree and $w = w_p(P) + 1$. There is an $O(n(mh^w)^{2hpk})$ algorithm to compute $N_{p^k}(P)$.*

4 Hardness Results for Counting

In all the results in this section, the polynomial is given in sparse representation. We refer the reader to the book by Papadimitriou [21] for the necessary complexity-theoretic definitions.

Theorem 4. *Let $\mathbb{F}_q$ be a finite field of characteristic p. Assume that r is not a power of p. Given a polynomial $P(\mathbf{X})$ over $\mathbb{F}_q$, the problem of computing $N_r(P)$ is* NP*-hard under randomized reductions.*

An instance of QE over $\mathbb{F}_q$ consists of m quadratic equations $Q_1(\mathbf{X}) = 0, \cdots, Q_m(\mathbf{X}) = 0$. It is well known that deciding if an instance of QE is feasible is NP-complete. An instance of UQE consists of a system of quadratic equation with the promise that in the Yes case, there is a unique solution. Similarly define U3SAT to be the unique version of 3SAT. We show that UQE is NP-complete under randomized reductions by a reduction from U3SAT.

Lemma 5. UQE *is* NP*-complete under randomized reductions over any field.*

Proof. We give a reduction from 3SAT to QE. The reduction itself is folklore, we just need to verify that it preserves the number of solutions. Assume that we have a 3SAT formula $\varphi(\mathbf{X})$ with clauses $C_1, \cdots, C_m$. We add auxiliary variables $Y_1, \cdots, Y_m$ and add the constraints $X_i^2 = X_i$, and $Y_i^2 = Y_i$, which ensures that all the variables need to be 0 or 1. Assume that $C_1 = X_1 \vee X_2 \vee X_3$. We replace this by $Y_1 = X_1 \vee X_2$ and $Y_1 \vee X_3 = 1$. This is done by the equations

$$Y_1 = X_1 + X_2 - X_1X_2, \quad Y_1 + X_3 - Y_1X_3 = 1$$

We perform a similar substitution for every clause. It is clear that this instance of QE is feasible iff $\varphi(\mathbf{X})$ is feasible. Further, this reduction preserves the number of solutions since the values of the auxiliary variables $Y_1, \cdots, Y_m$ are uniquely determined from the values assigned to $X_1, \cdots, X_n$.

Thus starting with an instance of U3SAT, we get an instance of UQE. It is known that U3SAT is NP-complete under randomized reductions [21], hence we infer the hardness of UQE for any $\mathbb{F}_q$. □

We now prove Theorem 4. The reduction used is the same reduction used by Ehrenfeucht and Karpsinski to show the #P-completeness of computing $N(P)$ [2], except that the uniqueness of the solution in the Yes case is crucial.

Proof. Given an instance $X_1, \cdots, X_n, Q_1 = 0, \cdots, Q_m = 0$ of UQE, we add new variables $Z_1, \cdots, Z_m$ and let $P(X_1, \cdots, X_n, Z_1, \cdots, Z_m)$ be the equation

$$\sum_{i=1,\cdots,m} Z_i Q_i(X_1, \cdots, X_n) = 1$$

Assume that $x_1, \cdots, x_n$ is a solution to the system of quadratic equations. Then the above equation reduced to $0 = 1$, so there is no solution. On the other hand if some equation say Q_m is unsatisfied, then we are left with a linear equation

$$\sum_{i=1,\cdots,m} c_i Z_i = 1, \qquad c_m \neq 0$$

Since $c_m \neq 0$, we can pick values for $Z_1, \cdots, Z_{m-1}$ arbitrarily, and then pick Z_m so that the above equation is satisfied. Thus when the instance of UQE is satisfiable, $N(P) = (q^n - 1)q^{m-1}$ whereas when it is unsatisfiable $N(P) = q^{n+m-1}$. Since $q = p^h$, if r is not a power of p, then $(q^n - 1)q^{m-1} \not\equiv q^{n+m-1} \mod r$. Hence an algorithm to compute $N_r(P)$ can be used to solve UQE. More precisely, deciding whether $N(P)$ lies in a particular congruence class modulo r is NP-hard. □

4.1 Hardness Results for Feasibility

The feasibility problem is, given a polynomial $P(\mathbf{X})$ over $\mathbb{F}_q$ does it have a root? When the field size q is constant, there is a simple algorithm for feasibility [4]. On the other hand, we show that the problem becomes NP-complete, when either the characteristic p or the extension degree h becomes large. A consequence of this is that exponential dependence on p and h in our algorithms is unavoidable, since the corresponding counting problems are also hard. To precisely quantify how large the field size needs to be, we parameterize an instance by the number of variables n. These results are proved by simple reductions from 3SAT, and their proofs can be found in the full version. By repeating the reductions starting with U3SAT, we can also show hardness for the related modular counting problems.

Theorem 5. *The problem of deciding whether a polynomial $P(\mathbf{X})$ over $\mathbb{Z}_p$ has a root is* NP*-complete for* $p \geq 2n$. *The problem of computing $N_p(P)$ given $P(\mathbf{X})$ over $\mathbb{Z}_p$ is* NP*-hard under randomized reductions for* $p \geq 2n$.

Theorem 6. *The problem of deciding whether a polynomial $P(\mathbf{X})$ over $\mathbb{F}_{2^t}$ has a root is* NP*-complete for $t \geq 2n$. The problem of computing $N_2(P)$ given $P(\mathbf{X})$ over $\mathbb{F}_{2^t}$ is* NP*-hard under randomized reductions for $t \geq 2n$.*

5 Maximum-Likelihood Reed-Solomon Decoding

Let $S_k(\mathbf{X})$ denote the kth elementary symmetric polynomial in $X_1, \cdots, X_n$. The polynomials $S_k(\mathbf{X})$ for $1 \leq k \leq n$ generate all symmetric polynomials [19]. If a symmetric polynomial is written as a sum of monomials in this basis, we say that it is sparsely represented. A natural question is what is the complexity of the feasibility problem for symmetric polynomials in the sparse representation. We show that maximum-likelihood decoding of Reed-Solomon codes is related to a variant of this problem.

An $[n, k]_q$ Reed Solomon codes consists of all univariate polynomials of degree at most k over $\mathbb{F}_q$ evaluated at a set of points $D = \{x_1, \cdots, x_n\} \subseteq \mathbb{F}_q$. The maximum likelihood decoding problem MLD-RS asks for the closest codeword to a vector $\mathbf{r} \in \mathbb{F}_q^n$. We will work with a different formulation of MLD-RS due to Guruswami and Vardy [17]. Given $D = \{x_1, \cdots, x_n\}$, define the matrix

$$H = \begin{pmatrix} 1 & 1 & \cdots & 1 \\ x_1 & x_2 & \cdots & x_n \\ x_1^2 & x_2^2 & \cdots & x_n^2 \\ \cdots & \cdots & \cdots & \cdots \\ x_1^w & x_2^w & \cdots & x_n^w \end{pmatrix}$$

We define the code $\mathbb{C} = \{\mathbf{z} \in \mathbb{F}_q^n \mid H\mathbf{z}^t = 0\}$, which is in fact a generalized Reed Solomon code. The problem MLD-RS is: *Given H and a syndrome $\mathbf{e} = (e_0, \cdots, e_w) \in \mathbb{F}_q^{w+1}$, is there a vector $\mathbf{z} \in \mathbb{F}_q^n$ with $wt(\mathbf{z}) \leq w$ satisfying $H\mathbf{z} = \mathbf{e}$?* Note that any $w + 1$ columns of H are linearly independent, so we can always find a vector $\mathbf{z}$ of weight $w + 1$ so that $H\mathbf{z} = \mathbf{e}$.

Theorem 7. *There exists a vector $\mathbf{z} \in \mathbb{F}_q^n$ with $wt(\mathbf{z}) \leq w$ so that $H\mathbf{z} = \mathbf{e}$ iff*

$$P(X_1, \cdots, X_w) = \sum_{i \leq w} (-1)^i e_{w-i} S_i(X_1, \cdots, X_w) = 0$$

has a root in D^w where $x_i \neq x_j$ for $i \neq j$.

Proof. We first prove the following identity:

$$\begin{vmatrix} 1 & \cdots & 1 & e_0 \\ x_1 & \cdots & x_w & e_1 \\ x_1^2 & \cdots & x_w^2 & e_2 \\ \cdots & \cdots & \cdots & \cdots \\ x_1^w & \cdots & x_w^w & e_w \end{vmatrix} = \prod_{i \neq j} (x_i - x_j) \sum_{i=0}^{w} (-1)^i e_{w-i} S_i(x_1, \cdots, x_w) \tag{6}$$

We evaluate the LHS by comparing it to the Vandermonde determinant. Let E denote a formal variable. Then

$$\begin{vmatrix} 1 & \cdots & 1 & E^0 \\ x_1 & \cdots & x_w & E^1 \\ x_1^2 & \cdots & x_w^2 & E^2 \\ \cdots & \cdots & \cdots & \cdots \\ x_1^w & \cdots & x_w^w & E^w \end{vmatrix} = \prod_{i \neq j}(x_i - x_j) \prod_{i \leq w}(E - x_i)$$

$$= \prod_{i \neq j}(x_i - x_j) \sum_{i \leq w}(-1)^i E^{w-i} S_i(x_1, \cdots, x_w)$$

Note that by expanding the determinant along the last column, we could derive the same formula without using fact that the various column entries are powers of E. They can be treated as formal symbols. Hence we deduce Equation (6).

Suppose that there exists $\mathbf{z}$ of weight w so that $H\mathbf{z}^t = \mathbf{e}$. Assume wlog that the first w co-ordinates of $\mathbf{z}$ are non-zero. Then $\mathbf{e}$ lies in the span of the first w columns of H, hence the LHS of Equation 6 vanishes. Since $x_i \neq x_j$, this implies that $P(x_1, \cdots, x_w) = 0$.

Conversely, given a root $(x_1, \cdots, x_w) \in D^w$ of P, where $x_i \neq x_j$, the determinant on the LHS of Equation (6) vanishes. Hence a non-trivial linear combination of its columns is 0. Since $x_i \neq x_j$, the columns corresponding to various x_is are linearly independent, so the column corresponding to $\mathbf{e}$ occurs in this combination with a non-zero multiplier. Hence we can write $\mathbf{e}$ as a linear combination of the other columns, which gives a solution to $H\mathbf{z}^t = \mathbf{e}$ of weight at most w. □

If we set $e_w = \gamma$, $e_{w-1} = 1$ and $e_i = 0$ for $i \leq w - 2$, the problem reduces to finding $(x_1, \cdots, x_w) \in D^w$ so that $\sum x_i = \gamma$. Guruswami and Vardy show this is NP-complete when the field size is exponential in n, which implies NP-completeness of MLD-RS over large fields [17]. However it is possible that the above feasibility problem and hence MLD-RS are intractable over $\mathbb{F}_q$ when q is polynomial in n, and when $D = \mathbb{F}_q$.

6 Open Problems

We recap some problems left unanswered by this work.

- Is there algorithm to compute $N_{p^k}(P)$ over $\mathbb{F}_q$ where $q = p^h$ which is singly exponential in p and h?
- Is the feasibility problem NP-complete for fields of characteristic 2 and size polynomial in n? Is it complete for polynomials of low degree?
- Is it possible to construct a family of modulus amplifying polynomials for a specific modulus p that have degree less than $2k - 1$?
- Is the feasibility problem hard for sparse symmetric polynomials when q is polynomial in n?

Acknowledgments. The first author would like to thank Matt Baker, Saugata Basu and Henry Cohn for useful discussions on this subject.

References

1. Lidl, R., Niederreiter, H.: Finite Fields, Encylopedia of Mathematics and Its Applications. Cambridge University Press (1997)
2. Ehrenfeucht, A., Karpinski, M.: The computational complexity of (*xor, and*)-counting problems. Technical Report 8543-CS, ICSI, Berkeley (1990)
3. Karpinski, M., Luby, M.: Approximating the number of zeroes of a GF[2] polynomial. Journal of Algorithms **14**(2) (1993) 280–287
4. Grigoriev, D., Karpinski, M.: An approximation algorithm for the number of zeroes of arbitrary polynomials over GF[q]. In: 32nd IEEE Symposium on Foundations of Computer Science. (1991) 662–669
5. Luby, M., Velicković, B., Wigderson, A.: Deterministic approximate counting of depth-2 circuits. In: Israel Symposium on Theory of Computing Systems. (1993) 18–24
6. Schoof, R.: Counting points on elliptic curves over finite fields. J. Th'eor. Nombres Bordeaux **7** (1995) 219–254.
7. Pila, J.: Frobenius maps of Abelian varieties and finding roots of unity in finite fields. Mathematics of Computation **55** (1990) 745–763
8. Adleman, L., Huang, M.D.: Counting rational points on curves and Abelian varieties over finite fields. In: Proceedings of the 1996 Algorithmic Number Theory Symposium, Springer-Verlag LNCS 1122. (1996) 1–16
9. Huang, M.D., Ierardi, D.: Counting rational points on curves over finite fields. In: Proceedings of the 34th IEEE Symposium on Foundations of Computer Science. (1993) 616–625
10. von zur Gathen, J., Karpinski, M., Shparlinski, I.: Counting curves and their projections. Computational Complexity **6** (1996) 64–99
11. Huang, M.D., Wong, Y.: Solvability of systems of polynomial congruences modulo a large prime. Journal of Computational Complexity **8** (1999) 227–257
12. Wan, D.: Computing Zeta functions over finite fields. Contemporary Mathematics **225** (1999) 135–141
13. Wan, D.: A Chevalley-Warning approach to p-adic estimate of character sums. Proceedings of the American Mathematical Society **123** (1995) 45–54
14. Toda, S.: PP is as hard as the polynomial-time hierarchy. SIAM Journal on Computing **20(5)** (1991) 865–877
15. Yao, A.C.: On ACC and threshold circuits. In: 31st IEEE Symposium on Foundations of Computer Science. (1990) 619–627
16. Beigel, R., Tarui, J.: On ACC. Computational Complexity **4** (1994) 350–366
17. Guruswami, V., Vardy, A.: Maximum-likelihood decoding of Reed-Solomon codes is NP-hard. In: Proceedings of the ACM-SIAM symposium on Discrete Algorithms. (2005) 470–478
18. Mason, R.C.: Diophantine Equations Over Function Fields. Cambridge University Press (1984)
19. Artin, M.: Algebra. Prentice-Hall (1991)
20. Moreno, O., Moreno, C.J.: Improvements of the Chevalley-Warning and the Ax-Katz theorems. American Journal of Mathematics **117(1)** (1995) 241–244
21. Papadimitriou, C.: Computational Complexity. Addison-Wesley (1994)

Hardness Amplification Via Space-Efficient Direct Products

Venkatesan Guruswami[1,*] and Valentine Kabanets[2,**]

[1] Department of Computer Science & Engineering,
University of Washington, Seattle, WA
venkat@cs.washington.edu

[2] School of Computing Science, Simon Fraser University,
Vancouver, BC, Canada
kabanets@cs.sfu.ca

Abstract. We prove a version of the derandomized Direct Product Lemma for deterministic space-bounded algorithms. Suppose a Boolean function $g : \{0,1\}^n \to \{0,1\}$ cannot be computed on more than $1 - \delta$ fraction of inputs by any deterministic time T and space S algorithm, where $\delta \leqslant 1/t$ for some t. Then, for t-step walks $w = (v_1, \ldots, v_t)$ in some explicit d-regular expander graph on 2^n vertices, the function $g'(w) \stackrel{\text{def}}{=} g(v_1) \ldots g(v_t)$ cannot be computed on more than $1 - \Omega(t\delta)$ fraction of inputs by any deterministic time $\approx T/d^t - \mathsf{poly}(n)$ and space $\approx S - O(t)$. As an application, by iterating this construction, we get a deterministic linear-space "worst-case to constant average-case" hardness amplification reduction, as well as a family of logspace encodable/decodable error-correcting codes that can correct up to a constant fraction of errors. Logspace encodable/decodable codes (with linear-time encoding and decoding) were previously constructed by Spielman [14]. Our codes have weaker parameters (encoding length is polynomial, rather than linear), but have a conceptually simpler construction. The proof of our Direct Product Lemma is inspired by Dinur's remarkable recent proof of the PCP theorem by gap amplification using expanders [4].

Keywords: Direct products, hardness amplification, error-correcting codes, expanders.

1 Introduction

1.1 Hardness Amplification Via Direct Products

Hardness amplification is, roughly, a procedure for converting a somewhat difficult computational problem into a much more difficult one. For example, one would like to convert a problem A that is worst-case hard (i.e., cannot be computed within certain restricted computational model) into a new problem B that is *average-case* hard (i.e., cannot be computed on a significant fraction of inputs).

* Supported in part by NSF grant CCF-0343672 and a Sloan Research Fellowship.
** Supported in part by an NSERC Discovery grant.

J.R. Correa, A. Hevia, and M. Kiwi (Eds.): LATIN 2006, LNCS 3887, pp. 556–568, 2006.

The first such "worst-case to average-case" reduction was given by Babai, Fortnow, Nisan, and Wigderson [2]. They used algebraic error-correcting codes to go from a worst-case hard function f to a weakly average-case hard function g. They further amplified the average-case hardness of g via the following Direct Product construction. Given $g : \{0,1\}^n \to \{0,1\}$, define $g^k : (\{0,1\}^n)^k \to \{0,1\}^k$ as $g^k(x_1, \dots, x_k) = g(x_1)\dots g(x_k)$. Intuitively, computing g on k independent inputs $x_1, \dots, x_k$ should be significantly harder than computing g on a single input. In particular, if g cannot be computed by circuits of certain size s on more than $1 - \delta$ fraction of inputs (i.e., g is δ-hard for circuit size s), then one would expect that g^k should not be computable (by circuits of approximately the same size s) on more than $(1-\delta)^k$ fraction of inputs. The result establishing the correctness of this intuition is known as Yao's Direct Product Lemma [16], and has a number of different proofs [11, 5, 9, 10].

1.2 Derandomized Direct Products and Error-Correcting Codes

Impagliazzo and Wigderson [9, 10] consider a "derandomized" version of the Direct Product lemma. Instead of evaluating a given n-variable Boolean function g on k *independent* inputs $x_1, \dots, x_k$, they generate the inputs using a certain deterministic function $F : \{0,1\}^r \to (\{0,1\}^n)^k$ such that the input size r of F is much smaller than the output size kn. They give several examples of the function F for which the function $g'(y)$ defined as $g(F(y)_1)\dots g(F(y)_k)$, where $F(y)_i$ denotes the ith n-bit string output by $F(y)$ for $y \in \{0,1\}^r$, has average-case hardness about the same as that of $g^k(x_1, \dots, x_k)$ for completely independent inputs x_i. In particular, Impagliazzo [9] shows that if g is δ-hard (for certain size circuits) for $\delta < 1/O(n)$, then, for a pairwise independent $F : \{0,1\}^{2n} \to (\{0,1\}^n)^n$, the function $g'(y) = g(F(y)_1)\dots g(F(y)_n)$ is $\Omega(\delta n)$-hard (for slightly smaller circuits).

Trevisan [15] observes that any Direct Product Lemma proved via "black-box" reductions can be interpreted as an error-correcting code mapping binary messages into codewords over a larger alphabet. Think of an $N = 2^n$-bit message *Msg* as a truth table of an n-variable Boolean function g. The encoding *Code* of this message will be the table of values of the direct-product function g^k. That is, the codeword *Code* is indexed by k-tuples of n-bit strings $(x_1, \dots, x_k)$, and the value of *Code* at position $(x_1, \dots, x_k)$ is the k-tuple $(g(x_1), \dots, g(x_k))$. The Direct Product Lemma says that if g is δ-hard, then g^k is $\epsilon \approx 1-(1-\delta)^k$-hard. In the language of codes, this means that given (oracle access to) a string w over the alphabet $\Sigma = \{0,1\}^k$ such that w and *Code* disagree in less than ϵ fraction of positions, we can construct an N-bit string Msg' such that *Msg* and Msg' disagree in less than δ fraction of positions.

Note that the error-correcting code derived from a Direct Product Lemma maps N-bit messages to N^k-symbol codewords over the larger alphabet $\Sigma = \{0,1\}^k$. A derandomized Direct Product Lemma, using a function $F : \{0,1\}^r \to (\{0,1\}^n)^k$ as described above, yields an error-correcting code with encoding length 2^r. For example, the pairwise-independent function F from Impagliazzo's derandomized Direct Product Lemma would yield codes with encoding length N^2, which is a significant improvement over the length N^k.

The complexity of the reduction used to prove a Direct Product Lemma determines the complexity of the decoding procedure for the corresponding error-correcting code.

In particular, if a reduction uses some non-uniformity (say, m bits of advice), then the corresponding error-correcting code will be only *list-decodable* with the list size at most 2^m. If one wants to get codes with ϵ being asymptotically close to 1, then list-decoding is indeed necessary. However, for a constant ϵ, unique decoding is possible, and so, in principle, there must be a proof of this weaker Direct Product Lemma that uses only uniform reductions (i.e., no advice).

1.3 Derandomized Direct Products Via Uniform Reductions

The derandomized Direct Product lemmas of [9, 10] are proved using nonuniform reductions. Using the graph-based construction of error-correcting codes of [6], Trevisan [15] proves a variant of a derandomized Direct Product lemma with a *uniform deterministic* reduction.

More precisely, for certain k-regular expander graphs G_n on 2^n vertices (labeled by n-bit strings), Trevisan [15] defines the function $F : \{0,1\}^n \to (\{0,1\}^n)^k$ as $F(y) = y_1, \dots, y_k$, where y_is are the neighbors of the vertex y in the graph G_n. He then argues that, for a Boolean function $g : \{0,1\}^n \to \{0,1\}$, if there is a deterministic algorithm running in time $t(n)$ that solves $g'(y) = g(F(y)_1) \dots g(F(y)_k)$ on $\Omega(1)$ fraction of inputs, then there is a deterministic algorithm running in time $O(t\mathsf{poly}(n,k))$ that solves g on $1-\delta$ fraction of inputs, for $\delta = O(1/k)$. That is, if g is δ-hard with respect to deterministic time algorithms, then g' is $\Omega(1)$-hard with respect to deterministic algorithms running in slightly less time. Note that the input size of g' is n, which is the same as the input size of g.

The given non-Boolean function $g' : \{0,1\}^n \to \{0,1\}^k$ can be converted into a Boolean function g'' on $n + O(\log k)$ input variables that has almost the same $\Omega(1)$ hardness with respect to deterministic algorithms. The idea is to use some binary error-correcting code $\mathcal{C}$ mapping k-bit messages to $O(k)$-bit codewords, and define $g''(x,i)$ to be the ith bit of $\mathcal{C}(g'(x))$.

1.4 Our Results

In this paper, we analyze a different derandomized Direct Product construction. Let G_n be a d-regular expander graph on 2^n vertices, for some constant d. Denote by $[d]$ the set $\{1, 2, \dots, d\}$. For any t and any given n-variable Boolean function g, we define g' to be the value of g along a t-step walk in G_n. That is, we define $g' : \{0,1\}^n \times [d]^t \to \{0,1\}^{t+1}$ as $g'(x, i_1, \dots, i_t) = g(x_0)g(x_1)\dots g(x_t)$, where $x_0 = x$, and each x_j (for $1 \leqslant j \leqslant t$) is the i_jth neighbor of x_{j-1} in the graph G_n. We show that if g is δ-hard to compute by deterministic uniform algorithms running in time T and space S for $\delta < 1/t$, then g' is $\Omega(t\delta)$-hard with respect to deterministic algorithms running in time $\approx T/d^t$ and space $\approx S - O(t)$.

Note that if g is δ-hard, then we expect $g^t(x_1, \dots, x_t) = g(x_1)\dots g(x_t)$ (on t independent inputs) to be $\delta' = 1-(1-\delta)^t$-hard. For $\delta \ll 1/t$, we have $\delta' \approx t\delta$, and so our derandomized Direct Product construction described above achieves asymptotically correct hardness amplification.

Combining the function g' with any linear error-correcting code $\mathcal{C}$ (with constant relative distance) mapping $(t+1)$-bit messages into $O(t)$-bit codewords, we can get from g' a Boolean function on $n+O(t)$ variables that also has hardness $\Omega(t\delta)$. Applying these two steps (our expander-walk Direct Product followed by an encoding using the error-correcting code $\mathcal{C}$) to a given δ-hard n-variable Boolean function g for $\log 1/\delta$ iterations, we can obtain a new Boolean function g'' on $n + O(t \log 1/\delta)$ variables that is $\Omega(1)$-hard. If g is δ-hard for deterministic time T and space S, then g'' is $\Omega(1)$-hard for deterministic time $\approx T\mathsf{poly}(\delta)$ and space $\approx S - O(\log 1/\delta)$.

In terms of running time, this iterated Direct Product construction matches the parameters of Trevisan's Direct Product construction described earlier. Both constructions are proved with uniform deterministic reductions. The main difference seems to be in the usage of space. Our reduction uses at most $O(n + \log 1/\delta)$ space, which is at most $O(n)$ even for $\delta = 2^{-n}$. Thus we get a deterministic uniform "worst-case to constant average-case" reduction computable in linear space. The space usage in Trevisan's construction is determined by the space complexity of encoding/decoding of the "inner" error correcting code $\mathcal{C}$ from k to $O(k)$ bits, for $k = O(1/\delta)$. A simple deterministically encodable/decodable code would use space $\Omega(k) = \Omega(1/\delta)$.

We also show that constant-degree expanders which have expansion better than degree$/2$ can be used to obtain a simple space-efficient hardness amplification. However, it is not known how to construct such expanders explicitly.

Related work. Impagliazzo and Wigderson [10] use expander walks in combination with the Nisan-Wigderson generator [12] to prove a different derandomized direct product lemma. They start with a Boolean function of constant average-case hardness (against circuits) and construct a new Boolean function of average-case hardness exponentially close to $1/2$. In contrast, (i) we analyze the hardness of a direct product using vertices of an expander walk only, (ii) our direct product lemma works for a different range of parameters (amplifying worst-case hardness to constant average-case hardness), and (iii) our reductions are completely uniform efficient *deterministic* algorithms.

Our deterministic linear-space hardness amplification result is not new. A deterministic linear-space "worst-case to constant average-case" reduction can be also achieved by using Spielman's expander-based error-correcting codes [14]. His codes have encoding/decoding algorithms of space complexity $O(\log N)$ for messages of length N, which translates into $O(n)$-space reductions for n-variable Boolean functions.

In light of the connection between Direct Product Lemmas and error-correcting codes, our iterated Direct Product construction also yields a deterministic logspace (in fact, uniform NC^1) encodable/decodable error-correcting code that corrects a constant fraction of errors. Spielman's NC^1 encodable/decodable codes also correct a constant fraction of errors, but they have much better other parameters. In particular, Spielman's encoding/decoding is in *linear* time, and so the length of the encoded message is linear in the size of the original message. In contrast, our encoding time and the length of the encoding are only polynomial in the size of the original message. We believe, however, that our codes have a conceptually simpler construction, which closely follows the "Direct Product Lemma" approach.

Finally, our proof method is inspired by Dinur's recent proof of the PCP Theorem [4]. She describes a procedure for increasing the unsatisfiability gap of a given unsatisfiable Boolean formula by a constant factor, at the cost of a constant-factor increase in the size of the new formula. Iterating this gap amplification for $O(\log n)$ steps, she converts any unsatisfiable formula with n clauses to a polynomially larger formula ϕ such that no assignment can satisfy more than a constant fraction of clauses in ϕ. A single step of gap amplification uses expanders to define a new, harder formula; intuitively, a new formula corresponds to a certain derandomized "direct product" of the old formula, where derandomization is done using constant-length expander walks. In the present paper, we also use constant-size expander walks to derandomize direct products, achieving a constant-factor hardness amplification at the cost of constant additive increase in the space complexity of the new function. Iterating this step $O(n)$ times, allows us to convert a Boolean function that is worst-case hard for linear space into one that is constant average-case hard for linear space.

Remainder of the paper. We give the necessary definitions in Section 2. In Section 3, we state and analyze our Direct Product Lemma. Applications of our Direct Product Lemma to linear-space hardness amplification and logspace encodable/decodable codes are given in Section 4. Section 5 proves a simpler version of the Direct Product Lemma, under the assumption that expanders with expansion better than degree/2 can be efficiently constructed.

2 Preliminaries

2.1 Worst-Case and Average-Case Hardness

Given a bound b on a computational resource $resource$ ($resource$ can be, e.g., deterministic time, space, circuit size, or some combination of such resources), we say that a function $f : A \to B$ (for some sets A and B) is *worst-case hard for b-bounded* $resource$ if every algorithm using at most b amount of $resource$ disagrees with the function f on at least one input $x \in A$.

For $0 \leqslant \delta \leqslant 1$ and a bound b on $resource$, a function $f : A \to B$ is called *average-case δ-hard (or, simply, δ-hard) for b-bounded* $resource$ if every algorithm using at most b amount of $resource$ disagrees with the function f on at least a fraction δ of inputs from A. Observe that for $\delta = 1/|A|$, the notion of δ-hardness coincides with that of worst-case hardness.

2.2 Expanders

Let $G = (V, E)$ be any d-regular undirected graph on n vertices. Let $A = \{a_{i,j}\}_{i,j=1}^{n}$ be the normalized adjacency matrix of G, i.e., $a_{i,j} = \frac{1}{d}*$(the number of edges between i and j). For a constant $\lambda < 1$, the graph $G = (V, E)$ is called a λ-expander if the second largest (in the absolute value) eigenvalue of the matrix A is at most λ.

Another (essentially equivalent) definition of expanders is the following. A d-regular graph $G = (V, E)$ is an (α, β)-expander if for every subset $W \subseteq V$ with $|W| \leqslant \alpha|V|$,

$$\Big|\{v \in V \mid \exists w \in W \text{ such that } (v, w) \in E\}\Big| \geqslant \beta|W| .$$

We will use the following lemma in the analysis of our Direct Product Lemma. A variant of this lemma (for edge sets rather than vertex sets) is proved in [4–Lemma 5.4]; the vertex-case is in fact simpler to argue.

Lemma 1. *Let $G = (V, E)$ be any d-regular λ-expander for some $\lambda < 1$, and let $S \subset V$ be any set. For any value t, let W_i, for $i \in [0..t]$, be the set of all t-step walks in G that pass through a vertex from S in step i. Then, for each $i \in [0..t]$, a random walk from the set W_i is expected to contain at most $t\frac{|S|}{|V|} + O(1)$ vertices from the set S.*

We will need an infinite family of d-regular λ-expanders $\{G_n = (V_n, E_n)\}_{n=1}^{\infty}$, where G_n is a graph on 2^n vertices; we assume that the vertices of G_n are identified with n-bit strings. We need that such a family of graphs be *efficiently constructible* in the sense that given the label of a vertex $v \in V_n$ and a number $i \in [d]$, the i'th neighbor of v in G_n can be computed efficiently by a deterministic polynomial-time and linear-space algorithm. We will spell out the exact constructibility requirement in Section 3.1.

2.3 Space Complexity

We review definitions concerning space complexity, since for our main Direct Product Lemma, we need to measure the space complexity of the algorithms very carefully.

Definition 2 (Standard Space Complexity). *An algorithm computes a function f in space S if given as input x on a* read-only *input tape, it uses a work tape of S cells and halts with $f(x)$ on the work tape. Such an algorithm is said to have* space complexity S.

Definition 3 (Total Space Complexity). *An algorithm A computes a function f with domain $\{0,1\}^n$ in* total space S *if on an n-bit input x,*

1. *A has read/write access to the input tape,*
2. *in addition to the n input tape cells, A is allowed another $S - n$ tape cells, and*
3. *at the end of its computation, the tape contains $f(x)$.*

Such an algorithm is then said to have total space complexity S.

Definition 4 (Input-Preserving Space Complexity). *An algorithm A computes a function f with domain $\{0,1\}^n$ in* input-preserving space S *if on an n-bit input x,*

1. *A has read/write access to the input tape,*
2. *in addition to the n input tape cells, A is allowed another $S - n$ tape cells, and*
3. *at the end of its computation, the tape contains $x; f(x)$.*

That is, we allow the algorithm to write on the input portion of the tape, provided it is restored to its original content at the end of the computation. Such an algorithm is then said to have input-preserving space complexity S. *(Note that the input-preserving space complexity of a function $f(x)$ is the same as the total space complexity of the function $f'(x) \stackrel{\text{def}}{=} x; f(x)$.).*

The following simple observation lets us pass between these models of space complexity with a linear additive difference.

Fact 5. *If there is an algorithm A with space complexity S to compute a function with domain $\{0,1\}^n$, then there is an algorithm A' with input-preserving (total) space complexity $S+n$ to compute f. Conversely, if there is an algorithm B' with input-preserving (total) space complexity S' to compute f, then there is an algorithm B with space complexity S' to compute f.*

We will use the input-preserving space complexity to analyze the efficacy of our Direct Product Lemma and its iterative application to amplify hardness. However, by Fact 5, our end result can be stated in terms of the standard space complexity of Definition 2.

3 A New Direct Product Lemma

3.1 Construction

We need the following two ingredients:

(i) Let $G = (V, E)$ be any efficiently constructible d-regular λ-expander on $|V| = 2^n$ vertices which are identified with n-bit strings (here d and $\lambda < 1$ are absolute constants, and we will typically hide factors depending on d in the O-notation). By efficient constructibility, we mean the following. There is an algorithm running in time $T_{expander} = \mathsf{poly}(n)$ and total space $S_{expander} = O(n)$, that given as input an n-bit string x and an index $i \in [d]$, outputs the pair $N_G(x, i) \stackrel{def}{=} (y, j)$, where $y \in \{0,1\}^n$ is the i'th neighbor in G of x, and $j \in [d]$ is such that x is the j'th neighbor of y. We can obtain such expanders from [13].

(ii) Let $\mathcal{C}$ be any polynomial-time and linear-space encodable (via Enc) and decodable (via Dec) linear binary error-correcting code with constant rate $1/c$ and constant relative distance ρ.

Our construction proceeds in two steps.

Step 1: Let $f : \{0,1\}^n \to \{0,1\}$ be any given Boolean function. For any $t \in \mathbb{N}$, define a new, non-Boolean function $g : \{0,1\}^n \times [d]^t \to \{0,1\}^{t+1}$ as follows:

$$g(v, i_1, \ldots, i_t) = f(v)f(v_1)\ldots f(v_t),$$

where for each $1 \leqslant j \leqslant t$, v_j is the i_jth neighbor of vertex v_{j-1} in the expander graph G (we identify v with v_0).

Step 2: Define a Boolean function $h : \{0,1\}^n \times [d]^t \times [c(t+1)] \to \{0,1\}$ as

$$h(v, i_1, \ldots, i_t, j) = Enc(g(v, i_1, \ldots, i_t))_j,$$

where $Enc(y)_j$ denotes the the jth bit in the encoding of the string y using the binary error-correcting code $\mathcal{C}$.

Complexity of the Encoding: Suppose that the n-variable Boolean function f is computable in deterministic time T and input-preserving space S. Then the non-Boolean function g obtained from f in Step 1 of the construction above will be computable in deterministic time $T_g = O(t(T + T_{expander})) = O(t(T + \mathsf{poly}(n)))$ and input-preserving space at most $S_g = \max\{S, S_{expander}\} + O(t)$. The claim about time complexity is obvious. For the space complexity, to compute $g(v, i_1, \ldots, i_t)$, we first compute $f(v)$ using input-preserving space S. We then re-use this space to compute $N_G(v, i_1) = (v_1, j_1)$ in total space $S_{expander}$. We remember i_1, j_1 (these take only $O(1)$ space) separately, but replace v by v_1, and compute $f(v_1)$ in input-preserving space S. We next likewise compute $N_G(v_1, i_2) = (v_2, j_2)$, and replace v_1 by v_2, compute $f(v_2)$, and so on. In the end, we would have computed $f(v)f(v_1)\dots f(v_t)$ in total space $\max\{S, S_{expander}\} + O(t)$. However, we need to restore the original input $v, i_1, i_2, \dots, i_t$. For this, we use the stored "back-indices" $j_t, j_{t-1}, \dots, j_1$ to walk back from v_t to v in a manner identical to the forward walk.

The Boolean function h obtained from g in Step 2 will be computable in time $T_g +$ poly(t) and input-preserving space $S_g + O(t)$. Note that, assuming $S \geqslant S_{expander}$, the input-preserving space complexity of h is at most an additive constant term $O(t)$ bigger than that of f.

3.2 Analysis

We will show that the "direct product construction" described above increases the hardness of a Boolean function f by a multiplicative factor $\Omega(t)$.

Lemma 6 (Direct Product Lemma). *Suppose an n-variable Boolean function f has hardness $\delta \leqslant 1/t$ for deterministic time T and input-preserving space $S \geqslant S_{expander} + \Omega(t)$. Let h be the Boolean function obtained from f using the direct product construction described above. Then h has hardness $\Omega(t\delta)$ for deterministic time $T' = \frac{T}{O(t^2 d^t)} - \mathsf{poly}(n)$ and input-preserving space $S' = S - O(t)$.*

The proof of the Direct Product Lemma above will consist of two parts (given by Lemmas 7 and 8 below). First we argue that the non-Boolean function g (obtained from f by evaluating f along t-step walks in the expander G) will have hardness $\Omega(t)$-factor larger than the hardness of f. Then we argue that turning the function g into the Boolean function h via encoding the outputs of g by a "good" error-correcting code will reduce its hardness by only a constant factor independent of t.

Lemma 7. *Suppose an n-variable Boolean function f has hardness $\delta \leqslant 1/t$ for deterministic time T and input-preserving space $S \geqslant S_{expander} + \Omega(t)$. Let g be the non-Boolean function obtained from f using the first step of the direct product construction described above. Then g has hardness $\delta' = \Omega(t\delta)$ for deterministic time $T' = \frac{T}{O(td^t)} - t\mathsf{poly}(n)$ and input-preserving space $S' = S - O(t)$.*

Proof: Let C' be a deterministic algorithm using time T' and input-preserving space S' that computes g correctly on $1 - \delta'$ fraction of inputs, for the least possible δ' that can be achieved by algorithms with these time/space bounds. We will define a new deterministic algorithm C using time at most T and input-preserving space S, and argue that δ' is at least $\Omega(t)$ times larger than the fraction of inputs computed incorrectly by C. Since the latter fraction must be at least δ (as f is assumed δ-hard for time T and input-preserving space S), we conclude that $\delta' \geqslant \Omega(t\delta)$.

We will compute f by an algorithm C defined as follows. On input $x \in \{0,1\}^n$, for each $i \in [0..t]$, record the majority value b_i taken over all values $C'(w)_i$, where w is a t-step walk in the graph G that passes through x at step i and $C'(w)_i$ is the ith bit in the $(t+1)$-tuple output by the circuit C' on input w. Output the majority over all the values b_i, for $0 \leqslant i \leqslant t$. A more formal description of the algorithm is given in the table Algorithm 1.

```
INPUT: x ∈ {0,1}^n.
GOAL: Compute f(x).

count_1 = 0
for each i = 0..t
    count_2 = 0
    for each t-tuple (k_1, k_2, ..., k_t) ∈ [d]^t
        Compute the vertex y reached from x in i steps by taking edges labeled
          k_1, k_2, ..., k_i, together with the "back-labels"
          ℓ_1, ℓ_2, ..., ℓ_i needed to get back from y to x.
        count_2 = count_2 + C'(y, ℓ_1, ℓ_2, ..., ℓ_i, k_{i+1}, ..., k_t)_i
        Restore x by walking from y for i steps using edge-labels ℓ_1, ℓ_2, ..., ℓ_i.
    end for
    if count_2 ⩾ d^t/2 then count_1 = count_1 + 1 end if
end for
if count_1 ⩾ t/2 then RETURN 1 else RETURN 0
end Algorithm
```

Algorithm 1. Algorithm C

It is straightforward to verify that the algorithm C can be implemented in deterministic time $O(td^t(T' + t\mathsf{poly}(n)))$. By choosing T' as in the statement of the lemma, we can ensure that the running time of C is at most T. It is also easy to argue that the input-preserving space complexity S of algorithm C is at most $\max\{S_{expander}, S'\} + O(t)$ (the argument goes along the lines of the one we used to argue about the complexity of the encoding at the end of Section 3.1). Hence by choosing $S' = S - O(t) \geqslant S_{expander}$ we get the input-preserving space complexity of C at most S.

We now analyze how many mistakes the algorithm C makes in computing f. Define the set $Bad = \{x \in \{0,1\}^n \mid C(x) \neq f(x)\}$. Pick a subset $B \subseteq Bad$ such that $|B|/|V| = \min\{|Bad|/|V|, 1/t\}$. By definition, if $x \in Bad$, then for each of at least $1/2$ values of $i \in [0..t]$, the algorithm C' is wrong on at least half of all t-step walks that pass through x in step i. Define a 0-1 matrix M with $|B|$ rows and $t+1$ columns such that, for $x \in B$ and $i \in [0..t]$, $M(x,i) = 0$ iff C' is wrong on at least half of all t-step walks that pass through x in step i. Then the fraction of 0s in the matrix M is at least $1/2$. By averaging, we conclude that there exists a subset $I \subseteq [0..t]$ of size at least $t/4$ such that, for each $i \in I$, the ith column of M contains at least $1/4$ fraction of 0s. This means that for each $i \in I$, the algorithm C' is wrong on at least $\frac{|B|}{4}\frac{d^t}{2} = \frac{1}{8}|B|d^t$ of all $|B|d^t$ walks of length t that pass through the set B at step i.

For $x \in B$ and $i \in [0..t]$, let us denote by $W_{i,x}$ the set of all t-step walks that pass through x in step i; observe that $|W_{i,x}| = d^t$. We define $W_i = \cup_{x \in B} W_{i,x}$. Since $W_{i,x}$ and $W_{i,y}$ are disjoint for $x \neq y$, we get $|W_i| = |B|d^t$. Also, for $x \in B$ and $i \in [0..t]$, denote by $W^*_{i,x}$ the set of all t-step walks $w \in W_{i,x}$ such that $C'(w) \neq g(w)$. Define $W^*_i = \cup_{x \in B} W^*_{i,x}$. Note that for each $i \in I$, $|W^*_i| \geqslant \frac{1}{8}|W_i|$. Finally, define $W^* = \cup_{i=0}^t W^*_i$ — by construction, for every $w \in W^*$, $C'(w) \neq g(w)$, so it suffices to give a lower bound on $|W^*|$ to argue that C' makes many mistakes.

For each $i \in [0..t]$, let $H_i \subseteq W_i$ be the set of all walks $w \in W_i$ that contain more than m elements from B. Using the properties of the expander G, we can choose

m to be a sufficiently large constant (independent of t) such that, for all i, $|H_i| < \frac{1}{16}|W_i|$. Indeed, by Lemma 1 above, for every i, a random walk $w \in W_i$ is expected to contain at most $t|B|/|V| + O(1)$ vertices from B. Since, by our choice of parameters, $|B|/|V| \leqslant 1/t$, a random $w \in W_i$ contains on average at most $b = O(1)$ vertices from B. By Markov's inequality, the probability that a random $w \in W_i$ contains more than $m = 16b$ vertices from B is at most $1/16$.

Thus we have

$$\sum_{i \in I} |W_i^* \setminus H_i| = \sum_{i \in I} (|W_i^*| - |H_i|) \geqslant |I|(\frac{1}{8} - \frac{1}{16})|W_i| \geqslant \frac{t}{64}|B|d^t. \quad (1)$$

On the other hand, we have

$$\sum_{i \in I} |W_i^* \setminus H_i| \leqslant m|W^* \setminus (\cup_{i=0}^t H_i)| \leqslant m|W^*|. \quad (2)$$

Combining Eqs. (1) and (2), we get $|W^*| \geqslant \frac{t}{64m}|B|d^t$. Dividing both sides by the number $|V|d^t$ of all possible t-step walks in G (which is the number of all possible inputs to the algorithm C'), we get that C' makes mistakes on at least $\frac{t}{64m}|B|/|V|$ fraction of inputs. Note that $|B|/|V| \geqslant \delta$ since f is assumed to be δ-hard for time T and input-preserving space S. It follows that the function g is $\Omega(t\delta)$-hard for time T' and input-preserving space S'. □

The analysis of the second step (Lemma 8) of our Direct Product construction uses the standard approach of "code concatenation".

Lemma 8. *Let $A = \{0,1\}^n \times [d]^t$. Suppose that a function $g : A \to \{0,1\}^{t+1}$ is δ-hard for deterministic time T and input-preserving space S. Let $h : A \times [c \cdot (t+1)] \to \{0,1\}$ be the Boolean function obtained from g as described in Step 2 of the Direct Product construction above (using the error-correcting code with relative distance ρ and rate $1/c$). Then the function h is $\delta \cdot \rho/2$-hard for deterministic time $T' = (T - \mathsf{poly}(t))/O(t)$ and input-preserving space $S' = S - O(t)$.*

3.3 Iteration

Our Direct Product Lemma (Lemma 6) can be applied repeatedly to increase the hardness of a given Boolean function at an exponential rate, as long as the current hardness is less than some universal constant. In particular, as shown in the corollary below, we can turn a δ-hard Boolean function into a $\Omega(1)$-hard Boolean function. Note that we state this result in terms of the usual space complexity, and not the input-preserving space complexity that we used to analyze a single Direct Product.

Corollary 9. *Let f be an n-variable Boolean function that is δ-hard for deterministic time T and space $S \geqslant \Omega(n + \log \frac{1}{\delta})$. Then there is a Boolean function f' on $n + O(\log \frac{1}{\delta})$ variables such that f' is $\Omega(1)$-hard for deterministic time $T' = T\mathsf{poly}(\delta) - \mathsf{poly}(n)$ and space $S' = S - n - O(\log \frac{1}{\delta})$. Moreover, if f is computable in time $\tilde{T}$ and space $\tilde{S}$, then f' is computable in time $(\tilde{T} + \mathsf{poly}(n))/\mathsf{poly}(\delta)$ and space $\tilde{S} + O(n + \log \frac{1}{\delta})$.*

Remark 10. The constant average-case hardness in Corollary 9 above can be boosted to any constant less than $1/4$ by one additional amplification with a suitable expander, as in [6, 15] (specifically, see Theorem 7 in [15]).

Remark 11. We want to point out that Spielman's logspace encodable/decodable codes can also be used for "worst-case to constant average-case" hardness amplification via deterministic linear-space reductions. So Corollary 9 is implicit in [14].

4 Applications

4.1 Hardness Amplification Via Deterministic Space-Efficient Reductions

The iterated Direct Product construction of Corollary 9 gives us a way to convert worst-case hard Boolean functions into constant-average-case hard ones, with space-efficient deterministic reductions. The following theorems are immediate consequences of Corollary 9 and Remark 10. Below we use standard notation for the complexity classes $\mathsf{E} = \mathsf{DTIME}(2^{O(n)})$ and $\mathsf{LINSPACE} = \mathsf{SPACE}(O(n))$.

Theorem 12. *Let $\alpha < 1/4$ be an arbitrary constant. If there is a language $L \in \mathsf{E} \setminus \mathsf{LINSPACE}$, then there is a language $L' \in \mathsf{E}$ that is α-hard for $\mathsf{LINSPACE}$.*

Theorem 13. *Let $\alpha < 1/4$ be an arbitrary constant. For every $c > 0$, there is a $c' > 0$ such that the following holds. If there is a language $L \in \mathsf{LINSPACE}$ that cannot be computed by any deterministic algorithm running in linear space and, simultaneously, time $2^{c'n}$, then there is a language $L' \in \mathsf{LINSPACE}$ that is α-hard for any deterministic algorithm running in linear space and, simultaneously, time 2^{cn}.*

4.2 Logspace Encodable/Decodable Error-Correcting Codes

Using the connection between error-correcting codes and hardness amplification mentioned in the Introduction (Section 1.2) , we get an alternative construction (with much weaker parameters) to Spielman's logspace encodable/decodable codes [14].

Theorem 14. *There is an explicit code C mapping n-bit messages to $\mathsf{poly}(n)$-bit codewords such that*

1. *C can correct a constant fraction of errors,*
2. *both encoding and decoding can be implemented in deterministic logspace (in fact, uniform NC^1).*

Remark 15. We are not aware of any logspace encodable/decodable asymptotically good codes other than Spielman's construction [14], and the recent improvements to its error-correction performance [6, 7]. Allowing NC^2 complexity seems to give several other choices of error-correcting codes.

5 A Simple Graph Based Amplification

Here we observe that the existence of efficiently constructible d-regular expanders with expansion factor better than $d/2$ would give us another deterministic linear-space hardness amplification. We recall Trevisan's derandomized Direct Product construction below. We note that a similar definition has been used in the construction of codes in several works beginning with [1] and more recently in [6, 8].

Definition 16. *Given a d-regular graph G on 2^n vertices, where each vertex is identified with an n-bit string, and a Boolean function $f : \{0,1\}^n \to \{0,1\}$, we define a function $g = G(f) : \{0,1\}^n \to \{0,1\}^d$ as follows. For $x \in \{0,1\}^n$, let $N_1(x), N_2(x), \ldots, N_d(x)$ denote the d neighbors of x in G (as per some fixed ordering). Then $g(x) \stackrel{def}{=} f(N_1(x))f(N_2(x))\ldots f(N_d(x))$.*

Lemma 17. *Let $G = (\{0,1\}^n, E)$ be an efficiently (in total space $S_{expander}$) constructible d-regular $(\delta, d/2 + \gamma_d)$-expander for some $\gamma_d > 0$. Let $f : \{0,1\}^n \to \{0,1\}$ be δ-hard for deterministic time T and input-preserving space $S \geqslant S_{expander} + \Omega(d)$. Then the function $g = G(f)$ from Definition 16 is $\gamma_d\delta$-hard for deterministic time $T' = \frac{T}{d} - \mathsf{poly}(n)$ and input-preserving space $S - O(d)$.*

Thus, provided explicit expanders with expansion better than $d/2$ are known, we can apply the above amplification repeatedly to get a deterministic linear-space "worst-case to constant average-case" hardness amplification. Unfortunately, we do not know explicit expanders with expansion factor better than $d/2$. The recent work of Capalbo *et al.* [3] applies only to bipartite graphs. Beating the $d/2$ barrier for general graphs remains a challenging open question.

References

1. N. Alon, J. Bruck, J. Naor, M. Naor, and R. Roth. Construction of asymptotically good low-rate error-correcting codes through pseudo-random graphs. *IEEE Transactions on Information Theory*, 38:509–516, 1992.
2. L. Babai, L. Fortnow, N. Nisan, and A. Wigderson. BPP has subexponential time simulations unless EXPTIME has publishable proofs. *Computational Complexity*, 3:307–318, 1993.
3. M.R. Capalbo, O. Reingold, S. Vadhan, and A. Wigderson. Randomness conductors and constant-degree lossless expanders. In *Proceedings of the Thirty-Fourth Annual ACM Symposium on Theory of Computing*, pages 659–668, 2002.
4. I. Dinur. The PCP theorem by gap amplification. *Electronic Colloquium on Computational Complexity*, TR05-046, 2005.
5. O. Goldreich, N. Nisan, and A. Wigderson. On Yao's XOR-Lemma. *Electronic Colloquium on Computational Complexity*, TR95-050, 1995.
6. V. Guruswami and P. Indyk. Expander-based constructions of efficiently decodable codes. In *Proceedings of the Forty-Second Annual IEEE Symposium on Foundations of Computer Science*, pages 658–667, 2001.
7. V. Guruswami and P. Indyk. Near-optimal linear-time codes for unique decoding and new list-decodable codes over smaller alphabets. In *Proceedings of the Thirty-Fourth Annual ACM Symposium on Theory of Computing*, pages 812–821, 2002.
8. V. Guruswami and P. Indyk. Linear-time encodable and list decodable codes. In *Proceedings of the Thirty-Fifth Annual ACM Symposium on Theory of Computing*, pages 126–135, 2003.
9. R. Impagliazzo. Hard-core distributions for somewhat hard problems. In *Proceedings of the Thirty-Sixth Annual IEEE Symposium on Foundations of Computer Science*, pages 538–545, 1995.

10. R. Impagliazzo and A. Wigderson. P=BPP if E requires exponential circuits: Derandomizing the XOR Lemma. In *Proceedings of the Twenty-Ninth Annual ACM Symposium on Theory of Computing*, pages 220–229, 1997.
11. L.A. Levin. One-way functions and pseudorandom generators. *Combinatorica*, 7(4):357–363, 1987.
12. N. Nisan and A. Wigderson. Hardness vs. randomness. *Journal of Computer and System Sciences*, 49:149–167, 1994.
13. O. Reingold, S. Vadhan, and A. Wigderson. Entropy waves, the zig-zag graph product, and new constant-degree expanders. *Annals of Mathematics*, 155(1):157–187, 2002.
14. D.A. Spielman. Linear-time encodable and decodable error-correcting codes. *IEEE Transactions on Information Theory*, 42(6):1723–1732, 1996.
15. L. Trevisan. List-decoding using the XOR lemma. In *Proceedings of the Forty-Fourth Annual IEEE Symposium on Foundations of Computer Science*, pages 126–135, 2003.
16. A.C. Yao. Theory and applications of trapdoor functions. In *Proceedings of the Twenty-Third Annual IEEE Symposium on Foundations of Computer Science*, pages 80–91, 1982.

The Online Freeze-Tag Problem

Mikael Hammar[1], Bengt J. Nilsson[2], and Mia Persson[2]

[1] Apptus Technologies AB, IDEON, SE-223 70 Lund, Sweden
mikael.hammar@apptus.com
[2] School of Technology and Society, Malmö University,
SE-205 06 Malmö, Sweden
Bengt.Nilsson@ts.mah.se,
mia@cs.lth.se

Abstract. We consider the following problem from swarm robotics: given one or more "awake" robots in some metric space M, wake up a set of "asleep" robots. A robot awakens a sleeping robot by moving to the sleeping robot's position. When a robot awakens, it is available to assist in awakening other slumbering robots. We investigate offline and online versions of this problem and give a 2-competitive strategy and a lower bound of 2 in the case when M is discrete and the objective is to minimize the total movement cost. We also study the case when M is continuous and show a lower bound of 7/3 when the objective is to minimize the time when the last robot awakens.

1 Introduction

The *Freeze-tag problem* has received some attention recently [1, 2, 7]: given a set of n robots located at points in some metric space. Initially, there is one or more *awake* or *active* robots and the other robots are *asleep*, i.e., in a *stand-by mode.* The objective is for the active robots to awaken the sleeping ones. An active robot is able to awaken a sleeping robot just by touching it. Once awake, this new robot is available to assist in awakening other sleeping robots. This problem was dubbed the Freeze-tag problem by the authors of [1, 2, 7].

In this paper, we consider online versions of this problem where the sleeping robots, also denoted as *requests*, occur in an online fashion, i.e., neither the total number of requests nor their exact location in space are known in advance. Therefore, decisions, i.e., which robots to move in order to wake up sleeping robots, has to be made without any knowledge about future requests. We consider the following two online problems. The *online step dependent Freeze-tag problem* (online $SDFT(k)$ for short) has k awake robots from the start and a request is released only when the previous request has been activated or served and the objective is to minimize the total distance travelled by all the robots. The *online time dependent Freeze-tag problem* (online $TDFT$ for short), closely models the Freeze-tag problem defined in [1] but in an online setting. The requests are released independently of how many of them have already been served and our objective is to minimize the makespan.

J.R. Correa, A. Hevia, and M. Kiwi (Eds.): LATIN 2006, LNCS 3887, pp. 569–579, 2006.

The Freeze-tag problem has many applications such as data distribution, network design, broadcasting, routing and scheduling [1]. Another feature is that *SDFT*(k) and *TDFT* exhibit some elements of the k-server problem and the Traveling Salesman Problem, respectively. In the well studied k-server problem (see e.g. [4, 5]) one aims to online serve a set of requests, positioned on points in some metric space M, by k mobile servers and simultaneously minimize the total movement cost. Here a request is a point in M and the servers are located on points of M. A request r is said to be served if one of the servers lies on r. Manasse *et al.*[5] have shown that k is a lower bound on the competitive ratio of any deterministic k-server algorithm in any metric space with at least $k+1$ points, whereas the work function algorithm for the k-server problem has been shown to have a competitive ratio of at most $(2k-1)$[4]. The online step dependent Freeze-tag problem closely models the online k-server problem but with the difference that now the number of servers, i.e., active robots, is not constant; a new server appears for each served request. In this paper, we provide a 2-competitive online algorithm for this variant of the k-server problem.

The online time dependent Freeze-tag problem can be viewed as a parallel online version of the Traveling Salesman Problem, in which the cities of the TSP instance correspond to the asleep robots initial locations in time dependent Freeze-tag, and the objective is to visit all cities as rapidly as possible. Furthermore, in *TDFT* there are many salesmen working in parallel since whenever a salesman visit a city, he recruits a new salesman to help visit other cities. The time dependent Freeze-tag problem was first introduced by Arkin *et al.*[1]. They showed that in the offline case, minimizing the time when the last robot awakens, even simple versions of the problem (e.g. in star metrics) are NP-complete and they provide a PTAS for geometric instances on a set of points in any constant dimension. Sztainberg *et al.*[7] further investigate offline Freeze-tag and prove that the greedy strategy gives a tight approximation bound for the case of points in the plane and that greedy yields a $\Theta((\log n)^{1-1/\delta})$-approximation for n points in R^{δ}. Arkin *et al.*[2] recently prove the NP-hardness and present an $O(1)$-approximation for offline Freeze-tag in unweighted graphs, in which there is one asleep robot at each node. They further generalize to the case when there are multiple robots at each node and edges are unweighted and they obtain a $\Theta(\sqrt{\log n})$-approximation. In the case of weighted edges for this problem, they provide an $O((L/d)\log n + 1)$-approximation, where L is the edge of heaviest weight and d is the diameter of the graph.

1.1 Our Results

In this paper we show that offline $SDFT(k)$ can be solved in polynomial time for all possible values of the parameter k. We further give a 2-competitive algorithm in the online case and also prove a lower bound of 2 on the competitive ratio for all k. For online *TDFT* we prove a lower bound of 7/3 on the competitive ratio.

Our paper is organized as follows. In section 2 we give some basic definitions and also formally define the different variants of the Freeze-tag problems. In section 3 we consider the offline version of $SDFT(k)$ and prove that this problem can be solved in polynomial time. In section 4 we present a strategy yielding a competitive ratio of 2 for online *SDFT*(k) and then we further prove that this is in fact optimal. In section 5 we prove a lower bound of 7/3 on the competitive ratio for the *TDFT* problem.

2 Preliminaries

We begin with some notation and basic definitions.

Given a metric space M, the robots are represented as points in M. A robot can be in two different states, the *awake* state and the *asleep* state. We also call sleeping robots *requests* and say that a request is *served* when a sleeping robot is awakened by an active robot. An active robot is also denoted as a *server*. The *release time* of robot r specifies the first point in time when r can be awakened and also the first point in time when the awake robots become aware of r.

The formal definition of the problems is as follows.

Definition 1. *In the online Step Dependent Freeze-tag problem,* (SDFT(k) *for short), one is given a set of k initially awake robots located on points in some discrete metric space M and the objective is to serve all requests in such a way that the total movement cost is minimized. A new request is released only when the previous request has been served.*

In the online Time Dependent Freeze-tag problem, (TDFT *for short*), *one is given one initially awake robot, located on a point in some continuous metric space L and the objective is to serve all requests in such a way that the makespan, i.e., the time when the last robot is awakened, is minimized. An awake robot can move with at most unit speed and associated to each robot is a release time.*

Note that in the *TDFT* problem a moving robot is allowed to reconsider its choice, i.e., aborting its motion when a new request occur. This is not possible in the *SDFT*(k) problem since a new request is released only when the previous request has been served.

The performance of deterministic online algorithms is measured in comparison with the optimal offline algorithm, denoted by *OPT*, using the standard competitive ratio, see e.g. [3]. It is assumed that *OPT* knows the entire input sequence, i.e., the total number of requests and their locations in M, and can hence achieve a lower cost. Furthermore, an online algorithm A is c-competitive for a constant c, if for all input sequences σ, the following holds: $A(\sigma) \leq cOPT(\sigma)$. The infimum of all such values c over all request sequences σ is called the competitive ratio of A. The input to an online algorithm is constructed by an *adversary*. For deterministic online algorithms, a *cruel adversary* knows exactly what the online algorithm's response will be to each input element and this adversary pays the optimal offline cost. A different kind of adversary is the *adaptive-online adversary*, who must serve each request it generates before the randomized online algorithm serves the request and this adversary knows its own strategy for generating requests as well as the description of the online algorithm and all its action taken thus far (see e.g. [3] for a more comprehensive survey on adversaries).

We refer to an algorithm A as being *lazy* if (1) A moves its robot along the shortest path to a sleeping robot and (2) A moves only one robot in order to serve a request.

The following lemma proves that any non-lazy online algorithm for *SDFT(k)* can be replaced by a lazy online algorithm without increasing the algorithms total movement cost.

Lemma 1. *Any online algorithm for the SDFT(k) problem can be replaced by a lazy online algorithm without increasing the total path travelled by the robots.*

Proof. Consider a feasible solution, i.e., a sequence $\sigma = (r_1, \ldots, r_n)$ of serving robots, generated by a non-lazy algorithm A for *SDFT(k)*. We show how to construct a new feasible solution by replacing all non-lazy movements by lazy movements without increasing the total distance travelled by the robots. The construction goes as follows. First, let us assume that A moves robot $r_i \in \sigma$ in order to wake up the sleeping robot $r(j)$ and furthermore, assume that A makes its first non-lazy movement in this step, denoted as the j:th step. This non-lazy movement is clearly (at least) one of the following kind: (1) A does not move along the shortest path to serve $r(j)$, or (2) A moves more than one robot to serve $r(j)$. Now replace this non-lazy movement by a lazy movement by moving only r_i along the shortest path to robot $r(j)$ and no further. It is clear that this replacement will not increase the total distance travelled by the robots up to step j. Repeat this process of exchanging non-lazy moves by lazy ones until all moves are lazy, using the same serving robot sequence, i.e., σ, as A. Since we only consider discrete metric spaces, the exchange of non-lazy moves to lazy ones will never result in a loss of cost. □

3 Offline Step Dependent Freeze-Tag

In this section we consider the offline version of *SDFT(k)* and we show that this problem has a polynomial time solution by reducing it to maximum cardinality minimum weight matching on bipartite graphs, which is known to have polynomial time algorithms (see e.g. [6]).

We reduce *SDFT(k)* to maximum cardinality minimum weight matching on bipartite graphs. Let $R = (r_1, \ldots, r_n)$ be a set of n requests where r_j denotes the j:th request in the sequence, where j goes from 1 to n. Let $S_0 = (s_{0_1}, \ldots, s_{0_k})$ be a set of k initially awake robots. We further define two new sets $S = (s_1, \ldots, s_n)$ and $S' = (s'_1, \ldots, s'_n)$ such that $s_j = s'_j = r_j$. Thus, S and S' are copies of R. Now define a graph $G = (S_0 \cup S \cup S', R, E)$, whose vertices represent the set of requests and the set of servers, i.e., active robots that can be used to awaken the sleeping robots. Clearly, G is bipartite because the set of vertices can be partitioned into two sets, one set containing the vertices corresponding to the servers (set $S_0 \cup S \cup S'$) and the other containing the vertices corresponding to the requests (set R). Two vertices $(s_i, r_j) \in G$ (or $(s'_i, r_j) \in G$), with $1 \leq i \leq n$, are adjacent if and only if $i < j$. Also, for each $s_{0_i} \in G$, there are edges connecting s_{0_i} to every $r_j \in R$. Furthermore, an edge (s_{0_i}, r_j)

(or (s_i, r_j) or (s'_i, r_j), respectively) is assigned weight equal to the distance between request r_j and server s_{0_i} (or s_i or s'_i, respectively). Note that if r_i and r_m, for $1 \leq m \leq n$, are two distinct requests with $i < m$, that happen to occur on the same point, then the corresponding vertices in G are considered as being distinct, although the edges (s_i, r_m) and (s'_i, r_m) have zero weights. Furthermore, the edges (s_i, r_i) and (s'_i, r_i) does not exist for any $r_i \in R$. It suffices to prove that the bipartite graph G has a maximum cardinality minimum weight matching of size l if and only if the total movement cost for the robots in *SDFT*(k) is l. Let $\{e_1, \ldots, e_n\}$ be a set of edges in a maximum cardinality minimum weight matching on G. Since $|S_0| > 1$ and an edge $(s_i, s_j) \in E$ for each $i < j$ it follows that a maximium cardinality minimum weight matching covers all vertices in R exactly once, whereas a vertex in set $S_0 \cup S \cup S'$ is covered at most once. Furthermore, note that the set of edges $E \in G$ simulates the precedence constraints in *SDFT*(k), i.e., a sleeping robot cannot be used for awaking other robots before it has been awaken. Hence, $\{e_1, \ldots, e_n\}$ is a set of movements costs with minimum cost for the serving robots in *SDFT*(k). Conversely, let $\{l_1, \ldots, l_n\}$ represent a set of movements costs of minimum cost for the serving robots generated from the *SDFT*(k) execution process on graph G. Since there is only one active robot located on each vertex in $S_0 \cup S \cup S'$ in G and since each request is served exactly once it follows that $\{l_1, \ldots, l_n\}$ forms a maximum cardinality minimum weight matching in G. See Fig. 1 for an illustrating example of the reduction, where the dashed edges indicate an optimal solution of cost 7 for *SDFT*(k).

We have proved the following theorem:

Theorem 1. *The problem SDFT(k) can be solved in polynomial time.*

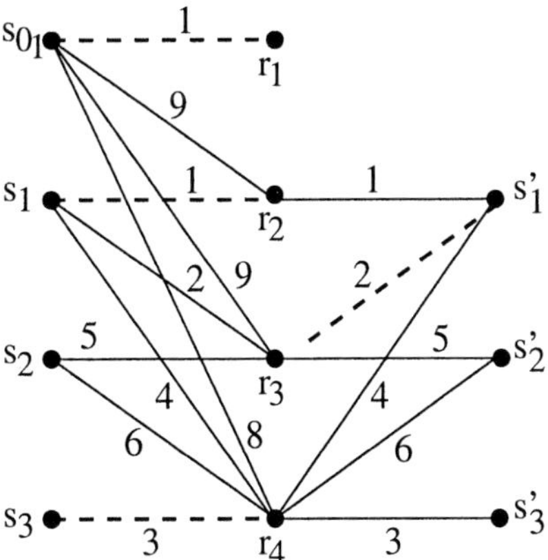

Fig. 1. Graph G for sequence of requests (r_1, r_2, r_3, r_4) with initially one active server s_0

4 Online Step Dependent Freeze-Tag

We now present a strategy, *Simple*, which achieves a competitive ratio of two for *SDFT*(k). This strategy works in arbitrary metric spaces and for all possible values of the parameter k. *Simple* works as follows. For each request in the sequence, the request is served with the closest robot which afterwards returns to its original position. Hence, *Simple* is a non-lazy strategy.

Theorem 2. *Simple is 2-competitive.*

Proof. To prove our claim about a competitive ratio of two for our strategy, we just need to consider that for any request in the request sequence, the optimal strategy must move at least the minimum distance from a previous request. *Simple* moves exactly this distance in order to serve the request and then back to its original position, thus proving the theorem. □

By using Lemma 1 *Simple* can be made lazy without increasing its competitive ratio.

We proceed by proving that *Simple* is in fact optimal. We start by giving an easy argument to why two is a lower bound for *SDFT*(k) when $k \geq 2$, i.e., there are at least two active robots initially.

Theorem 3. *There exist metric spaces for which the competitive ratio for the SDFT(k) is at least* 2*, when* $k \geq 2$.

Proof. Given a $k+1$ point unit-distance metric space M, i.e., M is a metric space with $k+1$ points such that for each pair of points p and q, distance(p, q)=1. Let A be any deterministic online algorithm for the *SDFT*(k) problem. We show that there exists a request sequence σ such that $A(\sigma) \geq 2OPT(\sigma)$. By Lemma 1, we assume that A is lazy. For the construction of σ, let the initial configuration of A consist of k active robots, occupying k distinct points in M, i.e., initially there exist one point p in M not occupied by an active robot. Now, the move for the cruel adversary is to put request r_1 on point p. Then, assume that A serves r_1 with one of its robots positioned at say point q, with q occupied by one active robot r. Hence, this movement of r leaves an empty point at position q and the next move for the cruel adversary strategy is to put the second request r_2 on position q, forcing A to move a total distance of two in order to serve σ. Clearly, $OPT(\sigma) = 1$ since OPT can move some other active robot than robot r to serve r_1 and this concludes the proof. □

Next, we prove a lower bound holding for all possible values of the parameter k, i.e, this lower bound holds also in the case when $k = 1$. We begin by defining a few concepts. A metric space is a *tree metric* if the underlying metric graph structure is a tree. The tree metric has *unit cost* if all tree edges have weight one. The distance between two points is the sum of the edge weights on the tree path between the points. Tree metrics generalize such metric spaces as star metrics and line metrics. An online algorithm for *SDFT*(k) in a tree metric is said to have the *nearness* property if it never serves a request using an active robot having some other active robot on the tree path to the request.

To show that we can restrict ourselves to consider only online algorithms having the nearness property, we first need to prove the following lemma.

Lemma 2. *For any online algorithm that solves SDFT(k) in tree metrics there is an online algorithm having the nearness property that solves the same instance with the same or better competitive ratio.*

Proof. Consider a request r that the algorithm serve with a robot r' without using nearness. This means that there is an active robot r'' on the tree path from r' to r and we can equivalently view the algorithm as moving r' to the position of r'' and let r'' serve the request r. However, this step is non-lazy and by Lemma 1 we can defer the movement of r' to the position of r'' to a later request, when it is needed. □

We are now in a position to prove our lower bound result.

Theorem 4. *There is a unit cost tree metric such that no randomized online algorithm for SDFT(k) has competitive ratio $2-\epsilon$, for any $\epsilon > 0$, against adaptive online adversaries.*

Proof. We let an adaptive online adversary construct a request sequence consisting of n requests in a unit cost tree metric where the tree initially has one or more active robots at a designated node that we denote the root r, i.e., the tree is a rooted tree. Each node in the tree has degree n and the distance from the root to each leaf node is n.

The adversary now places requests at different children of the root until the online algorithm has used all robots at the root to serve these requests. Note that by Lemma 2 we can assume that the online algorithm has the nearness property.

We denote a point that previously had active robots but now does not by a *hole* and a point that has a single active robot as a *single point*. Hence, once the initially active root robots all have served the requests at the children we have a hole at the root and an even number of active robots, two at each point having active robots. The adversary continues constructing requests in accordance with the following simple scheme S.

1. If the tree has a hole, then a request is placed at the hole.
2. If the tree has a single point, then a request is placed at a child of the single point being the root of a subtree having no active robots.

A simple induction proves that after the online algorithm has served each request, then the tree will either contain an even number of robots and a hole or an odd number of robots with one single point. All other points having active robots will have two active robots positioned at the same point. Again we rely on the fact that the online algorithm has the nearness property.

Let us divide the subsequent work of the online algorithm into stages. A *stage* begins with the online algorithm A and an adversarial online algorithm A^* both having a hole at the root of the tree and ends after a number of requests have been served when A and A^* again reach the same configuration, i.e., both have their hole at the root. The number of children of the root that have active robots we henceforth call the *r-degree*. Consider now an arbitrary stage i with r-degree $d_i \geq 2$. The adversary places a request at the root and A serves that request using an active robot at one of the children of the root, say p. A^* on the other hand serves that request using an active robot at some other child q of the root. The stage now continues with the adversary placing requests according to the simple scheme S to be served by both A and A^*. If A and A^* reach the same configuration, i.e., both have a single point at q, then the two subsequent requests will be handled by A^* so that it has a single point at the root. Without loss of generality we can assume that A does the same, thus one more request will generate a hole at the root ending the stage and starting the next stage. Figure 2 gives an example of a stage. The label of an edge is the sum of the movements cost along this edge during the stage. Note that there are always two active robots on each node (except for the root of the tree which has no active robot) after a stage.

Let us analyze the expected cost of A with respect to the expected cost of A^* during stage i. The adversary places a request at the root, which is served by A and A^*. Note that, independently of the probability distribution that A uses, the probability that A and A^* serve that request using the same active robot is $\frac{1}{d_i}$ if A^* chooses the robot to use with uniform distribution. Hence, the probability that A and A^* serve that request using active robots from two different children of the root is then $\frac{d_i-1}{d_i}$, where d_i denotes the r-degree.

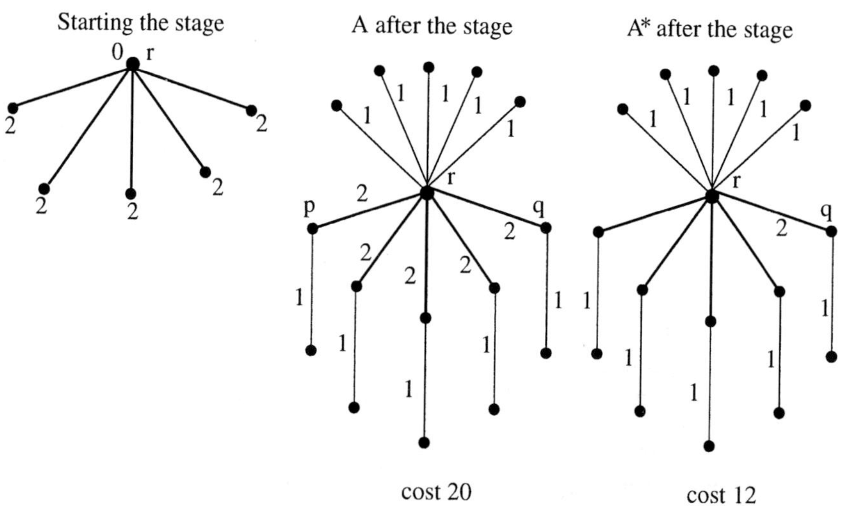

Fig. 2. Example illustrating the proof of Theorem 4

A *trial* is the sequence of steps that A does starting with a single point at the root until it has a single point at the root the next time. A trial consists of at least four steps so the expected cost of A in stage i can now be expressed as $E(A_i) \geq \sum_{j=1}^{d_i} E(A_i|j)$, where $E(A_i|j)$ denotes the conditional cost of A in stage i given that stage i consists of j trials. Then,

$$E(A_i|j) = k_{ij}\frac{1}{d_i - j + 1}\prod_{l=1}^{j-1}\frac{d_i - l}{d_i - l + 1} = \frac{k_{ij}}{d_i} , \tag{1}$$

where k_{ij} is the number of moves for A in stage i given that it consists of j trials. We know that $k_{ij} \geq j$ since each trial consists of at least four moves.

The corresponding expected cost of A^* in stage i given that A consists of j trials is

$$E(A_i^*|j) = \frac{k_{ij}/2 + 2}{d_i} , \tag{2}$$

since A_i^* will use only half the number of moves that A_i uses except in the last trial. In the last trial we can assume that both strategies use exactly four moves. In fact, we can let A_i know the moves of A_i^* when the last trial occurs so that A_i can follow A_i^*'s moves.

The ratio of the total expected costs for A and A^* is therefore bounded by

$$\begin{aligned}
\frac{E(A)}{E(A^*)} = \frac{\sum_{i=1}^{m} E(A_i)}{\sum_{i=1}^{m} E(A_i^*)} &= \frac{\sum_{i=1}^{m}\sum_{j=1}^{d_i} E(A_i|j)}{\sum_{i=1}^{m}\sum_{j=1}^{d_i} E(A_i^*|j)} = \\
\frac{\sum_{i=1}^{m}\sum_{j=1}^{d_i}\frac{k_{ij}}{d_i}}{\sum_{i=1}^{m}\sum_{j=1}^{d_i}\frac{\frac{k_{ij}}{2}+\frac{4}{2}}{d_i}} &\geq \frac{\sum_{i=1}^{m}\frac{1}{d_i}\sum_{j=1}^{d_i} k_{ij}}{\sum_{i=1}^{m}\frac{1}{2d_i}\sum_{j=1}^{d_i}(k_{ij}+4)} \geq \\
\frac{2\cdot\sum_{i=1}^{m}\frac{1}{d_i}\sum_{j=1}^{d_i} j}{\sum_{i=1}^{m}\frac{1}{d_i}\sum_{j=1}^{d_i}(j+1)} &\geq 2\cdot\frac{\sum_{i=1}^{m}\frac{1}{d_i}\cdot\frac{d_i(d_i+1)}{2}}{\sum_{i=1}^{m}\frac{1}{d_i}\cdot\frac{d_i(d_i+3)}{2}} = \\
2\cdot\frac{\sum_{i=1}^{m}(d_i+1)}{\sum_{i=1}^{m}(d_i+3)} &\geq 2\cdot\frac{\sum_{i=1}^{m}(i+1)}{\sum_{i=1}^{m}(i+3)} = \\
&2\cdot\frac{(m+3)}{(m+7)} \to 2 ,
\end{aligned} \tag{3}$$

as m increases, where m denote the number of stages. We have used the fact that $d_i \geq i$ since at each stage the r-degree increases by at least one.

Finally, note that there is one more case to handle, namely the case when m is constant. In this case, the ratio of the total expected costs for A and A^* depends on the cost of the last stage and is therefore bounded by

$$\begin{aligned}
\frac{E(A)}{E(A^*)} &\geq \frac{\sum_{i=1}^{m}\frac{1}{d_i}\sum_{j=1}^{d_i} k_{ij}}{\sum_{i=1}^{m}\frac{1}{2d_i}\sum_{j=1}^{d_i}(k_{ij}+4)} = \\
2\cdot &\frac{B + (\frac{1}{d_m})\cdot\sum_{j=1}^{d_i} k_{mj}}{C + (\frac{1}{d_m})\cdot\sum_{j=1}^{d_i}(k_{mj}+4)} \to 2 ,
\end{aligned} \tag{4}$$

as n increases, where n denotes the number of requests, and B and C are constants. □

5 Time Dependent Freeze-Tag

Let us change our problem specification for Freeze-tag. The robots are points in some general metric space and associated to each robot is a *release time*. The release time $t(r)$ specifies the first point in time when a robot r can be awakened and the awake robots become aware of r. An awake robot can move with at most unit speed to wake up sleeping robots. Hence, a request sequence is given by $\sigma = ((t(r_1), p(r_1)), \ldots, (t(r_n), p(r_n)))$, where $p(r_i)$ is the position of robot r_i. Initially one robot r_0 is awake and the objective is to find an awakening schedule that minimizes the time until all robots are awake, i.e., find the directed spanning tree of minimum height where the out degree of any point is at most two (except from the root point r_0 which has out degree one). We call such a tree an *optimal scheduling tree*. An optimal scheduling tree can be computed given a request sequence although the best known time complexity is exponential because of the NP-completeness of this problem [1].

The fact that the robots move in time requires us to use *continuous metric spaces* to accurately model the problem. This means that at any point in time a robot is positioned at some point in the metric space and robot motion is continuous.

We continue by providing a lower bound on the competitive ratio for *TDFT*.

Theorem 5. *There is a metric space such that no algorithm solves the time dependent Freeze-tag problem with competitive ratio* $7/3 - \epsilon$*, for any* $\epsilon > 0$.

Proof. Consider the graph structure of Fig. 3(a). It defines a continuous metric space where robots can move along the edges at unit speed. Points $p_1, \ldots, p_6$ are at distance three from the point p_0 having the initial active robot. The point p_7 is at distance six from p_0 and nine from the other ones.

The first request is $(0, p_0)$, i.e., at time zero a robot at p_0 is released thus giving us two active robots at p_0. These two robots are allowed to move along the edges to try to anticipate the release of subsequent robots. However, at time three the cruel adversary releases two robots, one on each of the points $p_1, \ldots, p_6$ that does not have an active robot on its corresponding edge. Without loss of generality we can assume that the requests are $(3, p_1)$ and $(3, p_2)$, thus giving us the situation depicted in Fig. 3(b)–(d).

The first active robot that serve a request will have to do so at the earliest at time six. Assume without loss of generality that this is p_1 and we look at the other robot r at time six. If r is further from p_2 than one, then nothing more happens giving a total time for the schedule of 7. The optimal schedule takes time three because the two robots can be at p_1 and p_2 respectively at time three.

On the other hand, if r is closer to p_2 than one, then a fourth request $(6, p_7)$ is generated by the cruel adversary. To serve this request, any of the robots will have to use at least a total of 14 time units. An optimal schedule will serve p_1

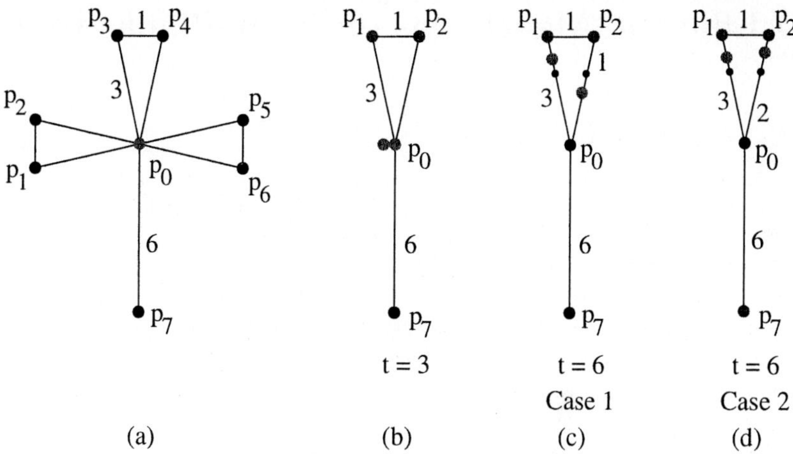

Fig. 3. Illustrating the proof of Theorem 5

and p_2 in time four with one robot and the other will serve p_7 in six time units giving a total optimal cost of six. In both cases the ratio for any algorithm is at least 7/3. □

6 Conclusions

An interesting open problem is to investigate online algorithms for time dependent Freeze-tag. Does there exists a strategy with constant competitive ratio for this problem?

References

1. E.M. Arkin, M.A. Bender, S.P. Fekete, J.S.B. Mitchell, and M. Skutella. The freeze-tag problem: How to wake up a swarm of robots. In *Proc. 13th ACM-SIAM Symp. on Discrete Algorithms*, pages 568–577, 2002.
2. E.M. Arkin, M.A. Bender, D. Ge, S. He, and J.S.B. Mitchell. Improved Approximation Algorithms for the Freeze-Tag Problem. In *Proc. 15th Annual ACM Symposium on Parallel Algorithms and Architectures*, pages 295–303, 2003.
3. A. Borodin and R. El-Yaniv. Online Computation and Competitive Analysis. Cambridge University Press, 1998.
4. E. Koutsoupias and C.H. Papadimitriou. On the k-server conjecture. *Journal of the ACM*, 42(5):971–983, 1995.
5. M. Manasse, L.A. McGeoch, D. Sleator. Competitive algorithms for server problems. *Journal of Algorithms*, 11:208–230, 1990.
6. C.H. Papadimitriou, K. Steiglitz. Combinatorial Optimization, Algorithms and Complexity. Dover Publications, Inc., Mineola, New York, 1998.
7. M.O. Sztainberg, E.M. Arkin, M.A. Bender, and J.S.B. Mitchell. Analysis of Heuristics for the Freeze-Tag problem. In *Proc. 8th Scandinavian Workshop on Algorithm Theory*, pages 270–279, 2002.

I/O-Efficient Algorithms on Near-Planar Graphs

Herman Haverkort[1,⋆] and Laura Toma[2]

[1] Department of Computing Science, Eindhoven University of Technology, PO Box 513, 5600 MB Eindhoven, The Netherlands
cs.herman@haverkort.net

[2] Department of Computer Science, Bowdoin College, 8650 College Station, Brunswick, ME 04011, United States of America
ltoma@bowdoin.edu

Abstract. Obtaining I/O-efficient algorithms for basic graph problems on sparse directed graphs is a long-standing open problem. While the best known upper bounds for most basic problems on such graphs with V vertices still require $\Omega(V)$ I/Os, optimal $O(sort(V))$ I/O algorithms are known for special classes of sparse graphs, like planar graphs and grid graphs. It is hard to accept that a problem becomes difficult as soon as the graph contains a few deviations from planarity. In this paper we extend the class of graphs on which basic graph problems can be solved I/O-efficiently. We give a characterization of near-planarity which covers a wide range of near-planar graphs, and obtain the first I/O-efficient algorithms for directed graphs that are near-planar.

1 Introduction

When working with massive graphs, only a fraction of the data can be held in the main memory of a computer. Thus, the transfer of blocks of data between main memory and disk, rather than the internal memory computation, is often the bottleneck. Therefore, developing *external-memory* or *I/O-efficient algorithms*—algorithms that specifically optimize the number of block transfers between main memory and disk, can lead to considerable runtime improvements.

I/O-efficient algorithms for graph problems has been an active area of research. Even though significant progress has been made, there is still a significant gap between the lower and the upper bound for all basic problems. Consider a directed graph (digraph) with non-negative real edge weights. A shortest path from vertex u to vertex v in G is a minimum-length path from u to v in G, where the length of a path is the sum of the weights of the edges on the path. The *single-source-shortest-paths (SSSP)* problem is to find shortest paths from a source

⋆ Part of this work was done while Herman Haverkort was at Karlsruhe University, supported by the European Commission, FET open project DELIS (IST-001907), and subsequently at Aarhus University, supported by a grant from the Danish National Science Research Council.

J.R. Correa, A. Hevia, and M. Kiwi (Eds.): LATIN 2006, LNCS 3887, pp. 580–591, 2006.

vertex s to all vertices in G. For *planar* digraphs (graphs that can be embedded in the plane such that no two edges intersect), there exist SSSP-algorithms with upper bounds on the number of block transfers that match proven lower bounds up to a constant factor. However, for general graphs, the SSSP problem is still open, as are other basic problems such as connected components (CC) and depth- and breadth-first search (DFS, BFS).

Both from a theoretical and from a practical point of view, it is hard to accept that SSSP should become extremely difficult as soon as a graph contains a few deviations from planarity. In practice, networks (e.g. transportation networks) may not be planar. However, when edges are expensive and junctions are cheap, such networks still have a strong tendency to planarity: there will be only relatively few links (e.g. motorways) that cross other edges without connecting to them. Other examples are networks in which each vertex is connected to a few nearby vertices. In such networks, there may be quite a number of crossings but they are all very 'local'. In this paper we give a characterization of near-planarity covering a wide range of near-planar graphs, and develop the first I/O-efficient algorithms for such graphs.

I/O-Model and related work. We develop I/O-efficient algorithms using the standard two-level I/O-model [2]. The model defines two parameters: M is the number of vertices/edges that fit into internal memory, and B the number of vertices/edges that fit into a disk block, where $B \leq M/2$. An *Input/Output* (or: *I/O*) is the operation of transferring a block of data between main memory and disk. The *I/O-complexity* of an algorithm is the number of I/Os it performs. The basic bounds in the I/O-model are those for scanning and sorting. The *scanning bound*, $scan(N) = \frac{N}{B}$, is the number of I/Os necessary to read N contiguous items from disk. The *sorting bound*, $sort(N) = \Theta(\frac{N}{B} \log_{M/B} \frac{N}{B})$, represents the number of I/Os required to sort N contiguous items on disk [2] when $N > M$. For all realistic values of B and $M < N$, we have $scan(N) < sort(N) \ll N$.

I/O-efficient graph algorithms have been considered by a number of authors; for a recent review see [23]. On general digraphs $G = (V, E)$ the best known algorithm for SSSP, as well as for the BFS and DFS traversal problems, use $\Omega(V)$ I/Os in the worst case[1]; their complexity is $O(\min\{(V + \frac{E}{B}) \cdot \log V + sort(E), V + \frac{V}{M}\frac{E}{B}\})$ [12, 13, 19]. On sparse graphs, which have $E = O(V)$, the best known bounds are thus $O(V)$ I/Os or worse, which is no better than just running the internal-memory algorithms in external memory. This is far from the currently best lower bound of $\Omega(\min\{V, sort(V)\} + E/B)$ I/Os, which on sparse graphs is practically $\Omega(sort(V))$.

The search for BFS, DFS and SSSP algorithms using $O(sort(E))$ I/Os on general (sparse) graphs has led to a number of improved results for special graph classes [5, 6, 7, 8, 10]. All these algorithms are based on the existence of small separators. For planar graphs, they exploit graph partitions, as introduced by Frederickson [16]. For any planar graph $\mathcal{K} = (V, E)$, given a parameter $R \leq V$, we can find a subset $V_S \subset V$ of $O(V/\sqrt{R})$ vertices, such that the removal of V_S

[1] We denote the size of a set by its name; the meaning will be clear from the context.

partitions $\mathcal{K}$ into subgraphs $\mathcal{K}_i$ such that: (1) there are $O(V/R)$ subgraphs; (2) each subgraph has size $O(R)$, and (3) (the vertices in) each $\mathcal{K}_i$ is (are) adjacent to $O(\sqrt{R})$ vertices of V_S. We call such a partition an *R-partition.* Assuming that $R \leq M/(c \log^2 B)$, for a sufficiently big constant c, an R-partition can be computed I/O-efficiently with $O(sort(V))$ I/Os [22]. On planar digraphs, using R-partitions, SSSP and BFS can be solved in $O(sort(V))$ I/Os [8], and DFS in $O(sort(V) \log \frac{V}{M})$ I/Os [10];

Our results. In this paper we extend the class of graphs that admit I/O-efficient algorithms. We introduce a class of near-planar graphs and show how to find small separators for planar subgraphs of such graphs that gracefully depend on the non-planarities. Using these separators, we develop the first I/O-efficient SSSP, BFS, DFS and topological sort algorithms for such near-planar graphs.

Our main result is the following. Let $G = (V, E \cup E_C)$ be a digraph that consists of a planar graph $\mathcal{K} = (V, E)$ and a given set of additional edges E_C; let $G_C = G - \mathcal{K} = (V_C, E_C)$ denote the non-planar part of G, where V_C is the set of vertices incident to edges in E_C. We show how to refine an R-partition of $\mathcal{K}$ to restrict the number of vertices of V_C per subgraph, while adding no more than $O(\sqrt{VV_C}/R^{1/4})$ vertices to the separator and increasing the number of subgraphs by no more than $O(V_C/\sqrt{R})$. Using refined R-partitions we show how to compute SSSP on G in $O(E_C + sort(V + E_C))$.

We generalize our result to graphs $G = (V, E \cup E_C)$ such that $\mathcal{K} = (V, E)$ can be drawn in the plane with T crossings. If we know for each edge (u, v) of $\mathcal{K}$ which edges it crosses, and in which order these crossings occur when traversing the edge from u to v, we can compute SSSP on such a graph G in $O(E_C + sort(V + T + E_C))$ I/Os.

When a graph is *near-planar* in the sense that $T = O(V)$ and $E_C = O(V/B)$, these bounds reduce to $O(sort(V))$, whereas the best known SSSP-algorithm for general graphs requires $O((V + \frac{E}{B}) \cdot \log \frac{V}{B} + sort(E)) \supset O(V)$ I/Os. If information about a suitable drawing of a graph is given, our results allow the computation of SSSP in $O(sort(E))$ I/Os on graphs with crossing number $O(E)$, on graphs that are k-embeddable in the plane for constant k, on graphs with skewness $O(E/B)$ and on graphs with splitting number $O(E/B)$. We obtain similar results for BFS, DFS, topological order and CC.

Outline. The paper is organized as follows. Sec. 2 presents refined R-partitions and Sec. 3 describes how to use these partitions to compute SSSP efficiently. Sec. 4 extends our approach to other basic graph problems. In Sec. 5 we explain how our technique could be used for problems on several types of graphs that are near-planar according to measures of planarity proposed in literature. We conclude in Sec. 6 and give directions for further research.

2 Partitioning a Near-Planar Graph

In this section we discuss how to compute small separators and extend Frederickson's R-partitions to graphs that are not planar. Consider a graph

(a) 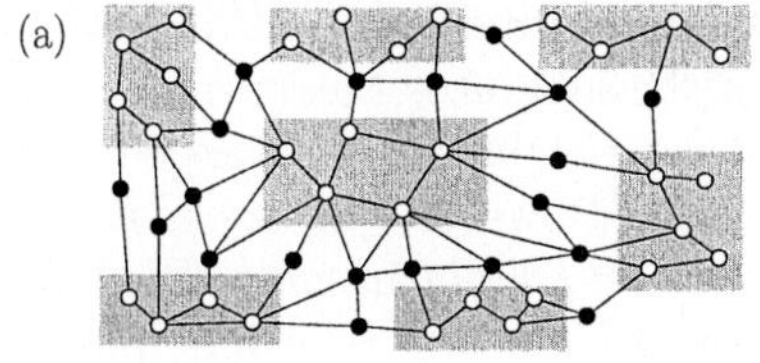 (b) 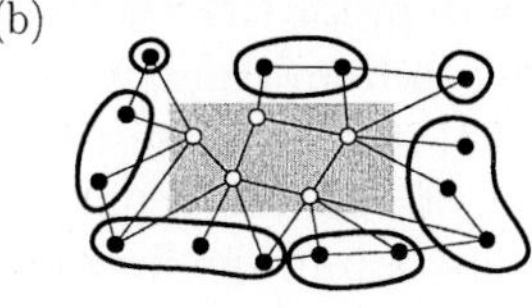

Fig. 1. (a) Partition of a planar graph into clusters (boxed) and separator vertices (black). (b) One cluster in the partition and its adjacent boundary sets.

$G = (V, E \cup E_C)$ that consists of planar subgraph $\mathcal{K} = (V, E)$ and a set of edges E_C. For this section we assume $\mathcal{K}$ to be known. Let $G_C = (V_C, E_C) = G - \mathcal{K}$ denote the non-planar part of G. We call the edges of G_C *cross-link edges*, and the vertices of G_C *cross-link vertices.* We assume that the vertices and edges in the *cross-link graph* G_C are labeled as such.

We start by computing an R-partition for $\mathcal{K} = (V, E)$, that is, a set $V_S \subset V$ of $O(V/\sqrt{R})$ vertices, such that the removal of V_S partitions $\mathcal{K}$ into subgraphs $\mathcal{K}_i$ such that there are $O(V/R)$ subgraphs, each subgraph has size $O(R)$, and is adjacent to $O(\sqrt{R})$ vertices of V_S. We use the following notation: the vertices in V_S are *separator vertices* and each of the subgraphs a *cluster*; the set of vertices in $\mathcal{K} - \mathcal{K}_i$ adjacent to $\mathcal{K}_i$ are the *boundary vertices* $\partial\mathcal{K}_i$ (or simply the *boundary*) of $\mathcal{K}_i$. We use $\overline{\mathcal{K}_i}$ to denote the graph consisting of $\mathcal{K}_i$, $\partial\mathcal{K}_i$ and the subset of edges of E connecting vertices in $\mathcal{K}_i \cup \partial\mathcal{K}_i$. The set of separator vertices can be partitioned into maximal subsets so that the vertices in each subset are adjacent to precisely the same set of clusters. These sets are the *boundary sets* of the partition. If the graph has bounded degree, which can be ensured for planar graphs using a simple transformation, there exists an R-partition with only $O(V/R)$ boundary sets [16] (Refer to Fig. 1).

The separator V_S is a separator for $\mathcal{K}$ but not necessarily for G, because any cluster in $\mathcal{K}$ may contain up to R cross-link vertices that are connected by cross-link edges to cross-link vertices in other clusters, by-passing the separator. Let G_i denote the clusters induced by $\mathcal{K}_i$ in G. A straightforward way to get a separator for G would be to add all cross-link vertices V_C to V_S; however, the SSSP algorithm of Sec. 3, run on the basis of such a separator, would use $O(E_C + V)$ I/Os.

We show how to refine the partition of $\mathcal{K}$ to incorporate the cross-link edges while ensuring that the total number of separator vertices and clusters is not too large and each cluster contains $O(\sqrt{R})$ cross-link vertices. Our approach is based on the following generalization of Lemma 2 from [16].

Lemma 1. *Given a subgraph* $G = (V, E)$ *of a planar graph with* $|\partial G| = O(\sqrt{V})$*, and a weight function* $w : V \to \mathbb{R}$ *such that* $\sum_{v \in V} w(v) = W$*, we can find a subset* $S \subset V$ *of size* $O(\sqrt{VW})$ *which separates* $G - S$ *into a set of* $O(W)$ *subgraphs (clusters)* G' *with the following properties:*

- *each cluster* $G' = (V', E')$ *has a total weight* $\sum_{v \in V'} w(v)$ *of at most 1.*
- *for each cluster* $G' = (V', E')$*, we have that* $\partial G'$ *has* $O(\sqrt{V})$ *vertices.*

Proof. The proof follows the proof of Lemma 1 and 2 from Frederickson [16], which is based on recursive application of the separator theorem by Lipton and Tarjan [21] in two phases: first with uniform weights on the vertices, and then with weights on the separator vertices only. However, we use a non-uniform weight function in the first phase. Note that we are not interested in low-weight separators: it is the weights of the clusters that count. The first phase of the recursive procedure is as follows. When G has weight $w(G)$ at most 1, we are done. Otherwise, applying Lipton and Tarjan's separator theorem, we find a subset S of at most $2\sqrt{2}\sqrt{V}$ vertices of V such that S separates $G - S$ into two clusters A and B that each have weight at most $\frac{2}{3}w(G)$. We partition the clusters A and B recursively.[2] This procedure results in a number of clusters. By construction each cluster $G' = (V', E')$ has weight at most 1, and the number of clusters is obviously $O(W)$. However, the boundary $\partial G'$ of a cluster G' may still have more than $O(\sqrt{V})$ vertices—this is solved by the second phase. But first we show that so far, the total number of vertices in the subsets S that were selected is $O(\sqrt{VW})$. Let $s(v, w)$ be the maximum number of separator vertices that may be selected while recursively partitioning a planar graph induced by a set of v vertices with weight w. Note that any of its subgraphs A and B may have total weight at most $\frac{2}{3}w$, and at least one of them has at most $v/2$ vertices. Therefore $s(v, w)$ is bounded by the following recursive expression: $s(v, w) \leq \max_{0<\alpha\leq 1/2, 1/3\leq\beta\leq 2/3} c\sqrt{v} + s(\alpha v, \beta w) + s((1-\alpha)v, (1-\beta)w)$ where $s(v, w) = 0$ if $w \leq 1$, and $c = 2\sqrt{2}$. This recursion solves to $s(V, W) = O(\sqrt{VW})$ (details in the full version of this paper). In the second phase of the procedure we recursively subdivide each cluster further until the size of its boundary is reduced to $O(\sqrt{V})$. It can be shown that this increases the number of separator vertices and the number of clusters by at most a constant factor; thus the lemma follows. □

Our algorithm first computes an R-partition of $\mathcal{K}$ in $O(sort(V))$ I/Os with the algorithm by Maheshwari and Zeh [22]; then we refine the partition by applying Lemma 1 to each cluster $\overline{G_i}$ that has more than $c\sqrt{R}$ cross-link vertices, for some fixed constant c. For each such cluster we assign weight $1/(c\sqrt{R})$ to every cross-link vertex in $\overline{G_i}$ and weight 0 to every other vertex. Thus each cluster that results from refining $\overline{G_i}$ has $O(\sqrt{R})$ cross-link vertices, $O(R)$ vertices in total, and $O(\sqrt{R})$ vertices on its boundary.

We use Lemma 1 to bound the number of separator vertices and number of clusters G' resulting from the refinement. Cluster $\overline{G_i}$ has total weight $W_i = \sum_{v\in\overline{G_i}} w(v) = |\overline{G_i} \cap V_C|/(c \cdot \sqrt{R})$. For each cluster G_i, the number of separator vertices obtained by refining it is $O((|\overline{G_i}| \cdot W_i)^{1/2}) = O(R^{1/4}(|\overline{G_i} \cap V_C|)^{1/2})$. Summed over all clusters G_i this adds $O(R^{1/4}\sum_{G_i}(|\overline{G_i} \cap V_C|)^{1/2})$ separator vertices in total. Since $2\sqrt{(a+b)/2} \geq \sqrt{a} + \sqrt{b}$, the worst case occurs if the cross-link vertices V_C are evenly distributed over the $O(V/R)$ subgraphs G_i, and we get: $R^{1/4}\sum_{G_i}(|\overline{G_i} \cap V_C|)^{1/2} \leq R^{1/4}O(V/R)O(\sqrt{V_C R/V}) = O(V/R^{3/4} + \sqrt{VV_C}/R^{1/4})$. Adding this to the $O(V/\sqrt{R})$ vertices that were already in V_S before we started refining the partition, we get a total of $O(V/\sqrt{R} + \sqrt{VV_C}/R^{1/4})$.

[2] Alternatively, one could apply the results of Aleksandrov et al. [3] for the first phase.

Similarly, the number of clusters obtained by refining each $\overline{G_i}$ is $O(W) = O(|\overline{G_i} \cap V_C|/(c \cdot \sqrt{R}))$ (by Lemma 1), and we can show that the total number of clusters is $O(V/R + V_C/\sqrt{R})$. Overall we have the following (due to space constraints the complete proof is omitted):

Theorem 1. *Let R be a parameter such that $R \leq M/(c \log^2 B)$, for a sufficiently big constant c. We can, with $sort(E)$ I/Os, find a subset $V_S \subset V$ whose removal separates $\mathcal{K}$ into a set of subgraphs G_i with the following properties:*

- *the total number of vertices in V_S is $O(V/\sqrt{R} + \sqrt{VV_C}/R^{1/4})$*
- *there are $O(V/R + V_C/\sqrt{R})$ subgraphs G_i in $\mathcal{K} - V_S$*
- *each subgraph contains $O(R)$ vertices, is adjacent to $O(\sqrt{R})$ separator vertices and contains $O(\sqrt{R})$ cross-link vertices.*

3 Computing SSSP Using the Refined *R*-Partition

We now show how to use the refined partition of a non-planar graph G obtained in Sec. 2 above to compute SSSP I/O-efficiently.

The standard approach used by I/O-efficient planar graph algorithms is as follows. Given an R-partition of a planar graph $\mathcal{K}$, we compute a substitute graph $\mathcal{K}^R$ defined on the separator vertices. The graph $\mathcal{K}^R$ is a *reduced* version of $\mathcal{K}$ (it has fewer vertices), and we construct it such that the lengths of the shortest paths in $\mathcal{K}^R$ are the same as in $\mathcal{K}$. The SSSP algorithm consists of three steps: (1) Compute $\mathcal{K}^R$; (2) Compute SSSP in $\mathcal{K}^R$ (by construction, we know these are the lengths of the shortest paths in $\mathcal{K}$); (3) Compute the shortest paths to vertices inside the clusters $\mathcal{K}_i$ of the R-partition.

To extend this approach to a non-planar graph G, we have to incorporate the cross-link (non-planar) edges E_C of G. We do this on the basis of a refined R-partition of G that divides G into subgraphs G_i, as explained in Sec. 2. Note that a shortest path between two arbitrary vertices in G enters and exits a subgraph $\overline{G_i}$ either through a boundary vertex or through a cross-link vertex. Therefore the substitute graph G^R will be defined on both the separator *and* the cross-link vertices and it contains an edge between each cross-link vertex and the boundary vertices of its cluster. Since this introduces $O(V_C\sqrt{R})$ edges in G^R, care must be taken so that the number of I/Os spent on them does not become $\Omega(V_C\sqrt{R})$.

Below we show how to exploit Theorem 1 to implement the substitute graph of a refined R-partition of a non-planar graph such that shortest paths can be computed efficiently. We will give details and prove this section's main result:

Theorem 2. *SSSP on a digraph $G = \mathcal{K} \cup G_C$ uses $O(E_C + sort(V + E_C))$ I/Os.*

3.1 The Substitute Graph

We obtain G^R as follows: First, it includes the edges between the separator vertices in the partition (that is, in G), and the edges between the cross-link vertices, i.e. the cross-link graph G_C. Second, it includes the union of all complete graphs G_i^R obtained by replacing each subgraph $\overline{G_i}$ as follows: the vertices of

G_i^R are the boundary vertices ∂G_i of G_i and the cross-link vertice $V_C \cap G_i$ of G_i, and there is an edge from u to v in G_i^R if there is a path from u to v in $\overline{G_i}$. The edge (u, v) has weight equal to the length of the shortest path from u to v in $\overline{G_i}$. Note that G_i^R contains edges between boundary vertices, between cross-link vertices and boundary vertices, and between cross-link vertices. Third, if the SSSP source vertex s is not a separator or a cross-link vertex, we add it to G^R and add edges from s to all the boundary vertices and all cross-link vertices of the subgraph G_i containing s; as above, the weight of an edge (s, v) is the length of the shortest path from s to v in $\overline{G_i}$.

Let $\delta_G(u, v)$ denote the shortest path from u to v in G. For any pair of vertices $u, v \in V_S \cup V_C \cup \{s\}$ we can show that $\delta_{G^R}(u, v) = \delta_G(u, v)$, that is, G^R maintains shortest paths between its vertices. The number of vertices in the substitute graph is $V_S + V_C + 1$, which, by Theorem 1, is $O(V/\sqrt{R} + \sqrt{VV_C}/R^{1/4} + V_C)$. By Theorem 1, there are $O(V/R + V_C/\sqrt{R})$ subgraphs in total, each of which has $O(\sqrt{R})$ boundary vertices, $O(\sqrt{R})$ cross-link vertices, and possibly a source vertex; thus each complete graph G_i^R has $O(R)$ edges in total. In total $\cup G_i^R$ has $O(V/R + V_C/\sqrt{R}) \cdot O(R) = O(V + V_C\sqrt{R})$ edges. Add the $O(V/\sqrt{R} + \sqrt{VV_C}/R^{1/4} + E_C)$ cross-link edges and edges between separator vertices in the partition, and we get:

Lemma 2. *The substitute graph G^R has $O(V/\sqrt{R} + \sqrt{VV_C}/R^{1/4} + V_C)$ vertices and $O(V + V_C\sqrt{R} + E_C)$ edges.*

We can also show that G^R can be computed in $O(scan(E) + sort(|G^R|))$. We defer the details to the full version of this paper.

3.2 Computing SSSP on G^R

To compute SSSP on G^R we use Dijkstra's algorithm, which we make I/O-efficient by modifying it to take advantage of the structure of G^R. In addition to a priority queue, we maintain a list L that stores the tentative distances from s to all the vertices in G^R, that is, in $V_S \cup V_C \cup \{s\}$. When extracting a vertex from the priority queue, we retrieve the tentative distances of its out-neighbors from L. For each out-neighbor w of v we check whether its tentative distance as stored in L is greater than $d(v)$ plus the weight of the edge (v, w); if it is, we update the distance of w in L, delete the old entry of w from the priority queue and insert a new entry for w with the updated distance in the queue.

In total, we perform $O(V(G^R)) = O(V_S + V_C)$ `ExtractMins`, and $O(E(G^R)) = O(V + V_C\sqrt{R} + E_C)$ `Deletes` and `Inserts` on the priority queue. These operations can be performed efficiently in $O(sort(V + V_C\sqrt{R} + E_C))$ I/Os using an I/O-efficient priority queue, e.g. [4]. We also perform $O(E(G^R)) = O(V + V_C\sqrt{R} + E_C)$ accesses to the list L; this is because every vertex in L is accessed once by each incoming edge in G^R. Of course, we cannot afford one I/O per edge. In order to perform the accesses to L efficiently, we store L in the following order: all vertices in V_S are at the front of L, grouped by boundary set, followed by the vertices in $V_C - V_S$, grouped by the index of the subgraph G_i that contains them. Note that with this order the vertices in the same boundary set, as well as cross-link vertices in the same cluster, are consecutive in L.

Lemma 3. *The accesses to the list L can be performed in $O(V_S+E_C+(V/\sqrt{R}+V_C)\cdot\lceil\sqrt{R}/B\rceil)$ I/Os.*

Proof. The accesses to the list L are of three types: (1) $O(E_C)$ accesses through the cross-link edges of G^R; (2) $O(V_S)$ accesses through edges between separator vertices; and (3) $O(V + V_C\sqrt{R})$ accesses through the edges in the substitute graphs G_i^R. The first two types of accesses clearly take $O(V_S+E_C)$ I/Os. We now analyze the third type of accesses to L by counting the number of accesses per boundary set (while ignoring the cross-link edges, which are counted separately in (1)). Recall that a boundary set is a maximal set of separator vertices which are adjacent to precisely the same subgraphs G_i. Every vertex $v \in V_S \cup V_C \cup \{s\}$ in G^R that is processed needs to access the tentative distances of its out-neighbors in L: that is, every separator vertex $v \in V_S$ needs to access all the boundary vertices and cross-link vertices of all subgraphs G_i adjacent to v; every vertex $v \in \{s\} \cup V_C \setminus V_S$ needs to access all the boundary vertices and all cross-link vertices in the subgraph G_i containing v. Every time a vertex in a boundary set needs to be accessed, the other vertices in the boundary set need to be accessed as well, since the vertices of a boundary set are adjacent to the same subgraphs. For simplicity, we can think of all the cross-link vertices in a subgraph G_i as an additional "boundary" set of that subgraph. Overall, each boundary set of G^R is accessed once by each of the vertices on the boundaries of the subgraphs adjacent to the boundary set, and by each of the cross-link vertices in these subgraphs. By Theorem 1, each subgraph G_i has $O(\sqrt{R})$ boundary and $O(\sqrt{R})$ cross-link vertices. Thus each boundary set is accessed $O(\sqrt{R})$ times for each adjacent subgraph.

By the planar graph argument [16] the number of boundary sets as well as the number of adjacencies between boundary sets and subgraphs G_i is asymptotically the same as the number of subgraphs G_i. Using Theorem 1 we get that the total number of accesses to boundary sets is $O(\sqrt{R}) \cdot O(V/R + V_C/\sqrt{R}) = O(V/\sqrt{R} + V_C)$. Since boundary sets are stored consecutively in L (including the "boundary" set consisting of the $O(\sqrt{R})$ cross-link vertices of a subgraph), each boundary set can be accessed in $\lceil\sqrt{R}/B\rceil$ I/Os.

Thus the accesses to boundary sets use in total $O(V/\sqrt{R}+V_C)\cdot\lceil\sqrt{R}/B\rceil$ I/Os. Adding the $O(V_S)$ accesses between separator vertices and the $O(E_C)$ I/Os to L caused by the cross-link edges (type (1) and (2)), we get a total of $O(V_S + E_C + (V/\sqrt{R}+V_C)\cdot\lceil\sqrt{R}/B\rceil)$ I/Os. □

Putting together the operations on the priority queue and the accesses to the list L (Lemma 3) we get that computing SSSP on G^R uses $O(V_S+E_C+(V/\sqrt{R}+V_C)\cdot\lceil\sqrt{R}/B\rceil + \mathit{sort}(V + V_C\sqrt{R}+E_C))$ I/Os.

The third step in the SSSP algorithm on G computes shortest paths to all vertices in $V-(V_S \cup V_C)$. In the full paper we show that this step is dominated by the previous two steps. From the above we get that the total number of I/Os to compute SSSP on G is $O(\mathit{sort}(V + V_C\sqrt{R}+E_C) + V_S + E_C + (V/\sqrt{R}$

$+V_C)\lceil\sqrt{R}/B\rceil)$, which is $O(V/\sqrt{R}+\sqrt{VV_C}/R^{1/4}+E_C+sort(V+V_C\sqrt{R}+E_C))$. Assume for simplicity that $M > B^2$. If $V_C < V/B$, we choose $R = B^2$ and the bound becomes $O(E_C + sort(V + E_C))$. If $V_C > V/B$, we choose $R = (V/V_C)^2 = O(M)$ and again get $O(E_C + sort(V + E_C))$. This concludes the proof of Theorem 2.

4 Other Graph Problems Using Refined Partitions

The ideas from the SSSP algorithm above can be extended to other algorithms on near-planar graphs. We mention results for connected components (CC), topological order and depth-first search (DFS) and leave details for the full version.

Theorem 3. *Let $G = \mathcal{K} \cup G_C$. A topological order (assuming G is a DAG) and the connected components of G (assuming G is undirected) can be computed with $O(E_C + sort(V + E_C))$ I/Os. A DFS ordering can be computed with $O(V/\sqrt{B} + E_C)$ I/Os.*

5 Planarizing Graphs

The question how close a given graph is to being planar, is much-studied and has obvious applications in, for example, graph drawing and in the manufacturing of VLSI circuits. Several generalizations of planarity and measures of planarity have been defined, including crossing number, k-embeddability in the plane, skewness, splitting number and thickness—for a survey, see Liebers [20]. The class of near-planar graphs studied in this paper includes graphs which have low crossing number, are k-embeddable for small k, have low skewness, or have low splitting number—provided information about a suitable drawing of the graph is given. We will now briefly review these measures of planarity and discuss how a near-planar graph can be preprocessed so that it can be operated on by the algorithms described in the previous sections of this paper.

Graphs with low crossing number. The crossing number of a graph $G = (V, E)$ is the minimum number of edge crossings needed in any drawing of a given graph in a plane. When a drawing with T crossings is given, it can be preprocessed so that our SSSP algorithm described in the previous sections uses $O(sort(E+T))$ I/O's. The idea is to represent each crossing i by a vertex $v(i)$, which is marked as a *crossing*. Each crossed edge (u, u'), with crossings $i_1, ..., i_n$ in order going from u towards u', is replaced by edges $(u, v(i_1)), (v(i_1), v(i_2)), ..., (v(i_{n-1}), v(i_n)), (v(i_n), u')$. The transformation can easily be carried out in $O(sort(E+T))$ I/O's.

The resulting graph is a planar graph with $O(V)$ original vertices and $O(T)$ crossing vertices, where the crossing vertices have the special property that shortest paths are not allowed to turn on such vertices. The partitioning scheme and SSSP algorithm described in the previous sections can easily be adapted to work on graphs in which some of the vertices represent such crossings. We start by applying the partitioning scheme as usual, ignoring the fact that some vertices represent crossings. After computing the refined separator V_S, we remove the crossing vertices and restore the original connectivity of the graph. When done carefully, this may make clusters and boundary sets non-planar, but it will not affect which boundary sets are adjacent to which clusters. Thus the SSSP algorithm will still work correctly within the claimed I/O-bounds, requiring $O(sort(V'))$ I/O's on such a graph, where $V' = O(V + T)$.

A graph is k-embeddable in the plane if it can be drawn in the plane so that each edge crosses at most k other edges [24]. Since a k-embeddable graph necessarily has small crossing number, the above approach can be taken.

Graphs with low skewness. The skewness of a graph $G = (V, E)$ is the minimum size of any set of edges E_C such that $G \setminus E_C$ is planar. When the skewness of a graph is $O(E/B)$ and E_C is given, our SSSP algorithm needs only $O(sort(E))$ I/Os, even if the edges and vertices in E_C form a clique with crossing number $\Theta(E^2/B^2)$.

When E_C is not given, it may be difficult to find it. Finding a minimum-size set E_C corresponds to finding a maximum-size planar subgraph of G. These are NP-complete problems [17]. When a drawing of the graph is given, we can define a *crossing graph* $G' = (V', E')$ in which V' has a vertex $v(e)$ for every edge e in G, and E' has an edge $(v(e), v(f))$ for every pair of crossing edges e and f in G. Finding a factor-two approximation of a minimum-size set E_C such that the drawing of $G \setminus E_C$ is intersection-free can be expressed as a maximal-matching problem in G', which can be solved with the randomized algorithm by Abello et al. [1]. This takes $O(sort(E')) = O(sort(T))$ I/Os (expected), where T is the number of crossings in the input graph.

Although theoretically, this transformation is not any cheaper than the one described in the previous section, it may still be advantageous because the resulting planar graph with added cross-links may be a lot smaller than a graph in which crossings are replaced by auxiliary vertices.

Graphs with small splitting number. Splitting a vertex is the process of replacing a vertex u by two vertices u_1, u_2, whereby some of the edges incident to u will be reconnected to u_1, while the remaining edges incident to u are reconnected to u_2. The splitting number of a graph is the minimum number of splittings that is needed to make the graph planar.

When the splitting number of a graph is $O(E/B)$ and the necessary splittings are given, we can solve the SSSP problem on such a graph in $O(sort(E))$ I/Os, using an approach similar to that for graphs with small skewness. Instead of running the shortest-paths algorithm on the original graph, we run it on the planar graph resulting from the splittings, augmented with a zero-weight bidirectional cross-link (u_1, u_2) for every vertex u split into u_1 and u_2.

Combining crossings and cross-links. Above we mentioned that graphs that have low crossing number can be handled efficiently by replacing crossings by special vertices, while graphs with small skewness or small splitting number can be handled efficiently by identifying a small number of cross-link edges. The two approaches can be combined: we can find shortest paths in $O(sort(E))$ I/Os on a graph that consists of $O(E/B)$ cross-links and a graph with crossing number $O(E)$, provided the cross-links and the intersections in the remaining graph are given. How to find a constant-factor approximation of a minimum-size set of cross-links such that the rest of the graph has crossing number $O(E)$, still remains as an open problem.

6 Discussion

In this paper we extended the class of graphs for which efficient SSSP computations are possible from planar graphs to several classes of near-planar graphs. Our approach yields efficient algorithms for graphs with low crossing number, low splitting number or low skewness, provided suitable drawings are given. In theory, creating suitable drawings is difficult, since identifying a maximum planar subgraph or computing the crossing number, splitting number or the skewness of a graph are NP-complete problems [15, 18, 25]. However, in many practical applications of graph algorithms, graphs are given with a drawing or suitable drawings can be produced by heuristic methods.

Even if a good drawing is given, the method to identify cross-links in a graph of low skewness as described in Sec. 5 needs to know all crossings in the drawing. The crossings would need to be given or would need to be computed: in the case of a rectilinear drawing[3] we could do so with the external-memory line segment intersection algorithm by Arge et al. [9] or the randomized algorithm by Crauser et al. [14]. One could hope to find an algorithm that can find an effective set of cross-links without computing all crossings in the drawing first. It would also be interesting to find a constant-factor approximation of a minimum-size set of cross-links such that the rest of the graph has crossing number $O(E)$, so that we may have only very few cross-links and handle the remaining crossings with auxiliary vertices as described in Sec. 5.

Furthermore, it would be interesting to look into more measures of planarity that may be exploited, for example thickness: the minimum number of planar subgraphs whose union is the original graph.

References

1. J. Abello, A. L. Buchsbaum, and J. R. Westbrook. A functional approach to external graph algorithms. *Algorithmica*, 32(3):437–458, 2002.
2. A. Aggarwal and J. S. Vitter. The Input/Output complexity of sorting and related problems. *Communications of the ACM*, 31(9):1116–1127, 1988.
3. L. Aleksandrov, H. Djidjev, H. Guo, and A. Maheshwari. Partitioning planar graphs with costs and weights. In *Proc. Workshop on Algorithm Engineering and Experimentation*, volume 2409 of *LNCS*, pages 98–110, 2002.

[3] A rectilinear drawing in itself already poses limitations. The minimum number of crossing required in a *rectilinear* drawing of the graph may be arbitrarily much greater than the crossing number for drawings with curves [11].

4. L. Arge. The buffer tree: A technique for designing batched external data structures. *Algorithmica*, 37(1):1–24, 2003.
5. L. Arge, G. S. Brodal, and L. Toma. On external memory MST, SSSP and multiway planar graph separation. *Journal of Algorithms*, 53(2):186–206, 2004.
6. L. Arge, U. Meyer, L. Toma, and N. Zeh. On external-memory planar depth first search. *Journal of Graph Algorithms*, 7(2):105–129, 2003.
7. L. Arge and L. Toma. Simplified external-memory algorithms for planar DAGs. In *Proc. Scandinavian Workshop on Algorithm Theory*, pages 493–503, 2004.
8. L. Arge, L. Toma, and N. Zeh. I/O-efficient topological sorting of planar DAGs. In *Proc. ACM Symposium on Parallel Algorithms and Architectures*, 2003.
9. L. Arge, D. E. Vengroff, and J. S. Vitter. External-memory algorithms for processing line segments in geographic information systems. In *Proc. Eur. Symp. Algorithms*, volume 979 of *LNCS*, pages 295–310, 1995.
10. L. Arge and N. Zeh. I/O-efficient strong connectivity and depth-first search for directed planar graphs. In *Proc. IEEE Symp. on Found. of Computer Sc.*, 2003.
11. D. Bienstock and N. Dean. Bounds for rectilinear crossing numbers. *Journal of Graph Theory*, 17:333–348, 1993.
12. A. L. Buchsbaum, M. Goldwasser, S. Venkatasubramanian, and J. R. Westbrook. On external memory graph traversal. In *Proc. Symposium on Discrete Algorithms*, pages 859–860, 2000.
13. Y.-J. Chiang, M. T. Goodrich, E. F. Grove, R. Tamassia, D. E. Vengroff, and J. S. Vitter. External-memory graph algorithms. In *Proc. Symposium on Discrete Algorithms*, pages 139–149, 1995.
14. A. Crauser, P. Ferragina, K. Mehlhorn, U. Meyer, and E. Ramos. Randomized external-memory algorithms for some geometric problems. In *Proc. ACM Symposium on Computational Geometry*, pages 259–268, 1998.
15. L. Faria, C. M. H. de Figueiredo, and C. F. X. de Mendon¸ca Neto. Splitting number is np-complete. *Discrete Applied Mathematics*, 108:65–83, 2001.
16. G. N. Frederickson. Fast algorithms for shortest paths in planar graphs, with applications. *SIAM Journal on Computing*, 16:1004–1022, 1987.
17. M. R. Garey and D. S. Johnson. *Computers and Intractability: A Guide to the Theory of NP-Completeness.* W H Freeman & Co, 1979.
18. M. R. Garey and D. S. Johnson. Crossing number is np-complete. *SIAM Journal on Algebraic and Discrete Methods*, 4:312–316, 1983.
19. V. Kumar and E. Schwabe. Improved algorithms and data structures for solving graph problems in external memory. In *Proc. IEEE Symposium on Parallel and Distributed Processing*, pages 169–177, 1996.
20. A. Liebers. Planarizing graphs—a survey and annotated bibliography. *Journal of Graph Algorithms and Applications*, 5(1):1–74, 2001.
21. R. J. Lipton and R. E. Tarjan. A separator theorem for planar graphs. *SIAM Journal of Applied Math.*, 36:177–189, 1979.
22. A. Maheshwari and N. Zeh. I/O-optimal algorithms for planar graphs using separators. In *Proc. Symposium on Discrete Algorithms*, pages 372–381, 2002.
23. U. Meyer, P. Sanders, and J. F. Sibeyn, editors. *Algorithms for Memory Hierarchies*, volume 2625 of *LNCS.* Springer, 2003.
24. J. Pach and G. Tóth. Graphs drawn with few crossings per edge. *Combinatorica*, 17:427–439, 1997.
25. T. Watanabe, T. Ae, and A. Nakamura. On the np-hardness of edge-deletion and -contraction problems. *Discrete Applied Mathematics*, 6:63–78, 1983.

Minimal Split Completions of Graphs*

Pinar Heggernes and Federico Mancini

Department of Informatics, University of Bergen, N-5020 Bergen, Norway
Pinar.Heggernes@ii.uib.no, Federico.Mancini@ii.uib.no

Abstract. We study the problem of adding edges to a given arbitrary graph so that the resulting graph is a split graph, called a split completion of the input graph. Our purpose is to add an inclusion minimal set of edges to obtain a minimal split completion, which means that no proper subset of the added edges is sufficient to create a split completion. Minimal completions of arbitrary graphs into chordal graphs have been studied previously, and new results have been added continuously. There is an increasing interest in minimal completion problems, and minimal completions of arbitrary graphs into interval graphs have been studied very recently. We extend these previous results to split graphs, and we give a characterization of minimal split completions, along with a linear time algorithm for computing a minimal split completion of an arbitrary input graph. Among our results is a new way of partitioning the vertices of a split graph uniquely into three subsets.

1 Introduction

Any graph can be embedded into a split graph by adding edges, and the resulting split graph is called a *split completion* of the input graph. A *minimum* split completion is a split completion with the minimum number of edges, and computing such split completions is an NP-hard problem [12]. A split completion H of a given graph G is *minimal* if no proper subgraph of H is a split completion of G. In this paper we show that a minimal split completion of a given graph can be computed in linear time.

Minimum and minimal chordal completions, also called *triangulations*, and minimum and minimal interval completions are defined analogously, by replacing split graphs with chordal graphs and with interval graphs. Computing a minimum triangulation and computing a minimum interval completion of a graph are NP-hard problems [4], [14], whereas it was shown already in 1976 that minimal triangulations can be computed in polynomial time [13]. Recently, there has been an increasing interest in minimal completion problems, which has led to faster algorithms for minimal triangulations [9], [10], [11], some of which were presented at recent years' SODA conferences, and a polynomial time algorithm for minimal interval completions [8] presented at this year's ESA conference. Minimal split completions have not been studied earlier, and with this paper we

* This work is supported by the Research Council of Norway through grant 166429/V30.

J.R. Correa, A. Hevia, and M. Kiwi (Eds.): LATIN 2006, LNCS 3887, pp. 592–604, 2006.

expand the knowledge about classes of graphs into which minimal completions of arbitrary graphs can be computed in polynomial time.

Minimal triangulations are well studied, and several characterizations of them have been given [7]. An algorithmically useful characterization is that a triangulation is minimal if and only if no single fill edge can be removed without destroying chordality of the triangulation [13] (fill edges are the edges added to the original graph to obtain a completion). This property does not hold for minimal interval completions. In this paper, we show that it holds for minimal split completions. Analogous to chordal graphs, we show that between a split graph $G_1 = (V, E_1)$ and a split graph $G_2 = (V, E_2)$ with $E_1 \subset E_2$, there is a sequence of split graphs that can be obtained by repeatedly removing one single edge from the previous split graph, starting from G_2. We characterize the fill edges that are candidates for removal when a non-minimal split completion H of an arbitrary graph G is given. Based on this, we give linear time algorithms both for computing minimal split completions, and for removing edges from a given split completion to obtain a minimal split completion.

2 Definitions and Background

All graphs in this paper are simple and undirected. For a graph $G = (V, E)$, we let $n = |V|$ and $m = |E|$. The set of neighbors of a vertex $v \in V$ is denoted by $N(v)$, and the degree of a vertex v is denoted by $d(v) = |N(v)|$. We distinguish between subgraphs and induced subgraphs. In this paper, a *subgraph* of $G = (V, E)$ is a graph $G_1 = (V, E_1)$ with $E_1 \subseteq E$, and a *supergraph* of G is a graph $G_2 = (V, E_2)$ with $E \subseteq E_2$. We will denote these relations informally by the notation $G_1 \subseteq G \subseteq G_2$ (proper subgraph relation is denoted by $G_1 \subset G$). The complement of G is denoted by $\bar{G}$.

A subset K of V is a *clique* if K induces a complete subgraph of G. A subset I of V is an *independent set* if no two vertices of I are adjacent in G. We use $\omega(G)$ to denote the size of a largest clique in G, and $\alpha(G)$ to denote the size of a largest independent set in G.

G is a *split graph* if there is a partition $V = I + K$ of its vertex set into an independent set I and a clique K. Such a partition is called a *split partition* of G. There is no restriction on the edges between vertices of I and vertices of K. The partition of a split graph into a clique and an independent set is not necessarily unique. The following theorem from [6] states the possible partition configurations.

Theorem 1. (Hammer and Simeone [6]) *Let G be a split graph whose vertices have been partitioned into an independent set I and a clique K. Exactly one of the following conditions holds:*

(i) $|I| = \alpha(G)$ *and* $|K| = \omega(G)$
(in this case the partition $I + K$ is unique),

(ii) $|I| = \alpha(G)$ *and* $|K| = \omega(G) - 1$
(in this case there exists a vertex $x \in I$ such that $K \cup \{x\}$ is a clique),

(iii) $|I| = \alpha(G) - 1$ *and* $|K| = \omega(G)$
(in this case there exists a vertex $y \in K$ *such that* $I \cup \{y\}$ *is independent).*

The following theorem characterizes split graphs, and we will use condition *(iii)* to characterize minimal split completions. For this result, note that a simple cycle on k vertices is denoted by C_k and that a complete graph on k vertices is denoted by K_k. Thus $2K_2$ is the graph that consists of 2 isolated edges.

Theorem 2. (Földes and Hammer [3]) *Let* G *be an undirected graph. The following conditions are equivalent:*

(i) G *is a split graph.*
(ii) G *and* $\bar{G}$ *are chordal graphs.*
(iii) G *contains no induced subgraph isomorphic to* $2K_2, C_4$ *or* C_5.

Note that every induced subgraph of a split graph is also a split graph. For a given arbitrary graph $G = (V, E)$, a split graph $H = (V, E \cup F)$, with $E \cap F = \emptyset$, is called a *split completion* of G. The edges in F are called *fill edges*. H is a *minimal* split completion of G if $(V, E \cup F')$ fails to be a split graph for every proper subset F' of F.

A graph is *chordal* if it contains no induced simple cycle of length at least 4. A graph is *interval* if sets of consecutive integers can be associated with its vertices such that two vertices are adjacent if and only if their associated sets intersect. (Minimal) chordal and interval completions of a given graph are defined analogously to (minimal) split completions. Chordal completions are also called *triangulations*. Both interval graphs and split graphs are chordal. For a chordal graph G, $\alpha(G)$ and $\omega(G)$ can be computed in linear time [5], whereas these are NP-hard problems for general graphs.

3 Sandwiching a Split Graph Between Two Given Split Graphs

Given two chordal graphs $G_1 = (V, E_1)$ and $G_2 = (V, E_2)$, such that $E_1 \subset E_2$, Rose, Tarjan, and Lueker [13] showed that there is an edge in $E_2 \setminus E_1$ whose removal from G_2 results in a chordal graph. A consequence of this result is that a given triangulation H of an arbitrary graph G is minimal if and only if it is impossible to obtain a chordal graph by removing a single fill edge from H. In this section, we show that analogous results hold for split graphs and minimal split completions.

The following two corollaries can be deduced directly from Theorem 1, and their proofs are omitted in this extended abstract.

Corollary 1. *Let* $G = (V, E)$ *and* $G' = (V, E')$ *be two split graphs with* $E \subseteq E'$, *and let* $V = I + K$ *and* $V = I' + K'$ *be two split partitions of* G *and* G', *respectively. Then* $|K' \cap K| \geq |K| - 1$.

Corollary 2. *Let $G = (V, E)$ and $G' = (V, E')$ be two split graphs with $E \subseteq E'$, and let $V = I + K$ and $V = I' + K'$ be two split partitions of G and G', respectively. Then $K' \setminus I \subseteq K$.*

Lemma 1. *Given two split graphs $G = (V, E)$ and $G' = (V, E \cup F)$ such that $E \cap F = \emptyset$, there is an edge $f \in F$ that can be removed from G' so that the result is a split graph.*

Proof. Let $V = I + K$ be a split partition of G, and let $V = I' + K'$ be a split partition of G'. If there is an edge $f \in F$ with one endpoint in I' and one endpoint in K', then f can be removed, and the resulting graph is split with split partition $V = I' + K'$. Assume for the rest of the proof that there is no fill edge between I' and K'.

We define the set $T = K' \cap I$, namely those vertices that belong to an independent set in the partition of G and to a clique in the partition of G'. According to our assumption, there is no fill edge between T and I'. Thus each edge in F has either both endpoints in T or is between a vertex of T and a vertex of $K' \setminus T$, since $K' \setminus T$ was already a clique in G, by Corollary 2. It follows that if $F \neq \emptyset$ then $T \neq \emptyset$, and all vertices in T must be incident to some edge of F in G'. Since T is a part of an independent set in G and a part of a clique in G', there are fill edges between each pair of vertices in T. If $|T| = 1$ the fill edges connect T to $K' \setminus T$.

By Corollary 1 we now have two possible situations: either $|K' \cap K| = |K|$ or $|K' \cap K| = |K| - 1$.

Assume first that $|K' \cap K| = |K|$. Then $K \subseteq K'$, and consequently $I' \subseteq I$. This means that no vertex of T is adjacent to a vertex of I' in G', because there can neither be original edges between these two sets since $T \cup I' = I$, nor edges from F since $T \subset K'$. In such a situation it is possible to pick a vertex $y \in T$ incident to one or more edges in F, and remove any of the fill edges incident to y. Doing this the graph will remain split because we can still partition it in an independent set $I' \cup \{y\}$ and a clique $K' \setminus \{y\}$. We proved above that such vertex y must exist.

Let now $|K' \cap K| = |K| - 1$. Then there must be a vertex x that in G belongs to K and in G' belongs to I', such that $(I' \setminus \{x\}) \subseteq I$. Now, each vertex of T can be adjacent to at most one vertex of I', namely x. If there is at least one vertex $y \in T$ which is not adjacent to x, then we can proceed as the previous case. If all vertices of T are adjacent to x, then $N(x) = K'$, so we can just swap x with any vertex in $y \in T$ incident to an edge of F, and remove this edge, since it now connects the independent set to the clique. Swapping the vertices we make a new partition where x is in the clique and y in the independent set, and thus the result is a split graph.

Note that if $F = \emptyset$ then it means that $G = G'$ and either $I = I'$ and $K = K'$, or $I + K$ and $I' + K'$ are two possible partitions of the same split graph G.

Corollary 3. *Given two split graphs $G = (V, E)$ and $G' = (V, E \cup F)$ with $E \cap F = \emptyset$, there is a sequence of split graphs G_0, G_1, G_2, ..., $G_{|F|}$ such that*

G_{i-1} is obtained by removing edge f_i from G_i, for $1 \leq i \leq |F|$, where $G_0 = G$, $G_{|F|} = G'$, and $F = \{f_1, f_2, ..., f_{|F|}\}$.

Theorem 3. *Given an arbitrary graph G and a split completion G' of G, G' is a minimal split completion if and only if no single fill edge can be removed from G' without destroying the split property.*

Proof. If G' is a minimal split completion then no subset of its fill edges can be removed, so no single fill edge can be removed either. If G' is not a minimal split completion, another split graph G'' exists between G and G'. Then by Lemma 1, there is a single fill edge that can be removed from G' while preserving the split property.

Thus we have a characterization of minimal split completions. We will use this to give another characterizations of minimal split completions and to describe the fill edges that can be removed from non-minimal split completions in Section 5. First, in the next section, we define a new way of partitioning the vertices of a split graph uniquely.

4 Unique 3-Partitions of Split Graphs

In this section, as an alternative to split partitions, we define another way of partitioning the vertices of a split graph that will be useful to decide whether a given split completion is minimal or not. We will call the new partition a *split 3-partition.*

When we are given a non-minimal split completion of an arbitrary graph, according to Lemma 1, the redundant fill edges can be removed one by one until we reach a minimal split completion. The edges that can be removed without problems are the ones connecting the independent set with the clique in the split partition of the completion. However, since this partition is not necessarily unique, and since we do not know the underlying minimal split completion, problems occur according to cases *(ii)* and *(iii)* of Theorem 1. To avoid this ambiguity we define a third set of vertices in the graph, that we will call Q. In case we do not have a unique split partition, this set will contain those vertices that can be chosen to be either in the independent set or in the clique, determining different partitions.

Definition 1. *Given a split graph $G = (V, E)$ that has no unique split partition, we define a* split 3-partition *$V = S + C + Q$ of G as follows:*

$$S = \{v \in V \mid d(v) < \omega(G) - 1\}$$
$$C = \{v \in V \mid d(v) > \omega(G) - 1\}$$
$$Q = \{v \in V \mid d(v) = \omega(G) - 1\}$$

If G has a unique split partition $V = I + K$, we do not need such a 3-partition, but for completeness, we define $S = I$, $C = K$, and $Q = \emptyset$ in this case, so that a split 3-partition is always defined. (Note that there can be vertices of degree $\omega(G) - 1$ in G also when its split partition is unique.) For a split graph

G, $\omega(G)$, $\alpha(G)$, and the corresponding maximum clique and independent set can be computed in linear time [5]. Thus it can be decided by Theorem 1 whether G has a unique split partition or not. Hence the 3-partition of a split graph is uniquely defined.

Lemma 2. *Let $G = (V, E)$ be a split graph with no unique split partition, and let $V = S + C + Q$ be the 3-partition of G. Then*

(i) $S \subseteq I$ and $C \subseteq K$, for every split partition $V = I + K$ of G.
(ii) $Q \neq \emptyset$.
(ii) Q is exactly the set of vertices each of which belongs to a clique and to an independent set in two different split partitions of G, respectively.

Proof. *(i)* Let $V = I + K$ be any split partition of G. By Theorem 1, each vertex of K belongs also to a clique of maximum size, and thus has degree at least $\omega(G) - 1$. Therefore, a vertex that has degree less than $\omega(G) - 1$ cannot belong to K, and it must belong to I. A vertex of I can be adjacent to at most $\omega(G) - 1$ vertices, because otherwise we have a clique of size $\omega(G) + 1$. Thus, a vertex that has degree more than $\omega(G) - 1$ must belong to K.

(ii) Let $V = I + K$ be any split partition of G. By Theorem 1, either there is a vertex x in I such that $K \cup \{x\}$ is a clique, or there is a vertex y in K such that $I \cup \{y\}$ is an independent set. In either case, each such vertex x or y is adjacent to all vertices of K and to no other vertex, and by Theorem 1, it has degree exactly $\omega(G) - 1$. Thus $Q \neq \emptyset$.

(iii) By the argument in *(ii)* every vertex that can be moved between an independent set and a clique in some split partition of G must have degree $\omega(G) - 1$. Let us show that each vertex of degree $\omega(G) - 1$ can indeed be moved between partitions. Let $V = I + K$ be any split partition of G, and let v be a vertex of degree $\omega(G) - 1$. Assume first that $v \in I$. If $|K| = \omega(G) - 1$, then by moving v from I to K, we get another split partition of G. If $|K| = \omega(G)$, then we know by Theorem 1 that a vertex x of K can be moved to I to give a different split partition. Thus x cannot be adjacent to v. We can swap v and x between I and K, and get a new split partition. Assume now that $v \in K$. If $|K| = \omega(G) - 1$, then there must be a vertex z in I that is adjacent to all vertices of K. Thus z is the only neighbor of v outside of K. We can swap z and v and get another partition. If $|K| = \omega(G)$ then v has no neighbors in I and we can move v from K to I and get a new partition.

Corollary 4. *A split graph $G = (V, E)$ has a unique partition $V = I + K$ if and only if there are exactly $\omega(G)$ vertices of degree $> \omega(G) - 1$.*

Corollary 5. *Let $G = (V, E)$ be a split graph with 3-partition $V = S + C + Q$. Then every vertex of Q is adjacent to all vertices of C and to no vertex of S.*

Lemma 3. *Let $G = (V, E)$ be a split graph with 3-partition $V = S + C + Q$. Then one of the following is true:*

(i) Q is a clique and $|C| + |Q| = \omega(G)$.
(ii) Q is an independent set, $|C| = \omega(G) - 1$, and $|Q| \geq 2$.

Proof. If Q is empty, there is nothing to prove. Assume that $|Q| \geq 1$, so that the split partition of G is not unique. If $\omega(G) > 1$ (otherwise the graph is a set of disconnected vertices and they would be all in Q) then $|C| + |Q| \geq \omega(G)$, so we can distinguish two situations: $|C| + |Q| = \omega(G)$ or $|C| + |Q| > \omega(G)$.

If $|C| + |Q| = \omega(G)$ then $|Q|$ is a clique, because the largest clique in G must have size $\omega(G)$ and it can only be obtained by adding to C all vertices of Q.

If $|C| + |Q| > \omega(G)$ then we will show that $|C| = \omega(G) - 1$ and Q is an independent set. If $|C| < \omega(G) - 1$ then $|Q| > 2$, and there must be at least a subset $Q' \subset Q$ that is a clique of size $\omega(G) - |C| \geq 2$. All vertices of Q' have degree $\omega(G)-1$, and since they make a clique of size $\omega(G)$ with C, they cannot be adjacent to any vertex in $Q \backslash Q'$. The vertices in $Q \backslash Q'$ must be an independent set, or it would not be possible to make a split partition $V = (C \cup Q') + (S \cup (Q \backslash Q'))$, so they are adjacent only to C (by Corollary 5) and consequently each of them has degree at most $\omega(G) - 2$, contradicting the fact that they belong to Q. So we must have $|C| = \omega(G) - 1$ and $|Q| \geq 2$, which implies that Q is an independent set, since otherwise we would get a too large clique in G by Corollary 5.

Corollary 6. *Let $G = (V, E)$ be a split graph with 3-partition $V = S + C + Q$ and $Q \neq \emptyset$. Then in any split partition $V = I + K$ of G, at least $\omega(G) - 1 - |C|$ vertices of Q belong to K.*

Lemma 4. *Let $G = (V, E)$ be a split graph with 3-partition $V = S + C + Q$ and $q \neq \emptyset$. If $|Q| = 1$, then $|C| = \omega(G) - 1$ and $|S| \geq 2$. If $|C| + |Q| = \omega(G)$ and $|Q| > 1$, then $|S| \geq 1$.*

Proof. Every vertex of C has degree greater than $\omega(G) - 1$, so every vertex of C has at least one neighbor in S (since $|C| + |Q| = \omega(G)$), but every vertex of S has degree at most $\omega(G) - 2$, so at least two of them are needed to be connected to C in the first case, and at least one in the last.

In the next section, we will use these results to characterize minimal split completions of arbitrary graphs.

5 Characterizing Minimal Split Completions

Assume that we are given an arbitrary graph $G = (V, E)$ and a split completion $H = (V, E \cup F)$ of G. We want to find a sufficient and necessary condition for H to be a minimal split completion of G. First we identify the fill edges that can be removed from any non-minimal split completion. Note that, when $V = S + C + Q$ is the 3-partition of H, any fill edge is either incident to a vertex of $S \cup Q$ or both of its endpoints belong to C.

Lemma 5. *Let $H = (V, E + F)$ be a split completion of an arbitrary graph $G = (V, E)$, and let $V = S + C + Q$ be the 3-partition of H. Then any fill edge incident to a vertex in $S \cup Q$ can be removed so that the resulting graph is split.*

Proof. We will prove that there is a split partition $V = I + K$ of H such that any fill edge incident to a vertex of $S \cup Q$ has one endpoint in I and one endpoint in K, and we know that such edges can be removed. We also know that all vertices of S belong to the independent set of any split partition of H, and all vertices of C belong to the clique of any split partition of H. This means that any edge between S and C can be removed. Let us then assume that there are no fill edges connecting S and C. Remember also that there are no edges between S and Q. Let us also assume that the partition is not unique, so that $Q \neq \emptyset$. Under these assumptions, each fill edge incident to a vertex of $S \cup Q$ can only be between two vertices of Q or between a vertex of Q and a vertex of C, and we have the following cases.

Case 1: Q is a clique and there is a fill edge between two vertices $x, y \in Q$. If Q is a clique, then we can make a partition where at most one of the vertices of Q is chosen to be in the independent set of a partition and all the others must be in the clique. If we put x (or y) in the independent set and y (or x) in the clique, there will be a fill edge (xy) between the independent set and the clique, that we can remove.

Case 2: There is a fill edge between Q and C. If a fill edge is between a vertex $x \in Q$ and a vertex $y \in C$, since we can always choose at least one vertex of Q to be in the independent set of a partition regardless of whether Q is a clique or an independent set, let us choose exactly x. Since y is in C it will always be in the clique of any partition of H, so we now have a fill edge connecting a vertex of the independent set (x) to a vertex of the clique (y), and we can remove it.

Note that Lemma 5 does not mean that all fill edges incident to $S \cup Q$ can be removed. We are guaranteed to be able to remove one such edge. After that the 3-partition of the resulting graph might change, and thus the set of fill edges that can be removed might also change. In the next section, we will describe precisely how the sets S, C, and Q might change after removing an unnecessary fill edge.

Lemma 6. *Let $H = (V, E + F)$ be a split completion of an arbitrary graph $G = (V, E)$, and let $V = S + C + Q$ be the 3-partition of H. If each fill edge has both its endpoints in C, then H is a minimal split completion of G.*

Proof. Assume that G, H, S, C, and Q are as in the premise of the lemma such that all fill edges of H have both their endpoints in C. Thus if $F \neq \emptyset$ then $|C| \geq 2$ and $\omega(H) \geq 2$. We show that removing any single fill edge from H results in a non-split graph.

If Q is empty and thus H has a unique split partition, then by Theorem 1 and Corollary 4, $|C| = \omega(H)$, no vertex of S is adjacent to whole C, and every vertex of C has a neighbor in S. Hence $|S| \geq 2$. If $|C| = 2$ then the single edge in C is a fill edge. Removing it we would get a $2K_2$, because we can pick two vertices in S adjacent each to only one vertex of C. If $|C| > 2$ then removing a fill edge we get two nonadjacent vertices $x, y \in C$. Now, x and y must each have a neighbor in S. If they have a common neighbor w, then we can find a vertex $v \in C$ which is not adjacent to w, since no vertex of S is adjacent to every vertex of C. But x and y are both adjacent to v, so this results in an induced cycle w, x, v, y, w of length 4. If they do not have a common neighbor, then there exist $w, z \in S$, where w is adjacent to x and not to y, and z is adjacent to y and not to x, So removing the edge between x and y we get a $2K_2$.

If $Q \neq \emptyset$ then S can be even empty or disconnected from C. Let us work on Q and C using Lemma 3 and 4.

In the case when $|C| + |Q| = \omega(H)$, we have that Q is a clique and $S \neq \emptyset$. If $|Q| = 1$, then $|C| = \omega(H) - 1$, so there must be at least 2 vertices in S adjacent to vertices of C, and no vertex of S is adjacent to every vertex of C, so we can use the same argument as above. If $|Q| > 1$, then $|C| < \omega(H) - 1$. Thus every vertex of C has a neighbor in S, and either we have the previous case, or there is a vertex $z \in S$ adjacent to all vertices in C. Recall that every vertex of Q is connected to all vertices of C and to no vertex of S. Let us now take any $x, y \in C$ and remove the edge xy. We can find a vertex $w \in Q$, adjacent to both x and y so that the subgraph induced by $\{z, x, w, y\}$ is a cycle of length 4.

In the case when $|C| + |Q| > \omega(H)$, we have that Q is an independent set, and $|C| = \omega(H) - 1$. In this case S can be empty. However, since $|C| \geq 2$, then $\omega(H) \geq 3$ and $|Q| \geq 2$. Since every vertex of Q is adjacent to all vertices of C, we can find two vertices w and z in Q, such that if we remove any fill edge xy from C, we get an induced cycle w, x, z, y, w of length 4.

Theorem 4. *Let $H = (V, E + F)$ be a split completion of an arbitrary graph $G = (V, E)$, and let $V = S + C + Q$ be the 3-partition of H. H is a minimal split completion of G if and only if all fill edges have both endpoints in C.*

Proof. One direction follows from Lemma 6. For the other direction assume that H is minimal. Then no single fill edge can be removed without destroying split property. Given the 3-partition $V = S + C + Q$, then each fill edge can have one endpoint in $S \cup Q$ and one endpoint in C, or both endpoints in C or in Q. This is because there cannot be fill edges between S and Q (by Corollary 5), and within S (it is an independent set). By Lemma 5 a fill edge incident to vertices in $S \cup Q$ can always be removed, so since the completion is minimal, the only possible fill edges are the ones in C.

6 Obtaining a Minimal Split Completion from a Given Split Completion

In the next section we will give an algorithm that computes a minimal split completion of any given graph. However, for some applications it might be desirable to compute a minimal split completion of that fits within an already given split completion. This problem has been studied and solved for minimal triangulations [1], [2], and we solve it for split completions in this section.

Assume that we are given an arbitrary graph G, a split completion H of G, and the 3-partition $S + C + Q$ of H. By the results of the previous section, we know how to decide whether H is a minimal split completion, and if not, we know that any single fill edge incident to $S \cup Q$ can be removed. After this removal, the sets S, C, and Q might change. So a straight forward algorithm to remove redundant fill edges from H to obtain a minimal split completion $M \subseteq H$ of G, would be to remove a fill edge incident to $S \cup Q$, recompute the 3-partition of the resulting split completion, and continue until a minimal split completion is reached. In this section, we will show that a new 3-partition of the intermediate graph does not have to be recomputed from scratch, and that a minimal split completion can be reached in time linear in the size of H.

Theorem 5. *Let $H = (V, E \cup F)$ be a split completion of an arbitrary graph $G = (V, E)$, with $F \cap E = \emptyset$. A minimal split completion M of G, such that $G \subseteq M \subseteq H$, can be computed in time $O(|V| + |E| + |F|)$.*

Proof. Here we give a sketch of the proof.

Let $V = S + C + Q$ be the 3-partition of H. Let pq be a fill edge that can be removed, H' the graph that is the result of removing pq from H, and $V = S' + C' + Q'$ the 3-partition of H'. We know that pq can be of three types. Based on this, for each fill edge that can be removed, we have to analyze all possible cases, and show that in each case, S', C', and Q' can be computed from S, C, and Q in constant time. Since we have a constant number of cases to check for each fill edge that can be removed, and since we can check in constant time whether a particular fill edge can be removed by the results of Section 5, the total number of steps will be at most F, and the total work we do will be linear in the size of H. To ease the notation, we let $w = \omega(H)$ and $w' = \omega(H')$.

The rest of this proof is a tedious case analysis of all the possibilities for the sets that p and q belong with the combination of whether Q is an independent set or a clique, and the number of vertices in Q. We leave this part to the full paper.

7 Computing a Minimal Split Completion Directly

In this section, we show that minimal split completion of a given graph can be computed in time linear in the size of the input graph. A simple and intuitive method to embed an arbitrary graph $G = (V, E)$ into a split graph, is to select a maximal independent set I of G and add edges to make $V \setminus I$ into a clique. Unfortunately, this procedure does not guarantee that the resulting split completion is minimal; it can add edges even to a graph that is already split, in particular if the graph does not have a unique partition. However, it can be modified to compute a minimal split completion, by choosing vertices of minimum degree first when computing the maximal independent set. We call this modified algorithm *MinimalSplit* and present it below.

Algorithm MinimalSplit
Input: An arbitrary graph $G = (V, E)$;
Output: A minimal split completion $H = (V, E \cup F)$ of G.
$I = \emptyset$; $K = \emptyset$; Unmark all vertices;
while there are unmarked vertices in V **do**

```
    Choose an unmarked vertex v with minimum degree in G;
    Mark and add v to I;
    Mark and add all neigbors of v to K;
  end-while
  Make K into a clique adding a set F of fill edges;
  H = (V, E ∪ F);
```

Lemma 7. *Given an arbitrary graph $G = (V, E)$, the graph $H = (V, E \cup F)$ computed by Algorithm MinimalSplit is a minimal split completion of G.*

Proof. Let $V = I + K$ be the split partition of H computed by the algorithm. By construction, I is an independent set, K is a clique, and no edges are added between I and K, so H is a split graph. It follows from Lemma 6 that if $V = I+K$ is a unique partition of H, then H is a minimal split completion.

Let us consider the case when $V = I+K$ is not a unique partition of H, and let $V = S+C+Q$ be the 3-partition of H. If $|K| < 2$ then no edges are added by the algorithm, so the completion is trivially minimal. Assume therefore that $|K| \geq 2$. By construction, $K = \bigcup_{v \in I} N(v)$, so every vertex of K has a neighbor in I. It follows that $|K| = \omega(H)-1$. Otherwise (if $|K|$ were $\omega(H)$), there would be $\omega(H)$ vertices in K with degree greater then $\omega(H)-1$, contradicting Corollary 4. Since the split partition is not unique, there is at least one vertex z of degree $\omega(H)-1$ in I, that can be moved to K by Theorem 1. Such vertices z belong to Q, but they are not adjacent to any fill edge. Consequently, the only possibility for the completion to be non-minimal is that a vertex $x \in K$ incident to a fill edge, has degree $\omega(H) - 1$ so that x belongs to Q. Thus x has exactly one neighbor in I. This means that there is exactly one vertex $y \in I$ of degree $\omega(H) - 1$, since vertices of degree $\omega(H) - 1$ in I must be adjacent to all vertices of K. So we have exactly one vertex $y \in I$ of degree $\omega(H) - 1$ and a vertex $x \in K$ of degree $\omega(H) - 1$, such that $N(x) \cap I = \{y\}$. The degree of x in the input graph G is actually less than $\omega(H) - 1$, because it is incident to at least one fill edge. But the degree of y is $\omega(G) - 1$ also in G since no edges are added to vertices in I. This means that $d(x) < d(y)$ in G, but a vertex can be in K only if one of its neighbors in G has been selected before it to be in I. In this case, since the only neighbor of x in G selected to be in I is y, it means that y has been processed by the algorithm before x, but that is a contradiction because $d(x) < d(y)$, and the algorithm always chooses the vertex with minimum degree among the unprocessed ones.

This means that any graph obtained by the algorithm is a minimal split completion of the input graph by Lemma 6.

Let us consider the time complexity of this algorithm. Since we add edges only between the vertices of K, we can actually skip the step of adding edges, because the resulting split partition will uniquely define the edges of H. Thus the algorithm can be modified to return just I and K (the edges between I and K are all edges of G). The degrees are computed only in the beginning of the algorithm, and need not be recomputed. This and the rest of the algorithm clearly require at most $\sum_{v \in V} d(v)$ steps, which sums up to time $O(|V| + |E|)$.

8 Conclusion

We have given a characterization of minimal split completions and we have shown how to compute minimal split completions in linear time. We have also given an algorithm for computing a minimal split completion between the input graph G and an already given non-minimal split completion H of G. To achieve these goals, we introduced a new way of uniquely partitioning the vertices of a split graph into three subsets instead of two.

With these results, polynomial time algorithms are now known for minimal triangulations, minimal interval completions, and minimal split completions of arbitrary graphs. There are other interesting graph classes into which any graph can be embedded by adding edges. We are interested in knowing whether minimal completions into these classes can be computed in polynomial time. Also, is there a graph class $\mathcal{C}$ that is recognizable in polynomial time, such that minimal $\mathcal{C}$ completion of arbitrary graphs is an NP-hard problem?

Acknowledgement

We thank Jan Arne Telle for an initial discussion on the topic.

References

1. A. Berry, J-P. Bordat, P. Heggernes, G. Simonet, and Y. Villanger. A wide-range algorithm for minimal triangulation from an arbitrary ordering. *J. Algorithms.* To appear.
2. J. R. S. Blair, P. Heggernes, and J. A. Telle. A practical algorithm for making filled graphs minimal. *Theor. Comput. Sci.*, 250:125–141, 2001.
3. S. Földes and P. L. Hammer. Split graphs. *Congressus Numerantium*, 19:311–315, 1977.
4. M. R. Garey and D. S. Johnson. *Computers and Intractability.* W. H. Freeman and Co., 1978.
5. M. C. Golumbic. *Algorithmic Graph Theory and Perfect Graphs.* Academic Press, 1980.
6. P. L. Hammer and B. Simeone. The splittance of a graph. *Combinatorica*, 1(3):275–284, 1981.
7. P. Heggernes. Minimal triangulations of graphs - a survey. *Dicrete Mathematics.* To appear.
8. P. Heggernes, K. Suchan, I. Todinca, and Y. Villanger. Minimal interval completions. In *Algorithms - ESA 2005.* Springer Verlag, 2005. LNCS to appear.
9. P. Heggernes, J. A. Telle, and Y. Villanger. Computing minimal triangulations in time $O(n^\alpha \log n) = o(n^{2.376})$. In *Proceedings of SODA 2005 - 16th Annual ACM-SIAM Symposium on Discrete Algorithms*, pages 907–916, 2005.
10. D. Kratsch and J. Spinrad. Between $O(nm)$ and $O(n^\alpha)$. In *Proceedings of SODA 2003 - 14th Annual ACM-SIAM Symposium on Discrete Algorithms*, pages 709–716, 2003.
11. D. Kratsch and J. Spinrad. Minimal fill in $o(n^3)$ time. 2004. Submitted.

12. A. Natanzon, R. Shamir, and R. Sharan. Complexity classification of some edge modification problems. *Disc. Appl. Math.*, 113:109–128, 2001.
13. D. Rose, R.E. Tarjan, and G. Lueker. Algorithmic aspects of vertex elimination on graphs. *SIAM J. Comput.*, 5:146–160, 1976.
14. M. Yannakakis. Computing the minimum fill-in is NP-complete. *SIAM J. Alg. Disc. Meth.*, 2:77–79, 1981.

Design and Analysis of Online Batching Systems

Regant Y.S. Hung and Hing-Fung Ting

Department of Computer Science,
The University of Hong Kong, Pokfulam, Hong Kong
{yshung, hfting}@cs.hku.hk

Abstract. In this paper, we study the design and analysis of online batching systems. In particular, we analyze the tradeoff relationship between the start-up delay and the efficient usage of resources in an online batching system, and analyze how the delay affects the performance of such system. We derive almost optimal upper and lower bounds on the competitive ratio of any deterministic scheduling algorithm for online batching systems. Our results cover in a general way many different batching systems and give interesting insights into the effect of start-up delay.

1 Introduction

A batching server is one that can process more than one job simultaneously. The jobs that are processed together form a batch, and all jobs in a batch share a server and start and complete at the same time. The model of a system of batching servers, or simply a batching system, has been proposed and studied extensively in the Operations Research community [2, 3, 9, 17] because many traditional production processes, such as burn-in process in VLSI manufacturing, diffusion process in semiconductor fabrication and heat treatment in metalworking can all be modeled as batching systems. This paper studies the design and analysis of batching systems that are online and non-preemptive. Online batching systems arise in many IT applications such as On-demand data broadcasting [11, 15, 16] and Video-on-demand systems [1, 8, 10].

Besides their practical importance, online batching systems are interesting theoretically because they embody a source of tension that does not exist in non-batching systems, namely, the tension between efficient usage of resources and timely completion of jobs. To take advantage of the sharing capability of the batching servers, the systems may ask their clients to be patient and wait for some moments so that different job requests can be aggregated in a batch and processed together. However, to maintain a good quality of services, the system also promises that clients do not need to wait longer than some specific *startup delay* [5]. Note that this delay has a profound effect on the cost-effectiveness of the systems. On one hand, a longer delay magnifies the sharing effect and also allows the scheduler to peep longer into the future to make better online scheduling decisions; this improves throughput and profits. On the other hand,

J.R. Correa, A. Hevia, and M. Kiwi (Eds.): LATIN 2006, LNCS 3887, pp. 605–616, 2006.

a shorter delay provides better quality of services and attracts more users to pay and use the systems.

In this paper, we study mathematically how the startup delay affects the performance of the systems. We also analyze the tradeoff relationship between the startup delay and the number of servers for maintaining some fixed level of performance of the systems. Our results are stated as a function of the laxity of the system, which is defined to be the ratio between the startup delay and the job length. Note that different batching systems may have different laxity; a Video-on-demand system has small laxity, while other traditional batching systems have larger laxity. Stating our results in terms of laxity not only allows us to study different batching systems in a unified way but also gives deeper insight into the effect of the start-up delay. In our study, we assume that all jobs have the same length. As pointed out in [15], this assumption is realistic for On-demand data broadcasting and Video-on-demand systems; most of the documents and movies are about the same length. The insights gained in our study can be used as a foundation for studying more complicated systems.

The Model. A batching system is specified by the tuple (d, S, c, F, ℓ) where $d > 0$ is the *start-up delay*, S is a set of identical *servers*, c is the *server capacity*, F is a set of *job families* and $\ell > 0$ is the *job length.* Only jobs from the same job family can be batched and processed by a server at the same time, and a batch can have no more than c jobs. Once processing of a batch has been initiated at some time t, no jobs can be removed or added to the batch until the server completes the processing at $t + \ell$. Every job request r can be specified by a tuple *(time , family, profit)* where *time* and *profit* are positive real numbers and *family* $\in F$. We denote the profit of r by $p(r)$. If r arrives at time t, then the system will gain a profit of $p(r)$ from r if it starts being processed within the startup delay, i.e., start being processed at or before $t + d$; the system cannot accept r after $t + d$. For example, an On-demand data broadcasting system is a (d, S, ∞, F, ℓ)-system where S is the set of broadcasting channels, F is the set of data pages, and ℓ is the page length.

Let $\mathcal{B}$ be a batching system. Given any sequence of job requests σ, we say that a schedule is a *$\mathcal{B}$-schedule* for σ if it specifies, for any server τ in $\mathcal{B}$, when τ will start processing the requests in σ in such a way that if we follow this schedule, τ is always available when it is needed in this schedule. The profit of the schedule is the total profit of the requests accepted by the schedule. An algorithm $\mathtt{A}$ is a *$\mathcal{B}$-scheduling algorithm* if it always produces $\mathcal{B}$-schedules, and it is online if its decision made at any time t depends only on the requests that arrive at or before t. We say that $\mathtt{A}$ has *competitive ratio* κ if for any request sequence σ, we have $\mathtt{A}(\sigma) \geq \frac{1}{\kappa}\mathtt{Opt}(\sigma)$, where $\mathtt{A}(\sigma)$ and $\mathtt{Opt}(\sigma)$ are respectively the profits of the schedules produced by $\mathtt{A}$ and an optimal offline $\mathcal{B}$-schedule. Let $\mathcal{B}'$ be another batching system. We say that $\mathtt{A}$ has *relative competitive ratio* κ *with respect to* $\mathcal{B}'$ if for any request sequence σ, we have $\mathtt{A}(\sigma) \geq \frac{1}{\kappa}\mathtt{Opt}_{\mathcal{B}'}(\sigma)$, where $\mathtt{A}(\sigma)$ and $\mathtt{Opt}_{\mathcal{B}'}(\sigma)$ are respectively the profits of the $\mathcal{B}$-schedule produced by $\mathtt{A}$ and an optimal offline $\mathcal{B}'$-schedule. Note that $\mathtt{A}$'s competitive ratio is just its relative competitive ratio with respect to $\mathcal{B}$.

Our Results. Let $\mathcal{B} = (d, S, c, F, \ell)$ be any batching system. We define the *laxity* of $\mathcal{B}$ to be the ratio $\alpha = d/\ell$. In this paper, we derive lower bounds on the competitive ratio κ of any $\mathcal{B}$-scheduling algorithm in terms of α. For general α, we prove that $\kappa \geq \frac{\lfloor\alpha\rfloor+2}{\lfloor\alpha\rfloor+1+\frac{1}{H_c}}$ where $H_c \approx \ln c$ is the cth Harmonic number. We have tightened our analysis for the case when $\alpha < 1$: we show that in this case, the lower bound can be improved to $\kappa \geq (\lceil 1/\alpha \rceil - 1)(1 - c^{\frac{-1}{\lceil 1/\alpha \rceil}})$.

For upper bounds, we design a $\mathcal{B}$-scheduling algorithm G and analyze its relative competitive ratio (and hence its competitive ratio) with respect to any batching system $\mathcal{B}' = (d', S', c, F, \ell)$. Let $\kappa_{\mathcal{B}'}$ be the relative competitive ratio of G with respect to $\mathcal{B}'$ and let $\alpha' = d'/\ell$ be the laxity of $\mathcal{B}'$. We prove that

$$\kappa_{\mathcal{B}'} \leq 1 + \left\lceil \frac{1 + d'/\Delta}{\lfloor d/\Delta \rfloor} \right\rceil \frac{|S'|}{\eta}$$

where $\Delta = \min\{d, \ell\}$ and $\eta = \left\lfloor \frac{|S|}{\lceil \ell/d \rceil} \right\rfloor = \left\lfloor \frac{|S|}{\lceil 1/\alpha \rceil} \right\rfloor$.[1]

	LB	UB
$0 < \alpha < 1$	$\max\{2, \lceil \frac{1}{\alpha} \rceil - 1\}$	$1{+}2 \lceil \frac{1}{\alpha} \rceil$
$\alpha \geq 1$	$1 + \frac{1}{1+\lfloor\alpha\rfloor}$	3

Fig. 1. Bounds on competitive ratios

$0 < \alpha < 1$	$1 + \lceil 1 + \frac{\alpha'}{\alpha} \rceil \lceil \frac{1}{\alpha} \rceil \frac{|S'|}{|S|}$
$\alpha \geq 1$	$1 + \left\lceil \frac{1+\alpha'}{\lfloor\alpha\rfloor} \right\rceil \frac{|S'|}{|S|}$

Fig. 2. G's relative competitive ratio

To help digest these bounds, we have summarized our results in Figures 1 and 2. Note that we have simplified the bounds by considering the cases $0 < \alpha < 1$ and $\alpha \geq 1$ separately, and assuming $c = \infty$ (for lower bounds) and ℓ/d divides $|S|$ (for upper bounds). Figure 1 shows that G is almost optimal. (Recall that G's competitive ratio is just its relative competitive ratio with respect to $\mathcal{B}$. Note that our formula gives an upper bound of 4 instead of 3 when $1 < \alpha < 2$. However, a more careful analysis on the boundary conditions for this case will reduce the bound back to 3. Details will be given in the full paper.) Furthermore, the bounds reveal the effect of the start-up delay on system performance. In particular, when $0 < \alpha < 1$, or equivalently, when $0 < d < \ell$, the competitive ratio is proportional to the inverse of the laxity; hence, the smaller the laxity, the poorer the performance guarantee.

Figure 2 reveals a tradeoff relationship between the number of servers and start-up delay that provides valuable information for a system designer to decide the right system configuration for a cost-effective system. For example,

[1] Note that $\eta \geq 1$ when $d \geq \ell$ and the bound becomes ∞ when $\eta = 0$, or equivalently, when $|S| < \ell/d$. It can be proved that for this case, there is no algorithm A that has finite competitive ratio. For example, consider the simplest case when $|S| = 1$ and $c = \infty$. Let σ be a request sequence with a request r at time 0 and an infinite number of requests at $d + \epsilon < \ell$. To ensure a finite competitive ratio, A must accept r lest there is no more request; then A cannot accept the following ones and has an infinite competitive ratio.

consider a system $\mathcal{B}'$ that has 5 servers and laxity 1/2. The top-right entry of Figure 1 guarantees that for any input σ, we can always produce a schedule with profit at least $\frac{1}{\kappa}\mathtt{Opt}_{\mathcal{B}'}(\sigma) = \frac{1}{5}\mathtt{Opt}_{\mathcal{B}'}(\sigma)$. On the other hand, the top entry of Figure 2 asserts that for another system $\mathcal{B}$ with 15 servers and laxity 1/4, the relative competitive ratio of the $\mathcal{B}$-scheduling algorithm G with respect to $\mathcal{B}'$ is 5. This suggests that we still gain a profit of at least $\frac{1}{5}\mathtt{Opt}_{\mathcal{B}'}(\sigma)$ if we reduce the laxity to 1/4 and increase the number of servers to 15. As another example, consider a system $\mathcal{B}'$ with laxity 10 and another system $\mathcal{B}$ with laxity 11. Suppose both systems have the same number of servers. Figure 1 suggests that using system $\mathcal{B}'$, we have a profit of at least $\frac{1}{3}\mathtt{Opt}_{\mathcal{B}'}(\sigma)$, while Figure 2 asserts that the $\mathcal{B}$-scheduling algorithm G has relative competitive ratio 2 with respect to $\mathcal{B}'$. In other words, by increasing the laxity from 10 to 11, we can improve the profit guarantee from $\frac{1}{3}\mathtt{Opt}_{\mathcal{B}'}(\sigma)$ to $\frac{1}{2}\mathtt{Opt}_{\mathcal{B}'}(\sigma)$. This improvement may give sufficient scientific justification for the service provider to reduce service charges to encourage the clients to be more patient.

Related Work. In [14], Goldwasser gave the first formal study on the effect of start-up delay on competitiveness. He focused on systems with a single non-batching server (i.e., $|S| = 1$ and $c = 1$). Unlike our model, he assumed that jobs may have different job lengths and start-up delays. He defined the *patience* τ to be the smallest number such that every job has a start-up delay at least $\tau\ell_J$. He proved that when all jobs have the same length, the competitive ratio for scheduling such system is exactly $1 + \frac{1}{\lfloor\tau\rfloor+1}$. Later, Bar-Noy, Garay and Herzberg [4] studied online scheduling of a Video-on-demand system, which is an example of an online batching system. Their results imply that if the laxity $\alpha \le \sqrt{s/u}$, then the competitive ratio $\kappa = \Theta(\ln u/s)$ where u is the total number of potential users and s is the total number of servers in the system, and if $\alpha = \Theta(1)$, then $\kappa = \Theta(1)$. Bar-Noy, Guha, Katz, Noar, Schieber and Shachnai [6] studied the scheduling of offline batching systems in which all the jobs belonging to the same family must have the same job length but jobs belonging to different families may have different job lengths. They designed an approximation algorithm that has approximation ratio 4. There were also studies on the relationship between start-up delay and resource requirements for other types of systems such as VOD systems that support stream-merging [5] and systems in which a job can be divided and processed by more than one server [7, 12, 13].

2 Definitions and Notations

For any server $\rho \in S$, we say that ρ is *dedicated to* some family $f \in F$ at time t if ρ starts to process a batch B of jobs belonging to the family f at t. For every job in B, we say that the request on that job *is accepted* at t. Let σ be any sequence of job requests. For any time interval I, let σ_I denote the sequence of requests in σ that arrive during I. For example, $\sigma_{[0,t)}$ is the sequence of requests in σ arriving before t, and $\sigma_{[t,t]}$ are those arriving at t. Note that there may be more than one request arriving at the same time and $\sigma_{[t,t]}$ may contain more than one request. For any family $f \in F$, we say that f *is in* σ_I if there is a request in σ_I

that is on some job belonging to f. Furthermore, we say that f has i requests in σ_I if there are exactly i requests in σ_I on jobs in f. For example, suppose that $\sigma_I = \{r_1, r_2, r_3, r_4\}$ and only r_1 and r_4 are requesting jobs in f. Then f is in σ_I and it has 2 requests in σ_I.

3 Lower Bounds on the Competitive Ratio

In this section, we derive two lower bounds on the competitive ratio of any online deterministic algorithm for scheduling (d, S, c, F, ℓ). We first derive a general lower bound, and then tighten our analysis to derive better lower bounds for some specific ranges of α. In our proofs, all jobs have a profit of 1 and hence the profit of a schedule is just the total number of jobs accepted by the schedule.

3.1 A General Lower Bound

In this section, we show that the competitive ratio of any online (d, S, c, F, ℓ)-scheduling algorithm is at least $\frac{\lfloor\alpha\rfloor+2}{\lfloor\alpha\rfloor+1+\frac{1}{H_c}}$ where $\alpha = d/\ell$ is the laxity of the batching system and $H_c = 1+1/2+\cdots+1/c$ is the cth Harmonic number. First, we give some definition.

Definition 1. *Let s, h be any positive integer. We say that a request sequence σ has an (s, h)-peak at time t_o if*

1. *there are exactly s different families in $\sigma_{[t_o,t_o]}$ and each of them has exactly h requests in $\sigma_{[t_o,t_o]}$, and*
2. *every family in $\sigma_{[0,t_o]}$ has no more than h requests.*

To prove the lower bound, we need to prove the following technical lemma, which reveals that given some input sequence σ with a peak at t_o, a competitive online algorithm has to accept many requests of σ at or before t_o.

Lemma 1. *Let $\mathtt{A}$ be an online (d, S, c, F, ℓ)-scheduling algorithm and σ be a request sequence that has an $(|S|, h)$-peak at time $t_o \geq 0$ where $h \leq c$. Suppose that $\mathtt{A}$ has competitive ratio κ. Then, given σ as input, the number m of requests accepted by $\mathtt{A}$ at or before t_o must be at least $\epsilon|S|h$ where $\epsilon = \frac{\lfloor d/\ell\rfloor+2}{\kappa} - (\lfloor d/\ell\rfloor+1)$.*

Proof. Let $\alpha = d/\ell$, $r = d - \lfloor\alpha\rfloor\ell$ and $t_1 = t_o + \ell - r$. Note that $0 \leq r < \ell$ and $t_o < t_1$. Consider the following request sequence δ, which is obtained by modifying σ as follows:

1. $\delta_{[0,t_o]} = \sigma_{[0,t_o]}$.
2. For any time $t > t_o$ and $t \neq t_1$, $\delta_{[t,t]}$ has no request.
3. $\delta_{[t_1,t_1]}$ has $(\lfloor\alpha\rfloor + 1)|S|$ different families, each with h requests.
4. $\delta_{[0,t_1)}$ and $\delta_{[t_1,t_1]}$ do not have any family in common.

It can be verified that every family has no more than h requests in δ.

First we estimate $\mathtt{A}(\delta)$, the number of requests in δ that $\mathtt{A}$ accepts. For any time interval I, let N_I be the number of requests in δ that $\mathtt{A}$ accepts during I.

Since A is online, it cannot distinguish σ and δ at or before t_o. Thus, A will also accept m requests of δ at or before t_o; in other words, $N_{[0,t_o]} = m$. Note that any server $s \in S$ can accept at most $(\lfloor\alpha\rfloor + 1)h$ requests in the time interval $(t_o, t_o + (\lfloor\alpha\rfloor + 1)\ell]$ because s can be dedicated at most $\lfloor\alpha\rfloor + 1$ times in the interval and each time s can accept at most h requests (recall that every family has at most h requests in δ). It follows that $N_{(t_o,t_o+(\lfloor\alpha\rfloor+1)\ell]} \le (\lfloor\alpha\rfloor + 1)|S|h$. Finally, note that $N_{(t_o+(\lfloor\alpha\rfloor+1)\ell,\infty)} = 0$ because all requests in δ arrive at or before t_1 and they cannot be accepted after $t_1 + d = t_o + (\lfloor\alpha\rfloor + 1)\ell$. Therefore,

$$\mathtt{A}(\delta) = N_{[0,t_o]} + N_{(t_o,t_o+(\lfloor\alpha\rfloor+1)\ell]} + N_{(t_o+(\lfloor\alpha\rfloor+1)\ell,\infty)} \le m + (\lfloor\alpha\rfloor + 1)|S|h. \quad (1)$$

Now, we estimate $\mathtt{Opt}(\delta)$, the maximum number of requests in δ that can be accepted. By the construction of δ and the fact that σ has an $(|S|, h)$-peak at t_o, we conclude that $\delta_{[t_o,t_o]}$ has $|S|$ families, each with h requests. Together with the requests in $\delta_{[t_1,t_1]}$, δ has at least $(\lfloor\alpha\rfloor + 2)|S|$ families, each with h requests. We can accept all these $(\lfloor\alpha\rfloor + 2)|S|h$ requests as follows. At time t_o, we dedicate all servers in S to serve the $|S|$ families in $\delta_{[t_o,t_o]}$. Since each of these families has h requests and $h \le c$, we can accept all the $h|S|$ requests in $\delta_{[t_o,t_o]}$. When we finish serving these requests at $t_o + \ell$, we start to accept the $(\lfloor\alpha\rfloor + 1)|S|h$ requests in $\delta_{[t_1,t_1]}$ during the time interval $[t_o + \ell, t_o + (\lfloor\alpha\rfloor + 1)\ell] = [t_1 + r, t_1 + d]$ by accepting $|S|h$ requests at $t_o + \ell$, another $|S|h$ requests at $t_o + 2\ell$, ..., and the last $|S|h$ requests at $t_o + (\lfloor\alpha\rfloor + 1)\ell = t_1 + d$. Hence, we have

$$\mathtt{Opt}(\delta) \ge (\lfloor\alpha\rfloor + 2)|S|h. \quad (2)$$

Since A has competitive ratio κ, (1) and (2) give us a lower bound on m as follows:

$$\frac{(\lfloor\alpha\rfloor + 2)|S|h}{m + (\lfloor\alpha\rfloor + 1)|S|h} \le \frac{\mathtt{Opt}(\delta)}{\mathtt{A}(\delta)} \le \kappa,$$

or equivalently,

$$m \ge \left(\frac{\lfloor\alpha\rfloor + 2}{\kappa} - (\lfloor\alpha\rfloor + 1)\right)|S|h = \epsilon|S|h. \qquad \square$$

Note that ϵ tends to 1 when κ tends to 1. This suggests that given any request sequence that has some peak at t_o, an online (d, S, c, F, ℓ)-scheduling algorithm A must accept many requests at or before t_o in order to be competitive. This gives us an idea to construct a difficult input sequence for A: we construct a request sequence with many peaks during some time interval I of length smaller than ℓ. Since a server that is dedicated at time t cannot be dedicated again before $t + \ell$, a server can be used at most once during I. It follows that A has at most $|S|$ servers to handle the peaks and thus cannot serve too many requests at these peaks. Together with Lemma 1, we can make the conclusion that A cannot be competitive. The proof of the following theorem formalizes this idea.

Theorem 1. *Any online (d, S, c, F, ℓ)-scheduling algorithm* A *has competitive ratio κ no smaller than $\frac{\lfloor\alpha\rfloor+2}{\lfloor\alpha\rfloor+1+\frac{1}{H_c}}$ where $\alpha = d/\ell$ and H_c is the cth Harmonic number.*

Proof. Let $\rho < \ell/c$ be any positive real number. Let σ be the request sequence in which

1. $\sigma_{[t,t]}$ is empty for all $t \notin \{\rho, 2\rho, \ldots, c\rho\}$, and
2. for each $1 \le h \le c$, σ has an $(|S|, h)$-peak at time $h\rho$, and
3. for any $1 \le i \ne j \le c$, $\sigma_{[i\rho,i\rho]}$ and $\sigma_{[j\rho,j\rho]}$ have no family in common.

Suppose that σ is given as input for A to schedule. For any time interval I, let N_I denote the number of requests of σ that A accepts during I, and for any integer $i > 0$, let s_i denote the number of times A dedicates a server to accept some family during time interval $((i-1)\rho, i\rho]$. Note that every family has no more than i requests in $\sigma_{[0,i\rho]}$ and this implies $N_{((i-1)\rho,i\rho]} \le is_i$. Furthermore, note that $N_{[0,0]} = 0$. Hence, for any $1 \le h \le c$,

$$N_{[0,h\rho]} = N_{[0,0]} + N_{(0,\rho]} + N_{(\rho,2\rho]} + \cdots + N_{((h-1)\rho,h\rho]} \le s_1 + 2s_2 + \cdots + hs_h.$$

Since σ has an $(|S|, h)$-peak at $h\rho$, we can apply Lemma 1 and further conclude that, for any $1 \le h \le c$,

$$s_1 + 2s_2 + \cdots + hs_h \ge N_{[0,h\rho]} \ge \epsilon|S|h. \tag{3}$$

where ϵ is as in Lemma 1. Below, we will use (3) to prove by induction that for all $1 \le i \le c$:

$$s_1 + s_2 + \cdots + s_i \ge \epsilon|S|(1 + 1/2 + \cdots + 1/i) = \epsilon|S|H_i. \tag{4}$$

Then, by noting that $s_1 + s_2 + \cdots + s_c$ is the total number of times that A dedicates some server to a family during the time interval $(0, c\rho] \subset (0, \ell)$, and that a server can be dedicated at most once in this interval, we have

$$|S| \ge s_1 + s_2 + \cdots + s_c \ge \epsilon|S|H_c = \left(\frac{\lfloor\alpha\rfloor + 2}{\kappa} - (\lfloor\alpha\rfloor + 1)\right)|S|H_c.$$

Simplify the above inequality and the theorem follows.

To prove (4) by induction, first note that from (3), we have $s_1 \ge \epsilon|S|$ and (4) holds for $i = 1$. Suppose it holds for all integers $1 \le i < h$ where $h \le c$. Then, we have

$$\sum_{1 \le i < h} (s_1 + s_2 + \cdots + s_i) \ge \epsilon|S| \sum_{1 \le i < h} H_i = \epsilon|S|(hH_h - h). \tag{5}$$

Adding (3) to (5), we get $h(s_1 + s_2 + \ldots + s_h) \ge \epsilon|S|h + \epsilon|S|(hH_h - h)$. It follows that (4) also holds for h, and hence it holds for all $1 \le i \le c$. □

3.2 A Better Lower Bound When $d < \ell$

Note that for the case when $\alpha < 1$, Theorem 1 asserts that the competitive ratio of any online algorithm cannot be smaller than $\frac{2}{1+1/H_c}$. In this section, we refine our analysis and improve this bound to $\Omega(\ell/d)$, which is much larger than 2 when d is much smaller than ℓ. The framework for proving this tighter lower bound is the same as that for the general case; we construct an input sequence which has many peaks within an interval I of length no greater than ℓ. Our observation for the improvement is that when $d < \ell$, a competitive online algorithm has to serve much more requests for these peaks because most of the requests will be expired after this interval I. To make this observation precise, we modify Lemma 1 as follows.

Lemma 2. *Let* $\mathtt{A}$ *be an online* (d, S, c, F, ℓ)*-scheduling algorithm. Let* $h \leq c$ *be any integer and* σ *be a request sequence such that (i) it has an* $(|S|, h)$*-peak at time* t_o*, and (ii) there is no request in* $\sigma_{(t_o,t_o+d]}$*. Suppose* $\mathtt{A}$ *has competitive ratio* κ*. Then, given* σ *as input, the number of requests accepted by* $\mathtt{A}$ *at or before* t_o+d *must be at least* $|S|h/\kappa$*.*

Proof. Suppose that $\mathtt{A}$ has accepted m requests of σ at or before $t_o + d$. Let τ be any input sequence where $\tau_{[0,t_o+d]} = \sigma_{[0,t_o+d]}$ and there is no request in $\tau_{(t_o+d,\infty)}$. Since $\mathtt{A}$ is online, it cannot distinguish τ and σ; it follows that $\mathtt{A}(\tau) = m$. On the other hand, note that we can accept all the $|S|h$ requests at t_o because $h \leq c$ and we have $|S|$ servers. Since $\mathtt{A}$ has competitive ratio κ, we have

$$m = \mathtt{A}(\tau) \geq \frac{1}{\kappa}\mathtt{Opt}(\tau) \geq |S|h/\kappa,$$

and the lemma follows. □

To make the basic ideas transparent, we first prove below a somewhat weaker lower bound on the competitive ratio. Then, we explain how to improve the bound.

Theorem 2. *Let* $\mathtt{A}$ *be an online* (d, S, c, F, ℓ)*-scheduling algorithm with competitive ratio* κ*. Suppose that* $d < \ell$ *and* $2^{\lceil \ell/d \rceil - 1} \leq c$*. Then, we have* $\kappa \geq \frac{1}{2}\lceil \ell/d \rceil$*.*

Proof. Let $\rho = \frac{\ell}{\lceil \ell/d \rceil - 1}$ and $m = \lceil \ell/d \rceil - 1$. Let σ be a request sequence in which

1. $\sigma_{[t,t]}$ is empty for all $t \notin \{\rho, 2\rho, \ldots, m\rho\}$, and
2. for each $1 \leq h \leq m$, σ has an $(|S|, 2^h)$-peak at time $h\rho$.

Suppose that σ is given to $\mathtt{A}$ for scheduling. For any time interval I, let N_I denote the number of requests $\mathtt{A}$ accepts during I. Note that $d < \rho$ and it follows that $N_{[0,d]} = 0$. For any $1 \leq i \leq m$, let s_i be the number of times that $\mathtt{A}$ dedicates a server to serve some family during $((i-1)\rho + d, i\rho + d]$. It can be verified that every family in $\sigma_{[0,i\rho+d]}$ has at most 2^i requests. It follows that $N_{((i-1)\rho+d,i\rho+d]} \leq 2^i s_i$. Therefore, for any $1 \leq h \leq m$, we have

$$\begin{aligned} N_{[0,h\rho+d]} &= N_{[0,d]} + N_{(d,\rho+d]} + N_{(\rho+d,2\rho+d]} + \cdots + N_{((h-1)\rho+d,h\rho+d]} \\ &\leq 2s_1 + 2^2 s_2 + \cdots + 2^h s_h. \end{aligned} \tag{6}$$

Furthermore, $d < \rho$ implies that for all $1 \leq h \leq m$, $h\rho + d < (h + 1)\rho$ and there is no request in $\sigma_{(h\rho,h\rho+d]}$. Together with the fact that σ has an $(|S|, 2^h)$-peak at $h\rho$ and $2^h \leq c$, we can apply Lemma 2 and conclude that

$$N_{[0,h\rho+d]} \geq 2^h|S|/\kappa. \tag{7}$$

Combining (6) and (7), we have, for all $1 \leq h \leq m$, $2s_1 + 2^2 s_2 + \cdots + 2^h s_h \geq 2^h|S|/\kappa$, which implies the following two inequalities:

$$\sum_{1\leq h\leq m} 2^{m-h}(2s_1 + 2^2 s_2 + \cdots + 2^h s_h) \geq \sum_{1\leq h\leq m} 2^m|S|/\kappa. \tag{8}$$

and

$$2s_1 + 2^2 s_2 + \cdots + 2^m s_m \geq 2^m|S|/\kappa. \tag{9}$$

Adding (8) to (9) and simplify, we have $2^{m+1}(s_1+s_2+\ldots+s_k) \geq (m+1)2^m|S|/\kappa$, or equivalently,

$$s_1 + s_2 + \cdots + s_m \geq \frac{(m+1)|S|}{2\kappa}.$$

Note that $m\rho = \ell$ and thus all the $s_1 + s_2 + \cdots + s_m$ servers are dedicated within the time interval $(d, d+\ell]$. As no server can be dedicated twice in this interval, we have

$$|S| \geq s_1 + s_2 + \cdots + s_k \geq \frac{(m+1)|S|}{2\kappa},$$

and the theorem follows. □

Note that we can remove the $2^{\lceil \ell/d \rceil} \leq c$ assumption in the above theorem by requiring σ to have an $(|S|, c^{\frac{h-1}{\lceil \ell/d \rceil}})$-peak instead of an $(|S|, 2^h)$-peak at $h\rho$. Then, a similar but more complicated analysis will give the following theorem.

Theorem 3. *Let* A *be an online (d, S, c, F, ℓ)-scheduling algorithm with competitive ratio κ. Suppose that $d < \ell$. Then we have $\kappa \geq (\lceil \ell/d \rceil - 1)(1 - c^{\frac{-1}{\lceil \ell/d \rceil}})$.*

4 A Competitive Algorithm for (d, S, c, F, ℓ)-Scheduling

In this section, we analyze a simple online algorithm G, which follows the Most-Profit-First heuristic for scheduling a (d, S, c, F, ℓ)-batching system. We derive an upper bound on its relative competitive ratio with respect to any batching system with the same set of job families, server capacity and job length, but with possibly different set of servers and start-up delay.

Let $\Delta = \min\{d, \ell\}$ and $\eta = \left\lfloor \frac{|S|}{\lceil \ell/d \rceil} \right\rfloor$. The online algorithm G decides a schedule as follows:

> Starting from time 0, G periodically dedicates η servers to accept requests at time 0, Δ, $2\Delta \ldots$ such that at time $i\Delta$, it schedules η servers to accept the set of requests with the largest possible profit that η servers can accept at $i\Delta$.

Lemma 3. *The online algorithm* G *is a (d, S, c, F, ℓ)-scheduling algorithm.*

Proof. By construction, G will only accept requests that are not expired. What needs to be proved is that at any time $i\Delta$, there are at least η free servers for G to dedicate.

Note that when a server is dedicated at $i\Delta$, it will be free at time $(i+\lceil \ell/d \rceil)\Delta$ because $\lceil \ell/d \rceil \Delta = \lceil \ell/d \rceil \min\{d,\ell\} \geq \ell$. Furthermore, during the interval $[i\Delta, (i+\lceil \ell/d \rceil)\Delta)$, G has dedicated $\lceil \ell/d \rceil \eta \leq |S|$ servers. This implies that the η servers dedicated at $i\Delta$ are free before G has no server to dedicate. □

Now, we derive an upper bound on the relative competitive ratio of G with respect to another system (d', S', c, F, ℓ). Note that this result also gives an upper bound on G's competitive ratio.

Theorem 4. *The competitive ratio of* G *relative to a* (d', S', c, F, ℓ)*-batching system is at most* $1 + \left\lceil \frac{1+d'/\Delta}{\lfloor d/\Delta \rfloor} \right\rceil \frac{|S'|}{\eta}$ *where* $\Delta = \min\{d, \ell\}$.

Proof. Let σ be any request sequence. Let O be the set of requests accepted in an optimal (d', S', c, F, ℓ)-schedule for σ, and D be the set of requests accepted in the (d, S, c, F, ℓ)-schedule constructed by the online algorithm G. Let $R = O - D$ be the set of requests that are accepted in the optimal schedule but not by G. It is obvious that the profits of O, R and D are related as follows:

$$p(O) \leq p(R) + p(D). \tag{10}$$

Now, we estimate $p(R)$ in terms of $p(D)$. For any server $\tau \in S'$ and integer i, let $R_{i,\tau}$ be the set of requests in R that are accepted by τ during the time interval $((i-1)\Delta, i\Delta]$. Let $m = \lfloor d/\Delta \rfloor$ and $k = \left\lceil \frac{1+d'/\Delta}{\lfloor d/\Delta \rfloor} \right\rceil$. Note that $(i-km)\Delta \leq (i-1)\Delta - d'$ and all requests in $\sigma_{[0,(i-km)\Delta]}$ are expired (with respect to the system (d', S', c, F, ℓ)) at or before $(i-1)\Delta$ and cannot be in $R_{i,\tau}$. Thus we have $R_{i,\tau} = R_{i,\tau} \cap \sigma_{((i-km)\Delta, i\Delta]}$, or equivalently,

$$\begin{aligned} R_{i,\tau} = \; & R_{i,\tau} \cap (\sigma_{((i-km)\Delta,(i-(k-1)m)\Delta]} \cup \sigma_{((i-(k-1)m)\Delta,(i-(k-2)m)\Delta]} \cup \cdots \\ & \cup \sigma_{((i-m)\Delta, i\Delta]}) \end{aligned} \tag{11}$$

Let us first estimate the profit of $R_{i,\tau} \cap \sigma_{((i-m)\Delta, i\Delta]}$, the set of requests in R that arrive during the interval $((i-m)\Delta, i\Delta]$, and are accepted by τ during the interval $((i-1)\Delta, i\Delta]$. Note that

- all requests in this set are of the same family type because $((i-1)\Delta, i\Delta]$ has length $\Delta = \min\{d, \ell\}$ and τ can be dedicated once to accept these requests, and
- they can still be accepted at time $i\Delta$ in the (d, S, c, F, ℓ) system because they all arrive after $(i-m)\Delta = (i - \lfloor d/\Delta \rfloor)\Delta \geq i\Delta - d$.

However, the fact that they are in R means they are not accepted by G. By the construction of G, we conclude that every of the η servers that G dedicates at $i\Delta$ must accept a set of requests at $i\Delta$ with profit no smaller than $p(R_{i,\tau} \cap \sigma_{((i-m)\Delta, i\Delta]})$. Hence, if we let D_i denote the set of requests accepted by G at time $i\Delta$, then we have $p\left(R_{i,\tau} \cap \sigma_{((i-m)\Delta, i\Delta]}\right) \leq p(D_i)/\eta$. For the other $R_{i,\tau} \cap \sigma_{((i-(j+1)m)\Delta,(i-jm)\Delta]}$, we can prove with a similar argument that $p\left(R_{i,\tau} \cap \sigma_{((i-(j+1)m)\Delta,(i-jm)\Delta]}\right) \leq p(D_{i-jm})/\eta$. Together with (11), we have

$$p(R_{i,\tau}) \leq \left(p(D_{i-(k-1)m}) + p(D_{i-(k-2)m}) + \cdots + p(D_i)\right)/\eta.$$

For any i, let $Q_i = p(D_{i-(k-1)m}) + p(D_{i-(k-2)m}) + \cdots + p(D_i)$. (To simplify notation, we let $p(D_j) = 0$ for all $j < 0$.) Then we have

$$p(R) = \sum_{i\geq 0}\sum_{\tau\in S'} p(R_{i,\tau}) \leq \sum_{i\geq 0}\sum_{\tau\in S'} Q_i/\eta = \frac{|S'|}{\eta}\sum_{i\geq 0} Q_i.$$

Note that for any j, D_j appears in at most k Q_i's (i.e. $Q_j, Q_{j+m}, \ldots, Q_{j+(k-1)m}$). Thus,

$$p(R) \leq \frac{|S'|}{\eta}\sum_{i\geq 0} Q_i \leq \frac{k|S'|}{\eta}(p(D_0) + p(D_1) + \ldots) = \frac{k|S'|}{\eta}p(D).$$

From (10), we have

$$p(O) \leq p(R) + p(D) \leq (\frac{k|S'|}{\eta} + 1)p(D) = \left(1 + \left\lceil \frac{1 + d'/\Delta}{\lfloor d/\Delta \rfloor} \right\rceil \frac{|S'|}{\eta}\right) p(D),$$

and the theorem is proved. □

5 Conclusion

We have analyzed the effect of the start-up delay on system performance in an online batching system and showed how different system configurations change this effect. Followings are some interesting unsolved problems.

1. Although we have derived almost optimal bounds on the competitive ratio, there are still gaps between the upper and lower bounds. We believe that a more careful analysis on the competitive ratio of G will give us a smaller upper bound.
2. In our analysis, we assume that all jobs have equal length. Although this assumption is reasonable for many applications such as On-demand data broadcasting, it is both practically and theoretically important to study systems with arbitrary job lengths, or systems in which the servers have different speeds or capacities.
3. Our lower bound proofs require the set F of job families has a fairly large size. However, it can be shown that when $|F| \leq \lfloor\alpha\rfloor|S|$, a simple round-robin scheduling always returns optimal schedules. This suggests the problem of studying the tradeoff between the start-up delay and the size of F. In particular, it is interesting to find out, for some fixed F, what is the smallest laxity that still allows us to find optimal schedules.
4. It is interesting to study how techniques such as randomization and preemption help to improve the performance guarantees.

References

1. C.C. Aggarwal, J.L. Wolf, and P.S. Yu. On optimal batching policies for video-on-demand storage servers. In *Proceedings of the IEEE International Conference on Multimedia Computing Multimedia Computing and Systems*, pages 200–209, 1996.
2. M. Azizoglu and S. Webster. Scheduling a batch processing machine with incompatible job families. *Computer and Industrial Engineering*, 39:325–335, 2001.
3. P. Baptiste. Batching identical jobs. *Mathematical Methods of Operations Research*, 53:355–367, 2000.
4. A. Bar-Noy, J.A. Garay, and A. Herzberg. Sharing video on demand. *Discrete Applied Mathematics*, 129:3–30, 2003.
5. A. Bar-Noy, J. Goshi, and R. Ladner. Off-line and on-line guaranteed start-up delay for media-on-demand with stream merging. In *Proceedings of the 15th Annual ACM Symposium on Parallel Algorithms and Architecture*, pages 164–173, 2003.
6. A. Bar-Noy, S. Guha, Y. Katz, J. Naor, B. Schieber, and H. Shachnai. Throughput maximization of real-time scheduling with batching. In *Proceedings of the 13th Symposium on Discrete Algorithms*, pages 742–751, 2002.
7. A. Bar-Noy, R.E. Ladner, and T. Tamir. Scheduling techniques for media-on-demand. In *Proceeding of the Annual ACM/SIAM Symposium on Discrete Algorithms*, pages 791–800, 2003.
8. C. Bouras, V. Kapoulas, G. Pantziou, and P. Spirakis. Competitive video on demand schedulers for popular movies. *Discrete Applied Mathematics*, 129:49–61, 2002.
9. P. Brucker, A. Gladky, H. Hoogeveen, M.Y. Kovalyov, C. Potts, T. Tauthenhahn, and S.L. Van de Velde. Scheduling a batching machine. *Journal of Scheduling*, 1:31–54, 1998.
10. A. Dan, D. Sitaram, and P. Shahabuddin. Dynamic batching policies for an on-demand video server. *ACM Multimedia Systems Journal*, 4(3):112–121, 1996.
11. J. Edmonds and K. Pruhs. Multicast pull scheduling: when fairness is fine. *Algorithmica*, 2003.
12. L. Engebretsen and M. Sudan. Harmonic broadcasting is optimal. In *Proceeding of the Annual ACM/SIAM Symposium on Discrete Algorithms*, pages 431–432, 2002.
13. W. Evans and D. Kirkpatrick. Optimally scheduling video-on-demand to minimize delay when server and receiver bandwidth may differ. In *Proceeding of the Annual ACM/SIAM Symposium on Discrete Algorithms*, pages 1041–1049, 2004.
14. M. Goldwasser. Patience is a virtue: The effect of slack on competitiveness for admission control. In *Proceedings of the 10th Annual Symposium on Discrete Algorithms*, pages pp. 396–405, 1999.
15. B. Kalyanasundaram and M. Velauthapillai. On-demand broadcasting under deadline. In *Proceedings of the 11th Annual European Symposium on Algorithms, volume 2832 of* Lecture Notes in Computer Science, pages 313–324, 2003.
16. J.H. Kim and K.Y. Chwa. Scheduling broadcasts with deadlines. In *Proceedings of the 9th Annual International Conference on Computing and Combinatorics, volume 2697 of* Lecture Notes in Computer Science, pages 415–424, 2003.
17. S.V. Mehta and R. Uzsoy. Minimizing total tardiness on a batch processing machine with incompatible jobs types. *IIE Transactions*, 32:165–175, 1998.

Competitive Analysis of Scheduling Algorithms for Aggregated Links

Wojciech Jawor[1,⋆], Marek Chrobak[1,⋆], and Christoph Dürr[2,⋆⋆,⋆⋆⋆]

[1] Department of Computer Science, University of California Riverside, USA
{wojtek, marek}@cs.ucr.edu

[2] CNRS, Laboratoire d'Informatique de l'Ecole Polytechnique, France
durr@lix.polytechnique.fr

Abstract. We study an online job scheduling problem arising in networks with aggregated links. The goal is to schedule n jobs, divided into k disjoint chains, on m identical machines, without preemption, so that the jobs within each chain complete in the order of release times and the maximum flow time is minimized.

We present a deterministic online algorithm Block with competitive ratio $O(\sqrt{n/m})$, and show a matching lower bound, even for randomized algorithms. The performance bound for Block we derive in the paper is, in fact, more subtle than a simple competitive analysis, and it shows that in overload conditions (when many jobs are released in a short amount of time), Block's performance is close to the optimum.

We also show efficient offline algorithms to minimize maximum flow time and makespan in our model for $k = 1$, and prove that minimizing the maximum flow time and makespan for $k, m \geq 2$ is $\mathcal{NP}$-hard.

1 Introduction

Link Aggregation is a method of grouping physical link segments (channels) between two network devices into a single logical link. The technology can be used to scale the bandwidth between the two devices, provide load balancing and improve system's fault-tolerance. Since Link Aggregation is also cost effective (it is often cheaper to add an additional channel to an existing link, then to replace the link with one of higher capacity) it is becoming very popular.

The traffic at the source of a multi-channel system (a network in which at least two nodes are interconnected with parallel links) is often divided into disjoint *conversations*, where a conversation is a distinguishable source-destination pair. In the system every output of a node is equipped with a scheduler. The scheduler receives packets from conversations that traverse the node, and chooses the transmission time and order of these packets over the output channels.

⋆ Supported by NSF grants OISE-0340752 and CCR-0208856.

⋆⋆ Supported by the CNRS/NSF grant 17171 and ANR Alpage.

⋆⋆⋆ Work conducted while being affiliated with Laboratoire de Recherche en Informatique, Université Paris-Sud.

J.R. Correa, A. Hevia, and M. Kiwi (Eds.): LATIN 2006, LNCS 3887, pp. 617–628, 2006.

Since packets of an individual conversation may be serviced concurrently by more than one channel, the order in which the data packets arrive at the receiver (i.e., the order in which the last bits of packets arrive) may be different from the order in which they arrived at the sender. Bennett et al. [2] argue that the parallelism of network components is one of the main reasons of packet reordering, contrary to the common belief that reordering is caused by malfunctioning network components. The inter-conversation reordering is an important issue for multi-channel systems, as it may negatively impact the performance of the transmission control protocol [2], and thus the performance of the whole system.

In practice, the requirement of maintaining packet ordering is met by transmitting all packets that compose a given conversation on a single channel. The distribution is achieved by using hash functions. This approach has several drawbacks: First, it does not fully utilize the capacity of a link if the number of conversations is smaller than the number of channels. Second, such scheduler does not provide load balancing, i.e., if traffic increases beyond a single channel's capacity, it is not distributed among additional channels. Third, it is hard or even not possible to design a hash function that would distribute the traffic well in all situations. And finally, in some (common) configurations the link aggregation algorithm must violate the layered architecture of network protocols by accessing higher layer information in order to compute useful hash functions.

Problem formulation. The above discussion raises the problem of designing appropriate schedulers to guide the packet transmission. Our goal is to optimize link utilization, under the constraint that packets complete their arrivals at the receiver in the order of their arrivals at the sender. Using scheduling terminology we state this problem as follows: We are given n jobs (packets) organized in k chains (conversations), with each job j specified by a triple (r_j, p_j, z_j) where r_j is a positive release (arrival) time, p_j is the processing time (length) of the job, and $1 \leq z_j \leq k$ is an index of the chain to which job j belongs. We assume that $\min_j r_j = 0$. In addition, the jobs are ordered so that if a job j precedes j' (we simply write $j < j'$) then $r_j \leq r_{j'}$. The ordering of job indices within a chain represents the ordering of packets in a conversation. We assume that at the node where scheduling takes places the packets arrived in a correct order, which justifies the ordering of the release times.

The jobs need to be scheduled on m identical machines and must satisfy the FRFC (**f**irst **r**eleased **f**irst **c**ompleted) order, i.e., if for jobs $j < j'$ we have $z_j = z_{j'}$ then $C_j^{\mathcal{A}} \leq C_{j'}^{\mathcal{A}}$, where $C_j^{\mathcal{A}}$ is the completion time (the time at which the packet arrives at the receiver) of j in schedule $\mathcal{A}$. Preemption is not allowed. Note that these constraints impose a fixed ordering even on jobs with a common release time, so these jobs must complete in the final schedule in this order.

In addition to the above constraints we want to optimize the machine utilization by constructing schedule which minimizes the *maximum flow time* $F_{\max}^{\mathcal{A}} = \max_j (C_j^{\mathcal{A}} - r_j)$, for schedule $\mathcal{A}$. The use of this function is motivated by *Quality of Service* applications in which, in order to provide the delay guarantees, an upper bound on the time *each* job spends in the system must be given. In the process, we will also construct schedules $\mathcal{A}$ to minimize *makespan* $C_{\max}^{\mathcal{A}} = \max_j C_j^{\mathcal{A}}$.

It is natural to require that the scheduling algorithms for this problem be *online*. In the online version of the problem jobs arrive at their release times, and the algorithm needs to schedule one of the pending jobs without the knowledge of the jobs that will arrive in the future. If the value of the objective function on a schedule produced by an online algorithm A is at most c times the value of an optimum schedule on the same instance, then we say that A is *c-competitive*. The smallest such value c is called the *competitive ratio* of A. If A is a randomized online algorithm, then the same definitions apply, except that we replace the value of the objective function with its expected value.

Our results. We give an online deterministic algorithm Block, which produces a schedule with maximum flow time at most $4\sqrt{2p_{\text{avg}}(n/m)(p_{\max} + F^*_{\max}(I))}$, where $F^*_{\max}(I)$ is the maximum flow time in an optimum schedule of instance I, $p_{\max} = \max_j p_j$, and $p_{\text{avg}} = \sum_j p_j/n$. Since $F^*_{\max}(I) \geq p_{\max} \geq p_{\text{avg}}$ this implies that Block's competitive ratio is at most $8\sqrt{n/m}$. We also show that for $m \geq 2$ there is no online randomized algorithm with expected competitive ratio better than $O(\sqrt{n/m})$ even for $p_{\max} = 3$. This means that Block is optimally competitive up to a constant factor, and that the asymptotic guarantee given by Block cannot be improved with randomization. Note that the performance bound of Block is stronger than what can be captured by classical competitive analysis. It implies that when $F^*_{\max}(I)$ is large (around $p_{\text{avg}}(n/m)$), the cost incurred by Block is no more than a constant times the optimum cost. We also give a series of simpler results: We show that for $k, m > 1$ minimizing maximum flow time or makespan is $\mathcal{NP}$-hard, and that for $k = 1$ it is possible to compute optimum solutions in polynomial time.

We would like to note that our main goal was to study the properties of the new FRFC scheduling constraint. The analysis of the algorithm Block, which we present in this paper, is a worst-case analysis. Whether this worst-case scenario reflects the properties of the real network traffic is a subject of further research.

Past work. Scheduling to minimize various functions of jobs' flow times has recently received a lot of attention. For recent surveys on online scheduling see [8, 7]. To the best of our knowledge we are the first to introduce the FRFC model. Therefore, in this section we briefly review only the results for related models.

In the context of online job scheduling with release times, the objective function $F_{\max}$ was first considered by Bender et al. [1] who give a deterministic lower bound of $\frac{4}{3}$ for $m = 2$. They also show that the FIFO algorithm is $(3 - 2/m)$-competitive. (The FIFO algorithm always schedules the job with the earliest release time on the next available machine; it does not necessarily create FRFC schedules.) Feuerstein et al. [5] study scheduling of jobs organized in a number of sequences called *threads*. Each job becomes available as soon as a scheduling decision has been made on all preceding jobs in the same thread. They show that Algorithm List [6] (a.k.a. Graham's algorithm), which always schedules the next job on the least loaded machine, is the best algorithm for the makespan problem if the number of machines is lower than the number of threads plus 2.

2 Preliminaries

FRFC scheduling. We are given m identical machines and n jobs, with each job j specified by a triple (r_j, p_j, z_j) where r_j is a positive release time, p_j is the processing time of the job, and z_j is an index of the chain to which job j belongs. The jobs are ordered so that for any two jobs $j < j'$ we have $r_j \le r_{j'}$.

A schedule $\mathcal{A}$ specifies when and where jobs are executed, i.e., for each job j it specifies a machine and an interval $[S_j^{\mathcal{A}}, C_j^{\mathcal{A}})$, such that (1) $S_j^{\mathcal{A}} \ge r_j$, (2) $C_j^{\mathcal{A}} - S_j^{\mathcal{A}} = p_j$, (3) all jobs assigned to the same machine are either completed before $S_j^{\mathcal{A}}$ or started after $C_j^{\mathcal{A}}$, and (4) for any two jobs $j < j'$ from the same chain, we have $C_j^{\mathcal{A}} \le C_{j'}^{\mathcal{A}}$. All jobs must be scheduled. Whenever the condition (4) is satisfied we say that jobs are scheduled in FRFC order. If $t \in [S_j^{\mathcal{A}}, C_j^{\mathcal{A}})$, then we say that job j is *running* at t in $\mathcal{A}$. Let $F_j^{\mathcal{A}} = C_j^{\mathcal{A}} - r_j$ denote the *flow time* of a job j in schedule $\mathcal{A}$. We set $F_{\max}^{\mathcal{A}} = \max_j F_j^{\mathcal{A}}$, and by $C_{\max}^{\mathcal{A}} = \max_j C_j^{\mathcal{A}}$ we denote the makespan of $\mathcal{A}$. Similarly, by $F_{\max}^*(I)$ we denote the maximum flow time of an offline schedule of I which minimizes maximum flow time, and by $C_{\max}^*(I)$ the makespan of an offline schedule of I which minimizes makespan. Our goal is to schedule the jobs in FRFC order so as to minimize the maximum flow time.

Online Algorithms. An algorithm A is called *online* if at each time t it decides which job to execute (if any) based only on the jobs released before or at time t.

Let $|\mathcal{A}|$ denote the value of the objective function on the schedule $\mathcal{A}$. Let $\mathsf{A}(I)$ denote the schedule computed by A on I. We say that an algorithm A is *c-competitive* if $|\mathsf{A}(I)| \le c|\mathcal{B}|$ for any schedule $\mathcal{B}$ of I. The smallest such value c is called the *competitive ratio* of A. If A is a randomized algorithm, then the same definition applies, except that $|\mathsf{A}(I)|$ is replaced with the expected value of the objective function on the schedule produced by A on I, where the expectation is taken over all random choices of the algorithm A.

3 Offline Algorithms

We now show how to compute schedules which minimize makespan or maximum flow for $k = 1$. We also give a 2-approximation algorithm for minimizing makespan for $k \ge 2$. These results are not difficult to prove, but some are used in our online algorithm in Section 4, and other are included for completeness.

In order to derive the algorithms that compute optimal schedules for $k = 1$, we consider a more general objective function, which includes both makespan and maximum flow as special cases. Consider an instance I and suppose that for each job $j \in I$ we define a *reference point*, denoted e_j, such that for any two jobs $j < j'$ we have $e_j \le e_{j'}$. We show an algorithm $\mathsf{RGreedy}$ which, when given an input instance I and a reference point for each job, computes a schedule $\mathcal{A}$ that minimizes $\max_j (C_j^{\mathcal{A}} - e_j)$. Observe that if we define $e_j = 0$ for all j, then the schedule $\mathcal{A} = \mathsf{RGreedy}(I)$ satisfies $C_{\max}^{\mathcal{A}} = C_{\max}^*(I)$; if we set $e_j = r_j$, then $F_{\max}^{\mathcal{A}} = F_{\max}^*(I)$.

```
for i ← 1 to m do start_i ← e_n

/* create auxiliary schedule X */
for j ← n downto 1 do
    l ← argmax{start_i : 1 ≤ i ≤ m}
    schedule j on machine l at time S_j^X ← min(e_j, start_l) − p_j
    start_l ← S_j^X

/* construct the final schedule A */
δ ← max_j(r_j − S_j^X)
for j ← 1 to n do S_j^A ← S_j^X + δ
```

Fig. 1. Algorithm RGreedy

The algorithm RGreedy is shown in Figure 1. It first computes an auxiliary schedule $\mathcal{X}$, and then shifts this schedule to meet all release times.

Theorem 1. *For $k = 1$ Algorithm* RGreedy *computes a schedule that minimizes* $\max_j(C_j^{\mathcal{A}} - e_j)$.

Proof. We first prove that the schedule computed by RGreedy, denoted $\mathcal{A}$, is feasible. Let $\mathcal{X}$ denote the auxiliary schedule constructed in the second loop. Consider a fixed iteration of this loop and observe that the job j, which is scheduled in this iteration, completes at $\min(e_j, \max_{i=1\ldots m} start_i)$, and that $start_i$ is equal to the minimum starting time of jobs scheduled on machine i in the previous iterations, or e_n if no jobs are scheduled on i. As j decreases from n to 1 the reference points e_j decrease, as well as $\max_{i=1,\ldots,m} start_i$. Thus $\mathcal{X}$ is an FRFC schedule. Shifting the jobs by δ in the last loop guarantees that all release times are met and does not change the order of completion times, so $\mathcal{A}$ is feasible.

Let $\mathcal{O}$ be any feasible schedule of jobs $1, 2, \ldots, n$ and let $\mathcal{Y}$ be a copy of $\mathcal{O}$ in which all jobs are shifted to the left by $\max_j(C_j^{\mathcal{O}} - e_j)$, i.e., $S_j^{\mathcal{Y}} = S_j^{\mathcal{O}} - \max_i(C_i^{\mathcal{O}} - e_i)$ for all $j = 1, 2, \ldots, n$. Observe that $C_j^{\mathcal{Y}} \leq e_j$ for all jobs j.

Claim 1. *For any $j = 1, 2, \ldots, n$, we have $C_j^{\mathcal{Y}} \leq C_j^{\mathcal{X}}$.*

Proof. The proof is by induction on $j = n, n-1, \ldots, 1$. For $j = n$ we have $C_n^{\mathcal{X}} = e_n \geq C_n^{\mathcal{Y}}$, so the claim holds. Now suppose that the claim holds for $j = n, n-1, \ldots, i+1$, where $i < n$. We show that it also holds for $j = i$. If $C_i^{\mathcal{X}} = e_i$ then the claim holds since $C_i^{\mathcal{Y}} \leq e_i$, so assume that $C_i^{\mathcal{X}} < e_i$. Let $B \subseteq \{i+1, i+2, \ldots, n\}$ be the set of jobs which are running at $C_i^{\mathcal{X}}$ in $\mathcal{X}$. Since $C_i^{\mathcal{X}} < e_i$ we have $|B| = m$ by the definition of RGreedy. We consider two cases.

If in $\mathcal{Y}$ all jobs from B are scheduled on different machines then $C_i^{\mathcal{Y}} \leq \max_{h\in B}(C_h^{\mathcal{Y}} - p_h) \leq \max_{h\in B}(C_h^{\mathcal{X}} - p_h) = C_i^{\mathcal{X}}$, where the second inequality follows from the inductive assumption, and the equality from the definition of RGreedy and the condition $C_i^{\mathcal{X}} < e_i$.

If there are two jobs $g < f$ such that $g, f \in B$ and both are scheduled on the same machine in $\mathcal{Y}$, then $C_i^{\mathcal{Y}} \leq C_g^{\mathcal{Y}} \leq C_f^{\mathcal{Y}} - p_f \leq C_f^{\mathcal{X}} - p_f \leq \max_{h\in B}(C_h^{\mathcal{X}} - p_h)$

$= C_i^{\mathcal{X}}$, where the first inequality follows from the FRFC assumption, the third inequality from the inductive assumption, and the equality from the definition of RGreedy and the condition $C_i^{\mathcal{X}} < e_i$. This ends the proof of Claim 1. □

Since $C_j^{\mathcal{X}} - e_j \le 0$ for all $j = 1, 2, \ldots, n$, and $C_n^{\mathcal{X}} = e_n$, we have $\max_j(C_j^{\mathcal{A}} - e_j) = \delta$. Let i be the job in $\mathcal{X}$ such that $r_i - S_i^{\mathcal{X}} = \delta$. We have $\max_j(C_j^{\mathcal{A}} - e_j) = \delta = r_i - S_i^{\mathcal{X}} = r_i + p_i - C_i^{\mathcal{X}} \le r_i + p_i - C_i^{\mathcal{Y}}$, where the last inequality follows from the above claim. We also have $\max_j(C_j^{\mathcal{O}} - e_j) = C_i^{\mathcal{O}} - C_i^{\mathcal{Y}} \ge r_i + p_i - C_i^{\mathcal{Y}}$, where the equality follows from the definition of $\mathcal{Y}$. Overall, $\max_j(C_j^{\mathcal{A}} - e_j) \le r_i + p_i - C_i^{\mathcal{Y}} \le \max_j(C_j^{\mathcal{O}} - e_j)$, ending the proof of Theorem 1. □

Consider an instance I and let $\widetilde{I}$ denote a copy of I in which all jobs are assigned to the same chain. Let RList denote Algorithm RGreedy when all reference points are set to 0. Consider an algorithm RList-M, which on any instance I returns the schedule produced by RList on $\widetilde{I}$.

Theorem 2. *Let $\mathcal{A}$ be the schedule produced by* RList-M *on some instance I. Then $\mathcal{A}$ is an FRFC schedule, and $C_{\max}^{\mathcal{A}} \le C_{\max}^*(I) + p_{\max}$.*

We omit the proof of the above theorem as it resembles the analysis of List from [6]. Since $C_{\max}^* \ge p_{\max}$, the above theorem yields the following corollary.

Corollary 1. *Algorithm* RList-M *is a 2-approximation algorithm for minimizing makespan.*

We conclude this section with the following theorem. Due to space constraints we omit the proof.

Theorem 3. *For $k > 1$, $m \ge 2$, minimizing maximum flow time or makespan is $\mathcal{NP}$-hard even if all jobs are released at time 0.*

4 Online Algorithms

4.1 An Upper Bound on the Competitive Ratio

Let A be any offline algorithm for minimizing makespan. We first show how to use algorithm A to create an online algorithm Block(A).

Algorithm Block(A)*:* The algorithm proceeds in phases numbered $1, 2, 3, \ldots$, where phase i starts at time β_i. First, let $\beta_1 = 0$. Consider phase i, and let Q_i be the set of jobs pending at time β_i. We apply algorithm A to schedule the jobs from Q_i. Suppose that the last job from Q_i completes at time $\beta_i + \delta_i$. Then $\beta_{i+1} \ge \beta_i + \delta_i$ is the first time when there is at least one pending job. (If no more jobs arrive, the computation completes.)

The above technique of applying an algorithm A to create an online algorithm Block(A) was first proposed by Shmoys et al. [9], who prove that if algorithm A produces a schedule with makespan at most $\rho C_{\max}^*(I)$ on any instance I, then algorithm Block(A) produces a schedule with makespan at most $2\rho C_{\max}^*(I')$ on any instance I'. Note that if A is an FRFC algorithm then so is Block(A). We therefore obtain the following corollary:

Corollary 2. Block(RList) *is a 2-competitive FRFC algorithm for minimizing makespan for* $k = 1$*, and* Block(RList-M) *is a 4-competitive FRFC algorithm to minimize makespan for* $k \geq 1$.

From now on we will denote Block(RList-M) simply by Block. In this section we show an upper bound on the maximum flow time of any job in the schedule produced by Block.

Theorem 4. *Let* $\mathcal{B}$ *be the schedule produced by* Block *on an instance* I*. Then* $F^{\mathcal{B}}_{\max} \leq 4\sqrt{2p_{\text{avg}}(n/m)(p_{\max} + F^*_{\max}(I))}$*, where* $p_{\max} = \max_j p_j$*, and* $p_{\text{avg}} = \sum_j p_j/n$.

Let B denote the number of phases. (Clearly, $B \leq n$.) For $i = 1, 2, \ldots, B$, let the schedule of Q_i created in i-th phase be called a *block*. By the definition of Algorithm RList-M in each block at least one machine is continuously busy processing jobs. In order to prove Theorem 4, we will need several lemmas.

Lemma 1. *Let* $\sigma_0 = 0$ *and let* $\sigma_1, \sigma_2, \ldots, \sigma_l \geq 0$ *be numbers such that* $\sigma_i - \sigma_{i-1} \leq \Delta$ *for* $i = 1, 2, \ldots, l$ *and some* $\Delta \geq 0$*. Then* $\sigma_l^2 \leq 2\Delta \sum_{i=1}^{l} \sigma_i$.

Proof. For any $i = 1, 2, \ldots, l$ we have $\sigma_i^2 - \sigma_{i-1}^2 \leq \Delta(\sigma_i + \sigma_{i-1})$. The lemma follows by adding these inequalities for all i. □

Consider any instance I and let $\mathcal{B} = \text{Block}(I)$. We show that in order to prove Theorem 4, we may assume that at all times $0 \leq t < C^{\mathcal{B}}_{\max}$ at least one machines is busy in the schedule $\mathcal{B}$.

Lemma 2. *Without loss of generality we may assume that at least one machine is busy in* $\mathcal{B}$ *at all times* $t \in [0, C^{\mathcal{B}}_{\max})$.

Proof. Assume that Theorem 4 holds for any instance I' such that at all times $t' \in [0, C^{\mathcal{B}'}_{\max})$ at least one machine is busy in $\mathcal{B}' = \text{Block}(I')$. We claim that then it also holds for I.

We divide the schedule into N segments that do not have idle times. Let $t_1 = \beta_1 = 0$. If t_i is defined, and if there is j such that $t_i < \beta_j + \delta_j < \beta_{j+1}$ then pick such smallest j and set $t'_i = \beta_j + \delta_j$ and $t_{i+1} = \beta_{j+1}$. Otherwise let $t'_i = \beta_B + \delta_B$ and $N = i$. Let I_i be the set of jobs released in the interval $[t_i, t'_i)$ for $i = 1, 2, \ldots, N$. Since Block is never idle when there are available jobs, all jobs from I_i are completed in $\mathcal{B}$ no later than at time t'_i. Let $\mathcal{B}_i$ be the schedule computed by Block on I_i. Note that the schedule $\mathcal{B}_i$ is identical to the schedule $\mathcal{B}$ in $[t_i, t'_i)$. Therefore, $F^{\mathcal{B}}_{\max} = \max_{i=1,\ldots,N} F^{\mathcal{B}_i}_{\max}$. Also, $F^*_{\max}(I) \geq \max_{i=1,\ldots,N} F^*_{\max}(I_i)$. Let $f(x, y) = 4\sqrt{2p_{\text{avg}}(y/m)(p_{\max} + x)}$. The function f is non-decreasing with x and y. Since we assumed that Theorem 4 holds for all instances I_i for $i = 1, 2, \ldots, N$, we obtain $F^{\mathcal{B}}_{\max} = \max_{i=1,\ldots,N} F^{\mathcal{B}_i}_{\max} \leq \max_{i=1,\ldots,N} f(F^*_{\max}(I_i), |I_i|) \leq f(F^*_{\max}(I), |I|)$. This means that Theorem 4 holds for schedule $\mathcal{B}$. This completes the proof. □

We now observe that the maximum flow time of any job in schedule $\mathcal{B}$ may be bounded by twice the length of a largest block in $\mathcal{B}$.

Lemma 3. $F^{\mathcal{B}}_{\max} \le 2\max_i \delta_i$.

Proof. Let j be the job with the maximum flow time in $\mathcal{B}$ and suppose $j \in Q_{i'}$ for some i'. Then $r_j > \beta_{i'-1}$ and $C^{\mathcal{B}}_j \le \beta_{i'} + \delta_{i'}$, therefore $F^{\mathcal{B}}_{\max} = F^{\mathcal{B}}_j \le \beta_{i'} + \delta_{i'} - \beta_{i'-1} = \delta_{i'-1} + \delta_{i'} \le 2\max_i \delta_i$, as $\beta_{i'} = \beta_{i'-1} + \delta_{i'-1}$ by Lemma 2. □

The idea behind the analysis of the algorithm is to estimate the size of a largest block in $\mathcal{B}$ in terms of $p_{\max}$ and $F^*_{\max}(I)$. In order to do this we derive an inequality that bounds the rate of growth of the blocks in $\mathcal{B}$.

Lemma 4. *For any* $i = 1, 2, \ldots, B$ *we have* $\delta_i \le \delta_{i-1} + F^*_{\max}(I) + p_{\max}$.

Proof. Let $\mathcal{O}$ denote an optimum schedule of I. Let $\alpha_i = \min_{j \in Q_i} S^{\mathcal{O}}_j$. We first observe that for all $i = 1, 2, \ldots, B$ we have

$$\alpha_i > \beta_{i-1}. \tag{1}$$

To justify this inequality, let $h = \operatorname{argmin}_{j \in Q_i} S^{\mathcal{O}}_j$. We have $S^{\mathcal{O}}_h = \alpha_i$, by the definition of α_i, and $r_h \le \alpha_i$. On the other hand, since the job $h \in Q_i$, we have $r_h > \beta_{i-1}$. So we obtain $\beta_{i-1} < r_h \le \alpha_i$.

Let $\lambda_i = \max_{j \in Q_i} C^{\mathcal{O}}_j - \alpha_i$. We now claim that for all $i = 1, 2, \ldots, B$ we have

$$\lambda_i \le \delta_{i-1} + F^*_{\max}(I). \tag{2}$$

Note that for all $j \in Q_i$ we have $r_j \le \beta_i = \beta_{i-1} + \delta_{i-1}$ by Lemma 2. Let $g \in Q_i$ be a job completed at $\alpha_i + \lambda_i$ in $\mathcal{O}$. We have $F^*_{\max}(I) \ge F^{\mathcal{O}}_g = \alpha_i + \lambda_i - r_g \ge \alpha_i + \lambda_i - (\beta_{i-1} + \delta_{i-1}) \ge \lambda_i - \delta_{i-1}$ by inequality (1) proving (2).

Theorem 2 implies that $\delta_i \le \lambda_i + p_{\max}$. Thus, using (2) we obtain $\delta_i \le \delta_{i-1} + F^*_{\max}(I) + p_{\max}$. This ends the proof. □

We now prove Theorem 4.

Proof. We may assume that $B \ge 2$. If for all $i = 1, 2, \ldots, B$, we have $\delta_i < 2p_{\max}$, then the theorem follows from Lemma 3. Otherwise let $l = \operatorname{argmax}_i \delta_i$. Let $1 \le f < l$ be the maximum index i such that $\delta_i < 2p_{\max}$ and $\delta_{i+1} \ge 2p_{\max}$. If such i does not exist, we set $f = 1$. Define $g = l - f + 1$. Let $\sigma_0 = 0$ and $\sigma_i = \delta_{f+i-1}$ for $i = 1, \ldots, g$.

Observe that for $i = 2, 3, \ldots, g$ we have $\sigma_i - \sigma_{i-1} \le p_{\max} + F^*_{\max}(I)$ by Lemma 4. If $f = 1$ we have $\sigma_1 = \delta_1 \le F^*_{\max}(I) \le p_{\max} + F^*_{\max}(I)$ by the definition of Block, and if $f > 1$ we have $\sigma_1 = \delta_f \le 2p_{\max} \le p_{\max} + F^*_{\max}(I)$.

We conclude that for $i = 1, 2, \ldots, g$, we have $\sigma_i - \sigma_{i-1} \le p_{\max} + F^*_{\max}(I)$, so the sequence σ_i satisfies the conditions of Lemma 1 with $\Delta = p_{\max} + F^*_{\max}(I)$. We now use this lemma to bound σ_g, the size of a largest block.

Observe that for $i = 2, 3, \ldots, g$ we have $\sigma_i \ge 2p_{\max}$, so the number of jobs in each block Q_{f+i-1} is more than m by the definition of Block. We therefore obtain $\sigma_i \le p_{\max} + \sum_{j \in Q_{f+i-1}} p_j/m$ by the definition of RList, and, since $\sigma_i \ge 2p_{\max}$, we have $\sigma_i \le 2\sum_{j \in Q_{f+i-1}} p_j/m$ for all $i = 2, 3, \ldots, g$. We now consider two cases. If $\sigma_1 \ge 2p_{\max}$ then the observation from the previous paragraph also applies to

$i = 1$, and $\sum_{i=1}^{g} \sigma_i \le \sum_{i=1}^{g} \left(2 \sum_{j \in Q_{f+i-1}} p_j/m\right) \le 2 \sum_{j=1}^{n} p_j/m$. If $\sigma_1 < 2p_{\max}$ then $\sigma_1 < \sigma_2$. Since we assumed that $\sigma_g \ge 2p$ we have $g > 1$, and $\sum_{i=1}^{g} \sigma_i \le \sigma_1 + \sum_{i=2}^{g}(2 \sum_{j \in Q_{f+i-1}} p_j/m) \le 2 \sum_{i=2}^{g}(2 \sum_{j \in Q_{f+i-1}} p_j/m) \le 4 \sum_{j=1}^{n} p_j/m$.

Overall we have $\sum_{i=1}^{g} \sigma_i \le 4 \sum_{j=1}^{n} p_j/m$. By Lemma 1 we obtain $(\sigma_g)^2 \le 2(p_{\max} + F^*_{\max}(I))4 \sum_{j=1}^{n} p_j/m \le 8(p_{\max} + F^*_{\max}(I))np_{\text{avg}}/m$, and, by Lemma 3, $F^{\mathcal{B}}_{\max} \le 2\sigma_g \le 4\sqrt{2p_{\text{avg}}(n/m)(p_{\max} + F^*_{\max}(I))}$, proving Theorem 4. □

4.2 A Lower Bound on the Competitive Ratio

We conclude the paper by proving that the algorithm Block has asymptotically the best possible competitive ratio.

Theorem 5. *No online randomized algorithm can be better than $O(\sqrt{n/m})$-competitive for $m \ge 2$.*

Proof. We use Yao's minimax principle [10, 3], and show a distribution on instances with $k = 1$ that forces each deterministic online algorithm to have expected ratio $\Omega(\sqrt{n/m})$.

Choose a large integer $a > 0$. For the sake of simplicity we assume that m is even. The construction is similar in the case when m is odd.

We generate each instance from a random binary string $\mu = \mu_1\mu_2 \ldots \mu_a$, where $\mu_i = 0$ or $\mu_i = 1$, each with probability $\frac{1}{2}$, independently. Suppose that $t_1 = 0$. For any $i = 1, 2, \ldots, a$, suppose that t_i and $I_1 \cup I_2 \cup \ldots \cup I_{i-1}$ have already been defined. Then let $t_{i+1} = t_i + 3(i+1) + 2\mu_i + 4$, and define instance I_i of $m(i+1) + 4m + m\mu_i$ jobs as follows:

- m jobs of length 1 released at time t_i;
- $i+1$ batches of m jobs of length 3, where batch j is released at time $t_i + 3j$ for $j = 0, 1, \ldots, i$;
- If $\mu_i = 1$, $m/2$ jobs of length 3 released at time $t_i + 3(i+1)$, and $m/2$ jobs of length 1 released at time $t_i + 3(i+1) + 2$;
- $3m$ jobs of length 1 released at time $t_i + 3(i+1) + 3\mu_i$.

The final instance I is obtained by concatenating sub-instances I_i for $i = 1, 2, \ldots, a$. Figure 2(a) shows sub-instance I_i for $\mu_i = 1$.

Since every sub-instance I_i contains at most $m(i+6)$ jobs, the total number of jobs in I is $n \le \sum_{i=1}^{a} m(i+6) = O(ma^2)$.

Claim 2. *There exists a schedule of the instance I with maximum flow time 5 and makespan at most t_{a+1}.*

Proof. Fix i. If $\mu_i = 0$, then we schedule the first m jobs from I_i at time t_i each on a separate machine. If $\mu_i = 1$, then we schedule the first $m/2$ jobs from I_i at time t_i on the first $m/2$ machines and the next $m/2$ jobs at time $t_i + 1$ on these same machines. The remaining jobs are always placed on the next available machine (ties broken arbitrarily). By routine inspection, this is a valid schedule, with the last job completed at time t_{i+1}. See Figures 2(b) and 2(c). It is easy to verify that in both cases the flow time of any job is at most 5. □

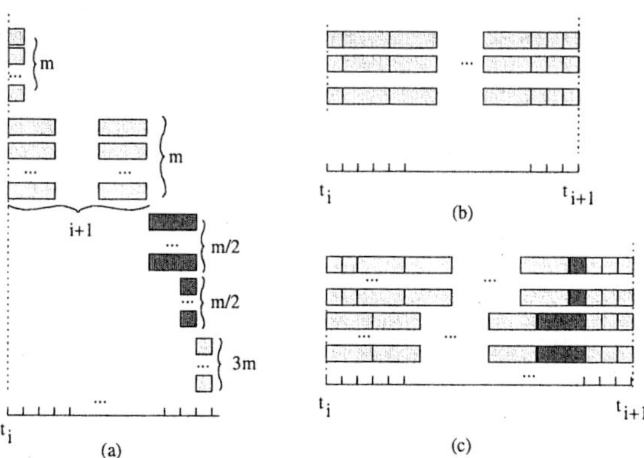

Fig. 2. (a) Sub-instance I_i for $\mu_i = 1$. When $\mu_i = 0$ the last $3m$ jobs are released three units earlier and the dark gray jobs are not released at all. (b) Optimum schedule of sub-instance I_i with $\mu_i = 0$. (c) Optimum schedule for sub-instance I_i with $\mu_i = 1$.

Since the total processing time of jobs from sub-instance I_i is exactly $m(t_{i+1}-t_i)$, the total processing time of jobs in I is $m(t_{a+1} - t_1)$, so to complete all jobs by time t_{a+1} all machines must be continuously busy in the interval $[t_1, t_{a+1})$.

Let A be an online algorithm. Since all release times and processing times of jobs in I are integral, we may restrict our attention to algorithms which start jobs at integral times only. This justifies the following claim:

Claim 3. *Without loss of generality we may assume that in any schedule of I algorithm* A *starts and completes jobs at integral times.*

For any schedule $\mathcal{X}'$ of the first n' jobs from I let an *idle slot at t on machine l* be a time unit such that l is idle at t, and a *stall slot at t on machine l* be an idle slot such that there either is a job which completes after time t on l in $\mathcal{X}'$, or $t < C^{\mathcal{X}'}_{\max} - 3$. Since a largest job in I has size 3, in any extension of $\mathcal{X}'$ to a full schedule of I no job will be running in any stall slot.

The main idea of the proof is to show that A creates a schedule with $\Omega(ma)$ idle slots for I, on average. The last completion time in the schedule created by A is, therefore, $t_{a+1} + \Omega(a)$. It follows that the flow time of the job that completes last in this schedule is $\Omega(a)$, as it is released before t_{a+1}. Finally, as the flow time of any job in an optimum schedule is bounded by a constant, we can conclude that the competitive ratio of A is $\Omega(a) = \Omega(\sqrt{n/m})$.

We now prove that A creates a schedule with $\Omega(ma)$ stall slots for I, on average. This follows from the following claim:

Claim 4. *For all $i = 1, 2, \ldots, a$ a schedule of $I_1 \cup I_2 \cup \ldots \cup I_i$ computed by* A *contains $\Omega(mi)$ stall time units on average, w.r.t. the random choice of $\mu_1 \ldots \mu_i$.*

Proof. Let $\mathcal{A} = \mathsf{A}(I)$. Fix i and let $\mathcal{B}$ denote the schedule $\mathcal{A}$ with all jobs removed except for $I_1 \cup I_2 \cup \ldots \cup I_{i-1}$ and the first m jobs from I_i.

If $C^{\mathcal{B}}_{\max} \geq t_i + i + 5$ then, since the total processing time of jobs in $\mathcal{B}$ is $m(t_i + 3)$ the algorithm must have already created mi stall slots and the claim follows. We can, therefore, assume that $C^{\mathcal{B}}_{\max} < t_i + i + 5$, which implies that $C^{\mathcal{B}}_{\max} < t_i + 3(i+1)$. So at $C^{\mathcal{B}}_{\max}$ the jobs which distinguish instance I_i for $\mu_i = 0$ from I_i for $\mu_i = 1$ are not released yet. We will prove that since at $C^{\mathcal{B}}_{\max}$ the algorithm cannot know the value of μ_i, it cannot avoid $\Omega(m)$ stall slots when scheduling jobs from I_i, on average.

Let $J = \{j \in I_i : p_j = 3\}$ and let $\mathcal{B}'$ denote the schedule $\mathcal{A}$ with all jobs removed except for jobs in $\mathcal{B}$ and J. We wish to count the number of stall slots in $[C^{\mathcal{B}}_{\max} - 3, C^{\mathcal{B}'}_{\max})$.

Observe that jobs in J are followed in I_i by at least $3m$ jobs of length 1. None of these jobs of length 1 may be started in $\mathcal{A}$ before $C^{\mathcal{B}'}_{\max} - 1$ (otherwise $\mathcal{A}$ would not be an FRFC schedule), and some of these jobs must complete at $C^{\mathcal{B}'}_{\max} + 3$ (since the total processing time of these jobs is at least $3m$). It follows that since $p_{\max} = 3$ these jobs force each idle slot in $[C^{\mathcal{B}'}_{\max} - 2, C^{\mathcal{B}'}_{\max} - 1)$ in $\mathcal{B}'$ to be a stall slot in $\mathcal{A}$. This allows us to include the idle slots in $[C^{\mathcal{B}'}_{\max} - 2, C^{\mathcal{B}'}_{\max} - 1)$ in $\mathcal{B}'$ in the total count of stall slots.

Let l_h denote the number of jobs from J scheduled on machine h. Let M denote the set of machines which complete all jobs in $\mathcal{B}$ at or before time $C^{\mathcal{B}}_{\max} - 2$. Let $\overline{M}$ denote the remaining machines. We now show that the number of new stall slots (i.e., stall slots which are in $\mathcal{B}'$ and not in $\mathcal{B}$) in $[C^{\mathcal{B}}_{\max} - 3, C^{\mathcal{B}'}_{\max} - 1)$ is at least $m/2$ for a certain value of μ_i. Distinguish the following cases.

Case 1: Suppose that $|M| \geq m/2$, and $\mu_i = 0$. First we claim that $C^{\mathcal{B}'}_{\max} \geq C^{\mathcal{B}}_{\max} + 3(i+1)$. Indeed, if $l_h = i + 1$ for $h = 1, 2, \ldots, m$, then $C^{\mathcal{B}'}_{\max} \geq C^{\mathcal{B}}_{\max} + 3(i+1)$. If for some h we have $l_h \geq i + 2$ then since the first job from J on h may not complete before $C^{\mathcal{B}}_{\max}$, we also have $C^{\mathcal{B}'}_{\max} \geq C^{\mathcal{B}}_{\max} + 3(i+1)$.

Any machine $h \in M$ completes all jobs in $\mathcal{B}$ by time $C^{\mathcal{B}}_{\max} - 2$, by definition. We claim that the number of new stall slots in $[C^{\mathcal{B}}_{\max} - 2, C^{\mathcal{B}'}_{\max} - 1)$ on any machine $h \in M$ is at least $i + 2 - l_h$. Indeed, comparing the processing power of h in that interval with the total processing time of jobs from J on h we see that there must be at least $(C^{\mathcal{B}'}_{\max} - 1) - (C^{\mathcal{B}}_{\max} - 2) - 3l_h \geq 3(i+1) + 1 - 3l_h \geq i + 2 - l_h$ stall slots on h. Using a similar argument we can show that every machine $h \in \overline{M}$ must contain at least $i + 1 - l_h$ stall slots. Thus, since $\sum_{h=1}^{m} l_h = m(i+1)$, the number of new stall slots is at least $\sum_{h \in M}(i + 2 - l_h) + \sum_{h \in \overline{M}}(i + 1 - l_h) = |M| \geq m/2$.

Case 2: Now assume that $|\overline{M}| > m/2$, and $\mu_i = 1$. We claim that $C^{\mathcal{B}'}_{\max} \geq C^{\mathcal{B}}_{\max} + 3(i+1) + 2$. Indeed, either for some $h \in \overline{M}$ we have $l_h \geq i + 2$, which implies that the last job on h completes in $\mathcal{B}'$ at least at $C^{\mathcal{B}}_{\max} - 1 + 3(i+2) = C^{\mathcal{B}}_{\max} + 3(i+1) + 2$, or, since $|M| < m/2$, for some $h \in M$ we have $l_h \geq i + 3$ and the last job on h completes in $\mathcal{B}'$ at least at $C^{\mathcal{B}}_{\max} - 3 + 3(i+3) > C^{\mathcal{B}}_{\max} + 3(i+1) + 2$.

We claim that the number of new stall slots in $[C^{\mathcal{B}}_{\max}, C^{\mathcal{B}'}_{\max} - 1)$ on any machine h is at least $i + 2 - l_h$. Indeed, comparing the processing power of h in that interval with the total processing time of jobs from J on h we see that there must be at least $(C^{\mathcal{B}'}_{\max} - 1) - C^{\mathcal{B}}_{\max} - 3l_h \geq 3(i+1) + 1 - 3l_h \geq i + 2 - l_h$ stall slots on h. Thus, since $\sum_{h=1}^{m} l_h = m(i+1) + m/2$, the number of new stall slots is at least $\sum_{h=1}^{m}(i + 2 - l_h) \geq m/2$.

Since the choice of i was arbitrary, it follows that each schedule of $I_1 \cup I_2 \cup \ldots \cup I_i$ computed by A contains $\Omega(mi)$ stall slots for each $i = 1, 2, \ldots, a$ on average. This ends the proof of Claim 4 and Theorem 5. □

We would like to point out that in the above construction we only use jobs with two different processing times: 1 and 3.

Acknowledgments. We would like to thank the anonymous referees for useful comments.

References

1. M. A. Bender, S. Chakrabarti, and S. Muthukrishnan. Flow and stretch metrics for scheduling continuous job streams. In *Proc. 9th Symp. on Discrete Algorithms (SODA)*, pages 270–279. ACM/SIAM, 1998.
2. J. C. R. Bennett, C. Partridge, and N. Shectman. Packet reordering is not pathological network behavior. *IEEE/ACM Transactions on networking*, 7(6):789–798, 1999.
3. A. Borodin and R. El-Yaniv. *Online Computation and Competitive Analysis.* Cambridge University Press, 1998.
4. J. A. Cobb and M. Lin. A theory of multi-channel schedulers for quality of service. *J. High Speed Netw.*, 12(1,2):61–86, 2003.
5. E. Feuerstein, M. Mydlarz, and L. Stougie. On-line multi-threaded scheduling. *J. Sched.*, 6(2):167–181, 2003.
6. R. L. Graham. Bounds for certain multiprocessing anomalies. *Bell System Technical J.*, 45:1563–1581, 1966.
7. W. Jawor. Three dozen papers on online algorithms. *SIGACT News*, 36(1):71–85, 2005.
8. K. Pruhs, E. Torng, and J. Sgall. Online scheduling. In J. Y.-T. Leung, editor, *Handbook of Scheduling: Algorithms, Models, and Performance Analysis*, chapter 15. CRC Press, 2004.
9. D. B. Shmoys, J. Wein, and D. P. Williamson. Scheduling parallel machines online. In L. A. McGeoch and D. D. Sleator, editors, *On-line Algorithms*, volume 7 of *DIMACS Series in Discrete Mathematics and Theoretical Computer Science*, pages 163–166. AMS/ACM, 1992.
10. A. C. C. Yao. Probabilistic computations: Towards a unified measure of complexity. In *Proc. 18th Symp. Foundations of Computer Science (FOCS)*, pages 222–227. IEEE, 1977.

A 4-Approximation Algorithm for Guarding 1.5-Dimensional Terrains

James King

School of Computer Science,
McGill University,
Montreal, H3A 2A7, Canada

Abstract. In the 1.5-dimensional terrain guarding problem we are given as input an x-monotone chain (the terrain) and asked for the minimum set of guards (points on the terrain) such that every point on the terrain is seen by at least one guard. It has recently been shown that the 1.5-dimensional terrain guarding problem is approximable to within a constant factor [3, 7], though no attempt has been made to minimize the approximation factor. We give a 4-approximation algorithm for the 1.5D terrain guarding problem that runs in quadratic time. Our algorithm is faster, simpler, and has a better worst-case approximation factor than previous algorithms.

1 Introduction

1.1 Problem Statement

In the 1.5-dimensional terrain guarding problem we are given as input a terrain T that is an x-monotone polygonal chain. An x-monotone polygonal chain is a polygonal chain that intersects any vertical line at most once. It can be thought of as an array of n vertices in 2-dimensional space sorted in ascending order of x-coordinate, where edges 'connect the dots' from left to right. Note that the x-monotonicity requires x-coordinates to be distinct.

We say that a point on the terrain sees another point on the terrain if there is a line of sight between them, *i.e.* the line segment connecting them is never strictly below T. A guard is simply a point on the terrain that we add to a 'guarding set'. Given a terrain T, we are asked for the smallest possible guarding set, *i.e.* the smallest set G of points on T such that every point on T is seen by some point in G.

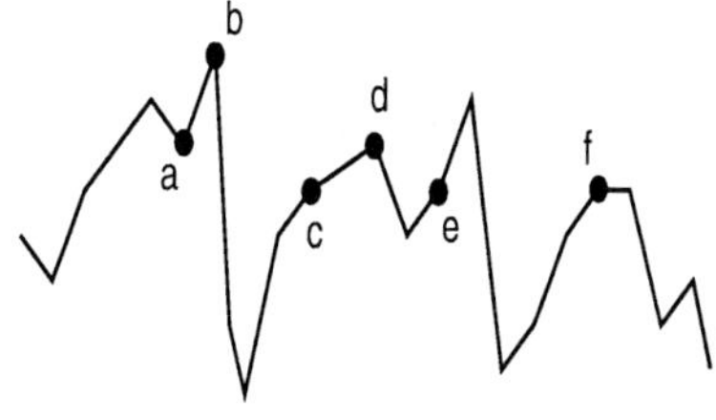

Fig. 1. An example of a 1.5D terrain. d can see b, c, and e but not a or f.

It is natural to consider two different versions of the terrain guarding problem: the discrete version and the continuous version. In the discrete version guards must be at vertices and only the vertices of the terrain need to be guarded. In the

J.R. Correa, A. Hevia, and M. Kiwi (Eds.): LATIN 2006, LNCS 3887, pp. 629–640, 2006.

continuous version guards may be anywhere on the terrain and every point on the terrain must be guarded. The discrete version is simpler but the continuous version is more natural to consider in a geometric context. For the rest of this paper we will use TG to denote the discrete version of the problem and TG-C to denote the continuous version.

Every instance of TG is an instance of SET COVER, but we know that SET COVER is NP-complete (see, *e.g.*, [13]) and no sub-logarithmic approximation factor can be obtained unless NP $\subseteq$ DTIME($n^{\log \log n}$) [12]. In general it is not particularly difficult to modify a TG algorithm to solve instances of TG-C, though this often involves some polynomial increase in time complexity.

1.2 Related Work

The 1.5D terrain guarding problem is very similar to the *art gallery problem* in which one must guard the interior of a simple polygon. The art gallery problem and its variants are well studied [1, 6, 11, 16, 17].

It is unknown whether or not TG is NP-hard. In 1995 Chen *et al.* [5] proposed an NP-hardness proof obtainable via a modification of Lee and Lin's proof that the art gallery problem is NP-complete [17]. However, the proof, whose details were omitted, was never completed successfully. Since then, attempts to find a polynomial-time algorithm for TG and attempts to prove that it is NP-hard have both been unsuccessful.

The first constant-factor approximation algorithm for the 1.5D terrain guarding problem was given by Ben-Moshe *et al.* [3]. Their algorithm works by first placing guards to divide the terrain into independent subterrains. Each subterrain has the property that it does not require internal guards, *i.e.* every unguarded vertex can be seen from outside the subterrain. For each such subterrain that is not completely guarded they then proceed with steps that either reduce the subterrain or split it into multiple independent subterrains. They made no attempt to minimize their algorithm's approximation factor; as such it is very large (at least 48). It could be brought down possibly as low as 6 with some minor modifications and careful accounting, but due to the inevitable cost incurred by repeatedly dividing the terrain it does not seem that it could be brought any lower than 6. Their algorithm runs in $O(n^2)$ time for the discrete version. They also provide a reduction from TG-C to TG that allows them to solve TG-C in $O(n^4)$ time, though the approximation factor can double in this case.

Another constant-factor approximation algorithm is given by Clarkson and Varadarajan [7]. Consider a partition of a 1.5D terrain into maximal intervals such that, for any two points p and p' in a given interval, the leftmost point that sees p and the leftmost point that sees p' are the same. If we label each interval with the leftmost point that sees it and read the labels from leftmost interval to rightmost interval, Clarkson and Varadarajan note that we end up with an $(n, 2)$ Davenport-Schinzel sequence [18]. Such a sequence must have length at most $2n$. This characterization of the lack of complexity in 1.5D terrains allows them to efficiently find appropriate ϵ-nets [14] for instances of TG. They then apply the SET COVER method of Brönnimann and Goodrich [4] to solve the problem using these ϵ-nets. The end result is a constant-factor approximation algorithm that runs in polynomial time. Using efficient derandomization their algorithm could probably be made to run deterministically in $O(n^2 \log n)$ time.

The 1.5D terrain guarding problem becomes easy if, instead of being placed on the terrain, all guards 'float' above the terrain at a fixed altitude that is above the highest vertex. Eidenbenz [8] gives a linear-time algorithm for finding an optimal set of guards in this case. The problem also becomes easy if guards can only look rightwards. Chen *et al.* [5] give a linear-time algorithm for this case.

A 2.5D terrain is a polyhedral surface that intersects every vertical line at most once and whose projection onto the x, y-plane is a simple polygon with no holes. The 2.5D Terrain Guarding Problem is therefore a natural extension of the 1.5D problem to the next dimension. Finding a minimum number of guards for a 2.5D terrain is NP-complete and Eidenbenz shows that it cannot be approximated within a sub-logarithmic factor unless NP $\subseteq$ DTIME($n^{\log \log n}$) [9]. Eidenbenz *et al.* show that the problem is also NP-complete and equally inapproximable when guards 'float' at a given altitude that is higher than the highest point in the terrain [10] (recall that this can be solved in linear time for 1.5D terrains).

1.3 Motivation

Naturally, the motivation for 1.5D terrain guarding comes from guarding or covering terrain. The 1.5D case appears, for example, when guarding or covering a road, perhaps with security cameras or street lights. The 2.5D case has more powerful applications, most notably for providing a wireless communication network that covers a given region. Its proven intractability and inapproximability, however, motivate us to look towards the 1.5D case for insight. The 1.5D case is also applicable, for example, if we only need to cover the path between two points on a polyhedral terrain. It has been pointed out [3] that the 1.5D terrain guarding problem can be utilized in heuristic methods for the 2.5D case.

The recent results of Ben-Moshe *et al.* [3] and Clarkson and Varadarajan [7] showed that constant-factor approximation algorithms exist for TG. Unfortunately they do not provide a small constant guaranteed approximation factor. Efforts to design an exact polynomial-time algorithm for TG have been unsuccessful and it is very possible that no such algorithm exists. If TG is NP-hard and P $\neq$ NP, then the best algorithm running in polynomial time will be the approximation algorithm with the lowest approximation factor. For this reason there is significant motivation to minimize the approximation factor.

The greedy algorithm for SET COVER, which achieves the optimal approximation factor of $O(\log n)$, repeatedly picks the set that contains the most uncovered elements. Similarly, the natural greedy algorithm for terrain guarding repeatedly picks the guard that sees the most unguarded vertices. There are terrains for which this method achieves a logarithmic approximation factor (such a terrain, provided by Ben-Moshe [2], is described in [15]). There are other natural greedy-like algorithms that one might consider. For example, one could repeatedly pick the guard that maximizes the leftmost unguarded vertex or the lowest unguarded vertex. Terrains exist for these algorithms that prove they do not achieve constant approximation factors. The apparent absence of simple algorithms that achieve constant approximation factors motivates us to consider more sophisticated techniques.

1.4 Our Contribution

Our result is a 4-approximation algorithm for the 1.5D terrain guarding problem. It runs in $O(n^2)$ time for TG and can be modified slightly to run in $O(n^2)$ time for TG-C.

1.5 Organization

The rest of the paper is organized as follows. In Section 2 we introduce notation and some small but fundamental lemmas. In Section 3 we give our 4-approximation algorithm for TG; the modifications required for TG-C are explained in Section 3.6. In Section 4 we discuss open problems and suggest directions for future work regarding 1.5D terrain guarding.

2 Preliminaries

2.1 Terminology and Notation

An instance of the 1.5D terrain guarding problem is simply an x-monotone chain T. This chain is a sequence of vertices $v_1, \ldots, v_n$ and edges $e_i = (v_i, v_{i+1})$, $i = 1 \ldots n-1$ such that the x-coordinate of v_i is smaller than that of v_j if $i < j$. Given two points p and q on T (not necessarily vertices of T), we say that $p < q$ if the x-coordinate of p is smaller than that of q.

For a point p we use $L(p)$ to denote the leftmost point that sees p and $R(p)$ to denote the rightmost point that sees p. It is not difficult to see that $L(p)$ and $R(p)$ will always be vertices, whether p is a vertex or not. We use $T_L(p)$ to denote the terrain restricted to the interval $[v_1, p]$ and use $T_R(p)$ to denote the terrain restricted to $[p, v_n]$. $CH(T)$ is the (upper) convex hull of T. We use $CH_L(p)$ to denote the convex hull of $T_L(p)$ and use $CH_R(p)$ for that of $T_R(p)$. If a point p sees every unguarded point that another point q sees we say that p *dominates* q. We can also say that a set S dominates a point p if every unguarded point seen by p is also seen by some vertex in S. We say that p dominates q with respect to a certain region of T if p sees every unguarded point in that region that q sees.

We consider a minimum guarding set G_{OPT} for the terrain T. We assume there is some mapping g of points of T to guards in G_{OPT} such that, for a point p, $g(p)$ is a guard in G_{OPT} that sees p. We say that $g(p)$ is the guard *responsible for* p. g is surjective but never injective (since $|G_{OPT}| < n$); we use it to simplify the explanation of our accounting scheme.

2.2 Elementary Lemmas

We will now state and prove several small but fundamental lemmas that we will use in the rest of the paper. These lemmas and corollaries can be used with left and right interchanged; this is stated explicitly for Corollary 1 as an example but is not stated for the others. Also note that these lemmas involve points on the terrain that need not necessarily be vertices.

Lemma 1 (Order Claim [3, 5]). *For points a, b, c, d such that $a \leq b < c \leq d$, if a sees c and b sees d then a sees d.*

Proof. This becomes quite clear with the help of a diagram (see Figure 2). It is trivially true if $a = b$ or $c = d$; otherwise we know that $a < b < c < d$. In this case b cannot be above $\overline{ac}$ and c cannot be above $\overline{bd}$ (otherwise the fact that a sees c and b sees d would be violated). This means that the two line segments must cross; we call their intersection point p. Considering the triangle formed by a, p, and d, we note that no point on the terrain can be above the lower hull and $\overline{ad}$ is the upper hull. Therefore no point on the terrain can be above $\overline{ad}$.

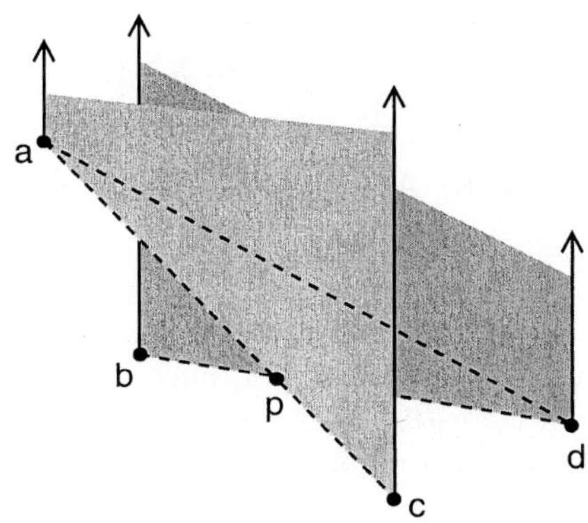

Fig. 2. The shaded areas are terrain free and their union contains $\overline{ad}$

Corollary 1. *For points u, v, w with $u \leq v < w$, if u and v can both be seen from $T_R(w)$ then $R(v) \leq R(u)$.*

Corollary 1 (Symmetric Version). *For points u, v, w with $u < v \leq w$, if v and w can both be seen from $T_L(u)$ then $L(w) \leq L(v)$.*

Lemma 2. *For an interval $[a, b]$ where a sees b, any guard in (a, b) is dominated with regard to $T_R(b)$ by a.*

Proof. Let p be a guard in (a, b) and let q be some point in $T_R(b)$ seen by p. If $q = b$ we know that a sees q. Otherwise the order claim, applied to a, p, b, q, states that a sees q.

Corollary 2. *For points p and q such that $p < q < R(p)$, we know that $R(q) \leq R(p)$.*

Lemma 3 (Lip Lemma). *For an interval $[a, b]$ where a sees b, if there are no unguarded points in (a, b) then $\{a, b\}$ dominates any guard in $[a, b]$.*

Proof. This follows from Lemma 2 since a and b see each other.

Lemma 4. *For a point q, any guard p in $T_L(q)$ is dominated with regard to $T_R(q)$ by a guard in $CH_L(q)$. In particular, p is dominated by the rightmost point in $T_L(p) \cap CH_L(q)$.*

Proof. Let u be the rightmost point in $T_L(p) \cap CH_L(q)$. If p is on $CH_L(q)$ then $p = q$ and the lemma clearly holds. Otherwise let w be the first point on $CH_L(q)$ to the right of u. Now u sees w, so u dominates p with regard to $T_R(q) \subseteq T_R(w)$ by Lemma 2.

Corollary 3. *For points p and q, if $L(q) \leq p \leq q$ then $L(q)$ is on $CH_L(p)$.*

3 The Algorithm

Our algorithm works by repeatedly finding an unguarded point u and a set S of up to 4 points such that S must dominate $g(u)$. By doing so, we achieve an approximation factor of 4, since we charge at most 4 guards to each guard in G_{OPT}. Our algorithm does not require any knowledge of previously placed guards other than which points are unguarded. The rest of this section basically deals with how to find an appropriate unguarded point. We first explain the algorithm as applied to TG, and in Section 3.6 we explain the minor modifications required for TG-C.

3.1 Introduction to GuardRight

Consider an unguarded vertex p not on $CH(T)$ along with a vertex c that can see every unguarded vertex in the range $[L(R(p)), p)$. c is like a good potential guard that lets us focus on the unguarded points in $[p, R(p)]$. Note that if we place a guard at c, no unguarded vertex in $[L(R(p)), R(p))$ can be seen from outside $[L(R(p)), R(p)]$. For this reason we say that the interval $[L(R(p)), R(p)]$ is *pseudo-independent.* Our algorithm repeatedly finds appropriate (p, c) pairs or advances trivially if such points are not available. If there is only one unguarded vertex s, we place a guard there that dominates $g(s)$. Otherwise consider the two leftmost unguarded vertices s and t with $s < t$. If $s \in CH(T)$ and $t \in CH(T)$ we just place a guard at $R(s)$ that must dominate $g(s)$. If $s \notin CH(T)$ then $p \leftarrow s$ and $c \leftarrow s$. If $s \in CH(T)$ but $t \notin CH(T)$ then $p \leftarrow t$ and $c \leftarrow s$.

If an appropriate (p, c) pair is found, our algorithm calls a recursive subroutine GuardRight(p, c). GuardRight will either find an unguarded vertex $u \in [p, R(p))$ for which $g(u)$ can be dominated by 4 guards or will find a pseudo-independent subinterval of $[p, R(p)]$, *i.e.* a pseudo-independent 'pocket' of the terrain that it can recurse into with new parameters p' and c'. At this point we introduce some new terminology and notation that depends on the parameters of GuardRight. It should be emphasized that this notation applies only to a particular call to GuardRight. We say that a *left vertex* is a vertex in $CH([L(R(p)), p]) - \{p\}$. A *right vertex* is a vertex in $[p, R(p)]$. An *exposed vertex* is an unguarded vertex in $[p, R(p))$ that can be seen by a left vertex. A *sheltered vertex* is an unguarded vertex in $[p, R(p))$ that cannot be seen by a left vertex. For an exposed vertex v we provide additional notation: $R'(v)$ is the rightmost left vertex that sees v and $L'(v)$ is the leftmost right vertex that sees v. $R'(v)$ and $L'(v)$ are undefined unless v is an exposed vertex.

Lemma 5. **(a)** *If v is a sheltered vertex then* $L(R(p)) \leq p \leq L(v) \leq v \leq R(v) \leq R(p)$. **(b)** *If v is an exposed vertex then* $L(R(p)) \leq L(v) \leq R'(v) < p \leq L'(v) \leq v \leq R(v) \leq R(p)$.

Proof. **(a)** $L(R(p)) \leq p$ since $R(p)$ sees p. $p \leq L(v)$ otherwise v would be an exposed vertex. $L(v) \leq v$ by definition. $R(v) \leq R(p)$ by Corollary 2. **(b)** $L(R(p)) \leq L(v)$ by Corollary 1 if $p < v$ and by the Order Claim otherwise. $L(v) \leq R'(v) < p \leq L'(v) \leq v$ by definition. $R(v) \leq R(p)$ by Corollary 2.

Lemma 6. *For an exposed vertex v, $L'(v)$ sees $R(v)$.*

Proof. If $v = p$ this is clearly true since $p = L'(p)$. Otherwise, it is easy to see that this is true as long as v is not above the line passing through $L'(v)$ and $R(v)$. $L'(v)$ cannot be below the line passing through p and v otherwise v would be seen by a vertex in $[p, L'(v))$ which contradicts the definition of $L'(v)$. Similarly, $R(v)$ cannot be below the line passing through v and $R(p)$. It should now be clear that v is not above the line passing through $L'(v)$ and $R(v)$ (see Figure 3). The rest follows trivially.

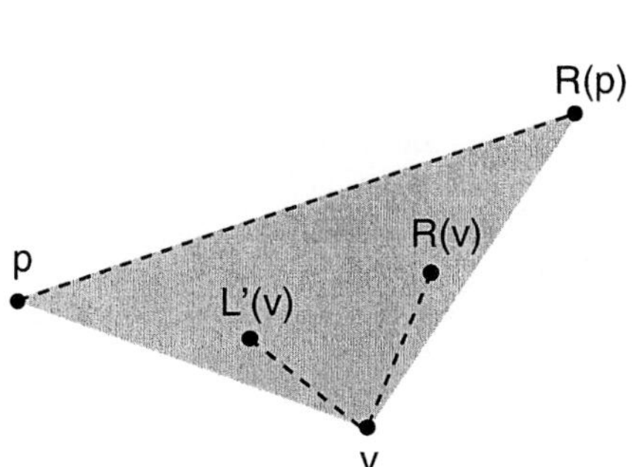

Fig. 3. $L'(v)$ and $R(v)$ must be in the shaded region

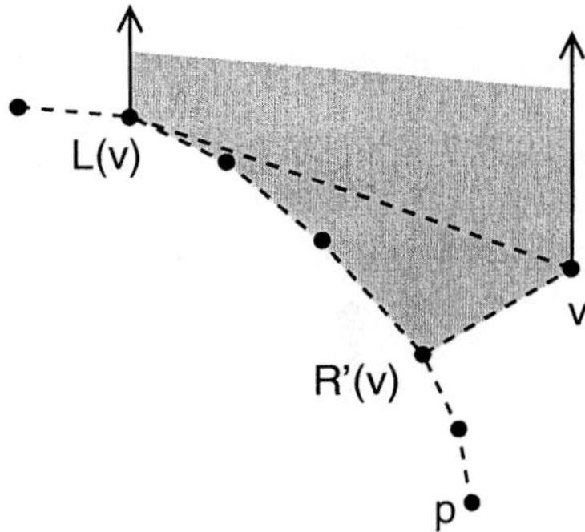

Fig. 4. The shaded region is terrain free, so every left vertex in $[L(v), R'(v)]$ must see v

Lemma 7. *For an exposed vertex v the set of left vertices that see v is contiguous,* i.e. *every left vertex in $[L(v), R'(v)]$ sees v.*

Proof. Consider $CH([L(v), R'(v)])$. This is a subset of $CH([L(R(p)), p]) - \{p\}$ since $L(v)$ and $R'(v)$ are both on $CH([L(R(p)), p])$. So we can see that $CH([L(v), R'(v)])$ is a set of left vertices and we know that no left vertex not in the set can see v. Now we will show that if w is a vertex in the set, $w \neq L(v)$, w sees v, and w' is the first vertex in the set to the left of w, then w' also sees v. w' must be above the line passing through v and w since $w \neq L(v)$. Since w' and w are consecutive points on the convex hull, w sees w'. Now we can see that $\overline{w'w}$ and $\overline{wv}$ are line segments that do not interfere with the terrain, so $\overline{w'v}$ cannot interfere with the terrain since it is above $\overline{w'w}$ and $\overline{wv}$. Therefore w' sees v. It is easy to extend this into an induction proof for the lemma. See Figure 4.

Lemma 8. *For any vertex v in $[L(R(p)), p)$, if w is the rightmost left vertex in $T_L(v)$ then $\{c, w\}$ dominates v.*

Proof. By Lemma 4 we know that w dominates v with regard to $T_R(p)$. If $w \neq v$ then v cannot see any vertex to the left of w so w dominates v with regard to $T_L(w)$. c can see every unguarded vertex in $[L(R(p)), p)$. Since $L(R(p)) \leq w$, we can see that $T_L(w) \cup [L(R(p)), p) \cup T_R(p) = T$. Therefore $\{c, w\}$ dominates v.

3.2 Finding a Good Left Vertex

Lemma 8 tells us that, as long as we place a guard at c when we place other guards, we needn't place any guard in $[L(R(p)), p)$ unless it is on a left vertex. The first thing we note is that there must be at least one exposed vertex in $[p, R(p))$, namely p. There may or may not be a sheltered vertex in $[p, R(p))$. We define b as the leftmost left vertex such that some exposed vertex v is seen by b but not by any left vertex to the right of b. In other words, b is the leftmost $R'(v)$ over all exposed vertices v. We define d as the leftmost exposed vertex for which $R'(d) = b$.

Lemma 9. *Every exposed vertex in $(L'(d), R(p))$ is seen by $L(d)$.*

Proof. If $d = p$ then the proof follows easily from the symmetric version of Corollary 1, so we will assume this is not the case. First we will prove that there are no exposed vertices in $(L'(d), d)$. Assume for the sake of contradiction that there is an exposed vertex v in $(L'(d), d)$. We can apply the order claim to $R'(v), L'(d), v, d$ to see that $R'(v)$ sees d. This tells us that $R'(v) \leq R'(d)$, which violates the definition of d, so there cannot be any such vertex v. Now we show that $L(d)$ sees every exposed vertex in $(d, R(p))$. Let w be an exposed vertex in $(d, R(p))$. We have $L(w) < d < w$ so by the symmetric version of Corollary 1 we know that $L(w) \leq L(d)$. By the definition of d we know that $R'(d) \leq R'(w)$. Therefore $L(d) \in [L(w), R'(w)]$, so by Lemma 7 we know that $L(d)$ sees w.

Lemma 10. *Any guard in $[L(R(p)), p)$ that sees d is dominated by $\{L(d), c\}$.*

Proof. Let v be a guard in $[L(R(p)), p)$ that sees d. Since c sees every unguarded vertex in $[L(R(p)), p)$ it suffices to prove that $L(d)$ dominates v with regard to $T_R(p)$. $L(d) \leq v$, so by Lemma 2 $L(d)$ dominates v with regard to $T_R(d)$. Now we show that no left vertex that sees d can see any exposed vertex to the left of d. It follows from the definition of $b = R'(d)$ and from Lemma 7 that any exposed vertex seen from the left of $R'(d)$ must be seen by $R'(d)$. However, d is the leftmost exposed vertex seen by $R'(d)$, so no exposed vertex to the left of d can be seen by $R'(d)$. In other words, no exposed vertex to the left of d can be seen by a left vertex that sees d. This, along with Lemma 4, tells us that v cannot see any unguarded vertices in $[p, d)$. Since v cannot see any sheltered vertices at all, this means that $L(d)$ dominates v with regard to $T_R(p)$. v cannot see anything left of $L(R(p))$ except possibly if $v = L(d)$, so $\{L(d), c\}$ dominates v over the entire terrain.

Recall that, while searching for a suitable vertex u for which we can dominate $g(u)$ with 4 guards, either we find one right away or we find some pseudo-independent pocket (*i.e.* a subinterval of $[p, R(p)))$ that we can recurse upon.

3.3 The Terminal Case

We first consider the case where there are no sheltered vertices in $(L'(d), R(d))$. We place guards at $\{c, L(d), L'(d), R(d)\}$ and claim that these guards dominate any guard that sees d. Lemma 9 tells us that every exposed vertex in $(L'(d), R(d))$ is seen by $L(d)$, and since there are no sheltered vertices in $(L'(d), R(d))$ there are no longer any unguarded vertices in $(L'(d), R(d))$. $L'(d)$ sees $R(d)$ by Lemma 6. Therefore, by Corollary 3, any guard in $[L'(d), R(d)]$ is dominated by $\{L(d), L'(d), R(d), c\}$. By Lemma 10 any guard in $[L(R(p)), p]$ that sees d is dominated by $\{L(d), L'(d), R(d), c\}$. Any guard that sees d must either be in $[L(R(p)), p]$ or in $[L'(d), R(d)]$, so $\{L(d), L'(d), R(d), c\}$ dominates any guard that can see d.

3.4 The Recursive Case

If there are sheltered vertices in $(L'(d), R(d))$ our job is slightly more complicated and requires recursion (this is where we find our pseudo-independent pocket). We require another subroutine, GuardLeft, that is simply a mirror image of GuardRight. For a call to GuardLeft(p', c') the condition that c' must satisfy is flipped horizontally: every unguarded vertex in $(p', R(L(p'))]$ must be seen by c'. Also, p' cannot be on $CH(T)$.

Let q be the rightmost sheltered vertex (note that q is not necessarily in the interval $(L'(d), R(d))$, but it must be in $(L'(d), R(p)))$. We will show that the preconditions are satisfied if we call GuardLeft$(q, L(d))$. By Corollary 2 $R(L(q)) \leq R(p)$ and by the definition of q any unguarded vertex in $(q, R(p))$ is an exposed vertex. Therefore by Lemma 9 every unguarded vertex in $(q, R(L(q)))$ is seen by $L(d)$. If $R(L(q)) < R(p)$ then either $R(L(q))$ is already guarded or it is an exposed vertex and is seen by $L(d)$. If $R(L(q)) = R(p)$ then $L(d)$ sees $R(L(q))$ since every vertex in $CH([L(R(p)), p])$ sees $R(p)$. Therefore every unguarded vertex in $(q, R(L(q))]$ is seen by $L(d)$. We know q is unguarded and $q \in (p, R(p))$ (and is therefore not on $CH(T)$), so the preconditions are satisfied.

In this way we can do a sort of recursive zig-zagging where each call to GuardRight will spawn a call to GuardLeft and each call to GuardLeft will spawn a call to GuardRight. It is not difficult to see that eventually, after at most a linear number of these zig-zagging steps, we will arrive at a terminal case. At this point we can simply place our 4 guards and, if we need to, start a brand new call to GuardRight.

3.5 Time Complexity

It is clear that at most $O(n)$ initial calls to GuardRight can be made. For TG it is also easy to see that an initial call to GuardRight will result in a number of guards being placed in $O(n^2)$ time. We can therefore give an upper bound of $O(n^3)$ for the running time of TG.

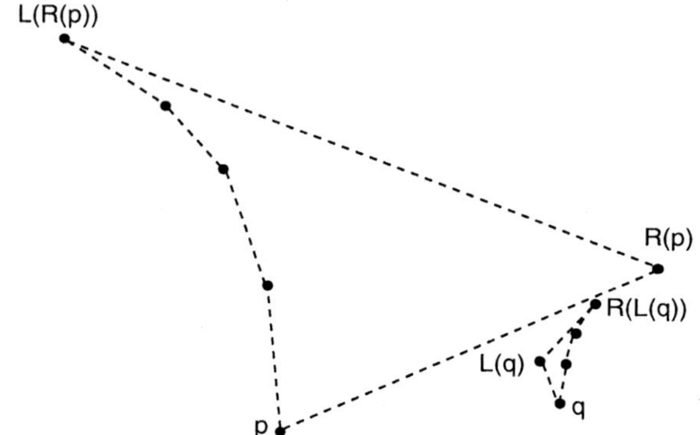

Fig. 5. The nested interval $[L(q), R(L(q))]$ can be handled independently with the help of a dominant outside vertex

If we want TG to be more efficient, we can make GUARDRIGHT(p, c) continue placing guards until $[p, R(p))$ has been completely guarded. This changes things slightly; on a given iteration, p is not necessarily unguarded so there is not necessarily an exposed vertex. If there is no exposed vertex, however, we can just recurse immediately by calling GUARDLEFT(q, c) so this is not a problem. To increase efficiency, we can sort the exposed vertices v by $R'(v)$ (breaking ties using the x-coordinates of exposed vertices) to find an appropriate b and d faster in each iteration. A call to GUARDRIGHT(p, c), ignoring all recursive calls that it spawns, can now run in $O(n + m \log m)$ time, where m is the number of exposed vertices in $[p, R(p))$. It is easy to see that the 'n' terms, added up over the entire course of the algorithm, will cost $O(n^2)$ time since there will be at most $O(n)$ calls to GUARDRIGHT. Any vertex will be an exposed vertex for at most one call to GUARDRIGHT, so the sum of all $m \log m$ factors encountered will actually be bounded by $O(n \log n)$. All other overhead incurred by the algorithm can be dealt with in $O(n^2)$ time, so the running time of TG is bounded by $O(n^2)$.

3.6 Modifications for TG-C

No real modifications need to be made to apply our TG algorithm to TG-C. However, we need to keep track of more information if we want our algorithm to run as efficiently as possible. When dealing with TG-C the only real problem is finding b and d at each iteration of a call to GUARDRIGHT. Instead of exposed vertices and sheltered vertices, we consider exposed edge sections and sheltered edge sections. It is not difficult to see that for each edge of the terrain, at most one contiguous section will be exposed and at most one will be sheltered. From left to right on an edge, we can have a guarded section, a sheltered section, an exposed section, and another guarded section, though not all of these sections will necessarily exist. For an exposed section, the leftmost point will have the leftmost R'.

$L(p)$ is always a vertex regardless of where p is. If we keep track of the transition points for the function $L(p)$ (there are only $O(n)$ of them [7]) then we can know where exposed sections end and sheltered sections begin. For every edge, our algorithm also keeps track of where the unguarded section starts and ends (it must be contiguous). After placing a guard, updating the unguarded section on every edge can be done quite easily in linear time. Assume we have just placed a guard at g. To the left of g call the first vertex v_1 and consider the edge e_1 whose left endpoint is v_1. Mark down that every point on e_1 is guarded. Now, moving left from v_1, find the first vertex above the line going through g and v_1; call this v_2, define e_2 appropriately and mark down that every point on e_2 above the line going through g and v_1 is guarded. It is easy to see how we can proceed to update the unguarded section of each edge in linear time. Since we place $O(n)$ guards the total cost of updating guarded edge sections of the terrain is $O(n^2)$.

If we do all of the aforementioned maintenance, we will only need to consider the leftmost point in each exposed section when looking for b and d. Therefore we do not need to worry about asymptotically more points in TG-C than in TG. The running time therefore remains $O(n^2)$.

4 Conclusions and Future Work

The 1.5D terrain guarding problem is not known to be in P. Constant-factor approximation algorithms for the problem have only recently been developed. We have provided an $O(n^2)$ time 4-approximation algorithm for both the discrete and continuous versions of the problem. Ours is the best known algorithm for the 1.5D terrain guarding problem.

The most pressing and obvious question regarding the 1.5D terrain guarding problem is whether or not it is NP-complete. All of our attempts at an NP-hardness proof have been stymied by the Order Claim. On the other hand, attempts at designing an exact polynomial-time algorithm have also been unsuccessful. If the problem is not NP-hard, we would be interested in a polynomial-time algorithm. If the problem is NP-hard, we would be interested in approximability thresholds, *e.g.* whether it is APX-complete or admits a PTAS or even an FPTAS. If the problem is APX-complete, the approximation factor should be lowered as much as possible.

Acknowledgements

This research was done under the supervision of Will Evans; I am extremely grateful for the help and guidance he gave me. I am also grateful to Boaz Ben-Moshe for the helpful discussions we had. Finally, I thank the anonymous referees for their helpful comments.

References

1. A. Aggarwal. *The art gallery problem: Its variations, applications, and algorithmic aspects.* PhD thesis, Johns Hopkins University, 1984.
2. B. Ben-Moshe. Personal communication, 2005.
3. B. Ben-Moshe, M. Katz, and J. Mitchell. A constant-factor approximation algorithm for optimal terrain guarding. In *Symposium on Discrete Algorithms*, 2005.
4. H. Brönnimann and M. T. Goodrich. Almost optimal set covers in finite VC-dimension. *Discrete & Computational Geometry*, 14, 1995.
5. D. Z. Chen, V. Estivill-Castro, and J. Urrutia. Optimal guarding of polygons and monotone chains (extended abstract), 1996.
6. V. Chvátal. A combinatorial theorem in plane geometry. *J. Comb. Theory Series B*, 18:39–41, 1975.
7. K. L. Clarkson and K. Varadarajan. Improved approximation algorithms for geometric set cover. In *Symposium on Computational Geometry*, 2005.
8. S. Eidenbenz. *(In-)Approximability of Visibility Problems on Polygons and Terrains.* PhD thesis, ETH Zurich, 2000.
9. S. Eidenbenz. Approximation algorithms for terrain guarding. *Information Processing Letters*, 82(2):99–105, April 2002.
10. S. Eidenbenz, C. Stamm, and P. Widmayer. Positioning guards at fixed height above a terrain — an optimum inapproximability result. *Lecture Notes in Computer Science*, 1461, 1998.
11. S. Eidenbenz, C. Stamm, and P. Widmayer. Inapproximability results for guarding polygons and terrains. *Algorithmica*, 31(1):79–113, 2001.
12. U. Feige. A threshold of $\ln n$ for approximating set cover. *Journal of the ACM*, 45(4):634–652, July 1998.
13. M. Garey and D. Johnson. *Computers and Intractibility: A Guide to the Theory of NP-Completeness.* W.H. Freeman and Co., 1979.
14. D. Haussler and E. Welzl. ϵ-Nets and Simplex Range Queries. *Discrete & Computational Geometry*, 2:127–151, 1987.
15. James King. Approximation algorithms for the 1.5 dimensional terrain guarding problem. Master's thesis, University of British Columbia, 2005.
16. A. A. Kooshesh and B. M. E. Moret. Three-coloring the vertices of a triangulated simple polygon. *Pattern Recognition*, 25, 1992.
17. D. T. Lee and A. K. Lin. Computational complexity of art gallery problems. *IEEE Transactions on Information Theory*, 32:276–282, 1986.
18. M. Sharir and P. K. Agarwal. Davenport-Schinzel sequences and their geometric applications. Technical report, 1995.

On Sampling in Higher-Dimensional Peer-to-Peer Systems

Goran Konjevod*, Andréa W. Richa**, and Donglin Xia*

Department of Computer Science and Engineering,
Arizona State University, Tempe AZ 85287, USA
{goran, aricha, dxia}@asu.edu

Abstract. We present fully distributed algorithms for random sampling of nodes in peer-to-peer systems, extending and generalizing the work of King and Saia [Proceedings of PODC 2004] from simple Chord-like distributed hash tables to systems based on higher-dimensional hierarchical constructions, like Content Addressable Networks (CAN). We also show preliminary results on the generalization of the problem to biased sampling. In addition, we provide an extension of CAN that requires only $O(1)$ space per node and achieves $O(\log n)$ lookup latency and message complexities.

1 Introduction

A distributed algorithm for random sampling of nodes in a peer-to-peer network provides a basic ingredient for the solution of several important problems, including load-balancing, Byzantine agreement and computing various statistics on data availability.

King and Saia [6] present a fully-distributed algorithm with expected logarithmic message complexity that with high probability (at least $1 - O(1/n)$) chooses a peer with *exactly* uniform distribution (i.e. with probability $1/n$). Both the expected latency (in terms of the number of links in the overlay network followed by the algorithm) and message complexity of their algorithm are logarithmic in the number of nodes. The algorithm does not assume any knowledge of the number of nodes n in the network. Their algorithm requires a peer-to-peer network with properties similar to those of Chord [11], in particular a one-dimensional circular keyspace.

In some applications, such as as peer-to-peer photo sharing and massively multiplayer games, multidimensional range-queries are critical, and the overlay network should support higher dimensional range queries [4]. Several structured overlay systems based on the geometry of two- or higher-dimensional space have been proposed, including CAN [9]. In this paper, we provide an efficient mechanism for *uniform random sampling in peer-to-peer overlay networks of higher*

* Supported in part by NSF grant CCR-0209138.
** Supported in part by NSF CAREER grant CCR-9985284.

J.R. Correa, A. Hevia, and M. Kiwi (Eds.): LATIN 2006, LNCS 3887, pp. 641–652, 2006.

dimensions. Our work applies to peer-to-peer systems with properties similar to those of CAN.

The algorithm of King and Saia is based on a procedure to associate with each peer a part of the keyspace so that **(a)** the parts assigned to different peers are disjoint, **(b)** the measure of the part assigned to each peer is exactly the same, and **(c)** a constant fraction of the total keyspace is assigned to the peers. These properties reduce the problem of peer sampling to that of sampling of points in the keyspace, because it ensures that only an expected constant number of random keys must be sampled before one is generated that lies in a region belonging to some peer.

It may be possible to generalize this algorithm directly to multiple dimensions by assigning to peers disjoint regions of the keyspace. A natural idea would be to assign to each peer its region in the Voronoi diagram of the set of peer nodes. However, even on a line, the distances between consecutive points in a set of n uniformly generated points vary with high probability from $\Theta(1/n^2)$ to $\Theta(\ln n/n)$ [6]. Thus the Voronoi regions need to be patched to give each peer an equal-area region and the resulting structure very quickly becomes prohibitively complex, even in two dimensions.

The reader may already have noticed a direct reduction of our problem to the one-dimensional case: concatenate the coordinates of a node's ID (assumed to be binary sequences) to get a single binary sequence. If each coordinate is uniformly distributed, then so is the resulting sequence. These sequences define coordinates of the nodes on a circle (ring). Building an overlay network (say, Chord) on top of this circle allows one to directly use the algorithm of King and Saia. This effectively creates two overlays: a multi-dimensional one based on the original node coordinates, which allows multi-dimensional queries, and a one-dimensional one that allows random sampling.

There are two main problems with the approach above. First, it creates unnecessary overhead by the need of building and maintaining a second one-dimensional overlay network to be used just for sampling. Second, since the random sampling will be done via the auxiliary one-dimensional overlay network, which does not take into account the proximity of the nodes in the multidimensional space, the expected latency of the algorithm *may no longer be logarithmic* with respect to the *original* multidimensional overlay network.

Hence, in contrast to this simplistic approach, we use the Hilbert space-filling curve [10] to map the multi-dimensional keyspace into a circle, and a single multi-dimensional overlay (say, CAN) to implement a routing table and other functions useful for the peer-to-peer network, including those needed for the sampling algorithm itself. Our goal is to assign to each peer a part of the keyspace of equal volume, while keeping these parts disjoint and large enough to jointly cover a constant fraction of the whole keyspace. Thanks to the properties of the Hilbert curve, contiguous segments of the circle correspond to connected regions in the original keyspace. This means that any basic step of the algorithm that searches linearly through a bounded number of peers along the circle will only have to consider a bounded connected region in the original keyspace, thereby ensuring low latency (logarithmic in the number of nodes) in terms of the number of links followed in the multi-dimensional overlay network, unlike the simplistic solution above. In addition to this advantage, the use of a "native" overlay allows for the implementation of basic functions in the overlay network using the stronger geometric properties of the multidimensional space.

1.1 Our Contributions

Given a fully decentralized peer-to-peer network in a multidimensional keyspace, we consider the problem of distributed random sampling of peer nodes. We assume that peers correspond to uniformly distributed points in a d-dimensional keyspace. We show that the one-dimensional sampling algorithm of King and Saia can be generalized to hierarchical systems with multidimensional keyspaces.

The sampling algorithm for the one-dimensional case relies on a basic function *next* that, given a peer, returns the next peer along the circle. In addition to this, the algorithm requires the ability to find the location of a given peer on the circle and to compute the measure of an interval between two given points. All we need assume is that the functions listed above can be computed efficiently by the peer-to-peer system, in other words, that the peer-to-peer system we use is compatible with the Hilbert curve. Our main result is stated in the theorem below:

Theorem 1. *Given a peer-to-peer d-dimensional keyspace satisfying the properties above, the algorithm* **Choose A Random Peer** *selects each peer with probability exactly* $1/n$*, with high probability. The algorithm has expected latency* $O(t_{lookup} + \log n)$ *and sends* $O(m_{lookup} + \log n)$ *messages, where* t_{lookup} *and* m_{lookup} *are the latency and message complexities of* lookup. *In particular, a CAN can be implemented so that for any dimension d,* $t_{lookup} = m_{lookup} = O(\log n)$*, and* $O(1)$ *routing items in each peer are maintained for* lookup. *Therefore, in CAN the latency and message complexity of our algorithm are both* $O(\log n)$ *in expectation.*

2 Background and Related Work

2.1 The Hilbert Curve

G. Peano discovered the first curve that passes through every point of a closed square [7], but one of the first graphical representations of such a space-filling curve was given by David Hilbert [5]. Space-filling curves are useful for reducing a multi-dimensional problem to a one-dimensional problem, and in this role have found uses in other contexts [3].

The Hilbert space-filling curve is a continuous mapping from the unit interval $[0, 1]$ onto the unit hypercube $[0, 1]^d$. It can be constructed recursively. First the d-dimensional cube is partitioned into 2^d congruent subcubes and accordingly, the unit interval into 2^d congruent subintervals. Then each subinterval is mapped onto a distinct subcube, and adjacent subintervals are mapped onto

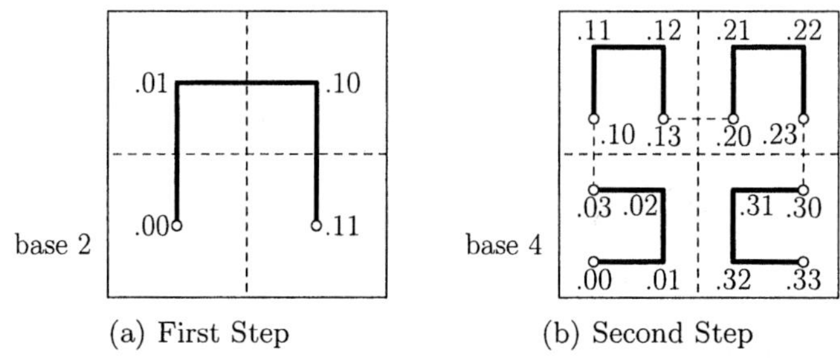

(a) First Step (b) Second Step

Fig. 1. The Generation of 2 Dimensional Hilbert Curve

adjacent subcubes with a common facet. The example for 2 dimensions are presented in Figure 1(a). The same algorithm is applied to each subcube and its corresponding subinterval. The subinterval in each subcube is rotated and reflected so that it can connect to the preceding one to form a single continuous unit interval. Figure 1(b) shows the second step of the construction in 2 dimensions. Albert [2] provides a mathematical formalism for the Hilbert curve in arbitrary dimension d.

We list several properties about the Hilbert curve in the following:

Proposition 1. *Each interval* $[0.b_1b_2..b_i, 0.b_1b_2..b_i + 2^{-i}]$ *of the curve fills a region of volume* 2^{-i}*, where* $i \in N$*, and* $b_j = 0, 1$*, for* $j = 1..i$*. We label this region by the i-digit binary sequence* $b_1b_2..b_i$*. Moreover, the region is a hyper-rectangle which can be split into two congruent sub-hyper-rectangles labeled by* $b_1b_2..b_i0$ *and* $b_1b_2..b_i1$*.*

In fact, we will think of the unit hypercube $[0, 1]^d$ as a wrapped d-torus, identifying for each coordinate i every pair of points $x, y \in [0, 1]^d$ such that $|x_i - y_i| = 1$ and $x_j = y_j$ for $i \neq j$, and the Hilbert curve $[0, 1]$ as a unit circle.

Proposition 2. *For any two regions labeled as* $A = a_1a_2..a_i$*,* $B = b_1b_2..b_j$*,* $(i \leq j)$*, if* A *is a prefix of* B*, then region* A *contains region* B*. If* A *is not a prefix of* B*, then the intersection of these two regions has volume of 0.*

Furthermore if $0.a_1a_2..a_i + 2^{-i} = 0.b_1b_2..b_j$*, or* $0.b_1b_2..b_j + 2^{-j} = 0.a_1a_2..a_i$*, i.e. their mapped intervals on the curve are connected, then regions* A *and* B *share a* $(d-1)$*-dimensional facet with positive* $(d-1)$*-dimensional volume.*

Proposition 3. *A random process of* n *points uniformly distributed on the Hilbert curve is equivalent to a random process of* n *points uniformly distributed in the unit hypercube according to the Hilbert curve mapping.*

2.2 CAN: A d-Dimensional Peer-to-Peer Network

CAN (Content-Addressable Network)[9] is a peer-to-peer system which takes a d-dimensional Cartesian coordinate space as the keyspace for its distributed hash

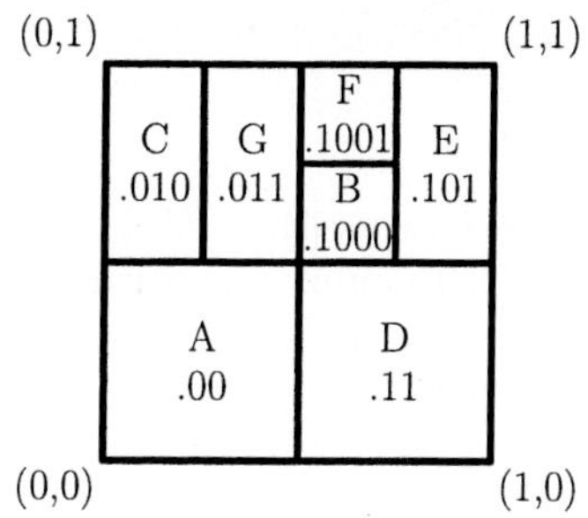

Fig. 2. A 2-Dimensional CAN with 6 Peers

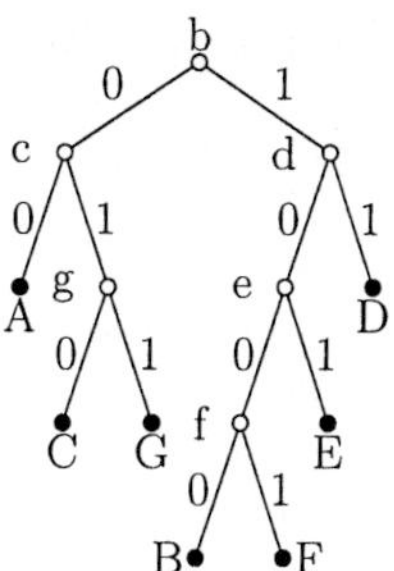

Fig. 3. The Binary Partition Tree for the CAN in Figure 2

table. The coordinate space is partitioned into hyper-rectangles, called zones, and each peer is responsible for a zone. Then a key is stored in the peer whose zone contains the key point. For example, Figure 2 shows a 2-dimensional $[0, 1] \times [0, 1]$ coordinate space with 6 peers.

Each peer maintains a routing table of all its neighbors in the coordinate space. Two peers are neighbors if they share a $(d-1)$-dimensional facet. Given a key, the lookup operation, or routing, in CAN is implemented by following the straight-line path through the coordinate space from the inquiring peer to the peer storing the key — that is, a peer will forward the lookup message through the peers responsible for the regions crossed by the respective straight-line path. Thus each peer maintains $2d$ neighbors and the average routing path length is $(d/4)(n^{1/d})$.

To join the network, a new peer first selects a random key point, and asks an existing peer to find the peer p who stores the key point. Then peer p will split its zone in half according to a given order of the dimensions, keeping one half of the zone assigned to itself and assigning the other half to the new peer.

We adapt the process by which CAN splits a zone to add a new peer to match the construction of the Hilbert curve. First, the whole key space $[0, 1]^d$ can be partitioned into two zones labeled by 0 and 1. Then by Proposition 1, a zone assigned to a peer p and labeled by $b_1b_2..b_i$ can be partitioned into two equally-sized zones labeled by $b_1b_2..b_i0$ and $b_1b_2..b_i1$. The peer p will keep the half zone labeled by $b_1b_2..b_i0$ and assign the other half labeled $b_1b_2..b_i1$ to the new peer. Therefore we have the following lemma:

Lemma 1. *The whole key space is partitioned into zones labeled by binary sequences. The zone labeled by $b_1b_2..b_i$ is filled by an interval $[0.b_1b_2..b_i, 0.b_1b_2..b_i + 2^{-i}]$ of the Hilbert curve. Therefore these intervals also partition the Hilbert curve.*

Thus each zone is labeled by a unique binary sequence. We will also use the zone's label to identify the peer responsible for the zone. By Lemma 1, the Hilbert curve is partitioned into intervals, each of which fills a distinct zone. Therefore, we have the following Lemma:

Lemma 2. *Given a peer labeled by $A = a_1a_2..a_i$, there exists one and only one peer labeled by $B = b_1b_2..b_j$ such that $0.a_1a_2..a_i + 2^{-i} = 0.b_1b_2..b_j$. Peers A and B are neighboring regions in the coordinate space.*

We call peer B the *next peer* of A, denoted by $B = next(A)$.

The routing scheme of CAN provided in [9] is a simple greedy routing in which the average routing path length is $(d/4)(n^{1/d})$. In fact, we can improve the expected routing path length to $O(\log n)$ by maintaining only $O(1)$ routing items at each node. We will maintain a binary tree that mimics the partitions performed by CAN[1].

We maintain a binary partition tree with n leaves according to the keyspace partition. Figure 3 illustrates the binary partition tree for the network shown in Figure 2. Each leaf of the binary tree corresponds to an existing zone (peer). Each inner node of the tree represents a zone that no longer exists, but was split at some previous time. The children of a tree node are the two zones into which it was split.

On the other hand, let each existing peer represent its corresponding leaf and the inner node that was split when the peer joined the network. Thus we have

Lemma 3. *The binary partition tree for CAN can be maintained with only $O(1)$ routing items in each peer.*

Now we show that the binary partition tree is well balanced.

Lemma 4. *The binary partition tree for CAN with n peers has $C_1 \log n \le d_1(n) \le d_2(n) \le C_2 \log n$ with high probability, where $C_1 < 1$ and $C_2 > 2$ are constants, $d_1(n)$ is the distance from the root to the closest leaf, and $d_2(n)$ is the depth of the tree.*

Proof. (1) $C_1 \log n \le d_1(n)$:Consider the full binary tree T with $C_1 \log n$ depth. The n^{C_1} leaves of tree T are considered as bins and the n peers as balls. Let X count the number of balls in a certain bin. Thus $E(X) = n^{1-C_1}$. By Chernoff bound, for any $0 < \delta < 1$, $Pr\{X < (1-\delta)E(X)\} < exp(-E(X)\delta^2/2)$. By setting $(1-\delta)E(X) = 1/2$, i.e. $\delta = 1 - 1/(2n^{1-C_1})$, we have

$$\begin{aligned} Pr\{X < 1/2\} &< exp(-n^{1-C_1}(1 - 1/(2n^{1-C_1}))^2/2) \\ &< exp(-n^{1-C_1}/8) \text{ (For } C_1 < 1,\ 1 - 1/(2n^{1-C_1}) > 1/2). \end{aligned} \tag{1}$$

Now the probability that there is a bin with less than $1/2$ ball is less than $n^{C_1} \cdot e^{-n^{1-C_1}/8} < e^{C_1 \ln n - n^{1-C_1}/8}$, which can be arbitrarily small for any constant $C_1 < 1$, and n large enough. Thus with high probability the number of balls in every bin is larger than or equal to $1/2$, thereby larger than or equal to 1. Thus $C_1 \log n \le d_1(n)$.

(2) $d_2(n) \le C_2 \log n$: Consider the full binary tree T with $C_2 \log n$ depth. The n^{C_2} leaves of tree T are considered as bins and the n peers as balls. Let X

[1] Note that this binary tree need not replace the standard greedy routing algorithm of CAN (depending on the application, this may not be desirable). This binary tree will be used for finding a suitable routing path during random sampling operations.

count the number of balls in a certain bin. Thus $E(X) = n^{1-C_2}$. By Chernoff bound, for any $\delta > 0$, we have $Pr\{X > (1+\delta)E(X)\} < (\frac{e}{1+\delta})^{(1+\delta)E(X)}$. By setting $(1+\delta) = 2n^{C_2-1}$, we have $Pr\{X \geq 2\} \leq (\frac{e}{2n^{C_2-1}})^2$. Thus the probability that there is a bin with more than or equal to 2 balls is less than or equal to $n^{C_2} \cdot (\frac{e}{2n^{C_2-1}})^2 = \frac{e^2}{4n^{C_2-2}}$, which can be arbitrarily small for any constant $C_2 > 2$ and n large enough. Therefore $d_2(n) \leq C_2 \log n$.

Therefore by maintaining such a binary partition tree we can achieve the average routing path length of $O(\log n)$ and $O(1)$ routing items in each peer.

3 Algorithm and Analysis

3.1 Estimating the Number of Peers

Since we want to choose each peer with the same probability, i.e., with probability $1/n$, where n is the number of peers in the network, it is clear that we must learn n in some sense. However, it is hard to count all the peers and keep n updated in a fully-distributed setting. Luckily, only an approximation to n is enough to sample peers uniformly. Before presenting the main algorithm, we first describe an algorithm by which a peer may estimate n to within a constant multiplicative factor. This is based on [6].

First the algorithm estimates n within a constant exponent by $\hat{n}_1$, the inverse of the volume of peer p's zone. Then it sums the volume of $c_1 \ln \hat{n}_1$ peers counting from p by the order of the *next* function. Finally n is estimated by the ratio of the number of these peers over their volume summation. The tightness of the estimation is determined by the constant c_1.

The algorithm is given as follows, where $vol(p)$ is the volume of peer p's zone, $next^{(s)}(p)$ means applying the *next* function s times starting from p, and $Vol(q,p)$ is the sum of the volumes of the peers from q to p, i.e.

$$Vol(q,p) = \sum_{\text{peer } r \text{ s.t. } 0.r \in [0.q, 0.p]} vol(r) \tag{2}$$

Estimate n

1. $\hat{n}_1 \leftarrow vol(p)^{-1}$
2. $s \leftarrow c_1 \ln \hat{n}_1$
3. $t \leftarrow Vol(p, next^{(s)}(p))$
4. Return $\hat{n}_2 \leftarrow s/t$.

We show that the above algorithm estimates n within a constant factor, basing our analysis on that of [6]. The main difference is that their result is based on the assumption that the peers are uniformly distributed in a unit circle, while in CAN each peer is responsible for a d-dimensional zone and cannot be abstracted as a point. Nevertheless, the peer in CAN is generated by a random key point, which allows us to generalize the analysis to work in our scenario. Note that by

Proposition 3 a random point process in the curve is equivalent to a random point process in d-dimensions according to the Hilbert curve mapping. while this observation is helpful, we point out that this equivalence only makes our proof simpler, and that in CAN a random key point is just selected directly in d-dimensions.

When a peer arrives, a random key point is generated by the random process on the Hilbert curve. Therefore for a peer p in CAN, we define its **original point** as the point on the Hilbert curve which corresponds to its random key point.

We list the notations used throughout the paper:

- For a peer p, let $x(p)$ denote its original point on the Hilbert curve.
- For x, y on the unit circle (Hilbert curve), define the distance from x to y as $d(x, y) = y - x$ if $y \geq x$ and $d(x, y) = (1 + y) - x$ otherwise.
- For any interval I of the unit circle, denote by $num_x(I)$ the number of random points in I.
- For any interval I in the unit circle, denote by $num_p(I)$ the number of peers s such that $0.s \in I$.
- For given functions $f(n)$ and $g(n)$, we say that $f(n)$ is a (γ_1, γ_2) *approximation* of $g(n)$ if $\gamma_1 g(n) \leq f(n) \leq \gamma_2 g(n)$.
- For any two peers p and q, the number of peers from q to p is given by $num(q, p) = |\{\text{peer } r \text{ s.t. } 0.r \in [0.q, 0.p]\}|$.

Since the original points of peers in CAN are considered as a random point process in the unit circle, we can relate the peers in CAN to their original points and generalize the original results.

Lemma 5. *For any two neighboring peers p and q, (e.g. $q = next(p)$), let X be the number of original points in $[0.p, 0.q]$. Then $E(X) = 1 - vol(p)$.*

Proof. Let $vol(p) = 2^{-t}$. Let Z_i be the zone with volume 2^{-i} that contains the zone p, for $i = 0, \ldots, t-1$. Note that zone Z_i is split into half when a new peer joins and the random point falls in the region of zone Z_i. Let x_i be a variable to indicate whether a random point falls in the region of zone p when zone Z_i is split. Then $E(x_i) = 2^{-t}/2^{-i} = 2^{i-t}$. Thus $E(X) = \sum_{i=0}^{t-1} E(x_i) = 1 - 2^{-t} = 1 - vol(p)$.

For reference, we list the lemma from [6] that we use in our proof:

Lemma 6. *[6] If n points are distributed uniformly at random in the unit circle, let $\alpha_1, \alpha_2, \epsilon$ be fixed positive constants with $\alpha_1 < \alpha_2$ and $0 \leq \epsilon \leq 1/2$. Let $C > 144/(\alpha_1 \epsilon^2)$. Then for two any interval I on the unit circle such that $C\alpha_1 \ln n \leq num_x(I) \leq C\alpha_2 \ln n$, we have $C(1-\epsilon)\alpha_1(\ln n/n) \leq |I| \leq C(1+\epsilon)\alpha_2 \ln n/n$ with probability at least $1 - 1/n$.*

Lemma 7. *Let $\alpha_1, \alpha_2, \epsilon$ be fixed positive constants with $\alpha_1 < \alpha_2$ and $0 \leq \epsilon \leq 1/2$. Let $C > 144/(\alpha_1 \epsilon^2)$. Then for any two peers p and q such that $C\alpha_1 \ln n \leq num(p, q) \leq C\alpha_2 \ln n$, we have $\frac{1}{2} \cdot C(1-\epsilon)\alpha_1(\ln n/n) \leq Vol(p, q) \leq 3 \cdot C(1 + \epsilon)\alpha_2 \ln n/n$ with probability at least $1 - 1/n$.*

Proof. Let $I = [0.p, 0.p + Vol(p,q)]$. Since the next peer t of q has $0.t = 0.q + vol(q) = 0.p + Vol(p,q)$, we have $I = [0.p, 0.t]$. Let $num_p(I)$ be the number of peer s such that $0.s \in I$, i.e. $num_p(I) = num(p,t)$. Then $num_p(I) = num(p,q) + 1$. Let peer r be the first peer laid down in the network such that $0.r \in [0.p, 0.t]$.

(1) $\frac{1}{2} \cdot C(1-\epsilon)\alpha_1(\ln n/n) \leq Vol(p,q)$

Let $I' = [2 \cdot 0.p - 0.r, 2 \cdot 0.t - 0.r]$. Then $|I'| = 2|I|$. For any peer s such that $0.s \in [0.p, 0.t]$, $s \neq r$, we have $|x(s) - 0.r| \leq 2|0.s - 0.r|$. Then $x(s) \in I'$ since $0.s \in [0.p, 0.t]$. Thus $num_x(I') \geq num(p,q) + 1 - 1 \geq C\alpha_1 \ln n$. Thus by Lemma 6, $|I'| \geq C(1-\epsilon)\alpha_1(\ln n/n)$. Therefore $Vol(p,q) = |I| \geq \frac{1}{2} \cdot C(1-\epsilon)\alpha_1(\ln n/n)$.

(2) $Vol(p,q) \leq 3 \cdot C(1+\epsilon)\alpha_2 \ln n/n$

Let $I' = [0.p', 0.t']$ where p' and t' be the first peer laid down in the network such that $0.p' \in [0.p, 0.r)$ and $0.t' \in (0.r, 0.t]$ respectively. Then $|I'| \geq \frac{|I|}{2}$.

Let $num_x(I') = x_1 + x_2$, where x_1 is the number of such peers s that $x(s) \in I'$ and $0.s \in I'$, and x_2 is the number of such peers s that $x(s) \in I'$ but $0.s \notin I'$. First we have $x_1 \leq num_p(I') \leq num_p(I) \leq C\alpha_2 \ln n$. Then since all the peers s such that $0.s \in (0.p', 0.r)$ or $0.s \in (0.r, 0.t')$ should have $x(s) \in [0.p', 0.r]$ or $x(s) \in [0.r, 0.t']$ respectively, by Lemma 5 we have $E(x_2) < (1 - (0.r - 0.p')) + (1 - (0.t' - 0.r)) = 2 - |I'| < 2$. Thus by Chernoff bound, $Pr\{x_2 \geq \frac{1}{2} \cdot C\alpha_2 \ln n\} \leq 2^{-\frac{1}{2} \cdot C\alpha_2 \ln n} = n^{-\frac{1}{2} \cdot C\alpha_2 \ln 2}$. Thus with probability $1 - O(1/n)$, we have $num_x(I') = x_1 + x_2 \leq \frac{3}{2} \cdot C\alpha_2 \ln n$. Thus by Lemma 6, $|I'| \leq \frac{3}{2} \cdot C(1+\epsilon)\alpha_2 \ln n/n$. Therefore $Vol(p,q) = |I| \leq 3 \cdot C(1-\epsilon)\alpha_1(\ln n/n)$.

Lemma 8. *With probability at least $1 - 2/n$, the algorithm 'Estimate n' ensures that $(1/6 - \epsilon_1)n \leq \hat{n}_2 \leq 6 + \epsilon_1$, for any positive constant ϵ_1 and n sufficiently large.*

Proof. Since for a peer p, $\log(vol(p)^{-1})$ is the depth of p's leaf in the binary partition tree, by Lemma 4, $\log(vol(p)^{-1})$ is an (α, β) approximation to $\log n$ for any fixed constants $\alpha < 1$ and $\beta > 2$. Thus $s = c_1 \ln vol(p)^{-1}$ is an (α, β) approximation to $c_1 \ln n$. Similarly, Lemma 7 shows that t in our algorithm is a $(\alpha/2 - \epsilon, 3\beta + \epsilon)$ approximation to $(c_1 \ln n)/n$ for any $\epsilon > 0$, for n and c_1 sufficiently large. Thus, $\hat{n}$ is a $(\frac{\alpha}{3\beta+\epsilon}, \frac{\beta}{\alpha/2-\epsilon})$ approximation to n for c_1 and n sufficiently large.

3.2 Choosing a Random Peer

By Lemma 8, we can estimate the number of peers as $\hat{n}_2$, a (γ_1, γ_2)-approximation to n, for constants γ_1, γ_2. Then let $n' = \hat{n}_2/\gamma_1$, and $\lambda = 1/(13n')$. Thus $n' \geq n$, $\lambda \leq 1/(13n)$ and $\lambda = \Theta(1/n)$.

Our algorithm for choosing a random peer works as follows. First it randomly selects a key x in the key space $[0,1]^d$, looks up the peer p who stores the key and picks up a random number T in $[0, vol(p)]$. Then if there is a peer p such that $Vol(lookup(x), p) - (vol(lookup(x)) - T) \leq \lambda \cdot num(lookup(x), p)$ with $num(lookup(x), p) \leq 12 \ln n'$, the algorithm returns the first such peer. Else it repeats until a peer is returned. We will show that the expected number of repetitions of the *while* loop is $O(1)$ with high probability. The algorithm is a direct generalization of the one given by King and Saia [6].

Choose Random Peer

```
While TRUE do:
    x ← random number in [0,1]^d;
    p = lookup(x); T ← random number in [0, vol(p)];
    For (i = 1; i ≤ 12 ln n'; i++)
        If (T ≤ i · λ) return p;
        Else
            p = next(p);
            T = T + vol(p);
```

Definition 1. *For any peer labeled as p, let* ***first(p)*** *be the first peer such that the sum of the volumes of the peers from $first(p)$ to p including $first(p)$ and p is larger than or equal to the number of these peers multiplied by λ, i.e. $Vol(first(p), p) \geq \lambda num(first(p), p)$.*

Lemma 9. *For any peer p, if $num(first(p), p) \leq 12\ln(n')$, the algorithm will choose p with probability λ in each iteration of the* ***while*** *loop.*

Proof. Let $q = first(p)$. Note that if peer $s = lookup(x)$ is not in $[0.q, 0.p]$, then p couldn't be returned by the algorithm. Otherwise assume $0.s \notin [0.q, 0.p]$ and p is returned. Denote t the previous peer of $first(p)$, i.e. $next(t) = first(p)$. Since t is visited by algorithm before p, then $T_1 > \lambda num(s,t)$, where T_1 is the value of T at the time t is visited. Then when p is visited, $T = T_1 + Vol(first(p), p) > \lambda num(s,t) + \lambda num(first(p), p) = \lambda num(s,p)$, which contradicts the condition to return p.

Then we argue by induction on $k = num(first(p), p)$.

Base: $k=1$, i.e. $vol(p) \geq \lambda$. The probability that p is selected is $\frac{vol(p)}{1} \cdot \frac{\lambda}{vol(p)} = \lambda$.

Induction Step: For $k > 1$, assume peer p will be chosen if $num(first(p), p) < k$. Now consider the case $num(first(p), p) = k$.

Denote $q = first(p)$. Then for any peer s such that $0.s \in [0.q, 0.p)$, $first(s)$ should also be in $[0.q, 0.p)$. Otherwise by the definition of $first()$, $Vol(q, s) < \lambda num(q, s), Vol(next(s), p) < \lambda num(next(s), p)$. Thus $Vol(q, p) < \lambda num(q, p)$, which contradicts $q = first(p)$. Therefore all such s have $first(s)$ in $[0.q, 0.p)$ and by the induction hypothesis they will be selected with probability λ. According to our algorithm, if the peer $s = lookup(x)$ is in $(0.q, 0.p]$, then one of the peers in $(0.q, 0.p]$ must be selected since p is a candidate due to $first(p) = q$. And if the peer $s = lookup(x)$ is equal to q, then one of the peers in $[0.q, 0.p]$ will be selected with probability $(\lambda num(q,p) - Vol(next(q), p))/vol(q)$.

Therefore the probability that one of the peers in $[0.q, 0.p]$ will be selected is given as

$$\frac{Vol(q,p)}{1} \cdot \frac{Vol(next(q), p) + \frac{\lambda num(q,p) - Vol(next(q),p)}{vol(q)} \cdot vol(q)}{Vol(q,p)} = \lambda num(q,p) \quad (3)$$

Since every peer $s \in [0.q, 0.p]$ other than p has probability λ of being selected by the induction hypothesis, peer p must be selected with probability λ.

Lemma 10. *With probability greater than* $1 - 1/n$*, for any peer* p *and* q *such that* $num(p,q) > 12\ln n$*, we have* $Vol(p,q) > \ln n/n$*.*

Proof. Let $I = [0.p, 0.p + Vol(p,q)]$. Denote $t = next(q)$. Then $I = [0.p, 0.t]$ and $num(p,t) = num(p,q) + 1$. Let peer r be the first peer laid down in the network such that $0.r \in [0.p, 0.t]$. Let $I' = [2 \cdot 0.p - 0.r, 2 \cdot 0.t - 0.r]$. Then $|I'| = 2|I|$. Note that for any peer s such that $0.s \in [0.p, 0.t]$, $s \neq r$, its original point will fall in the interval I'. Thus $num_x(I') \geq num(p,q) + 1 - 1 > 12\ln n$. By a simply application of Chernoff bound, we have any interval containing more than $6\ln n$ points has length greater than $\ln n/n$[6]. Thus $|I'| > 2\ln n/n$. Therefore $Vol(p,q) = |I| > \ln n/n$.

Lemma 11. *With probability greater than* $1 - 1/n$*, for any peer* p*,* $num(first(p), p) \leq 12\ln n'$*.*

Proof. Let $q = first(p)$. By contradiction, assume $num(first(p),p) > 12\ln n'$. Let s be the peer such that $0.s \in (0.q, 0.p]$ and $num(s,p) = 12\ln n'$. Then by Lemma 10, $Vol(s,p) \geq \ln n/n \geq \ln n'/n' \geq 12\ln n'/(13n') = \lambda \cdot num(s,p)$. Then $first(p)$ must be in $[0.s, 0.p]$, which contradicts $q = first(p)$ and $0.s \in (0.q, 0.p]$. Therefore $num(first(p),p) \leq 12\ln n'$.

Theorem 2. *With probability at least* $1 - 3/n$*, each peer is chosen with probability exactly* $1/n$*.*

Proof. By Lemma 11, for any peer p, $num(first(p),p) \leq 12\ln n'$. Then by Lemma 9, our algorithm will choose p with probability λ in each iteration of the *while* loop. Since $\lambda = \Theta(1/n)$, the expected number of the repetitions of the *while* loop will be $\Theta(1)$. Therefore each peer will be chosen with the same probability.

3.3 Latency and Message Complexity

Proof of Theorem 1. By Theorem 2, our algorithm selects each peer with probability $1/n$, with high probability. Now consider its latency and message complexity.

Since our algorithm for estimating n takes $O(\log n)$ *next* operations in expectation, it has expected latency $O(\log n)$ and message complexity $O(\log n)$.

For each iteration of the *while* loop of the algorithm, there is one *lookup* operation and at most $O(\log n)$ *next* operations. By Lemma 4 and 3, the lookup operations in CAN implemented with the aid of the partition tree have expected complexity $O(\log n)$, with $O(1)$ routing items being maintained at each peer.

Therefore, in CAN, our algorithm has expected latency $O(\log n)$ and sends $O(\log n)$ messages in expectation. ∎

4 Future Work: Biased Distributions and More General Networks

In this section, we propose two natural generalizations of the uniform random sampling problem which pose interesting lines of future work.

First we consider the problem of handling more general sampling distributions and to choose peers with a biased probability. In other words, we would like to choose a peer p with probability proportional to $f(p)$, thereby with probability $f(p)/\sum f(x)$, where $f(p)$ can be any positive function on peer p. A case that is easy to solve is when $\max(f(x))/\min(f(x)) = \Theta(1)$. We can estimate $\sum f(x)$ as σ to a constant factor with a technique similar to that we used for estimating n. We assign each peer a region of volume $\lambda(p) = c \cdot f(p)/\sigma$, where c is a constant. Since $\max(f(x))/\min(f(x)) = \Theta(1)$ and σ approximates $\sum f(x)$ within a constant factor, we have $\lambda(p) = \Theta(1/n)$ for all peers p. Then the sampling algorithm still works if Line 5 is replaced by '**If** $(T \leq \Lambda_i)$ **return** p;', where $\Lambda_i = \sum_{k=0}^{i-1} \lambda(next^{(k)}(p))$. For brevity, we omit the proof of correctness. Dealing with distributions that are not "almost uniform" as described above seems to be rather more difficult.

The second generalization of the random sampling problem that we consider is to devise efficient algorithms for selecting a peer uniformly at random in other overlay peer-to-peer systems, such as the locality-aware systems of Plaxton et al. [8] and Abraham et al. [1], or in systems that provide even less structure than Chord or CAN. Our approach of using a space-filling curve can be used in most systems based on peers embedded in geometric space.

References

1. I. Abraham, D. Malkhi, O. Dobzinski: *LAND: Stretch* $(1+\epsilon)$ *Locality Aware Networks for DHTs.* In ACM-SIAM SODA, 2004.
2. J. Alber, R. Niedermeier: *On Multi-Dimensional Hilbert Indexings.* In 4th COCOON, 1998.
3. L. K. Platzmann, J. J. Bartholdi: *Spacefilling curves and the planar travelling salesman problem.* In Journal of the ACM, 36(4):719-737, Oct 1989.
4. P. Ganesan, B. Yang, H. Garcia-Molina: *One torus to rule them all: multi-dimensional queries in P2P systems.* In 7th WebDB, 2004.
5. D. Hilbert: *Über stetige Abbildung einer Linie auf ein Flachenstuck.* Math. Annln., 38:459-460, 1891.
6. V. King, J. Saia: *Choosing a random peer.* In Proc. of the ACM PODC, 2004.
7. G. Peano: *Sur une courbe qui remplit toute une aire plane.* Math. Annln., 36:157-160, 1890.
8. C. Plaxton, R. Rajaraman, A. Richa: *Accessing nearby copies of replicated objects in a distributed environment.* In Proc. of the ACM SPAA, 1997.
9. S. Ratnasamy, P. Francis, M. Handley, R. Karp, S. Schenker: *A scalable content-addressable network.* In Proc. of the ACM SIGCOMM, 2001.
10. H. Sagan: *Space-filling curves*, Springer Verlag 1994.
11. I. Stoica, R. Morris, D. Karger, F. Kaashoek, H. Balakrishnan: *Chord: A Scalable Peer-to-peer Lookup Service for Internet Applications.* In ACM SIGCOMM, 2001.

Mobile Agent Rendezvous in a Synchronous Torus

Evangelos Kranakis[1,⋆], Danny Krizanc[2], and Euripides Markou[3,⋆⋆]

[1] School of Computer Science, Carleton University, Ottawa, Ontario, Canada
kranakis@scs.carleton.ca

[2] Department of Mathematics, Wesleyan University, Middletown, Connecticut 06459, USA
dkrizanc@caucus.cs.wesleyan.edu

[3] Department of Computer Science, National Kapodistrian University of Athens, Athens, Greece
emarkou@cs.ntua.gr

Abstract. We consider the rendezvous problem for identical mobile agents (i.e., running the same deterministic algorithm) with tokens in a synchronous torus with a sense of direction and show that there is a striking *computational difference* between one and more tokens. More specifically, we show that 1) two agents with a constant number of unmovable tokens, or with one movable token, each cannot rendezvous if they have $o(\log n)$ memory, while they can perform rendezvous with detection as long as they have one unmovable token and $O(\log n)$ memory; in contrast, 2) when two agents have two movable tokens each then rendezvous (respectively, rendezvous with detection) is possible with constant memory in an arbitrary $n \times m$ (respectively, $n \times n$) torus; and finally, 3) two agents with three movable tokens each and constant memory can perform rendezvous with detection in a $n \times m$ torus. This is the first publication in the literature that studies tradeoffs between the number of tokens, memory and knowledge the agents need in order to meet in such a network.

Keywords: Mobile agent, rendezvous, rendezvous with detection, tokens, torus, synchronous.

1 Introduction

We study the following problem: how should two mobile agents move along the nodes of a network so as to ensure that they meet or rendezvous?

⋆ Research supported in part by NSERC (Natural Sciences and Engineering Research Council of Canada) and MITACS (Mathematics of Information Technology and Complex Systems) grants.

⋆⋆ Work done partly while visiting Carleton University (April - May 2005). Research supported in part by NSERC grant and by PYTHAGORAS project 70/3/7392 under the EPEAEK program of the Greek Ministry of Educational and Religious Affairs.

J.R. Correa, A. Hevia, and M. Kiwi (Eds.): LATIN 2006, LNCS 3887, pp. 653–664, 2006.

The problem is well studied for several settings. When the nodes of the network are uniquely numbered, solving the rendezvous problem is easy (the two agents can move to a node with a specific label). However even in that case the agents need enough memory in order to remember and distinguish node labels. Symmetry in the rendezvous problem is usually broken by using randomized algorithms or by having the mobile agents use different deterministic algorithms. (See the surveys by Alpern [1] and [2], as well as the recent book by Alpern and Gal [3]). Yu and Yung [13] prove that the rendezvous problem cannot be solved on a general graph as long as the mobile agents use the same deterministic algorithm. While Baston and Gal [5] mark the starting points of the agents, they still rely on randomized algorithms or different deterministic algorithms to solve the rendezvous problem.

Research has focused on the power, memory and knowledge the agents need, to rendezvous in a network. In particular what is the 'weakest' possible condition which makes rendezvous possible? For example Yu and Yung [13] have considered attaching unique identifiers to the agents while Dessmark, Fraigniaud and Pelc [6] added unbounded memory; note that having different identities allows each agent to execute a different algorithm. Other researchers (Barriere et al [4] and Dobrev et al [7]) have given the agents the ability to leave notes in each node they visit. In another approach each agent has a stationary token placed at the initial position of the agent. This model is much less powerful than distinct identities or than the ability to write in every node. Assuming that the agents have enough memory, the tokens can be used to break symmetries. This is the approach introduced in [11] and studied in Kranakis et al [10] and Flocchini et al [8] for the ring topology. In particular the authors proved in [10] that two agents with one unmovable token each in a synchronous, n-node oriented ring need at least $\Omega(\log\log n)$ memory in order to do rendezvous with detection. They also proved that if the token is movable then rendezvous without detection is possible with constant memory.

We are interested here in the following scenario: there are two identical agents running the same deterministic algorithm in an anonymous oriented torus. In particular we are interested in answering the following questions. What memory do the agents need to solve rendezvous using unmovable tokens? What is the situation if they can move the tokens? What is the tradeoff between memory and the number of tokens?

1.1 Model and Terminology

Our model consists of two identical mobile agents that are placed in an anonymous, synchronous and oriented torus. The torus consists of n rings and each of these rings consists of m nodes. Since the torus is oriented we can say that it consists of n vertical rings. A horizontal ring of the torus consists of n nodes while a vertical ring consists of m nodes. We call such a torus a $n \times m$ torus. The mobile agents share a common orientation of the torus, i.e., they agree on any direction (clockwise vertical or horizontal). Each mobile agent owns a number of identical tokens, i.e. the tokens are indistinguishable. A token or an agent at a given node is visible to all agents on the same node, but is not visible to any other agents. The agents follow the **same** deterministic algorithm and begin execution at the same time.

At any single time unit, the mobile agent occupies a node of the torus and may 1) stay there or move to an adjacent node, 2) detect the presence of one or more tokens at the node it is occupying and 3) release/take one or more tokens to/from the node it is occupying. We call a token *movable* if it can be moved by any mobile agent to any node of the network, otherwise we call the token *unmovable* in the sense that it can occupy only the node in which it has been released.

More formally we consider a mobile agent as a finite Moore automaton[1] $\mathcal{A} = (X, Y, \mathcal{S}, \delta, \lambda, S_0)$, where $X \subseteq \mathcal{D} \times \mathcal{C}_v \times \mathcal{C}_{MA}$, $Y \subseteq \mathcal{D} \times \{\texttt{drop}, \texttt{take}\}$, $\mathcal{S}$ is a set of σ states among which there is a specified state S_0 called the *initial* state, $\delta : \mathcal{S} \times X \rightarrow \mathcal{S}$, and $\lambda : \mathcal{S} \rightarrow Y$. $\mathcal{D}$ is the set of possible directions that an agent could follow in the torus. Since the torus is oriented, the direction port labels are globally consistent. We assume labels *up, down, left, right.* Therefore $\mathcal{D} = \{\texttt{up}, \texttt{down}, \texttt{left}, \texttt{right}, \texttt{stay}\}$ ($\texttt{stay}$ represents the situation where the agent does not move). $\mathcal{C}_v = \{\texttt{agent}, \texttt{token}, \texttt{empty}\}$ is the set of possible configurations of a node (if there is an agent and a token in a node then its configuration is $\texttt{agent}$). Finally, $\mathcal{C}_{MA} = \{\texttt{token}, \texttt{no}-\texttt{token}\}$ is the set of possible configurations of the agent according to whether it carries a token or not.

Initially the agent is at some node u_0 in the initial state $S_0 \in \mathcal{S}$. S_0 determines an action (drop token or nothing) and a direction by which the agent leaves u_0, $\lambda(S_0) \in Y$. When incoming to a node v, the behavior of the agent is as follows. It reads the direction i of the port through which it entered v, the configuration $c_v \in \mathcal{C}_v$ of node v (i.e., whether there is a token or an agent in v) and of course the configuration $c_{MA} \in \mathcal{C}_{MA}$ of the agent itself (i.e., whether the agent carries a token or not). The triple $(i, c_v, c_{MA}) \in X$ is an input symbol that causes the transition from state S to state $S' = \delta(S, (i, c_v, c_{MA}))$. S' determines an action (such as release or take a token or nothing) and a port direction $\lambda(S')$, by which the agent leaves v. The agent continues moving in this way, possibly infinitely.

We assume that the memory required by an agent is at least proportional to the number of bits required to encode its state which we take to be $\Theta(\log(|S|))$ bits. Memory permitting, an agent can count the number of nodes between tokens, or the total number of nodes of the torus, etc. In addition, an agent might already know the number of nodes of the torus, or some other network parameter such as a relation between n and m. Since the agents are identical they face the same limitations on their knowledge of the network.

Let $U = \{(n_1/2, 0, ..., 0), (0, n_2/2, ..., 0), ..., (0, 0, ..., n_d/2)\}$, where each n_i is even, be a set consisting of d vectors in d-dimension. The *distance* between two nodes on a d-dimensional torus is a d-vector the ith of element of which is $\min\{(x_i - y_i), (n_i + y_i - x_i)\}$ where wlog $x_i \geq y_i$ are the ith co-ordinates of the nodes.

[1] The first known algorithm designed for graph exploration by a mobile agent, modeled as a finite automaton, was introduced by Shannon [12] in 1951.

Theorem 1. *Consider two agents placed in a d-dimensional oriented torus* $(n_1 \times n_2 \times \cdots \times n_d)$ *so that their distance is the sum of vectors contained in any nonempty subset* S *of* U. *Assume that for any non-zero element of the distance, the number of nodes of that dimension of the torus is even. Then, no matter how many tokens or how much memory the agents have, it is impossible for the agents to rendezvous.*

Corollary 1. *Two agents placed in a* $n \times m$ *torus are incapable of meeting each other (no matter how many tokens, movable or unmovable they have) if their initial distance is either* $(n/2, 0)$ *or* $(0, m/2)$ *or* $(n/2, m/2)$.

Theorem 1 is a generalization of Theorem 1 in [10] which states that it is impossible for two agents equipped with one unmovable token each, to rendezvous in a ring with n nodes if their initial distance is $n/2$, where n is even.

Definition 1. *We call rendezvous with detection (RVD) the problem in which the agents meet each other if their distance is not the sum of vectors contained in any nonempty subset* S *of* U, *otherwise they stop moving and declare that is impossible to meet each other.*

We say that an algorithm $\mathcal{A}$ solves RVD (or is an RVD algorithm) if the agents rendezvous when their initial distance is not the sum of vectors contained in any nonempty subset S of U. If, the distance is indeed the sum of vectors contained in a subset S of U then $\mathcal{A}$ halts after a finite number of steps and the agents declare that rendezvous is impossible.

Definition 2. *We call rendezvous without detection (RV) the problem in which the agents meet each other if their distance is not the sum of vectors contained in any nonempty subset* S *of* U.

Therefore we say that an algorithm $\mathcal{A}$ solves RV (or is an RV algorithm) if the agents rendezvous when their initial distance is not the sum of vectors contained in any nonempty subset S of U. If, however, the distance is indeed the sum of vectors contained in a subset S of U then $\mathcal{A}$ may run forever.

We assume that at any single time unit an agent can traverse one edge of the network or wait at a node (we assume that taking or leaving a token can be done instantly). For a given torus G and starting positions s, s' of the agents we define as cost $\mathcal{CT}_{RVD}(A, G, s, s')$ of an RVD algorithm A, the maximum time (number of steps plus waiting time) needed either to rendezvous or to decide that rendezvous is impossible. The cost $\mathcal{CT}_{RV}(A', G, s, s')$ of an RV algorithm A', is the time needed to rendezvous (when it is possible of course). Finally, the time complexity of an RVD or RV algorithm is its maximum cost overall possible starting positions of the agents.

1.2 Our Results

In the study of the rendezvous problem this paper shows that there is a striking *computational difference* between one and more tokens. More specifically, we show that

1. Two agents with a constant number of unmovable tokens each cannot rendezvous if they have $o(\log n)$ memory.
2. Two agents with one movable token each cannot rendezvous if they have $o(\log n)$ memory.
3. Two agents with one unmovable token each can perform rendezvous with detection as long as they have $O(\log n)$ memory.
4. When two agents have two movable tokens each then rendezvous (respectively, rendezvous with detection) is possible with constant memory in an arbitrary $n \times m$ (respectively, $n \times n$) torus.
5. Two agents with three movable tokens each and constant memory can perform rendezvous with detection in an arbitrary $n \times m$ torus.

This is the first publication in the literature that studies tradeoffs between the number of tokens, memory and knowledge the agents need in order to meet in such a network.

1.3 Outline of the Paper

In Section 2 we first give some preliminary results concerning possible ways that an agent can move in a torus using either no tokens or a constant number of unmovable tokens. Then we prove that rendezvous without detection in a torus cannot be solved by two agents with one movable token each, or with a constant number of unmovable tokens unless their memory is $\Omega(\log n)$ bits.

In Section 3 we give an algorithm for rendezvous with detection in a $n \times n$ torus using one unmovable token and $O(\log n)$ memory. We also give an algorithm for rendezvous with detection in a $n \times n$ torus using two movable tokens and constant memory. Next we give an algorithm for rendezvous without detection in an arbitrary $n \times m$ torus using two movable tokens and constant memory, stating the relation that m and n must have in order to do rendezvous with detection following that algorithm. Finally we give an algorithm for rendezvous with detection in a $n \times m$ torus using three movable tokens and constant memory.

In Section 4 we discuss the results and state some open problems. Due to space limitations, the proofs, formal algorithms and figures have been omitted in this extended abstract.

2 Memory Lower Bounds of Rendezvous

2.1 Preliminary Results

Lemma 1. *Consider one mobile agent with σ states and no tokens. We can always (for any configuration of the automaton, i.e. states and transition function) select a $n \times n$ oriented torus, where $n = \Omega(\sigma)$ so that no matter what is the starting position of the agent, it cannot visit all nodes of the torus. In fact, the agent will visit at most $(\sigma + 1)n$ nodes.*

Lemma 2. *Consider one mobile agent with σ states and one unmovable token. We can always (for any configuration of the automaton, i.e. states and transition function) select an oriented $n \times n$ torus, where $n = \Omega(\sigma^2)$ so that no matter what is the starting position of the agent, it cannot visit all nodes of the torus. In fact, the agent will visit at most $(\sigma + 1)(1 + \sigma n) = O(\sigma^2 n)$ nodes.*

Theorem 2. *Consider one mobile agent with σ states and a constant number of identical unmovable tokens. We can always (for any configuration of the automaton, i.e. states and transition function) select a $n \times n$ oriented torus, where $n = \Omega(\sigma^2)$ so that no matter what is the starting position of the agent, it cannot visit all nodes of the torus. In fact, the agent will visit at most $O(\sigma^2 n)$ nodes.*

2.2 An $\Omega(\log n)$ Memory Lower Bound for Rendezvous Using One Movable Token

Lemma 3. *Consider two mobile agents with σ states. They each have a token (identical to each other). Then we can always (for any configuration of the automatons, i.e. states and transition function) find an oriented $n \times n$ torus, where $n = \Omega(\sigma^2)$ and place the agents so that if they can not move tokens then they cannot rendezvous.*

Proof. (**Sketch**) We place the first agent A in a node. If A can meet only its token, then by Lemma 2, the agent would visit at most $(\sigma + 1)(1 + \sigma n)$ nodes before it repeats everything. We prove that we can initially choose a node to place the other agent B so that anyone's token is out of reach of the other. In other words we place the second agent B so that

- it releases its token t_B at a node different from at most $(\sigma+1)(1+\sigma n)$ nodes visited by the first agent A and
- to avoid to visit the node where the first agent A released its token t_A □

Notice that in the previous scenario, where the two agents cannot move the tokens, there are still unvisited nodes (from the same agent) in the torus. In fact we proved Lemma 3 by describing a way to 'hide' token t_A in a node not visited by agent B and token t_B in a node not visited by agent A.

If there are two starting nodes s, s' for the agents A and B so that agent A drops its token t_A in a node not visited by agent B and agent B drops its token t_B in a node not visited by agent A then we say that s, s' satisfy property π.

If the agents could move tokens, then it is easy to think of an algorithm where all nodes of any torus are visited by the same agent. For example consider the following algorithm:

- 1: go right until you meet the second token;
- 2: move the token down;
- 3: repeat from step 1;

Nevertheless the goal is again to place the agents in a way that they could meet only their own token. To achieve this we place the agents so that in a phase which starts when the agents move their tokens, up to the moment that they move their tokens again they do not meet each other's token.

Lemma 4. *Consider two mobile agents with σ states. They each have a token (identical to each other). Then we can always (for any configuration of the automatons, i.e. states and transition function) find an oriented $n \times n$ torus, where $n > 8\sigma^3 = \Omega(\sigma^3)$ and place the agents so that even if they can move tokens they cannot rendezvous.*

This implies the following theorem:

Theorem 3. *Two agents in a $n \times n$ torus with one movable token need at least $\Omega(\log n)$ memory to solve the RV problem.*

Proof. Suppose that the agents have a memory of r bits. Hence they can have at most 2^r states. By Lemma 4 as long as $n > 8\sigma^3$ the agents cannot perform rendezvous. Hence, the agents need at least $r = \Omega(\log n)$ memory in order to perform rendezvous. □

2.3 An $\Omega(\log n)$ Memory Lower Bound for Rendezvous Using $O(1)$ Unmovable Tokens

Lemma 5. *Consider two mobile agents with σ states. They each have two tokens (identical to each other). Then we can always (for any configuration of the automatons, i.e. states and transition function) find a $n \times n$ oriented torus, where $n = \Omega(\sigma^2)$ and place the agents so that if they cannot move tokens they cannot rendezvous.*

Proof. (**Sketch**) In view of Lemmas 2, 3 we can select the torus and the starting positions so that an agent will visit at most $(\sigma+1)(1+\sigma n)$ nodes until it decides to release its second token and up to that point does not meet the other's token. Its second token will have to be released at a 'short' distance from the first one since an agent cannot count more than σ. Using similar arguments as in the proof of Lemma 3 one can show that there are at least $n^2 - 5(\sigma+1)(1+\sigma n)$ pairs of starting nodes that satisfy property π. □

This implies the following theorem:

Theorem 4. *Two agents in a $n \times n$ torus with two identical unmovable tokens each, need at least $\Omega(\log n)$ memory to solve the RV problem.*

Applying similar arguments we can prove the following lemma and theorem:

Lemma 6. *Consider two mobile agents with σ states. They each have a constant number of k identical tokens. Then we can always (for any configuration of the automatons, i.e. states and transition function) find an oriented $n \times n$ torus, where $n = \Omega(\sigma^2)$ and place the agents so that if they cannot move tokens they cannot rendezvous.*

Proof. (**Sketch**) Using similar arguments as in the proof of Lemma 3 one can show that there are at least $n^2 - \frac{k(k+2)}{2}(\sigma+1)(1+\sigma n)$ pairs of starting nodes that satisfy property π. □

Theorem 5. *Two agents in a $n \times n$ torus with a constant number of unmovable tokens need at least $\Omega(\log n)$ memory to solve RV problem.*

3 Rendezvous

3.1 Rendezvous with Detection (RVD) in a $n \times n$ Torus Using One Token and $O(\log n)$ Memory

We describe an algorithm which solves the RVD problem of two agents in a $n \times n$ torus, equipped with one unmovable token and $O(\log n)$ memory each. Below is a high level description of the algorithm.

First the agent (both agents run the same algorithm) moves in the initial horizontal ring; it releases its token and it counts steps until it meets a token twice. If its counters differ, then it can meet the other agent. Otherwise it does the same in the initial vertical ring. If it does not meet the other or decide that rendezvous is impossible (which means that the agents must have started in different rings), then it searches one by one the horizontal rings of the torus counting its steps. If at least one of its counters (representing horizontal or vertical distances) is different than $n/2$ then it can meet the other agent. Otherwise it stops and declares rendezvous impossible. The formal description of the algorithm will appear in the full version of the paper.

Theorem 6. *The Rendezvous with Detection problem on a $n \times n$ torus can be solved by two agents using one unmovable token and $O(\log n)$ memory each, in time $O(n^2)$.*

The above result can be extended for the case of an arbitrary $n \times m$ torus. The main difference in that case is how the agents decide if they have started on the same ring or not: they again explore one by one the horizontal rings. They will meet a token while going down (passing from one horizontal ring to the next) if and only if they have started on the same ring. Otherwise, they will meet a token while going right (before finishing the exploration of a horizontal ring). They can solve the Rendezvous with Detection problem in $O(nm)$ steps as long as they have $O(\log n + \log m)$ memory each.

3.2 Rendezvous with Detection in a $n \times n$ Torus Using Two Movable Tokens and Constant Memory

We define Procedure HorScan which will be used in our algorithms.

In this procedure the agent stops immediately after it meets a token. So for example, if after it goes right it meets a token then it stops immediately; it does not go up.

Procedure HorScan

```
repeat
    go down, right, up
until you meet a token
```

Procedure FindTokenHor

```
1: repeat
2:     HorScan
3:     if you meet token up then
4:         HorScan
5:         go one step down and drop (or move) the second token
6:     end if
7: until you meet a token down or right
```

We also use Procedure FindTokenHor.

An agent following Procedure FindTokenHor, scans one by one the horizontal rings of the torus until it meets a token while moving down or right. Below we explain procedure FindTokenHor and prove some of its properties.

Let the agents release their first token and execute procedure FindTokenHor. During execution of HorScan (step 2 of Procedure FindTokenHor), the agent has to meet a token for the first time either after it moved down in the first step, or up or right (he can not meet a token while going down at a later step of Horscan since it would have met the token while going right earlier).

If it meets a token after it moved up, then this can be any token: its or the other's first token (or its or the other's second token when it scans a later horizontal ring). However, if it executes Horscan again (step 4 of Procedure FindTokenHor), then no matter what was the case, it is easy to see that the first token it meets now is its token (first or second) and it meets it after it moved up[2]. Furthermore in this case it is sure that the down ring had no tokens.

If it meets a token right then it is clear that it is the other's first token and that the two agents have started in different rings.

If it meets a token while it goes down then either it is its first token or the other's first token. In both cases this means that they have started in the same ring: if it is its first token it means that it has searched the whole torus and did not meet any other token while it was moving right.

Therefore the agent exits procedure FindTokenHor knowing that it has started either in the same ring with the other agent (if it met a token after it moved down) or in different rings (if it met a token after it moved right). Procedure FindTokenHor needs $O(n^2)$ time units.

Below is a high level description of the algorithm RVD2n which solves the RVD problem in a $n \times n$ torus. The two agents search one by one the horizontal rings of the torus (using Procedure FindTokenHor) to discover whether they have started in the same ring. If so, then they execute a procedure which appeared

[2] Supposing that there are at most two tokens in the same horizontal ring.

in [9], for rendezvous with detection in a ring using two tokens and constant memory. Otherwise they try to 'catch' each other on the torus using a path, marked by their tokens. If they do not rendezvous then they search one by one the vertical rings of the torus (using a procedure similar to FindTokenHor for searching one by one the vertical rings). They again try to 'catch' each other on the torus. If they do not meet this time they declare rendezvous impossible. The algorithm takes $O(n^2)$ time.

Theorem 7. *The Rendezvous with Detection problem on a $n \times n$ torus can be solved by two agents using two movable tokens and constant memory each, in time $O(n^2)$.*

Another possible algorithm could be if after discovering that the agents started on different rings, **first** to search whether they are at distance $(n/2, n/2)$ and if not, then searching one by one the horizontal rings of the torus. We have chosen to present here the first approach since it is expandable to a $n \times m$ torus.

3.3 Rendezvous Without Detection in a $n \times m$ Torus Using Two Movable Tokens and Constant Memory

We give now algorithm RV2mn which is a RV algorithm for two agents with constant memory in a $n \times m$ torus. Algorithm RV2mn, at first, copies algorithm RVD2n. If no rendezvous occurs and no decision is made about its impossibility (i.e. the agents have started in different rings), the algorithm instructs the agents to mark a rectangle with their tokens on the torus and then execute Procedure Pendulum: they try to shrink the rectangle and eventually meet which will happen unless they had started at distance $(n/2, m/2)$ (in that case the algorithm runs forever).

In fact one of the following things could happen: either the agents rendezvous, or they detect that they are in the same ring and their distance is half the size of the ring or the algorithm runs forever (in that case they are at horizontal distance $n/2$ and vertical distance $m/2$). Algorithm RV2mn needs $O(n^4 + m^4)$ time.

Theorem 8. *The Rendezvous without Detection problem on an arbitrary $n \times m$ torus can be solved by two agents using two movable tokens and constant memory each, in time $O(n^4 + m^4)$.*

An interesting question which naturally follows is: what is the relation of n and m for which algorithm RV2mn is indeed a RVD algorithm? The answer is given by the following lemma.

Lemma 7. *If after the horizontal and vertical scanning of Algorithm RV2mn the agents do not rendezvous and $\frac{n-1}{10} \leq m \leq 2n + 17$ then their distance is $(n/2, m/2)$ and therefore rendezvous is impossible.*

Hence by Lemma 7 if we knew that $\frac{n-1}{10} \leq m \leq 2n + 17$ then algorithm RV2mn would be a RVD algorithm for the $n \times m$ torus.

3.4 Rendezvous with Detection in a $n \times m$ Torus Using Three Movable Tokens and Constant Memory

If the agents have 3 tokens then we can extend our RVD2n algorithm to get a RVD algorithm for a $n \times m$ torus. The idea is the following: If the agents do not meet while they copy Algorithm RVD2n then they mark a rectangle on the torus using their two tokens each. Next they release their third token to the right of their starting position. They travel on this rectangle (one agent from inside and the other from outside), each time moving one step the fifth token they meet: first they move it to the right until it hits another token and then down until it touches a token. Next they go left until they meet a token and then up until they meet a token. If at that point they see two tokens adjacent then they declare rendezvous impossible. Otherwise they wait until rendezvous which will occur in less than $n + m$ time. Algorithm RVD3mn takes $O(n^2 + m^2)$ time.

Theorem 9. *The Rendezvous with Detection problem on an arbitrary $n \times m$ torus can be solved by two agents using three movable tokens and constant memory each, in time $O(n^2 + m^2)$.*

4 Conclusions

In this paper we investigated on the number of tokens and memory that two agents need in order to rendezvous in an anonymous oriented torus.

It appears that there is a strict hierarchy on the power of tokens and memory with respect to rendezvous: a constant number of unmovable tokens are less powerful than two movable tokens. While the hierarchy collapses on three tokens (we gave an algorithm for rendezvous with detection in a $n \times m$ torus when the agents have constant memory each), it remains an open question if three tokens are strictly more powerful than two with respect to rendezvous with detection.

It is also interesting that although a movable token is more powerful than an unmovable one (we showed that an agent with one unmovable token cannot visit all the nodes of a torus with a properly selected size unless it has $\Omega(\log n)$ memory, while it could do it with a constant memory if it could move its token) it appears that this power is not enough with respect to rendezvous; the agents with one movable token each, still require $\Omega(\log n)$ memory to rendezvous in the torus.

As this is the first publication in the literature that studies tradeoffs between the number of tokens, memory, knowledge and power the agents need in order to meet on a torus network, a lot of interesting questions remain open:

- Can we improve the time complexity for rendezvous without detection on a $n \times m$ torus using constant memory? Can we improve the time complexity for rendezvous with detection on a $n \times n$ torus using constant memory?

- What is the lower memory bound for two agents with two movable tokens each in order to do rendezvous with detection in a $n \times m$ torus? In particular, can they do it with constant memory?

- What is the situation in a d-dimensional torus? Is it the case that with $d - 1$ movable tokens, rendezvous needs $\Omega(\log n)$ memory while with d movable tokens and constant memory rendezvous with detection can be done? How does this change if the size of the torus is not the same in every dimension?

- What are the results if the torus is not oriented? If the torus is asynchronous?
- Finally, an interesting problem is that of many agents trying to rendezvous (or gathering) in a torus network.

References

1. S. Alpern, The Rendezvous Search Problem, SIAM Journal of Control and Optimization, 33, pp. 673-683, 1995. (Earlier version: LSE CDAM Research Report, 53, 1993.)
2. S. Alpern, Rendezvous Search: A Personal Perspective, Operations Research, 50, No. 5, pp. 772-795, 2002.
3. S. Alpern and S. Gal, The Theory of Search Games and Rendezvous, Kluwer Academic Publishers, Norwell, Massachusetts, 2003.
4. L. Barriere, P. Flocchini, P. Fraigniaud, and N. Santoro, Election and Rendezvous of Anonymous Mobile Agents in Anonymous Networks with Sense of Direction, Proceedings of the 9th International Colloquium on Structural Information and Communication Complexity (SIROCCO), pp. 17-32, 2003.
5. V. Baston and S. Gal, Rendezvous Search When Marks Are Left at the Starting Points, Naval Research Logistics, 47, No. 6, pp. 722-731, 2001.
6. A. Dessmark, P. Fraigniaud, and A. Pelc, Deterministic Rendezvous in Graphs, 11th Annual European Symposium on Algorithms (ESA), pp. 184-195, 2003.
7. S. Dobrev, P. Flocchini, G. Prencipe, and N. Santoro, Multiple agents rendezvous in a ring in spite of a black hole, Symposium on Principles of Distributed Systems (OPODIS '03), LNCS 3144, pp. 34-46, 2004.
8. P. Flocchini, E. Kranakis, D. Krizanc, N. Santoro, and C. Sawchuk, Multiple Mobile Agent Rendezvous in the Ring, LATIN 2004, LNCS 2976, pp. 599-608, 2004.
9. L. Gasieniec, E. Kranakis, D. Krizanc, X. Zhang, Optimal Memory Rendezvous of Anonymous Mobile Agents in a Uni-directional Ring. In proceedings of SOFSEM 2006, 32nd International Conference on Current Trends in Theory and Practice of Computer Science January 21 - 27, 2006 Merin, Czech Republic, SVLNCS, 2006, to appear.
10. E. Kranakis, D. Krizanc, N. Santoro, and C. Sawchuk, Mobile Agent Rendezvous Search Problem in the Ring, International Conference on Distributed Computing Systems (ICDCS), pp. 592-599, 2003.
11. C. Sawchuk, Mobile Agent Rendezvous in the Ring, PhD thesis, Carleton University, School of Computer Science, Ottawa, Canada, 2004.
12. CL. E. Shannon, Presentation of a Maze-Solving Machine, in 8th Conf. of the Josiah Macy Jr. Found. (Cybernetics), pp. 173-180, 1951.
13. X. Yu and M. Yung, Agent Rendezvous: A Dynamic Symmetry-Breaking Problem, in Proceedings of ICALP '96, LNCS 1099, pp. 610-621, 1996.

Randomly Colouring Graphs with Girth Five and Large Maximum Degree

Lap Chi Lau and Michael Molloy

Department of Computer Science,
University of Toronto,
Toronto, Ontario, Canada
{chi, molloy}@cs.toronto.edu

Abstract. We prove that the Glauber dynamics on the k-colourings of a graph G on n vertices with girth 5 and maximum degree $\Delta \geq 1000 \log^3 n$ mixes rapidly if $k = q\Delta$ and $q > \beta$ where $\beta = 1.645...$ is the root of $2 - (1 - e^{-1/\beta})^2 - 2\beta e^{-1/\beta} = 0$.

1 Introduction

The Glauber dynamics is a Markov chain on the proper colourings of a graph that has been widely studied in both computer science and statistical physics. For a given graph G and integer k which is at least the chromatic number of G, the Markov chain is described as follows: We start with an arbitrary k-colouring, and at each step we choose a uniformly random vertex v, and a uniformly random colour c from $L(v)$, the list of colours which do not appear on any neighbours of v. Then we change the colour of v to c.

This chain is of great interest for a number of reasons. For one, it is the most natural chain on the colourings of a graph, and so is an obvious attempt at a procedure to approximately count the colourings of a graph and to generate such a colouring nearly uniformly at random. It is also of interest in the statistical physics community, in part because of its relation to the Potts model.

The main question in this area is: For what values of k does this Markov chain mix in polytime? Usually this is studied in terms of Δ, the maximum degree of G. It is well known that for some graphs, the chain does not mix for $k \leq \Delta + 1$. It is conjectured that for every graph, the chain mixes in polytime for $k \geq \Delta + 2$, or at least for $k \geq \Delta + o(\Delta)$, but this appears to be a very difficult conjecture. Jerrum[11], and independently Salas and Sokal[14], showed that for all graphs the chain mixes in polytime for $k \geq 2\Delta$. Vigoda[15] showed that for all graphs, a different chain mixes in optimal time for $k \geq \frac{11}{6}\Delta$ and this implies that for the same values of k, the Glauber dynamics mixes in polytime. This is the best progress to date for general graphs.

A recent trend has been to study the performance of the Glauber dynamics on graphs with restrictions on the girth and maximum degree. At first, these restrictions were rather severe, and the number of colours remained far from Δ: Dyer and Frieze[4] showed that if Δ is at least $O(\log n)$ and the girth is at least

J.R. Correa, A. Hevia, and M. Kiwi (Eds.): LATIN 2006, LNCS 3887, pp. 665–676, 2006.

$O(\log \Delta)$ then we have rapid mixing for roughly $k = 1.763\Delta$ colours (note that $1.763 < 11/6$). Since then, several improvements[12, 7, 8, 9, 10, 5, 6] have reduced these restrictions substantially, and this line of research is producing surprisingly strong results and shedding much insight on the general conjecture. Some notable results are that we obtain rapid mixing for $\Delta = O(\log n)$, girth at least 9, and $k \geq (1+\epsilon)\Delta$[8] and for Δ at least a particular large constant, girth at least 6 and k roughly 1.489Δ[5].

Recently, Hayes and Vigoda[10] introduced "coupling from the stationary distribution" (described below) with which they managed to improve the girth requirement from five to four in one of these results (from Hayes[7]). (An improvement of 1 may not seem like much at first glance, but when the numbers are this small, each such improvement can be a huge gain.) They showed that we have rapid mixing with $\Delta = O(\log n)$, girth at least 4 and with k roughly 1.763Δ. This value of k is one that is often obtained by using a particular property that we call the *first local uniformity condition* (defined below). Hayes[7] had also proved rapid mixing $\Delta = O(\log n)$, girth at least 6 and with k roughly 1.489Δ, a value that is often obtained by using the *second local uniformity condition.* The main result of this paper, is to incorporate that second local uniformity condition into a coupling from the stationary distribution argument, and reduce the girth requirement from the latter result to 5. In doing so, difficulties cost us in two ways: (i) we must increase the restriction on Δ somewhat, and (ii) we obtain a number larger than the usual 1.489....

Define $\beta = 1.645...$ to be the solution to

$$2 - (1 - e^{-1/\beta})^2 - 2\beta e^{-1/\beta} = 0.$$

Theorem 1. *The Glauber dynamics mixes in $O(n \log n)$ time on all graphs on n vertices with maximum degree $\Delta \geq 1000 \log^3 n$, when the number of colours is $k \geq (\beta + \epsilon)\Delta$ for any constant $\epsilon > 0$.*

Remark. We made no attempt to optimize the exponent "3" in the lower bound on Δ. It is not hard to reduce it somewhat.

1.1 Outline

The proof of our main result uses the framework of "coupling with the stationary distribution" developed by Hayes and Vigoda[10] to prove their aforementioned result. Here is the basic idea: To analyse the mixing time via a coupling argument, we can assume that one Markov chain X is distributed according to the uniform distribution. Given a graph of girth at least 4 and maximum degree $\Delta = \Omega(\log n)$, one can show that with high probability, X_t has the first local uniformity condition. Hayes and Vigoda then show that, given an *arbitrary* colouring Y_t, the Hamming distance between X_t and Y_t decreases in expectation for k roughly 1.763Δ so long as X_t has the first local uniformity. So, with high probability, X_t and Y_t tend to drift together, and their theorem follows.

The main advantage of using the coupling with the stationary distribution is that one only needs to prove that a uniformly random colouring has local uniformity properties rather than a colouring generated by the Markov chain. This allows one to skip the analysis of the burn-in period, which is the most technical part of many previous papers. In addition, short cycles are a bit less harmful in uniform colourings than in "burn-in" colourings; this allowed the girth requirement to be reduced by one in [10]. One substantial drawback to this technique is that it does not accommodate path-coupling, a very useful technique introduced in [2]. This means that one needs to analyse the expected change in Hamming distance between two colourings with *arbitrary* Hamming distance, rather than just analyzing the much simpler case where the Hamming distance is one. Carrying out that analysis turned out to be manageable in [10] where they were able to adopt the original coupling argument from Jerrum[11], which predated path-coupling.

The main thrust of this paper is to incorporate the second local uniformity into the framework of "coupling with the uniform condition". In this case, it is much more difficult to extend the path coupling analysis to the case where two colourings have arbitrary distance. In fact, we are unable to do so without some loss, and this is why our result requires 1.645Δ colours rather than 1.489Δ. This portion of our analysis makes use of a novel "charging" argument (Lemma 1). That argument does not make use of any special structure of G (such as its girth or maximum degree) and so it might be useful in other settings. This argument appears in Section 2.

A second difficulty that arises in this paper is in proving that the second uniformity condition holds for a uniformly random colouring when the girth requirement is reduced from 6 (in [7]) to 5. The main problem is that the second local uniformity condition is defined in terms of vertices of distance two from a specific vertex v. Every previous paper that established a uniformity condition made crucial use of the fact that the vertices which defined the condition were very close to being an independent set. This is true in our setting for girth 6 graphs, but girth 5 graphs can have many edges between those vertices. The difficulties caused by these edges are what require us to increase the bound on Δ from $O(\log n)$ to $O(\log^3 n)$. We present this part of the proof in Section 3.

Remark. Our main theorem applies to graphs with maximum degree Δ. However, for brevity and ease of presentation, we only present the proof for the case where the graph is Δ-regular. For the most part, it is straightforward to extend the proof to non-regular graphs. The material in Section 3 is not as straightforward to extend, but the arguments used in [12] will suffice.

1.2 Definitions

In a graph G, we define $N(v)$ to be the set of neighbours of vertex v.

For a colouring X of G, we define $X(v)$ to be the colour at vertex v. We denote by $L_X(v)$ the list of available colours at v in X; i.e. the colours that do not appear on any neighbours of v. We denote by L_X the minimum of $|L_X(v)|$ over all possible v. Given two colourings X, Y, $P_v(X,Y) := L_Y(v) - L_X(v)$ and $P_v(Y,X) := L_X(v) - L_Y(v)$. In other words, $P_v(X,Y)$ is the set of colours appearing in the neighbourhood of v in X but not appearing in the neighbourhood of v in Y. Given a colouring X, suppose we recolour v by colour c and denote the resulting colouring by X'; then $R_X(v,c)$ is defined to be the set of neighbours w of v such that $L_X(w) = L_{X'}(w)$. In other words, $R_X(v,c)$ is the set of vertices $w \in N(v)$ such that $X(v)$ and c both appear in $N_G(w) - v$. We further define R_X to be the minimum of $|R_X(v,c)|$ over all possible v and c.

For the purposes of this paper, the local uniformity conditions are defined as follows. Set $q = k/\Delta$.

First Local Uniformity Condition[4]
For every $\zeta > 0$, $(qe^{-1/q} - \zeta)\Delta < L_X < (qe^{-1/q} + \zeta)\Delta$.

Second Local Uniformity Condition[12]
For every $\zeta > 0$, $((1 - e^{-1/q})^2 - \zeta)\Delta < R_X < ((1 - e^{-1/q})^2 + \zeta)\Delta$.

Given a particular value of k, we define Ω to be the set of k-colourings of G.

1.3 A Concentration Tool

We will make use of the following inequality, which is particularly useful in this paper because it can be applied to random trials that are not independent. The version that we use is from [13] and is a distillation of Azuma's original statement[1].

Azuma's Inequality. *Let X be a random variable determined by n trials $T_1, ..., T_n$, such that for each i, and any two possible initial sequences of outcomes $t_1, ..., t_i$ and $t_1, ..., t_{i-1}, t_i'$ that differ only on the ith outcome:*

$$|\exp(X|T_1 = t_1, ..., T_i = t_i) - \exp(X|T_1 = t_1, ..., T_i = t_i')| \le \gamma_i$$

then

$$\mathbf{Pr}(|X - \exp(X)| > \tau) \le 2e^{-\tau^2/(2\sum_{i=1}^n \gamma_i^2)},$$

for every $\tau > 0$.

2 Distance Decreasing with the Local Uniformities

Consider two colourings X, Y of G. We use $d(X,Y)$ to denote the Hamming distance of X, Y; i.e. the number of vertices on which they differ. The key lemma in this paper is the following:

Lemma 1. *For any two colourings X, Y of a Δ-regular graph,*

$$\sum_{w \in V} \max\{P_w(X,Y), P_w(Y,X)\} \le (1 - \frac{R_X}{2})\Delta d(X,Y).$$

We defer its proof until the end of the section.

Let X', Y' denote random colourings generated by applying one step of the Glauber dynamics to X, Y respectively. Following the notation in [10], we say that X, Y are δ*-distance decreasing* if there exists a coupling of X', Y' under which the expected value of $d(X', Y')$ is at most $(1 - \delta)d(X, Y)$.

Recall that $\beta = 1.645...$ is defined in the introduction. Using Lemma 1, it is fairly straightforward to prove the following:

Lemma 2. *Suppose that $k \geq (\beta+\epsilon)\Delta$ for some $\epsilon > 0$. Then there exists $\zeta, \delta > 0$ such that if $X \in \Omega$ satisfies the first and the second local uniformity conditions for $\zeta > 0$, then for every $Y \in \Omega$, (X, Y) is δ-distance-decreasing.*

Proof. We need to prove, for every $Y \in \Omega$,

$$\mathbf{E}(d(X', Y')) \leq (1 - \frac{\delta}{n})d(X, Y)$$

for some $\delta > 0$. Let v be the vertex selected for recolouring at the first time step. First we bound the probability that the chains recolour v to different colours. For a colour c available to both chains, c will be chosen in X with probability $\frac{1}{|L_X(v)|}$ and in Y with probability $\frac{1}{|L_Y(v)|}$, and hence c will be chosen in both chains with probability $\frac{1}{\max\{|L_X(v)|,|L_Y(v)|\}}$ if we use, as usual, Jerrum's coupling. Therefore, the probability that v will be coloured differently is:

$$\begin{aligned}\mathbf{Pr}(X'(v) \neq Y'(v) \mid v) = 1 - \frac{|L_X(v) \cap L_Y(v)|}{\max\{|L_X(v)|, |L_Y(v)|\}} \\ = \frac{\max\{|L_X(v)|, |L_Y(v)|\} - |L_X(v) \cap L_Y(v)|}{\max\{|L_X(v)|, |L_Y(v)|\}} \\ = \frac{\max\{|P_v(X,Y)|, |P_v(Y,X)|\}}{\max\{L_X(v), L_Y(v)\}}, \qquad (1)\end{aligned}$$

recall that $P_v(X,Y) := L_Y(v) - L_X(v)$ and $P_v(Y,X) := L_X(v) - L_Y(v)$. Now we bound the expected distance after one step.

$$\begin{aligned}&\mathbf{E}(d(X', Y')) \\ &= \sum_{w \in V} \mathbf{Pr}(X'(w) \neq Y'(w)) \\ &= \sum_{w \in V} \mathbf{Pr}(v \neq w \wedge X(w) \neq Y(w)) + \sum_{w \in V} \mathbf{Pr}(v = w \wedge X'(w) \neq Y'(w)) \\ &= \frac{n-1}{n} d(X,Y) + \frac{1}{n} \sum_{w \in V} \mathbf{Pr}(X'(w) \neq Y'(w) \mid v = w) \\ &= \frac{n-1}{n} d(X,Y) + \frac{1}{n} \sum_{w \in V} \frac{\max\{|P_w(X,Y)|, |P_w(Y,X)|\}}{\max\{|L_Y(w)|, |L_X(w)|\}} \quad \text{(by (1))} \\ &\leq \frac{n-1}{n} d(X,Y) + \frac{1}{nL_X} \sum_{w \in V} \max\{P_w(X,Y), P_w(Y,X)\} \\ &\leq \frac{n-1}{n} d(X,Y) + \frac{1}{nL_X}(1 - \frac{R_X}{2})\Delta d(X,Y) \quad \text{(by Lemma 1)}\end{aligned}$$

$$\leq \frac{n-1}{n}d(X,Y) + \frac{1}{n}\frac{(2-(1-e^{-\frac{\Delta}{k}})^2+\zeta)\Delta}{2ke^{-\frac{\Delta}{k}}-\zeta}d(X,Y)$$

$$\leq \frac{n-1}{n}d(X,Y) + \frac{1}{n}(1-\delta)d(X,Y) = (1-\frac{\delta}{n})d(X,Y) \quad \text{(since } k \geq (\beta+\epsilon)\Delta\text{)}$$

for some $\delta > 0$, if we take ζ to be sufficiently small in terms of ϵ. The second last inequality follows from the local uniformity properties. ∎

The following theorem about couplings which "usually" decrease distances is by Hayes and Vigoda[10].

Theorem 2. *Let $X_0, \ldots, X_T$, $Y_0, \ldots, Y_T$ be coupled Markov chains such that, for every $0 \leq t \leq T-1$,*

$$\mathbf{Pr}((X_t, Y_t) \textit{ is not } \delta \textit{ distance decreasing}) \leq \epsilon.$$

Then

$$\mathbf{Pr}(X_t \neq Y_t) \leq ((1-\delta)^T + \epsilon/\delta)\, diam(\Omega).$$

In Section 3, we will prove (Lemma 3) that if G is a graph of girth 5 and maximum degree $\Delta \geq 1000\log^3 n$, and if X is a uniformly random k-colouring of G where $k \geq (1+\epsilon)\Delta$ for some $\epsilon > 0$ then X satisfies the first and second local uniformity properties. That will allow us to apply the preceding lemmas to such graphs, and thus prove the main result of this paper, which we do now.

Proof (**Proof of Theorem 1**). The proof is along the same line as in Hayes and Vigoda[10]. Here we just give a quick sketch. For ease of exposition, we assume that G is Δ-regular.

Let X_0 be distributed according to π (the uniform distribution) and Y_0 be arbitrary. Generate $X_1, \ldots, X_T$, $Y_1, \ldots, Y_T$ using Jerrum's coupling with initial states X_0, Y_0. For every $t \geq 0$, X_t is distributed according to π. By Lemma 3, X_t has the first and the second local uniformity properties with sufficiently high probability. Hence, by Lemma 2, X_t and Y_t are δ-distance decreasing with high probability. Now, applying Theorem 2 gives the theorem. ∎

Finally, we close this section by proving the key lemma.

Proof (**Proof of Lemma 1**). Let $d := d(X,Y)$ be the Hamming distance between X and Y, and $v_1, \ldots, v_d$ be the d vertices with different colours in X and Y. Let Z_i be a colouring equal to X except $Z_j(v_j) = Y(v_j)$ for $1 \leq j \leq i$. Let $P_w(i) := \max\{P_w(X, Z_i), P_w(Z_i, X)\}$. So, $Z_d = Y$ and $\sum_{w \in V} P_w(d)$ is the value we would like to bound. To bound $\sum_{w \in V} P_w(d)$, we consider $\sum_{w \in V} P_w(i)$ for $1 \leq i \leq d$. Intuitively, we consider the colour changes one-at-a-time.

Note that $P_w(i) > P_w(i-1)$ only when w is a neighbour of v_i, and note also that $P_w(i) \leq P_w(i-1)+1$ by definition. Since the maximum degree in G is Δ, it follows that $\sum_{w \in V} P_w(d) \leq d\Delta$. With the second local uniformity, however, we can give a better bound. For example, when the colour of v_1 is changed from $X(v_1)$ to $Y(v_1)$, for each vertex w in $R_X(v_1, Y(v_1))$, both colours $X(v_1)$

and $Y(v_1)$ appear in $N_G(w) - v$ and thus $L_X(w)$ and $L_{Z_1}(w)$ are the same. Hence, $P_w(1) = P_w(0)$ for $w \in R_X(v_1, Y(v_1))$. Since $|R_X(v_1, Y(v_1))| \geq R_X \Delta$ by definition, we have $\sum_{w \in V} P_w(1) \leq (1 - R_X)\Delta$. Notice that the above argument does not hold in general at time i for $i > 1$, since the colours have been changed at $v_1, \ldots, v_{i-1}$. But one may still hope that $P_w(i) = P_w(i-1)$ for "many" vertices in $R_X(v_i, Y(v_i))$. In light of this, we say that a *good* event happens at w at time i if $P_w(i) = P_w(i-1)$ when the colour of v_i is changed from $X(v_i)$ to $Y(v_i)$; otherwise a *bad* event if $P_w(i) = P_w(i-1)+1$ at w at time i when the colour of v_i is changed from $X(v_i)$ to $Y(v_i)$. In the following, we focus on a bad event at w at time i where $w \in R_X(v_i, Y(v_i))$.

Let $a := X(v_i)$ and $b := Y(v_i)$. Consider a vertex w in $R_X(v_i, b)$ when the colour of v_i is changed from a to b. Suppose that a bad event happens at w (i.e. $P_w(i) = P_w(i-1)+1$). Since w is in $R_X(v_i, b)$, by definition, there are two vertices $u_a, u_b \in N_G(w) - v_i$ so that $X(u_a) = a$ and $X(u_b) = b$. Recall that $P_w(i) := \max\{P_w(X, Z_i), P_w(Z_i, W)\}$. Since both colours a and b appear in the neighbourhood of w in X, we have $P_w(Z_i, X) = P_w(Z_{i-1}, X)$. Since we assume $P_w(i) = P_w(i-1)+1$, it must be the case that $P_w(X, Z_i) = P_w(X, Z_{i-1}) + 1$. This can only happen when the colour a disappears in the neighbourhood of w at time i (i.e. $a \notin L_{Z_{i-1}}(w)$ and $a \in L_{Z_i}(w)$). In particular, this implies that the colour of u_a had been changed from a to some other colour in some j-th step where $j < i$. Consider that colour change of $u_a = v_j$ at the j-th step. At the j-th step, $Z_j(v_i)$ is still of colour a. Therefore, by changing the colour of u_a from a to some other colour, we have $P_w(X, Z_j) = P_w(X, Z_{j-1})$. Notice that $P_w(X, Y) \leq \sum_{i=1}^{d} |P_w(X, Z_i) - P_w(X, Z_{i-1})|$ and similarly $P_w(Y, X) \leq \sum_{i=1}^{d} |P_w(Z_i, X) - P_w(Z_{i-1}, X)|$. From the above argument, if a bad event happens at w at the i-th step, we have $|P_w(X, Z_i) - P_w(X, Z_{i-1})| = 0$ and $|P_w(Z_j, X) - P_w(Z_{j-1}, X)| = 0$ for some $j < i$. And thus the bad event at w at time i and the event at w at time j combine to contribute at most 1 to $P_w(d)$, in particular this implies that $P_w(d)$ is at most $d-1$. Formally, we map a bad event at w at time i to another event at w at time j where $j < i$ and $X(v_i) = X(v_j) = a$, so that they combine to contribute at most 1 to $P_w(d)$. We call the events in this mapping a *couple*. We can do the mapping for each bad event at w at time j for each $w \in R_X(v_j, Y(v_j))$ by the above argument. If there are T_1 disjoint couples and T_2 distinct good events such that no event appears therein more than once, then $\sum_{w \in V} P_w(d) \leq d\Delta - T_1 - T_2$. Suppose for now that each bad event at w at time j for $w \in R_X(v_j, Y(v_j))$ maps to a distinct good event (we will prove this claim in the next paragraph). Since $\sum_{j=1}^{d} |R_X(v_j, Y(v_j))| \geq R_X \Delta d$ and each event therein is either good or is in a distinct couple, we have $\sum_{w \in V} P_w(d) \leq d\Delta - (R_X \Delta d)/2 = ((1 - R_X/2)\Delta)d$, as desired. (The worst case is that there are $(R_X \Delta d)/2$ disjoint couples where each couple contains two distinct events therein).

To finish the proof, it remains to show that each bad event at w at time j for $w \in R_X(v_j, Y(v_j))$ maps to a distinct good event. To see this, we need to review the mapping process. As argued previously, a bad event at w at time i for $w \in R_X(v_i, Y(v_i))$ happens only if the colour $X(v_i)$ disappears in the neighbourhood of w at time i. Then we map this bad event to another event at w at time j where $j < i$ and $X(v_i) = X(v_j)$. This implies that $Z_j(v_i) = X(v_i)$ and thus the colour $X(v_j)$ does not disappear in the neighbourhood of w at time j, and hence the event at w at time j is not a bad event. So, a bad event does not map to another bad event. Also, two bad events cannot map to the same event since a colour can disappear at most once in the neighbourhood of a vertex, as each vertex is recoloured at most once. This proves the claim and completes the proof. ∎

3 Uniform Colourings of Graphs with Girth 5

In this section, we establish that the uniform random colourings we consider satisfy the first and second uniformity properties. Recall that our setting is: G is a graph on n vertices with maximum degree Δ and with girth at least 5. We consider k-colourings of G where $k = q\Delta$ for some $q > 1$. For ease of exposition, we assume that G is Δ-regular.

Lemma 3. *Consider a uniform random k-colouring X of G and consider any $\zeta > 0$. With probability at least $1 - n^{-3}$ we have:*

(a) $(qe^{-1/q} - \zeta)\Delta < L_X < (qe^{-1/q} + \zeta)\Delta$;
(b) $((1 - e^{-1/q})^2 - \zeta)\Delta < R_X < ((1 - e^{-1/q})^2 + \zeta)\Delta$.

The lower bound in part (a) was proven in [10] for graphs of girth 4. The rest of Lemma 3 was (essentially) proven in [7] to hold for graphs of girth 6. Roughly speaking, having girth 6 is very helpful as follows: Define $N_2(v)$ to be the vertices of distance 2 from v. Note that $R_X(v, c)$ is a function of the colours appearing on $v \cup N_2(v)$ which, if G has girth at least 6, is an independent set. If we pretend that the colours on those vertices are independent uniformly random colours, then Lemma 3(b) follows easily. Of course they aren't independent: some dependency is induced by the edges joining the independent set to the rest of G; this is a bit difficult to deal with, but the techniques from [12] will suffice. When we reduce the girth requirement to 5, $N_2(v)$ can now have many edges, and these edges bring dependencies that are not straightforward to deal with. Fortunately, there are some restrictions - for example, no two vertices of $N_2(v)$ with a common "parent" in $N(v)$ can be joined. This allows us to overcome the dependency. The way we do so comprises the new ideas in the proof; the rest just follows techniques from [12].

In order to better control the edges within $v \cup N(v) \cup N_2(v)$, we partition it into smaller subgraphs. In particular, for each vertex v, we partition $N(v)$ into sets $U_1(v), ..., U_{\Delta^{2/3}}(v)$ each of size $\Delta^{1/3}$. (For convenience, we treat $\Delta^{1/3}$ as an integer; it is trivial to extend the argument to the non-integer case.) Instead of analyzing the colour distribution of colours on all of $N_2(v)$, we will sometimes consider, for each i, the distribution on the subset of $N_2(v)$ that is adjacent to $U_i(v)$.

Given a colouring of G, for any vertex v and colours c, c' we define:

- $T_{v,c} = \sum \frac{1}{|L(u)|}$ summed over all $u \in N(v)$ with c not appearing on $N(u) - v$;
- $T^j_{v,c} = \sum \frac{1}{|L(u)|}$ summed over all $u \in U_j(v)$ with c not appearing on $N(u) - v$.

We define $\alpha_0 = 0, \beta_0 = 1, \lambda_0 = 1$, and

- $\alpha_{i+1} = e^{-\beta_i}/\lambda_i$
- $\beta_{i+1} = e^{-\alpha_i}/qe^{-1/q}$
- $\lambda_{i+1} = \frac{q\beta_i - 1}{\beta_i - \alpha_i} e^{-\alpha_i} + \frac{1 - q\alpha_i}{\beta_i - \alpha_i} e^{-\beta_i}$

As shown in [12], $\lim_{i\to\infty} \alpha_i = \lim_{i\to\infty} \beta_i = 1/q$ and $\lim_{i\to\infty} \lambda_i = qe^{-1/q}$.

We will prove:

Lemma 4. *In a uniformly random k-colouring X of G, for every v, c, j, i we have: with probability at least $1 - (\Delta^3)^i \exp(-\Delta^{1/3})$:*

(a) $qe^{-1/q}\Delta - o(\Delta) < |L(v)| < \lambda_i \Delta + o(\Delta)$;
(b) $\alpha_i \Delta^{-2/3} - o(\Delta^{-2/3}) < |T^j_{v,c}| < \beta_i \Delta^{-2/3} + o(\Delta^{-2/3})$.

Proof. Many details are the same as those that have appeared already in several papers. So we gloss over those, focusing more on the details that are new.

The lower bound in (a) was proven in [10]. For the rest, we use induction on i. The base case $i = 0$ is trivial. So suppose it holds for i; we will prove that it holds for $i + 1$.

Expose the colours of every vertex except for those in $N(v)$. This yields a list $L(u)$ for each $u \in N(v)$.

By induction, and multiplying by Δ^2 vertices u in $N(v) \cup N_2(v)$ plus less than $2\Delta^{8/3}$ triples j, c, u with $u \in N(v)$, we see that with probability at least $1 - (\Delta^2 + 2\Delta^{8/3})(\Delta^3)^i \exp(-\Delta^{1/3})$, each $u \in N(v) \cup N_2(v)$ has $qe^{-1/q}\Delta - o(\Delta) < |L(u)| < \lambda_i \Delta + o(\Delta)$ and for every $u \in N(v)$ and c, $T_{u,c} = \sum_{j=1}^{\Delta^{2/3}} T^j_{u,c}$ is between $\alpha_i + o(1)$ and $\beta_i + o(1)$.

We will show that, if the exposed colours behave as described in the previous paragraph, then the probability that (a) is violated is at most $\exp(-\Delta^{1/3})$ and so is the probability that (b) is violated. This yields an overall bound of at most $(\Delta^2 + 2\Delta^{8/3})(\Delta^3)^i \exp(-\Delta^{1/3}) + 2\exp(-\Delta^{1/3}) < (\Delta^3)^{i+1} \exp(-\Delta^{1/3})$.

Part (a): Straightforward calculations, as in [12], show that $\mathbf{Exp}(|L(v)|) \le \lambda_{i+1}\Delta + o(\Delta)$. Because G is triangle-free, we can regard the assignments of colours to $N(v)$ as independent uniform choices from the lists $L(u)$. Thus standard concentration bounds (such as Azuma's Inequality) easily yield that the probability of $|L(u)|$ differing from its mean by more than $\Delta^{2/3}$ is at most $\exp(-\theta(\Delta^{2/3})) < \exp(-\Delta^{1/3})$.

Part (b): Let $t^j_{v,c}$ denote the number of neighbours $u \in U_j(v)$ with $c \in L(u)$. We will show that, with sufficiently high probability, $\alpha_i \Delta^{1/3} + o(\Delta^{1/3}) < t^j_{v,c} < \alpha_i \Delta^{1/3} + o(\Delta^{1/3})$. This, along with the inductive bound on $|L(u)|$ will establish (b).

Let H be the subgraph induced by $\cup_{u \in U_j} N(u) - v$. Note that, since the girth of G is at least 5, no $w \in H$ can be adjacent to more than one neighbour of any $u \in U_i(v)$. Thus the maximum degree in H is at most $|U_j(v)| = \Delta^{1/3}$.

Expose the colours of every vertex except those in H. This yields a list $L(u)$ for each $u \in U_j(v)$. With probability at least $1-2ie^{-\Delta^{1/20}}$, each $w \in N(U_j(v))-v$ has $qe^{-1/q}\Delta - o(\Delta) < |L(w)| < \lambda_i\Delta + o(\Delta)$ and for every c and $u \in U_j(v)$, $T_{u,c}$ is between $\alpha_i + o(1)$ and $\beta_i + o(1)$.

Let Ω be the set of colourings of H in which each $w \in H$ has a colour from $L(w)$. Thus, the unexposed colours form a uniformly random member of Ω. Since H is not an independent set, we can't simply treat those colours as independent uniform choices from their lists, as we did in part (a). How we deal with this complication is the main new idea required for this section.

Suppose that the vertices of H are $w_1, ..., w_t$; we will colour the vertices one-at-a-time, in order. After r vertices have been coloured, we use Ω_r to denote the set of completions of the partial colouring to a colouring of H taken from the lists $L(w)$; thus $\Omega_0 = \Omega$. When we colour w_r, we choose each colour c with probability $p_r(c)$ which is equal to the proportion of colourings in Ω_{r-1} in which w_r has c. Thus, the resultant colouring is a uniformly random member of Ω, as required.

Claim. For each r, c such that c is not assigned to a neighbour $w_{r'}$ of w_r with $r' < r$, we have: $p_r(c) = |L(w_r)|^{-1}(1 \pm O(\Delta^{-2/3}))$.

Proof. First note that since the maximum degree of H is at most $\Delta^{1/3}$, at any point in the colouring process, every w has at least $|L(w)| - \Delta^{1/3} = \Theta(\Delta)$ colours not appearing on any of its neighbours.

We prove the claim by induction on r. The base case is $r = t$; i.e. the last vertex coloured in H. Here, each colour in $L(w_t)$ that does not yet appear on a neighbour is equally likely, so $|L(w_r)|^{-1} \leq p_t(c) \leq (|L(w_r)| - \Delta^{1/3})^{-1} < |L(w_r)|^{-1}(1 + O(\Delta^{-2/3})$. Now assume that the claim holds for every $r' > r$; we will prove it for r.

Consider two colours $c, c' \in L(w_r)$ that don't appear on any neighbour of w_r. Let $\Omega(c), \Omega(c')$ denote the sets of colourings in Ω_{r-1} in which w_r gets c, c' respectively. Suppose that we set $w_r = c'$ and then continue our colouring process. By induction, the probability that at least one neighbour of w_r receives c is at most $\Delta^{1/3} \times O(\Delta^{-1}) = O(\Delta^{-2/3})$. Since this process yields a uniform member of $\Omega(c')$, this implies that at least $(1 - O(\Delta^{-2/3}))|\Omega(c')|$ of the colourings of $\Omega(c')$ can be mapped to a colouring in $\Omega(c)$ by switching the colour of w_r to c. Therefore, $|\Omega(c)| \geq |\Omega(c')|(1 - O(\Delta^{-2/3}))$. Since this is true for every pair c, c' the claim follows. ∎

Having proven the Claim, we now consider any $u \in U_j$; we will bound the probability that c is not assigned to any $w \in N(u) - v$. Since our colouring procedure yields a uniformly random member of Ω, this probability is not affected by the actual order in which we colour the vertices. So we can take $w_1, ...,$ the first vertices to be coloured, to be $N(u) - v$. Since $N(u) - v$ is an independent set, if $c \in L(w)$ for some $w \in N(u) - v$, c will still be eligible to be assigned to w when we come to choose the colour for w. Therefore by our claim, the desired probability is $\prod(1 - |L_w|^{-1} + O(\Delta^{-5/3}))$ over all $w \in N(u) - v$ with $c \in L(w)$, and this is between $\alpha_i + o(1)$ and $\beta_i + o(1)$ by the same calculations as in [12]. Thus, $\alpha_i \Delta^{1/3} + o(\Delta^{1/3}) < \mathbf{Exp}(t^j_{v,c}) < \beta_i \Delta^{1/3} + o(\Delta^{1/3})$.

Next we will use Azuma's Inequality to show that $t^j_{v,c}$ is concentrated.

Consider a particular w_r adjacent to $u_\ell \in U_j$. We want to measure how much the colour chosen for w_r can affect the expected value of $t^j_{v,c}$, where the expectation is over the remaining $t - r$ random colour assignments. The extreme case is when we choose to assign c to w_r (we omit the straightforward dispensation of the other cases). This will cause u_ℓ to not have $c \in L(u_\ell)$ and thus might reduce the conditional expectation of $t^j_{v,c}$ by 1. Possibly it will also have a further effect on the conditional expectation because it changes the probability that other members of U_j will have c in their lists; we bound that effect as follows: For each of the $\Delta^{1/3}$ neighbours w of w_r, the assignment of c to w_r drops the probability of w receiving c to zero; for every other w, by reasoning similar to that in our claim, this affects the probability of w receiving c by a negligible amount. Each $u \in N(v)$ has at most one neighbour adjacent to w_r and so the effect on the probability of c not appearing in $N(u)$ is at most a factor of $(1-1/\Theta(\Delta))$; since this probability is $\Theta(1)$, by the previous paragraph, the effect is an additive term of at most $O(1/\Delta)$. Thus, the overall affect on $\mathbf{Exp}(t_{v,c})$ is $1 + |U_i| \times O(1/\Delta) = 1 + O(\Delta^{-2/3}) < 2$.

Note that $t_{v,c}$ is determined by $\Delta^{4/3}$ trials - the colour choice for each neighbour of every member of U_i. If we try to apply Azuma's Inequality directly with each $\gamma_i = 2$ and with $\Delta^{4/3}$ trials, we fail. So we reduce the number of trials to $\Delta^{1/3}$ as follows: For each $u \in N(v)$, we treat the assignments to all of $N(u) - v$ as a single random choice. A simple concentration argument (details omitted) implies that with probability at least $1 - \exp(-\Theta(\Delta))$, no u has more than $\sqrt{\Delta}$ neighbours that receive c. Standard arguments (details omitted) allow us to assume that no u has more than $\Delta^{1/10}$ such neighbours, as far as the remainder of the argument is concerned. Thus, by the same calculations as in the previous paragraph, the maximum effect that any one of these random choices can have is $1+\sqrt{\Delta}\times O(\Delta^{-2/3}) < 2$. Thus we can apply Azuma's Inequality with $\Delta^{1/3}$ trials, each $\gamma_i = 2$ and with $\tau = O(\Delta^{9/30})$ to show that the probability of $t_{v,c}$ differing from its mean by more than $O(\Delta^{9/30}) = o(\Delta^{1/3})$ is at most $\exp(-\Theta(\Delta^{4/15}))$. This yields an overall probability of $t^j_{v,c}$ differing from its mean by more than $O(\Delta^{9/30})$ of less than $\exp(-\Theta(\Delta)) + \exp(-\Theta(\Delta^{-4/15})) < \exp(-\Delta^{1/3})$. ∎

Now we finish the proof of Lemma 3.

*Proof (***Proof of Lemma 3***).* We will show that for every v, c, the probability that $L_X(v)$ violates part (a) or that $R_X(v,c)$ violates part (b) is at most $\exp(-\frac{1}{2}\Delta^{1/3})$. If $\Delta \geq 1000 \log^3 n$ then this is at most $1/n^5$. Thus, after multiplying by the fewer than n^2 choices for v, c, we obtain that conditions (a,b) hold for every n, c with probability at least $1 - n^{-3}$, as required.

The bound on the probability that $L_X(v)$ is in violation follows immediately from Lemma 4 by taking i to be a large enough constant that λ_i differs from its limit by at most $\zeta/2$ and noting that $(\Delta^3)^i \exp(-\Delta^{1/3}) < \exp(-\frac{1}{2}\Delta^{1/3})$.

For $R_X(v,c)$, we also apply Lemma 4 for a particular large value of i. Then we carry out an argument nearly identical to that in the proof of Lemma 4(b) to bound, for every v, c the number of neighbours $u \in N(v)$ with $X(v), c$ both in $N(u)-v$. Straightforward calculations, as in [12], show that the expected number is within $\zeta\Delta/2$ of $(1-e^{-1/q})^2\Delta$ so long as i is large enough that $\alpha_i, \beta_i, \lambda_i$ are sufficiently close to their limits. A concentration proof nearly identical to that in the proof of Lemma 4 shows that the probability that this number differs from its mean by at least $\zeta\Delta/2$ is less than $\exp(-\frac{1}{2}\Delta^{1/3})$; we omit the repetitive details. ∎

References

1. K. Azuma. *Weighted sums of certain dependent random variables.* Tokuku. Math. Journal, 19, pp. 357-367, 1967.
2. R. Bubley and M. Dyer. *Path coupling: a technique for proving rapid mixing in Markov chains.* Proceedings of FOCS 1997, pp. 223-231.
3. M. Dyer, A. Flaxman, A. Frieze and E. Vigoda. *Randomly coloring sparse random graphs with fewer colors than the maximum degree.* Manuscript, 2004.
4. M. Dyer and A. Frieze. *Randomly colouring graphs with lower bounds on girth and maximum degree.* Random Structures and Algorithms, 23, pp. 167-179, 2003.
5. M. Dyer, A. Frieze, T. Hayes, E. Vigoda. *Randomly coloring constant degree graphs.* Proceedings of FOCS 2004, pp. 582-589.
6. A. Frieze and J. Vera. *On randomly colouring locally sparse graphs.* Manuscript, 2004.
7. T. Hayes. *Randomly coloring graphs of girth at least five.* Proceedings of STOC 2003, pp. 269-278.
8. T. Hayes and E. Vigoda. *A non-Markovian coupling for randomly sampling colorings.* Proceedings of FOCS 2003, pp. 618-627.
9. T. Hayes and E. Vigoda. *Variable length path coupling.* Proceedings of SODA 2004, pp. 103-110.
10. T. Hayes and E. Vigoda. *Coupling with the stationary distribution and improved sampling for colorings and independent sets.* Proceedings of SODA 2005, pp. 971-979.
11. M. Jerrum. *A very simple algorithm for estimating the number of k-colourings of a low-degree graph.* Random Structures and Algorithms, 7, pp. 157-165, 1995.
12. M. Molloy. *The Glauber dynamics on colorings of a graph with high girth and maximum degree.* SIAM Journal on Computing, 33, pp. 721-737, 2004.
13. M. Molloy and B. Reed. *Graph colourings and the probabilistic method.* Springer, 2002.
14. J. Salas and A. Sokal. *Absence of phase transition for antiferromagnetic Potts models via the Dobrushin uniqueness theorem.* J. Statist. Phys., 86, pp. 551-579, 1997.
15. E. Vigoda. *Improved bounds for sampling colorings.* Journal of Mathematical Physics, 41, pp. 1555-1569, 2000.

Packing Dicycle Covers in Planar Graphs with No $K_5 - e$ Minor

Orlando Lee[1,*] and Aaron Williams[2,**]

[1] Instituto de Computao Universidade Estadual de Campinas (UNICAMP)
lee@ic.unicamp.br
[2] Dept. of Computer Science, University of Victoria
haron@uvic.ca

Abstract. We prove that the minimum weight of a dicycle is equal to the maximum number of disjoint dicycle covers, for every weighted digraph whose underlying graph is planar and does not have $K_5 - e$ as a minor ($K_5 - e$ is the complete graph on five vertices, minus one edge). Equality was previously known when forbidding K_4 as a minor, while an infinite number of weighted digraphs show that planarity does not guarantee equality. The result also improves upon results known for Woodall's Conjecture and the Edmonds-Giles Conjecture for packing dijoins. Our proof uses Wagner's characterization of planar 3-connected graphs that do not have $K_5 - e$ as a minor.

1 Introduction

Min-max theorems are fundamental to directed graph theory. For example, Menger's Theorem [7] proves that the minimum number of arcs separating node s from node t equals the maximum number of arc-disjoint dipaths from s to t. Reversing the roles of these objects gives another min-max theorem: the minimum number of arcs in a dipath from s to t equals the maximum number of arc-disjoint cuts separating s from t. Similarly, the celebrated Lucchesi-Younger Theorem [6] proves that the minimum number of arcs in a dijoin equals the maximum number of arc-disjoint dicuts. In all three cases, the min-max theorems can be extended from digraphs to weighted digraphs.

Still, many important min-max questions remain open or are untrue. Woodall's Conjecture [13] reverses the roles of the Lucchesi-Younger Theorem and asks if the minimum number of arcs in a dicut equals to the maximum number of arc-disjoint dijoins. Although Woodall's Conjecture remains one of the biggest open conjectures in graph theory, its weighted version (the Edmonds-Giles Conjecture [2]) is not true [9], [5], [12], [1]. In particular, Figure 1 shows that the Edmonds-Giles Conjecture is not true for planar digraphs. On the other hand, the conjecture was verified for series-parallel digraphs [8] (see also [10], [3], [4] which proves the conjecture for source-sink connected digraphs). In this

* Research supported by FAPESP/CNPq (Pronex Proc. 2003/09925-5), CNPq (Edital Universal Proc. 471460/2004-4) and CNPq (PROSUL Proc. 490333/04-4).

** Research supported by NSERC PGS-D.

J.R. Correa, A. Hevia, and M. Kiwi (Eds.): LATIN 2006, LNCS 3887, pp. 677–688, 2006.

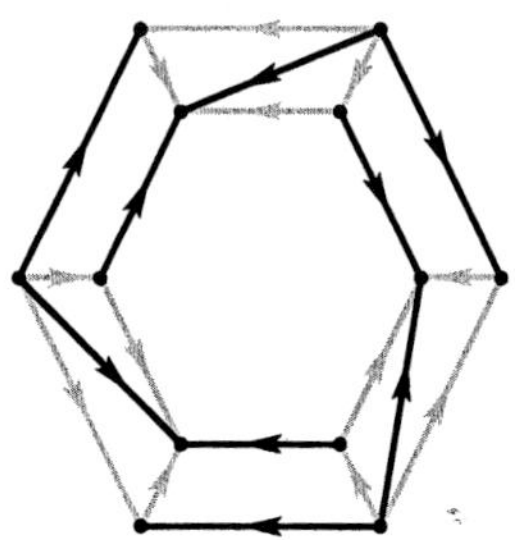

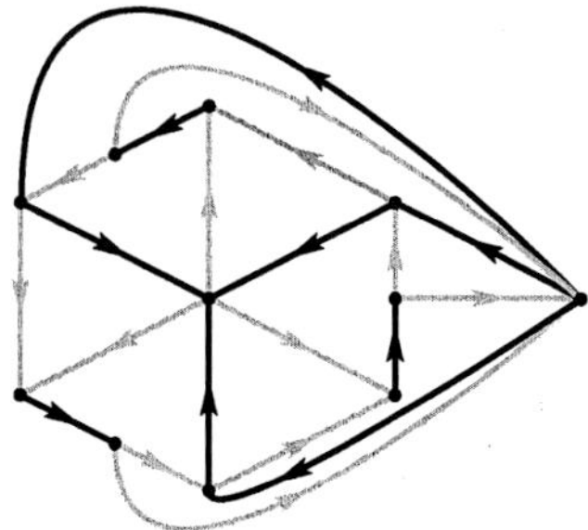

Fig. 1. A counterexample to the Edmonds-Giles Conjecture (left), and its planar dual (right). Light arcs have weight zero and dark arcs have weight one.

paper we narrow the gap between these two results by working on the planar dual problem. Along the way we introduces new techniques and lemmas that hold promise for future results in this challenging and important area.

Claim. If digraph D is planar and has no $K_5 - e$ minor, then for any arc weights, the minimum weight of a dicycle is equal to the maximum number of disjoint dicycle covers.

Our proof relies upon Wagner's characterization of 3-connected graphs that have no $K_5 - e$ minor. In particular, we show that dicycle covers can be glued across vertex cuts of size 1 and 2. Then, despite the global nature of dicycle covers, we are able to reduce the problem of finding dicycle covers locally. We redistribute weight around individual nodes and then eliminate arcs with zero weight or large weight. Furthermore, we employ a wye-delta reduction which removes a vertex of degree 3 and replaces it with edges between the vertex's neighbours.

Theorem 1 (Wagner). *If planar digraph D is 3-connected and has no $K_5 - e$ minor, then D is either a small complete graph, the envelope graph, or a wheel [11].*

2 Definitions, Notation, and Terminology

In this section we group together definitions, notation, and terminology necessary for the remainder of the paper. Graph-theoretic concepts that are open to different interpretations will be formally defined, while more standardized concepts will not. Included in this section are ideas that are common to many packing and covering theorems, so experienced readers may wish to skim this portion of the text. At the end of the section we introduce the notion of pushing weight into a cut, and point out its use in Remarks 1 and 2.

A *graph* $G = (V, E)$ is a set of vertices V and a set of edges E, where each edge is an unordered distinct pair of vertices. A *digraph* $D = (N, A)$ is a set of nodes N and a set of arcs A, where each arc is an ordered distinct pair of

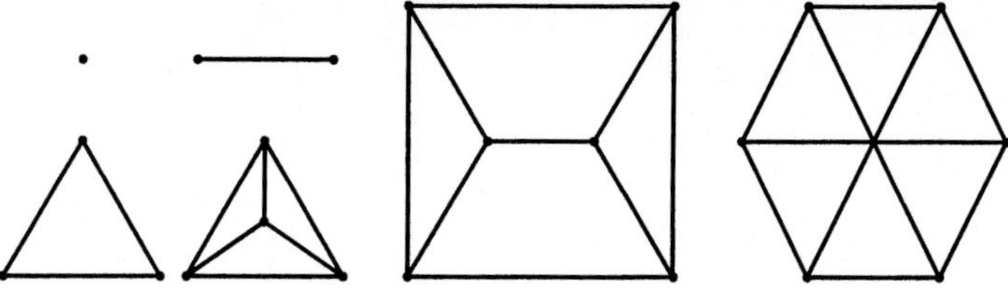

Fig. 2. From left to right: complete graphs K_1 through K_4, the envelope graph, and a wheel graph with seven vertices

nodes. Given a digraph $D = (N, A)$, its *underlying graph* is equal to (V, E), where $V = N$ and $E = \{xy : xy \in A \text{ or } yx \in A\}$. A *weighted digraph* (D, ω) is a digraph $D = (N, A)$ together with non-negative arc weights $\omega \in \{0, 1, 2, \ldots\}^A$.

Let C be a dicycle in D with arcs $A(C)$. The *weight* of C is denoted $\omega(C)$ and is equal to $\sum_{a \in A(C)} \omega(a)$. The minimum weight of a dicycle in (D, ω) is denoted $\tau(D, \omega)$. Let $J \subseteq A$ be a subset of arcs. J *covers* C if $J \cap A(C) \neq \emptyset$. J is a *(dicycle) cover* of D if J covers every dicycle in D. A cover is *minimal* if every proper subset of it is not a cover. A collection of arc subsets $J_1, J_2, \ldots, J_k \subseteq A$ are *disjoint* in (D, ω) if at most $\omega(a)$ of the covers use a, for all $a \in A$. The maximum number of disjoint covers in (D, ω) is denoted $\nu(D, \omega)$. Notice that $\nu(D, \omega) \leq \tau(D, \omega)$. If equality holds then we say that (D, ω) *packs*; otherwise (D, ω) does not pack. Finally, a collection of $\tau(D, \omega)$ disjoint covers is called a *packing* of covers. Central to finding a packing of covers is the pursuit of special covers which we will call valid and accommodating. Let $v_J \in \{0, 1\}^A$ be an incidence weighting for $J \subseteq A$, where $v_J(a) = 1$ if $a \in J$, and $v_J(a) = 0$ if $a \notin J$. We say that J is *valid* in (D, ω) if $\omega - v_J$ has only non-negative entries; that is, if $\omega(a) > 0$ for each $a \in J$. J *accommodates* dicycle C in (D, ω) if $\omega'(C) \geq \tau(D, \omega) - 1$, where $\omega' = \omega - v_J$. Furthermore, J is *accommodating* in (D, ω) if every dicycle in C is accommodated by J in (D, ω). Notice that if J is accommodating in (D, ω) then $\tau(D, \omega - v_J) = \tau(D, \omega) - 1$; that is, J leaves enough room for the possibility of finding $\tau(D, \omega) - 1$ disjoint covers after its *removal* forms $(D, \omega - v_J)$. Notice that every cover in a packing is valid and accommodating, and that the ability to always find a valid and accommodating cover allows one to construct a packing of covers.

Let $X \subseteq N$ be a set of nodes in digraph (N, A). We let $\overline{X} = N - X$, and the *cut induced by* X is represented by $\delta(X)$ and is equal to $\delta^{\text{in}}(X) \cup \delta^{\text{out}}(X)$, where $\delta^{\text{in}}(X) = \{xy \in A : x \in \overline{X} \text{ and } y \in X\}$ and $\delta^{\text{out}}(X) = \{xy \in A : x \in X \text{ and } y \in \overline{X}\}$. If $\delta^{\text{out}}(X) = \emptyset$ then we say that $\delta^{\text{in}}(X)$ is a *dicut*.

A *digon* is a dicycle of length 2, and any arc that is in a digon is called a *digon arc*.

Given a graph $G = (V, E)$ and $e \in E$, we let $G \backslash e$ represent the result of deleting edge e, and we let G/e represent the result of contracting edge e. Likewise, given $v \in V$, we let $G \backslash v$ be the result of deleting vertex v and every edge that is incident with v. For a subset of vertices $X \subseteq V$, we let $G[X]$ be the result of deleting every vertex in $\overline{X}$. In graph $G = (V, E)$, a *k-separation* is a pair of subgraphs (G_1, G_2) where $G_1 = (V_1, E_1)$ and $G_2 = (V_2, E_2)$, such that

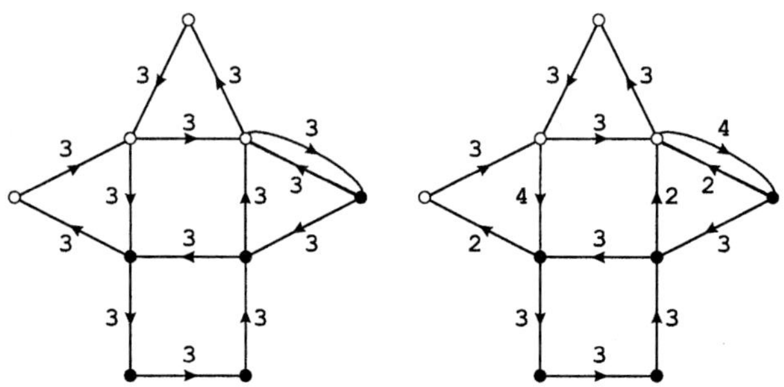

Fig. 3. Before and after pushing into $\delta(X)$, where X is given by the black nodes

$E = E_1 \cup E_2$, $E_1 \cap E_2 = \emptyset$, $V_1 \cup V_2 = V$, and $|V_1 \cap V_2| = k$. If a graph does not have an i-separation for any $i \leq k-1$, then we say that it is *k-connected.* We use analogous notation for digraphs and weighted digraphs, except that for weighted digraphs we implicitly assume that contraction and deletion will result in weights that are restricted to the remaining arcs.

Given a weighted digraph (D, ω) with $D = (N, A)$, and $X \subset N$, then *pushing into* $\delta(X)$ results in a new weighting for D, denoted by $\rho(\omega, X) = \omega'$ where

$$\omega'(a) = \begin{cases} \omega(a) + 1 & \text{if } a \in \delta^{\text{in}}(X) \\ \omega(a) - 1 & \text{if } a \in \delta^{\text{out}}(X) \\ \omega(a) & \text{otherwise} \end{cases}$$

Remark 1. If C is a dicycle in (D, ω), then $\omega(C) = \omega'(C)$ where $\omega' = \rho(\omega, X)$, for $X \subseteq N$. In particular, $\tau(D, \omega) = \tau(D, \omega')$.

Remark 2. J is accommodating in (D, ω) $\iff$ J is accommodating in $(D, \rho(\omega, X))$, for $X \subseteq N$.

Remarks 1 and 2 follow from the fact that $|A(C) \cap \delta^{\text{in}}(X)| = |A(C) \cap \delta^{\text{out}}(X)|$ for any dicycle C, and any subset of nodes X. It is worth noting that Remarks 1 and 2 hold regardless of how many cuts we push into, and whether or not we push into the same cut more than once. To perform successive pushes, let us define

$$\rho^0(\omega, X) = \omega$$
$$\rho^i(\omega, X) = \rho(\rho^{i-1}(\omega, X), X)$$

Often we want to push as much weight into a cut as possible, and we also want to avoid making an arc have negative weight, so we are constrained by the minimum weight of an outgoing arc in the cut. For this reason we introduce the following notation: let $\rho^*(\omega, X)$ be shorthand for $\rho^i(\omega, X)$ where

$$i = \min\{\omega(a) : a \in \delta^{\text{out}}(X)\}$$

and $i = \tau(D, \omega)$ if $\delta^{\text{out}}(X) = \emptyset$. To aid in the readability of this document, we suggest that $\rho(\omega, \overline{X})$ be verbalized as "pushing out of $\delta(X)$" as opposed to "pushing into $\delta(\overline{X})$". Finally, we generally wish to push into or out of cuts surrounding a single node, and so we will use the notation $\rho(\omega, x)$ as a short-form for $\rho(\omega, \{x\})$, and $\rho(\omega, \overline{x})$ as a short-form for $\rho(\omega, \overline{\{x\}})$, for node x.

3 Connectivity Lemmas

In this section we show that packings can be combined across dicuts, 1-separations, and 2-separations whose overlapping vertices form a digon. For this entire section we let (D, ω) be a weighted digraph with $D = (N, A)$.

Lemma 1 (Dicut). *Suppose that $\delta^{in}(S)$ is a dicut in D. Let $D_1 = D[S]$, let $D_2 = D[\overline{S}]$, and let ω_i be ω restricted to D_i for each $i = 1, 2$. If (D_1, ω_1) packs and (D_2, ω_2) packs, then (D, ω) packs.*

Proof. Note that $\tau(D_i, \omega_i) \geq \tau(D, \omega)$ for each $i = 1, 2$. Suppose that (D_i, ω_i) packs for each $i = 1, 2$. Then there exists a packing including $J^i_1, \ldots, J^i_{\tau(D,\omega)}$ in (D_i, ω_i) for each $i = 1, 2$. Let $J_j = J^1_j \cup J^2_j$, for $1 \leq j \leq \tau(D, \omega)$. Clearly J_j is a cover of D, for $1 \leq j \leq \tau(D, \omega)$. Thus, $J_1, \ldots, J_{\tau(D,\omega)}$ is a packing for (D, ω).

Lemma 2 (1-separation). *Let (D_1, D_2) be a 1-separation of D. Let ω_i be ω restricted to D_i for each $i = 1, 2$. If (D_1, ω_1) packs and (D_2, ω_2) packs, then (D, ω) packs.*

Proof. The proof is identical to the proof of Lemma 1.

Lemma 3 (2-separation). *Let (D'_1, D'_2) be a 2-separation of D such that D'_1 and D'_2 share vertices x and y. Let ω_i be ω restricted to D_i for each $i = 1, 2$. Let α_i be the minimum weight of a dipath from x to y in (D'_i, ω_i) for each $i = 1, 2$. Assume that $\alpha_1 \leq \alpha_2$, and let $\alpha = \min\{\tau(D, \omega), \alpha_1\}$. Let $e = xy, f = yx$ be new arcs. For each $i = 1, 2$, let $D_i = D'_i \cup \{e, f\}$, let $\omega_i(e) = \alpha$, and let $\omega_i(f) = \tau(D, \omega) - \alpha$. If (D_1, ω_1) packs and (D_2, ω_2) packs, then (D, ω) packs.*

Proof. Let $\tau = \tau(D, \omega)$. We claim that $\tau(D_i, \omega_i) = \tau$ for each $i = 1, 2$. We prove it for D_1; the proof is analogous for D_2. Suppose there exists a dicycle C in D'_1 such that $\omega_1(C) < \tau$. Clearly, C must contain e or f. If $f \in A(C)$ then $C - f$ gives a dipath from x to y in D_1 such that $\omega_1(C - f) < \alpha \leq \alpha_1$, which contradicts the choice of α_1. If $e \in A(C)$ then $\alpha_1 = \alpha < \tau$, and $C - e$ gives a path in D_1 such that $\omega(C - e) = \omega_1(C - e) < \omega - \alpha_1$. Let Q be a minimum length dipath from x to y in $(D'_1, \omega_1[D'_1])$, that is, $\omega(Q) = \alpha_1$. Then $\omega((C - e) \cup Q) < \tau - \alpha_1 + \alpha_1 = \tau$, and hence $(C - e) \cup Q$ contains a dicycle Z in D such that $\omega(Z) < \tau$, which is a contradiction. Hence, $\tau(D'_1, \omega_1) = \tau$.

If $\tau(D_i, \omega_i) = \nu(D_i, \omega_i)$ for each $i = 1, 2$, then there exists a packing including τ covers of D'_i, say $\{J^i_1, \ldots, J^i_\tau\}$, for each $i = 1, 2$.; We may assume that for each $i = 1, 2$, we have $e = xy \in J^i_j$, for $1 \leq j \leq \alpha$, and $f = yx \in J^i_j$, for $\alpha + 1 \leq j \leq \tau$. Let $J_j = (J^1_j \cup J^2_j) - \{e, f\}$, $1 \leq j \leq \tau$. We claim that each J_j is a cover of D, for $1 \leq j \leq \tau$. In fact, if C is a dicycle contained in D_1 or in D_2 then J_j clearly intersects C. Otherwise, x and y are vertices in C, and C can be partitioned in two paths P and Q from x to y and from y to x, respectively. Without loss of generality, say that P is contained within the arcs of D_1, and Q is contained within the arcs of D_2. We have two cases to consider.

(a) $1 \leq j \leq \alpha$: note that $P \cup f$ is a dicycle in D'_1 and recall that J_j^1 is a cover of D'_1. Since $f \notin J_j^1$, then J_j^1 (and hence, J_j) intersects P.
(b) $\alpha + 1 \leq j \leq \tau$: note that $Q \cup e$ is a dicycle of D'_2 and recall that J_j^2 is a cover of D'_2. Since $e \notin J_j^2$, then J_j^2 (and hence, J_j) intersects Q.

Thus, in both cases each J_j is a cover of D, and hence, $\{J_1, \ldots, J_\tau$ is a packing in (D, ω).

4 Contraction and Deletion Lemmas

In this section, we continue to show how packings in a smaller weighted digraph can be extended to packings in a larger original weighted digraph called (D, ω) with $D = (N, A)$. However, in this section there is a single smaller weighted digraph, and it arises not from dicuts or separations, but instead from individual arcs that are deleted or contracted. In particular, we associate deletion with arcs of weight at least $\tau(D, \omega)$, and contraction with non-transitive arcs of weight 0. We require non-transitive arcs for contracting since we do not wish to introduce new dicycles. On the other hand, we are not concerned with removing dicycles when deleting an arc of weight at least $\tau(D, \omega)$ since the arc can be added to every cover of the smaller weighted digraph.

We also extend these results by pushing weight into a cut $\delta(X)$ to bring an arbitrary arc to weight $\tau(D, \omega)$, or a non-transitive arc to weight 0. We point out that it is always possible to create arcs of weight 0 in a cut, however it is not always possible to create arcs of weight $\tau(D, \omega)$ in a cut. In particular, if we are pushing into $\delta(X)$ with $\delta^{\text{in}}(X) \neq \emptyset$, then it must be that $\max_{a \in \delta^{\text{in}}(X)} \omega(a) + \min_{a \in \delta^{\text{out}}(X)} \omega(a) \geq \tau(D, \omega)$. After manipulating weights and forming a smaller weighted digraph, we are not interested in revealing an entire packing for (D, ω), but merely a single valid accommodating cover. Because of Remarks 1 and 2, our challenge is ensuring that at least one cover in the smaller weighted digraph is valid in (D, ω). For this reason, $|\delta^{\text{out}}(X) \cap \{a \in A : \omega(a) = 0\}|$ becomes important. If the value is strictly less than $\tau(D, \omega)$, then we can ensure that at least one of the $\tau(D, \omega)$ covers found in the smaller weighted digraph will be valid in (D, ω); otherwise, we can think of $\delta(X)$ as being *protected* against such an argument.

Lemma 4 (Contract). *(a) Suppose $\exists$ non-transitive $a \in A$ with $\omega(a) = 0$. If $(D, \omega)/a$ packs, then (D, ω) packs.*
(b) Suppose $\exists$ non-transitive $a \in \delta^{out}(X)$ and $X \subseteq N$, such that $\omega(a) = \min\{\omega(b) : b \in \delta^{out}(X)\}$ and $|\delta^{in}(X) \cap \{b \in A : \omega(b) = 0\}| < \tau(D, \omega)$. If $(D, \rho^*(\omega, X))/a$ packs, then (D, ω) has a valid accommodating cover.

Proof. (a) Since a is non-transitive, it means that the dicycles in D and D/a are identical (except that some dicycles in D/a no longer include a). Therefore,

by Remark 1 and since $\omega(a) = 0$, we have that $\tau(D,\omega) = \tau(D,\omega)/a$, and that a cover in D/a is a cover in D. Therefore, the packing in $(D,\omega)/a$ is also a packing in (D,ω).
(b) Let $\omega' = \rho^*(\omega, X)$. Since a is non-transitive, it means that the dicycles in D and D/a are identical (except that some dicycles in D/a no longer include a). Therefore, by Remark 1 and since $\omega'(a) = 0$, we have that $\tau(D,\omega) = \tau(D,\omega') = \tau((D,\omega')/a)$, and that a cover in D/a is a cover in D. Let $J_1, \dots, J_{\tau(D,\omega)}$ be a packing in (D,ω'). From Remark 2, each J_i is accommodating in (D,ω). Furthermore, since the only arcs that have $\omega(b) = 0$ and $\omega'(b) > 0$ are contained in $\delta^{\text{in}}(X)$, and since $|\delta^{\text{in}}(X) \cap \{b \in A : \omega(b) = 0\}| < \tau(D,\omega)$, it must be that one of the J_i covers is also valid in (D,ω).

Lemma 5 (Delete). *(a) Suppose $\exists a \in A$ with $\omega(a) \geq \tau(D,\omega)$. If $(D,\omega)\backslash a$ packs, then (D,ω) packs.*
(b) Suppose $\exists a \in A$ and $X \subseteq N$, such that $a \in \delta^{in}(X)$ and $\max_{b\in\delta^{in}(X)} \omega(b) + \min_{b\in\delta^{out}(X)} \omega(b) \geq \tau(D,\omega)$ and $|\delta^{in}(X) \cap \{b \in A : \omega(b) = 0\}| < \tau(D,\omega)$. If $(D,\rho^*(\omega,X))\backslash a$ packs, then (D,ω) has a valid accommodating cover.

Proof. (a) Deleting a does not decrease the minimum weight of a dicycle, so there is a packing of covers in $(D,\omega)\backslash a$ that includes $J_1, \dots, J_{\tau(D,\omega)}$. Notice that each J_i covers every dicycle in D, except possibly for some dicycles containing a. Therefore, since $\omega(a) \geq \tau(D,\omega)$, we have that $J_1 \cup \{a\}, \dots, J_{\tau(D,\omega)}$ is a packing for (D,ω).
(b) Let $\omega' = \rho^*(\omega, X)$. Notice that $\omega'(a) \geq \tau(D,\omega)$. By Remark 1, and since deleting a does not decrease the minimum weight of a dicycle, we have that $\tau(D,\omega) = \tau(D,\omega') \geq \tau((D,\omega')\backslash a)$. Therefore, there is a packing of covers in $(D,\omega)\backslash a$ that includes $J_1, \dots, J_{\tau(D,\omega)}$. Notice that each J_i covers every dicycle in D, except possibly for some dicycles containing a. Therefore, $J_1 \cup \{a\}, \dots, J_{\tau(D,\omega)}$ are all covers in (D,ω). From Remark 2, each J_i is accommodating in (D,ω). Furthermore, since the only arcs that have $\omega(a) = 0$ and $\omega'(a) > 0$ are contained in $\delta^{\text{in}}(X)$, and since $|\delta^{\text{in}}(X) \cap \{b \in A : \omega(b) = 0\}| < \tau(D,\omega)$, it must be that one of the J_i covers is also valid in (D,ω).

5 Proof of Claim

Now we are ready to use the results of the two previous sections in order to prove Claim 1, which is restated below as Claim 5.

Claim. (D,ω) packs whenever the underlying graph of D has no $K_5 - e$ minor.

To prove Claim 5, we show that a smallest counterexample cannot exist. We call such a counterexample (D_m, ω_m), where $D_m = (N_m, A_m)$. For notational convenience, let $\tau_m = \tau(D_m, \omega_m)$. We choose (D_m, ω_m) to be smallest in the sense that it first minimizes τ, then the number of nodes, and finally the number of arcs. Hence, (D,ω) packs whenever one of the following holds:

M1: $\tau(D, \omega) < \tau_m$

M2: $\tau(D, \omega) = \tau_m$ and $|N| < |N_m|$

M3: $\tau(D, \omega) = \tau_m$ and $|N| = |N_m|$ and $|A| < |A_m|$

Remark 3. D_m is 3-connected.

From these choices and the results in Section 3, we have proven that D_m is 3-connected (Remark 3). Therefore, by Theorem 1 it must be that the underlying graph of D_m is a small complete graph, the envelope graph, or a wheel. Without too much difficulty, we can eliminate the possibility of K_2 and K_3. Furthermore, a wheel with three vertices is K_3, and a wheel with four vertices is K_4. Therefore, we have the following:

Remark 4. The underlying graph of D_m is a member of the set $S = \{ K_4, E, W_5, W_6, W_7, \ldots\}$, where E is the envelope graph, and W_i is a wheel with i vertices.

Fortunately, each member of S contains vertices of degree 3. In fact, every vertex has degree 3 except for the middle vertex in the wheels. Even more fortuitously, if we perform a wye-delta reduction on any of the graphs in S, then we either get K_3, or a graph in S with one less vertex. (If v is a vertex of degree 3 with neighbours x, y, z, then a wye-delta reduction is the result of removing vertex v and adding edges xy, xz, yz.)

Remark 5. D_m has no dicuts.

Since D_m has no dicuts, there is at least one arc entering and leaving each node in D_m. Therefore, Figure 4 shows the possible configurations for the directed versions of its degree 3 vertices (without distinguishing between neighbours).

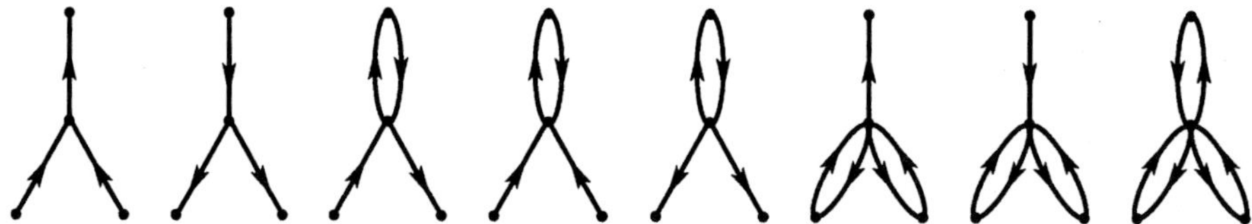

Fig. 4. From left to right: Configurations 1 through 8

By using the ideas from Section 4, let us now eliminate all but the first three configurations. From (M2) and Lemma 5, we cannot push any arc in A_m to weight τ_m, unless the cut we are pushing on is protected by $|\delta^{\text{out}}(X) \cap \{a \in A : \omega_m(a) = 0\}| \geq \tau_m$. Notice that pushing a digon arc to weight zero is equivalent to pushing its partnered digon arc to weight τ_m. Therefore, Configurations 6, 7, 8 cannot appear in (D_m, ω_m). For Configuration 4 and 5, notice that in both cases there is a digon arc a that would appear in no other dicycle except with the digon arc that it is partnered with, and we will call b. Therefore, we can delete a, find τ_m covers by minimality, and then extend these to covers of (D_m, ω_m) simply by adding a to the covers that b is not included in. Hence, Configurations 4 and 5 cannot appear in (D_m, ω_m).

Configuration 3 is slightly more difficult to eliminate. Let us label the nodes and arc weights of Configuration 3 as in the left portion of Figure 5. As in the previous paragraph, we cannot push a digon arc to weight 0. Therefore, we have the following conditions:

$$s < r$$
$$t < q$$

Our strategy is now to replace this configuration by using a wye-delta reduction which is illustrated in the right portion of Figure 5. We will call the newly formed weighted digraph (D, ω). In particular, we construct (D, ω) so that every dicycle in (D_m, ω_m) has a corresponding dicycle in (D, ω) with the same weight, except for the digon, which is not transferred to (D, ω). From the previous discussion on wye-delta reductions on the set of graphs S, and from (M2), (D, ω) packs. From the arc weights shown in Figure 5, it is clear that $\tau(D, \omega) = \tau_m$, and so we have a packing $J_1, \ldots, J_{\tau_m}$ for (D, ω). We now wish to show that this packing gives a valid accommodating cover for (D_m, ω_m). Before proceeding, we must constrain which sets of arcs the individual covers can contain by introducing Lemma 6.

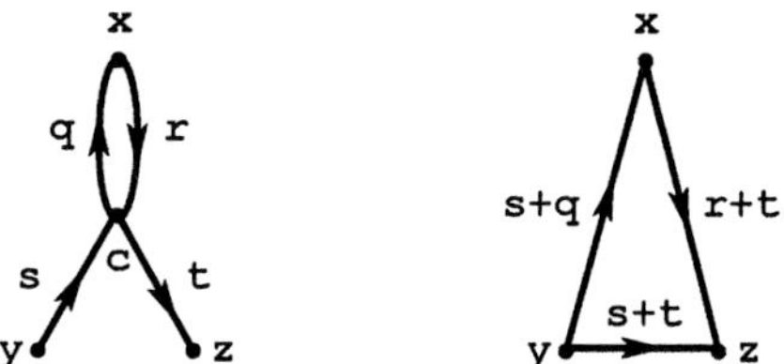

Fig. 5. Delta-wye reduction for Configuration 3

Lemma 6. *Let C be a cycle that is not a dicycle, and of the two directions of arcs within C, let one direction be called* forward *and the other* backward. *If J is a minimal cover and includes every forward arc of C, then J includes at least one backward arc of C.*

Proof. Otherwise, since J is minimal, for every forward arc a in C, there is a dicycle C_a such that $A(C_a) \cap J = \{a\}$. However, if J does not include any backward arc of C, then we contradict that J is a cover, since taking C_a for each forward arc a in C, together with the backward arcs of C, gives a dicycle that is not covered by J.

From Lemma 6, we know that the possible sets of arcs contained in each J_i are given by Figure 6. If J_i is of the first type, then it can be converted into a

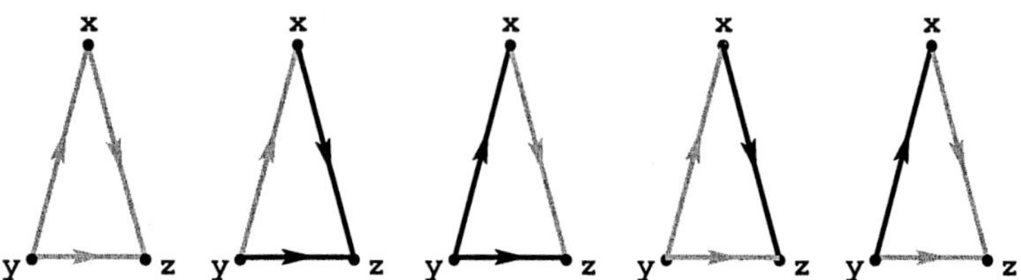

Fig. 6. From left to right, the five possible arc sets included in each J_i are given by the light arcs

valid accommodating cover for (D_m, ω_m) by replacing arcs yx, yz, xz by cx, cz (Figure 7 (left)). If J_i is of the second type, then it can be converted into a valid accommodating cover for (D_m, ω_m) by replacing arc yx with cx (Figure 7 (center)). If J_i is of the third type, then it can be converted into a valid accommodating cover for (D_m, ω_m) by replacing arc xz with xc (Figure 7 (right)). Therefore, we can eliminate Configuration 3 by guaranteeing that at least one of $J_1, \ldots, J_\tau$ is of one of the first three types given by Figure 6. Fortunately, the last two types given by Figure 6 can be used at most $s+t$ times by the weight given in Figure 5. From the above discussion on pushing, $s < r$ and $t < q$, so $s+t < r+q$. However, $r+q$ are the weights of a digon, so $s+t < \tau_m$, and therefore, at least one of $J_1, \ldots, J_{\tau_m}$ can be converted into a valid accommodating cover for (D_m, ω_m).

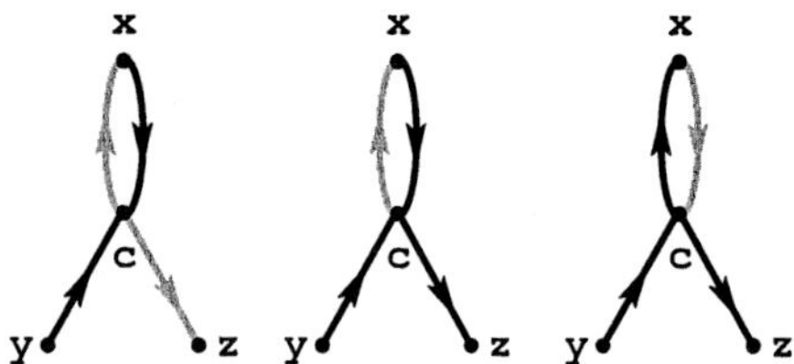

Fig. 7. From left to right, converting the first three possible arc sets into valid accommodating covers for (D_m, ω_m)

Therefore, we have now shown that every degree 3 vertex in the underlying graph of D_m is of Configuration 1 or Configuration 2 in Figure 4. Furthermore, from the results of Section 4, we know that at least one of the cuts surrounding these degree 3 vertices must be protected by $|\delta^{\text{out}}(X) \cap \{a \in A : \omega_m(a) = 0\}| \geq \tau_m$. Therefore, we must have that $\tau_m = 2$ and the arc weights must be as shown in Figure 8.

The endgame for our proof now consists of showing that when the underlying graph of (D_m, ω_m) is the envelope graph, or a wheel, and $\tau_m = 2$, and the degree 3 nodes of (D_m, ω_m) are given by Figure 8, then (D_m, ω_m) packs. It is not difficult to verify by hand that the envelope graph cannot possibly have nodes of degree 3 that are consistent with Figure 8. Therefore, we turn our attention to the infinite class of wheels.

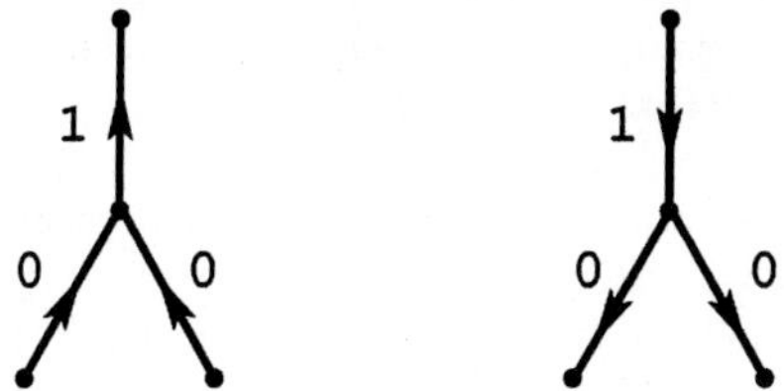

Fig. 8. Every degree 3 node in (D_m, ω_m) must be isomorphic to one of the above

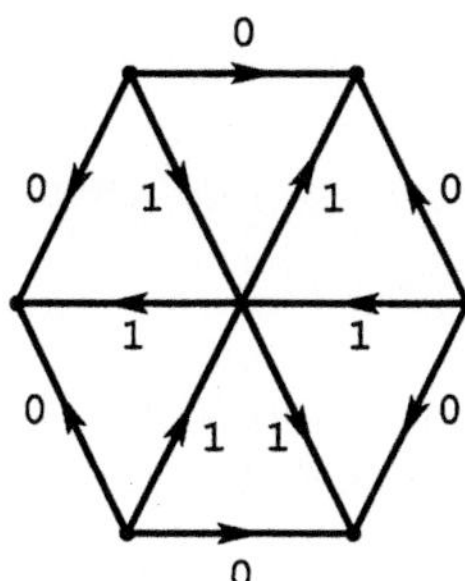

Fig. 9. If (D_m, ω_m) is a wheel, then it must have an odd number of nodes as above

Fortunately, the heavy restrictions force the wheels to have an odd number of vertices, with the outer cycle alternating in direction with weight zero arcs, and the inner arcs alternating in direction around the wheel with weight one. Figure 9 shows an example with seven nodes. Given such a weighted digraph, it is easy to find a packing of two covers, since we can let $J_1 = \{xc : xc \in A_m\}$ and $J_2 = \{cx : cx \in A_m\}$, where c is the center node in the wheel.

Therefore, we have show that (D_m, ω_m) does in fact pack. Therefore, there is no minimal counterexample, and we have verified our original claim. We conclude the paper by pointing out the corollary that it has for the Edmonds-Giles Conjecture, and less generally, Woodall's Conjecture.

Corollary 1. *Let (D, ω) be a weighted digraph, and let P be the planar dual of the $K_5 - e$ graph. If the underlying graph for D does not contain P as a minor, then the maximum number of disjoint dijoins is equal to the minimum weight of a dicut.*

References

1. B. Guenin A. Williams. Advances in packing directed joins. In *Second Brazilian Symposium on Graphs, Algorithms and Combinatorics,* volume 7, pages 212218. Electronic Notes in Discrete Mathematics, 2005.
2. J. Edmonds and R. Giles. A min-max relation for submodular functions on graphs. *Annals of Discrete Math.,* 1:185204, 1977.
3. P. Feofiloff. *Disjoint Transversals of Directed Coboundaries.* PhD thesis, University of Waterloo, Ontario, Canada, 1983.

4. P. Feofiloff and D. H. Younger. Directed cut transversal packing for source-sink connected graphs. *Combinatorica,* 7(3):255263, 1987.
5. B. Guenin G. Cornuejols. A note on dijoins. *Discrete Math.,* 243:213216, 2002.
6. C.L. Lucchesi and D.H. Younger. A minimax relation for directed graphs. *J. London Math. Soc.,* 17(2):369374, 1978.
7. K. Menger. Zur allgemeinen kurventheoric. *Fund. Math,* 10:96115, 1927.
8. Y. Wakabayashi O. Lee. A note on a min-max conjecture of Woodall. *J. Graph Theory,* 38:3641, 2001.
9. A. Schrijver. A counterexample to a conjecture of Edmonds and Giles. *Discrete Math.,* 32:213214, 1980.
10. A. Schrijver. Min-max relations for directed graphs. *Annals of Discrete Math.,* 16:261280, 1982.
11. K.Wagner. Bemerkungen zu hadwigers vermutung. *ath. Ann.,* 141 (1960):433451, 1960.
12. A. M. Williams. Packing directed joins. Masters thesis, University of Waterloo, Ontario, Canada, 2003.
13. D. R. Woodall. Menger and Konig systems. In *Theory and Applications of Graphs,* volume 642, pages 620635. Springer-Verlag Lecture notes in Mathematics, 1978.

Sharp Estimates for the Main Parameters of the Euclid Algorithm

Loïck Lhote and Brigitte Vallée

GREYC, University of Caen, France
{loick.lhote, brigitte.vallee}@info.unicaen.fr

Abstract. We provide sharp estimates for the probabilistic behaviour of the main parameters of the Euclid algorithm, and we study in particular the distribution of the bit-complexity which involves two main parameters : digit–costs and length of continuants. We perform a "dynamical analysis" which heavily uses the dynamical system underlying the Euclidean algorithm. Baladi and Vallée [2] have recently designed a general framework for "distributional dynamical analysis", where they have exhibited asymptotic gaussian laws for a large class of digit–costs. However, this family contains neither the bit–complexity cost nor the length of continuants. We first show here that an asymptotic gaussian law also holds for the length of continuants at a fraction of the execution. There exist two gcd algorithms, the standard one which only computes the gcd, and the extended one which also computes the Bezout pair, and is widely used for computing modular inverses. The extended algorithm is more regular than the standard one, and this explains that our results are more precise for the extended algorithm. We prove that the bit–complexity of the extended Euclid algorithm asymptotically follows a gaussian law, and we exhibit the speed of convergence towards the normal law. We describe also conjectures [quite plausible], under which we can obtain an asymptotic gaussian law for the plain bit-complexity, or a sharper estimate of the speed of convergence towards the gaussian law.

1 Introduction

The Euclid algorithm computes the greatest common divisor (in short gcd) of u and v, with Euclidean divisions of the form $v = m \cdot u + r$ with $0 \le r < u$. On an input (u, v), with $v_0 := v, v_1 := u$, it performs a sequence of Euclidean divisions

$$v_0 = m_1 \cdot v_1 + v_2, \quad \dots \quad v_i = m_{i+1} \cdot v_{i+1} + v_{i+2} \cdots \quad v_{p-1} = v_p \cdot m_p + 0. \quad (1)$$

It stops when the remainder v_{p+1} is zero, and the last non-zero remainder v_p is the greatest common divisor d of u and v.

We wish to study the bit–complexity of the Euclid algorithm, i.e., the total number of binary operations performed during the execution of the Euclid algorithm. The (naive) bit–complexity of a Euclidean division $v = m \cdot u + r$ is $\ell(u) \cdot \ell(m)$, where $\ell(v)$ is the binary length of the integer v; it equals $\lfloor \lg v \rfloor + 1$,

J.R. Correa, A. Hevia, and M. Kiwi (Eds.): LATIN 2006, LNCS 3887, pp. 689–702, 2006.

where lg denotes the logarithm in base 2. Then, the bit–complexity of the Euclid algorithm on the input (u, v) is

$$B(u,v) = \sum_{i=1}^{p} \ell(m_i) \cdot \ell(v_i), \qquad [\, p := P(u,v) \text{ is the number of iterations}] \quad (2)$$

The extended Euclid algorithm ouputs, at the same time, the Bezout pair (r, s) for which $d = rv + su$. It computes the sequence s_i defined by $s_0 = 0, \quad s_1 = 1,$ $s_i = s_{i-2} - s_{i-1} \cdot m_{i-1}, \quad 2 \le i < p$. The last element s_p is the Bezout coefficient s. The bit–complexity D of this algorithm on (u, v) is

$$D(u,v) = \ell(m_p) \cdot \ell(v_p) + \sum_{i=1}^{p-1} \ell(m_i) \cdot [\ell(v_i) + \ell(s_i)]. \quad (3)$$

We introduce also a so–called "smoothed" version $\widetilde{D}, \widetilde{B}$ of costs D, B, where we replace the size $\ell(s_i), \ell(v_i)$ of s_i, v_i by their logarithms $\lg s_i, \lg v_i$,

$$\widetilde{D}(u,v) = \ell(m_p) {\cdot} \lg v_p + \sum_{i=1}^{p-1} \ell(m_i) {\cdot} [\lg v_i + \lg s_i], \qquad \widetilde{B}(u,v) = \sum_{i=1}^{p} \ell(m_i) {\cdot} \lg v_i. \quad (4)$$

We observe that all the costs of interest can be expressed as a sum of terms, each of them being a product of two factors: the first one involves the (bit-)size of digits, and the second one involves the size of the so-called continuants v_i, s_i.

1.1 Distributional Analysis

We are interested here in studying the probabilistic behaviour of the gcd algorithm. The set Ω of inputs for the Euclid algorithm is $\Omega := \{(u,v) \in \mathbb{N}^2; \quad 0 \le u < v\}$. For any (u, v) of Ω, the size of pair (u, v), denoted by $L(u, v)$, is just the binary length (or the size) of v, i.e., $L(u, v) := \ell(v)$. The subset Ω_n of inputs (u, v) with a fixed size n,

$$\Omega_n := \{(u,v) \in \Omega; \quad L(u,v) = n\}, \quad (5)$$

is endowed with the uniform probability $\mathbb{P}_n$. For a random variable R defined on Ω, its restriction to Ω_n is denoted by R_n, and we wish to analyze the asymptotic behaviour of R, i.e., the evolution of variables R_n when n becomes large.

The evolution of the mean values $\mathbb{E}[R_n]$ is of great interest and, more generally, the study of all moments $\mathbb{E}[R_n^\ell]$ provides a first understanding of the probabilistic behaviour of the algorithm: this is the average–case analysis. However, the distributional analysis, which describes the evolution of the distribution of variable R_n, provides a much more precise analysis of the algorithm: this is the ultimate purpose in analysis of algorithms. Very often, variables R_n have a distribution which tends to the gaussian law: this phenomenon is easily proved as soon as cost R_n is the sum of n elementary costs, which are independent, and possess the same distribution. However, in the "Euclidean world", the steps of (1) are not independent, and the distribution of numbers may evolve with the evolution of the algorithm. This is why asymptotic gaussian laws, even if they are widely expected, are often difficult to prove in this context.

We provide here such a distributional analysis, for the most precise parameter of the extended Euclid algorithm, its bit–complexity D. We are also interested in describing the evolution of the size of remainders v_i. There exist now many well-known results about the probabilistic behaviour of the Euclid algorithm, even if the last ones have been obtained recently. The first results on probabilistic analysis of Euclid's algorithm are due to Heilbronn and Dixon who have shown, around 1970, that the average number of iterations is linear with respect to the size. In 1994, Hensley [6] performed the first distributional analysis, and proved that the number of steps has an asymptotic gaussian behaviour. However, his proof is not easily extended to other parameters of the algorithm. During the last ten years, the research team in Caen has designed a complete framework for analyzing an entire class of Euclidean algorithms, with a large class of parameters (see [10]). It is possible to obtain precise results on the average behaviour of the main parameters of the algorithm : the digits m_i, and the size of continuants v_i and s_i. Akhavi and Vallée have also analyzed the average bit–complexity [1]. These methods consider the underlying dynamical systems, and make a deep use of dynamical tools, like the transfer operator. However, all the analyses were "average–case analyses". There was a breakthrough two years ago, when Baladi and Vallée [2] extended the previous method for obtaining limit distributions, for a large class of costs, the so-called additive costs of moderate growth; they consider costs C defined on Ω and associated to an elementary cost c on digits,

$$C(u,v) := \sum_{i=1}^{p} c(m_i). \tag{6}$$

When $c(m)$ is $O(\log m)$, the cost c, and the cost C are said to be of moderate growth. This class of costs contains quite natural parameters, as the number of steps (for $c = 1$), the number of occurrences of a given digit m_0 (for $c(m) := [[m = m_0]]$), the total encoding length (when c equals the binary length ℓ), but NOT the bit–complexity. These bit–complexity costs are more difficult to deal with, because they involve both continuants and digits, in a multiplicative way. Here, we aim to study the distribution of the bit–complexity, and we wish to extend both the results of Akhavi and Vallée, about the average bit–complexity, and the distributional methods of Baladi and Vallée. We wish also to study the evolution of the size of remainders v_i.

As in previous works [2, 3], we make a deep use of the weighted transfer operator relative to an elementary cost c and which depends on two parameters (s, w),

$$\mathbf{G}_{s,w,[c]}[f](x) := \sum_{m\geq 1} \frac{1}{(m+x)^{2s}} \cdot \exp[wc(m)] \cdot f\left(\frac{1}{m+x}\right). \tag{7}$$

When c is of moderate growth, the operator $\mathbf{G}_{s,w,[c]}$ admits (on a convenient functional space) a unique dominant eigenvalue, for (s,w) near $(1,0)$. The logarithm of the dominant eigenvalue, called the pressure, and denoted by $\Lambda_{[c]}(s,w)$, plays a central work in [2], and also in the present paper. The particular case when c equals the binary length ℓ is crucial in study of bit–complexities.

1.2 Asymptotic Gaussian Laws

We prove here that many variables R defined on Ω follow asymptotically a gaussian law. We first provide a precise definition:

Definition. [Asymptotic gaussian law.] *Consider a cost R defined on Ω and its restriction R_n to Ω_n. The cost R asymptotically follows a gaussian law if there exist three sequences a_n, b_n, r_n, with $r_n \to 0$, for which*

$$\mathbb{P}\left[(u,v) \in \Omega_n \ \Big| \ \frac{R_n(u,v) - a_n}{\sqrt{b_n}} \leq y\right] = \frac{1}{\sqrt{2\pi}} \int_{-\infty}^{y} e^{-t^2/2}\, dt + O(r_n)\,.$$

The sequence r_n defines the speed of convergence, denoted also by $r[R_n]$. The expectation $\mathbb{E}[R_n]$ and the variance $\mathbb{V}[R_n]$ satisfy $\mathbb{E}[R_n] \sim a_n, \quad \mathbb{V}[R_n] \sim b_n$. We say that the triple (a_n, b_n, r_n) is a characteristic triple for the gaussian law of R.

For instance, the result of Baladi and Vallée can be stated as follows.

Theorem 0. [Asymptotic gaussian Law for additive costs of moderate growth] (Baladi and Vallée). *Consider an additive cost C relative to an elementary cost c of moderate growth [defined in (6)].*
(i) On the set of integer inputs of size n, the cost C asymptotically follows a gaussian law, with a characteristic triple given by: $r[C_n] = O(n^{-1/2})$,

$$\mathbb{E}[C_n] = \mu(c)\cdot n + \mu_1(c) + O(2^{-n\gamma}), \qquad \mathbb{V}[C_n] = \rho(c)\cdot n + \rho_1(c) + O(2^{-n\gamma}),$$

Here γ is a strictly positive constant which does not depend on cost c.
(ii) The constants $\mu(c)$ and $\rho(c)$ involve the first five derivatives of order 1 and 2 of the pressure function $\Lambda(s,w) = \Lambda_{[c]}(s,w)$ of $\mathbf{G}_{s,w,[c]}$ at $(s,w) = (1,0)$.

In the case when $c = \ell$, the constant $\rho(\ell)$ is (only) polynomial–time computable (see [8]) while $\mu(\ell)$ admits a closed form

$$\mu(\ell) = \frac{12}{\pi^2} \log \prod_{i=0}^{\infty} (1 + \frac{1}{2^i}).$$

1.3 Our Main Results

The "extended" cost D defined in (3) is easier to analyze because it is, in a sense, more regular than cost B. We prove here that the cost D follows asymptotically a gaussian law, with a characteristics triple which involves constants $\mu(\ell), \rho(\ell)$ of Thm 0 relative to the binary–length ℓ.

Theorem 1. [Asymptotic gaussian law for the extended bit–complexity.] *(i) On the set of integer inputs of size n, the bit complexity D of the extended Euclid algorithm follows asymptotically a gaussian law, with the characteristic triple*

$$\mathbb{E}[D_n] = \mu(\ell)\cdot n^2\,[1+O\left(\frac{1}{n}\right)],\ \ \mathbb{V}[D_n] = \rho(\ell)\cdot n^3\,[1+O\left(\frac{1}{n}\right)],\ \ r[D_n] = O(n^{-1/3}).$$

The smoothed bit–complexity $\widetilde{D}$ asymptotically follows a gaussian law with the same characteristic triple $[\mu(\ell)\cdot n^2,\ \rho(\ell)\cdot n^3,\ O(n^{-1/3})]$.

(ii) Under conjecture (C1), the speed of convergence $r[\widetilde{D}_n]$ is $O(n^{-1/2})$.

Conjecture $(C1)$ is described in 2.3. For the standard bit–cost B, defined in (2), we exhibit a precise estimate for the variance, and propose a conjecture $(C2)$, described in 2.4, under which we prove an asymptotic gaussian law.

Theorem 2. [Standard integer bit–complexity.] *(i) On the set of integer inputs of size n, the mean and the variance of the bit-complexity B satisfy*

$$\mathbb{E}[B_n] = \frac{1}{2}\mu(\ell)\cdot n^2\,[1+O\left(\frac{1}{n}\right)], \qquad \mathbb{V}[B_n] = \tau\cdot n^3\,[1+O\left(\frac{1}{n}\right)]. \qquad (8)$$

Here τ is a strictly positive constant, which involves spectral objects of the operator $\mathbf{G}_{s,w,[c]}$. The same holds for the smoothed version $\widetilde{B}$.

(ii) Under Conjecture $(C2)$, the speed of convergence $r[\widetilde{B}_n]$ is $O(n^{-1/3})$, and the equality $4\tau = \rho(\ell)$ holds.

We are also interested in describing the evolution of the size of remainders v_i during the execution of the algorithm, and we consider the size of the remainder v_i at "a fraction of the depth". More precisely, for a real $\delta\in]0,1[$, we denote by $\ell^{[\delta]}$ the logarithm of v_i when i equals $\lfloor\delta P\rfloor$, [P is the number of iterations of the Euclid algorithm]. The following result shows that the remainders at a fraction of the depth asymptotically follow a log-normal law, and that the evolution of the sizes of continuants is very regular. This result constitutes a "discrete version" of the well-known result of [9] (sharpened by Vallée in [11]) who shows that the n-th continuant of a real $x\in\mathcal{I}$ asymptotically follows a gaussian law, when $\mathcal{I}$ is endowed with any density of class C^1.

This result also plays a central rôle in the analysis of the so–called Interrupted Euclidean algorithm which stops as soon as the remainder v_i is less than v_0^δ. An average–case analysis of the Interrupted algorithm is provided in [4], and the present results are a first [but crucial] step towards the distributional analysis of the algorithm. And the Interrupted algorithm is itself a basic procedure of the Lehmer Euclid algorithm [7], or the recursive Euclidean algorithms.

Theorem 3. [gaussian limit law for sizes of continuants at a fraction of the depth.] *Consider a rational δ of $]0,1[$. On the set of integer inputs of size n, the length $\ell^{[\delta]}$ follows asymptotically a gaussian law, with mean, variance and speed of convergence given by* $r[\ell_n^{[\delta]}] = O(n^{-1/2})$,

$$\mathbb{E}[\ell_n^{[\delta]}] = \mu_{[\delta]}\cdot n + \mu_1(\delta) + O(2^{-n\gamma}), \qquad \mathbb{V}[\ell_n^{[\delta]}] = \rho_{[\delta]}\cdot n + \rho_1(\delta) + O(2^{-n\gamma}).$$

Here γ is a strictly positive constant which depends on δ, and the constants $\mu_{[\delta]}$ and $\rho_{[\delta]}$ are related to the derivatives of the pressure function $\Lambda(s)$ at $s = 1$,

$$\mu_{[\delta]} = (1 - \delta), \qquad \rho_{[\delta]} = \delta(1 - \delta)\frac{|\Lambda''(1)|}{|\Lambda'(1)|} > 0 .$$

1.4 Plan of the Paper

Section 2 provides a description of the main steps for proving Theorems 1 and 2 and states Theorem 4, which will be a main tool in these proofs. Section 3 presents the transfer operators and explains their generating rôle. Then, it describes the main principles of the analytical study which provides a proof of Theorems 3 and 4. Finally, we describe the two conjectures and provide some hints towards a possible proof.

2 Main Steps for Theorems 1 and 2

Here, we explain how to obtain asymptotic gaussian laws for the bit–complexities. We prove Theorem 1, Assertion (i), describe conjectures $(C1)$ and $(C2)$ and explain how to prove Theorem 1 (ii) and Theorem 2 (ii) under these conjectures.

2.1 Expressions for Continuants

Each division–step of the Euclid algorithm $v = m \cdot u + r$ uses a digit m and changes the old pair (u, v) into a new pair (r, u). Instead of integers, we consider rationals [the old rational $x = u/v$, and the new rational $y = r/u$] which both belong to the unit interval, and we look for a relation between y and x. One has

$$\frac{r}{u} = \frac{v - mu}{u} = \frac{v}{u} - \lfloor\frac{v}{u}\rfloor \qquad \text{so that} \quad y = T(x) \qquad \text{with} \quad T(x) := \frac{1}{x} - \lfloor\frac{1}{x}\rfloor.$$

With $T(0) = 0$, the map $T : [0, 1] \to [0, 1]$ is called the Gauss map and plays a fundamental rôle in the study of the Euclid algorithm. When the quotient is m, there exists also a linear fractional transformation (LFT) $h_{[m]}$ for which

$$x = h_{[m]}(y) \qquad \text{with} \quad h_{[m]}(y) = 1/(m + y) .$$

Of course, the LFT's $h_{[m]}$ are the inverse branches of T. On an input (u, v), the execution (1) creates a continued fraction of the form

$$\frac{u}{v} = h_{[m_1]} \circ h_{[m_2]} \circ \dots h_{[m_p]} = h(0). \tag{9}$$

When the algorithm performs p iterations, it gives rise to a continued fraction of depth p. Here, we show that the main parameters of the Euclid algorithm on the input (u, v) (quotients m_i, remainders v_i and continuants s_i) can be read on the continued fraction of the rational u/v. When the CFE of u/v is split at depth i, the LFT h defines three LFT's, the beginning LFT b_i, the middle LFT h_i and the ending LFT e_i, respectively defined as

$$b_i := h_{[m_1]} \circ h_{[m_2]} \circ \ldots \circ h_{[m_{i-1}]}, \qquad h_i := h_{[m_i]} \qquad e_i := h_{[m_{i+1}]} \circ \ldots \circ h_{[m_p]}.$$

Then, the i-th continuants admit expressions which involve LFT's e_i and b_i,

$$s_i^{-2} = |b_i'(0)|, \qquad v_i^{-2} = v_p^{-2} \cdot |e_i'(0)|. \tag{10}$$

2.2 Bit-Complexity Cost

This entails the following decompositions,

Proposition 1. *The bit–complexity costs $D, \widetilde{D}$ of the extended Euclidean algorithm decompose as* $\widetilde{D} = (L-1) \cdot Z + \widetilde{Y}, \qquad D = (L+1) \cdot Z + Y,$

$$\text{with} \quad \widetilde{Y} = -Y^{(1)} + O(Y^{(2)}) + Y^{(3)} + Y^{(4)}, \qquad Y = \widetilde{Y} + Y^{(5)}.$$

Here L is the size of the input, defined by $L(u,v) = \ell(v) = \ell(v_0)$ and

$$Z = \sum_{i=1}^{p-1} \ell(m_i), \qquad Y^{(1)} = \sum_{i=1}^{p-1} \ell(m_i) \cdot \lg m_i, \qquad Y^{(2)} = (\ell(m_p) + \lg v_p)^2, \tag{11}$$

$$Y^{(3)} = f \cdot \sum_{i=1}^{p-1} \ell(m_i), \qquad Y^{(4)} = \sum_{i=1}^{p-1} d_i \cdot \ell(m_i) \qquad Y^{(5)} = \sum_{i=1}^{p-1} f_i \cdot \ell(m_i), \tag{12}$$

$$\text{with} \quad d_i =: \lg \left| \frac{h_i'(0)}{h_i'(e_i(0))} \right| + \lg \left| \frac{b_i'(e_{i-1}(0))}{b_i'(0)} \right|, \quad f := \{\lg v_0\}, \ f_i := -\{\lg v_i\} - \{\lg s_i\}.$$

Moreover, the so-called distortions d_i admit uniform lower and upper bounds.

We have then "splitted" the extended cost D into two costs: the "main" cost $X := L \cdot Z$ which will be proven to be (asymptotically) gaussian, and a "remainder" cost Y, which will be proven to be (asymptotically) more concentrated that the main cost. Then, the total cost $X + Y$ will be (asymptotically) gaussian: .

Proposition 2. *Consider two costs X and Y, defined on Ω and their restrictions X_n, Y_n to Ω_n. Suppose that X admits a gaussian limit law with speed of convergence $r[X_n]$ and the variances of X_n and Y_n satisfy $\mathbb{V}[Y_n] = \alpha_n \cdot \mathbb{V}[X_n]$, with $\alpha_n \to 0$. Then, the random variable $X + Y$ follows asymptotically a gaussian limit law with a characteristic triple given by:* $r[X_n + Y_n] = r[X_n] + O(\alpha_n^{1/3})$,

$$\mathbb{E}[X_n + Y_n] = \mathbb{E}_n[X] + \mathbb{E}_n[Y], \quad \mathbb{V}[X_n + Y_n] = \mathbb{V}[X_n] \cdot [1 + O(\alpha_n)].$$

2.3 Proof of Theorem 1

Theorem 1 (i) is easily deduced from Propositions 1 and 2. The "main" cost is $X := L \cdot Z$, where $L(u,v)$ is the size of pair (u,v), equal to $\ell(v)$. With results of Baladi and Vallée [Theorem 0], the cost Z follows an asymptotic gaussian law. Since $X_n = n \cdot Z_n$, the cost X follows itself an asymptotic gaussian law with the characteristic triple

$$\mathbb{E}[X_n] = n \cdot \mathbb{E}[Z_n] = O(n^2), \quad \mathbb{V}[X_n] = n^2 \cdot \mathbb{V}[Z_n] = O(n^3), \quad r[X_n] = O(n^{-1/2}).$$

In Proposition 1, there appear three different kinds of costs: – (i) cost $Y^{(1)}$ – (ii) cost $Y^{(2)}$ which is an end-cost, [i.e., it depends only on variables used in the last step $\ell(m_p), \ell(v_p)$, and in a polynomial way.] – (iii) The other costs R [the distortion cost $Y^{(4)}$ and the two fractional costs $Y^{(3)}$, $Y^{(5)}$] deal with bounded sequence f_i, d_i. For these costs R, one has:

$$\mathbb{E}[R_n] = O(\mathbb{E}[Z_n]) = O(n), \qquad \mathbb{V}[R_n] \le \mathbb{E}[R_n^2] = O(\mathbb{E}[Z_n^2]) = O(n^2).$$

In the following Theorem 4, we will prove that the cost $Y = Y^{(1)}$ fulfills the concentration property, and that end-costs R are negligible i.e.,

$$\mathbb{E}[Y_n] = O(n), \quad \mathbb{V}[Y_n] = O(n), \qquad \mathbb{E}[R_n] = O(1), \quad \mathbb{V}[R_n] = O(1).$$

This leads to Theorem 1 [Assertion (i)], with a speed of convergence $O(n^{-1/3})$.

If we wish to obtain a speed of convergence of order $n^{-1/2}$, we must study more carefully costs $Y^{(i)}$ for $j = 3, 4, 5$. The fractional cost $Y^{(5)}$ is clearly very difficult to study: this is why we have introduced the smoothed cost $\widetilde{D}$, which no longer involves $Y^{(5)}$. It is possible to generate the distortion cost $Y^{(4)}$ and the fractional cost $Y^{(3)}$ with some convenient transfer operator. However, we do not succeed in proving that the concentration property holds for them.

Conjecture $(C1)$: *The costs $Y^{(3)}$ and $Y^{(4)}$ satisfy the concentration property.*

Under this conjecture, Theorem 1 (ii) is proven. ∎

2.4 An Asymptotic Gaussian Law for $\widetilde{B}$?

For proving Theorem 2 (ii), we relate $w_i := v_i/v_p = (e_i'(0))^{1/2}$ to the approximate continuant $\overline{s}_i := b_i'(e_{i-1}(0))^{-1/2}$ and we introduce two (new) costs

$$A(u,v) := \sum_{i=1}^{p} \ell(m_i) \cdot \lg w_i, \qquad \overline{A}(u,v) := \sum_{i=1}^{p} \ell(m_i) \cdot \lg \overline{s}_i.$$

First, as in 2.3, the cost $(A + \overline{A})$ will be asymptotically gaussian with the same characteristic triple as $\widetilde{D}$. Second, since the cost A is close to costs $\widetilde{B}, B$, it is sufficient for Theorem 2 (ii) to prove that the decomposition $A = (1/2)(A + \overline{A}) - (1/2)(A - \overline{A})$ provides a new instance of application of Propositions 1 and 2.

This is possible if the second cost $(A_n - \overline{A}_n)$ has a variance of order $O(n^2)$. Since $\mathbb{E}[A_n - \overline{A}_n]$ is of order $O(n)$ [see Proposition 3, Section 3], we study

$$\mathbb{E}[(A_n - \overline{A}_n)^2] = \mathbb{E}[A_n^2] + \mathbb{E}[\overline{A}_n^2] - 2\mathbb{E}[A_n \cdot \overline{A}_n],$$

where each term is of order $O(n^4)$. Proposition 3 proves a cancellation between the dominant terms, and entails for α_n an order of $O(1/n)$.

Conjecture $(C2)$: $\mathbb{E}[A_n^2]$, $\mathbb{E}[\overline{A}_n^2]$, $\mathbb{E}[A_n \cdot \overline{A}_n]$ *have the same terms of order* n^3.

This conjecture is plausible since it is based on a property of "semi-commutativity" which generalizes Proposition 3. Under $(C2)$, it is easy to prove Theorem 2 (ii).

2.5 Various Kinds of Costs

We are then led to study various costs C, and the behaviour of additive costs C heavily depends on the behaviour of cost c. We then introduce the Dirichlet series $A_c(s, w)$,

$$A_c(s,w) := \sum_{m \in \mathcal{M}} \frac{1}{m^{2s}} \exp[wc(m)],$$

closely related to the operator $\mathbf{G}_{s,w,[c]}$, which helps to define the behaviour of c.

Definition 1. (a) *A cost R is an end–cost if it depends only on variables used in the last step $\ell(m_p), \ell(v_p)$, and in a polynomial way.*
(b) *An elementary cost c and its associated additive cost C are of moderate growth if –$(b1)$ the bivariate generating function $A_c(s,w)$ is convergent for $\Re s > \sigma_0$ and $\Re w < \nu_0$ with $\sigma_0 < 1$ and $\nu_0 > 0$ – $(b2)$ it is analytic at $(1, 0)$,*
(c) *An elementary cost c and its associated additive cost C are of intermediate growth if –$(c1)$ its generating function $A_c(s,w)$ is convergent for $\Re s > \sigma_0$ with $\sigma_0 < 1$ and $\Re w \le 0$, – $(c2)$ it is not analytic at $(s,w) = (1,0)$, but, as a function of the real variable w, it admits derivatives of any order wrt w, at $w = 0^-$.*

Remark. The size cost $c = \ell$ is of moderate growth, while any power of the size of the form $c = \ell^\alpha$ (with $\alpha > 1$) defines a cost of intermediate growth.

The following theorem is one of the basic results of our paper. Note that Assertion (b) is already proven by Baladi and Vallée [2].

Theorem 4. *The following holds:*

(a) *An end cost R is negligible, i.e., the expectation $\mathbb{E}[R_n]$ and the variance $\mathbb{V}[R_n]$ are $O(1)$.*

(b) *An additive cost C of moderate growth is asymptotically gaussian with a characteristic triple of the form $[O(n), O(n), O(n^{-1/2})]$.*

(c) *An additive cost C of intermediate growth satisfies the concentration property, i.e., the expectation $\mathbb{E}[C_n]$ and the variance $\mathbb{V}[C_n]$ are $O(n)$.*

3 Dynamical Systems and Generating Operators

We explain here how dynamical systems allow to derive alternative forms for generating functions. This will be done via various extensions of the transfer operator, which plays here the rôle of a "generating operator".

3.1 Dynamical Systems and Transfer Operators

A continuous extension of one step of the Euclid algorithm to real numbers x of $\mathcal{I} := [0,1]$ is provided by the Gauss map $T : \mathcal{I} \to \mathcal{I}$, together with the set $\mathcal{H} := \{h_{[m]}; m \in \mathbb{N}\}$ of the branches of T^{-1}. The pair $(\mathcal{I}, T)$ defines a dynamical system. The set $\mathcal{H}^k$ is the set of the inverse branches of the iterate T^k, and the set $\mathcal{H}^\star := \cup_k \mathcal{H}^k$ is the semi-group generated by $\mathcal{H}$.

If $\mathcal{I}$ is endowed with some initial density $f = f_0$, the time evolution governed by the map T modifies the density. The successive densities $f_1, f_2, \ldots, f_n, \ldots$ describe the global evolution of the system, and there exists an operator, the density transformer $\mathbf{G}$ which transforms f_0 into f_1. The weighted transfer operator $\mathbf{G}_{s,w,[c]}$ relative to some digit cost c,

$$\mathbf{G}_{s,w,[c]}[f](x) = \sum_{h \in \mathcal{H}} \exp[wc(h)] \cdot |h'(x)|^s \cdot f \circ h(x),$$

is a perturbation of the density transformer $\mathbf{G}$ [obtained for $(s,w) = (1,0)$]. When $w = 0$, we omit the variable w and the cost c, so that $\mathbf{G}_s := \mathbf{G}_{s,0,[c]}$. Now, if we extend cost c on $\mathcal{H}^\star$ by additivity, the quasi–inverse is of the form

$$(I - \mathbf{G}_{s,w,[c]})^{-1}[f](x) = \sum_{h \in \mathcal{H}^\star} \exp[wc(h)] \cdot |h'(x)|^s \cdot f \circ h(x)\ .$$

3.2 Generating Functions

We consider a general parameter R defined on Ω, and we wish to study its distribution on Ω_n, when endowed with the uniform probability. Our final probabilistic tool [for distributional analyses] is the sequence of moment generating functions $\mathbb{E}[\exp(wR_n)]$,

$$\mathbb{E}[\exp(wR_n)] = \frac{R(n,w)}{R(n,0)}, \qquad \text{with} \quad R(n,w) := \sum_{(u,v) \in \Omega_n} \exp[wR(u,v)]. \tag{13}$$

We first consider the whole set Ω of inputs and our strategy consists in encapsulating all the moment generating functions $\mathbb{E}[\exp(wR_n)]$ in a Dirichlet series

$$S_R(s,w) := \sum_{(u,v) \in \Omega} \frac{1}{v^{2s}} \exp[wR(u,v)] = \sum_{m \geq 1} \frac{1}{m^{2s}} r_m(w)\,, \tag{14}$$

where $r_m(w)$ is the cumulative value of $\exp[wR]$ on inputs (u,v) for which $v = m$. The series $S_R(s,w)$ is a bivariate generating function which depends on two parameters, s "marks" the input size, and w "marks" the cost of interest. This is a Dirichlet series with respect to s. The study of moments of order k,

$$\mathbb{E}[R_n^k] = \frac{R(n)^{[k]}}{R(n)^{[0]}}, \qquad \text{with} \quad R(n)^{[k]} := \sum_{(u,v)\in\Omega_n} R^k(u,v), \tag{15}$$

deals with a Dirichlet Series $S_R^{[k]}(s)$

$$S_R^{[k]}(s) := \frac{\partial^k}{\partial w^k} S_R(s,w)|_{w=0} = \sum_{(u,v)\in\Omega} \frac{1}{v^{2s}} R^k(u,v) = \sum_{m\geq 1} \frac{1}{m^{2s}} r_m^{[k]}, \tag{16}$$

where $r_m^{[k]}$ is the cumulative value of R^k on inputs (u,v) for which $v = m$.

In both cases, the plain moment generating function, , and the plain moment of order k of R_n can be recovered from series $S_R(s,w)$ or $S_R^{[k]}(s)$ with (13,15), and relations

$$R(n,w) = \sum_{m=2^{n-1}}^{2^n-1} r_m(w), \qquad R(n)^{[k]} = \sum_{m=2^{n-1}}^{2^n-1} r_m^{[k]}. \tag{17}$$

We first look for an alternative expression for series $S_R(s,w)$ [defined in (14)] from which the position and the nature of its dominant singularity become apparent. With taking derivatives, we also obtain alternative expressions for $S_R^{[k]}(s)$ [defined in (16)]. Then, we transfer these informations on the asymptotic behaviour of coefficients of $S_R(s,w)$ or $S_R^{[k]}(s)$, which are closely related via (13,15,17) to our prime objects of interest $\mathbb{E}[\exp(wR_n)]$, $\mathbb{E}[R_n^k]$.

3.3 Alternative Expressions for Bivariate Dirichlet Series

We will use transfer operators $\mathbf{G}_s$, $\mathbf{G}_{s,w,[c]}$ (or some of their extensions) as "generating" operators: Bivariate generating functions $S_R(s,w)$ can be expressed via quasi–inverses of these operators.

Additive costs. If C is the total cost relative to c, the quasi-inverse $(I-\mathbf{G}_{s,w,[c]})^{-1}$ "generates" the bivariate generating function of cost C (relative to coprime inputs). Furthermore, the Zeta function defined as $\zeta(2s) := \sum_{d\geq 1} d^{-2s}$ allows to deal with general inputs [not only coprime inputs]. Finally,

$$S_C(s,w) = \zeta(2s)\cdot(I-\mathbf{G}_{s,w,[c]})^{-1}[1](0). \tag{18}$$

Continuant at a fraction of the depth. We study the parameter $\ell^{[\delta]}$ which equals the logarithm of remainder v_i for $i = \lfloor \delta P \rfloor$. For an input (u,v) of Ω on which the algorithm performs p iterations, there exists LFT h of depth p such that $u/v = h(0)$. One decomposes h in two LFT's g and r of depth $\lfloor \delta p \rfloor$ and $p - \lfloor \delta p \rfloor$ such that $h = g \circ r$, and if δ is a rational of the form $\delta = c/(c+d)$, then

$$S_{\ell^{[\delta]}}(s,w) = \zeta(2s-2w)\cdot \sum_{j=0}^{c+d-1} \mathbf{G}_{s-w}^{j-\lfloor\delta j\rfloor} \circ \Big(\sum_{k\geq 0} \mathbf{G}_{s-w}^{dk} \circ \mathbf{G}_s^{ck}\Big) \circ \mathbf{G}_s^{\lfloor \delta j\rfloor}[1](0). \tag{19}$$

The operator $\mathbb{G}_{s,w} := \sum_{k\geq 0} \mathbf{G}_{s-w}^{dk} \circ \mathbf{G}_s^{ck}$ is called a pseudo–quasi-inverse; of course, since $\mathbf{G}_s$ and $\mathbf{G}_{s,w}$ do not commute, this is not a "true" quasi-inverse. However, we study this operator when w is near to 0, and we can hope that the properties of $\mathbb{G}_{s,w}$ will be close to properties of a true quasi-inverse.

3.4 Alternative Expressions for Dirichlet Series $S_R^{[j]}(s)$

In other cases of cost R, we look for alternative expressions for the series $S_R^{[i]}(s)$ for $i = 1, 2$. We denote by $W_{[c]}$ the derivation wrt w (at $w = 0$), and by Δ the derivation wrt s,

$$W_{[c]}\mathbf{G}_s = \frac{\partial}{\partial w}\mathbf{G}_{s,w,[c]}|_{w=0}, \qquad \Delta := \frac{1}{\log 2}\frac{d}{ds}\mathbf{G}_s.$$

Then, the operators $W_{[c]}$ or Δ operate themselves on transfer operators. Our Dirichlet series of interest can be written as a sequence of occurrences of the quasi–inverse $(I - \mathbf{G}_s)^{-1}$, separated by occurrences of the form $A\mathbf{G}_s$ where A is a monomial of the (commutative) algebra $\mathcal{A}$ generated by $\{\Delta, W_{[c]}\}$. Then, we adopt shorthand notations where we omit the quasi-inverses, the zeta function, the function 1, and the point 0: we only take into account the operators "between" the quasi inverses.

Additive costs C of intermediate growth. In this case, it is not possible to deal directly with the transfer operator $\mathbf{G}_{s,w,[c]}$. However, the univariate series $S_C^{[j]}(s)$ admit alternative expressions of the form

$$S_C^{[1]} = [W_{[c]}] \qquad S_C^{[2]} = [W_{[c]}^2] + 2[W_{[c]}, W_{[c]}]. \tag{20}$$

Bit-Complexity Costs $A, \overline{A}$. Here, we omit also the index $[\ell]$ in $W_{[\ell]}$ and we obtain, in the same vein $\quad S_A^{[1]} = [\Delta, W], \quad S_{\overline{A}}^{[1]} = [W, \Delta],$

$$(1/2)\, S_A^{[2]} \approx 2[\Delta,\Delta,W,W]+[\Delta,W,\Delta,W]+[\Delta^2,W,W]+[\Delta,\Delta W,W]+[\Delta,\Delta,W^2],$$

$$(1/2)\, S_{\overline{A}}^{[2]} \approx 2[W,W,\Delta,\Delta]+[W,\Delta,W,\Delta]+[W,W,\Delta^2]+[W,\Delta W,\Delta]+[W^2,\Delta,\Delta],$$

$$S_{A\overline{A}}^{[1]} \approx 2[W,\Delta,\Delta,W]+2[\Delta,W,W,\Delta]+[W,\Delta,W,\Delta]+[\Delta,W,\Delta,W]+$$
$$+[W,\Delta^2,W]+[\Delta,W,\Delta W]+[W,\Delta,\Delta W]+[\Delta W,\Delta,W]+[\Delta W,W,\Delta]+[\Delta,W^2,\Delta].$$

3.5 Analysis of Costs

With alternative expressions of Dirichlet series provided in Sections 3.3 and 3.4 at hand, we now perform the second step: we find the dominant singularities of these Dirichlet series and their nature, and then transfer these informations towards coefficients and obtain asymptotic expressions for their coefficients. We use, as a main tool, convenient "extractors" which express coefficients of series as a function of the series itself. There are two main "extractors" for Dirichlet series: Tauberian Theorems [which do not provide remainder terms] are well-adapted for the average–case analysis or the study of all the (non centered) moments [Thm 4 (a)] — the Perron Formula [which may provide remainder terms] constitutes an essential step, both in the studies of the variance [Thms 1, 2, 4(c)] and in distributional analyses [Thm 3].

Both extractors need informations on the quasi-inverse (QI), closely related to the dominant spectral properties of the transfer operator on the Banach space $\mathcal{C}^1(\mathcal{I})$. However, Tauberian Theorems "only" need informations on the QI on the domain $\Re s \geq 1$. For using with some success the Perron Formula, we need a more precise knowledge of the QI on vertical strips on the left of the vertical line $\Re s = 1$. Properties of the same vein are very often difficult to prove and intervene for instance in the proof of the Prime Number Theorem. The *US* Property [Uniformity on Strips] describes a convenient behaviour of the QI and informally says: "there exists a vertical strip $|\Re(s) - 1| < \alpha$ which contains only one pole of the QI; moreover, on the left line $\Re(s) = 1 - \alpha$, an adequate norm of the QI is bounded by $M \cdot |\Im s|^\xi$ (with $\xi > 0$ small). Baladi and Vallée [2] have generalized ideas due to Dolgopyat [5] and prove that the $US(s)$ Property holds for $(I - \mathbf{G}_s)^{-1}$, and that a uniform $US(s,w)$ Property (uniform wrt w) holds for $(I - \mathbf{G}_{s,w,[c]})^{-1}$ (when c is of moderate growth). Here, for Thm 3, we prove that a uniform $US(s,w)$ Property also holds for the "pseudo quasi–inverse".

For $(s,w) = (1,0)$, and for any cost c, the operator $\mathbf{G}_{s,w,[c]}$ is just the density transformer $\mathbf{G}$, which possesses a unique dominant eigenvalue equal to 1 and an invariant function $\Psi(x) = (1/\log 2)(1/1+x)$. Then, each occurrence of the quasi–inverse $(I - \mathbf{G}_s)^{-1}$ brings a pole at $s = 1$, with an explicit residue:

Proposition 3. *Any Dirichlet series denoted by an expression $[A_1, A_2, \ldots A_k]$ [see Section 3.4], where each A_i is a monomial of the algebra generated by $\{\Delta, W_{[c]}\}$, has a pôle of order $k+1$ at $s = 1$, with an expansion of the form*

$$[A_1, A_2, \ldots, A_k](s) = \frac{1}{\log 2} \sum_{i \geq 0} a_i \cdot (|\lambda'(1)|(s-1))^{i-k-1},$$

with $a_0 = \prod_{i=1}^{k} I[A_i \mathbf{G}],\ \ I[\mathbf{H}] := \int_I \mathbf{H}[\Psi](t)dt,\ \ \Psi(x) := (1/\log 2)(1/1+x).$

Since the dominant constant a_0 depends only on the subset $\{A_1, A_2, \ldots A_k\}$, this proves that, for additive costs $R = C$ or bit-complexities $R = A, \overline{A}$, there exists a relation of the form $b_0 = 2a_0^2$ between the dominant constant a_0 of $S_R^{[1]}$ and the dominant constant b_0 of $S_R^{[2]}$, which entails a cancellation in the variance.

Conjecture $(C2)$**.** It is based on a similar property which involves the Porter operator $\mathbf{Q}$ defined as the constant term in the expansion of $(I - \mathbf{G}_s)^{-1}$ near $s = 1$. Conjecture $(C2)$ says: *The following equality holds:*

$$\sum_{\substack{X,Y,X',Y' \in \{\Delta, W\} \\ X' \neq X, Y' \neq Y}} (-1)^{[X=Y]} \cdot I[X\mathbf{G}] \cdot I[Y\mathbf{G}] \cdot (I[X'\mathbf{G} \circ \mathbf{Q} \circ Y'\mathbf{G}] - I[X'Y'\mathbf{G}]) = 0.$$

Conjecture $(C1)$**.** It deals with two families of costs.
Costs $R = f \cdot C$. To prove that $\mathbb{V}R_n$ is $O(n)$, we use generating functions relative to moments of $\overline{R} := \lg v_0 \cdot C$. They can be expressed with $[\cdot, \ldots \cdot, \cdot]$, and we have to prove cancellations between the constants, as in Conjecture $(C2)$.

Distortion costs. Generating functions for the distortion costs involve generalized transfer operators acting on functions with two variables as in [11]. The US properties are not yet proven to hold for such operators, and proving cancellations between constants needs to deal with their dominant spectral objects.

References

1. A. Akhavi, B. Vallée. Average bit–complexity of Euclidean Algorithms, Proceedings of ICALP'2000, Lecture Notes in Computer Science 1853, pp 373–387, Springer.
2. V. Baladi, B. Vallée. Euclidean Algorithms are Gaussian, Journal of Number Theory, Volume 110, Issue 2 (2005) pp 331–386
3. E. Cesaratto, B. Vallée. Reals with bounded digit averages, Proceedings of the Colloquium on Mathematics and Computer Science: Algorithms, Trees, Combinatorics and Probability, pp 473–490, M. Drmota et al., ed., Birkhauser Verlag, Trends in Mathematics, 2004.
4. B. Daireaux, B. Vallée. Dynamical analysis of the parameterized Lehmer-Euclid Algorithm, Combinatorics, Probability, Computing, pp 499–536 (2004).
5. D. Dolgopyat. On decay of correlations in Anosov flows, *Ann. of Math.* 147 (1998) 357–390.
6. D. Hensley. The number of steps in the Euclidean algorithm, Journal of Number Theory, 49, 2 (1994), 142-182
7. D.H. Lehmer. Euclid's algorithm for large numbers. Am. Math. Mon. (1938) 45 pp 227–233.
8. L. Lhote. Computation of a Class of Continued Fraction Constants Proceedings of Alenex–ANALCO04, pp 199–210
9. W. Philipp. Some metrical theorems in number theory II, Duke Math. J. 37 (1970) pp 447–488. Errata, ibid, 788.
10. B. Vallée. Euclidean Dynamics, to appear in Discrete and Continuous Dynamical Systems, 2005, web page: `www.info.unicaen.fr/~brigitte`
11. B. Vallée. Opérateurs de Ruelle-Mayer généralisés et analyse en moyenne des algorithmes de Gauss et d'Euclide, Acta Arithmetica 81.2 (1997), pp 101–144

Position-Restricted Substring Searching

Veli Mäkinen[1,⋆] and Gonzalo Navarro[2,⋆⋆]

[1] Department of Computer Science, University of Helsinki, Finland
vmakinen@cs.helsinki.fi
[2] Center for Web Research,
Dept. of Computer Science, University of Chile
gnavarro@dcc.uchile.cl

Abstract. A full-text index is a data structure built over a text string $T[1,n]$. The most basic functionality provided is (a) counting how many times a pattern string $P[1,m]$ appears in T and (b) locating all those occ positions. There exist several indexes that solve (a) in $O(m)$ time and (b) in $O(occ)$ time. In this paper we propose two new queries, (c) counting how many times $P[1,m]$ appears in $T[l,r]$ and (d) locating all those $occ_{l,r}$ positions. These can be solved using (a) and (b) but this requires $O(occ)$ time. We present two solutions to (c) and (d) in this paper. The first is an index that requires $O(n \log n)$ bits of space and answers (c) in $O(m + \log n)$ time and (d) in $O(\log n)$ time per occurrence (that is, $O(occ_{l,r} \log n)$ time overall). A variant of the first solution answers (c) in $O(m + \log\log n)$ time and (d) in constant time per occurrence, but requires $O(n \log^{1+\epsilon} n)$ bits of space for any constant $\epsilon > 0$. The second solution requires $O(nm \log \sigma)$ bits of space, solving (c) in $O(m\lceil \log \sigma / \log\log n \rceil)$ time and (d) in $O(m\lceil \log \sigma / \log\log n \rceil)$ time per occurrence, where σ is the alphabet size. This second structure takes less space when the text is compressible.

Our solutions can be seen as a generalization of *rank* and *select* dictionaries, which allow computing how many times a given character c appears in a prefix $T[1,i]$ and also locate the i-th occurrence of c in T. Our solution to (c) extends character *rank* queries to *substring rank* queries, and our solution to (d) extends character *select* to *substring select* queries.

As a byproduct, we show how *rank* queries can be used to implement fractional cascading in little space, so as to obtain an alternative implementation of a well-known two-dimensional range search data structure by Chazelle. We also show how Grossi et al.'s *wavelet trees* are suitable for two-dimensional range searching, and their connection with Chazelle's data structure.

1 Introduction and Related Work

The indexed string matching problem is that of, given a long text $T[1,n]$ over an alphabet Σ of size σ, build a data structure called *full-text index* on it, to solve

⋆ Funded by the Academy of Finland under grant 108219.

⋆⋆ Funded by Millennium Nucleus Center for Web Research, Grant P04-067-F, Mideplan, Chile.

J.R. Correa, A. Hevia, and M. Kiwi (Eds.): LATIN 2006, LNCS 3887, pp. 703–714, 2006.

two types of queries: (*a*) Given a short pattern $P[1,m]$ over Σ, *count* the occurrences of P in T; (*b*) *locate* those *occ* positions in T. There are several classical full-text indexes requiring $O(n \log n)$ bits of space which can answer counting queries in $O(m)$ time (like suffix trees [2]) or $O(m + \log n)$ time (like suffix arrays [14]). Both locate each occurrence in constant time once the counting is done. Similar complexities are obtained with modern compressed data structures [5, 10, 7], requiring space $nH_k(T) + o(n \log \sigma)$ bits, where $H_k(T) \leq \log \sigma$ is the k-th order empirical entropy of T.[1]

In this paper we introduce a new problem, *position restricted* substring searching, which consists of two new queries: (*c*) Given $P[1,m]$ and two integers $1 \leq l \leq r \leq n$, *count* all the occurrences of P in $T[l,r]$, and (*d*) *locate* those $occ_{l,r}$ occurrences. These queries are fundamental in many text search situations where one wants to search only a part of the text collection, e.g. restricting the search to a subset of dynamically chosen documents in a document database, restricting the search to only parts of a long DNA sequence, and so on. Curiously, there seem to be no solutions to this problem apart from locating all the occurrences and then filter those in the range $[l,r]$. This costs at least $O(m+occ)$ for (*c*) and (*d*) together.

We present several alternative structures to solve this problem. The best complexities are summarized in Table 1.

Table 1. Simplified complexities achieved for queries (*c*) and (*d*). Locating time is given per occurrence.

Section	Bits of space	Counting time	Locating time
4	$O(n \log^{1+\epsilon} n)$	$O(m + \log \log n)$	$O(1)$
4	$n \log n(1 + o(1)) + O(nH_k(T) \log^\gamma n)$	$O(m + \log n)$	$O(\log n)$
4	$n \log n(1 + o(1)) + nH_k(T)$	$O(m\lceil \frac{\log \sigma}{\log \log n} \rceil + \log n)$	$O(\log n)$
5	$n \sum_{k=0}^{m-1} H_k(T)$	$O(m\lceil \frac{\log \sigma}{\log \log n} \rceil)$	$O(m\lceil \frac{\log \sigma}{\log \log n} \rceil)$

Interestingly, our solutions are also useful to solve a generalization of another well-studied problem. Given a sequence S over an alphabet Σ of size σ and a character $c \in \Sigma$, query $rank_c(S,i)$ returns the number of occurrences of c in $S[1,i]$, while $select_c(S,j)$ returns the position of the j-th occurrence of c in S. Both queries can be answered in constant time using data structures that require $nH_0(S)+o(n)$ bits of space if the alphabet of the sequence is $\sigma = O(\text{polylog}(n))$, or in $O(\log \sigma / \log \log n)$ time in general [9, 8]. They can also be solved in $O(\log \sigma)$ time using wavelet trees [10, 11]. For the case of binary sequences, apart from the simple $n + o(n)$ bits data structures [12, 4, 16], there are others that answer *rank* and *select* in constant time using $nH_0(S) + o(n)$ bits [18].

A natural generalization of the above problem is *substring rank and select.* For a string s, $rank_s(S,i)$ is the number of occurrences of s in $S[1,i]$, and $select_s(S,j)$ is the starting position of the j-th occurrence of s in S. We can use the indexes for position-restricted substring searching to answer $rank_s$ in the same time of

[1] In this paper log stands for $\log_2$.

a counting query (type (c)), and $select_s$ in the same time of a counting query plus the time to locate one occurrence (type (d)).

As a byproduct, we present a space-efficient implementation of a well-known two-dimensional range search data structure by Chazelle [3]. We show in particular how the fractional cascading information (which is simulated rather than stored in Chazelle's data structure) can be represented by constant-time *rank* queries on bit arrays. We also show that Grossi et al.'s wavelet trees [10, 11] are suitable for two-dimensional range searching, pointing out in particular their connection with Chazelle's data structure.

2 Two-Dimensional Range Searching

In this section we describe a range search data structure to query by rectangular areas. The structure is a succinct variant of one from Chazelle [3, 13] where we have completely removed binary searching and fractional cascading and have replaced them by constant-time *rank* queries over bit arrays. Given a set of points in $[1, n] \times [1, n]$, the data structure permits determining the number of points that lie in a range $[i, i'] \times [j, j']$ in time $O(\log n)$, as well as retrieving each of those points in $O(\log n)$ time. The structure can be implemented using $n \log n(1 + o(1))$ bits.

Structure. We describe a slightly simpler version of the original structure [3], which is sufficient for our problem. The simplification is that our set of points come from pairing two permutations of $[1, n]$. Therefore, no two different points share their same first or second coordinates, that is, for every pair of points $(i, j) \neq (i', j')$ it holds $i \neq i'$ and $j \neq j'$. Moreover, there is a point with first coordinate i for any $1 \leq i \leq n$ and a point with second coordinate j for any $1 \leq j \leq n$.

The structure is built as follows. First, sort the points by their j coordinate. Then, form a perfect binary tree where each node handles an interval of the first coordinate i, and thus knows only the points whose first coordinate falls in the interval. The root handles the interval $[1, n]$, and the children of a node handling interval $[i, i']$ are associated to $[i, \lfloor (i+i')/2 \rfloor]$ and $[\lfloor (i+i')/2 \rfloor + 1, i']$. The leaves handle intervals for the form $[i, i]$. All those intervals will be called *tree intervals*.

Each node v contains a bitmap B_v so that $B_v[r] = 0$ iff the r-th point handled by node v (in the order given by the initial sorting by j coordinate) belongs to the left child. Each of those bitmaps B_v is preprocessed for constant-time *rank* queries [12, 4, 16]). The bitmaps with rank functionality give a space-efficient way to implement fractional cascading, and also avoid any need of binary searching.

Querying. We first show how to *track* a particular point (i, j) as we go down the tree. In the root, the position given by the sorting of coordinates is precisely j, because there is exactly one point with second coordinate j for any $j \in [1, n]$. Then, if $B_{root}[j] = 0$, this means that point (i, j) is in the left subtree, otherwise it is in the right subtree. In the first case, the new position of (i, j) in the left

Algorithm. RangeCount$(v, [i, i'], [j, j'], [ti, ti'])$
if $j > j'$ **then return** 0;
if $[ti, ti'] \cap [i, i'] = \emptyset$ **then return** 0;
if $[ti, ti'] \subseteq [i, i']$ **then return** $j' - j + 1$;
$tm \leftarrow \lfloor (ti + ti')/2 \rfloor$;
$[j_l, j'_l] \leftarrow [rank_0(B_v, j-1) + 1, rank_0(B_v, j')]$;
$[j_r, j'_r] \leftarrow [rank_1(B_v, j-1) + 1, rank_1(B_v, j')]$;
return RangeCount$(left(v), [i, i'], [j_l, j'_l], [ti, tm])$ +
RangeCount$(right(v), [i, i'], [j_r, j'_r], [tm + 1, ti'])$;

Fig. 1. Algorithm for counting the number of points in $[i, i'] \times [j, j']$ on a tree structure rooted by v with nodes $left(v)$ and $right(v)$. The last argument is the tree interval handled by node v. The first invocation is RangeCount$(root, [i, i'], [j, j'], [1, n])$.

subtree is $j \leftarrow rank_0(B_{root}, j)$, which is the number of points preceding (i, j) in B_{root} which chose the left subtree. Similarly, the new position on the right subtree it is $j \leftarrow rank_1(B_{root}, j)$.

Range searching for $[i, i'] \times [j, j']$ is carried out as follows. Find in the tree the $O(\log n)$ maximal tree intervals that cover $[i, i']$. The answer is then the set of points in those intervals whose second coordinate is in $[j, j']$. Those points form an interval in the B array of each of the nodes that form the cover of $[i, i']$. However, we need to track those j and j' coordinates as we descend by the tree. Every time we descend to the left child of a node v, we update $[j, j'] \leftarrow [rank_0(B_v, j-1) + 1, rank_0(B_v, j')]$, and similarly with $rank_1$ for a right child. When we arrive at a node whose interval is contained in $[i, i']$, the number of qualifying points is just $j' - j + 1$. Thus the whole procedure takes $O(\log n)$ time. Figure 1 shows the pseudocode.

For retrieving the points, we start from each of the tree nodes that cover $[i, i']$. Each point in the node whose second coordinate is in $[j, j']$ is tracked down in the tree until the leaves, so as to find its first coordinate i. This can be done in $O(\log n)$ time per retrieved element. (For our application, we do not describe how to associate the proper j value to each i coordinate found, but it can be done by traversing the tree upwards from each leaf using *select*.) We traverse the whole subtree of each node included in $[i, i']$, as long as it has some point in $[j, j']$. The leaves found in this process are reported. Figure 2 gives the pseudocode.

Space. We do not need any pointer for this tree. We only need $1 + \lceil \log n \rceil$ bit streams, one per tree level. All the bit streams at level h of the tree are concatenated into a single one, of length exactly n. A single *rank* structure is computed for each whole level, totalizing $n \log n(1 + O(\log\log n / \log n))$ bits. Maintaining the initial position p of the sequence corresponding to node v at level h is easy. There is only one sequence at the root, so $p = 1$ at level $h = 1$. Now, assume that the sequence for v starts at position p (in level h), and we move to a child (in level $h + 1$). Then the left child starts at the same position

Algorithm. RangeLocate(v, $[j, j']$, $[ti, ti']$)
(1) **if** $ti = ti'$ **then** { output ti; **return**; }
(2) **if** $j > j'$ **then return**;
(3) $tm \leftarrow \lfloor (ti + ti')/2 \rfloor$;
(4) $[j_l, j_l'] \leftarrow [rank_0(B_v, j-1)+1, rank_0(B_v, j')]$;
(5) $[j_r, j_r'] \leftarrow [rank_1(B_v, j-1)+1, rank_1(B_v, j')]$;
(6) RangeLocate($left(v)$, $[j_l, j_l']$, $[ti, tm]$);
(7) RangeLocate($right(v)$, $[j_r, j_r']$, $[tm+1, ti']$);

Fig. 2. Algorithm to invoke instead of returning $j' - j + 1$ in line (3) of RangeCount, so as to locate occurrences instead of just counting them

p, while the right child starts at $p + rank_0(B_v, |B_v|)$. The length of the current sequence $|B_v|$ is also easy to maintain. The root sequence is of length n. Then the left child of v is of length $rank_0(B_v, |B_v|)$ and the right child is of length $rank_1(B_v, |B_v|)$. Finally, if we know that v starts at position p and we have the whole-level sequence B^h instead of B_v, then $rank_b(B_v, j) = rank_b(B^h, p-1+j) - rank_b(B^h, p-1)$. Figure 3 shows again the counting algorithm, this time over the real data structure.

Algorithm. RangeCount(B, $[i, i']$, $[j, j']$, h, p, ℓ, $[ti, ti']$)
(1) **if** $j > j'$ **then return** 0;
(2) **if** $[ti, ti'] \cap [i, i'] = \emptyset$ **then return** 0;
(3) **if** $[ti, ti'] \subseteq [i, i']$ **then return** $j' - j + 1$;
(4) $tm \leftarrow \lfloor (ti + ti')/2 \rfloor$;
(5) $[j_l, j_l'] \leftarrow [rank_0(B^h, p, p-1+j-1)+1, rank_0(B^h, p, p-1+j')]$;
(6) $[j_r, j_r'] \leftarrow [rank_1(B^h, p, p-1+j-1)+1, rank_1(B^h, p, p-1+j')]$;
(7) $[\ell_l, \ell_r] \leftarrow [rank_0(B^h, p, p-1+\ell), rank_1(B^h, p, p-1+\ell)]$
(8) $p' \leftarrow p + rank_0(B^h, \ell)$
(9) **return** RangeCount(B, $[i, i']$, $[j_l, j_l']$, $h+1, p, \ell_l$, $[ti, tm]$) +
RangeCount(B, $[i, i']$, $[j_r, j_r']$, $h+1, p', \ell_r$, $[tm+1, ti']$);

Fig. 3. Algorithm for counting the number of points in $[i, i'] \times [j, j']$ on the real, levelwise, structure. The first invocation is RangeCount(B, $[i, i']$, $[j, j']$, $1, 1, n$, $[1, n]$). We use $rank_b(B^h, a, b)$ as shorthand for $rank_b(B^h, b) - rank_b(B^h, a-1)$.

Wavelet Trees. Wavelet trees [10, 11] are data structures for text indexing introduced by Grossi et al. The wavelet tree is a perfectly balanced tree of height $\lceil \log \sigma \rceil$. Each tree node corresponds to a subinterval of $[1, \sigma]$ and represents the text subsequence of characters in that subinterval. At each node, the current alphabet range is partitioned into two halves, and the corresponding alphabet subintervals are assigned to the left and right child of the node. The only data stored at a node is a bitmap where, for each character of the text it represents, it is indicated whether that character went left or right.

Each bitmap is processed for *rank* and *select* queries. If one uses basic techniques for those queries [12, 4, 16] the wavelet tree takes $n\lceil\log\sigma\rceil + O(n\log\log n/\log_\sigma n)$ bits of space for a text $T[1,n]$, that is, the same text size. With more advanced techniques [18], the size of the wavelet tree achieves $nH_0(T) + O(n\log\log n/\log_\sigma n)$ bits of space, where $H_0(T)$ is the zero-order entropy of T. In both cases, the wavelet tree solves in $O(\log\sigma)$ time the following queries: (a) $T[i]$, that is, finding the i-th character of T; (b) $rank_c(T,i)$, that is, finding the number of occurrences of c in $T[1,i]$; and (c) $select_c(T,j)$, that is, finding the position in T of the j-th occurrence of c.

We note now that wavelet trees have yet other applications not considered before. Assume we have a set of points $(i,j) \in [1,n]\times[1,n]$ which is the product of two permutations of $[1,n]$ as explained in the beginning of this section. Call $i(j)$ the unique i value such that (i,j) is a point in the set. Then consider the text $T[1,n] = i(1)i(2)i(3)\ldots i(n)$. Then, *the wavelet tree of T is exactly the data structure we have described in this section.* This text has alphabet of size n and its zero-order entropy is also $\log n$, thus this wavelet tree takes $n\log n(1+o(1))$ bits as expected. Although the original wavelet-tree queries are not especially interesting in this range search scenario, we have shown in this section that the wavelet tree structure can indeed be used to solve two-dimensional range search queries in $O(\log n)$ time, and report each occurrence in $O(\log n)$ time as well.

3 A Simple $O(m + \log n)$ Time Solution

Our first solution is composed of two data structures. The first is the familiar suffix array $\mathcal{A}[1,n]$ of T, enriched with longest common prefix (lcp) information [14]. This structure needs $2n\lceil\log n\rceil$ bits and permits determining the interval $\mathcal{A}[sp,ep]$ of suffixes that start with $P[1,m]$ in $O(m+\log n)$ time [14]. The second is the range search data structure $\mathcal{R}$ described in Section 2, indexing the points $(i,\mathcal{A}[i])$. Both structures together require $3n\log n(1+o(1))$ bits, or $3n+o(n)$ words.

To find the number of occurrences of $P[1,m]$ in $T[l,r]$, we first find the interval $\mathcal{A}[sp,ep]$ of the occurrences of P in T, and then count the number of points in the range $[l, r-m+1]\times[sp,ep]$ using $\mathcal{R}$. This takes overall $O(m+\log n)$ time. Additionally, each first coordinate (that is, text position $l \le i \le r-m+1$) of an occurrence can be retrieved in $O(\log n)$ time, that is, the $occ_{l,r}$ occurrences can be located in $O(occ_{l,r}\log n)$ time.

A plus of the index is that, unlike plain suffix arrays, this structure locates the occurrences in text position order, not in suffix array order. In order to find them in suffix array order, we should rather index points $(\mathcal{A}[i], i)$ and search for the interval $[sp,ep]\times[l, r-m+1]$. Then $\mathcal{R}$ would retrieve the suffix array positions i (in increasing order in $\mathcal{A}$) such that $\mathcal{A}[i]$ is an occurrence.

Larger and faster. It is possible to improve the locating time to $O(1)$ by using slightly more space. Instead of the structure of Section 2, that of Alstrup et al. [1] can be used to index the points $(i,\mathcal{A}[i])$. This structure retrieves the $occ_{l,r}$

occurrences of a range query in $O(\log\log n + occ_{l,r})$ time. In exchange, it needs $O(n \log^{1+\epsilon} n)$ bits of space, for any constant $0 < \epsilon < 1$. Thus, by using slightly more space, we achieve $O(m + \log n)$ counting time and $O(1)$ locating time per occurrence.

Given the complexity $O(\log\log n)$ for the range-search part of the counting query, it makes sense to replace the suffix array by a suffix tree, so that we still have $O(n \log^{1+\epsilon} n)$ bits of space and can solve the counting query in $O(m + \log\log n)$ time, and the locating query in constant time per occurrence.

Smaller and slower. Alternatively, it is possible to replace the suffix array $\mathcal{A}$ and its lcp information by any of the wealth of existing compressed data structures [17]. For example, by using the LZ-index of Ferragina and Manzini [6] we obtain $n \log n(1 + o(1)) + O(nH_k(T) \log^\gamma n)$ bits of space (for any $\gamma > 0$ and any $k = O(\log_\sigma \log n))$ and the same time complexities. On the other hand, we can use the alphabet-friendly FM-index of Ferragina et al. [7, 8] to obtain $n \log n(1 + o(1)) + nH_k(T)$ bits of space (for any $\sigma = o(n/\log\log n)$ and any $k \le \alpha \log_\sigma n$ for any constant $0 < \alpha < 1$). In this case the counting time raises to $O(m\lceil \log \sigma / \log\log n\rceil + \log n)$. This is still $O(m+\log n)$ if $\sigma = O(\text{polylog}(n))$.

4 An $O(m \log \sigma)$ Time Solution

We present now a solution that, given a construction parameter t, requires $nt \log \sigma(1+o(1))$ bits of space and achieves $O(m\lceil \log \sigma / \log\log n\rceil)$ time for counting the occurrences of any pattern of length $m \le t$. Likewise, each such occurrence can be located in $O(m\lceil \log \sigma / \log\log n\rceil)$ time. For example, choosing $t = \log_\sigma n$ gives a structure using $n \log n(1 + o(1))$ bits of space able to search for patterns of length $m \le \log_\sigma n$.

Actually, we show that this structure can be smaller for compressible texts, taking $n \sum_{k=0}^{t-1} H_k(T)$ instead of $nt \log \sigma$, where $H_k(T)$ is the k-th order empirical entropy of T [15, 10]. This is a lower bound to the number of bits per character achievable by any compressor that considers contexts of length k to model T.

Structure. Our structure indexes the positions of all the t-grams (substrings of length t) of T. It can be tought of as an extension of the wavelet tree [10, 11] to t-grams.

The structure is a perfectly balanced binary tree, which indexes the binary representation of all the t-grams of T, and searches for the binary representation of P. The binary representation $b(s)$ of a string s over an alphabet σ is obtained by expanding each character of s to the $\lceil \log \sigma \rceil$ bits necessary to code it. We index n t-grams of T, namely $b(T[1,t])$, $b(T[2,t+1])$, $\ldots$, $b(T[n, n+t-1])$. The text T is padded with $t-1$ dummy characters at the end.

The binary tree has $\ell = t\lceil \log \sigma \rceil$ levels. Each tree node v is associated a binary string $s(v)$ according to the path from the root to v. That is, $s(root) = \varepsilon$ and, if v_l and v_r are the left and right children of v, respectively, then $s(v_l) = s(v)0$ and $s(v_r) = s(v)1$. To each node v we also associate a subsequence of text positions $S_v = \{i,\ s(v) \text{ is a prefix of } b(T[i, i+t-1])\}$.

Note that each $i \in S_v$ will belong exactly to one of its two children, v_l or v_r. At each internal node v we store a bitmap B_v of length $n_v = |S_v|$, such that $B_v[i] = 0$ iff $i \in S_{v_l}$. Neither $s(v)$ nor S_v are explicitly stored, only B_v is.

Querying. Given a text position i at the root node, we can track its corresponding position in B_v for any node v such that $i \in S_v$. At the root, we start with $i_{root} = i$. When we descend to the left child v_l of a node v in the path, we set $i_{v_l} = rank_0(B_v, i)$, and if we descend to the right child v_r we set $i_{v_r} = rank_1(B_v, i)$. Then we arrive with the proper i_v value at any node v.

In order to search for P in the interval $[l, r]$, we start at the root with $l_{root} = l$ and $r_{root} = r - m + 1$, and find the tree node v such that $s(v) = b(P)$ (following the bits of $b(P)$ to choose the path from the root). At the same time we obtain the proper values l_v and r_v. Then the answer to the counting query is $r_v - l_v + 1$. The process requires $O(m \log \sigma)$ time.

To locate each such occurrence $l_v \leq i_v \leq r_v$, we must do the inverse tracking upwards. If v is the left child of its parent v_p, then the corresponding position in v_p is $i_{v_p} = select_0(B_{v_p}, i_v)$. If v is a right child, then $i_{v_p} = select_1(B_{v_p}, i_v)$. The final position in T is thus i_{root}. This takes $O(m \log \sigma)$ time for each occurrence.

Space. The bulk of the space requirement corresponds to the overall size of bit arrays B_v. Vectors B_v could be represented using the technique of Clark and Munro [4, 16], which permits answering *rank* and *select* queries in constant time over the bit arrays B_v using $n_v(1 + o(1))$ bits. All the n_v values at any depth add up n, and since the tree height is ℓ, we have $nt\lceil \log \sigma \rceil (1 + o(1))$ bits overall. The same technique used before to concatenate all the bitmaps at each level is used here to ensure that $o(1)$ is sublinear in n.

We show now that, by using more sophisticated techniques [18], the space requirement may be reduced on compressible texts T. In that work they represent bit array B_v using $n_v H_0(B_v) + o(n_v)$ bits, and answer *rank* and *select* queries in constant time. As we already know that the $o(n_v)$ parts add up $o(nm \log \sigma)$ bits (more precisely, $O(nm \log \sigma \log \log n / \log n)$ bits), we focus on the entropy-related part. Let us assume for simplicity that σ is a power of 2.

Let us analyze all the $n_v H_0(B_v)$ terms together. For a binary string s, let us define $n_s = |\{i,\ s \text{ is a prefix of } b(T[i, i+t-1])\}|$. Thus, if we consider vector B_{root}, its representation takes $nH_0(B_{root}) = -n_0 \log \frac{n_0}{n} - n_1 \log \frac{n_1}{n}$.

Consider now the vectors B for the two children of the root. The entropy part of their representations add up $-n_{00} \log \frac{n_{00}}{n_0} - n_{01} \log \frac{n_{01}}{n_0} - n_{10} \log \frac{n_{10}}{n_1} - n_{11} \log \frac{n_{11}}{n_1}$. We notice that $n_0 = n_{00} + n_{01}$ and $n_1 = n_{10} + n_{11}$. By adding up the size of representations of the root and its two children, we get $-n_{00} \log \frac{n_{00}}{n} - n_{01} \log \frac{n_{01}}{n} - n_{10} \log \frac{n_{10}}{n} - n_{11} \log \frac{n_{11}}{n}$ bits. This can be extended inductively to $\log \sigma$ levels, so that the sum of all the representations from the root to level $\log \sigma - 1$ is

$$- \sum_{s \in \{0,1\}^{\log \sigma}} n_s \log \frac{n_s}{n} = nH_0(T)$$

where $0 \log 0 = 0$.

Similarly, starting from each node v such that $s(v) \in \{0,1\}^{\log\sigma}$, we have that $nH_0(B_v) = -n_{s(v)0}\log\frac{n_{s(v)0}}{n_{s(v)}} - n_{s(v)1}\log\frac{n_{s(v)1}}{n_{s(v)}}$, and all the B vectors in the next $\log\sigma$ levels of its subtree add up

$$-\sum_{s\in\{0,1\}^{\log\sigma}} n_{s(v)s}\log\frac{n_{s(v)s}}{n_{s(v)}} .$$

Summing this for all the nodes representing all the possible $s(v) \in \{0,1\}^{\log\sigma}$, we have

$$-\sum_{s,s'\in\{0,1\}^{\log\sigma}} n_{ss'}\log\frac{n_{ss'}}{n_s} = nH_1(T).$$

This can be continued inductively until level $t\log\sigma$, to show that the overall space is

$$n\sum_{k=0}^{t-1} H_k(T) + O(nt\log\sigma\log\log n/\log n)$$

bits. For incompressible texts this is $nt\log\sigma(1+o(1))$, but for compressible texts it may be significantly less.

Higher arity trees. A generalization of the rank/select data structures [18] permit handling sequences with alphabets of size up to $O(\mathrm{polylog}(n))$ with constant time $rank_c$ and $select_c$ [9, 8]. Instead of handling one bit of $b(T[i,i+t-1])$ at a time, we could handle a bits at a time. This way, our binary tree would be 2^a-ary instead of binary. Instead of a sequence of bits B_v at each node, we would store a sequence B_v of integers in $[0,a-1]$. As long as $2^a = O(\mathrm{polylog}(n))$ (that is, $a = O(\log\log n)$), we can index those sequences B_v with the generalized data structure so as to answer in constant time the rank/select queries we need to navigate the tree.

The search algorithm is adapted in the obvious way. When going down to the d-th child of node v, $0 \le d < a$, we update i_v to $i_{v_d} = rank_d(B_v, i_v)$ and, similarly, when going up to v from child d, $i_v = select_d(B_v, i_{v_d})$. Note that a must divide $\log\sigma$ to ensure that any pattern search will arrive exactly at a tree node. The overall time is $O(m\log(\sigma)/a) = O(m\lceil\log\sigma/\log\log n\rceil)$, either for counting or for locating an occurrence. This is $O(m)$ whenever $\sigma = O(\mathrm{polylog}(n))$.

We note that it is necessary, again, to concatenate all sequences at each tree level, so that the limit $a = O(\log\log n)$ remains constant as we descend in the tree. For space occupancy related to entropy, the analysis is very similar; we just consider a bits at once.

Compared to the solution of the previous section requiring $O(n\log n)$ bits of space and $O(m+\log n)$ counting time, we can use $t = O(\log_\sigma n)$ to achieve the same space complexity, so that any query of length up to t can be answered. The structure of this section is faster than that of the previous section in this range of m values. Compared to the faster structure requiring $O(n\log^{1+\epsilon} n)$ bits and $O(m)$ counting time, our structure could answer in the same space counting queries on patterns of length up to $O(\log_\sigma n\log^\epsilon n)$. The time for counting is better than the previous structure for $m = O(\log\log n)$.

5 Substring Rank and Select

The techniques developed for the problem of counting and locating the occurrences of a pattern P in $T[l,r]$ can be used to solve the *substring rank* and *substring select* problem. As far as we know, this problem has not been addressed before. It extends the $rank_c$ and $select_c$ queries, $c \in \Sigma$, to strings over Σ. That is, given $s \in \Sigma^*$, $rank_s(T,i)$ is the number of occurrences of s in $T[1,i]$, while $select_s(T,j)$ is the initial position of the j-th occurrence of s in T.

Note that $rank_s(T,i)$ is just the number of occurrences of s in $T[1,i]$, and therefore it corresponds directly to a particular case of our counting queries. On the other hand, $select_s(T,j)$ is solved by using the locating mechanism. We detail this query now.

With the structure of Section 3 we must start with a counting query for s in the interval $[1,n]$. Therefore, we end up at the unique interval $[sp, ep]$ at the tree root. Then, to solve $select_s(T,j)$ we must track down in the tree the position $sp+j-1$ at the tree root. Therefore, we need overall $O(|s| + \log n)$ time for $select_s(T,j)$ (just as for $rank_s(T,i)$), yet ℓ calls to $select_s$ cost $O(|s| + \ell \log n)$. It is not clear whether the more complicated $O(n \log^{1+\epsilon} n)$ bits structure can extract random occurrences in constant time.

Let us now consider the structure of Section 4. Once we search for s in the tree starting with range $[l,r] = [1,n]$, we end up at some node v (such that $s(v) = b(s)$) with $[l_v, r_v]$. To solve $select_s(T,j)$ we take entry $l_v + j - 1$ at node v and walk the tree upwards until finding the position in the root node, and that position is the answer. This takes overall time $O(|s| \lceil \log \sigma / \log\log n \rceil)$ (just as for $rank_s$), and requires $O(n|s| \log \sigma)$ bits of space (or less if T is compressible).

6 A Small Experiment

We have implemented the simplest mechanism described in Section 3, and compared it against a brute-force solution, that is, use the plain suffix array to discover the *occ* occurrences and then pass over those determining which are in the right text range.

As the suffix array search is identical in both cases, we have focused on the time to complete the process once the suffix array range is known. For counting, the brute-force method has complexity $O(occ)$, whereas our method in Section 3 requires $O(\log n)$ time. For locating the occurrences, the brute-force method is still $O(occ)$ time, while our method requires $O(occ_{l,r} \log n)$.

We tested over different English texts, ranging from 1 to 100 megabytes. We randomly generated subintervals of the suffix array of a fixed length and compared the time to pass over it counting/reporting the text positions within some range, against using the generated suffix array interval as a key for the two-dimensional search of our method. Note that the fact that the suffix array ranges generated do not come from an actual search is irrelevant for the performance of the process, and it permits us tight control over *occ*.

According to the experiments, our counting method becomes faster than brute force approximately for $occ > 1{,}000$. This did not depend significatively on the text size nor on the text interval $[l, r]$ chosen. The two-dimensional search part takes, in our method, time similar to the suffix array search.

For locating queries, on the other hand, our method was superior for $\frac{occ_{l,r}}{occ} < 0.004$. This is obtained when *occ* exceeds 1,000 by a sufficient margin (say, 10 times). The reason is that our method has a constant overhead that is independent of $occ_{l,r}$, so that even for $occ_{l,r} = 0$ the brute force method is faster for $occ < 1,000$.

The ranges of values obtained show that our method is reasonably practical, and it wins when the query is sufficiently selective, as expected.

7 Conclusions

We have addressed several important generalizations of well-studied problems in string matching and succinct data structures. First, we generalized the indexed string matching problem to *position-restricted searching*, where the search can be done inside any text substring. We have obtained space and time complexities close to those obtained for the basic problem. Second, we generalized *rank* and *select* queries on sequences to substring *rank* and *select*, where the occurrences of any substring s can be tracked instead of only characters. Our time complexities are slightly over the ideal $O(|s|)$.

It is an interesting open question whether we can close those small gaps, that is (1) whether we can answer position-restricted counting queries in $O(m)$ time and locating each result in $O(1)$ time with structures taking $O(n \log n)$ bits of space, or even better, compressed data structures requiring $O(nH_k)$ bits of space; and (2) whether we can answer *rank* and *select* queries for substring s in $O(|s|)$ time.

In addition, we have shown some interesting connections between well-known two-dimensional range search data structures by Chazelle and recent data structures for compressed text indexing (the wavelet trees by Grossi et al.). We also showed how *rank* queries permit implement Chazelle's structure without using any fractional cascading information nor binary searches.

References

1. S. Alstrup, G. Brodal, and T. Rahue. New data structures for orthogonal range searching. In *Proc. 41st IEEE Symposium on Foundations of Computer Science (FOCS)*, pages 198–207, 2000.
2. A. Apostolico. The myriad virtues of subword trees. In *Combinatorial Algorithms on Words*, NATO ISI Series, pages 85–96. Springer-Verlag, 1985.
3. B. Chazelle. A functional approach to data structures and its use in multidimensional searching. *SIAM Journal on Computing*, 17(3):427–462, 1988.
4. D. Clark. *Compact Pat Trees.* PhD thesis, University of Waterloo, 1996.
5. P. Ferragina and G. Manzini. Opportunistic data structures with applications. In *Proc. 41st IEEE Symposium on Foundations of Computer Science (FOCS)*, pages 390–398, 2000.

6. P. Ferragina and G. Manzini. Indexing compressed texts. *Journal of the ACM*, 52(4):552–581, 2005.
7. P. Ferragina, G. Manzini, V. Mäkinen, and G. Navarro. An alphabet-friendly FM-index. In *Proc. 11th International Symposium on String Processing and Information Retrieval (SPIRE)*, LNCS v. 3246, pages 150–160, 2004.
8. P. Ferragina, G. Manzini, V. Mäkinen, and G. Navarro. Compressed representation of sequences and full-text indexes. Technical Report 2004-05, Technische Fakultät, Universität Bielefeld, Germany, December 2004. Submitted to a journal.
9. P. Ferragina, G. Manzini, V. Mäkinen, and G. Navarro. Succinct representation of sequences. Technical Report TR/DCC-2004-5, Department of Computer Science, University of Chile, Chile, August 2004. `ftp://ftp.dcc.uchile.cl/pub/users/gnavarro/sequences.ps.gz`.
10. R. Grossi, A. Gupta, and J. Vitter. High-order entropy-compressed text indexes. In *Proc. 14th Annual ACM-SIAM Symposium on Discrete Algorithms (SODA)*, pages 841–850, 2003.
11. R. Grossi, A. Gupta, and J. Vitter. When indexing equals compression: Experiments with compressing suffix arrays and applications. In *Proc. 15th Annual ACM-SIAM Symposium on Discrete Algorithms (SODA)*, pages 636–645, 2004.
12. G. Jacobson. Space-efficient static trees and graphs. In *Proc. 30th IEEE Symp. Foundations of Computer Science (FOCS'89)*, pages 549–554, 1989.
13. J. Kärkkäinen. *Repetition-based text indexes.* PhD thesis, Dept. of Computer Science, University of Helsinki, Finland, 1999. Also available as Report A-1999-4, Series A.
14. U. Manber and G. Myers. Suffix arrays: a new method for on-line string searches. *SIAM Journal on Computing*, pages 935–948, 1993.
15. G. Manzini. An analysis of the Burrows-Wheeler transform. *Journal of the ACM*, 48(3):407–430, 2001.
16. I. Munro. Tables. In *Proc. 16th Foundations of Software Technology and Theoretical Computer Science (FSTTCS'96)*, LNCS 1180, pages 37–42, 1996.
17. G. Navarro and V. Mäkinen. Compressed full-text indexes. Technical Report TR/DCC-2005-7, Department of Computer Science, University of Chile, Chile, June 2005. `ftp://ftp.dcc.uchile.cl/pub/users/gnavarro/survcompr.ps.gz`. Submitted to a journal.
18. R. Raman, V. Raman, and S. Srinivasa Rao. Succinct indexable dictionaries with applications to encoding k-ary trees and multisets. In *Proc. 13th Annual ACM-SIAM Symposium on Discrete Algorithms (SODA'02)*, pages 233–242, 2002.

Rectilinear Approximation of a Set of Points in the Plane

Yan Mayster and Mario A. Lopez

University of Denver, Department of Computer Science, 2360 S. Gaylord St., Denver, CO 80208, USA
{ymayster, mlopez}@cs.du.edu

Abstract. We derive algorithms for approximating a set S of n points in the plane by an x-monotone rectilinear polyline with k horizontal segments. The quality of the approximation is measured by the maximum distance from a point in S to the segment above or below it. We consider two types of problems: min -ε, where the goal is to minimize the error for k horizontal segments and min -#, where the goal is to minimize the number of segments for error ε. After $O(n)$ preprocessing time, we solve the latter in $O(\min\{k \log n, n\})$ time per instance. We then solve the former in $O(\min\{n^2, nk \log n\})$ time. We also describe an approximation algorithm for the min -ε problem that computes a solution within a factor of 3 of the optimal error for k segments, or with at most the same error as the k-optimal but using $2k - 1$ segments. Both approximations run in $O(n \log n)$ time.

1 Introduction

The problem of approximating a set of two-dimensional points by a polygonal line has been studied extensively in the literature. Different error metrics and constraints on the nature of the approximating curve result in many variants of this problem (see [1, 2, 4, 5, 6, 7, 8, 9, 10] for an assortment of these). For each variant two problem types, min -ε and min -#, are often considered. In min -ε approximation the goal is to minimize the error for a given complexity of the approximating curve $\mathcal{R}$. Conversely, in min -# approximation the objective is to minimize the complexity of $\mathcal{R}$ for a given allowed error ε.

We consider both min -ε and min -# problems for the special case of rectilinear approximation, i.e., for the case where the approximating curve $\mathcal{R}$ is a step function. In this case, the complexity (size) of $\mathcal{R}$ is given by the number of its horizontal segments. This problem has been studied in [3, 11]. Díaz-Báñez and Mesa derive a simple $O(n)$ time algorithm to solve the rectilinear min -# problem and use it as a subroutine to solve the min -ε problem in $O(n^2 \log n)$ time. This result was improved by Wang, who proposed an algorithm that runs in $O(n^2)$ time. His algorithm also makes use of the min -# algorithm of [3] but achieves better performance by reducing the number of subroutine calls.

Our contribution to the rectilinear approximation is twofold. First, we develop a new min -# algorithm which, after a preprocessing cost of $O(n)$, can solve

J.R. Correa, A. Hevia, and M. Kiwi (Eds.): LATIN 2006, LNCS 3887, pp. 715–726, 2006.

multiple instances (with different target errors) at a cost of $O(\min\{k \log n, n\})$ time per instance. When coupled with the approach proposed by Wang, this results in an algorithm for the min-ε problem than runs in $O(\min\{n^2, nk \log n\})$ time. This is asymptotically faster than the result in [11] when $k = o(n/\log n)$. Our second algorithm has $O(n \log n)$ running time and yields an approximation curve with error within a factor of 3 of the optimal and with the same number of segments. Furthermore, a curve generated by our algorithm of size $2k - 1$ achieves an error no more than that of an optimal curve of size k.

We now define the problem formally as well as introduce useful notation. Let $S = \{p_i = (x_i, y_i), i = 1, \ldots, n\}, x_1 < x_2 < \ldots < x_n$, be a set of points in the plane. For $1 \leq i \leq j \leq n$, define $S_{ij} := \{p_i, p_{i+1}, \ldots, p_j\}$. A curve $\mathcal{R}$ is rectilinear if it consists only of alternating horizontal and vertical segments. It is x-monotone if any two neighboring horizontal segments share exactly one x-coordinate, namely that of the vertical segment joining them. The error function used in our method is based on *vertical distance*. First, for $p = (x, y)$ and $p' = (x', y')$, the vertical distance $d_v(p, p')$ between p and p' is $|y - y'|$ if $x = x'$, and ∞ otherwise. Then, the vertical distance between a point p and a curve $\mathcal{R}$ is simply $d_v(p, \mathcal{R}) = \min_{q \in \mathcal{R}} d_v(p, q)$. Following the notation of [3], the eccentricity of $\mathcal{R}$ with respect to S is the maximum vertical error between the points of S and $\mathcal{R}$, i.e., $e(S, \mathcal{R}) = \max_{1 \leq i \leq n} d_v(p_i, \mathcal{R})$.

Because of our definition of distance, it suffices to restrict our attention to rectilinear x-monotone curves whose domain in x includes the interval $[x_1, x_n]$ and all of whose vertical segments fall strictly within it. We also assume that the first and the last segments of an approximation curve are horizontal and, if needed, can be extended arbitrarily far in the negative or positive x-directions. Given a horizontal segment s of a curve $\mathcal{R}$ with x-domain $[a, b]$, we say that a point p_i of S is "covered" by s if $x_i \in [a, b]$. Obviously, every point of S falls in the x-domain of some horizontal segment of $\mathcal{R}$ and the set of all points covered by s is some S_{ij} (which, as in [3], we call the *allocation set* of s). Furthermore, since we are not interested in increasing the complexity of $\mathcal{R}$ without changing its eccentricity, all allocation sets are nonempty, i.e. there is no horizontal segment that does not cover at least one point of S.

Let us call the y-span of points in the allocation set of s, the *range* of s. Then, it is clear (and shown in [3]) that in the search for an optimal solution to the min-ε problem we may restrict our attention to curves whose horizontal segments are centered with respect to their range. We denote by $\epsilon_{ij} = \frac{1}{2} \max_{p_r, p_s \in S_{ij}} |y_r - y_s|$ the error of a segment covering S_{ij} and note that $\mathcal{E} = \{\epsilon_{12}, \epsilon_{13}, \ldots, \epsilon_{1n}, \epsilon_{23}, \ldots, \epsilon_{n-1n}\}$, the set of all possible errors for the segments in a candidate curve $\mathcal{R}$, must include the eccentricity of $\mathcal{R}$.

The min-# problem can be solved in $O(n)$ time [3] by sweeping the points from left to right and adding them to the current allocation set provided that its range is no more than twice the allowed eccentricity. Then, the min-ε problem can be reduced to a binary search on $\mathcal{E}$, resulting in an $O(n^2 \log n)$ algorithm. See [3] for details.

Wang [11] improves the running time for min-ε by avoiding the generation and sorting of all possible eccentricities. Instead he uses a double sweepline scan to determine which instances of the min-# problem to solve. Every step of the algorithm advances one of the two sweeplines based on the comparison between the number of segments in the curve from the previous step and k. As a result, [11] tries no more than $2n$ values of ϵ and, thus, solves the min-ε problem in $O(n^2)$ time. See [11] for details.

2 An Optimal min-ε Algorithm

In our solution to the min-ε problem, we use Wang's approach to generate min-# problem instances but solve each of them using a different algorithm. Let $k = \lceil \log_2 n \rceil$, where $n = |S|$, and let A be an array of size $2^{k+1} - 1$. The last 2^k elements of A contain the points of S sorted by x-coordinate, padded on the right with $2^k - n$ copies of (x_n, y_n) for $n < 2^k$. The entries of A can be interpreted as a full binary tree whose leaves are the (padded) elements of S and whose internal nodes occupy locations $A[1..2^k - 1]$. We adopt the convention that the parent of $A[i]$ is stored at location $\lfloor i/2 \rfloor$, as it is normally done for heaps. Each of the first $2^k - 1$ elements of A stores the range of the y values for all points in its subtree. This information can be generated in $O(1)$ time per node by proceeding bottom-up, one level at a time. Thus, the tree can be built in linear time.

The algorithm creates an optimal k-curve with eccentricity $\leq \epsilon$ one segment at a time but does not necessarily investigate every point. Rather, it first expands a segment by covering progressively larger groups of consecutive points and then contracts its allocation set when (and if) its error exceeds ϵ. The expansion and contraction phases correspond to following upward and downward node sequences through the tree, which we shall now describe. The construction of the first segment starts at the first leaf node (i.e., $A[1] = p_1$) and proceeds upward as long as the union of y-ranges of the nodes examined (i.e., the error of the segment being built) is no more than ϵ. Each new node in the upward sequence is the parent of the previous node's right neighbor, i.e. the parent of the node whose array index in A is one greater than that of the previous node. Note that this relationship means that the new node may not necessarily be linked to the previous one by an edge, i.e. generally we do not traverse a connected path. If the previous node had no right neighbor, then the upward sequence must terminate since it means that the last leaf node is already covered and so the segment extends to include the nth point. Once the algorithm reaches a node that causes the union of y-ranges to exceed ϵ, the upward sequence is finished and to determine the rightmost point of the segment under construction we begin the downward sequence. Now, every time the algorithm goes down one level of the tree it first examines the left child of a node followed (if necessary) by the right child to see which of the y-ranges of the two children causes the error of the segment to be greater than ϵ. It then descends to that child. Once the algorithm reaches a leaf element, the segment is finalized and a new segment is started with its allocation set beginning at this element. Note that in each upward sequence no more than 1 node and in each downward one no more than 2 nodes from the same tree level are examined. Therefore, the total cost of creating a segment is proportional to $3 \log n$. Since in order to advance either of the sweeplines in Wang's min-ε algorithm all we need to know is how the number of segments k' in the candidate curve compares to the target number k, our min-# algorithm may finish and simply return that $k' > k$ whenever any points are not covered after k segments have been finalized. This ensures that we spend no more than $O(k \log n)$ time building a single candidate curve.

The other upper bound for this min-# algorithm can be shown by observing that for each segment the length of upward and downward node sequences depends on the number of points it would be extended to cover. Letting $n_i, 1 \leq i \leq k$, be the number of points in the final allocation set of the ith segment, we know that no more than $3 \log 4n_i$ nodes were in both sequences for segment i since the highest node in the upward sequence could have at most $4n_i$ nodes in its subtree (since the subtree sizes in the upward sequence double every time we move up a level and the last node in this sequence may contribute no points to the allocation set, it may have $2n_i$ leaves in its subtree and therefore have height $\log 4n_i$). Hence, the total number of nodes visited by the algorithm is $\sum_{i=1}^{k} 3 \log 4n_i \leq 3(2k + n) = O(n)$ since $n_1 + \ldots + n_k \leq n$. This implies that no matter what k is, our min-# algorithm takes no more than linear time and so our solution to the min-ε problem, which uses Wang's method of solving $2n$ min-# problems, takes total time $O(\min\{n^2, nk \log n\})$ and is asymptotically faster than Wang's algorithm when $k < n/\log n$.

We note that the above algorithm can be implemented without padding the leaf point data to bring its size to a power of 2. Then, the number of the non-leaf elements preceding S in A is $n - 1$ if n is even and n otherwise. Now, the point with the leftmost x-coordinate may not necessarily be the leftmost leaf due to the fact that leaves may reside at two different levels of the tree. Therefore, we have to allow some upward sequence that reached the rightmost node of some level to cross over to the leftmost node on that same level. This situation may occur exactly once during the execution of the algorithm and to detect it it is enough to keep track of the largest leaf index reachable from the currently inspected node.

3 An $O(n \log n)$ Approximation Algorithm

Let $k > 0$ and S be a set of n points in the plane. Let us denote an optimal k-curve (a curve with at most k segments) for S by C^* and its eccentricity by ϵ^*. The algorithm in [11] finds C^* and ϵ^* in $\Theta(n^2)$ time. We sought a faster approximation algorithm to generate curves with eccentricity no more than ϵ^* and at most αk horizontal segments, where $\alpha > 1$ is a small constant. In this section we first describe our simple greedy algorithm and then show that $\alpha = 2$ always suffices. Furthermore, we shall prove that the eccentricity of a k-curve generated by our algorithm is within a factor of 3 from optimal.

```
GCSA(A, m)
 1  Sort A in ascending order by x-coordinate
 2  L ← BuildSegmentList(A)
 3  if size[A] ≤ m
 4     then return L
 5  H ← BuildCostHeap(L)
 6  while size[H] > m
 7  do s ← ExtractMin(H)
 8     low[left[s]] ← Min(low[left[s]], low[s])
 9     high[left[s]] ← Max(high[left[s]], high[s])
10     right[left[s]] ← right[s]
11     if cost[left[s]] ≠ ∞
12        then h ← Max(high[left[s]], high[left[left[s]]])
13             l ← Min(low[left[s]], low[left[left[s]]])
14             cost[left[s]] ← h − l
15             Heapify(H, left[s])
16     if right[s] ≠ NIL
17        then h ← Max(high[left[s]], high[right[s]])
18             l ← Min(low[left[s]], low[right[s]])
19             cost[right[s]] ← h − l
20             left[right[s]] ← left[s]
21             Heapify(H, right[s])
22     delete s
23  return L
```

Fig. 1. Pseudocode for the GCSA Algorithm

We start with a set of n points stored in an array A. The *Greedy Combine Segment Approximation* (GCSA) Algorithm, described in Fig. 1, is called on A with a single additional parameter m that specifies the number of segments in the output curve C. The first step of the GCSA algorithm is to sort the points in A by x-coordinate. Then, GCSA creates a doubly-linked list L of n segments (line 2), each going through a different point and linked in the order of appearance in A of the corresponding points. Thus, we begin with an n-curve of eccentricity 0 that consists of singleton segments ordered from left to right. In order to produce an m-curve, adjacent allocation sets will be merged and new longer segments created. GCSA follows a greedy approach minimizing the eccentricity of the resulting curve at every merge step. GCSA represents each segment as a structure with 6 fields: *start*, which is the index in A of the leftmost point in the allocation set, *low* and *high*, which are the lowest and highest y-coordinates of the points in the allocation set, *left* and *right*, which point to the left and right neighboring segments, and *cost* that records the eccentricity of the curve resulting from merging the segment with its left neighbor. Initialization of these fields is done inside of the function BuildSegmentList. The fields *low* and *high* are set to the y-coordinate of the only point covered by the segment. The only field whose initialization requires a computation is *cost*, which is set to the length of the y-range of the points covered by the segment and its left neighbor. Since the leftmost segment has no left neighbor, its cost is ∞. Note that if $n \leq m$, the algorithm simply returns the curve obtained in this first step.

The next step is to prepare the list of segments for processing (merging) by prioritizing them according to cost. GCSA repeatedly extracts the segment with the least cost from the heap (line 7) and merges it with its left neighbor. Merging is carried out by updating the neighbor's fields (lines 8-14), re-heapifying (line 15), and then updating the fields of and re-heapifying on its right neighbor (lines 16-20). As part of this operation, the algorithm also maintains the adjacency pointers between segments (lines 10, 20). At every point in the execution of this loop, the eccentricity of the curve is dominated by the cost of the segment at the top of the heap and equals the error of the last generated segment. Another important observation is that the costs of the neighbors of an extracted segment can only increase, which justifies the calls to HEAPIFY in lines 15 and 21. Note that the leftmost segment is never extracted from the heap and as a result the splicing of nodes never changes the head node of L. The GCSA algorithm stops after $n - m$ iterations and L stores the segments of the curve sorted by x-range. The following theorem follows easily from Fig. 1 and the complexity of heap operations.

Theorem 1. *The GCSA Algorithm produces the curve C in $O(n \log n)$ time.*

Let us now introduce some terminology which will facilitate our discussion of the properties of curves constructed with GCSA.

Definition 1. *The cardinality of a rectilinear x-monotone curve C, denoted $|C|$, is taken to be the number of horizontal segments contained in C (including the possibly semi-infinite beginning and ending segments).*

We shall call a curve constructed by the GCSA algorithm (after any number of iterations) a *GCSA curve* and classify its segments through their relationship with the segments of some (fixed) optimal k-curve.

Definition 2. *Let S be a set of points and C^* an optimal k-curve for S. A horizontal segment s of a GCSA curve C is called an* inside *segment* with respect to *C^* if its allocation set is a contiguous subset of the allocation set of a horizontal segment s^* of C^*. In that case, we say that s^* contains s and denote it $s \subseteq s^*$. Also, if $s \subseteq s^*$ and s^* covers a point not covered by s, we say that s^* properly contains s. Any segment of C which is not an inside segment with respect to C^* is a* straddling *segment (or a* straddler*)* with respect to *C^*.*

Since we shall always keep the choice of C^* fixed throughout every argument, we shall drop the reference to C^* when qualifying the segments of C and simply speak of them as *inside* or *straddling* segments. When the allocation sets of a segment s of C and a segment s^* of C^* intersect, we shall simply say that s intersects s^*. Clearly, any segment of C can only intersect a subset of the segments of C^* that are adjacent. We shall also expand the last definition and differentiate between the straddlers that intersect exactly two segments of C^* and call these *simple* straddlers and the straddlers that intersect exactly three segments of C^* and name those *double* straddlers. Obviously, no GCSA curve C may contain more than $k - 1$ straddlers with respect to an optimal k-curve C^*

(and only have that many when they are all simple) because there are exactly $k-1$ pairs of adjacent segments in C^*.

We now prove one of the two main properties of the GCSA algorithm.

Theorem 2. *If $n \geq 2k$, the GCSA algorithm with $m = 2k-1$ produces a curve C with eccentricity $\epsilon \leq \epsilon^*$.*

Proof. Let S be a set of n points and $k > 0$. Let C^* be an optimal k-curve with eccentricity ϵ^*. Suppose that the curve C returned by GCSA and consisting of $2k-1$ segments has eccentricity $\epsilon > \epsilon^*$. Then, C contains at least one segment s, whose error is equal to ϵ. Consider the situation just before s was created (i.e., before its cost is extracted from the heap) and call the curve constructed up to that point C'. Then, clearly, $|C'| > 2k-1$, since s resulted from merging two segments. Now, let s^* be a segment of C^* and consider how many inside segments of C' may be contained by s^*. Assume there are two or more such segments of C'. Then, there must be at least one adjacent pair of these inside segments and the cost of merging any such pair is $\leq error(s^*) \leq \epsilon^*$ and so is $< \epsilon$ by hypothesis. Therefore, merging these pairs of segments has smaller cost than creating s contradicting the fact that the cost of s is at the top of the heap. Thus, we infer that all such pairs have been merged before and there can be at most one segment of C' completely inside of s^*. Since s^* was arbitrary, it follows each segment of C^* may contain at most one segment of C'. Then, since there can be no more than $k-1$ straddling segments, C' has no more than $k+k-1 = 2k-1$ segments. However, earlier we established that $|C'| > 2k-1$, a contradiction. We conclude that the curve C consisting of $2k-1$ segments has eccentricity $\epsilon \leq \epsilon^*$. Since GCSA starts with n segments and with each iteration decreases their number by one, C is obtained after $n-2k+1$ iterations.

Construction 1. We can construct an arbitrarily large set of points S whose GCSA $(2k-2)$-curve has eccentricity bigger than ϵ^*. For $k=2$, Fig. 2a shows a set of $n=4$ points along with an optimal 2-curve (dotted lines) and the GCSA curve with $2k-2=2$ segments (solid lines). If δ is very small, this shows that the GCSA 2-curve is almost twice as bad as an optimal one. Figure 2b shows how we can build a bigger example with three copies of the same point set in a V-shaped arrangement (with point D shared) and this construction carries over to arbitrarily large numbers of points and segments. In general, an optimal k-curve may have eccentricity almost twice as small as that of the GCSA $(2k-2)$-curve. To see this, let $k = h+1$ and so $2k-2 = 2h$. With h copies of the point set in Fig. 2a we have $3h+1$ points and after merging in each copy the middle pair of points we obtain h 2-point segments and $h+1$ singleton segments. This necessitates the creation of one more segment with error strictly bigger than the eccentricity of an optimal $(h+1)$-curve.

Before we proceed with our next result, we need one more definition.

Definition 3. *For $0 < i < n$ and the same set of points S, the GCSA curves produced after $0, \ldots, i-1$ merge steps are called the* ancestor curves *of the GCSA curve C obtained after the ith step. Furthermore, any segment in an ancestor*

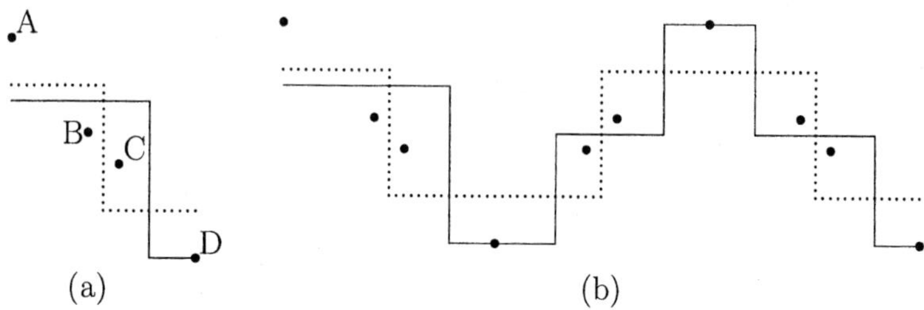

Fig. 2. (a) A set of 4 points with the optimal 2-curve (dotted) and the GCSA curve approximating it (solid). The distance between the points A and B as well as between C and D is 1 unit, while the distance between B and C is $1-\delta$, (b) 3 copies of the set in (a) adjoined in a V-shaped arrangement with the optimal 4-curve (dotted) and the GCSA 6-curve (solid) whose eccentricity is almost twice as bad.

curve of C contained by a segment s of C is an ancestor segment *of s (i.e., the ancestors of s are all those segments whose merging eventually produced s).*

In the next two theorems, we shall make use of the following easy to prove lemma.

Lemma 1. *Let s and t be neighboring segments in a GCSA curve C. Suppose that with respect to some optimal curve C^*, s is an inside segment and t is a straddler with error $error(t)$. Then, $error(s \cup t) \leq \epsilon^* + error(t)$.*

We now prove an intermediate result that determines the minimum number of segments for a GCSA curve C to have guaranteed eccentricity within a factor of 2 of optimal.

Theorem 3. *The GCSA curve C with at least $k + \lfloor \frac{k}{3} \rfloor$ segments achieves eccentricity at most $2\epsilon^*$.*

Proof. Consider the curve C generated by GCSA just prior to creating a segment with error $\epsilon > 2\epsilon^*$. Clearly, $|C| \leq 2k-1$ (Theorem 2). First, suppose that C has no segment properly contained by a segment of C^*. Then, the theorem follows since in that case C is made up only of straddling segments or segments that also belong to C^*, and thus $|C| \leq k$.

Therefore, we assume that C has at least one segment properly contained by a segment of C^*. Let s be the leftmost such segment of C, inside of some segment s^* of C^*, that does not extend all the way to the right end of s^*. We observe that, if s exists, the number of segments in C to the left of s is no more than the number of segments in C^* to the left of s^*. If s does not exist, then $|C| \leq |C^*| = k$ and the theorem follows. We shall compute the greatest possible number of segments in C between s and the next inside segment r to the right of s and compare it to the number of segments in C^* between s^* and the segment containing r (note that r may or may not extend to the right end of that segment). Let C' be the last ancestor curve of C with eccentricity $\epsilon' \leq \epsilon^*$

(soon we shall see that because of the existence of s, C' must be different from C). Then, since $error(s) \leq \epsilon^*$, $s \in C'$ and from the proof of Theorem 2 we know that s has no neighbors in C' and, consequently, in C that are properly inside of s^*. Therefore, the right neighbor of s in C, call it t, "straddles" s^* and its right neighbor t^*.

Now, can $t \in C'$? Clearly not, as otherwise $error(t) \leq \epsilon^*$, and, therefore, merging s with t results in a segment whose error is less than ϵ (by Lemma 1, $error(s \cup t) \leq error(t) + \epsilon^* \leq 2\epsilon^*$), but by hypothesis ϵ resides at the top of the heap after C was created. Therefore, $t \notin C'$ and so t has an ancestor segment $t' \in C'$ that also straddles s^* and t^* (since t covers points of s^* that s does not, C' cannot contain an ancestor of t that covers only these points, otherwise merging s with that ancestor creates a segment of eccentricity $\leq \epsilon^*$). Now, t' covers all points of s^* that are covered by t and $error(t') \leq \epsilon^*$. Since both t and t' cover points of t^* and $t' \subseteq t$, it follows that t was obtained from t' by merging it on the right with another segment of C' and so C cannot have any segment contained in t^* (since at most one existed in C').

If the rightmost point covered by t is the same as the rightmost point covered by t^*, then C has no more segments up to that point than C^* (same number of segments prior to s^* and then $\{s, t\}$ in C and $\{s^*, t^*\}$ in C^*). Now, suppose that t does not cover at least one point of t^*. Then, the right neighbor of t^*, u^*, cannot contain a segment of C for otherwise the segment of C that straddles t^* and u^* is also in C' (this is because its ancestor in C' must also be a straddler and cannot be merged afterwards as its only inside neighbor on the left has to be merged with t' to produce t and, by assumption, its inside neighbor on the right survives) and we have a contradiction - it and the segment inside of u^* must be merged before we get to C (again by Lemma 1)! Thus, there are at most two segments of C that cover points of u^*, one that straddles t^* and u^* (call it u) and a segment (call it v) that straddles u^* and its right neighbor r^*. Therefore, the next inside segment r of C to the right of s must be beyond u^* (the first segment of C^* that can contain it is r^*). Consequently, since every segment of C from s to r is straddling except s itself, there is at most one more segment in C up to r than there are segments in C^* up to the segment containing r. Since there are at least 3 such segments in C^* (namely, $\{s^*, t^*, u^*\}$ versus $\{s, t, u, v\}$ in C), it follows that C has at most $\lfloor \frac{k}{3} \rfloor$ more segments than C^*. It remains to consider the case when t ends in u^* or beyond. In this case t is a double-straddler and C has no more segments spanning the range from s to the left endpoint of r than C^* (again, since all segments other than s are straddling and t is a double straddler). Thus, it is always true that $|C| \leq k + \lfloor \frac{k}{3} \rfloor$.

We are now ready to prove that not only does GCSA produce a good approximation curve with $2k - 1$ segments but that a curve with as few as k segments has eccentricity no more than 3 times greater than optimal.

Theorem 4. *A GCSA curve C with k segments has eccentricity at most $3\epsilon^*$.*

Proof. Let C be the GCSA curve prior to creating a segment with error $> 3\epsilon^*$. Previously, we have shown that no two segments of C can be inside of the same segment of C^* (Theorem 2). Then, we have argued that no two inside segments of C can be adjacent to the same simple straddler (Theorem 3 proves a stronger claim). Now, we show that in our case no simple straddler can be a neighbor of an inside segment in C. This is easy. Suppose that s is such a simple straddler next to an inside segment t of C. Let C' be the last ancestor curve of C whose eccentricity is no more than ϵ^*. Since $t \in C'$, the proof of Theorem 2 implies that there is an ancestor of s in C', call it s', which is also a straddler. Moreover, s' is a neighbor of t in C' and, therefore, s was obtained from s' by merging on (at most) one side. Therefore, $error(s) \leq 2\epsilon^*$. It follows by Lemma 1 that $error(s \cup t) \leq error(s) + \epsilon^* = 3\epsilon^*$. Hence, s and t cannot both be present in C and we have a contradiction. Thus, we have shown that for every inside segment C must contain a non-simple straddler. Otherwise, if C has no inside segments, it consists only of straddlers and segments that are also in C^*. Therefore, we are done and $|C| \leq k$.

Construction 2. We shall illustrate with this example that situations when the GCSA algorithm produces k-curves with eccentricity arbitrarily close to $3\epsilon^*$ do arise. Figure 3 shows one such situation with 12 points, numbered in the order of increasing x-coordinate, and Table 1 displays the y-coordinates and the vertical distance between consecutive points (δ is an arbitrarily small positive number). We let $k = 4$ and display the optimal 4-curve with eccentricity $\epsilon^* = 1 + 3\delta$ in Fig. 3b.

In Fig. 3c, d, e we show the construction of an approximating 4-curve with our algorithm, divided into 3 stages. The first stage (3c) merges 5 pairs of singleton segments, the second stage (3d) merges two pairs of neighboring segments created in the first stage and separated by $1 + \delta$, and in the final stage there is no choice but to increase the eccentricity to $3 + 3\delta$ and one possibility for it is Fig. 3e.

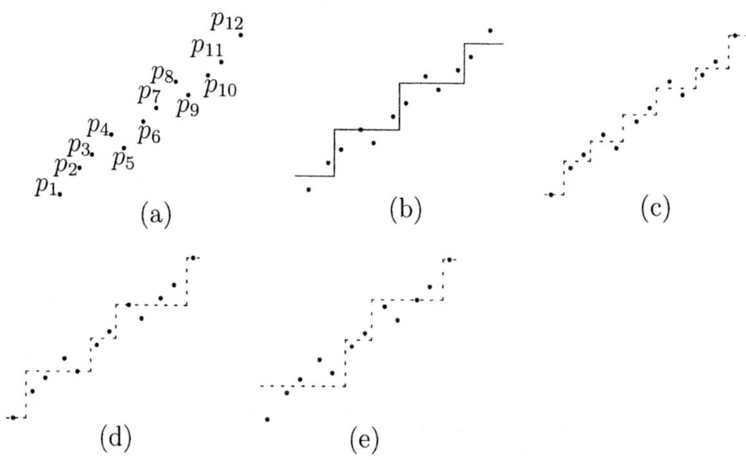

Fig. 3. (a) A set of 12 points, (b) An optimal approximation, (c) Stage 1 of GCSA ($\epsilon = 1$), (d) Stage 2 of GCSA ($\epsilon = 2 + \delta$), (e) Stage 3 of GCSA ($\epsilon = 3 + 3\delta$)

Table 1. Vertical distances between neighboring points in Fig. 3

Point	y_i	$\lvert y_i - y_{i-1}\rvert$	Point	y_i	$\lvert y_i - y_{i-1}\rvert$
p_1	0	--	p_7	$4+5\delta$	1
p_2	$1+2\delta$	$1+2\delta$	p_8	$5+7\delta$	$1+2\delta$
p_3	$2+2\delta$	1	p_9	$4+7\delta$	1
p_4	$3+3\delta$	$1+\delta$	p_{10}	$5+8\delta$	$1+\delta$
p_5	$2+3\delta$	1	p_{11}	$6+8\delta$	1
p_6	$3+5\delta$	$1+2\delta$	p_{12}	$7+10\delta$	$1+2\delta$

This example carries over to arbitrarily large n and k. Imagine putting together h copies of the 12 points in Fig. 3a, with p_1 of the $(i+1)$st copy having x-coordinate greater and y-coordinate 1 less than those of p_{12} of the ith copy. Then, it is an easy exercise to verify that the optimal $(3h+1)$-curve has eccentricity $1+4\delta$ while the GCSA curve of the same complexity will have eccentricity $3+3\delta$. Therefore, for arbitrarily large n and k, as $\delta \to 0$, we have the ratio

$$\frac{\epsilon}{\epsilon^*} = \frac{3+3\delta}{1+4\delta} \to 3.$$

The worst-case curves used to prove the tightness of the bounds in Theorems 2 and 4 are artificial constructions which are unlikely to occur in practice. The GCSA algorithm performs very well on both synthetic (correlated and uncorrelated) and real data and achieves approximation factors well below these upper bounds. In fact, the experiments we have conducted on real datasets taken from websites such as the Time Series Data Library[1] and The Financial Data Finder[2] have produced near-optimal or optimal k-curves. The specific results are displayed in Table 2 for four data files and for several representative values of k. The contents of the files used in our experiments are described below.

`arosa.dat`[1] Ozone concentrations in Arosa, Switzerland, 1932-1972
`daily.asc`[2] Daily Dow Jones Industrial avg. stock price index, 1915-1989
`pphil.dat`[1] Monthly precipitation in mm in Philadelphia, 1820-1950
`tpmon.dat`[1] Monthly temperature in England (F), 1723-1970

For the datasets in Table 2 GCSA produced highly accurate curves with k segments and always significantly outperformed k-optimal curves with $2k-1$ segments. The GCSA approximation is extremely accurate for meteorological data and is within 11% of optimal for the Dow Jones data with just $\sqrt{n} = 137$ segments. Furthermore, we note that on this dataset GCSA computed its approximation $\sqrt{n}$-curve 1000 times faster than Wang's algorithm took to find an optimal curve of the same complexity. Overall, we conclude that with real data the GCSA algorithm quickly produces remarkably good k-curve approximations with eccentricity within 15% of optimal while the eccentricity of $(2k-1)$-curves always remains significantly below k-optimal.

[1] *http://www-personal.buseco.monash.edu.au/~hyndman/TSDL/*
[2] *http://fisher.osu.edu/fin/osudown.htm*

Table 2. Results of GCSA test runs on financial and meteorological data shown as ratios of GCSA k-curve eccentricities ϵ to those (ϵ^*) of optimal k-curves and ratios of optimal k-curve eccentricities ϵ^* to those (ϵ') of GCSA $(2k-1)$-curves (in parentheses)

File	n	$k = \log n$	$k = \sqrt{n}$	$k = n/20$	$k = n/10$	$k = n/5$
Ozone	481	1.00 (1.07)	1.00 (1.16)	1.00 (1.16)	1.00 (1.67)	1.10 (1.94)
Precipitation	1572	1.00 (1.15)	1.00 (1.18)	1.00 (1.24)	1.00 (1.33)	1.00 (1.62)
Temperature	2976	1.00 (1.07)	1.00 (1.09)	1.00 (1.14)	1.00 (1.69)	1.14 (1.89)
Dow Jones	18840	1.13 (1.38)	1.11 (1.39)	1.07 (1.82)	1.05 (1.96)	1.04 (2.56)

References

1. Chan, S., Chin, F.: Approximation of polygonal curves with minimum number of line segments or minimum error. International Journal of Computational Geometry and Applications. **6** (1996) 59–77
2. Díaz-Bánez, J.M., Gomez, F., Hurtado, F.: Approximation of point sets by 1-corner polygonal chains. INFORMS Journal on Computing. **12** (2000) 317–323
3. Díaz-Bánez, J.M., Mesa, J.A.: Fitting rectilinear polygonal curves to a set of points in the plane. European Journal of Operations Research. **130** (2001) 214–222
4. Eu, D., Toussaint, G.T.: On approximating polygonal curves in two and three dimensions. CVGIP: Graphical Models and Image Processing. **56**(3) (1994) 231–246
5. Hakimi, S.L., Schmeichel, E.F.: Fitting polygonal functions to a set of points in the plane. CVGIP: Graphical Models and Image Processing. **53**(2) (1991) 132–136
6. Imai, H., Iri, M.: Computational-geometric methods for polygonal approximations of a curve. Computer Vision, Graphics and Image Processing. **36**(1) (1986) 31–41
7. Imai, H., Iri, M.: An optimal algorithm for approximating a piecewise linear function. Journal of Information Processing. **9**(3) (1986) 159–162
8. Imai, H., Iri, M.: Polygonal approximations of a curve – formulations and algorithms. In: Toussaint, G.T. (ed.): Computational Morphology. North-Holland, Amsterdam, Netherlands (1988) 71–86
9. Melkman, A., O'Rourke, J.: On polygonal chain approximation. In: Toussaint, G.T. (ed.): Computational Morphology. North-Holland, Amsterdam, Netherlands (1988) 87–95
10. Varadarajan, K.R.: Approximating Monotone Polygonal Curves Using the Uniform Metric. In: SCG '94 Proceedings of the 12th annual symposium on Computational geometry. (1996) 311–318
11. Wang, D.P.: A new algorithms for fitting a rectilinear x-monotone curve to a set of points in the plane. Pattern Recognition Letters. **23** (2002) 329–334

The Branch-Width of Circular-Arc Graphs

Frédéric Mazoit

LIF, Université de Provence,
13453 Marseille Cedex 13, France

Abstract. We prove that the branch-width of circular-arc graphs can be computed in polynomial time.

Keywords: Branch-width, circular-arc graphs, algorithm.

1 Introduction

The notions of tree-width and tree-decomposition of a graph have been introduced by Robertson and Seymour [1] for their graph minor project. These notions have been intensively investigated for algorithmic purposes and it is well known that many intractable problems can be solved in polynomial (and very often in linear) time when the input is restricted to graphs with bounded tree-width (see [2] for a comprehensive survey). While working on their graph minor project, Robertson and Seymour defined, in connection with tree-width, the notion of branch-width [3]. They proved that for any graph G, $\mathrm{bw}(G) \leq \mathrm{tw}(G)+1 \leq 1.5\mathrm{bw}(G)$. Both bounds are tight and achievable on trees and complete graphs. Branch-width appeared to be an even more appropriate tool than tree-width for the graph minor theory. Since both parameters are so close, one can expect the algorithmic behaviour of these problems to be quite similar. However, this is not true. For example, on planar graphs branch-width can be computed in polynomial time [4] while computing the tree-width of a planar graph in polynomial time is a long standing open problem. An even more striking example was observed by Kloks et al. [5]: deciding the branch-width of a split-graph is NP-hard while deciding the tree-width of a split-graph can be done in linear time.

In [6], the author studied the relation between both tree-decompositions and branch-decompositions and, in particular, how they can be associated to triangulations in a similar way. Using his techniques, Fomin et al. [7] describe the analogue of minimal triangulations and potential maximal cliques for branch-width: *efficient triangulations* and *blocks*. They also note that, using a large enough family of blocks together with their *block branch-width*, it is possible to compute the branch-width of any graph in exponential time. The algorithm is essentially the same as the one Fomin et al. use [8] to compute the tree-width.

In this article, we use the same framework to show that it is possible to compute the branch-width of circular-arc graphs in polynomial time. Section 3 is devoted to a proof of the *efficient triangulation* theorem which is simpler than the original one [6]. Section 4 presents some results of [7] and Sect. 5 shows how to use these tools to compute the branch-width of circular-arc graphs.

J.R. Correa, A. Hevia, and M. Kiwi (Eds.): LATIN 2006, LNCS 3887, pp. 727–736, 2006.

2 Preliminaries

Throughout this paper, G is a graph with vertex set V and edge set E, $n = |V|$ and $m = |E|$. The *neighbourhood* $N(x)$ of a vertex x is the set of vertices adjacent to x. The *neighbourhood* of a set of vertices C is the set of vertices not in C that are adjacent to at least one vertex of C.

A set of vertices S is a *separator* if $G \setminus S$ has at least two connected components, an a, b*-separator* if a and b are in different connected components of $G \setminus S$, an a, b*-minimal separator* if no proper subset of S is an a, b-separator. The connected component of a in $G \setminus S$ is $C_a(S)$. The component $C_a(S)$ *is a full connected component* if S is the neighbourhood of $C_a(S)$. For an a, b-minimal separator S, both $C_a(S)$ and $C_b(S)$ are full. A set S is a *minimal separator* if there exist a and b such that S is an a, b-minimal separator or, which is equivalent, if $G \setminus S$ has at least two full connected components.

A *clique* of a graph G is a set of vertices of G that are pairwise adjacent in G. The maximum clique size of G is denoted by $\omega(G)$. A graph is *chordal* (or *triangulated*) if every cycle of length at least four has a chord, that is an edge between two non-consecutive vertices of the cycle.

Theorem 1 ([9]). *A graph is chordal if and only if it is the intersection graph of a family of sub-trees of a tree.*

If H is the intersection graph of a family $\{T_x \mid x \in V(H)\}$ of sub-trees of a tree T, every maximal clique Ω of H can be associated to a vertex v_Ω of T such that the set of vertices whose sub-tree contains v_Ω is exactly Ω. The minimal separators of H can be associated to edges of T in a similar way.

A graph H is a *super-graph* of a graph G if H and G have the same vertices and every edge of G is an edge of H. A *triangulation* of a graph G is a chordal super-graph of G. A triangulation H of G is *minimal* if no strict sub-graph of H is a triangulation of G.

Definition 1 (Efficient triangulation). *A triangulation H of G is* efficient *if*

1. *each minimal separator of H is also a minimal separator of G;*
2. *for each minimal separator S of H, the connected components of $H \setminus S$ are exactly the connected components of $G \setminus S$.*

Note that according to a result of Parra and Scheffler [10], minimal triangulations are efficient.

If X is a set of edges, $V(X)$ (the vertices of X) denotes the set of vertices incident to X. The *border* of X, $\delta(X)$, is the set of vertices $V(X) \cap V(E \setminus X)$. A *pack* of a set of vertices S is either an edge whose ends belong to S or the set of edges incident to a connected component of $G \setminus S$. Note that if X is a set of edges, a pack of $\delta(X)$ is either a subset of X or disjoint from X. We can thus define the *packs* of a set of edges X as being the packs of $\delta(X)$ that are subsets of X.

The notion of branch-width is due to Robertson and Seymour [3]. A *branch-decomposition* T of a graph G is a pair (T, τ) with T a ternary tree and τ a bijective mapping from the leaves of T to the edges of G. The vertices of T are its *nodes*. For any edge e of $\mathcal{T}$, the two connected components $T_1^*(e)$ and $T_2^*(e)$ of $T \backslash e$ are the *e-branches* of $\mathcal{T}$. A *branch* of $\mathcal{T}$ is a e-branch for some edge e. The set of edges of G mapped on the leaves of a branch T^* is its *ground*. By using the ground, we can define the packs, the border and the set $V(T^*)$ of a branch T^*. We also extend the definition of packs and border to the edges of a branch-decomposition. A branch-decomposition $\mathcal{T}$ is *compatible* with a set of vertices S if S is a subset of at least one border of $\mathcal{T}$. The maximum size of the border of an edge of $\mathcal{T}$ denoted by $\mathrm{bw}(\mathcal{T})$ is called the *width* of the branch-decomposition. The *branch-width* ($\mathrm{bw}(G)$) of a graph G is the minimum width of one of its branch-decompositions. Note that the definitions of branch-decomposition and branch-width also apply to hyper-graphs.

The notion of branch-width is closely related to the well-known notion of tree-width. The *tree-width* of a graph G ($\mathrm{tw}(G)$) is the minimum of $\omega(H)$ over the triangulations H of G. In particular, Robertson and Seymour [3] showed that $\mathrm{bw}(G) \leq \mathrm{tw}(G) + 1 \leq \lfloor 2\,\mathrm{bw}(G)/3 \rfloor$. The branch-decompositions of a graph can also be associated to triangulations. Indeed, given a branch-decomposition $\mathcal{T}$ of a graph G, we can associate to each vertex x of G the subtree T_x of T covering all the leaves of T containing edges incident to x. The border of a branch e of $\mathcal{T}$ is exactly the set of vertices x of G such that e belongs to T_x. The intersection graph of the subtrees T_x is the triangulation $H_{\mathcal{T}}$ of G *associated* with $\mathcal{T}$.

Proposition 1. *Let G be a graph, $\mathcal{T}$ be any branch-decomposition of G of optimal width and $H_{\mathcal{T}}$ be the triangulation of G associated with $\mathcal{T}$.*

$$\mathrm{bw}(G) = \mathrm{bw}(H_{\mathcal{T}}).$$

Proof. By induction on the number p of edges in $H_{\mathcal{T}} \backslash G$. If G and $H_{\mathcal{T}}$ are equal, the result is obvious.

Otherwise, let (x, y) be an edge of $H_{\mathcal{T}} \setminus G$ and G' be the graph $G \cup \{(x, y)\}$. Since (x, y) is an edge of $H_{\mathcal{T}}$, T_x and T_y have a non empty intersection and since (x, y) does not belong to G, this intersection contains an edge e of $\mathcal{T}$. Add to T a new vertex in the middle of e and a new leaf u attached to that vertex and map the edge (x, y) to u. The branch-decomposition $\mathcal{T}'$ obtained is a branch-decomposition of G' such that $\mathrm{bw}(\mathcal{T}) = \mathrm{bw}(\mathcal{T}')$. Since $\mathcal{T}$ is optimal for G, we have $\mathrm{bw}(G) \geq \mathrm{bw}(G')$ and since G is a sub-graph of G', $\mathrm{bw}(G) = \mathrm{bw}(G')$. By construction, the triangulation $H_{\mathcal{T}'}$ of G' associated to $\mathcal{T}'$ is equal to $H_{\mathcal{T}}$. By induction, $\mathrm{bw}(H_{\mathcal{T}}) = \mathrm{bw}(G')$ which finishes the proof. □

3 Tight Branch-Decompositions

Proposition 1 implies that the branch-width of G is the minimal branch-width of a triangulation of G. This latter result is also true for tree-width. It is also known that while dealing with tree-width, we can only consider minimal triangulations but the same restriction is not possible for branch-width for there are examples of chordal graphs H such that H is never the triangulation associated to an optimal branch-decomposition of H. However we show that we can restrict ourselves to efficient triangulations. To prove this, we order the branch-decompositions of a given graph and then show that the triangulations arising from the minimal branch-decompositions are efficient.

Definition 2. *A branch-decomposition $\mathcal{T}$ of a graph G is* tighter *than $\mathcal{T}'$ if* $\mathrm{bw}(\mathcal{T})$ *is at most* $\mathrm{bw}(\mathcal{T}')$ *and $H_\mathcal{T}$ is a subgraph of $H_{\mathcal{T}'}$.*

The following theorem defines a "cleaning" process that does not increase the tightness and that we can use to perform local optimisations to a branch-decomposition. Figure 1 gives an idea of how the proof works.

Theorem 2. *Let $\mathcal{T}$ be a branch-decomposition of a graph G and T^* a branch of $\mathcal{T}$.*

There exists a branch-decomposition $\mathcal{T}'$ obtained from $\mathcal{T}$ by replacing T^* with another branch such that $\mathcal{T}'$ is tighter than $\mathcal{T}$ and such that every pack of T'^* is the ground of a sub-branch of T'^*.

Moreover, if a border S of T^* is neither a subset of $\delta(T^*)$, nor a subset of $V(X)$ with X a pack of T^*, then $\mathcal{T}'$ is strictly tighter than $\mathcal{T}$.

Proof. For every pack X_i of T^*, consider the rooted sub-tree of T^* covering the leaves containing an edge of X_i. By contracting edges, we can "remove" the nodes of degree 2 and thus obtain a branch $T_i'^*$ of ground X_i. By linking these branches by there root, we can obtain a branch T'^* of ground X.

We claim that the branch-decomposition $\mathcal{T}'$ obtained by replacing T^* by T'^* in $\mathcal{T}$ is tighter than $\mathcal{T}$. Indeed, by construction, every sub-branch B_i' of $T_i'^*$ is a pruned sub-branch B of T^*. Since X_i is a pack of T^*, $\delta(B_i')$ is a subset of $\delta(B)$. Finally, since the border of all the edges added to link the branches $T_i'^*$ is

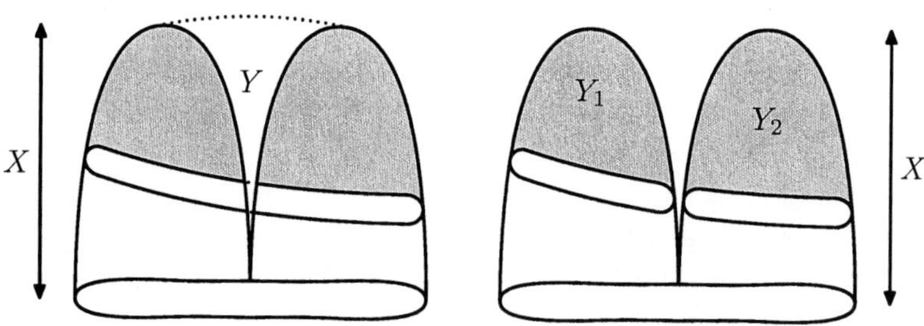

On the left, we consider a branch T_X^* of a decomposition $\mathcal{T}$ with a ground X that has two packs and a sub-branch T_Y^* of ground Y that "crosses" the two packs.

If we prune T_X^* to each pack and rearrange the new branches, the sub-branch T_Y^* and $\delta(Y)$ will be split in two.

The new decomposition is strictly tighter than $\mathcal{T}$.

Fig. 1. A trimmed branch

a subset of $\delta(T^*)$, every border of T' is a subset of a border of T. This proves that $\mathcal{T}'$ is tighter than $\mathcal{T}$.

Now suppose that S is a border of T^* which is not a subset of $\delta(T^*)$ and not a subset of any $V(X_i)$. Let T^*_S be a sub-branch of T^* of border S. Let u be a vertex of $S \setminus \delta(T^*)$ (such a vertex exists by hypothesis) and X_u be the pack of T^* such that $V(X_u)$ contains u. Since S is not a subset of $V(X_u)$, there exists a vertex v of S in $S \setminus V(X_u)$. By construction, T'_u and T'_v do not meet and $\mathcal{T}'$ is strictly tighter than $\mathcal{T}$. □

The branch-decomposition built in Th. 2 is *trimmed* along T^*. We can now easily prove Th. 3.

Theorem 3. *The triangulation associated to a tightest branch-decomposition of a graph G is an efficient triangulation of G.*

Proof. Let $\mathcal{T}$ be a tightest branch-decomposition of G and $H_{\mathcal{T}}$ the triangulation associated to $\mathcal{T}$. Let S be a minimal separator of $H_{\mathcal{T}}$ and Ω_1 and Ω_2 be two maximal cliques of $H_{\mathcal{T}}$ contaning S.

The triangulation $H_{\mathcal{T}}$ is the intersection graph of the sub-trees T_x, thus Ω_1 and Ω_2 correspond to vertices v_{Ω_1} and v_{Ω_2} of T and there is an edge e_S on the path from v_{Ω_1} to v_{Ω_2} that corresponds to S. Let T^*_1 and T^*_2 be the two e_S-branches with v_{Ω_1} in T^*_1 and v_{Ω_2} in T^*_2.

Suppose now that S is not a minimal separator of G, it has at most one full connected component. We can thus suppose that S is the border of no pack of T^*_1. By Th. 2, we can trim $\mathcal{T}$ along T^*_1 and suppose that all the packs of S in G are the grounds of some sub-branches of the two e-branches. In the resulting decomposition $\mathcal{T}'$, no border of a sub-branch of T'^*_1 contains S which proves that Ω_1 is not a maximal clique of $H_{\mathcal{T}'}$. Thus $\mathcal{T}'$ is strictly tighter than $\mathcal{T}$ which is absurd. The minimal separator S of $H_{\mathcal{T}}$ is also a minimal separator of G.

Using the same techniques, we can also deduce the fact that the connected components of $G \setminus S$ and $H_{\mathcal{T}} \setminus S$ are the same. □

4 Block Branch-Width

Suppose that $\mathrm{bw}(G) = \mathrm{bw}(H_{\mathcal{T}})$. We can see the maximal cliques of $H_{\mathcal{T}}$ as pieces of a puzzle that match along minimal separators. Since we can suppose that $H_{\mathcal{T}}$ is efficient, we can characterize its maximal cliques as *blocks*.

Definition 3 (Block). *A set of vertices B of G is called a* block *if, for each connected component C_i of $G \setminus B$,*

- *its neighbourhood $S_i = N(C_i)$ is a minimal separator;*
- *$B \setminus S_i$ is non empty and contained in a connected component of $G \setminus S_i$.*

We say that the minimal separators S_i border *the block B and we denote by $s(B)$ the number of these separators.*

Let Ω be a maximal clique of $H_{\mathcal{T}}$. The branch-decomposition $\mathcal{T}$ induces a branch-decomposition $\mathcal{T}_\Omega$ of the complete graph $K(\Omega)$ on Ω. This branch-decomposition *respects* $K(\Omega)$ in that $\mathcal{T}_\Omega$ is compatible with the minimal separators of $H_{\mathcal{T}}$ included in Ω. The branch-width of $H_{\mathcal{T}}$ is the maximum width of the branch-decompositions $\mathcal{T}_\Omega$.

Definition 4 (Block branch-width). *The* block branch-width *of a block* Ω *(*$\mathrm{bbw}(\Omega)$*) is the minimal width of a branch-decomposition of* $K(\Omega)$ *respecting* Ω.

Conversely, if we have optimal respectful branch-decompositions of the maximal cliques of $H_{\mathcal{T}}$, we can construct an optimal branch-decomposition of $H_{\mathcal{T}}$ which leads to the following theorem.

Theorem 4 ([7])

$$\mathrm{bw}(G) = \min_{H \textit{ efficient triangulation of } G} \max\{\mathrm{bbw}(\Omega) \mid \Omega \textit{ maximal clique of } H\}.$$

If we have "enough" blocks of a graph G and if we know their block branch-width, computing the branch-width of G is indeed a large puzzle in which we try to match blocks of low block branch-width along minimal separators bordering them to construct a chordal graph. This puzzle can be solved in linear time in the number of blocks.

Theorem 5 ([7]). *Given a graph* G *and a complete list* $\mathcal{B}_G$ *of blocks together with their block branch-widths, the branch-width of* G *can be computed in* $\mathcal{O}(nm|\mathcal{B}_G|)$ *time.*

The last tool we need to be able to compute the branch-width of a graph is to be able to compute the block branch-width of a block. Unfortunately, deciding the block branch-width of a block is strictly equivalent to deciding the branch-width of a split-graph which is NP-complete [5] as already stated. Fortunately, if a block Ω is bordered by "few" minimal separators, we can still compute $\mathrm{bbw}(\Omega)$.

Theorem 6 ([7]). *The block branch-width of any block* B *can be computed in* $\mathcal{O}(3^{s(B)})$ *time.*

We can sketch the proof as follows. In a decomposition $\mathcal{T}$ of Ω, there is a node v_Ω corresponding to Ω and three branches T_i^* attached to v_Ω. If $\mathcal{T}$ respects Ω, then the minimal separators bordering Ω are partitioned in three according to the branch in which they appear as a border. If we choose a 3-partition of the minimal separators, we can compute the optimal width of a decomposition leading to this 3-partition in constant time. To compute the block branch-width, we only have to try all the possible 3-partitions.

5 Circular- Arc Graphs

In this section, we apply Th. 5 to circular-arc graphs to prove that their branch-width can be computed in polynomial time.

A *circular-arc* graph is the intersection graph of the arcs of a circle. The tree-width of circular-arc graphs can be computed in polynomial time as it is shown in [11]. To prove this, the authors use a circular interpretation of the graph (which can be obtained in linear time [12]) and give a geometrical interpretation of maximal potential cliques which allows them to prove that a tree-decomposition corresponds to a planar triangulation of some polygon. We will follow exactly the same path to prove that the branch-width of circular-arc graphs can be computed in polynomial time.

From now on, G is a circular-arc graph. By shifting them a little, we can suppose that ends of two distinct arcs of an intersection model $\mathcal{I}$ of G are also distinct. We will only consider such representations. Between two such ends, we put a *scan-point*. A *scan-line* is a chord of the disk Σ between two distinct scan-points; these chords are different from the chords of a chordal graph. It is easy to see that there are $2n$ scan-points and $n(2n-1)$ scan-lines. The arcs inside of which lie the ends of a scan-line are *cut* by the scan-line. A scan-line λ or a family of scan-lines Λ *realises* the set $V(\lambda)$ or $V(\Lambda)$ of vertices whose arcs they cut.

Let S be a minimal separator of G and C a full component of S. The union of the arcs of the connected component C is also an arc μ_C which is bordered by two scan-points. The scan-line defined by these scan-points is *close to* C. It is easy to see that the scan-line close to C realises S. Since blocks of a graph are characterised by the minimal separators that border them, we can realise blocks with scan-lines. More precisely, a block Ω is characterised by the connected components of $G \setminus \Omega$. The scan-lines close to these connected components define a *block-realiser* of Ω:

Definition 5 (Block-realiser). *A* realiser *of a block Ω is a family of scan-lines Λ such that:*

1. *Λ realises the minimal separators bordering Ω;*
2. *no two scan-lines of Λ cross;*
3. *there is a connected component $\Sigma \setminus \Lambda$ which is incident to all the scan-lines of Λ (the* domain *of Λ);*
4. *every minimal separator bordering Ω is realised by at least one scan-line of Λ.*

Figure 2 shows two realisers. The block-realiser we have just described is the *loose* realiser. In this realiser, distinct minimal separators correspond to distinct scan-lines. It may be possible to use less scan-lines. For example, if S is a subset of S', we only need to realise S'. By grouping some minimal separators under a common scan-line, we can hope to bound the number of scan-lines of a realiser. This would prove that there is a polynomial number of blocks.

More precisely, let $\mathcal{T}$ be a branch-decomposition of G and Ω a block of G associated with $\mathcal{T}$. Let v_Ω be a vertex of T corresponding to Ω and T_1^*, T_2^* and T_3^* be the three connected components of $T \setminus \{v_\Omega\}$.

Definition 6 (Respectful realiser). *A scan-line λ* respects *a branch T^* if $V(\lambda)$ is a subset of $V(T^*)$.*

A realiser Λ of Ω respects *the three branches T_i^* if it is a realiser and if all its scan-lines respect one of the branches.*

By construction, the loose realiser is respectful.

Proposition 2. *Let Ω be a block of a tightest branch-decomposition $\mathcal{T}$ of G and v_Ω and T_i^* be defined as above.*

A realiser Λ of Ω using as few scan-points as possible respecting T_i^* has at most three scan-lines.

Proof. Suppose for a contradiction that Λ has at least four scan-lines. At least two scan-lines λ_1 and λ_2 of Λ respect the same branch $T_{i_0}^*$. The ends of λ_1 and λ_2 define a chord λ_3 of the domain of Λ which respects $T_{i_0}^*$ (see Fig. 2). This chord partitions Λ in Λ_1 and Λ_2. If $V(\Lambda_1)$ is a subset of $V(\lambda_3)$, then $\Lambda_2 \cup \{\lambda_3\}$ is a realiser of Ω respecting T_i^* which uses strictly less scan-points than Λ which is absurd. For the same reason, $V(\Lambda_2)$ is not a subset of $V(\lambda_3)$.

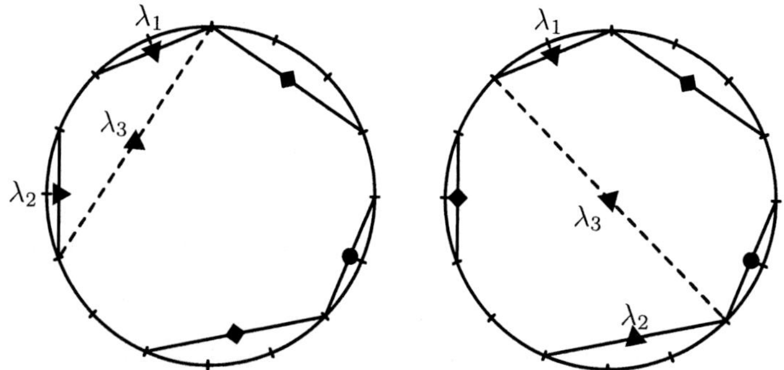

Two chords of a same branch have the same symbol on them.
In the first case, we can either reduce the size of the realiser or produce a strictly tighter decomposition than $\mathcal{T}$.
In the second case, we produce a strictly tighter decomposition than $\mathcal{T}$.

Fig. 2. Two realisers of a block and two scan-lines λ_3

Let $\mathcal{T}'$ be the decomposition $\mathcal{T}$ trimmed along T_1^*, T_2^* and T_3^*.

By rearranging the sub-branches of $T_i'^*$ corresponding the packs of T_i^*, we can build two branches $T_{i1}'^*$ and $T_{i2}'^*$ such that $V(T_{i1}'^*)$ contains $V(\Lambda_1)$ and $V(T_{i2}'^*)$ contains $V(\Lambda_2)$. By rearranging the three branches $T_{i1}'^*$ and the three branches $T_{i2}'^*$ in two branches $T_1''^*$ and $T_2''^*$ that we link, we can define a new branch-decomposition $\mathcal{T}''$ which is tighter than $\mathcal{T}$. Moreover, by construction $V(\lambda_1)$ and $V(\Lambda_2)$ are cliques of $H_{\mathcal{T}''}$ but Ω is not one because λ_3 separates two vertices of Ω. This implies that $\mathcal{T}''$ is strictly tighter than $\mathcal{T}$ which is absurd and finishes the proof. □

Since there are $O(n^2)$ scan-lines, Prop. 2 implies that there are at most $\mathcal{O}(n^6)$ blocks that can appear in a tightest branch-decomposition of G. Moreover, since the realiser of Ω gives the "good" three-partition of the minimal separators bordering Ω, we can compute bbw(Ω) in constant time using Th. 6. These last two results show that we can use Th. 5 to prove:

Theorem 7. *There is a polynomial time algorithm to compute the branch-width of a circular-arc graph.*

6 Conclusion and Open Problems

Theorem 5 can be used for any class of graphs. If we can bound the number of "interesting" blocks in a class of graphs $\mathcal{C}$ and if we can compute the block branch-width of these blocks, it shows that we can compute the branch-width of the graphs in $\mathcal{C}$ efficiently. This can easily be done with graphs of bounded asteroidal number with a polynomial number of minimal separators for which we can also compute the tree-width in polynomial time. The specific ideas used for the circular-arc graph rely on the existence of scan-lines that can realise minimal separators. A scan-line λ_3 using the ends points of two other scan-lines λ_1 and λ_2 must realise a subset of $V(\lambda_1) \cup V(\lambda_2)$. Such a notion exists for circular permutation graphs and more generally for d-trapezoid circular graphs.

The work we have conducted seems to show that the branch-width problem is more difficult that the tree-width problem. The only class we know for which this might not be the case is the class of planar graph. Otherwise, if we can compute the branch-width of a class of graphs, then we can compute the tree-width for this same class and with a more efficient algorithm. We feel that this is because tree-decompositions cannot decompose cliques whereas branch-decompositions can. Indeed, in our Th. 5, we not only need to be able to compute the blocks but we need to compute their block branch-width. This second point has no equivalent in the tree-width version of the algorithm. We feel that there could be some theorem stating that if we only use minimal separators, triangulations and blocks it is more difficult to compute branch-width than tree-width.

References

1. Robertson, N., Seymour, P.: Graphs minors. III. Planar tree-width. Journal of Combinatorial Theory Series B **36** (1984) 49–64
2. Bodlaender, H.: A partial k-arboretum of graphs with bounded treewidth. Theoretical computer science **209** (1998) 1–45
3. Robertson, N., Seymour, P.: Graphs minors. X. Obstruction to tree-decomposition. Journal of Combinatorial Theory Series B **52** (1991) 153–190
4. Seymour, P., Thomas, R.: Call routing and the ratcatcher. Combinatorica **14**(2) (1994) 217–241
5. Kloks, T., Kratochvíl, J., Müller, H.: New branchwidth territories. In: Proceedings 16th Annual Symposium on Theoretical Aspects of Computer Science (STACS'99). Volume 1563 of Lecture Notes in Computer Science., Springer-Verlag (1999) 173–183
6. Mazoit, F.: Décomposition algorithmique des graphes. PhD thesis, École Normale Supérieure de Lyon (2004)

7. Fomin, F., Mazoit, F., Todinca, I.: Computing branchwidth via efficient triangulations and blocks. In: Proceedings 31th International Workshop on Graphs, WG 2005. Volume 3787 of Lecture Notes in Computer Science., Springer-Verlag (2005) 374 – 384
8. Fomin, F., Kratsch, D., Todinca, I.: Exact (exponential) algorithms for treewidth and minimum fill-in. In: Proceedings 31th International Colloquium on Automata, Languages, and Programming (ICALP'04). Volume 3142 of Lecture Notes in Computer Science., Springer-Verlag (2004) 568–580
9. Gavril, F.: The intersection graphs of a path in a tree are exactly the chordal graphs. Journal of Combinatorial Theory **16** (1974) 47–56
10. Parra, A., Scheffler, P.: Characterizations and algorithmic applications of chordal graph embeddings. Discrete Applied Mathematics **79**(1-3) (1997) 171–188
11. Sundaram, R., Sher Singh, K., Pandu Rangan, C.: Treewidth of circular-arc graphs. SIAM Journal Discrete Mathematics **7** (1994) 647–655
12. McConnell, R.M.: Linear-time recognition of circular-arc graphs. Algorithmica **37** (2003) 93–147

Minimal Eulerian Circuit in a Labeled Digraph*

Eduardo Moreno and Martín Matamala

Departamento de Ingeniería Matemática, Centro de Modelamiento Matemático,
Universidad de Chile, UMR 2071, UCHILE-CNRS,
Casilla 170-3, Correo 3, Santiago, Chile
emoreno@dim.uchile.cl, mmatamal@dim.uchile.cl

Abstract. Let $G = (V, A)$ be an Eulerian directed graph with an arc-labeling. In this work we study the problem of finding an Eulerian circuit of lexicographically minimal label among all Eulerian circuits of the graph. We prove that this problem is NP-hard by showing a reduction from the DIRECTED-HAMILTONIAN-CIRCUIT problem.

If the labeling of the arcs is such that arcs going out from the same vertex have different labels, the problem can be solved in polynomial time. We present an algorithm to construct the unique Eulerian circuit of lexicographically minimal label starting at a fixed vertex. Our algorithm is a recursive greedy algorithm which runs in $\mathcal{O}(|A|)$ steps.

We also show an application of this algorithm to construct the minimal De Bruijn sequence of a language.

1 Introduction

Eulerian graphs were an important concept in the beginning of graph theory. The "Königsberg bridge problem" and its solution given by Euler in 1736 is considered the first paper of what is nowadays called *graph theory*.

In this work, we consider Eulerian digraphs with an arc-labeling into a finite alphabet, and we study the problem of finding the Eulerian circuit of lexicographically minimal label among all Eulerian circuits in the digraph.

By the BEST theorem (see [1]), we can compute the number of Eulerian circuits in a graph. This number is usually exponential in the number of vertices of the graph (at least $((\gamma - 1)!)^{|V|}$ where V is the set of vertices and γ is the minimum degree of vertices in V). Therefore, finding the Eulerian circuit of lexicographically minimal label can be costly.

This problem can be stated as a Chinese postman problem with a kind of priority over the streets: The postman must deliver mail in a network of streets and return to his depot without walking any street more than once (minimizing the walked distance) and at each corner he wants to choose the street of minimal slope. Therefore, the post office needs to give an itinerary to the postman such that at each corner he will choose the unvisited street of minimal slope unless it produces an unfeasible itinerary.

* Partially supported by Programa Iniciativa Científica Milenio P01-005 and Fundación Andes.

J.R. Correa, A. Hevia, and M. Kiwi (Eds.): LATIN 2006, LNCS 3887, pp. 737–744, 2006.

To find an Eulerian circuit of lexicographically minimal label is also interesting with respect to the problem of finding optimal encodings for DRAM address bus. In this model, an address space of size 2^{2n} is represented as labels of arcs in a complete digraph with 2^n vertices. An Eulerian circuit over this digraph produces an optimal multiplexed code (see [2]). If we want to give priority to some address in particular, an Eulerian circuit of lexicographically minimal label give us this code.

Eulerian digraphs with an arc-labeling are commonly employed in automata theory: a labeled digraph represents deterministic automata where vertices are the states of the automata, and arcs represent the transitions from one state to another, depending on the label of the arc. Eulerian circuits over these digraphs are related with synchronization of automata (see [3]).

Eulerian digraphs with an arc-labeling are also used in the study of DNA. By DNA sequencing we can obtain fragments of DNA which need to be assembled in the correct way. To solve this problem, we can simply construct a *DNA graph* (see [4]) and find an Eulerian circuit over this digraph. This strategy is already implemented and it is now one of the most promising algorithms for DNA sequencing (see [5, 6]).

Another interesting application of these digraphs is to find *De Bruijn sequences* of a language. De Bruijn sequences are also known as "shift register sequences" and were originally studied in [7] by N. G. De Bruijn for the binary alphabet. These sequences have many different applications, such as memory wheels in computers and other technological device, network models, DNA algorithms, pseudo-random number generation and modern public-key cryptographic schemes, to mention a few (see [8, 9, 10]). More details about this application are discussed in Section 3.

Note that these last applications consider digraphs with an arc-labeling with a particular property: Arcs going out from the same vertex have different labels.

In Section 2, we define the problem and we study its complexity. We prove that the problem is NP-hard. In Section 3 we study the problem when the arc-labeling has different labels for arcs going out from the same vertex. We show that in this case the problem can be polynomially solved: we give a recursive greedy algorithm that runs in linear time in the number of arcs of the digraph. Finally, in Section 4 we show an application of this algorithm to construct the minimal De Bruijn sequence of a language.

2 The Problem and Its Complexity

Let G be a digraph and let $l : A(G) \to N$ be a labeling of the arcs of G over an alphabet N such that arcs going out from the same vertex have different labels.

A *trail* is a sequence $T = a_1 a_2 \ldots a_k$ of arcs a_j such that the tail of a_i is the head of a_{i-1} for every $i = 1, 2, \ldots, k$ and all arcs are distinct. If the tail of a_1 is equal to the head of a_k then T is a closed trail or *circuit*. A circuit is an Eulerian circuit if the arcs of T are all the arcs of G. An Eulerian digraph is a digraph with an Eulerian circuit. The label of T, $l(T)$, is the word $l(a_1) \ldots l(a_k)$.

Our problem is the following: given an Eulerian digraph and a vertex r, we intend to find the Eulerian circuit starting in r with the lexicographically minimal label. We note that is important to fix a starting vertex r so as to define an order in which vertices are visited, which allows us to define a lexicographical order among Eulerian circuits.

First, we prove that this problem is NP-hard. We define the decision problem:

> MIN-LEX-EULERIAN-CIRCUIT
> *Instance*: An Eulerian digraph G, a labeling $l : A(G) \to N$ of its arcs, a starting vertex r and a word $X \in N^{|A(G)|}$.
> *Question*: Is there an Eulerian circuit T starting at r such that $l(T) \leq X$?

Theorem 1. MIN-LEX-EULERIAN-CIRCUIT *is NP-complete.*

Proof. We present a transformation of a DIRECTED-HAMILTONIAN-CIRCUIT instance (see [11]) into a MIN-LEX-EULERIAN-CIRCUIT instance, polynomially bounded in the size of the input graph.

Let G be a digraph. We want to verify if G contains a directed Hamiltonian circuit. We construct a digraph H in the following way: for each vertex $v \in G$, we include two vertices v_1 and v_2 and an arc v_1v_2 in H. Additionally, for each arc $vw \in G$ we include the arc v_2w_1 in H (see Figure 1). Finally, we label all arcs in H with the label 0.

It is easy to see that G has a Hamiltonian circuit if and only if H has a circuit with label $0^{2|V(G)|}$.

We can complete the digraph H to an Eulerian digraph $\bar{H}$ with additional vertices and arcs with label 1 in the following way: we add a vertex c and we

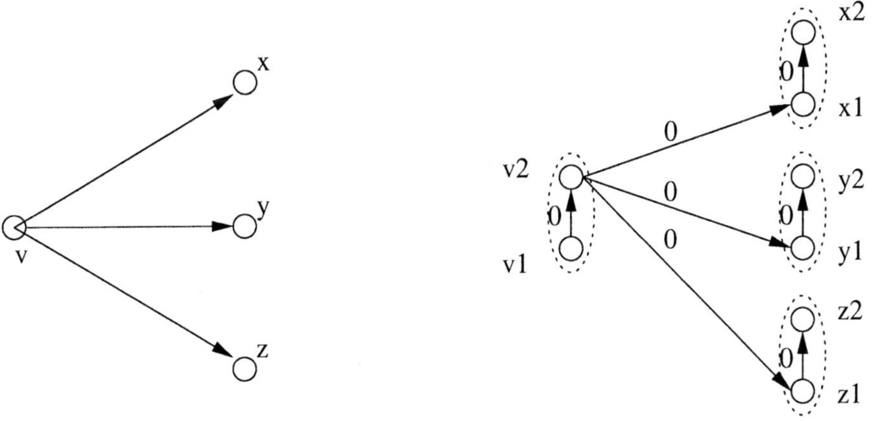

Fig. 1. Transformation of a digraph G (DIRECTED-HAMILTONIAN-CIRCUIT instance) into a labeled digraph H (MIN-LEX-EULERIAN-CIRCUIT instance)

connect every vertex v in H to c with two arcs vc and cv of label 1. With these connections, the resulting digraph is strongly connected even if we remove all arcs in H. Finally, to each arc xy in H we add an arc yx with label 1. These arcs provide the equality between the in-degree and the out-degree of each vertex in $\bar{H}$. Hence, the resulting digraph is Eulerian. Moreover, if G has a Hamiltonian circuit then we can remove the arcs of its associated circuit of label $0^{2|V(G)|}$ in $\bar{H}$ and the remaining graph is still Eulerian.

Therefore, G has a Hamiltonian circuit if and only if $\bar{H}$ has an Eulerian circuit starting at any vertex $r \in H$ with label smaller or equal to $0^{2|V(G)|}1^{|A(\bar{H})|-2|V(G)|}$.

3 A Linear Algorithm

In this section we assume that the arc-labeling gives a different label to each arc going out from the same vertex. We note that if we fix an initial vertex r, then there is a bijection between the trails starting at r and its labels.

We define the following greedy strategy to construct a circuit: Starting at a given vertex r, follow the unvisited arc (if exists) of minimal label. This strategy finishes with a trail, and this trail exhausts the vertex r. A trail constructed by this strategy is called an *alphabetic trail* starting at r.

Let U be a subset of vertices in G. A *cut* defined by U is the set of arcs with one end in U and the other in $V(G)\setminus U$, and is denoted by $\delta_G(U)$. For simplicity, for a trail T we write $\delta_G(T)$ instead of $\delta_G(V(T))$, where $V(T)$ is the set of the tail and head vertices of the arcs in T.

A vertex v is *exhausted* by a trail T if $\delta_{G\setminus A(T)}(v) = \emptyset$. We note that an alphabetic trail starting at r is the trail of lexicographically minimal label among all trails starting at r and exhausting r. We denote by $LastNotEx(T)$ the last vertex visited by T among all vertices not exhausted by T.

Let $T = e_1 \dots e_M$ be a trail and let e_i be an arc in the trail T. We denote by Te_i the subtrail $e_1 \dots e_i$, by e_iT the subtrail $e_i \dots e_M$ and by e_iTe_j the subtrail $e_i \dots e_j$ for $i < j$.

Lemma 2. *Let T be a circuit starting at r and exhausting r and let $v = LastNotEx(T)$. If e_i is the arc in T after the last visit to v then*

$$\delta_{G\setminus A(Te_{i-1})}(e_{i+1}T) = \{e_i\}$$

Proof. Let e be an arc of $\delta_G(e_{i+1}T)$. Since all vertices of $e_{i+1}T$ are exhausted by T, $e \in T$. Hence either $e \in Te_{i-1}$ or $e \in e_iT$. Therefore $e \in \delta_{G\setminus A(Te_{i-1})}(e_{i+1}T)$ if and only if $e = e_i$.

We note that these properties are valid for any trail starting at r and exhausting r, it does not need to be an Eulerian circuit.

For a trail $T = e_1 \dots e_M$ over G, we define a *failure* of T as a pair of arcs $e_i = vw, e_j = vx$ in the trail such that $i < j$ but $l(e_i) > l(e_j)$ and such that $\forall k < i$ with $e_k = vy$, $l(e_k) < l(e_j)$. The vertex v is called a *failure vertex.* Note that an alphabetic trail is a trail with no failures.

Our strategy to construct the Eulerian circuit starting at r of lexicographically minimal label is the following: Start at r with an alphabetic trail T_0. Let $v = LastNotEx(T_0)$ and let e_i be the arc in T after the last visit to v. Start at v an alphabetic trail T_1 over $G \setminus A(T_0)$. If the trail T_1 exhaust all its vertices, merge both trails obtaining $T_2 = (T_0 e_{i-1}) T_1 (e_i T_0)$ and repeat the process over $LastNotEx(T_2)$. If T_1 does not exhaust all its vertices, repeat the strategy recursively.

We note that if T_1 exhausts all its vertices, the trail $T_2 = (T_0 e_{i-1}) T_1 (e_i T_0)$ is the trail of minimal label exhausting r and having one failure.

This strategy can be stated as in Algorithm MinLex. Note that we include a global counter s and a bound $MaxSteps$ in order to count the number of failures of the resulting trail.

MinLex(A,r) : Compute the lexicographically minimal Eulerian circuit starting at r on $(V(G), A)$

Require: A an arc-subset of $A(G)$, r a vertex of G.
$s \leftarrow s + 1$
$T \leftarrow$ AlphabeticTrail(A, r)
while $NotEx(T) \neq \emptyset$ and $s \leq MaxSteps$
    $v \leftarrow LastNotEx(T)$
    $e_i \leftarrow$ the arc in T after the last visit to v
    $T \leftarrow (T e_{i-1})(\text{MinLex}(A \setminus A(T), v))(e_i T)$
end while
Return: T

Where AlphabeticTrail(A, r) returns the alphabetic trail starting at r over $(V(G), A)$, $NotEx(T)$ is the set of not exhausted vertices in T, $MaxSteps$ is a fixed integer and s is a global counter initialized in 0.

In order to prove the correctness of our algorithm, we define O^k as the circuit of minimal label starting at r, having k failures and exhausting r and its failures vertices. In the following, we will define the trail O^{k-1} in terms of O^k.

First, we study the position of failures over O^k. The following lemma can be proved:

Lemma 3. *Let $\langle e_i, e_j \rangle$ and $\langle e_{i'}, e_{j'} \rangle$ be two different failures of O^k. Then it is not possible that $e_{i'} \in e_i O^k e_j$ and $e_{j'} \in e_j O^k$.*

The previous lemma state that two failures of O^k are either nested or in different subtrails of O^k. A failure $\langle e_i, e_j \rangle$ will be called simplicial if and only if there is not another failure in $e_i O^k e_j$. Therefore, there exist simplicial failures of O^k.

Let $\langle e_i, e_j \rangle$ be the first simplicial failure of O^k. We define $\tilde{O} = (O^k e_{i-1})(e_j O^k)$. We will prove that $\tilde{O} = O^{k-1}$. In order to prove this equality, we will prove some intermediate lemmas. All the proofs have the same idea: if the statement of the lemma is not fulfilled, then we can construct a trail with k failures with a label smaller than the label of O^k.

Lemma 4. *Let $\langle e_i, e_j \rangle$ be the first simplicial failure of O^k and let v be the vertex $LastNotEx(O^k)$. Then v is last visited in $O^k e_j$.*

Proof. (sketch) By contradiction, suppose that v is last visited by an arc $e_m \in e_{j+1}O^k$ and let W be the alphabetic trail starting at v over $G \setminus A(\tilde{O})$. Hence, the trail $(\tilde{O}e_{m-1})(W)(e_m\tilde{O})$ is a trail with k failures and a label smaller than the label of O^k, which is a contradiction.

Corollary 5. *Let $\langle e_i, e_j \rangle$ be the first simplicial failure of O^k and let v be the vertex $LastNotEx(\tilde{O})$. Then v is the failure vertex of $\langle e_i, e_j \rangle$.*

Proof. By contradiction, let x be a vertex in $e_j\tilde{O}$ not exhausted by $\tilde{O}$ and let e_n be the last visit of $\tilde{O}$ to x. By Lemma 4, the vertex x is exhausted by O^k, then there exists an arc $e_m \in e_iO^ke_j$ such that x is the head of e_{m-1} and the tail of e_m.

Therefore, the trail $(\tilde{O}e_{n-1})(e_mO^ke_{j-1})(e_iO^ke_{m-1})(e_n\tilde{O})$ uses the same arcs that O^k but its label is smaller than the label of O^k.

Lemma 6. *The trail $\tilde{O}$ is the trail O^{k-1}.*

Proof. (sketch) The trail $\tilde{O}$ starts and exhausts the vertex r and it has $k-1$ failures. By Corollary 5, $\tilde{O}$ exhausts all its failure. We only need to compare the label of $\tilde{O}$ with the label of O^{k-1}.

If the label of O^{k-1} is smaller than the label of $\tilde{O}$ then the arc producing this difference needs to be in a failure of $\tilde{O}$. Its corresponding failure vertex cannot occur before v, then $\tilde{O}e_{i-1} = O^{k-1}e_{i-1} = O^ke_{i-1}$.

Let $x = LastNotEx(O^{k-1})$. If x is last visited after e_j then we can merge an alphabetic trail starting at x over $G \setminus A(O^{k-1})$ and we obtain a trail with k failures and with a label smaller than the label of O^k, therefore $LastNotEx(O^{k-1}) = v$.

By Lemma 2, there is no arc between the vertices of subtrails $e_iO^ke_{j-1}$ and e_jO^{k-1}. Hence, if the label of e_jO^{k-1} is smaller than the label of $e_j\tilde{O}$, the trail $(O^{k-1}e_{i-1})(e_iO^ke_{j-1})(e_jO^{k-1})$ has k failures and a label smaller than the label of O^k. Therefore, $\tilde{O} = O^{k-1}$.

Now we are ready to prove the correctness of our algorithm.

Theorem 7. *The trail O^k is the trail obtained by the* MinLex*$(A(G), r)$ algorithm after k steps ($MaxSteps = k$).*

Proof. We prove this result by induction on the number of steps.

For $k = 0$, the algorithm produces the alphabetic trail starting at r, which is the minimal trail exhausting r with no failures.

Now we assume that $k > 0$. By Lemma 6, O^k is equal to merge O^{k-1} with the alphabetic trail over $G \setminus A(O^{k-1})$ starting at v, where $v = LastNotEx(O^{k-1})$. By induction hypothesis, O^{k-1} is obtained by the algorithm after $k-1$ steps. Moreover, the next step in the algorithm is an alphabetic trail starting at v over $G \setminus A(O^{k-1})$. Therefore, the resulting trail of the algorithm after k steps is exactly O^k.

Corollary 8. *The algorithm* MINLEX*($A(G), r$) with a sufficient large integer $MaxSteps$ finishes with the Eulerian circuit of minimal label starting at* r

The complexity of our algorithm is linear in $|A(G)|$. We can use an adjacency list (see [12]) to represent the digraph, where each vertex v has a list with the head of each arc starting at v in the alphabetical order of its labels. Knowing this structure of a digraph, the algorithm can easily construct the alphabetic trails over $G \setminus O^i$ for every i, removing the visited arcs from the list and keeping track of exhausted vertices. Since this algorithm visits each arc at most twice, it can be implemented in $\mathcal{O}(|A(G)|)$, which is best possible.

4 An Application: Minimal De Bruijn Sequence

Given a set $\mathcal{D}$ of words of length n, a De Bruijn sequence of span n is a periodic sequence B such that every word in $\mathcal{D}$ (and no other n-tuple) appears exactly once in B. Historically, De Bruijn sequence was studied in an arbitrary alphabet considering the language of all the n-tuples. In [13] the concept of De Bruijn sequences was generalized to restricted languages with a finite set of forbidden substrings and it was proved the existence of these sequences and presented an algorithm to generate one of them. Nevertheless, it remained to find the minimal De Bruijn sequence in this general case.

In [14] was studied some particular cases where an alphabetic trail obtains the minimal De Bruijn sequence. Using Algorithm MINLEX we can solve this problem efficiently in all cases.

A word p is said to be a *factor* of a word w if there exist words $u, v \in N^*$ such that $w = upv$. If u is the empty word (denoted by ε), then p is called a *prefix* of w, and if v is empty then is called a *suffix* of w.

Let $\mathcal{D}$ be a set of words of length $n+1$. We call this set a *dictionary.* A *De Bruijn sequence of span* $n+1$ for $\mathcal{D}$ is a (circular) word $B^{\mathcal{D},n+1}$ of length $|\mathcal{D}|$ such that all the words in $\mathcal{D}$ are factors of $B^{\mathcal{D},n+1}$. In other words,

$$\{(B^{\mathcal{D},n+1})_i \dots (B^{\mathcal{D},n+1})_{i+n \bmod (|\mathcal{D}|)} | i = 0 \dots |\mathcal{D}| - 1\} = \mathcal{D}$$

De Bruijn sequences are closely related to De Bruijn digraphs. The *De Bruijn graph of span* n, denoted by $G^{\mathcal{D},n}$, is the digraph with vertex set

$$V(G^{\mathcal{D},n}) = \{u \in N^n | u \text{ is a prefix or a suffix of a word in } \mathcal{D}\}$$

and arc set

$$A(G^{\mathcal{D},n}) = \{(\alpha v, v\beta) | \alpha, \beta \in N, \alpha v \beta \in \mathcal{D}\}$$

Note that the original definitions of De Bruijn sequences and De Bruijn graph given in [7] are the particular case of $\mathcal{D} = N^{n+1}$.

We label the arcs of the digraph $G^{\mathcal{D},n}$ using the following function l: if $e = (\alpha u, u\beta)$ then $l(e) = \beta$. This labeling has an interesting property: Let $T = v_0 e_0 \dots e_m v_{m+1}$ be a trail over $G^{\mathcal{D},n}$ of length $m \geq n$. Then T finishes in a vertex u if and only if u is a suffix of $l(T) = l(e_0) \dots l(e_m)$. This property explains the relation between De Bruijn graphs and De Bruijn sequence: $B^{\mathcal{D},n+1}$ is the label of an Eulerian circuit of $G^{\mathcal{D},n}$. Therefore, given a dictionary $\mathcal{D}$, the existence of a De Bruijn sequence of span $n+1$ is characterized by the existence of an Eulerian circuit on $G^{\mathcal{D},n}$.

Let D be a dictionary such that $G^{\mathcal{D},n}$ is an Eulerian digraph. Let z be the vertex of minimum label among all vertices. Clearly, the minimal De Bruijn sequence has z as prefix. Hence, the minimal Eulerian circuit on $G^{\mathcal{D},n}$ starts at an (unknown) vertex and after n steps it arrives to z. Therefore if we start our Algorithm MinLex in the vertex z we obtain the Eulerian circuit of minimal label starting at z which have label $B = B' \cdot z$. Hence $z \cdot B'$ is the minimal De Bruijn sequence of span $n+1$ for $\mathcal{D}$.

References

1. Tutte, W.T.: Graph theory. Volume 21 of Encyclopedia of Mathematics and its Applications. Addison-Wesley Publishing Company Advanced Book Program, Reading, MA (1984)
2. Cheng, W.C., Pedram, M.: Power-optimal encoding fod DRAM address bus. In: ISLPED, ACM (2000) 250–252
3. Kari, J.: Synchronizing finite automata on Eulerian digraphs. Theoret. Comput. Sci. **295**(1-3) (2003) 223–232 Mathematical foundations of computer science (Mariánské Lázně, 2001).
4. Blazewicz, J., Hertz, A., Kobler, D., de Werra, D.: On some properties of DNA graphs. Discrete Appl. Math. **98**(1-2) (1999) 1–19
5. Pevzner, P.A.: L-tuple DNA sequencing: computer analysis. J. Biomol. Struct. Dyn. **7** (1989) 63–73
6. Pevzner, P.A., Tang, H., Waterman, M.S.: An eulerian path approach to DNA fragment assembly. Proceedings of the National Academy of Sciences **98**(17) (2001) 9748–9753
7. de Bruijn, N.G.: A combinatorial problem. Nederl. Akad. Wetensch., Proc. **49** (1946) 758–764
8. Stein, S.K.: The mathematician as an explorer. Sci. Amer. **204**(5) (1961) 148–158
9. Bermond, J.C., Dawes, R.W., Ergincan, F.Ö.: De Bruijn and Kautz bus networks. Networks **30**(3) (1997) 205–218
10. Chung, F., Diaconis, P., Graham, R.: Universal cycles for combinatorial structures. Discrete Math. **110**(1-3) (1992) 43–59
11. Garey, M.R., Johnson, D.S.: Computers and intractability. W. H. Freeman and Co., San Francisco, Calif. (1979) A guide to the theory of NP-completeness, A Series of Books in the Mathematical Sciences.
12. Gibbons, A.: Algorithmic graph theory. Cambridge University Press, Cambridge (1985)
13. Moreno, E.: De Bruijn sequences and de Bruijn graphs for a general language. Inf. Process. Lett. **96** (2005) 214–219
14. Moreno, E., Matamala, M.: Minimal de Bruijn sequence in a language with forbidden subtrings. In Hromkovic, J., Nagl, M., Westfechtel, B., eds.: Graph-Theoretic Concepts in Computer Science. Volume 3353 of Lect. Notes in Comp. Sci., Springer-Verlag Heidelberg (2004) 168–176

Speeding up Approximation Algorithms for NP-Hard Spanning Forest Problems by Multi-objective Optimization

Frank Neumann[1] and Marco Laumanns[2]

[1] Institute of Computer Science,
Christian-Albrechts-Univ. of Kiel, 24098 Kiel, Germany
[2] Institute for Operations Research,
ETH Zürich, CH-8092 Zürich, Switzerland
fne@informatik.uni-kiel.de, laumanns@ifor.math.ethz.ch

Abstract. We give faster approximation algorithms for the generalization of two NP-hard spanning tree problems. First, we investigate the problem of minimizing the degree of minimum spanning forests. Fischer [3] has shown how to compute a minimum spanning tree of degree at most $b \cdot \Delta^* + \lceil \log_b n \rceil$ in time $O(n^{4+1/\ln b})$ for any $b > 1$, where Δ^* is the value of an optimal solution. We model our generalization as a multi-objective optimization problem and give a deterministic algorithm that computes for each number of connected components a solution with the same approximation quality as the algorithm of Fischer and runs in time $O(n^{3+1/\ln b})$. After that, we take a multi-objective view on the problem of computing minimum spanning trees with nonuniform degree bounds, which has been examined by Könemann and Ravi [7]. Given degree bounds B_v for each vertex $v \in V$, we construct an algorithm that computes for each number of connected components a spanning forest in which each vertex v has degree $O(B_v + \log n)$ and whose weight is at most a constant times the weight of a minimum spanning forest obeying the degree bounds. The total runtime of our algorithm is $O(n^{3+2/\ln b})$ for an arbitrary constant $b > 1$. Setting $b = e^k$, $k > 2/3$ an arbitrary constant, the runtime is by a factor $n^{3-2/k} \log n$ less than the given bound by Könemann and Ravi.

1 Introduction

Many problems in computer science ask for solutions with certain attributes or properties that can be expressed as functions mapping potential solutions to scalar numeric values. The usual optimization approach is to take these attributes as constraints to determine the feasibility of a solution, while one of them is chosen as the objective function to determine the preference order of the feasible solutions. In the minimum spanning tree problem, for example, constraints are imposed on the number of connected components (one) and the number of cycles (zero) of the chosen subgraph, while the total weight of its edges is the objective to be minimized. A more general approach is multi-objective optimization, where several attributes are employed as objective functions and used to define a partial preference order of the solutions, with respect to which the set of minimal (maximal) elements is sought (see, e.g., Ehrgott [2]

J.R. Correa, A. Hevia, and M. Kiwi (Eds.): LATIN 2006, LNCS 3887, pp. 745–756, 2006.

or Papadimitriou and Yannakakis [9]). Most of the best known single-objective polynomial solvable problems like shortest path or minimum spanning tree become NP-hard when at least two weight functions have to be optimized at the same time. Moreover, the set of minimal elements always contains the solution of the corresponding constrained single-objective problem. In this sense, multi-objective optimization is considered (at least) as difficult as single-objective optimization.

The question arises whether working in this more general framework can lead to better understanding of the problem and sometimes also to the design of more efficient algorithms. This is indeed the case for the average case analysis of a well-known algorithm for the 0/1 knapsack problem. Beier and Vöcking [1] have considered different input distributions for this problem and shown that the number of minimal elements in objective space is polynomially bounded. This implies that the well-known Nemhauser and Ullmann algorithm has an expected polynomial runtime for these distributions. Neumann and Wegener [8] have analyzed the runtime of randomized search heuristics for minimum spanning trees. Their results show that randomized search heuristics find minimum spanning trees easier in a multi-objective model than in a single-objective one. A further of a successful multi-objective approach is to obtain more information (a set of minimal elements instead of only one specific element in it) with the same or — as for the example of this paper — even less computational effort.

In this paper we consider two NP-hard spanning forest problems and use a multi-objective formulation to obtain faster approximation algorithms. Given a connected graph $G = (V, E)$ with n vertices and m edges and positive integer weights $w(e)$ for each edge $e \in E$, we are searching (i) for minimum spanning forests of minimum degree, and (ii) for minimum spanning forests obeying given degree bounds on the vertices. A forest with i connected components is a cycle-free subgraph of G that contains exactly $n - i$ edges. A minimum spanning forest with i connected components is a forest where the sum over all edge weights is minimal among all spanning forests with i connected components. In a minimum spanning forest of minimum degree, the largest vertex degree is as small as possible. This generalizes the problem of computing minimum spanning trees of minimum degree. For our algorithms it is, in contrast to previous work, not necessary to assume the graph to be connected. For simplicity we work under this assumption, but in the case that G is not connected our algorithms would produce solutions for each possible value of i. Note that solutions for the different number of connected components may be of additional interest when each spanning tree has a weight that is not acceptable in practical applications such that the graph has to be partitioned into different clusters. Having a solution for each number of connected components the designer of a network can decide how to build these clusters.

1.1 Previous Work and Our Results

Let Δ^* be the degree of an optimal solution. When edge weights are not considered, or assumed uniform, a $\Delta^* + 1$ approximation algorithm for minimizing the degree of spanning trees has been obtained by Fürer and Raghavachari [6]. For the weighted case, Fischer [3] has presented an approximation algorithm that computes a minimum spanning tree of degree at most $b \cdot \Delta^* + \lceil \log_b n \rceil$ in time $O(n^{4+1/\ln b})$ for any $b > 1$, which is the best-known algorithm for this problem up to now. His algorithm is an adaptation of a local search algorithm of Fürer and Raghavachari [5] to the weighted case. The idea of the local search is to perform edge exchanges until the spanning tree is locally optimal.

We model the problem of computing minimum spanning forests of minimum degree as a multi-objective optimization problem where one objective is to minimize the number of connected components and the other to minimize the weight and degree. Our aim is to compute for each i, $1 \leq i \leq n$, a minimum spanning forest with i connected components with the same approximation quality as the algorithm of Fischer. Our algorithm can be seen as extension of Kruskal's algorithm for the computation of minimum spanning trees and runs in time $O(n^{3+1/\ln b})$. Note that during the run, Kruskal's algorithm computes solutions that are minimum spanning forests for each possible number of connected components. The working principle of our algorithm is also to start with an empty graph and to compute the minimum spanning forests as in the run of Kruskal's algorithm one after another. After inclusion of a new edge that leads to a minimum spanning forest with smaller number of connected components, its degree is improved by edge exchanges as long as we cannot guarantee our desired approximation quality.

Könemann and Ravi [7] have considered the problem of approximating minimum spanning trees with nonuniform degree bounds. Given degree bounds B_v for all vertices, they have presented an algorithm that runs in time $O(n^6 \log n)$ to compute a spanning tree in which the degree of each vertex v is $O(B_v + \log n)$ and the weight is by at most a constant factor higher than the weight of any spanning tree that obeys the given degree constraints, if such a tree exists. Note, that the question whether there exists a spanning tree obeying the given degree bounds is already NP-complete.

Starting with the empty edge set, we compute in time $O(n^{3+2/\ln b})$ for each number of connected components a minimum spanning forest with the same approximation quality given by Könemann and Ravi [7]. For $b > 1$ a large constant, the runtime of our algorithm approaches the upper bound $O(n^3)$. An additional advantage of our algorithm is that we do not need the assumption that there is a spanning tree obeying the given degree bounds. If this is not the case our algorithm will produce a set of solutions that contains for each i where a spanning forest exists that respects the degree constraints a solution that has the stated approximation quality.

The paper is organized as follows. In Section 2, we introduce our model for the computation of minimum spanning forests with minimum degree and give a new algorithm for minimizing the degree of minimum spanning forests that runs in time $O(n^{3+1/\ln b})$. Section 3 applies our technique in combination with an extension of the primal-dual method for minimum spanning trees [7] to the problem of computing minimum spanning forests with nonuniform degree bounds. We finish with some conclusions.

2 Minimizing the Degree of Minimum Spanning Forests

2.1 A Multi-objective Formulation

We take a multi-objective view on the computation of minimum spanning forests with minimum degree. Let $X = \{0,1\}^m$ be the search space. A search point $x \in X$ describes the set of all edges e_i where $x_i = 1$ holds. Let $c(x)$ be the number of connected components of the solution x, $w(x)$ be the weight of the chosen edges, $d_j(x)$

be the number of vertices of degree j in x, and $\Delta(x)$ the maximum vertex degree of x. The objective function value of x is given by the vector $f(x) = (f_1(x), f_2(x))$, where $f_1(x) = c(x)$ and $f_2(x) = (w(x), d_{n-1}(x), \ldots, d_0(x))$. Both objectives f_1 and f_2 have to be minimized. Minimizing the second objective means minimization with respect to the lexicographic order. This model generalizes the function $g(x) = (c(x), w(x))$ that has been examined by Neumann and Wegener [8] for the computation of minimum spanning trees by randomized search heuristics. They have shown that a multi-objective model leads to a more efficient optimization process than in the case of a single objective.

Let $f(X)$ be the image of the search space under the objective function f defined above. By intersecting the canonic order on $f_1(X)$ with the lexicographic order on $f_2(X)$, both of which are total orders, a partial order on $f(X)$ can be defined as

$$f(x) \preceq f(x') :\Leftrightarrow f_1(x) \leq f_1(x') \wedge f_2(x) \leq_{\text{lex}} f_2(x') \tag{1}$$

for all $x, x' \in X$. This partial order represents our preference relation regarding the solutions. The aim is to identify all minimal elements of $(f(X), \preceq)$, and with each minimal element one of its pre-images from X.

As the edge weights are positive, a minimum spanning forest with $i+1$ connected components has a smaller weight than a minimum spanning forest with i connected components. Therefore, $(f(X), \preceq)$ has n minimal elements, representing for each i, $1 \leq i \leq n$, a minimum spanning forest with i connected components and minimum degree. Our goal is to approximate the set of minimal elements as good as possible. We want to compute for each i a minimum spanning forest with i connected components that has degree at most $b \cdot \Delta_i^* + \lceil \log_b n \rceil$, where Δ_i^* is the smallest maximum degree of any minimum spanning forest with i connected components.

2.2 Local Improvements

Fischer's algorithm [3] for the computation of a minimum spanning tree with degree at most $b \cdot \Delta^* + \lceil \log_b n \rceil$ starts with an arbitrary minimum spanning tree and improves the degree of high-degree vertices. The number of these improvements is bounded by $O(n^{2+1/\ln b})$. A better bound on the number of necessary improvements would yield a better upper bound for the runtime of Fischer's algorithm. We consider the number of necessary improvements for our multi-objective model and start with some general properties of minimum spanning forests with i connected components.

Lemma 1. *Let Δ_i^*, $1 \leq i \leq n$, be the minimum degree of a minimum spanning forest with i connected components. Then $\Delta_n^* \leq \Delta_{n-1}^* \leq \ldots \leq \Delta_1^*$ holds.*

Let s_i be a solution with i connected components and minimal weight. We call s_i locally optimal if there is no solution s_i' with $c(s_i') = c(s_i)$ and hamming distance 2 that is better than s_i with respect to $f_2(s_i)$ when disregarding all $d_j(s_i)$ with $j < \Delta(s_i) - \lceil \log_b n \rceil$. Otherwise, we say that s_i' improves s_i. For the case $i = 1$ Fischer has shown in [3] that if there is no improvement for s_1 then the minimum spanning tree has already degree at most $b \cdot \Delta_1^* + \lceil \log_b n \rceil$ for any $b > 1$. We generalize this approximation guarantee of local optimal minimum spanning trees given by Fischer to locally optimal minimum spanning forests.

Lemma 2. *Let s_i be a solution that is locally optimal and Δ_i be its maximum degree. Then $\Delta_i \leq b \cdot \Delta_i^* + \lceil \log_b n \rceil$ holds for any constant $b > 1$.*

Lemma 3. *The total number of local improvements until a minimum spanning forest of degree at most $b \cdot \Delta_i^* + \lceil \log_b n \rceil$ has been computed for each i, $1 \leq i \leq n$, can be bounded by $O(n^{1+1/\ln b})$.*

Proof. Consider a situation where a minimum spanning forest with j connected components and degree at most $b \cdot \Delta_j^* + \lceil \log_b n \rceil$ has been computed for each $j, i \leq j \leq n$. We want to show that no more than $3 \cdot (n-i) \cdot \mu$ local improvements are necessary to reach this state, where $\mu = O(n^{1/\ln b})$. Setting $i = 1$ then proves the lemma.

Let a potential function be defined as $p(s) := \sum_{j=0}^{\lceil \log_b n \rceil + 1} d_{r+j}(s) \cdot e^j$, where $r = \max\{\Delta(s_i) - \lceil \log_b n \rceil, 0\}$. The empty edge set $s_n = \emptyset$ is obviously a minimum spanning forest with n components and minimum degree, and no improvements are necessary to reach this. In addition $p(s_n) = 0$ holds. For going from i to $i-1$ we introduce into s_i a lightest edge e that does not create a cycle. This yields a minimum spanning forest with $i-1$ components, denoted as s'_{i-1}. Introducing an arbitrary edge into s_i increases the potential value $p(s_i)$ by at most $2e^{\lceil \log_b n \rceil + 1} \leq 2e^2 \cdot e^{\log_b n} = 2e^2 \cdot e^{\ln(n)/\ln(b)} =: \mu$, hence $p(s'_{i-1}) - p(s_i) \leq \mu$, where $\mu = O(n^{1/\ln b})$. Now, s'_{i-1} can undergo a number ν_{i-1} of local improvements to arrive at a new solution s_{i-1} that is locally optimal or satisfies $\Delta(s_{i-1}) = \Delta(s_i)$, which can be achieved by reducing the whole potential $p(s'_{i-1})$. Due to Lemma 1 and 2 in both cases the claimed approximation holds.

In a local improvement $d_k(s)$ decreases by at least 1 for some $k \geq r+2$. The potential reduces by the smallest amount if $d_k(s)$ reduces by one, $d_{k-1}(s)$ increases by three and $d_{k-2}(s)$ decreases by 2. This means that one local improvement step reduces the potential by at least $e^2 - 3e + 2 > 1/3$, i.e., a constant amount.

Therefore, and because r cannot increase in this process, the relation $p(s_i) \leq p(s_{i+1}) - \nu_{i+1}/3 + \mu$ holds for each i, $1 \leq i \leq n-1$. Using this, the potential value of a solution s'_i that has been created by the introduction of a new edge into s_{i+1} can be bounded with respect to the cumulated number of all ν_j, $i < j \leq n$, previous improvement steps by $p(s'_i) \leq (n-i)\mu - \sum_{j=i+1}^{n} \nu_j/3$. As $\nu_i \leq 3p(s'_i)$, $\sum_{j=i}^{n} \nu_j \leq 3\mu(n-i)$. □

2.3 The Algorithm

The analysis in Section 2.2 has shown that the number of improvements in the multi-objective model can be bounded by $O(n^{1+1/\ln b})$, which is by a factor n smaller than then the number of improvements in the algorithm of Fischer. Based on this observation we give a deterministic algorithm that computes for each i a minimum spanning forest with i connected components and degree at most $b \cdot \Delta_i^* + \lceil \log_b n \rceil$ in time $O(n^{3+1/\ln b})$ for any $b > 1$.

Let s_i be a minimum spanning forest of degree at most $b \cdot \Delta_i^* + \lceil \log_b n \rceil$. Then we can produce a minimum spanning forest s_{i-1} with $i-1$ connected components by introducing a lightest edge that does not create a cycle. If $\Delta(s_{i-1}) = \Delta(s_i)$ holds, s_{i-1} has the desired approximation quality. Otherwise we have to improve s_{i-1}. The pseudo-code of our algorithm called Minimum Spanning Forest Optimizer (MSFO) is given in Fig. 1.

1. Initialize: let $i := n$, $s_i := \emptyset$, $S := \{s_i\}$.
2. Create s_{i-1} by introducing into s_i the lightest edge that does not create a cycle.
3. Improve the solution s_{i-1} until it is locally optimal or $\Delta(s_{i-1}) = \Delta(s_i)$ holds.
4. $S := S \cup \{s_{i-1}\}$.
5. $i := i - 1$.
6. If $i > 1$ continue at 2., otherwise output S and stop.

Fig. 1. Minimum Spanning Forest Optimizer (MSFO)

MSFO can be seen as a variant of Kruskal's algorithm where after each insertion of an edge the degree of the current solution s_{i-1} is improved as long as we cannot guarantee the desired approximation quality. The algorithm of Kruskal can be implemented in time $O((m+n)\log n)$. Hence, to bound the runtime of MSFO it is necessary to bound the number of local improvements (as done in Section 2.2) and the time to achieve such an improvement.

Lemma 4. *Let s_i be a solution with i connected components that is not locally optimal. Then an improvement can be found in time $O(n^2)$.*

Using the bound on the number of necessary improvements and the time bound to achieve such an improvement, we can give an upper bound on the runtime of MSFO.

Theorem 1. *The algorithm MSFO computes for any $b > 1$ in time $O(n^{3+1/\ln b})$ a set of solutions that includes for each i, $1 \leq i \leq n$, a minimum spanning forest with i connected components and degree at most $b \cdot \Delta_i^* + \lceil \log_b n \rceil$.*

Proof. Consider the time the solutions $\{s_n, s_{n-1}, \ldots, s_i\} \subset S$ have been produced. These solutions have the following properties. Each s_j, $i \leq j \leq n$, is a minimum spanning forest with j connected components. In addition, s_j is locally optimal or $\Delta(s_j) = \Delta(s_{j+1})$ holds. Obviously, s_n is a locally optimal solution. We introduce into s_i an edge e of minimal weight that does not create a cycle. This can be easily done by checking each remaining edge in time $O(m)$. Note that the whole computation in step 2 in the run of the algorithm can be implemented in time $O((m+n)\log n)$ using the ideas of Kruskal's algorithm. After step 3, the solution s_{i-1} has minimal weight among all solutions with $i-1$ components. If e is not incident to at least one edge of degree $\Delta(s_i)$, $\Delta(s_{i-1}) = \Delta(s_i)$ holds. Otherwise, the number of vertices with degree $\Delta(s_i)+1$ is at most 2 and we have to improve s_{i-1} to reach a locally optimal solution or to achieve $\Delta(s_{i-1}) = \Delta(s_i)$.

The number of local improvements in the run of MSFO is bounded by $O(n^{1+1/\ln b})$ as shown in Lemma 3 and an improvement of a non locally optimal solution can be found in time $O(n^2)$ due to Lemma 4. Hence, the time until MSFO has achieved the desired approximation can be bounded by $O(n^{3+1/\ln b})$. □

3 Minimum Spanning Forests with Nonuniform Degree Bounds

Könemann and Ravi [7] have examined the case of non-uniform degree bounds B_v for all vertices $v \in V$. They presented an algorithm that finds, in time $O(n^6 \log n)$, a spanning tree where the degree of each vertex is $O(B_v + \log n)$ and whose total edge weight is at most a constant times the weight of any tree that satisfies the degree constraints. There it is necessary to assume that a tree obeying the given degree bounds exists. Note that the problem to decide this question is already NP-complete. The algorithm uses a combination of primal-dual methods and local search, where in each local search step the normalized degree of the high-degree vertices in a current spanning tree is reduced. We generalize the primal-dual idea of Könemann and Ravi to the approximation of minimum spanning forests with nonuniform degree bounds. The task is to find for each i, $1 \leq i \leq n$, a spanning forest with i connected components and minimum total edge weight such that the maximum degree of each vertex v is at most B_v. The algorithm presented here runs in time $O(n^{3+2/\ln b})$, $b > 1$ an arbitrary constant, and outputs for each i, $1 \leq i \leq n$, a spanning forest F_i of i connected components whose vertex degrees are $O(B_v + \log n)$ and whose total weight is at most a constant times the total weight of any minimum spanning forest with i connected components. Here we do not need the assumption that there exists a spanning tree obeying the given degree bounds. For each value of i where a spanning forest respecting the degree constraints exists a solution with the stated approximation quality is produced.

We first adapt some results of [7] to the case of spanning forests. A feasible partition of V is a set $\pi = \{V_1, \ldots, V_k\}$ where $V_i \cap V_j = \emptyset$ for all $i \neq j$, $V = V_1 \cup \ldots \cup V_k$, and the induced subgraphs $G[V_i]$ are connected. Let G_π be the multigraph obtained from G by contracting each V_i into a single vertex, Π be the set of all feasible partitions of V, and $x(e)$ be the variable indicating whether edge e is included in the current solution, i.e., $x(e_i) = x_i$. We consider the following integer linear program (IP) formulation for the problem of computing the minimum spanning forest with i, $1 \leq i \leq n$, connected components that obeys all degree bounds B_v.

$$\min \sum_{e \in E} w(e)x(e) \tag{2}$$

$$\text{s.t.} \quad \sum_{e \in E[G_\pi]} x(e) \geq |\pi| - i \quad \forall \pi \in \Pi \tag{3}$$

$$\sum_{e \in E: v \in e} x(e) \leq B_v \quad \forall v \in V \tag{4}$$

$$x(e) \in \{0,1\} \quad \forall e \in E \tag{5}$$

The dual of the linear programming relaxation of (IP) is given by

$$\max \sum_{\pi \in \Pi} (|\pi| - i) \cdot y_\pi - \sum_{v \in V} \lambda_v B_v \tag{6}$$

$$\text{s.t.} \quad \sum_{\pi : e \in E[G_\pi]} y_\pi \leq w(e) + \lambda_u + \lambda_v \quad \forall e = (u,v) \in E \tag{7}$$

$$y, \lambda \geq 0 \tag{8}$$

Könemann and Ravi have given a primal-dual interpretation of Kruskal's algorithm. Let (IP-SP) denote (IP) without constraints of type (4), its LP relaxation denoted by (LP-SP) and its dual be (D-SP). Kruskal's algorithm can be seen as a continuous process over time that starts with an empty edge set at time 0 and ends with a minimum spanning tree at time t^*. At any time t, $0 \leq t \leq t^*$, a pair (x^t, y^t) is kept, where x^t is a partial primal solution for (LP-SP) and y^t is a feasible solution for (D-SP). In the initialization step $x(e)^0 = 0$ is set for all $e \in E$, and $y^t_\pi = 0$ for all $\pi \in \Pi$. Consider the forest F^t that corresponds to the partial solution x^t and let π^t be the partition induced by the connected components of F^t. At time t the algorithm increases y^t_π until a constraint of type (7) becomes tight. If this happens for edge e, this edge e is included into the primal solution. If more than one edge becomes tight, the edges are processed in arbitrary order. We denote by MSF_i a variant of this algorithm that stops when a minimum spanning forest with i connected components has been computed in the continuous process.

Let $deg_F(v)$ be the degree of vertex v in the spanning forest F with i connected components. The normalized degree of a vertex v is denoted by $ndeg_F(v) = \max\{0, deg_F(v) - 1 - b\alpha \cdot B_v\}$, where b and α are constants depending on the desired approximation quality. Let Δ^t the maximum normalized degree of any vertex in the current spanning forest F^t_i at a given time t and denote by U^t_j the set of vertices whose normalized degree is at least j at time t. The following lemma was shown in [7].

Lemma 5. *There is a* $d^t \in \{\Delta^t - 2\log_b n, \ldots, \Delta^t\}$ *such that*

$$\sum_{v \in U_{d^t-1}} B_v \leq b \cdot \sum_{v \in U_{d^t}} B_v$$

for any constant $b > 1$.

Our algorithm called Primal Dual Forest Optimizer (PDFO) is given in Fig. 2. The idea is to start with an empty edge set and compute the solutions with the desired approximation quality one after another. If we maintain a solution x^t with i connected components that does not have the desired approximation quality with respect to the degree bounds, we compute a new solution x^{t+1} which improves x^t with respect to the normalized degree. Let F^t_i be the forest corresponding to x^t. We increase the weight of an edge $e \in E$ by ϵ^t if it is either in F^t_i and adjacent to vertices of U_{d^t}, or in $E \setminus F^t_i$ and adjacent to vertices of U_{d^t-1}. The weight increment ϵ^t is defined as the smallest weight increase when deleting an edge adjacent to a vertex of U_{d^t} and inserting an edge

1. $t := 0$; $\lambda^t_v := 0, \forall v \in V$; $w^t(e) = w(e), \forall e \in E$;
2. $i := n$; $(x^t, y^t) := \mathrm{MSF}_i(G, w^t)$; $S := \{x^t\}$;
3. while $i > 1$ do
   (a) $i := i - 1$; $w^{t+1}(e) = w^t(e)$; $(x^{t+1}, y^{t+1}) := \mathrm{MSF}_i(G, w^{t+1})$; $t := t + 1$;
   (b) while $\Delta^t > 2\log_b n$ do
      i. Choose $d^t \in \{\Delta^t - 2\log_b n, \ldots, \Delta^t\}$ s.t. $\sum_{v \in U_{d^t-1}} B_v \leq b \cdot \sum_{v \in U_{d^t}} B_v$
      ii. Choose ϵ^t and let $\lambda^{t+1}_v := \lambda^t_v + \epsilon^t$ if $v \in U^t_{d^t-1}$ and $\lambda^{t+1}_v := \lambda^t_v$ otherwise
      iii. $w^{t+1}(e) := w^t(s) + \epsilon^t$ if $((e \in F^t_i) \wedge (e \cap U_{d^t} \neq \emptyset) \vee ((e \notin F^t_i) \wedge (e \cap U_{d^t-1} \neq \emptyset))$ and $w^{t+1}(e) := w^t(e)$ otherwise
      iv. $(x^{t+1}, y^{t+1}) := \mathrm{MSF}_i(G, w^{t+1})$; $t := t + 1$;
   (c) $S := S \cup \{x^t\}$;

Fig. 2. Primal Dual Forest Optimizer (PDFO)

adjacent to vertices that are not contained in U_{d^t-1} such that a new cyclefree subgraph of G is constructed. After that, x^{t+1} is a minimum spanning forest with i connected components with respect to the updated weight function w^{t+1}. We have also stated the computation of the dual variables corresponding to the primal solutions in Fig. 2 using MSF_i. The dual variables will be used later to show the approximation quality of our algorithm, but it is not necessary to compute these values during in the run.

We want to show that the algorithm computes in time $O(n^{3+2/\ln b})$ an approximation of the set of minimal elements that contains for each i, $1 \leq i \leq n$, a spanning forest with i connected components in which each vertex v has degree $O(B_v + \log n)$ and weight at most a constant times the weight of an optimal solution obeying the degree bounds. First, we consider the approximation quality of the solutions that are introduced into the set S in step 3c. Here we use an extension of the arguments given in [7] to the case of minimum spanning forest with given degree bounds.

Lemma 6. *For all iterations $t \geq 0$ where we are considering solutions with i connected components in the algorithm PDFO, the relation*

$$\sum_{\pi \in \Pi} (|\pi| - j) y_\pi^{t+1} \geq \sum_{\pi \in \Pi} (|\pi| - j) y_\pi^t + \epsilon^t \alpha \sum_{v \in U_{d^t-1}} B_v \tag{9}$$

holds for and all j, $1 \leq j \leq i$.

Lemma 7. *Let $\omega > 1$ be a constant and $\alpha = \max\{\omega/(\omega-1), \omega\}$. For all iterations $t \geq 0$ where we are considering solutions with i connected components in the algorithm PDFO, the relation*

$$\omega \sum_{v \in V} B_v \lambda_v^t \leq (\omega - 1) \sum_{\pi \in \Pi} (|\pi| - j) \cdot y_\pi^t \tag{10}$$

holds for each j, $1 \leq j \leq i$.

Lemma 8. *For all iterations $t \geq 0$ where we are considering solutions with i connected components in the algorithm PDFO, the relation*

$$\sum_{e \in F_j^t} w(e) \leq \omega \left[\sum_{\pi \in \Pi} ((|\pi| - j) \cdot y_\pi^t) - \sum_{v \in V} (B_v \cdot \lambda_v^t) \right] \tag{11}$$

holds for each j, $1 \leq j \leq i$.

Lemma 8 shows that in each iteration the weight of a spanning forest with j, $1 \leq j \leq i$, is only a constant times the weight of an optimal solution. It remains to show an upper bound on the runtime of PDFO. To do this we first consider the time to produce a new solution x^{t+1} from the current solution x^t.

Lemma 9. *The solution x^{t+1} can be computed from x^t in time $O(n^2)$.*

Proof. If the computation of x^{t+1} is carried out in step 3a of the algorithm introducing the lightest edge for the weight function w^{t+1} into x^t that does not create a cycle yields

x^{t+1}. This can be done in time $O(n^2)$ by inspecting every edge at most once. In the other case x^t is a minimum spanning forest with i connected components with respect to w^t and x^{t+1} is a minimum spanning forest with i connected components with respect to the updated weight function w^{t+1}. To determine x^{t+1} we have to compute ϵ^t and execute the resulting exchange operation. For the computation of d^t we use an integer array of size n and store at position j, $0 \leq j \leq n-1$, the sum over the B_v-values with vertices of normalized degree j in F_i^t. This can be done in time $O(n)$ using a breath first search traversal on F_i^t in which we compute the B_v value for the current vertex v in the traversal and add the value to the value of the corresponding position in the array. After that we determine the values $\sum_{v \in U_j} B_v$, $0 \leq j \leq n-1$, one after another starting with U_0. Note that $\sum_{v \in U_0} B_v = \sum_{v \in V} B_v$. The value $\sum_{v \in U_{j+1}} B_v$ can be computed by subtracting from $\sum_{v \in U_j} B_v$ the entry at position j in the array. Each computation can be done in constant time based on the corresponding array values. Hence, the value d^t of Lemma 5 can be determined in time $O(n)$. To compute the ϵ^t value we determine the exchange operation that leads to a primal solution of $\mathrm{MSF}_i(G, w^{t+1})$. Note that ϵ^t is the smallest weight increase such that deleting an edge adjacent to at least one vertex of U_{d^t} and inserting an edge adjacent to vertices that are not contained in U_{d^t-1} yields a minimum spanning forest with i components for the weight function w^{t+1}. We consider two possibilities for the exchange operation.

First we investigate the case where introducing an edge e connects two components of the current forest F_i^t. Then another edge from the resulting forest has to be removed to create a solution with i connected components. Introducing the edge e with smallest weight that is not incident to vertices of U_{d^t-1} and deleting the edge $e' \in F_i^t$ with the largest weight of all edges incident to vertices of U_{d^t} in F_i^t leads to the desired exchange operation with the smallest weight increase. Each edge of G has to be examined once, which gives an upper bound of $O(n^2)$ on the runtime in this case. Let $\epsilon^{t'}$ be the value obtained by this exchange operation.

The other possibility to get a smaller value than $\epsilon^{t'}$ is to introduce into F_i^t an edge that creates a cycle. Then we have to delete one edge of this cycle. We use a depth first search traversal of F_i^t from every vertex $v \in V$. Let w be the current vertex in this traversal. Assume that there is an edge $e = (v, w)$ in E and that v and w are not contained in U_{d^t-1}. Otherwise we can continue the traversal since the pair (v, w) does not fulfill the properties for the exchange operation. Let w_i be the largest weight of an edge e' in the path from v to w that is incident to vertices of U_{d^t}. If no such edge e' exists then e can not participate in the exchange operation we are looking for. The weight increase of introducing e and deleting e' can be computed in constant time, and the variables w_i can be maintained in constant time per step of the traversal using stacks. Hence, we can determine the exchange operation with the smallest weight increase in the second case in time $O(n^2)$. Let $\epsilon^{t''}$ be the weight increase of this exchange operation. Choosing $\epsilon^t = \min\{\epsilon^{t'}, \epsilon^{t''}\}$ and computing a primal solution of MSF_i^{t+1} by executing the corresponding exchange operation gives the stated upper bound. In addition we update the weight with respect to w^{t+1} for the next iteration which can be done in time $O(n^2)$. □

Theorem 2. *The algorithm PDFO computes, for any $b > 1$ and $\omega > 1$, in time $O(n^{3+2/\ln b})$ a set of solutions that includes a minimum spanning forest with*

i connected components for each i, $1 \leq i \leq n$, in which each vertex degree is at most $b \cdot \alpha \cdot B_v + 2\log_b n + 1$ and the total weight is at most $\omega \cdot w(F_i^*)$, where $\alpha = \max\{\omega/(\omega-1), \omega\}$ and $w(F_i^*)$ is the minimum weight of any spanning forest with i connected components satisfying the degree bounds.

Proof. As long as the algorithm has not achieved a solution with i connected components such that the vertex degree is at most $b \cdot \alpha \cdot B_v + 2\log_b n + 1$, the right hand side of (11) is ω times the optimal value of the dual objective function. This implies that the weight of a minimum spanning forest with i connected components for the weight function w^t is at most ω times the value of an optimal solution obeying the degree bounds for each j, $1 \leq j \leq i$. Hence, the solutions introduced into the set S in step 3c of PDFO have the stated approximation quality. As each possible value of i is considered in the run of PDFO, the set S includes after termination for each i, $1 \leq i \leq n$, a solution with i connected components with the desired approximation quality.

In the following we give an upper bound of $O(n^{3+2/\ln b})$ on the runtime until the algorithm terminates. A new solution x^{t+1} can be computed from the current solution in time $O(n^2)$ due to Lemma 9. It remains to bound the number of primal solutions x^t that have to be computed until the algorithm terminates. The number of solutions that are computed in step 3a of the algorithm is at most $n-1$ as the number of connected components is bound by n. In the inner while-loop we compute for the solution with i connected components new primal solutions as long as we have not reached a solution with i connected components that has the desired approximation quality.

Consider a modification of the potential function p introduced in Lemma 3. For a solution s, let $\hat{\Delta}(s)$ be the maximum normalized degree of s, and let $\hat{d}_i$, $0 \leq i \leq n-1$, be the number of vertices with normalized degree i in this solution. The potential of a solution s is given by $p'(s) := \sum_{j=0}^{\lceil 2\log_b n \rceil + 1} \hat{d}_{r+j}(s) \cdot e^j$, where $r = \hat{\Delta}(s) - \lceil 2\log_b n \rceil$. Assume that a minimum spanning forest with i, $2 \leq i \leq n$, has been computed. After initialization this is true for $i = n$. The solution of $\mathrm{MSF}_{i-1}(G, w^t)$ differs from $\mathrm{MSF}_i(G, w^t)$ by one single edge that is additionally introduced into $\mathrm{MSF}_{i-1}(G, w^t)$ at any time t. Introducing this edge into s_i, a solution s'_{i-1} with $p'(s'_{i-1}) - p'(s_i) = O(n^{2/\ln b})$ is created. Each iteration of the inner while-loop reduces the potential by at least $1/3$. Hence, we can upper bound the number of improvements in the run of PDFO by $O(n^{1+2/\ln b})$ using the ideas of Lemma 3. □

Note that choosing b as a constant large enough the runtime of PDFO approximates the upper bound $O(n^3)$. For $b = e$, $b = e^2$, ..., $b = e^k$, where $e = 2.71...$ and k is a constant, we get runtimes $O(n^5)$, $O(n^4)$, ..., $O(n^{3+2/k})$. The degrees of the produced solutions are bounded by $O(B_v + \log n)$, and the weight of a solution with i connected components introduced into the set S in step 3c is at most a constant times the weight of any minimum spanning forest with i connected components obeying the degree bounds.

4 Conclusions

In this paper we have shown that a multi-objective formulation can help to design faster approximation algorithms for the generalization of two NP-hard spanning tree problems. Our algorithms can be seen as an incremental construction procedure starting with the empty edge set and producing the solutions for the different number of connected components after another. Nevertheless we point out that our results have been obtained by the multi-objective formulation and this new view on the problem. We also think that this approach may be helpful to get a better understanding of other problems that have additional constraints. Based on our observations we have given an algorithm that computes for each i, $1 \leq i \leq n$, a minimum spanning forest with i components and degree at most $b \cdot \Delta_i^* + \lceil \log_b n \rceil$ in a total time of $O(n^{3+1/\ln b})$. In the case of nonuniform degree bounds we have presented an algorithm that runs in time $O(n^{3+2/\ln b})$ and computes for each i, $1 \leq i \leq n$, a spanning forest in which each vertex has degree $O(B_v + \log n)$ and the weight is a most a constant times the weight of a minimum spanning forest with i components obeying the given degree bounds.

References

1. Beier, R., and Vöcking, B. (2003). Random Knapsack in Expected Polynomial Time. In Proc, ACM Symposium on Theory of Computing (STOC), 306 - 329
2. Ehrgott, M. (2005). Multicriteria Optimization. 2nd Edition, Springer
3. Fischer, T. (1993). Optimizing the Degree of Minimum Weight Spanning Trees. Technical Report 93-1338, Department of Computer Science, Cornell University, Ithaca, NY, USA.
4. Fürer, M., and Raghavachari, B. (1990). An NC approximation algorithm for the minimum degree spanning tree problem. In Proc. of the 28th Annual Allerton Conf. on Communication, Control and Computing, 274-281.
5. Fürer, M., and Raghavachari, B. (1992). Approximating the Minimum-Degree Spanning Tree to within One from the Optimal Degree. In Proc. of the third annual ACM-SIAM symposium on Discrete algorithms (SODA), 317-324.
6. Fürer, M., and Raghavachari, B. (1994). Approximating the Minimum-Degree Steiner Tree to within One of Optimal. Journal of Algorithms 17, 409-423.
7. Könemann, J., and Ravi, R. (2003). Primal-dual meets local search: approximating MST's with nonuniform degree bounds. In Proc, ACM Symposium on Theory of Computing (STOC), 389-395
8. Neumann, F. and Wegener, I. (2005). Minimum spanning trees made easier via multi-objective optimization. In: Beyer et al. (Eds.): Genetic and Evolutionary Computation Conference - GECCO 2005, Volume 1, ACM Press, New York, 763-770
9. Papadimitriou, C. H., Yannakakis, M. (2000). The complexity of tradeoffs, and optimal access of web sources. 41st Annual Symposium on Foundations of Computer Science (FOCS 2000)

RISOTTO: Fast Extraction of Motifs with Mismatches

Nadia Pisanti[1,*], Alexandra M. Carvalho[2,**], Laurent Marsan[3], and Marie-France Sagot[4]

[1] Dipartimento di Informatica, University of Pisa, Italy
[2] INESC-ID, Lisbon
[3] PRiSM, Université de Versailles St Quentin, France
[4] INRIA Rhône-Alpes, France

Abstract. We present in this paper an exact algorithm for motif extraction. Efficiency is achieved by means of an improvement in the algorithm and data structures that applies to the whole class of motif inference algorithms based on suffix trees. An average case complexity analysis shows a gain over the best known exact algorithm for motif extraction. A full implementation was developed and made available online. Experimental results show that the proposed algorithm is more than two times faster than the best known exact algorithm for motif extraction.

1 Introduction

Patterns appearing repeated either inside a same string or over a set of strings are important objects to identify. Such repeated patterns are called *motifs* and their identification is called *motif inference* or *motif extraction*. The area has many potential applications, namely to data compression, natural languages, databases, basically, any activity or research requiring text mining [4]. The field of application that concerns us is molecular biology. The motifs in this case may correspond to functional elements in DNA, RNA or protein molecules, or to whole genes whose sequences are strongly similar. In biological applications, it is mandatory to allow for some mismatches between different occurrences of the same motif. In fact point mutations might have taken place, as well as errors in the sequencing procedure, so that molecules that have the same or related function(s), have no longer identical sequences. This is what makes the problem difficult from the computational point of view. In this paper we propose an exact algorithm for the extraction of motifs with mismatches. In particular, we consider *single* and *structured motifs*, which are motifs composed of several disjoint single motifs placed at given distances from each other. The extraction of structured motifs appears particularly interesting because of its application to the detection of binding sites ([3]). Given a text s, the problem is to find repeated

* Partially supported by PRIN project ALGONEXT.
** Partially supported by FCT grant SFRH/BD/18660/2004 and FCT Project FEDER POSI/SRI/47778/2002 BioGrid, and by the FCT grant SFRH/BD/18660/2004.

J.R. Correa, A. Hevia, and M. Kiwi (Eds.): LATIN 2006, LNCS 3887, pp. 757–768, 2006.

patterns in s according to some parameters that specify the frequency and the structure required for the motifs. In molecular biology, the text is in general a set of DNA sequence. Several exact, heuristic, and probabilistic algorithms for extracting structured motifs exist. Up to date, the best known exact algorithms for the extraction of single [7] and structured [2] motifs perform well when searching for short motifs. In this paper, we propose an improvement to such algorithms in order to deal with long motifs. The problem of extracting long motifs was first adressed by Pevzner and Sze [6]. They considered a precise version of the motif discovery problem: find all single motifs of length 15 with at most 4 mismatches in 20 sequences of size 600. A general set for this problem deserves attention from the algorithmic point of view because its computational complexity is in the worst case exponential with respect to the number e of mismatches allowed among different occurrences of the same motif. The reason is that, to identify motifs of the required length, there can be an explosion of the number of candidates of intermediate length whose extension has to be attempted. This imposes in practice a limit to the length of the motifs themselves, as in many applications the value of e depends on this length. The improvement introduced in this paper acts exactly in these cases, and hence applies to relatively long motifs, being a way to increase the length of motifs that are detectable in practice.

2 Single Motif Extraction

A *single motif* is a word over an alphabet Σ. Given an error rate e, a motif is said to *e-occur* in a sequence if it occurs with at most e letters substitution. The *single motif extraction problem* takes as input N sequences, a quorum $q \leq N$, a maximal number e of mismatches allowed, and a minimal and maximal length for the motifs, k_{min} and k_{max}, respectively. The problem consists in determining all motifs that e-occur in at least q input sequences. Such motifs are called *valid.* An efficient exact algorithm for the extraction of single motifs with mismatches has been introduced in [7] and is based on a suffix tree: motifs are considered in lexicographical order starting from the empty word, and they are extended to the right as long as the quorum is satisfied until either a valid motif of maximal length is found (if the k_{max} length is reached), or the quorum is no longer satisfied. In both cases, a new motif is attempted. Formally, the algorithm ([7]) we refer to is sketched in Algorithm 1.. At the beginning **ExtractSingleMotif** is called on the empty word. The algorithm recursively calls itself for longer motifs built by adding letters (step 4), and considers new ones (step 1) when the extension fails (step 2). A valid motif is spelled out whenever a motif whose

Algorithm 1. Single motif extraction

ExtractSingleMotif(motif m)
1. **for all** $\alpha \in \Sigma$ **do**
2.   **if** $m\alpha$ is valid **then**
3.   **if** $|m\alpha| \geq k_{min}$ **then** spell out the valid motif $m\alpha$
4.   **if** $|m\alpha| < k_{max}$ **then** **ExtractSingleMotif**($m\alpha$)

length lies within the required minimal and maximal length is detected (step 3). The order in which motifs are generated corresponds to a depth-first visit of a complete trie $\mathcal{M}$ (the *motif tree*) of all words of length k_{max} on the alphabet Σ. The algorithm does not need to allocate the motif tree. The only memory requirement is for the suffix tree $\mathcal{T}$. Assuming that the required length of the motif is k (that is $k_{min} = k_{max} = k$), and that at most e mismatches are allowed, the algorithm has worst case time complexity in $O(Nn_k\nu(e,k))$, where n_k is the number of tree nodes at depth k, and $\nu(e,k)$ is the number of words that differ in at most e letters from a word m of length k. This value does not depend on m, and it holds that $\nu(e,k) \leq k^e|\Sigma|^e$. This upper bound is in practice not tight. Nevertheless, no better bound can be given and therefore the time complexity is linear in the input size, but possibly exponential in the number e of mismatches. Since reasonable values for e are proportional to the value of k, this actually places a practical bound on the length required for the motifs. The goal of this paper is to move this bound.

2.1 Using Maximal Extensibility of Factors

The modification we suggest consists in storing information concerning maximal extensibility in order to avoid trying to extend hopeless motifs. For instance (see Fig. 1), assume that in our virtual depth-first visit of the motif tree, we have found out that motif m can be further extended without losing the quorum up to a length of $MaxExt(m)$ only, the latter representing its maximal extensibility. If later on, we are processing a motif m' that has m as a suffix, then the $MaxExt(m)$ information could be useful, as it applies to m' as well because m' can also be extended with at most $MaxExt(m)$ symbols (and possibly less). In particular, we have that if $|m'| + MaxExt(m) < k_{min}$, then we can avoid any further attempt to extend m' as there is no hope to reach length k_{min} for motifs that have m' as prefix. In Algorithm 1., motifs are considered in lexicographical order by a depth-first (virtual) visit of the motif tree $\mathcal{M}$. Every time we stop extending a motif, that is, when we (virtually) backtrack in $\mathcal{M}$, it is either because we found a valid motif of the maximal length, or because the quorum is no longer satisfied ($m\alpha$ does not satisfy the condition at step 2, and we start to consider the next one in lexicographical order). In the first case, m is valid, as are all its prefixes, and $|m| = k_{max}$. No information on the maximal extension of m nor of

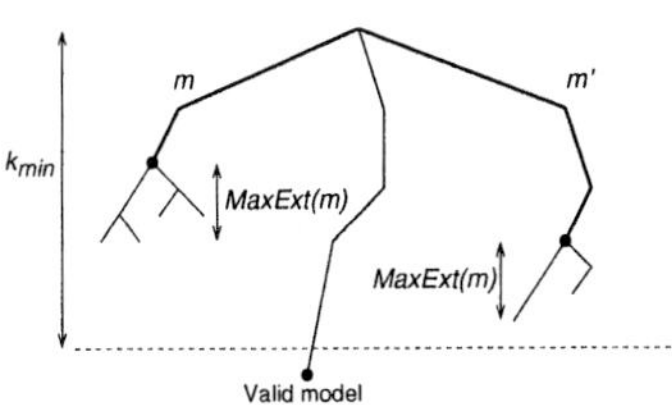

Fig. 1. Example where the extension of m' can be avoided, using $MaxExt(m)$, where m is a suffix of m', because $|m'| + MaxExt(m) < k_{min}$

its prefixes can be of any use because all motifs having a prefix of m as suffix can in general still be extended as much as necessary to reach at least the length k_{min}. For this reason, we set $MaxExt(m) = +\infty$. In the second case, m does not satisfy the quorum while all its prefixes do. For reasons that will be clearer later, we chose to only use the maximal extensibility information of motifs of length up to $k_{min} - 1$, hence this case can be subdivided into two subcases. When a motif m cannot be extended anymore and it has not reached the length $k_{min} - 1$, we set $MaxExt(m) = 0$. If the motif has reached a length h between $k_{min} - 1$ and k_{max}, we set $MaxExt(\langle m\alpha|_{k_{min}-1}) = h - (k_{min} - 1)$, where $\langle m\alpha|_{k_{min}-1}$ is the prefix of length $k_{min} - 1$ of $m\alpha$. Since it can be that $MaxExt(\langle m\alpha|_{k_{min}-1})$ had already received some value because a previous extension of $\langle m\alpha|_{k_{min}-1}$ was interrupted, then we change the value of $MaxExt(\langle m\alpha|_{k_{min}-1})$ only if we are increasing it, as maximal extensibility of a motif refers to its longest extension. All maximal extensibility values are initially set to -1, hence the first attribution to $MaxExt(\langle m\alpha|_{k_{min}-1})$ will always increase its value.

In all the cases above, the algorithm does not consider any further extension of m, and backtracks. This backtracking consists in either replacing the last letter $\sigma_{|m|}$ of m (line 1), or considering a shorter motif which in general shares a prefix with m, if $\sigma_{|m|}$ was the last letter of the alphabet Σ. In this latter case, the whole subtree rooted at the node spelling $\sigma_1 \dots \sigma_{|m|-1}$ has been (virtually) completely visited. Thus, we have all the information necessary to set the value of $MaxExt(\sigma_1 \dots \sigma_{|m|-1})$ according to $MaxExt(x) = 1 + \max_{\alpha \in \Sigma} MaxExt(x\alpha)$, for all valid motifs x such that $|x| < k_{min} - 1$. If the letter $\sigma_{|m|-1}$ was the last of the alphabet, then the backtracking goes further. In that case, also the $MaxExt$ information concerning the word $\sigma_1 \dots \sigma_{|m|-2}$ can be filled in in the same way, and so on. As mentioned before, maximal extensibility information can be used for motifs whose extension is being considered and for which this information could actually prevent some useless attempts. Namely, assume we are trying to extend the motif $m = \sigma_1, \sigma_2 \dots, \sigma_{|m|}$. Obviously, we do not know the value of $MaxExt(m)$ yet, and we know $MaxExt(\sigma_2, \dots, \sigma_{|m|})$ only if it lexicographically precedes m, that is, it has already been virtually visited in the motif tree. If this is not the case, we check whether $MaxExt(\sigma_3, \dots, \sigma_{|m|})$ is already known, and so on, possibly until the singleton $\sigma_{|m|}$. If they are all lexicographically greater than m, then no maximal extension information can be used for m, but if for any of them $MaxExt$ is known and it holds that the maximal possible extension is not enough to reach k_{min}, then the information is useful as it guarantees that attempting to further extend m is useless.

Lemma 1. Let $w \in \Sigma^*$. We have $MaxExt(w) \leq MaxExt(v)$ for each v which is a suffix of w.

A consequence of Lemma 1 is that longer suffixes of m can give us more tight bounds on the maximal extensibility information with respect to shorter ones. Therefore, since we start by checking the longest one, as soon as we find a suffix of m that enables us to state that m is not worth further attempts, then we can stop checking the other (shorter) suffixes. That is, if we find a suffix $|m\rangle_j = \sigma_j, \dots, \sigma_{|m|}$ of m, with $1 < j \leq |m|$, such that $MaxExt(|m\rangle_j)$ is not enough for m

to reach k_{min} because $MaxExt(|m\rangle_j)+|m| < k_{min}$, then we can quit attempting m and all its extensions, and we can consequently update $MaxExt(m)$. On the other hand, if no suffix $|m\rangle_j$ of m is such that $MaxExt(|m\rangle_j)+|m| < k_{min}$, then the maximal extension does not disallow to reach k_{min}. In this case, we have to go on trying to extend m even if it might be the case that it will never reach the minimal length. The algorithm for single motif extraction using the maximal extensibility information is presented in Algorithm 2.. For simplicity, we denote in the same way a node x and the word spelled by the path from the root to x. Recall that we use $\langle m\alpha|_{k_{min}-1}$ to denote the prefix of $m\alpha$ of length $k_{min}-1$, and $|x\rangle_{|x|-1}$ to denote the suffix of x of length $|x|-1$. Finally, for step 3, recall that we assumed that all maximal extensibility values are initially set to -1.

Algorithm 2. Single motif extraction with maximal extensibility information

ExtractSingleMotif(motif m)
1. **for all** $\alpha \in \Sigma$ **do**
2. $\quad x := m\alpha$
3. $\quad$**repeat** $x := |x\rangle_{|x|-1}$ **until** ($x = root$ **or** $MaxExt(x) \neq -1$)
4. $\quad$**if** $x \neq root$ **and** $MaxExt(x) + |m\alpha| < k_{min}$ **then**
5. $\quad\quad MaxExt(m\alpha) := MaxExt(x)$
6. $\quad\quad$stop spelling $m\alpha$ and **continue**
7. $\quad$**if** $m\alpha$ is valid **then**
8. $\quad\quad$**if** $|m\alpha| \geq k_{min}$ **then** spell out the valid motif
9. $\quad\quad$**if** $|m\alpha| < k_{max}$ **then** **ExtractSingleMotif**($m\alpha$)
10. $\quad\quad$**else** $MaxExt(\langle m\alpha|_{k_{min}-1}) := +\infty$
11. $\quad$**else**
12. $\quad\quad$**if** $|m\alpha| < k_{min}$ **then** $MaxExt(m\alpha) := 0$
13. $\quad\quad$**else if** $MaxExt(\langle m\alpha|_{k_{min}-1}) < |m\alpha|-(k_{min}-1)$ **then** $MaxExt(\langle m\alpha|_{k_{min}-1}) := |m\alpha|-(k_{min}-1)$
14. **if** $|m| < (k_{min}-1)$ **then** $MaxExt(m) := 1 + \max_{\alpha\in\Sigma} MaxExt(m\alpha)$

2.2 Complexity Analysis

The time complexity of Algorithm 2. remains the same as for Algorithm 1. in the worst case. Nevertheless, the proposed improvement has (very positive) effects on the average case. Next we compute the average ratio between the number of attempted extensions by RISO and RISOTTO for single motif extraction and compute the limit from which RISOTTO performs better than RISO.

Assume that the dataset has r planted random motifs of size t, where each motif can be extracted with at most e mismatches, and that the remaining text is uniformly random. This assumption captures the fact that we want to analyze the ratio between the number of attempted extensions by RISO and RISOTTO in the context of a dataset with highly correlated sequences (meeting the application requirements to biological datasets).

Let M_i be the random variable that gives the number of extracted motifs of size i with at most e mismatches for the assumed dataset, where $0 \leq i \leq t$. Clearly, we have that $P(M_0 = 1) = 1$ and $P(M_t \geq r) = 1$. The number of attempted extensions by RISO at level $i > 0$ (when the recursion step is at level i) is given by the random variable $E_i = M_{i-1}|\Sigma|$, and the total number of attempted extensions for the extraction of a single motif of size k is given by

$R_k = \sum_{i=1}^{k} E_i$. On the other hand, RISOTTO will only extend words at level i if they fulfill the maximum extensibility requirement. Therefore the number of attempted extensions by RISOTTO at level i is given by $E'_i = M_{i-1}|\Sigma|(1-p(i))$, where $p(i)$ is the probability of a i-word having maximal extensibility information to avoid its extension. Furthermore, the total number of attempted extensions by RISOTTO for the extraction of a single motif of size k is $R'_k = \sum_{i=1}^{k} E'_i$.

We conclude that to compute the average value of $\frac{R'_k}{R_k}$ we need to determine the average of the random variables M_i and the values $p(i)$, for $i = 1, \ldots, k$. We proceed by computing the average values of M_i. Clearly, a planted motif of size t has $t - i + 1$ segments of size i (considering overlapping). Observe that the average number of mismatches of the e-occurrences of a motif of size t is

$$\overline{e} = \sum_{j=0}^{e} j \frac{\binom{t}{j} (|\Sigma| - 1)^j}{\nu(e, t)}.$$

Hence, if we assume the mismatches to distribute uniformly over the segments, the average number of mismatches of the segments of size i of the e-occurrences is $\overline{e}_i = \frac{i}{t}\overline{e}$. Thus, the motifs extracted at level i due to the planted motifs are all the neighbors differing at most $(e - \overline{e}_i)$ letters from the segments of size i of the planted motifs. Since there are $r(t - i + 1)$ segments of size i, the average number of extracted motifs of size i with at most e mismatches, due to the planted motifs, is

$$\overline{T}_i = |\Sigma|^i \left(\sum_{j=0}^{r(t-i+1)-1} \left(1 - \frac{\nu(e - \overline{e}_i, i)}{|\Sigma|^i}\right)^j \frac{\nu(e - \overline{e}_i, i)}{|\Sigma|^i} \right).$$

Finally, to determine the average value of M_i, we need to take into account the motifs extracted from the random part of the text, and so, we have $\overline{M}_i = \overline{T}_i + (|\Sigma|^i - \overline{T}_i)(1 - \pi_i)$, where π_i is the probability of a random word of size i not being extracted with quorum q from a set of N sequences. Given that the probability of an e-neighbor of a word of size i not appearing in a random text of size n is

$$\delta(i, e, n) = (1 - 1/|\Sigma|^i)^{(n-i+1)\nu(e,i)} \approx (1 - 1/|\Sigma|^i)^{n i^e |\Sigma|^e},$$

the value of π_i can be computed by the following binomial

$$\pi_i = \sum_{j=0}^{q-1} \binom{N}{j} \delta(i, e, n)^{N-j} (1 - \delta(i, e, n))^j.$$

We finalize by computing the probability $p(i)$. Since the probability of a suffix of a random word being lexicographically smaller than the random word is $\frac{1}{2}$, we have that

$$p(i) = \sum_{j=1}^{i} \frac{1}{2^j} \gamma_{k-i}$$

where γ_{k-i} is the probability of the suffix of size $k-i$ to have information to avoid the extension. Notice that γ_{k-i} is the probability of the suffix of size $k-i$ not being extended to a size greater than $k-1$, and is given by

$$\begin{aligned}\gamma_{k-i} &= \pi_{k-i} + (1-\pi_{k-i})\pi_{k-i+1}^{|\Sigma|} + (1-\pi_{k-i})(1-\pi_{k-i+1}^{|\Sigma|})\pi_{k-i+2}^{|\Sigma|^2} + \dots \\ &= \sum_{j=0}^{i-1} \pi_{k-i+j}^{|\Sigma|^j} \prod_{\ell=1}^{j} (1-\pi_{k-i+j-\ell}^{|\Sigma|^{j-\ell}}) .\end{aligned}$$

To understand when RISOTTO starts to provides a gain over RISO, it is important to look to E_i' and E_i. Note that E_i' will be much smaller than E_i if $p(i)$ is close to 1. Moreover, as soon as random motifs start to disappear, M_{i-1} will be larger than M_i, which happens when π_i is close to 1. Both π_i and $p(i)$ depend tightly of $\delta(i,e,n)$, that is, if $\delta(i,e,n)$ is close to 0, so are π_i and $p(i)$, and if $\delta(i,e,n)$ is close to 1, so are π_i and $p(i)$. Since $\delta(i,e,n)$ behaves like a Dirac cumulative function for large values of n, that is, it jumps very fast from 0 to 1, we just need to solve the equation $\delta(i,e,n) = 1/2$ for the variable i to grasp when RISOTTO starts to be faster than RISO, which is just slightly before the solution. The solution of that equation is the fixed point of the following function

$$f(x) = -\log(1 - \frac{1}{2^{nx^e|\Sigma|^e}})/\log(|\Sigma|).$$

Given that $f(x)$ is contractive, that is, its derivate function takes values in the interval $(-1+\epsilon, 1-\epsilon)$, the fixed point can be computed by iterating f over an initial value. Finally, notice that the fixed point increases with the values of e, n and Σ. With the previous analysis, we have all the machinery necessary for computing the ratio between the expected number of attempted extensions between RISO and RISOTTO, as well as, from which point RISOTTO performs

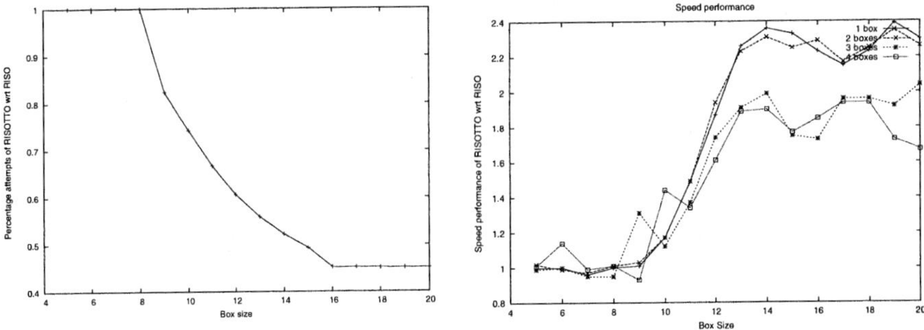

Fig. 2. Left: Ratio between the expected number of extensions attempted by RISOTTO and RISO (cf Fig. 3 to compare theoretical with experimental results obtained in the same set). **Right:** Ratio between performance of RISOTTO and RISO.

better than RISO. As an example, the ratio between the expected number of extensions attempted by RISOTTO and RISO for a dataset consisting of $N = 100$ sequences of size $n = 1000$ where we planted $r = 1$ motif of size $t = k = 5..20$, with up to $e = 2$ mismatches, and quorum $q = 100$, is given in Fig. 2(**left**). For the dataset considered, the fixed point for $f(x)$ is $x = 10.6616$.

3 Structured Motif Extraction

A *structured motif* is a pair (m, d) where $m = (m_i)_{1 \le i \le p}$ is a p-tuple of single motifs and $d = (d_{min_i}, d_{max_i})_{1 \le i < p}$ is a $(p-1)$-tuple of pairs, denoting $p-1$ intervals of distance between the p single motifs. Each element m_i of a structured motif is called a *box* and its minimal and maximal length denoted by k_{min_i} and k_{max_i}, respectively. The *structured motif extraction problem* takes as parameters N input sequences, a quorum $q \le N$, p maximal error rates $(e_i)_{i \le 1 \le p}$ (one for each of the p boxes), p minimal and maximal lengths $(k_{min_i})_{i \le 1 \le p}$ and $(k_{max_i})_{i \le 1 \le p}$ (one for each of the p boxes), and $p - 1$ intervals of distance $(d_{min_i}, d_{max_i})_{i \le 1 \le p-1}$ (one for each pair of consecutive boxes). Given these parameters, the problem consists in searching for the contents of the boxes, that is the motifs, that have the structure defined by the parameters above and that satisfy the quorum. The algorithm for single motif extraction introduced in [7] is the ancestor of others [2, 5] that infer structured motifs. The optimisation introduced in this paper can be applied to any of them. In a few words, the algorithm first builds the factor tree $\mathcal{T}$ of the input sequences, then it searches for all valid motifs of length at least k_{min} and up to k_{max} (as in [7]) and, after updating the data structure (see [2] for details), checks whether there is a second valid motif (again as in [7]) with the required interval between them. The algorithm is described by Algorithm 3. for $p = 2$, where i indicates whether we are dealing with the first or the second box, and λ is the empty word.

Algorithm 3. Structured motif extraction

```
ExtractStructuredMotif(motif m, box i)
for all α ∈ Σ do
   if mα is valid then
      if |mα| ≥ k_min_i then
         if i = 2 then spell out the valid motif
         else update T to ExtractStructuredMotif(λ, 2)
      if |mα| < k_max_i then ExtractStructuredMotif(mα, i)
```

3.1 Using Maximal Extensibility of Factors

In the case of structured motifs, the maximal extensibility information for the first box of a motif should be updated as described in Sect. 2.1. However, any failure in attempting to extend a motif during the search of a second box cannot update any value of $MaxExt$ because it refers only to parts of the text that follow a specific first box at a specific distance. In fact, when a first box m_1 of a structured motif is fixed at any given step, the maximal extensibility information

Algorithm 4. Structured motif extraction with maximal extensibility information

```
ExtractStructuredMotif(motif m, box i)
for all α ∈ Σ do
  if i = 1 or e_2 ≤ e_1 then
    x := mα
    while (x ≠ root or MaxExt(x) = −1) x := |x⟩_{|x|−1}
    if x ≠ root and MaxExt(x) + |mα| < k_{min_i} then
      if i = 1 then MaxExt(mα) := MaxExt(x)
      stop spelling mα and continue
  if mα is valid then
    if |mα| ≥ k_{min_i} then
      if i = 2 then spell out the valid motif
      else follow box-links and update T to ExtractStructuredMotif(λ, 2)
    if |mα| < k_{max_i} then ExtractStructuredMotif(mα, i)
    else if i = 1 then MaxExt(⟨mα|_{k_{min_1}−1}) := +∞
  else if i = 1 then
    if |mα| < k_{min_1} then MaxExt(mα) := 0
    else if MaxExt(⟨mα|_{k_{min_1}−1}) < |mα| − (k_{min_1} − 1) then MaxExt(⟨mα|_{k_{min_1}−1}) :=
  |mα| − (k_{min_1} − 1)
if i = 1 and |m| < (k_{min_1} − 1) then MaxExt(m) := 1 + max_{α∈Σ} MaxExt(mα)
```

that concerns the whole sequence is in general an upper bound on the maximal extensibility of fragments of the sequence that are at a given distance from the occurrences of m_1. Given this observation, a possibility is to use the maximal extensibility information of the first box when searching and trying to extend a second box. Another possibility, while attempting to find a motif for the second box, is to compute and store tighter maximal extensibility information which we can use for the second box being attempted as long as the first box is fixed. In the following, we only address the first alternative, that is, only the first box stores extensibility information. The conditions needed for our optimisation to be applicable in the case of structured motifs may hold even more frequently than in the case of single motifs. In fact, since the search for a valid motif as second box is made after a valid motif for the first box is found, maximal extensibility information may be known also for the whole motif whose extension is attempted and not just for its prefixes. In other words, it may happen that when Algorithm 3. is called with parameters m and 2, the value of $MaxExt(m)$ is already known. Proper suffixes are thus not the only candidates to give useful information when we are trying to find a motif for the second box. The extensibility information can be used as for the case of single motifs except that one has to deal with different error rates among boxes. Indeed, e_2 must be less than or equal to e_1 in order for the extensibility information to be useful for the second box. Otherwise, the maximal extensibility information stored for the first box may be too restrictive, and if it is used, it may cancel the extension of valid motifs. The algorithm for structured motif extraction using the maximal extensibility information is presented in Algorithm 4. Similarly to the case of single motif extraction, the time complexity of Algorithm 4. remains the same as for Algorithm 3. in the worst case, and the improvement proposed accounts only for the average case, as we shall verify in the next section.

4 Implementation and Experimental Results

In order to verify the improvement proposed in this paper, a C implementation of the maximal extensibility algorithm, called RISOTTO[1], was made. The new implementation was tested against a C implementation of the algorithm presented in [2] and called RISO. The results of the experiments we made show a sensible improvement for both single and structured motif extraction when using maximal extensibility information. As we shall see in this section, maximal extensibility may cost some extra space, which is a delicate issue for large datasets, but it can definitely save some hopeless visits, and in general it results very efficient.

We start with some considerations concerning the storage of extensibility information. As we have seen in Sect. 2.1, due to the order in which motifs are considered, we have that only certain subwords of motifs can give useful information concerning maximal extensibility, namely, those that are lexicographically smaller. Since no motif is smaller than itself, we actually only use the $MaxExt$ information of motifs that are shorter than the current one, that is, they are proper suffixes. Therefore, since the condition to check is whether or not we can hope to reach the k_{min} length, then we make use of the $MaxExt$ data only for strings of length at most $k_{min} - 1$. Hence, it is not necessary to store this information for motifs that have length k_{min} or more for the purpose mentioned above. Let us now discuss how much space is required to store the extensibility information until level $k_{min} - 1$. We say that a tree is *uncompact complete* if it is a trie where all possible nodes are present. There is thus no arc whose label contains more than one letter. A previous result [1] makes use of some statistical analysis for stating that a suffix tree of a text of length n is expected to be uncompact complete at the $log_{|\Sigma|}(n)$ top levels, where Σ is the alphabet of the text. This fact suggests a model to store extensibility information: a static data structure to keep the $MaxExt$ values until level $log_{|\Sigma|}(n)$, and a dynamic structure for deeper levels. Since we are interested in the DNA alphabet (composed of the four nucleotides A, C, G, and T), then we have that our suffix tree is uncompact complete at the top $log_4(n)$ levels where n is the size of the input sequence s. The function $log_4(n)$ reaches 10 for $n \approx 10^6$, it is greater than 11 for $n = 10^7$, it is more than 13 for $n = 10^8$, and nearly 15 for $n = 10^9$. These values correspond to reasonable values for the minimal length k_{min} of the motif, and they are reached for values n of the text size corresponding to quite big datasets. In the RISOTTO implementation, we took all the observations above into consideration. Since k_{min} has to be relatively small for our approach to be tractable spacewise, we considered only 1 byte (a char in C) to store $MaxExt$ values. In this case, extensibility values must be less than 256, which is quite reasonable. To build a static data structure to store such values until level z, we need $z + 1$ 1-byte arrays, where the j-th array has size $|\Sigma|^j$ with $0 \leq j \leq z$. Therefore, for the case of DNA, the total amount of memory required is $\frac{4^{z+1}-1}{3}$ bytes. This function gives us values of 1.3MB for $z = 10$, 5.3MB for $z = 11$,

[1] RISOTTO is available at `http://algos.inesc-id.pt/~asmc/software/riso.html`.

85.3MB for $z = 13$, and 1.3GB for $z = 15$. In our experiments, we achieved an optimum trade-off between memory allocation/management and maximal extensibility gain when $z = 10$. Taking this observation into account, we only allocate values for $MaxExt$ until level $z = \min\{10, k_{min} - 1\}$, even for large values of k_{min}, and disregard deeper levels as well as the dynamic data structure mentioned above. Nevertheless, we allowed this z level to be an implementation parameter. In the end, considering $z = \min\{10, k_{min} - 1\}$, RISOTTO requires at most 1.3MB more that RISO for DNA databases, being more than twice faster as we shall see next.

To test maximal extensibility performance we used several randomly generated (with a uniform distribution over the four letters size DNA alphabet) synthetic datasets with planted structured motifs. Each dataset consists of 100 sequences of size 1000 where we planted one motif, possibly structured into several boxes, with 2 mismatches per box. We ran both RISO and RISOTTO requiring a quorum $q = 100$ and at most 2 mismatches per box so that the output contains at least the planted motif. For each dataset, we made several runs for increasing lengths of the motifs. In particular, given the number of boxes of the structured motifs (in our tests there are p boxes for $p = 1, \ldots, 4$), we have increased the size of the boxes ranging from 5 to 20. As a result, the total motifs size (without counting the gaps) ranges from 5 to 80. For each p (number of boxes), we have plotted in Fig. 3, against the size of the motif (x axis), the

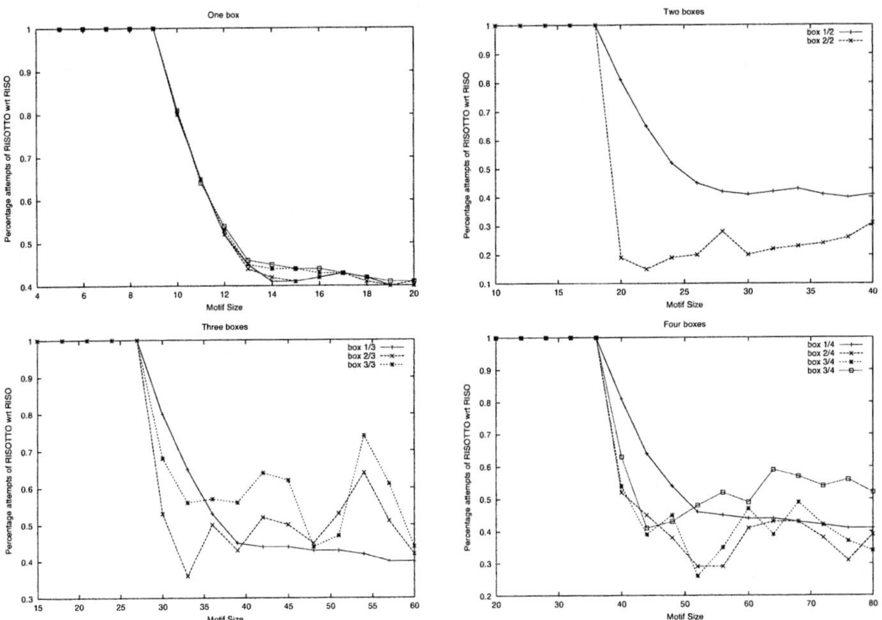

Fig. 3. Ratio between the number of extensions attempted by RISOTTO and RISO (cf Fig. 2(**left**) to compare theoretical with experimental results obtained in the same set)

ratio between the number of extensions attempted by RISOTTO and those by RISO (y axis). Given than RISOTTO only saves useless attempts, this equals the percentage of saved calls of the recursive procedure. For one box (Fig. 3 top left) we have depicted the results for several runs, while for two, three and four boxes (Fig. 3 top right and bottom) there are one curve for the inference of each box of the structured model. As one would expect, the attempts saved are more when the length of the motif increases and, in particular, the improvement starts when the length of the box is about 10 (this value depends in general from the input sequence and the alphabet size). For one box (see Fig. 3 top left), the number of attempted extension of RISOTTO decreases fast to 40% with respect to RISO (for growing values of the length of the motifs). Even better results, getting as good as attempting only 20% of the extensions of RISO, were achieved when extracting an i-th box with $2 \leq i \leq p$ (see Fig. 3 (top right and bottom)). Moreover, we present the ratio of speed performance of the computation of RISOTTO with respect to that of RISO. This is shown for all tests together in Fig. 2(**right**) for all possible sizes of the boxes. One can see that the best relative performance is achieved for the first boxes (that is where it is more needed because the search space is very large and noisy), where RISOTTO is up to 2.4 faster than RISO. Finally, in [6] a challenging problem was launched that concerned finding all single motifs of length 15 with at most 4 mismatches in 20 texts of size 600. We ran both RISO and RISOTTO on such instances. We observe a speedup of 1.6 of RISOTTO over RISO. We actually believe that a true challenge should involve texts of larger size. Therefore, we ran tests with the same parameters (length 15 and at most 4 mismatches) on larger input sequences. The results confirm the 1.6 speedup for sequences of length 700 and 800, 1.3 speedup for length 900, and then the speedup decreases, but the time required by RISOTTO is always lower than for RISO.

References

1. J. Allali. Structures d'indexation: les arbres des facteurs. Memoire de maitrise, University of Marne-la-Vallée, 2000.
2. A. M. Carvalho, A. T. Freitas, A. L. Oliveira, and M.-F. Sagot. A highly scalable algorithm for the extraction of cis-regulatory regions. In *Proc. APBC'05*, pages 273–282. Imperial College Press, 2005.
3. D. Guha-Thakurta and G. D. Stormo. Identifying target sites for cooperatively binding factors. *Bioinformatics*, 17:608–621, 2001.
4. M. Lothaire. *Applied Combinatorics on words.* Cambridge University Press, 2005.
5. L. Marsan and M.-F. Sagot. Algorithms for extracting structured motifs using a suffix tree with application to promoter and regulatory consensus identification. *J. Comp. Bio.*, 7:345–360, 2001.
6. P. A. Pevzner and S. H. Sze. Combinatorial algorithm for finding subtle signals in dna sequences. In *Proc. ISMB'00*, pages 269–278, 2000.
7. M.-F. Sagot. Spelling Approximate repeated or common motifs using a suffix tree. In C.L. Lucchesi and A.V. Moura, editors, *Proc. Latin'98*, volume 1380, pages 111–127, 1998. LNCS.

Minimum Cost Source Location Problems with Flow Requirements

Mariko Sakashita[1], Kazuhisa Makino[2], and Satoru Fujishige[3]

[1] Graduate School of Informatics,
Kyoto University, Kyoto 606-8501, Japan
sakasita@amp.i.kyoto-u.ac.jp

[2] Graduate School of Information Science and Technology,
University of Tokyo, Tokyo 113-8656, Japan
makino@mist.i.u-tokyo.ac.jp

[3] Research Institute for Mathematical Sciences,
Kyoto University, Kyoto 606-8502, Japan
fujishig@kurims.kyoto-u.ac.jp

Abstract. In this paper, we consider source location problems and their generalizations with three connectivity requirements λ, κ and $\hat{\kappa}$. We show that the source location problem with edge-connectivity requirement λ in undirected networks is strongly NP-hard, and that no source location problems with three connectivity requirements in undirected/directed networks are approximable within a ratio of $O(\ln D)$, unless NP has an $O(N^{\log \log N})$-time deterministic algorithm. Here D denotes the sum of given demands. We also devise $(1 + \ln D)$-approximation algorithms for all the extended source location problems if we have the integral capacity and demand functions.

Furthermore, we study the extended source location problems when a given graph is a tree. Our algorithms for all the extended source location problems run in pseudo-polynomial time and the ones for the source location problem with vertex-connectivity requirements κ and $\hat{\kappa}$ run in polynomial time, where pseudo-polynomiality for the source location problem with the arc-connectivity requirement λ is best possible unless P=NP, since it is known to be weakly NP-hard, even if a given graph is a tree.

Keywords: Connectivity, location problem, combinatorial optimization, and approximation algorithm.

1 Introduction

There is vast literature on location problems in the fields of operations research, computer science, etc. (see, e.g., [12]). Location problems in networks are often formulated as optimization problems to determine the best location of facilities such as industrial plants or warehouses in given networks to satisfy a certain property. Location problems based on flow (i.e., connectivity) requirements, called *source location problems*, were introduced by Tamura *et al.* [17, 18], and have recently received much attention from many authors (e.g., [1, 2, 3, 8, 10, 11, 14]).

J.R. Correa, A. Hevia, and M. Kiwi (Eds.): LATIN 2006, LNCS 3887, pp. 769–780, 2006.

Connectivity is one of the most important factors in applications to control and design of multimedia networks. Suppose that we are asked to locate a set S of multiple servers which can provide a certain service in a multimedia network $\mathcal{N}$. A user at vertex v can receive a service by connecting to a server in S through a path in $\mathcal{N}$. To ensure the quality of the service to v even if certain number $d-1$ of links and/or vertices become out of order, we should select S so that the arc- and/or vertex-connectivity between S and v is at least d. Therefore, such a kind of fault-tolerant settings can be formulated as a source location problem.

(Extended) Source Location Problems

Formally, source location problems can be described as follows. Let $\mathcal{N} = (G = (V, A), u)$ be a network with a vertex set V of cardinality n, an arc set A of cardinality m, and a capacity function $u : A \to \mathbb{R}_+$, where $\mathbb{R}_+$ denotes the set of all nonnegative reals. It has two demand functions $d^-, d^+ : V \to \mathbb{R}_+$, and a cost function $c : V \to \mathbb{R}_+$. Then the problem is given as

$$\begin{aligned} &\text{Minimize} \quad \sum_{v \in S} c(v) \\ &\text{subject to} \quad \psi^-(S, v) \geq d^-(v) \quad \text{and} \quad \psi^+(v, S) \geq d^+(v) \quad (v \in V), \\ &\qquad\qquad\quad S \subseteq V. \end{aligned} \tag{1.1}$$

Here $\psi^{\pm}(X, Y)$ denote certain measurements based on the connectivity from vertex set X to vertex set Y in $\mathcal{N}$. For any $v \in V$, we simply write $\psi^-(S, v)$ and $\psi^+(v, S)$ instead of $\psi^-(S, \{v\})$ and $\psi^+(\{v\}, S)$, respectively. For such measurements $\psi^{\pm}$, the present paper studies three basic connectivity requirements: arc-connectivity λ, and two kinds of vertex-connectivity κ and $\hat{\kappa}$. Note that Problem (1.1) sometimes adopts a single measurement ψ $(= \psi^- = \psi^+)$. Let us further note that in a more general problem setting we consider multiple constraints in (1.1) as $\psi_i^-(S, v) \geq d_i^-(v)$, $\psi_i^+(v, S) \geq d_i^+(v)$ for all $v \in V$ and $i = 1, 2, \cdots, \ell$. For example, we can consider the arc- and vertex-connectivity simultaneously.

We note that the cost function of the source location problem depends only on the fixed setup costs of the facilities at vertices. It is natural to consider the cost functions which depend not only on the setup costs but also on the supply values. This kind of generalization was introduced in [15], where we deal with source location problems with supply values $x(v)$ of facilities at $v \in V$, each of whose cost functions c_v is the sum of the opening cost and the monotone concave running cost for a facility at $v \in V$. We remark that monotonicity and concavity are natural assumptions on the cost, and are required in many network design problems (see, e.g., [6]).

Previous Work

We briefly survey the developments in (extended) source location problems.

Arc Connectivity Requirements. Tamura *et al.* [17] first considered the source location problem with the edge-connectivity (or arc-connectivity) requirement ψ $(= \psi^- = \psi^+) = \lambda$, when $\mathcal{N}$ is undirected and both the cost c and demand

d are uniform (i.e., $c(v) = 1$ and $d(v) = k$ for all $v \in V$), and gave a polynomial time algorithm for it. Since then, Tamura *et al.* [17], Ito *et al.* [10] and Arata *et al.* [2] have investigated the source location problem with edge-connectivity requirement λ in undirected networks. They provided polynomial time algorithms when the cost function or the demand function is uniform. On the other hand, it was shown that the problem is in general weakly NP-hard [2]. But, it remains open to prove the NP-hardness in the strong sense or to devise a pseudo-polynomial time algorithm.

For directed networks, Ito *et al.* [11] showed that the problem is strongly NP-hard, even if either the cost function c or the demand functions d^- and d^+ is uniform, and Bárász *et al.* [3] and Heuvel and Johnson [8] provided a polynomial time algorithm if c, d^- and d^+ are uniform.

Vertex Connectivity Requirements. The source location problem with vertex-connectivity requirement κ (i.e., $\psi^- = \psi^+ = \kappa$) was investigated by Ito *et al.* [10]. They considered the case in which G is undirected, and the cost function c and the demand function d ($= d^- = d^+$) are uniform. They showed that the problem is polynomially solvable for $k \leq 2$, but NP-hard for $k \geq 3$. They also showed that the positive result for $k \leq 2$ can be extended to the case in which the edge-connectivity $\lambda(S, v) \geq \ell$ is required simultaneously.

Let us note that the vertex-connectivity requirement, say, $\kappa(S, v)$ ensures that there exists at least $\kappa(S, v)$ *internally* vertex-disjoint paths from S to v. This means that any source in S never has a breakdown. To take possible breakdowns of sources into consideration, Nagamochi *et al.* [14] introduced another kinds of vertex-connectivity requirements $\hat{\kappa}^-(S, v)$ and $\hat{\kappa}^+(v, S)$, where $\hat{\kappa}^-(S, v)$ (resp., $\hat{\kappa}^+(v, S)$) is the maximum number of paths from S to v (resp., from v to S) which are vertex-disjoint except at v. They presented a polynomial time algorithm for the source location problem when d^- and d^+ are uniform and $\psi^{\pm} = \hat{\kappa}^{\pm}$. Ishii *ct al.* [9] considered the problem with a non-uniform demand function in undirected networks, gave a linear time algorithm if $d(v)$ ($= d^-(v) = d^+(v)$) ≤ 3, and showed that the problem is NP-hard if there exists a vertex $v \in V$ with $d(v) \geq 4$.

Some other types of the source location problems were also studied. These include single cover problem [17, 18], simultaneous location [1], and compound requirements [10].

For the extended source location problems, [15] investigated when it has uniform edge-connectivity requirement λ in undirected networks. By modeling this as a laminar cover problem, it can be shown that it is solvable in $O(nm + n^2(q + \log n))$ time, where q is the time required to compute $c_v(x)$ for each $x \in \mathbb{R}_+$ and $v \in V$.

The Results in this Paper

We address the (extended) source location problems. Briefly, the results obtained in this paper can be summarized as follows.

- We show that the source location problem with edge-connectivity requirement λ in undirected networks is strongly NP-hard. This solves an open

problem posed in [2], and gives us a complete picture of the tractability-intractability of the source location problems. Moreover, we show that no source location problems with three connectivity requirements in undirected/directed networks are approximable within a ratio of $O(\ln \sum_{v \in V}(d^+(v) + d^-(v)))$, unless NP has an $O(N^{\log \log N})$-time deterministic algorithm.
- We devise $(1 + \ln \sum_{v \in V}(d^+(v) + d^-(v)))$-approximation algorithms for the extended source location problems if we have integral capacity and demand functions. We remark that our approximation algorithm is applicable to all connectivity requirements. By combining the hardness result described above, we can say that our algorithm is *optimal* for all the extended source location problems, i.e., it is $\Theta(\ln \sum_{v \in V}(d^+(v) + d^-(v)))$-approximable.
- We also develop algorithms for the (extended) source location problems when a given graph is a tree, where we say that a directed graph G is a tree if G becomes a tree by ignoring the orientation of arcs and then regarding parallel edges as a single one. Our algorithms for the integral versions of the extended source location problems run in pseudo-polynomial time and the ones for the source location problems with vertex-connectivity requirements κ and $\hat{\kappa}$ run in polynomial time. We remark that pseudo-polynomiality for the arc-connectivity requirement λ is *best possible* unless P=NP, since it is known [2] that the source location problem with arc-connectivity requirement λ is weakly NP-hard, even if the underlying graph is a tree. Our positive results for the source location problems with the vertex-connectivity requirements κ and $\hat{\kappa}$ reveal for the first time a tractable subclass of the problems.

The rest of the paper is organized as follows. Section 2 introduces some notation and definitions. Section 3 shows the intractability of the source location problems, and Section 4 presents approximation algorithms for the (extended) source location problems. Section 5 considers the source location problem in tree networks.

Due to the space limitation, we skip the proofs, which can be found in [16].

2 Definitions and Preliminaries

For a network $\mathcal{N} = (G = (V, A), u)$, let us define three connectivities λ, κ, and $\hat{\kappa}$. For vertex subsets $X, Y \subseteq V$, we say that X is *k-arc-connected* to Y if there exists a feasible flow φ from X to Y whose value is at least k, where a flow $\varphi : A \to \mathbb{R}_+$ is feasible from X to Y if it satisfies the following conditions:

$$\partial\varphi(v) \stackrel{\text{def}}{=} \sum_{(v,w) \in A} \varphi(v, w) - \sum_{(w,v) \in A} \varphi(w, v) = 0 \quad (v \in V - (X \cup Y)), \tag{2.1}$$

$$0 \leq \varphi(a) \leq u(a) \quad (a \in A), \tag{2.2}$$

and the value of φ is defined by $\sum_{v \in X} \partial\varphi(v)$. The *arc-connectivity* from X to Y, denoted by $\lambda(X, Y)$, is the maximum number k such that X is k-arc-connected to Y. Here we define $\lambda(X, Y) = +\infty$ if $X \cap Y \neq \emptyset$.

For two sets $X, Y \subseteq V$, we say that X is *k-vertex-connected* to Y if there exists k internally vertex-disjoint paths from X to Y. The vertex-connectivity from X to Y, denoted by $\kappa(X,Y)$, is the maximum number k such that X is k-vertex-connected to Y. We define $\kappa(X,Y) = +\infty$ if $X \cap Y \neq \emptyset$ or there exists an arc from X to Y. For two sets $X, Y \subseteq V$, $\hat{\kappa}^-(X,Y)$ (resp., $\hat{\kappa}^+(X,Y)$) denotes the maximum number of paths from X to Y such that no pair of paths contains a common vertex in $V \setminus Y$ (resp., $V \setminus X$). We define $\hat{\kappa}^-(X,Y) = \hat{\kappa}^+(X,Y) = +\infty$, if $X \cap Y \neq \emptyset$.

This paper studies the source location problem given by (1.1) with three basic connectivity requirements λ, κ and $\hat{\kappa}$, i.e., the constraints $\psi^-(S,v) \geq d^-(v)$ and $\psi^+(S,v) \geq d^+(v)$ for λ, κ, and $\hat{\kappa}$ are, respectively, given as follows.

$$\lambda(S,v) \geq d^-(v) \text{ and } \lambda(v,S) \geq d^+(v), \tag{2.3}$$

$$\kappa(S,v) \geq d^-(v) \text{ and } \kappa(v,S) \geq d^+(v), \tag{2.4}$$

$$\hat{\kappa}^-(S,v) \geq d^-(v) \text{ and } \hat{\kappa}^+(v,S) \geq d^+(v). \tag{2.5}$$

For any $X, Y \subseteq V$, we denote $A(X,Y) = \{(x,y) \in A \mid x \in X,\ y \in Y\}$. For $X \subseteq V$, let $u^-(X)$ (resp., $u^+(X)$) denote the sum of the capacities of arcs entering (resp., leaving) X, i.e.,

$$u^-(X) = \sum_{a \in A(V \setminus X, X)} u(a), \quad u^+(X) = \sum_{a \in A(X, V \setminus X)} u(a),$$

and, for an undirected network, let us define $u(X)$ by $u(X) = u^-(X) (= u^+(X))$, where we note that $u(X) = u(V \setminus X)$. For every $X \subseteq V$, a vertex $v \in V \setminus X$ is called an *in-neighbor* (resp., *out-neighbor*) of X if there is an arc $(v,x) \in A$ (resp., $(x,v) \in A$) for some $x \in X$, and v is simply called a *neighbor* of X if it is an in- or out-neighbor of X. The set of all in-neighbors (resp., out-neighbors) of X is denoted by $N^-(X)$ (resp., $N^+(X)$). For a vertex set $X \subseteq V$, $d^-(X)$ (resp., $d^+(X)$) denotes the maximum in-demand (resp., out-demand) among all vertices in X, i.e.,

$$d^-(X) = \max_{v \in X} d^-(v) \quad (\text{resp., } d^+(X) = \max_{v \in X} d^+(v)).$$

A set $W \subseteq V$ is, respectively, called *deficient with respect to* (1) λ, (2) $\hat{\kappa}$, *and* (3) κ if (1) $u^-(X) < d^-(X)$ or $u^+(X) < d^+(X)$, (2) $|N^-(W)| < d^-(W)$ or $|N^+(W)| < d^+(W)$, and (3) W can be represented by $W = X \cup N^-(X)$ with $|N^-(X)| < d^-(X)$ or $W = X \cup N^+(X)$ with $|N^+(X)| < d^+(X)$. A deficient set W is called *minimal* if no nonempty proper subset $X \subsetneq W$ is deficient. We denote by $\mathcal{W}_\lambda$, $\mathcal{W}_{\hat{\kappa}}$ and $\mathcal{W}_\kappa$ the families of all minimal deficient sets with respect to λ, $\hat{\kappa}$ and κ, respectively.

From the max-flow and min-cut theorem, we have the following lemma, which is frequently used in the proofs given in the sequel.

Lemma 2.1. *A set $S \subseteq V$ is a feasible solution of the source location problem with requirement $\psi = \lambda, \hat{\kappa}, \kappa$ if and only if $S \cap W \neq \emptyset$ holds for every $W \in \mathcal{W}_\psi$.* $\square$

We next define the extended source location problems. Due to the space limitation, we only generalizes the arc-connectivity requirement.

A flow $\varphi : A \to \mathbb{R}_+$ is *feasible with a supply* $x : V \to \mathbb{R}_+$ if it satisfies the following conditions:

$$-x(v) \leq \partial\varphi(v) \leq x(v) \quad (v \in V), \tag{2.6}$$
$$0 \leq \varphi(a) \leq u(a) \qquad (a \in A), \tag{2.7}$$

where $\partial\varphi(v)$ is the net out-flow at vertex $v \in V$ for φ defined by (2.1). (2.6) means that the net out-flow $\partial\varphi(v)$ and the net in-flow $-\partial\varphi(v)$ at v is at most the supply at v. For a vertex $v \in V$ and a supply $x \in \mathbb{R}_+^V$, let $\lambda^-(x; v)$ (resp., $\lambda^+(x; v)$) denote the sum of the supply $x(v)$ and the maximum net in-flow $-\partial\varphi(v)$ (resp., net out-flow $\partial\varphi(v)$) at v among all feasible flows with a supply x. In other words, $\lambda^-(x; v)$ (resp., $\lambda^+(x; v)$) denotes the maximum (s, v)-flow (resp., (v, s)-flow) value in the augmented network $\mathcal{N}^* = (G^* = (V^*, E^*), u^*)$ defined by

$$\begin{aligned} V^* &= V \cup \{s\}, \\ A^* &= A \cup \{(s, v), (v, s) \mid v \in V\}, \\ u^*(a) &= \begin{cases} u(a) & \text{if } a \in A, \\ x(v) & \text{if } a = (s, v), (v, s). \end{cases} \end{aligned} \tag{2.8}$$

The extended source location problem with arc-connectivity requirement λ asks for a minimum-cost supply $x \in \mathbb{R}_+^V$, i.e.,

$$\begin{aligned} &\text{Minimize} \quad \sum_{v \in V} c_v(x(v)) \\ &\text{subject to} \quad \lambda^-(x; v) \geq d^-(v) \ \text{ and } \ \lambda^+(x; v) \geq d^+(v) \quad (v \in V), \\ &\qquad\qquad\quad x(v) \geq 0 \quad (v \in V). \end{aligned} \tag{2.9}$$

Here we assume that the cost $c_v : \mathbb{R}_+ \to \mathbb{R}_+$ is monotone (i.e., $f(x) \leq f(y)$ holds for arbitrary two reals $x, y \in \mathbb{R}$ with $x \leq y$) and concave (i.e., $f(\alpha x + (1 - \alpha)y) \geq \alpha f(x) + (1 - \alpha)f(y)$ for arbitrary two reals $x, y \in \mathbb{R}$ and α with $0 \leq \alpha \leq 1$).

Note that the flows φ_v^- and φ_v^+ in (2.9) may depend on v. It is not difficult to see that (2.9) is a generalization of the source location problem. In fact, for all $v \in V$, let $c_v(x(v)) = 0$ if $x(v) = 0$, and $= c(v)$ otherwise. Then we can see that (2.9) represents the source location problem.

3 Intractability of the Source Location Problems

In this section, we show that the source location problem with edge-connectivity requirement λ in undirected networks is strongly NP-hard. Recall that the problem can be described as follows. Given an undirected network $\mathcal{N} = (G = (V, E), u)$ with a capacity function $u : E \to \mathbb{R}_+$, a demand function $d : V \to \mathbb{R}_+$, a cost function $c : V \to \mathbb{R}_+$,

$$\begin{aligned} \text{Minimize} \ & \sum\{c(v) \mid v \in S\} \\ \text{subject to} \ & \lambda(S,v) \geq d(v) \quad (v \in V), \\ & S \subseteq V, \end{aligned} \tag{3.1}$$

where $\lambda(S,v)$ denotes the edge-connectivity between S and v.

Theorem 3.1. *Problem (3.1) is strongly NP-hard.*

Proof. We show the present theorem by reducing to Problem (3.1) the set cover problem, which is known to be strongly NP-hard [7].

Problem SET COVER
Input. A set $U = \{1, 2, \cdots, p\}$ and a family $\mathcal{S} = \{S_1, \cdots, S_q\} \subseteq 2^U$.
Output. A subfamily $\mathcal{X} \subseteq \mathcal{S}$ such that $\bigcup_{S_i \in \mathcal{X}} S_i = U$ and $|\mathcal{X}|$ is minimum.

Given a problem instance I of SET COVER, we construct the corresponding instance J $(\mathcal{N} = ((G,u), d, c))$ of Problem (3.1) as follows.

$$V = \{t_1, t_2\} \cup \{s_1, \cdots, s_q\} \cup \{x_1, \cdots, x_p\},$$
$$E = \{(t_1, s_i) \mid S_i \in \mathcal{S}\} \cup \{(s_i, x_j) \mid j \in S_i \in \mathcal{S}\} \cup \{(x_j, t_2) \mid j \in U\},$$
$$u(v,w) = \begin{cases} \ell_i & \text{if } v = t_1 \text{ and } w = s_i \ (S_i \in \mathcal{S}), \\ 1 & \text{if } v = s_i, w = x_j \text{ and } j \in S_i \in \mathcal{S}, \\ k_j - 1 & \text{if } v = x_j \ (j \in U) \text{ and } w = t_2, \end{cases}$$
$$d(v) = \begin{cases} \sum_{i=1}^{q} \ell_i & \text{if } v = t_1, t_2, \\ 0 & \text{otherwise,} \end{cases} \qquad c(v) = \begin{cases} 0 & \text{if } v = t_2, \\ 1 & \text{if } v \in \{s_1, \cdots, s_q\}, \\ q+1 & \text{otherwise,} \end{cases}$$

where $\ell_i = |S_i|$ and $k_j = |\{S_i \mid S_i \ni j\}|$. We denote $S = \{s_i, \cdots, s_q\}$ and $X = \{x_1, \cdots, x_p\}$. Note that $\sum_{i=1}^{q} \ell_i = \sum_{j=1}^{p} k_j$.

Although we skip the details, we can show that any optimal solution Y of J can be represented as

$$Y = \{t_2\} \cup \{s_i \mid S_i \in \mathcal{X}\} \tag{3.2}$$

for an optimal solution $\mathcal{X}$ of I. This completes the proof. □

We remark that the reduction above is *gap-preserving*. In fact, any feasible solution $\mathcal{X}$ of I produces a feasible solution Y of J given by (3.2), whose cost is $\sum\{c(v) \mid c \in Y\} = |\mathcal{X}|$. Any feasible solution Y of J that cannot be represented by (3.2) has $\sum\{c(v) \mid v \in Y\} \geq q+1 > |S|$. It is known [13, 4] that problem SET COVER is not approximable within a ratio of $O(\log p)$, unless NP has an $O(N^{\log\log N})$ time deterministic algorithm. Here we can assume that q is bounded by a polynomial in p. Since we have $\sum\{d(v) \mid v \in V\} \leq 2pq$, which is polynomial in p, the reduction above implies the following inapproximable result.

Theorem 3.2. *The source location problem with edge-connectivity requirement λ in (un)directed networks is not approximable within a ratio of* $O(\ln \sum\{d^+(v) + d^-(v) \mid v \in V\})$, *unless* NP *has an* $O(N^{\log\log N})$ *time deterministic algorithm.*

Similarly, we can show the inapproximability of the problems with vertex-connectivity requirements κ and $\hat{\kappa}$.

Theorem 3.3. *The source location problem with vertex-connectivity requirement κ (or $\hat{\kappa}$) in (un)directed networks is not approximable within a ratio of* $\mathrm{O}(\ln \sum\{d^+(v) + d^-(v) \mid v \in V\})$*, unless* NP *has an* $\mathrm{O}(N^{\log \log N})$ *time deterministic algorithm.*

Before concluding this section, we remark that the bounds in Theorems 3.2 and 3.3 are tight. This can be shown in the next section by constructing $(1+\ln \sum_{v \in V}(d^+(v)+d^-(v)))$-approximation algorithms for the extended source location problems.

4 Approximation Algorithms for the Extended Source Location Problems

In this section, we introduce the submodular cover problem as a natural generalization of the set cover problem [4, 13] and the submodular set cover problem [5, 19], and show that the extended source location problems can be regarded as a special case of the submodular cover problem. We then show that the extended source location problems are all approximable within a ratio of $(1 + \ln \sum_{v \in V}(d^-(v) + d^+(v)))$ by producing a simple greedy algorithm for the submodular cover problem.

A function $f : \mathbb{R}_+^V \to \mathbb{R}$ is *submodular* if

$$f(x) + f(y) \geq f(x \vee y) + f(x \wedge y) \tag{4.1}$$

holds for arbitrary two vectors $x, y \in \mathbb{R}_+^V$. Here $(x \vee y)(v) = \max\{x(v), y(v)\}$ and $(x \wedge y)(v) = \min\{x(v), y(v)\}$. The *submodular cover problem* is described as follows. Given a finite set V, monotone concave cost functions $c_v : \mathbb{R}_+ \to \mathbb{R}_+$ $(v \in V)$, a monotone submodular function $f : \mathbb{R}_+^V \to \mathbb{R}_+$ and a real M, the problem is to find a minimum-cost vector $x \in \mathbb{R}_+^V$ that satisfies $f(x) \geq M$, i.e.,

$$\begin{array}{lll} \text{(SC)} & \text{Minimize} & \displaystyle\sum_{v \in V} c_v(x(v)) \\ & \text{subject to} & f(x) \geq M, \\ & & x \in \mathbb{R}_+^V. \end{array} \tag{4.2}$$

Here, by putting $f(x) \leftarrow \min\{f(x) - f(0),\ M - f(0)\}$ if necessary, we can assume without loss of generality that $f(0) = 0$ and $f(x) \leq M$ for all $x \in \mathbb{R}_+^V$. This paper also considers its integral version, which is obtained from (4.2) by replacing $x \in \mathbb{R}_+^V$ with $x \in \mathbb{Z}_+^V$, where $\mathbb{Z}_+$ denotes the set of all nonnegative integers. We remark that Wolsey [19] considered a similar generalization, but he assumed that the cost function c is linear and f is piecewise linear and concave as well as monotone and submodular.

We show that all the extended source location problems can be formulated as submodular cover problems. Because of space limitations we only consider the extended source location problem (2.9) with arc-connectivity requirement λ.

Given a directed network $\mathcal{N} = (G = (V, A), u)$ and two demand functions d^- and d^+, we define a function f and a real M by

$$f(x) = \sum_{v \in V} \left(\min\{\lambda^-(x; v), d^-(v)\} + \min\{\lambda^+(x; v), d^+(v)\} \right), \qquad (4.3)$$

$$M = \sum_{v \in V} (d^-(v) + d^+(v)). \qquad (4.4)$$

Then we have

Lemma 4.1. *A function f defined by (4.3) is monotone and submodular.* □

Hence we see that the extended source location problem can be formulated as a submodular cover problem given by (4.2).

Similarly to the case of arc-connectivity requirement λ, we can show that the extended source location problems with κ and $\hat{\kappa}$ can also be formulated as (4.2).

We now propose a simple greedy algorithm for the submodular cover problem, which can be seen as a natural generalization of the one for the set cover problem. The algorithm starts with $x = \mathbf{0}$ and repeatedly increases x until it becomes a feasible solution of the problem. In each iteration, it finds an element $v^* \in V$ and a positive real $\delta^* > 0$ that is *the most cost-effective*, i.e., that attains

$$g_x(v^*; \delta^*) = \min_{v \in V, \delta > 0} \{g_x(v; \delta)\}, \qquad (4.5)$$

where

$$g_x(v; \delta) = \frac{c_v(x(v) + \delta) - c_v(x(v))}{f(x + \delta\chi_v) - f(x)}, \qquad (4.6)$$

and χ_v is the vth unit vector, i.e., $\chi_v(w) = 1$ if $w = v$, and $= 0$ otherwise. Here we assume the existence of the minimum in (4.5).

The algorithm is formally described as follows.

Algorithm. Greedy_SC
Input: A finite set V, a monotone submodular function $f : \mathbb{R}_+^V \to \mathbb{R}_+$, monotone concave cost functions $c_v : \mathbb{R}_+ \to \mathbb{R}_+$ $(v \in V)$, and a real M (> 0).
/* Let us assume that $f(0) = 0$ and $f(x) \leq M$ for all $x \in \mathbb{R}_+^V$ */
Output: A feasible solution $x \in \mathbb{R}_+^V$ for Problem SC.
Step 0. $x(v) := 0$ for all $v \in V$.
Step 1. While $f(x) < M$ **do**
(I) Find an element $v^* \in V$ and a real $\delta^* > 0$ that satisfies (4.5).
(II) $x(v^*) := x(v^*) + \delta^*$.
Step 2. Output x and halt. □

We remark that the algorithm might not halt in general since for v^* and δ^* computed in Step 1, $f(x+\delta^*\chi_{v^*})-f(x)$ might converge to $+0$. However, this does not hold for many problem instances, including the ones constructed from the (extended) source location problem.

As for the integral version of the problem, we modify algorithm GREEDY_SC by replacing " a real $\delta^*>0$ " in Step 1 with " an integer $\delta^*>0$ ".

Theorem 4.2. *Let ε be a nonnegative real such that v^* and δ^* in Step 1 always satisfy $f(x+\delta^*\chi_{v^*})-f(x)\geq\varepsilon$. Then* GREEDY_SC *computes a solution whose cost is at most $\left(1+\ln\frac{M}{\varepsilon}\right)$ times the optimum. Moreover, it is polynomial if*

(i) *v^* and δ^* in Step 1 can be computed in polynomial time, and*

(ii) *the number of iterations is bounded by a polynomial in the input size.* □

For requirement λ, we can see that $M=\sum_{v\in V}(d^-(v)+d^+(v))$ by (4.4), and $\varepsilon\geq 1$ if d^-, d^+ and u are integral. Moreover (i) and (ii) in Theorem 4.2 are satisfied for requirement λ. For example, we can prove that GREEDY_SC has at most $n(2n+1)$ iterations. This implies the following results, where we have similar results for κ and $\hat{\kappa}$.

Theorem 4.3. *For (the integral version of) the extended source location problems with λ, κ and $\hat{\kappa}$,* GREEDY_SC *computes in polynomial time a feasible solution whose cost is at most $1+\ln\sum_{v\in V}(d^-(v)+d^+(v))$ times the optimum if d^-, d^+ and u are integral.*

Corollary 4.4. *For the source location problems with κ and $\hat{\kappa}$,* GREEDY_SC *computes in polynomial time a feasible solution whose cost is at most $1+\ln 2n$ times the optimum.*

5 Extended Source Location Problems in Tree Networks

In this section, we present algorithms for the (extended) source location problems when the underlying graph $G=(V,E)$ is a tree, where we say that a directed graph G is a tree if G becomes a tree by ignoring the orientation of arcs and then regarding parallel edges as a single one. Our algorithms are based on dynamic programming. The algorithms for the extended source location problems run in pseudo-polynomial time and the ones for the source location problems with vertex-connectivity requirements κ and $\hat{\kappa}$ run in polynomial time.

Due to space limitations, we briefly discuss the source location problems in undirected tree networks.

Let $\mathcal{N}=(T=(V,E),u)$ be an undirected tree network with integer capacity u, let $d:V\to\mathbb{Z}_+$ be a demand function and let $c:V\to\mathbb{R}_+$ be a cost function. Then our source location problem can be represented by

$$\begin{aligned} &\text{Minimize} && \sum_{v\in S} c(v) \\ &\text{subject to} && \psi(S,v)\geq d(v) \quad (v\in V) \qquad (5.1)\\ & && S\subseteq V. \end{aligned}$$

Here $\psi = \lambda$, κ or $\hat{\kappa}$. Let us arbitrarily take a vertex r as a root of T. For a $v \in V$, let $CH(v)$ be the set of all children of v, and let $V(v)$ be the subset of V consisting of all descendants of v (including v). For example, if v is a leaf, we have $CH(v) = \emptyset$ and $D(v) = \{v\}$. For a vertex $v \in V$ and $X \subseteq CH(v)$, let $V(v, X)$ be the subset of V consisting of v and all descendants of X. We denote by $\mathcal{N}(v)$ (resp., $\mathcal{N}(v, X)$) the network induced by $V(v)$ (resp., $V(v, X)$).

Let $D = \max\{d(v) \mid v \in V\}$ and $U = \max\{u(e) \mid e \in E\}$. For a $v \in V$, let Y be a subset of $CH(v)$. We define a tree network $\mathcal{N}'(v, Y) = ((V', E'), u')$ by

$$V' = V(v, Y) \cup \{t_v\} \quad \text{and} \quad E' = E(v, Y) \cup \{(t_v, v)\}, \tag{5.2}$$

and $u'(e) = D$ if $e = (t_v, v)$, and $= u(e)$ otherwise, where t_v is a new vertex that is not in V. Here $E(v, Y)$ denotes the edge set of $\mathcal{N}(v, Y)$.

Then our dynamic programming solves the following problems in the networks $\mathcal{N}'(v, Y)$.

$$\begin{array}{lll} \mathbf{P}(\mathcal{N}'(v,Y),k,p) & \text{Minimize} & \sum\{c(v) \mid v \in S\} \\ & \text{subject to} & \psi(S,t) \geq d(t) - k \quad (t \in V(v,Y)), \qquad (5.3) \\ & & \psi(S \cup \{t_v\}, t) \geq d(t) \ (t \in V(v,Y) \setminus \{v\}), \ (5.4) \\ & & \psi(S, t_v) \geq p, \qquad (5.5) \\ & & S \subseteq V(v,Y), \end{array}$$

where k is an integer with $0 \leq k \leq D$, and p is an integer with $0 \leq p \leq \min\{D, U\}$ if $\psi = \lambda$; $p = 0, 1, D$ if $\psi = \kappa$; and $p = 0, 1$ if $\psi = \hat{\kappa}$. Constraint (5.3) means that the deficiency at each vertex $t \in V(v, Y)$ for S is at most k, (5.4) takes into account the possibility that sources outside $\mathcal{N}(v, Y)$ may cancel this deficiency, and (5.5) the possibility that S may satisfy at least p deficiency outside $\mathcal{N}(v, Y)$.

Although we skip the details, we have the following positive results.

Theorem 5.1. *The integral versions of the extended source location problems for $\psi = \lambda, \kappa, \hat{\kappa}$ can be solved in pseudo-polynomial time when the underlying graph is a tree, and the capacity and demand functions are integral.*

Theorem 5.2. *The source location problems for $\psi = \hat{\kappa}$ and κ can be solved in polynomial time when the underlying graph is a tree.*

References

1. K. Andreev, C. Garrod, and B. Maggs: Simultaneous source location, *SCS Technical Report* CMU-CS-03-162, Carnegie Mellon University, Pittsburgh.
2. K. Arata, S. Iwata, K. Makino, and S. Fujishige: Locating sources to meet flow demands in undirected networks, *J. Algorithms*, **42** (2002), 54–68.
3. M. Bárász, J. Becker, and A. Frank: An algorithm for source location in directed graphs, *Operations Research Letters*, **33** (2005), 221–230.
4. U. Feige: A threshold of $\ln n$ for approximating set cover, *J. ACM*, **45** (1998), 634–652.

5. T. Fujito: Approximation algorithms for submodular set cover with applications, *IEICE Trans.*, **E83-D** (2000), 480–487.
6. A. Goel and D. Estrin: Simultaneous optimization for concave costs: Single sink aggregation or single source buy-at-bulk, *Proceedings of 14th Annual ACM-SIAM Symposium on Discrete Algorithms* (2003), 499–505.
7. M. R. Garey and D. S. Johnson: *Computers and Intractability: A Guide to the Theory NP-Completeness*, Freeman, New York, 1979.
8. J. van den Heuvel and M. Johnson: Transversals of subtree hypergraphs and the source location problem in digraphs, *CDAM Research Report*, LSE-CDAM-2004-10, London School of Economics.
9. T. Ishii, H. Fujita, and H. Nagamochi: Minimum cost source location problem with local 3-vertex-connectivity requirements, *Computing Theory: The Australian Theory Symposium*, (2005), 97-105.
10. H. Ito, M. Ito, Y. Itatsu, H. Uehara, and M. Yokoyama: Source location problems considering vertex-connectivity and edge-connectivity simultaneously, *Networks*, **40** (2002), 63–70.
11. H. Ito, K. Makino, K. Arata, S. Honami, Y. Itatsu, and S. Fujishige: Source location problem with flow requirements in directed networks, *Optimization Methods and Software*, **18** (2003), 427–435.
12. M. Labbe, D. Peeters, and J.-F. Thisse: Location on networks, In M. O. Ball et al. (eds.), *Handbooks in OR & MS*, **8**, North-Holland (1995), 551–624.
13. C. Lund and M. Yannakakis: On the hardness of approximating minimization problems, *J. ACM*, **41** (1994), 960–981.
14. H. Nagamochi, T. Ishii, and H. Ito: Minimum cost source location problem with vertex-connectivity requirements in digraphs, *Information Processing Letters*, **80** (2001), 287–294.
15. M. Sakashita, K. Makino, and S. Fujishige: Minimizing a monotone concave function with laminar covering constraints, *ISAAC2005*, LNCS **3827** (2005), 71–81.
16. M. Sakashita, K. Makino, and S. Fujishige: Minimum cost source location problems with flow requirements, *unpublished manuscript.*
17. H. Tamura, M. Sengoku, S. Shinoda, and T. Abe: Location problems on undirected flow networks, *IEICE Trans.*, **E73** (1990), 1989–1993.
18. H. Tamura, M. Sengoku, S. Shinoda, and T. Abe: Some covering problems in location theory on flow networks, *IEICE Trans.*, **E75-A** (1992), 678–683.
19. L.A. Wolsey: An analysis of the greedy algorithm for the submodular set covering problem, *Combinatorica*, **2** (1982), 385–393.

Exponential Lower Bounds on the Space Complexity of OBDD-Based Graph Algorithms*

Daniel Sawitzki**

University of Dortmund, Computer Science 2, D-44221 Dortmund, Germany
daniel.sawitzki@cs.uni-dortmund.de

Abstract. Ordered Binary Decision Diagrams (OBDDs) are a data structure for Boolean functions which is successfully applied in many areas like Integer Programming, Model Checking, and Relational Algebra. Nevertheless, many basic graph problems like Connectivity, Reachability, Single-Source Shortest-Paths, and Flow Maximization are known to be PSPACE-hard if their input graphs are represented by OBDDs. This holds even for input OBDDs of constant width. We extend these results by concrete exponential lower bounds on the space complexity of OBDD-based algorithms for the Reachability Problem, the Single-Source Shortest-Paths Problem, and the Maximum Flow Problem. This involves the first exponential lower bound on the OBDD size for the highest bit of Integer Multiplication w. r. t. the natural interleaved variable order.

1 Introduction

Algorithms on (weighted) graphs G with node set V and edge set $E \subseteq V^2$ typically work on adjacency lists of size $\Theta(|V| + |E|)$ or on adjacency matrices of size $\Theta(|V|^2)$. But in areas like CAD, Model Checking, and Relational Algebra graphs arise whose size does not allow an explicit enumeration of all their elements. There [2, 10] and in further areas like Algorithmic Learning [7] and Integer Programming [1], the implicit representation of data by Ordered Binary Decision Diagrams (OBDDs) [3, 4, 18] is well-established as a succinct alternative. Their convenient algorithmic properties help to save time and space through solving problems by efficient logical operations. So OBDDs are applied in heuristic methods with hopefully sublinear resource usage.

Though each single OBDD manipulation is always efficient, a short sequence of them may suffice to cause an exponential blow-up in the OBDD size. Most algorithms on OBDD-represented graphs have only been analyzed experimentally [11, 12, 21] or w. r. t. rough measures like the number of OBDD manipulations [8, 9, 13]. Feigenbaum et al. [6] showed that even the very basic problem of s–t-Connectivity is PSPACE-complete on OBDD-represented graphs. That is, the success of OBDD-based approaches has to be explained by means of advantageous properties of real-world instances causing an essentially better behavior

* An extended version of this paper is available at http://ls2-www.cs.uni-dortmund.de/~sawitzki/.

** Supported by DFG grant We 1066/10-2.

J.R. Correa, A. Hevia, and M. Kiwi (Eds.): LATIN 2006, LNCS 3887, pp. 781–792, 2006.

than in the worst case. Recent research tries to build theoretical foundations for the analysis of the over-all runtime of OBDD-based algorithms [15, 16, 20]. This includes the investigation of the parameterized complexity of graph problems when structural properties of input and/or output OBDDs are considered as fixed parameters [17]. So basic graph problems are known to be fixed-parameter intractable w. r. t. a fixed input OBDD width (unless P=PSPACE). (The book of Downey and Fellows [5] gives a comprehensive introduction to the field of parameterized complexity.)

Despite these hardness results, there is no nontrivial lower bound for the complexity of any problem on OBDD-represented instances so far. The challenge is to prove both an upper bound on the input's OBDD size and an exponentially larger lower bound on the size of some OBDD occurring during the computation. We present such bounds for the Single-Source Shortest-Paths Problem and the Maximum Flow Problem. For the Reachability Problem this succeeds only for a certain class of OBDD-based algorithms. However, all existing algorithms known to the author belong to this class. We do not assume a separate output tape because the separation of working space and output size is not reasonable in practical applications.

The paper is organized as follows: After giving foundations on OBDDs in Section 2, we sum up both known and some trivial new results on the complexity of graph problems on OBDD-represented instances in Section 3. In Section 4, we introduce a construction method for functions with constant OBDD width. With these preliminaries we are able to construct pathological instances for all three considered graph problems in Sections 5, 6, and 7 giving us the desired exponential lower bounds. Due to space limitations, the technical lower bound on the OBDD size of the highest bit of multiplication has been shifted into the extended version of the paper. Finally, Section 8 gives conclusions on the work.

2 Ordered Binary Decision Diagrams

For $\mathbb{B} := \{0,1\}$, let us denote the ith character of a binary string $x \in \mathbb{B}^n$ by x_i and let $|x| := \sum_{i=0}^{n-1} x_i 2^i$ identify its value. The class of Boolean functions $f\colon \{0,1\}^n \to \{0,1\}$ is denoted by B_n. The set of all permutations of n elements is denoted by Σ_n.

A function $f \in B_n$ defined on variables $x_0, \ldots, x_{n-1}$ can be represented by an *Ordered Binary Decision Diagram (OBDD)* [3, 4]. An OBDD $\mathcal{G}$ is a directed acyclic graph consisting of *internal nodes* and *sink nodes*. Each internal node is labeled with a Boolean variable x_i, while each sink node is labeled with a Boolean constant. Each internal node is left by two edges one labeled 0 and the other 1. A *function pointer* p marks a special node that represents f. Moreover, a permutation $\pi \in \Sigma_n$ called *variable order* must be respected by the internal nodes' labels on every path from p to a sink. For a given variable assignment $\alpha \in \mathbb{B}^n$, we compute the function value $f(\alpha)$ by traversing $\mathcal{G}$ from p to a sink labeled $f(\alpha)$ while leaving each node labeled x_i via its α_i-edge. An OBDD with variable order π is called π-OBDD. The minimal-size π-OBDD for a function

$f \in B_n$ is known to be canonical and will be denoted by π-OBDD$[f]$. Its *size* size(π-OBDD$[f]$) is measured by the number of its nodes; its *width* is the maximum number of inner nodes labeled with the same variable. Finding an optimal variable order leading to the minimum size OBDD for a given function is known to be NP-hard. Independent of π, it is size(π-OBDD$[f]$) $\leq \big(2 + o(1)\big)2^n/n$ for any $f \in B_n$. The book of Wegener [18] gives a comprehensive survey on different types of Binary Decision Diagrams.

Efficient Algorithms on OBDDs. Functional operations on OBDD-represented functions can be implemented by efficient algorithms called *OBDD operations* in the following. The satisfiability of f can be decided in time $\mathcal{O}(1)$. The negation $\overline{f}$, the replacement of a variable x_i by some constant c (i. e., $f_{|x_i=c}$), and computing $|f^{-1}(1)|$ are possible in time $\mathcal{O}\big(\text{size}(\pi\text{-OBDD}[f])\big)$. The set $f^{-1}(1)$ of f's minterms can be obtained in time $\mathcal{O}\big(n \cdot |f^{-1}(1)|\big)$. Whether two functions f and g are equivalent (i. e., $f = g$) can be decided in time $\mathcal{O}\big(\text{size}(\pi\text{-OBDD}[f]) + \text{size}(\pi\text{-OBDD}[g])\big)$. The most important OBDD operation is the *binary synthesis* $f \otimes g$ for $f, g \in B_n$, $\otimes \in B_2$ (e. g., $\wedge$, $\vee$); in general, it produces the result π-OBDD$[f \otimes g]$ in time and space $\mathcal{O}\big(\text{size}(\pi\text{-OBDD}[f]) \cdot \text{size}(\pi\text{-OBDD}[g])\big)$. The synthesis is also used to implement *quantifications* $(\mathcal{Q}x_i)f$ for $\mathcal{Q} \in \{\exists, \forall\}$. Hence, computing π-OBDD$\big[(\mathcal{Q}x_i)f\big]$ takes time $\mathcal{O}\big(\text{size}^2(\pi\text{-OBDD}[f])\big)$ in general. All operations produce minimum size π-OBDDs.

Representing Graphs by OBDDs. One canonical way to represent data implicitly by an OBDD is to express it in terms of a subset $S \subseteq \{0, \dots, N-1\}^k$, where N depends on the data size and k is constant. Assuming w. l. o. g. $N = 2^n$, S can be represented by an OBDD for the *characteristic function* $\chi_S \in B_{kn}$ of S defined by $\chi_S(x^{(1)}, \dots, x^{(k)}) = 1 :\Leftrightarrow (|x^{(1)}|, \dots, |x^{(k)}|) \in S$, where $x^{(1)}, \dots, x^{(k)} \in \mathbb{B}^n$. Correspondingly, a digraph $G = (V, E)$ with nodes $v_0, \dots, v_{N-1}$ can be represented by $\chi_G \in B_{2n}$ with $\chi_G(x, y) = 1 :\Leftrightarrow (v_{|x|}, v_{|y|}) \in E$ for $x, y \in \mathbb{B}^n$. Undirected graphs are then considered as digraphs with symmetric edges. If G's edges are weighted by $c\colon E \to \{0, \dots, B\}$ with maximum weight B, we extend the definition to $\chi_G(x, y, a) = 1 :\Leftrightarrow (v_{|x|}, v_{|y|}) \in E \wedge c(v_{|x|}, v_{|y|}) = |a|$. In the context of characteristic functions, one further functional operation is of interest: the argument reordering.

Definition 1. *Let $\rho \in \Sigma_k$ and $f \in B_{kn}$ be defined on variable vectors $x^{(1)}, \dots, x^{(k)} \in \mathbb{B}^n$. The* argument reordering *$\mathcal{R}_\rho(f) \in B_{kn}$ w. r. t. ρ is defined by $\mathcal{R}_\rho(f)\big(x^{(1)}, \dots, x^{(k)}\big) = f\big(x^{(\rho(1))}, \dots, x^{(\rho(k))}\big)$.*

In order to enable efficient argument reorderings (see Lemma 3 and Theorem 1), it is common to use k-interleaved variable orders.

Definition 2. *The* k-interleaved *variable order $\pi^\tau_{k,n} \in \Sigma_{kn}$ of k variable vectors $x^{(1)}, \dots, x^{(k)} \in \mathbb{B}^n$ reads bits of same significance index en bloc:*

$$\pi^\tau_{k,n} := \left(x^{(1)}_{\tau(0)}, \dots, x^{(k)}_{\tau(0)}, x^{(1)}_{\tau(1)}, \dots, x^{(k)}_{\tau(1)}, \dots\dots, x^{(k)}_{\tau(n-1)}\right),$$

where τ is the local order of each $x^{(1)}, \dots, x^{(k)}$. The order $\pi^{\text{id}}_{k,n}$ is called natural.

3 Survey of Previous and New Results

Even the very basic problem of deciding whether two nodes s and t are connected in a digraph $G = (V, E)$ is known to be PSPACE-complete if the input graph is represented by π-OBDD$[\chi_G]$ (see [6]). In [17], this result is extended to fixed-parameter intractability results for a variety of fundamental graph problems including Connectivity, Bipartiteness, Planarity, Acyclicity, Single-Source Shortest-Paths, and Flow Maximization. This is for the fixed parameter of the input graph's OBDD width and under the assumption $\mathrm{P} \neq \mathrm{PSPACE}$. Unless stated otherwise, we assume in the following that input and, if required, output are represented by OBDDs.

With similar techniques, it can be proved that these problems as well as computing minimum spanning trees on OBDD-represented graphs remain fixed parameter intractable when the maximum of input and output OBDD width is considered as fixed parameter. That is, even if the characteristic functions of input *and* output have OBDDs of constant width and, therefore, size $\mathcal{O}(\log|V|)$, intermediately generated OBDDs may still be superpolynomially larger, unless P=PSPACE. Because these negative results rely on the compactness of OBDD-represented configuration graphs of Turing machines, they directly carry over to more general branching program models. Interestingly, the situation changes when the edge weight zero is forbidden. Then, the All-Pairs Shortest-Paths Problem can be solved in polynomial time if certain width restrictions apply to input and output OBDD [15]. The prefix "PW-" indicates the restriction to positive weights (PW-APSP and PW-SSSP).

On the other hand, all decision problems mentioned so far can be trivially solved in space poly$(\log|V|)$ by a nondeterministic Turing machine using χ_G as oracle. Each oracle request can be implemented by an OBDD evaluation operation. Together with the reasonable assumption size$(\pi\text{-OBDD}[\chi_G]) \geq \log_2|V|$ and the fact NPSPACE=PSPACE, we conclude that these problems can be solved in polynomial space w. r. t. size$(\pi\text{-OBDD}[\chi_G])$. But what about search problems?

If the output OBDD has polynomial size w. r. t. the input size size$(\pi\text{-OBDD}[\chi_G])$, we can enumerate all potential output OBDDs of polynomial size and verify the result in polynomial space. Without bounding the output OBDD size, this is not possible: We prove in this paper that constant input OBDD width does not suffice for polynomial space complexity of the Single-Source Shortest-Paths Problem, the Maximum Flow Problem, and the restricted reachability problem Reachability*. That is, their space complexity is not fixed-parameter tractable w. r. t. the parameter of input OBDD width.

Table 1 gives an overview of the state of affairs w. r. t. eight complexity classes named α-β-γ for $\alpha \in \{\mathrm{I}, \mathrm{IO}\}$, $\beta \in \{\mathrm{FPT}, \mathrm{P}\}$, and $\gamma \in \{\mathrm{T}, \mathrm{S}\}$. Component α indicates whether the complexity is related to input (I) or input *and* output OBDDs (IO); β separates fixed-parameter (FPT) from polynomial (P) complexities; γ separates time (T) from space (S) complexity. The classes are related as follows:

$$\text{I-}\beta\text{-T} \begin{array}{c} \subseteq \text{IO-}\beta\text{-T} \subsetneq \\ \subseteq \text{I-}\beta\text{-S} \subsetneq \end{array} \text{IO-}\beta\text{-S}, \qquad \alpha\text{-P-}\gamma \subseteq \alpha\text{-FPT-}\gamma \ .$$

Table 1. The complexity of graph problems on OBDD-represented inputs, unless P=PSPACE. The IO case is left out for decision problems. Results from [15, 17] are marked with daggers. The main contributions of this paper are starred.

	I-FPT-T	IO-FPT-T	I-FPT-S	IO-FPT-S	I-P-T	IO-P-T	I-P-S	IO-P-S
MaxFlow	no †	no	no ⋆	yes	no †	no	no ⋆	yes
APSP	no †	no	no ⋆	yes	no †	no	no ⋆	yes
PW-APSP	no †	yes †	no ⋆	yes †	no †	?	no ⋆	yes
SSSP	no †	no	no ⋆	yes	no †	no	no ⋆	yes
PW-SSSP	no †	?	no ⋆	yes	no †	?	no ⋆	yes
Reachability*	no	no	no ⋆	yes	no	no	no ⋆	yes
TransClos*	no	no	no ⋆	yes	no	no	no ⋆	yes
MST	no	no	?	yes	no	no	?	yes
s–t-Conn.	no †	-	yes	-	no †	-	yes	-
Connected	no †	-	yes	-	no †	-	yes	-
Bipartite	no †	-	yes	-	no †	-	yes	-
Acyclic	no †	-	yes	-	no †	-	yes	-
Euler Cycle	no †	-	yes	-	no †	-	yes	-

4 Constructing Functions with Constant OBDD Width

The pathological graph instances constructed in the following sections will have constant OBDD width w. r. t. natural interleaved variable orders. This section supplies a convenient construction method for functions with constant OBDD width. Actually we generate OBDDs with constant complete-OBDD width.

Definition 3. *An OBDD for $f \in B_n$ is called* complete *if every path from its function pointer to a sink has length n.*

That is, complete OBDDs are not allowed to skip variable tests. The minimal-size complete π-OBDD for $f \in B_n$ is also known to be canonical [18] and will be denoted by $\pi\text{-OBDD}_\text{c}[f]$ in the following.

Definition 4. *The* complete-OBDD width *of a function $f \in B_n$ w. r. t. a variable order $\pi \in \Sigma_n$ is the width of $\pi\text{-OBDD}_\text{c}[f]$.*

So it is $\text{size}(\pi\text{-OBDD}[f]) \leq \text{size}(\pi\text{-OBDD}_\text{c}[f]) = \mathcal{O}(nw)$ for any $f \in B_n$ with complete-OBDD width w and variable order π. On the other hand, it is $\text{size}(\pi\text{-OBDD}_\text{c}[f]) \leq n \cdot \text{size}(\pi\text{-OBDD}[f])$ (see, e. g., [18]).

The basic building blocks of the construction technique are multivariate threshold functions [20].

Definition 5. *Let $f \in B_{kn}$ be defined on variable vectors $x^{(1)}, \dots, x^{(k)} \in \mathbb{B}^n$. A function f is called* k-variate threshold function *iff there are $W \in \mathbb{N}$, $T \in \mathbb{Z}$, and $\delta_1, \dots, \delta_k \in \{-W, \dots, W\}$ such that*

$$f\left(x^{(1)}, \dots, x^{(k)}\right) = \left(\sum_{i=1}^{k} \delta_i \cdot \left|x^{(i)}\right| \geq T\right) .$$

The corresponding class of functions is denoted by $\mathbb{T}_{k,n}^W$.

Clearly, each of the relations $>$, $\leq$, $<$, and $=$ can be composed by binary syntheses of a constant number of multivariate threshold functions.

Lemma 1 ([20]). *Functions* $f \in \mathbb{T}_{k,n}^W$ *have complete OBDDs of width* $\mathcal{O}(k^2W)$ *using the variable order* $\pi_{k,n}^{\mathrm{id}} \in \Sigma_{kn}$.

That is, for $k, W = \mathcal{O}(1)$ multivariate threshold functions have constant complete-OBDD width. Moreover, both critical OBDD operations that may increase the OBDD size preserve constant complete-OBDD width (proved in the paper's extended version):

Let $f_1, f_2 \in B_n$ be defined on variables $x_0, \ldots, x_{n-1} \in \mathbb{B}$; assume f_1 resp. f_2 has complete-OBDD width w_1 resp. w_2 w. r. t. $\pi \in \Sigma_n$.

Lemma 2. *The binary synthesis result* π-OBDD$[f_1 \otimes f_2]$, $\otimes \in B_2$, *has a complete-OBDD width of at most* $w_1 w_2$.

Let $f_3 \in B_{kn}$ be defined on variable vectors $x^{(1)}, \ldots, x^{(k)} \in \mathbb{B}^n$; assume f_3 has complete-OBDD width w_3 w. r. t. $\pi_{k,n}^{\tau}$, $\tau \in \Sigma_n$. Let $\rho \in \Sigma_k$.

Lemma 3. *The argument reordering result* $\mathcal{R}_\rho(f_3)$ *of* f_3 *w. r. t.* ρ *has a complete-OBDD width of at most* $w_3 3^k$.

We conclude that a constant number of operations increases the complete-OBDD width independently of n.

Theorem 1. *Let* $x^{(1)}, \ldots, x^{(k)} \in \mathbb{B}^n$, *k constant. Let* $\mathcal{S}$ *be a sequence of* $\mathcal{O}(1)$ *operations as introduced in Section 2 applied to functions from* $\mathbb{T}_{k,n}^W$ *defined on* $x^{(1)}, \ldots, x^{(k)}$ *and to intermediate results generated by the current prefix of* $\mathcal{S}$.

Each function generated by $\mathcal{S}$ *has complete-OBDD width* $\beta(W)$ *w. r. t.* $\pi_{k,n}^{\mathrm{id}}$ *for some appropriate function* $\beta\colon \mathbb{N} \to \mathbb{N}$ *independent of* n.

5 Single-Source Shortest-Paths

The previous section has enabled us to construct functions with constant OBDD width starting from simply structured multivariate threshold functions. We now have to use this framework to generate instances for graph problems with an exponential gap between the input and output OBDD size. At first, we consider the *Single-Source Shortest-Paths Problem* on a weighted graph $G = (V, E, c)$ as introduced in Section 2. Let D be the set of all solution pairs $(v, d) \in V \times \mathbb{N}$ such that a shortest s–v-path $(s, \ldots, v) =: P$ has weight $d := \sum_{e \in P} c(e)$. The input for the problem's OBDD-based version consists of π-OBDD$[\chi_G]$ for some $\pi \in \Sigma_{kn}$ and a source node $s \in V$; the output is π-OBDD$[\chi_D]$ for D's characteristic function χ_D. That is, we even restrict ourselves to computing only the *costs* of shortest s–v-paths.

We define a sequence $(G_m)_m$ of pathological graph instances with solution sets D_m. $G_m := (V_m, E_m, c_m)$ consists of 2^m components H_i, $0 \leq i < 2^m$. Each

H_i is a path of 2^m nodes $w_{i,0}, \ldots, w_{i,2^m-1}$ with edges $(w_{i,j}, w_{i,j+1})$ of weight i for $0 \le i < 2^m$. Moreover, a common source node s is connected to all H_is by edges $(s, w_{i,0})$ of weight 0. So the path $(s, \ldots, w_{i,j})$ has weight ij and it is $(w_{i,j}, ij) \in D_m$. We add $2^{2(m+1)} - (2^{2m}+1)$ dummy singletons and number the nodes in $V_m := \{v_0, \ldots, v_{2^{2(m+1)}-1}\}$ by $w_{i,j} := v_{i2^{m+1}+j}$ and $s := v_{2^{2m+1}}$.

Claim. The function χ_{G_m} has a complete $\pi^{\mathrm{id}}_{5,m+1}$-OBDD of size $\mathcal{O}(m)$ with constant width independent of m.

Proof. We express $\chi_{G_m} \in B_{5(m+1)}$ in terms of Theorem 1. For $x \in \mathbb{B}^{2(m+1)}$ let $i(x) := x_{2m+1} \ldots x_{m+1}$, and $j(x) := x_m \ldots x_0$. We begin with a tentative function $\chi^*_{G_m}$.

$$\begin{aligned}\chi^*_{G_m}(x,y,a) := {} & \big[(|x| = 2^{2m+1} \neq |y|) \wedge (|j(y)| = |a| = 0)\big] \\ & \vee \big[(|x| \neq 2^{2m+1} \neq |y|) \wedge (|i(x)| = |i(y)| = |a|) \wedge (|j(y)| = |j(x)| + 1)\big] \quad (1)\end{aligned}$$

for node numbers $x, y \in \mathbb{B}^{2(m+1)}$ and a weight encoding $a \in \mathbb{B}^{m+1}$. This definition does not take care yet of dummy singletons occurring due to the node numbering. Hence, let $\chi_{V_m} \in B_{2(m+1)}$ be defined by

$$\chi_{V_m}(x) := (|x| = 2^{2m+1}) \vee (|i(x)|, |j(x)| < 2^m) \quad (2)$$

which is 1 exactly for all nondummy nodes. We finally have χ_{G_m} with

$$\chi_{G_m}(x,y,a) := \chi_{V_m}(x) \wedge \chi_{V_m}(y) \wedge \chi^*_{G_m}(x,y,a) \ . \quad (3)$$

Each comparison in (1)–(3) can be realized by $\mathcal{O}(1)$ functions from $\mathbb{T}^{\mathcal{O}(1)}_{5,m+1}$. So Theorem 1 applies and $\pi^{\mathrm{id}}_{5,m+1}$-OBDD$_{\mathrm{c}}[\chi_{G_m}]$ has constant width. □

Having proved that G_m has compact complete OBDDs of constant width, it remains to show that the output χ_{D_m} has exponential OBDD size.

Claim. Every OBDD for χ_{D_m} has exponential size w. r. t. m.

Proof. Assume w. l. o. g. that m is even. Replacing some variables of a Boolean function by constant values does not enlarge the corresponding π-OBDD. So we show a lower bound on a subfunction $f_m \in B_{2m}$ of χ_{D_m} which is obtained by replacing $x_{2m+1}, \ldots, x_{m+m/2+1}$ and $x_m, \ldots, x_{m/2}$ by 0. Hence, argument x represents a $w_{i,j}$ node with $i, j < 2^{m/2}$ and it is $f_m(x,a) = 1 \Leftrightarrow |i(x)| \cdot |j(x)| = |a|$, where $|a| < 2^m$. So f_m is the *Graph of $m/2$-bit Integer Multiplication* whose π-OBDD size is bounded below by $2^{m/1536-1}$ for any variable order π (proved in [16]). □

The claims in this section imply the result on the space complexity of SSSP.

Theorem 2. *The Single-Source Shortest-Paths Problem on OBDD-represented graphs has exponential space complexity, even for instances with constant complete-OBDD width. This implies* SSSP $\notin$ I-FPT-S.

By further variable replacements, the single-source variant can be trivially reduced to the all-pairs variant as defined in [15]. In Section 3, the special role of the edge weight zero has been mentioned. Though G_m contains such edges they can be avoided by constructions which are a little more complicated.

Proposition 1. PW-SSSP, PW-APSP $\notin$ I-FPT-S.

6 Maximum Flow

We continue with an exponential lower bound on the space complexity of the OBDD-based *Maximum Flow Problem.* Again, the input is an OBDD π-OBDD$[\chi_G]$ representing a weighted graph instance G. We want to compute a maximum flow $\phi\colon E \to \{0,\dots,B\}$ from a source $s \in V$ to a terminal $t \in V$. This time, the edge weights represent capacities. The flow ϕ must respect $\phi(e) \le c(e)$ as well as $\sum_{e=(u,v)} \phi(e) = \sum_{e=(v,w)} \phi(e)$ for each $v \in V$. The final output is π-OBDD$[\chi_F]$ for the solution set $F := \{(v,w,d) : \phi(v,w) = d\}$.

We define a sequence $(G_m)_m$ of pathological graph instances with unique maximum flows ϕ_m and solution sets F_m. $G_m := (V_m, E_m, c_m)$ consists of 2^{2m} components $H_{i,j}$, $0 \le i,j < 2^m$. Each $H_{i,j}$ consists of $j+2$ nodes $s_{i,j}$, $t_{i,j}$, and $w_{i,j,\ell}$ for $0 \le \ell < j$. $H_{i,j}$ contains $2j$ edges $(s_{i,j}, w_{i,j,\ell})$ and $(w_{i,j,\ell}, t_{i,j})$ with capacity i. The global source s and terminal t are connected to all components $H_{i,j}$ by edges $(s, s_{i,j})$ and $(t_{i,j}, t)$ of capacity 2^{2m}. Obviously, ϕ_m sends $i \cdot j$ units of flow through each $H_{i,j}$ and it is $(s, s_{i,j}, ij) \in F_m$.

The nodes in $V_m := \{v_0, \dots, v_{2^{3(m+1)}-1}\}$ are numbered by $w_{i,j,\ell} := v_{i2^{2(m+1)}+j2^{m+1}+\ell}$, $s_{i,j} := v_{i2^{2(m+1)}+j2^{m+1}+2^m}$, $t_{i,j} := v_{i2^{2(m+1)}+j2^{m+1}+2^m+1}$, $s := v_{2^{2m+1}}$, and $t := v_{2^{2m+1}+1}$.

Claim. Function χ_{G_m} has a complete $\pi^{\mathrm{id}}_{8,m+1}$-OBDD of size $\mathcal{O}(m)$ with constant width independent of m.

Proof. We express $\chi_{G_m} \in B_{8(m+1)}$ in terms of Theorem 1. For $x \in \mathbb{B}^{3(m+1)}$ let $i(x) := x_{3m+2} \dots x_{2m+2}$, $j(x) := x_{2m+1} \dots x_{m+1}$, and $\ell(x) := x_m \dots x_0$. Again we begin with a tentative function $\chi^*_{G_m}$ reflecting all four types of edges.

$$\begin{aligned}
\chi^*_{G_m}(x,y,a,b) := {} & \left[(|x| = 2^{2m+1}) \wedge (v_{|y|} \in \{s_{i,j} : i,j\}) \wedge (|ab| = 2^{2m})\right] \\
& \vee \left[(v_{|x|} \in \{t_{i,j} : i,j\}) \wedge (|y| = 2^{2m+1}+1) \wedge (|ab| = 2^{2m})\right] \\
& \vee \left[(v_{|x|} \in \{s_{i,j}\}_{i,j}) \wedge (v_{|y|} \in \{w_{i,j,\ell}\}_{i,j,\ell}) \right.\\
& \left. \wedge (|i(x)| = |i(y)| = |ab|) \wedge (|\ell(y)| < |j(x)| = |j(y)|)\right] \\
& \vee \left[(v_{|x|} \in \{w_{i,j,\ell}\}_{i,j,\ell}) \wedge (v_{|y|} \in \{t_{i,j}\}_{i,j}) \right.\\
& \left. \wedge (|i(x)| = |i(y)| = |ab|) \wedge (|\ell(x)| < |j(x)| = |j(y)|)\right] \qquad (4)
\end{aligned}$$

for node numbers $x, y \in \mathbb{B}^{3(m+1)}$ and a weight encoding ab consisting of concatenated components $a, b \in \mathbb{B}^{m+1}$. This definition does not take care yet of dummy singletons occurring due to the node numbering. Hence, let $\chi_{V_m} \in B_{3(m+1)}$ be defined by

$$\chi_{V_m}(x) := (v_{|x|} \in \{s, t, s_{i,j}, w_{i,j,\ell}, t_{i,j} : i, j, \ell\}) \ . \tag{5}$$

We finally have χ_{G_m} with

$$\chi_{G_m}(x, y, a) := \chi_{V_m}(x) \wedge \chi_{V_m}(y) \wedge \chi^*_{G_m}(x, y, a) \ . \tag{6}$$

Each comparison in (4)–(6) can be realized by $\mathcal{O}(1)$ functions from $\mathbb{T}^{\mathcal{O}(1)}_{8,m+1}$. This holds also for type checks like $v_{|x|} \in \{s_{i,j} : i, j\} \Leftrightarrow (|i(x)|, |j(x)| < 2^m) \wedge (|\ell(m)| = 2^m)$. Comparisons with the concatenation $|ab|$ can be broken down to $\mathcal{O}(1)$ comparisons with both parts $|a|$ and $|b|$. So Theorem 1 applies and $\pi^{\mathrm{id}}_{8,m+1}$-OBDD$_{\mathrm{c}}[\chi_{G_m}]$ has constant width. □

Claim. Every OBDD for χ_{F_m} has exponential size w. r. t. m.

Proof. We show a lower bound on a subfunction $f_m \in B_{4m}$ of the solution χ_{F_m} which is obtained by replacing x by the source number 2^{2m+1}, y_{3m+2} and y_{2m+1} by 0, and $|\ell(y)|$ by 2^m. Hence, argument y represents an $s_{i,j}$ node with $i, j < 2^m$. The maximum flow ϕ sends $i \cdot j$ flow units through $(s, s_{i,j})$ and it is $f_m(y, a, b) = 1 \Leftrightarrow |i(y)| \cdot |j(y)| = |ab|$. So f_m is the Graph of m-bit Integer Multiplication whose π-OBDD size is bounded below by $2^{m/768-1}$ for any variable order π (see [16]). □

The claims in this section imply the result on the space complexity of MaxFlow.

Theorem 3. MaxFlow $\notin$ I-FPT-S.

7 Reachability

Computing the set R of nodes that are reachable from some source $s \in V$ in a digraph $G = (V, E)$ is an important problem in CAD and Model Checking (see, e. g., [18–Chapters 13.2 and 13.3]). Let G be defined as in Section 2. In the OBDD-based setting, we want to compute the characteristic function χ_R of the solution set $R \subseteq V$. There are both BFS-like approaches with $\Omega(|V|)$ OBDD operations [11] as well as iterative squaring methods with $\mathcal{O}(\log^2 |V|)$ operations [14]. All popular algorithms known to the author achieve this by iteratively increasing the length of considered paths. This involves computing intermediate subfunctions χ_{R_p} with $\chi_{R_p}(x) = 1$ iff s and $v_{|x|}$ are connected by a directed path not longer than 2^p for $p \in \{1, \ldots, \lfloor \log_2 |V| \rfloor\}$. So we denote the problem of computing $\{\chi_{R_p}, \chi_R\}_p$ by *Reachability**. Moreover, we assume that the variable order is not changed during the algorithm.

We construct instances $(G_m)_m$ with constant complete-OBDD width whose intermediate result $R_{m,p}$ has exponential OBDD size for some maximum path length 2^p. Each $G_m := (V_m, E_m)$ consists of 2^{2m} components $H_{i,j}$, $0 \leq i, j < 2^m$. Each $H_{i,j}$ is the concatenation $P_{i-1} \ldots P_0$ of paths $P_\ell := (w_{i,j,\ell,j-1}, \ldots, w_{i,j,\ell,0})$. P_ℓ is concatenated to $P_{\ell-1}$ by $(w_{i,j,\ell,0}, w_{i,j,\ell-1,j-1})$. Moreover, a common source node s is connected to all $H_{i,j}$s by edges $(s, w_{i,j,i-1,j-1})$. The nodes in $V_m := \{v_0, \ldots, v_{2^{4(m+1)}-1}\}$ are numbered by $w_{i,j,\ell,r} := v_{i2^{3(m+1)}+j2^{2(m+1)}+\ell 2^{m+1}+r}$ and $s := v_{2^{4m+3}}$.

Claim. Function χ_{G_m} has a complete $\pi^{\mathrm{id}}_{8,m+1}$-OBDD of size $\mathcal{O}(m)$ with constant width independent of m.

Proof. We express $\chi_{G_m} \in B_{8(m+1)}$ in terms of Theorem 1. For $x \in \mathbb{B}^{4(m+1)}$ let $i(x) := x_{4m+3} \ldots x_{3m+3}$, $j(x) := x_{3m+2} \ldots x_{2m+2}$, $\ell(x) := x_{2m+1} \ldots x_{m+1}$, and $r(x) := x_m \ldots x_0$. Again we begin with a tentative function $\chi^*_{G_m}$.

$$\begin{aligned}\chi^*_{G_m}(x,y) := {} & [(|x| = 2^{4m+3} \neq |y|) \wedge (|\ell(y)| = |i(y)| - 1) \wedge (|r(y)| = |j(y)| - 1)] \\ & \vee [(|x| \neq 2^{2m+1} \neq |y|) \wedge (|i(x)| = |i(y)|) \wedge (|j(x)| = |j(y)|) \\ & \wedge (|\ell(x)| = |\ell(y)| < |i(x)|) \wedge (|j(x)| > |r(x)| = |r(y)| + 1)] \\ & \vee [(|x| \neq 2^{2m+1} \neq |y|) \wedge (|i(x)| = |i(y)|) \wedge (|j(x)| = |j(y)| \\ & \wedge (|i(x)| > |\ell(x)| = |\ell(y)| + 1) \wedge (|r(x)| = 0) \wedge (|r(y)| = |j(y)| - 1)] \end{aligned} \qquad (7)$$

for node numbers $x, y \in \mathbb{B}^{4(m+1)}$. This definition does not take care yet of dummy singletons occurring due to the node numbering. Hence, let $\chi_{V_m} \in B_{4(m+1)}$ be defined by

$$\chi_{V_m}(x) := (v_{|x|} = 2^{4m+3}) \vee (|i(x)|, |j(x)|, |\ell(x)|, |r(x)| < 2^m) \qquad (8)$$

which is 1 exactly for all nondummy nodes. We finally have χ_{G_m} with

$$\chi_{G_m}(x,y,a) := \chi_{V_m}(x) \wedge \chi_{V_m}(y) \wedge \chi^*_{G_m}(x,y,a) \ . \qquad (9)$$

Each comparison in (7)–(9) can be realized by $\mathcal{O}(1)$ functions from $\mathbb{T}^{\mathcal{O}(1)}_{8,m+1}$. So Theorem 1 applies and $\pi^{\mathrm{id}}_{8,m+1}$-OBDD$_c[\chi_{G_m}]$ has constant width. □

Having proved that G_m has compact complete OBDDs of constant width, it remains to show that for some appropriate p the function $\chi_{R_{m,p}}$ has exponential OBDD size. We first consider the ith bit of n-bit Integer Multiplication.

Definition 6. *Let $x, y \in \mathbb{B}^n$. The ith bit of n-bit Integer Multiplication $\mathrm{MULT}_{n,i} \in B_{2n}$ on variables x, y is defined to be the ith bit of $|x| \cdot |y|$.*

There are well-known exponential lower bounds on the OBDD-size of the middle bit $\mathrm{MULT}_{n,n-1}$ (see [19]). The π-OBDD size of the highest bit $\mathrm{MULT}_{n,2n-1}$ for any nontrivial variable order $\pi \in \Sigma_{2n}$ has been open so far [18–Problem 4.12].

Theorem 4. *The size of $\pi^{\mathrm{id}}_{2,n}$-OBDD$[\mathrm{MULT}_{n,2n-1}]$ is at least $2^{(n-5)/6}$.*

The proof of this theorem uses techniques from analytic number theory; it has been shifted into the extended version of this paper.

Claim. The size of $\pi^{\mathrm{id}}_{4,m+1}$-OBDD$[\chi_{R_{m,2m-1}}]$ is exponential w. r. t. m.

Proof. Replacing some variables of a Boolean function by constant values does not enlarge the corresponding π-OBDD. So we show a lower bound on a subfunction $f_m \in B_{2m}$ of $\chi_{R_{m,2m-1}}$ which is obtained by replacing x_{4m+3}, x_{3m+2}, and $x_{2m+1}, \ldots, x_0$ by 0. Hence, argument x represents a $w_{i,j,0,0}$ node which is reachable from s via at most 2^{2m-1} edges iff the s–$w_{i,j,0,0}$-path length $i \cdot j$ is not larger than 2^{2m-1}. So it is $f_m(x) = 1 \Leftrightarrow |i(x)| \cdot |j(x)| \leq 2^{2m-1}$.

Let $g_m(x) = 1 :\Leftrightarrow |i(x)| \cdot |j(x)| = 2^{2m-1}$. It is easy to see that the $\pi_{2,m}^{\mathrm{id}}$-OBDD size of g_m is $\mathcal{O}(m^2)$. Hence, the $\pi_{2,m}^{\mathrm{id}}$-OBDD for

$$h_m(x) := f_m(x) \wedge \overline{g(x)} = (|i(x)| \cdot |j(x)| < 2^{2m-1})$$

is at most polynomially larger than for f_m. Due to $\mathrm{MULT}_{m,2m-1}(x, y) = 1 :\Leftrightarrow |x| \cdot |y| \geq 2^{2m-1}$ for $x, y \in \mathbb{B}^m$, it is $\overline{h_m} = \mathrm{MULT}_{m,2m-1}$.

Altogether, we showed that the $\pi_{2,m}^{\mathrm{id}}$-OBDD size of $\mathrm{MULT}_{m,2m-1}$ is at most polynomially larger than of $\chi_{R_{m,2m-1}}$ implying the claim's statement. □

Both claims imply the result on the space complexity of Reachability*.

Theorem 5. Reachability* $\notin$ I-FPT-S.

By further variable replacements, the Reachability Problem can be trivially reduced to computing the OBDD of all connected node pairs—the *transitive closure*. It is an important submodule of many OBDD-based graph algorithms [11, 14, 20]. So it follows for the analogous restricted variant *TransClos**:

Proposition 2. TransClos* $\notin$ I-FPT-S.

These results rely on the assumption that the output OBDDs of the starred problem variants use the same variable order as the input OBDDs. In contrast, practical algorithms usually run variable reordering heuristics on intermediate OBDD results in order to minimize their size. However, we conjecture that $\mathrm{MULT}_{n,2n-1}$ has exponential OBDD size for every variable order.

8 Conclusions

None of the graph problems that have been considered on OBDD-represented instances has an FPT algorithm w. r. t. a fixed input OBDD width, unless P=PSPACE. Except restricted shortest paths problems, this holds also if both input and output OBDD width are fixed parameters. On the other hand, a polynomially bounded output OBDD size allows to solve all considered problems in polynomial space. We contributed exponential lower bounds on the space (and, therewith, time) complexity of general OBDD-based shortest paths and maximum flow algorithms and a common class of reachability algorithms. Consequently, a restriction of the input OBDD width does not suffice to guarantee polynomial space for these problems. The analyses include the first nontrivial exponential lower bound on the OBDD size for the highest bit of Integer Multiplication w. r. t. the natural interleaved variable order.

It remains an open question whether the OBDD-based Minimum Spanning Tree Problem is in I-FPT-S.

Acknowledgments. Thanks to Oliver Giel, Martin Sauerhoff, and Ingo Wegener for proofreading and discussions.

References

[1] B. Becker, M. Behle, F. Eisenbrand, and R. Wimmer. BDDs in a branch and cut framework. In *WEA 2005*, volume 3503 of *LNCS*, pages 452–463. Springer, 2005.
[2] R. Berghammer and F. Neumann. RELVIEW - An OBDD-based computer algebra system for relations. In *CASC 2005*, volume 3718 of *LNCS*, pages 40–51. Springer, 2005.
[3] R.E. Bryant. Symbolic manipulation of Boolean functions using a graphical representation. In *DAC 1985*, pages 688–694. ACM Press, 1985.
[4] R.E. Bryant. Graph-based algorithms for Boolean function manipulation. *IEEE Transactions on Computers*, 35:677–691, 1986.
[5] R.G. Downey and M.R. Fellows. *Parameterized Complexity.* Springer, Berlin Heidelberg New-York, 1999.
[6] J. Feigenbaum, S. Kannan, M.Y. Vardi, and M. Viswanathan. Complexity of problems on graphs represented as OBDDs. In *STACS 1998*, volume 1373 of *LNCS*, pages 216–226. Springer, 1998.
[7] R. Gavalda and D. Guijarro. Learning Ordered Binary Decision Diagrams. In *Algorithmic Learning Theory 1995*, volume 997 of *LNCS*, pages 228–238. Springer, 1995.
[8] R. Gentilini, C. Piazza, and A. Policriti. Computing strongly connected components in a linear number of symbolic steps. In *SODA 2003*, pages 573–582. ACM Press, 2003.
[9] R. Gentilini and A. Policriti. Biconnectivity on symbolically represented graphs: A linear solution. In *ISAAC 2003*, volume 2906 of *LNCS*, pages 554–564. Springer, 2003.
[10] G.D. Hachtel and F. Somenzi. *Logic Synthesis and Verification Algorithms.* Kluwer Academic Publishers, Boston, 1996.
[11] G.D. Hachtel and F. Somenzi. A symbolic algorithm for maximum flow in 0–1 networks. *Formal Methods in System Design*, 10:207–219, 1997.
[12] H. Jin, A. Kuehlmann, and F. Somenzi. Fine-grain conjunction scheduling for symbolic reachability analysis. In *TACAS 2002*, volume 2280 of *LNCS*, pages 312–326. Springer, 2002.
[13] K. Ravi, R. Bloem, and F. Somenzi. A comparative study of symbolic algorithms for the computation of fair cycles. In *FMCAD 2000*, volume 1954 of *LNCS*, pages 143–160. Springer, 2000.
[14] D. Sawitzki. Implicit flow maximization by iterative squaring. In *SOFSEM 2004*, volume 2932 of *LNCS*, pages 301–313. Springer, 2004.
[15] D. Sawitzki. A symbolic approach to the all-pairs shortest-paths problem. In *WG 2004*, volume 3353 of *LNCS*, pages 154–167. Springer, 2004.
[16] D. Sawitzki. Lower bounds on the OBDD size of graphs of some popular functions. In *SOFSEM 2005*, volume 3381 of *LNCS*, pages 298–309. Springer, 2005.
[17] D. Sawitzki. The complexity of problems on implicitly represented inputs. To appear in *SOFSEM 2006*, 2006.
[18] I. Wegener. *Branching Programs and Binary Decision Diagrams.* SIAM, Philadelphia, 2000.
[19] P. Woelfel. New bounds on the OBDD-size of integer multiplication via universal hashing. In *STACS 2001*, volume 2010 of *LNCS*, pages 563–574. Springer, 2001.
[20] P. Woelfel. Symbolic topological sorting with OBDDs. In *MFCS 2003*, volume 2747 of *LNCS*, pages 671–680. Springer, 2003.
[21] A. Xie and P.A. Beerel. Implicit enumeration of strongly connected components. In *ICCAD 1999*, pages 37–40. ACM Press, 1999.

Constructions of Approximately Mutually Unbiased Bases

Igor E. Shparlinski[1] and Arne Winterhof[2]

[1] Department of Computing, Macquarie University,
Sydney, NSW 2109, Australia
igor@ics.mq.edu.au

[2] Johann Radon Institute for Computational and Applied Mathematics,
Austrian Academy of Sciences,
Altenberger Straße 69, A-4040 Linz, Austria
arne.winterhof@oeaw.ac.at

Abstract. We construct systems of bases of $\mathbb{C}^n$ which are mutually almost orthogonal and which might turn out to be useful for quantum computation. Our constructions are based on bounds of classical exponential sums and exponential sums over elliptic curves.

1 Introduction

A maximal set of *mutually unbiased bases*, for short MUBs, is given by a set of n^2+n vectors in $\mathbb{C}^n$, the n-dimensional vector space over the complex numbers $\mathbb{C}$, which are the elements of $n+1$ orthonormal bases $\mathcal{B}_h = \{\mathbf{w}_{h,1}, \dots, \mathbf{w}_{h,n}\}$ of $\mathbb{C}^n$ where $h = 0, \dots, n$. Hence,

$$\langle \mathbf{w}_{h,i}, \mathbf{w}_{h,j} \rangle = \delta_{i,j}, \tag{1}$$

where

$$\delta_{i,j} = \begin{cases} 1, & i = j, \\ 0, & i \neq j, \end{cases}$$

and the defining property is the mutual unbiasness, given by

$$|\langle \mathbf{w}_{f,i}, \mathbf{w}_{g,j} \rangle| = \frac{1}{\sqrt{n}} \tag{2}$$

for $0 \le f, g \le n$, $f \neq g$, and $1 \le i, j \le n$, where

$$\langle \mathbf{a}, \mathbf{b} \rangle = \sum_{u=1}^{n} \overline{a_u} b_u$$

denotes the standard inner product of two vectors

$$\mathbf{a} = (a_1, \dots, a_n), \ \mathbf{b} = (b_1, \dots, b_n) \in \mathbb{C}^n.$$

However, so far maximally sets of $n+1$ MUBs in dimension n are only known to exist in any dimension $n = p^r$ which is a power of a prime p, see [5, 10] for

J.R. Correa, A. Hevia, and M. Kiwi (Eds.): LATIN 2006, LNCS 3887, pp. 793–799, 2006.

an overview and some of such constructions. For $p \geq 3$, one of the most elegant constructions, see [10], is based on Gaussian sums and in the case of prime $n = p$ can be described as

$$\mathbf{w}_{h,k} = \frac{1}{\sqrt{p}} \left(\mathbf{e}_p(hu^2 + ku)\right)_{u=1}^{p}, \qquad 1 \leq h, k \leq p,$$

where $\mathbf{e}_m(x) = \exp(2\pi\sqrt{-1}x/m)$, and also $\mathcal{B}_0$ being a standard orthonormal basis, that is, $\mathbf{w}_{0,j} = (\delta_{j,u})_{u=1}^{p}$. One can use additive characters over an arbitrary finite field to extend this construction to an arbitrary prime power $n = p^r$. Very interesting links between constructing MUBs and some classical problems of Lie algebras have recently been discovered in [5].

However the condition that $n = p^r$ is a prime power is still somewhat too restrictive and unnatural for quantum computation. So a natural question arises whether MUBs exist for every positive integer n. Unfortunately, despite efforts of many researchers, no example of MUBs in other dimensions is known, and in fact most it is believed that indeed MUBs may exist only in prime power dimensions, see [1, 2, 3, 5, 9, 10, 14, 15, 19] and references therein.

Here, following the proposed in [11] approach, we consider vector systems where we relax the condition (2). In [11] exponential sums have been used to construct vector systems for any dimension n which

- satisfy (1) but instead of (2) all other inner products are $O(n^{-1/3})$;
- satisfy (1) and assuming some natural and widely believed conjecture on the distribution of primes in arithmetic progressions all other inner products are $O(n^{-1/2} \log n)$.

Here we improve the bound for the first construction from [11] from $O(n^{-1/3})$ to $O(n^{-1/2}(\log n)^{1/2})$. We also present a new construction, based on elliptic curves, satisfying (1) and all other inner products are $O(n^{-1/2})$. This construction applies to almost all dimensions n, and under some widely accepted conjectures about the gaps between consecutive prime numbers, it in fact applies to all dimensions.

2 Finite Field Construction

We need the following lemma which could be of independent interest.

For positive integers h, k, n and a prime $p \geq n$ we put

$$S_{h,k}(p, n) = \sum_{u=1}^{n} \mathbf{e}_p(hu^2)\mathbf{e}_n(ku).$$

Lemma 1. *For a positive integer n and a prime $p \geq n$ we have*

$$\max_{h=1,\ldots,p-1} \max_{k=0,\ldots,n-1} |S_{h,k}(p, n)| \leq \left(2\pi^{-1/2} + O\left(\frac{1}{\log p}\right)\right) p^{1/2}(\log p)^{1/2}.$$

Proof. We have

$$|S_{h,k}(p,n)|^2 = \sum_{u,v=1}^{n} \mathbf{e}_p\left(h\left(u^2-v^2\right)\right)\mathbf{e}_n(k(u-v))$$

$$\le 2\left|\sum_{1\le v<u\le n} \mathbf{e}_p\left(h\left(u^2-v^2\right)\right)\mathbf{e}_n(k(u-v))\right| + n.$$

Making the change of variable $u = v + w$ we derive

$$\begin{aligned}|S_{h,k}(p,n)|^2 &\le 2\left|\sum_{v=1}^{n-1}\sum_{w=1}^{n-v} \mathbf{e}_p\left(h\left((v+w)^2-v^2\right)\right)\mathbf{e}_n(kw)\right| + n\\
&= 2\left|\sum_{v=1}^{n-1}\sum_{w=1}^{n-v} \mathbf{e}_p\left(h\left(2vw+w^2\right)\right)\mathbf{e}_n(kw)\right| + n\\
&= 2\left|\sum_{w=1}^{n-1} \mathbf{e}_p\left(hw^2\right)\mathbf{e}_n(kw)\sum_{v=1}^{n-w}\mathbf{e}_p\left(2hvw\right)\right| + n\\
&\le 2\sum_{w=1}^{n-1}\left|\sum_{v=1}^{n-w}\mathbf{e}_p\left(2hvw\right)\right| + n\\
&\le 2\sum_{w=1}^{p-1}\left|\sum_{v=1}^{n-w}\mathbf{e}_p\left(2hvw\right)\right| + n\\
&= 2\sum_{w=1}^{p-1}\left|\sum_{v=1}^{n-w}\mathbf{e}_p\left(vw\right)\right| + n.\end{aligned}$$

Since the inner sum is the sum of a geometric progression, we obtain

$$\begin{aligned}|S_{h,k}(p,n)|^2 &\le 2\sum_{w=1}^{p-1}\left|\frac{\mathbf{e}_p\left((n-w+1)w\right)-\mathbf{e}_p\left(vw\right)}{\mathbf{e}_p\left(w\right)-1}\right| + n\\
&\le 4\sum_{w=1}^{p-1}\frac{1}{|\mathbf{e}_p\left(w\right)-1|} + n.\end{aligned}$$

Finally, since

$$\mathbf{e}_p\left(w\right) - 1 = 2\sqrt{-1}\mathbf{e}_p\left(w/2\right)\sin(\pi w/p),$$

we conclude

$$|S_{h,k}(p,n)|^2 \le 2\sum_{w=1}^{p-1}\frac{1}{|\sin(\pi w/p)|} + n.$$

Now the bound of Lemma 2 of [6] immediately implies the desired result. □

Let p be the smallest prime with $p \ge n$ and note that, in particular $p \sim n$ by the Prime Number Theorem. For each $f = 1, \ldots, n$ we consider the basis

$$\mathcal{B}_f = \{\mathbf{u}_{f,1}, \ldots, \mathbf{u}_{f,n}\}, \text{ where } \mathbf{u}_{f,i} = \frac{1}{\sqrt{n}}\left(\mathbf{e}_p\left(fu^2\right)\mathbf{e}_n\left(iu\right)\right)_{u=1}^{n}. \tag{3}$$

Theorem 1. *The standard basis $\mathcal{B}_0$ and the n bases $\mathcal{B}_f$, $f = 1,\ldots,n$, given by* (3) *are orthonormal and also satisfy*

$$|\langle \mathbf{u}_{f,i}, \mathbf{u}_{g,j}\rangle| \le \left(2\pi^{-1/2} + O\left(\frac{1}{\log n}\right)\right) n^{-1/2}(\log n)^{1/2}$$

where $f,g = 0,\ldots,n$, $f \ne g$ and $1 \le i,j \le n$.

Proof. The orthonormality of $\mathcal{B}_j$, $j = 0,\ldots,n$, is obvious (see also [11]).
If $f \ne g$ and $f = 0$ or $g = 0$ then obviously

$$|\langle \mathbf{u}_{f,i}, \mathbf{u}_{g,j}\rangle| = \frac{1}{\sqrt{n}}.$$

Finally, if $fg \ne 0$, we also have

$$\langle \mathbf{u}_{f,i}, \mathbf{u}_{g,j}\rangle = \sum_{u=1}^{n} \mathbf{e}_p((f-g)u^2)\mathbf{e}_n((i-j)u)$$

and the result now follows from Lemma 1. □

3 Elliptic Curve Construction

Let $\mathcal{E}$ be an elliptic curve over a finite field $\mathbb{F}_p$ of prime order $p > 3$ defined by the *affine Weierstrass equation*

$$Y^2 = X^3 + aX + b, \quad a,b \in \mathbb{F}_p,$$

such that $4a^3 + 27b^2 \ne 0$.

We recall that the set of $\mathbb{F}_p$ rational points on $\mathcal{E}$ forms an Abelian group (with the point at infinity $\mathcal{O}$ as the neutral element). The cardinality $n = \#\mathcal{E}(\mathbb{F}_p)$ of this group satisfies the *Hasse–Weil inequality*

$$|n - p - 1| \le 2p^{1/2}$$

or equivalently

$$n^{1/2} - 1 \le p^{1/2} \le n^{1/2} + 1 \tag{4}$$

(see [17] for this, and other general properties of elliptic curves).

Each polynomial $f \in \mathbb{F}_p(\mathcal{E})$ in the function field $\mathbb{F}_p(\mathcal{E})$ over $\mathcal{E}$, can be uniquely written in the form $f(X,Y) = u(X) + v(X)Y$ with some polynomials $u(X), v(X) \in \mathbb{F}_p[X]$. The degree of f is defined as

$$\deg(f) = \max(2\deg(u), 3 + 2\deg(v))$$

with the convention that the degree of the zero polynomial is $-\infty$.

For $2 \le d \le n-1$ denote by $\mathcal{F}_d$ the set of polynomials of degree at most d with $f(0,0) = 0$.

Lemma 2. *The cardinality of* $\mathcal{F}_d$ *is* $|\mathcal{F}_d| = p^{d-1}$.

Proof. If d is even then $f = u + vy$ has degree at most d if and only if $\deg(u) \le d/2$ and $\deg(v) \le d/2 - 2$. So we have $p^{d/2+1+d/2-1-1}$ polynomials f of degree at most d with $f(0,0) = u(0) = 0$. If d is odd the result follows similarly. □

As we have noticed, $\mathcal{E}(\mathbb{F}_p)$ is an Abelian group. Let $\mathcal{X}$ denote the corresponding character group.

For $f \in \mathbb{F}_p[\mathcal{E}]$ we define the set

$$\mathcal{B}_f = \{\mathbf{v}_{f,\chi} \ : \ \chi \in \mathcal{X}\},$$

where for a character $\chi \in \mathcal{X}$, the vector $\mathbf{v}_{f,\chi}$ is given by

$$\mathbf{v}_{f,\chi} = \frac{1}{\sqrt{n}} \left(\mathbf{e}_p(f(P))\chi(P)\right)_{P \in \mathcal{E}}$$

where we also define $f(\mathcal{O}) = 0$.

Theorem 2. *For* $2 \le d \le n-1$ *the standard basis and the* p^{d-1} *sets* $\mathcal{B}_f = \{\mathbf{v}_{f,\chi} \ : \ \chi \in \mathcal{X}\}$, *with* $f \in \mathcal{F}_d$, *are orthonormal and satisfy*

$$|\langle \mathbf{v}_{f,\chi}, \mathbf{v}_{g,\psi}\rangle| \le \frac{2d + (2d+1)n^{-1/2}}{n^{1/2}},$$

where $f, g \in \mathcal{F}_d$, $f \ne g$, *and* $\chi, \psi \in \mathcal{X}$.

Proof. The orthonormality follows immediately from the property of group characters,

$$\langle \mathbf{v}_{f,\chi}, \mathbf{v}_{f,\psi}\rangle = \frac{1}{n} \sum_{P \in \mathcal{E}} \chi(P)\psi^{-1}(P) = \begin{cases} 1, \chi = \psi, \\ 0, \chi \ne \psi. \end{cases}$$

For $f \ne g$ we have

$$\begin{aligned} |\langle \mathbf{v}_{f,\chi}, \mathbf{v}_{g,\psi}\rangle| &\le \frac{1}{n}\left(\left|\sum_{\substack{P \in \mathcal{E} \\ P \ne \mathcal{O}}} \mathbf{e}_p(f(P) - g(P))\chi(P)\psi^{-1}(P)\right| + 1\right) \\ &\le \frac{1}{n}(2dp^{1/2} + 1) \le \frac{2dn^{1/2} + 2d + 1}{n} \end{aligned}$$

by [12, Theorem 1] and the bound (4). □

4 Remarks

By the classical results of Deuring [8], see also [4, 16, 18], for every prime $p > 3$, each integer n in the so-called *Hasse-Weil interval* $[p + 1 - 2p^{1/2}, p + 1 + 2p^{1/2}]$ is a cardinality of some elliptic curve over $\mathbb{F}_p$, see also [4, 16, 18].

Note that the probability that an integer n is not in such an interval for some prime p is very small. More precisely, for any fixed $\varepsilon > 0$, the number of such integers $n \le x$ is $O(x^{25/36+\varepsilon})$ by a result of [13]. By the Cramer conjecture [7] the distance of the nth and $(n+1)$th primes p_n and p_{n+1} is

$$p_{n+1} - p_n = O\left((\log p_n)^2\right).$$

Certainly under this conjecture every positive integer n presents a cardinality of some elliptic curve over a finite field. On the other hand, the Riemann Hypothesis seems not to be strong enough to derive this statement.

Acknowledgments

During the preparation of this paper, I. S. was supported in part by ARC grant DP0556431, A. W. was supported in part by the Austrian Academy of Sciences and by FWF grant S8313.

References

1. M. Aschbacher, A. M. Childs, and P. Wocjan, 'The limitations of nice mutually unbiased bases', *Preprint*, 2004, available from `http://arxiv.org/abs/quant-ph/0412066`.
2. S. Bandyopadhyay, P. O. Boykin, V. Roychowdhury, and F. Vatan, 'A new proof for the existence of mutually unbiased bases', *Algorithmica*, **34** (2001), 512–528.
3. I. Bengtsson and A. Ericsson, 'Mutually unbiased bases and the complementarity polytope', *Open Syst. Inform. Dynamics*, **12** (2005), 107–120.
4. B. J. Birch, 'How the number of points of an elliptic curve over a fixed prime field varies', *J. Lond. Math. Soc.*, **43** (1968), 57–60.
5. P. O. Boykin, M. Sitharam, P. H. Tiep and P. Wocjan, 'Mutually unbiased bases and orthogonal decompositions of Lie algebras', *Preprint*, 2005, available from `http://arxiv.org/abs/quant-ph/0506089`
6. T. Cochrane and J. C. Peral, 'An asymptotic formula for a trigonometric sum of Vinogradov', *J. Number Theory*, **91** (2001), 1–19.
7. H. Cramer, 'On the order of magnitude of the difference between consecutive prime numbers', *Acta Arith.*, **2** (1936), 23–46.
8. M. Deuring, 'Die Typen der Multiplikatorenringe elliptischer Funktionenkörper', *Abh. Math. Sem. Hansischen Univ.*, **14** (1941), 197–272.
9. S. Chaturvedi, 'Aspects of mutually unbiased bases in odd-prime-power dimensions', *Phys. Rev. A*, **65**, 2002, 044301.
10. A. Klappenecker and M. Rötteler, 'Constructions of mutually unbiased bases', *Lect. Notes in Comp. Sci.*, Springer-Verlag, Berlin, **2948** (2004), 137–144.
11. A. Klappenecker, M. Rötteler, I. Shparlinski and A. Winterhof, 'On approximately symmetric informationally complete positive operator-valued measures and related systems of quantum states', *J. Math. Physics*, **46** (2005), 082104.
12. D. R. Kohel and I. E. Shparlinski, 'Exponential sums and group generators for elliptic curves over finite fields', *Lect. Notes in Comp. Sci.*, Springer-Verlag, Berlin, **1838** (2000), 395–404.

13. A. S. Peck, 'Differences between consecutive primes', *Proc. London Math. Soc.*, **76** (1998), 33–69.
14. M. Planat and H. Rosu, 'Mutually unbiased bases, phase uncertainties and Gauss sums', *European Physical J. D*, **36** (2005), 133–139.
15. M. Planat and M. Saniga, 'Ovals in finite projective planes and complete sets of mutually unbiased bases', *Intern. J. of Modern Physics B*, (to appear).
16. R. Schoof, 'Nonsingular plane cubic curves over finite fields', *J. Combin. Theory, Ser.A*, **47** (1987), 183–211.
17. J. H. Silverman, *The arithmetic of elliptic curves*, Springer-Verlag, Berlin, 1995.
18. W. C. Waterhouse, 'Abelian varieties over finite fields', *Ann. Sci. Ecole Norm. Sup.*, **2** (1969), 521–560.
19. P. Wocjan and T. Beth, 'New construction of mutually unbiased bases in square dimensions', *Quantum Infor. and Comp.*, **5** (2005), 93–101.

Improved Exponential-Time Algorithms for Treewidth and Minimum Fill-In

Yngve Villanger

Department of Informatics,
University of Bergen N-5020 Bergen,Norway
yngvev@ii.uib.no

Abstract. Exact exponential-time algorithms for NP-hard problems is an emerging field, and an increasing number of new results are being added continuously. Two important NP-hard problems that have been studied for decades are the treewidth and the minimum fill problems. Recently, an exact algorithm was presented by Fomin, Kratsch, and Todinca to solve both of these problems in time $\mathcal{O}^*(1.9601^n)$. Their algorithm uses the notion of potential maximal cliques, and is able to list these in time $\mathcal{O}^*(1.9601^n)$, which gives the running time for the above mentioned problems. We show that the number of potential maximal cliques for an arbitrary graph G on n vertices is $\mathcal{O}^*(1.8135^n)$, and that all potential maximal cliques can be listed in $\mathcal{O}^*(1.8899^n)$ time. As a consequence of this results, treewidth and minimum fill-in can be computed in $\mathcal{O}^*(1.8899^n)$ time.

1 Introduction

Recently there has been a growing interest for exact exponential time algorithms for NP-hard problems. There are several reasons for this. One is the need for exact solutions, which approaches like approximation algorithms, randomized algorithms, and heuristics, cannot deal with.

An exhaustive search is a trivial way to cope with the problem of finding an exact solution. In the recent yeas it has been shown that it is possible to design algorithms which are significantly faster than exhaustive search, though still not in polynomial time. Nice examples of this type of algorithms are a $\mathcal{O}^*(1.4802^n)$ time algorithm for 3-SAT [9] and Eppstein's algorithm for graph coloring in $\mathcal{O}^*(2.4150^n)$ time [10]. (In this paper we use a modified big-Oh notation that suppresses all other (polynomial bounded) terms. For functions f and g we write $f(n) = \mathcal{O}^*(g(n))$ if $f(n) = \mathcal{O}(g(n) \cdot poly(|n|))$, where $poly(|n|)$ is a polynomial. This modification may be justified by the exponential growth of $f(n)$.) An overview of applied techniques used for exact algorithms can be found in [17].

The *treewidth* of a graph, introduced by Robertson and Seymour [14], has been intensively investigated in the last years, mainly because many NP-hard problems become solvable in polynomial time when restricted to graphs with small treewidth. These algorithms use a *tree-decomposition* (or a triangulation)

J.R. Correa, A. Hevia, and M. Kiwi (Eds.): LATIN 2006, LNCS 3887, pp. 800–811, 2006.

of small width of the graph. In recent years [8] it has bee shown that the results on graphs of bounded treewidth (branchwidth) are not only of theoretical interest but can successfully be applied to find optimal solutions of lower time bounds for various optimization problems. Finding a small treewidth is useful and important in areas like artificial intelligence, databases, and logical-circuit design. See [1] for further references.

The *minimum fill-in* problem asks to find a triangulation (equivalently a tree-decomposition) with the minimum number of edges. This problem has applications in sparse matrix computations [15], database management [16], and knowledge base systems [13].

Computing the treewidth and minimum fill-in are NP-hard problems [2, 18]. Treewidth is known to be fixed parameter tractable, moreover, for a fixed k, the treewidth of size k can be computed in linear time (with a huge hidden constant) [4]. There exists also an approximation algorithm for treewidth, with a factor $logOPT$ [1, 5], and it is an open question if there exists a constant factor approximation.

Both treewidth and minimum fill-in can be computed exactly in $\mathcal{O}^*(2^n)$ time by reformulating the problems to finding a special vertex ordering and using the technique proposed by Held and Karp [12] for the traveling salesman problem, or by using the algorithm of Arnborg et al.[2]. In 2004 Fomin, Kratsch, and Todinca [11] improved this bound to $\mathcal{O}^*(1.9601^n)$ by listing all the minimal separators and potential maximal cliques of the graph, and then using these to compute the treewidth and minimum fill-in for the graph. The most expensive operation used in [11] to obtain the $\mathcal{O}^*(1.9601^n)$ time bound is listing the potential maximal cliques. It is actually known from [7] that the number of potential maximal cliques in a graph is bounded by the number of *nice* potential maximal cliques in the graph.

In this paper we find a new theoretical bound ($\mathcal{O}^*(1.8135^n)$) for the number of nice potential maximal cliques in a graph, and thus also a new bound for the number of potential maximal cliques in the graph. This is obtained using a non constructive proof, and cannot be used directly to create faster algorithms. The second result in this paper is a new way of partitioning the graph, such that any nice potential maximal clique can be represented by a vertex set of size $n/3$ or less, which is less than the $2n/5$ bound used in [11]. This new bound improves the time required to list all the potential maximal cliques to $\mathcal{O}^*(1.8899^n)$, and thus also the bound for computing the treewidth and minimum fill-in.

2 Basic Definitions

We consider finite, simple, undirected, and connected graphs. Given a graph $G = (V, E)$, we denote the number of vertices as $n = |V|$ and the number of edges as $m = |E|$. For any non empty subset $W \subseteq V$, the subgraph of G induced by W is denoted by $G[W]$, and the subgraph induced by $G[V \setminus W]$ is denoted by $G \setminus W$. The *neighborhood* of a vertex $u \in V$ is denoted by $N_G(u) = \{v$ for $uv \in E\}$, and $N_G[u] = N_G(u) \cup \{u\}$. In the same way we define the neighborhood of a set $A \subseteq V$ of vertices by $N_G(A) = \cup_{u \in A} N_G(u) \backslash A$, and $N_G[A] = N_G(A) \cup A$. A sequence $v_1 - v_2 - ... - v_k$ of distinct vertices describes a *path* if $v_i v_{i+1}$ is an edge for $1 \leq i < k$. The *length* of a path is the number of edges in the path. A *cycle* is defined as a path except that it starts and ends with the same vertex. If there is an edge between every pair of vertices in a set $A \subseteq V$, then the set A is called a *clique.*

The notion of treewidth is due to Robertson and Seymour [14]. A *tree decomposition* of a graph $G = (V, E)$, denoted by $TD(G)$, is a pair (X, T) such that $T = (V_T, E_T)$ is a tree and $X = \{X_i \mid i \in V_T\}$ is a family of subsets of V such that:

1. $\bigcup_{i \in V_T} X_i = V$;
2. for each edge $uv \in E$ there exists an $i \in V_T$ such that both u and v belong to X_i;
3. for all $v \in V$, the set of nodes $\{i \in V_T \mid v \in X_i\}$ induces a connected subtree of T.

The *width* of a tree decomposition is defined as maximum of $|X_i| - 1$ where $i \in V_T$, and the *treewidth* of the graph G is the minimum width over all tree decompositions of G.

A *chord* of a cycle is an edge connecting two non-consecutive vertices of the cycle. A graph H is *chordal*, or equivalently *triangulated*, if it contains no induced chordless cycle of length ≥ 4. A graph $H = (V, E \cup F)$ is called a *triangulation* of $G = (V, E)$ if H is chordal. The edges in F are called *fill edges.* H is a *minimal triangulation* if $(V, E \cup F')$ is non-chordal for every proper subset F' of F. H is a *minimum triangulation* if there exists no edge set F' such that $|F'| < |F|$ and $(V, E \cup F')$ is chordal. The problem of finding the smallest value of $|F|$, such that $H = (V, E \cup F)$ is chordal is called the *minimum fill-in* problem for the graph $G = (V, E)$.

A vertex set $S \subset V$ is a *separator* if $G \backslash S$ is disconnected. Given two vertices u and v, S is a *u, v-separator* if u and v belong to different connected components of $G \setminus S$, and S is then said to *separate* u and v. A u, v-separator S is *minimal* if no proper subset of S separates u and v. In general, S is a *minimal separator* of G if there exist two vertices u and v in G such that S is a minimal u, v-separator. We denote by Δ_G the set of all minimal separators of G. The following two results will be used to list all minimal separators, and give an upper bound for the number of minimal separators.

Theorem 1. ([3]) *There is an algorithm listing all minimal separators of an input graph G in $\mathcal{O}(n^3 |\Delta_G|)$ time.*

Theorem 2. ([11]) *For any graph G, $|\Delta_G| = \mathcal{O}(n \cdot 1.7087^n)$.*

For a set $K \subseteq V$, a connected component C of $G \setminus K$ is a *full component associated* to K if $N(C) = K$. A vertex set $\Omega \subset V$ is called a *potential maximal clique* of G if there is a minimal triangulation H of G, such that Ω is a maximal clique in H. We denote by Π_G the set of all potential maximal cliques of G.

Theorem 3. (Bouchitté and Todinca [6]) *Let $K \subseteq V$ be a set of vertices and let $\mathcal{C}(K) = \{C_1, ..., C_p\}$ be the set of connected components of $G \setminus K$. Let $\mathcal{S}(K) = \{S_1, S_2, ..., S_p\}$ where $S_i(K)$ is the set of vertices of K adjacent to at least one vertex of $C_i(K)$. Then K is a potential maximal clique if and only if:*

1. *$G \setminus K$ has no full component associated to K, and*
2. *the graph on the vertex set K obtained from $G[K]$ by turning each $S_i \in \mathcal{S}(K)$ into a clique, is a complete graph.*

The following result is an easy consequence of Theorem 3.

Theorem 4. ([6]) *There is an algorithm that, given a graph $G = (V, E)$ and a set of vertices $K \subseteq V$, verifies if K is a potential maximal clique of G. The time complexity of the algorithm is $\mathcal{O}(nm)$.*

Three different ways of representing a potential maximal clique is given in the next lemma. We will see that potential maximal cliques that can be represented by the two first of these already can be found and listed within a good time bound.

Lemma 1. (Fomin, Kratsch, and Todinca [11]) *Let Ω be a potential maximal clique of G, S be a minimal separator contained in Ω and C be the component of $G \setminus S$ intersecting Ω. Then one of the following holds:*

1. *there is $a \in \Omega \setminus S$ such that $\Omega = N[a]$;*
2. *there is $a \in S$ such that $\Omega = S \cup (N(a) \cap C)$;*
3. *$\Omega = N(C \setminus \Omega)$.*

The number of potential maximal cliques covered by the first case is clearly bounded by n, since only one such potential maximal clique can exist for each vertex in the graph.

From [11] we have the following statement covering the second case. Let Ω be a potential maximal clique of G. The triple (S, a, b) is called a *separator representation* of Ω if S is a minimal separator of G, $a \in S$, $b \in V \setminus S$, and $\Omega = S \cup (N(a) \cap C_b(S))$, where $C_b(S)$ is the component of $G \setminus S$ containing b. Note that for a given triple (S, a, b) one can check in polynomial time if (S, a, b) is the separator representation of a (unique) potential maximal clique Ω.

The number of unique potential maximal cliques in a graph, that have a separator representation is bounded by $n^2|\Delta_G|$, since there are $\mathcal{O}(n^2)$ triples for each separator. From Theorem 2 we have that $|\Delta_G| = \mathcal{O}(n \cdot 1.7087^n)$, thus the number of unique potential maximal cliques with a separator representation is of order $\mathcal{O}(n^3 \cdot 1.7087^n)$.

Let Ω be a potential maximal clique of a graph G, and let $S \subset \Omega$ be a minimal separator of G. We say that S is an *active separator* for Ω, if Ω is not a clique in the graph $G_{S(\Omega)\setminus\{S\}}$, obtained from G by completing all the minimal separators contained in Ω, except S. If S is active, a pair of vertices $x, y \in S$ non adjacent in $G_{S(\Omega)\setminus\{S\}}$ is called an *active pair*.

Theorem 5. (Bouchitté and Todinca [6]) *Let Ω be a potential maximal clique of G, S be a minimal separator contained in Ω and C be the component of $G \setminus S$ intersecting Ω, and let $x, y \in S$ be an active pair. Then $\Omega \setminus S$ is a minimal x, y-separator in $G[C \cup \{x, y\}]$.*

We say that a potential maximal clique Ω is *nice* if at least one of the minimal separators contained in Ω is active for Ω.

Theorem 6. (Bouchitté and Todinca [7]) *Let Ω be a potential maximal clique of G, let u be a vertex of G, and let $G' = G \setminus \{u\}$. Then one of the following holds:*

1. *Ω or $\Omega \setminus \{u\}$ is a potential maximal clique of G'.*
2. *$\Omega = S \cup \{u\}$, where S is a minimal separator of G.*
3. *Ω is nice.*

The following result can be found using Theorem 6.

Corollary 1. [11] *A graph G on n vertices has at most $n^2|\Delta_G| + n \cdot \Pi_{NG} = n^2 \cdot 1.701^n + n \cdot \Pi_{NG}$ potential maximal cliques, where Π_{NG} is the number of nice potential maximal cliques in the graph.*

Proof. This follows from the Theorems 2 and 6, and the proof of Theorem 16 of [11].

Finally we can relate the upper bound for listing all potential maximal cliques of G to computing the treewidth and minimum fill-in of G. Theorem 7 is the tool we need to obtain this.

Theorem 7. (Fomin, Kratsch, and Todinca [11]) *There is an algorithm that, given a graph G together with the list of its minimal separators Δ_G and the list of its potential maximal cliques Π_G, computes the threewidth and the minimum fill-in of G in $\mathcal{O}^*(|\Pi_G|)$ time. The algorithm also constructs optimal triangulations for the threewidth and the minimum fill-in.*

3 Theoretical Upper Bound for the Number of Potential Maximal Cliques

In this section we show that the upper bound for the number of potential maximal cliques in a graph is $\mathcal{O}(n^3 \cdot 1.8135^n)$. This bound is obtained by finding a new upper bound for the number of nice potential maximal cliques. We do this by computing two numbers: the number of potential maximal cliques of size less than αn and the number of potential maximal cliques of size at least αn, for $0 < \alpha < 1$.

Let Ω be a potential maximal clique of G and let x be a vertex in Ω. Let $C_{\Omega x}$ be the connected component of $G \setminus (\Omega \setminus \{x\})$ containing x. Notice that $G[C_{\Omega x}]$ is connected, and that every component C of $G \setminus \Omega$ such that $x \in N(C)$ is contained in $C_{\Omega x}$.

Corollary 2. *Let Ω be a potential maximal clique of G and let x be a vertex in Ω. Then $\Omega = N(C_{\Omega x}) \cup \{x\}$.*

Proof. This follows directly from Theorem 3, which gives a definition of a potential maximal clique.

Definition 1. *We will say that the pair (Z, z) is a* vertex representation *of Ω if $Z = C_{\Omega z} \setminus \{z\}$, $z \in \Omega$, and $\Omega = N(Z \cup \{z\}) \cup \{z\}$.*

Lemma 2. *Let Ω be a nice potential maximal clique, α be a constant such that $\alpha n = |\Omega|$. Then there exists a* vertex representation *(U, u) of Ω such that $|U| \leq \lceil 2n(1-\alpha)/3 \rceil$.*

Proof. Let Ω be a nice potential maximal clique of G, S be a minimal separator active for Ω, $x, y \in S$ be an active pair, and z be a vertex contained in $\Omega \setminus S$. Let us now prove that there exists a vertex u such that $|C_{\Omega u} \setminus \{u\}| \leq \lceil 2n(1-\alpha)/3 \rceil$. Partition the connected components of $G \setminus \Omega$ into three sets: $A_1 = C_{\Omega x} \cap C_{\Omega y}$; $A_2 = C_{\Omega x} \setminus (C_{\Omega y} \cup \{x\})$; $A_3 = (V \setminus \Omega) \setminus (A_1 \cup A_2)$. Notice the following: $|A_1 \cup A_2 \cup A_3| = n(1-\alpha)$ since $A_1 \cup A_2 \cup A_3 = V \setminus \Omega$, A_1, A_2, A_3 are pairwise non intersecting, and most important: $C_{\Omega x} \setminus \{x\} = A_1 \cup A_2$; $C_{\Omega y} \setminus \{y\} \subseteq A_1 \cup A_3$; $C_{\Omega z} \setminus \{z\} \subseteq A_2 \cup A_3$.

One of the vertex sets A_1, A_2, A_3 will be of size at least $n(1-\alpha)/3$, thus the remaining two are of size at most $\lceil 2n(1-\alpha)/3 \rceil$. Let us without loss of generality assume that $|A_1| \geq n(1-\alpha)/3$, then $|A_2| + |A_3| \leq \lceil 2n(1-\alpha)/3 \rceil$. It follows that $|C_{\Omega z} \setminus \{z\}| \leq \lceil 2n(1-\alpha)/3 \rceil$ since $C_{\Omega z} \setminus \{z\} \subseteq A_2 \cup A_3$, and thus there exists a vertex representation (U, u) of Ω as claimed by the lemma.

Lemma 3. *For a constant $0 < \alpha < 1$, and a graph G, the number of nice potential maximal cliques of size at least αn vertices is not more than $n \sum_{i=1}^{\lceil 2n(1-\alpha)/3 \rceil} \binom{n}{i}$.*

Proof. It follows from Lemma 2 that every potential maximal clique Ω of size at least αn has a vertex representation (X, x) such that $|X| \leq \lceil 2n(1-\alpha)/3 \rceil$. The idea of the proof is to give a bound for the number of such pairs. The number of unique vertex sets of size $\lceil 2n(1-\alpha)/3 \rceil$ or less is $\sum_{i=1}^{\lceil 2n(1-\alpha)/3 \rceil} \binom{n}{i}$. For each such vertex set X we create a pair (X, x) for each vertex $x \in V \setminus S$, which give us the multiplication by n.

Lemma 4. *For a constant $0 < \alpha < 1$, and a graph G, the number of nice potential maximal cliques of size less than αn vertices is not more than $2^{n(2+\alpha)/3}$.*

Proof. We know from Lemma 2 that every potential maximal clique Ω of size less than αn has a vertex representation (U, u) such that $|V \setminus (\Omega \cup U)| \geq n(1-\alpha)/3$. We say that (x, X) is a *bad pair* associated to Ω if $\Omega = N(C_x) \cup \{x\}$, where C_x is the connected component of $G[X \cup \{x\}]$ containing x.

Let (x, X) be a bad pair associated to Ω_x and let (y, Y) be associated to Ω_y, where $\Omega_x \neq \Omega_y$. We want to prove that $(x, X) \neq (y, Y)$. Suppose that $x = y$ and that $X = Y$. From the definition of bad pair we know that $N(C_x) \cup$

$\{x\} = N(C_y) \cup \{y\}$. Now we have a contradiction since $N(C_x) \cup \{x\} = \Omega_x$, $N(C_y) \cup \{y\} = \Omega_y$, and $\Omega_x \neq \Omega_y$.

Since (U, u) is a vertex representation of Ω, then $U \cup \{u\} = C_{\Omega u}$. Remember that $C_{\Omega u}$ is connected and that $|V \setminus N[C_{\Omega u}]| \geq n(1-\alpha)/3$. Thus we can create $2^{n(1-\alpha)/3}$ unique bad pairs u, X for Ω, by selecting $X = C_{\Omega u} \cup Z$, where Z is any of the $2^{n(1-\alpha)/3}$ subset of $V \setminus N[C_{\Omega u}]$.

It follows that $2^n \geq |\Pi_{NGs\alpha}| \cdot 2^{n(1-\alpha)/3}$, which can be restated as $|\Pi_{NGs\alpha}| \leq 2^{n(2+\alpha)/3}$, where $|\Pi_{NGs\alpha}|$ is the number of nice potential maximal cliques of size less than αn.

Lemma 5. *The number of nice potential maximal cliques in a graph G with n vertices is $\mathcal{O}(n^2 \cdot 1.8135^n)$.*

Proof. Let Π_{NG} be the set of nice potential maximal cliques, $\Pi_{NGl\alpha}$ be the set of potential maximal cliques of size at least αn, and $\Pi_{NGs\alpha}$ be the set of potential maximal cliques of size less than αn. Then $|\Pi_{NG}| = |\Pi_{NGl\alpha}| + |\Pi_{NGs\alpha}| \leq n \cdot \sum_{i=1}^{\lceil 2n(1-\alpha)/3 \rceil} \binom{n}{i} + 2^{n(2+\alpha)/3}$. By making use of Stirling's formula and using $\alpha = 0.5763$ we obtain the bound $\mathcal{O}(n^2 \cdot 1.8135^n)$.

Theorem 8. *For any graph G, $\Pi_G = \mathcal{O}(n^3 \cdot 1.8135^n)$.*

Proof. From Corollary 1 we have that the number of potential maximal cliques in G is less than $n^2|\Delta_G| + n \cdot \Pi_{NG} = n^3 \cdot 1.701^n + n \cdot \Pi_{NG}$ potential maximal cliques, where Π_{NG} is the number of nice potential maximal cliques in the graph. By inserting the result from Lemma 5 we get the new result that $\Pi_G = \mathcal{O}(n^3 \cdot 1.8135^n)$.

4 Listing All the Potential Maximal Cliques

In this section we show that any potential maximal clique of a graph with n vertices can be represented with $n/3$ vertices or less, thus it follows that all potential maximal cliques of the graph can be listed in $\mathcal{O}^*(\binom{n}{n/3})$, or equivalent $\mathcal{O}^*(1.8899^n)$ time.

The idea is to show that every nice potential maximal clique which is not covered by the two first cases of Lemma 1 can be represented by a vertex set of size $n/3$ or less. From the results of [11] we know that the number of nice potential maximal cliques covered by the two first cases of Lemma 1 is bounded by $n + n^2|\Delta_G|$ and from [7] it follows that the potential maximal cliques which is not nice can be generated from the nice potential maximal cliques.

To describe these different representations of a potential maximal clique we need to partition the graph into different vertex set. The first step towards this partitioning is given in Lemma 6 which is a slightly refinement of similar lemma and proof given in [11].

Lemma 6. *Let Ω be a nice potential maximal clique, S be a minimal separator active for Ω, $x, y \in S$ be an active pair, and C be the component of $G \setminus S$ containing $\Omega \setminus S$. There is a partition (D_x, D_y, D_r) of $C \setminus \Omega$ such that $N(D_x \cup \{x\}) \cap C = N(D_y \cup \{y\}) \cap C = \Omega \setminus S$.*

Proof. By Theorem 5, $\Omega \setminus S$ is a minimal x, y-separator in $G[C \cup \{x, y\}]$. Let C_x be the full component associated to $\Omega \setminus S$ in $G[C \cup \{x, y\}]$ containing x, $D_x = C_x \setminus \{x\}$, and let C_y be the full component associated to $\Omega \setminus S$ in $G[C \cup \{x, y\}]$ containing y, $D_y = C_y \setminus \{y\}$, and $D_r = C \setminus (\Omega \cup D_x \cup D_y)$. Since $D_x \cup \{x\}$ and $D_y \cup \{y\}$ are full components of $\Omega \setminus S$, we have that $N(D_x \cup \{x\}) \cap C = N(D_y \cup \{y\}) \cap C = \Omega \setminus S$.

Definition 2. *For a potential maximal clique Ω of G, we say that a pair (X, c), where $X \subset V$ and $c \in X$ is a* partial representation *of Ω if $\Omega = N(C_c) \cup (X \setminus C_c)$, where C_c is the connected component of $G[X]$ containing c.*

Definition 3. *For a potential maximal clique Ω of G, we say that a triple (X, x, c), where $X \subset V$ and $x, c \notin X$ is an* indirect representation *of Ω if $\Omega = N(C_c \cup D_x \cup \{x\}) \cup \{x\}$, where*

- *C_c is the connected component of $G \setminus N[X]$ containing c;*
- *D_x is the vertex set of the union of all connected components C' of $G[X]$ such that $x \in N(C')$.*

Let us note that for a given vertex set X and two vertices x, c one can check in polynomial time whether the pair (X, c) is a partial representation or if the triple (X, x, c) is a separator representation or indirect representation of a (unique) potential maximal clique Ω.

We state now the main tool for upper bounding the number of nice potential maximal cliques.

Lemma 7. *Let Ω be a nice potential maximal clique of G. Then one of the following holds:*

1. *There is a vertex a such that $\Omega = N[a]$;*
2. *Ω has a separator representation;*
3. *Ω has a partial representation (X, c) such that $|X| \leq n/3$;*
4. *Ω has a indirect representation (X, x, c) such that $|X| \leq n/3$.*

Proof. Let S be a minimal separator active for Ω, $x, y \in S$ be an active pair, and C be the component of $G \setminus S$ containing $\Omega \setminus S$. By Lemma 6, there is a partition (D_x, D_y, D_r) of $C \setminus \Omega$ such that $N(D_x \cup \{x\}) \cap C = N(D_y \cup \{y\}) \cap C = \Omega \setminus S$. If one of the sets D_x, D_y, say D_x, is equal the emptyset, then $N(D_x \cup \{x\}) \cap C = N(x) \cap C = \Omega \setminus S$, and thus the triple (S, x, z) is a separator representation of Ω.

Suppose that none of the first two conditions of the lemma holds. Then D_x and D_y are nonempty. In order to argue that Ω has a partial representation (X, c) or a indirect representation (X, x, c) such that $|X| \leq n/3$, we partition the graph further. Let $R = \Omega \setminus S$ and let D_S be the set of full components associated to S in $G \setminus \Omega$. The vertex set D_x is the union of vertex sets of all connected components C' of $G \setminus (\Omega \cup D_S)$ such that x is contained in the neighborhood of C'. Thus a connected component C' of $G \setminus (\Omega \cup D_S)$ is contained in D_x if and only if $x \in N(C')$. Similarly, a connected component C' of $G \setminus (\Omega \cup D_S)$ is contained in D_y if and only if $y \in N(C')$. We also define $D_r = V \setminus (\Omega \cup D_S \cup D_x \cup D_y)$, which is the set of vertices of the components of $G \setminus (\Omega \cup D_S)$ which are not in D_x and D_y.

We partition S in the following sets

- $S_{\overline{x}} = (S \setminus N(D_x)) \cap N(D_y)$;
- $S_{\overline{y}} = (S \setminus N(D_y)) \cap N(D_x)$;
- $S_{\overline{xy}} = S \setminus (N(D_y) \cup N(D_x))$;
- $S_{xy} = S \cap N(D_y) \cap N(D_x)$.

Thus $S_{\overline{x}}$ is the set of vertices in S with no neighbor in D_x and with at least one neighbor in D_y, $S_{\overline{y}}$ is the set of vertices in S with no neighbor in D_y and with at least one neighbor in D_x, $S_{\overline{xy}}$ is the set of vertices in S with neighbors neither in D_x or D_y, and finally S_{xy} is the set of vertices in S with neighbors both in D_x and D_y. Notice that the vertex sets D_S, D_x, D_y, D_r, R,$S_{\overline{x}}$, $S_{\overline{y}}$, $S_{\overline{xy}}$, and S_{xy} are pairwise disjoint. The set S_{xy} is only mentioned to complete the partition of S, and will not be used in the rest of the proof.

Both for size requirements and because of the definition of indirect representation we can not use the sets $S_{\overline{x}}, S_{\overline{y}}$, and $S_{\overline{xy}}$ directly, they have to be represented by the sets $Z_{\overline{x}}, Z_{\overline{y}}$, and $Z_{\overline{r}}$, which are subsets of the vertex sets D_y, D_x, and D_r. By the definition of $S_{\overline{x}}$ and $S_{\overline{y}}$ it follows that there exists two vertex sets $Z_{\overline{x}} \subseteq D_y$ and $Z_{\overline{y}} \subseteq D_x$ such that $S_{\overline{x}} \subseteq N(Z_{\overline{x}})$ and $S_{\overline{y}} \subseteq N(Z_{\overline{y}})$, let $Z_{\overline{x}}$ and $Z_{\overline{y}}$ be the smallest such sets. By Lemma 1, $\Omega = N(D_x \cup D_y \cup D_r)$, thus it follows that there exists a vertex set $Z_{\overline{r}} \subseteq D_r$ such that $S_{\overline{xy}} \subseteq N(Z_{\overline{r}})$, let $Z_{\overline{r}}$ be the smallest such set.

Let C^* be a connected component of $G[D_S]$, remember that $N(C^*) = S$. We define the following sets

- $X_1 = C^* \cup R$;
- $X_2 = D_x \cup Z_{\overline{x}} \cup Z_{\overline{r}}$;
- $X_3 = D_y \cup Z_{\overline{y}} \cup Z_{\overline{r}}$.

First we claim that

- the pair (X_1, c), where $c \in C^*$, is a partial representation of Ω;
- the triple (X_2, x, c), where $c \in C^*$ is an indirect representation of Ω;
- the triple (X_3, x, c), where $c \in C^*$ is an indirect representation of Ω.

In fact, the pair $(X_1, c) = (C^* \cup R, c)$ is a partial representation of Ω because $N(C^*) \cap R = \emptyset$, C^* induces a connected graph, and $\Omega = N(C^*) \cup R$. Thus (X_1, c) is a partial representation of Ω.

To prove that $(X_2, x, c) = (D_x \cup Z_{\overline{x}} \cup Z_{\overline{r}}, x, c)$ is an indirect representation of Ω, we have to show that $\Omega = N(C_c \cup D'_x \cup \{x\}) \cup \{x\}$ where C_c is the connected component of $G \setminus N[X_2]$ containing c, and D'_x is the vertex set of the union of all connected components C' of $G[X_2]$ such that $x \in N(C')$. Notice that $(S \cup C^*) \cap X_2 = \emptyset$ and that $S \subseteq N(X_2)$ since $S \subseteq N(D_x \cup Z_{\overline{x}} \cup Z_{\overline{r}})$ and $X_2 = D_x \cup Z_{\overline{x}} \cup Z_{\overline{r}}$. Hence the connected component C_c of $G \setminus N[X_2]$ containing c is C^*.

Every connected component C' of $G[X_2]$ is contained in D_x, $Z_{\overline{x}}$, or $Z_{\overline{r}}$ since $\Omega \cap (D_x \cup Z_{\overline{x}} \cup Z_{\overline{r}}) = \emptyset$ and Ω separates D_x, $Z_{\overline{x}}$, and $Z_{\overline{r}}$. From the definition of D_x

it follows that $x \in N(C')$ for every component C' of $G[D_x]$, and from the definition of D_y and D_r follows that $x \notin N(C')$ for every component C' of $G[Z_{\overline{x}} \cup Z_{\overline{r}}]$. We can now conclude that D_x is the vertex set of the union of all connected components C' of $G[X_2]$ such that $x \in N(C')$. It remains to prove that $\Omega = N(C^* \cup D_x \cup \{x\}) \cup \{x\}$. By Lemma 6, we have that $\Omega \setminus S = R$ is subset of $N(D_x \cup \{x\})$ and $N(D_y \cup \{y\})$, and remember that $N(C^*) = S$. From this observations it follows that $\Omega = N(C^* \cup D_x \cup \{x\}) \cup \{x\}$ since $N(C^* \cup D_x \cup \{x\}) = (S \cup R) \setminus \{x\}$.

By similar arguments, (X_3, x, c) is an indirect representation of Ω.

To conclude the proof of Lemma, we argue that at least one of the vertex sets X_1, X_2, or X_3 used to represent Ω, contains at most $n/3$ vertices.

We partition the graph in the following three sets:

- $V_1 = D_S \cup R$;
- $V_2 = D_x \cup S_{\overline{x}} \cup S_{\overline{xy}}$;
- $V_3 = D_y \cup S_{\overline{y}} \cup D_r$.

These sets are pairwise disjoint and at least one of them is of size at most $n/3$ and to prove the Lemma we show that $|X_1| \le |V_1|$, $|X_2| \le |V_2|$, and $|X_3| \le |V_3|$.

$|X_1| \le |V_1|$. Since $C^* \subseteq D_S$, we have that $X_1 = C^* \cup R \subseteq V_1 = D_S \cup R$.

$|X_2| \le |V_2|$. To prove the inequality we need an additional result

$$|Z_{\overline{x}}| \le |S_{\overline{x}}|, |Z_{\overline{y}}| \le |S_{\overline{y}}|, \text{ and } |Z_{\overline{r}}| \le |S_{\overline{xy}}|. \tag{1}$$

In fact, since $Z_{\overline{x}}$ is the smallest subset of D_y such that $S_{\overline{x}} \subseteq N(Z_{\overline{x}})$, we have that for any vertex $u \in Z_{\overline{x}}$, $S_{\overline{x}} \not\subseteq N(Z_{\overline{x}} \setminus \{u\})$. Thus u has a private neighbor in $S_{\overline{x}}$, or in other words there exists $v \in S_{\overline{x}}$ such that $\{u\} = N(v) \cap Z_{\overline{x}}$. Therefore $S_{\overline{x}}$ contains at least one vertex for every vertex in $Z_{\overline{x}}$, which yields $|Z_{\overline{x}}| \le |S_{\overline{x}}|$. The proof of inequalities $|Z_{\overline{y}}| \le |S_{\overline{y}}|$, and $|Z_{\overline{r}}| \le |S_{\overline{xy}}|$ is similar.

Now the proof of $|X_2| \le |V_2|$, which is equivalent to $|D_x \cup Z_{\overline{x}} \cup Z_{\overline{r}}| \le |D_x \cup S_{\overline{x}} \cup S_{\overline{xy}}|$, follows from (1) and the fact that all subsets on each side of inequality are pairwise disjoint.

$|X_3| \le |V_3|$. This inequality is equivalent to $|D_y \cup Z_{\overline{y}} \cup Z_{\overline{r}}| \le |D_y \cup S_{\overline{y}} \cup D_r|$. Again, the sets on each side of inequality are pairwise disjoint. $|Z_{\overline{r}}| \le |D_r|$ because $Z_{\overline{r}} \subseteq D_r$, and $|Z_{\overline{y}}| \le |S_{\overline{y}}|$ by (1).

Thus $\min\{|X_1|, |X_2|, |X_3|\} \le n/3$ which concludes the proof of the lemma.

Lemma 8. *Every graph on n vertices has at most $2n^2 \sum_{i=1}^{n/3} \binom{n}{i}$ nice potential maximal cliques which can be listed in $\mathcal{O}^*(\binom{n}{n/3})$ time.*

Proof. By Lemma 7, the number of the number of possible partial representations (X, c) and indirect representations (X, x, c) with $|X| \le n/3$ is at most $2n^2 \sum_{i=1}^{n/3} \binom{n}{i}$. By Theorem 2, the number of all possible separator representations is at most $n^2 |\Delta_G| \le n^2 \binom{n}{n/3}$ and we deduce that the number of nice potential maximal cliques is at most $2n^2 \sum_{i=1}^{n/3} \binom{n}{i}$. Moreover, these potential maximal cliques can be computed in $\mathcal{O}^*(\binom{n}{n/3})$ time as follows. We enumerate all the triples (S,a,b) where S is a minimal separator and a,b are vertices, and check if the triple is the separator representation of a potential maximal clique Ω; if so, we store this potential maximal clique. We also enumerate all the potential maximal cliques of type $N[a], a \in V(G)$ in polynomial time. Finally, by listing all the sets X of at most $n/3$ vertices and all the couples of vertices (x,c), we compute all the nice potential maximal cliques with a partial representation (X,c) or a indirect representation (X,x,c).

Not all potential maximal cliques of a graph are necessarily nice (see [7] for an example). These non nice potential maximal cliques can be found as shown in the proof of Theorem 9, by using Theorem 6 and an algorithm to find nice potential maximal cliques.

Theorem 9. *A graph G on n vertices has at most $2n^3\sum_{i=1}^{n/3}\binom{n}{i} = \mathcal{O}(n^4 \cdot 1.8899^n)$ potential maximal cliques. There is an algorithm to list all potential maximal cliques of a graph in time $\mathcal{O}^*(1.8899^n)$.*

Proof. Let $x_1, x_2, \ldots, x_n$ be the vertices of G and $G_i = G[\{x_1,\ldots,x_i\}]$, for all $i \in \{1,2,\ldots,n\}$. Theorem 6 and Lemma 8 imply that $|\Pi_{G_i}| \leq |\Pi_{G_{i-1}}| + n|\Delta_{G_i}| + 2n^2\sum_{i=1}^{n/3}\binom{n}{i}$, for all $i \in \{2,3,\ldots,n\}$. By Theorem 2, $|\Pi_G| \leq 2n^3\sum_{i=1}^{n/3}\binom{n}{i}$.

Clearly, if we have the potential maximal cliques of G_{i-1}, the potential maximal cliques of G_i can be computed in $\mathcal{O}^*(|\Pi_{G_{i-1}}| + \binom{n}{n/3})$ time by making use of Theorems 2, 6, and Lemma 8. The graph G_1 has a unique potential maximal clique, namely $\{x_1\}$. Therefore Π_G can be listed in time $\mathcal{O}^*(\binom{n}{n/3})$ time which is approximately $\mathcal{O}^*(1.8899^n)$.

Theorem 10. *For a graph G on n vertices, the treewidth and the minimum fill-in of G can be computed in $\mathcal{O}^*(1.8899^n)$ time.*

Proof. The result follows from the Theorems 1, 2, 7, and 9.

5 Concluding Remarks

It is still an open question whether or not it is possible to list all potential maximal cliques in a graph in less than $\mathcal{O}^*(1.8899^n)$ time. The fact that the theoretical bound for the number of potential maximal cliques is $\mathcal{O}^*(1.8135^n)$ points in the direction of a better bound. Unfortunately there exits no nice algorithm for listing the potential maximal cliques of G in $\mathcal{O}^*(|\Pi_G|)$ time, like there exists for minimal separators [3].

Acknowledgments

The author would like to thank Ioan Todinca for useful discussions on the subject and for helpful comments to the text.

References

1. E. Amir. Efficient approximation for triangulation of minimum treewidth. In *UAI '01: Proceedings of the 17th Conference in Uncertainty in Artificial Intelligence*, pages 7–15, San Francisco, CA, USA, 2001. Morgan Kaufmann Publishers Inc.
2. S. Arnborg, D. G. Corneil, and A. Proskurowski. Complexity of finding embeddings in a k-tree. *SIAM J. Alg. Disc. Meth.*, 8:277–284, 1987.
3. A. Berry, J. P. Bordat, and O. Cogis. Generating all the minimal separators of a graph. *Int. J. Foundations Comp. Sci.*, 11(3):397–403, 2000.
4. H. L. Bodlaender. A linear time algorithm for finding tree-decompositions of small treewidth. *SIAM Journal on Computing*, 25:1305–1317, 1996.
5. V. Bouchitté, D. Kratsch, H. Müller, and I. Todinca. On treewidth approximations. *Discrete Appl. Math.*, 136(2-3):183–196, 2004.
6. V. Bouchitté and I. Todinca. Treewidth and minimum fill-in: Grouping the minimal separators. *SIAM J. Comput.*, 31:212–232, 2001.
7. V. Bouchitté and I. Todinca. Listing all potential maximal cliques of a graph. *Theor. Comput. Sci.*, 276(1-2):17–32, 2002.
8. W. Cook and P. Seymour. Tour merging via branch-decomposition. *INFORMS J. Comput.*, 15(3):233–248, 2003.
9. E. Dantsin, A. Goerdt, E. A. Hirsch, R. Kannan, J. Kleinberg, C. Papadimitriou, P. Raghavan, and U. Schöning. A deterministic (2 - 2/(k+ 1))n algorithm for k-sat based on local search. *Theor. Comput. Sci.*, 289(1):69–83, 2002.
10. D. Eppstein. Small maximal independent sets and faster exact graph coloring. In *WADS '01: Proceedings of the 7th International Workshop on Algorithms and Data Structures*, pages 462–470, London, UK, 2001. Springer-Verlag.
11. F. V. Fomin, D. Kratsch, and I. Todinca. Exact (exponential) algorithms for treewidth and minimum fill-in. In *ICALP*, pages 568–580, 2004.
12. M. Held and R. M. Karp. A dynamic programming approach to sequencing problems. *J. Soc. Indust. Appl. Math.*, 10:196–210, 1962.
13. S. L. Lauritzen and D. J. Spiegelhalter. Local computations with probabilities on graphical structures and their applications to expert systems. *J. Royal Statist. Soc., ser B*, 50:157–224, 1988.
14. N. Robertson and P. Seymour. Graph minors II. Algorithmic aspects of tree-width. *J. Algorithms*, 7:309–322, 1986.
15. D. J. Rose. A graph-theoretic study of the numerical solution of sparse positive definite systems of linear equations. In R. C. Read, editor, *Graph Theory and Computing*, pages 183–217. Academic Press, New York, 1972.
16. R. E. Tarjan and M. Yannakakis. Simple linear-time algorithms to test chordality of graphs, test acyclicity of hypergraphs, and selectively reduce acyclic hypergraphs. *SIAM J. Comput.*, 13:566–579, 1984.
17. Gerhard J. Woeginger. Exact algorithms for np-hard problems: A survey. In Michael Jünger, Gerhard Reinelt, and Giovanni Rinaldi, editors, *Combinatorial Optimization*, volume 2570 of *Lecture Notes in Computer Science*, pages 185–208. Springer, 2001.
18. M. Yannakakis. Computing the minimum fill-in is NP-complete. *SIAM J. Alg. Disc. Meth.*, 2:77–79, 1981.

Author Index

Lecture Notes in Computer Science

For information about Vols. 1–3792

please contact your bookseller or Springer

Vol. 3901: P.M. Hill (Ed.), Logic Based Program Synthesis and Transformation. X, 179 pages. 2006.

Vol. 3899: S. Frintrop, VOCUS: A Visual Attention System for Object Detection and Goal-Directed Search. XIV, 216 pages. 2006. (Sublibrary LNAI).

Vol. 3894: W. Grass, B. Sick, K. Waldschmidt (Eds.), Architecture of Computing Systems - ARCS 2006. XII, 496 pages. 2006.

Vol. 3889: J. Rosca, D. Erdogmus, J.C. Príncipe, S. Haykin (Eds.), Independent Component Analysis and Blind Signal Separation. XXI, 980 pages. 2006.

Vol. 3887: J. R. Correa, A. Hevia, M. Kiwi (Eds.), LATIN 2006: Theoretical Informatics. XVI, 814 pages. 2006.

Vol. 3884: B. Durand, W. Thomas (Eds.), STACS 2006. XIV, 714 pages. 2006.

Vol. 3881: S. Gibet, N. Courty, J.-F. Kamp (Eds.), Gesture in Human-Computer Interaction and Simulation. XIII, 344 pages. 2006. (Sublibrary LNAI).

Vol. 3879: T. Erlebach, G. Persinao (Eds.), Approximation and Online Algorithms. X, 349 pages. 2006.

Vol. 3878: A. Gelbukh (Ed.), Computational Linguistics and Intelligent Text Processing. XVII, 589 pages. 2006.

Vol. 3877: M. Detyniecki, J.M. Jose, A. Nürnberger, C. J. '. van Rijsbergen (Eds.), Adaptive Multimedia Retrieval: User, Context, and Feedback. XI, 279 pages. 2006.

Vol. 3876: S. Halevi, T. Rabin (Eds.), Theory of Cryptography. XI, 617 pages. 2006.

Vol. 3875: S. Ur, E. Bin, Y. Wolfsthal (Eds.), Haifa Verification Conference. X, 265 pages. 2006.

Vol. 3874: R. Missaoui, J. Schmidt (Eds.), Formal Concept Analysis. X, 309 pages. 2006. (Sublibrary LNAI).

Vol. 3873: L. Maicher, J. Park (Eds.), Charting the Topic Maps Research and Applications Landscape. VIII, 281 pages. 2006. (Sublibrary LNAI).

Vol. 3872: H. Bunke, A. L. Spitz (Eds.), Document Analysis Systems VII. XIII, 630 pages. 2006.

Vol. 3870: S. Spaccapietra, P. Atzeni, W.W. Chu, T. Catarci, K.P. Sycara (Eds.), Journal on Data Semantics V. XIII, 237 pages. 2006.

Vol. 3869: S. Renals, S. Bengio (Eds.), Machine Learning for Multimodal Interaction. XIII, 490 pages. 2006.

Vol. 3868: K. Römer, H. Karl, F. Mattern (Eds.), Wireless Sensor Networks. XI, 342 pages. 2006.

Vol. 3863: M. Kohlhase (Ed.), Mathematical Knowledge Management. XI, 405 pages. 2006. (Sublibrary LNAI).

Vol. 3861: J. Dix, S.J. Hegner (Eds.), Foundations of Information and Knowledge Systems. X, 331 pages. 2006.

Vol. 3860: D. Pointcheval (Ed.), Topics in Cryptology – CT-RSA 2006. XI, 365 pages. 2006.

Vol. 3858: A. Valdes, D. Zamboni (Eds.), Recent Advances in Intrusion Detection. X, 351 pages. 2006.

Vol. 3857: M.P.C. Fossorier, H. Imai, S. Lin, A. Poli (Eds.), Applied Algebra, Algebraic Algorithms and Error-Correcting Codes. XI, 350 pages. 2006.

Vol. 3855: E. A. Emerson, K.S. Namjoshi (Eds.), Verification, Model Checking, and Abstract Interpretation. XI, 443 pages. 2005.

Vol. 3853: A.J. Ijspeert, T. Masuzawa, S. Kusumoto (Eds.), Biologically Inspired Approaches to Advanced Information Technology. XIV, 388 pages. 2006.

Vol. 3852: P.J. Narayanan, S.K. Nayar, H.-Y. Shum (Eds.), Computer Vision - ACCV 2006, Part II. XXXI, 977 pages. 2005.

Vol. 3851: P.J. Narayanan, S.K. Nayar, H.-Y. Shum (Eds.), Computer Vision - ACCV 2006, Part I. XXXI, 973 pages. 2006.

Vol. 3850: R. Freund, G. Păun, G. Rozenberg, A. Salomaa (Eds.), Membrane Computing. IX, 371 pages. 2006.

Vol. 3849: I. Bloch, A. Petrosino, A.G.B. Tettamanzi (Eds.), Fuzzy Logic and Applications. XIV, 438 pages. 2006. (Sublibrary LNAI).

Vol. 3848: J.-F. Boulicaut, L. De Raedt, H. Mannila (Eds.), Constraint-Based Mining and Inductive Databases. X, 401 pages. 2006. (Sublibrary LNAI).

Vol. 3847: K.P. Jantke, A. Lunzer, N. Spyratos, Y. Tanaka (Eds.), Federation over the Web. X, 215 pages. 2006. (Sublibrary LNAI).

Vol. 3846: H. J. van den Herik, Y. Björnsson, N.S. Netanyahu (Eds.), Computers and Games. XIV, 333 pages. 2006.

Vol. 3845: J. Farré, I. Litovsky, S. Schmitz (Eds.), Implementation and Application of Automata. XIII, 360 pages. 2006.

Vol. 3844: J.-M. Bruel (Ed.), Satellite Events at the MoDELS 2005 Conference. XIII, 360 pages. 2006.

Vol. 3843: P. Healy, N.S. Nikolov (Eds.), Graph Drawing. XVII, 536 pages. 2006.

Vol. 3842: H.T. Shen, J. Li, M. Li, J. Ni, W. Wang (Eds.), Advanced Web and Network Technologies, and Applications. XXVII, 1057 pages. 2006.

Vol. 3841: X. Zhou, J. Li, H.T. Shen, M. Kitsuregawa, Y. Zhang (Eds.), Frontiers of WWW Research and Development - APWeb 2006. XXIV, 1223 pages. 2006.

Vol. 3840: M. Li, B. Boehm, L.J. Osterweil (Eds.), Unifying the Software Process Spectrum. XVI, 522 pages. 2006.

Vol. 3839: J.-C. Filliâtre, C. Paulin-Mohring, B. Werner (Eds.), Types for Proofs and Programs. VIII, 275 pages. 2006.

Vol. 3838: A. Middeldorp, V. van Oostrom, F. van Raamsdonk, R. de Vrijer (Eds.), Processes, Terms and Cycles: Steps on the Road to Infinity. XVIII, 639 pages. 2005.

Vol. 3837: K. Cho, P. Jacquet (Eds.), Technologies for Advanced Heterogeneous Networks. IX, 307 pages. 2005.

Vol. 3836: J.-M. Pierson (Ed.), Data Management in Grids. X, 143 pages. 2006.

Vol. 3835: G. Sutcliffe, A. Voronkov (Eds.), Logic for Programming, Artificial Intelligence, and Reasoning. XIV, 744 pages. 2005. (Sublibrary LNAI).

Vol. 3834: D.G. Feitelson, E. Frachtenberg, L. Rudolph, U. Schwiegelshohn (Eds.), Job Scheduling Strategies for Parallel Processing. VIII, 283 pages. 2005.

Vol. 3833: K.-J. Li, C. Vangenot (Eds.), Web and Wireless Geographical Information Systems. XI, 309 pages. 2005.

Vol. 3832: D. Zhang, A.K. Jain (Eds.), Advances in Biometrics. XX, 796 pages. 2005.

Vol. 3831: J. Wiedermann, G. Tel, J. Pokorný, M. Bieliková, J. Štuller (Eds.), SOFSEM 2006: Theory and Practice of Computer Science. XV, 576 pages. 2006.

Vol. 3829: P. Pettersson, W. Yi (Eds.), Formal Modeling and Analysis of Timed Systems. IX, 305 pages. 2005.

Vol. 3828: X. Deng, Y. Ye (Eds.), Internet and Network Economics. XVII, 1106 pages. 2005.

Vol. 3827: X. Deng, D.-Z. Du (Eds.), Algorithms and Computation. XX, 1190 pages. 2005.

Vol. 3826: B. Benatallah, F. Casati, P. Traverso (Eds.), Service-Oriented Computing - ICSOC 2005. XVIII, 597 pages. 2005.

Vol. 3824: L.T. Yang, M. Amamiya, Z. Liu, M. Guo, F.J. Rammig (Eds.), Embedded and Ubiquitous Computing – EUC 2005. XXIII, 1204 pages. 2005.

Vol. 3823: T. Enokido, L. Yan, B. Xiao, D. Kim, Y. Dai, L.T. Yang (Eds.), Embedded and Ubiquitous Computing – EUC 2005 Workshops. XXXII, 1317 pages. 2005.

Vol. 3822: D. Feng, D. Lin, M. Yung (Eds.), Information Security and Cryptology. XII, 420 pages. 2005.

Vol. 3821: R. Ramanujam, S. Sen (Eds.), FSTTCS 2005: Foundations of Software Technology and Theoretical Computer Science. XIV, 566 pages. 2005.

Vol. 3820: L.T. Yang, X.-s. Zhou, W. Zhao, Z. Wu, Y. Zhu, M. Lin (Eds.), Embedded Software and Systems. XXVIII, 779 pages. 2005.

Vol. 3819: P. Van Hentenryck (Ed.), Practical Aspects of Declarative Languages. X, 231 pages. 2005.

Vol. 3818: S. Grumbach, L. Sui, V. Vianu (Eds.), Advances in Computer Science – ASIAN 2005. XIII, 294 pages. 2005.

Vol. 3817: M. Faundez-Zanuy, L. Janer, A. Esposito, A. Satue-Villar, J. Roure, V. Espinosa-Duro (Eds.), Nonlinear Analyses and Algorithms for Speech Processing. XII, 380 pages. 2006. (Sublibrary LNAI).

Vol. 3816: G. Chakraborty (Ed.), Distributed Computing and Internet Technology. XXI, 606 pages. 2005.

Vol. 3815: E.A. Fox, E.J. Neuhold, P. Premsmit, V. Wuwongse (Eds.), Digital Libraries: Implementing Strategies and Sharing Experiences. XVII, 529 pages. 2005.

Vol. 3814: M. Maybury, O. Stock, W. Wahlster (Eds.), Intelligent Technologies for Interactive Entertainment. XV, 342 pages. 2005. (Sublibrary LNAI).

Vol. 3813: R. Molva, G. Tsudik, D. Westhoff (Eds.), Security and Privacy in Ad-hoc and Sensor Networks. VIII, 219 pages. 2005.

Vol. 3812: C. Bussler, A. Haller (Eds.), Business Process Management Workshops. XIII, 520 pages. 2006.

Vol. 3811: C. Bussler, M.-C. Shan (Eds.), Technologies for E-Services. VIII, 127 pages. 2006.

Vol. 3810: Y.G. Desmedt, H. Wang, Y. Mu, Y. Li (Eds.), Cryptology and Network Security. XI, 349 pages. 2005.

Vol. 3809: S. Zhang, R. Jarvis (Eds.), AI 2005: Advances in Artificial Intelligence. XXVII, 1344 pages. 2005. (Sublibrary LNAI).

Vol. 3808: C. Bento, A. Cardoso, G. Dias (Eds.), Progress in Artificial Intelligence. XVIII, 704 pages. 2005. (Sublibrary LNAI).

Vol. 3807: M. Dean, Y. Guo, W. Jun, R. Kaschek, S. Krishnaswamy, Z. Pan, Q.Z. Sheng (Eds.), Web Information Systems Engineering – WISE 2005 Workshops. XV, 275 pages. 2005.

Vol. 3806: A.H. H. Ngu, M. Kitsuregawa, E.J. Neuhold, J.-Y. Chung, Q.Z. Sheng (Eds.), Web Information Systems Engineering – WISE 2005. XXI, 771 pages. 2005.

Vol. 3805: G. Subsol (Ed.), Virtual Storytelling. XII, 289 pages. 2005.

Vol. 3804: G. Bebis, R. Boyle, D. Koracin, B. Parvin (Eds.), Advances in Visual Computing. XX, 755 pages. 2005.

Vol. 3803: S. Jajodia, C. Mazumdar (Eds.), Information Systems Security. XI, 342 pages. 2005.

Vol. 3802: Y. Hao, J. Liu, Y.-P. Wang, Y.-m. Cheung, H. Yin, L. Jiao, J. Ma, Y.-C. Jiao (Eds.), Computational Intelligence and Security, Part II. XLII, 1166 pages. 2005. (Sublibrary LNAI).

Vol. 3801: Y. Hao, J. Liu, Y.-P. Wang, Y.-m. Cheung, H. Yin, L. Jiao, J. Ma, Y.-C. Jiao (Eds.), Computational Intelligence and Security, Part I. XLI, 1122 pages. 2005. (Sublibrary LNAI).

Vol. 3799: M. A. Rodríguez, I.F. Cruz, S. Levashkin, M.J. Egenhofer (Eds.), GeoSpatial Semantics. X, 259 pages. 2005.

Vol. 3798: A. Dearle, S. Eisenbach (Eds.), Component Deployment. X, 197 pages. 2005.

Vol. 3797: S. Maitra, C. E. V. Madhavan, R. Venkatesan (Eds.), Progress in Cryptology - INDOCRYPT 2005. XIV, 417 pages. 2005.

Vol. 3796: N.P. Smart (Ed.), Cryptography and Coding. XI, 461 pages. 2005.

Vol. 3795: H. Zhuge, G.C. Fox (Eds.), Grid and Cooperative Computing - GCC 2005. XXI, 1203 pages. 2005.

Vol. 3794: X. Jia, J. Wu, Y. He (Eds.), Mobile Ad-hoc and Sensor Networks. XX, 1136 pages. 2005.

Vol. 3793: T. Conte, N. Navarro, W.-m.W. Hwu, M. Valero, T. Ungerer (Eds.), High Performance Embedded Architectures and Compilers. XIII, 317 pages. 2005.

Lecture Notes in Computer Science 3887

Commenced Publication in 1973
Founding and Former Series Editors:
Gerhard Goos, Juris Hartmanis, and Jan van Leeuwen

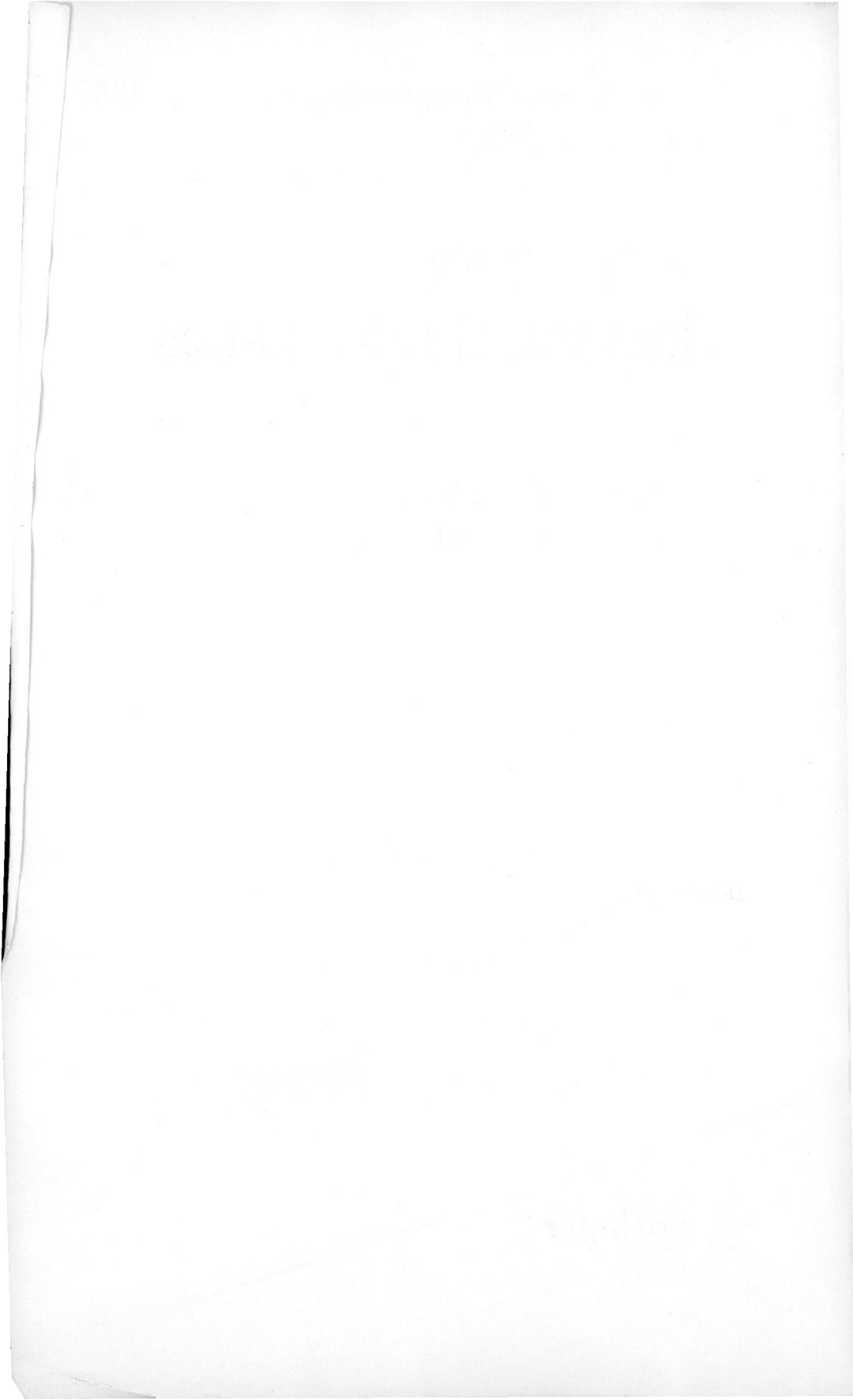